Prefixes for Powers of 10[†]

Multiple	Prefix	Abbreviation
10^{24}	yotta	Y
10^{21}	zetta	Z
10^{18}	exa	E
10^{15}	peta	P
10^{12}	tera	T
10^{9}	giga	G
10^{6}	mega	M
10^{3}	kilo	k
10^{2}	hecto	h
10^{1}	deka	da
10^{-1}	deci	d
10^{-2}	centi	c
10^{-3}	milli	m
10^{-6}	micro	μ
10^{-9}	nano	n
10^{-12}	pico	p
10^{-15}	femto	f
10^{-18}	atto	a
10^{-21}	zepto	z
10^{-24}	yocto	y

† Commonly used prefixes are in bold. All prefixes are pronounced with the accent on the first syllable.

The Greek Alphabet

Alpha	A	α	Nu	N	ν
Beta	B	β	Xi	Ξ	ξ
Gamma	Γ	γ	Omicron	O	o
Delta	Δ	δ	Pi	Π	π
Epsilon	E	ϵ, ε	Rho	P	ρ
Zeta	Z	ζ	Sigma	Σ	σ
Eta	H	η	Tau	T	τ
Theta	Θ	θ	Upsilon	Y	υ
Iota	I	ι	Phi	Φ	ϕ
Kappa	K	κ	Chi	X	χ
Lambda	Λ	λ	Psi	Ψ	ψ
Mu	M	μ	Omega	Ω	ω

Mathematical Symbols

$=$	is equal to
$\equiv$	is defined by
$\neq$	is not equal to
$\approx$	is approximately equal to
$\sim$	is of the order of
$\propto$	is proportional to
$>$	is greater than
$\geq$	is greater than or equal to
$>>$	is much greater than
$<$	is less than
$\leq$	is less than or equal to
$<<$	is much less than
Δx	change in x
dx	differential change in x
$|x|$	absolute value of x
$|\vec{v}|$	magnitude of $\vec{v}$
$n!$	$n(n-1)(n-2)...1$
Σ	sum
$\lim$	limit
$\Delta t \to 0$	Δt approaches zero
$\dfrac{dx}{dt}$	derivative of x with respect to t
$\dfrac{\partial x}{\partial t}$	partial derivative of x with respect to t
$\displaystyle\int_{x_1}^{x_2} f(x)dx$	definite integral

$$= F(x)\Big|_{x_1}^{x_2} = F(x_2) - F(x_1)$$

Terrestrial and Astronomical Data[†]

Acceleration of gravity at earth's surface	g	$9.81 \text{ m/s}^2 = 32.2 \text{ ft/s}^2$
Radius of earth	R_E	$6370 \text{ km} = 3960 \text{ mi}$
Mass of earth	M_E	$5.98 \times 10^{24} \text{ kg}$
Mass of sun		$1.99 \times 10^{30} \text{ kg}$
Mass of moon		$7.36 \times 10^{22} \text{ kg}$
Escape speed at earth's surface		$11.2 \text{ km/s} = 6.95 \text{ mi/s}$
Standard temperature and pressure (STP)		$0°C = 273.15 \text{ K}$ $1 \text{ atm} = 101.3 \text{ kPa}$
Earth–moon distance[‡]		$3.84 \times 10^8 \text{ m} = 2.39 \times 10^5 \text{ mi}$
Earth–sun distance (mean)[‡]		$1.50 \times 10^{11} \text{ m} = 9.30 \times 10^7 \text{ mi}$
Speed of sound in dry air (at STP)		331 m/s
Speed of sound in dry air (20°C, 1 atm)		343 m/s
Density of air (STP)		1.29 kg/m^3
Density of water (4°C, 1 atm)		1000 kg/m^3
Heat of fusion of water (0°C, 1 atm)	L_f	333.5 kJ/kg
Heat of vaporization of water (100°C, 1 atm)	L_v	2.257 MJ/kg

† Additional data on the solar system can be found in Appendix B and at http://nssdc.gsfc.nasa.gov/planetary/planetfact.html.
‡ Center to center.

Abbreviations for Units

A	ampere	H	henry	nm	nanometer (10^{-9} m)
Å	angstrom (10^{-10} m)	h	hour	pt	pint
atm	atmosphere	Hz	hertz	qt	quart
Btu	British thermal unit	in	inch	rev	revolution
Bq	becquerel	J	joule	R	roentgen
C	coulomb	K	kelvin	Sv	seivert
°C	degree Celsius	kg	kilogram	s	second
cal	calorie	km	kilometer	T	tesla
Ci	curie	keV	kilo-electron volt	u	unified mass unit
cm	centimeter	lb	pound	V	volt
dyn	dyne	L	liter	W	watt
eV	electron volt	m	meter	Wb	weber
°F	degree Fahrenheit	MeV	mega-electron volt	y	year
fm	femtometer, fermi (10^{-15} m)	Mm	megameter (10^6 m)	yd	yard
ft	foot	mi	mile	μm	micrometer (10^{-6} m)
Gm	gigameter (10^9 m)	min	minute	μs	microsecond
G	gauss	mm	millimeter	μC	microcoulomb
Gy	gray	ms	millisecond	Ω	ohm
g	gram	N	newton		

Some Conversion Factors

Length

1 m = 39.37 in = 3.281 ft = 1.094 yd

1 m = 10^{15} fm = 10^{10} Å = 10^9 nm

1 km = 0.6215 mi

1 mi = 5280 ft = 1.609 km

1 lightyear = 1 $c \cdot y$ = 9.461 $\times$ 10^{15} m

1 in = 2.540 cm

Volume

1 L = 10^3 cm^3 = 10^{-3} m^3 = 1.057 qt

Time

1 h = 3600 s = 3.6 ks

1 y = 365.24 d = 3.156 $\times$ 10^7 s

Speed

1 km/h = 0.278 m/s = 0.6215 mi/h

1 ft/s = 0.3048 m/s = 0.6818 mi/h

Angle–angular speed

1 rev = 2π rad = 360°

1 rad = 57.30°

1 rev/min = 0.1047 rad/s

Force–pressure

1 N = 10^5 dyn = 0.2248 lb

1 lb = 4.448 N

1 atm = 101.3 kPa = 1.013 bar = 76.00 cmHg = 14.70 lb/in^2

Mass

1 u = [(10^{-3} mol^{-1})/N_A] kg = 1.661 $\times$ 10^{-27} kg

1 tonne = 10^3 kg = 1 Mg

1 slug = 14.59 kg

1 kg weighs about 2.205 lb

Energy–power

1 J = 10^7 erg = 0.7373 ft·lb = 9.869 $\times$ 10^{-3} L·atm

1 kW·h = 3.6 MJ

1 cal = 4.184 J = 4.129 $\times$ 10^{-2} L·atm

1 L·atm = 101.325 J = 24.22 cal

1 eV = 1.602 $\times$ 10^{-19} J

1 Btu = 778 ft·lb = 252 cal = 1054 J

1 horsepower = 550 ft·lb/s = 746 W

Thermal conductivity

1 W/(m·K) = 6.938 Btu·in/(h·ft^2·F°)

Magnetic field

1 T = 10^4 G

Viscosity

1 Pa·s = 10 poise

fifth edition

PHYSICS

FOR SCIENTISTS AND ENGINEERS

Volume 2
Electricity, Magnetism, Light &
Elementary Modern Physics

W. H. Freeman and Company
New York

Publisher:	Susan Finnemore Brennan
Senior Development Editors:	Kathleen Civetta/Jennifer Van Hove
Assistant Editors:	Rebecca Pearce/Amanda McCorquodale/Eileen McGinnis
Marketing Manager:	Mark Santee
Project Editors:	Georgia L. Hadler/Cathy Townsend, PreMediaONE, A Black Dot Group Company
Text Designer:	Marsha Cohen
Cover Designer:	Blake Logan
Illustrations:	Network Graphics/PreMediaONE, A Black Dot Group Company
Photo Editors:	Patricia Marx/Dena Betz
Production Manager:	Julia DeRosa
Media and Supplements Editor:	Brian Donnellan
Composition:	PreMediaONE, A Black Dot Group Company
Manufacturing:	RR Donnelley & Sons Company

Cover image: Digital Vision

Library of Congress Cataloging-in-Publication Data
Physics for Scientists and Engineers. - 5th ed.
    p. cm.
    By Paul A. Tipler and Gene Mosca
    Includes index.
    ISBN: 0-7167-0809-4 (Vol. 1 Hardback Ch. 1-20, R)
    ISBN: 0-7167-0900-7 (Vol. 1A Softcover Ch. 1-13, R)
    ISBN: 0-7167-0903-1 (Vol. 1B Softcover Ch. 14-20)
    ISBN: 0-7167-0810-8 (Vol. 2 Hardback Ch. 21-41)
    ISBN: 0-7167-0902-3 (Vol. 2A Softcover Ch. 21-25)
    ISBN: 0-7167-0901-5 (Vol. 2B Softcover Ch. 26-33)
    ISBN: 0-7167-0906-6 (Vol. 2C Softcover Ch. 34-41)
    ISBN: 0-7167-8339-8 (Standard Hardback Ch. 1-33, R)
    ISBN: 0-7167-4389-2 (Extended Hardback Ch. 1-41)

Printed in the United States of America

First printing 2003

PT: For Claudia

GM: For Vivian

CONTENTS IN BRIEF

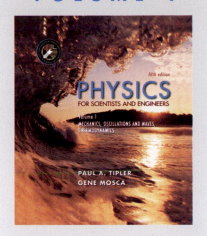

PART V   LIGHT

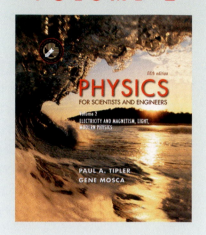

PART VI   MODERN PHYSICS: QUANTUM MECHANICS, RELATIVITY, AND THE STRUCTURE OF MATTER

APPENDIX

CONTENTS

CHAPTER 10

CONSERVATION OF ANGULAR MOMENTUM / 309

CHAPTER R

SPECIAL RELATIVITY / R-1

CHAPTER 11

GRAVITY / 339

CHAPTER 12 *

STATIC EQUILIBRIUM AND ELASTICITY / 370

VOLUME 2

PART IV ELECTRICITY AND MAGNETISM/651

CHAPTER 21

THE ELECTRIC FIELD I: DISCRETE CHARGE DISTRIBUTIONS / 651

CHAPTER 22

THE ELECTRIC FIELD II: CONTINUOUS CHARGE DISTRIBUTIONS / 682

CHAPTER 23

ELECTRIC POTENTIAL / 717

CHAPTER 24

ELECTROSTATIC ENERGY AND CAPACITANCE / 748

CHAPTER 25

ELECTRIC CURRENT AND DIRECT-CURRENT CIRCUITS / 786

CHAPTER 26

THE MAGNETIC FIELD / 829

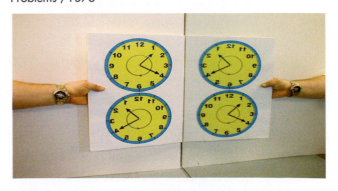

PART VI   MODERN PHYSICS: QUANTUM MECHANICS, RELATIVITY, AND THE STRUCTURE OF MATTER

CHAPTER 34

WAVE-PARTICLE DUALITY AND QUANTUM PHYSICS / 1117

CHAPTER 35

APPLICATIONS OF THE SCHRÖDINGER EQUATION / 1149

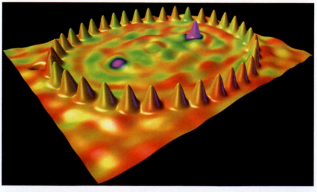

CHAPTER 36

ATOMS / 1171

CHAPTER 40

NUCLEAR PHYSICS / 1306

CHAPTER 41

ELEMENTARY PARTICLES AND THE BEGINNING OF THE UNIVERSE / 1335

APPENDIX A

SI UNITS AND CONVERSION FACTORS / AP-1

APPENDIX B

NUMERICAL DATA / AP-3

APPENDIX C

PERIODIC TABLE OF ELEMENTS / AP-6

APPENDIX D

REVIEW OF MATHEMATICS / AP-8

PREFACE

We are exceptionally pleased to present the fifth edition of *Physics for Scientists and Engineers*. Over the course of this revision, we have built upon the strengths of the fourth edition so that the new text is an even more reliable, engaging and motivating learning tool for the calculus-based introductory physics course. With the help of reviewers and the many users of the fourth edition we have carefully scrutinized and refined every aspect of the book, with an eye toward improving student comprehension and success. Our goals included helping students to increase their problem-solving ability, making the text more accessible and fun to read, and keeping the text flexible for the instructor.

Examples

One of the most important ways we've addressed our goals was to add some new features to the side-by-side worked examples that were introduced in the fourth edition. These examples juxtapose the problem-solving steps with the necessary equations so that it's easier for students to watch the problem unfold.

The side-by-side format for the worked examples came from a student suggestion; we've just added a few finishing touches:

- After each problem statement, students are asked to *Picture the Problem*. Here, the problem is analyzed both conceptually and visually, with students frequently directed to draw a free-body diagram. Each step of the solution is then presented with a written statement in the left-hand column and the corresponding mathematical equations in the right-hand column.

- *Remarks* at the end of the example point out the importance or relevance of the example, or suggest a different way to approach it.

- NEW *Plausibility Checks* remind students to check their results for mathematical accuracy, and for reasonableness as well.

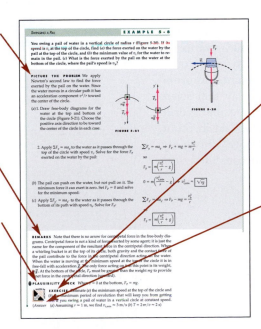

- An *Exercise* often follows the solution of the example, allowing students to check their understanding by solving a similar problem without help. Answers are included with the Exercise to provide immediate feedback and alternative solutions.

- NEW *Master the Concept Exercises* appear at least once in each chapter and help build students' problem-solving skills online.

Every example has been scrutinized, with additional steps added wherever an assumption might have been made, new Remarks included, and new follow-up exercises, free-body diagrams added where appropriate. The answers are now boxed to make them easier to find. Our new features include the Plausibility Check, which offers quick tests that help students learn to evaluate their answers with logic. We've also added interactive Master the Concept exercises to help students work through key problems. The exercises follow examples in the textbook and are marked with a Master the Concept icon that directs students to our Web site. There, the exercise is set up with algorithmically generated variables and students work the problem with step-by-step guidance and immediate feedback.

This edition also includes two types of specialized examples that provide unique problem-solving opportunities for students. The Try it Yourself examples prompt students to take an active role in solving the problem, and the Put It in Context examples approximate the real life scenarios they might encounter as scientists.

Try It Yourself examples

Like the regular worked example, these use the side-by-side format, but here the Picture the Problem section is sometimes missing, and the descriptions in the left-hand column are more terse. These examples take students step-by-step through the solution without doing the math for them. Students find it helpful to cover the right-hand column and attempt to perform the calculations on their own before looking at the equations. In this way, students can think through the steps as they fill in the answers.

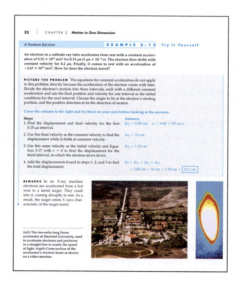

New Put It in Context examples

Each chapter now identifies at least one worked example as "context rich." These examples may include information not needed to solve the problem, or may require the student to find additional information in tables or to draw from experience or previously obtained information. Context-rich examples reflect the way that scientists and engineers solve problems in the real world. Laura McCullough of the University of Wisconsin, Stout, and Thomas Foster of Southern Illinois University, Edwardsville, initiated this feature and consulted with us in creating many of these examples.

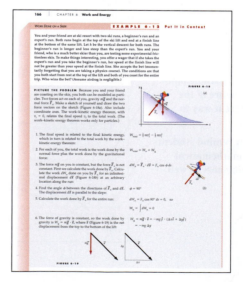

Practice Problems

Care has been taken to improve the quality and clarity of the end-of-chapter problems. About twenty percent of the 4,500 problems are new, written by Charles Adler of St. Mary's College of Maryland. Conceptual problems have been grouped together at the beginning of each problem set, and a new category of Estimation and Approximation problems have been added to encourage students to think more like scientists or engineers. Answers to odd-numbered problems appear at the back of the text. Solutions to approximately twenty-five percent of the problems appear in the newly revised Student Solutions Manual. This was written by David Mills of the College of the Redwoods to provide detailed solutions and to mirror the popular side-by-side format of the textbook examples.

About 1,100 of the text's problems are included in the new **iSOLVE** homework service. These problems can be accessed at www.whfreeman.com/tipler5e. About a third of the iSOLVE problems are Checkpoint Problems, which ask students to note the key principles and equations they're using and indicate their confidence level.

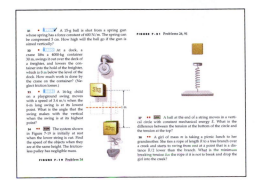

Each problem is marked with:

• a series of one, two, or three bullets, which identify its level of difficulty

• a **SSM** icon if the answer is in the Student Solutions Manual

• an **iSOLVE** icon if the problem is part of the isolve homework service and a **iSOLVE ✓** icon if the problem is a Checkpoint problem.

Features

This new edition of *Physics* has a number of textual features that make the book a valuable teaching tool. Key aspects of the last edition have been revised, and some new features have been added to make the book more engaging, inviting and up to date.

New chapter-opening pedagogy

• Each chapter now begins with a photograph and a question that is answered in a worked example within the chapter. These draw students into the material and provide motivation for problem solving.

• Chapter outlines list the major section headings, giving students a "road map" to the chapter.

• Chapter goal statements highlight the main ideas of the chapter.

PART I **MECHANICS**

CHAPTER **2**

Motion in One Dimension

MOTION IN ONE DIMENSION IS MOTION ALONG A STRAIGHT LINE, LIKE THAT OF A CAR ON A STRAIGHT ROAD. THIS DRIVER ENCOUNTERS STOPLIGHTS AND DIFFERENT SPEED LIMITS ON HER COMMUTE ALONG A STRAIGHT HIGHWAY TO SCHOOL.

? How can she estimate her arrival time? (See Example 2-2.)

2-1 Displacement, Velocity, and Speed
2-2 Acceleration
2-3 Motion With Constant Acceleration
2-4 Integration

Content improvements

Chapter R, an optional "mini" chapter in Volume 1, brief enough to be covered in a lecture or two, allows instructors to include this popular modern topic early in the course. The chapter avoids the abstraction associated with the Lorentz transformations and focuses on the basic concepts of length contraction, time dilation, and simultaneity, using thought experiments involving meter sticks and light clocks. The relation between relativistic momentum and relativistic energy is also developed.

Quantum Theory: Chapters 17, "Wave-Particle Duality and Quantum Physics," and 27, "The Microscopic Theory of Electrical Conduction" of the fourth edition have been moved to their more traditional location in Volume II of the fifth edition as Chapters 37 and 38. Should instructors wish to include these chapters earlier in the course, both chapters are available on the web at www.whfreeman.com/tipler5e.

Changes in Approach: Dozens of smaller, yet significant improvements in content have been made throughout the book. For example:

- Motion-diagrams are introduced in Section 3-3 and used to estimate the direction of the acceleration vector using the definition of acceleration.

- In Section 4-4, frictional forces are now introduced qualitatively, allowing for free-body diagrams that include frictional forces. A quantitative treatment of frictional forces appears in Section 5-1.

- Section 4-7 introduces problems with two or more objects. Selecting a separate set of coordinate axes for each object is a robust problem-solving practice when using Newton's laws with systems consisting of two or more objects. The value of this practice is revealed in the example where Steve is sliding down the glacier while Paul has already fallen over its edge.

- In Section 8-8, "Systems With Variable Mass," the basic equation of motion for an object with continuously varying mass (the rocket equation) is developed using an object that is acquiring mass—like an open boxcar in the rain—rather than one that is losing mass—like a rocket spewing exhaust gasses. This approach facilitates both the development of the basic equation of motion and the application of it to certain situations.

- In Chapter 9, "Rotation," there is a new section that provides problem-solving guidelines for applying Newton's Second Law to rotation.

- In Section 13-3 the discussion of buoyancy now includes the buoyant force on objects supported by a submerged surface.

- In Chapter 18, work-energy relations are expressed in terms of the work done on the system. The first law of thermodynamics is now expressed in terms of the work done on the system also. (The Educational Testing Service has adopted the convention that the work term in first law of thermodynamics be the work done on the system. This will be adhered to on all Advanced Placement physics exams.)

More engineering and biological applications

Additional applications emphasize the relevance of physics to students' experiences, further studies, and future careers.

New focus on common pitfalls

Topics that commonly cause confusion are identified with a new ❶ icon where the difficulty is addressed. For example, in Section 3-4 the icon is used to identify the discussion pointing out that the horizontal and vertical motions are independent in projectile motion.

For instructor and student convenience, the fifth edition of *Physics for Scientists and Engineers* is available in five paperback volumes—

Vol. 1A Mechanics (Ch. 1-13, plus a mini-chapter on relativity, Ch. R) 0-7167-0900-7

Vol. 1B Oscillations & Waves; Thermodynamics (Ch. 14-20) 0-7167-0903-1

Vol. 2A Electricity (Ch. 21-25) 0-7167-0902-3

Vol. 2B Electrodynamics, Light (Ch. 26-33) 0-7167-0901-5

Vol. 2C Elementary Modern Physics (Ch. 34-41) 0-7167-0906-6

or in four hardcover versions—

Vol. 1 Mechanics, Oscillations and Waves; Thermodynamics (Ch. 1-20, R) 0-7167-0809-4

Vol. 2 Electricity, Magnetism, Light & Modern Physics (Ch. 21-41) 0-7167-0810-8

Standard Version (Vol 1A-2B) 0-7167-8339-8

Extended Version (Vol 1A-2C) 0-7167-4389-2

New design and improved illustrations

The book has a warmer, more colorful look. Each piece of art has been carefully considered and many have been revised to increase clarity. Approximately 245 new figures have been added, including many new free-body diagrams within the worked examples. New photos bring to life the many real-world applications of physics.

Optional sections

The book was designed to allow professors to be flexible by designating certain sections "optional." These sections are marked with an *, and professors who choose to skip this section can do so knowing that their students won't be missing any material they will need in later chapters.

Summary

End of chapter summaries are organized with important topics on the left and relevant remarks and equations on the right. Here the key equations from the chapter appear together for easy reference.

Exploring essays

Students are invited to examine interesting extensions of the chapter concepts in Exploring sections, which are now found on the Web. These short pieces relate the chapter concepts to everything from the weather to transducers.

Media and Print Supplements

The supplements package has been updated and improved in response to reviewer suggestions and those from users of the fourth edition.

For the Student:

Student Solutions Manual: *Vol. 1, 0-7167-8333-9; Vol. 2, 0-7167-8334-7.* The new manual prepared by David Mills of College of the Redwoods, Charles Adler of St. Mary's College of Maryland, Ed Whittaker of Stevens Institute of Technology, George Zober of Yough Senior High School and Patricia Zober of Ringgold High School provides solutions for about twenty-five percent of the problems in the textbook, using the same side-by-side format and level of detail as the textbook's worked examples.

Study Guide: *Vol. 1, 0-7167-8332-0; Vol. 2, 0-7167-8331-2.* Prepared by Gene Mosca of the United States Naval Academy and Todd Ruskell of Colorado School of Mines, the Study Guide describes the key ideas and potential pitfalls of each chapter, and also includes true and false questions that test essential definitions and relations, questions and answers that require qualitative reasoning, and problems and solutions.

Student Web Site: Robin Jordan of Florida Atlantic University has put together a site designed to make studying and testing easier for both students and professors. The Web site includes:

- **On-line quizzing:** Multiple choice quizzes are available for each chapter. Students will receive immediate feedback, and the quiz results are collected for the instructor in a grade book.

- **iSOLVE homework service**: *0-7167-5802-4.* About one-fourth of the book's end-of-chapter problems, 1,100 altogether, are available on-line in W.H. Freeman's iSOLVE homework service. This service will offer each student a different version of every problem similar to CAPA and WebAssign, and the iSOLVE problems will be marked with an icon in the textbook. Homework scores can be collected in a grade book. Students may purchase access to iSOLVE for three semesters at a time.

• **iSOLVE Checkpoint problems:** A third of our iSOLVE questions are Checkpoint problems, which prompt students to describe how they arrived at their answer and to indicate their confidence level. All student responses will be gathered and included in the instructor's grade book report. Rolf Enger of the U.S. Air Force Academy inspired the development of Checkpoints to help professors gauge their student's understanding of the material.

• **Master the Concept exercises:** For each chapter, one or more exercises from the book will be available on-line so students can practice working the problem with randomized variables and step-by-step guidance. The on-line exercise will walk the student slowly through the problem-solving process and use interactive animations, simulations, video, and other graphic aids to help students visualize the problem. Teachers can collect grade book information on their progress. These premium examples are called out in the book with a Master the Concept icon.

Homework services: In addition to the iSOLVE network, there are three other homework services that are compatible with this textbook. End of chapter problems are available in WebAssign as well as CAPA: A Computer-Assisted Personalized Approach. A list of all the fifth edition problems included in WebAssign and CAPA is posted on the instructor's section of the *Physics* Web site. Our text is also compatible with the University of Texas Interactive Homework Service.

> **The iSolve homework service is available at**
> *www.whfreeman.com/tipler5e*
>
> **For more information about WebAssign, CAPA or UTX homework services, find their Web sites at:**
> *http://webassign.net/info*
> *http://www.pa.msu.edu/educ/CAPA/*
> *http://hw.utexas.edu/hw.html*

For the Instructor:

Instructor's Resource CD-ROM: *0-7167-9839-5.* This multi-faceted resource will give instructors the tools to make their own Web sites and presentations. The CD contains illustrations from the text in .jpg format, Powerpoint Lecture Slides for each chapter of the book, Lab Demonstration Videos, and Applied Physics videos in QuickTime format, and Presentation Manager Pro v.2.0, as well as all of the solutions to the end-of-chapter problems in editable Microsoft Word format.

Instructor's Resource Manual: The updated IRM contains Classroom Demonstrations for each chapter, a film and video guide with suggestions for each chapter, links to valuable Web sites, and links to free sources for Physlets, animations, and other teaching tools. This manual will be available on the book's Web site at www.whfreeman.com/tipler5e.

Instructor's Solutions Manual: *Vol. 1, 0-7167-9640-6; Vol. 2, 0-7167-9639-2.* This guide contains fully worked solutions for all of the problems in the textbook, using the side-by-side format wherever possible. It is available in print and is also included in editable Word files on the Instructor's CD-ROM.

Test Bank: *In print, 0-7167-9652-X; CD-ROM, 0-7167-9653-8.* Prepared by Mark Riley of Florida State University and David Mills of College of the Redwoods, this set of more than 4,000 multiple choice questions is available both in print and on a CD-ROM for Windows and Macintosh users. All questions refer to specific sections in the book. The CD-ROM version of the Test Bank makes it easy to add, edit and re-sequence questions to suit your needs.

Transparencies: *0-7167-9664-3.* Approximately 150 full color acetates of figures and tables from the text are included, with type enlarged for projection.

Acknowledgments

We are grateful to the many instructors, students, colleagues, and friends who have contributed to this, and to earlier editions.

Charles Adler of St. Mary's College of Maryland authored the excellent new problems. David Mills of the College of the Redwoods extensively revised the solutions manual. Robin Jordan of Florida Atlantic University created the innovative Master the Concept exercises and iSOLVE Checkpoint problems. Laura McCullough of the University of Wisconsin, Stout, and Thomas Foster of Southern Illinois University, Edwardsville, drawing from their background in Physics Education Research, were instrumental in providing context-rich examples in every chapter as well as our new Estimation and Approximation problems. We received invaluable help in accuracy checking of text and problems from professors:

Karamjeet Arya,
San Jose State University

Michael Crivello,
San Diego Mesa College

David Faust,
Mt. Hood Community College

Jerome Licini,
Lehigh University

Dan Lucas,
University of Wisconsin

Jeannette Myers,
Clemson University

Marian Peters,
Appalachian State University

Paul Quinn,
Kutztown University

Michael G. Strauss,
University of Oklahoma

George Zober,
Yough Senior High School

Patricia Zober,
Ringgold High School

Many instructors and students have provided extensive and helpful reviews of one or more chapters. They have each made a fundamental contribution to the quality of this revision, and deserve our gratitude. We would like to thank the following reviewers:

Edward Adelson,
The Ohio State University

Todd Averett,
The College of William and Mary

Yildirim M. Aktas,
University of North Carolina at Charlotte

Karamjeet Arya,
San Jose State University

Alison Baski,
Virginia Commonwealth University

Gary Stephen Blanpied,
University of South Carolina

Ronald Brown,
California Polytechnic State University

Robert Coakley,
University of Southern Maine

Robert Coleman,
Emory University

Andrew Cornelius,
University of Nevada at Las Vegas

Peter P. Crooker,
University of Hawaii

N. John DiNardo,
Drexel University

William Ellis,
University of Technology - Sydney

John W. Farley,
University of Nevada at Las Vegas

David Flammer,
Colorado School of Mines

Tom Furtak,
Colorado School of Mines

Patrick C. Gibbons,
Washington University

John B. Gruber,
San Jose State University

Christopher Gould,
University of Southern California

Phuoc Ha,
Creighton University

Theresa Peggy Hartsell,
Clark College

James W. Johnson,
Tallahassee Community College

Thomas O. Krause,
Towson University

Donald C. Larson,
Drexel University

Paul L. Lee,
California State University, Northridge

Peter M. Levy,
New York University

Jerome Licini,
Lehigh University

Edward McCliment,
University of Iowa

Robert R. Marchini,
The University of Memphis

Pete E.C. Markowitz,
Florida International University

Fernando Medina,
Florida Atlantic University

Laura McCullough,
University of Wisconsin at Stout

John W. Norbury,
University of Wisconsin at Milwaukee

Melvyn Jay Oremland,
Pace University

Antonio Pagnamenta,
University of Illinois at Chicago

John Parsons,
Columbia University

Dinko Pocanic,
University of Virginia

Bernard G. Pope,
Michigan State University

Yong-Zhong Qian,
University of Minnesota

Ajit S. Rupaal,
Western Washington University

Todd G. Ruskell,
Colorado School of Mines

Mesgun Sebhatu,
Winthrop University

Marllin L. Simon,
Auburn University

Zbigniew M. Stadnik,
University of Ottawa

G. R. Stewart,
University of Florida

Michael G. Strauss,
University of Oklahoma

Chin-Che Tin,
Auburn University

Stephen Weppner,
Eckerd College

Suzanne E. Willis,
Northern Illinois University

Ron Zammit,
California Polytechnic State University

Problems/solutions reviewers

Lay Nam Chang,
Virginia Polytechnic Institute

Mark W. Coffey,
Colorado School of Mines

Brent A. Corbin,
UCLA

Alan Cresswell,
Shippensburg University

Ricardo S. Decca,
Indiana University-Purdue University

Michael Dubson,
University of Colorado at Boulder

David Faust,
Mount Hood Community College

Philip Fraundorf,
University of Missouri, Saint Louis

Clint Harper,
Moorpark College

Kristi R.G. Hendrickson,
University of Puget Sound

Michael Hildreth,
University of Notre Dame

David Ingram,
Ohio University

James J. Kolata,
University of Notre Dame

Eric Lane,
University of Tennessee, Chattanooga

Jerome Licini,
Lehigh University

Daniel Marlow,
Princeton University

Laura McCullough,
University of Wisconsin at Stout

Carl Mungan,
United States Naval Academy

Jeffry S. Olafsen,
University of Kansas

Robert Pompi,
The State University of New York
at Binghamton

R. J. Rollefson,
Wesleyan University

Andrew Scherbakov,
Georgia Institute of Technology

Bruce A. Schumm,
University of California, Santa Cruz

Dan Styer,
Oberlin College

Jeffrey Sundquist,
Palm Beach Community College - South

Cyrus Taylor,
Case Western Reserve University

Fulin Zuo,
University of Miami

Study Guide & Test Bank reviewers

Anthony J. Buffa,
California Polytechnic State University

Mirela S. Fetea,
University of Richmond

James Garner,
University of North Florida

Tina Harriott,
Mount Saint Vincent, Canada

Roger King,
City College of San Francisco

John A. McClelland,
University of Richmond

Chun Fu Su,
Mississippi State University

John A. Underwood,
Austin Community College

Media reviewers

Mick Arnett,
Kirkwood Community College

Colonel Rolf Enger,
U.S. Air Force Academy

John W. Farley,
The University of Nevada at Las Vegas

David Ingram,
The Ohio State University

Shawn Jackson,
The University of Tulsa

Dan MacIsaac,
Northern Arizona University

Peter E.C. Markowitz,
Florida International University

Dean Zollman,
Kansas State University

Media focus group participants

Edwin R. Jones,
University of South Carolina

William C. Kerr,
Wake Forest University

Taha Mzoughi,
Mississippi State University

Charles Niederriter,
Gustavus Adolphus College

Cindy Schwarz,
Vassar College

Dave Smith,
University of the Virgin Islands

D.J. Wagner,
Grove City College

George Watson,
University of Delaware

Frank Wolfs,
University of Rochester

We also remain indebted to the reviewers of past editions. We would therefore like to thank the following reviewers, who provided immeasurable support as we developed the fourth edition:

Michael Arnett,
Iowa State University

William Bassichis,
Texas A&M

Joel C. Berlinghieri,
The Citadel

Frank Blatt,
Michigan State University

John E. Byrne,
Gonzaga University

Wayne Carr,
Stevens Institute of Technology

George Cassidy,
University of Utah

I.V. Chivets,
Trinity College, University of Dublin

Harry T. Chu,
University of Akron

Jeff Culbert,
London, Ontario

Paul Debevec,
University of Illinois

Robert W. Detenbeck,
University of Vermont

Bruce Doak,
Arizona State University

John Elliott,
University of Manchester, England

James Garland,
Retired

Ian Gatland,
Georgia Institute of Technology

Ron Gautreau,
New Jersey Institute of Technology

David Gavenda,
University of Texas at Austin

Newton Greenberg,
SUNY Binghamton

Huidong Guo,
Columbia University

Richard Haracz,
Drexel University

Michael Harris,
University of Washington

Randy Harris,
University of California at Davis

Dieter Hartmann,
Clemson University

Robert Hollebeek,
University of Pennsylvania

Madya Jalil,
University of Malaya

Monwhea Jeng,
University of California – Santa Barbara

Ilon Joseph,
Columbia University

David Kaplan,
University of California – Santa Barbara

John Kidder,
Dartmouth College

Boris Korsunsky,
Northfield Mt. Hermon School

Andrew Lang (graduate student),
University of Missouri

David Lange,
University of California – Santa Barbara

Isaac Leichter,
Jerusalem College of Technology

William Lichten,
Yale University

Robert Lieberman,
Cornell University

Fred Lipschultz,
University of Connecticut

Graeme Luke,
Columbia University

Howard McAllister,
University of Hawaii

M. Howard Miles,
Washington State University

Matthew Moelter,
University of Puget Sound

Eugene Mosca,
United States Naval Academy

Aileen O'Donughue,
St. Lawrence University

Jack Ord,
University of Waterloo

Richard Packard,
University of California

George W. Parker,
North Carolina State University

Edward Pollack,
University of Connecticut

John M. Pratte,
Clayton College and State University

Brooke Pridmore,
Clayton State College

David Roberts,
Brandeis University

Lyle D. Roelofs,
Haverford College

Larry Rowan,
University of North Carolina
 at Chapel Hill

Lewis H. Ryder,
University of Kent, Canterbury

Bernd Schuttler,
University of Georgia

Cindy Schwarz,
Vassar College

Murray Scureman,
Amdahl Corporation

Scott Sinawi,
Columbia University

Wesley H. Smith,
University of Wisconsin

Kevork Spartalian,
University of Vermont

Kaare Stegavik,
University of Trondheim, Norway

Jay D. Strieb,
Villanova University

Martin Tiersten,
City College of New York

Oscar Vilches,
University of Washington

Fred Watts,
College of Charleston

John Weinstein,
University of Mississippi

David Gordon,
Wilson, MIT

David Winter,
Columbia University

Frank L.H. Wolfe,
University of Rochester

Roy C. Wood,
New Mexico State University

Yuriy Zhestkov,
Columbia University

Of course, our work is never done. We hope to continue to receive comments and suggestions from our readers so that we can improve the text and correct any errors. If you believe you have found an error, or have any other comments, suggestions, or questions, send us a note at asktipler@whfreeman.com. We will incorporate corrections into the text during subsequent reprinting.

Finally, we would like to thank our friends at W. H. Freeman and Company for their help and encouragement. Susan Brennan, Kathleen Civetta, Georgia Lee Hadler, Julia DeRosa, Margaret Comaskey, Dena Betz, Rebecca Pearce, Brian Donnellan, Jennifer Van Hove, Patricia Marx, and Mark Santee were extremely generous with their creativity and hard work at every stage of the process. We are also grateful for the contributions of Cathy Townsend and Denise Kadlubowski at PreMediaONE and the help of our colleagues Larry Tankersley, John Ertel, Steve Montgomery, and Don Treacy.

Paul Tipler
Alameda, California

Gene Mosca
Annapolis, Maryland

ABOUT THE AUTHORS

PAUL A TIPLER

Paul Tipler was born in the small farming town of Antigo, Wisconsin, in 1933. He graduated from high school in Oshkosh, Wisconsin, where his father was superintendent of the Public Schools. He received his B.S. from Purdue University in 1955 and his Ph.D. at the University of Illinois in 1962, where he studied the structure of nuclei. He taught for one year at Wesleyan University in Connecticut while writing his thesis, then moved to Oakland University in Michigan, where he was one of the original members of the Physics department, playing a major role in developing the physics curriculum. During the next 20 years, he taught nearly all the physics courses and wrote the first and second editions of his widely used textbooks *Modern Physics* (1969, 1978) and *Physics* (1976, 1982). In 1982, he moved to Berkeley, California, where he now resides, and where he wrote *College Physics* (1987) and the third edition of *Physics* (1991). In addition to physics, his interests include music, hiking, and camping, and he is an accomplished jazz pianist and poker player.

GENE MOSCA

Gene Mosca was born in New York City and grew up on Shelter Island, New York. His undergraduate studies were at Villanova University and his graduate studies were at the University of Michigan and the University of Vermont, where he received his Ph.D. in 1974. He taught at Southampton High School, the University of South Dakota, and Emporia State University. Since 1986 Gene has been teaching at the U.S. Naval Academy. There he coordinated the core physics course for 16 semesters, and instituted numerous enhancements to both the laboratory and classroom. Proclaimed by Paul Tipler as, "the best reviewer I ever had," Mosca authored the popular Study Guide for the third and fourth editions of the text.

PART IV ELECTRICITY AND MAGNETISM

The Electric Field I: Discrete Charge Distributions

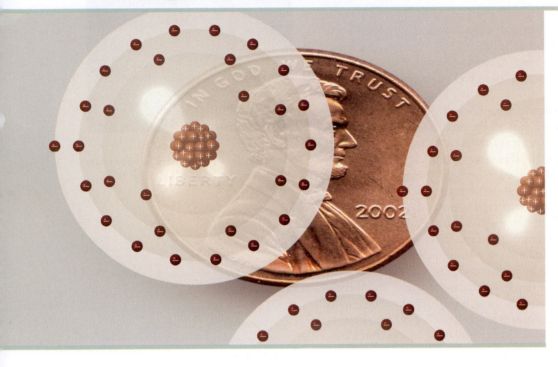

COPPER IS A CONDUCTOR, A MATERIAL WITH SPECIFIC PROPERTIES WE FIND USEFUL BECAUSE THESE PROPERTIES MAKE IT POSSIBLE TO TRANSPORT ELECTRICITY. THE ELECTRICITY WE HARNESS TO POWER MACHINES IS ALSO RESPONSIBLE FOR THE COPPER ATOM ITSELF: ATOMS ARE HELD TOGETHER BY ELECTRICAL FORCES.

? **What is the total charge of all the electrons in a penny? (See Example 21-1.)**

While just a century ago we had nothing more than a few electric lights, we are now extremely dependent on electricity in our daily lives. Yet, although the use of electricity has only recently become widespread, the study of electricity has a history reaching long before the first electric lamp glowed. Observations of electrical attraction can be traced back to the ancient Greeks, who noticed that after amber was rubbed, it attracted small objects such as straw or feathers. Indeed, the word *electric* comes from the Greek word for amber, *elektron*.

➤ In this chapter, we begin our study of electricity with *electrostatics*, the study of electrical charges at rest. After introducing the concept of electric charge, we briefly look at conductors and insulators and how conductors can be given a net charge. We then study Coulomb's law, which describes the force exerted by one electric charge on another. Next, we introduce the electric field and show how it can be visualized by electric field lines that indicate the magnitude and direction of the field, just as we visualized the velocity field of a flowing fluid using streamlines (Chapter 13). Finally, we discuss the behavior of point charges and electric dipoles in electric fields.

21-1 Electric Charge

Suppose we rub a hard rubber rod with fur and then suspend the rod from a string so that it is free to rotate. Now we bring a second similarly rubbed hard rubber rod near it. The rods repel each other (Figure 21-1). We get the same results if we use two glass rods that have been rubbed with silk. But, when we place a hard rubber rod rubbed with fur near a glass rod rubbed with silk they attract each other.

Rubbing a rod causes the rod to become electrically charged. If we repeat the experiment with various materials, we find that all charged objects fall into one of just two groups—those like the hard rubber rod rubbed with fur and those like the glass rod rubbed with silk. Objects from the same group repel each other, while objects from different groups attract each other. Benjamin Franklin explained this by proposing a model in which every object has a *normal* amount of electricity that can be transferred from one object to the other when two objects are in close contact, as when they are rubbed together. This leaves one object with an excess charge and the other with a deficiency of charge in the same amount as the excess. Franklin described the resulting charges with plus and minus signs, choosing positive to be the charge acquired by a glass rod when it is rubbed with a piece of silk. The piece of silk acquires a negative charge of equal magnitude during the procedure. Based on Franklin's convention, hard rubber rubbed with fur acquires a negative charge and the fur acquires a positive charge. Two objects that carry the same type of charge repel each other, and two objects that carry opposite charges attract each other (Figure 21-2).

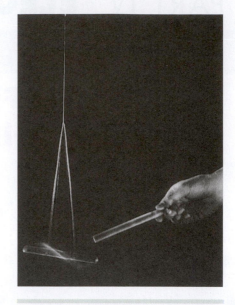

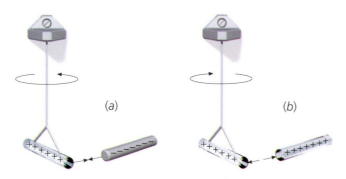

(a)   (b)

FIGURE 21-2
(*a*) Objects carrying charges of opposite sign attract each other. (*b*) Objects carrying charges of the same sign repel each other.

Today, we know that when glass is rubbed with silk, electrons are transferred from the glass to the silk. Because the silk is negatively charged (according to Franklin's convention, which we still use) electrons are said to carry a negative charge. Table 21-1 is a short version of the **triboelectric series.** (In Greek *tribos* means "a rubbing.") The further down the series a material is, the greater its affinity for electrons. If two of the materials are brought in contact, electrons are transferred from the material higher in the table to the one further down the table. For example, if Teflon is rubbed with nylon, electrons are transferred from the nylon to the Teflon.

Charge Quantization

Matter consists of atoms that are electrically neutral. Each atom has a tiny but massive nucleus that contains protons and neutrons. Protons are positively charged, whereas neutrons are uncharged. The number of protons in the nucleus

TABLE 21-1

The Triboelectric Series

+ Positive End of Series
Asbestos
Glass
Nylon
Wool
Lead
Silk
Aluminum
Paper
Cotton
Steel
Hard rubber
Nickel and copper
Brass and silver
Synthetic rubber
Orlon
Saran
Polyethylene
Teflon
Silicone rubber
– Negative End of Series

is the atomic number Z of the element. Surrounding the nucleus is an equal number of negatively charged electrons, leaving the atom with zero net charge. The electron is about 2000 times less massive than the proton, yet the charges of these two particles are exactly equal in magnitude. The charge of the proton is e and that of the electron is $-e$, where e is called the **fundamental unit of charge**. The charge of an electron or proton is an intrinsic property of the particle, just as mass and spin are intrinsic properties of these particles.

All observable charges occur in integral amounts of the fundamental unit of charge e; that is, *charge is quantized*. Any charge Q occurring in nature can be written $Q = \pm Ne$, where N is an integer.[†] For ordinary objects, however, N is usually very large and charge appears to be continuous, just as air appears to be continuous even though air consists of many discrete molecules. To give an everyday example of N, charging a plastic rod by rubbing it with a piece of fur typically transfers 10^{10} or more electrons to the rod.

Charge Conservation

When objects are rubbed together, one object is left with an excess number of electrons and is therefore negatively charged; the other object is left lacking electrons and is therefore positively charged. The net charge of the two objects remains constant; that is, *charge is conserved*. The **law of conservation of charge** is a fundamental law of nature. In certain interactions among elementary particles, charged particles such as electrons are created or annihilated. However, in these processes, equal amounts of positive and negative charge are produced or destroyed, so the net charge of the universe is unchanged.

The SI unit of charge is the coulomb, which is defined in terms of the unit of electric current, the ampere (A).[‡] The **coulomb** (C) is the amount of charge flowing through a wire in one second when the current in the wire is one ampere. The fundamental unit of electric charge e is related to the coulomb by

$$e = 1.602177 \times 10^{-19}\,\text{C} \approx 1.60 \times 10^{-19}\,\text{C} \qquad \text{21-1}$$

FUNDAMENTAL UNIT OF CHARGE

EXERCISE A charge of magnitude 50 nC (1 nC $= 10^{-9}$ C) can be produced in the laboratory by simply rubbing two objects together. How many electrons must be transferred to produce this charge?
(*Answer* $N = Q/e = (50 \times 10^{-9}\,\text{C})/(1.6 \times 10^{-19}\,\text{C}) = 3.12 \times 10^{11}$. Charge quantization cannot be detected in a charge of this size; even adding or subtracting a million electrons produces a negligibly small effect.)

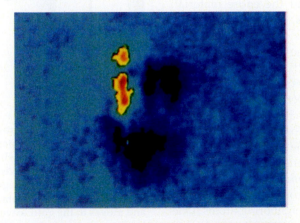

Charging by contact. A piece of plastic about 0.02 mm wide was charged by contact with a piece of nickel. Although the plastic carries a net positive charge, regions of negative charge (dark) as well as regions of positive charge (yellow) are indicated. The photograph was taken by sweeping a charged needle of width 10^{-7} m over the sample and recording the electrostatic force on the needle.

---

† In the standard model of elementary particles, protons, neutrons, and some other elementary particles are made up of more fundamental particles called quarks that carry charges of $\pm\frac{1}{3}e$ or $\pm\frac{2}{3}e$. Only combinations that result in a net charge of $\pm Ne$ or 0 are known.
‡ The ampere (A) is the unit of current used in everyday electrical work.

E X A M P L E   2 1 - 1

A copper penny† (Z = 29) has a mass of 3 grams. What is the total charge of all the electrons in the penny?

PICTURE THE PROBLEM The electrons have a total charge given by the number of electrons in the penny, N_e, times the charge of an electron, $-e$. The number of electrons is 29 (the atomic number of copper) times the number of copper atoms N. To find N, we use the fact that one mole of any substance has Avogadro's number ($N_A = 6.02 \times 10^{23}$) of molecules, and the number of grams in a mole is the molecular mass M, which is 63.5 g/mol for copper. Since each molecule of copper is just one copper atom, we find the number of atoms per gram by dividing N_A (atoms/mole) by M (grams/mole).

1. The total charge is the number of electrons times the electronic charge:

$$Q = N_e(-e)$$

2. The number of electrons is Z times the number of copper atoms N_a:

$$N_e = ZN_a$$

3. Compute the number of copper atoms in 3 g of copper:

$$N_a = (3 \text{ g})\frac{6.02 \times 10^{23} \text{ atoms/mol}}{63.5 \text{ g/mol}} = 2.84 \times 10^{22} \text{ atoms}$$

4. Compute the number of electrons N_e:

$$N_e = ZN_a = (29 \text{ electrons/atom})(2.84 \times 10^{22} \text{ atoms})$$
$$= 8.24 \times 10^{23} \text{ electrons}$$

5. Use this value of N_e to find the total charge:

$$Q = N_e(-e)$$
$$= (8.24 \times 10^{23} \text{ electrons})(-1.6 \times 10^{-19} \text{ C/electron})$$
$$= \boxed{-1.32 \times 10^5 \text{ C}}$$

EXERCISE If one million electrons were given to each man, woman, and child in the United States (about 285 million people), what percentage of the number of electrons in a penny would this represent? (*Answer*   About 35×10^{-9} percent)

21-2 Conductors and Insulators

In many materials, such as copper and other metals, some of the electrons are free to move about the entire material. Such materials are called **conductors**. In other materials, such as wood or glass, all the electrons are bound to nearby atoms and none can move freely. These materials are called **insulators**.

In a single atom of copper, 29 electrons are bound to the nucleus by the electrostatic attraction between the negatively charged electrons and the positively charged nucleus. The outer electrons are more weakly bound than the inner electrons because of their greater distance from the nucleus and because of the repulsive force exerted by the inner electrons. When a large number of copper atoms are combined in a piece of metallic copper, the binding of the electrons of each individual atom is reduced by interactions with neighboring atoms. One or more of the outer electrons in each atom is no longer bound

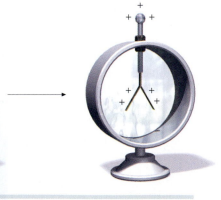

FIGURE 21-3 An electroscope. Two gold leaves are attached to a conducting post that has a conducting ball on top. The leaves are otherwise insulated from the container. When uncharged, the leaves hang together vertically. When the ball is touched by a negatively charged plastic rod, some of the negative charge from the rod is transferred to the ball and moves to the gold leaves, which then spread apart because of electrical repulsion between their negative charges. Touching the ball with a positively charged glass rod also causes the leaves to spread apart. In this case, the positively charged glass rod attracts electrons from the metal ball, leaving a net positive charge on the leaves.

---

† The penny was composed of 100 percent copper from 1793 to 1837. In 1982, the composition changed from 95 percent copper and 5 percent zinc to 2.5 percent copper and 97.5 percent zinc.

as a gas molecule is free to move about in a box. The number of free electrons depends on the particular metal, but it is typically about one per atom. An atom with an electron removed or added, resulting in a net charge on the atom, is called an **ion.** In metallic copper, the copper ions are arranged in a regular array called a *lattice.* A conductor is electrically neutral if for each lattice ion carrying a positive charge $+e$ there is a free electron carrying a negative charge $-e$. The net charge of the conductor can be changed by adding or removing electrons. A conductor with a negative net charge has an excess of free electrons, while a conductor with a positive net charge has a deficit of free electrons.

Charging by Induction

The conservation of charge is illustrated by a simple method of charging a conductor called **charging by induction,** as shown in Figure 21-4. Two uncharged metal spheres are in contact. When a charged rod is brought near one of the spheres, free electrons flow from one sphere to the other, toward a positively charged rod or away from a negatively charged rod. The positively charged rod in Figure 21-4a attracts the negatively charged electrons, and the sphere nearest the rod acquires electrons from the sphere farther away. This leaves the near sphere with a net negative charge and the far sphere with an equal net positive charge. A conductor that has *separated* equal and opposite charges is said to be **polarized.** If the spheres are separated before the rod is removed, they will be left with equal amounts of opposite charges (Figure 21-4b). A similar result would be obtained with a negatively charged rod, which would drive electrons from the near sphere to the far sphere.

EXERCISE Two identical conducting spheres, one with an initial charge $+Q$, the other initially uncharged, are brought into contact. (a) What is the new charge on each sphere? (b) While the spheres are in contact, a negatively charged rod is moved close to one sphere, causing it to have a charge of $+2Q$. What is the charge on the other sphere? (*Answer*   (a) $+\frac{1}{2}Q$. Since the spheres are identical, they must share the total charge equally. (b) $-Q$, which is necessary to satisfy the conservation of charge)

EXERCISE Two identical spheres are charged by induction and then separated; sphere 1 has charge $+Q$ and sphere 2 has charge $-Q$. A third identical sphere is initially uncharged. If sphere 3 is touched to sphere 1 and separated, then touched to sphere 2 and separated, what is the final charge on each of the three spheres? (*Answer* $Q_1 = +Q/2$, $Q_2 = -Q/4$, $Q_3 = -Q/4$)

For many purposes, the earth itself can be considered to be an infinitely large conductor with an abundant supply of free charge. If a conductor is connected to the earth, it is said to be **grounded** (indicated schematically in Figure 21-5b by a connecting wire ending in parallel horizontal lines). Figure 21-5 demonstrates

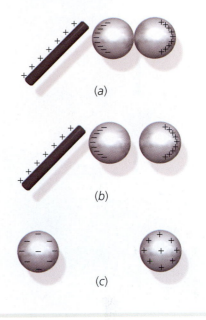

FIGURE 21-4 Charging by induction. (*a*) Conductors in contact become oppositely charged when a charged rod attracts electrons to the left sphere. (*b*) If the spheres are separated before the rod is removed, they will retain their equal and opposite charges. (*c*) When the rod is removed and the spheres are far apart, the distribution of charge on each sphere approaches uniformity.

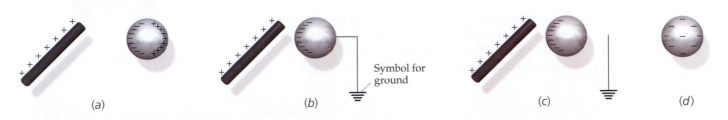

FIGURE 21-5 Induction via grounding. (*a*) The free charge on the single conducting sphere is polarized by the positively charged rod, which attracts negative charges on the sphere. (*b*) When the conductor is grounded by connecting it with a wire to a very large conductor, such as the earth, electrons from the ground neutralize the positive charge on the far face. The conductor is then negatively charged. (*c*) The negative charge remains if the connection to the ground is broken before the rod is removed. (*d*) After the rod is removed, the sphere has a uniform negative charge.

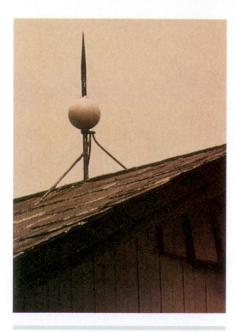

The lightning rod on this building is grounded so that it can conduct electrons from the ground to the positively charged clouds, thus neutralizing them.

These fashionable ladies are wearing hats with metal chains that drag along the ground, which were supposed to protect them from lightning.

how we can induce a charge in a single conductor by transferring charge from the earth through the ground wire and then breaking the connection to the ground.

21-3 Coulomb's Law

Charles Coulomb (1736–1806) studied the force exerted by one charge on another using a torsion balance of his own invention.[†] In Coulomb's experiment, the charged spheres were much smaller than the distance between them so that the charges could be treated as point charges. Coulomb used the method of charging by induction to produce equally charged spheres and to vary the amount of charge on the spheres. For example, beginning with charge q_0 on each sphere, he could reduce the charge to $\frac{1}{2}q_0$ by temporarily grounding one sphere to discharge it and then placing the two spheres in contact. The results of the experiments of Coulomb and others are summarized in **Coulomb's law:**

The force exerted by one point charge on another acts along the line between the charges. It varies inversely as the square of the distance separating the charges and is proportional to the product of the charges. The force is repulsive if the charges have the same sign and attractive if the charges have opposite signs.

COULOMB'S LAW

Coulomb's torsion balance.

---

† Coulomb's experimental apparatus was essentially the same as that described for the Cavendish experiment in Chapter 11, with the masses replaced by small charged spheres. For the magnitudes of charges easily transferred by rubbing, the gravitational attraction of the spheres is completely negligible compared with their electric attraction or repulsion.

The *magnitude* of the electric force exerted by a charge q_1 on another charge q_2 a distance r away is thus given by

$$F = \frac{k|q_1 q_2|}{r^2} \qquad \text{21-2}$$

where k is an experimentally determined constant called the **Coulomb constant,** which has the value

$$k = 8.99 \times 10^9 \, \text{N} \cdot \text{m}^2 / \text{C}^2 \qquad \text{21-3}$$

If q_1 is at position $\vec{r}_1$ and q_2 is at $\vec{r}_2$ (Figure 21-6), the force $\vec{F}_{1,2}$ exerted by q_1 on q_2 is

$$\vec{F}_{1,2} = \frac{k q_1 q_2}{r_{1,2}^2} \hat{r}_{1,2} \qquad \text{21-4}$$

COULOMB'S LAW FOR THE FORCE EXERTED BY q_1 ON q_2

where $\vec{r}_{1,2} = \vec{r}_2 - \vec{r}_1$ is the vector pointing from q_1 to q_2, and $\hat{r}_{1,2} = \vec{r}_{1,2}/r_{1,2}$ is a unit vector pointing from q_1 to q_2.

By Newton's third law, the force $\vec{F}_{2,1}$ exerted by q_2 on q_1 is the negative of $\vec{F}_{1,2}$. Note the similarity between Coulomb's law and Newton's law of gravity. (See Equation 11-3.) Both are inverse-square laws. But the gravitational force between two particles is proportional to the masses of the particles and is always attractive, whereas the electric force is proportional to the charges of the particles and is repulsive if the charges have the same sign and attractive if they have opposite signs.

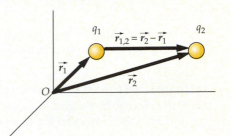

FIGURE 21-6 Charge q_1 at position $\vec{r}_1$ and charge q_2 at $\vec{r}_2$ relative to the origin O. The force exerted by q_1 on q_2 is in the direction of the vector $\vec{r}_{1,2} = \vec{r}_2 - \vec{r}_1$ if both charges have the same sign, and in the opposite direction if they have opposite signs.

ELECTRIC FORCE IN HYDROGEN   **EXAMPLE 21-2**

In a hydrogen atom, the electron is separated from the proton by an average distance of about 5.3×10^{-11} m. Calculate the magnitude of the electrostatic force of attraction exerted by the proton on the electron.

PICTURE THE PROBLEM Substitute the given values into Coulomb's law:

$$F = \frac{k|q_1 q_2|}{r^2} = \frac{k e^2}{r^2} = \frac{(8.99 \times 10^9 \, \text{N} \cdot \text{m}^2)(1.6 \times 10^{-19} \, \text{C})^2}{(5.3 \times 10^{-11} \, \text{m})^2}$$

$$= \boxed{8.19 \times 10^{-8} \, \text{N}}$$

REMARKS Compared with macroscopic interactions, this is a very small force. However, since the mass of the electron is only about 10^{-30} kg, this force produces an enormous acceleration of $F/m = 8 \times 10^{22}$ m/s^2.

EXERCISE Two point charges of 0.05 μC each are separated by 10 cm. Find the magnitude of the force exerted by one point charge on the other. (*Answer* 2.25×10^{-3} N)

Since the electrical force and the gravitational force between any two particles both vary inversely with the square of the separation between the particles, the ratio of these forces is independent of separation. We can therefore compare the relative strengths of the electrical and gravitational forces for elementary particles such as the electron and proton.

RATIO OF ELECTRIC AND GRAVITATIONAL FORCES   **E X A M P L E   21 - 3**

Compute the ratio of the electric force to the gravitational force exerted by a proton on an electron in a hydrogen atom.

PICTURE THE PROBLEM We use Coulomb's law with $q_1 = e$ and $q_2 = -e$ to find the electric force, and Newton's law of gravity with the mass of the proton, $m_p = 1.67 \times 10^{-27}$ kg, and the mass of the electron, $m_e = 9.11 \times 10^{-31}$ kg.

1. Express the magnitudes of the electric force F_e and the gravitational force F_g in terms of the charges, masses, separation distance r, and electrical and gravitational constants:

$$F_e = \frac{ke^2}{r^2}; F_g = \frac{Gm_p m_e}{r^2}$$

2. Take the ratio. Note that the separation distance r cancels:

$$\frac{F_e}{F_g} = \frac{ke^2}{Gm_p m_e}$$

3. Substitute numerical values:

$$\frac{F_e}{F_g} = \frac{(8.99 \times 10^9 \text{ N·m}^2/\text{C}^2)(1.6 \times 10^{-19} \text{ C})^2}{(6.67 \times 10^{-11} \text{ N·m}^2/\text{kg}^2)(1.67 \times 10^{-27} \text{ kg})(9.11 \times 10^{-31} \text{ kg})}$$

$$= \boxed{2.27 \times 10^{39}}$$

REMARKS This result shows why the effects of gravity are not considered when discussing atomic or molecular interactions.

Although the gravitational force is incredibly weak compared with the electric force and plays essentially no role at the atomic level, it is the dominant force between large objects such as planets and stars. Because large objects contain almost equal numbers of positive and negative charges, the attractive and repulsive electrical forces cancel. The net force between astronomical objects is therefore essentially the force of gravitational attraction alone.

Force Exerted by a System of Charges

In a system of charges, each charge exerts a force given by Equation 21-4 on every other charge. The net force on any charge is the vector sum of the individual forces exerted on that charge by all the other charges in the system. This follows from the principle of superposition of forces.

NET FORCE   **E X A M P L E   21 - 4**   **Try It Yourself**

Three point charges lie on the x axis; q_1 is at the origin, q_2 is at $x = 2$ m, and q_0 is at position x ($x > 2$m).

(a) Find the net force on q_0 due to q_1 and q_2 if $q_1 = +25$ nC, $q_2 = -10$ nC and $x = 3.5$ m (Figure 21-7).

(b) Find an expression for the net force on q_0 due to q_1 and q_2 throughout the region 2 m $< x < \infty$ (Figure 21-8).

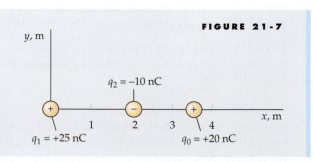

FIGURE 21-7

PICTURE THE PROBLEM The net force on q_0 is the vector sum of the force $\vec{F}_{1,0}$ exerted by q_1, and the force $\vec{F}_{2,0}$ exerted by q_2. The individual forces are found using Coulomb's law. Note that $\hat{r}_{1,0} = \hat{r}_{2,0} = \hat{i}$ because both $\vec{r}_{1,0}$ and $\vec{r}_{2,0}$ are in the positive x direction.

Cover the column to the right and try these on your own before looking at the answers.

Steps   **Answers**

(a) 1. Draw a sketch of the system of charges.
Label the distances $r_{1,0}$ and $r_{2,0}$.

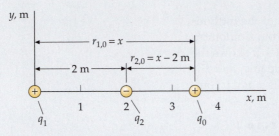

FIGURE 21-8

2. Find the force $\vec{F}_{1,0}$ due to q_1. $\vec{F}_{1,0} = (0.367\ \mu\text{N})\,\hat{i}$

3. Find the force $\vec{F}_{2,0}$ due to q_2. $\vec{F}_{2,0} = (-0.799\ \mu\text{N})\,\hat{i}$

4. Combine your results to obtain the net force. $\vec{F}_{net} = \vec{F}_{1,0} + \vec{F}_{2,0} = \boxed{-(0.432\ \mu\text{N}\hat{i})}$

(b) 1. Find an expression for the force due to q_1. $\vec{F}_{1,0} = \dfrac{kq_1q_0}{x^2}\,\hat{i}$

2. Find an expression for the force due to q_2. $\vec{F}_{1,0} = \dfrac{kq_2q_0}{(x-2\text{ m})^2}\,\hat{i}$

3. Combine your results to obtain an expression for the net force. $\vec{F}_{net} = \vec{F}_{1,0} + \vec{F}_{2,0} = \left(\dfrac{kq_1q_0}{x^2} + \dfrac{kq_2q_0}{(x-2\text{ m})^2}\right)\hat{i}$

REMARKS Figure 21-9 shows the x component of the force F_x on q_0 as a function of the position x of q_0 throughout the region $2\text{ m} < x < \infty$. Near q_2 the force due to q_2 dominates, and because opposite charges attract the force on q_2 is in the negative x direction. For $x \gg 2\text{m}$ the force is in the positive x direction. This is because for large x the distance between q_1 and q_2 is negligible so the force due to the two charges is almost the same as that for a single charge of $+15$ nC.

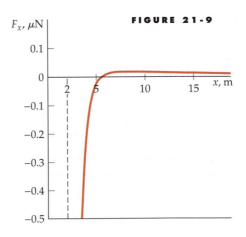

FIGURE 21-9

EXERCISE If q_0 is at $x = 1$ m, find (a) $\hat{r}_{1,0}$, (b) $\hat{r}_{2,0}$, and (c) the net force acting on q_0. (*Answer*   (a) $\hat{i}$, (b) $-\hat{i}$, (c) $(6.29\ \mu\text{N})\hat{i}$)

If a system of charges is to remain stationary, then there must be other forces acting on the charges so that the net force from all sources acting on each charge is zero. In the preceding example, and those that follow throughout the book, we assume that there are such forces so that all the charges remain stationary.

NET FORCE IN TWO DIMENSIONS   **EXAMPLE   21-5**

Charge $q_1 = +25$ nC is at the origin, charge $q_2 = -15$ nC is on the x axis at $x = 2$ m, and charge $q_0 = +20$ nC is at the point $x = 2$ m, $y = 2$ m as shown in Figure 21-10. Find the magnitude and direction of the resultant force $\Sigma \vec{F}$ on q_0.

PICTURE THE PROBLEM The resultant force is the vector sum of the individual forces exerted by each charge on q_0. We compute each force from Coulomb's law and write it in terms of its rectangular components. Figure 21-10*a* shows the resultant force on charge q_0 as the vector sum of the forces $\vec{F}_{1,0}$ due to q_1 and $\vec{F}_{2,0}$ due to q_2. Figure 21-10*b* shows the net force in Figure 21-10*a* and its x and y components.

FIGURE 21-10

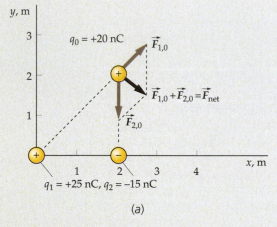

(a)

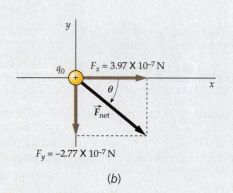

(b)

1. Draw the coordinate axes showing the positions of the three charges. Show the resultant force on charge q_0 as the vector sum of the forces $\vec{F}_{1,0}$ due to q_1 and $\vec{F}_{2,0}$ due to q_2.

2. The resultant force $\Sigma \vec{F}$ on q_0 is the sum of the individual forces:

$$\Sigma \vec{F} = \vec{F}_{1,0} + \vec{F}_{2,0}$$

$$\Sigma F_x = F_{1,0\,x} + F_{2,0\,x}$$

$$\Sigma F_y = F_{1,0\,y} + F_{2,0\,y}$$

3. The force $\vec{F}_{1,0}$ is directed along the line from q_1 to q_0. Use $r_{1,0} = 2\sqrt{2}$ for the distance between q_1 and q_0 to calculate its magnitude:

$$F_{1,0} = \frac{k|q_1 q_0|}{r_{1,0}^2}$$

$$= \frac{(8.99 \times 10^9 \text{ N·m}^2/\text{C}^2)(25 \times 10^{-9} \text{ C})(20 \times 10^{-9} \text{ C})}{(2\sqrt{2} \text{ m})^2}$$

$$= 5.62 \times 10^{-7} \text{ N}$$

4. Since $\vec{F}_{1,0}$ makes an angle of 45° with the x and y axes, its x and y components are equal to each other:

$$F_{1,0x} = F_{1,0y} = F_{1,0}\cos 45° = \frac{5.62 \times 10^{-7} \text{ N}}{\sqrt{2}} = 3.97 \times 10^{-7} \text{ N}$$

5. The force $\vec{F}_{2,0}$ exerted by q_2 on q_0 is attractive and in the negative y direction as shown in Figure 21-10*a*:

$$\vec{F}_{2,0} = \frac{kq_2 q_0}{r_{2,0}^2}\hat{r}_{2,0}$$

$$= \frac{(8.99 \times 10^9 \text{ N·m}^2/\text{C}^2)(-15 \times 10^{-9} \text{ C})(20 \times 10^{-9} \text{ C})}{(2 \text{ m})^2}\hat{j}$$

$$= (-6.74 \times 10^{-7} \text{ N})\hat{j}$$

6. Calculate the components of the resultant force:

$$\Sigma F_x = F_{1,0\,x} + F_{2,0\,x} = (3.97 \times 10^{-7} \text{ N}) + 0 = 3.97 \times 10^{-7} \text{ N}$$

$$\Sigma F_y = F_{1,0\,y} + F_{2,0\,y} = (3.97 \times 10^{-7} \text{ N}) + (-6.74 \times 10^{-7} \text{ N})$$

$$= -2.77 \times 10^{-7} \text{ N}$$

7. Draw the resultant force along with its two components:

8. The magnitude of the resultant force is found from its components:

$$F = \sqrt{F_x^2 + F_y^2} = \sqrt{(3.97 \times 10^{-7}\,\text{N})^2 + (-2.77 \times 10^{-7}\,\text{N})^2}$$

$$= \boxed{4.84 \times 10^{-7}\,\text{N}}$$

9. The resultant force points to the right and downward as shown in Figure 21-10b, making an angle θ with the x axis given by:

$$\tan\theta = \frac{F_y}{F_x} = \frac{-2.77}{3.97} = -0.698$$

$$\theta = \boxed{-34.9°}$$

EXERCISE Express $\hat{r}_{1,0}$ in Example 21-5 in terms of $\hat{i}$ and $\hat{j}$. [*Answer* $\hat{r}_{1,0} = (\hat{i} + \hat{j})/\sqrt{2}$]

21-4 The Electric Field

The electric force exerted by one charge on another is an example of an action-at-a-distance force, similar to the gravitational force exerted by one mass on another. The idea of action at a distance presents a difficult conceptual problem. What is the mechanism by which one particle can exert a force on another across the empty space between the particles? Suppose that a charged particle at some point is suddenly moved. Does the force exerted on the second particle some distance r away change instantaneously? To avoid the problem of action at a distance, the concept of the **electric field** is introduced. One charge produces an electric field $\vec{E}$ everywhere in space, and this field exerts the force on the second charge. Thus, it is the *field* $\vec{E}$ at the position of the second charge that exerts the force on it, not the first charge itself which is some distance away. Changes in the field propagate through space at the speed of light, c. Thus, if a charge is suddenly moved, the force it exerts on a second charge a distance r away does not change until a time r/c later.

Figure 21-11 shows a set of point charges, q_1, q_2, and q_3, arbitrarily arranged in space. These charges produce an electric field $\vec{E}$ everywhere in space. If we place a small positive **test charge** q_0 at some point near the three charges, there will be a force exerted on q_0 due to the other charges.[†] The net force on q_0 is the vector sum of the individual forces exerted on q_0 by each of the other charges in the system. Because each of these forces is proportional to q_0, the net force will be proportional to q_0. The electric field $\vec{E}$ at a point is this force divided by q_0:[‡]

$$\vec{E} = \frac{\vec{F}}{q_0} \quad (q_0 \text{ small})$$

21-5

DEFINITION—ELECTRIC FIELD

The SI unit of the electric field is the newton per coulomb (N/C). Table 21-2 lists the magnitudes of some of the electric fields found in nature.

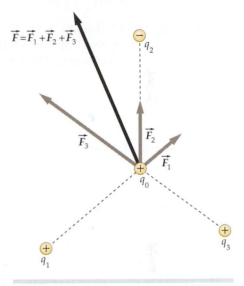

FIGURE 21-11 A small test charge q_0 in the vicinity of a system of charges q_1, q_2, q_3, ... experiences a force $\vec{F}$ that is proportional to q_0. The ratio $\vec{F}/q_0$ is the electric field at that point.

TABLE 21-2

Some Electric Fields in Nature

	E, N/C
In household wires	10^{-2}
In radio waves	10^{-1}
In the atmosphere	10^2
In sunlight	10^3
Under a thundercloud	10^4
In a lightning bolt	10^4
In an X-ray tube	10^6
At the electron in a hydrogen atom	6×10^{11}
At the surface of a uranium nucleus	2×10^{21}

---

† The presence of the charge q_0 will generally change the original distribution of the other charges, particularly if the charges are on conductors. However, we may choose q_0 to be small enough so that its effect on the original charge distribution is negligible.

‡ This definition is similar to that for the gravitational field of the earth, which was defined in Section 4-3 as the force per unit mass exerted by the earth on an object.

The electric field describes the condition in space set up by the system of point charges. By moving a test charge q_0 from point to point, we can find $\vec{E}$ at all points in space (except at any point occupied by a charge q). The electric field $\vec{E}$ is thus a vector function of position. The force exerted on a test charge q_0 at any point is related to the electric field at that point by

$$\vec{F} = q_0\vec{E}$$ 21-6

EXERCISE When a 5-nC test charge is placed at a certain point, it experiences a force of 2×10^{-4} N in the direction of increasing x. What is the electric field $\vec{E}$ at that point? [*Answer* $\vec{E} = \vec{F}/q_0 = (4 \times 10^4 \text{ N/C})\hat{i}$]

EXERCISE What is the force on an electron placed at a point where the electric field is $\vec{E} = (4 \times 10^4 \text{ N/C})\hat{i}$? [*Answer* $(-6.4 \times 10^{-15} \text{ N})\hat{i}$]

The electric field due to a single point charge can be calculated from Coulomb's law. Consider a small, positive test charge q_0 at some point P a distance $r_{i,P}$ away from a charge q_i. The force on it is

$$\vec{F}_{i,0} = \frac{kq_iq_0}{r_{i,P}^2}\hat{r}_{i,P}$$

The electric field at point P due to charge q_i (Figure 21-12) is thus

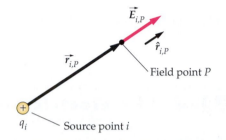

$$\vec{E}_{i,P} = \frac{kq_i}{r_{i,P}^2}\hat{r}_{i,P}$$ 21-7

COULOMB'S LAW FOR $\vec{E}$ DUE TO A POINT CHARGE

FIGURE 21-12 The electric field $\vec{E}$ at a field point P due to charge q_i at a source point i.

where $\hat{r}_{i,P}$ is the unit vector pointing from the **source point** i to the **field point** P. The net electric field due to a distribution of point charges is found by summing the fields due to each charge separately:

$$\vec{E}_P = \sum_i\vec{E}_{i,P} = \sum_i\frac{kq_i}{r_{i,P}^2}\hat{r}_{i,P}$$ 21-8

ELECTRIC FIELD $\vec{E}$ DUE TO A SYSTEM OF POINT CHARGES

ELECTRIC FIELD ON A LINE THROUGH TWO POSITIVE CHARGES **EXAMPLE 21-6**

A positive charge $q_1 = +8$ nC is at the origin, and a second positive charge $q_2 = +12$ nC is on the x axis at $a = 4$ m (Figure 21-13). Find the net electric field (*a*) at point P_1 on the x axis at $x = 7$ m, and (*b*) at point P_2 on the x axis at $x = 3$ m.

FIGURE 21-13

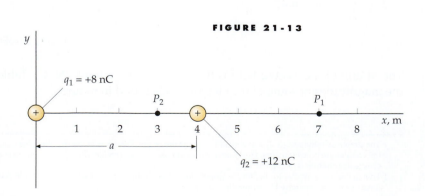

PICTURE THE PROBLEM Because point P_1 is to the right of both charges, each charge produces a field to the right at that point. At point P_2, which is between the charges, the 5-nC charge gives a field to the right and the 12-nC charge gives a field to the left. We calculate each field using

$$\vec{E} = \sum_i \frac{kq_i}{r_{i,P}^2}\hat{r}_{i,P}$$

At point P_1, both unit vectors point along the x axis in the positive direction, so $\hat{r}_{1,P_1} = \hat{r}_{2,P_1} = \hat{i}$. At point P_2, $\hat{r}_{1,P_2} = \hat{i}$, but the unit vector from the 12-nC charge points along the negative x direction, so $\hat{r}_{2,P_2} = -\hat{i}$.

1. Calculate $\vec{E}$ at point P_1, using $r_{1,P_1} = x = 7$ m and $r_{2,P_1} = (x - a) = 7\,\text{m} - 4\,\text{m} = 3\,\text{m}$:

$$\vec{E} = \frac{kq_1}{r_{1,P_1}^2}\hat{r}_{1,P_1} + \frac{kq_2}{r_{2,P_1}^2}\hat{r}_{2,P_1} = \frac{kq_1}{x^2}\hat{i} + \frac{kq_2}{(x-a)^2}\hat{i}$$

$$= \frac{(8.99 \times 10^9\,\text{N·m}^2/\text{C}^2)(8 \times 10^{-9}\text{C})}{(7\,\text{m})^2}\hat{i}$$

$$+ \frac{(8.99 \times 10^9\,\text{N·m}^2/\text{C}^2)(12 \times 10^{-9}\text{C})}{(3\,\text{m})^2}\hat{i}$$

$$= (1.47\,\text{N/C})\hat{i} + (12.0\,\text{N/C})\hat{i} = \boxed{(13.5\,\text{N/C})\hat{i}}$$

2. Calculate $\vec{E}$ at point P_2, where $r_{1,P_2} = x = 3$ m and $r_{2,P_2} = a - x = 4\,\text{m} - 3\,\text{m} = 1\,\text{m}$:

$$\vec{E} = \frac{kq_1}{r_{1,P_2}^2}\hat{r}_{1,P_2} + \frac{kq_2}{r_{2,P_2}^2}\hat{r}_{2,P_2} = \frac{kq_1}{x^2}\hat{i} + \frac{kq_2}{(a-x)^2}(-\hat{i})$$

$$= \frac{(8.99 \times 10^9\,\text{N·m}^2/\text{C}^2)(8 \times 10^{-9}\text{C})}{(3\,\text{m})^2}\hat{i}$$

$$+ \frac{(8.99 \times 10^9\,\text{N·m}^2/\text{C}^2)(12 \times 10^{-9}\text{C})}{(1\,\text{m})^2}(-\hat{i})$$

$$= (7.99\,\text{N/C})\hat{i} - (108\,\text{N/C})\hat{i} = \boxed{(-100\,\text{N/C})\hat{i}}$$

REMARKS The electric field at point P_2 is in the negative x direction because the field due to the +12-nC charge, which is 1 m away, is larger than that due to the +8-nC charge, which is 3 m away. The electric field at source points close to the +8-nC charge is dominated by the field due to the +8-nC charge. There is one point between the charges where the net electric field is zero. At this point, a test charge would experience no net force. A sketch of E_x versus x for this system is shown in Figure 21-14.

EXERCISE Find the point on the x axis where the electric field is zero. (*Answer* $x = 1.80$ m)

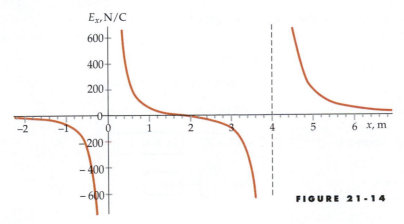

FIGURE 21-14

EXAMPLE 21-7 Try It Yourself

Find the electric field on the y axis at $y = 3$ m for the charges in Example 21-6.

PICTURE THE PROBLEM On the y axis, the electric field $\vec{E}_1$ due to charge q_1 is directed along the y axis, and the field $\vec{E}_2$ due to charge q_2 makes an angle θ with the y axis (Figure 21-15a). To find the resultant field, we first find the x and y components of these fields, as shown in Figure 21-15b.

FIGURE 21-15

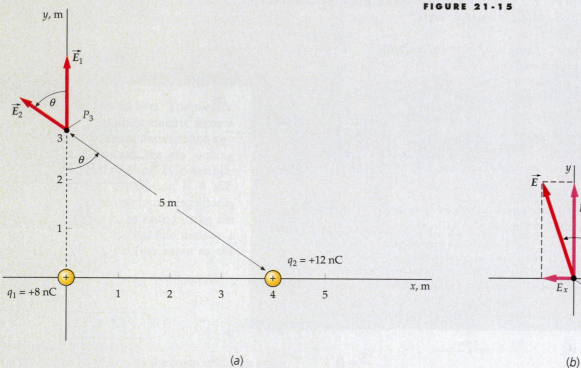

(a)

(b)

Cover the column to the right and try these on your own before looking at the answers.

Steps	Answers
1. Calculate the magnitude of the field $\vec{E}_1$ due to q_1. Find the x and y components of $\vec{E}_1$.	$E_1 = kq_1/y_2 = 7.99$ N/C $E_{1x} = 0, E_{1y} = 7.99$ N/C
2. Calculate the magnitude of the field $\vec{E}_2$ due to q_2.	$E_2 = 4.32$ N/C
3. Write the x and y components of $\vec{E}_2$ in terms of the angle θ.	$E_x = -E_2 \sin \theta; E_y = E_2 \cos \theta$
4. Compute $\sin \theta$ and $\cos \theta$.	$\sin \theta = 0.8; \cos \theta = 0.6$
5. Calculate E_{2x} and E_{2y}.	$E_{2x} = -3.46$ N/C; $E_{2y} = 2.59$ N/C
6. Find the x and y components of the resultant field $\vec{E}$.	$E_x = -3.46$ N/C; $E_y = 10.6$ N/C
7. Calculate the magnitude of $\vec{E}$ from its components.	$E = \sqrt{E_x^2 + E_y^2} = \boxed{11.2 \text{ N/C}}$
8. Find the angle θ_1 made by $\vec{E}$ with the x axis.	$\theta_1 = \tan^{-1}\left(\dfrac{E_y}{E_x}\right) = \boxed{108°}$

ELECTRIC FIELD DUE TO TWO EQUAL AND OPPOSITE CHARGES **EXAMPLE 21-8**

A charge $+q$ is at $x = a$ and a second charge $-q$ is at $x = -a$ (Figure 21-16). (a) Find the electric field on the x axis at an arbitrary point $x > a$. (b) Find the limiting form of the electric field for $x \gg a$.

PICTURE THE PROBLEM   We calculate the electric field using

$$\vec{E} = \sum_i \frac{kq_i}{r_{i,P}^2} \hat{r}_{i,P}$$

(Equation 21-8). For $x > a$, the unit vector for each charge is $\hat{i}$. The distances are $x - a$ to the plus charge and $x - (-a) = x + a$ to the minus charge.

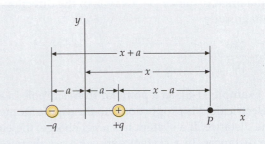

FIGURE 21-16

(a) 1. Draw the charge configuration on a coordinate axis and label the distances from each charge to the field point:

2. Calculate $\vec{E}$ due to the two charges for $x > a$: (*Note:* The equation on the right holds only for $x > a$. For $x < a$, the signs of the two terms are reversed. For $-a < x < a$, both terms have negative signs.)

$$\vec{E} = \frac{kq}{(x-a)^2}\hat{i} + \frac{k(-q)}{(x+a)^2}\hat{i}$$

$$= kq\left[\frac{1}{(x-a)^2} - \frac{1}{(x+a)^2}\right]\hat{i}$$

3. Put the terms in square brackets under a common denominator and simplify:

$$\vec{E} = kq\left[\frac{(x+a)^2 - (x-a)^2}{(x+a)^2(x-a)^2}\right]\hat{i} = \boxed{kq\frac{4ax}{(x^2-a^2)^2}\hat{i}}$$

(b) In the limit $x \gg a$, we can neglect a^2 compared with x^2 in the denominator:

$$\vec{E} = kq\frac{4ax}{(x^2-a^2)^2}\hat{i} \approx kq\frac{4ax}{x^4}\hat{i} = \boxed{\frac{4kqa}{x^3}\hat{i}}$$

REMARKS   Figure 21-17 shows E_x versus x for all x, for $q = 1$ nC and $a = 1$ m. Far from the charges, the field is given by

$$\vec{E} = \frac{4kqa}{|x|^3}\hat{i}$$

Between the charges, the contribution from each charge is in the negative direction. An expression that holds for all x is

$$\vec{E} = \frac{kq}{(x-a)^2}\left[\frac{(x-a)\hat{i}}{|x-a|}\right] + \frac{k(-q)}{(x+a)^2}\left[\frac{(x+a)\hat{i}}{|x+a|}\right]$$

Note that the unit vectors (quantities in square brackets in this expression) point in the proper direction for all x.

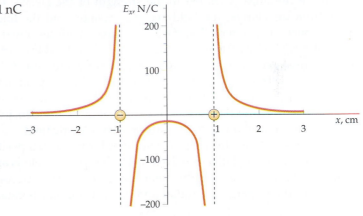

FIGURE 21-17   A plot of E_x versus x on the x axis for the charge distribution in Example 21-8.

Electric Dipoles

A system of two equal and opposite charges q separated by a small distance L is called an **electric dipole**. Its strength and orientation are described by the **electric dipole moment** $\vec{p}$, which is a vector that points from the negative charge to the positive charge and has the magnitude $q\vec{L}$ (Figure 21-18).

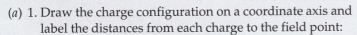

$$\boxed{\vec{p} = q\vec{L}} \qquad\qquad 21\text{-}9$$

DEFINITION—ELECTRIC DIPOLE MOMENT

where $\vec{L}$ is the vector from the negative charge to the positive charge.

FIGURE 21-18   An electric dipole consists of a pair of equal and opposite charges. The dipole moment is $\vec{p} = q\vec{L}$, where q is the magnitude of one of the charges and $\vec{L}$ is the relative position vector from the negative to the positive charge.

For the system of charges in Figure 21-16, $\vec{L} = 2a\hat{i}$ and the electric dipole moment is

$$\vec{p} = 2aq\hat{i}$$

In terms of the dipole moment, the electric field on the axis of the dipole at a point a great distance $|x|$ away is in the direction of the dipole moment and has the magnitude

$$E = \frac{2kp}{|x|^3} \qquad\qquad 21\text{-}10$$

(See Example 21-8). At a point far from a dipole in any direction, the magnitude of the electric field is proportional to the dipole moment and decreases with the cube of the distance. If a system has a net charge, the electric field decreases as $1/r^2$ at large distances. In a system with zero net charge, the electric field falls off more rapidly with distance. In the case of an electric dipole, the field falls off as $1/r^3$.

21-5 Electric Field Lines

We can picture the electric field by drawing lines to indicate its direction. At any given point, the field vector $\vec{E}$ is tangent to the lines. Electric field lines are also called **lines of force** because they show the direction of the force exerted on a positive test charge. At any point near a positive point charge, the electric field $\vec{E}$ points radially away from the charge. Consequently, the electric field lines near a positive charge also point away from the charge. Similarly, near a negative point charge the electric field lines point toward the negative charge.

Figure 21-19 shows the electric field lines of a single positive point charge. The spacing of the lines is related to the strength of the electric field. As we move away from the charge, the field becomes weaker and the lines become farther apart. Consider a spherical surface of radius r with its center at the charge. Its area is $4\pi r^2$. Thus, as r increases, the density of the field lines (the number of lines per unit area) decreases as $1/r^2$, the same rate of decrease as E. So, if we adopt the convention of drawing a fixed number of lines from a point charge, the number being proportional to the charge q, and if we draw the lines symmetrically about the point charge, the field strength is indicated by the density of the lines. The more closely spaced the lines, the stronger the electric field.

Figure 21-20 shows the electric field lines for two equal positive point charges q separated by a small distance. Near each charge, the field is approximately due to that charge alone because the other charge is far away. Consequently, the field lines near either charge are radial and equally spaced. Because the charges are

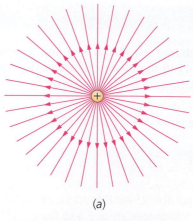

(a)

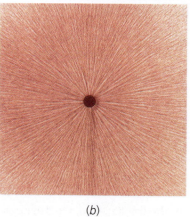

(b)

FIGURE 21-19 (*a*) Electric field lines of a single positive point charge. If the charge were negative, the arrows would be reversed. (*b*) The same electric field lines shown by bits of thread suspended in oil. The electric field of the charged object in the center induces opposite charges on the ends of each bit of thread, causing the threads to align themselves parallel to the field.

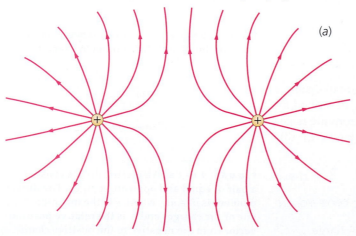

(a)

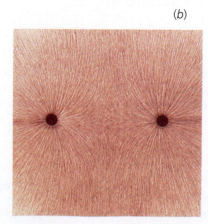

(b)

FIGURE 21-20 (*a*) Electric field lines due to two positive point charges. The arrows would be reversed if both charges were negative. (*b*) The same electric field lines shown by bits of thread in oil.

equal, we draw an equal number of lines originating from each charge. At very large distances, the details of the charge configuration are not important and the system looks like a point charge of magnitude $2q$. (For example, if the two charges were 1 mm apart and we were looking at them from a point 100 km away, they would look like a single charge.) So at a large distance from the charges, the field is approximately the same as that due to a point charge $2q$ and the lines are approximately equally spaced. Looking at Figure 21-20, we see that the density of field lines in the region between the two charges is small compared to the density of lines in the region just to the left and just to the right of the charges. This indicates that the magnitude of the electric field is weaker in the region between the charges than it is in the region just to the right or left of the charges, where the lines are more closely spaced. This information can also be obtained by direct calculation of the field at points in these regions.

We can apply this reasoning to draw the electric field lines for any system of point charges. Very near each charge, the field lines are equally spaced and leave or enter the charge radially, depending on the sign of the charge. Very far from all the charges, the detailed structure of the system is not important so the field lines are just like those of a single point charge carrying the net charge of the system. The rules for drawing electric field lines can be summarized as follows:

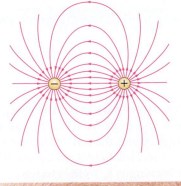

(a)

(b)

FIGURE 21-21 (*a*) Electric field lines for an electric dipole. (*b*) The same field lines shown by bits of thread in oil.

1. Electric field lines begin on positive charges (or at infinity) and end on negative charges (or at infinity).

2. The lines are drawn uniformly spaced entering or leaving an isolated point charge.

3. The number of lines leaving a positive charge or entering a negative charge is proportional to the magnitude of the charge.

4. The density of the lines (the number of lines per unit area perpendicular to the lines) at any point is proportional to the magnitude of the field at that point.

5. At large distances from a system of charges with a net charge, the field lines are equally spaced and radial, as if they came from a single point charge equal to the net charge of the system.

6. Field lines do not cross. (If two field lines crossed, that would indicate two directions for $\vec{E}$ at the point of intersection.)

RULES FOR DRAWING ELECTRIC FIELD LINES

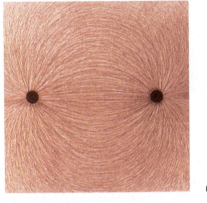

(a)

Figure 21-21 shows the electric field lines due to an electric dipole. Very near the positive charge, the lines are directed radially outward. Very near the negative charge, the lines are directed radially inward. Because the charges have equal magnitudes, the number of lines that begin at the positive charge equals the number that end at the negative charge. In this case, the field is strong in the region between the charges, as indicated by the high density of field lines in this region in the field.

Figure 21-22a shows the electric field lines for a negative charge $-q$ at a small distance from a positive charge $+2q$. Twice as many lines leave the positive charge as enter the negative charge. Thus, half the lines beginning on the positive charge $+2q$ enter the negative charge $-q$; the rest leave the system. Very far from the charges (Figure 21-22b), the lines leaving the system are approximately symmetrically spaced and point radially outward, just as they would for a single positive charge $+q$.

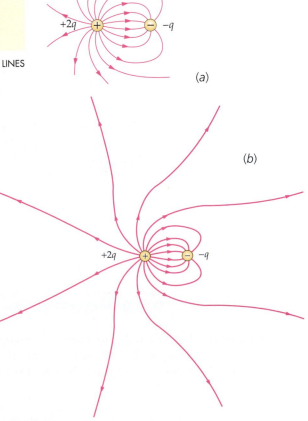

(b)

FIGURE 21-22 (*a*) Electric field lines for a point charge $+2q$ and a second point charge $-q$. (*b*) At great distances from the charges, the field lines approach those for a single point charge $+q$ located at the center of charge.

EXAMPLE 21-9

The electric field lines for two conducting spheres are shown in Figure 21-23. What is the relative sign and magnitude of the charges on the two spheres?

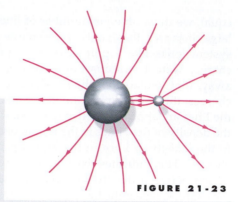

FIGURE 21-23

PICTURE THE PROBLEM The charge on a sphere is positive if more lines leave than enter and negative if more enter than leave. The ratio of the magnitudes of the charges equals the ratio of the net number of lines entering or leaving.

Since 11 electric field lines leave the large sphere on the left and 3 enter, the net number leaving is 8, so the charge on the large sphere is positive. For the small sphere on the right, 8 lines leave and none enter, so its charge is also positive. Since the net number of lines leaving each sphere is 8, the spheres carry equal positive charges. The charge on the small sphere creates an intense field at the nearby surface of the large sphere that causes a local accumulation of negative charge on the large sphere—indicated by the three entering field lines. Most of the large sphere's surface has positive charge, however, so its total charge is positive.

The convention relating the electric field strength to the electric field lines works because the electric field varies inversely as the square of the distance from a point charge. Because the gravitational field of a point mass also varies inversely as the square of the distance, field-line drawings are also useful for picturing the gravitational field. Near a point mass, the gravitational field lines converge on the mass just as electric field lines converge on a negative charge. However, unlike electric field lines near a positive charge, there are no points in space from which gravitational field lines diverge. That's because the gravitational force is always attractive, never repulsive.

21-6 Motion of Point Charges in Electric Fields

When a particle with a charge q is placed in an electric field $\vec{E}$, it experiences a force $q\vec{E}$. If the electric force is the only significant force acting on the particle, the particle has acceleration

$$\vec{a} = \frac{\Sigma \vec{F}}{m} = \frac{q}{m}\vec{E}$$

where m is the mass of the particle. (If the particle is an electron, its speed in an electric field is often a significant fraction of the speed of light. In such cases, Newton's laws of motion must be modified by Einstein's special theory of relativity.) If the electric field is known, the charge-to-mass ratio of the particle can be determined from the measured acceleration. J. J. Thomson used the deflection of electrons in a uniform electric field in 1897 to demonstrate the existence of electrons and to measure their charge-to-mass ratio. Familiar examples of devices that rely on the motion of electrons in electric fields are oscilloscopes, computer monitors, and television picture tubes.

Schematic drawing of a cathode-ray tube used for color television. The beams of electrons from the electron gun on the right activate phosphors on the screen at the left, giving rise to bright spots whose colors depend on the relative intensity of each beam. Electric fields between deflection plates in the gun (or magnetic fields from coils surrounding the gun) deflect the beams. The beams sweep across the screen in a horizontal line, are deflected downward, then sweep across again. The entire screen is covered in this way 30 times per second.

FIGURE 21-24

$\vec{E}$

$\vec{v}_0$

e

EXAMPLE 21-10

An electron is projected into a uniform electric field $\vec{E} = (1000 \text{ N/C})\hat{i}$ with an initial velocity $\vec{v}_0 = (2 \times 10^6 \text{ m/s})\hat{i}$ in the direction of the field (Figure 21-24). How far does the electron travel before it is brought momentarily to rest?

PICTURE THE PROBLEM Since the charge of the electron is negative, the force $\vec{F} = -e\vec{E}$ acting on the electron is in the direction opposite that of the field. Since $\vec{E}$ is constant, the force is constant and we can use constant acceleration formulas from Chapter 2. We choose the field to be in the positive x direction.

1. The displacement Δx is related to the initial and final velocities:

$$v_x^2 = v_{0x}^2 + 2a_x\,\Delta x$$

2. The acceleration is obtained from Newton's second law:

$$a_x = \frac{F_x}{m} = \frac{-eE}{m}$$

3. When $v_x = 0$, the displacement is:

$$\Delta x = \frac{v_x^2 - v_{0x}^2}{2a_x} = \frac{0 - v_{0x}^2}{2(-eE/m)} = \frac{mv_0^2}{2eE}$$

$$= \frac{(9.11 \times 10^{-31}\,\text{kg})(2 \times 10^6\,\text{m/s})^2}{2(1.6 \times 10^{-19}\,\text{C})(1000\,\text{N/C})}$$

$$= 1.14 \times 10^{-2}\,\text{m} = \boxed{1.14\,\text{cm}}$$

ELECTRON MOVING PERPENDICULAR TO A UNIFORM
ELECTRIC FIELD

EXAMPLE 21-11

An electron enters a uniform electric field $\vec{E} = (-2000\,\text{N/C})\hat{j}$ with an initial velocity $\vec{v}_0 = (10^6\,\text{m/s})\hat{i}$ perpendicular to the field (Figure 2125). (*a*) Compare the gravitational force acting on the electron to the electric force acting on it. (*b*) By how much has the electron been deflected after it has traveled 1 cm in the x direction?

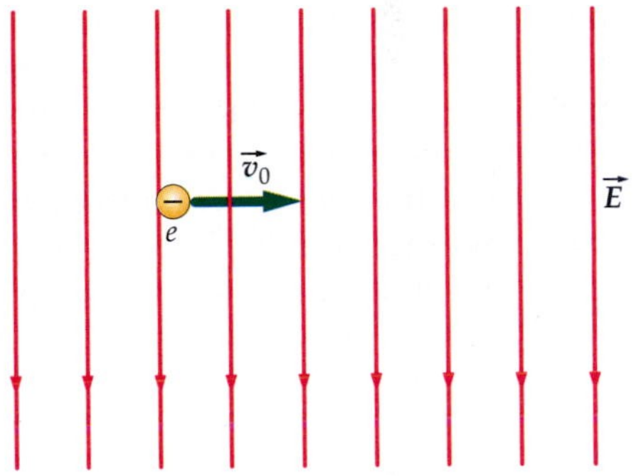

FIGURE 21-25

PICTURE THE PROBLEM (*a*) Calculate the ratio of the electric force $qE = -eE$ to the gravitational force mg. (*b*) Since mg is negligible, the force on the electron is $-eE$ vertically upward. The electron thus moves with constant horizontal velocity v_x and is deflected upward by an amount $y = \frac{1}{2}at^2$, where t is the time to travel 1 cm in the x direction.

(*a*) Calculate the ratio of the magnitude of the electric force, F_e, to the magnitude of the gravitational force, F_g:

$$\frac{F_e}{F_g} = \frac{eE}{mg} = \frac{(1.6 \times 10^{-19}\,\text{C})(2000\,\text{N/C})}{(9.11 \times 10^{-31}\,\text{kg})(9.81\,\text{N/kg})} = \boxed{3.6 \times 10^{13}}$$

(*b*) 1. Express the vertical deflection in terms of the acceleration a and time t:

$$y = \frac{1}{2}a_y t^2$$

2. Express the time required for the electron to travel a horizontal distance x with constant horizontal velocity v_0:

$$t = \frac{x}{v_0}$$

3. Use this result for t and eE/m for a_y to calculate y:

$$y = \frac{1}{2}\frac{eE}{m}\left(\frac{x}{v_0}\right)^2$$

$$= \frac{1}{2}\frac{(1.6 \times 10^{-19}\,\text{C})(2000\,\text{N/C})}{9.11 \times 10^{-31}\,\text{kg}}\left(\frac{0.01\,\text{m}}{10^6\,\text{m/s}}\right)^2$$

$$= \boxed{1.76\,\text{cm}}$$

REMARKS (*a*) As is usually the case, the electric force is huge compared with the gravitational force. Thus, it is not necessary to consider gravity when designing a cathode-ray tube, for example, or when calculating the deflection in the problem above. In fact, a television picture tube works equally well upside down and right side up, as if gravity were not even present. (*b*) The path of an electron moving in a uniform electric field is a parabola, the same as the path of a neutron moving in a uniform gravitational field.

THE ELECTRIC FIELD IN AN INK-JET PRINTER   **EXAMPLE 21-12** **Put It in Context**

You've just finished printing out a long essay for your English professor, and you get to wondering about how the ink-jet printer knows where to place the ink. You search the Internet and find a picture (Figure 21-26) that shows that the ink drops are given a charge and pass between a pair of oppositely charged metal plates that provide a uniform electric field in the region between the plates. Since you've been studying the electric field in physics class, you wonder if you can determine how large a field is used in this type of printer. You do a bit more searching and find that the 40-μm-diameter ink drops have an initial velocity of 40 m/s, and that a drop with a 2-nC charge is deflected upward a distance of 3 mm as the drop transits the 1-cm-long region between the plates. Find the magnitude of the electric field. (Neglect any effects of gravity on the motion of the drops.)

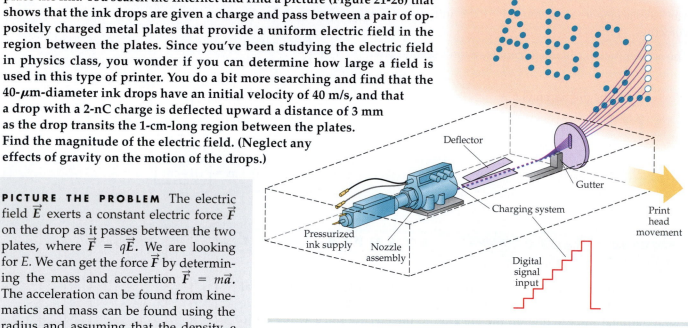

PICTURE THE PROBLEM The electric field $\vec{E}$ exerts a constant electric force $\vec{F}$ on the drop as it passes between the two plates, where $\vec{F} = q\vec{E}$. We are looking for E. We can get the force $\vec{F}$ by determining the mass and accelertion $\vec{F} = m\vec{a}$. The acceleration can be found from kinematics and mass can be found using the radius and assuming that the density ρ of ink is 1000 kg/m³ (the same as the density of water).

FIGURE 21-26 An ink-jet used for printing. The ink exits the nozzle in discrete droplets. Any droplet destined to form a dot on the image is given a charge. The deflector consists of a pair of oppositely charged plates. The greater the charge a drop receives, the higher the drop is deflected as it passes between the deflector plates. Drops that do not receive a charge are not deflected upward. These drops end up in the gutter, and the ink is returned to the ink reservoir.

1. The electric field equals the force to charge ratio:

$$E = \frac{F}{q}$$

2. The force, which is in the $+y$ direction (upward), equals the mass times the acceleration:

$$F = ma$$

3. The vertical displacement is obtained using a constant-acceleration kinematic formula with $v_{0y} = 0$:

$$\Delta y = v_{0y}t + \tfrac{1}{2}at^2$$
$$= 0 + \tfrac{1}{2}at^2$$

4. The time is how long it takes for the drop to travel the $\Delta x = 1$ cm at $v_0 = 40$ m/s:

$$\Delta x = v_{0x}t = v_0 t, \text{ so } t = \Delta x/v_0$$

5. Solving for a gives:

$$a = \frac{2\Delta y}{t^2} = \frac{2\Delta y}{(\Delta x/v_0)^2} = \frac{2v_0^2\Delta y}{(\Delta x)^2}$$

6. The mass equals the density times the volume:

$$m = \rho V = \rho \tfrac{4}{3}\pi r^3$$

7. Solve for E:

$$E = \frac{F}{q} = \frac{ma}{q} = \frac{\rho \tfrac{4}{3}\pi r^3}{q}\frac{2v_0^2\Delta y}{(\Delta x)^2}$$

$$= \frac{8\pi}{3}\frac{\rho r^3 v_0^2 \Delta y}{q(\Delta x)^2}$$

$$= \frac{8\pi}{3}\frac{(1000 \text{ kg/m}^3)(20 \times 10^{-6} \text{ m})^3(40 \text{ m/s})^2(3 \times 10^{-3} \text{ m})}{(2 \times 10^{-9} \text{ C})(0.01 \text{ m})^2}$$

$$= \boxed{1610 \text{ N/C}}$$

REMARKS The ink jet in this example is called a multiple-deflection continuous ink jet. It is used in some industrial printers. The ink-jet printers sold for use with home computers do not use charged droplets deflected by an electric field.

21-7 Electric Dipoles in Electric Fields

In Example 21-6 we found the electric field produced by a dipole, a system of two equal and opposite point charges that are close together. Here we consider the behavior of an electric dipole in an external electric field. Some molecules have permanent electric dipole moments due to a nonuniform distribution of charge within the molecule. Such molecules are called **polar molecules.** An example is HCl, which is essentially a positive hydrogen ion of charge $+e$ combined with a negative chlorine ion of charge $-e$. The center of charge of the positive ion does not coincide with the center of charge for the negative ion, so the molecule has a permanent dipole moment. Another example is water (Figure 21-27).

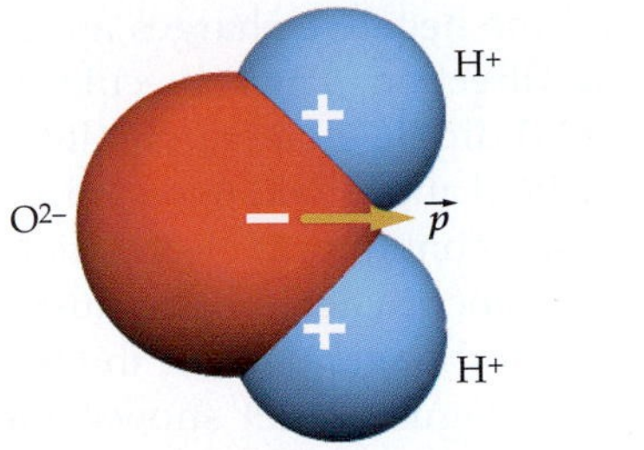

A uniform external electric field exerts no net force on a dipole, but it does exert a torque that tends to rotate the dipole into the direction of the field. We see in Figure 21-28 that the torque calculated about the position of either charge has the magnitude $F_1 L \sin \theta = qEL \sin \theta = pE \sin \theta$.[†] The direction of the torque is into the paper such that it rotates the dipole moment $\vec{p}$ into the direction of $\vec{E}$. The torque can be conveniently written as the cross product of the dipole moment $\vec{p}$ and the electric field $\vec{E}$.

$$\vec{\tau} = \vec{p} \times \vec{E} \qquad\qquad 21\text{-}11$$

When the dipole rotates through $d\theta$, the electric field does work:

$$dW = -\tau d\theta = -pE \sin \theta \, d\theta$$

(The minus sign arises because the torque opposes any increase in θ.) Setting the negative of this work equal to the change in potential energy, we have

$$dU = -dW = +pE \sin \theta \, d\theta$$

Integrating, we obtain

$$U = -pE \cos \theta + U_0$$

If we choose the potential energy U_0 to be zero when $\theta = 90°$, then the potential energy of the dipole is

$$U = -pE \cos \theta = -\vec{p} \cdot \vec{E} \qquad\qquad 21\text{-}12$$

POTENTIAL ENERGY OF A DIPOLE IN AN ELECTRIC FIELD

Microwave ovens take advantage of the electric dipole moment of water molecules to cook food. Like all electromagnetic waves, microwaves have oscillating electric fields that exert torques on electric dipoles, torques that cause the water molecules to rotate with significant rotational kinetic energy. In this manner, energy is transferred from the microwave radiation to the water molecules throughout the food at a high rate, accounting for the rapid cooking times that make microwave ovens so convenient.

---

† The torque produced by two equal and opposite forces (an arrangement called a couple) is the same about any point in space.

FIGURE 21-27 An H_2O molecule has a permanent electric dipole moment that points in the direction from the center of negative charge to the center of positive charge.

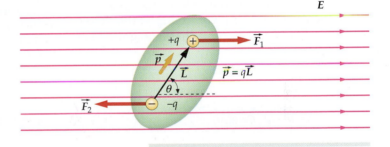

FIGURE 21-28 A dipole in a uniform electric field experiences equal and opposite forces that tend to rotate the dipole so that its dipole moment is aligned with the electric field.

Nonpolar molecules have no permanent electric dipole movement. However, all neutral molecules contain equal amounts of positive and negative charge. In the presence of an external electric field $\vec{E}$, the charges become separated in space. The positive charges are pushed in the direction of $\vec{E}$ and the negative charges are pushed in the opposite direction. The molecule thus acquires an induced dipole moment parallel to the external electric field and is said to be **polarized.**

In a nonuniform electric field, an electric dipole experiences a net force because the electric field has different magnitudes at the positive and negative poles. Figure 21-29 shows how a positive point charge polarizes a nonpolar molecule and then attracts it. A familiar example is the attraction that holds an electrostatically charged balloon against a wall. The nonuniform field produced by the charge on the balloon polarizes molecules in the wall and attracts them. An equal and opposite force is exerted by the wall molecules on the balloon.

The diameter of an atom or molecule is of the order of 10^{-10} m = 0.1 nm. A convenient unit for the electric dipole moment of atoms and molecules is the fundamental electronic charge e times the distance 1 nm. For example, the dipole moment of H_2O in these units has a magnitude of about $0.04\,e \cdot$nm.

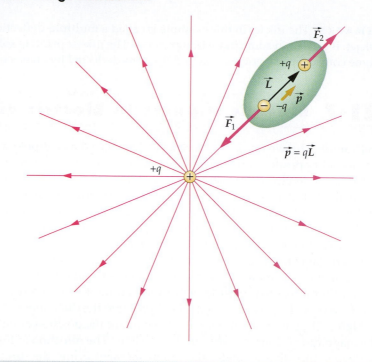

FIGURE 21-29 A nonpolar molecule in the nonuniform electric field of a positive point charge. The induced electric dipole moment $\vec{p}$ is parallel to the field of the point charge. Because the point charge is closer to the center of negative charge than to the center of positive charge, there is a net force of attraction between the dipole and the point charge. If the point charge were negative, the induced dipole moment would be reversed, and the molecule would again be attracted to the point charge.

TORQUE AND POTENTIAL ENERGY **EXAMPLE 21-13**

A dipole with a moment of magnitude 0.02 $e \cdot$nm makes an angle of 20° with a uniform electric field of magnitude 3×10^3 N/C (Figure 21-30). Find (a) the magnitude of the torque on the dipole, and (b) the potential energy of the system.

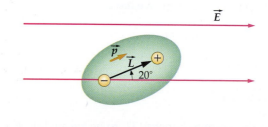

FIGURE 21-30

PICTURE THE PROBLEM The torque is found from $\vec{\tau} = \vec{p} \times \vec{E}$ and the potential energy is found from $U = -\vec{p} \cdot \vec{E}$.

1. Calculate the magnitude of the torque:

$$\tau = |\vec{p} \times \vec{E}| = pE\sin\theta = (0.02\,e \cdot \text{nm})(3 \times 10^3\,\text{N/C})(\sin 20°)$$

$$= (0.02)(1.6 \times 10^{-19}\,\text{C})(10^{-9}\,\text{m})(3 \times 10^3\,\text{N/C})(\sin 20°)$$

$$\boxed{= 3.28 \times 10^{-27}\,\text{N·m}}$$

2. Calculate the potential energy:

$$U = -\vec{p} \cdot \vec{E} = -pE\cos\theta$$

$$= -(0.02)(1.6 \times 10^{-19}\,\text{C})(10^{-9}\,\text{m})(3 \times 10^3\,\text{N/C})\cos 20°$$

$$\boxed{= -9.02 \times 10^{-27}\,\text{J}}$$

SUMMARY

1. Quantization and conservation are fundamental properties of electric charge.
2. Coulomb's law is the fundamental law of interaction between charges at rest.
3. The electric field describes the condition in space set up by a charge distribution.

Topic	Relevant Equations and Remarks
1. Electric Charge	There are two kinds of electric charge, positive and negative.
Quantization	Electric charge is quantized—it always occurs in integral multiples of the fundamental unit of charge e. The charge of the electron is $-e$ and that of the proton is $+e$.
Magnitude	$e = 1.60 \times 10^{-19}\,\text{C}$ 21-1
Conservation	Charge is conserved. It is neither created nor destroyed in any process, but is merely transferred.
2. Conductors and Insulators	In conductors, about one electron per atom is free to move about the entire material. In insulators, all the electrons are bound to nearby atoms.
Ground	A very large conductor that can supply an unlimited amount of charge (such as the earth) is called a ground.
3. Charging by Induction	A conductor can be charged by holding a charge near the conductor to attract or repel the free electrons and then grounding the conductor to drain off the faraway charges.
4. Coulomb's Law	The force exerted by a charge q_1 on q_2 is given by $$\vec{F}_{1,2} = \frac{kq_1q_2}{r_{1,2}^2}\hat{r}_{1,2} \qquad 21\text{-}2$$ where $\hat{r}_{1,2}$ is a unit vector that points from q_1 to q_2.
Coulomb constant	$k = 8.99 \times 10^9\,\text{N·m}^2/\text{C}^2$ 21-3
5. Electric Field	The electric field due to a system of charges at a point is defined as the net force exerted by those charges on a very small positive test charge q_0 divided by q_0: $$\vec{E} = \frac{\vec{F}}{q_0} \qquad 21\text{-}5$$
Due to a point charge	$$\vec{E}_{i,P} = \frac{kq_i}{r_{i,P}^2}\hat{r}_{i,P} \qquad 21\text{-}7$$
Due to a system of point charges	The electric field due to several charges is the vector sum of the fields due to the individual charges: $$\vec{E}_{i,P} = \sum_i \vec{E}_i = \sum_i \frac{kq_i}{r_{i,P}^2}\hat{r}_{i,P} \qquad 21\text{-}8$$
6. Electric Field Lines	The electric field can be represented by electric field lines that originate on positive charges and end on negative charges. The strength of the electric field is indicated by the density of the electric field lines.

7. **Electric Dipole**	An electric dipole is a system of two equal but opposite charges separated by a small distance.	
Dipole moment	$$\vec{p} = q\vec{L}$$ where $\vec{L}$ points from the negative charge to the positive charge.	**21-9**
Field due to dipole	The electric field far from a dipole is proportional to the dipole moment and decreases with the cube of the distance.	
Torque on a dipole	In a uniform electric field, the net force on a dipole is zero, but there is a torque that tends to align the dipole in the direction of the field. $$\vec{\tau} = \vec{p} \times \vec{E}$$	**21-11**
Potential energy of a dipole	$$U = -\vec{p} \cdot \vec{E}$$	**21-12**
8. **Polar and Nonpolar Molecules**	Polar molecules, such as H_2O, have permanent dipole moments because their centers of positive and negative charge do not coincide. They behave like simple dipoles in an electric field. Nonpolar molecules do not have permanent dipole moments, but they acquire induced dipole moments in the presence of an electric field.	

PROBLEMS

- Single-concept, single-step, relatively easy
- ● Intermediate-level, may require synthesis of concepts
- ●●● Challenging

SSM Solution is in the *Student Solutions Manual*

iSOLVE Problems available on iSOLVE online homework service

 ✓ These "Checkpoint" online homework service problems ask students additional questions about their confidence level, and how they arrived at their answer.

In a few problems, you are given more data than you actually need; in a few other problems, you are required to supply data from your general knowledge, outside sources, or informed estimates.

Conceptual Problems

1 ●● **SSM** Discuss the similarities and differences in the properties of electric charge and gravitational mass.

2 ● Can insulators be charged by induction?

3 ●● A metal rectangle *B* is connected to ground through a switch *S* that is initially closed (Figure 21-31). While the charge $+Q$ is near *B*, switch *S* is opened. The charge $+Q$ is then removed. Afterward, what is the charge state of the metal rectangle *B*? (*a*) It is positively charged. (*b*) It is uncharged. (*c*) It is negatively charged. (*d*) It may be any of the above depending on the charge on *B* before the charge $+Q$ was placed nearby.

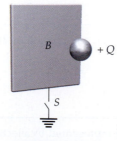

FIGURE 21-31
Problem 3

4 ●● Explain, giving each step, how a positively charged insulating rod can be used to give a metal sphere (*a*) a negative charge, and (*b*) a positive charge. (*c*) Can the same rod be used to simultaneously give one sphere a positive charge and another sphere a negative charge without the rod having to be recharged?

5 ●● **SSM** Two uncharged conducting spheres with their conducting surfaces in contact are supported on a large wooden table by insulated stands. A positively charged rod is brought up close to the surface of one of the spheres on the side opposite its point of contact with the other sphere. (*a*) Describe the induced charges on the two conducting spheres, and sketch the charge distributions on them. (*b*) The two spheres are separated far apart and the charged rod is removed. Sketch the charge distributions on the separated spheres.

6 • Three charges, $+q, +Q,$ and $-Q,$ are placed at the corners of an equilateral triangle as shown in Figure 21-32. The net force on charge $+q$ due to the other two charges is (a) vertically up. (b) vertically down. (c) zero. (d) horizontal to the left. (e) horizontal to the right.

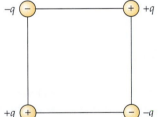

FIGURE 21-32 Problem 6

7 • **SSM** A positive charge that is free to move but is at rest in an electric field $\vec{E}$ will

(a) accelerate in the direction perpendicular to $\vec{E}$.
(b) remain at rest.
(c) accelerate in the direction opposite to $\vec{E}$.
(d) accelerate in the same direction as $\vec{E}$.
(e) do none of the above.

8 • **SSM** If four charges are placed at the corners of a square as shown in Figure 21-33, the field $\vec{E}$ is zero at

(a) all points along the sides of the square midway between two charges.
(b) the midpoint of the square.
(c) midway between the top two charges and midway between the bottom two charges.
(d) none of the above.

FIGURE 21-33
Problem 8

9 •• At a particular point in space, a charge Q experiences no net force. It follows that

(a) there are no charges nearby.
(b) if charges are nearby, they have the opposite sign of Q.
(c) if charges are nearby, the total positive charge must equal the total negative charge.
(d) none of the above need be true.

10 • Two charges $+4q$ and $-3q$ are separated by a small distance. Draw the electric field lines for this system.

11 • **SSM** Two charges $+q$ and $-3q$ are separated by a small distance. Draw the electric field lines for this system.

12 • **SSM** Three equal positive point charges are situated at the corners of an equilateral triangle. Sketch the electric field lines in the plane of the triangle.

13 • Which of the following statements are true?

(a) A positive charge experiences an attractive electrostatic force toward a nearby neutral conductor.
(b) A positive charge experiences no electrostatic force near a neutral conductor.
(c) A positive charge experiences a repulsive force, away from a nearby conductor.
(d) Whatever the force on a positive charge near a neutral conductor, the force on a negative charge is then oppositely directed.
(e) None of the above is correct.

14 • **SSM** The electric field lines around an electrical dipole are best represented by which, if any, of the diagrams in Figure 21-34?

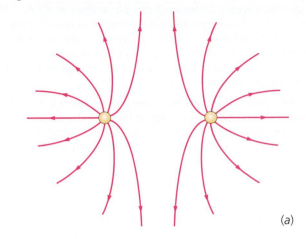

(a)

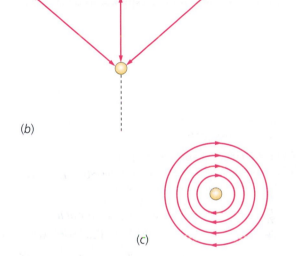

(b)

(c)

(d)

FIGURE 21-34 Problem 14

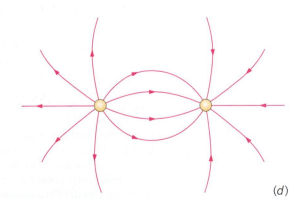

15 •• **SSM** A molecule with electric dipole moment $\vec{p}$ is oriented so that $\vec{p}$ makes an angle θ with a uniform electric field $\vec{E}$. The dipole is free to move in response to the force from the field. Describe the motion of the dipole. Suppose the electric field is nonuniform and is larger in the x direction. How will the motion be changed?

16 •• True or false:

(a) The electric field of a point charge always points away from the charge.

(b) All macroscopic charges Q can be written as $Q = \pm Ne$, where N is an integer and e is the charge of the electron.

(c) Electric field lines never diverge from a point in space.

(d) Electric field lines never cross at a point in space.

(e) All molecules have electric dipole moments in the presence of an external electric field.

17 •• Two metal balls have charges $+q$ and $-q$. How will the force on one of them change if (a) the balls are placed in water, the distance between them being unchanged, and (b) a third uncharged metal ball is placed between the first two? Explain.

18 •• [SSM] A metal ball is positively charged. Is it possible for it to attract another positively charged ball? Explain.

19 •• [SSM] A simple demonstration of electrostatic attraction can be done simply by tying a small ball of tinfoil on a hanging string, and bringing a charged wand near it. Initially, the ball will be attracted to the wand, but once they touch, the ball will be repelled violently from it. Explain this behavior.

Estimation and Approximation

20 •• Two small spheres are connected to opposite ends of a steel cable of length 1 m and cross-sectional area 1.5 cm². A positive charge Q is placed on each sphere. Estimate the largest possible value Q can have before the cable breaks, given that the tensile strength of steel is 5.2×10^8 N/m².

21 •• The net charge on any object is the result of the surplus or deficit of only an extremely small fraction of the electrons in the object. In fact, a charge imbalance greater than this would result in the destruction of the object. (a) Estimate the force acting on a 0.5 cm × 0.5 cm × 4 cm rod of copper if the electrons in the copper outnumbered the protons by 0.0001%. Assume that half of the excess electrons migrate to opposite ends of the rod of the copper. (b) Calculate the largest possible imbalance, given that copper has a tensile strength of 2.3×10^8 N/m².

22 ••• Electrical discharge (sparks) in air occur when free ions in the air are accelerated to a high enough velocity by an electric field to ionize other gas molecules on impact. (a) Assuming that the ion moves, on average, 1 mean free path through the gas before hitting a molecule, and that it needs to acquire an energy of approximately 1 eV to ionize it, estimate the field strength required for electrical breakdown in air at a pressure and temperature of 1×10^5 N/m² and 300 K. Assume that the cross-sectional area of a nitrogen molecule is about 0.1 nm². (b) How should the breakdown potential depend on temperature (all other things being equal)? On pressure?

23 •• [SSM] A popular classroom demonstration consists of rubbing a "magic wand" made of plastic with fur to charge it, and then placing it near an empty soda can on its side (Figure 21-35.) The can will roll toward the wand, as it acquires a charge on the side nearest the wand by induction. Typically, if the wand is held about 10 cm away from the can, the can will have an initial acceleration of about 1 m/s². If the mass of the can is 0.018 kg, estimate the charge on the rod.

FIGURE 21-35
Problem 23

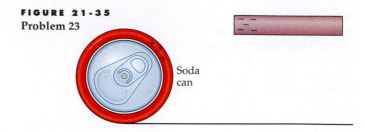

Soda can

24 •• Estimate the force required to bind the He nucleus together, given that the extent of the nucleus is about 10^{-15} m and contains 2 protons.

Electric Charge

25 • [SOLVE] A plastic rod is rubbed against a wool shirt, thereby acquiring a charge of $-0.8\ \mu C$. How many electrons are transferred from the wool shirt to the plastic rod?

26 • A charge equal to the charge of Avogadro's number of protons ($N_A = 6.02 \times 10^{23}$) is called a *faraday*. Calculate the number of coulombs in a faraday.

27 • [SSM] [SOLVE] How many coulombs of positive charge are there in 1 kg of carbon? Twelve grams of carbon contain Avogadro's number of atoms, with each atom having six protons and six electrons.

Coulomb's Law

28 • [SOLVE] A charge $q_1 = 4.0\ \mu C$ is at the origin, and a charge $q_2 = 6.0\ \mu C$ is on the x axis at $x = 3.0$ m. (a) Find the force on charge q_2. (b) Find the force on q_1. (c) How would your answers for Parts (a) and (b) differ if q_2 were $-6.0\ \mu C$?

29 • [SOLVE] Three point charges are on the x axis: $q_1 = -6.0\ \mu C$ is at $x = -3.0$ m, $q_2 = 4.0\ \mu C$ is at the origin, and $q_3 = -6.0\ \mu C$ is at $x = 3.0$ m. Find the force on q_1.

30 •• Three charges, each of magnitude 3 nC, are at separate corners of a square of edge length 5 cm. The two charges at opposite corners are positive, and the other charge is negative. Find the force exerted by these charges on a fourth charge $q = +3$ nC at the remaining corner.

31 •• [SOLVE] A charge of 5 μC is on the y axis at $y = 3$ cm, and a second charge of $-5\ \mu C$ is on the y axis at $y = -3$ cm. Find the force on a charge of 2 μC on the x axis at $x = 8$ cm.

32 •• [SSM] A point charge of $-2.5\ \mu C$ is located at the origin. A second point charge of 6 μC is at $x = 1$ m, $y = 0.5$ m. Find the x and y coordinates of the position at which an electron would be in equilibrium.

33 •• [SSM] A charge of $-1.0\ \mu C$ is located at the origin; a second charge of 2.0 μC is located at $x = 0$, $y = 0.1$ m; and a third charge of 4.0 μC is located at $x = 0.2$ m, $y = 0$. Find the forces that act on each of the three charges.

34 •• A charge of 5.0 μC is located at $x = 0$, $y = 0$ and a charge Q_2 is located at $x = 4.0$ cm, $y = 0$. The force on a 2-μC charge at $x = 8.0$ cm, $y = 0$ is 19.7 N, pointing in the negative x direction. When this 2-μC charge is positioned at $x = 17.75$ cm, $y = 0$, the force on it is zero. Determine the charge Q_2.

35 •• Five equal charges Q are equally spaced on a semi-circle of radius R as shown in Figure 21-36. Find the force on a charge q located at the center of the semicircle.

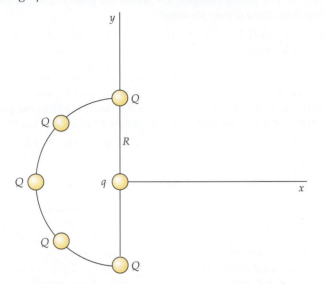

FIGURE 21-36 Problem 35

36 ••• The configuration of the NH_3 molecule is approximately that of a regular tetrahedron, with three H^+ ions forming the base and an N^{3-} ion at the apex of the tetrahedron. The length of each side is 1.64×10^{-10} m. Calculate the force that acts on each ion.

The Electric Field

37 • SSM iSOLVE A charge of 4.0 μC is at the origin. What is the magnitude and direction of the electric field on the x axis at (a) $x = 6$ m, and (b) $x = -10$ m? (c) Sketch the function E_x versus x for both positive and negative values of x. (Remember that E_x is negative when E points in the negative x direction.)

38 • SSM iSOLVE Two charges, each $+4$ μC, are on the x axis, one at the origin and the other at $x = 8$ m. Find the electric field on the x axis at (a) $x = -2$ m, (b) $x = 2$ m, (c) $x = 6$ m, and (d) $x = 10$ m. (e) At what point on the x axis is the electric field zero? (f) Sketch E_x versus x.

39 • When a test charge $q_0 = 2$ nC is placed at the origin, it experiences a force of 8.0×10^{-4} N in the positive y direction. (a) What is the electric field at the origin? (b) What would be the force on a charge of -4 nC placed at the origin? (c) If this force is due to a charge on the y axis at $y = 3$ cm, what is the value of that charge?

40 • iSOLVE The electric field near the surface of the earth points downward and has a magnitude of 150 N/C. (a) Compare the upward electric force on an electron with the downward gravitational force. (b) What charge should be placed on a penny of mass 3 g so that the electric force balances the weight of the penny near the earth's surface?

41 •• iSOLVE Two equal positive charges of magnitude $q_1 = q_2 = 6.0$ nC are on the y axis at $y_1 = +3$ cm and $y_2 = -3$ cm. (a) What is the magnitude and direction of the electric field on the x axis at $x = 4$ cm? (b) What is the force exerted on a third charge $q_0 = 2$ nC when it is placed on the x axis at $x = 4$ cm?

42 •• SSM iSOLVE A point charge of $+5.0$ μC is located at $x = -3.0$ cm, and a second point charge of -8.0 μC is located at $x = +4.0$ cm. Where should a third charge of $+6.0$ μC be placed so that the electric field at $x = 0$ is zero?

43 •• A point charge of -5 μC is located at $x = 4$ m, $y = -2$ m. A second point charge of 12 μC is located at $x = 1$ m, $y = 2$ m. (a) Find the magnitude and direction of the electric field at $x = -1$ m, $y = 0$. (b) Calculate the magnitude and direction of the force on an electron at $x = -1$ m, $y = 0$.

44 •• Two equal positive charges q are on the y axis, one at $y = +a$ and the other at $y = -a$. (a) Show that the electric field on the x axis is along the x axis with $E_x = 2kqx(x^2 + a^2)^{-3/2}$. (b) Show that near the origin, when x is much smaller than a, E_x is approximately $2kqx/a^3$. (c) Show that for values of x much larger than a, E_x is approximately $2kq/x^2$. Explain why you would expect this result even before calculating it.

45 •• SSM A 5-μC point charge is located at $x = 1$ m, $y = 3$ m; and a -4-μC point charge is located at $x = 2$ m, $y = -2$ m. (a) Find the magnitude and direction of the electric field at $x = -3$ m, $y = 1$ m. (b) Find the magnitude and direction of the force on a proton at $x = -3$ m, $y = 1$ m.

46 •• (a) Show that the electric field for the charge distribution in Problem 44 has its greatest magnitude at the points $x = a/\sqrt{2}$ and $x = -a/\sqrt{2}$ by computing dE_x/dx and setting the derivative equal to zero. (b) Sketch the function E_x versus x using your results for Part (a) of this problem and Parts (b) and (c) of Problem 44.

47 ••• For the charge distribution in Problem 44, the electric field at the origin is zero. A test charge q_0 placed at the origin will therefore be in equilibrium. (a) Discuss the stability of the equilibrium for a positive test charge by considering small displacements from equilibrium along the x axis and small displacements along the y axis. (b) Repeat Part (a) for a negative test charge. (c) Find the magnitude and sign of a charge q_0 that when placed at the origin results in a net force of zero on each of the three charges. (d) What will happen if any of the charges is displaced slightly from equilibrium?

48 ••• SSM Two positive point charges $+q$ are on the y axis at $y = +a$ and $y = -a$ as in Problem 44. A bead of mass m carrying a negative charge $-q$ slides without friction along a thread that runs along the x axis. (a) Show that for small displacements of $x \ll a$, the bead experiences a restoring force that is proportional to x and therefore undergoes simple harmonic motion. (b) Find the period of the motion.

Motion of Point Charges in Electric Fields

49 • iSOLVE The acceleration of a particle in an electric field depends on the ratio of the charge to the mass of the particle. (a) Compute e/m for an electron. (b) What is the magnitude and direction of the acceleration of an electron in a uniform electric field with a magnitude of 100 N/C? (c) When the speed of an electron approaches the speed of light c, relativistic mechanics must be used to calculate its motion, but at speeds significantly less than c, Newtonian mechanics applies. Using Newtonian mechanics, compute the time it takes for an electron placed at rest in an electric field with a magnitude of 100 N/C to reach a speed of $0.01c$. (d) How far does the electron travel in that time?

50 • **SSM** **iSOLVE** (a) Compute e/m for a proton, and find its acceleration in a uniform electric field with a magnitude of 100 N/C. (b) Find the time it takes for a proton initially at rest in such a field to reach a speed of $0.01c$ (where c is the speed of light).

51 • **iSOLVE** An electron has an initial velocity of 2×10^6 m/s in the x direction. It enters a uniform electric field $\vec{E} = (400 \text{ N/C})\hat{j}$; which is in the y direction. (a) Find the acceleration of the electron. (b) How long does it take for the electron to travel 10 cm in the x direction in the field? (c) By how much, and in what direction, is the electron deflected after traveling 10 cm in the x direction in the field?

52 •• **iSOLVE** An electron, starting from rest, is accelerated by a uniform electric field of 8×10^4 N/C that extends over a distance of 5.0 cm. Find the speed of the electron after it leaves the region of uniform electric field.

53 •• A 2-g object, located in a region of uniform electric field $\vec{E} = (300 \text{ N/C})\hat{i}$, carries a charge Q. The object, released from rest at $x = 0$, has a kinetic energy of 0.12 J at $x = 0.50$ m. Determine the charge Q.

54 •• **SSM** **iSOLVE** A particle leaves the origin with a speed of 3×10^6 m/s at 35° to the x axis. It moves in a constant electric field $\vec{E} = E_y\hat{j}$. Find E_y such that the particle will cross the x axis at $x = 1.5$ cm if the particle is (a) an electron, and (b) a proton.

55 •• An electron starts at the position shown in Figure 21-37 with an initial speed $v_0 = 5 \times 10^6$ m/s at 45° to the x axis. The electric field is in the positive y direction and has a magnitude of 3.5×10^3 N/C. On which plate and at what location will the electron strike?

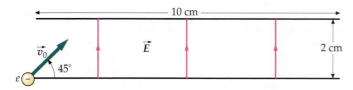

FIGURE 21-37 Problem 55

56 •• An electron with kinetic energy of 2×10^{-16} J is moving to the right along the axis of a cathode-ray tube as shown in Figure 21-38. There is an electric field $\vec{E} = (2 \times 10^4 \text{ N/C})\hat{j}$ in the region between the deflection plates. Everywhere else, $\vec{E} = 0$. (a) How far is the electron from the axis of the tube when it reaches the end of the plates? (b) At what angle is the electron moving with respect to the axis? (c) At what distance from the axis will the electron strike the fluorescent screen?

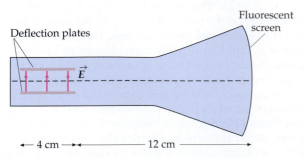

FIGURE 21-38 Problem 56

57 • **iSOLVE** Two point charges, $q_1 = 2.0$ pC and $q_2 = -2.0$ pC, are separated by 4 μm. (a) What is the dipole moment of this pair of charges? (b) Sketch the pair, and show the direction of the dipole moment.

58 • **SSM** **iSOLVE** A dipole of moment 0.5 e·nm is placed in a uniform electric field with a magnitude of 4.0×10^4 N/C. What is the magnitude of the torque on the dipole when (a) the dipole is parallel to the electric field, (b) the dipole is perpendicular to the electric field, and (c) the dipole makes an angle of 30° with the electric field? (d) Find the potential energy of the dipole in the electric field for each case.

59 •• **SSM** For a dipole oriented along the x axis, the electric field falls off as $1/x^3$ in the x direction and $1/y^3$ in the y direction. Use dimensional analysis to prove that, in any direction, the field far from the dipole falls off as $1/r^3$.

60 •• A water molecule has its oxygen atom at the origin, one hydrogen nucleus at $x = 0.077$ nm, $y = 0.058$ nm and the other hydrogen nucleus at $x = -0.077$ nm, $y = 0.058$ nm. If the hydrogen electrons are transferred completely to the oxygen atom so that it has a charge of $-2e$, what is the dipole moment of the water molecule? (Note that this characterization of the chemical bonds of water as totally ionic is simply an approximation that overestimates the dipole moment of a water molecule.)

61 •• An electric dipole consists of two charges $+q$ and $-q$ separated by a very small distance $2a$. Its center is on the x axis at $x = x_1$, and it points along the x axis in the positive x direction. The dipole is in a nonuniform electric field, which is also in the x direction, given by $\vec{E} = Cx\hat{i}$, where C is a constant. (a) Find the force on the positive charge and that on the negative charge, and show that the net force on the dipole is $Cp\hat{i}$. (b) Show that, in general, if a dipole of moment $\vec{p}$ lies along the x axis in an electric field in the x direction, the net force on the dipole is given approximately by $(dE_x/dx)p\hat{i}$.

62 ••• A positive point charge $+Q$ is at the origin, and a dipole of moment $\vec{p}$ is a distance r away ($r \gg L$) and in the radial direction as shown in Figure 21-29. (a) Show that the force exerted on the dipole by the point charge is attractive and has a magnitude $\approx 2kQp/r^3$ (see Problem 61). (b) Now assume that the dipole is centered at the origin and that a point charge Q is a distance r away along the line of the dipole. Using Newton's third law and your result for part (a), show that at the location of the positive point charge the electric field $\vec{E}$ due to the dipole is toward the dipole and has a magnitude of $\approx 2kp/r^3$.

General Problems

63 • **SSM** (a) What mass would a proton have if its gravitational attraction to another proton exactly balanced out the electrostatic repulsion between them? (b) What is the true ratio of these two forces?

64 •• Point charges of -5.0 μC, $+3.0$ μC, and $+5.0$ μC are located along the x axis at $x = -1.0$ cm, $x = 0$, and $x = +1.0$ cm, respectively. Calculate the electric field at $x = 3.0$ cm and at $x = 15.0$ cm. Is there some point on the x axis where the magnitude of the electric field is zero? Locate that point.

65 •• For the charge distribution of Problem 64, find the electric field at $x = 15.0$ cm as the vector sum of the electric field due to a dipole formed by the two 5.0-μC charges and a point charge of 3.0 μC, both located at the origin. Compare your result with the result obtained in Problem 64, and explain any difference between these two.

66 •• **SSM** **iSOLVE** In copper, about one electron per atom is free to move about. A copper penny has a mass of 3 g. (a) What percentage of the free charge would have to be removed to give the penny a charge of 15 μC? (b) What would be the force of repulsion between two pennies carrying this charge if they were 25 cm apart? Assume that the pennies are point charges.

67 •• Two charges q_1 and q_2 have a total charge of 6 μC. When they are separated by 3 m, the force exerted by one charge on the other has a magnitude of 8 mN. Find q_1 and q_2 if (a) both are positive so that they repel each other, and (b) one is positive and the other is negative so that they attract each other.

68 •• Three charges, $+q$, $+2q$, and $+4q$, are connected by strings as shown in Figure 21-39. Find the tensions T_1 and T_2.

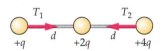

FIGURE 21-39 Problem 68

69 •• **SSM** A positive charge Q is to be divided into two positive charges q_1 and q_2. Show that, for a given separation D, the force exerted by one charge on the other is greatest if $q_1 = q_2 = \frac{1}{2}Q$.

70 •• **SSM** A charge Q is located at $x = 0$, and a charge $4Q$ is at $x = 12.0$ cm. The force on a charge of -2 μC is zero if that charge is placed at $x = 4.0$ cm, and is 126.4 N in the positive x direction if placed at $x = 8.0$ cm. Determine the charge Q.

71 •• Two small spheres (point charges) separated by 0.60 m carry a total charge of 200 μC. (a) If the two spheres repel each other with a force of 80 N, what are the charges on each of the two spheres? (b) If the two spheres attract each other with a force of 80 N, what are the charges on the two spheres?

72 •• **iSOLVE** A ball of known charge q and unknown mass m, initially at rest, falls freely from a height h in a uniform electric field $\vec{E}$ that is directed vertically downward. The ball hits the ground at a speed $v = 2\sqrt{gh}$. Find m in terms of E, q, and g.

73 •• **SSM** A rigid stick one meter long is pivoted about its center (Figure 21-40). A charge $q_1 = 5 \times 10^{-7}$ C is placed on one end of the rod, and an equal but opposite charge q_2 is placed a distance $d = 10$ cm directly below it. (a) What is the net force between the two charges? (b) What is the torque (measured from the center of the rod) due to that force? (c) To counterbalance the attraction between the two charges, we hang a block 25 cm from the pivot on the *opposite* side of the balance point. What value should we choose for the mass m of the block? (See Figure 21-40.) (d) We now move the block and hang it a distance of 25 cm from the balance point on the *same* side of the balance as the charge. Keeping q_1 the same, and d the same, what value should we choose for q_2 to keep this apparatus in balance?

FIGURE 21-40 Problem 73

74 •• Charges of 3.0 μC are located at $x = 0$, $y = 2.0$ m, and at $x = 0$, $y = -2.0$ m. Charges Q are located at $x = 4.0$ m, $y = 2.0$ m, and at $x = 4.0$ m, $y = -2.0$ m (Figure 21-41). The electric field at $x = 0$, $y = 0$ is $(4.0 \times 10^3$ N/C$)\hat{i}$. Determine Q.

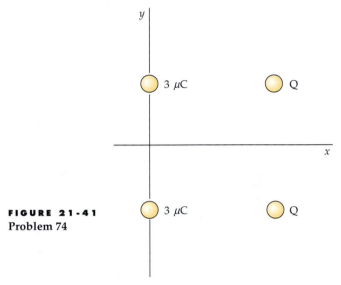

FIGURE 21-41
Problem 74

75 •• Two identical small spherical conductors (point charges), separated by 0.60 m, carry a total charge of 200 μC. They repel one another with a force of 120 N. (a) Find the charge on each sphere. (b) The two spheres are placed in electrical contact and then separated so that each carries 100 μC. Determine the force exerted by one sphere on the other when they are 0.60 m apart.

76 •• Repeat Problem 75 if the two spheres initially attract one another with a force of 120 N.

77 •• A charge of -3.0 μC is located at the origin; a charge of 4.0 μC is located at $x = 0.2$ m, $y = 0$; a third charge Q is located at $x = 0.32$ m, $y = 0$. The force on the 4.0-μC charge is 240 N, directed in the positive x direction. (a) Determine the charge Q. (b) With this configuration of three charges, where, along the x direction, is the electric field zero?

78 •• **SSM** Two small spheres of mass m are suspended from a common point by threads of length L. When each sphere carries a charge q, each thread makes an angle θ with the vertical as shown in Figure 21-42. (a) Show that the charge q is given by

$$q = 2L \sin\theta \sqrt{\frac{mg \tan\theta}{k}}$$

where k is the Coulomb constant. (b) Find q if $m = 10$ g, $L = 50$ cm, and $\theta = 10°$.

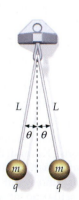

FIGURE 21-42 Problem 78

79 •• **iSOLVE** (*a*) Suppose that in Problem 78 $L = 1.5$ m, $m = 0.01$ kg, and $q = 0.75$ μC. What is the angle that each string makes with the vertical? (*b*) Find the angle that each string makes with the vertical if one mass carries a charge of 0.50 μC, the other a charge of 1.0 μC.

80 •• Four charges of equal magnitude are arranged at the corners of a square of side L as shown in Figure 21-43. (*a*) Find the magnitude and direction of the force exerted on the charge in the lower left corner by the other charges. (*b*) Show that the electric field at the midpoint of one of the sides of the square is directed along that side toward the negative charge and has a magnitude E given by

$$E = k\frac{8q}{L^2}\left(1 - \frac{\sqrt{5}}{25}\right)$$

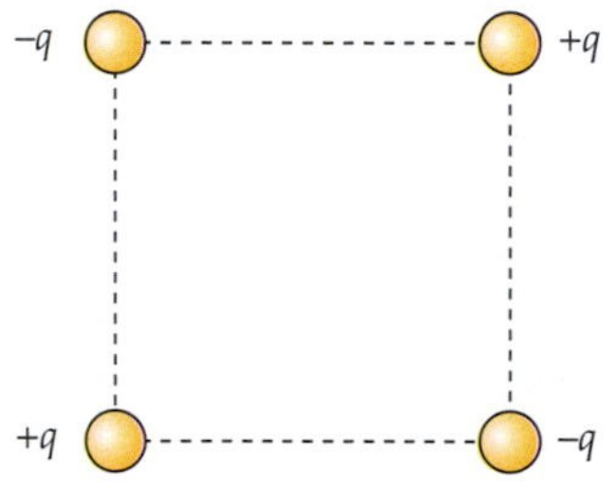

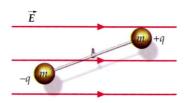

FIGURE 21-43
Problem 80

81 •• Figure 21-44 shows a dumbbell consisting of two identical masses m attached to the ends of a thin (massless) rod of length a that is pivoted at its center. The masses carry charges of $+q$ and $-q$, and the system is located in a uniform electric field $\vec{E}$. Show that for small values of the angle θ between the direction of the dipole and the electric field, the system displays simple harmonic motion, and obtain an expression for the period of that motion.

FIGURE 21-44 **Problems 81 and 82**

82 •• For the dumbbell in Figure 21-44, let $m = 0.02$ kg, $a = 0.3$ m, and $\vec{E} = (600$ N/C$)\hat{i}$. Initially the dumbbell is at rest and makes an angle of 60° with the x axis. The dumbbell is then released, and when it is momentarily aligned with the electric field, its kinetic energy is 5×10^{-3} J. Determine the magnitude of q.

83 •• **SSM** An electron (charge $-e$, mass m) and a positron (charge $+e$, mass m) revolve around their common center of mass under the influence of their attractive coulomb force. Find the speed of each particle v in terms of e, m, k, and their separation r.

84 •• The equilibrium separation between the nuclei of the ionic molecule KBr is 0.282 nm. The masses of the two ions, K$^+$ and Br$^-$, are very nearly the same, 1.4×10^{-25} kg and each of the two ions carries a charge of magnitude e. Use the result of Problem 81 to determine the frequency of oscillation of a KBr molecule in a uniform electric field of 1000 N/C.

85 ••• A small (point) mass m, which carries a charge q, is constrained to move vertically inside a narrow, frictionless cylinder (Figure 21-45). At the bottom of the cylinder is a point mass of charge Q having the same sign as q. (*a*) Show that the mass m will be in equilibrium at a height $y_0 = (kqQ/mg)^{1/2}$. (*b*) Show that if the mass m is displaced by a small amount from its equilibrium position and released, it will exhibit simple harmonic motion with angular frequency $\omega = (2g/y_0)^{1/2}$.

FIGURE 21-45
Problem 85

86 ••• A small bead of mass m and carrying a negative charge $-q$ is constrained to move along a thin, frictionless rod (Figure 21-46). A distance L from this rod is a positive charge Q. Show that if the bead is displaced a distance x, where $x \ll L$, and released, it will exhibit simple harmonic motion. Obtain an expression for the period of this motion in terms of the parameters L, Q, q, and m.

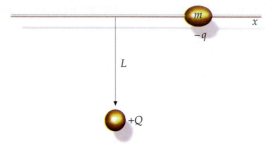

FIGURE 21-46 **Problem 86**

87 ••• Repeat Problem 79 with the system located in a uniform electric field of 1.0×10^5 N/C that points vertically downward.

88 ••• Suppose that the two spheres of mass in Problem 78 are not equal. One mass is 0.01 kg, the other is 0.02 kg. The charges on the two masses are 2.0 μC and 1.0 μC, respectively. Determine the angle that each of the strings supporting the masses makes with the vertical.

89 ••• **iSOLVE** A simple pendulum of length $L = 1.0$ m and mass $M = 5.0 \times 10^{-3}$ kg is placed in a uniform, vertically directed electric field $\vec{E}$. The bob carries a charge of -8.0 μC. The period of the pendulum is 1.2 s. What is the magnitude and direction of $\vec{E}$?

90 ••• **SSM** Two neutral polar molecules attract each other. Suppose that each molecule has a dipole moment $\vec{p}$, and that these dipoles are aligned along the x axis and separated by a distance d. Derive an expression for the force of attraction in terms of p and d.

91 ••• Two equal positive charges Q are on the x axis at $x = \frac{1}{2}L$ and $x = -\frac{1}{2}L$. (*a*) Obtain an expression for the electric field as a function of y on the y axis. (*b*) A ring of mass m, which carries a charge q, moves on a thin, frictionless rod along the y axis. Find the force that acts on the charge q as a function of y; determine the sign of q such that this force always points toward $y = 0$. (*c*) Show that for small values of y the ring exhibits simple harmonic motion. (*d*) If $Q = 5$ μC, $|q| = 2$ μC, $L = 24$ cm, and $m = 0.03$ kg, what is the frequency of the oscillation for small amplitudes?

92 ••• In the Millikan experiment used to determine the charge on the electron, a charged polystyrene microsphere is released in still air in a known vertical electric field. The charged microsphere will accelerate in the direction of the net force until it reaches terminal speed. The charge on the microsphere is determined by measuring the terminal speed. In one such experiment, the bead has radius $r = 5.5 \times 10^7$ m, and the field has a magnitude $E = 6 \times 10^4$ N/C. The magnitude of the drag force on the sphere is $F_D = 6\pi\eta r v$, where v is the speed of the sphere and η is the viscosity of air ($\eta = 1.8 \times 10^{-5}$ N·s/m²). The polystyrene has density 1.05×10^3 kg/m³. (a) If the electric field is pointing down so that the polystyrene microsphere rises with a terminal speed $v = 1.16 \times 10^{-4}$ m/s, what is the charge on the sphere? (b) How many excess electrons are on the sphere? (c) If the direction of the electric field is reversed but its magnitude remains the same, what is the terminal speed?

93 ••• **SSM** In Problem 92, there was a description of the Millikan experiment used to determine the charge on the electron. In the experiment, a switchable power supply is used so that the electrical field can point both up and down, but with the same magnitude, so that one can measure the terminal speed of the microsphere as it is pushed up (against the force of gravity) and down. Let v_u represent the terminal speed when the particle is moving up, and v_d the terminal speed when moving down. (a) If we let $v = v_u + v_d$, show that $v = \dfrac{qE}{3\pi\eta r}$, where q is the microsphere's net charge. What advantage does measuring both v_u and v_d give over measuring only one? (b) Because charge is quantized, v can only change by steps of magnitude Δv. Using the data from Problem 92, calculate Δv.

The Electric Field II: Continuous Charge Distributions

BY DESCRIBING CHARGE IN TERMS OF CONTINUOUS CHARGE DENSITY, IT BECOMES POSSIBLE TO CALCULATE THE CHARGE ON THE SURFACE OF OBJECTS AS LARGE AS CELESTIAL BODIES.

? **How would you calculate the charge on the surface of the Earth? (See Example 22-10.)**

On a microscopic scale, electric charge is quantized. However, there are often situations in which many charges are so close together that they can be thought of as continuously distributed. The use of a continuous charge density to describe a large number of discrete charges is similar to the use of a continuous mass density to describe air, which actually consists of a large number of discrete molecules. In both cases, it is usually easy to find a volume element ΔV that is large enough to contain a multitude of individual charges or molecules and yet is small enough that replacing ΔV with a differential dV and using calculus introduces negligible error.

We describe the charge per unit volume by the **volume charge density** ρ:

$$\rho = \frac{\Delta Q}{\Delta V} \qquad\qquad 22\text{-}1$$

Often charge is distributed in a very thin layer on the surface of an object. We define the **surface charge density** σ as the charge per unit area:

$$\sigma = \frac{\Delta Q}{\Delta A} \qquad\qquad 22\text{-}2$$

Similarly, we sometimes encounter charge distributed along a line in space. We define the **linear charge density** λ as the charge per unit length:

$$\lambda = \frac{\Delta Q}{\Delta L} \qquad\qquad 22\text{-}3$$

➤ In this chapter, we show how Coulomb's law is used to calculate the electric field produced by various types of continuous charge distributions. We then introduce Gauss's law, which relates the electric field on a closed surface to the net charge within the surface, and we use this relation to calculate the electric field for symmetric charge distributions.

22-1   Calculating $\vec{E}$ From Coulomb's Law

Figure 22-1 shows an element of charge $dq = \rho\,dV$ that is small enough to be considered a point charge. Coulomb's law gives the electric field $d\vec{E}$ at a field point P due to this element of charge as:

$$d\vec{E} = \frac{k\,dq}{r^2}\hat{r}$$

where $\hat{r}$ is a unit vector that points from the source point to the field point P. The total field at P is found by integrating this expression over the entire charge distribution. That is,

$$\vec{E} = \int_V \frac{k\,dq}{r^2}\hat{r} \qquad\qquad 22\text{-}4$$

ELECTRIC FIELD DUE TO A CONTINUOUS CHARGE DISTRIBUTION

where $dq = \rho\,dV$. If the charge is distributed on a surface or line, we use $dq = \sigma\,dA$ or $dq = \lambda\,dL$ and integrate over the surface or line.

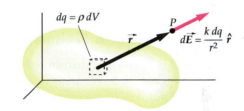

FIGURE 22-1 An element of charge dq produces a field $d\vec{E} = (k\,dq/r^2)\,\hat{r}$ at point P. The field at P is found by integrating over the entire charge distribution.

$\vec{E}$ on the Axis of a Finite Line Charge

A charge Q is uniformly distributed along the x axis from $x = -\frac{1}{2}L$ to $x = +\frac{1}{2}L$, as shown in Figure 22-2. The linear charge density for this charge is $\lambda = Q/L$. We wish to find the electric field produced by this line charge at some field point P on the x axis at $x = x_P$, where $x_P > \frac{1}{2}L$. In the figure, we have chosen the element of charge dq to be the charge on a small element of length dx at position x. Point P is a distance $r = x_P - x$ from dx. Coulomb's law gives the electric field at P due to the charge dq on this length dx. It is directed along the x axis and is given by

$$dE_x\hat{i} = \frac{k\,dq}{(x_P - x)^2}\hat{i} = \frac{k\lambda\,dx}{(x_P - x)^2}\hat{i}$$

We find the total field $\vec{E}$ by integrating over the entire line charge in the direction of increasing x (from $x_1 = -\frac{1}{2}L$ to $x_2 = +\frac{1}{2}L$):

$$E_x = k\lambda \int_{-L/2}^{+L/2} \frac{dx}{(x_P - x)^2} = -k\lambda \int_{x_P+(L/2)}^{x_P-(L/2)} \frac{du}{u^2}$$

where $u = x_P - x$ (so $du = -dx$). Note that if $x = -\frac{1}{2}L$, $u = x_P + \frac{1}{2}L$, and if $x = +\frac{1}{2}L$, $u = x_P - \frac{1}{2}L$. Evaluating the integral gives

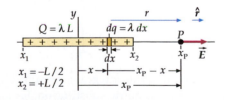

FIGURE 22-2 Geometry for the calculation of the electric field on the axis of a uniform line charge of length L, charge Q, and linear charge density $\lambda = Q/L$. An element $dq = \lambda\,dx$ is treated as a point charge.

$$E_x = +k\lambda \frac{1}{u}\Big|_{x_P+(L/2)}^{x_P-(L/2)} = k\lambda \left\{ \frac{1}{x_P - \frac{1}{2}L} - \frac{1}{x_P + \frac{1}{2}L} \right\} = \frac{k\lambda L}{x_P^2 - (\frac{1}{2}L)^2}$$

Substituting Q for λL, we obtain

$$E_x = \frac{kQ}{x_P^2 - (\frac{1}{2}L)^2}, \qquad x_P > \frac{1}{2}L \qquad\qquad 22\text{-}5$$

We can see that if x_P is much larger than L, the electric field at x_P is approximately kQ/x_P^2. That is, if we are sufficiently far away from the line charge, it approaches that of a point charge Q at the origin.

EXERCISE The validity of Equation 22-5 is established for the region $x_P > \frac{1}{2}L$. Is it also valid in the region $-\frac{1}{2}L \le x_P \le \frac{1}{2}L$? Explain. (*Answer* No. Symmetry dictates that E_x is zero at $x_P = 0$. However, Equation 22-5 gives a negative value for E_x at $x_P = 0$. These contradictory results cannot both be valid.)

$\vec{E}$ off the Axis of a Finite Line Charge

A charge Q is uniformly distributed on a straight-line segment of length L, as shown in Figure 22-3. We wish to find the electric field at an arbitrarily positioned field point P. To calculate the electric field at P we first choose coordinate axes. We choose the x axis through the line charge and the y axis through point P as shown. The ends of the charged line segment are labeled x_1 and x_2. A typical charge element $dq = \lambda\,dx$ that produces a field $d\vec{E}$ is shown in the figure. The field at P has both an x and a y component. Only the y component is computed here. (The x component is to be computed in Problem 22-27.)

The magnitude of the field produced by an element of charge $dq = \lambda\,dx$ is

$$|d\vec{E}| = \frac{k\,dq}{r^2} = \frac{k\lambda\,dx}{r^2}$$

and the y component is

$$dE_y = |d\vec{E}|\cos\theta = \frac{k\lambda\,dx}{r^2}\frac{y}{r} = \frac{k\lambda y\,dx}{r^3} \qquad\qquad 22\text{-}6$$

where $\cos\theta = y/r$ and $r = \sqrt{x^2 + y^2}$. The total y component E_y is computed by integrating from $x = x_1$ to $x = x_2$.

$$E_y = \int_{x=x_1}^{x=x_2} dE_y = k\lambda y \int_{x_1}^{x_2} \frac{dx}{r^3} \qquad\qquad 22\text{-}7$$

In calculating this integral y remains fixed. One way to execute this calculation is to use trigonometric substitution. From the figure we can see that $x = y\tan\theta$, so $dx = y\sec^2\theta\,d\theta$.[†] We also can see that $y = r\cos\theta$, so $1/r = \cos\theta/y$. Substituting these into Equation 22-7 gives

$$E_y = k\lambda y\frac{1}{y^2}\int_{\theta_1}^{\theta_2}\cos\theta\,d\theta = \frac{k\lambda}{y}(\sin\theta_2 - \sin\theta_1) = \frac{kQ}{Ly}(\sin\theta_2 - \sin\theta_1) \qquad 22\text{-}8a$$

E_y DUE TO A UNIFORMLY CHARGED LINE SEGMENT

EXERCISE Show that for the line charge shown in Figure 22-3 $dE_x = -k\lambda x\,dx/r^3$.

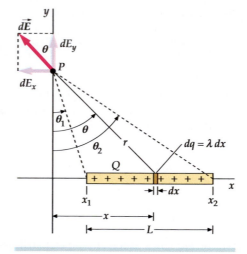

FIGURE 22-3 Geometry for the calculation of the electric field at field point P due to a uniform finite line charge.

[†] We have used the relation $d(\tan\theta)/d\theta = \sec^2\theta$.

The x component for the finite line charge shown in Figure 22-3 (and computed in Problem 22-27) is

$$E_x = \frac{k\lambda}{y}(\cos\theta_2 - \cos\theta_1) \qquad 22\text{-}8b$$

E_x DUE TO A UNIFORMLY CHARGED LINE SEGMENT

$\vec{E}$ Due to an Infinite Line Charge

A line charge may be considered infinite if for any field point of interest P (see Figure 22-3), $x_1 \to -\infty$ and $x_2 \to +\infty$. We compute E_x and E_y for an infinite line charge using Equations 22-8a and b in the limit that $\theta_1 \to -\pi/2$ and $\theta_2 \to \pi/2$. (From Figure 22-3 we can see that this is the same as the limit that $x_1 \to -\infty$ and $x_2 \to +\infty$.) Substituting $\theta_1 = -\pi/2$ and $\theta_2 = \pi/2$ into Equations 22-8a and b gives

$E_x = 0$ and $E_y = \dfrac{2k\lambda}{y}$, where y is the perpendicular distance from the line charge to the field point. Thus,

Electric field lines near a long wire. The electric field near a high-voltage power line can be large enough to ionize air, making the air a conductor. The glow resulting from the recombination of free electrons with the ions is called corona discharge.

$$E_R = 2k\frac{\lambda}{R} \qquad 22\text{-}9$$

$\vec{E}$ AT A DISTANCE R FROM AN INFINITE LINE CHARGE

where R is the perpendicular distance from the line charge to the field point.

EXERCISE Show that Equation 22-9 has the correct units for the electric field.

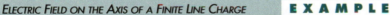

ELECTRIC FIELD ON THE AXIS OF A FINITE LINE CHARGE   **E X A M P L E   2 2 - 1**

Using Equations 22-8a and b, obtain an expression for the electric field on the perpendicular bisector of a uniformly charged line segment with linear charge density λ and length L.

FIGURE 22-4

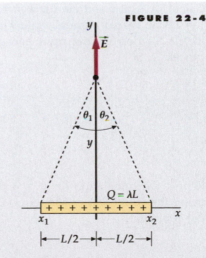

PICTURE THE PROBLEM Sketch the line charge on the x axis with the y axis as its perpendicular bisector. According to Figure 22-4 this means choosing $x_1 = -\frac{1}{2}L$ and $x_2 = \frac{1}{2}L$ so $\theta_1 = -\theta_2$. Then use Equations 22-8a and 22-8b to find the electric field.

1. Sketch the charge configuration with the line charge on the x axis with the y axis as its perpendicular bisector. Show the field point on the positive y axis a distance y from the origin:

2. Use Equation 22-8a to find an expression for E_y. Simplify using $\theta_2 = -\theta_1 = \theta$:

$$E_y = \frac{k\lambda}{y}(\sin\theta_2 - \sin\theta_1) = \frac{k\lambda}{y}[\sin\theta - \sin(-\theta)]$$

$$= \frac{2k\lambda}{y}\sin\theta$$

3. Express $\sin\theta$ in terms of y and L and substitute into the step 2 result:

$$\sin\theta = \frac{\frac{1}{2}L}{\sqrt{(\frac{1}{2}L)^2 + y^2}}$$

so

$$E_y = \frac{2k\lambda}{y}\frac{\frac{1}{2}L}{\sqrt{(\frac{1}{2}L)^2 + y^2}}$$

4. Use Equation 22-8b to determine E_x:

$$E_x = \frac{k\lambda}{y}(\cos\theta_2 - \cos\theta_1) = \frac{k\lambda}{y}\left[\cos\theta - \cos(-\theta)\right]$$

$$= \frac{k\lambda}{y}(\cos\theta - \cos\theta) = 0$$

5. Express the vector $\vec{E}$:

$$\vec{E} = E_x\hat{i} + E_y\hat{j} = \boxed{\frac{2k\lambda}{y}\frac{\frac{1}{2}L}{\sqrt{(\frac{1}{2}L)^2 + y^2}}\hat{j}}$$

ELECTRIC FIELD NEAR AND FAR FROM A FINITE LINE CHARGE   **EXAMPLE 22-2**

A line charge of linear density $\lambda = 4.5$ nC/m lies on the x axis and extends from $x = -5$ cm to $x = 5$ cm. Using the expression for E_y obtained in Example 22-1, calculate the electric field on the y axis at (*a*) $y = 1$ cm, (*b*) $y = 4$ cm, and (*c*) $y = 40$ cm. (*d*) Estimate the electric field on the y axis at $y = 1$ cm, assuming the line charge to be infinite. (*e*) Find the total charge and estimate the field at $y = 40$ cm, assuming the line charge to be a point charge.

PICTURE THE PROBLEM Use the result of Example 22-1 to obtain the electric field on the y axis. In the expression for $\sin\theta_0$, we can express L and y in centimeters because the units cancel. (*d*) To find the field very near the line charge, we use $E_y = 2k\lambda/y$. (*e*) To find the field very far from the charge, we use $E_y = kQ/y^2$ with $Q = \lambda L$.

1. Calculate E_y at $y = 1$ cm for $\lambda = 4.5$ nC/m and $L = 10$ cm. We can express L and y in centimeters in the fraction on the right because the units cancel.

$$E_y = \frac{2k\lambda}{y}\frac{\frac{1}{2}L}{\sqrt{(\frac{1}{2}L)^2 + y^2}}$$

$$= \frac{2(8.99 \times 10^9 \text{ N·m}^2/\text{C}^2)(4.5 \times 10^{-9}\text{ C/m})}{0.01\text{ m}}\frac{5\text{ cm}}{\sqrt{(5\text{ cm})^2 + (1\text{ cm})^2}}$$

$$= \frac{80.9\text{ N·m/C}}{0.01\text{ m}}\frac{5\text{ cm}}{\sqrt{(5\text{ cm})^2 + (1\text{ cm})^2}} = 7.93 \times 10^3 \text{ N/C}$$

$$= \boxed{7.93\text{ kN/C}}$$

2. Repeat the calculation for $y = 4$ cm $= 0.04$ m using the result $2k\lambda = 80.9$ N·m/C to simplify the notation:

$$E_y = \frac{80.9\text{ N·m/C}}{0.04\text{ m}}\frac{5\text{ cm}}{\sqrt{(5\text{ cm})^2 + (4\text{ cm})^2}} = 1.58 \times 10^3 \text{ N/C}$$

$$= \boxed{1.58\text{ kN/C}}$$

3. Repeat the calculation for $y = 40$ cm:

$$E_y = \frac{80.9\text{ N·m/C}}{0.40\text{ m}}\frac{5\text{ cm}}{\sqrt{(5\text{ cm})^2 + (40\text{ cm})^2}} = \boxed{25.1\text{ N/C}}$$

4. Calculate the field at $y = 1$ cm $= 0.01$ m due to an infinite line charge:

$$E_y \approx \frac{2k\lambda}{y} = \frac{80.9\text{ N·m/C}}{0.01\text{ m}} = \boxed{8.09\text{ kN/m}}$$

5. Calculate the total charge λL for $L = 0.1$ m and use it to find the field of a point charge at $y = 0.4$ m:

$$Q = \lambda L = (4.5\text{ nC/m})(0.1\text{ m}) = 0.45\text{ nC}$$

$$E_y \approx \frac{k\lambda L}{y^2} = \frac{kQ}{y^2} = \frac{(8.99 \times 10^9 \text{ N·m}^2/\text{C}^2)(0.45 \times 10^{-9}\text{ C})}{(0.40\text{ m})^2}$$

$$= \boxed{25.3\text{ N/C}}$$

REMARKS At 1 cm from the 10-cm-long line charge, the estimated value of 8.09 kN/C obtained by assuming an infinite line charge differs from the exact value of 7.93 calculated in (*a*) by about 2 percent. At 40 cm from the line charge, the approximate value of 25.3 N/C obtained by assuming the line charge to be a point charge differs from the exact value of 25.1 N/C obtained in (*c*) by about 1 percent. Figure 22-5 shows the exact result for this line segment of length 10 cm and charge density 4.5 nC/m, and for the limiting cases of an infinite line charge of the same charge density, and a point charge $Q = \lambda L$.

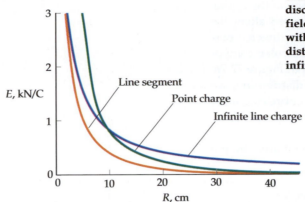

FIGURE 22-5 The magnitude of the electric field is plotted versus distance for the 10-cm-long line charge, the point charge, and the infinite line charge discussed in Example 22-2. Note that the field of the finite line segment converges with the field of the point charge at large distances, and with the field of the infinite line charge at small distances.

FIELD DUE TO A LINE CHARGE AND A POINT CHARGE   **EXAMPLE 22-3** **Try It Yourself**

An infinitely long line charge of linear charge density $\lambda = 0.6 \ \mu\text{C/m}$ lies along the z axis, and a point charge $q = 8 \ \mu\text{C}$ lies on the y axis at $y = 3$ m. Find the electric field at the point P on the x axis at $x = 4$ m.

PICTURE THE PROBLEM The electric field for this system is the superposition of the fields due to the infinite line charge and the point charge. The field of the line charge, $\vec{E}_L$, points radially away from the z axis (Figure 22-6). Thus, at point P on the x axis, $\vec{E}_L$ is in the positive x direction. The point charge produces a field $\vec{E}_P$ along the line connecting q and the point P. The distance from q to P is
$$r = \sqrt{(3 \text{ m})^2 + (4 \text{ m})^2} = 5 \text{ m}.$$

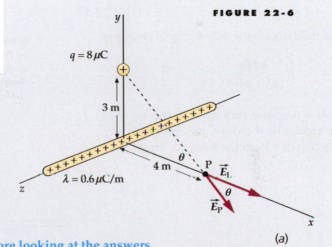

FIGURE 22-6

(a)

Cover the column to the right and try these on your own before looking at the answers.

Steps	Answers
1. Calculate the field $\vec{E}_L$ at point P due to the infinite line charge.	$\vec{E}_L = 2.70 \text{ kN/C}\,\hat{i}$
2. Find the field $\vec{E}_P$ at point P due to the point charge. Express $\vec{E}_P$ in terms of the unit vector $\hat{r}$ that points from q toward P.	$\vec{E}_P = 2.88 \text{ kN/C}\,\hat{r}$
3. Find the x and y components of $\vec{E}_P$.	$E_{Px} = E_P(0.8) = 2.30 \text{ kN/C}$ $E_{Py} = E_P(-0.6) = -1.73 \text{ kN/C}$
4. Find the x and y components of the total field at point P.	$E_x = \boxed{5.00 \text{ kN/C}}$, $E_y = \boxed{-1.73 \text{ kN/C}}$
5. Use your result in step 4 to calculate the magnitude of the total field.	$E = \sqrt{E_x^2 + E_y^2} = \boxed{5.29 \text{ kN/C}}$
6. Use your results in step 4 to find the angle ϕ between the field and the direction of increasing x.	$\phi = \tan^{-1}\dfrac{E_y}{E_x} = \boxed{-19.1°}$

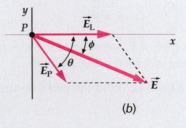

(b)

$\vec{E}$ on the Axis of a Ring Charge

Figure 22-7a shows a uniform ring charge of radius a and total charge Q. The field $d\vec{E}$ at point P on the axis due to the charge element dq is shown in the figure. This field has a component dE_x directed along the axis of the ring and a component $dE_\perp$ directed perpendicular to the axis. The perpendicular components cancel in pairs, as can be seen in Figure 22-7b. From the symmetry of the charge distribution, we can see that the net field due to the entire ring must lie along the axis of the ring; that is, the perpendicular components sum to zero.

The axial component of the field due to the charge element shown is

$$dE_x = \frac{k\,dq}{r^2}\cos\theta = \frac{k\,dq}{r^2}\frac{x}{r} = \frac{k\,dq\,x}{(x^2+a^2)^{3/2}}$$

where

$$r^2 = x^2 + a^2 \quad \text{and} \quad \cos\theta = \frac{x}{r} = \frac{x}{\sqrt{x^2+a^2}}$$

The field due to the entire ring of charge is

$$E_x = \int \frac{kx\,dq}{(x^2+a^2)^{3/2}}$$

Since x does not vary as we integrate over the elements of charge, we can factor any function of x from the integral. Then

$$E_x = \frac{kx}{(x^2+a^2)^{3/2}}\int dq$$

or

$$E_x = \frac{kQx}{(x^2+a^2)^{3/2}} \qquad \text{22-10}$$

A plot of E_x versus x along the axis of the ring is shown in Figure 22-8.

EXERCISE Find the point on the axis of the ring where E_x is maximum. (*Answer* $x = a/\sqrt{2}$)

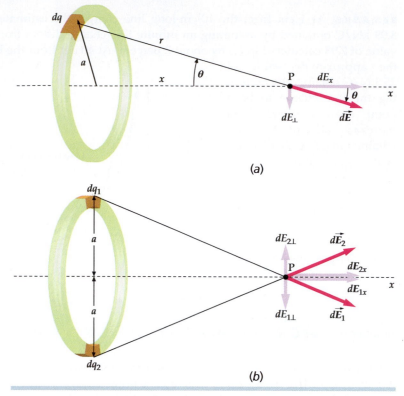

(a)

(b)

FIGURE 22-7 (*a*) A ring charge of radius a. The electric field at point P on the x axis due to the charge element dq shown has one component along the x axis and one perpendicular to the x axis. (*b*) For any charge element dq_1 there is an equal charge element dq_2 opposite it, and the electric-field components perpendicular to the x axis sum to zero.

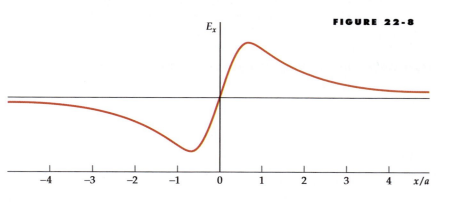

FIGURE 22-8

$\vec{E}$ on the Axis of a Uniformly Charged Disk

Figure 22-9 shows a uniformly charged disk of radius R and total charge Q. We can calculate the field on the axis of the disk by treating the disk as a set of concentric ring charges. Let the axis of the disk be the x axis. $\vec{E}$ due to the charge on each ring is along the x axis. A ring of radius a and width da is shown in the figure. The area of this ring is $dA = 2\pi a\,da$, and its charge is $dq = \sigma\,dA = 2\pi\sigma a\,da$, where $\sigma = Q/\pi R^2$ is the surface charge density (the charge per unit area). The field produced by this ring is given by Equation 22-10 if we replace Q with $dq = 2\pi\sigma a\,da$.

$$dE_x = \frac{kx2\pi\sigma a\,da}{(x^2+a^2)^{3/2}}$$

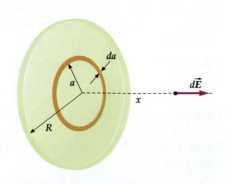

FIGURE 22-9 A uniform disk of charge can be treated as a set of ring charges, each of radius a.

The total field is found by integrating from $a = 0$ to $a = R$:

$$E_x = \int_0^R \frac{kx2\pi\sigma a\,da}{(x^2 + a^2)^{3/2}} = kx\pi\sigma \int_0^R (x^2 + a^2)^{-3/2}2a\,da = kx\pi\sigma \int_{x^2+0^2}^{x^2+R^2} u^{-3/2}du$$

where $u = x^2 + a^2$, so $du = 2a\,da$. The integration thus gives

$$E_x = kx\pi\sigma \frac{u^{-1/2}}{-1/2}\bigg|_{x^2}^{x^2+R^2} = -2kx\pi\sigma\left(\frac{1}{\sqrt{x^2 + R^2}} - \frac{1}{\sqrt{x^2}}\right)$$

This can be expressed

$$E_x = 2\pi k\sigma\left(1 - \frac{1}{\sqrt{1 + \dfrac{R^2}{x^2}}}\right), \qquad x > 0 \qquad\qquad \text{22-11}$$

$\vec{E}$ ON THE AXIS OF A DISK CHARGE

EXERCISE Find an expression for E_x on the negative x axis. (*Answer* $E_x = -2\pi k\sigma\left(1 - \dfrac{1}{\sqrt{1 + \dfrac{R^2}{x^2}}}\right)$ for $x < 0$)

For $x \gg R$ (on the positive x axis far from the disk) we expect it to look like a point charge. If we merely replace R^2/x^2 with 0 for $x \gg R$, we get $E_x \to 0$. Although this is correct, it does not tell us anything about how E_x depends on x for large x. We can find this dependence by using the binomial expansion, $(1 + \epsilon)^n \approx 1 + n\epsilon$, for $|\epsilon| \ll 1$. Using this approximation on the second term in Equation 22-11, we obtain

$$\frac{1}{\left(1 + \dfrac{R^2}{x^2}\right)^{1/2}} = \left(1 + \frac{R^2}{x^2}\right)^{-1/2} \approx 1 - \frac{R^2}{2x^2}$$

Substituting this into Equation 22-11 we obtain

$$E_x \approx 2\pi k\sigma\left(1 - 1 + \frac{R^2}{2x^2}\right) = \frac{2k\pi R^2\sigma}{2x^2} = \frac{kQ}{x^2}, \qquad x \gg R \qquad\qquad \text{22-12}$$

where $Q = \sigma\pi R^2$ is the total charge on the disk. For large x, the electric field of the charged disk approaches that of a point charge Q at the origin.

$\vec{E}$ Due to an Infinite Plane of Charge

The field of an infinite plane of charge can be obtained from Equation 22-11 by letting the ratio R/x go to infinity. Then

$$E_x = 2\pi k\sigma, \qquad x > 0 \qquad\qquad \text{22-13a}$$

$\vec{E}$ NEAR AN INFINITE PLANE OF CHARGE

Thus, the field due to an infinite-plane charge distribution is uniform; that is, the field does not depend on x. On the other side of the infinite plane, for negative values of x, the field points in the negative x direction, so

$$E_x = -2\pi k\sigma, \qquad x < 0 \qquad\qquad \text{22-13b}$$

As we move along the x axis, the electric field jumps from $-2\pi k\sigma\,\hat{\imath}$ to $+2\pi k\sigma\,\hat{\imath}$ when we pass through an infinite plane of charge (Figure 22-10). There is thus a discontinuity in E_x in the amount $4\pi k\sigma$.

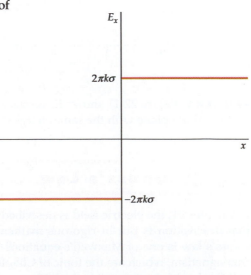

FIGURE 22-10 Graph showing the discontinuity of $\vec{E}$ at a plane charge.

ELECTRIC FIELD ON THE AXIS OF A DISK

EXAMPLE 22-4

A disk of radius 5 cm carries a uniform surface charge density of 4 $\mu C/m^2$. Using appropriate approximations, find the electric field on the axis of the disk at distances of (a) 0.01 cm, (b) 0.03 cm, and (c) 6 m. (d) Compare the results for (a), (b), and (c) with the exact values arrived at by using Equation 22-11.

PICTURE THE PROBLEM For the comparisons in Part (d), we will carry out all calculations to five-figure accuracy. For (a) and (b), the field point is very near the disk compared with its radius, so we can approximate the disk as an infinite plane. For (c), the field point is sufficiently far from the disk ($x/R = 120$) that we can approximate the disk as a point charge. (d) To compare, we find the percentage difference between the approximate values and the exact values.

(a) The electric field near the disk is approximately that due to an infinite plane charge:

$$E_x \approx 2\pi k\sigma$$
$$= 2\pi (8.98755 \times 10^9 \, N\cdot m^2/C^2)(4 \times 10^{-6} \, C/m^2)$$
$$= \boxed{225.88 \, kN/C}$$

(b) Since 0.03 cm is still very near the disk, the disk still looks like an infinite plane charge:

$$E_x \approx 2\pi k\sigma = \boxed{225.88 \, kN/C}$$

(c) Far from the disk, the field is approximately that due to a point charge:

$$E_x \approx \frac{kQ}{x^2} = \frac{k\sigma\pi R^2}{x^2} = 2\pi k\sigma \frac{R^2}{2x^2}$$
$$= (225.88 \, kN/C)\frac{(0.05 \, m)^2}{2(6 \, m)^2} = \boxed{7.8431 \, N/C}$$

(d) Using the exact expression (Equation 22-11) for E_x, we calculate the exact values at the specified points:

$$E_x(\text{exact}) = 2\pi k\sigma \left(1 - \frac{1}{\sqrt{1 + \dfrac{R^2}{x^2}}}\right)$$

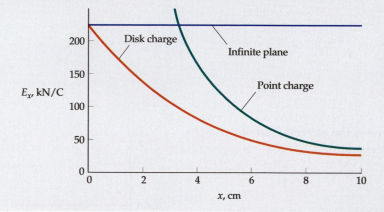

x (cm)	E_x (exact) (N/C)	E_x (approx) (N/C)	% diff
0.01	225,430	225,880	0.2
0.03	224,530	225,880	0.6
600	7.8427	7.8431	0.005

FIGURE 22-11 Note that the field of the disk charge converges with the field of the point charge at large distances, and equals the field of the infinite plane charge in the limit that x approaches zero.

REMARKS Figure 22-11 shows E_x versus x for the disk charge in this example, for an infinite plane with the same charge density, and for a point charge.

22-2 Gauss's Law

In Chapter 21, the electric field is described visually via electric field lines. Here that description is put in rigorous mathematical language called Gauss's law. Gauss's law is one of Maxwell's equations—the fundamental equations of electromagnetism, which are the topic of Chapter 31. For static charges, Gauss's law and Coulomb's law are equivalent. Electric fields arising from some symmetrical charge distributions, such as a spherical shell of charge or an infinite line of charge, can be easily calculated using Gauss's law. In this section, we give an argument for the validity of Gauss's law based on the properties of electric field lines. A rigorous derivation of Gauss's law is presented in Section 22-6.

A closed surface is one that divides the universe into two distinct regions, the region inside the surface and the region outside the surface. Figure 22-12 shows a closed surface of arbitrary shape enclosing a dipole. The number of electric field lines beginning on the positive charge and penetrating the surface from the inside depends on where the surface is drawn, but any line penetrating the surface from the inside also penetrates it from the outside. To count the net number of lines out of any closed surface, count any line that penetrates from the inside as +1, and any penetration from the outside as −1. Thus, for the surface shown (Figure 22-12), the net number of lines out of the surface is zero. For surfaces enclosing other types of

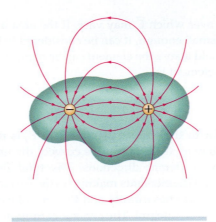

FIGURE 22-12 A surface of arbitrary shape enclosing an electric dipole. As long as the surface encloses both charges, the number of lines penetrating the surface from the inside is exactly equal to the number of lines penetrating the surface from the outside no matter where the surface is drawn.

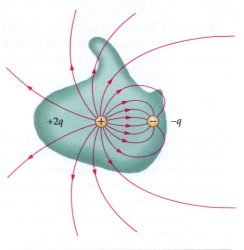

FIGURE 22-13 A surface of arbitrary shape enclosing the charges +2q and −q. Either the field lines that end on −q do not pass through the surface or they penetrate it from the inside the same number of times as from the outside. The net number that exit, the same as that for a single charge of +q, is equal to the net charge enclosed by the surface.

charge distributions, such as that shown in Figure 22-13, *the net number of lines out of any surface enclosing the charges is proportional to the net charge enclosed by the surface.* This rule is a qualitative statement of Gauss's law.

Electric Flux

The mathematical quantity that corresponds to the number of field lines penetrating a surface is called the **electric flux** ϕ. For a surface perpendicular to $\vec{E}$ (Figure 22-14), the electric flux is the product of the magnitude of the field E and the area A:

$$\phi = EA$$

The units of flux are N·m²/C. Because E is proportional to the number of field lines per unit area, the flux is proportional to the number of field lines penetrating the surface.

In Figure 22-15, the surface of area A_2 is not perpendicular to the electric field $\vec{E}$. However, the number of lines that penetrate the surface of area A_2 is the same as the number that penetrate the surface of area A_1, which is perpendicular to $\vec{E}$. These areas are related by

$$A_2 \cos \theta = A_1$$

where θ is the angle between $\vec{E}$ and the unit vector $\hat{n}$ that is normal to the surface A_2, as shown in the figure. The electric flux through a surface is defined to be

$$\phi = \vec{E} \cdot \hat{n} A = EA \cos \theta = E_n A \qquad 22\text{-}15$$

where $E_n = \vec{E} \cdot \hat{n}$ is the component of $\vec{E}$ normal (perpendicular) to the surface.

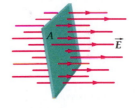

FIGURE 22-14 Electric field lines of a uniform field penetrating a surface of area A that is oriented perpendicular to the field. The product EA is the electric flux through the surface.

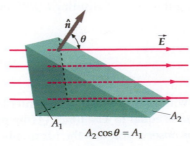

22-14

FIGURE 22-15 Electric field lines of a uniform electric field that is perpendicular to the surface of area A_1 but makes an angle θ with the unit vector $\hat{n}$ that is normal to the surface of area A_2. Where $\vec{E}$ is not perpendicular to the surface, the flux is $E_n A$, where $E_n = E \cos \theta$ is the component of $\vec{E}$ that is perpendicular to the surface. The flux through the surface of area A_2 is the same as that through the surface of area A_1.

Figure 22-16 shows a curved surface over which $\vec{E}$ may vary. If the area ΔA_i of the surface element that we choose is small enough, it can be considered to be a plane, and the variation of the electric field across the element can be neglected. The flux of the electric field through this element is

$$\Delta\phi_i = E_{ni}\,\Delta A_i = \vec{E}_i \cdot \hat{n}_i\,\Delta A_i$$

where $\hat{n}_i$ is the unit vector perpendicular to the surface element and $\vec{E}_i$ is the electric field anywhere on the surface element. If the surface is curved, the unit vectors for different elements will have different directions. The total flux through the surface is the sum of $\Delta\phi_i$ over all the elements making up the surface. In the limit, as the number of elements approaches infinity and the area of each element approaches zero, this sum becomes an integral. The general definition of electric flux is thus:

$$\phi = \lim_{\Delta A_i \to 0} \sum_i \vec{E}_i \cdot \hat{n}_i\,\Delta A_i = \int_S \vec{E} \cdot \hat{n}\,dA \qquad \text{22-16}$$

DEFINITION—ELECTRIC FLUX

where the S stands for the surface we are integrating over.

On a *closed* surface we are interested in the electric flux out of the surface, so we choose the unit vector $\hat{n}$ to be outward at each point. The integral over a closed surface is indicated by the symbol $\oint$. The total or net flux out of a closed surface is therefore written

$$\phi_{\text{net}} = \oint_S \vec{E} \cdot \hat{n}\,dA = \oint_S E_n\,dA \qquad \text{22-17}$$

The net flux ϕ_{net} through the closed surface is positive or negative, depending on whether $\vec{E}$ is predominantly outward or inward at the surface. At points on the surface where $\vec{E}$ is inward, E_n is negative.

Quantitative Statement of Gauss's Law

Figure 22-17 shows a spherical surface of radius R with a point charge Q at its center. The electric field everywhere on this surface is normal to the surface and has the magnitude

$$E_n = \frac{kQ}{R^2}$$

The net flux of $\vec{E}$ out of this spherical surface is

$$\phi_{\text{net}} = \oint_S E_n\,dA = E_n \oint_S dA$$

where we have taken E_n out of the integral because it is constant everywhere on the surface. The integral of dA over the surface is just the total area of the surface, which for a sphere of radius R is $4\pi R^2$. Using this and substituting kQ/R^2 for E_n, we obtain

$$\phi_{\text{net}} = \frac{kQ}{R^2}\,4\pi R^2 = 4\pi kQ \qquad \text{22-18}$$

Thus, the net flux out of a spherical surface with a point charge at its center is independent of the radius R of the sphere and is equal to $4\pi k$ times Q (the point charge). This is consistent with our previous observation that the net number of

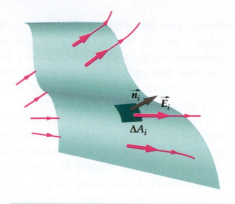

FIGURE 22-16 If E_n varies from place to place on a surface, either because E varies or because the angle between $\vec{E}$ and $\hat{n}$ varies, the area of the surface is divided into small elements of area ΔA_i. The flux through the surface is computed by summing $\vec{E}_i \cdot \hat{n}_i\,\Delta A_i$ over all the area elements.

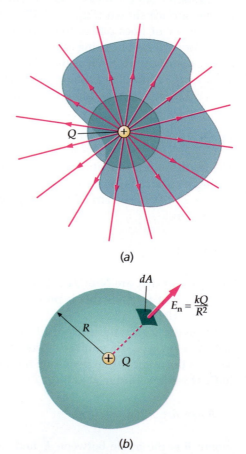

FIGURE 22-17 A spherical surface enclosing a point charge Q. (*a*) The net number of electric field lines out of this surface and the net number out of any surface that also encloses Q is the same. (*b*) The net flux is easily calculated for a spherical surface. It equals E_n times the surface area, or $E_n 4\pi R^2$.

lines going out of a closed surface is proportional to the net charge inside the surface. *This number of lines is the same for all closed surfaces surrounding the charge, independent of the shape of the surface.* Thus, the net flux out of *any surface* surrounding a point charge Q equals $4\pi kQ$.

We can extend this result to systems containing multiple charges. In Figure 22-18, the surface encloses two point charges, q_1 and q_2, and there is a third point charge q_3 outside the surface. Since the electric field at any point on the surface is the vector sum of the electric fields produced by each of the three charges, the net flux $\phi_{net} = \oint_s \vec{E} \cdot \hat{n}\, dA$ out of the surface is just the sum of the fluxes due to the individual charges. The flux due to charge q_3, which is outside the surface, is zero because every field line from q_3 that enters the surface at one point leaves the surface at some other point. The flux out of the surface due to charge q_1 is $4\pi kq_1$ and that due to charge q_2 is $4\pi kq_2$. The net flux out of the surface therefore equals $4\pi k(q_1 + q_2)$, which may be positive, negative, or zero depending on the signs and magnitudes of q_1 and q_2.

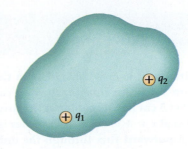

FIGURE 22-18 A surface enclosing point charges q_1 and q_2, but not q_3. The net flux out of this surface is $4\pi k(q_1 + q_2)$.

The net outward flux through any closed surface equals $4\pi k$ times the net charge inside the surface:

$$\phi_{net} = \int_S E_n\, dA = 4\pi kQ_{inside} \qquad 22\text{-}19$$

GAUSS'S LAW

This is **Gauss's law.** Its validity depends on the fact that the electric field due to a single point charge varies inversely with the square of the distance from the charge. It was this property of the electric field that made it possible to draw a fixed number of electric field lines from a charge and have the density of lines be proportional to the field strength.

It is customary to write the Coulomb constant k in terms of another constant ϵ_0, which is called the **permittivity of free space:**

$$k = \frac{1}{4\pi\,\epsilon_0} \qquad 22\text{-}20$$

Using this notation, Coulomb's law for $\vec{E}$ is written

$$\vec{E} = \frac{1}{4\pi\,\epsilon_0}\frac{q}{r^2}\hat{r} \qquad 22\text{-}21$$

and Gauss's law is written

$$\phi_{net} = \oint_S E_n\, dA = \frac{Q_{inside}}{\epsilon_0} \qquad 22\text{-}22$$

The value of ϵ_0 in SI units is

$$\epsilon_0 = \frac{1}{4\pi k} = \frac{1}{4\pi(8.99 \times 10^9\ \text{N·m}^2/\text{C}^2)} = 8.85 \times 10^{-12}\ \text{C}^2/\text{N·m}^2 \qquad 22\text{-}23$$

Gauss's law is valid for all surfaces and all charge distributions. For charge distributions that have high degrees of symmetry, it can be used to calculate the electric field, as we illustrate in the next section. For static charge distributions, Gauss's law and Coulomb's law are equivalent. However, Gauss's law is more general in that it is always valid and Coulomb's law is valid only for static charge distributions.

EXAMPLE 22-5

An electric field is $\vec{E} = (200 \text{ N/C})\hat{i}$ in the region $x > 0$ and $\vec{E} = (-200 \text{ N/C})\hat{i}$ in the region $x < 0$. An imaginary soup-can shaped surface of length 20 cm and radius $R = 5$ cm has its center at the origin and its axis along the x axis, so that one end is at $x = +10$ cm and the other is at $x = -10$ cm (Figure 22-19). (*a*) What is the net outward flux through the entire closed surface? (*b*) What is the net charge inside the closed surface?

PICTURE THE PROBLEM The closed surface described, which is piecewise continuous, consists of three pieces—two flat ends and a curved side. Separately calculate the flux of $\vec{E}$ out of each piece of the surface. To calculate the flux out of a piece draw the outward normal $\hat{n}$ at a randomly chosen point on the piece and draw the vector $\vec{E}$ at the same point. If $E_n = \vec{E} \cdot \hat{n}$ is the same everywhere on the piece, then the outward flux through it is $\phi = \vec{E} \cdot \hat{n}A$ (Equation 22-15). The net outward flux through the entire closed surface is obtained by summing the fluxes through the individual pieces. The net outward flux is related to the charge inside by Gauss's law (Equation 22-19).

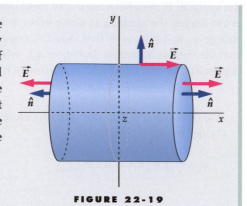

FIGURE 22-19

(*a*) 1. Sketch the soup-can shaped surface. On each piece of the surface draw the outward normal $\hat{n}$ and the vector $\vec{E}$:

2. Calculate the outward flux through the right circular flat surface where $\hat{n} = \hat{i}$:

$$\phi_{\text{right}} = \vec{E}_{\text{right}} \cdot \hat{n}_{\text{right}}A = \vec{E}_{\text{right}} \cdot \hat{i}\,\pi R^2$$
$$= (200 \text{ N/C})\hat{i} \cdot \hat{i}(\pi)(0.05 \text{ m})^2$$
$$= 1.57 \text{ N·m}^2/\text{C}$$

3. Calculate the outward flux through the left circular surface where $\hat{n} = -\hat{i}$:

$$\phi_{\text{left}} = \vec{E}_{\text{left}} \cdot \hat{n}_{\text{left}}A = \vec{E}_{\text{left}} \cdot (-\hat{i})\,\pi R^2$$
$$= (-200 \text{ N/C})\hat{i} \cdot (-\hat{i})(\pi)(0.05 \text{ m})^2$$
$$= 1.57 \text{ N·m}^2/\text{C}$$

4. Calculate the outward flux through the curved surface where $\vec{E}$ is perpendicular to $\hat{n}$:

$$\phi_{\text{curved}} = \vec{E}_{\text{curved}} \cdot \hat{n}_{\text{curved}}A = 0$$

5. The net outward flux is the sum through all the individual surfaces:

$$\phi_{\text{net}} = \phi_{\text{right}} + \phi_{\text{left}} + \phi_{\text{curved}}$$
$$= 1.57 \text{ N·m}^2/\text{C} + 1.57 \text{ N·m}^2/\text{C} + 0$$
$$= \boxed{3.14 \text{ N·m}^2/\text{C}}$$

(*b*) Gauss's law relates the charge inside to the net flux:

$$Q_{\text{inside}} = \epsilon_0\,\phi_{\text{net}}$$
$$= (8.85 \times 10^{-12} \text{ C}^2/\text{N·m}^2)(3.14 \times \text{N·m}^2/\text{C})$$
$$= \boxed{2.78 \times 10^{-11} \text{ C} = 27.8 \text{ pC}}$$

REMARKS The flux does not depend on the length of the can. This means the charge inside resides entirely on the yz plane.

22-3 Calculating $\vec{E}$ From Gauss's Law

Given a highly symmetrical charge distribution, the electric field can often be calculated more easily using Gauss's law than it can be using Coulomb's law. We first find an imaginary closed surface, called a **Gaussian surface** (the soup can in Example 22-5). Optimally, this surface is chosen so that on each of its pieces $\vec{E}$ is

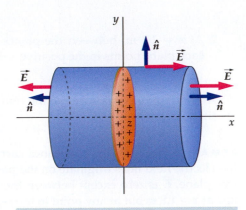

either zero, perpendicular to $\hat{n}$, or parallel to $\hat{n}$ with E_n constant. Then the flux through each piece equals $E_n A$ and Gauss's law is used to relate the field to the charges inside the closed surface.

Plane Symmetry

A charge distribution has **plane symmetry** if the views of it from all points on an infinite plain surface are the same. Figure 22-20 shows an infinite plane of charge of uniform surface charge density σ. By symmetry, $\vec{E}$ must be perpendicular to the plane and can depend only on the distance from it. Also, $\vec{E}$ must have the same magnitude but the opposite direction at points the same distance from the charged plane on either side of the plane. For our Gaussian surface, we choose a soup-can shaped cylinder as shown, with the charged plane bisecting the cylinder. On each piece of the cylinder is drawn both $\hat{n}$ and $\vec{E}$. Since $\vec{E} \cdot \hat{n}$ is zero everywhere on the curved piece of the Gaussian surface, there is no flux through it. The flux through each flat piece of the surface is $E_n A$, where A is the area of each flat piece. Thus, the total outward flux through the closed surface is $2E_n A$. The net charge inside the surface is σA. Gauss's law then gives

FIGURE 22-20 Gaussian surface for the calculation of $\vec{E}$ due to an infinite plane of charge. (Only the part of the plane that is inside the Gaussian surface is shown.) On the flat faces of this soup can, $\vec{E}$ is perpendicular to the surface and constant in magnitude. On the curved surface $\vec{E}$ is parallel with the surface.

$$Q_{\text{inside}} = \epsilon_0 \, \phi_{\text{net}}$$

$$\sigma A = \epsilon_0 \, 2E_n A$$

(Can you see why $Q_{\text{inside}} = \sigma A$?) Solving for E_n gives

$$E_n = \frac{\sigma}{2\epsilon_0} = 2\pi k \sigma \qquad\qquad\qquad \text{22-24}$$

$\vec{E}$ FOR AN INFINITE PLANE OF CHARGE

E_n is positive if σ is positive, and E_n is negative if σ is negative. This means if σ is positive $\vec{E}$ is directed away from the charged plane, and if σ is negative $\vec{E}$ points toward it. This is the same result that we obtained, with much more difficulty, using Coulomb's law (Equations 22-13a and b). Note that the field is discontinuous at the charged plane. If the charged plane is the yz plane, the field is $\vec{E} = \sigma/(2\epsilon_0)\hat{i}$ in the region $x > 0$ and $\vec{E} = -\sigma/(2\epsilon_0)\hat{i}$ in the region $x < 0$. Thus, the field is discontinuous by $\Delta\vec{E} = \sigma/(2\epsilon_0)\hat{i} - [-\sigma/(2\epsilon_0)\hat{i}] = (\sigma/\epsilon_0)\hat{i}$.

ELECTRIC FIELD DUE TO TWO INFINITE PLANES   **EXAMPLE 22-6**

FIGURE 22-21

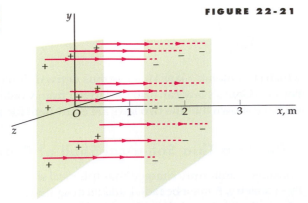

In Figure 22-21, an infinite plane of surface charge density $\sigma = +4.5$ nC/m^2 lies in the $x = 0$ plane, and a second infinite plane of surface charge density $\sigma = -4.5$ nC/m^2 lies in a plane parallel to the $x = 0$ plane at $x = 2$ m. Find the electric field at (a) $x = 1.8$ m and (b) $x = 5$ m.

PICTURE THE PROBLEM Each plane produces a uniform electric field of magnitude $E = \sigma/(2\epsilon_0)$. We use superposition to find the resultant field. Between the planes the fields add, producing a net field of magnitude σ/ϵ_0 in the positive x direction. For $x > 2$ m and for $x < 0$, the fields point in opposite directions and cancel.

(a) 1. Calculate the magnitude of the field E produced by each plane:

$$E = \frac{\sigma}{2\,\epsilon_0} = \frac{4.5 \times 10^{-9}\,\text{C/m}^2}{2(8.85 \times 10^{-12}\,\text{C}^2/\text{N·m}^2)}$$

$$= 254 \text{ N/C}$$

2. At $x = 1.8$ m, between the planes, the field due to each plane points in the positive x direction:

$$E_{x,net} = E_1 + E_2 = 254 \text{ N/C} + 254 \text{ N/C}$$

$$= \boxed{508 \text{ N/C}}$$

(b) At $x = 5$ m, the fields due to the two planes are oppositely directed:

$$E_{x,net} = E_1 - E_2 = \boxed{0}$$

REMARKS Because the two planes carry equal and opposite charge densities, the electric field lines originate on the positive plane and terminate on the negative plane. $\vec{E}$ is zero except between the planes. Note that $E_{x,net} = 508$ N/C not just at $x = 1.8$ m but at any point in the region between the charged planes.

Spherical Symmetry

Assume a charge distribution is concentric within a spherical surface. The charge distribution has **spherical symmetry** if the views of it from all points on the spherical surface are the same. To calculate the electric field due to spherically symmetric charge distributions, we use a spherical surface for our Gaussian surface. We illustrate this by first finding the electric field at a distance r from a point charge q. We choose a spherical surface of radius r, centered at the point charge, for our Gaussian surface. By symmetry, $\vec{E}$ must be directed either radially outward or radially inward. It follows that the component of $\vec{E}$ normal to the surface equals the radial component of E at each point on the surface. That is, $E_n = \vec{E} \cdot \hat{n} = E_r$, where $\hat{n}$ is the outward normal, has the same value everywhere on the spherical surface. Also, the magnitude of $\vec{E}$ can depend on the distance from the charge but not on the direction from the charge. The net flux through the spherical surface of radius r is thus

$$\phi_{net} = \oint_S \vec{E} \cdot \hat{n} \, dA = \oint_S E_r \, dA = E_r \oint_S dA = E_r 4\pi r^2$$

where $\oint_S dA = 4\pi r^2$ the total area of the spherical surface. Since the total charge inside the surface is just the point charge q, Gauss's law gives

$$E_r 4\pi r^2 = \frac{q}{\epsilon_0}$$

Solving for E_r gives

$$E_r = \frac{1}{4\pi\epsilon_0} \frac{q}{r^2}$$

which is Coulomb's law. We have thus derived Coulomb's law from Gauss's law. Because Gauss's law can also be derived from Coulomb's law (see Section 22-6), we have shown that the two laws are equivalent for static charges.

$\vec{E}$ Due to a Thin Spherical Shell of Charge

Consider a uniformly charged thin spherical shell of radius R and total charge Q. By symmetry, $\vec{E}$ must be radial, and its magnitude can depend only on the distance r from the center of the sphere. In Figure 22-22, we have chosen a spherical Gaussian surface of radius $r > R$. Since $\vec{E}$ is normal to this surface, and has the same magnitude everywhere on the surface, the flux through the surface is

$$\phi_{net} = \oint_S E_r \, dA = E_r \oint_S dA = E_r 4\pi r^2$$

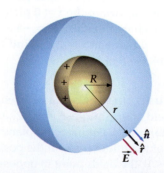

FIGURE 22-22 Spherical Gaussian surface of radius $r > R$ for the calculation of the electric field outside a uniformly charged thin spherical shell of radius R.

Since the total charge inside the Gaussian surface is the total charge on the shell Q, Gauss's law gives

$$E_r 4\pi r^2 = \frac{Q}{\epsilon_0}$$

or

$$E_r = \frac{1}{4\pi\epsilon_0}\frac{Q}{r^2}, \qquad r > R \qquad\qquad\qquad 22\text{-}25a$$

$\vec{E}$ OUTSIDE A SPHERICAL SHELL OF CHARGE

Thus, the electric field outside a uniformly charged spherical shell is the same as if all the charge were at the center of the shell.

If we choose a spherical Gaussian surface inside the shell, where $r < R$, the net flux is again $E_r\, 4\pi r^2$, but the total charge inside the surface is zero. Therefore, for $r < R$, Gauss's law gives

$$\phi_{\text{net}} = E_r 4\pi r^2 = 0$$

so

$$E_r = 0, \qquad r < R \qquad\qquad\qquad\qquad 22\text{-}25b$$

$\vec{E}$ INSIDE A SPHERICAL SHELL OF CHARGE

These results can also be obtained by direct integration of Coulomb's law, but that calculation is much more difficult.

Figure 22-23 shows E_r versus r for a spherical-shell charge distribution. Again, note that the electric field is discontinuous at $r = R$, where the surface charge density is $\sigma = Q/4\pi R^2$. Just outside the shell at $r \approx R$, the electric field is $E_r = Q/4\pi\epsilon_0 R^2 = \sigma/\epsilon_0$, since $\sigma = Q/4\pi R^2$. Because the field just inside the shell is zero, the electric field is discontinuous by the amount σ/ϵ_0 as we pass through the shell.

(a)

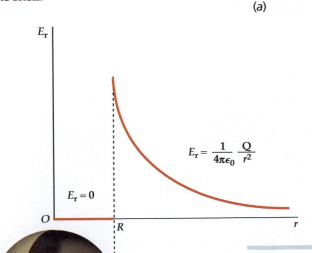

E_r

$E_r = \frac{1}{4\pi\epsilon_0}\frac{Q}{r^2}$

$E_r = 0$

O

R

r

(b)

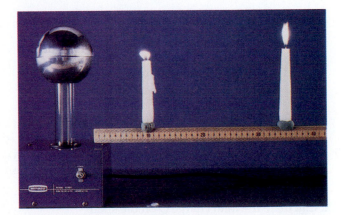

FIGURE 22-23 (a) A plot of E_r versus r for a spherical-shell charge distribution. The electric field is discontinuous at $r = R$, where there is a surface charge of density σ. (b) The decrease in E_r over distance due to a charged spherical shell is evident by the effect of the field on the flames of these two candles. The spherical shell at the left (part of a Van de Graaff generator, a device that is discussed in Chapter 24) carries a large negative charge that attracts the positive ions in the nearby candle flame. The flame at right, which is much farther away, is not noticeably affected.

E X A M P L E   2 2 - 7

A spherical shell of radius $R = 3$ m has its center at the origin and carries a
surface charge density of $\sigma = 3$ nC/m². A point charge $q = 250$ nC is on the y
axis at $y = 2$ m. Find the electric field on the x axis at (a) $x = 2$ m and (b) $x = 4$ m.

PICTURE THE PROBLEM We find the field due to the point charge and that
due to the spherical shell and sum the field vectors. For (a), the field point is inside the shell, so the field is due only to the point charge (Figure 22-24a). For (b),
the field point is outside the shell, so the shell can be considered as a point charge
at the origin. We then find the field due to two point charges (Figure 22-24b).

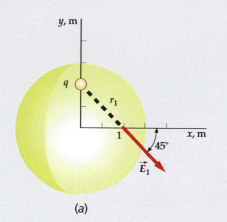

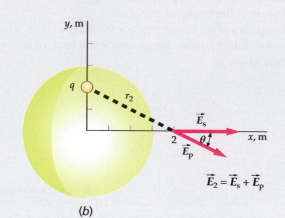

$$\vec{E}_2 = \vec{E}_s + \vec{E}_p$$

FIGURE 22-24

(a)   (b)

(a) 1. Inside the shell, $\vec{E}_1$ is due only to the point charge:

$$\vec{E}_1 = \frac{kq}{r_1^2}\hat{r}_1$$

2. Calculate the square of the distance r_1:

$$r_1^2 = (2\text{ m})^2 + (2\text{ m})^2 = 8\text{ m}^2$$

3. Use r_1^2 to calculate the magnitude of the field:

$$E_1 = \frac{kq}{r_1^2} = \frac{(8.99 \times 10^9 \text{ N·m}^2/\text{C}^2)(250 \times 10^{-9}\text{ C})}{8\text{ m}^2}$$

$$= 281\text{ N/C}$$

4. From Figure 22-24a, we can see that the field makes
an angle of 45° with the x axis:

$$\theta_1 = 45°$$

5. Express $\vec{E}_1$ in terms of its components:

$$\vec{E}_1 = E_{1x}\hat{i} + E_{1y}\hat{j} = E_1\cos 45°\hat{i} - E_1\sin 45°\hat{j}$$

$$= (281\text{ N/C})\cos 45°\hat{i} - (281\text{ N/C})\sin 45°\hat{j}$$

$$= \boxed{199\,(\hat{i} - \hat{j})\text{ N/C}}$$

(b) 1. Outside of its perimeter, the shell can be treated as a
point charge at the origin, and the field due to the
shell $\vec{E}_s$ is therefore along the x axis:

$$\vec{E}_s = \frac{kQ}{x_2^2}\hat{i}$$

2. Calculate the total charge Q on the shell:

$$Q = \sigma 4\pi R^2 = (3\text{ nC/m}^2)4\pi(3\text{ m})^2 = 339\text{ nC}$$

3. Use Q to calculate the field due to the shell:

$$E_s = \frac{kQ}{x_2^2} = \frac{(8.99 \times 10^9 \text{ N·m}^2/\text{C}^2)(339 \times 10^{-9}\text{ C})}{(4\text{ m})^2}$$

$$= 190\text{ N/C}$$

4. The field due to the point charge is:

$$\vec{E}_p = \frac{kq}{r_2^2}\hat{r}_2$$

5. Calculate the square of the distance from the point charge q on the y axis to the field point at $x = 4$ m:

$$r_2^2 = (2 \text{ m})^2 + (4 \text{ m})^2 = 20 \text{ m}^2$$

6. Calculate the magnitude of the field due to the point charge:

$$E_p = \frac{kq}{r_2^2} = \frac{(8.99 \times 10^9 \text{ N·m}^2/\text{C}^2)(250 \times 10^{-9} \text{ C})}{20 \text{ m}^2}$$

$$= 112 \text{ N/C}$$

7. This field makes an angle θ with the x axis, where:

$$\tan \theta = \frac{2 \text{ m}}{4 \text{ m}} = \frac{1}{2} \Rightarrow \theta = \tan^{-1}\frac{1}{2} = 26.6°$$

8. The x and y components of the net electric field are thus:

$$E_x = E_{px} + E_{sx} = E_p \cos \theta + E_s$$

$$= (112 \text{ N/C}) \cos 26.6° + 190 \text{ N/C} = 290 \text{ N/C}$$

$$E_y = E_{py} + E_{sy} = -E_p \sin \theta + 0$$

$$= -(112 \text{ N/C}) \sin 26.6° = -50.0 \text{ N/C}$$

$$\vec{E} = \boxed{(290\hat{i} - 50.0\hat{j})\text{N/C}}$$

REMARKS Giving the x, y, and z components of a vector completely specifies the vector. In these cases, the z component is zero.

$\vec{E}$ Due to a Uniformly Charged Sphere

ELECTRIC FIELD DUE TO A CHARGED SOLID SPHERE   **EXAMPLE 22-8**

Find the electric field (a) outside and (b) inside a uniformly charged solid sphere of radius R carrying a total charge Q that is uniformly distributed throughout the volume of the sphere with charge density $\rho = Q/V$, where $V = \frac{4}{3}\pi R^3$ is the volume of the sphere.

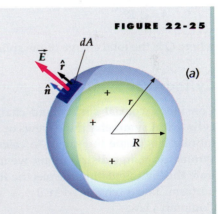

FIGURE 22-25

PICTURE THE PROBLEM By symmetry, the electric field must be radial. (a) To find E_r outside the charged sphere, we choose a spherical Gaussian surface of radius $r > R$ (Figure 22-25a). (b) To find E_r inside the charge we choose a spherical Gaussian surface of radius $r > R$ (Figure 22-25b). On each of these surfaces, E_r is constant. Gauss's law then relates E_r to the total charge inside the Gaussian surface.

(a) 1. (Outside) Draw a charged sphere of radius R and draw a spherical Gaussian surface with radius $r > R$:

2. Relate the flux through the Gaussian surface to the electric field E_r on it. At every point on this surface $\hat{n} = \hat{r}$ and E_r has the same value:

$$\phi_{\text{net}} = \vec{E} \cdot \hat{n}A = \vec{E} \cdot \hat{r}A = E_r4\pi r^2$$

3. Apply Gauss's law to relate the field to the total charge inside the surface, which is Q:

$$E_r4\pi r^2 = \frac{Q_{\text{inside}}}{\epsilon_0} = \frac{Q}{\epsilon_0}$$

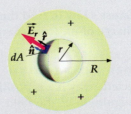

(b)

4. Solve for E_r:

$$E_r = \boxed{\frac{1}{4\pi \epsilon_0}\frac{Q}{r^2}, \quad r > R}$$

(b) 1. (Inside) Again draw the charged sphere of radius R. This time draw a spherical Gaussian surface with radius $r < R$:

2. Relate the flux through the Gaussian surface to the electric field E_r on it. At every point on this surface $\hat{n} = \hat{r}$ and E_r has the same value:

$$\phi_{\text{net}} = \vec{E} \cdot \hat{n}A = \vec{E} \cdot \hat{r}A = E_r4\pi r^2$$

3. Apply Gauss's law to relate the field to the total charge inside the surface Q_{inside}:

$$E_r 4\pi r^2 = \frac{Q_{inside}}{\epsilon_0}$$

4. The total charge inside the surface is $\rho V'$, where $\rho = Q/V$, $V = \frac{4}{3}\pi R^3$ and $V' = \frac{4}{3}\pi r^3$. V is the volume of the solid sphere and V' is the volume inside the Gaussian surface:

$$Q_{inside} = \rho V' = \left(\frac{Q}{V}\right)V' = \left(\frac{Q}{\frac{4}{3}\pi R^3}\right)\left(\frac{4}{3}\pi r^3\right) = Q\frac{r^3}{R^3}$$

5. Substitute this value for Q_{inside} and solve for E_r:

$$E_r 4\pi r^2 = \frac{Q_{inside}}{\epsilon_0} = \frac{1}{\epsilon_0}Q\frac{r^3}{R^3}$$

$$\boxed{E_r = \frac{1}{4\pi\epsilon_0}\frac{Q}{R^3}r, \quad r \leq R}$$

REMARKS Figure 22-26 shows E_r versus r for the charge distribution in this example. Inside a sphere of charge, E_r increases with r. Note that E_r is continuous at $r = R$. A uniformly charged sphere is sometimes used as a model to describe the electric field of an atomic nucleus.

We see from Example 22-8 that the electric field a distance r from the center of a uniformly charged sphere of radius R is given by

$$E_r = \frac{1}{4\pi\epsilon_0}\frac{Q}{r^2}, \quad r \geq R \qquad 22\text{-}26a$$

$$E_r = \frac{1}{4\pi\epsilon_0}\frac{Q}{R^3}r, \quad r \leq R \qquad 22\text{-}26b$$

where Q is the total charge of the sphere.

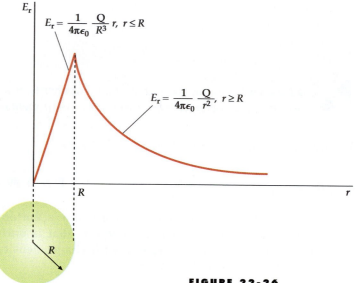

FIGURE 22-26

Cylindrical Symmetry

Consider a coaxial surface and charge distribution. A charge distribution has **cylindrical symmetry** if the views of it from all points on a cylindrical surface of infinite length are the same. To calculate the electric field due to cylindrically symmetric charge distributions, we use a cylindrical Gaussian surface. We illustrate this by calculating the electric field due to an infinitely long line charge of uniform linear charge density, a problem we have already solved using Coulomb's law.

ELECTRIC FIELD DUE TO INFINITE LINE CHARGE **EXAMPLE 22-9**

Use Gauss's law to find the electric field everywhere due to an infinitely long line charge of uniform charge density λ.

PICTURE THE PROBLEM Because of the symmetry, we know the electric field is directed away if λ is positive (directly toward it if λ is negative), and we know the magnitude of the field depends only on the radial distance from the line charge. We therefore choose a soup-can shaped Gaussian surface coaxial with the line. This surface consists of three pieces, the two flat ends and the curved side. We calculate the outward flux of $\vec{E}$ through each piece and, using Gauss's law, relate the net outward flux to the charge density λ.

1. Sketch the wire and a coaxial soup-can shaped Gaussian surface (Figure 22-27) with length L and radius R. The closed surface consists of three pieces, the two flat ends and the curved side. At a randomly chosen point on each piece, draw the vectors $\vec{E}$ and $\hat{n}$. Because of the symmetry, we know that the direction of $\vec{E}$ is directly away from the line charge if λ is positive (directly toward it if λ is negative), and we know that the magnitude of E depends only on the radial distance from the line charge.

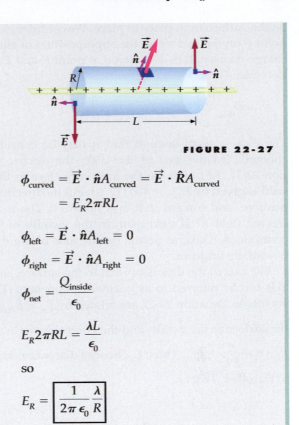

FIGURE 22-27

2. Calculate the outward flux through the curved piece of the Gaussian surface. At each point on the curved piece $\hat{R} = \hat{n}$, where $\hat{R}$ is the unit vector in the radial direction.

$$\phi_{\text{curved}} = \vec{E} \cdot \hat{n} A_{\text{curved}} = \vec{E} \cdot \hat{R} A_{\text{curved}}$$
$$= E_R 2\pi RL$$

3. Calculate the outward flux through each of the flat ends of the Gaussian surface. On these pieces the direction of $\hat{n}$ is parallel with the line charge (and thus perpendicular to $\vec{E}$):

$$\phi_{\text{left}} = \vec{E} \cdot \hat{n} A_{\text{left}} = 0$$
$$\phi_{\text{right}} = \vec{E} \cdot \hat{n} A_{\text{right}} = 0$$

4. Apply Gauss's law to relate the field to the total charge inside the surface Q_{inside}. The net flux out of the Gaussian surface is the sum of the fluxes out of the three pieces of the surface, and Q_{inside} is the charge on a length L of the line charge:

$$\phi_{\text{net}} = \frac{Q_{\text{inside}}}{\epsilon_0}$$

$$E_R 2\pi RL = \frac{\lambda L}{\epsilon_0}$$

so

$$\boxed{E_R = \frac{1}{2\pi\,\epsilon_0} \frac{\lambda}{R}}$$

REMARKS Since $1/(2\pi\epsilon_0) = 2k$, the field is $2k\lambda/R$, the same as Equation 22-9.

It is important to realize that although Gauss's law holds for any surface surrounding any charge distribution, it is very useful for calculating the electric fields of charge distributions that are highly symmetric. It is also useful doing calculations involving conductors in electrostatic equilibrium, as we shall see in Section 22.5. In the calculation of Example 22-9, we needed to assume that the field point was very far from the ends of the line charge so that E_n would be constant everywhere on the cylindrical Gaussian surface. (This is equivalent to assuming that, at the distance R from the line, the line charge appears to be infinitely long.) If we are near the end of a finite line charge, we cannot assume that $\vec{E}$ is perpendicular to the curved surface of the soup can, or that E_n is constant everywhere on it, so we cannot use Gauss's law to calculate the electric field.

FIGURE 22-28 (a) A surface carrying surface-charge. (b) The electric field $\vec{E}_{\text{disk}}$ due to the charge on a circular disk, plus the electric field $\vec{E}'$ due to all other charges. The right side of the disk is the + side, the left side the − side.

22-4 Discontinuity of E_n

We have seen that the electric field for an infinite plane of charge and a thin spherical shell of charge is discontinuous by the amount σ/ϵ_0 on either side of a surface carrying charge density σ. We now show that this is a general result for the component of the electric field that is perpendicular to a surface carrying a charge density of σ.

Figure 22-28 shows an arbitrary surface carrying a surface charge density σ. The surface is arbitrary in that it is arbitrarily curved, although it does not have any sharp folds, and σ may vary continuously

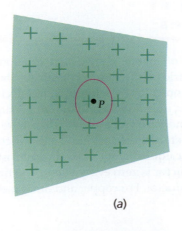

(a)

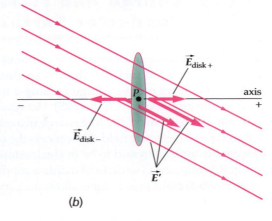

(b)

on the surface from place to place. We consider electric field $\vec{E}$ in the vicinity of a point P on the surface as the superposition of electric field $\vec{E}_{disk}$, due just to the charge on a small disk centered at point P, and $\vec{E}$ due to all other charges in the universe. Thus,

$$\vec{E} = \vec{E}_{disk} + \vec{E}'$$

22-27

The disk is small enough that it may be considered both flat and uniformly charged. On the axis of the disk, the electric field $\vec{E}_{disk}$ is given by Equation 22-11. At points on the axis very close to the disk, the magnitude of this field is given by $E_{disk} = |\sigma|/(2\,\epsilon_0)$ and its direction is away from the disk if σ is positive, and toward it if σ is negative. The magnitude and direction of the electric field $\vec{E}'$ is unknown. In the vicinity of point P, however, this field is continuous. Thus, at points on the axis of the disk and very close to it, $\vec{E}'$ is essentially uniform.

The axis of the disk is normal to the surface, so vector components along this axis can be referred to as normal components. The normal components of the vectors in Equation 22-27 are related by $E_n = E_{disk\,n} + E'_n$. If we refer one side of the surface as the $+$ side, and the other side the $-$ side, then $E_{n+} = \dfrac{\sigma}{2\,\epsilon_0} + E'_{n+}$ and $E_{n-} = -\dfrac{\sigma}{2\,\epsilon_0} + E'_{n+}$. Thus, E_n changes discontinuously from one side of the surface to the other. That is:

$$\Delta E_n = E_{n+} - E_{n-} = \frac{\sigma}{2\,\epsilon_0} - \left(-\frac{\sigma}{2\,\epsilon_0}\right) = \frac{\sigma}{\epsilon_0}$$

22-28

DISCONTINUITY OF E_n AT A SURFACE CHARGE

where we have made use of the fact that near the disk $E'_{n+} = E'_{n-}$ (since $\vec{E}'$ is continuous and uniform).

Note that the discontinuity of E_n occurs at a finite disk of charge, an infinite plane of charge (refer to Figure 22-10), and a thin spherical shell of charge (see Figure 22-23). However, it does not occur at the perimeter of a solid sphere of charge (see Figure 22-26). The electric field is discontinuous at any location with an infinite volume-charge density. These include locations with a finite point charge, locations with a finite line-charge density, and locations with a finite surface-charge density. At all locations with a finite surface-charge density, the normal component of the electric field is discontinuous—in accord with Equation 22-28.

22-5 Charge and Field at Conductor Surfaces

A conductor contains an enormous amount of mobile charge that can move freely within the conductor. If there is an electric field within a conductor, there will be a net force on this charge causing a momentary electric current (electric currents are discussed in Chapter 25). However, unless there is a source of energy to maintain this current, the free charge in a conductor will merely redistribute itself to create an electric field that cancels the external field within the conductor. The conductor is then said to be in **electrostatic equilibrium.** Thus, in electrostatic equilibrium, the electric field inside a conductor is zero everywhere. The time taken to reach equilibrium depends on the conductor. For copper and other metal conductors, the time is so small that in most cases electrostatic equilibrium is reached in a few nanoseconds.

We can use Gauss's law to show that any net electric charge on a conductor resides on the surface of the conductor. Consider a Gaussian surface completely inside the material of a conductor in electrostatic equilibrium (Figure 22-29). The size and shape of the Gaussian surface doesn't matter, as long as the entire surface is within the material of the conductor. The electric field is zero everywhere on the Gaussian surface because the surface is completely within the conductor where the field is everywhere zero. The net flux of the electric field through the surface must therefore be zero, and, by Gauss's law, the net charge inside the surface must be zero. Thus, there can be no net charge inside any surface lying completely within the material of the conductor. If a conductor carries a net charge, it must reside on the conductor's surface. At the surface of a conductor in electrostatic equilibrium, $\vec{E}$ must be perpendicular to the surface. We conclude this by reasoning that if the electric field had a tangential component at the surface, the free charge would be accelerated tangential to the surface until electrostatic equilibrium was reestablished.

Since E_n is discontinuous at any charged surface by the amount σ/ϵ_0, and since $\vec{E}$ is zero inside the material of a conductor, the field just outside the surface of a conductor is given by

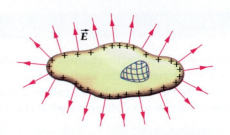

FIGURE 22-29 A Gaussian surface completely within the material of a conductor. Since the electric field is zero inside a conductor in electrostatic equilibrium, the net flux through this surface must also be zero. Therefore, the net charge density ρ within the material of a conductor must be zero.

$$E_n = \frac{\sigma}{\epsilon_0}$$ 22-29

E_n JUST OUTSIDE THE SURFACE OF A CONDUCTOR

This result is exactly twice the field produced by a uniform disk of charge. We can understand this result from Figure 22-30. The charge on the conductor consists of two parts: (1) the charge near point P and (2) all the rest of the charge. The charge near point P looks like a small, uniformly charged circular disk centered at P that produces a field near P of magnitude $\sigma/(2\epsilon_0)$ just inside and just outside the conductor. The rest of the charges in the universe must produce a field of magnitude $\sigma/(2\epsilon_0)$ that exactly cancels the field inside the conductor. This field due to the rest of the charge adds to the field due to the small charged disk just outside the conductor to give a total field of σ/ϵ_0.

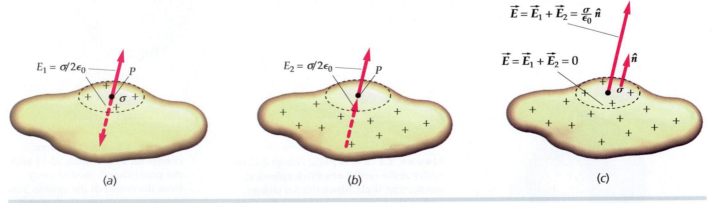

FIGURE 22-30 An arbitrarily shaped conductor carrying a charge on its surface. (a) The charge in the vicinity of point P near the surface looks like a small uniformly charged circular disk centered at P, giving an electric field of magnitude $\sigma/(2\epsilon_0)$ pointing away from the surface both inside and outside the surface. Inside the conductor, this field points away from point P in the opposite direction. (b) Since the net field inside the conductor is zero, the rest of the charges in the universe must produce a field of magnitude $\sigma/(2\epsilon_0)$ in the outward direction. The field due to this charge is the same just inside the surface as it is just outside the surface. (c) Inside the surface, the fields shown in (a) and (b) cancel, but outside at point P they add to give $E_n = \sigma/\epsilon_0$.

While watching a science show on the atmosphere, you find out that on average the electric field of the Earth is about 100 N/C directed vertically downwards. Given that you have been studying electric fields in your physics class, you wonder if you can determine what the total charge on the Earth's surface is.

PICTURE THE PROBLEM The earth is a conductor, so any charge it carries resides on the surface of the earth. The surface charge density σ is related to the normal component of the electric field E_n by Equation 22-29. The total charge Q equals the charge density σ times the surface area A.

1. The surface charge density σ is related to the normal component of the electric field E_n by Equation 22-29:

$$E_n = \frac{\sigma}{\epsilon_0}$$

2. On the surface of the earth $\hat{n}$ is upward and $\vec{E}$ is downward, so E_n is negative:

$$E_n = \vec{E} \cdot \hat{n} = E \times 1 \times \cos 180° = -E = -100 \text{ n/C}$$

3. The charge Q is the charge per unit area. Combine this with the step 1 and 2 results to obtain an expression for Q:

$$Q = \sigma A = \epsilon_0 E_n A = -\epsilon_0 EA$$

4. The surface area of a sphere of radius r is given by $A = 4\pi r^2$.

$$Q = -\epsilon_0 EA = -\epsilon_0 E 4\pi R_E^2 = -4\pi \epsilon_0 E R_E^2$$

5. The radius of the earth is 6.38×10^6 m:

$$Q = -4\pi \epsilon_0 E R_E^2$$
$$= -4\pi (8.85 \times 10^{-12} \text{ C}^2/\text{N·m}^2)(100 \text{ N/C})(6.38 \times 10^6 \text{ m})^2$$
$$= \boxed{-4.53 \times 10^5 \text{ C}}$$

Figure 22-31 shows a positive point charge q at the center of a spherical cavity inside a spherical conductor. Since the net charge must be zero within any Gaussian surface drawn within the conductor, there must be a negative charge $-q$ induced in the inside surface. In Figure 22-32, the point charge has been moved so that it is no longer at the center of the cavity. The field lines in the cavity are altered, and the surface charge density of the induced negative charge on the inner surface is no longer uniform. However, the positive surface charge density on the outside surface is not disturbed—it is still uniform—because it is electrically shielded from the cavity by the conducting material.

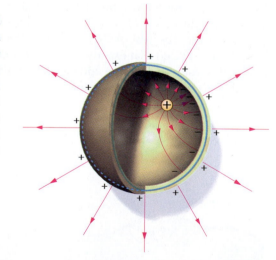

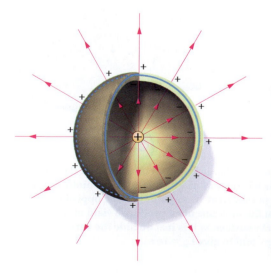

FIGURE 22-31 A point charge q in the cavity at the center of a thick spherical conducting shell. Since the net charge within the Gaussian surface (indicated in blue) must be zero, a surface charge $-q$ is induced on the inner surface of the shell, and since the conductor is neutral, an equal but opposite charge $+q$ is induced on the outer surface. Electric field lines begin on the point charge and end on the inner surface. Field lines begin again on the outer surface.

FIGURE 22-32 The same conductor as in Figure 22-31 with the point charge moved away from the center of the sphere. The charge on the outer surface and the electric field lines outside the sphere are not affected.

An infinite, nonconducting, uniformly charged plane is located in the $x = -a$ plane, and a second such plane is located in the $x = +a$ plane (Figure 22-33a). The plane at $x = -a$ carries a positive charge density whereas the plane at $x = +a$ carries a negative charge density of the same magnitude. The electric field due to the charges on both planes is $\vec{E}_{\text{applied}} = (450 \text{ kN/C})\hat{i}$ in the region between them. A thin, uncharged 2-m diameter conducting disk is placed in the $x = 0$ plane and centered at the origin (Figure 22-33b). (a) Find the charge density on each face of the disk. Also, find the electric field just outside the disk at each face. (Assume that any charge on either face is uniformly distributed.) (b) A net charge of 96 μC is placed on the disk. Find the new charge density on each face and the electric field just outside each face but far from the edges of the sheet.

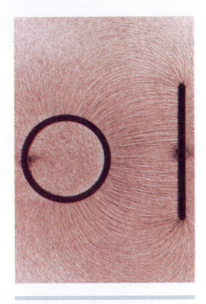

Electric field lines for an oppositely charged cylinder and plate, shown by bits of fine thread suspended in oil. Note that the field lines are perpendicular to the conductors and that there are no lines inside the cylinder.

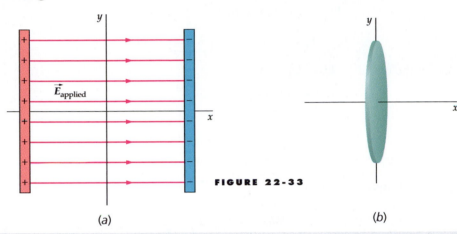

FIGURE 22-33

(a)   (b)

PICTURE THE PROBLEM (a) We find the charge density by using the fact that the total charge on the disk is zero and that there is no electric field inside the conducting material of the disk. The surface charges on the disk must produce an electric field inside it that exactly cancels $\vec{E}_{\text{applied}}$. (b) The additional charge of 96 μC must be distributed so that the electric field inside the conducting disk remains zero.

(a) 1. Let σ_R and σ_L be the charge densities on the right and left faces on the conducting sheet, respectively. Since the disk is uncharged, these densities must add to zero.

$$\sigma_R + \sigma_L = 0$$

so

$$\sigma_L = -\sigma_R$$

2. Inside the conducting sheet the electric field due to the charges on its surface must cancel $\vec{E}_{\text{applied}}$. Let $\vec{E}_R$ and $\vec{E}_L$ be the electric field due to the charges on the right and left faces, respectively.

$$\vec{E}_R + \vec{E}_L + \vec{E}_{\text{applied}} = 0$$

3. Using Equations 22-13a and b we can express the electric field due to the charge on each surface of the disk by the corresponding surface charge density. The field due to a disk of surface charge σ next to the disk is given by $[\sigma/(2\,\epsilon_0)]\,\hat{u}$, where $\hat{u}$ is a unit vector directed away from the surface charge.

$$\vec{E}_R + \vec{E}_L + \vec{E}_{\text{applied}} = 0$$

$$\frac{\sigma_R}{2\,\epsilon_0}(-\hat{i}) + \frac{\sigma_L}{2\,\epsilon_0}\,\hat{i} + \vec{E}_{\text{applied}} = 0$$

4. Substituting $-\sigma_R$ for σ_L and solving for the surface charge densities gives:

$$\frac{\sigma_R}{2\,\epsilon_0}(-\hat{i}) + \frac{-\sigma_R}{2\,\epsilon_0}\,\hat{i} + \vec{E}_{\text{applied}} = 0$$

$$-\frac{\sigma_R}{\epsilon_0}\,\hat{i} + \vec{E}_{\text{applied}} = 0$$

$$\sigma_R \hat{i} = \epsilon_0 \vec{E}_{applied}$$

$$= (8.85 \times 10^{-12} \, C^2/N \cdot m^2)(450 \, kN/C) \hat{i}$$

$$\sigma_R = 3.98 \times 10^{-6} \, C/m^2 = \boxed{3.98 \, \mu C/m^2}$$

$$\sigma_L = -\sigma_R = \boxed{-3.98 \, \mu C/m^2}$$

5. Use Equation 22-29 ($E_n = \sigma/\epsilon_0$) to relate the electric field just outside a conductor to the surface charge density on it. Just outside the right side of the disk $\hat{n} = \hat{i}$, and just outside the left side $\hat{n} = -\hat{i}$:

$$E_{Rn} = \frac{\sigma_R}{\epsilon_0} = \frac{3.98 \, \mu C/m^2}{8.85 \times 10^{-12} \, C^2/N \cdot m^2}$$

$$= 450 \, kN/C$$

$$\vec{E}_R = E_{Rn}\hat{n} = E_{Rn}\hat{i} = \boxed{450 \, kN/C \hat{i}}$$

$$E_{Ln} = \frac{\sigma_L}{\epsilon_0} = \frac{-3.98 \, \mu C/m^2}{8.85 \times 10^{-12} \, C^2/N \cdot m^2}$$

$$\vec{E}_L = E_{Ln}\hat{n} = E_{Ln}(-\hat{i}) = \boxed{450 \, kN/C \hat{i}}$$

(b) 1. The sum of the charges on the two faces of the disk must equal the net charge on the disk.

$$Q_R + Q_L = Q_{net}$$

$$\sigma_R A + \sigma_L A = Q_{net}$$

or

$$\sigma_L = \frac{Q_{net}}{A} - \sigma_R$$

2. Substitute for σ_L in the Part (a), step 2 result and solve for the surface charge densities:

$$\frac{\sigma_R}{2\epsilon_0}(-\hat{i}) + \frac{(Q_{net}/A) - \sigma_R}{2\epsilon_0}\hat{i} + \vec{E}_{applied} = 0$$

$$\frac{(Q_{net}/A) - 2\sigma_R}{2\epsilon_0}\hat{i} + \vec{E}_{applied} = 0$$

$$\sigma_R \hat{i} = \epsilon_0 \vec{E}_{applied} + \frac{Q_{net}}{2A}\hat{i} = \epsilon_0 (450 \, kN/C)\hat{i} + \frac{Q_{net}}{2A}\hat{i}$$

$$\sigma_R = (8.85 \times 10^{-12} \, C^2/N \cdot m^2)(450 \, kN/C) + \frac{Q_{net}}{2A}$$

$$= 3.98 \, \mu C/m^2 + \frac{96 \, \mu C}{2\pi(1 \, m)^2} = \boxed{19.3 \, \mu C/m^2}$$

$$\sigma_L = \frac{Q_{net}}{A} - \sigma_R = \frac{Q_{net}}{A} - \left(\epsilon_0 (450 \, kN/C) + \frac{Q_{net}}{2A}\right)$$

$$= -\epsilon_0 (450 \, kN/C) + \frac{Q_{net}}{2A}$$

$$= -3.98 \, \mu C/m^2 + \frac{96 \, \mu C}{2\pi(1 \, m)^2} = \boxed{11.3 \, \mu C/m^2}$$

3. Using Equation 22-29 ($E_n = \epsilon_0\sigma$), relate the electric field just outside a conductor to the surface charge density on it.

$$E_{Rn} = \frac{\sigma_R}{\epsilon_0} = \frac{19.3 \, \mu C/m^2}{8.85 \times 10^{12} \, C^2/N \cdot m^2}$$

$$= 2.17 \times 10^6 \, N/C$$

$$\vec{E}_R = E_{Rn}\hat{n} = E_{Rn}\hat{i} = \boxed{+2.17 \, MN/C \hat{i}}$$

$$E_{Ln} = \frac{\sigma_L}{\epsilon_0} = \frac{11.3 \, \mu C/m^2}{8.85 \times 10^{12} \, C^2/N \cdot m^2}$$

$$\vec{E}_L = E_{Ln}\hat{n} = E_{Ln}(-\hat{i}) = \boxed{-1.28 \, MN/C \hat{i}}$$

REMARKS The charge added to the disk was distributed equally, half on one side and half on the other. The electric field inside the disk due to this added charge is exactly zero. On each side of a real charged conducting thin disk the magnitude of the charge density is greatest near the edge of the disk.

EXERCISE The electric field just outside the surface of a certain conductor points away from the conductor and has a magnitude of 2000 N/C. What is the surface charge density on the surface of the conductor? (*Answer* 17.7 nC/m^2)

*22-6 Derivation of Gauss's Law From Coulomb's Law

Gauss's law can be derived mathematically using the concept of the **solid angle.** Consider an area element ΔA on a spherical surface. The solid angle $\Delta \Omega$ subtended by ΔA at the center of the sphere is defined to be

$$\Delta \Omega = \frac{\Delta A}{r^2}$$

where r is the radius of the sphere. Since ΔA and r^2 both have dimensions of length squared, the solid angle is dimensionless. The SI unit of the solid angle is the **steradian** (sr). Since the total area of a sphere is $4\pi r^2$, the total solid angle subtended by a sphere is

$$\frac{4\pi r^2}{r^2} = 4\pi \text{ steradians}$$

There is a close analogy between the solid angle and the ordinary plane angle $\Delta \theta$, which is defined to be the ratio of an element of arc length of a circle Δs to the radius of the circle:

$$\Delta \theta = \frac{\Delta s}{r} \text{ radians}$$

The total plane angle subtended by a circle is 2π radians.

In Figure 22-34, the area element ΔA is not perpendicular to the radial lines from point O. The unit vector $\hat{n}$ normal to the area element makes an angle θ with the radial unit vector $\hat{r}$. In this case, the solid angle subtended by ΔA at point O is

FIGURE 22-34 An area element ΔA whose normal is not parallel to the radial line from O to the center of the element. The solid angle subtended by this element at O is defined to be $(\Delta A \cos \theta)/r^2$.

$$\Delta \Omega = \frac{\Delta A\, \hat{n} \cdot \hat{r}}{r^2} = \frac{\Delta A \cos \theta}{r^2} \qquad \text{22-30}$$

Figure 22-35 shows a point charge q surrounded by a surface S of arbitrary shape. To calculate the flux of $\vec{E}$ through this surface, we want to find $\vec{E} \cdot \hat{n}\Delta A$ for each element of area on the surface and sum over the entire surface. The electric field at the area element shown is given by

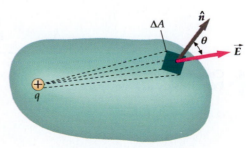

FIGURE 22-35 A point charge enclosed by an arbitrary surface S. The flux through an area element ΔA is proportional to the solid angle subtended by the area element at the charge. The net flux through the surface, found by summing over all the area elements, is proportional to the total solid angle 4π at the charge, which is independent of the shape of the surface.

$$\vec{E} = \frac{kq}{r^2}\hat{r}$$

so the flux through the element is

$$\Delta\phi = \vec{E} \cdot \hat{n}\,\Delta A = \frac{kq}{r^2}\hat{r} \cdot \hat{n}\,\Delta A = kq\,\Delta\Omega$$

The solid angle $\Delta\Omega$ is the same as that subtended by the corresponding area element of a spherical surface of any radius. The sum of the fluxes through the entire surface is kq times the total solid angle subtended by the closed surface, which is 4π steradians:

$$\phi_{net} = \oint_S \vec{E} \cdot \hat{n}\,dA = kq \oint d\Omega = kq4\pi = 4\pi kq = \frac{q}{\epsilon_0} \qquad \text{22-31}$$

which is Gauss's law.

SUMMARY

1. Gauss's law is a fundamental law of physics that is equivalent to Coulomb's law for static charges.

2. For highly symmetric charge distributions, Gauss's law can be used to calculate the electric field.

Topic	Relevant Equations and Remarks	
1. Electric Field for a Continuous Charge Distribution	$\vec{E} = \int_V \frac{k\,dq}{r^2}\hat{r} = \frac{1}{4\pi\epsilon_0}\int_V \frac{dq}{r^2}\hat{r}$ (Coulomb's law)	22-4
	where $dq = \rho\,dV$ for a charge distributed throughout a volume, $dq = \sigma\,dA$ for a charge distributed on a surface, and $dq = \lambda\,dL$ for a charge distributed along a line.	
2. Electric Flux	$\phi = \lim_{\Delta A_i \to 0}\sum_i \vec{E}_i \cdot \hat{n}_i\,\Delta A_i = \int_S \vec{E} \cdot \hat{n}\,dA$	22-16
3. Gauss's Law	$\phi_{net} = \int_S E_n\,dA = 4\pi k\,Q_{inside} = \frac{Q_{inside}}{\epsilon_0}$	22-19
	The net outward flux through a closed surface equals $4\pi k$ times the net charge within the surface.	
4. Coulomb Constant k and Permittivity of Free Space ϵ_0	$k = \frac{1}{4\pi\epsilon_0} = 8.99 \times 10^9 \text{ N·m}^2/\text{C}^2$	
	$\epsilon_0 = \frac{1}{4\pi k} = 8.85 \times 10^{-12} \text{ C}^2/\text{N·m}^2$	22-23

5. **Coulomb's Law and Gauss's Law**	$$\vec{E} = \frac{1}{4\pi\epsilon_0}\frac{q}{r^2}\hat{r}$$	22-21
	$$\phi_{net} = \oint_S E_n\, dA = \frac{Q_{inside}}{\epsilon_0}$$	22-22

6. **Discontinuity of E_n**

At a surface carrying a surface charge density σ, the component of the electric field perpendicular to the surface is discontinuous by σ/ϵ_0.

$$E_{n+} - E_{n-} = \frac{\sigma}{\epsilon_0}$$ 22-28

7. **Charge on a Conductor**

In electrostatic equilibrium, the net electric charge on a conductor resides on the surface of the conductor.

8. **$\vec{E}$ Just Outside a Conductor**

The resultant electric field just outside the surface of a conductor is perpendicular to the surface and has the magnitude σ/ϵ_0, where σ is the local surface charge density at that point on the conductor:

$$E_n = \frac{\sigma}{\epsilon_0}$$ 22-29

The force per unit area exerted on the charge on the surface of a conductor by all the other charges is called the electrostatic stress.

9. **Electric Fields for Various Uniform Charge Distributions**

Of a line charge	$$E_y = \frac{k\lambda}{y}(\sin\theta_2 - \sin\theta_1); \quad E_x = \frac{k\lambda}{y}(\cos\theta_2 - \cos\theta_1)$$	22-8
Of a line charge of infinite length	$$E_R = 2k\frac{\lambda}{R} = \frac{1}{2\pi\epsilon_0}\frac{\lambda}{R}$$	22-9
On the axis of a charged ring	$$E_x = \frac{kQx}{(x^2 + a^2)^{3/2}}$$	22-10
On the axis of a charged disk	$$E_x = \frac{\sigma}{2\epsilon_0}\left(1 - \frac{1}{\sqrt{1 + \dfrac{R^2}{x^2}}}\right), \quad x > 0$$	22-11
Of a charged plane	$$E_x = \frac{\sigma}{2\epsilon_0}, \quad x > 0$$	22-24
Of a charged spherical shell	$$E_r = \frac{1}{4\pi\epsilon_0}\frac{Q}{r^2}, \quad r > R$$	22-25a
	$$E_r = 0, \quad r < R$$	22-25b
Of a charged solid sphere	$$E_r = \frac{1}{4\pi\epsilon_0}\frac{Q}{r^2}, \quad r \geq R$$	22-26a
	$$E_r = \frac{1}{4\pi\epsilon_0}\frac{Q}{R^3}r, \quad r \leq R$$	22-26b

PROBLEMS

- Single-concept, single-step, relatively easy
- •• Intermediate-level, may require synthesis of concepts
- ••• Challenging
- **SSM** Solution is in the *Student Solutions Manual*
- **iSOLVE** Problems available on iSOLVE online homework service
- **iSOLVE✓** These "Checkpoint" online homework service problems ask students additional questions about their confidence level, and how they arrived at their answer.

In a few problems, you are given more data than you actually need; in a few other problems, you are required to supply data from your general knowledge, outside sources, or informed estimates.

Conceptual Problems

1 •• **SSM** True or false:

(a) Gauss's law holds only for symmetric charge distributions.
(b) The result that $E = 0$ inside a conductor can be derived from Gauss's law.

2 •• What information, in addition to the total charge inside a surface, is needed to use Gauss's law to find the electric field?

3 ••• Is the electric field E in Gauss's law only that part of the electric field due to the charge inside a surface, or is it the total electric field due to all charges both inside and outside the surface?

4 •• Explain why the electric field increases with r rather than decreasing as $1/r^2$ as one moves out from the center inside a spherical charge distribution of constant volume charge density.

5 • **SSM** True or false:

(a) If there is no charge in a region of space, the electric field on a surface surrounding the region must be zero everywhere.
(b) The electric field inside a uniformly charged spherical shell is zero.
(c) In electrostatic equilibrium, the electric field inside a conductor is zero.
(d) If the net charge on a conductor is zero, the charge density must be zero at every point on the surface of the conductor.

6 • If the electric field E is zero everywhere on a closed surface, is the net flux through the surface necessarily zero? What, then, is the net charge inside the surface?

7 • A point charge $-Q$ is at the center of a spherical conducting shell of inner radius R_1 and outer radius R_2, as shown in Figure 22-36. The charge on the inner surface of the shell is (a) $+Q$. (b) zero. (c) $-Q$. (d) dependent on the total charge carried by the shell.

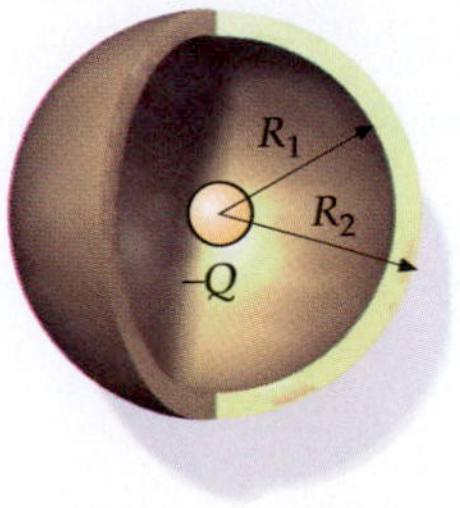

FIGURE 22-36 Problem 7

8 • For the configuration of Figure 22-36, the charge on the outer surface of the shell is (a) $+Q$. (b) zero. (c) $-Q$. (d) dependent on the total charge carried by the shell.

9 •• **SSM** Suppose that the total charge on the conducting shell of Figure 22-36 is zero. It follows that the electric field for $r < R_1$ and $r > R_2$ points

(a) away from the center of the shell in both regions.
(b) toward the center of the shell in both regions.
(c) toward the center of the shell for $r < R_1$ and is zero for $r > R_2$.
(d) away from the center of the shell for $r < R_1$ and is zero for $r > R_2$.

10 •• **SSM** If the conducting shell in Figure 22-36 is grounded, which of the following statements is then correct?

(a) The charge on the inner surface of the shell is $+Q$ and that on the outer surface is $-Q$.
(b) The charge on the inner surface of the shell is $+Q$ and that on the outer surface is zero.
(c) The charge on both surfaces of the shell is $+Q$.
(d) The charge on both surfaces of the shell is zero.

11 •• For the configuration described in Problem 10, in which the conducting shell is grounded, the electric field for $r < R_1$ and $r > R_2$ points

(a) away from the center of the shell in both regions.
(b) toward the center of the shell in both regions.
(c) toward the center of the shell for $r < R_1$ and is zero for $r > R_2$.
(d) toward the center of the shell for $r < R_1$ and is zero for $r > R_1$.

12 •• If the net flux through a closed surface is zero, does it follow that the electric field E is zero everywhere on the surface? Does it follow that the net charge inside the surface is zero?

13 •• True or false: The electric field is discontinuous at all points at which the charge density is discontinuous.

Estimation and Approximation

14 •• **SSM** Given that the maximum field sustainable in air without electrical discharge is approximately 3×10^6 N/C, estimate the total charge of a thundercloud. Make any assumptions that seem reasonable.

15 •• If you rub a rubber balloon against dry hair, the resulting static charge will be enough to make the hair stand on end. Estimate the surface charge density on the balloon and its electric field.

16 • A disk of radius 2.5 cm carries a uniform surface charge density of 3.6 $\mu C/m^2$. Using reasonable approximations, find the electric field on the axis at distances of (a) 0.01 cm, (b) 0.04 cm, (c) 5 m, and (d) 5 cm.

Calculating $\vec{E}$ From Coulomb's Law

17 • **SSM** **iSOLVE✓** A uniform line charge of linear charge density $\lambda = 3.5$ nC/m extends from $x = 0$ to $x = 5$ m. (a) What is the total charge? Find the electric field on the x axis at (b) $x = 6$ m, (c) $x = 9$ m, and (d) $x = 250$ m. (e) Find the field at $x = 250$ m, using the approximation that the charge is a point charge at the origin, and compare your result with that for the exact calculation in Part (d).

18 • Two infinite vertical planes of charge are parallel to each other and are separated by a distance $d = 4$ m. Find the electric field to the left of the planes, to the right of the planes, and between the planes (a) when each plane has a uniform surface charge density $\sigma = +3$ $\mu C/m^2$ and (b) when the left plane has a uniform surface charge density $\sigma = +3$ $\mu C/m^2$ and that of the right plane is $\sigma = -3$ $\mu C/m^2$. Draw the electric field lines for each case.

19 • **iSOLVE✓** A 2.75-μC charge is uniformly distributed on a ring of radius 8.5 cm. Find the electric field on the axis at (a) 1.2 cm, (b) 3.6 cm, and (c) 4.0 m from the center of the ring. (d) Find the field at 4.0 m using the approximation that the ring is a point charge at the origin, and compare your results with that for Part (c).

20 • For the disk charge of Problem 16, calculate exactly the electric field on the axis at distances of (a) 0.04 cm and (b) 5 m, and compare your results with those for Parts (b) and (c) of Problem 16.

21 • A uniform line charge extends from $x = -2.5$ cm to $x = +2.5$ cm and has a linear charge density of $\lambda = 6.0$ nC/m. (a) Find the total charge. Find the electric field on the y axis at (b) $y = 4$ cm, (c) $y = 12$ cm, and (d) $y = 4.5$ m. (e) Find the field at $y = 4.5$ m, assuming the charge to be a point charge, and compare your result with that for Part (d).

22 • **iSOLVE✓** A disk of radius a lies in the yz plane with its axis along the x axis and carries a uniform surface charge density σ. Find the value of x for which $E_x = \frac{1}{2}\sigma/2\epsilon_0$.

23 • A ring of radius a with its center at the origin and its axis along the x axis carries a total charge Q. Find E_x at (a) $x = 0.2a$, (b) $x = 0.5a$, (c) $x = 0.7a$, (d) $x = a$, and (e) $x = 2a$. (f) Use your results to plot E_x versus x for both positive and negative values of x.

24 • Repeat Problem 23 for a disk of uniform surface charge density σ.

25 •• **SSM** (a) Using a spreadsheet program or graphing calculator, make a graph of the electric field on the axis of a disk of radius r = 30 cm carrying a surface charge density $\sigma = 0.5$ nC/m². (b) Compare the field to the approximation $E = 2\pi k\sigma$. At what distance does the approximation differ from the exact solution by 10 percent?

26 •• Show that E_x on the axis of a ring charge of radius a has its maximum and minimum values at $x = +a/\sqrt{2}$ and $x = -a/\sqrt{2}$. Sketch E_x versus x for both positive and negative values of x.

27 •• A line charge of uniform linear charge density λ lies along the x axis from $x = x_1$ to $x = x_2$ where $x_1 < x_2$. Show the x component of the electric field at a point on the y axis is given by

$$E_x = \frac{k\lambda}{y}(\cos\theta_2 - \cos\theta_1)$$

where $(\theta_1 = \tan^{-1}(x_1/y)$ and $\theta_2 = \tan^{-1}(x_2/y)$.

28 •• A ring of radius R has a charge distribution on it that goes as $\lambda(\theta) = \lambda_0 \sin\theta$, as shown in the figure below. (a) In what direction does the field at the center of the ring point? (b) What is the magnitude of the field at the center of the ring?

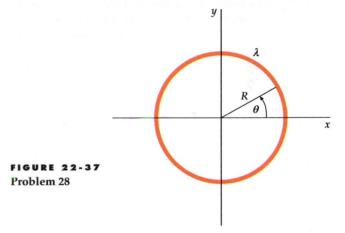

FIGURE 22-37
Problem 28

29 •• A finite line charge of uniform linear charge density λ lies on the x axis from $x = 0$ to $x = a$. Show that the y component of the electric field at a point on the y axis is given by

$$E_y = \frac{k\lambda}{y}\frac{a}{\sqrt{y^2 + a^2}}$$

30 ••• **SSM** A hemispherical thin shell of radius R carries a uniform surface charge σ. Find the electric field at the center of the hemispherical shell ($r = 0$).

Gauss's Law

31 • **iSOLVE✓** Consider a uniform electric field $\vec{E} = 2$ kN/C$\hat{i}$. (a) What is the flux of this field through a square of side 10 cm in a plane parallel to the yz plane? (b) What is the flux through the same square if the normal to its plane makes a 30° angle with the x axis?

32 • **SSM** A single point charge $q = +2$ μC is at the origin. A spherical surface of radius 3.0 m has its center on the x axis at $x = 5$ m. (a) Sketch electric field lines for the point charge. Do any lines enter the spherical surface? (b) What is the net number of lines that cross the spherical surface, counting those that enter as negative? (c) What is the net flux of the electric field due to the point charge through the spherical surface?

33 • An electric field is $\vec{E} = 300$ N/C$\hat{i}$ for $x > 0$ and $\vec{E} = -300$ N/C$\hat{i}$ for $x < 0$. A cylinder of length 20 cm and radius 4 cm has its center at the origin and its axis along the x axis such that one end is at $x = +10$ cm and the other is at $x = -10$ cm. (*a*) What is the flux through each end? (*b*) What is the flux through the curved surface of the cylinder? (*c*) What is the net outward flux through the entire cylindrical surface? (*d*) What is the net charge inside the cylinder?

34 • Careful measurement of the electric field at the surface of a black box indicates that the net outward flux through the surface of the box is 6.0 kN·m²/C. (*a*) What is the net charge inside the box? (*b*) If the net outward flux through the surface of the box were zero, could you conclude that there were no charges inside the box? Why or why not?

35 • A point charge $q = +2$ μC is at the center of a sphere of radius 0.5 m. (*a*) Find the surface area of the sphere. (*b*) Find the magnitude of the electric field at points on the surface of the sphere. (*c*) What is the flux of the electric field due to the point charge through the surface of the sphere? (*d*) Would your answer to Part (*c*) change if the point charge were moved so that it was inside the sphere but not at its center? (*e*) What is the net flux through a cube of side 1 m that encloses the sphere?

36 • **SSM** Since Newton's law of gravity and Coulomb's law have the same inverse-square dependence on distance, an expression analogous in form to Gauss's law can be found for gravity. The gravitational field $\vec{g}$ is the force per unit mass on a test mass m_0. Then, for a point mass m at the origin, the gravitational field g at some position r is

$$\vec{g} = -\frac{Gm}{r^2}\hat{r}$$

Compute the flux of the gravitational field through a spherical surface of radius R centered at the origin, and show that the gravitational analog of Gauss's law is $\phi_{net} = -4\pi Gm_{inside}$.

37 •• **SOLVE** A charge of 2 μC is 20 cm above the center of a square of side length 40 cm. Find the flux through the square. (*Hint: Don't integrate.*)

38 •• **SOLVE** ✓ In a particular region of the earth's atmosphere, the electric field above the earth's surface has been measured to be 150 N/C downward at an altitude of 250 m and 170 N/C downward at an altitude of 400 m. Calculate the volume charge density of the atmosphere assuming it to be uniform between 250 and 400 m. (You may neglect the curvature of the earth. Why?)

Spherical Symmetry

39 • A spherical shell of radius R_1 carries a total charge q_1 that is uniformly distributed on its surface. A second, larger spherical shell of radius R_2 that is concentric with the first carries a charge q_2 that is uniformly distributed on its surface. (*a*) Use Gauss's law to find the electric field in the regions $r < R_1$, $R_1 < r < R_2$, and $r > R_2$. (*b*) What should be the ratio of the charges q_1/q_2 and their relative signs be for the electric field to be zero for $r > R_2$? (*c*) Sketch the electric field lines for the situation in Part (*b*) when q_1 is positive.

40 • **SOLVE** ✓ A spherical shell of radius 6 cm carries a uniform surface charge density $\sigma = 9$ nC/m². (*a*) What is the total charge on the shell? Find the electric field at (*b*) $r = 2$ cm, (*c*) $r = 5.9$ cm, (*d*) $r = 6.1$ cm, and (*e*) $r = 10$ cm.

41 •• A sphere of radius 6 cm carries a uniform volume charge density $\rho = 450$ nC/m³. (*a*) What is the total charge of the sphere? Find the electric field at (*b*) $r = 2$ cm, (*c*) $r = 5.9$ cm, (*d*) $r = 6.1$ cm, and (*e*) $r = 10$ cm. Compare your answers with Problem 40.

42 •• **SSM** Consider two concentric conducting spheres (Figure 22-38). The outer sphere is hollow and initially has a charge $-7Q$ deposited on it. The inner sphere is solid and has a charge $+2Q$ on it. (*a*) How is the charge distributed on the outer sphere? That is, how much charge is on the outer surface and how much charge is on the inner surface? (*b*) Suppose a wire is connected between the inner and outer spheres. After electrostatic equilibrium is established, how much total charge is on the outside sphere? How much charge is on the outer surface of the outside sphere, and how much charge is on the inner surface? Does the electric field at the surface of the inside sphere change when the wire is connected? If so, how? (*c*) Suppose we return to the original conditions in Part (*a*), with $+2Q$ on the inner sphere and $-7Q$ on the outer. We now connect the outer sphere to ground with a wire and then disconnect it. How much total charge will be on the outer sphere? How much charge will be on the inner surface of the outer sphere and how much will be on the outer surface?

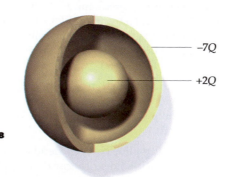

FIGURE 22-38
Problem 42

43 •• **SOLVE** ✓ A nonconducting sphere of radius $R = 0.1$ m carries a uniform volume charge of charge density $\rho = 2.0$ nC/m³. The magnitude of the electric field at $r = 2R$ is 1883 N/C. Find the magnitude of the electric field at $r = 0.5R$.

44 •• A nonconducting sphere of radius R carries a volume charge density that is proportional to the distance from the center: $\rho = Ar$ for $r \leq R$, where A is a constant; $\rho = 0$ for $r > R$. (*a*) Find the total charge on the sphere by summing the charges on shells of thickness dr and volume $4\pi r^2\, dr$. (*b*) Find the electric field E_r both inside and outside the charge distribution, and sketch E_r versus r.

45 •• Repeat Problem 44 for a sphere with volume charge density $\rho = B/r$ for $r < R$; $\rho = 0$ for $r > R$.

46 •• **SSM** Repeat Problem 44 for a sphere with volume charge density $\rho = C/r^2$ for $r < R$; $\rho = 0$ for $r > R$.

47 ••• A thick, nonconducting spherical shell of inner radius a and outer radius b has a uniform volume charge density ρ. Find (*a*) the total charge and (*b*) the electric field everywhere.

Cylindrical Symmetry

48 •• Show that the electric field due to an infinitely long, uniformly charged cylindrical shell of radius R carrying a surface charge density σ is given by

$$E_r = 0, \qquad r < R$$

$$E_r = \frac{\sigma R}{\epsilon_0 r} = \frac{\lambda}{2\pi\epsilon_0 r} \qquad r > R$$

where $\lambda = 2\pi R\sigma$ is the charge per unit length on the shell.

49 •• **SOLVE** A cylindrical shell of length 200 m and radius 6 cm carries a uniform surface charge density of $\sigma = 9 \text{ nC/m}^2$. (a) What is the total charge on the shell? Find the electric field at (b) $r = 2$ cm, (c) $r = 5.9$ cm, (d) $r = 6.1$ cm, and (e) $r = 10$ cm. (Use the results of Problem 48.)

50 •• An infinitely long nonconducting cylinder of radius R carries a uniform volume charge density of $\rho(r) = \rho_0$. Show that the electric field is given by

$$E_r = \frac{\rho R^2}{2\epsilon_0 r} = \frac{1}{2\pi\epsilon_0}\frac{\lambda}{r} \qquad r > R$$

$$E_r = \frac{\rho}{2\epsilon_0}r = \frac{\lambda}{2\pi\epsilon_0 R^2}r \qquad r < R$$

where $\lambda = \rho\pi R^2$ is the charge per unit length.

51 •• A cylinder of length 200 m and radius 6 cm carries a uniform volume charge density of $\rho = 300 \text{ nC/m}^3$. (a) What is the total charge of the cylinder? Use the formulas given in Problem 50 to calculate the electric field at a point equidistant from the ends at (b) $r = 2$ cm, (c) $r = 5.9$ cm, (d) $r = 6.1$ cm, and (e) $r = 10$ cm. Compare your results with those in Problem 49.

52 •• **SSM** Consider two infinitely long, concentric cylindrical shells. The inner shell has a radius R_1 and carries a uniform surface charge density of σ_1, and the outer shell has a radius R_2 and carries a uniform surface charge density of σ_2. (a) Use Gauss's law to find the electric field in the regions $r < R_1$, $R_1 < r < R_2$, and $r > R_2$. (b) What is the ratio of the surface charge densities σ_2/σ_1 and their relative signs if the electric field is zero at $r > R_2$? What would the electric field between the shells be in this case? (c) Sketch the electric field lines for the situation in Part (b) if σ_1 is positive.

53 •• **SOLVE** Figure 22-39 shows a portion of an infinitely long, concentric cable in cross section. The inner conductor carries a charge of 6 nC/m; the outer conductor is uncharged. (a) Find the electric field for all values of r, where r is the distance from the axis of the cylindrical system. (b) What are the surface charge densities on the inside and the outside surfaces of the outer conductor?

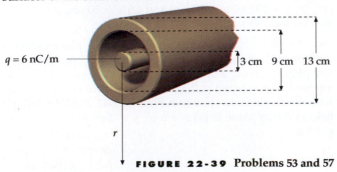

$q = 6 \text{ nC/m}$ — 3 cm 9 cm 13 cm

r

FIGURE 22-39 Problems 53 and 57

54 •• An infinitely long nonconducting cylinder of radius R and carrying a nonuniform volume charge density of $\rho(r) = ar$. (a) Show that the charge per unit length of the cylinder is $\lambda = 2\pi aR^3/3$. (b) Find the expressions for the electric field due to this charged cylinder. You should find one expression for the electric field in the region $r < R$ and a second expression for the field in the region $r > R$, as in Problem 50.

55 •• Repeat Problem 54 for a nonuniform volume charge density of $\rho = br^2$. In Part (a) show $\lambda = \pi bR^4/2$ (instead of the expression given for λ in Problem 54).

56 ••• An infinitely long, thick, nonconducting cylindrical shell of inner radius a and outer radius b has a uniform volume charge density ρ. Find the electric field everywhere.

57 ••• Suppose that the inner cylinder of Figure 22-39 is made of nonconducting material and carries a volume charge distribution given by $\rho(r) = C/r$, where $C = 200 \text{ nC/m}^2$. The outer cylinder is metallic. (a) Find the charge per meter carried by the inner cylinder. (b) Calculate the electric field for all values of r.

Charge and Field at Conductor Surfaces

58 • **SSM** **SOLVE** ✓ A penny is in an external electric field of magnitude 1.6 kN/C directed perpendicular to its faces. (a) Find the charge density on each face of the penny, assuming the faces are planes. (b) If the radius of the penny is 1 cm, find the total charge on one face.

59 • **SOLVE** ✓ An uncharged metal slab has square faces with 12-cm sides. It is placed in an external electric field that is perpendicular to its faces. The total charge induced on one of the faces is 1.2 nC. What is the magnitude of the electric field?

60 • **SOLVE** A charge of 6 nC is placed uniformly on a square sheet of nonconducting material of side 20 cm in the yz plane. (a) What is the surface charge density σ? (b) What is the magnitude of the electric field just to the right and just to the left of the sheet? (c) The same charge is placed on a square conducting slab of side 20 cm and thickness 1 mm. What is the surface charge density σ? (Assume that the charge distributes itself uniformly on the large square surfaces.) (d) What is the magnitude of the electric field just to the right and just to the left of each face of the slab?

61 • A spherical conducting shell with zero net charge has an inner radius a and an outer radius b. A point charge q is placed at the center of the shell. (a) Use Gauss's law and the properties of conductors in equilibrium to find the electric field in the regions $r < a$, $a < r < b$, and $b < r$. (b) Draw the electric field lines for this situation. (c) Find the charge density on the inner surface ($r = a$) and on the outer surface ($r = b$) of the shell.

62 •• **SOLVE** The electric field just above the surface of the earth has been measured to be 150 N/C downward. What total charge on the earth is implied by this measurement?

63 •• **SSM** A positive point charge of magnitude 2.5 μC is at the center of an uncharged spherical conducting shell of inner radius 60 cm and outer radius 90 cm. (a) Find the charge densities on the inner and outer surfaces of the shell and the total charge on each surface. (b) Find the electric field everywhere. (c) Repeat Part (a) and Part (b) with a net charge of +3.5 μC placed on the shell.

64 •• ⚡SOLVE✔ If the magnitude of an electric field in air is as great as 3×10^6 N/C, the air becomes ionized and begins to conduct electricity. This phenomenon is called dielectric breakdown. A charge of 18 μC is to be placed on a conducting sphere. What is the minimum radius of a sphere that can hold this charge without breakdown?

65 •• A square conducting slab with 5-m sides carries a net charge of 80 μC. (a) Find the charge density on each face of the slab and the electric field just outside one face of the slab. (b) The slab is placed to the right of an infinite charged nonconducting plane with charge density 2.0 μC/m² so that the faces of the slab are parallel to the plane. Find the electric field on each side of the slab far from its edges and the charge density on each face.

General Problems

66 •• Consider the three concentric metal spheres shown in Figure 22-40. Sphere one is solid, with radius R_1. Sphere two is hollow, with inner radius R_2 and outer radius R_3. Sphere three is hollow, with inner radius R_4 and outer radius R_5. Initially, all three spheres have zero excess charge. Then a negative charge $-Q_0$ is placed on sphere one and a positive charge $+Q_0$ is placed on sphere three. (a) After the charges have reached equilibrium, will the electric field in the space between spheres one and two point *toward* the center, *away* from the center, or neither? (b) How much charge will be on the inner surface of sphere two? Give the correct sign. (c) How much charge will be on the outer surface of sphere two? (d) How much charge will be on the inner surface of sphere three? (e) How much charge will be on the outer surface of sphere three? (f) Plot E versus r.

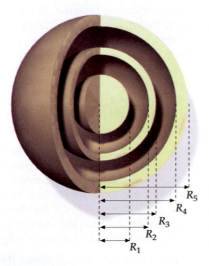

FIGURE 22-40 Problem 66

67 •• ⚡SOLVE A nonuniform surface charge lies in the yz plane. At the origin, the surface charge density is $\sigma = 3.10$ μC/m². Other charged objects are present as well. Just to the right of the origin, the x component of the electric field is $E_x = 4.65 \times 10^5$ N/C. What is E_x just to the left of the origin?

68 •• An infinite line charge of uniform linear charge density $\lambda = -1.5$ μC/m lies parallel to the y axis at $x = -2$ m. A point charge of 1.3 μC is located at $x = 1$ m, $y = 2$ m. Find the electric field at $x = 2$ m, $y = 1.5$ m.

69 •• SSM A thin nonconducting uniformly charged spherical shell of radius r (Figure 22-41a) has a total charge of Q. A small circular plug is removed from the surface. (a) What is the magnitude and direction of the electric field at the center of the hole? (b) The plug is put back in the hole (Figure 22-41b). Using the result of part a, calculate the force acting on the plug. (c) From this, calculate the "electrostatic pressure" (force/unit area) tending to expand the sphere.

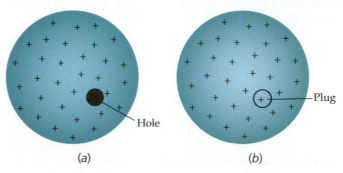

(a)   (b)

FIGURE 22-41 Problem 69

70 •• A soap bubble of radius $R_1 = 10$ cm has a charge of 3 nC uniformly spread over it. Because of electrostatic repulsion, the soap bubble expands until it bursts at a radius $R_2 = 20$ cm. From the results of Problem 69, calculate the work done by the electrostatic force in expanding the soap bubble.

71 •• If the soap bubble of Problem 70 collapses into a spherical water droplet, estimate the electric field at its surface.

72 •• Two infinite planes of charge lie parallel to each other and to the yz plane. One is at $x = -2$ m and has a surface charge density of $\sigma = -3.5$ μC/m². The other is at $x = 2$ m and has a surface charge density of $\sigma = 6.0$ μC/m². Find the electric field for (a) $x < -2$ m, (b) -2 m $< x < 2$ m, and (c) $x > 2$ m.

73 •• SSM An infinitely long cylindrical shell is coaxial with the y axis and has a radius of 15 cm. It carries a uniform surface charge density $\sigma = 6$ μC/m². A spherical shell of radius 25 cm is centered on the x axis at $x = 50$ cm and carries a uniform surface charge density $\sigma = -12$ μC/m². Calculate the magnitude and direction of the electric field at (a) the origin; (b) $x = 20$ cm, $y = 10$ cm; and (c) $x = 50$ cm, $y = 20$ cm. (See Problem 48.)

74 •• ⚡SOLVE An infinite plane in the xz plane carries a uniform surface charge density $\sigma_1 = 65$ nC/m². A second infinite plane carrying a uniform charge density $\sigma_2 = 45$ nC/m² intersects the xz plane at the z axis and makes an angle of 30° with the xz plane, as shown in Figure 22-42. Find the electric field in the xy plane at (a) $x = 6$ m, $y = 2$ m and (b) $x = 6$ m, $y = 5$ m.

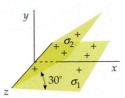

FIGURE 22-42 Problem 74

75 •• A quantum-mechanical treatment of the hydrogen atom shows that the electron in the atom can be treated as a smeared-out distribution of charge, which has the form: $\rho(r) = \rho_0 e^{-2r/a}$, where r is the distance from the nucleus, and a is the Bohr radius ($a = 0.0529$ nm). (a) Calculate ρ_0, from the fact that the atom is uncharged. (b) Calculate the electric field at any distance r from the nucleus. Treat the proton as a point charge.

76 •• SSM Using the results of Problem 75, if we placed a proton above the nucleus of a hydrogen atom, at what distance r would the electric force on the proton balance the gravitational force mg acting on it? From this result, explain why even though the electrostatic force is enormously stronger than the gravitational force, it is the gravitational force we notice more.

77 •• A ring of radius R carries a uniform, positive, linear charge density λ. Figure 22-43 shows a point P in the plane of the ring but not at the center. Consider the two elements of the ring of lengths s_1 and s_2 shown in the figure at distances r_1 and r_2, respectively, from point P. (a) What is the ratio of the charges of these elements? Which produces the greater field at point P? (b) What is the direction of the field at point P due to each element? What is the direction of the total electric field at point P? (c) Suppose that the electric field due to a point charge varied as $1/r$ rather than $1/r^2$. What would the electric field be at point P due to the elements shown? (d) How would your answers to Parts (a), (b), and (c) differ if point P were inside a spherical shell of uniform charge and the elements were of areas s_1 and s_2?

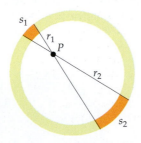

FIGURE 22-43 Problem 77

78 •• A uniformly charged ring of radius R that lies in a horizontal plane carries a charge Q. A particle of mass m carries a charge q, whose sign is opposite that of Q, is on the axis of the ring. (a) What is the minimum value of $|q|/m$ such that the particle will be in equilibrium under the action of gravity and the electrostatic force? (b) If $|q|/m$ is twice that calculated in Part (a), where will the particle be when it is in equilibrium?

79 •• A long, thin, nonconducting plastic rod is bent into a loop with radius R. Between the ends of the rod, a small gap of length l ($l \ll R$) remains. A charge Q is equally distributed on the rod. (a) Indicate the direction of the electric field at the center of the loop. (b) Find the magnitude of the electric field at the center of the loop.

80 •• A nonconducting sphere 1.2 m in diameter with its center on the x axis at $x = 4$ m carries a uniform volume charge of density $\rho = 5\ \mu C/m^3$. Surrounding the sphere is a spherical shell with a diameter of 2.4 m and a uniform surface charge density $\sigma = -1.5\ \mu C/m^2$. Calculate the magnitude and direction of the electric field at (a) $x = 4.5$ m, $y = 0$; (b) $x = 4.0$ m, $y = 1.1$ m; and (c) $x = 2.0$ m, $y = 3.0$ m.

81 •• An infinite plane of charge with surface charge density $\sigma_1 = 3\ \mu C/m^2$ is parallel to the xz plane at $y = -0.6$ m. A second infinite plane of charge with surface charge density $\sigma_2 = -2\ \mu C/m^2$ is parallel to the yz plane at $x = 1$ m. A sphere of radius 1 m with its center in the xy plane at the intersection of the two charged planes ($x = 1$ m, $y = -0.6$ m) has a surface charge density $\sigma_3 = -3\ \mu C/m^2$. Find the magnitude and direction of the electric field on the x axis at (a) $x = 0.4$ m and (b) $x = 2.5$ m.

82 •• An infinite plane lies parallel to the yz plane at $x = 2$ m and carries a uniform surface charge density $\sigma = 2\ \mu C/m^2$. An infinite line charge of uniform linear charge density $\lambda = 4\ \mu C/m$ passes through the origin at an angle of $45°$ with the x axis in the xy plane. A sphere of volume charge density $\rho = -6\ \mu C/m^3$ and radius 0.8 m is centered on the x axis at $x = 1$ m. Calculate the magnitude and direction of the electric field in the xy plane at $x = 1.5$ m, $y = 0.5$ m.

83 •• SOLVE ✓ An infinite line charge λ is located along the z axis. A particle of mass m that carries a charge q whose sign is opposite to that of λ is in a circular orbit in the xy plane about the line charge. Obtain an expression for the period of the orbit in terms of m, q, R, and λ, where R is the radius of the orbit.

84 •• SSM A ring of radius R that lies in the yz plane carries a positive charge Q uniformly distributed over its length. A particle of mass m that carries a negative charge of magnitude q is at the center of the ring. (a) Show that if $x \ll R$, the electric field along the axis of the ring is proportional to x. (b) Find the force on the particle of mass m as a function of x. (c) Show that if m is given a small displacement in the x direction, it will perform simple harmonic motion. Calculate the period of that motion.

85 •• SOLVE When the charges Q and q of Problem 84 are 5 μC and $-5\ \mu C$, respectively, and the radius of the ring is 8.0 cm, the mass m oscillates about its equilibrium position with an angular frequency of 21 rad/s. Find the angular frequency of oscillation of the mass if the radius of the ring is doubled to 16 cm and all other parameters remain unchanged.

86 •• SOLVE Given the initial conditions of Problem 85, find the angular frequency of oscillation of the mass if the radius of the ring is doubled to 16 cm while keeping the linear charge density on the ring constant.

87 •• A uniformly charged nonconducting sphere of radius a with center at the origin has volume charge density ρ. (a) Show that at a point within the sphere a distance r from the center $\vec{E} = \dfrac{\rho}{3\,\epsilon_0} r\hat{r}$. (b) Material is removed from the sphere leaving a spherical cavity of radius $b = a/2$ with its center at $x = b$ on the x axis (Figure 22-44). Calculate the electric field at points 1 and 2 shown in Figure 22-44. (*Hint: Replace the sphere-with-cavity with two uniform spheres of equal positive and negative charge densities.*)

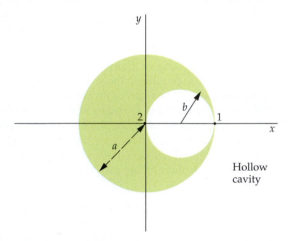

FIGURE 22-44 Problem 87

88 ••• Show that the electric field throughout the cavity of Problem 87 is uniform and is given by

$$\vec{E} = \frac{\rho}{3\,\epsilon_0} b\hat{i}$$

89 •• Repeat Problem 87 assuming that the cavity is filled with a uniformly charged material wth a total charge of Q.

90 •• A nonconducting cylinder of radius 1.2 m and length 2.0 m carries a charge of 50 μC uniformly distributed throughout the cylinder. Find the electric field *on the cylinder axis* at a distance of (a) 0.5 m, (b) 2.0 m, and (c) 20 m from the center of the cylinder.

91 •• **iSOLVE** A uniform line charge of density λ lies on the x axis between $x = 0$ and $x = L$. Its total charge is $Q = 8$ nC. The electric field at $x = 2L$ is 600 N/C$\hat{i}$. Find the electric field at $x = 3L$.

92 ••• A *small* gaussian surface in the shape of a cube with faces parallel to the xy, xz, and yz planes (Figure 22-45) is in a region in which the electric field remains parallel with the x axis. Using the Taylor series (and neglecting terms higher than first order), show that the net flux of the electric field out of the gaussian surface is given by

$$\phi_{\text{net}} = \frac{\partial E_x}{\partial x} \Delta V$$

where ΔV is the volume enclosed by the gaussian surface.

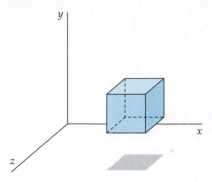

FIGURE 22-45 Problem 92

Remark: The corresponding result for situations for which the direction of the electric field is not restricted to one dimension is

$$\phi_{\text{net}} = \left(\frac{\partial E_x}{\partial x} + \frac{\partial E_y}{\partial y} + \frac{\partial E_z}{\partial z} \right) \Delta V$$

where the combination of derivatives in the parentheses is commonly written $\vec{\nabla} \cdot \vec{E}$ and is called the *divergence* of $\vec{E}$.

93 •• Using Gauss's law and the results of Problem 92 show that

$$\vec{\nabla} \cdot \vec{E} = \frac{\rho}{\epsilon_0}$$

where ρ is the volume charge density. (This equation is known as the point form of Gauss's law.)

94 ••• **SSM** A dipole $\vec{p}$ is located at a distance r from an infinitely long line charge with a uniform linear charge density λ. Assume that the dipole is aligned with the field due to the line charge. Determine the force that acts on the dipole.

95 •• Consider a simple but surprisingly accurate model for the Hydrogen molecule: two positive point charges, each with charge $+e$, are placed inside a sphere of radius R, which has uniform charge density $-2e$. The two point charges are placed symmetrically (Figure 22-46). Find the distance from the center, a, where the net force on either charge is 0.

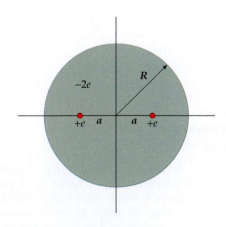

FIGURE 22-46
Problem 95

Electric Potential

Did you know that the maximum potential that the dome of a Van de Graaff generator can be raised to is determined by the radius of the dome? For a discussion of this, see Example 23-14.

The electric force between two charges is directed along the line joining the charges and varies inversely with the square of their separation, the same dependence as the gravitational force between two masses. Like the gravitational force, the electric force is conservative, so there is a potential energy function U associated with it. If we place a test charge q_0 in an electric field, its potential energy is proportional to q_0. The potential energy per unit charge is a function of the position in space of the charge and is called the electric potential. As it is a scalar field, it is easier to manipulate than the electric field in many circumstances. ➤ **In this chapter, we will establish the relationship between the electric field and electric potential and calculate the electric potential of various continuous charge distributions. Then, we can use the electric potential to determine the electric field of these regions.**

23-1 Potential Difference

In general, when the point of application of a conservative force $\vec{F}$ undergoes a displacement $d\vec{\ell}$, the change in the potential energy function dU is given by

$$dU = -\vec{F} \cdot d\vec{\ell}$$

The force exerted by an electric field $\vec{E}$ on a point charge q_0 is

$$\vec{F} = q_0\vec{E}$$

Thus, when a charge undergoes a displacement $d\vec{\ell}$, the change in the electrostatic potential energy is

$$dU = -q_0\vec{E} \cdot d\vec{\ell} \qquad\qquad 23\text{-}1$$

The potential energy change is proportional to the charge q_0. The potential energy change *per unit charge* is called the **potential difference** dV:

$$dV = \frac{dU}{q_0} = -\vec{E} \cdot d\vec{\ell} \qquad\qquad 23\text{-}2a$$

DEFINITION—POTENTIAL DIFFERENCE

For a finite displacement from point a to point b, the change in potential is

$$\Delta V = V_b - V_a = \frac{\Delta U}{q_0} = -\int_a^b \vec{E} \cdot d\vec{\ell} \qquad\qquad 23\text{-}2b$$

DEFINITION—FINITE POTENTIAL DIFFERENCE

The potential difference $V_b - V_a$ is the negative of the work per unit charge done by the electric field on a small positive test charge when the test charge moves from point a to point b. During this calculation, the positions of any and all other charges remain fixed.

The function V is called the **electric potential,** often it is shortened to the **potential.** Like the electric field, the potential V is a function of position. Unlike the electric field, V is a scalar function, whereas $\vec{E}$ is a vector function. As with potential energy U, only *differences* in the potential V are important. We are free to choose the potential to be zero at any convenient point, just as we are when dealing with potential energy. For convenience, the electric potential and the potential energy of a test charge are chosen to be zero at the same point. Under these conditions they are related by

$$U = q_0V \qquad\qquad 23\text{-}3$$

RELATION BETWEEN POTENTIAL ENERGY AND POTENTIAL

Continuity of V

In Chapter 22, we saw that the electric field is discontinuous by σ/ϵ_0 at points where there is a surface charge density σ. The potential function, on the other hand, is continuous everywhere, except at points where the electric field is infinite (points where there is a point charge or a line charge). We can see this from its definition. Consider a region occupied by an electric field $\vec{E}$. The difference in potential between two nearby points separated by displacement $d\vec{\ell}$ is related to the electric field by $dV = -\vec{E} \cdot d\vec{\ell}$ (Equation 23-2a). The dot product can be expressed $\vec{E} \cdot d\vec{\ell} = E_\parallel d\ell$, where $E_\parallel$ is the component of $\vec{E}$ in the direction of $d\vec{\ell}$ and $d\ell$ is the magnitude of $d\vec{\ell}$. Substituting into Equation 23-2a gives $dV = -E_\parallel d\ell$. If $\vec{E}$ is finite at each of the two points and along the line segment of infinitesimal length $d\ell$ joining them, then dV is infinitesimal. Thus, the potential function V is continuous at any point not occupied by a point charge or a line charge.

Units

Since electric potential is the potential energy per unit charge, the SI unit for potential and potential difference is the joule per coulomb, called the **volt** (V):

$$1\text{ V} = 1\text{ J/C} \qquad\qquad 23\text{-}4$$

The potential difference between two points (measured in volts) is sometimes called the **voltage.** In a 12-V car battery, the positive terminal has a potential 12 V higher than the negative terminal. If we attach an external circuit to the battery and one coulomb of charge is transferred from the positive terminal through the circuit to the negative terminal, the potential energy of the charge decreases by $Q\,\Delta V = (1\text{ C})(12\text{ V}) = 12\text{ J}$.

We can see from Equation 23-2 that the dimensions of potential are also those of electric field times distance. Thus, the unit of the electric field is equal to one volt per meter:

$$1\text{ N/C} = 1\text{ V/m} \qquad\qquad 23\text{-}5$$

so we may think of the electric field strength as either a force per unit charge or as a rate of change of V with respect to distance. In atomic and nuclear physics, we often have elementary particles with charges of magnitude e, such as electrons and protons, moving through potential differences of several to thousands or even millions of volts. Since energy has dimensions of electric charge times electric potential, a unit of energy is the product of the fundamental charge unit e times a volt. This unit is called an **electron volt** (eV). Energies in atomic and molecular physics are typically a few eV, making the electron volt a convenient-sized unit for atomic and molecular processes. The conversion between electron volts and joules is obtained by expressing the electronic charge in coulombs:

$$1\text{ eV} = 1.60 \times 10^{-19}\text{ C} \cdot \text{V} = 1.60 \times 10^{-19}\text{ J} \qquad\qquad 23\text{-}6$$

THE ELECTRON VOLT

For example, an electron moving from the negative terminal to the positive terminal of a 12-V car battery loses 12 eV of potential energy.

Potential and Electric Fields

If we place a positive test charge q_0 in an electric field $\vec{E}$ and release it, it accelerates in the direction of $\vec{E}$. As the kinetic energy of the charge increases, its potential energy decreases. The charge therefore accelerates toward a region of lower potential energy, just as a mass accelerates toward a region of lower gravitational potential energy (Figure 23-1). Thus, as illustrated in Figure 23-2,

The electric field points in the direction in which the potential decreases most rapidly.

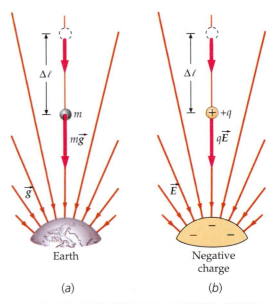

FIGURE 23-1 (a) The work done by the gravitational field $\vec{g}$ on a mass m is equal to the decrease in the gravitational potential energy. (b) The work done by the electric field $\vec{E}$ on a charge q is equal to the decrease in the electric potential energy.

FIGURE 23-2 The electric field points in the direction in which the potential decreases most rapidly. If a positive test charge q_0 is in an electric field, it accelerates in the direction of the field. If it is released from rest, its kinetic energy increases and its potential energy decreases.

E X A M P L E   2 3 - 1

An electric field points in the positive x direction and has a constant magnitude of $E = 10$ N/C $= 10$ V/m. Find the potential as a function of x, assuming that $V = 0$ at $x = 0$.

PICTURE THE PROBLEM

1. By definition, the change in potential dV is related to the displacement $d\vec{\ell}$ and the electric field $\vec{E}$:

$$dV = -\vec{E} \cdot d\vec{\ell} = -E\hat{i} \cdot (dx\,\hat{i} + dy\,\hat{j} + dz\,\hat{k}) = -E\,dx$$

2. Integrate dV:

$$V(x) = \int dV = \int -E\,dx = -Ex + C$$

3. The constant of integration C is found by setting $V = 0$ at $x = 0$:

$$V(0) = C \implies 0 = C$$

4. The potential is then:

$$V(x) = -Ex = \boxed{-(10\text{ V/m})x}$$

REMARKS The potential is zero at $x = 0$ and decreases by 10 V for every 1-m increase in x.

EXERCISE Repeat this example for the electric field $\vec{E} = (10\text{ V/m}^2)x\hat{i}$ [*Answer* $V(x) = -(5\text{ V/m}^2)x^2$]

23-2 Potential Due to a System of Point Charges

The electric potential at a distance r from a point charge q at the origin can be calculated from the electric field:

$$\vec{E} = \frac{kq}{r^2}\hat{r}$$

For an infinitesimal displacement $d\vec{\ell}$ where we have replaced r_P (the distance to the field point) with r (Figure 23-3), the change in potential is

$$dV = -\vec{E} \cdot d\vec{\ell} = -\frac{kq}{r^2}\hat{r} \cdot d\vec{\ell} = -\frac{kq}{r^2}\,dr$$

Integrating along a path from an arbitrary reference point to an arbitrary field point gives

$$\int_{\text{ref}}^{P} dV = -\int_{\text{ref}}^{P} \vec{E} \cdot d\vec{\ell} = -kq\int_{r_{\text{ref}}}^{r_P} r^{-2}\,dr = -kq\frac{r^{-1}}{-1}\Big|_{r_{\text{ref}}}^{r_P} = \frac{kq}{r_P} - \frac{kq}{r_{\text{ref}}}$$

or

$$V = \frac{kq}{r} - \frac{kq}{r_{\text{ref}}} \qquad\qquad 23\text{-}7$$

POTENTIAL DUE TO A POINT CHARGE

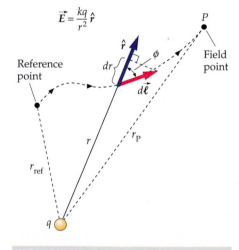

FIGURE 23-3 The change in r is dr. It is the component of $d\vec{\ell}$ in the direction of $\hat{r}$. It can be seen from the figure that $|d\vec{\ell}|\cos\phi = dr$. Since $\hat{r} \cdot d\vec{\ell} = |d\vec{\ell}|\cos\phi$, it follows that $dr = \hat{r} \cdot d\vec{\ell}$.

where we let $r_P = r$. We are free to choose the reference point, so we choose it to give the potential the simplest algebraic form. Choosing the reference point infinitely far from the point charge ($r_{\text{ref}} = \infty$) accomplishes this. Thus,

$$V = \frac{kq}{r} \qquad 23\text{-}8$$

COULOMB POTENTIAL

where we have replaced r_P (the distance to the field point) with r. The potential given by Equation 23-8 is called the **Coulomb potential.** It is positive or negative depending on whether q is positive or negative.

The potential energy U of a test charge q_0 placed a distance r from the point charge q is

$$U = q_0 V = \frac{kq_0 q}{r} \qquad 23\text{-}9$$

ELECTROSTATIC POTENTIAL ENERGY OF A TWO-CHARGE SYSTEM

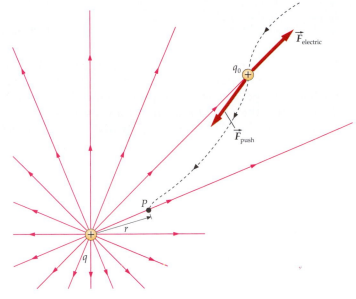

FIGURE 23-4 The work required to bring a test charge q_0 from infinity to a point P is kq_0q/r, where r is the distance from P to a charge q. The work per unit charge is kq/r, the electric potential at point P relative to zero potential at infinity. If the test charge is released from point P, the electric field does work kq_0q/r on the charge as the charge moves out to infinity.

This is the electric potential energy of the two-charge system relative to $U = 0$ at infinite separation. If we release a test charge q_0 from rest at a distance r from q (and hold q fixed), the test charge will be accelerated away from q (assuming that q has the same sign as q_0). At a very great distance from q, its potential energy will be zero so its kinetic energy will be kq_0q/r. Alternatively, if we move a test charge q_0 initially at rest at infinity to rest at a point a distance r from q, the work we must do is kq_0q/r (Figure 23-4). The work per unit charge is kq/r, the electric potential at point P relative to zero potential at infinity.

Choosing the electrostatic potential energy of two charges to be zero at an infinite separation is analogous to the choice we made in Chapter 11 when we chose the gravitational potential energy of two point masses to be zero at an infinite separation. If two charges (or two masses) are at infinite separation, we think of them as not interacting. It has a certain appeal that the potential energy is zero if the particles are not interacting.

POTENTIAL ENERGY OF A HYDROGEN ATOM   **EXAMPLE   23-2**

(a) What is the electric potential at a distance $r = 0.529 \times 10^{-10}$ m from a proton? (This is the average distance between the proton and electron in a hydrogen atom.) (b) What is the electric potential energy of the electron and the proton at this separation?

PICTURE THE PROBLEM

(a) Use $V = kq/r$ to calculate the potential V due to the proton:

$$V = \frac{kq}{r} = \frac{ke}{r} = \frac{(8.99 \times 10^9 \text{ N·m}^2/\text{C}^2)(1.6 \times 10^{-19} \text{ C})}{0.529 \times 10^{-10} \text{ m}}$$

$$= 27.2 \text{ N·m/C} = \boxed{27.2 \text{ V}}$$

(b) Use $U = q_0 V$, with $q_0 = -e$ to calculate the potential energy:

$$U = q_0 V = (-e)(27.2 \text{ V}) = \boxed{-27.2 \text{ eV}}$$

REMARKS If the electron were at rest at this distance from the proton, it would take a minimum of 27.2 eV to remove it from the atom. However, the electron has kinetic energy equal to 13.6 eV, so its total energy in the atom is 13.6 eV − 27.2 eV = −13.6 eV. The minimum energy needed to remove the electron from the atom is thus 13.6 eV. This energy is called the ionization energy.

EXERCISE What is the potential energy of the electron and proton in Example 23-2 in SI units? (*Answer* −4.35 × 10⁻¹⁸ J)

POTENTIAL ENERGY OF NUCLEAR-FISSION PRODUCTS   **E X A M P L E   2 3 - 3**

In nuclear fission, a uranium-235 nucleus captures a neutron and splits apart into two lighter nuclei. Sometimes the two fission products are a barium nucleus (charge 56e) and a krypton nucleus (charge 36e). Assume that immediately after the split these nuclei are positive point charges separated by $r = 14.6 \times 10^{-15}$ m. Calculate the potential energy of this two-charge system in electron volts.

PICTURE THE PROBLEM The potential energy for two point charges separated by a distance r is $U = kq_1q_2/r$. To find this energy in electron volts, we calculate the potential due to one of the charges kq_1/r in volts and multiply by the other charge.

1. Equation 23-9 gives the potential energy of the two charges:

$$U = \frac{kq_1q_2}{r} = \frac{k(56e)(36e)}{r}$$

2. Factor out e and substitute the given values:

$$U = \frac{k(56e)(36e)}{r} = e\frac{56 \cdot 36ke}{r}$$

$$= e\frac{56 \cdot 36 \cdot (8.99 \times 10^9 \, \text{N·m}^2/\text{C}^2)(1.6 \times 10^{-19} \, \text{C})}{14.6 \times 10^{-15} \, \text{m}}$$

$$= e(1.99 \times 10^8 \, \text{V}) = \boxed{199 \, \text{MeV}}$$

REMARKS The separation distance r was chosen to be the sum of the radii of the two nuclei. After the fission, the two nuclei repel because of their electrostatic repulsion. Their potential energy of 199 MeV is converted into kinetic energy and, upon colliding with surrounding atoms, thermal energy. Two or three neutrons are also released in the fission process. In a chain reaction, one or more of these neutrons produces a fission of another uranium nucleus. The average energy given off in chain reactions of this type is about 200 MeV per nucleus, as calculated in this example.

The potential at some point due to several point charges is the sum of the potentials due to each charge separately. (This follows from the superposition principle for the electric field.) The potential due to a system of point charges q_i is thus given by

$$V = \sum_i \frac{kq_i}{r_i} \qquad\qquad 23\text{-}10$$

POTENTIAL DUE TO A SYSTEM OF POINT CHARGES

where the sum is over all the charges, and r_i is the distance from the ith charge to the field point at which the potential is to be found.

POTENTIAL DUE TO TWO POINT CHARGES   **E X A M P L E   2 3 - 4**

Two +5 nC point charges are on the x-axis, one at the origin and the other at $x = 8$ cm. Find the potential at (a) point P_1 on the x axis at $x = 4$ cm and (b) point P_2 on the y-axis at $y = 6$ cm.

PICTURE THE PROBLEM The two positive point charges on the x-axis are shown in Figure 23-5, and the potential is to be found at points P_1 and P_2.

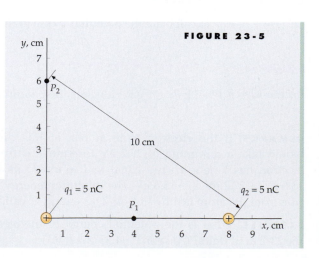

FIGURE 23-5

(a) 1. Use Equation 23-10 to write V as a function of the distances r_1 and r_2 to the charges:

$$V = \sum_i \frac{kq_i}{r_i} = \frac{kq_1}{r_1} + \frac{kq_2}{r_2}$$

2. Point P_1 is 4 cm from each charge, and the charges are equal:

$$r_1 = r_2 = r = 0.04 \text{ m}$$
$$q_1 = q_2 = q = 5 \times 10^{-9} \text{ C}$$

3. Use these to find the potential at point P_1:

$$V = \frac{kq_1}{r_1} + \frac{kq_2}{r_2} = \frac{2kq}{r}$$

$$= \frac{2 \times (8.99 \times 10^9 \text{ N·m}^2/\text{C}^2)(5 \times 10^{-9} \text{ C})}{0.04 \text{ m}}$$

$$= 2250 \text{ V} = \boxed{2.25 \text{ kV}}$$

(b) Point P_2 is 6 cm from one charge and 10 cm from the other. Use these to find the potential at point P_2:

$$V = \frac{(8.99 \times 10^9 \text{ N·m}^2/\text{C}^2)(5 \times 10^{-9} \text{ C})}{0.06 \text{ m}}$$

$$+ \frac{(8.99 \times 10^9 \text{ N·m}^2/\text{C}^2)(5 \times 10^{-9} \text{ C})}{0.10 \text{ m}}$$

$$= 749 \text{ V} + 450 \text{ V} \approx \boxed{1.20 \text{ kV}}$$

REMARKS Note that in Part (a), the electric field is zero at the point midway between the charges but the potential is not. It takes work to bring a test charge to this point from a long distance away, because the electric field is zero only at the final position.

FIGURE 23-6

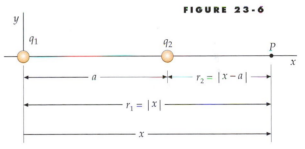

POTENTIAL THROUGHOUT THE X-AXIS   **E X A M P L E   2 3 - 5**

In Figure 23-6, a point charge q_1 is at the origin, and a second point charge q_2 is on the x-axis at $x = a$. Find the potential everywhere on the x-axis.

PICTURE THE PROBLEM The total potential is the sum of the potential due to each charge separately. The distance r_1 from q_1 to an arbitrary field point P is $r_1 = |x|$, and the distance r_2 from q_2 to P is $r_2 = |x - a|$.

Write the potential as a function of the distances to the two charges:

$$V = \frac{kq_1}{r_1} + \frac{kq_2}{r_2} = \boxed{\frac{kq_1}{|x|} + \frac{kq_2}{|x - a|}}$$

$$x \neq 0, \quad x \neq a$$

REMARKS Figure 23-7 shows V versus x for $q_1 = q_2 > 0$. The potential becomes infinite at each charge.

FIGURE 23-7

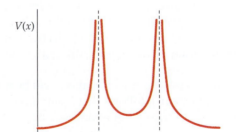

An electric dipole consists of a positive charge $+q$ on the x-axis at $x = +a$ and a negative charge $-q$ on the x-axis at $x = -a$, as shown in Figure 23-8. Find the potential on the x-axis for $x \gg a$ in terms of the dipole moment $p = 2qa$.

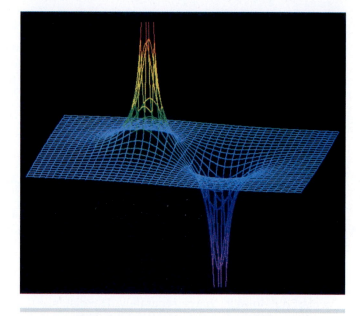

PICTURE THE PROBLEM The potential is the sum of the potentials for each charge. For $x > a$, the distance from the field point P to the positive charge is $x - a$ and the distance to the negative charge is $x + a$.

FIGURE 23-8

1. For $x > a$, the potential due to the two charges is:

$$V = \frac{kq}{x-a} + \frac{k(-q)}{x+a} = \frac{2kqa}{x^2 - a^2}$$

2. For $x \gg a$, we can neglect a^2 compared with x^2 in the denominator. We then have:

$$\boxed{V \approx \frac{2kqa}{x^2} = \frac{kp}{x^2}, \quad x \gg a}$$ 23-11

REMARKS Far from the dipole, the potential decreases as $1/r^2$ (compared to the potential of a point charge, which decreases as $1/r$).

23-3 Computing the Electric Field From the Potential

If we know the potential, we can use the potential to calculate the electric field. Consider a small displacement $d\vec{\ell}$ in an arbitrary electric field $\vec{E}$. The change in potential is

$$dV = -\vec{E} \cdot d\vec{\ell} = -E \cos \theta \, d\ell = -E_t \, d\ell \qquad 23\text{-}12$$

where $E_t = E \cos \theta$ is the component of $\vec{E}$ in the direction of $d\vec{\ell}$. Then

$$E_t = -\frac{dV}{d\ell} \qquad 23\text{-}13$$

The elecrostatic potential in the plane of an electric dipole. The potential due to each charge is proportional to the charge and inversely proportional to the distance from the charge.

If the displacement $d\vec{\ell}$ is perpendicular to the electric field, then $dV = 0$ (the potential does not change). For a given $d\ell$, the maximum increase in V occurs when the displacement $d\vec{\ell}$ is in the same direction as $-\vec{E}$. A vector that points in the direction of the greatest change in a scalar function and that has a magnitude equal to the derivative of that function with respect to the distance in that direction is called the **gradient** of the function. Thus, the electric field $\vec{E}$ is the negative gradient of the potential V. The electric field lines point in the direction of the greatest rate of decrease with respect to distance in the potential function.

If the potential V depends only on x, there will be no change in V for displacements in the y or z direction, it follows that E_y and E_z equal zero. For a displacement in the x direction, $d\vec{\ell} = dx\hat{i}$, and Equation 23-12 becomes

$$dV(x) = -\vec{E} \cdot d\vec{\ell} = -\vec{E} \cdot dx\hat{i} = -\vec{E} \cdot \hat{i} \, dx = -E_x dx$$

Then

$$E_x = -\frac{dV(x)}{dx} \qquad\qquad\qquad 23\text{-}14$$

Similarly, for a spherically symmetric charge distribution, the potential can be a function only of the radial distance r. Displacements perpendicular to the radial direction give no change in $V(r)$, so the electric field must be radial. A displacement in the radial direction is written $d\vec{\ell} = dr\hat{r}$. Equation 23-12 is then

$$dV(r) = -\vec{E} \cdot d\vec{\ell} = -\vec{E} \cdot dr\hat{r} = -E_r dr$$

and

$$E_r = -\frac{dV(r)}{dr} \qquad\qquad\qquad 23\text{-}15$$

If we know either the potential or the electric field over some region of space, we can use one to calculate the other. The potential is often easier to calculate because it is a scalar function, whereas the electric field is a vector function. Note that we cannot calculate $\vec{E}$ if we know the potential V at just a single point— we must know V over a region of space to compute the derivative necessary to obtain $\vec{E}$.

$\vec{E}$ FOR A POTENTIAL THAT VARIES WITH X	**EXAMPLE 23 - 7**

Find the electric field for the electric potential function V given by $V = 100\ \text{V} - (25\ \text{V/m})x$.

PICTURE THE PROBLEM This potential function depends only on x. The electric field is found from Equation 23-14:

$$\vec{E} = -\frac{dV}{dx}\hat{i} = \boxed{+(25\ \text{V/m})\hat{i}}$$

REMARKS This electric field is uniform and in the x direction. Note that the constant 100 V in the expression for $V(x)$ has no effect on the electric field. The electric field does not depend on the choice of zero for the potential function.

EXERCISE (a) At what point does V equal zero in this example? (b) Write the potential function corresponding to the same electric field with $V = 0$ at $x = 0$. [Answer (a) $x = 4$ m, (b) $V = -(25\ \text{V/m})x$]

*General Relation Between $\vec{E}$ and V

In vector notation, the gradient of V is written as either $\vec{grad}\ V$ or $\vec{\nabla}V$. Then

$$\vec{E} = -\vec{\nabla}V \qquad\qquad\qquad 23\text{-}16$$

In general, the potential function can depend on x, y, and z. The rectangular components of the electric field are related to the partial derivatives of the potential with respect to x, y, or z, while the other variables are held constant. For example, the x component of the electric field is given by

$$E_x = -\frac{\partial V}{\partial x} \qqud\qquad\qquad 23\text{-}17a$$

Similarly, the y and z components of the electric field are related to the potential by

$$E_y = -\frac{\partial V}{\partial y}$$ 23-17b

and

$$E_z = -\frac{\partial V}{\partial z}$$ 23-17c

Thus, Equation 23-16 in rectangular coordinates is

$$\vec{E} = -\vec{\nabla}V = -\left(\frac{\partial V}{\partial x}\hat{i} + \frac{\partial V}{\partial y}\hat{j} + \frac{\partial V}{\partial z}\hat{k}\right)$$ 23-18

23-4 Calculations of V for Continuous Charge Distributions

The potential due to a continuous distribution of charge can be calculated by choosing an element of charge dq, which we treat as a point charge, and, invoking superposition, changing the sum in Equation 23-10 to an integral:

$$V = \int \frac{k\,dq}{r}$$ 23-19

POTENTIAL DUE TO A CONTINUOUS CHARGE DISTRIBUTION

This equation assumes that $V = 0$ at an infinite distance from the charges, so we cannot use it when there is charge at infinity, as is the case for artificial charge distributions like an infinite line charge or an infinite plane charge.

V on the Axis of a Charged Ring

Figure 23-9 shows a uniformly charged ring of radius a and charge Q. The distance from an element of charge dq to the field point P on the axis of the ring is $r = \sqrt{x^2 + a^2}$. Since this distance is the same for all elements of charge on the ring, we can remove this term from the integral in Equation 23-19. The potential at point P due to the ring is thus

$$V = \int_0^Q \frac{k\,dq}{r} = \frac{k}{r}\int_0^Q dq = \frac{kQ}{r}$$

or

$$V = \frac{kQ}{\sqrt{x^2 + a^2}} = \frac{kQ}{|x|}\frac{1}{\sqrt{1 + (a^2/x^2)}}$$ 23-20

POTENTIAL ON THE AXIS OF A CHARGED RING

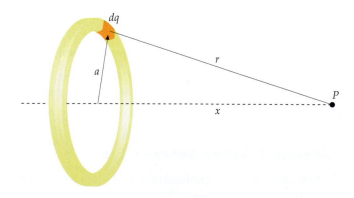

FIGURE 23-9 Geometry for the calculation of the electric potential at a point on the axis of a charged ring of radius a.

Note that when $|x|$ is much greater than a, the potential approaches $kQ/|x|$, the same as for a point charge Q at the origin.

A RING AND A PARTICLE   **EXAMPLE 23-8**   **Try It Yourself**

A ring of radius 4 cm is in the yz plane with its center at the origin. The ring carries a uniform charge of 8 nC. A small particle of mass $m = 6$ mg $= 6 \times 10^{-6}$ kg and charge $q_0 = 5$ nC is placed at $x = 3$ cm and released. Find the speed of the particle when it is a great distance from the ring. Assume gravitational effects are negligible.

PICTURE THE PROBLEM The particle is repelled by the ring. As the particle moves along the x-axis, its potential energy decreases and its kinetic energy increases. Use conservation of mechanical energy to find the kinetic energy of the particle when it is far from the ring. The final speed is found from the final kinetic energy.

Cover the column to the right and try these on your own before looking at the answers.

Steps

1. Write down the relation between the kinetic energy and the speed.

2. Use $U = q_0 V$, with V given by Equation 23-20, to obtain an expression for the potential energy of the point charge q_0 as a function of its distance x from the center of the ring.

3. Use conservation of mechanical energy to relate the speed to the position x and solve for the speed when x approaches infinity.

Answers

$K = \frac{1}{2}mv^2$

$U = q_0 V = \dfrac{kq_0 Q}{\sqrt{x^2 + a^2}}$

$U_f + K_f = U_i + K_i$

$$\dfrac{kq_0 Q}{\sqrt{x_f^2 + a^2}} + \dfrac{1}{2}mv_f^2 = \dfrac{kq_0 Q}{\sqrt{x_i^2 + a^2}} + \dfrac{1}{2}mv_i^2$$

so

$v_f = \boxed{1.55 \text{ m/s}}$

EXERCISE What is the potential energy of the particle when it is at $x = 9$ cm? (*Answer* 3.65×10^{-6} J)

V on the Axis of a Uniformly Charged Disk

We can use our result for the potential on the axis of a ring charge to calculate the potential on the axis of a uniformly charged disk.

FIND V FOR A CHARGED DISK   **EXAMPLE 23-9**

Find the potential on the axis of a disk of radius R that carries a total charge Q distributed uniformly on its surface.

PICTURE THE PROBLEM We take the axis of the disk to be the x-axis, and we treat the disk as a set of ring charges. The ring of radius a and thickness da in Figure 23-10 has an area of $2\pi a\, da$, and its charge is $dq = \sigma\, dA = \sigma 2\pi a\, da$ where $\sigma = Q/(\pi R^2)$, the surface charge density. The potential due to the charge on this ring at point P is given by Equation 23-20. We then integrate from $a = 0$ to $a = R$ to find the total potential due to the charge on the disk.

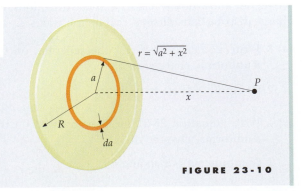

FIGURE 23-10

1. Write the potential dV at point P due to the charged ring of radius a:

$$dV = \frac{k\,dq}{(x^2 + a^2)^{1/2}} = \frac{k\sigma\,2\pi a\,da}{(x^2 + a^2)^{1/2}}$$

2. Integrate from $a = 0$ to $a = R$:

$$V = \int_0^R \frac{k\sigma\,2\pi a\,da}{(x^2 + a^2)^{1/2}} = k\sigma\pi \int_0^R (x^2 + a^2)^{-1/2}2a\,da$$

3. The integral is of the form $\int u^n\,du$, with $u = x^2 + a^2$, $du = 2x\,dx$, and $n = -\frac{1}{2}$. When $a = 0$, $u = x^2$ and when $a = R$, $u = x^2 + R^2$:

$$V = k\sigma\pi \int_{x^2+0^2}^{x^2+R^2} u^{-1/2}\,du = k\sigma\pi \left.\frac{u^{1/2}}{\frac{1}{2}}\right|_{x^2}^{x^2+R^2} = 2k\sigma\pi\left(\sqrt{x^2 + R^2} - \sqrt{x^2}\right)$$

4. Rearranging this result to find V gives:

$$\boxed{V = 2\pi k\sigma|x|\left(\sqrt{1 + \frac{R^2}{x^2}} - 1\right)}$$

PLAUSIBILITY CHECK For $|x| \gg R$, the potential function V should approach that of a point charge Q at the origin. We expect that for large $|x|$, $V \approx kQ/|x|$. To approximate our result for $|x| \gg R$, we use the binomial expansion:

$$\left(1 + \frac{R^2}{x^2}\right)^{1/2} \approx 1 + \frac{1}{2}\frac{R^2}{x^2} + \;\ldots$$

Then

$$V \approx 2\pi k\sigma|x|\left[\left(1 + \frac{1}{2}\frac{R^2}{x^2} + \;\ldots\;\right) - 1\right] = \frac{k(\sigma\pi R^2)}{|x|} = \frac{kQ}{|x|}$$

From Example 23-9, we see that the potential on the axis of a uniformly charged disk is

$$V = 2\pi k\sigma|x|\left(\sqrt{1 + \frac{R^2}{x^2}} - 1\right) \qquad\qquad 23\text{-}21$$

POTENTIAL ON THE AXIS OF A UNIFORMLY CHARGED DISK

Find $\vec{E}$ *Given* V

EXAMPLE 23-10

Calculate the electric field on the axis of (a) a uniformly charged ring and (b) a uniformly charged disk using the potential functions previously given for these charge distributions.

PICTURE THE PROBLEM Using $E_x = -dV/dx$, we can evaluate E_x by direct differentiation. We cannot evaluate either E_y or E_z by direct differentiation because we do not know how V varies in those directions. However, the symmetry of the charge distributions dictates that on the x-axis, $E_y = E_z = 0$.

(a) 1. Write Equation 23-20 for the potential on the axis of a uniformly charged ring:

$$V = \frac{kQ}{\sqrt{x^2 + a^2}} = kQ(x^2 + a^2)^{-1/2}$$

2. Compute $-dV/dx$ to find E_x:

$$E_x = -\frac{dV}{dx} = +\frac{1}{2}kQ(x^2 + a^2)^{-3/2}(2x) = \boxed{\frac{kQx}{(x^2 + a^2)^{3/2}}}$$

(b) 1. Write Equation 23-21 for the potential on the axis of a uniformly charged disk:

$$V = 2\pi k\sigma\left[(x^2 + a^2)^{1/2} - |x|\right]$$

2. Compute $-dV/dx$ to find E_x:

$$E_x = -\frac{dV}{dx} = -2\pi k\sigma\left[\frac{1}{2}(x^2 + a^2)^{-1/2}\,2x - \frac{d|x|}{dx}\right]$$

3. Evaluate $d|x|/dx$. It is the slope of a graph of $|x|$ versus x (see Figure 23-11):

$$\frac{d|x|}{dx} = +1, \quad x > 0; \frac{d|x|}{dx} = -1, \quad x < 0$$

4. Substituting for $d|x|/dx$ in the Part (b), step 2 result gives:

$$E_x = -2\pi k\sigma\left(\frac{x}{\sqrt{x^2 + a^2}} - 1\right), \quad x > 0$$

and

$$E_x = -2\pi k\sigma\left(\frac{x}{\sqrt{x^2 + a^2}} + 1\right), \quad x < 0$$

5. A little rearranging puts these expressions in a form that better reveals that E_x is an odd function (Figure 23-12). [A function f is odd if $f(-x) = f(x)$ for all values of x]:

$$\boxed{E_x = +2\pi k\sigma\left(1 - \frac{1}{\sqrt{1 + (a^2/x^2)}}\right), \quad x > 0}$$

and

$$\boxed{E_x = -2\pi k\sigma\left(1 - \frac{1}{\sqrt{1 + (a^2/x^2)}}\right), \quad x < 0}$$

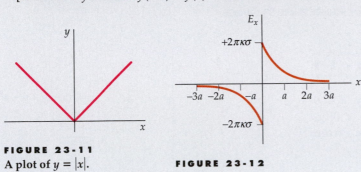

FIGURE 23-11
A plot of $y = |x|$.

FIGURE 23-12

REMARKS The results for Parts (a) and (b) are the same as Equations 22-10 and 22-11, which were calculated directly from Coulomb's law.

V Due to an Infinite Plane of Charge

If we let R become very large, our disk approaches an infinite plane. As R approaches infinity, the potential function (Equation 23-21) approaches infinity. However, we obtained Equation 23-21 from Equation 23-19, which assumes that $V = 0$ at infinity, so Equation 23-21 cannot be used. For infinite charge distributions, we must choose $V = 0$ at some finite point rather than at infinity. For such cases, we first find the electric field $\vec{E}$ (by direct integration or from Gauss's law) and then calculate the potential function V from its definition $dV = -\vec{E} \cdot d\vec{\ell}$. For an infinite plane of uniform charge of density σ in the yz plane, the electric field for positive x is given by

$$\vec{E} = \frac{\sigma}{2\,\epsilon_0}\,\hat{i} = 2\pi k\sigma\hat{i}$$

The potential is then

$$dV = -\vec{E} \cdot d\vec{\ell} = -(2\pi k\sigma\hat{i}) \cdot (dx\hat{i} + dy\hat{j} + dz\hat{k}) = -2\pi k\sigma\,dx$$

where we have used $d\vec{\ell} = dx\hat{i} + dy\hat{j} + dz\hat{k}$. Integrating, we obtain

$$V = -2\pi k\sigma x + V_0$$

where the arbitrary integration constant V_0 is the potential at $x = 0$. Note that the potential decreases with distance from the plane and approaches $-\infty$ as x approaches $+\infty$. Therefore, at $x = +\infty$ the potential equals negative infinity.
    For negative x, the electric field is

$$\vec{E} = -2\pi k\sigma\hat{i}$$

so

$$dV = -\vec{E} \cdot d\vec{\ell} = +2\pi k\sigma \, dx$$

and the potential is

$$V = V_0 + 2\pi k\sigma x$$

Since x is negative, the potential again decreases with distance from the plane and approaches $-\infty$ as x approaches $-\infty$. For either positive or negative x, the potential can be written

$$V = V_0 - 2\pi k\sigma |x| \qquad\qquad 23\text{-}22$$

POTENTIAL NEAR AN INFINITE PLANE OF CHARGE

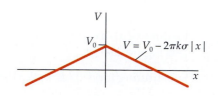

FIGURE 23-13 Plot of V versus x for an infinite plane of charge in the yz plane. Note that the potential is continuous at $x = 0$ even though $E_x = dV/dx$ is not.

A PLANE AND A POINT CHARGE   **EXAMPLE 23-11**

An infinite plane of uniform charge density σ is in the $x = 0$ plane, and a point charge q is on the x axis at $x = a$ (Figure 23-14). Find the potential at some point P a distance r from the point charge.

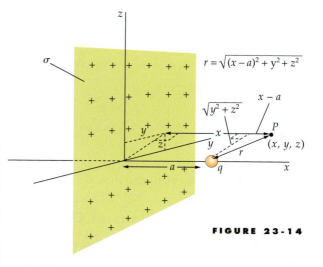

FIGURE 23-14

PICTURE THE PROBLEM We can use the principle of superposition. The total potential V is the sum of the individual potentials due to the plane and the point charge. We must add an arbitrary constant in our expression for V, which is determined by our choice of the reference point, where $V = 0$. We are free to choose the reference point, except at $x = \pm\infty$ or at $x = a$ on the x-axis. For this solution, we choose $V = 0$ at the origin.

1. The potential due to the charged plane is given by Equation 23-22:

$$V_{\text{plane}} = -2\pi k\sigma |x|$$

2. Equation 23-7 gives the potential due to a point charge. The distance r from the point charge to the field point equals $\sqrt{(x-a)^2 + y^2 + z^2}$:

$$V_{\text{point}} = \frac{kq}{r} = \frac{kq}{\sqrt{(x-a)^2 + y^2 + z^2}}$$

3. Sum the above results to find the total potential V. A constant is added to the sum so we can set the potential at the reference point to zero:

$$V = V_{\text{plane}} + V_{\text{point}}$$
$$= -2\pi k\sigma |x| + \frac{kq}{\sqrt{(x-a)^2 + y^2 + z^2}} + C$$

4. We choose to let $V = 0$ at the origin. To do that, set $V = 0$ at $x = y = z = 0$ and solve for the constant C:

$$0 = 0 + \frac{kq}{a} + C, \quad \text{so } C = -\frac{kq}{a}$$

5. Substitute $-kq/a$ for C in the step 3 result:

$$V = -2\pi k\sigma |x| + \frac{kq}{\sqrt{(x-a)^2 + y^2 + z^2}} - \frac{kq}{a}$$

$$\boxed{= -2\pi k\sigma |x| + \frac{kq}{r} - \frac{kq}{a}}$$

REMARKS The answer is not unique. We could have specified the potential at any point other than at $x = a$ or at $x = \pm\infty$.

V Inside and Outside a Spherical Shell of Charge

We find the potential due to a thin spherical shell of radius R with charge Q uniformly distributed on its surface next. We are interested in the potential at all points inside, outside, and on the shell. Unlike the infinite plane of charge, this charge distribution is confined to a finite region of space, so, in principle, we could calculate the potential by direct integration of Equation 23-19. However, there is a simpler way. Since the electric field for this charge distribution is easily obtained from Gauss's law, we will calculate the potential from the known electric field using $dV = -\vec{E} \cdot d\vec{\ell}$.

Outside the spherical shell, the electric field is radial and is the same as if all the charge Q were a point charge at the origin:

$$\vec{E} = \frac{kQ}{r^2}\hat{r}$$

The change in the potential for some displacement $d\vec{r}$ outside the shell is then

$$dV = -\vec{E} \cdot d\vec{\ell} = -\frac{kQ}{r^2}\hat{r} \cdot d\vec{\ell}$$

The product $\hat{r} \cdot d\vec{\ell}$ is dr (the component of $d\vec{\ell}$ in the direction of $\hat{r}$). Integrating along a path from the reference point at infinity, we obtain

$$V_P = -\int_{\infty}^{\vec{r}_P} \vec{E} \cdot d\vec{\ell} = -\int_{\infty}^{r_P} \frac{kQ}{r^2}\,dr = -kQ\int_{\infty}^{r_P} r^{-2}\,dr = \frac{kQ}{r_P}$$

where P is an arbitrary field point in the region $r \geq R$, and r_P is the distance from the center of the shell to the field point P. The potential is chosen to be zero at infinity. Since P is arbitrary, we let $r_P = r$ to obtain

$$V = \frac{kQ}{r}, \qquad r \geq R$$

Inside the spherical shell, the electric field is zero everywhere. Again integrating from the reference point at infinity, we obtain

$$V_P = -\int_{\infty}^{\vec{r}_P} \vec{E} \cdot d\vec{r} = -\int_{\infty}^{R} \frac{kQ}{r^2}\,dr - \int_{R}^{r_P} (0)dr = \frac{kQ}{R}$$

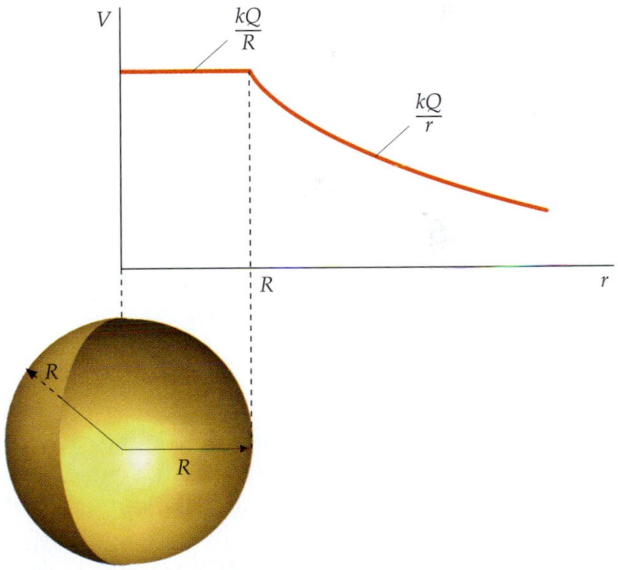

where P is an arbitrary field point in the region $r < R$, and r_P is the distance from the center of the shell to the field point P. The potential at an arbitrary point inside the shell is kQ/R, where R is the radius of the shell. Inside the shell V is the same everywhere. It is the work per unit charge to bring a test charge from infinity to the shell. No additional work is required to bring it from the shell to any point inside the shell. Thus,

$$V = \frac{kQ}{r}, \qquad r \geq R$$

$$V = \frac{kQ}{R}, \qquad r \leq R \qquad\qquad \text{23-23}$$

POTENTIAL DUE TO A SPHERICAL SHELL

FIGURE 23-15 Electric potential of a uniformly charged spherical shell of radius R as a function of the distance r from the center of the shell. Inside the shell, the potential has the constant value kQ/R. Outside the shell, the potential is the same as that due to a point charge at the center of the sphere.

This potential function is plotted in Figure 23-15.

❶ A common mistake is to think that the potential must be zero inside a spherical shell because the electric field is zero throughout that region. But a region of zero electric field merely implies that the potential is uniform. Consider a spherical shell with a small hole so that we can move a test charge in and out of the shell. If we move the test charge from an infinite distance to the shell, the work per charge we must do is kQ/R. Inside the shell there is no electric field, so it takes no work to move the test charge around inside the shell. The total amount of work per unit charge it takes to bring the test charge from infinity to any point inside the shell is just the work per charge it takes to bring the test charge up to the shell radius R, which is kQ/R. The potential is therefore kQ/R everywhere inside the shell.

EXERCISE What is the potential of a spherical shell of radius 10 cm carrying a charge of 6 μC? (*Answer* 5.39×10^5 V = 539 kV)

| FIND V FOR A UNIFORMLY CHARGED SPHERE | **EXAMPLE 23-12** **Try It Yourself** |

In one model, a proton is considered to be a spherical ball of charge of uniform volume charge density with radius R and total charge Q. The electric field inside the sphere is given by Equation 22-26b,

$$E_r = k \frac{Q}{R^3} r$$

Find the potential V both inside and outside the sphere.

PICTURE THE PROBLEM Outside the sphere, the charge looks like a point charge, so the potential is given by $V = kQ/r$. Inside the sphere, V can be found by integrating $dV = -\vec{E} \cdot d\vec{\ell}$.

Cover the column to the right and try these on your own before looking at the answers.

Steps   **Answers**

1. Outside the sphere, the electric field is the same as that of a point charge. If we set the potential equal to zero at infinity, the potential there is also the same as that of a point charge.

$$V(r) = \boxed{\frac{kQ}{r}}, \quad r \geq R$$

2. For $r \leq R$, find dV from $dV = -\vec{E} \cdot d\vec{\ell}$.

$$dV = -\vec{E} \cdot d\vec{\ell} = -E_r\, dr = -\frac{kQ}{R^3} r\, dr, \quad r \leq R$$

3. Find the definite integral using your expression in step 2. Find the change in potential from infinity to an arbitrary field point P in the region $r_P < R$, where r_P is the distance of point P from the center of the sphere:

$$V_P = -\int_\infty^{r_P} E_r\, dr = -\int_\infty^R \frac{kQ}{r^2}\, dr - \int_R^{r_P} \frac{kQ}{R^3} r\, dr$$

$$= \frac{kQ}{R} - \frac{kQ}{2R^3}(r_P^2 - R^2) = \frac{kQ}{2R}\left(3 - \frac{r_P^2}{R^2}\right), \quad r \leq R$$

4. Since the field point position is arbitrary, express the result in terms of $r = r_P$:

$$V(r) = \boxed{\frac{kQ}{2R}\left(3 - \frac{r^2}{R^2}\right), \quad r \leq R}$$

❶ **PLAUSIBILITY CHECK** Substituting $r = R$ in the step 4 result gives $V(R) = kQ/R$ as required. At $r = 0$, $V(0) = 3kQ/2R = 1.5\, kQ/R$, which is greater than $V(R)$, as it should be, because the electric field is in the positive radial direction for $r < R$, so positive work must be done to move a positive test charge against the field from $r = R$ to $r = 0$.

REMARKS Figure 23-16 shows $V(r)$ as a function of r. Note that both $V(r)$ and $E_r = -dV/dr$ are continuous everywhere.

EXERCISE What is $V(r)$, if we choose $V(R) = 0$? [*Answer* $V(r) = kQ/r - kQ/R$ for $r \geq R$; $V(r) = \frac{1}{2}(kQ/R)(1 - r^2/R^2)$ for $r \leq R$]

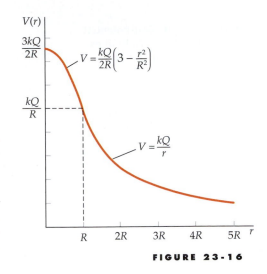

FIGURE 23-16

V Due to an Infinite Line Charge

We will now calculate the potential due to a uniform infinite line charge. Let the charge per unit length be λ. Like the infinite plane of charge, this charge distribution is not confined to a finite region of space, so, in principle, we cannot calculate the potential by direct integration of $dV = kdq/r$ (Equation 23-19). Instead, we will find the potential by integrating the electric field directly. First, we must obtain the electric field of a uniformly charged infinite line. The field, a cylindrically symmetric charge distribution like this one, can be obtained using Gauss's law ($\phi_{net} = 4\pi kQ_{inside}$). The outward flux through a coaxial soup-can-shaped Gaussian surface of radius R and length L is E_R $(2\pi RL)$, and the charge inside is λL. Substituting these expressions into the Gauss's-law equation and solving for E_R gives $E_R = 2k\lambda/R$. The change in potential for a displacement $d\vec{\ell}$ is given by

$$dV = -\vec{E} \cdot d\vec{\ell} = -E_R \hat{R} \cdot d\vec{\ell}$$

where $\hat{R}$ is the radial direction. The product $\hat{R} \cdot d\vec{\ell}$ is dR (the component of $d\vec{\ell}$ in the direction of $\hat{R}$), so $dV = -E_R dR$. Integrating from an arbitrary reference point to an arbitrary field point P (Figure 23-17) gives

$$V_P - V_{ref} = -\int_{R_{ref}}^{R_P} E_R dR = -2k\lambda \int_{R_{ref}}^{R_P} \frac{dR}{R} = -2k\lambda \ln \frac{R_P}{R_{ref}}$$

where R_P and R_{ref} are the radial distances of the field point P and the reference point from the line charge, respectively. For convenience, we choose the potential to equal zero at the reference point ($V_{ref} = 0$). We cannot choose R_{ref} to be zero because $\ln(0) = -\infty$, and we cannot choose R_{ref} to be infinity because $\ln(\infty) = +\infty$. However, any other choice in the interval $0 < R_{ref} < \infty$ is acceptable, and the potential function is given by

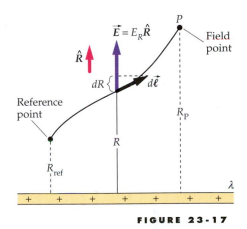

FIGURE 23-17

$$V = 2k\lambda \ln \frac{R_{ref}}{R} \qquad\qquad 23\text{-}24$$

POTENTIAL DUE TO A LINE CHARGE

We do not encounter infinite planes or lines of charge, but these distributions make excellent models for some real situations. For example, the potential near a 500-m-long, nearly straight, high-voltage transmission power line.

23-5 Equipotential Surfaces

Since there is no electric field inside the material of a conductor that is in static equilibrium, the change in potential as we move about the region occupied by the conducting material is zero. Thus, the electric potential is the same throughout the material of the conductor; that is, the conductor is a three-dimensional **equipotential region** and its surface is an **equipotential surface**. Because the potential V has the same value everywhere on an equipotential surface, the change in V is zero. If a test charge on the surface is given a small displacement $d\vec{\ell}$ parallel to the surface, $dV = -\vec{E} \cdot d\vec{\ell} = 0$. Since $\vec{E} \cdot d\vec{\ell}$ is zero for *any* $d\vec{\ell}$ parallel to the surface, $\vec{E}$ must be perpendicular to any and every $d\vec{\ell}$ parallel to

FIGURE 23-18 Equipotential surfaces and electric field lines outside a uniformly charged spherical conductor. The equipotential surfaces are spherical and the field lines are radial and perpendicular to the equipotential surfaces.

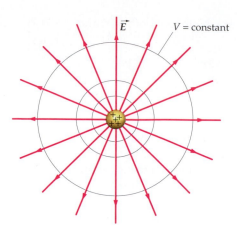

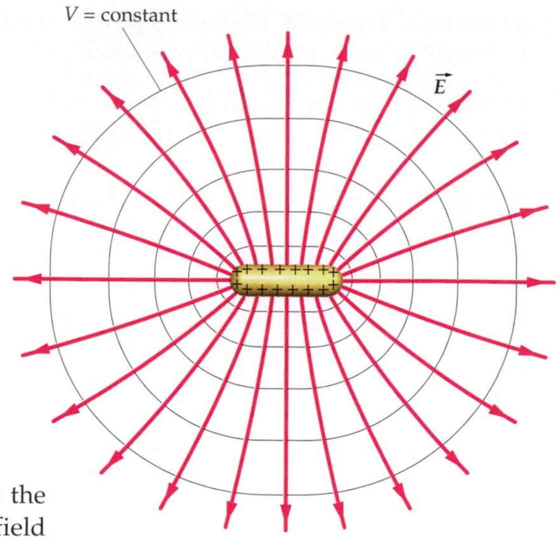

FIGURE 23-19 Equipotential surfaces and electric field lines outside a nonspherical conductor. Electric field lines are always intersect equipotential surfaces at right angles.

the surface. The only way $\vec{E}$ can be perpendicular to every $d\vec{\ell}$ parallel to the surface, however, is for $\vec{E}$ to be normal to the surface. Therefore, any electric field lines beginning or terminating on the equipotential surface must be normal to it. Figures 23-18 and 23-19 show equipotential surfaces near a spherical conductor and a nonspherical conductor. Note that anywhere a field line meets or penetrates an equipotential surface, shown in blue, the field line is normal to the equipotential surface. If we go from one equipotential surface to a neighboring equipotential surface by undergoing a displacement $d\vec{\ell}$ along a field line in the direction of the field, the potential changes by $dV = -\vec{E} \cdot d\vec{\ell} = -E\,d\ell$. It follows that equipotential surfaces that have a fixed potential difference between them are more closely spaced where the electric field strength E is greater.

A HOLLOW SPHERICAL SHELL **EXAMPLE 23-13**

A hollow, uncharged spherical conducting shell has an inner radius a and an outer radius b. A positive point charge $+q$ is in the cavity, at the center of the sphere. (*a*) Find the charge on each surface of the conductor. (*b*) Find the potential $V(r)$ everywhere, assuming that $V = 0$ at $r = \infty$.

PICTURE THE PROBLEM (*a*) The charge distribution has spherical symmetry, so applying Gauss's law should be a good method for finding the charges on the inner and outer surface of the shell. (*b*) Sum the individual potentials for the individual charges to obtain the resultant potential. The potential for a point charge and for a uniform thin spherical shell of charge have already been established (Equations 23-8 and 23-23).

(*a*) 1. The charge inside a closed surface is proportional to the outward flux of $\vec{E}$ through the surface:

$$\phi_{net} = 4\pi k Q_{inside}, \text{ where } \phi_{net} = \oint_S E_n\,dA$$

2. Sketch the point charge and the spherical shell. On a conductor, charge resides only on its surface. Label the charge on each surface of the shell. Include a Gaussian surface completely inside the material of the conductor:

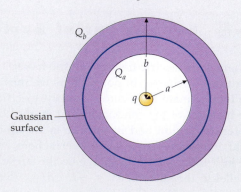

FIGURE 23-20

3. Apply Gauss's law (the step 1 result) to the Gaussian surface and solve for the charge on the inner surface of the shell:

$$E_n = 0 \quad \Rightarrow \quad Q_{\text{inside}} = q + Q_a = 0$$

so

$$\boxed{Q_a = -q}$$

4. The shell is neutral, so solve for the charge on its outer surface:

$$Q_a + Q_b = 0$$

so

$$\boxed{Q_b = +q}$$

(b) 1. The potential is the sum of the potentials due to the individual charges:

$$V = V_q + V_{Q_a} + V_{Q_b}$$

2. Add the potentials in the region outside the shell. The potential for a thin charged spherical shell is given in Equation 23-23:

$$V = \frac{kq}{r} - \frac{kq}{r} + \frac{kq}{r} = \boxed{\frac{kq}{r}, \quad r \geq b}$$

3. Add the potentials in the region inside the material of the conducting shell:

$$V = \frac{kq}{r} - \frac{kq}{r} + \frac{kq}{b} = \boxed{\frac{kq}{b}, \quad a \leq r \leq b}$$

4. Add the potentials in the region between the point charge and the shell:

$$V = \boxed{\frac{kq}{r} + \frac{kq}{b} - \frac{kq}{a}, \quad 0 < r \leq a}$$

REMARKS Each of the individual potential functions has its zero-potential reference point at $r = \infty$. Thus, the sum of these functions also has its zero-potential reference point at $r = \infty$. The potential arrived at in the example can be obtained by directly evaluating $-\int_{\infty}^{P} \vec{E} \cdot d\vec{\ell} = -\int_{\infty}^{r_P} E_r\, dr$. Yet a third way to obtain the potential is by evaluating the indefinite integral $-\int E_r\, dr$ in each region to find the integration constants by matching the potential functions at the boundaries. Matching the potential functions at the boundaries is valid because the potential must be continuous.

Figure 23-21 shows the electric potential as a function of the distance from the center of the cavity. Inside the conducting material, where $a \leq r \leq b$, the potential has the constant value kq/b. Outside the shell, the potential is the same as that of a point charge q at the center of the shell. Note that $V(r)$ is continuous everywhere. The electric field is discontinuous at the conductor surfaces, as reflected in the discontinuous slope of $V(r)$ at $r = a$ and $r = b$.

In general, two conductors that are separated in space will not be at the same potential. The potential difference between such conductors depends on their geometrical shapes, their separation, and the net charge on each. When two conductors are brought into contact, the charge on the conductors distributes itself so that electrostatic equilibrium is established, and the electric field is zero inside both conductors. While in contact, the two conductors may be considered to be a single conductor with a single equipotential surface. If we put a spherical charged conductor in contact with a second spherical conductor that is uncharged, charge will flow between them until both conductors are at the same potential. If the spherical conductors are identical, they share the original charge equally. If the identical spherical conductors are now separated, each carries half the original charge.

FIGURE 23-21

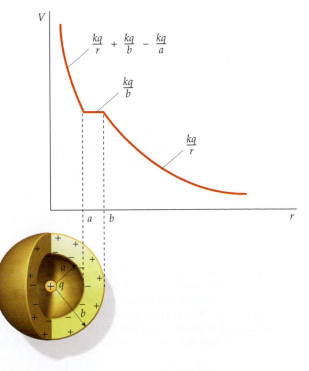

The Van de Graaff Generator

In Figure 23-22, a small conductor carrying a positive charge q is inside the cavity of a larger conductor. In equilibrium, the electric field is zero inside the conducting material of both conductors. The electric field lines that begin on the positive charge q must terminate on the inner surface of the large conductor. This must occur no matter what the charge may be on the outside surface of the large conductor. Regardless of the charge on the large conductor, the small conductor in the cavity is at a greater potential because the electric field lines go from this conductor to the larger conductor. If the conductors are now connected, say, with a fine conducting wire, *all* the charge originally on the smaller conductor will flow to the larger conductor. When the connection is broken, there is no charge on the small conductor in the cavity, and there are no field lines between the conductors. The positive charge transferred from the smaller conductor resides completely on the outside surface of the larger conductor. If we put more positive charge on the small conductor in the cavity and again connect the conductors with a fine wire, all of the charge on the inner conductor will again flow to the outer conductor. The procedure can be repeated indefinitely. This method is used to produce large potentials in a device called the Van de Graaff generator, in which the charge is brought to the inner surface of a larger spherical conductor by a continuous charged belt (Figure 23-23). Work must be done by the motor driving the belt to bring the charge from the bottom to the top of the belt, where

FIGURE 23-22 Small conductor carrying a positive charge inside a larger hollow conductor.

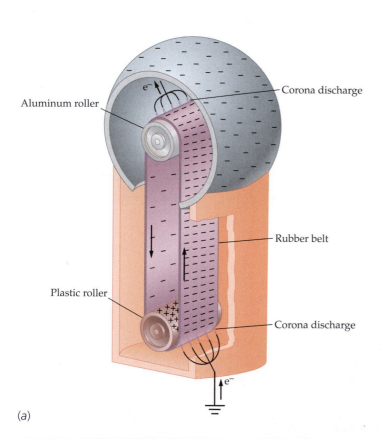

(a)

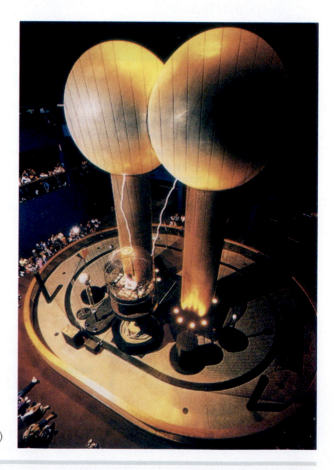

(b)

FIGURE 23-23 (*a*) Schematic diagram of a Van de Graaff generator. The lower roller becomes positively charged due to contact with the moving belt. (The inner surface of the belt acquires an equal amount of negative charge that is distributed over a larger area.) The dense positive charge on the roller attracts electrons to the tips of the lower comb where dielectric breakdown takes place and negative charge is transported to the belt via corona discharge. At the top roller the negatively charged belt repels electrons from the tips of the comb and negative charge is transferred from the belt to the comb. The charge is then transferred to the outer surface of the dome. (*b*) These large demonstration Van de Graaff generators in the Boston Science Museum are discharging to the grounded wire cage housing the operator.

the potential is very high. One can often hear the motor speed decrease as the sphere charges. The greater the charge on the outer conductor, the greater its potential, and the greater the electric field just outside its outer surface. A Van de Graaff accelerator is a device that uses the intense electric field produced by a Van de Graaff generator to accelerate charged ions particles, such as protons.

Dielectric Breakdown

Many nonconducting materials become ionized in very high electric fields and become conductors. This phenomenon, called **dielectric breakdown,** occurs in air at an electric field strength of $E_{max} \approx 3 \times 10^6$ V/m = 3 MN/C. In air, some of the existing ions are accelerated to greater kinetic energies before they collide with neighboring molecules. Dielectric breakdown occurs when these ions are accelerated to kinetic energies sufficient to result in a growth in ion concentration due to the collisions with neighboring molecules. The maximum potential that can be obtained in a Van de Graaff generator is limited by the dielecric break- down of the air. In a vacuum, Van de Graaff generators can achieve much higher potentials. The magnitude of the electric field for which dielectric breakdown occurs in a material is called the **dielectric strength** of that material. The dielec- tric strength of air is about 3 MV/m. The discharge through the conducting air resulting from dielectric breakdown is called **arc discharge.** The electric shock you receive when you touch a metal doorknob after walking across a rug on a dry day is a familiar example of arc discharge. These breakdowns occur more often on dry days because moist air can conduct the charge away before the breakdown condition is reached. Lightning is an example of arc discharge on a large scale.

DIELECTRIC BREAKDOWN FOR A CHARGED SPHERE　　　**EXAMPLE 23-14**

A spherical conductor has a radius of 30 cm (≈ 1 ft). (*a*) **What is the maximum charge that can be placed on the sphere before dielectric breakdown of the surrounding air occurs?** (*b*) **What is the maximum potential of the sphere?**

PICTURE THE PROBLEM (*a*) We find the maximum charge by relating the charge to the electric field and setting the field equal to the dielectric strength of air, E_{max}. (*b*) The maximum potential is then found from the maximum charge calculated in Part (*a*).

(*a*) 1. The surface charge density on the conductor σ is re- lated to the electric field just outside the conductor:

$$E = \frac{\sigma}{\epsilon_0} = 4\pi k\sigma$$

2. Set this field equal to E_{max}:

$$E_{max} = 4\pi k\sigma_{max}$$

3. The maximum charge Q_{max} is found from σ_{max}:

$$\sigma_{max} = \frac{Q_{max}}{4\pi R^2}$$

4. Solving for Q_{max} gives:

$$Q_{max} = 4\pi R^2 \sigma_{max} = 4\pi R^2 \frac{E_{max}}{4\pi k} = \frac{R^2 E_{max}}{k}$$

$$= \frac{(0.3\text{ m})^2 (3 \times 10^6\text{ N/C})}{(8.99 \times 10^9\text{ N·m}^2/\text{C}^2)} = \boxed{3.00 \times 10^{-5}\text{ C}}$$

(*b*) Use the expression for the maximum charge to calcu- late the maximum potential of the sphere:

$$V_{max} = \frac{kQ_{max}}{R} = \frac{k}{R}\left(\frac{R^2 E_{max}}{k}\right) = RE_{max}$$

$$= (0.3\text{ m})(3 \times 10^6\text{ N/C}) = \boxed{9.00 \times 10^5\text{ V}}$$

TWO CHARGED SPHERICAL CONDUCTORS   **EXAMPLE 23-15**

Two charged spherical conductors of radius R_1 = 6 cm and R_2 = 2 cm (Figure 23-24) are separated by a distance much greater than 6 cm and are connected by a long, thin conducting wire. A total charge $Q = +80$ nC is placed on one of the spheres. (*a*) What is the charge on each sphere? (*b*) What is the electric field near the surface of each sphere? (*c*) What is the electric potential of each sphere? (Assume that the charge on the connecting wire is negligible.)

FIGURE 23-24

PICTURE THE PROBLEM The total charge will be distributed with Q_1 on sphere 1 and Q_2 on sphere 2 so that the spheres will be at the same potential. We can use $V = kQ/R$ for the potential of each sphere because they are far apart.

(*a*) 1. Conservation of charge gives us one relation between the charges Q_1 and Q_2:

$$Q_1 + Q_2 = Q$$

2. Equating the potential of the spheres gives us a second relation for the charges Q_1 and Q_2:

$$\frac{kQ_1}{R_1} = \frac{kQ_2}{R_2} \Rightarrow Q_2 = \frac{R_2}{R_1}Q_1$$

3. Combine the results from step 1 and step 2 and solve for Q_1 and Q_2:

$$Q_1 + \frac{R_2}{R_1}Q_1 = Q, \quad \text{so}$$

$$Q_1 = \frac{R_1}{R_1 + R_2}Q = \frac{6 \text{ cm}}{8 \text{ cm}}(80 \text{ nC}) = \boxed{60 \text{ nC}}$$

$$Q_2 = Q - Q_1 = \boxed{20 \text{ nC}}$$

(*b*) Use these results to calculate the electric fields at the surface of the spheres:

$$E_1 = \frac{kQ_1}{R_1^2} = \frac{(8.99 \times 10^9 \text{ N·m}^2/\text{C}^2)(60 \times 10^{-9} \text{ C})}{(0.06 \text{ m})^2}$$

$$= \boxed{150 \text{ kN/C}}$$

$$E_2 = \frac{kQ_2}{R_2^2} = \frac{(8.99 \times 10^9 \text{ N·m}^2/\text{C}^2)(20 \times 10^{-9} \text{ C})}{(0.02 \text{ m})^2}$$

$$= \boxed{450 \text{ kN/C}}$$

(*c*) Calculate the common potential from kQ/R for either sphere:

$$V_1 = \frac{kQ_1}{R_1} = \frac{(8.99 \times 10^9 \text{ N·m}^2/\text{C}^2)(60 \times 10^{-9} \text{ C})}{0.06 \text{ m}}$$

$$= \boxed{8.99 \text{ kV}}$$

PLAUSIBILITY CHECK If we use sphere 2 to calculate V, we obtain $V_2 = kQ_2/R_2 = (8.99 \times 10^9 \text{ N·m}^2/\text{C}^2)(20 \times 10^{-9} \text{ C})/0.02 \text{ m} = 8.99 \times 10^3$ V. An additional check is available, since the electric field at the surface of each sphere is proportional to its charge density. The radius of sphere 1 is three times that of sphere 2, so its surface area is nine times that of sphere 2. And since it carries three times the charge, its charge density is $\frac{1}{3}$ that of sphere 2. Therefore, the field of sphere 1 should be $\frac{1}{3}$ that of sphere 2, which is what we found.

When a charge is placed on a conductor of nonspherical shape, like that in Figure 23-25a, the surface of the conductor will be an equipotential surface, but the surface charge density and the electric field just outside the conductor will vary from point to point. Near a point where the radius of curvature is small, such as point A in the figure, the surface charge density and electric field will be large, whereas near a point where the radius of curvature is large, such as point B in the figure, the field and surface charge density will be small. We can understand this qualitatively by considering the ends of the conductor to be spheres of different radii. Let σ be the surface charge density.

The potential of a sphere of radius R is

$$V = \frac{kQ}{R} = \frac{1}{4\pi\epsilon_0}\frac{Q}{R}$$ 23-25

Since the area of a sphere is $4\pi R^2$, the charge on a sphere is related to the charge density by $Q = 4\pi R^2\sigma$. Substituting this expression for Q into Equation 23-25 we have

$$V = \frac{1}{4\pi\epsilon_0}\frac{4\pi R^2\sigma}{R} = \frac{R\sigma}{\epsilon_0}$$

Solving for σ, we obtain

$$\sigma = \frac{\epsilon_0 V}{R}$$ 23-26

Since both *spheres* are at the same potential, the sphere with the smaller radius must have the greater surface charge density. And since $E = \sigma/\epsilon_0$ near the surface of a conductor, the electric field is greatest at points on the conductor where the radius of curvature is least.

For an arbitrarily shaped conductor, the potential at which dielectric breakdown occurs depends on the smallest radius of curvature of any part of the conductor. If the conductor has sharp points of very small radius of curvature, dielectric breakdown will occur at relatively low potentials. In the Van de Graaff generator (see Figure 23-23a), the charge is transferred onto the belt by sharp-edged conductors near the bottom of the belt. The charge is removed from the belt by sharp-edged conductors near the top of the belt. Lightning rods at the top of a tall building draw the charge off a nearby cloud before the potential of the cloud can build up to a destructively large value.

(a)

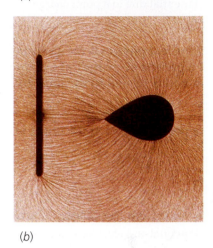

(b)

FIGURE 23-25 (a) A nonspherical conductor. If a charge is placed on such a conductor, it will produce an electric field that is stronger near point A, where the radius of curvature is small, than near point B, where the radius of curvature is large. (b) Electric field lines near a nonspherical conductor and plate carrying equal and opposite charges. The lines are shown by small bits of thread suspended in oil. Note that the electric field is strongest near points of small radius of curvature, such as at the ends of the plate and at the pointed left side of the conductor. The equipotential surfaces are more closely spaced where the field strength is greater.

SUMMARY

1. Electric potential, which is defined as the electrostatic potential energy per unit charge, is an important derived physical concept that is related to the electric field.

2. Because potential is a scalar quantity, it is often easier to calculate than the vector electric field. Once V is known, $\vec{E}$ can be calculated from V.

Topic	Relevant Equations and Remarks
1. Potential Difference	The potential difference $V_b - V_a$ is defined as the negative of the work per unit charge done by the electric field when a test charge moves from point a to point b:

$$\Delta V = V_b - V_a = \frac{\Delta U}{q_0} = -\int_a^b \vec{E}\cdot d\vec{\ell}$$ 23-2b

Potential difference for infinitesimal displacements	$dV = -\vec{E} \cdot d\vec{\ell}$	**23-2a**

2. Electric Potential

Potential due to a point charge	$V = \dfrac{kq}{r} - \dfrac{kq}{r_{\text{ref}}}, \quad (V = 0 \text{ at } r = r_{\text{ref}})$	**23-7**
Coulomb potential	$V = \dfrac{kq}{r}, \quad (V = 0 \text{ at } r = \infty)$	**23-8**
Potential due to a system of point charges	$V = \sum_i \dfrac{kq_i}{r_i}, \quad (V = 0 \text{ at } r_i = \infty, i = 1, 2, \dots)$	**23-10**
Potential due to a continuous charge distribution	$V = \displaystyle\int \dfrac{k\,dq}{r}, \quad (V = 0 \text{ at } r = \infty)$	**23-19**
	This expression can be used only if the charge distribution is contained in a finite volume so that the potential can be chosen to be zero at infinity.	
Potential and electric field lines	Electric field lines point in the direction of decreasing electric potential.	
Continuity of electric potential	The potential function V is continuous everywhere in space.	

3. Computing the Electric Field From the Potential

	The electric field points in the direction of the greatest decrease in the potential. $$E_t = \dfrac{dV}{d\ell}$$	**23-13**
Gradient	A vector that points in the direction of the greatest rate of change in a scalar function and that has a magnitude equal to the derivative of that function, with respect to the distance in that direction, is called the gradient of the function. $\vec{E}$ is the negative gradient of V.	
Potential a function of x alone	$E_x = -\dfrac{dV(x)}{dx}$	**23-14**
Potential a function of r alone	$E_r = -\dfrac{dV(r)}{dr}$	**23-15**

4. *General Relation Between $\vec{E}$ and V

	$\vec{E} = -\vec{\nabla} V = -\left(\dfrac{\partial V}{\partial x}\hat{i} + \dfrac{\partial V}{\partial y}\hat{j} + \dfrac{\partial V}{\partial z}\hat{k}\right)$ or $V_b - V_a = -\displaystyle\int_a^b \vec{E} \cdot d\vec{\ell}$	**23-18**

5. Units

V and ΔV	The SI unit of potential and potential difference is the volt (V): $1\,\text{V} = 1\,\text{J/C}$	**23-4**
Electric field	$1\,\text{N/C} = 1\,\text{V/m}$	**23-5**
Electron volt	The electron volt (eV) is the change in potential energy of a particle of charge e as it moves from a to b, where $\Delta V = V_b - V_a = 1$ volt: $1\,\text{eV} = 1.60 \times 10^{-19}\,\text{C} \cdot \text{V} = 1.60 \times 10^{-19}\,\text{J}$	**23-6**

6. Potential Energy of Two Point Charges

	$U = q_0 V = \dfrac{kq_0 q}{r}, \quad (U = 0 \text{ at } r = \infty)$	**23-9**

7. Potential Functions

On the axis of a uniformly charged ring	$V = \dfrac{kQ}{\sqrt{x^2 + a^2}}$, $(V = 0 \text{ at } |x| = \infty)$	23-20
On the axis of a uniformly charged disk	$V = 2\pi k\sigma|x|\left(\sqrt{1 + \dfrac{R^2}{x^2}} - 1\right)$, $(V = 0 \text{ at } |x| = \infty)$	23-21
Near an infinite plane of charge	$V = V_0 - 2\pi k\sigma|x|$, $(V = V_0 \text{ at } x = 0)$	23-22
For a spherical shell of charge	$V = \dfrac{kQ}{r}$, $r \geq R$ $(V = 0 \text{ at } r = \infty)$	
	$V = \dfrac{kQ}{R}$, $r \leq R$ $(V = 0 \text{ at } r = \infty)$	23-23
For an infinite line charge	$V = 2k\lambda \ln\dfrac{R_{\text{ref}}}{R}$, $V = 0 \text{ at } r = R_{\text{ref}}$	23-24

8. Charge on a Nonspherical Conductor

On a conductor of arbitrary shape, the surface charge density σ is greatest at points where the radius of curvature is smallest.

9. Dielectric Breakdown

The amount of charge that can be placed on a conductor is limited by the fact that molecules of the surrounding medium undergo dielectric breakdown at very high electric fields, causing the medium to become a conductor.

Dielectric strength

The dielectric strength is the magnitude of the electric field at which dielectric breakdown occurs. The dielectric strength of air is

$$E_{\text{max}} \approx 3 \times 10^6 \text{ V/m} = 3 \text{ MV/m}$$

PROBLEMS

- Single-concept, single-step, relatively easy
- •• Intermediate-level, may require synthesis of concepts
- ••• Challenging, for advanced students
- SSM Solution is in the *Student Solutions Manual*
- ISOLVE Problems available on iSOLVE online homework service
- ISOLVE✓ These "Checkpoint" online homework service problems ask students additional questions about their confidence level, and how they arrived at their answer.

In a few problems, you are given more data than you actually need; in a few other problems, you are required to supply data from your general knowledge, outside sources, or informed estimates.

Conceptual Problems

1 • SSM A positive charge is released from rest in an electric field. Will it move toward a region of greater or smaller electric potential?

2 •• A lithium nucleus and an α particle are at rest. The lithium nucleus has a charge of $+3e$ and a mass of $7\ u$; the α particle has a charge of $+2e$ and a mass of $4\ u$. Which of the following methods would accelerate them both to the same kinetic energy? (a) Accelerate them through the same electrical potential difference. (b) Accelerate the α particle through potential V_1 and the lithium nucleus through $(2/3)V_1$. (c) Accelerate the α particle through potential V_1 and the lithium nucleus through $(7/4)V_1$. (d) Accelerate the α particle through potential V_1 and the lithium nucleus through $(2 \times 7)/(3 \times 4)V$. (e) None of the answers are correct.

3 • If the electric potential is constant throughout a region of space, what can you say about the electric field in that region?

4 • If E is known at just one point, can V be found at that point?

5 • In what direction can you move relative to an electric field so that the electric potential does not change?

6 •• In the calculation of V at a point x on the axis of a ring of charge, does it matter whether the charge Q is uniformly distributed around the ring? Would either V or E_x be different if it were not?

7 •• **SSM** Figure 23-26 shows a metal sphere carrying a charge $-Q$ and a point charge $+Q$. Sketch the electric field lines and equipotential surfaces in the vicinity of this charge system.

$-Q$

$+Q$

FIGURE 23-26 Problems 7 and 8

8 •• Repeat Problem 7 with the charge on the metal sphere changed to $+Q$.

9 •• Sketch the electric field lines and the equipotential surfaces, both near and far, from the conductor shown in Figure 23-25a, assuming that the conductor carries some charge Q.

10 •• Two equal positive charges are separated by a small distance. Sketch the electric field lines and the equipotential surfaces for this system.

11 • **SSM** Two equal positive point charges $+Q$ are on the x-axis. One is at $x = -a$ and the other is at $x = +a$. At the origin, (a) $E = 0$ and $V = 0$, (b) $E = 0$ and $V = 2kQ/a$, (c) $\vec{E} = (2kQ^2/a^2)\hat{i}$ and $V = 0$, (d) $\vec{E} = (2kQ^2/a^2)\hat{i}$ and $V = 2kQ/a$, or (e) none of the answers are correct.

12 • The electrostatic potential is measured to be $V(x, y, z) = 4|x| + V_0$, where V_0 is a constant. The charge distribution responsible for this potential is (a) a uniformly charged thread in the xy plane, (b) a point charge at the origin, (c) a uniformly charged sheet in the yz plane, or (d) a uniformly charged sphere of radius $1/\pi$ at the origin.

13 • Two point charges of equal magnitude but opposite sign are on the x axis; $+Q$ is at $x = -a$ and $-Q$ is at $x = +a$. At the origin, (a) $E = 0$ and $V = 0$, (b) $E = 0$ and $V = 2kQ/a$, (c) $\vec{E} = (2kQ^2/a^2)\hat{i}$ and $V = 0$, (d) $\vec{E} = (2kQ^2/a^2)\hat{i}$ and $V = 2kQ/a$, or (e) none of the answers are correct.

14 •• True or false:

(a) If the electric field is zero in some region of space, the electric potential must also be zero in that region.

(b) If the electric potential is zero in some region of space, the electric field must also be zero in that region.

(c) If the electric potential is zero at a point, the electric field must also be zero at that point.

(d) Electric field lines always point toward regions of lower potential.

(e) The value of the electric potential can be chosen to be zero at any convenient point.

(f) In electrostatics, the surface of a conductor is an equipotential surface.

(g) Dielectric breakdown occurs in air when the potential is 3×10^6 V.

15 •• (a) V is constant on a conductor surface. Does this mean that σ is constant? (b) If E is constant on a conductor surface, does this mean that σ is constant? Does it mean that V is constant?

16 • **SSM** Two charged metal spheres are connected by a wire, and sphere A is larger than sphere B (Figure 23-27). The magnitude of the electric potential of sphere A is (a) greater than that at the surface of sphere B; (b) less than that at the surface of sphere B; (c) the same as that at the surface of sphere B; (d) greater than or less than that at the surface of sphere B, depending on the radii of the spheres; or (e) greater than or less than that at the surface of sphere B, depending on the charge on the spheres.

A

B

FIGURE 23-27 Problem 16

Estimation and Approximation

17 • Estimate the potential difference between a thundercloud and the earth, given that the electrical breakdown of air occurs at fields of roughly 3×10^6 V/m.

18 • **SSM** Estimate the potential difference across the spark gap in a typical automobile spark plug. Because of the high compression of the gas in the piston, the electric field at which the gas sparks is roughly 2×10^7 V/m.

19 •• A proton can be thought of as having a "radius" of approximately 10^{-15} m. Two protons have a head-on collision with equal and opposite momenta. (a) Estimate the minimum kinetic energy (in MeV) required by each proton to allow the protons to overcome electrostatic repulsion and collide. Do this estimate without using relativity. (b) The rest energy of the proton is 938 MeV. If your value for the kinetic energy is much less than this, then a nonrelativistic calculation was justified. What fraction of the rest energy of the proton is the kinetic energy you calculated in Part (a)?

20 •• **iSOLVE** When you touch a friend after walking across a rug on a dry day, you typically draw a spark of about 2 mm. Estimate the potential difference between you and your friend before the spark.

Potential Difference

21 • **iSOLVE** A uniform electric field of 2 kN/C is in the x direction. A positive point charge $Q = 3 \mu C$ is released from rest at the origin. (a) What is the potential difference $V(4 \text{ m}) - V(0)$? (b) What is the change in the potential energy of the charge from $x = 0$ to $x = 4$ m? (c) What is the kinetic energy of the charge when it is at $x = 4$ m? (d) Find the potential $V(x)$ if $V(x)$ is chosen to be zero at $x = 0$, (e) 4 kV at $x = 0$, and (f) zero at $x = 1$ m.

22 • Two large parallel conducting plates separated by 10 cm carry equal and opposite surface charge densities so that the electric field between them is uniform. The difference in potential between the plates is 500 V. An electron is released from rest at the negative plate. (a) What is the magnitude of the electric field between the plates? Is the positive or negative plate at the higher potential? (b) Find the work done by the electric field on the electron as the electron moves from the negative plate to the positive plate. Express your answer in both electron volts and joules. (c) What is the change in potential energy of the electron when it moves from the negative plate to the positive plate? What is its kinetic energy when it reaches the positive plate?

23 • A positive charge of magnitude 2 μC is at the origin. (a) What is the electric potential V at a point 4 m from the origin relative to $V = 0$ at infinity? (b) How much work must be done by an outside agent to bring a 3-μC charge from infinity to $r = 4$ m, assuming that the 2-μC charge is held fixed at the origin? (c) How much work must be done by an outside agent to bring the 2-μC charge from infinity to the origin if the 3-μC charge is first placed at $r = 4$ m and is then held fixed?

24 •• ISOLVE✓ The distance between the K⁺ and Cl⁻ ions in KCl is 2.80×10^{-10} m. Calculate the energy required to separate the two ions to an infinite distance apart, assuming them to be point charges initially at rest. Express your answer in eV.

25 •• ISOLVE Protons from a Van de Graaff accelerator are released from rest at a potential of 5 MV and travel through a vacuum to a region at zero potential. (a) Find the final speed of the 5-MeV protons. (b) Find the accelerating electric field if the same potential change occurred *uniformly* over a distance of 2.0 m.

26 •• SSM ISOLVE An electron gun fires electrons at the screen of a television tube. The electrons start from rest and are accelerated through a potential difference of 30,000 V. What is the energy of the electrons when they hit the screen (a) in electron volts and (b) in joules? (c) What is the speed of impact of electrons with the screen of the picture tube?

27 •• (a) Derive an expression for the distance of closest approach of an α particle with kinetic energy E to a massive nucleus of charge Ze. Assume that the nucleus is fixed in space. (b) Find the distance of closest approach of a 5.0- and a 9.0-MeV α particle to a gold nucleus; the charge of the gold nucleus is $79e$. (Neglect the recoil of the gold nucleus.)

Potential Due to a System of Point Charges

28 • Four 2-μC point charges are at the corners of a square of side 4 m. Find the potential at the center of the square (relative to zero potential at infinity) if (a) all the charges are positive, (b) three of the charges are positive and one is negative, and (c) two are positive and two are negative.

29 • ISOLVE✓ Three point charges are on the x-axis: q_1 is at the origin, q_2 is at $x = 3$ m, and q_3 is at $x = 6$ m. Find the potential at the point $x = 0$, $y = 3$ m if (a) $q_1 = q_2 = q_3 = 2$ μC, (b) $q_1 = q_2 = 2$ μC and $q_3 = -2$ μC, and (c) $q_1 = q_3 = 2$ μC and $q_2 = -2$ μC.

30 • Points a, b, and c are at the corners of an equilateral triangle of side 3 m. Equal positive charges of 2 μC are at a and b. (a) What is the potential at point c? (b) How much work is required to bring a positive charge of 5 μC from infinity to point c if the other charges are held fixed? (c) Answer Parts (a) and (b) if the charge at b is replaced by a charge of -2 μC.

31 • ISOLVE✓ A sphere with radius 60 cm has its center at the origin. Equal charges of 3 μC are placed at 60° intervals along the equator of the sphere. (a) What is the electric potential at the origin? (b) What is the electric potential at the north pole?

32 • SSM Two point charges q and q' are separated by a distance a. At a point $a/3$ from q and along the line joining the two charges the potential is zero. Find the ratio q/q'.

33 •• Two positive charges $+q$ are on the x-axis at $x = +a$ and $x = -a$. (a) Find the potential $V(x)$ as a function of x for points on the x-axis. (b) Sketch $V(x)$ versus x. (c) What is the significance of the minimum on your curve?

34 •• SSM A point charge of $+3e$ is at the origin and a second point charge of $-2e$ is on the x-axis at $x = a$. (a) Sketch the potential function $V(x)$ versus x for all x. (b) At what point or points is $V(x)$ zero? (c) How much work is needed to bring a third charge $+e$ to the point $x = \frac{1}{2}a$ on the x-axis?

Computing the Electric Field From the Potential

35 • ISOLVE✓ A uniform electric field is in the negative x direction. Points a and b are on the x-axis, a at $x = 2$ m and b at $x = 6$ m. (a) Is the potential difference $V_b - V_a$ positive or negative? (b) If the magnitude of $V_b - V_a$ is 10^5 V, what is the magnitude E of the electric field?

36 • SSM The potential due to a particular charge distribution is measured at several points along the x-axis, as shown in Figure 23-28. For what value(s) in the range $0 < x < 10$ m is $E_x = 0$?

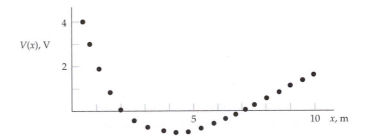

FIGURE 23-28 Problem 36

37 • A point charge $q = 3.00$ μC is at the origin. (a) Find the potential V on the x-axis at $x = 3.00$ m and at $x = 3.01$ m. (b) Does the potential increase or decrease as x increases? Compute $-\Delta V / \Delta x$, where ΔV is the change in potential from $x = 3.00$ m to $x = 3.01$ m and $x = 0.01$ m. (c) Find the electric field at $x = 3.00$ m, and compare its magnitude with $-\Delta V / \Delta x$ found in Part (b). (d) Find the potential (to three significant figures) at the point $x = 3.00$ m, $y = 0.01$ m, and compare your result with the potential on the x-axis at $x = 3.00$ m. Discuss the significance of this result.

38 • A charge of $+3.00 \ \mu C$ is at the origin, and a charge of $-3.00 \ \mu C$ is on the x-axis at $x = 6.00$ m. (a) Find the potential on the x-axis at $x = 3.00$ m. (b) Find the electric field on the x-axis at $x = 3.00$ m. (c) Find the potential on the x-axis at $x = 3.01$ m, and compute $-\Delta V/\Delta x$, where ΔV is the change in potential from $x = 3.00$ m to $x = 3.01$ m and $x = 0.01$ m. Compare your result with your answer to Part (b).

39 • A uniform electric field is in the positive y direction. Points a and b are on the y-axis, a at $y = 2$ m and b at $x = 6$ m. (a) Is the potential difference $V_b - V_a$ positive or negative? (b) If the magnitude of $V_b - V_a$ is 2×10^4 V, what is the magnitude E of the electric field?

40 • In the following, V is in volts and x is in meters. Find E_x when (a) $V(x) = 2000 + 3000x$, (b) $V(x) = 4000 + 3000x$, (c) $V(x) = 2000 - 3000x$, and (d) $V(x) = -2000$, independent of x.

41 •• A charge q is at $x = 0$ and a charge $-3q$ is at $x = 1$ m. (a) Find $V(x)$ for a general point on the x-axis. (b) Find the points on the x-axis where the potential is zero. (c) What is the electric field at these points? (d) Sketch $V(x)$ versus x.

42 •• **SSM** **SOLVE** An electric field is given by $E_x = 2.0x^3$ kN/C. Find the potential difference between the points on the x-axis at $x = 1$ m and $x = 2$ m.

43 •• Three equal charges lie in the xy plane. Two are on the y-axis at $y = -a$ and $y = +a$, and the third is on the x-axis at $x = a$. (a) What is the potential $V(x)$ due to these charges at a point on the x-axis? (b) Find E_x along the x-axis from the potential function $V(x)$. Evaluate your answers to Parts (a) and (b) at the origin and at $x = \infty$ to see if they yield the expected results.

Calculations of V for Continuous Charge Distributions

44 • A charge of $q = +10^{-8}$ C is uniformly distributed on a spherical shell of radius 12 cm. (a) What is the magnitude of the electric field just outside and just inside the shell? (b) What is the magnitude of the electric potential just outside and just inside the shell? (c) What is the electric potential at the center of the shell? What is the electric field at that point?

45 • An infinite line charge of linear charge density $\lambda = 1.5 \ \mu C/m$ lies on the z-axis. Find the potential at distances from the line charge of (a) 2.0 m, (b) 4.0 m, and (c) 12 m, assuming that $V = 0$ at 2.5 m.

46 •• Derive Equation 23-21 by integrating the electric field E_x along the axis of the disk. (See Equation 22-11.)

47 •• **SSM** A rod of length L carries a charge Q uniformly distributed along its length. The rod lies along the y-axis with its center at the origin. (a) Find the potential as a function of position along the x-axis. (b) Show that the result obtained in Part (a) reduces to $V = kQ/x$ for $x \gg L$.

48 •• A disk of radius R carries a surface charge distribution of $\sigma = \sigma_0 R/r$. (a) Find the total charge on the disk. (b) Find the potential on the axis of the disk a distance x from its center.

49 •• Repeat Problem 48 if the surface charge density is $\sigma = \sigma_0 r^2/R^2$.

50 •• A rod of length L carries a charge Q uniformly distributed along its length. The rod lies along the y-axis with one end at the origin. Find the potential as a function of position along the x-axis.

51 •• **SSM** A disk of radius R carries a charge density $+\sigma_0$ for $r < a$ and an equal but opposite charge density $-\sigma_0$ for $a < r < R$. The total charge carried by the disk is zero. (a) Find the potential a distance x along the axis of the disk. (b) Obtain an approximate expression for $V(x)$ when $x \gg R$.

52 •• Use the result obtained in Problem 51a to calculate the electric field along the axis of the disk. Then calculate the electric field by direct integration using Coulomb's law.

53 •• A rod of length L has a charge Q uniformly distributed along its length. The rod lies along the x-axis with its center at the origin. (a) What is the electric potential as a function of position along the x-axis for $x > L/2$? (b) Show that for $x \gg L/2$, your result reduces to that due to a point charge Q.

54 •• A conducting spherical shell of inner radius b and outer radius c is concentric with a small metal sphere of radius $a < b$. The metal sphere has a positive charge Q. The total charge on the conducting spherical shell is $-Q$. (a) What is the potential of the spherical shell? (b) What is the potential of the metal sphere?

55 •• Two very long, coaxial cylindrical shell conductors carry equal and opposite charges. The inner shell has radius a and charge $+q$; the other shell has radius b and charge $-q$. The length of each cylindrical shell is L. Find the potential difference between the shells.

56 •• **SOLVE✓** A uniformly charged sphere has a potential on its surface of 450 V. At a radial distance of 20 cm from this surface, the potential is 150 V. What is the radius of the sphere, and what is the charge of the sphere?

57 •• Consider two infinite parallel planes of charge, one in the yz plane and the other at distance $x = a$. (a) Find the potential everywhere in space when $V = 0$ at $x = 0$ if the planes carry equal positive charge densities $+\sigma$. (b) Repeat the problem with charge densities equal and opposite, and the charge in the yz plane positive.

58 •• **SSM** **SOLVE** Show that for $x \gg R$ the potential on the axis of a disk charge approaches kQ/x, where $Q = \sigma \pi R^2$ is the total charge on the disk. [Hint: Write $(x^2 + R^2)^{1/2} = x(1 + R^2/x^2)^{1/2}$ and use the binomial expression.]

59 •• In Example 23-12, you derived the expression

$$V(r) = \frac{kQ}{2R}\left(3 - \frac{r^2}{R^2}\right)$$

for the potential inside a solid sphere of constant charge density by first finding the electric field. In this problem you derive the same expression by direct integration. Consider a sphere of radius R containing a charge Q uniformly distributed. You wish to find V at some point $r < R$. (a) Find the charge q' inside a sphere of radius r and the potential V_1 at r due to this part of the charge. (b) Find the potential dV_2 at r due to the charge in a shell of radius r' and thickness dr' at $r' > r$. (c) Integrate your expression in Part (b) from $r' = r$ to $r' = R$ to find V_2. (d) Find the total potential V at r from $V = V_1 + V_2$.

60 • **iSOLVE✔** An infinite plane of charge has a surface charge density 3.5 $\mu C/m^2$. How far apart are the equipotential surfaces whose potentials differ by 100 V?

61 • A point charge $q = +\frac{1}{9} \times 10^{-8}$ C is at the origin. Taking the potential to be zero at $r = \infty$, locate the equipotential surfaces at 20-V intervals from 20 to 100 V, and sketch them to scale. Are these surfaces equally spaced?

62 • **iSOLVE✔** (a) Find the maximum net charge that can be placed on a spherical conductor of radius 16 cm before dielectric breakdown of the air occurs. (b) What is the potential of the sphere when it carries this maximum charge?

63 • **SSM** **iSOLVE** Find the greatest surface charge density σ_{max} that can exist on a conductor before dielectric breakdown of the air occurs.

64 •• Charge is placed on two conducting spheres that are very far apart and connected by a long thin wire. The radius of the smaller sphere is 5 cm and that of the larger sphere is 12 cm. The electric field at the surface of the larger sphere is 200 kV/m. Find the surface charge density on each sphere.

65 •• Two concentric spherical shell conductors carry equal and opposite charges. The inner shell has radius a and charge $+q$; the outer shell has radius b and charge $-q$. Find the potential difference between the shells, $V_a - V_b$.

66 ••• **SSM** Calculate the potential relative to infinity at the point a distance $R/2$ from the center of a uniformly charged thin spherical shell of radius R and charge Q.

67 •• Two identical uncharged metal spheres connected by a wire are placed close by two similar conducting spheres with equal and opposite charges, as shown in Figure 23-29. (a) Sketch the electric field lines between spheres 1 and 3 and between spheres 2 and 4. (b) What can be said about the potentials V_1, V_2, V_3, and V_4 of the spheres? (c) If spheres 3 and 4 are also connected by a wire, show that the final charge on each must be zero.

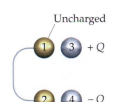

FIGURE 23-29 Problem 67

General Problems

68 • An electric dipole has a positive charge of 4.8×10^{-19} C separated from a negative charge of the same magnitude by 6.4×10^{-10} m. What is the electric potential at a point 9.2×10^{-10} m from each of the two charges? (a) 9.4 V. (b) Zero. (c) 4.2 V. (d) 5.1×10^9 V. (e) 1.7 V.

69 • Two positive charges $+q$ are on the y-axis at $y = +a$ and $y = -a$. (a) Find the potential V for any point on the x-axis. (b) Use your result in Part (a) to find the electric field at any point on the x-axis.

70 • **iSOLVE✔** If a conducting sphere is to be charged to a potential of 10,000 V, what is the smallest possible radius of the sphere so that the electric field will not exceed the dielectric strength of air?

71 •• **SSM** Two infinitely long parallel wires carry a uniform charge per unit length λ and $-\lambda$ respectively. The wires are in the xz plane, parallel with the z axis. The positively charged wire intersects the x axis at $x = -a$, and the negatively charged wire intersects the x axis at $x = +a$. (a) Choose the origin as the reference point where the potential is zero, and express the potential at an arbitrary point (x, y) in the xy plane in terms of x, y, λ, and a. Use this expression to solve for the potential everywhere on the y axis. (b) Use a spreadsheet program to plot the equipotential curve in the xy plane that passes through the point $x = \frac{1}{4}a$, $y = 0$. Use $a = 5$ cm and $\lambda = 5$ nC/m.

72 •• The equipotential curve graphed in Problem 71 looks like a circle. (a) Show explicitly that it is a circle. (b) The equipotential circle in the xy plane is the intersection of a three-dimensional equipotential surface and the xy plane. Describe the three-dimensional surface in a sentence or two.

73 •• The hydrogen atom can be modeled as a positive point charge of magnitude $+e$ (the proton) surrounded by a charge density (the electron) which has the formula $\rho = \rho_0 e^{-2r/a}$ (from quantum mechanics), where $a = 0.523$ nm. (a) Calculate the value of ρ_0 needed for charge neutrality. (b) Calculate the electrostatic potential (relative to infinity) at any distance r from the proton.

74 • **iSOLVE** An isolated aluminum sphere of radius 5.0 cm is at a potential of 400 V. How many electrons have been removed from the sphere to raise it to this potential?

75 • **iSOLVE** A point charge Q resides at the origin. A particle of mass $m = 0.002$ kg carries a charge of 4.0 μC. The particle is released from rest at $x = 1.5$ m. Its kinetic energy as it passes $x = 1.0$ m is 0.24 J. Find the charge Q.

76 •• **SSM** **iSOLVE** A Van de Graaff generator has a potential difference of 1.25 MV between the belt and the outer shell. Charge is supplied at the rate of 200 $\mu C/s$. What minimum power is needed to drive the moving belt?

77 •• A positive point charge $+Q$ is located at $x = -a$. (a) How much work is required to bring a second equal positive point charge $+Q$ from infinity to $x = +a$? (b) With the two equal positive point charges at $x = -a$ and $x = +a$, how much work is required to bring a third charge $-Q$ from infinity to the origin? (c) How much work is required to move the charge $-Q$ from the origin to the point $x = 2a$ along the semicircular path shown (Figure 23-30)?

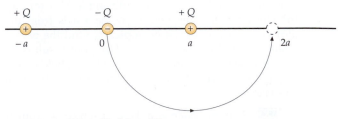

FIGURE 23-30 Problem 77

78 •• A charge of 2 nC is uniformly distributed around a ring of radius 10 cm that has its center at the origin and its axis along the x-axis. A point charge of 1 nC is located at $x = 50$ cm. Find the work required to move the point charge to the origin. Give your answer in both joules and electron volts.

79 •• **ISOLVE**✓ The centers of two metal spheres of radius 10 cm are 50 cm apart on the x-axis. The spheres are initially neutral, but a charge Q is transferred from one sphere to the other, creating a potential difference between the spheres of 100 V. A proton is released from rest at the surface of the positively charged sphere and travels to the negatively charged sphere. At what speed does it strike the negatively charged sphere?

80 •• (a) Using a spreadsheet program, graph $V(x)$ versus x for the uniformly charged ring in the yz plane given by Equation 23-20. (b) At what point is $V(x)$ a maximum? (c) What is E_x at this point?

81 •• A spherical conductor of radius R_1 is charged to 20 kV. When it is connected by a long fine wire to a second conducting sphere far away, its potential drops to 12 kV. What is the radius of the second sphere?

82 •• **SSM** **ISOLVE**✓ A metal sphere centered at the origin carries a surface charge of charge density $\sigma = 24.6$ nC/m². At $r = 2.0$ m, the potential is 500 V and the magnitude of the electric field is 250 V/m. Determine the radius of the metal sphere.

83 •• **ISOLVE** Along the axis of a uniformly charged disk, at a point 0.6 m from the center of the disk, the potential is 80 V and the magnitude of the electric field is 80 V/m; at a distance of 1.5 m, the potential is 40 V and the magnitude of the electric field is 23.5 V/m. Find the total charge residing on the disk.

84 •• **ISOLVE** A radioactive ²¹⁰Po nucleus emits an α particle of charge $+2e$ and energy 5.30 MeV. Assume that just after the α particle is formed and escapes from the nucleus, it is a distance R from the center of the daughter nucleus ²⁰⁶Pb, which has a charge $+82e$. Calculate R by setting the electrostatic potential energy of the two particles at this separation equal to 5.30 MeV. (Neglect the size of the α particle.)

85 •• Two large, parallel, nonconducting planes carry equal and opposite charge densities of magnitude σ. The planes have area A and are separated by a distance d. (a) Find the potential difference between the planes. (b) A conducting slab having thickness a and area A, the same area as the planes, is inserted between the original two planes. The slab carries no net charge. Find the potential difference between the original two planes and sketch the electric field lines in the region between the original two planes.

86 ••• A point charge q_1 is at the origin and a second point charge q_2 is on the x-axis at $x = a$, as in Example 23-5. (a) Calculate the electric field everywhere on the x-axis from the potential function given in that example. (b) Find the potential at a general point on the y-axis. (c) Use your result from Part (b) to calculate the y component of the electric field on the y-axis. Compare your result with that obtained directly from Coulomb's law.

87 ••• **SSM** A point charge q is a distance d from a grounded conducting plane of infinite extent (Figure 23-31a). For this configuration the potential V is zero, both at all points infinitely far from the particle in all directions, and at all points on the conducting plane. Consider a set of coordinate axes with the particle located on the x axis at $x = d$. A second configuraton (Figure 23-31b) has the conducting plane replaced by a particle of charge $-q$ located on the x axis at $x = -d$. (a) Show

that for the second configuration the potential function is zero at all points infinitely far from the particle in all directions, and at all points on the yz plane—just as was the case for the first configuration. (b) A theorem, called the uniqueness theorem, shows that throughout the half-space $x > 0$ the potential function V—and thus the electric field $\vec{E}$—for the two configurations are identical. Using this result, obtain the electric field $\vec{E}$ at every point in the yz plane in the second configuration. (The uniqueness theorem tells us that in the first configuration the electric field at each point in the yz plane is the same as it is in the second configuration.) Use this result to find the surface charge density σ at each point in the conducting plane (in the first configuration).

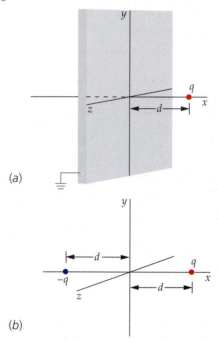

(a)

(b)

FIGURE 23-31 Problem 87

88 ••• A particle of mass m carrying a positive charge q is constrained to move along the x-axis. At $x = -L$ and $x = L$ are two ring charges of radius L (Figure 23-32). Each ring is centered on the x-axis and lies in a plane perpendicular to it. Each carries a positive charge Q. (a) Obtain an expression for the potential due to the ring charges as a function of x. (b) Show that $V(x)$ is a minimum at $x = 0$. (c) Show that for $x \ll L$, the potential is of the form $V(x) = V(0) + \alpha x^2$. (d) Derive an expression for the angular frequency of oscillation of the mass m if it is displaced slightly from the origin and released.

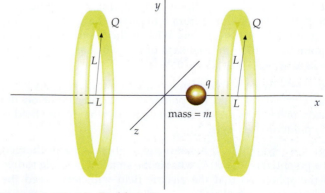

FIGURE 23-32 Problem 88

89 ••• Three concentric conducting spherical shells have radii a, b, and c so that $a < b < c$. Initially, the inner shell is uncharged, the middle shell has a positive charge Q, and the outer shell has a negative charge $-Q$. (a) Find the electric potential of the three shells. (b) If the inner and outer shells are now connected by a wire that is insulated as it passes through the middle shell, what is the electric potential of each of the three shells, and what is the final charge on each shell?

90 ••• **SSM** Consider two concentric spherical metal shells of radii a and b, where $b > a$. The outer shell has a charge Q, but the inner shell is grounded. This means that the inner shell is at zero potential and that electric field lines leave the outer shell and go to infinity, but other electric field lines leave the outer shell and end on the inner shell. Find the charge on the inner shell.

91 ••• Show that the total work needed to assemble a uniformly charged sphere with charge Q and radius R is given by $W = U = \dfrac{3}{5}\dfrac{Q^2}{4\pi\epsilon_0 R}$, where U is the electrostatic potential energy of the sphere. *Hint: Let ρ be the charge density of the sphere with charge Q and radius R. Calculate the work dW to bring in charge dq from infinity to the surface of a uniformly charged sphere of radius r $(r < R)$ and charge density ρ. (No additional work is required to smear dq throughout a spherical shell of radius r, thickness dr, and charge density ρ.)*

92 •• Use the result of Problem 91 to calculate the *classical electron radius*, the radius of a uniform sphere of charge $-e$ that has electrostatic potential energy equal to the rest energy of the electron (5.11×10^5 eV). Comment on the shortcomings of this model for the electron.

93 •• (a) Consider a uniformly charged sphere of radius R and total charge Q which is composed of an incompressible fluid, such as water. If the sphere fissions (splits) into two halves of equal volume and equal charge, and if these halves stabilize into spheres, what is the radius R' of each? (b) Using the expression for potential energy shown in Problem 91, calculate the change in the total electrostatic potential energy of the charged fluid. Assume that the spheres are separated by a large distance.

94 ••• **SSM** Problem 93 can be modified to be used as a very simple model for nuclear fission. When a ^{235}U nucleus absorbs a neutron, it can fission into the fragments ^{140}Xe and ^{94}Sr, plus 2 neutrons ejected. The ^{235}U has 92 protons, while ^{140}Xe has 54 and ^{94}Sr has 38. Estimate the energy liberated by this fission process (in MeV), assuming that the mass density of the nucleus is constant and has a value $\rho \sim 4 \times 10^{17}$ kg/m^3.

95 ••• (a) Consider an imaginary spherical surface and a point charge q that is located outside the surface. Show by direct integration that the potential at the center of the spherical surface due to the presence of point charge is the average of the potential over the surface of the sphere. (b) Argue from the superposition principle that this result must hold for any spherical surface and any configuration of charges outside the surface.

Electrostatic Energy and Capacitance

THE ENERGY FOR THE ELECTRONIC FLASH OF THE CAMERA WAS STORED IN A CAPACITOR IN THE FLASH UNIT.

? **How is energy stored in a capacitor? (See Section 24-3.)**

When we bring a point charge q from far away to a region where other charges are present, we must do work qV, where V is the potential at the final position due to the other charges in the vicinity. The work done is stored as electrostatic potential energy. The electrostatic potential energy of a system of charges is the total work needed to assemble the system.

When positive charge is placed on an isolated conductor, the potential of the conductor increases. The ratio of the charge to the potential is called the **capacitance** of the conductor. A useful device for storing charge and energy is the capacitor, which consists of two conductors, closely spaced but insulated from each other. When attached to a source of potential difference, such as a battery, the conductors acquire equal and opposite charges. The ratio of the magnitude of the charge on either conductor to the potential difference between the conductors is the capacitance of the capacitor. Capacitors have many uses. The flash attachment for your camera uses a capacitor to store the energy needed to provide the sudden flash of light. Capacitors are also used in the tuning circuits of devices such as radios, televisions, and cellular phones, allowing them to operate at specific frequencies. ➤ **Circuits containing batteries and capacitors are presented in this chapter. In the next few chapters, these techniques and concepts will be further developed in circuits containing resistors, inductors, and other devices.**

The first capacitor was the Leyden jar, a glass container lined inside and out with gold foil. It was invented at the University of Leyden in the Netherlands by eighteenth-century experimenters who, while studying the effects of electric charges on people and animals, got the idea of trying to store a large amount of charge in a bottle of water. An experimenter held up a jar of water in one hand while charge was conducted to the water by a chain from a static electric generator. When the experimenter reached over to lift the chain out of the water with his other hand, he was knocked unconscious. Benjamin Franklin realized that the device for storing charge did not have to be jar shaped and used foil-covered window glass, called Franklin panes. With several of these connected in parallel, Franklin stored a large charge and attempted to kill a turkey with it. Instead, he knocked himself out. Franklin later wrote, "I tried to kill a turkey but nearly succeeded in killing a goose."

Capacitors are used in large numbers in common electronic devices such as television sets. Some capacitors are used to store energy, but the majority of them are used to filter unwanted electrical frequencies.

24-1 Electrostatic Potential Energy

If we have a point charge q_1 at point 1, the potential V_2 at point 2 a distance $r_{1,2}$ away is given by

$$V_2 = \frac{kq_1}{r_{1,2}}$$

To bring a second point charge q_2 in from rest at infinity to rest at point 2 requires that we do work:

$$W_2 = q_2 V_2 = \frac{kq_2 q_1}{r_{1,2}}$$

The potential at point 3, a distance $r_{1,3}$ from q_1 and a distance $r_{2,3}$ from q_2, is given by

$$V_3 = \frac{kq_1}{r_{1,3}} + \frac{kq_2}{r_{2,3}}$$

To bring in an additional point charge q_3 from rest at infinity to rest at point 3 requires that we must do additional work:

$$W_3 = q_3 V_3 = \frac{kq_3 q_1}{r_{1,3}} + \frac{kq_3 q_2}{r_{2,3}}$$

The total work required to assemble the three charges is the **electrostatic potential energy** U of the system of three point charges:

$$U = \frac{kq_2 q_1}{r_{1,2}} + \frac{kq_3 q_1}{r_{1,3}} + \frac{kq_3 q_2}{r_{2,3}} \qquad 24\text{-}1$$

This quantity of work is independent of the order in which the charges are brought to their final positions. In general,

The electrostatic potential energy of a system of point charges is the work needed to bring the charges from an infinite separation to their final positions.

ELECTROSTATIC POTENTIAL ENERGY OF A SYSTEM

The first two terms on the right-hand side of Equation 24-1 can be written

$$\frac{kq_2 q_1}{r_{1,2}} + \frac{kq_3 q_1}{r_{1,3}} = q_1 \left(\frac{kq_2}{r_{1,2}} + \frac{kq_3}{r_{1,3}} \right) = q_1 V_1$$

where V_1 is the potential at the location of q_1 due to charges q_2 and q_3. Similarly, the second and third terms represent the charge q_3 times the potential due to charges q_1 and q_2, and the first and third terms equal the charge q_2 times the potential due to charges q_1 and q_2. We can thus rewrite Equation 24-1 as

$$U = \frac{kq_2 q_1}{r_{1,2}} + \frac{kq_3 q_1}{r_{1,3}} + \frac{kq_3 q_2}{r_{2,3}} = \frac{1}{2}(U + U)$$

$$= \frac{1}{2}\left(\frac{kq_2 q_1}{r_{1,2}} + \frac{kq_3 q_1}{r_{1,3}} + \frac{kq_3 q_2}{r_{2,3}} + \frac{kq_2 q_1}{r_{1,2}} + \frac{kq_3 q_1}{r_{1,3}} + \frac{kq_3 q_2}{r_{2,3}} \right)$$

$$= \frac{1}{2}\left[q_1 \left(\frac{kq_2}{r_{1,2}} + \frac{kq_3}{r_{1,3}} \right) + q_2 \left(\frac{kq_3}{r_{2,3}} + \frac{kq_1}{r_{1,2}} \right) + q_3 \left(\frac{kq_1}{r_{1,3}} + \frac{kq_2}{r_{2,3}} \right) \right]$$

The electrostatic potential energy U of a system of n point charges is thus

$$U = \frac{1}{2} \sum_{i=1}^{n} q_i V_i \qquad\qquad 24\text{-}2$$

ELECTROSTATIC POTENTIAL ENERGY OF A SYSTEM OF POINT CHARGES

where V_i is the potential at the location of the ith charge due to all of the other charges.

Equation 24-2 also describes the electrostatic potential energy of a continuous charge distribution. Consider a spherical conductor of radius R. When the sphere carries a charge q, its potential relative to $V = 0$ at infinity is

$$V = \frac{kq}{R}$$

The work we must do to bring an additional amount of charge dq from infinity to the conductor is $V\,dq$. This work equals the increase in the potential energy of the conductor:

$$dU = V\,dq = \frac{kq}{R}\,dq$$

The total potential energy U is the integral of dU as q increases from zero to its final value Q. Integrating, we obtain

$$U = \frac{k}{R} \int_0^Q q\,dq = \frac{kQ^2}{2R} = \frac{1}{2}QV \qquad\qquad 24\text{-}3$$

where $V = kQ/R$ is the potential on the surface of the fully charged sphere. We can interpret Equation 24-3 as $U = Q \times \frac{1}{2}V$ where $\frac{1}{2}V$ is the average potential of the sphere during the charging process. During the charging process, bringing the first element of charge in from infinity to the uncharged sphere requires no work because the sphere is initially uncharged. Therefore, the charge being brought in is not repelled by the charge on the sphere. As the charge on the sphere accumulates, bringing in additional elements of charge to the sphere requires more and more work; when the sphere is almost fully charged, bringing in the last element of charge in against the repulsive force of the charge on the sphere requires the most work. The average potential of the sphere during the

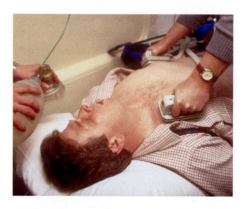

In about two thirds of the people that go into cardiac arrest the heart goes into a state called ventricular fribrillation. In this state the heart quivers, spasms chaotically, and does not pump. To defibrillate the heart a significant current is passed through it, which causes it to stop. Then the pacemaker cells in the heart can again establish a regular heartbeat. An external defibrillator applies a large voltage across the chest.

charging process is one-half its final potential V, so the total work required to bring in the entire charge equals $\frac{1}{2}QV$. Although we derived Equation 24-3 for a spherical conductor, it holds for any conductor. The potential of any conductor is proportional to its charge q, so we can write $V = \alpha q$, where α is a constant. The work needed to bring an additional charge dq from infinity to the conductor is $V\,dq = \alpha q\,dq$, and the total work needed to put a charge Q on the conductor is $\frac{1}{2}\alpha Q^2 = \frac{1}{2}QV$. If we have a set of n conductors with the ith conductor at potential V_i and carrying a charge Q_i, the electrostatic potential energy is

$$U = \frac{1}{2}\sum_{i=1}^{n} Q_i V_i \qquad\qquad 24\text{-}4$$

ELECTROSTATIC POTENTIAL ENERGY OF A SYSTEM OF CONDUCTORS

WORK REQUIRED TO MOVE POINT CHARGES   **EXAMPLE 24-1**

Points A, B, C, and D are at the corners of a square of side a, as shown in Figure 24-1. Four identical positive point charges, each with charge q, are initially at rest at infinite separation. (*a*) Calculate the total work required to place the point charges at each corner of the square by separately calculating the work required to move each charge to its final position. (*b*) Show that Equation 24-2 gives the total work.

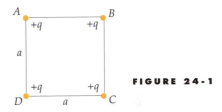

FIGURE 24-1

PICTURE THE PROBLEM No work is needed to place the first charge at point A because the potential is zero when the other three charges are at infinity. As each additional charge is brought into place, work must be done because of the presence of the previous charges.

(*a*) 1. Place the first charge at point A. To accomplish this step, the work W_A that is needed is zero:

$$W_A = 0$$

2. Bring the second charge to point B. The work required is $W_B = qV_A$, where V_A is the potential at point B due to the first charge at point A a distance a away:

$$W_B = qV_A = q\left(\frac{kq}{a}\right) = \frac{kq^2}{a}$$

3. $W_C = qV_C$, where V_C is the potential at point C due to q at point A a distance $\sqrt{2}a$ away and q at point B a distance a away:

$$W_C = qV_C = q\left(\frac{kq}{a} + \frac{kq}{\sqrt{2}a}\right) = \left(1 + \frac{1}{\sqrt{2}}\right)\frac{kq^2}{a}$$

4. Similar considerations give W_D, the work needed to bring the fourth charge to point D:

$$W_D = qV_D = q\left(\frac{kq}{a} + \frac{kq}{\sqrt{2}a} + \frac{kq}{a}\right)$$

$$= \left(2 + \frac{1}{\sqrt{2}}\right)\frac{kq^2}{a}$$

5. Summing the individual contributions gives the total work required to assemble the four charges:

$$W_{total} = W_A + W_B + W_C + W_D = \boxed{\left(4 + \sqrt{2}\right)\frac{kq^2}{a}}$$

(*b*) 1. Calculate W_{total} from Equation 24-2. Use V_D from Part (*a*), step 4 for the potential at the location of each charge. There are four identical terms, one from each charge:

$$W_{total} = U = \frac{1}{2}\sum_{i=1}^{4} q_i V_i$$

2. The potential at the location of each charge is V_D from step 4. Substitute V_D for V_i and solve for W_{total}:

$$W_{total} = \frac{1}{2}\sum_{i=1}^{4}\left[q_i\left(2 + \frac{1}{\sqrt{2}}\right)\frac{kq}{a}\right] = \frac{1}{2}\left(2 + \frac{1}{\sqrt{2}}\right)\frac{kq}{a}\sum_{i=1}^{4} q_i$$

$$= \frac{1}{2}\left(2 + \frac{1}{\sqrt{2}}\right)\frac{kq}{a}\,4q = \boxed{\left(4 + \sqrt{2}\right)\frac{kq^2}{a}}$$

REMARKS W_{total} equals the total electrostatic energy of the charge distribution.

EXERCISE (a) How much additional work is required to bring a fifth positive charge q from infinity to the center of the square? (b) What is the total work required to assemble the five-charge system? [*Answer* (a) $4\sqrt{2}\,kq^2/a$, (b) $(4 + 5\sqrt{2})\,kq^2/a$]

24-2 Capacitance

The potential V due to the charge Q on a single isolated conductor is proportional to Q and depends on the size and shape of the conductor. Typically, the larger the surface area of a conductor the more charge it can carry for a given potential. For example, the potential of a spherical conductor of radius R carrying a charge Q is

$$V = \frac{kQ}{R}$$

The ratio of charge Q to the potential V of an isolated conductor is called its capacitance C:

$$C = \frac{Q}{V}$$

24-5

DEFINITION—CAPACITANCE

Capacitance is a measure of the capacity to store charge for a given potential difference. Since the potential is proportional to the charge, this ratio does not depend on either Q or V, but only on the size and shape of the conductor. The self-capacitance of a spherical conductor is

$$C = \frac{Q}{V} = \frac{Q}{kQ/R} = \frac{R}{k} = 4\pi\,\epsilon_0\,R$$

24-6

The SI unit of capacitance is the coulomb per volt, which is called a **farad** (F) after the great English experimentalist Michael Faraday:

$$1\,F = 1\,C/V$$

24-7

The farad is a rather large unit, so submultiples such as the microfarad ($1\,\mu F = 10^{-6}\,F$) or the picofarad ($1\,pF = 10^{-12}\,F$) are often used. Since capacitance is in farads and R is in meters, we can see from Equation 24-6 that the SI unit for the permittivity of free space, ϵ_0, can also be written as a farad per meter:

$$\epsilon_0 = 8.85 \times 10^{-12}\,F/m = 8.85\,pF/m$$

24-8

EXERCISE Find the radius of a spherical conductor that has a capacitance of 1 F. (*Answer* 8.99×10^9 m, which is about 1400 times the radius of the earth)

We see from the above exercise that the farad is indeed a very large unit.

EXERCISE A sphere of capacitance C_1 carries a charge of 20 μC. If the charge is increased to 60 μC, what is the new capacitance C_2? (*Answer* $C_2 = C_1$. The capacitance does not depend on the charge. If the charge is tripled, the potential of the sphere will be tripled and the ratio Q/V, which depends only on the radius of the sphere, remains unchanged.)

Capacitors

A device consisting of two conductors carrying equal but opposite charges is called a **capacitor.** A capacitor is usually charged by transferring a charge Q from one conductor to the other conductor, which leaves one of the conductors with a charge $+Q$ and the other conductor with a charge $-Q$. The capacitance of the device is defined to be Q/V, where Q is the magnitude of the charge on either conductor and V is the magnitude of the potential difference between the conductors.[†] To calculate the capacitance, we place equal and opposite charges on the conductors and then find the potential difference V by first finding the electric field $\vec{E}$ due to the charges.

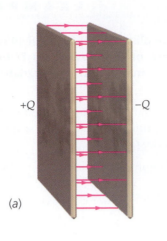

(a)

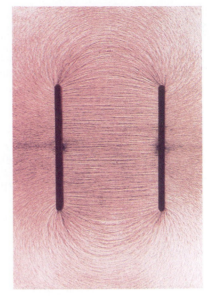

(b)

FIGURE 24-2 (*a*) Electric field lines between the plates of a parallel-plate capacitor. The lines are equally spaced between the plates, indicating that the field is uniform. (*b*) Electric field lines in a parallel-plate capacitor shown by small bits of thread suspended in oil.

Parallel-Plate Capacitors

A common capacitor is the **parallel-plate capacitor,** which utilizes two parallel conducting plates. In practice, the plates are often thin metallic foils that are separated and insulated from one another by a thin plastic film. This "sandwich" is then rolled up, which allows for a large surface area in a relatively small space. Let A be the area of the surface (the area of the side of each plate that faces the other plate), and let d be the separation distance, which is small compared to the length and width of the plates. We place a charge $+Q$ on one plate and $-Q$ on the other plate. These charges attract each other and become uniformly distributed on the inside surfaces of the plates. Since the plates are close together, the electric field between them is approximately the same as the field between two infinite planes of equal and opposite charge. Each plate contributes a uniform field of magnitude $E = \sigma/(2\epsilon_0)$; Equation 22-24 giving a total field strength $E = \sigma/\epsilon_0$, where $\sigma = Q/A$ is the magnitude of the charge per unit area on either plate. Since $\vec{E}$ is uniform between the plates (Figure 24-2), the potential difference between the plates equals the field strength E times the plate separation d:

$$V = Ed = \frac{\sigma}{\epsilon_0} d = \frac{Qd}{\epsilon_0 A} \qquad 24\text{-}9$$

The capacitance of the parallel-plate capacitor is thus

$$C = \frac{Q}{V} = \frac{\epsilon_0 A}{d} \qquad 24\text{-}10$$

CAPACITANCE OF A PARALLEL-PLATE CAPACITOR

Note that because V is proportional to Q, the capacitance does not depend on either Q or V. For a parallel-plate capacitor, the capacitance is proportional to the area of the plates and is inversely proportional to the gap width (separation distance). In general, capacitance depends on the size, shape, and geometrical arrangement of the conductors and capacitance also depends on the properties of the insulating medium between the conductors.

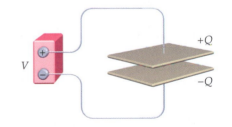

FIGURE 24-3 When the conductors of an uncharged capacitor are connected to the terminals of a battery, the battery "pumps" charge from one conductor to the other until the potential difference between the conductors equals that between the battery terminals.[‡] The amount of charge transferred through the battery is $Q = CV$.

[†] When we speak of the charge on a capacitor, we mean the magnitude of the charge on either conductor. The use of V rather than ΔV for the magnitude of the potential difference between the plates is standard and simplifies many of the equations relating to capacitance.

[‡] We will discuss batteries more fully in Chapter 25. Here, all we need to know is that a battery is a device that stores energy, supplies electrical energy, and maintains a constant potential difference V between its terminals.

THE CAPACITANCE OF A PARALLEL-PLATE CAPACITOR **EXAMPLE 24-2**

A parallel-plate capacitor has square plates of edge length 10 cm separated by 1 mm. (*a*) Calculate the capacitance of this device. (*b*) If this capacitor is charged to 12 V, how much charge is transferred from one plate to another?

PICTURE THE PROBLEM The capacitance C is determined by the area and the separation of the plates. Once C is found, the charge for a given voltage V is found from the definition of capacitance $C = Q/V$.

1. We find the capacitance using Equation 24-10:

$$C = \frac{\epsilon_0 A}{d} = \frac{(8.85\ \text{pF/m})(0.1\ \text{m})^2}{0.001\ \text{m}} = \boxed{88.5\ \text{pF}}$$

2. The charge transferred is found from the definition of capacitance:

$$Q = CV = (88.5\ \text{pF})(12\ \text{V}) = 1.06 \times 10^{-9}\ \text{C}$$

$$= \boxed{1.06\ \text{nC}}$$

REMARKS Q is the magnitude of the charge on each plate of the capacitor. In this case, Q corresponds to roughly 6.6×10^9 electrons.

EXERCISE How large would the plate area have to be for the capacitance to be 1 F? (*Answer* $A = 1.13 \times 10^8\ \text{m}^2$, which corresponds to a square 10.6 km on a side)

Cylindrical Capacitors

A cylindrical capacitor consists of a small conducting cylinder or wire of radius R_1 and a larger, concentric cylindrical conducting shell of radius R_2. A coaxial cable, such as that used for cable television, can be thought of as a cylindrical capacitor. The capacitance per unit length of a coaxial cable is important in determining the transmission characteristics of the cable.

AN EXPRESSION FOR THE CAPACITANCE OF A CYLINDRICAL CAPACITOR **EXAMPLE 24-3**

FIGURE 24-4

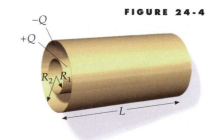

Find an expression for the capacitance of a cylindrical capacitor consisting of two conductors, each of length L. One conductor is a cylinder of radius R_1 and the second conductor is a coaxial cylindrical shell of inner radius R_2, with $R_1 < R_2 \ll L$ as shown in Figure 24-4.

PICTURE THE PROBLEM We place charge $+Q$ on the inner conductor and charge $-Q$ on the outer conductor and calculate the potential difference $V = V_b - V_a$ from the electric field between the conductors, which is found from Gauss's law. Since the electric field is not uniform (it depends on R) we must integrate to find the potential difference.

1. The capacitance is defined as the ratio Q/V:

$$C = Q/V$$

2. V is related to the electric field:

$$dV = -\vec{E} \cdot d\vec{\ell}$$

3. To find E_R we choose a soup-can shaped Gaussian surface of radius R and length ℓ, where $(R_1 < R < R_2)$ and $\ell \ll L$. The Gaussian surface is located far from the ends of the cylindrical shells (Figure 24-5):

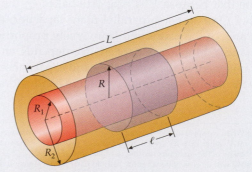

FIGURE 24-5

4. Far from the ends of the shells $\vec{E}$ is radial, so there is no flux of $\vec{E}$ through the flat ends of the can. The area of the curved part of the can is $2\pi R\ell$, so Gauss's law gives:

$$\phi_{net} = \oint_S E_n \, dA = \frac{1}{\epsilon_0} Q_{inside}$$

$$= E_R 2\pi R_1 = \frac{1}{\epsilon_0} Q_{inside}$$

5. Assuming the charge per unit length on the inner shell is uniformly distributed, find Q_{inside}:

$$Q_{inside} = \frac{\ell}{L} Q$$

6. Substitute for Q_{inside} and solve for E_R:

$$E_R 2\pi R\ell = \frac{1}{\epsilon_0} \frac{\ell}{L} Q$$

so

$$E_R = \frac{Q}{2\pi L \epsilon_0 R}$$

7. Integrate to find $V = |V_{R_2} - V_{R_1}|$:

$$V_{R_2} - V_{R_1} = \int_{V_{R_1}}^{V_{R_2}} dV = -\int_{R_1}^{R_2} E_R \, dR$$

$$= -\frac{Q}{2\pi L \epsilon_0} \int_{R_1}^{R_2} \frac{dR}{R} = -\frac{Q}{2\pi L \epsilon_0} \ln \frac{R_2}{R_1}$$

so

$$V = |V_{R_2} - V_{R_1}| = \frac{Q}{2\pi L \epsilon_0} \ln \frac{R_2}{R_1}$$

8. Substitute this result to find C:

$$C = \frac{Q}{V} = \boxed{\frac{2\pi \epsilon_0 L}{\ln(R_2/R_1)}}$$

REMARKS The capacitance of a cylindrical capacitor is proportional to the length of the conductors.

EXERCISE How is the capacitance affected if the potential across a cylindrical capacitor is increased from 20 V to 80 V? (*Answer* The capacitance of any capacitor does not depend on the potential. To increase V, you must increase the charge Q. The ratio Q/V depends only on the geometry of the capacitor and the nature of the insulators.)

From Example 24-3 we see that the capacitance of a cylindrical capacitor is given by

$$C = \frac{2\pi \epsilon_0 L}{\ln(R_2/R_1)} \qquad 24\text{-}11$$

CAPACITANCE OF A CYLINDRICAL CAPACITOR

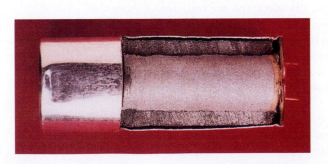

A coaxial cable is a long cylindrical capacitor with a solid wire for the inner conductor and a braided-wire shield for the outer conductor. The outer rubber coating has been peeled back from the cable to show the conductors and the white plastic insulator that separates the conductors.

Cutaway of a 200-μF capacitor used in an electronic strobe light.

A variable air-gap capacitor like those that were used in the tuning circuits of old radios. The semicircular plates rotate through the fixed plates, which changes the amount of surface area between the plates, and hence the capacitance.

Cross section of a foil-wound capacitor.

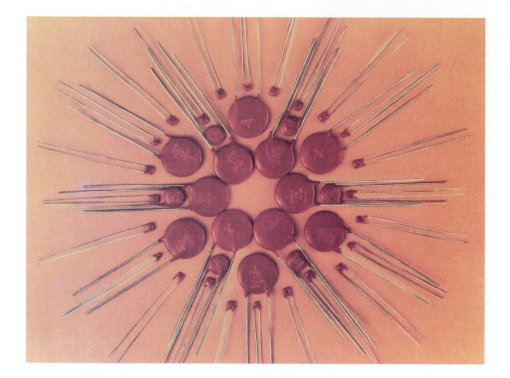

Ceramic capacitors for use in electronic circuits.

24-3 The Storage of Electrical Energy

When a capacitor is being charged, typically electrons are transferred from the positively charged conductor to the negatively charged conductor. This leaves the positive conductor with an electron deficit and the negative conductor with an electron surplus. Alternatively, transferring positive charges from the negative to the positive conductor can also charge capacitors. Either way, work must be done to charge a capacitor, and at least some of this work is stored as electrostatic potential energy.

Let q be the positive charge that has been transferred at some time during the charging process. The potential difference is then $V = q/C$. If a small amount of additional positive charge dq is now transferred from the negative conductor to the positive conductor through a potential increase of V (Figure 24-6), the potential energy of the charge, and thus the capacitor, is increased by

$$dU = V\,dq = \frac{q}{C}\,dq$$

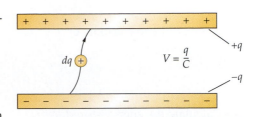

The total increase in potential energy U is the integral of dU as q increases from zero to its final value Q (Figure 24-7):

$$U = \int dU = \int_0^Q \frac{q}{C}\,dq = \frac{1}{2}\frac{Q^2}{C}$$

This potential energy is the energy stored in the capacitor. Using $C = Q/V$, we can express this energy in a variety of ways:

$$U = \frac{1}{2}\frac{Q^2}{C} = \frac{1}{2}QV = \frac{1}{2}CV^2 \qquad\qquad \text{24-12}$$

ENERGY STORED IN A CAPACITOR

EXERCISE A 15-μF capacitor is charged to 60 V. How much energy is stored in the capacitor? (*Answer* 0.027 J)

EXERCISE Obtain the expression for the electrostatic energy stored in a capacitor (Equation 24-12) from Equation 24-4, using $Q_1 = +Q$, $Q_2 = -Q$, $n = 2$, and $V = V_1 - V_2$.

Suppose we charge a capacitor by connecting it to a battery. The potential difference V when the capacitor is fully charged with charge Q is just the potential difference between the terminals of the battery before they were connected to the capacitor. The total work done *by the battery* in charging the capacitor is QV, which is twice the energy stored in the capacitor. The additional work done by the battery is either dissipated as thermal energy in the battery and in the connecting wires[†] or radiated as electromagnetic energy via an electromagnetic wave.[‡]

FIGURE 24-6 When a small amount of positive charge dq is moved from the negative conductor to the positive conductor, its potential energy is increased by $dU = V\,dq$, where V is the potential difference between the conductors.

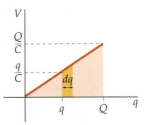

FIGURE 24-7 The work needed to charge a capacitor is the integral of $V\,dq$ from the original charge of $q = 0$ to the final charge of $q = Q$. This work is the triangular area under the curve $\frac{1}{2}(Q/C)Q$.

---

† We will show in Section 25-6 that if the capacitor is connected to a battery by wires of some resistance R, half the energy supplied by the battery in charging the capacitor is dissipated as thermal energy in the wires.
‡ We will show in Section 30-3 that under certain circumstances the circuit will act as a broadcast antenna and a significant portion of the work will be broadcast as electromagnetic radiation.

CHARGING A PARALLEL-PLATE CAPACITOR WITH A BATTERY **E X A M P L E 2 4 - 4**

A parallel-plate capacitor with square plates 14 cm on a side and separated by 2.0 mm is connected to a battery and charged to 12 V. (*a*) What is the charge on the capacitor? (*b*) How much energy is stored in the capacitor? (*c*) The battery is then disconnected from the capacitor and the plate separation is then increased to 3.5 mm. By how much is the energy increased when the plate separation is changed?

PICTURE THE PROBLEM (*a*) The charge on the capacitor can be calculated from the capacitance and then used to calculate the energy stored in Part (*b*). (*c*) Since the capacitor is no longer connected to the battery, the charge remains constant as the plates are separated. The energy increase is found by using the charge and new potential to calculate the new energy, from which we subtract the original energy.

(*a*) 1. The charge Q on the capacitor equals the product of C_0 and V_0, where C_0 is the capacitance and $V_0 = 12$ V is the battery voltage: $Q = C_0 V_0$

2. Calculate the capacitance of the parallel-plate capacitor: $C_0 = \dfrac{\epsilon_0 A}{d_0}$

3. Substitute to calculate Q:

$$Q = C_0 V_0 = \frac{\epsilon_0 A}{d_0} V_0$$

$$= \frac{(8.85 \text{ pF/m})(0.14 \text{ m})^2}{0.002 \text{ m}} (12 \text{ V}) = \boxed{1.04 \text{ nC}}$$

(b) Calculate the energy stored:

$$U_0 = \tfrac{1}{2} Q V_0 = \tfrac{1}{2}(1.04 \text{ nC})(12 \text{ V}) = \boxed{6.24 \text{ nJ}}$$

(c) 1. The battery is disconnected. The potential difference V between the plates is the field strength E times the separation distance d:

$$V = Ed$$

2. At the surface of a conductor, E is proportional to the surface charge density $\sigma = Q/A$. Since Q is constant, so is σ, and thus E:

$$E = \frac{\sigma}{\epsilon_0} = \frac{Q}{A \epsilon_0}$$

3. Combining the last two steps reveals that V is proportional to d:

$$V = Ed = \frac{Q}{A \epsilon_0} d$$

so

$$\frac{V}{d} = \frac{V_0}{d_0}, \quad \text{or} \quad \left(V = \frac{d}{d_0} V_0 \right)$$

4. Calculate U and ΔU, obtaining U_0 from Part (b):

$$U = \frac{1}{2} QV = \frac{1}{2} Q \frac{d}{d_0} V_0 = \frac{d}{d_0} \frac{1}{2} QV_0 = \frac{d}{d_0} U_0$$

so

$$\Delta U = U - U_0 = \frac{d}{d_0} U_0 - U_0 = \left(\frac{d}{d_0} - 1 \right) U_0$$

$$= \left(\frac{3.5 \text{ mm}}{2.0 \text{ mm}} - 1 \right)(6.24 \text{ nJ}) = \boxed{4.68 \text{ nJ}}$$

REMARKS The additional energy calculated in Part (c) comes from work done by the agent responsible for increasing the separation between the plates, which attract each other. An application of the dependence of capacitance on separation distance is shown in Figure 24-8.

EXERCISE Find the final voltage V between the capacitor plates. (*Answer* 21.0 V)

EXERCISE (a) Find the initial capacitance C_0 in this example when separation of the plates is 2.0 mm. (b) Find the final capacitance C when separation of the plates is 3.5 mm. (*Answer* (a) $C_0 = 86.7$ pF (b) $C = 49.6$ pF)

It is instructive to work Part (c) of Example 24-4 in another way. The oppositely charged plates of a capacitor exert attractive forces on one another. Work must be done against these forces to increase the plate separation. Assume that the lower plate is held fixed and the upper plate is moved. The force on the upper plate is the charge Q on the plate times the electric field $\vec{E}'$ *due to the charge* $-Q$ *on the lower plate.* This field is half the total field $\vec{E}$ between the plates (because the charge on the upper plate and the charge on the lower plate contribute equally to the field). When the potential difference is 12 V and the separation is 2 mm, the total field strength between the plates is

$$E = \frac{V}{d} = \frac{12 \text{ V}}{2 \text{ mm}} = 6 \text{ V/mm} = 6 \text{ kV/m}$$

The magnitude of the force exerted on the upper plate by the bottom plate is thus

$$F = QE' = Q(\tfrac{1}{2}E) = (1.04 \text{ nC})(3 \text{ kV/m}) = 3.12 \ \mu\text{N}$$

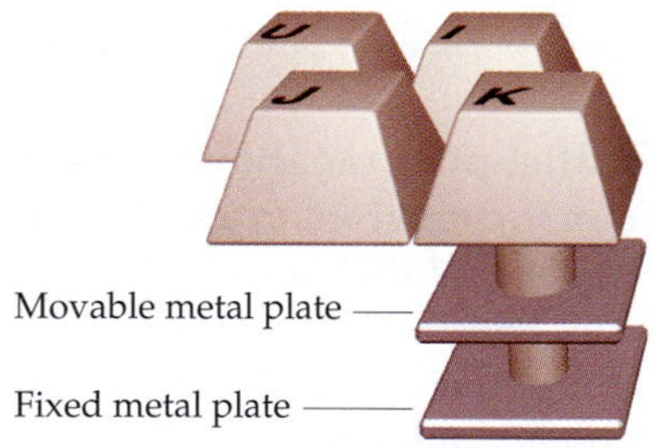

FIGURE 24-8 Capacitance switching in computer keyboards. A metal plate attached to each key acts as the top plate of a capacitor. Depressing the key decreases the separation between the top and bottom plates and increases the capacitance, which triggers the electronic circuitry of the computer to acknowledge the keystroke.

The work that must be done to move the upper plate a distance of $\Delta d = 1.5$ mm is then

$$W = F\,\Delta d = (3.12\ \mu\text{N})(1.5\ \text{mm}) = 4.68\ \text{nJ}$$

This is the same number of joules calculated in Part (c) of Example 24-4. This work equals the increase in the energy stored.

Electrostatic Field Energy

In the process of charging a capacitor, an electric field is produced between the plates. The work required to charge the capacitor can be thought of as the work required to create the electric field. That is, we can think of the energy stored in a capacitor as energy stored in the electric field, called **electrostatic field energy.**

Consider a parallel-plate capacitor. We can relate the energy stored in the capacitor to the electric field strength E between the plates. The potential difference between the plates is related to the electric field by $V = Ed$, where d is the plate separation distance. The capacitance is given by $C = \epsilon_0 A/d$ (Equation 24-10). The energy stored is

$$U = \frac{1}{2}CV^2 = \frac{1}{2}\left(\frac{\epsilon_0 A}{d}\right)(Ed)^2 = \frac{1}{2}\epsilon_0 E^2(Ad)$$

The quantity Ad is the volume of the space between the plates of the capacitor containing the electric field. The energy-per-unit volume is called the **energy density** u_e. The energy density in an electric field strength E is thus

$$u_e = \frac{energy}{volume} = \frac{1}{2}\epsilon_0 E^2 \qquad\qquad 24\text{-}13$$

ENERGY DENSITY OF AN ELECTROSTATIC FIELD

Thus, the energy per unit volume of the electrostatic field is proportional to the square of the electric field strength. *Although we obtained Equation 24-13 by considering the electric field between the plates of a parallel-plate capacitor, the result applies to any electric field.* Whenever there is an electric field in space, the electrostatic energy per unit volume is given by Equation 24-13.

EXERCISE (a) Calculate the energy density u_e for Example 24-4 when the plate separation is 2.0 mm. (b) Show that the increase in energy in Example 24-4 is equal to u_e times the increase in volume (Δ vol) between the plates. (*Answer* (a) $u_e = \frac{1}{2}\epsilon_0 E^2 = 159.3\ \mu\text{J}/\text{m}^3$, (b) $\Delta\text{vol} = A\,\Delta d = 2.94 \times 10^{-5}\ \text{m}^3$, $u_e\,\Delta\text{vol} = 4.68\ \text{nJ}$, in agreement with Example 24-4)

We can illustrate the generality of Equation 24-13 by calculating the electrostatic field energy of a spherical conductor of radius R that carries a charge Q. The electrostatic potential energy in terms of the charge Q and potential V is given by Equation 24-12:

$$U = \frac{kQ^2}{2R} = \frac{1}{2}QV \qquad\qquad 24\text{-}14$$

We now obtain the same result by considering the energy density of an electric field given by Equation 24-13. When the conductor carries a charge Q, the electric field is radial and is given by

$$E_r = 0, \quad r < R \text{ (inside the conductor)}$$

$$E_r = \frac{kQ}{r^2}, \quad r > R \text{ (outside the conductor)}$$

Since the electric field is spherically symmetric, we choose a spherical shell for our volume element. If the radius of the shell is r and its thickness is dr, the volume is $d\mathcal{V} = 4\pi r^2\, dr$ (Figure 24-9). The energy dU in this volume element is

$$dU = u_e\, d\mathcal{V} = \frac{1}{2}(\epsilon_0 E^2)4\pi r^2 dr$$

$$= \frac{1}{2}\epsilon_0 \left(\frac{kQ}{r^2}\right)^2 (4\pi r^2\, dr) = \frac{1}{2}(4\pi\epsilon_0 k^2)Q^2\frac{dr}{r^2} = \frac{1}{2}kQ^2 r^{-2}\, dr$$

where we have used $4\pi\epsilon_0 = 1/k$. Since the electric field is zero for $r < R$, we obtain the total energy in the electric field by integrating from $r = R$ to $r = \infty$:

$$U = \int u_e\, d\mathcal{V} = \frac{1}{2}kQ^2 \int_R^\infty r^{-2} dr = \frac{1}{2}k\frac{Q^2}{R} = \frac{1}{2}Q\left(\frac{kQ}{R}\right) = \frac{1}{2}QV \qquad \text{24-15}$$

which is the same as Equation 24-12.

24-4 Capacitors, Batteries, and Circuits

Next, we examine what happens when an initially uncharged capacitor is connected to the terminals of a battery. The potential difference between the two terminals of a battery is called its **terminal voltage.** Typically, one terminal of a battery is positively charged and the other terminal is negatively charged; this charge separation is maintained by chemical action within the battery. Within the battery, there is an electric field directed away from the positive terminal toward the negative terminal.[†] When a plate of an uncharged capacitor is connected to the negative terminal of the battery, the negative charge on that terminal is shared with the plate. This gives the plate a small negative charge and momentarily reduces the amount of negative charge on that battery terminal. If the second capacitor plate is then connected to the positive battery terminal, the charge on the positive battery terminal is then shared with it—momentarily reducing the positive charge on that battery terminal. These charge reductions on the battery terminals result in a decrease in the terminal voltage of the battery. This decrease in terminal voltage triggers the chemical activity within the battery that transfers charge from one terminal to the other terminal in an effort to maintain the terminal voltage at its initial level, which is called the **open-circuit terminal voltage.** This chemical action ceases when the battery has transferred sufficient charge from one capacitor plate to the other capacitor plate to raise the potential difference between the plates to the open-circuit terminal voltage of the battery.

It is useful to think of a battery as a charge pump. When we connect the plates of an uncharged capacitor to the terminals of a battery, the terminal voltage drops causing the battery to pump charge from one plate to the other plate until the open circuit terminal voltage is again reached.

In electric circuit diagrams, the symbol representing a battery is ⊣⊢, where the longer, thinner vertical line represents the positive terminal and the shorter, thicker vertical line represents the negative terminal. The symbol representing a capacitor is ⊣⊢.

EXERCISE A 6-μF capacitor, initially uncharged, is connected to the terminals of a 9-V battery. What total amount of charge flows through the battery? (*Answer* 54 μC)

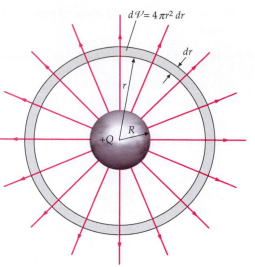

FIGURE 24-9 Geometry for the calculation of the electrostatic energy of a spherical conductor carrying a charge Q. The volume of the space between r and $r + dr$ is $d\mathcal{V} = 4\pi r^2\, dr$. The electrostatic field energy in this volume element is $u_e d\mathcal{V}$, where $u_e = \frac{1}{2}\epsilon_0 E^2$ is the energy density.

---

† This electric field from the positive to the negative terminal exists outside the battery also.

Combinations of Capacitors

FIGURE 24-10

CAPACITORS CONNECTED IN PARALLEL   E X A M P L E   2 4 - 5

A circuit consists of a 6-μF capacitor, a 12-μF capacitor, a 12-V battery, and a switch, connected as shown in Figure 24-10. Initially, the switch is open and the capacitors are uncharged. The switch is then closed and the capacitors charge. When the capacitors are fully charged and open-circuit terminal voltage is restored (*a*) what is the potential of each conductor in the circuit? (Choose the zero-potential reference point on the negative battery terminal.) (*b*) What is the charge on each capacitor plate? (*c*) What total charge passed through the battery?

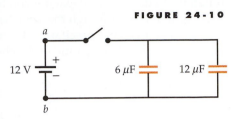

PICTURE THE PROBLEM The potential is the same throughout a conductor in electrostatic equilibrium. After the charges stop moving, all of the conductors connected by a conducting wire are at the same potential. The charge on a capacitor (step 2 and step 3) is related to the potential difference across the capacitor by $Q = CV$. The charges on the plates of a single capacitor are equal but opposite.

FIGURE 24-11

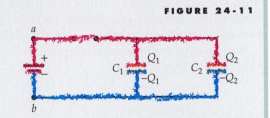

(*a*) Use a red marker to color the positive ($+$) battery terminal and all the conductors connected to it (Figure 24-11), and use a blue marker to color the negative ($-$) battery terminal and all the conductors connected to it:

All points colored red are at potential $\boxed{V_a = 12 \text{ V}}$

All points colored blue are at potential $\boxed{V_b = 0}$

(*b*) Use $Q = CV$ to find the magnitude of the charge on the plates. The capacitor plate at the higher potential carries a positive charge:

$Q_1 = C_1V = (6 \ \mu\text{F})(12 \text{ V}) = \boxed{72 \ \mu\text{C}}$

$Q_2 = C_2V = (12 \ \mu\text{F})(12 \text{ V}) = \boxed{144 \ \mu\text{C}}$

(*c*) The plates become charged because the battery acts as a charge pump:

$Q = Q_1 + Q_2 = \boxed{216 \ \mu\text{C}}$

REMARKS The equivalent capacitance of the two-capacitor combination is Q/V, where Q is the charge passing through the battery and V is the open-circuit terminal voltage of the battery. For this example $C_{eq} = (216 \ \mu\text{C})/(12 \text{ V}) = 18 \ \mu\text{F}$.

When two capacitors are connected, as shown in Figure 24-12, so that the upper plates of the two capacitors are connected by a conducting wire and are therefore at a common potential, and the lower plates are also connected together and are at a common potential, just like the capacitors in Example 24-5, the capacitors are said to be connected in **parallel**. Devices connected in parallel share a common potential difference across each device *due solely to the way they are connected*.

In Figure 24-12, assume that points *a* and *b* are connected to a battery or some other device that provides a potential difference $V = V_a - V_b$ between the plates of each capacitor. If the capacitances are C_1 and C_2, the charges Q_1 and Q_2 stored on the plates are given by

$$Q_1 = C_1V$$

and

$$Q_2 = C_2V$$

The total charge stored is

$$Q = Q_1 + Q_2 = C_1V + C_2V = (C_1 + C_2)V$$

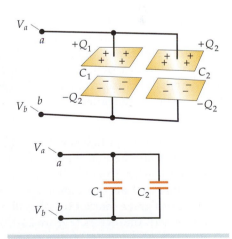

FIGURE 24-12 Two capacitors in parallel. The upper plates are connected together and are therefore at a common potential V_a; the lower plates are similarly connected together and therefore at a common potential V_b.

A combination of capacitors in a circuit can sometimes be substituted with a single capacitor that is operationally equivalent to the combination. The substitute capacitor is said to have an **equivalent capacitance.** That is, if a combination of initially uncharged capacitors is connected to a battery, the charge Q that flows through the battery as the capacitor combination becomes charged is the same as the charge that flows through the same battery if connected to a single uncharged capacitor of equivalent capacitance. Therefore, the equivalent capacitance of two capacitors in parallel is the ratio of the charge $Q_1 + Q_2$ to the potential difference:

$$C_{eq} = \frac{Q}{V} = \frac{Q_1 + Q_2}{V} = \frac{Q_1}{V} + \frac{Q_2}{V} = C_1 + C_2 \qquad 24\text{-}16$$

Thus, for two capacitors in parallel, C_{eq} is the sum of the individual capacitances. When we add a second capacitor in parallel, we increase the capacitance of the combination. The area that the charge is distributed on is effectively increased, allowing more charge to be stored for the same potential difference.

The same reasoning can be extended to three or more capacitors connected in parallel, as in Figure 24-13:

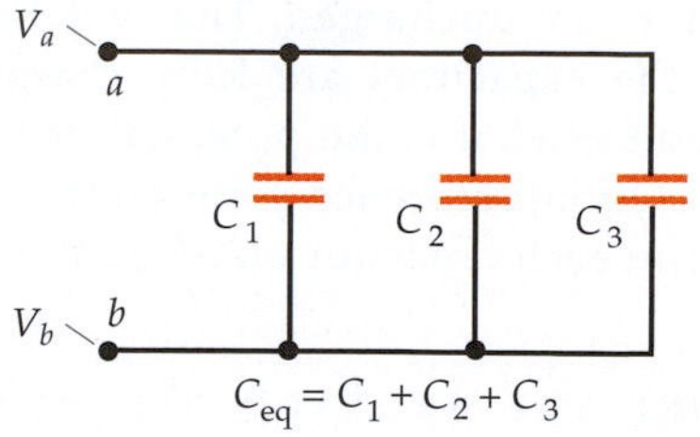

$$C_{eq} = C_1 + C_2 + C_3 + \dots \qquad 24\text{-}17$$

EQUIVALENT CAPACITANCE FOR CAPACITORS IN PARALLEL

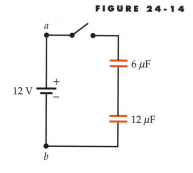

FIGURE 24-13 Three capacitors in parallel. The effect of adding a parallel capacitor to a circuit is an increase in the equivalent capacitance.

CAPACITORS CONNECTED IN SERIES   **EXAMPLE 24-6**

FIGURE 24-14

A circuit consists of a 6-μF capacitor, a 12 μ-F capacitor, a 12-V battery, and a switch, connected as shown in Figure 24-14. Initially, the switch is open and the capacitors are uncharged. The switch is then closed and the capacitors charge. When the capacitors are fully charged and open-circuit terminal voltage is restored, (a) what is the potential of each conductor in the circuit? (Choose the zero-potential reference point on the negative battery terminal.) If the potential of a conductor is not known, represent its potential symbolically. (b) What is the charge on each capacitor plate? (c) What total charge passed through the battery?

PICTURE THE PROBLEM (a) The potential is the same throughout a conductor in electrostatic equilibrium. After the charges stop moving, all of the conductors connected by a conducting wire are at the same potential. The charge on a capacitor, Parts (b) and (c), is related to the potential difference across the capacitor by $Q = CV$. Charge does not travel from one plate of a capacitor to the other.

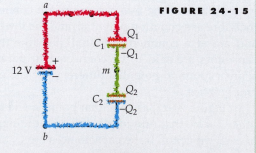

FIGURE 24-15

(a) Use a red marker to color the positive (+) battery terminal and all conductors connected to it, use a blue marker to color the negative (−) battery terminal and all the conductors connected to it, and use a green marker to color all other mutually connected conductors (Figure 24-15):

All points colored red are at potential $\boxed{V_a = 12\ \text{V}}$

All points colored blue are at potential $\boxed{V_b = 0}$

All points colored green are at the yet unknown potential $\boxed{V_m}$

(b) 1. Express the potential difference across each capacitor in terms of the Part (a) results:

$V_1 = V_a - V_m$

and

$V_2 = V_m - V_b$

2. Use $Q = CV$ to relate the charge on each capacitor to the potential difference:

$$Q_1 = C_1 V_1 = C_1(V_a - V_m)$$

and

$$Q_2 = C_2 V_2 = C_2(V_m - V_b)$$

3. Eliminating V_m gives:

$$\left.\begin{array}{l} V_a - V_m = \dfrac{Q_1}{C_1} \\[2ex] V_m - V_b = \dfrac{Q_2}{C_2} \end{array}\right\} \Rightarrow V_a - V_b = \dfrac{Q_1}{C_1} + \dfrac{Q_2}{C_2}$$

4. During charging, there is no charge transferred either to or from the green region in Figure 24-15, so its net charge remains zero:

$$(-Q_1) + Q_2 = 0$$

so

$$Q_1 = Q_2$$

5. Let $Q = Q_1 = Q_2$. Substitute Q for Q_1 and Q_2 and solve for Q:

$$V_a - V_b = \dfrac{Q}{C_1} + \dfrac{Q}{C_2}$$

so

$$Q = \dfrac{V_a - V_b}{\dfrac{1}{C_1} + \dfrac{1}{C_2}} = \dfrac{12\text{ V} - 0}{\dfrac{1}{6\ \mu\text{F}} + \dfrac{1}{12\ \mu\text{F}}} = 48\ \mu\text{C}$$

$$Q_1 = Q_2 = \boxed{48\ \mu\text{C}}$$

(c) All the charge passing through the battery ends up on the upper plate of C_1:

$$Q_1 = Q = \boxed{48\ \mu\text{C}}$$

REMARKS The equivalent capacitance of the two-capacitor combination is Q/V, where Q is the charge passing through the battery and V is the open-circuit terminal voltage of the battery. For this example $C_{eq} = (48\ \mu\text{C})/(12\text{ V}) = 4\ \mu\text{F}$.

EXERCISE Find the potential V_m on the conductors colored green in Figure 24-15. (*Answer* 4.0 V)

In Figure 24-16, two capacitors are connected so that the potential difference across the pair is the sum of the potential differences across the individual capacitors, just like those in Example 24-6. Devices connected in this manner are connected in **series**.

Capacitors C_1 and C_2 in Figure 24-16 are connected in series and initially they are without charge. If points a and b are then connected to the terminals of a battery, electrons will be pumped from the upper plate of C_1 to the lower plate of C_2. This leaves the upper plate of C_1 with a charge $+Q$ and the lower plate of C_2 with a charge $-Q$. When a charge $+Q$ appears on the upper plate of C_1, the electric field produced by that charge induces an equal negative charge, $-Q$, on the lower plate of C_1. This charge comes from electrons drawn from the upper plate of C_2. Thus, there will be an equal charge $+Q$ on the upper plate of the second capacitor and a corresponding charge $-Q$ on its lower plate. The potential difference across the first capacitor is

$$V_1 = \frac{Q}{C_1}$$

Similarly, the potential difference across the second capacitor is

$$V_2 = \frac{Q}{C_2}$$

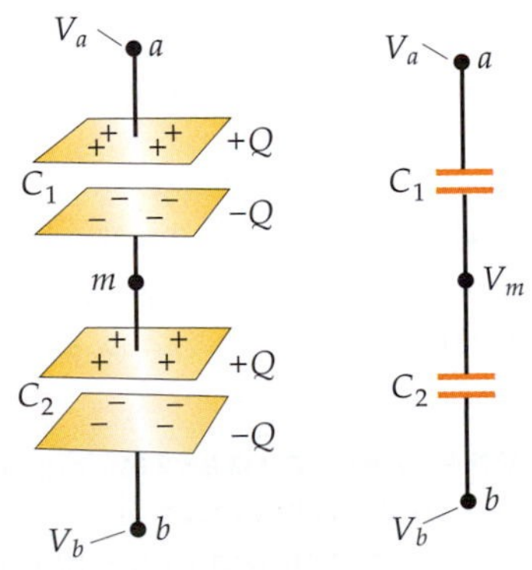

FIGURE 24-16 The total charge on the two interconnected capacitor plates equals zero. The potential difference across the pair equals the sum of the potential differences across the individual capacitors. The two capacitors are connected in series.

The potential difference across the two capacitors in series is the sum of these potential differences:

$$V = V_a - V_b = V_1 + V_2 = \frac{Q}{C_1} + \frac{Q}{C_2} = Q\left(\frac{1}{C_1} + \frac{1}{C_2}\right) \qquad 24\text{-}18$$

The equivalent capacitance of the two capacitors in series is defined as

$$C_{eq} = \frac{Q}{V} \qquad 24\text{-}19$$

Substituting Q/C_{eq} for V in Equation 24-18, and then dividing both sides by Q, gives

$$\frac{1}{C_{eq}} = \frac{1}{C_1} + \frac{1}{C_2} \qquad 24\text{-}20$$

Note that in the preceding exercise, the equivalent capacitance of the two capacitors in series is less than the capacitance of either capacitor. Adding a capacitor in series increases $1/C_{eq}$, which means the equivalent capacitance C_{eq} decreases. When we add a second capacitor in series, we decrease the capacitance of the combination. The plate separation is essentially increased, requiring a greater potential difference to store the same charge.

Equation 24-20 can be generalized to three or more capacitors connected in series:

$$\frac{1}{C_{eq}} = \frac{1}{C_1} + \frac{1}{C_2} + \frac{1}{C_3} + \dots \qquad 24\text{-}21$$

EQUIVALENT CAPACITANCE FOR EQUALLY CHARGED CAPACITORS IN SERIES

❗ **This formula is valid only if the capacitors are in series *and* the total charge on each pair of capacitor plates connected by a wire is zero.**

EXERCISE Two capacitors have capacitances of 20 μF and 30 μF. Find the equivalent capacitance if the capacitors are connected (*a*) in parallel and (*b*) in series. (*Answer* (*a*) 50 μF, (*b*) 12 μF)

A capacitor bank for storing energy to be used by the pulsed Nova laser at Lawrence Livermore Laboratories. The laser is used in fusion studies.

USING THE EQUIVALENCE FORMULA   **EXAMPLE 24-7**

A 6-μF capacitor and a 12-μF capacitor, each initially uncharged, are connected in series across a 12-V battery. Using the equivalence formula for capacitors in series, find the charge on each capacitor and the potential difference across each.

PICTURE THE PROBLEM Figure 24-17*a* shows the circuit in this example and Figure 24-17*b* shows an equivalent capacitor that carries the same charge $Q = C_{eq}V$. After finding the charge, we can find the potential drop across each capacitor.

1. The charge on each capacitor equals the charge on the equivalent capacitor: $Q = C_{eq}V$

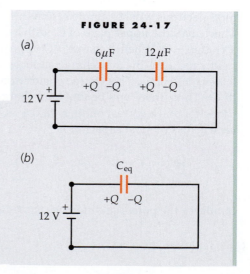

FIGURE 24-17

2. The equivalent capacitance of the series combination is found from:

$$\frac{1}{C_{eq}} = \frac{1}{C_1} + \frac{1}{C_2} = \frac{1}{6\ \mu F} + \frac{1}{12\ \mu F} = \frac{3}{12\ \mu F}$$

$$C_{eq} = 4\ \mu F$$

3. Use this value to find the charge Q. This is the charge that went through the battery. It is the charge on each capacitor:

$$Q = C_{eq} V = (4\ \mu F)(12\ V) = \boxed{48\ \mu C}$$

4. Use the result for Q to find the potential across the 6-μF capacitor:

$$V_1 = \frac{Q}{C_1} = \frac{48\ \mu C}{6\ \mu F} = \boxed{8\ V}$$

5. Again, use the result for Q to find the potential across the 12-μF capacitor:

$$V_2 = \frac{Q}{C_2} = \frac{48\ \mu C}{12\ \mu F} = \boxed{4\ V}$$

PLAUSIBILITY CHECK The sum of these potential differences is 12 V, as required.

REMARKS The results are the same as those obtained in Example 24-6.

CAPACITORS IN SERIES REARRANGED IN PARALLEL   **EXAMPLE 24-8**   **Try It Yourself**

The two capacitors in Example 24-7 are removed from the battery and carefully disconnected from each other so that the charge on the plates is not disturbed (Figure 24-18a). They are then reconnected in a circuit containing open switches, positive plate to positive plate and negative plate to negative plate (Figure 24-18b). Find the potential difference across the capacitors and the charge on each capacitor after the switches are closed and the charges have stopped flowing.

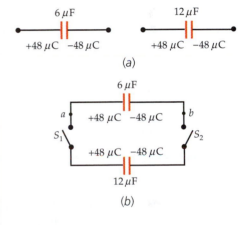

(a)

(b)

FIGURE 24-18

PICTURE THE PROBLEM Just after the two capacitors are disconnected from the battery, they carry equal charges of 48 μC. After switches S_1 and S_2 in the new circuit are closed, the capacitors are in parallel between points a and b. The potential across each of them is the same, and the equivalent capacitance of the system is $C_{eq} = C_1 + C_2$. The two positive plates form a single conductor with charge $Q = 48\ \mu C$, and the negative plates form a conductor with charge $-Q = -48\ \mu C$. Therefore, the potential difference is $V = Q/C_{eq}$, and the charges on the two capacitors are $Q_1 = C_1 V$ and $Q_2 = C_2 V$.

Cover the column to the right and try these on your own before looking at the answers.

Steps	Answers
1. The wiring is such that after the switches are closed the potential difference is the same across each capacitor.	$V = V_1 = V_2$
2. For each capacitor $V = Q/C$. Substitute this into the step 1 result. Let C_1 be the 2-μF capacitor.	$\dfrac{Q_1}{C_1} = \dfrac{Q_2}{C_2}$
3. The sum of the charges on the two capacitor plates on the left remains 96 μC.	$Q_1 + Q_2 = 96\ \mu C$
4. Solve for the charge on each capacitor.	$Q_1 = \boxed{32\ \mu C}$, $Q_2 = \boxed{64\ \mu C}$
5. Calculate the potential difference.	$V = \dfrac{Q_1}{C_1} = \boxed{5.33\ V}$

PLAUSIBILITY CHECK Note that $Q = Q_1 + Q_2 = 96\ \mu C$, and that $Q_2/C_2 = 5.33\ V$ as required.

REMARKS After the switches are closed, the two capacitors are connected in parallel with the potential difference between point a and point b being the potential difference across the pair. Thus, $C_{eq} = C_1 + C_2 = 18\ \mu F$, $Q = Q_1 + Q_2 = 96\ \mu C$, and $V = Q/C_{eq} = 5.33\ V$.

EXERCISE Find the energy stored in the capacitors before and after they are connected. [*Answer* $U_i = q^2/(2C_1) + q^2/(2C_2)$, where $q = 48\ \mu C$. Thus, $U_i = 288\ \mu J$. $U_f = Q_1^2/(2C_1) + Q_2^2/(2C_2) = 256\ \mu J$. Note that 32 μJ is *lost* to thermal energy in the wires or radiated away.]

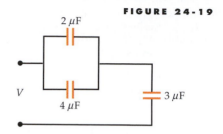

FIGURE 24-19

CAPACITORS IN SERIES AND IN PARALLEL   **E X A M P L E   2 4 - 9**

(a) Find the equivalent capacitance of the network of three capacitors in Figure 24-19. (b) The capacitors are initially uncharged. Find the charge on each capacitor and the voltage drop across it after the capacitor combination is connected to a 6-V battery.

PICTURE THE PROBLEM (a) The 2-μF capacitor and the 4-μF capacitor are connected in parallel, and the parallel combination is connected in series with the 3-μF capacitor. We first find the equivalent capacitance of the 2-μF capacitor and the 4-μF capacitor (Figure 24-20a), then combine this equivalent capacitance with the 3-μF capacitor to reach a final equivalent capacitance (Figure 24-20b). (b) The charge on the 3-μF capacitor is the charge passing through the battery $Q = C_{eq}V$ as shown in Figure 24-20a.

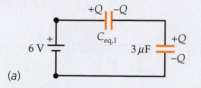

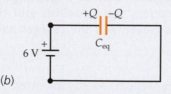

FIGURE 24-20

(a)   (b)

(a) 1. The equivalent capacitance of the two capacitors in parallel is the sum of the capacitances:

$$C_{eq,1} = C_1 + C_2 = 2\ \mu F + 4\ \mu F = 6\ \mu F$$

2. Find the equivalent capacitance of a 6-μF capacitor in series with a 3-μF capacitor:

$$\frac{1}{C_{eq}} = \frac{1}{C_{eq,1}} + \frac{1}{C_3} = \frac{1}{6\ \mu F} + \frac{1}{3\ \mu F} = \frac{1}{2\ \mu F}$$

$$C_{eq} = \boxed{2\ \mu F}$$

(b) 1. Calculate the charge Q delivered by the battery. This is also the charge on the 3-μF capacitor:

$$Q = C_{eq}V = (2\ \mu F)(6\ V) = 12\ \mu C$$

2. The potential drop across the 3-μF capacitor is Q/C_3:

$$V_3 = \frac{Q_3}{C_3} = \frac{Q}{C_3} = \frac{12\ \mu C}{3\ \mu F} = \boxed{4\ V}$$

3. The potential drop across the parallel combination $V_{2,4}$ is $Q/C_{eq,1}$:

$$V_{2,4} = \frac{Q}{C_{eq,1}} = \frac{12\ \mu C}{6\ \mu F} = \boxed{2\ V}$$

4. The charge on each of the parallel capacitors is found from $Q_i = C_i V_{2,4}$, where $V_{2,4} = 2\ V$:

$$Q_2 = C_2 V_{2,4} = (2\ \mu F)(2\ V) = \boxed{4\ \mu C}$$

$$Q_4 = C_4 V_{2,4} = (4\ \mu F)(2\ V) = \boxed{8\ \mu C}$$

PLAUSIBILITY CHECK The voltage drop across the parallel combination (2 V) plus that across the 3-μF capacitor (4 V) equals the voltage of the battery. Also, the sum of the charges on the parallel capacitors (4 μC + 8 μC) equals the total charge (12 μC) on the 3-μF capacitor.

EXERCISE Find the energy stored in each capacitor. (*Answer* $U_2 = 4\ \mu J$, $U_3 = 24\ \mu J$, $U_4 = 8\ \mu J$. Note that $U_2 + U_3 + U_4 = 36\ \mu J = \frac{1}{2}QV = \frac{1}{2}Q^2/C_{eq} = \frac{1}{2}C_{eq}V^2$.)

24-5 Dielectrics

A nonconducting material (e.g., air, glass, paper, or wood) is called a **dielectric**. When the space between the two conductors of a capacitor is occupied by a dielectric, the capacitance is increased by a factor κ that is characteristic of the dielectric, a fact discovered experimentally by Michael Faraday. The reason for this increase is that the electric field between the plates of a capacitor is weakened by the dielectric. Thus, for a given charge on the plates, the potential difference is reduced and the capacitance (Q/V) is increased.

Consider an isolated charged capacitor without a dielectric between its plates. A dielectric slab is then inserted between the plates, completely filling the space between the plates. If the electric field is E_0 before the dielectric slab is inserted, after the dielectric slab is inserted between the plates the field is

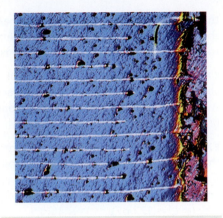

A cut section of a multilayer capacitor with a ceramic dielectric. The white lines are the edges of the conducting plates.

$$E = \frac{E_0}{\kappa} \qquad\qquad 24\text{-}22$$

ELECTRIC FIELD INSIDE A DIELECTRIC

where κ (kappa) is called the **dielectric constant**. For a parallel-plate capacitor of separation d, the potential difference V between the plates is

$$V = Ed = \frac{E_0 d}{\kappa} = \frac{V_0}{\kappa}$$

where V is the potential difference with the dielectric and $V_0 = E_0 d$ is the original potential difference without the dielectric. The new capacitance is

$$C = \frac{Q}{V} = \frac{Q}{V_0/\kappa} = \kappa \frac{Q}{V_0}$$

or

$$C = \kappa C_0 \qquad\qquad 24\text{-}23$$

EFFECT OF A DIELECTRIC ON CAPACITANCE

where $C_0 = Q/V_0$ is the capacitance without the dielectric. The capacitance of a parallel-plate capacitor filled with a dielectric of constant κ is thus

$$C = \frac{\kappa \epsilon_0 A}{d} = \frac{\epsilon A}{d} \qquad\qquad 24\text{-}24$$

where

$$\epsilon = \kappa \epsilon_0 \qquad\qquad 24\text{-}25$$

is called the **permittivity** of the dielectric.

In the preceding discussion, the capacitor was isolated so we assumed that the charge on its plates did not change as the dielectric was inserted. This is the case if the capacitor is charged and then removed from the charging source (the battery) before the insertion of the dielectric. If the dielectric is inserted while the battery remains connected, the battery pumps additional charge to maintain the original potential difference. The total charge on the plates is then $Q = \kappa Q_0$. In either case, the capacitance (Q/V) is increased by the factor κ.

EXERCISE The 88.5-pF capacitor of Example 24-2 is filled with a dielectric of constant $\kappa = 2$. (*a*) Find the new capacitance. (*b*) Find the charge on the capacitor with the dielectric in place if the capacitor is attached to a 12-V battery. (*Answer* (*a*) 177 pF, (*b*) 2.12 nC)

EXERCISE The capacitor in the previous exercise is charged to 12 V without the dielectric and is then disconnected from the battery. The dielectric of constant $\kappa = 2$ is then inserted. Find the new values for (*a*) the charge Q, (*b*) the voltage V, and (*c*) the capacitance C. (*Answer* (*a*) $Q = 1.06$ nC, which is unchanged; (*b*) $V = 6$ V; (*c*) $C = 177$ pF)

Dielectrics not only increase the capacitance of a capacitor, they also provide a means for keeping parallel conducting plates apart and they raise the potential difference at which dielectric breakdown occurs.[†] Consider a parallel-plate capacitor made from two sheets of metal foil that are separated by a thin plastic sheet. The plastic sheet allows the metal sheets to be very close together without actually being in electrical contact, and because the dielectric strength of plastic is greater than that of air, a greater potential difference can be attained before dielectric breakdown occurs. Table 24-1 lists the dielectric constants and dielectric strengths of some dielectrics. Note that for air $\kappa \approx 1$; so, for most situations we do not need to distinguish between air and a vacuum.

TABLE 24-1

Dielectric Constants and Dielectric Strengths of Various Materials

Material	Dielectric Constant κ	Dielectric Strength, kV/mm
Air	1.00059	3
Bakelite	4.9	24
Glass (Pyrex)	5.6	14
Mica	5.4	10–100
Neoprene	6.9	12
Paper	3.7	16
Paraffin	2.1–2.5	10
Plexiglas	3.4	40
Polystyrene	2.55	24
Porcelain	7	5.7
Transformer oil	2.24	12

USING A DIELECTRIC IN A PARALLEL-PLATE CAPACITOR **EXAMPLE 24-10**

A parallel-plate capacitor has square plates of edge length 10 cm and a separation of $d = 4$ mm. A dielectric slab of constant $\kappa = 2$ has dimensions 10 cm × 10 cm × 4 mm. (*a*) What is the capacitance without the dielectric? (*b*) What is the capacitance if the dielectric slab fills the space between the plates? (*c*) What is the capacitance if a dielectric slab with dimensions 10 cm × 10 cm × 3 mm is inserted into the 4-mm gap?

[†] Recall from Chapter 23 that for electric fields greater than about 3×10^6 V/m, air breaks down; that is, it becomes ionized and begins to conduct.

PICTURE THE PROBLEM The capacitance without the dielectric, C_0, is found from the area and spacing of the plates (Figure 24-21a). When the capacitor is filled with a dielectric κ, (Figure 24-21b), the capacitance is $C = \kappa C_0$ (Equation 24-23). If the dielectric only partially fills the capacitor (Figure 24-21c), we calculate the potential difference V for a given charge Q, then apply the definition of capacitance, $C = Q/V$.

FIGURE 24-21

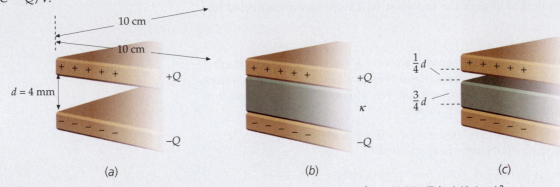

(a)   (b)   (c)

(a) If there is no dielectric, the capacitance C_0 is given by Equation 24-10:

$$C_0 = \frac{\epsilon_0 A}{d} = \frac{(8.85\ \text{pF/m})(0.1\ \text{m})^2}{0.004\ \text{m}} = \boxed{22.1\ \text{pF}}$$

(b) When the capacitor is filled with a dielectric κ, its capacitance C is increased by the factor κ:

$$C = \kappa C_0 = (2)(22.1\ \text{pF}) = \boxed{44.2\ \text{pF}}$$

(c) 1. The new capacitance is related to the original charge Q and the new potential difference V:

$$C = \frac{Q}{V}$$

2. The potential difference V between the plates is the sum of the potential difference for the empty gap plus the potential difference for the dielectric slab:

$$V = V_{\text{gap}} + V_{\text{slab}} = E_{\text{gap}}(\tfrac{1}{4}d) + E_{\text{slab}}(\tfrac{3}{4}d)$$

3. The field in the gap just outside the conductor is the original field E_0:

$$E_{\text{gap}} = E_0 = \frac{Q}{\epsilon_0 A}$$

4. The field in the dielectric slab is reduced by the factor κ:

$$E_{\text{slab}} = \frac{E_0}{\kappa}$$

5. Combining the previous two results yields V in terms of κ. Note that the original potential difference is $V_0 = E_0 d$:

$$V = E_0\left(\frac{1}{4}d\right) + \frac{E_0}{\kappa}\left(\frac{3}{4}d\right) = E_0 d\left(\frac{1}{4} + \frac{3}{4\kappa}\right)$$

$$= V_0\left(\frac{\kappa + 3}{4\kappa}\right)$$

6. Using $C = Q/V$, we find the new capacitance in terms of the original capacitance, $C_0 = Q/V_0$:

$$C = \frac{Q}{V} = \frac{Q}{V_0\dfrac{\kappa + 3}{4\kappa}} = \frac{Q}{V_0}\left(\frac{4\kappa}{\kappa + 3}\right) = C_0\left(\frac{4\kappa}{\kappa + 3}\right)$$

$$= (22.1\ \text{pF})\left(\frac{8}{5}\right) = \boxed{35.4\ \text{pF}}$$

⊘ PLAUSIBILITY CHECK The absence of a dielectric corresponds to $\kappa = 1$. In this case, our result for the final step in Part (c) would reduce to $C = C_0$ as expected. Suppose that the dielectric slab were a conducting slab. In a conductor, $E = 0$; so, according to Equation 24-22, κ for a conductor would equal infinity. As κ approaches infinity, the quantity $4\kappa/(\kappa + 3)$ approaches 4, so the result for the final step in Part (c) approaches $4C_0$. A conducting slab simply extends the capacitor plate, hence the plate separation with the conducting dielectric in place would be $\frac{1}{4}d$. This means that C should be $4C_0$, as it is for very large κ.

REMARKS Note that the results of this example are independent of the vertical position of the dielectric (or conducting) slab in the space between the plates.

EXAMPLE 24-11 Put It in Context

When studying capacitors in physics class, your professor claims that you could build a parallel-plate capacitor from waxed paper and aluminum foil. You decide to try it, and build one about the size of a piece of notebook paper. Before testing its charge-storing power on your gullible roommate, you decide to calculate the amount of charge the capacitor will store when connected to a 9-V battery.

PICTURE THE PROBLEM We want charge, which we can get from the definition $C = Q/V$ if we know the capacitance. We can get the capacitance from the parallel-plate capacitor formula $C = \epsilon_0 A/d$. We will need to either measure or to estimate the thickness of the waxed paper.

1. The charge on a capacitor is related to the voltage and the capacitance by the definition of capacitance:

$$Q = CV$$

2. The capacitance is obtained from the parallel-plate capacitance formula:

$$C = \frac{\kappa \epsilon_0 A}{d}$$

3. Substituting for C and solving for Q gives:

$$Q = CV = \frac{\kappa \epsilon_0 VA}{d}$$

4. A sheet of notebook paper is approximately 8.5-by-11 in.:

$$A = 8.5 \text{ in.} \times 11 \text{ in.} = 93.5 \text{ in.}^2 = 0.0603 \text{ m}^2$$

5. We assume a sheet of wax paper is the same thickness as a sheet of the paper your physics textbook is made of. Measure the thickness of 300 sheets of paper in your book (from page 1 through page 600):

300 sheets of paper are 2.0 cm (0.020 m) thick. So, the thickness of a single sheet of paper is

$$0.020 \text{ m}/300 = 66.7 \ \mu\text{m}$$

6. Using the step 3 result, solve for the charge. Assume the dielectric constant of wax paper is 2.3 (the same as that of paraffin):

$$Q = \frac{\kappa \epsilon_0 AV}{d} = \frac{2.3(88.6 \text{ pF/m})(0.0603 \text{ m}^2)(9 \text{ V})}{66.7 \times 10^{-6} \text{ m}}$$

$$= 1.66 \times 10^6 \text{ pC} = \boxed{1.66 \ \mu\text{C}}$$

Energy Stored in the Presence of a Dielectric

The energy stored in a parallel-plate capacitor with dielectric is

$$U = \tfrac{1}{2}QV = \tfrac{1}{2}CV^2$$

We can express the capacitance C in terms of the area and separation of the plates, and the voltage difference V in terms of the electric field and plate separation, to obtain

$$U = \frac{1}{2}CV^2 = \frac{1}{2}\left(\frac{\epsilon A}{d}\right)(Ed)^2 = \frac{1}{2}\epsilon E^2 (Ad)$$

The quantity Ad is the volume between the plates containing the electric field. The energy per unit volume is thus

$$u_e = \tfrac{1}{2}\epsilon E^2 = \tfrac{1}{2}\kappa \epsilon_0 E^2 \qquad\qquad 24\text{-}26$$

Part of this energy is the energy associated with the electric field (Equation 24-13) and the rest is the energy associated with the polarization of the dielectric (discussed in Section 24-6).

INSERTING THE DIELECTRIC—BATTERY DISCONNECTED   **EXAMPLE 24-12**

Two parallel-plate capacitors, each having a capacitance of $C_1 = C_2 = 2 \mu F$, are connected in parallel across a 12-V battery. (a) Find the charge on each capacitor. (b) Find the total energy stored in the capacitors.

The parallel combination is then disconnected from the battery and a dielectric slab of constant $\kappa = 2.5$ is inserted between the plates of the capacitor C_2, completely filling the gap. After the dielectric is inserted, find (c) the potential difference across each capacitor, (d) the charge on each capacitor, and (e) the total energy stored in the capacitors.

PICTURE THE PROBLEM (a) The charge Q and (b) total energy U can be found for each capacitor from its capacitance C and voltage V. (c) After the capacitors are removed from the battery, the total charge on the pair remains the same. When the dielectric is inserted into one of the capacitors, its capacitance C_2 changes. The potential across the parallel combination can be found from the total charge and the equivalent capacitance.

(a) The charge on each capacitor is found from its capacitance C and voltage V:

$$Q = CV = (2 \mu F)(12 \text{ V}) = \boxed{24 \mu C}$$

(b) 1. The energy stored in each capacitor is found from its charge Q and its voltage V:

$$U = \tfrac{1}{2}QV = \tfrac{1}{2}(24 \mu C)(12 \text{ V}) = 144 \mu J$$

2. The total energy is twice that stored in each capacitor:

$$U_{total} = 2U = \boxed{288 \mu J}$$

(c) 1. The potential across the parallel combination is related to the total charge Q_{total} and the equivalent capacitance C_{eq}:

$$V = \frac{Q_{total}}{C_{eq}}$$

2. The capacitance C_2 of the capacitor with the dielectric is increased by the factor κ. The equivalent capacitance is the sum of the capacitances:

$$C_{eq} = C_1 + C_2 = C_1 + \kappa C_2 = (2 \mu F) + (2.5)(2 \mu F)$$
$$= 2 \mu F + 5 \mu F = 7 \mu F$$

3. The total charge remains 48 μC. Substitute for Q_{total} and C_{eq} to calculate V:

$$V = \frac{Q_{total}}{C_{eq}} = \frac{48 \mu C}{7 \mu F} = \boxed{6.86 \text{ V}}$$

(d) The charge on each capacitor is again derived from its capacitance and the voltage V:

$$Q_1 = C_1 V = (2 \mu F)(6.86 \text{ V}) = \boxed{13.7 \mu C}$$

$$Q_2 = C_2 V = (5 \mu F)(6.86 \text{ V}) = \boxed{34.3 \mu C}$$

(e) The energy stored in each capacitor is found from its new charge and new voltage:

$$U = U_1 + U_2 = \tfrac{1}{2}Q_1 V + \tfrac{1}{2}Q_2 V = \tfrac{1}{2}(Q_1 + Q_2)V$$
$$= \tfrac{1}{2}(13.7 \mu C + 34.3 \mu C)(6.86 \text{ V}) = \boxed{165 \mu J}$$

PLAUSIBILITY CHECK When the dielectric is inserted into one of the capacitors, the field is weakened and the potential difference is lowered. Since the two capacitors are connected in parallel, charge must flow from the other capacitor so that the potential difference is the same across both capacitors. Note that the capacitor with the dielectric has the greater charge, and that when the charges calculated for each capacitor in Part (d) are added, $Q_1 + Q_2 = 13.7 \mu C + 34.3 \mu C = 48 \mu C$, the result is the same as the original sum.

REMARKS The total energy of 165 μJ is less than the original energy of 288 μJ. When the dielectric is inserted, it is pulled in and work is done on whatever was holding it. To remove the dielectric, work $W = 288 \mu J - 165 \mu J = 123 \mu J$ must be done, and this work is stored as electrostatic potential energy.

Find (a) the charge on each capacitor and (b) the total energy stored in the capacitors of Example 24-12, if the dielectric is inserted into one of the capacitors while the battery is still connected.

PICTURE THE PROBLEM Since the battery is still connected, the potential difference across the capacitors remains 12 V. This condition determines the charge and energy stored in each capacitor. Let subscript 1 refer to the capacitor without the dielectric and subscript 2 refer to the capacitor with the dielectric.

Cover the column to the right and try these on your own before looking at the answers.

Steps	Answers
(a) Calculate the charge on each capacitor from $Q = CV$ using the result that $C_1 = 2 \ \mu F$ and $C_2 = 5 \ \mu F$ as found in Example 24-12.	$Q_1 = C_1 V = \boxed{24 \ \mu C}$ $Q_2 = C_2 V = \boxed{60 \ \mu C}$
(b) 1. Calculate the energy stored in each capacitor from $U = \frac{1}{2} CV^2$. Check your results by using $U = \frac{1}{2} QV$.	$U_1 = 144 \ \mu J, \ U_2 = 360 \ \mu J$
2. Add your results for U_1 and U_2 to obtain the final energy.	$U_{\text{total}} = \boxed{504 \ \mu J}$

REMARKS Note that Q_2 is two and a half times its value before the dielectric was inserted (since $\kappa = 2.5$). The battery supplies this additional charge in order to maintain a fixed potential difference. Because of the work done by the battery to supply this charge, the total energy of the system is higher with the dielectric in place (504 μJ) than without the dielectric (288 μJ).

24-6 Molecular View of a Dielectric

A dielectric weakens the electric field between the plates of a capacitor, because the molecules in the dielectric produce an electric field within the dielectric in a direction opposite to the field produced by the charges on the plates. The electric field produced by the dielectric is due to the electric dipole moments of the molecules of the dielectric.

Although atoms and molecules are electrically neutral, they are affected by electric fields because they contain positive and negative charges that can respond to external fields. We can think of an atom as a very small, positively charged nucleus surrounded by a negatively charged electron cloud. In some atoms and molecules, the electron cloud is spherically symmetric, so its "center of negative charge" is at the center of the atom or molecule, coinciding with the center of positive charge. An atom or molecule like this has zero dipole moment and is said to be nonpolar. But in the presence of an external electric field, the positive and negative charges experience forces in opposite directions, so the positive and negative charges then separate until the attractive force they exert on each other balances the forces due to the external electric field (Figure 24-22). The molecule is then said to be polarized and it behaves like an electric dipole.

In some molecules (e.g., HCl and H_2O), the centers of positive and negative charge do not coincide, even in the absence of an external electric field. As we noted in Chapter 21, these polar molecules have a permanent electric dipole moment.

When a dielectric is placed in the field of a capacitor, its molecules are polarized in such a way that there is a net dipole moment parallel to the field. If the

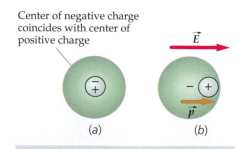

Center of negative charge coincides with center of positive charge

(a)   (b)

FIGURE 24-22 Schematic diagrams of the charge distributions of an atom or nonpolar molecule. (a) In the absence of an external electric field, the center of positive charge coincides with the center of negative charge. (b) In the presence of an external electric field, the centers of positive and negative charge are displaced, producing an induced dipole moment in the direction of the external field.

molecules are polar their dipole moments, originally oriented at random, tend to become aligned due to the torque exerted by the field.† If the molecules are nonpolar, the field induces dipole moments that are parallel to the field. In either case, the molecules in the dielectric are polarized in the direction of the external field (Figure 24-23).

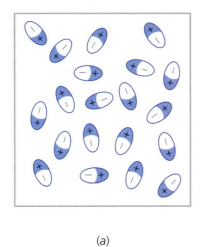

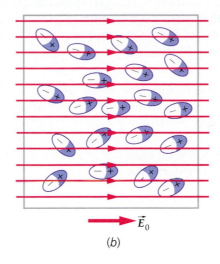

FIGURE 24-23 (*a*) The randomly oriented electric dipoles of a polar dielectric in the absence of an external electric field. (*b*) In the presence of an external electric field, the dipoles are partially aligned parallel to the field.

(*a*)   (*b*)

The net effect of the polarization of a homogeneous dielectric in a parallel-plate capacitor is the creation of a surface charge on the dielectric faces near the plates, as shown in Figure 24-24. The surface charge on the dielectric is called a **bound charge**, because the surface charge is bound to the molecules of the dielectric and cannot move about like the free charge on the conducting capacitor plates. This bound charge produces an electric field opposite in direction to the electric field produced by the free charge on the conductors. Thus, the net electric field between the plates is reduced, as illustrated in Figure 24-25.

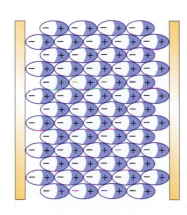

FIGURE 24-24 When a dielectric is placed between the plates of a capacitor, the electric field of the capacitor polarizes the molecules of the dielectric. The result is a bound charge on the surface of the dielectric that produces its own electric field; this field opposes the external field. The field of the bound surface charges thus weakens the electric field within the dielectric.

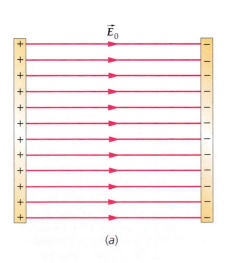

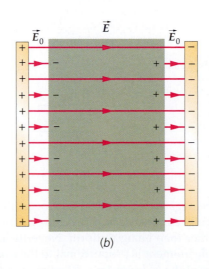

FIGURE 24-25 The electric field between the plates of a capacitor (*a*) with no dielectric and (*b*) with a dielectric. The surface charge on the dielectric weakens the original field between the plates.

(*a*)   (*b*)

† The degree of alignment depends on the external field and on the temperature. It is approximately proportional to pE/kT, where pE is the maximum energy of a dipole in a field E, and kT is the characteristic thermal energy.

E X A M P L E   2 4 - 1 4

A hydrogen atom consists of a proton nucleus of charge $+e$ and an electron of charge $-e$. The charge distribution of the atom is spherically symmetric, so the atom is nonpolar. Consider a model in which the hydrogen atom consists of a positive point charge $+e$ at the center of a uniformly charged spherical cloud of radius R and total charge $-e$. Show that when such an atom is placed in a uniform external electric field $\vec{E}$, the induced dipole moment is proportional to $\vec{E}$; that is, $\vec{p} = \alpha\vec{E}$, where α is called the *polarizability*.

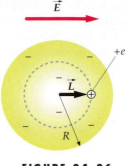

FIGURE 24-26

PICTURE THE PROBLEM In the external field, the center of the uniform negative cloud is displaced from the positive charge by an amount L so that the force exerted by the field $e\vec{E}$ is balanced by the force exerted by the negative cloud $e\vec{E}'$, where $\vec{E}'$ is the field due to the cloud (Figure 24-26). We use Gauss's law to find E', and then we calculate the induced dipole moment $\vec{p} = e\vec{L}$, where $\vec{L}$ is the position of the positive charge relative to the center of the cloud.

1. Write the magnitude of the induced dipole moment in terms of e and L:

$$p = eL$$

2. We can find L by calculating the field E_n' due to the negatively charged cloud at a distance L from the center. We use Gauss's law to compute E_n'. Choose a spherical Gaussian surface of radius L concentric with the cloud. Then E_n' is constant on this surface:

$$\phi_{\text{net}} = \oint E_n \, dA = \frac{Q_{\text{inside}}}{\epsilon_0}$$

$$E_n' (4\pi L^2) = \frac{Q_{\text{inside}}}{\epsilon_0}$$

$$E_n' = \frac{Q_{\text{inside}}}{4\pi \epsilon_0 L^2}$$

3. The charge inside the sphere of radius L equals the charge density times the volume:

$$Q_{\text{inside}} = \rho \frac{4}{3}\pi L^3 = \frac{-e}{\frac{4}{3}\pi R^3}\frac{4}{3}\pi L^3 = -e\frac{L^3}{R^3}$$

4. Substitute this value of Q_{inside} to calculate E_n':

$$E_n' = \frac{Q_{\text{inside}}}{4\pi \epsilon_0 L^2} = \frac{-eL^3/R^3}{4\pi \epsilon_0 L^2} = -\frac{e}{4\pi \epsilon_0 R^3}L$$

5. Solve for L:

$$L = -\frac{4\pi \epsilon_0 R^3}{e}E_n'$$

6. E_n' is negative because it points inward on the Gaussian surface. At the positive charge, E_n' points to the left, so $E_n' = -E$:

$$E_n' = -E$$

so

$$L = \frac{4\pi \epsilon_0 R^3}{e}E$$

7. Substitute these results for L and E_n' to express p in terms of the external field E:

$$p = eL = 4\pi \epsilon_0 R^3 E = \alpha E$$

so

$$\boxed{\vec{p} = \alpha\vec{L}}$$

where

$$\alpha = 4\pi \epsilon_0 R^3$$

REMARKS The charge distribution of the negative charge in a hydrogen atom, obtained from quantum theory, is spherically symmetric, but the charge density decreases exponentially with distance rather than being uniform. Nevertheless, the above calculation shows that the dipole moment is proportional to the external field $p = \alpha E$, and the polarizability α is of the order of $4\pi \epsilon_0 R^3$ where R is the radius of the atom or molecule. The dielectric constant κ can be related to the polarizability and to the number of molecules per unit volume.

Magnitude of the Bound Charge

The bound charge density σ_b on the surfaces of the dielectric is related to the dielectric constant κ and to the free charge density σ_f on the plates. Consider a dielectric slab between the plates of a parallel-plate capacitor, as shown in Figure 24-27. If the dielectric is a very thin slab between plates that are close together, the electric field inside the dielectric slab due to the bound charge densities, $+\sigma_b$ on the right and $-\sigma_b$ on the left, is just the field due to two infinite-plane charge densities. Thus, the field E_b has the magnitude

$$E_b = \frac{\sigma_b}{\epsilon_0}$$

This field is directed to the left and subtracts from the electric field E_0 due to the free charge density on the capacitor plates, which has the magnitude

$$E_0 = \frac{\sigma_f}{\epsilon_0}$$

The magnitude of the net field $E = E_0/\kappa$ is the difference between these magnitudes:

$$E = E_0 - E_b = \frac{E_0}{\kappa}$$

or

$$E_b = \left(1 - \frac{1}{\kappa}\right)E_0$$

Writing σ_b/ϵ_0 for E_b and σ_f/ϵ_0 for E_0, we obtain

$$\sigma_b = \left(1 - \frac{1}{\kappa}\right)\sigma_f \qquad \text{24-27}$$

The bound charge density σ_b is always less than the free charge density σ_f on the capacitor plates, and it is zero if $\kappa = 1$, which is the case when there is no dielectric. For a conducting slab, $\kappa = \infty$ and $\sigma_b = \sigma_f$.

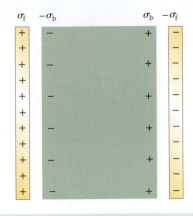

FIGURE 24-27 A parallel-plate capacitor with a dielectric slab between the plates. If the plates are closely spaced, each of the surface charges can be considered an infinite plane charge. The electric field due to the free charge on the plates is directed to the right and has a magnitude $E_0 = \sigma_f/\epsilon_0$. That due to the bound charge is directed to the left and has a magnitude $E_b = \sigma_b/\epsilon_0$.

*The Piezoelectric Effect

In certain crystals that contain polar molecules (e.g., quartz, tourmaline, and topaz), a mechanical stress applied to the crystal produces polarization of the molecules. This is known as the **piezoelectric effect**. The polarization of the stressed crystal causes a potential difference across the crystal, which can be used to produce an electric current. Piezoelectric crystals are used in transducers (e.g., microphones, phonograph pickups, and vibration-sensing devices) to convert mechanical strain into electrical signals. The converse piezoelectric effect, in which a voltage applied to such a crystal induces mechanical strain (deformation), is used in headphones and many other devices.

Because the natural frequency of vibration of quartz is in the range of radio frequencies, and because its resonance curve is very sharp,[†] quartz is used extensively to stabilize radio-frequency oscillators and to make accurate clocks.

---

† Resonance in AC circuits, which will be discussed in Chapter 29, is analogous to mechanical resonance, which was discussed in Chapter 14.

1. Capacitance is an important defined quantity that relates charge to potential difference.
2. Devices connected in *parallel* share a common potential difference across each device *due solely to the way they are connected.*

Topic	Relevant Equations and Remarks	
1. Electrostatic Potential Energy	The electrostatic potential energy of a system of point charges is the work needed to bring the charges from an infinite separation to their final positions.	
Of point charges	$$U = \frac{1}{2} \sum_{i=1}^{n} q_i V_i$$	24-2
Of a conductor with charge Q at potential V	$$U = \frac{1}{2} QV$$	24-3
Of a system of conductors	$$U = \frac{1}{2} \sum_{i=1}^{n} Q_i V_i$$	24-4
Energy stored in a capacitor	$$U = \frac{1}{2} \frac{Q^2}{C} = \frac{1}{2} QV = \frac{1}{2} CV^2$$	24-12
Energy density of an electric field	$$u_e = \frac{1}{2} \epsilon_0 E^2$$	24-13
2. Capacitor	A capacitor is a device for storing charge and energy. It consists of two conductors insulated from each other that carry equal and opposite charges.	
3. Capacitance	Definition of capacitance.	
	$$C = \frac{Q}{V}$$	24-5
Isolated conductor	Q is the conductor's total charge, V is the conductor's potential relative to infinity.	
Capacitor	Q is the magnitude of the charge on either conductor, V is the magnitude of the potential difference between the conductors.	
Of an isolated spherical conductor	$$C = 4\pi \epsilon_0 R$$	24-6
Of a parallel-plate capacitor	$$C = \frac{\epsilon_0 A}{d}$$	24-10
Of a cylindrical capacitor	$$C = \frac{2\pi \epsilon_0 L}{\ln(R_2/R_1)}$$	24-11
4. Equivalent Capacitance		
Parallel capacitors	When devices are connected in parallel, the voltage drop is the same across each.	
	$$C_{eq} = C_1 + C_2 + C_3 + \dots$$	24-17

Series capacitors	When capacitors are in series, the voltage drops add. If the net charge on each connected pair of plates is zero, then:

$$\frac{1}{C_{eq}} = \frac{1}{C_1} + \frac{1}{C_2} + \frac{1}{C_3} + \dots \qquad \text{24-21}$$

5. Dielectrics

Macroscopic behavior	A nonconducting material is called a dielectric. When a dielectric is inserted between the plates of a capacitor, the electric field within the dielectric is weakened and the capacitance is thereby increased by the factor κ, which is the dielectric constant.
Microscopic view	The field in the dielectric of a capacitor is weakened because the dipole moments of the molecules (either preexisting or induced) tend to align with the field and thereby produce an electric field inside the dielectric that opposes the applied field. The aligned dipole moment of the dielectric is proportional to the applied field.
Electric field inside	$E = \dfrac{E_0}{\kappa}$ 24-22
Effect on capacitance	$C = \kappa C_0$ 24-23
Permittivity ϵ	$\epsilon = \kappa\,\epsilon_0$ 24-25
Uses of a dielectric	1. Increases capacitance 2. Increases dielectric strength 3. Physically separates conductors

*6. Piezoelectric Effect

In certain crystals containing polar molecules, a mechanical stress polarizes the molecules, which induces a voltage across the crystal. Conversely, an applied voltage induces mechanical strain (deformation) in the crystal.

PROBLEMS

- • Single-concept, single-step, relatively easy
- •• Intermediate-level, may require synthesis of concepts
- ••• Challenging, for advanced students
- **SSM** Solution is in the *Student Solutions Manual*
- **iSOLVE** Problems available on iSOLVE online homework service
- **iSOLVE✓** These "Checkpoint" online homework service problems ask students additional questions about their confidence level, and how they arrived at their answer.

In a few problems, you are given more data than you actually need; in a few other problems, you are required to supply data from your general knowledge, outside sources, or informed estimates.

Conceptual Problems

1 • **SSM** If the voltage across a parallel-plate capacitor is doubled, its capacitance (a) doubles. (b) drops by half. (c) remains the same.

2 • If the charge on an isolated spherical conductor is doubled, its capacitance (a) doubles. (b) drops by half. (c) remains the same.

3 • True or false: The electrostatic energy per unit volume at some point is proportional to the square of the electric field at that point.

4 • If the potential difference of a parallel-plate capacitor is doubled by changing the plate separation without changing the charge, by what factor does its stored electric energy change?

5 •• SSM A parallel-plate air capacitor is connected to a constant-voltage battery. If the separation between the capacitor plates is doubled while the capacitor remains connected to the battery, the energy stored in the capacitor (a) quadruples. (b) doubles. (c) remains unchanged. (d) drops to half its previous value. (e) drops to one-fourth its previous value.

6 •• If the capacitor of Problem 5 is disconnected from the battery before the separation between the plates is doubled, the energy stored in the capacitor upon separation of the plates (a) quadruples. (b) doubles. (c) remains unchanged. (d) drops to half its previous value. (e) drops to one-fourth its previous value.

7 • True or false:

(a) The equivalent capacitance of two capacitors in parallel equals the sum of the individual capacitances.
(b) The equivalent capacitance of two capacitors in series is less than the capacitance of either capacitor alone.

8 •• Two initially uncharged capacitors of capacitance C_0 and $2C_0$, respectively, are connected in series across a battery. Which of the following is true?

(a) The capacitor $2C_0$ carries twice the charge of the other capacitor.
(b) The voltage across each capacitor is the same.
(c) The energy stored by each capacitor is the same.
(d) None of the above statements is correct.

9 • True or false: A dielectric inserted into a capacitor increases the capacitance.

10 •• SSM Two capacitors half-filled with a dielectric are shown in Figure 24-28. The area and separation of each capacitor is the same. Which has the higher capacitance, that shown in Figure (a) or in Figure (b)?

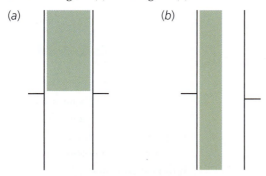

FIGURE 24-28 Problem 10

11 • True or false:

(a) The capacitance of a capacitor is defined as the total amount of charge the capacitor can hold.
(b) The capacitance of a parallel-plate capacitor depends on the voltage difference between the plates.
(c) The capacitance of a parallel-plate capacitor is proportional to the charge on its plates.

12 •• Two identical capacitors are connected in series to a 100-V battery. When only one capacitor is connected to this battery, the energy stored is U_0. What is the total energy stored in the two capacitors when the series combination is connected to the battery? (a) $4U_0$. (b) $2U_0$. (c) U_0. (d) $U_0/2$. (e) $U_0/4$.

Estimation and Approximation

13 •• Disconnect the coaxial cable from a television or other device and measure (estimate) the diameter of the center conductor and the braided conductor, shown in the photo on page 755. Assume a plausible value (see Table 24-1) for the dielectric constant of the material separating the two conductors and estimate the capacitance per unit length of the cable.

14 •• SSM To create the high-energy densities needed to operate a pulsed nitrogen laser, the discharge from a high-capacitance capacitor is used. Typically, the energy requirement per pulse (i.e., per discharge) is 100 J. Estimate the capacitance required if the discharge is applied through a spark gap of 1 cm width. Assume that the dielectric breakdown of nitrogen occurs at $E \approx 3 \times 10^6$ V/m.

15 •• Measurements reveal that the earth's electric field extends upward for 1000 m and has an average magnitude of 200 V/m. Estimate the electrical energy stored in the atmosphere. (Hint: You may treat the atmosphere as a flat slab with an area equal to the surface area of the earth. Why?)

16 •• Estimate the capacitance of a typical hot-air balloon.

Electrostatic Potential Energy

17 • Three point charges are on the x axis: q_1 at the origin, q_2 at $x = 3$ m, and q_3 at $x = 6$ m. Find the electrostatic potential energy for (a) $q_1 = q_2 = q_3 = 2$ μC; (b) $q_1 = q_2 = 2$ μC, and $q_3 = -2$ μC; and (c) $q_1 = q_3 = 2$ μC, and $q_2 = -2$ μC.

18 • Point charges q_1, q_2, and q_3 are at the corners of an equilateral triangle of side 2.5 m. Find the electrostatic potential energy of this charge distribution if (a) $q_1 = q_2 = q_3 = 4.2$ μC; (b) $q_1 = q_2 = 4.2$ μC, and $q_3 = -4.2$ μC; and (c) $q_1 = q_2 = -4.2$ μC, and $q_3 = +4.2$ μC.

19 • SSM iSOLVE What is the electrostatic potential energy of an isolated spherical conductor of 10 cm radius that is charged to 2 kV?

20 •• iSOLVE Four point charges of magnitude 2 μC are at the corners of a square of side 4 m. Find the electrostatic potential energy if (a) all of the charges are negative, (b) three of the charges are positive and one of the charges is negative, and (c) two of the charges are positive and two of the charges are negative.

21 •• iSOLVE Four charges are at the corners of a square centered at the origin as follows: q at $(-a, +a)$; $2q$ at $(+a, +a)$; $-3q$ at $(+a, -a)$; and $6q$ at $(-a, -a)$. A fifth charge $+q$ is placed at the origin and released from rest. Find its speed when it is a great distance from the origin.

Capacitance

22 • SSM An isolated spherical conductor of 10 cm radius is charged to 2 kV. (a) How much charge is on the conductor? (b) What is the capacitance of the sphere? (c) How does the capacitance change if the sphere is charged to 6 kV?

23 • **iSOLVE** A capacitor has a charge of 30 μC. The potential difference between the conductors is 400 V. What is the capacitance?

24 •• Two isolated conducting spheres of equal radius have charges $+Q$ and $-Q$, respectively. If they are separated by a large distance compared to their radius, what is the capacitance of this unusual capacitor?

The Storage of Electrical Energy

25 • **iSOLVE** (a) A 3-μF capacitor is charged to 100 V. How much energy is stored in the capacitor? (b) How much additional energy is required to charge the capacitor from 100 V to 200 V?

26 • **iSOLVE** A 10-μF capacitor is charged to $Q = 4$ μC. (a) How much energy is stored in the capacitor? (b) If half the charge is removed from the capacitor, how much energy remains?

27 • **iSOLVE** (a) Find the energy stored in a 20-pF capacitor when it is charged to 5 μC. (b) How much additional energy is required to increase the charge from 5 μC to 10 μC?

28 • **SSM** Find the energy per unit volume in an electric field that is equal to 3 MV/m, which is the dielectric strength of air.

29 • A parallel-plate capacitor with a plate area of 2 m² and a separation of 1.0 mm is charged to 100 V. (a) What is the electric field between the plates? (b) What is the energy per unit volume in the space between the plates? (c) Find the total energy by multiplying your answer from Part (b) by the total volume between the plates. (d) Find the capacitance C. (e) Calculate the total energy from $U = \frac{1}{2} CV^2$, and compare your answer with your result from Part (c).

30 •• **iSOLVE** Two concentric metal spheres have radii $r_1 = 10$ cm and $r_2 = 10.5$ cm, respectively. The inner sphere has a charge $Q = 5$ nC spread uniformly on its surface, and the outer sphere has charge $-Q$ on its surface. (a) Calculate the total energy stored in the electric field inside the spheres. *Hint: You can treat the spheres essentially as parallel flat slabs separated by 0.5 cm—why?* (b) Find the capacitance of this two-sphere system and show that the total energy stored in the field is equal to $\frac{1}{2}Q^2/C$.

31 •• **SSM** **iSOLVE** A parallel-plate capacitor with plates of area 500 cm² is charged to a potential difference V and is then disconnected from the voltage source. When the plates are moved 0.4 cm farther apart, the voltage between the plates increases by 100 V. (a) What is the charge Q on the positive plate of the capacitor? (b) How much does the energy stored in the capacitor increase due to the movement of the plates?

32 ••• A ball of charge of radius R has a uniform charge density ρ and a total charge $Q = \frac{4}{3}\pi R^3 \rho$. (a) Find the electrostatic energy density at a distance r from the center of the ball for $r < R$ and for $r > R$. (b) Find the energy in a spherical shell of volume $4\pi r^2\, dr$ for $r < R$ and for $r > R$. (c) Compute the total electrostatic energy by integrating your expressions from Part (b), and show that your result can be written $U = kQ^2/R$. Explain why this result is greater than that for a spherical conductor of radius R carrying a total charge Q.

Combinations of Capacitors

33 • (a) How many 1-μF capacitors connected in parallel would it take to store a total charge of 1 mC with a potential difference of 10 V across each capacitor? (b) What would be the potential difference across the combination? (c) If the number of 1-μF capacitors found in Part (a) is connected in series and the potential difference across each is 10 V, find the charge on each capacitor and the potential difference across the combination.

34 • **iSOLVE** A 3-μF capacitor and a 6-μF capacitor are connected in series, and the combination is connected in parallel with an 8-μF capacitor. What is the equivalent capacitance of this combination?

35 • **SSM** Three capacitors are connected in a triangle as shown in Figure 24-29. Find the equivalent capacitance between points a and c.

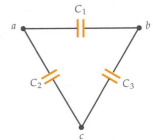

FIGURE 24-29 Problem 35

36 • A 10-μF capacitor and a 20-μF capacitor are connected in parallel across a 6-V battery. (a) What is the equivalent capacitance of this combination? (b) What is the potential difference across each capacitor? (c) Find the charge on each capacitor.

37 •• A 10-μF capacitor is connected in series with a 20-μF capacitor across a 6-V battery. (a) Find the charge on each capacitor. (b) Find the potential difference across each capacitor.

38 •• **SSM** Three identical capacitors are connected so that their maximum equivalent capacitance is 15 μF. (a) Describe how the capacitors are combined. (b) There are three other ways to combine all three capacitors in a circuit. What are the equivalent capacitances for each arrangement?

39 •• For the circuit shown in Figure 24-30, find (a) the total equivalent capacitance between the terminals, (b) the charge stored on each capacitor, and (c) the total stored energy.

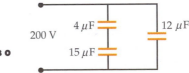

FIGURE 24-30
Problem 39

40 •• (a) Show that the equivalent capacitance of two capacitors in series can be written

$$C_{eq} = \frac{C_1 C_2}{C_1 + C_2}$$

(b) Use this expression to show that $C_{eq} < C_1$ and $C_{eq} < C_2$.
(c) Show that the correct expression for the equivalent capacitance of three capacitors in series is

$$C_{eq} = \frac{C_1 C_2 C_3}{C_1 C_2 + C_2 C_3 + C_1 C_3}$$

41 •• For the circuit shown in Figure 24-31, find (a) the total equivalent capacitance between the terminals, (b) the charge stored on each capacitor, and (c) the total stored energy.

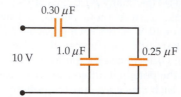

FIGURE 24-31
Problem 41

42 •• Five identical capacitors of capacitance C_0 are connected in a bridge network, as shown in Figure 24-32. (a) What is the equivalent capacitance between points a and b? (b) Find the equivalent capacitance between points a and b if the capacitor at the center is replaced by a capacitor with a capacitance of $10\,C_0$.

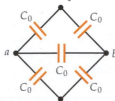

FIGURE 24-32 Problem 42

43 •• Design a network of capacitors that has a capacitance of 2 μF and breakdown voltage of 400 V, using only 2-μF capacitors that have individual breakdown voltages of 100 V.

44 •• **SSM** Find all the different possible equivalent capacitances that can be obtained using a 1-μF, a 2-μF, and a 4-μF capacitor in any combination that includes all three, or any two, of the capacitors.

45 ••• (a) What is the capacitance of the infinite ladder of capacitors shown in Figure 24-33a? (b) If we were to replace the ladder with a single capacitor (as shown in Figure 24-33b), what capacitance C would we need so that the combination had the same capacitance as the infinite ladder?

(a)

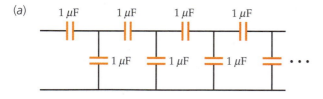

(b)

FIGURE 24-33 Problem 45

Parallel-Plate Capacitors

46 • **ISOLVE✓** A parallel-plate capacitor has a capacitance of 2 μF and a plate separation of 1.6 mm. (a) What is the maximum potential difference between the plates, so that dielectric breakdown of the air between the plates does not occur? (Use E_{max} = 3 MV/m.) (b) How much charge is stored at this maximum potential difference?

47 • **ISOLVE✓** An electric field of 2×10^4 V/m exists between the plates of a circular parallel-plate capacitor that has a plate separation of 2 mm. (a) What is the voltage across the capacitor? (b) What plate radius is required if the stored charge is 10 μC?

48 •• **ISOLVE** A parallel-plate, air-gap capacitor has a capacitance of 0.14 μF. The plates are 0.5 mm apart. (a) What is the area of each plate? (b) What is the potential difference if the capacitor is charged to 3.2 μC? (c) What is the stored energy? (d) How much charge can the capacitor carry before dielectric breakdown of the air between the plates occurs?

49 •• **SSM** **ISOLVE✓** Design a 0.1-μF parallel-plate capacitor with air between the plates that can be charged to a maximum potential difference of 1000 V. (a) What is the minimum possible separation between the plates? (b) What minimum area must the plates of the capacitor have?

Cylindrical Capacitors

50 • A Geiger tube consists of a wire of radius R = 0.2 mm, length L = 12 cm, and a coaxial cylindrical shell conductor of the same length L = 12 cm with a radius of 1.5 cm. (a) Find the capacitance, assuming that the gas in the tube has a dielectric constant of κ = 1. (b) Find the charge per unit length on the wire, when the potential difference between the wire and shell is 1.2 kV.

51 •• A cylindrical capacitor consists of a long wire of radius R_1 and length L with a charge $+Q$ and a concentric outer cylindrical shell of radius R_2, length L, and charge $-Q$. (a) Find the electric field and energy density at any point in space. (b) How much energy resides in a cylindrical shell between the conductors of radius R, thickness dr, and volume $2\pi rL\,dr$? (c) Integrate your expression from Part (b) to find the total energy stored in the capacitor and compare your result with that obtained, using $U = \frac{1}{2}CV^2$.

52 ••• Three concentric, thin conducting cylindrical shells have radii of 0.2, 0.5, and 0.8 cm. The space between the shells is filled with air. The innermost and outermost cylinders are connected at one end by a conducting wire. Find the capacitance per unit length of this system.

53 •• **SSM** A goniometer is a precise instrument for measuring angles. A capacitive goniometer is shown in Figure 24-34a. Each plate of the variable capacitor (Figure 24-34b) consists of a flat metal semicircle with inner radius R_1 and the outer radius R_2. The plates share a common rotation axis, and the width of the air gap separating the plates is d. Calculate the capacitance as a function of the angle θ and the parameters given.

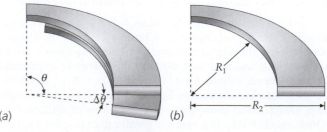

FIGURE 24-34 Problem 53

54 •• A capacitive pressure gauge is shown in Figure 24-35. Two plates of area A are separated by a material with dielectric constant κ, thickness d, and Young's modulus Y. If a pressure increase ΔP is applied to the plates, what is the change in their capacitance?

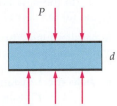

FIGURE 24-35 Problem 54

Spherical Capacitors

55 •• SSM A spherical capacitor consists of two thin, concentric spherical shells of radii R_1 and R_2. (a) Show that the capacitance is given by $C = 4\pi\varepsilon_0 R_1 R_2/(R_2 - R_1)$. (b) Show that when the radii of the shells are nearly equal, the capacitance is given approximately by the expression for the capacitance of a parallel-plate capacitor, $C = \varepsilon_0 A/d$, where A is the area of the sphere and $d = R_2 - R_1$.

56 •• A spherical capacitor has an inner sphere of radius R_1 with a charge of $+Q$ and an outer concentric spherical thin shell of radius R_2 with a charge of $-Q$. (a) Find the electric field and the energy density at any point in space. (b) Calculate the energy in the electrostatic field in a spherical shell of radius r, thickness dr, and volume $4\pi r^2\, dr$ between the conductors? (c) Integrate your expression from Part (b) to find the total energy stored in the capacitor, and compare your result with that obtained using $U = \frac{1}{2}QV$.

57 ••• A spherical shell of radius R carries a charge Q distributed uniformly over its surface. Find the distance from the center of the sphere such that half the total electrostatic field energy of the system is within that distance.

Disconnected and Reconnected Capacitors

58 •• ISOLVE✓ A 2-μF capacitor is charged to a potential difference of 12 V. The wires connecting the capacitor to the battery are then disconnected from the battery and connected across a second, initially uncharged, capacitor. The potential difference across the 2-μF capacitor then drops to 4 V. What is the capacitance of the second capacitor?

59 •• ISOLVE A 100-pF capacitor and a 400-pF capacitor are both charged to 2 kV. They are then disconnected from the voltage source and are connected together, positive plate to positive plate and negative plate to negative plate. (a) Find the resulting potential difference across each capacitor. (b) Find the energy lost when the connections are made.

60 •• SSM ISOLVE Two capacitors, $C_1 = 4\ \mu$F and $C_2 = 12\ \mu$F, are connected in series across a 12-V battery. They are carefully disconnected so that they are not discharged and they are then reconnected to each other, with positive plate to positive plate and negative plate to negative plate. (a) Find the potential difference across each capacitor after they are connected. (b) Find the initial energy stored and the final energy stored in the capacitors.

61 •• A 1.2-μF capacitor is charged to 30 V. After charging, the capacitor is disconnected from the voltage source and is connected to another uncharged capacitor. The final voltage is 10 V. (a) What is the capacitance of the first capacitor? (b) How much energy was lost when the connection was made?

62 •• Rework Problem 59, imagining that the capacitors are connected positive plate to negative plate, after they have been charged to 2 kV.

63 •• Rework Problem 60, imagining that the two capacitors are first connected in parallel across the 12-V battery and are then connected, with the positive plate of each capacitor connected to the negative plate of the other.

64 •• SSM ISOLVE A 20-pF capacitor is charged to 3 kV and then removed from the battery and connected to an uncharged 50-pF capacitor. (a) What is the new charge on each capacitor? (b) Find the initial energy stored in the 20-pF capacitor, and find the final energy stored in the two capacitors. Is electrostatic potential energy gained or lost when the two capacitors are connected?

65 ••• A parallel combination of three capacitors, $C_1 = 2\ \mu$F, $C_2 = 4\ \mu$F, and $C_3 = 6\ \mu$F is charged with a 200-V source. The capacitors are then disconnected from the voltage source and from each other and are reconnected positive plates to negative plates, as shown in Figure 24-36. (a) What is the voltage across each capacitor with switches S_1 and S_2 closed but switch S_3 open? (b) After switch S_3 is closed, what is the final charge on each capacitor? (c) Give the voltage across each capacitor after switch S_3 is closed.

FIGURE 24-36
Problem 65

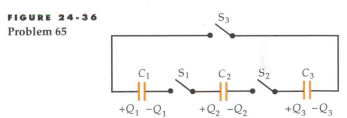

66 •• SSM A capacitor of capacitance C has a charge Q. A student connects one terminal of the capacitor to a terminal of an identical uncharged capacitor. When the remaining two terminals are connected, charge flows until electrostatic equilibrium is reestablished and both capacitors have charge $Q/2$ on them. Compare the total energy initially stored in the one capacitor to the total energy stored in the two after the second electrostatic equilibrium. Where did the missing energy go? This energy was dissipated in the connecting wires via Joule heating, which is discussed in Chapter 25.

Dielectrics

67 • ISOLVE✓ A parallel-plate capacitor is made by placing polyethylene ($\kappa = 2.3$) between two sheets of aluminum foil. The area of each sheet of aluminum foil is 400 cm^2, and the thickness of the polyethylene is 0.3 mm. Find the capacitance.

68 •• Suppose the Geiger tube of Problem 50 is filled with a gas of dielectric constant $\kappa = 1.8$ and breakdown field of 2×10^6 V/m. (a) What is the maximum potential difference that can be maintained between the wire and shell? (b) What is the charge per unit length on the wire?

69 •• Repeat Problem 56, with the space between the two spherical shells filled with a dielectric of dielectric constant κ.

70 •• **iSOLVE✓** A certain dielectric, with a dielectric constant $\kappa = 24$, can withstand an electric field of 4×10^7 V/m. Suppose we want to use this dielectric to construct a 0.1-μF capacitor that can withstand a potential difference of 2000 V. (*a*) What is the minimum plate separation? (*b*) What must the area of the plates be?

71 •• A parallel-plate capacitor has plates separated by a distance d. The space between the plates is filled with two dielectrics, one of thickness $\frac{1}{4}d$ and dielectric constant κ_1, and the other with thickness $\frac{3}{4}d$ and dielectric constant κ_2. Find the capacitance of this capacitor in terms of C_0, the capacitance with no dielectrics.

72 •• **SSM** Two capacitors, each consisting of two conducting plates of surface area A, with an air gap of width d. They are connected in parallel, as shown in Figure 24-37, and each has a charge Q. A slab of width d and area A with dielectric constant κ is inserted between the plates of one of the capacitors. Calculate the new charge Q' on that capacitor.

FIGURE 24-37
Problem 72

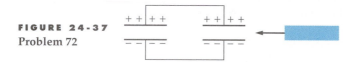

73 •• A parallel-plate capacitor with no dielectric has a capacitance C_0. If the separation distance between the plates is d, and a slab with dielectric constant κ and thickness $t < d$ is placed in the capacitor, find the new capacitance.

74 •• The membrane of the axon of a nerve cell is a thin cylindrical shell of radius $R = 10^{-5}$ m, length $L = 0.1$ m, and thickness $d = 10^{-8}$ m. The membrane has a positive charge on one side and a negative charge on the other, and the membrane acts as a parallel-plate capacitor of area $A = 2\pi r L$ and separation d. The membrane's dielectric constant is approximately $\kappa = 3$. (*a*) Find the capacitance of the membrane. If the potential difference across the membrane is 70 mV, find (*b*) the charge on each side of the membrane, and (*c*) the electric field through the membrane.

75 •• **SSM** What is the dielectric constant of a dielectric on which the induced bound charge density is (*a*) 80 percent of the free-charge density on the plates of a capacitor filled by the dielectric, (*b*) 20 percent of the free charge density, and (*c*) 98 percent of the free charge density?

76 •• Two parallel plates have charges Q and $-Q$. When the space between the plates is devoid of matter, the electric field is 2.5×10^5 V/m. When the space is filled with a certain dielectric, the field is reduced to 1.2×10^5 V/m. (*a*) What is the dielectric constant of the dielectric? (*b*) If $Q = 10$ nC, what is the area of the plates? (*c*) What is the total induced charge on either face of the dielectric?

77 •• **SSM** Find the capacitance of the parallel-plate capacitor shown in Figure 24-38.

78 •• A parallel-plate capacitor has plates of area 600 cm² and a separation of 4 mm. The capacitor is charged to 100 V and is then disconnected from the battery. (*a*) Find the electric field E_0 and the electrostatic energy U. A dielectric of constant $\kappa = 4$ is then inserted, completely filling the space between the plates. Find (*b*) the new electric field E, (*c*) the potential difference V, and (*d*) the new electrostatic energy.

79 ••• A parallel-plate capacitor is constructed using a dielectric whose constant varies with position. The plates have area A. The bottom plate is at $y = 0$ and the top plate is at $y = y_0$. The dielectric constant is given as a function of y according to $\kappa = 1 + (3/y_0)y$. (*a*) What is the capacitance? (*b*) Find σ_b/σ_f on the surfaces of the dielectric. (*c*) Use Gauss's law to find the induced volume charge density $\rho(y)$ within this dielectric. (*d*) Integrate the expression for the volume charge density found in Part (*c*) over the dielectric, and show that the total induced bound charge, including that on the surfaces, is zero.

General Problems

80 •• You are given 4 identical capacitors and a 100-V battery. When only one capacitor is connected to this battery the energy stored is U_0. Can you find a combination of the four capacitors so that the total energy stored in all four capacitors is U_0?

81 • **SSM** Three capacitors have capacitances of 2 μF, 4 μF, and 8 μF. Find the equivalent capacitance if (*a*) the capacitors are connected in parallel and (*b*) if the capacitors are connected in series.

82 • A 1-μF capacitor is connected in parallel with a 2-μF capacitor, and the combination is connected in series with a 6-μF capacitor. What is the equivalent capacitance of this combination?

83 • The voltage across a parallel-plate capacitor with plate separation 0.5 mm is 1200 V. The capacitor is disconnected from the voltage source and the separation between the plates is increased until the energy stored in the capacitor has been doubled. Determine the final separation between the plates.

84 •• **iSOLVE✓** Determine the capacitance of each of the networks shown in Figure 24-39.

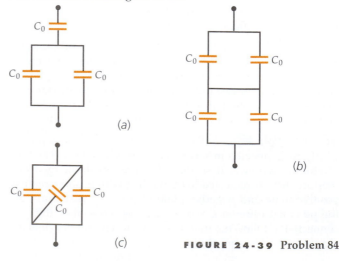

(a)

(b)

(c)

FIGURE 24-39 Problem 84

FIGURE 24-38 Problem 77

85 •• [SSM] Figure 24-40 shows four capacitors connected in the arrangement known as a capacitance bridge. The capacitors are initially uncharged. What must the relation between the four capacitances be so that the potential between points c and d is zero when a voltage V is applied between points a and b?

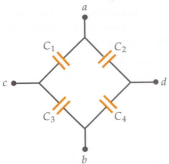

FIGURE 24-40 Problem 85

86 •• Two conducting spheres of radius R are separated by a large distance, compared to their size. One sphere initially has a charge Q, and the other sphere is uncharged. A thin wire is then connected between the spheres. What fraction of the initial energy is dissipated?

87 •• A parallel-plate capacitor of area A and separation distance d is charged to a potential difference V and then disconnected from the charging source. The plates are then pulled apart until the separation is $2d$. find expressions in terms of A, d, and V for (a) the new capacitance, (b) the new potential difference, and (c) the new stored energy. (d) How much work was required to change the plate separation from d to $2d$?

88 •• A parallel-plate capacitor has capacitance C_0 with no dielectric. The capacitor is then filled with dielectric of constant κ. When a second capacitor of capacitance C' is connected in series with the first capacitor, the capacitance of the series combination is C_0. Find C'.

89 •• A Leyden jar, the earliest type of capacitor, is a glass jar coated inside and out with metal foil. Suppose that a Leyden jar is a cylinder 40-cm high, with 2.0-mm-thick walls, and an inner diameter of 8 cm. Ignore any field fringing. (a) Find the capacitance of this Leyden jar, if the dielectric constant κ of the glass is 5. (b) If the dielectric strength of the glass is 15 MV/m, what maximum charge can the Leyden jar carry without undergoing dielectric breakdown? (*Hint: Treat the device as a parallel-plate capacitor.*)

90 •• [SSM] [ISOLVE✓] A parallel-plate capacitor is constructed from a layer of silicon dioxide of thickness 5×10^{-6} m between two conducting films. The dielectric constant of silicon dioxide is 3.8 and its dielectric strength is 8×10^6 V/m. (a) What voltage can be applied across this capacitor without dielectric breakdown? (b) What should the surface area of the layer of silicon dioxide be for a 10-pF capacitor? (c) Estimate the number of these capacitors that can fit into a square 1 cm by 1 cm.

91 •• A parallel combination of two identical 2-μF parallel-plate capacitors is connected to a 100-V battery. The battery is then removed and the separation between the plates of one of the capacitors is doubled. Find the charge on each of the capacitors.

92 •• A parallel-plate capacitor has a capacitance C_0 and a plate separation d. Two dielectric slabs of constants κ_1 and κ_2, each of thickness $\frac{1}{2}d$ and having the same area as the plates, are inserted between the plates as shown in Figure 24-41. When the charge on the plates is Q, find (a) the electric field in each dielectric, and (b) the potential difference between the plates. (c) Show that the new capacitance is given by $C = 2\kappa_1\kappa_2/(\kappa_1 + \kappa_2)C_0$. (d) Show that this system can be considered to be a series combination of two capacitors of thickness $\frac{1}{2}d$ filled with dielectrics of constant κ_1 and κ_2.

FIGURE 24-41 Problem 92

93 •• A parallel-plate capacitor has a plate area A and a separation distance d. A metal slab of thickness t and area A is inserted between the plates. (a) Show that the capacitance is given by $C = \epsilon_0 A/(d - t)$, regardless of where the metal slab is placed. (b) Show that this arrangement can be considered to be a capacitor of separation a in series with one of separation b, where $a + b + t = d$.

94 •• [SSM] A parallel-plate capacitor is filled with two dielectrics of equal size, as shown in Figure 24-42. (a) Show that this system can be considered to be two capacitors of area $\frac{1}{2}A$ connected in parallel. (b) Show that the capacitance is increased by the factor $(\kappa_1 + \kappa_2)/2$.

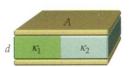

FIGURE 24-42 Problem 94

95 •• A parallel-plate capacitor of plate area A and separation x is given a charge Q and is then removed from the charging source. (a) Find the stored electrostatic energy as a function of x. (b) Find the increase in energy dU due to an increase in plate separation dx from $dU = (dU/dx)\,dx$. (c) If F is the force exerted by one plate on the other, the work needed to move one plate a distance dx is $F\,dx = dU$. Show that $F = Q^2/2\epsilon_0 A$. (d) Show that the force in Part (c) equals $\frac{1}{2}EQ$, where Q is the charge on one plate and E is the electric field between the plates. Discuss the reason for the factor $\frac{1}{2}$ in this result.

96 •• A rectangular parallel-plate capacitor of length a and width b has a dielectric of width b partially inserted a distance x between the plates, as shown in Figure 24-43. (a) Find the capacitance as a function of x. Neglect edge effects. (b) Show that your answer gives the expected results for $x = 0$ and $x = a$.

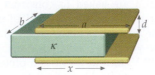

FIGURE 24-43 Problem 96

97 ••• SSM An electrically isolated capacitor with charge Q is partly filled with a dielectric substance as shown in Figure 24-43. The capacitor consists of two rectangular plates of edge lengths a and b separated by distance d. The distance which the dielectric is inserted is x. (*a*) What is the energy stored in the capacitor? (*Hint: the capacitor can be thought of as two capacitors connected in parallel.*) (*b*) Because the energy of the capacitor decreases as x increases, the electric field must be doing positive work on the dielectric, meaning that there must be an electric force pulling it in. Calculate the force by examining how the stored energy varies with x. (*c*) Express the force in terms of the capacitance and voltage. (*d*) Where does this force originate from?

98 •• Two identical, 4-μF parallel-plate capacitors are connected in series across a 24-V battery. (*a*) What is the charge on each capacitor? (*b*) What is the total stored energy of the capacitors? A dielectric that has a dielectric constant of 4.2 is inserted between the plates of one of the capacitors, while the battery is still connected. (*c*) After the dielectric is inserted, what is the charge on each capacitor? (*d*) What is the potential difference across each capacitor? (*e*) What is the total stored energy of the capacitors?

99 •• A parallel-plate capacitor has a plate area A of 1 m² and a plate separation distance d of 0.5 cm. Completely filling the space between the conducting plates is a glass plate that has a dielectric constant of $\kappa = 5$. The capacitor is charged to a potential difference of 12 V and the capacitor is then removed from its charging source. How much work is required to pull the glass plate out of the capacitor?

100 •• A capacitor carries a charge of 15 μC, when the potential between its plates is V. When the charge on the capacitor is increased to 18 μC, the potential between the plates increases by 6 V. Find the capacitance of the capacitor and the initial and final voltages.

101 •• A capacitance balance is shown in Figure 24-44. On one side of the balance, a weight is attached, while on the other side is a capacitor whose two plates are separated by a gap of variable width. When the capacitor is charged to a voltage V, the attractive force between the plates balances the weight of the hanging mass. (*a*) Is the balance stable? That is, if we balance it out, and then move the plates a little closer together, will they snap shut or move back to the equilibrium point? (*b*) Calculate the voltage required to balance a mass M, assuming the plates are separated by distance d and have area A. The force between the plates is given by the derivative of the stored energy with respect to the plate separation. Why is this?

FIGURE 24-44
Problem 101

102 ••• SSM You are asked to construct a parallel-plate, air-gap capacitor that will store 100 kJ of energy. (*a*) What minimum volume is required between the plates of the capacitor? (*b*) Suppose you have developed a dielectric that can withstand 3×10^8 V/m and that has a dielectric constant of $\kappa = 5$. What volume of this dielectric, between the plates of the capacitor, is required for it to be able to store 100 kJ of energy?

103 ••• Consider two parallel-plate capacitors, C_1 and C_2, that are connected in parallel. The capacitors are identical except that C_2 has a dielectric inserted between its plates. A voltage source of 200 V is connected across the capacitors to charge them and, the voltage source is then disconnected. (*a*) What is the charge on each capacitor? (*b*) What is the total stored energy of the capacitors? (*c*) The dielectric is removed from C_2. What is the final stored energy of the capacitors? (*d*) What is the final voltage across the two capacitors?

104 ••• A capacitor is constructed of two concentric cylinders of radii a and b ($b > a$), which has a length $L \gg b$. A charge of $+Q$ is on the inner cylinder, and a charge of $-Q$ is on the outer cylinder. The region between the two cylinders is filled with a dielectric that has a dielectric constant κ. (*a*) Find the potential difference between the cylinders. (*b*) Find the density of the free charge σ_f on the inner cylinder and the outer cylinder. (*c*) Find the bound charge density σ_b on the inner cylindrical surface of the dielectric and on the outer cylindrical surface of the dielectric. (*d*) Find the total stored electrostatic energy. (*e*) If the dielectric will move without friction, how much mechanical work is required to remove the dielectric cylindrical shell?

105 ••• Two parallel-plate capacitors have the same separation distance and plate area. The capacitance of each is initially 10 μF. When a dielectric is inserted, so that it completely fills the space between the plates of one of the capacitors, the capacitance of that capacitor increases to 35 μF. The 35-μF and 10-μF capacitors are connected in parallel and are charged to a potential difference of 100 V. The voltage source is then disconnected. (*a*) What is the stored energy of this system? (*b*) What are the charges on the two capacitors? (*c*) The dielectric is removed from the capacitor. What are the new charges on the plates of the capacitors? (*d*) What is the final stored energy of the system?

106 ••• SSM The two capacitors shown in Figure 24-45 have capacitances $C_1 = 0.4$ μF and $C_2 = 1.2$ μF. The voltages across the two capacitors are V_1 and V_2, respectively, and the total stored energy in the two capacitors is 1.14 mJ. If terminals b and c are connected together, the voltage is $V_a - V_d = 80$ V; if terminal a is connected to terminal b, and terminal c is connected to terminal d, the voltage $V_a - V_d = 20$ V. Find the initial voltages V_1 and V_2.

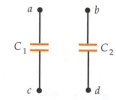

FIGURE 24-45 Problem 106

107 ••• Before Switch S is closed, as shown in Figure 24-46, the voltage across the terminals of the switch is 120 V and the voltage across the 0.2 μF capacitor is 40 V. The total energy stored in the two capacitors is 1440 μJ. After closing the switch, the voltage across each capacitor is 80 V, and the energy stored by the two capacitors has dropped to 960 μJ. Determine the capacitance of C_2 and the charge on that capacitor before the switch was closed.

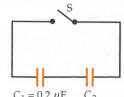

FIGURE 24-46 Problem 107 $C_1 = 0.2$ μF C_2

108 ••• A parallel-plate capacitor of area A and separation distance d is charged to a potential difference V and is then removed from the charging source. A dielectric slab of constant $\kappa = 2$, thickness d, and area $\frac{1}{2}A$ is inserted, as shown in Figure 24-47. Let σ_1 be the free charge density at the conductor–dielectric surface, and let σ_2 be the free charge density at the conductor–air surface. (a) Why must the electric field have the same value inside the dielectric as in the free space between the plates? (b) Show that $\sigma_1 = 2\sigma_2$. (c) Show that the new capacitance is $3\epsilon_0 A/2d$, and that the new potential difference is $\frac{2}{3}V$.

FIGURE 24-47 Problem 108

109 ••• Two identical, $10\text{-}\mu\text{F}$ parallel-plate capacitors are given equal charges of $100\ \mu\text{C}$ each and are then removed from the charging source. The charged capacitors are connected by a wire between their positive plates and by another wire between their negative plates. (a) What is the stored energy of the system? A dielectric that has a dielectric constant of $\kappa = 3.2$ is inserted between the plates of one of the capacitors, so that it completely fills the region between the plates. (b) What is the final charge on each capacitor? (c) What is the final stored energy of the system?

110 ••• **SSM** A capacitor has rectangular plates of length a and width b. The top plate is inclined at a small angle, as shown in Figure 24-48. The plate separation varies from $d = y_0$ at the left to $d = 2y_0$ at the right, where y_0 is much less than a or b. Calculate the capacitance using strips of width dx and length b to approximate differential capacitors of area $b\ dx$ and separation $d = y_0 + (y_0/a)x$ that are connected in parallel.

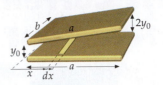

FIGURE 24-48 Problem 110

111 ••• Not all dielectrics that separate the plates of a capacitor are rigid. For example, the membrane of a nerve axon is a bilipid layer that has a finite compressibility. Consider a parallel-plate capacitor whose plate separation is maintained by a dielectric of dielectric constant $\kappa = 3.0$ and thickness $d = 0.2$ mm, when the potential across the capacitor is zero. The dielectric, which has a dielectric strength of 40 kV/mm, is highly compressible, with a Young's modulus[†] for compressive stress of $5 \times 10^6\ \text{N/m}^2$. The capacitance of the capacitor in the limit $V \to 0$ is C_0. (a) Derive an expression for the capacitance, as a function of voltage across the capacitor. (b) What is the maximum voltage that can be applied to the capacitor? (Assume that κ does not change under compression.) (c) What fraction of the total energy of the capacitor is electrostatic field energy and what fraction is mechanical stress energy stored in the compressed dielectric when the voltage across the capacitor is just below the breakdown voltage?

112 ••• A conducting sphere of radius R_1 is given a free charge Q. The sphere is surrounded by an uncharged, concentric spherical dielectric shell that has an inner radius R_1, an outer radius R_2, and a dielectric constant κ. The system is far removed from other objects. (a) Find the electric field everywhere in space. (b) What is the potential of the conducting sphere relative to $V = 0$ at infinity? (c) Find the total electrostatic potential energy of the system.

---

† Young's modulus is discussed in Section 12-8.

Electric Current and Direct-Current Circuits

UNDERSTANDING DIRECT CURRENT CIRCUITS CAN HELP YOU PERFORM POTENTIALLY DANGEROUS TASKS LIKE JUMP-STARTING A VEHICLE.

 **When jump-starting your car, which battery terminals should be connected? (See Example 25-15.)**

When we turn on a light, we connect the wire filament in the lightbulb across a potential difference that causes electric charge to flow through the wire, which is similar to the way a pressure difference in a garden hose causes water to flow through the hose. The flow of electric charge constitutes an electric current. Usually we think of currents as being in conducting wires, but the electron beam in a video monitor and a beam of charged ions from a particle accelerator also constitute electric currents.

➤ **In Chapter 25, we will look at direct current (dc) circuits, which are circuits where the direction of the current in a circuit element does not vary. Direct currents can be produced by batteries connected to resistors and capacitors. In Chapter 29, we discuss alternating current (ac) circuits, in which the direction of the current alternates.**

When a switch is thrown to turn on a circuit, a very small amount of charge accumulates along the surfaces of the wires and other conducting elements of the circuit and these charges produce an electric field that, within the material of the conductors, drives the motion of mobile charges throughout the conducting materials in the circuit. Many complicated changes take place as the current builds up and small charges accumulate at various points in the circuit, but an equilibrium or steady state is quickly established. The time for steady state to be established depends on the size and the conductivity of the elements in the circuit, but the time is practically instantaneous as far as our perceptions are concerned. In steady state, charge no longer continues to accumulate at points along the circuit and the current is steady. (For circuits containing capacitors and resistors, the current may increase or decrease slowly, but appreciable changes occur only over a period of time that is much longer than the time needed to reach the steady state.)

25-1 Current and the Motion of Charges

Electric **current** is defined as the rate of flow of electric charge through a cross-sectional area. Figure 25-1 shows a segment of a current-carrying wire in which charge carriers are moving. If ΔQ is the charge that flows through the cross-sectional area A in time Δt, the current I is

$$I = \frac{\Delta Q}{\Delta t}$$

25-1

DEFINITION—ELECTRIC CURRENT

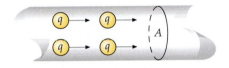

FIGURE 25-1 A segment of a current-carrying wire. If ΔQ is the amount of charge that flows through the cross-sectional area A in time Δt, the current through A is $I = \Delta Q/\Delta t$.

The SI unit of current is the **ampere** (A)[†]:

$1\,A = 1\,C/s$

25-2

By convention, the direction of current is considered to be the direction of flow of positive charge. This convention was established before it was known that free electrons are the particles that actually travel in current-carrying metal wires. Thus, electrons move in the direction *opposite* to the direction of the conventional current. (In an accelerator that produces a proton beam, both the direction of the current and the direction of motion of the positively charged protons are the same.)

In a conducting metal wire, the motion of negatively charged free electrons is quite complex. When there is no electric field in the wire, the free electrons move in random directions with relatively large speeds of the order of 10^6 m/s.[‡] In addition, the electrons collide repeatedly with the lattice ions in the wire. Since the velocity vectors of the electrons are randomly oriented, the *average* velocity is zero. When an electric field is applied, the field exerts a force $-e\vec{E}$ on each free electron, giving it a change in velocity in the direction opposite the field. However, any additional kinetic energy acquired is quickly dissipated by collisions with the lattice ions in the wire. During the time between collisions with the lattice ions, the free electrons, on average, acquire an additional velocity in the direction opposite to the field. The net result of this repeated acceleration and dissipation of energy is that the electrons drift along the wire with a small average velocity, directed opposite to the electric-field direction, called their **drift velocity**. The **drift speed** is the magnitude of the drift velocity.

The motion of the free electrons in a metal is similar to the motion of the molecules of a gas, such as air. In still air, the gas molecules move with large instantaneous velocities (due to their thermal energy), but their average velocity is zero. When there is a breeze, the air molecules have a small average velocity or drift velocity in the direction of the breeze superimposed on their much larger instantaneous velocities. Similarly, when there is no applied electric field, the *electron gas* in a metal has a zero average velocity, but when there is an applied electric field, the electron gas acquires a small drift velocity.

---

† The ampere is operationally defined (see Chapter 26) in terms of the magnetic force that current-carrying wires exert on one another. The coulomb is then defined as the ampere·second.

‡ The average energy of the free electrons in a metal is quite large, even at very low temperatures. These electrons do not have the classical Maxwell–Boltzmann energy distribution and do not obey the classical equipartition theorem. We discuss the energy distribution of these electrons and calculate their average speed in Chapter 38.

Let n be the number of free charge-carrying particles per unit volume in a conducting wire of cross-sectional area A. We call n the **number density** of charge carriers. Assume that each particle carries a charge q and moves with a drift velocity v_d. In a time Δt, all the particles in the volume $Av_d \Delta t$, shown in Figure 25-2 as a shaded region, pass through the area element. The number of particles in this volume is $nAv_d \Delta t$, and the total free charge is

$$\Delta Q = qnAv_d \Delta t$$

The current is thus

$$I = \frac{\Delta Q}{\Delta t} = qnAv_d \qquad \text{25-3}$$

RELATION BETWEEN CURRENT AND DRIFT VELOCITY

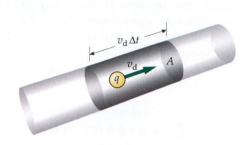

FIGURE 25-2 In time Δt, all the free charges in the shaded volume pass through A. If there are n charge carriers per unit volume, each with charge q, the total free charge in this volume is $\Delta Q = qnAv_d \Delta t$, where v_d is the drift velocity of the charge carriers.

Equation 25-3 can be used to find the current due to the flow of any species of charged particle, simply by substituting the average velocity of the particle species for the drift velocity v_d.

The number density of charge carriers in a conductor can be measured by the Hall effect, which is discussed in Chapter 26. The result is that, in most metals, there is about one free electron per atom.

FINDING THE DRIFT SPEED **EXAMPLE 25-1**

A typical wire for laboratory experiments is made of copper and has a radius 0.815 mm. Calculate the drift speed of electrons in such a wire carrying a current of 1 A, assuming one free electron per atom.

PICTURE THE PROBLEM Equation 25-3 relates the drift speed to the number density of charge carriers, which equals the number density of copper atoms n_a. We can find n_a from the mass density of copper, its molecular mass, and Avogadro's number.

1. The drift velocity is related to the current and number density of charge carriers:

$$I = nqv_d A$$

2. If there is one free electron per atom, the number density of free electrons equals the number density of atoms n_a:

$$n = n_a$$

3. The number density of atoms n_a is related to the mass density ρ_m, Avogadro's number N_A, and the molar mass M. For copper, $\rho = 8.93 \text{ g/cm}^3$ and $M = 63.5 \text{ g/mol}$:

$$n_a = \frac{\rho_m N_A}{M}$$

$$= \frac{(8.93 \text{ g/cm}^3)(6.02 \times 10^{23} \text{ atoms/mol})}{63.5 \text{ g/mol}}$$

$$= 8.47 \times 10^{22} \text{ atoms/cm}^3 = 84.7 \text{ atoms/nm}^3$$

$$= 8.47 \times 10^{28} \text{ atoms/m}^3$$

4. The magnitude of the charge is e, and the area is related to the radius r of the wire:

$$q = e$$

$$A = \pi r^2$$

5. Substituting numerical values yields v_d:

$$v_d = \frac{I}{nqA} = \frac{I}{n_a e \pi r^2}$$

$$= \frac{1 \text{ C/s}}{(8.47 \times 10^{28} \text{ m}^{-3})(1.6 \times 10^{-19} \text{ C})\pi(8.15 \times 10^{-4} \text{ m})^2}$$

$$= 3.54 \times 10^{-5} \text{ m/s} = \boxed{3.54 \times 10^{-2} \text{ mm/s}}$$

REMARKS Typical drift speeds are of the order of a few hundredths of a millimeter per second, quite small by macroscopic standards.

EXERCISE How long would it take for an electron to drift from your car battery to the starter motor, a distance of about 1 m, if its drift speed is 3.5×10^{-5} m/s? (*Answer*   7.9 h)

If electrons drift down a wire at such low speeds, why does an electric light come on instantly when the switch is thrown? A comparison with water in a hose may prove useful. If you attach an empty 100-ft-long hose to a water faucet and turn on the water, it typically takes several seconds for the water to travel the length of the hose to the nozzle. However, if the hose is already full of water when the faucet is opened, the water emerges from the nozzle almost instantaneously. Because of the water pressure at the faucet, the segment of water near the faucet pushes on the water immediately next to it, which pushes on the next segment of water, and so on, until the last segment of water is pushed out the nozzle. This pressure wave moves down the hose at the speed of sound in water, and the water quickly reaches a steady flow rate.

Unlike a water hose, a metal wire is never empty. That is, there are always a very large number of conduction electrons throughout the metal wire. Thus, charge starts moving along the entire length of the wire (including the wire inside the lightbulb) almost immediately after the light switch is thrown. The transport of a significant amount of charge in a wire is accomplished not by a few charges moving rapidly down the wire, but by a very large number of charges slowly drifting down the wire. Surface charges on the wires produce an electric field, and this electric field drives the conduction electrons through the wire.

FINDING THE NUMBER DENSITY   **EXAMPLE   25-2**

In a certain particle accelerator, a current of 0.5 mA is carried by a 5-MeV proton beam that has a radius of 1.5 mm. (*a*) Find the number density of protons in the beam. (*b*) If the beam hits a target, how many protons hit the target in 1 s?

PICTURE THE PROBLEM To find the number density, we use the relation $I = qnAv$ (Equation 25-3), where v is the drift speed of the charge carriers. (The drift speed is the magnitude of the average velocity.) We can find v from the energy. The amount of charge Q that hits the target in time Δt is $I\Delta t$, and the number N of protons that hits the target is Q divided by the charge per proton.

(*a*) 1. The number density is related to the current, the charge, the cross-sectional area, and the speed:

$$I = qnAv$$

2. We find the speed of the protons from their kinetic energy:

$$K = \tfrac{1}{2}mv^2 = 5\,\text{MeV}$$

3. Use $m = 1.67 \times 10^{-27}$ kg for the mass of a proton, and solve for the speed:

$$v = \sqrt{\frac{2K}{m}} = \sqrt{\frac{(2)(5 \times 10^6\,\text{eV})}{1.6 \times 10^{-27}\,\text{kg}} \times \frac{1.6 \times 10^{-19}\,\text{J}}{1\,\text{eV}}}$$

$$= \boxed{3.10 \times 10^7\,\text{m/s}}$$

4. Substitute to calculate n:

$$n = \frac{I}{qAv}$$

$$= \frac{0.5 \times 10^{-3}\,\text{A}}{(1.6 \times 10^{-19}\,\text{C/proton})\,\pi(1.5 \times 10^{-3}\,\text{m})^2(3.10 \times 10^7\,\text{m/s})}$$

$$= \boxed{1.43 \times 10^{13}\,\text{protons/m}^3}$$

(b) 1. The number of protons N that hit the target in 1 s is related to the total charge ΔQ that hits in 1 s and the proton charge q:

$$\Delta Q = Nq$$

2. The charge ΔQ that strikes the target in time Δt is the current times the time:

$$\Delta Q = I\Delta t$$

3. The number of protons is then:

$$N = \frac{\Delta Q}{q} = \frac{I\Delta t}{q} = \frac{(0.5 \times 10^{-3}\,\text{A})(1\,\text{s})}{1.6 \times 10^{-19}\,\text{C/proton}}$$

$$= \boxed{3.13 \times 10^{15}\,\text{protons}}$$

PLAUSIBILITY CHECK The number of protons N hitting the target in time Δt is the number in the volume $Av\,\Delta t$. Then $N = nAv\,\Delta t$. Substituting $n = I/(qAv)$ then gives $N = nAv\,\Delta t = [I/(qAv)](Av)\,\Delta t = I\,\Delta t/q = \Delta Q/q$, which is what we used in Part (b).

REMARKS We were able to use the classical expression for kinetic energy in step 2 without taking relativity into consideration, because the proton kinetic energy of 5 MeV is much less than the proton rest energy (about 931 MeV). The speed found, 3.1×10^7 m/s, is about one-tenth the speed of light.

EXERCISE Using the number density found in Part (a), how many protons are there in a volume of 1 mm³ of the space containing the beam? (*Answer* 14,300)

25-2 Resistance and Ohm's Law

Current in a conductor is driven by an electric field $\vec{E}$ inside the conductor that exerts a force $q\vec{E}$ on the free charges. (In electrostatic equilibrium, the electric field must be zero inside a conductor, but when a conductor carries a current, it is no longer in electrostatic equilibrium and the free charge drifts down the conductor, driven by the electric field.) Since the direction of the force on a positive charge is also the direction of the electric field, $\vec{E}$ is in the direction of the current.

Figure 25-3 shows a wire segment of length ΔL and cross-sectional area A carrying a current I. Since the electric field points in the direction of decreasing potential, the potential at point a is greater than the potential at point b. If we think of the current as the flow of positive charge, these positive charges move in the direction of decreasing potential. Assuming the electric field $\vec{E}$ to be uniform throughout the segment, the **potential drop** V between points a and b is

$$V = V_a - V_b = E\,\Delta L \qquad\qquad 25\text{-}4$$

The ratio of the potential drop to the current is called the **resistance** of the segment.

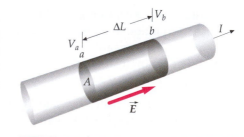

FIGURE 25-3 A segment of wire carrying a current I. The potential drop is related to the electric field by $V_a - V_b = E\,\Delta L$.

$$R = \frac{V}{I} \qquad\qquad 25\text{-}5$$

DEFINITION—RESISTANCE

The SI unit of resistance, the volt per ampere, is called an **ohm** (Ω):

$$1\,\Omega = 1\,\text{V/A} \qquad\qquad 25\text{-}6$$

For many materials, the resistance does not depend on the potential drop or the current. Such materials, which include most metals, are called **ohmic materials.**

For ohmic materials, the potential drop across a segment is proportional to the current:

$$V = IR, \quad R \text{ constant} \qquad 25\text{-}7$$

OHM'S LAW

For **nonohmic materials,** the resistance depends on the current I, so V is not proportional to I. Figure 25-4 shows the potential difference V versus the current I for ohmic and nonohmic materials. For ohmic materials (Figure 25-4a), the relation is linear, but for nonohmic materials (Figure 25-4b), the relation is not linear. Ohm's law is not a fundamental law of nature, like Newton's laws or the laws of thermodynamics, but rather is an empirical description of a property shared by many materials.

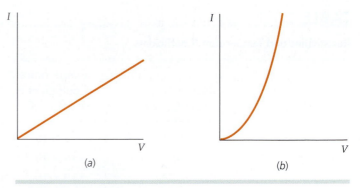

(a)   (b)

FIGURE 25-4 Plots of V versus I for (a) ohmic and (b) nonohmic materials. The resistance $R = V/I$ is independent of I for ohmic materials, as indicated by the constant slope of the line in Figure 25-4a.

EXERCISE A wire of resistance 3 Ω carries a current of 1.5 A. What is the potential drop across the wire? (*Answer* 4.5 V)

The resistance of a conducting wire is found to be proportional to the length of the wire and inversely proportional to its cross-sectional area:

$$R = \rho\frac{L}{A} \qquad 25\text{-}8$$

where the proportionality constant ρ is called the **resistivity** of the conducting material.[†] The unit of resistivity is the ohm-meter ($\Omega\cdot$m). Note that Equation 25-7 and Equation 25-8 for electrical conduction and electrical resistance are of the same form as Equation 20-9 ($\Delta T = IR$) and Equation 20-10 [$R = \Delta x/(kA)$] for thermal conduction and thermal resistance. For the electrical equations, the potential difference V replaces the temperature difference ΔT and $1/\rho$ replaces the thermal conductivity k. (In fact, $1/\rho$ is called the electrical conductivity.[‡]) Ohm was led to his law by the similarity between the conduction of electricity and the conduction of heat.

THE LENGTH OF A 2-Ω RESISTOR   **EXAMPLE 25-3**

A Nichrome wire ($\rho = 10^{-6}$ $\Omega\cdot$m) has a radius of 0.65 mm. What length of wire is needed to obtain a resistance of 2.0 Ω?

Solve $R = \rho L/A$ (Equation 25-8) for L:
$$L = \frac{RA}{\rho} = \frac{(2\ \Omega)\pi(6.5 \times 10^{-4}\ \text{m})^2}{10^{-6}\ \Omega\cdot\text{m}} = \boxed{2.65\ \text{m}}$$

The resistivity of any given metal depends on the temperature. Figure 25-5 shows the temperature dependence of the resistivity of copper. This graph is nearly a straight line, which means that the resistivity varies nearly linearly with temperature.[§] In tables, the resistivity is usually given in terms of its value at 20°C, ρ_{20}, along with the **temperature coefficient of resistivity,** α, which is defined by

$$\alpha = \frac{(\rho - \rho_{20})/\rho_{20}}{t_\text{C} - 20°\text{C}} \qquad 25\text{-}9$$

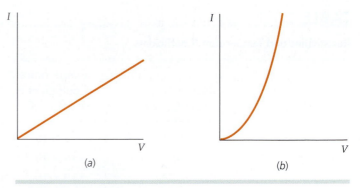

FIGURE 25-5 Plot of resistivity ρ versus temperature for copper. Since the Celsius and absolute temperatures differ only in the choice of zero, the resistivity has the same slope whether it is plotted against t or T.

---

† The symbol ρ used here for the resistivity was used in previous chapters for volume charge density. Care must be taken to distinguish what quantity ρ refers to. Usually this will be clear from the context.
‡ The unit of conductivity is the siemens (S), 1 siemens $= 1\ \Omega^{-1}\cdot$m^{-1}.
§ There is a breakdown in this linearity for all metals at very low temperatures that is not shown in Figure 25-5.

TABLE 25-1

Resistivities and Temperature Coefficients

Material	Resistivity ρ at 20°C, $\Omega \cdot$m	Temperature Coefficient α at 20°C, K^{-1}
Silver	1.6×10^{-8}	3.8×10^{-3}
Copper	1.7×10^{-8}	3.9×10^{-3}
Aluminum	2.8×10^{-8}	3.9×10^{-3}
Tungsten	5.5×10^{-8}	4.5×10^{-3}
Iron	10×10^{-8}	5.0×10^{-3}
Lead	22×10^{-8}	4.3×10^{-3}
Mercury	96×10^{-8}	0.9×10^{-3}
Nichrome	100×10^{-8}	0.4×10^{-3}
Carbon	3500×10^{-8}	-0.5×10^{-3}
Germanium	0.45	-4.8×10^{-2}
Silicon	640	-7.5×10^{-2}
Wood	$10^8 - 10^{14}$	
Glass	$10^{10} - 10^{14}$	
Hard rubber	$10^{13} - 10^{16}$	
Amber	5×10^{14}	
Sulfur	1×10^{15}	

TABLE 25-2

Wire Diameters and Cross-Sectional Areas for Commonly Used Copper Wires

Gauge Number	Diameter at 20°C, mm	Area, mm^2
4	5.189	21.15
6	4.115	13.30
8	3.264	8.366
10	2.588	5.261
12	2.053	3.309
14	1.628	2.081
16	1.291	1.309
18	1.024	0.8235
20	0.8118	0.5176
22	0.6438	0.3255

Table 25-1 gives the resistivity ρ at 20°C and the temperature coefficient α at 20°C for various materials. Note the tremendous range of values for ρ.

Electrical wires are manufactured in standard sizes. The diameter of the circular cross section is indicated by a *gauge number,* with higher numbers corresponding to smaller diameters, as can be seen from Table 25-2.

RESISTANCE PER UNIT LENGTH　　　　　　　**E X A M P L E   2 5 - 4**

Calculate the resistance per unit length of a 14-gauge copper wire.

1. From Equation 25-8, the resistance per unit length equals the resistivity per unit area:

$$R = \rho \frac{L}{A}$$

so

$$\frac{R}{L} = \frac{\rho}{A}$$

2. Find the resistivity of copper from Table 25-1 and the area from Table 25-2:

$$\rho = 1.7 \times 10^{-8} \; \Omega \cdot m$$

$$A = 2.08 \; mm^2$$

3. Use these values to find R/L:

$$\frac{R}{L} = \frac{\rho}{A} = \frac{1.7 \times 10^{-8} \; \Omega \cdot m}{2.08 \times 10^{-6} \; m^2} = \boxed{8.17 \times 10^{-3} \; \Omega/m}$$

REMARKS 14-gauge copper wire is commonly used for household lighting circuits. The resistance of a 100-W, 120-V lightbulb filament is 144 Ω and the resistance of a 100 m of the wire is 0.817 Ω, so the resistance of the wire is negligible compared to the resistance of the lightbulb filament.

Carbon, which has a relatively high resistivity, is used in resistors found in electronic equipment. Resistors are often marked with colored stripes that indicate their resistance value. The code for interpreting these colors is given in Table 25-3.

TABLE 25-3

The Color Code for Resistors and Other Devices

Colors		Numeral	Tolerance		
Black	=	0	Brown	=	1 %
Brown	=	1	Red	=	2 %
Red	=	2	Gold	=	5 %
Orange	=	3	Silver	=	10 %
Yellow	=	4	None	=	20 %
Green	=	5			
Blue	=	6			
Violet	=	7			
Gray	=	8			
White	=	9			

The color bands are read starting with the band closest to the end of the resistor. The first two bands represent an integer between 1 and 99. The third band represents the number of zeros that follow. For the resistor shown, the colors of the first three bands are, respectively, orange, black, and blue. Thus, the number is 30,000,000 and the resistance is 30 MΩ. The fourth band is the tolerance band. If the fourth band is silver, as shown here, the tolerance is 10 percent. Ten percent of 30 is 3, so the resistance is (30 ± 3) MΩ.

Color-coded carbon resistors on a circuit board.

THE ELECTRIC FIELD THAT DRIVES THE CURRENT   **EXAMPLE 25-5**

Find the electric field strength E in the 14-gauge copper wire of Example 25-4 when the wire is carrying a current of 1.3 A.

PICTURE THE PROBLEM We find the electric field strength as the potential drop for a given length of wire, $E = V/L$. The potential drop is found using Ohm's law, $V = IR$, and the resistance per length is given in Example 25-4.

1. The electric field strength equals the potential drop per unit length:

$$E = \frac{V}{L}$$

2. Write Ohm's law for the potential drop:

$$V = IR$$

3. Substitute this expression into the equation for E:

$$E = \frac{V}{L} = \frac{IR}{L} = I\frac{R}{L}$$

4. Substitute the value of R/L found in Example 25-4 to calculate E:

$$E = I\frac{R}{L} = (1.3 \text{ A})(8.17 \times 10^{-3} \ \Omega/\text{m}) = \boxed{1.06 \times 10^{-2} \text{ V/m}}$$

REMARKS Since $R/L = \rho/A$, $E = I\rho/A$, which is the same throughout the length of the wire. Thus, E is uniform throughout the length of the wire.

25-3 Energy in Electric Circuits

When there is an electric field in a conductor, the *electron gas* gains kinetic energy due to the work done on the free electrons by the field. However, steady state is soon achieved as the kinetic energy gain is continuously dissipated into the thermal energy of the conductor by collisions between the electrons and the lattice ions of the conductor. This mechanism for increasing the thermal energy of a conductor is called **Joule heating.**

Consider the segment of wire of length L and cross-sectional area A shown in Figure 25-6a. The wire is carrying a steady current to the right. Consider the free charge Q initially in the segment. During time Δt, this free charge undergoes a small displacement to the right (Figure 25-6b). This displacement is equivalent to an amount of charge ΔQ (Figure 25-6c) being moved from its left end, where it had potential energy $\Delta Q\, V_a$, to its right end, where it has potential energy $\Delta Q\, V_b$. The net change in the potential energy of Q is thus

$$\Delta U = \Delta Q(V_b - V_a)$$

since $V_a > V_b$, this represents a net loss in the potential energy of Q. The potential energy lost is then

$$-\Delta U = \Delta Q\, V$$

where $V = V_a - V_b$ is the potential drop across the segment. The rate of potential energy loss is

$$-\frac{\Delta U}{\Delta t} = \frac{\Delta Q}{\Delta t}V = IV$$

where $I = \Delta Q/\Delta t$ is the current. The potential energy loss per unit time is the power P dissipated in the conducting segment:

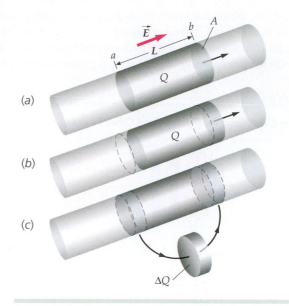

FIGURE 25-6 During a time Δt, an amount of charge ΔQ passes point a, where the potential is V_a. During the same time interval, an equal amount of charge leaves the segment, passing point b, where the potential is V_b. The net effect during time Δt is: the charge Q that was initially in the segment both loses an amount of potential energy equal to $\Delta Q\, V_a$, and gains an amount equal to $\Delta Q\, V_b$. This amounts to a net decrease in potential energy since $V_a > V_b$.

$$P = IV \qquad\qquad\qquad 25\text{-}10$$

POTENTIAL ENERGY LOSS PER UNIT TIME

If V is in volts and I is in amperes, the power is in watts. The power loss is the product of the decrease in potential energy per unit charge, V, and the charge flowing per unit time, I. Equation 25-10 applies to any device in a circuit. The rate at which potential energy is delivered to the device is the product of the potential drop across the device and the current through the device. In a conductor, the potential energy is dissipated as thermal energy in the conductor. Using $V = IR$, or $I = V/R$, we can write Equation 25-10 in other useful forms

$$P = IV = I^2R = \frac{V^2}{R} \qquad\qquad 25\text{-}11$$

POWER DISSIPATED IN A RESISTOR

POWER DISSIPATED IN A RESISTOR   **EXAMPLE 25-6**

A 12-Ω resistor carries a current of 3 A. Find the power dissipated in this resistor.

PICTURE THE PROBLEM Since we are given the current and the resistance, but not the potential drop, $P = I^2R$ is the most convenient equation to use. Alternatively, we could find the potential drop from $V = IR$, then use $P = IV$.

Compute I^2R: $P = I^2R = (3\text{ A})^2(12\text{ Ω}) = \boxed{108\text{ W}}$

PLAUSIBILITY CHECK The potential drop across the resistor is $V = IR = (3\text{ A})(12\text{ Ω}) = 36$ V. We can use this to find the power from $P = IV = (3\text{ A})(36\text{ V}) = 108$ W.

EXERCISE A wire of resistance 5 Ω carries a current of 3 A for 6 s. (*a*) How much power is put into the wire? (*b*) How much thermal energy is produced? (*Answer* (*a*) 45 W, (*b*) 270 J)

EMF and Batteries

To maintain a steady current in a conductor, we need a constant supply of electrical energy. A device that supplies electrical energy to a circuit is called a **source of emf.** (The letters *emf* stand for *electromotive force*, a term that is now rarely used. The term is something of a misnomer because it is definitely not a force.) Examples of emf sources are a battery, which converts chemical energy into electrical energy, and a generator, which converts mechanical energy into electrical energy. A source of emf does work on the charge passing through it, raising the potential energy of the charge. The work per unit charge is called the **emf** $\mathcal{E}$ of the source. The unit of emf is the volt, the same as the unit of potential difference. An **ideal battery** is a source of emf that maintains a constant potential difference between its two terminals, independent of the current through the battery. The potential difference between the terminals of an ideal battery is equal in magnitude to the emf of the battery.

Figure 25-7 shows a simple circuit consisting of a resistance R connected to an ideal battery. The resistance is indicated by the symbol -W-. The straight lines indicate connecting wires of negligible resistance. The source of emf ideally maintains a constant potential difference equal to $\mathcal{E}$ between points a and b, with point a being at the higher potential. There is negligible potential difference between points a and c and between points d and b, because the connecting wire is assumed to have negligible resistance. The potential drop from points c to d is therefore equal in magnitude to the emf $\mathcal{E}$, and the current through the resistor is given by $I = \mathcal{E}/R$. The direction of the current in this circuit is clockwise, as shown in the figure.

Note that *inside* the source of emf, the charge flows from a region of low potential to a region of high potential, so it gains potential energy.[†] When charge ΔQ flows through the source of emf $\mathcal{E}$, its potential energy is increased by the amount $\Delta Q\,\mathcal{E}$. The charge then flows through the resistor, where this potential energy is

The electric ray has two large electric organs on each side of its head, where current passes from the lower to the upper surface of the body. These organs are composed of columns, with each column consisting of one hundred forty to half a million gelatinous plates. In saltwater fish, these batteries are connected in parallel, whereas in freshwater fish the batteries are connected in series, transmitting discharges of higher voltage. Fresh water has a higher resistivity than salt water, so to be effective a higher voltage is required. It is with such a battery that an average electric ray can electrocute a fish, delivering 50 A at 50 V.

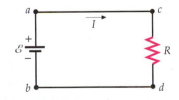

FIGURE 25-7 A simple circuit consisting of an ideal battery of emf $\mathcal{E}$, a resistance R, and connecting wires that are assumed to be of negligible resistance.

---

† When a battery is being charged by a generator or by another battery, the charge flows from a high-potential to a low-potential region within the battery being charged, thus losing electrostatic potential energy. The energy lost is converted to chemical energy and stored in the battery being charged.

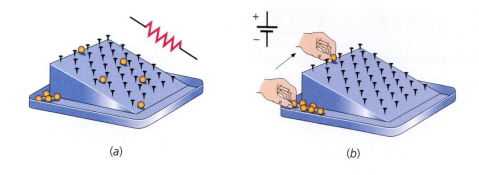

(a)   (b)

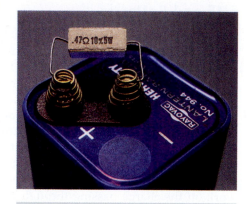

FIGURE 25-8 A mechanical analog of a simple circuit consisting of a resistance and source of emf. (*a*) The marbles start at some height *h* above the bottom and are accelerated between collisions with the nails by the gravitational field. The nails are analogous to the lattice ions in the resistor. During the collisions, the marbles transfer the kinetic energy they obtained between collisions to the nails. Because of the many collisions, the marbles have only a small, approximately constant, drift velocity toward the bottom. (*b*) When the marbles reach the bottom, a child picks them up, lifts them to their original height *h*, and starts them again. The child, who does work *mgh* on each marble, is analogous to the source of emf. The energy source in this case is the internal chemical energy of the child.

dissipated as thermal energy. The rate at which energy is supplied by the source of emf is the power output:

$$P = \frac{\Delta Q \mathcal{E}}{\Delta t} = I\mathcal{E}$$

25-12

POWER SUPPLIED BY AN EMF SOURCE

In the simple circuit of Figure 25-7, the power output by the source of emf equals that dissipated in the resistor.

A source of emf can be thought of as a charge pump that pumps the charge from a region of low potential energy to a region of higher potential energy. Figure 25-8 shows a mechanical analog of the simple electric circuit just discussed.

In a **real battery,** the potential difference across the battery terminals, called the **terminal voltage,** is not simply equal to the emf of the battery. Consider the circuit consisting of a real battery and a resistor in Figure 25-9. If the current is varied by varying the resistance *R* and the terminal voltage is measured, the terminal voltage is found to decrease slightly as the current increases (Figure 25-10), just as if there were a small resistance within the battery.

Thus, we can consider a real battery to consist of an ideal battery of emf $\mathcal{E}$ plus a small resistance *r*, called the **internal resistance** of the battery.

The circuit diagram for a real battery and resistor is shown in Figure 25-11. If the current in the circuit is *I*, the potential at point *a* is related to the potential at point *b* by

$$V_a = V_b + \mathcal{E} - Ir$$

The terminal voltage is thus

$$V_a - V_b = \mathcal{E} - Ir$$ 25-13

The terminal voltage of the battery decreases linearly with current, as we saw in Figure 25-10. The potential drop across the resistor *R* is *IR* and is equal to the terminal voltage:

$$IR = V_a - V_b = \mathcal{E} - Ir$$

Solving for the current *I*, we obtain

$$I = \frac{\mathcal{E}}{R + r}$$ 25-14

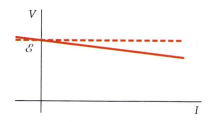

FIGURE 25-9 A simple circuit consisting of a real battery, a resistor, and connecting wires.

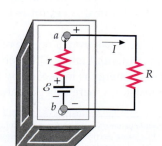

FIGURE 25-10 Terminal voltage *V* versus *I* for a real battery. The dashed line shows the terminal voltage of an ideal battery, which has the same magnitude as $\mathcal{E}$.

FIGURE 25-11 Circuit diagram for the circuit shown in Figure 25-9. A real battery can be represented by an ideal battery of emf $\mathcal{E}$ and a small resistance *r*.

If a battery is connected as shown in Figure 25-11, the terminal voltage given by Equation 25-13 is less than the emf of the battery because of the potential drop across the internal resistance of the battery. Real batteries, such as a good car battery, usually have an internal resistance of the order of a few hundredths of an ohm, so the terminal voltage is nearly equal to the emf unless the current is very large. One sign of a bad battery is an unusually high internal resistance. If you suspect that your car battery is bad, checking the terminal voltage with a voltmeter, which draws very little current, is not always sufficient. You need to check the terminal voltage while current is being drawn from the battery, such as while you are trying to start your car. Then the terminal voltage may drop considerably, indicating a high internal resistance and a bad battery.

Batteries are often rated in ampere-hours (A·h), which is the total charge that batteries can deliver:

$$1 \text{ A·h} = (1 \text{ C/s})(3600 \text{ s}) = 3600 \text{ C}$$

The total energy stored in the battery is the product of the emf and the total charge it can deliver:

$$W = Q\mathcal{E} \qquad\qquad\qquad 25\text{-}15$$

TERMINAL VOLTAGE, POWER, AND STORED ENERGY **EXAMPLE 25-7**

An 11-Ω resistor is connected across a battery of emf 6 V and internal resistance 1 Ω. Find (a) the current, (b) the terminal voltage of the battery, (c) the power delivered by the emf source, (d) the power delivered to the external resistor, and (e) the power dissipated by the battery's internal resistance. (f) If the battery is rated at 150 A·h, how much energy does the battery store?

PICTURE THE PROBLEM The circuit diagram is the same as the circuit diagram shown in Figure 25-11. We find the current from Equation 25-14 and then use it to find the terminal voltage and power delivered to the resistors.

1. Equation 25-14 gives the current: $\qquad I = \dfrac{\mathcal{E}}{R + r} = \dfrac{6 \text{ V}}{11 \ \Omega + 1 \ \Omega} = \boxed{0.5 \text{ A}}$

2. Use the current to calculate the terminal voltage of the battery: $\qquad V_a - V_b = \mathcal{E} - Ir = 6 \text{ V} - (0.5 \text{ A})(1 \ \Omega) = \boxed{5.5 \text{ V}}$

3. The power delivered by the source of emf equals $\mathcal{E}I$: $\qquad P = \mathcal{E}I = (6 \text{ V})(0.5 \text{ A}) = \boxed{3 \text{ W}}$

4. The power delivered to and dissipated by the external resistance equals I^2R: $\qquad I^2R = (0.5 \text{ A})^2(11 \ \Omega) = \boxed{2.75 \text{ W}}$

5. The power dissipated in the internal resistance is I^2r. $\qquad I^2r = (0.5 \text{ A})^2(1 \ \Omega) = \boxed{0.25 \text{ W}}$

6. The total energy stored is the emf times the total charge it can deliver: $\qquad W = Q\mathcal{E} = 150 \text{ A·h} \times \dfrac{3600 \text{ C}}{1 \text{ A·h}} \times 6 \text{ V} = \boxed{3.24 \text{ MJ}}$

REMARKS The value of the internal resistance is exaggerated in this example to simplify calculations. In other examples, we may simply ignore the internal resistance. Of the 3 W of power delivered by the battery, 2.75 W is dissipated in the resistor and 0.25 W is dissipated in the internal resistance of the battery.

MAXIMUM POWER DELIVERED | **EXAMPLE 25-8** **Try It Yourself**

For a battery of given emf $\mathcal{E}$ and internal resistance r, what value of external resistance R should be placed across the terminals to obtain the maximum power delivered to the resistor?

PICTURE THE PROBLEM The circuit diagram is the same as the circuit diagram shown in Figure 25-11. The power input to R is I^2R, where $I = \mathcal{E}/(R + r)$. To find the maximum power, we compute dP/dR and set it equal to zero.

Cover the column to the right and try these on your own before looking at the answers.

Steps

Answers

1. Use Equation 25-14 to eliminate I from $P = I^2R$ so that P is written as a function of R and the constants $\mathcal{E}$ and r only.

$$P = \frac{\mathcal{E}^2 R}{(R + r)^2} = \mathcal{E}^2 R(R + r)^{-2}$$

2. Calculate the derivative dP/dR. Use the product rule.

$$\frac{dP}{dR} = \mathcal{E}^2(R + r)^{-2} - 2\mathcal{E}^2 R(R + r)^{-3}$$

3. Set $dP/dR = 0$ and solve for R in terms of r.

$$R = r$$

REMARKS The maximum value of P occurs when $R = r$, that is, when the load resistance equals the internal resistance. A similar result holds for alternating current circuits. Choosing $R = r$ to maximize the power delivered to the load is known as *impedance matching*. A graph of P versus R is shown in Figure 25-12.

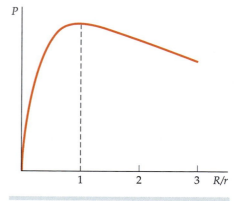

25-4 Combinations of Resistors

The analysis of a circuit can often be simplified by replacing two or more resistors with a single equivalent resistor that carries the same current with the same potential drop as the original resistors. The replacement of a set of resistors by an equivalent resistor is similar to the replacement of a set of capacitors by an equivalent capacitor, discussed in Chapter 24.

FIGURE 25-12 The power delivered to the external resistor is maximum if $R = r$.

Resistors in Series

When two or more resistors are connected like R_1 and R_2 in Figure 25-13 so that they carry the same current I, the resistors are said to be connected in series. The potential drop across R_1 is IR_1 and the potential drop across R_2 is IR_2. The potential drop across the two resistors is the sum of the potential drops across the individual resistors:

$$V = IR_1 + IR_2 = I(R_1 + R_2) \qquad \text{25-16}$$

The single equivalent resistance R_{eq} that gives the same total potential drop V when carrying the same current I is found by setting V equal to IR_{eq} (Figure 25-13b). Then R_{eq} is given by

$$R_{eq} = R_1 + R_2$$

FIGURE 25-13 (a) Two resistors in series carry the same current. (b) The resistors in Figure 25-13a can be replaced by a single equivalent resistance $R_{eq} = R_1 + R_2$ that gives the same total potential drop when carrying the same current as in Figure 25-13a.

(a)

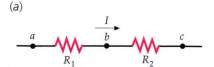

(b)

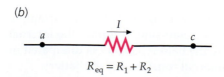

When there are more than two resistors in series, the equivalent resistance is

$$R_{eq} = R_1 + R_2 + R_3 + \ldots \qquad \text{25-17}$$

EQUIVALENT RESISTANCE FOR RESISTORS IN SERIES

Resistors in Parallel

Two resistors that are connected, as in Figure 25-14a, so that they have the same potential difference across them, are in parallel. Note that the resistors are connected at both ends by wires. Let I be the current leading to point a. At point a the circuit branches out into two branches, and the current I divides into two parts, with current I_1 in the upper branch containing resistor R_1, and with current I_2 in the lower branch containing R_2. The two **branch currents** sum to the current in the wire leading into point a:

$$I = I_1 + I_2 \qquad \text{25-18}$$

At point b the branch currents recombine so the current in the wire following point b is also equal to $I = I_1 + I_2$. The potential drop across either resistor, $V = V_a - V_b$, is related to the currents by

$$V = I_1 R_1 = I_2 R_2 \qquad \text{25-19}$$

The equivalent resistance for parallel resistors is the resistance R_{eq} for which the same total current I requires the same potential drop V (Figure 25-14b):

$$R_{eq} = \frac{V}{I}$$

Solving this equation for I and using $I = I_1 + I_2$, we have

$$I = \frac{V}{R_{eq}} = I_1 + I_2 = \frac{V}{R_1} + \frac{V}{R_2} = V\left(\frac{1}{R_1} + \frac{1}{R_2}\right) \qquad \text{25-20}$$

where we have used Equation 25-19 for I_1 and I_2. The equivalent resistance for two resistors in parallel is therefore given by

$$\frac{1}{R_{eq}} = \frac{1}{R_1} + \frac{1}{R_2}$$

This result can be generalized for combinations, such as that in Figure 25-15, in which three or more resistors are connected in parallel:

$$\frac{1}{R_{eq}} = \frac{1}{R_1} + \frac{1}{R_2} + \frac{1}{R_3} + \ldots \qquad \text{25-21}$$

EQUIVALENT RESISTANCE FOR RESISTORS IN PARALLEL

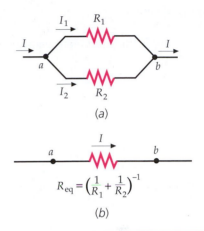

$$R_{eq} = \left(\frac{1}{R_1} + \frac{1}{R_2}\right)^{-1}$$

(b)

FIGURE 25-14 (a) Two resistors are in parallel when they are connected together at both ends so that the potential drop is the same across each. (b) The two resistors in Figure 25-14a can be replaced by an equivalent resistance R_{eq} that is related to R_1 and R_2 by $1/R_{eq} = 1/R_1 + 1/R_2$.

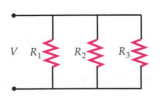

FIGURE 25-15 Three resistors in parallel.

EXERCISE A 2-Ω resistor and a 4-Ω resistor are connected (a) in series and (b) in parallel. Find the equivalent resistances for both cases. (*Answer* (a) 6 Ω, (b) 1.33 Ω)

RESISTORS IN PARALLEL

EXAMPLE 25-9

A battery applies a potential difference of 12 V across the parallel combination of 4-Ω and 6-Ω resistors shown in Figure 25-16. Find (*a*) the equivalent resistance, (*b*) the total current, (*c*) the current through each resistor, (*d*) the power dissipated in each resistor, and (*e*) the power delivered by the battery.

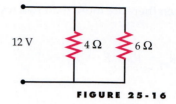

FIGURE 25-16

PICTURE THE PROBLEM Choose symbols and directions for the currents in Figure 25-17.

(*a*) Calculate the equivalent resistance:

$$\frac{1}{R_{eq}} = \frac{1}{4\,\Omega} + \frac{1}{6\,\Omega} = \frac{3}{12\,\Omega} + \frac{2}{12\,\Omega} = \frac{5}{12\,\Omega}$$

$$R_{eq} = \frac{12\,\Omega}{5} = \boxed{2.4\,\Omega}$$

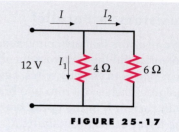

FIGURE 25-17

(*b*) The total current is the potential drop divided by the equivalent resistance:

$$I = \frac{V}{R_{eq}} = \frac{12\,V}{2.4\,\Omega} = \boxed{5\,A}$$

(*c*) We obtain the current through each resistor using Equation 25-19 and the fact that the potential drop is 12 V across the parallel combination:

$$I_1 = \frac{12\,V}{4\,\Omega} = \boxed{3\,A}$$

$$I_2 = \frac{12\,V}{6\,\Omega} = \boxed{2\,A}$$

(*d*) Use these currents to find the power dissipated in each resistor:

$$P_1 = I_1^2 R = (3\,A)^2(4\,\Omega) = \boxed{36\,W}$$

$$P_2 = I_2^2 R = (2\,A)^2(6\,\Omega) = \boxed{24\,W}$$

(*e*) Use $P = VI$ to find the power delivered by the battery:

$$P = VI = (12\,V)(5\,A) = \boxed{60\,W}$$

PLAUSIBILITY CHECK The power delivered by the battery equals the power dissipated in the two resistors $P = 60\,W = 36\,W + 24\,W$. In step 4, we could have calculated the power dissipated in each resistor from $P_4 = VI_4 = (12\,V)(3\,A) = 36\,W$ and $P_6 = VI_6 = (12\,V)(2\,A) = 24\,W$.

RESISTORS IN SERIES

EXAMPLE 25-10 Try It Yourself

A 4-Ω resistor and a 6-Ω resistor are connected in series to a battery of emf 12 V with negligible internal resistance. Find (*a*) the equivalent resistance of the two resistors, (*b*) the current in the circuit, (*c*) the potential drop across each resistor, (*d*) the power dissipated in each resistor, and (*e*) the total power dissipated.

Cover the column to the right and try these on your own before looking at the answers.

Steps	Answers
(*a*) 1. Draw a circuit diagram (Figure 25-18).	
2. Calculate R_{eq} for the two series resistors.	$R_{eq} = \boxed{10\,\Omega}$
(*b*) Use $V = IR_{eq}$ to find the current through the battery.	$I = \boxed{1.2\,A}$

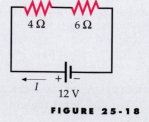

FIGURE 25-18

(c) Use Ohm's law to find the potential drop across each resistor. $V_4 =$ ☐ 4.8 V , $V_6 =$ ☐ 7.2 V

(d) Find the power dissipated in each resistor using $P = I^2R$. Check your result using $P = IV$ for each resistor. $P_4 =$ ☐ 5.76 W , $P_6 =$ ☐ 8.64 W

(e) Add your results from Part (d) to find the total power. Check your result, using $P = IV$ and $P = I^2R_{eq}$. $P =$ ☐ 14.4 W

REMARKS Note that much less power is dissipated in the series circuit than in the corresponding parallel circuit of Example 25-9.

Note from Example 25-9 that the equivalent resistance of two parallel resistances is less than the resistance of either resistor alone. This is a general result. Suppose we have a single resistor R_1 carrying current I_1 with potential drop $V = I_1R_1$. If we add a second resistor in parallel, it will carry some additional current I_2 without affecting I_1. The equivalent resistance is $V/(I_1 + I_2)$, which is less than $R_1 = V/I_1$. Note also from Example 25-9 that the ratio of the currents in the two parallel resistors equals the inverse ratio of the resistances. This general result follows from Equation 25-19:

$$I_1R_1 = I_2R_2$$

$$\frac{I_1}{I_2} = \frac{R_2}{R_1} \text{ (parallel resistors)}$$ 25-22

SERIES AND PARALLEL COMBINATIONS   **E X A M P L E   2 5 - 1 1**   **Try It Yourself**

Consider the circuit in Figure 25-19. When the switch S_1 is open and switch S_2 is closed, find (a) the equivalent resistance of the circuit, (b) the total current in the source of emf, (c) the potential drop across each resistor, and (d) the current carried by each resistor. (e) If switch S_1 is now closed, find the current in the 2-Ω resistor. (f) If switch S_2 is now opened (while switch S_1 remains closed), find the potential drops across the 6-Ω resistor and across switch S_2.

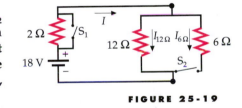

FIGURE 25-19

PICTURE THE PROBLEM (a) To find the equivalent resistance of the circuit, first replace the two parallel resistors by their equivalent resistance. Ohm's law can then be used to find the current and potential drops. For Part (b) and Part (c), use Ohm's law.

Cover the column to the right and try these on your own before looking at the answers.

Steps   **Answers**

(a) 1. Find the equivalent resistance of the 6- and 12-Ω parallel combination. $R_{eq} = 4\,\Omega$

   2. Combine your result in step 1 with the 2-Ω resistor in series to find the total equivalent resistance of the circuit. $R'_{eq} =$ ☐ 6 Ω

(b) Find the total current using Ohm's law. This is the current in both the battery and in the 2-Ω resistor. $I =$ ☐ 3 A

(c) 1. Find the potential drop across the 2-Ω resistor from $V_2 = IR$. $V_{2\Omega} =$ ☐ 6 V

   2. Find the potential drop across each resistor in the parallel combination using $V_p = IR_{eq}$. $V_{6\Omega} = V_{12\Omega} =$ ☐ 12 V

(d) Find the current in the 6-Ω and 12-Ω resistors from $I = V_p/R$.

$I_{6\Omega} = \boxed{2\ \text{A}}$, $I_{12\Omega} = \boxed{1\ \text{A}}$

(e) With S_1 closed the potential drop across the 2-Ω resistor is zero. Using Ohm's law, calculate the current through the 2-Ω resistor.

$I_{2\Omega} = \boxed{0}$

(f) With S_2 open, the current through the 6-Ω resistor is zero. Using Ohm's law, calculate the potential drop across the 6-Ω resistor. The potential drop across the 6-Ω resistor plus the potential drop across switch S_2 equals the potential drop across the 12-Ω resistor.

$V_{6\Omega} = \boxed{0}$, $V_{S_2} = \boxed{18\ \text{V}}$

PLAUSIBILITY CHECK The current in the 6-Ω resistor is twice that in the 12-Ω resistor, as we should expect. Also, these two currents sum to give I, the total current in the circuit, as they must. Finally, note that the potential drops across the 2-Ω resistor and the parallel combination sum to the emf of the battery; $V_2 + V_p = 6\ \text{V} + 12\ \text{V} = 18\ \text{V}$.

EXERCISE Repeat Part (a) through Part (d) of this example with the 6-Ω resistor replaced by a wire of negligible resistance. (*Answer* (a) $R'_{eq} = 2\ \Omega$; (b) $I = 9\ \text{A}$; (c) $V_2 = 18\ \text{V}$, $V_0 = 0$, $V_{12} = 0$; (d) $I_2 = 9\ \text{A}$, $I_0 = 9\ \text{A}$, $I_{12} = 0$)

COMBINATIONS OF COMBINATIONS　　　　**EXAMPLE 25-12** **Try It Yourself**

Find the equivalent resistance of the combination of resistors shown in Figure 25-20.

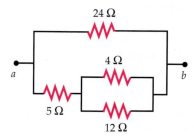

24 Ω

4 Ω

a　　　　　　b

5 Ω

12 Ω　　　　**FIGURE 25-20**

PICTURE THE PROBLEM You can analyze this complicated combination step by step. First, find the equivalent resistance R_{eq} of the 4-Ω and 12-Ω parallel combination; next, find the equivalent resistance R'_{eq} of the series combination of the 5-Ω resistor and R_{eq}; and finally, find the equivalent resistance R''_{eq} of the parallel combination of the 24-Ω resistor and R'_{eq}.

Cover the column to the right and try these on your own before looking at the answers.

Steps　　　　　　　　　　　　　　　　**Answers**

1. Find the equivalent resistance R_{eq} of the 4-Ω and 12-Ω resistors in parallel.　　　$R_{eq} = 3\ \Omega$

2. Find the equivalent resistance R'_{eq} of R_{eq} in series with the 5-Ω resistor.　　　$R'_{eq} = 8\ \Omega$

3. Find the equivalent resistance of R'_{eq} in parallel with the 24-Ω resistor.　　　$R''_{eq} = \boxed{6\ \Omega}$

BLOWING THE FUSE **EXAMPLE 25-13** **Put It in Context**

You are making a snack for some friends to help you get ready for a full night of studying. You decide that coffee, toast, and popcorn would be a good start. You start the toaster and get some popcorn going in the microwave. Since your apartment is in an older building, you know you have problems with the fuse blowing when you turn too many things on. Should you start the coffeemaker? You look on the appliances and find that the toaster has a rating of 900 W, the microwave is rated at 1200 W, and the coffeemaker is rated at 600 W. Past experience with replacing fuses has shown that your house has 20-A fuses.

PICTURE THE PROBLEM We can assume that household circuits are wired in parallel, since plugging in one device usually does not affect others that are in the circuit. Household voltage in the United States is 120 V. (We can neglect the fact that it is not dc.) If we can determine the current through each device, we can add up the total current in the circuit and see how it compares to the fuse current.

1. The power delivered to a device is the current times the potential drop. That is, $P = IV$. Solve for the current for each device:

$$I_{\text{toaster}} = \frac{P_{\text{toaster}}}{V} = \frac{900\ \text{W}}{120\ \text{V}} = 7.5\ \text{A}$$

$$I_{\text{m-wave}} = \frac{P_{\text{m-wave}}}{V} = \frac{1200\ \text{W}}{120\ \text{V}} = 10\ \text{A}$$

$$I_{\text{c-maker}} = \frac{P_{\text{c-maker}}}{V} = \frac{600\ \text{W}}{120\ \text{V}} = 5\ \text{A}$$

2. The current through the fuse is the sum of these currents: $I_{\text{fuse}} = 22.5\ \text{A}$

3. A current this large is above the 20-A rating of the fuse: | Your guests will have to wait on the coffee. |

REMARKS We have assumed that the apartment has only one circuit, and thus only one fuse. Typically, there are several circuits, each fused separately. The coffeemaker can be plugged into an outlet that is on a different circuit than the outlet for the toaster and microwave without a fuse blowing.

25-5 Kirchhoff's Rules

There are many simple circuits, such as the simple circuit shown in Figure 25-21, that cannot be analyzed by merely replacing combinations of resistors by an equivalent resistance. The two resistors R_1 and R_2 in this circuit look as if they might be in parallel, but they are not. The potential drop is not the same across both resistors because of the presence of the emf source $\mathcal{E}_2$ in series with R_2. Nor are R_1 and R_2 in series, because they do not carry the same current.

Two rules, called **Kirchhoff's rules,** apply to this circuit and to any other circuit:

1. When any closed-circuit loop is traversed, the algebraic sum of the changes in potential must equal zero.
2. At any junction (branch point) in a circuit where the current can divide, the sum of the currents into the junction must equal the sum of the currents out of the junction.

KIRCHHOFF'S RULES

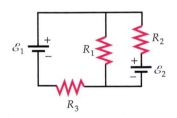

FIGURE 25-21 An example of a simple circuit that cannot be analyzed by replacing combinations of resistors in series or parallel with their equivalent resistances. The potential drops across R_1 and R_2 are not equal because of the emf source $\mathcal{E}_2$, so these resistors are not in parallel. (Parallel resistors would be connected together at both ends.) The resistors do not carry the same current, so they are not in series.

Kirchhoff's first rule, called the **loop rule,** follows directly from the presence of a conservative field $\vec{E}$.[†] To say $\vec{E}$ is conservative means that

$$\oint_C \vec{E} \cdot d\vec{r} = 0 \qquad \text{25-23}$$

where the integral is taken around any closed curve C. Changes in potential ΔV and $\vec{E}$ are related by $\Delta V = V_b - V_a = -\int_a^b \vec{E} \cdot d\vec{r}$. Thus, Equation 25-23 implies that the sum of the changes in potential (the sum of the ΔVs) around any closed path equals zero.

Kirchhoff's second rule, called the **junction rule,** follows from the conservation of charge. Figure 25-22 shows the junction of three wires carrying currents I_1, I_2, and I_3. Since charge does not originate or accumulate at this point, the conservation of charge implies the junction rule, which for this case gives

$$I_1 = I_2 + I_3 \qquad \text{25-24}$$

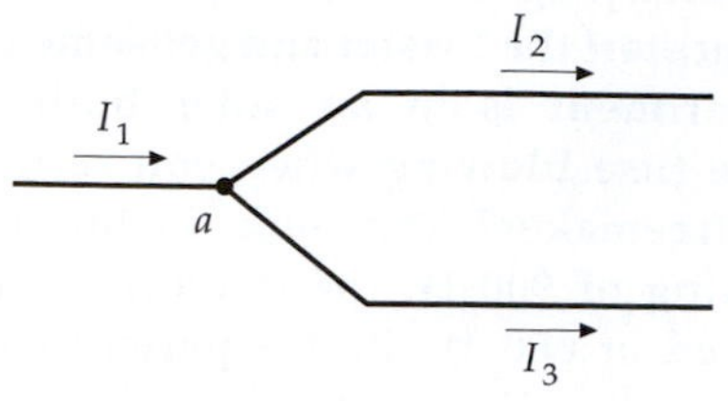

FIGURE 25-22 Illustration of Kirchhoff's junction rule. The current I_1 into point a equals the sum $I_2 + I_3$ of the currents out of point a.

Single-Loop Circuits

As an example of using Kirchhoff's loop rule, consider the circuit shown in Figure 25-23, which contains two batteries with internal resistances r_1 and r_2 and three external resistors. We wish to find the current in terms of the emfs and resistances.

We choose clockwise as positive, as indicated in Figure 25-23. We then apply Kirchhoff's loop rule as we traverse the circuit in the positive direction, beginning at point a. Note that we encounter a potential drop as we traverse the source of emf between points c and d and we encounter a potential increase as we traverse the source of emf between e and a. Assuming that I is positive, we encounter a potential drop as we traverse each resistor. Beginning at point a, we obtain from Kirchhoff's loop rule

$$-IR_1 - IR_2 - \mathcal{E}_2 - Ir_2 - IR_3 + \mathcal{E}_1 - Ir_1 = 0$$

Solving for the current I, we obtain

$$I = \frac{\mathcal{E}_1 - \mathcal{E}_2}{R_1 + R_2 + R_3 + r_1 + r_2} \qquad \text{25-25}$$

If $\mathcal{E}_2$ is greater than $\mathcal{E}_1$, we get a negative value for the current I, indicating that the current is in the negative direction (counterclockwise).

For this example, suppose that $\mathcal{E}_1$ is the greater emf. In battery 2, the charge flows from high potential to low potential. Therefore, a charge ΔQ moving through battery 2 from point c to point d loses potential energy $\Delta Q \, \mathcal{E}_2$ (plus any energy dissipated within the battery via Joule heating). If battery 2 is a rechargeable battery, much of this potential energy is stored in the battery as chemical energy, which means that battery 2 is *charging*.

The analysis of a circuit is usually simplified if we choose one point to be at zero potential and then find the potentials of the other points relative to it. Since only potential differences are important, any point in a circuit can be chosen to have zero potential. In the following example, we choose point e in the figure to be at zero potential. This is indicated by the ground symbol $\perp$ at point e.[‡]

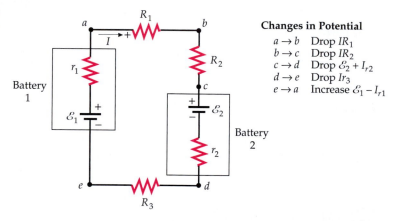

Changes in Potential

$a \rightarrow b$	Drop IR_1
$b \rightarrow c$	Drop IR_2
$c \rightarrow d$	Drop $\mathcal{E}_2 + Ir_2$
$d \rightarrow e$	Drop Ir_3
$e \rightarrow a$	Increase $\mathcal{E}_1 - Ir_1$

FIGURE 25-23 Circuit containing two batteries and three external resistors.

[†] There is also a nonconservative electric field that is discussed in Chapter 28. The resultant electric field is the superposition of the conservative electric field and the nonconservative electric field.

[‡] As we saw in Section 21-2, the earth can be considered to be a very large conductor with a nearly unlimited supply of charge, which means that the potential of the earth remains essentially constant. In practice, electrical circuits are often grounded by connecting one point to the earth. The outside metal case of a washing machine, for example, is usually grounded by connecting it by a wire to a water pipe that is in contact with the earth. Since everything so grounded is at the same potential, it is convenient to designate this potential as zero.

E X A M P L E   2 5 - 1 4

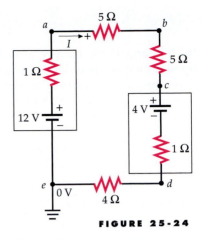

FIGURE 25-24

Suppose the elements in the circuit in Figure 25-23 have the values $\mathcal{E}_1 = 12$ V, $\mathcal{E}_2 = 4$ V, $r_1 = r_2 = 1\ \Omega$, $R_1 = R_2 = 5\ \Omega$, and $R_3 = 4\ \Omega$, as shown in Figure 25-24. (*a*) Find the potentials at points *a* through *e* in the figure, assuming that the potential at point *e* is zero. (*b*) Find the power input and output in the circuit.

PICTURE THE PROBLEM To find the potential differences, we first need to find the current *I* in the circuit. The potential drop across each resistor is then *IR*. To discuss the energy balance, we calculate the power into or out of each element using Equations 25-11 and 25-12.

(*a*) 1. The current *I* in the circuit is found using Equation 25-25:

$$I = \frac{12\ \text{V} - 4\ \text{V}}{5\ \Omega + 5\ \Omega + 4\ \Omega + 1\ \Omega + 1\ \Omega} = \frac{8\ \text{V}}{16\ \Omega} = 0.5\ \text{A}$$

2. We now find the potential at each labeled point in the circuit:

$$V_a = V_e + \mathcal{E}_1 - Ir_1 = 0 + 12\ \text{V} - (0.5\ \text{A})(1\ \Omega) = \boxed{11.5\ \text{V}}$$

$$V_b = V_a - IR_1 = 11.5\ \text{V} - (0.5\ \text{A})(5\ \Omega) = \boxed{9\ \text{V}}$$

$$V_c = V_b - IR_2 = 9\ \text{V} - (0.5\ \text{A})(5\ \Omega) = \boxed{6.5\ \text{V}}$$

$$V_d = V_c - \mathcal{E}_2 - Ir_2 = 6.5\ \text{V} - 4\ \text{V} - (0.5\ \text{A})(1\ \Omega) = \boxed{2.0\ \text{V}}$$

$$V_e = V_d - IR_3 = 2.0\ \text{V} - (0.5\ \text{A})(4\ \Omega) = \boxed{0}$$

(*b*) 1. First, calculate the power supplied by the emf source $\mathcal{E}_1$:

$$P_{\mathcal{E}_1} = \mathcal{E}_1 I = (12\ \text{V})(0.5\ \text{A}) = \boxed{6\ \text{W}}$$

2. Part of this power is dissipated in the resistors, both internal and external:

$$P_R = I^2 R_1 + I^2 R_2 + I^2 R_3 + I^2 r_1 + I^2 r_2$$

$$= (0.5\ \text{A})^2 (5\ \Omega + 5\ \Omega + 4\ \Omega + 1\ \Omega + 1\ \Omega) = 4.0\ \text{W}$$

3. The remaining 2 W of power goes into charging battery 2:

$$P_{\mathcal{E}_2} = \mathcal{E}_2 I = (4\ \text{V})(0.5\ \text{A}) = 2\ \text{W}$$

4. The rate at which potential energy being taken out of the circuit is:

$$P = P_R + P_{\mathcal{E}_1} = \boxed{6\ \text{W}}$$

Note that the terminal voltage of the battery that is being charged in Example 25-14 is $V_c - V_d = 4.5$ V, which is greater than the emf of the battery. If the same 4-V battery were to deliver 0.5 A to an external circuit, its terminal voltage would be 3.5 V (again assuming that its internal resistance is 1 Ω). If the internal resistance is very small, the terminal voltage of a battery is nearly equal to its emf, whether the battery is delivering energy to an external circuit or is being charged. Some real batteries, such as those used in automobiles, are nearly reversible and can easily be recharged. Other types of batteries are not reversible. If you attempt to recharge one of these by driving current from its positive to its negative terminal, most, if not all, of the energy will be dissipated into thermal energy rather than being transformed into the chemical energy of the battery.

EXAMPLE 25-15

A fully charged[†] car battery is to be connected by jumper cables to a discharged car battery in order to charge it. (*a*) To which terminal of the discharged battery should the positive terminal of the charged battery be connected? (*b*) Assume that the charged battery has an emf of $\mathcal{E}_1 = 12$ V and the discharged battery has an emf of $\mathcal{E}_2 = 11$ V, that the internal resistances of the batteries are $r_1 = r_2 = 0.02\ \Omega$, and that the resistance of the jumper cables is $R = 0.01\ \Omega$. What will the charging current be? (*c*) What will the current be if the batteries are connected incorrectly?

PICTURE THE PROBLEM

FIGURE 25-25

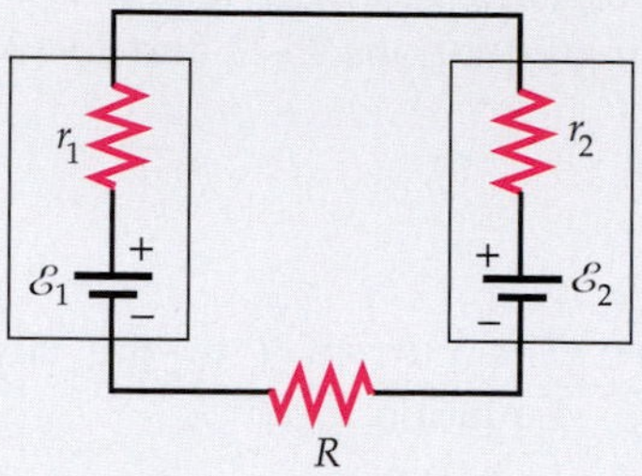

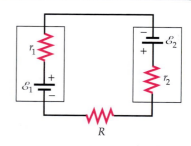

1. To charge the discharged battery, we connect the terminals positive to positive and negative to negative, to drive current through the discharged battery from the positive terminal to the negative terminal (Figure 25-25):

2. Use Kirchhoff's loop rule to find the charging current:

$$\mathcal{E}_1 - Ir_1 - Ir_2 - \mathcal{E}_2 - IR = 0$$

so

$$I = \frac{\mathcal{E}_1 - \mathcal{E}_2}{R + r_1 + r_2} = \frac{12\ \text{V} - 11\ \text{V}}{0.05\ \Omega} = \boxed{20\ \text{A}}$$

3. When the batteries are connected incorrectly, positive terminals to negative terminals, the emfs add:

$$\mathcal{E}_1 - Ir_1 + \mathcal{E}_2 - Ir_2 - IR = 0$$

so

$$I = \frac{\mathcal{E}_1 + \mathcal{E}_2}{R + r_1 + r_2} = \frac{12\ \text{V} + 11\ \text{V}}{0.05\ \Omega} = \boxed{460\ \text{A}}$$

REMARKS If the batteries are connected incorrectly, as shown in Figure 25-26, the total resistance of the circuit is of the order of hundredths of an ohm, the current is very large, and the batteries could explode in a shower of boiling battery acid.

Multiloop Circuits

In multiloop circuits, often the directions of the currents in the different branches of the circuit are unknown. Fortunately, Kirchhoff's rules do not require that we know the directions of the current initially. In fact, these rules allow us to solve for the directions of the currents. To accomplish this, for each branch we arbitrarily assign a positive direction along the branch, and we indicate this assignment by placing a corresponding arrow on the circuit diagram (Figure 25-27). If the actual current in the branch is in the positive direction, when we solve for it we will get a positive value, and if the actual current is opposite to the positive direction, when we solve for it we will get a negative value. The current through a resistor always goes from high potential to low potential. Therefore, any time we traverse a resistor in the direction of the current, the change in potential is negative, and vice versa. Here is the rule:

> For each branch of a circuit, we draw an arrow to indicate the positive direction for that branch. Then, if we traverse a resistor in the direction of the arrow, the change in potential ΔV is equal to $-IR$ (and if we traverse a resistor in the opposite direction, ΔV is equal to $+IR$).

SIGN RULE FOR THE CHANGE IN POTENTIAL ACROSS A RESISTOR

FIGURE 25-26 Two batteries connected incorrectly—dangerous!

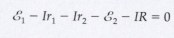

FIGURE 25-27 It is not known whether or not the current I has a positive or a negative value. Whether it is positive or negative, $V_b - V_a = -IR$. If the current is upward, then I is positive and $-IR$ is negative. However, if the current is downward, then I is negative and $-IR$ is positive.

[†] Batteries do not store charge. A *fully charged* battery is one with a maximum amount of stored chemical energy.

If we traverse a resistor in the positive direction, and if I is positive, then $-IR$ is negative. This is as expected, since the current is always in the direction of decreasing potential. If we traverse a resistor in the positive direction, and if I is negative, then $-IR$ is positive. Similarly, if we traverse a resistor in the negative direction, and if I is positive, then $+IR$ is positive, and if we traverse a resistor in the negative direction and if I is negative, then $+IR$ is negative.

To analyze circuits containing more than one loop, we need to use both of Kirchhoff's rules, with Kirchhoff's junction rule applied to points where the current splits into two or more parts.

FIGURE 25-28

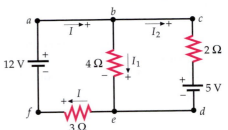

APPLYING KIRCHHOFF'S RULES

E X A M P L E 2 5 - 1 6

(a) Find the current in each branch of the circuit shown in Figure 25-28.
(b) Find the energy dissipated in the 4-Ω resistor in 3 s.

PICTURE THE PROBLEM There are three branch currents, I, I_1, and I_2, to be determined, so we need three relations. One relation comes from applying the junction rule to point b. (We can also apply the junction rule to point e, the only other junction in the circuit, but it gives exactly the same information.) The other two relations are obtained by applying the loop rule. There are three loops in the circuit: the two interior loops, $abefa$ and $bcdeb$, and the exterior loop, $abcdefa$. We can use any two of these loops—the third will give redundant information. There is a direction arrow on each branch in Figure 25-28. Each direction arrow indicates the positive direction for that branch. If our analysis results in a negative value for a branch current, then that current is in the direction opposite to the direction arrow for that branch.

(a) 1. Apply the junction rule to point b:

$I = I_1 + I_2$

2. Apply the loop rule to the outer loop, $abcdefa$:

$12\,\text{V} - (2\,\Omega)I_2 - 5\,\text{V} - (3\,\Omega)(I_1 + I_2) = 0$

3. Divide the above equation by 1 Ω, recalling that $(1\,\text{V})/(1\,\Omega) = 1\,\text{A}$, then simplify:

$7\,\text{A} - 3I_1 - 5I_2 = 0$

4. For the third condition, apply the loop rule to the loop on the right, $bcdeb$:

$-(2\,\Omega)I_2 - 5\,\text{V} + (4\,\Omega)I_1 = 0$

$-5\,\text{A} + 4I_1 - 2I_2 = 0$

5. The results for steps 3 and 4 can be combined to solve for I_1 and I_2. To do so, first multiply the result for step 3 by 2, and then multiply the result for step 4 by -5:

$14\,\text{A} - 6I_1 - 10I_2 = 0$

$25\,\text{A} - 20I_1 + 10I_2 = 0$

6. Add the equations in step 5 to eliminate I_2, then solve for I_1:

$39\,\text{A} - 26I_1 = 0$

$I_1 = \dfrac{39\,\text{A}}{26} = \boxed{1.5\,\text{A}}$

7. Substitute I_1 in the results for step 3 or 4 to solve for I_2:

$7\,\text{A} - 3(1.5\,\text{A}) - 5I_2 = 0$

$I_2 = \dfrac{2.5\,\text{A}}{5} = \boxed{0.5\,\text{A}}$

8. Finally, I_1 and I_2 determine I using the equation in step 1:

$I = I_1 + I_2 = 1.5\,\text{A} + 0.5\,\text{A} = \boxed{2.0\,\text{A}}$

(b) 1. The power dissipated in the 4-Ω resistor is found using $P = I_1^2 R$:

$P = I_1^2 R = (1.5\,\text{A})^2 (4\,\Omega) = 9\,\text{W}$

2. The total energy dissipated in a time Δt is $W = P\Delta t$. In this case, $t = 3\,\text{s}$:

$W = P\Delta t = (9\,\text{W})(3\,\text{s}) = \boxed{27\,\text{J}}$

❶**PLAUSIBILITY CHECK** In Figure 25-29, we have chosen the potential to be zero at point f, and we have labeled the currents and the potentials at the other points. Note that $V_b - V_e = 6$ V and $V_e - V_f = 6$ V.

REMARKS Applying the loop rule to the loop on the left, $abefa$, gives $12 \text{ V} - (4 \, \Omega) I_1 - (3 \, \Omega)(I_1 + I_2) = 0$, or $12 \text{ A} - 7I_1 - 3I_2 = 0$. Note that this is just the result for step 3 minus the result for step 4 and hence contains no new information, as expected.

EXERCISE Find I_1 for the case in which the 3-Ω resistor approaches (*a*) zero resistance and (*b*) infinite resistance. [*Answer* (*a*) The potential drop across the 4-Ω resistor is 12 V; thus, $I_1 = 3$ A. (*b*) In this case, the loop on the left is an open circuit, so $I = 0$ and $I_2 = -I_1$. Thus, $I_1 = (5 \text{ V})/(2 \, \Omega + 4 \, \Omega) = 0.833$ A.]

FIGURE 25-29

Example 25-16 illustrates the general methods for the analysis of multiloop circuits:

1. Draw a sketch of the circuit.

2. Replace any series or parallel resistor combinations or capacitor combinations by their equivalent values.

3. Choose the positive direction for each branch of the circuit and indicate the positive direction with a direction arrow. Label the current in each branch. Add plus and minus signs to indicate the high-potential terminal and low-potential terminal of each source of emf.

4. Apply the junction rule to all but one of the branch points (junctions).

5. Apply the loop rule to each loop until you obtain as many independent equations as there are unknowns. When traversing a resistor in the positive direction, the change in potential equals $-IR$. When traversing a battery from the negative terminal to the positive terminal, the change in potential equals $\mathcal{E} - IR$.

6. Solve the equations to obtain the desired values.

7. Check your results by assigning a potential of zero to one point in the circuit and use the values of the currents found to determine the potentials at other points in the circuit.

GENERAL METHOD FOR ANALYZING MULTILOOP CIRCUITS

A THREE-BRANCH CIRCUIT   **EXAMPLE 25-17**   **Try It Yourself**

(*a*) **Find the current in each part of the circuit shown in Figure 25-30. Draw the circuit diagram with the correct magnitudes and directions for the current in each part.** (*b*) **Assign $V = 0$ to point c and then label the potential at each other point a through f.**

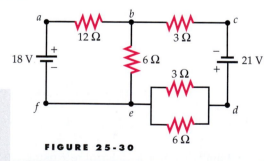

FIGURE 25-30

PICTURE THE PROBLEM First, replace the two parallel resistors by an equivalent resistance. Let I be the current through the 18-V battery, and let I_1 be the current from point b to point e. The currents can then be found by applying the junction rule at branch points b and c and by applying the loop rule to each loop.

Cover the column to the right and try these on your own before looking at the answers.

Steps	Answers
(*a*) 1. Find the equivalent resistance of the 3-Ω and 6-Ω parallel resistors.	$R_{eq} = 2 \, \Omega$

FIGURE 25-31

2. Apply the junction rule at points b and e and redraw the circuit diagram with the positive branch directions indicated (Figure 25-31).

$I = I_1 + I_2$ or $(I_1 = I - I_2)$

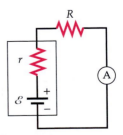

3. Apply Kirchhoff's loop rule to loop $abefa$ to obtain an equation involving I and I_2.

$18\text{ V} - (12\ \Omega)I - (6\ \Omega)(I - I_2) = 0$

4. Simplify your equation from step 3.

$3\text{ A} - 3I + I_2 = 0$

5. Apply Kirchhoff's loop rule to loop $bcdeb$ to obtain an equation involving I and I_2.

$-(3\ \Omega)I_2 + 21\text{ V} - (2\ \Omega)I_2 + (6\ \Omega)(I - I_2) = 0$

6. Simplify your equation in step 5.

$21\text{ A} + 6I - 11I_2 = 0$

7. Solve your simultaneous equations from step 4 and step 6 for I and I_2. One way to do this is to multiply the equation in step 4 by 11 and then add the equations to eliminate I_2.

$I = \boxed{2\text{ A}}$, $I_2 = \boxed{3\text{ A}}$

8. Find the current through the 6-Ω resistor.

$I_1 = I - I_2 = \boxed{-1\text{ A}}$

9. Use $V = I_2 R_{eq}$ to find the potential drop across the parallel 3-Ω and 6-Ω resistors.

$V = 6\text{ V}$

10. Use the result of step 9 to find the current in each of the parallel resistors.

$I_{3\Omega} = \boxed{2\text{ A}}$, $I_{6\Omega} = \boxed{1\text{ A}}$

(b) Redraw Figure 25-31 showing the current through each part of the circuit (Figure 25-32). Begin with $V = 0$ at point c and calculate the potential at points d, e, f, a, and b.

$V_d = V_c + 21\text{ V} = 0 + 21\text{ V} = \boxed{21\text{ V}}$

$V_e = V_d - (3\text{ A})(2\ \Omega) = 21\text{ V} - 6\text{ V} = \boxed{15\text{ V}}$

$V_f = V_e = \boxed{15\text{ V}}$

$V_a = V_f + 18\text{ V} = 15\text{ V} + 18\text{ V} = \boxed{33\text{ V}}$

$V_b = V_a - (2\text{ A})(12\ \Omega) = 33\text{ V} - 24\text{ V} = \boxed{9\text{ V}}$

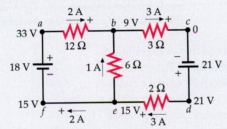

FIGURE 25-32

PLAUSIBILITY CHECK From point b to point c the potential drops by $(3\text{ A})(3\ \Omega) = 9\text{ V}$, which gives $V_c = 0$, as assumed. From point e to point b the potential drops by $(1\text{ A})(6\ \Omega) = 6\text{ V}$, so $V_b = V_e - 6\text{ V} = 15\text{ V} - 6\text{ V} = 9\text{ V}$.

Ammeters, Voltmeters, and Ohmmeters

The devices that measure current, potential difference, and resistance are called **ammeters**, **voltmeters**, and **ohmmeters**, respectively. Often, all three of these meters are included in a single *multimeter* that can be switched from one use to another. You might use a voltmeter to measure the terminal voltage of your car battery and an ohmmeter to measure the resistance of some electrical device at home (e.g., a toaster or lightbulb) when you suspect a short circuit or a broken wire.

To measure the current through a resistor in a simple circuit, we place an ammeter in series with the resistor, as shown in Figure 25-33, so that the ammeter and the resistor carry the same current. Since the ammeter has a very low (but finite) resistance, the current in the circuit decreases very slightly when the ammeter is inserted. Ideally, the ammeter should have a negligibly small resistance so that the current to be measured is only negligibly affected.

FIGURE 25-33 To measure the current in a resistor R, an ammeter A (circled) is placed in series with the resistor so that it carries the same current as the resistor.

The potential difference across a resistor is measured by placing a voltmeter across the resistor (in parallel with it), as shown in Figure 25-34, so that the potential drop across the voltmeter is the same as that across the resistor. The voltmeter reduces the resistance between points a and b, thus increasing the total current in the circuit and changing the potential drop across the resistor. A good voltmeter has an extremely large resistance so that its effect on the current in the circuit is negligible.

The principal component of many common ammeters and voltmeters is a **galvanometer,** a device that detects small currents passing through it. The galvanometer is designed so that the scale reading is proportional to the current passing through. A typical galvanometer used in many student laboratories consists of a coil of wire in the magnetic field of a permanent magnet. When the coil carries a current, the magnetic field exerts a torque on the coil, which causes the coil to rotate. A pointer attached to the coil indicates the reading on a scale. The coil itself contributes a small amount of resistance when the galvanometer is placed within a circuit.

To construct an ammeter from a galvanometer, we place a small resistor called a **shunt resistor** in *parallel* with the galvanometer. The shunt resistance is usually much smaller than the resistance of the galvanometer so that most of the current is carried by the shunt resistor. The equivalent resistance of the ammeter is then approximately equal to the shunt resistance, which is much smaller than the internal resistance of the galvanometer alone. To construct a voltmeter, we place a resistor with a large resistance in *series* with the galvanometer so that the equivalent resistance of the voltmeter is much larger than that of the galvanometer alone. Figure 25-35 illustrates the construction of an ammeter and voltmeter from a galvanometer. The resistance of the galvanometer R_g is shown separately in these schematic drawings, but it is actually part of the galvanometer.

A simple ohmmeter consists of a battery connected in series with a galvanometer and a resistor, as shown in Figure 25-36*a*. The resistance R_s is chosen so that when the terminals a and b are shorted (put in electrical contact, with negligible resistance between them), the current through the galvanometer gives a full-scale deflection. Thus, a full-scale deflection indicates no resistance between terminals a and b. A zero deflection indicates an infinite resistance between the terminals. When the terminals are connected across an unknown resistance R, the current through the galvanometer depends on R, so the scale can be calibrated to give a direct reading of R, as shown in Figure 25-36*b*. Because an ohmmeter sends a current through the resistance to be measured, some caution must be exercised when using this instrument. For example, you would not want to try to measure the resistance of a sensitive galvanometer with an ohmmeter, because the current provided by the battery in the ohmmeter would probably damage the galvanometer.

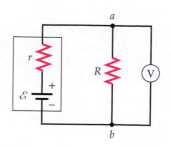

FIGURE 25-34 To measure the potential drop across a resistor, a voltmeter V (circled) is placed in parallel with the resistor so that the potential drops across the voltmeter and the resistor are the same.

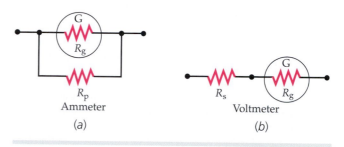

(a)   (b)

FIGURE 25-35 (*a*) An ammeter consists of a galvanometer G (circled) whose resistance is R_g and a small parallel resistance R_p. (*b*) A voltmeter consists of a galvanometer G (circled) and a large series resistance R_s.

(a)

(b)

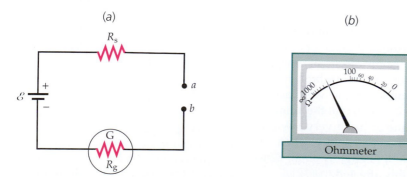

Ohmmeter

FIGURE 25-36 (*a*) An ohmmeter consists of a battery connected in series with a galvanometer and a resistor R_s, which is chosen so that the galvanometer gives full-scale deflection when points a and b are shorted. (*b*) When a resistor R is placed across a and b, the galvanometer needle deflects by an amount that depends on the value of R. The galvanometer scale is calibrated to give a readout in ohms.

25-6 *RC Circuits*

A circuit containing a resistor and a capacitor is called an **RC circuit**. The current in an *RC* circuit flows in a single direction, as in all dc circuits, but the magnitude of the current varies with time. A practical example of an *RC* circuit is the circuit in the flash attachment of a camera. Before a flash photograph is taken, a battery in the flash attachment charges the capacitor through a resistor. When the charge is accomplished, the flash is ready. When the picture is taken, the capacitor discharges through the flashbulb. The battery then recharges the capacitor, and a short time later the flash is ready for another picture. Using Kirchhoff's rules, we can obtain equations for the charge Q and the current I as functions of time for both the charging and discharging of a capacitor through a resistor.

Discharging a Capacitor

Figure 25-37 shows a capacitor with initial charges of $+Q_0$ on the upper plate and $-Q_0$ on the lower plate. The capacitor is connected to a resistor R and a switch S, which is initially open. The potential difference across the capacitor is initially $V_0 = Q_0/C$, where C is the capacitance.

We close the switch at time $t = 0$. Since there is now a potential difference across the resistor, there must be a current in it. The initial current is

$$I_0 = \frac{V_0}{R} = \frac{Q_0}{RC} \qquad 25\text{-}26$$

The current is due to the flow of charge from the positive plate of the capacitor to the negative plate through the resistor. After a time, the charge on the capacitor is reduced. If we choose the positive direction to be clockwise, then the current equals the rate of decrease of that charge. If Q is the charge on the upper plate of the capacitor at time t, the current at that time is

$$I = -\frac{dQ}{dt} \qquad 25\text{-}27$$

(The minus sign is needed because while Q decreases, dQ/dt is negative.)[†] Traversing the circuit in the clockwise direction, we encounter a potential drop IR across the resistor and a potential increase Q/C across the capacitor. Thus, Kirchhoff's loop rule gives

$$\frac{Q}{C} - IR = 0 \qquad 25\text{-}28$$

where Q and I, both functions of time, are related by Equation 25-27. Substituting $-dQ/dt$ for I in Equation 25-28, we have

$$\frac{Q}{C} + R\frac{dQ}{dt} = 0$$

or

$$\frac{dQ}{dt} = -\frac{1}{RC}Q \qquad 25\text{-}29$$

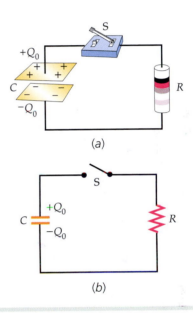

(a)

(b)

FIGURE 25-37 (*a*) A parallel-plate capacitor in series with a switch *S* and a resistor *R*. (*b*) A circuit diagram for Figure 25-37*a*.

[†] If the positive direction were chosen to be counterclockwise, then the sign in Equation 25-27 would be a positive sign.

To solve this equation, we first separate the variables Q and t by multiplying both sides by dt/Q, and then integrate. Multiplying both sides by dt/Q, we obtain

$$\frac{dQ}{Q} = -\frac{1}{RC} dt \qquad\qquad 25\text{-}30$$

The variables Q and t are now in separate terms. Integrating from Q_0 at $t = 0$ to Q' at time t' gives

$$\int_{Q_0}^{Q'} \frac{dQ}{Q} = -\frac{1}{RC} \int_0^{t'} dt$$

so

$$\ln\frac{Q'}{Q_0} = -\frac{t'}{RC}$$

Since t' is arbitrary, we can replace t' with t, and then $Q' = Q(t)$. Solving for $Q(t)$ gives

$$Q(t) = Q_0 e^{-t/(RC)} = Q_0 e^{-t/\tau} \qquad\qquad 25\text{-}31$$

where τ, called the **time constant,** is the time it takes for the charge to decrease by a factor of e^{-1}:

$$\tau = RC \qquad\qquad 25\text{-}32$$

DEFINITION—TIME CONSTANT

Figure 25-38 shows the charge on the capacitor in the circuit of Figure 25-37 as a function of time. After a time $t = \tau$, the charge is $Q = e^{-1}Q_0 = 0.37\,Q_0$, after a time $t = 2\tau$, the charge is $Q = e^{-2}Q_0 = 0.135Q_0$, and so forth. After a time equal to several time constants, the charge Q is negligible. This type of decrease, which is called an **exponential decrease,** is very common in nature. It occurs whenever the rate at which a quantity decreases is proportional to the quantity itself.[†]

The decrease in the charge on a capacitor can be likened to the decrease in the amount of water in a bucket with vertical sides that has a small hole in the bottom. The rate at which the water flows out of the bucket is proportional to the pressure of the water, which is in turn proportional to the amount of water still in the bucket.

The current is obtained by differentiating Equation 25-31

$$I = -\frac{dQ}{dt} = \frac{Q_0}{RC} e^{-t/(RC)}$$

Substituting, using Equation 25-26, we obtain

$$I = I_0 e^{-t/\tau} \qquad\qquad 25\text{-}33$$

where $I_0 = V_0/R = Q_0/(RC)$ is the initial current. The current as a function of time is shown in Figure 25-39. As with the charge, the current decreases exponentially with time constant $\tau = RC$.

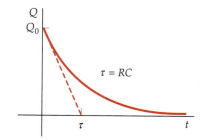

FIGURE 25-38 Plot of the charge on the capacitor versus time for the circuit shown in Figure 25-37 when the switch is closed at time $t = 0$. The time constant $\tau = RC$ is the time it takes for the charge to decrease by a factor of e^{-1}. (The time constant is also the time it would take the capacitor to discharge fully if its discharge rate remains constant, as indicated by the dashed line.)

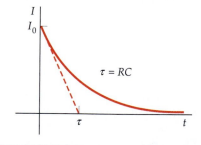

FIGURE 25-39 Plot of the current versus time for the circuit in Figure 25-37. The curve has the same shape as that in Figure 25-38. If the rate of decrease of the current remains constant, the current would reach zero after one time constant, as indicated by the dashed line.

---

† We encountered exponential decreases in Chapter 14 when we studied the damped oscillator.

DISCHARGING A CAPACITOR **EXAMPLE 25-18**

A 4-μF capacitor is charged to 24 V and then connected across a 200-Ω resistor. Find (*a*) the initial charge on the capacitor, (*b*) the initial current through the 200-Ω resistor, (*c*) the time constant, and (*d*) the charge on the capacitor after 4 ms.

PICTURE THE PROBLEM The circuit diagram is the same as the circuit diagram shown in Figure 25-37.

(*a*) The initial charge is related to the capacitance and voltage:

$$Q_0 = CV_0 = (4\ \mu\text{F})(24\ \text{V}) = \boxed{96\ \mu\text{C}}$$

(*b*) The initial current is the initial voltage divided by the resistance:

$$I_0 = \frac{V_0}{R} = \frac{24\ \text{V}}{200\ \Omega} = \boxed{0.12\ \text{A}}$$

(*c*) The time constant is *RC*:

$$\tau = RC = (200\ \Omega)(4\ \mu\text{F}) = 800\ \mu\text{s} = \boxed{0.8\ \text{ms}}$$

(*d*) Substitute $t = 4$ ms into Equation 25-31 to find the charge on the capacitor at that time:

$$Q = Q_0 e^{-t/\tau} = (96\ \mu\text{C})e^{-(4\ \text{ms})/(0.8\ \text{ms})}$$

$$= (96\ \mu\text{C})e^{-5} = \boxed{0.647\ \mu\text{C}}$$

REMARKS After five time constants, the *Q* is less than 1 percent of its initial value.

EXERCISE Find the current through the 200-Ω resistor at $t = 4$ ms. (*Answer* 0.809 mA)

Charging a Capacitor

Figure 25-40*a* shows a circuit for charging a capacitor. The capacitor is initially uncharged. The switch S, originally open, is closed at time $t = 0$. Charge immediately begins to flow through the battery (Figure 25-40*b*). If the charge on the rightmost plate of the capacitor at time *t* is *Q*, the current in the circuit is *I*, and clockwise is positive, then Kirchhoff's loop rule gives

$$\mathcal{E} - IR - \frac{Q}{C} = 0 \qquad\qquad 25\text{-}34$$

By inspecting this equation we can see that at time $t = 0$, the charge on the capacitor is zero and the current is $I_0 = \mathcal{E}/R$. The charge then increases and the current decreases. The charge reaches a maximum value of $Q_f = C\mathcal{E}$ when the current *I* equals zero, as can also be seen from Equation 25-34.

In this circuit, we have chosen the positive direction so if *I* is positive *Q* is increasing. Thus,

$$I = +\frac{dQ}{dt}$$

Substituting *dQ/dt* for *I* in Equation 25-34 gives

$$\mathcal{E} - R\frac{dQ}{dt} - \frac{Q}{C} = 0 \qquad\qquad 25\text{-}35$$

Equation 25-35 can be solved in the same way as Equation 25-29. The details are left as a problem (see Problem 119). The result is

$$Q = C\mathcal{E}(1 - e^{-t/(RC)}) = Q_f(1 - e^{-t/\tau}) \qquad\qquad 25\text{-}36$$

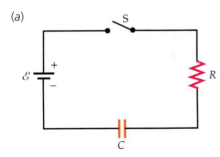

(*a*)

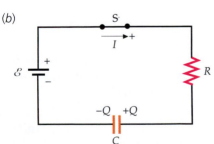

(*b*)

FIGURE 25-40 (*a*) A circuit for charging a capacitor to a potential difference $\mathcal{E}$. (*b*) After the switch is closed, there is current through and a potential drop across the resistor and a charge on and a potential drop across the capacitor.

where $Q_f = C\mathcal{E}$ is the final charge. The current is obtained from $I = dQ/dt$:

$$I = \frac{dQ}{dt} = C\mathcal{E}\left(-\frac{-1}{RC}e^{-t/(RC)}\right) = \frac{\mathcal{E}}{R}e^{-t/(RC)}$$

or

$$I = \frac{\mathcal{E}}{R}e^{-t/(RC)} = I_0 e^{-t/\tau} \qquad\qquad 25\text{-}37$$

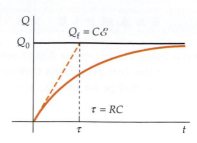

FIGURE 25-41 Plot of the charge on the capacitor versus time for the charging circuit of Figure 25-40 after the switch is closed (at $t = 0$). After a time $t = \tau = RC$, the charge on the capacitor is 0.63 $C\mathcal{E}$, where $C\mathcal{E}$ is its final charge. If the charging rate were constant, the capacitor would be fully charged after a time $t = \tau$.

where the initial current in this case is $I_0 = \mathcal{E}/R$.

Figure 25-41 and Figure 25-42 show the charge and the current as functions of time.

EXERCISE Show that Equation 25-36 does indeed satisfy Equation 25-35 by substituting $Q(t)$ and dQ/dt into Equation 25-35.

EXERCISE What fraction of the maximum charge is on the charging capacitor after a time $t = 2\tau$? (*Answer* 0.86)

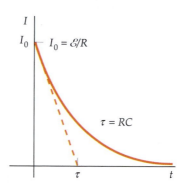

FIGURE 25-42 Plot of the current versus time for the charging circuit of Figure 25-40. The current is initially $\mathcal{E}/R$, and the current decreases exponentially with time.

CHARGING A CAPACITOR   **EXAMPLE 25-19 Try It Yourself**

A 6-V battery of negligible internal resistance is used to charge a 2-μF capacitor through a 100-Ω resistor. Find (*a*) the initial current, (*b*) the final charge on the capacitor, (*c*) the time required for the charge to reach 90 percent of its final value, and (*d*) the charge when the current is half its initial value.

PICTURE THE PROBLEM

Cover the column to the right and try these on your own before looking at the answers.

Steps	Answers
(*a*) Find the initial current from $I_0 = \mathcal{E}/R$.	$I_0 = \boxed{0.06\text{ A}}$
(*b*) Find the final charge from $Q = C\mathcal{E}$.	$Q_f = \boxed{12\ \mu C}$
(*c*) Set $Q = 0.9\ Q_f$ in Equation 25-36 and solve for t. (First solve for $e^{t/\tau}$, then take the natural log of both sides, then solve for t.)	$t = 2.3\ \tau = \boxed{460\ \mu s}$
(*d*) 1. Apply Kirchhoff's loop rule to the circuit using Figure 25-40*b*.	$\mathcal{E} - IR - \dfrac{Q}{C} = 0$
2. Set $I = I_0/2$ and solve for Q.	$Q = \dfrac{Q_f}{2} = \boxed{6\ \mu C}$

REMARKS The answer to Part (*d*) can be obtained by first solving for t using Equation 25-37, then substituting that time into Equation 25-36 and solving for Q. However, using the loop rule is certainly the more direct approach.

FINDING VALUES AT SHORT AND LONG TIMES   **EXAMPLE   25-20**

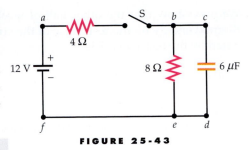

FIGURE 25-43

The 6-μF capacitor in the circuit shown in Figure 25-43 is initially uncharged. Find the current through the 4-Ω resistor and the current through the 8-Ω resistor (a) immediately after the switch is closed, and (b) a long time after the switch is closed. (c) Find the charge on the capacitor a long time after the switch is closed.

PICTURE THE PROBLEM Since the capacitor is initially uncharged, and since the 4-Ω resistor limits the current through the battery, the initial potential difference across the capacitor is zero. The capacitor and the 8-Ω resistor are connected in parallel, and the difference in potential across each is the same. Thus, the initial potential difference across the 8-Ω resistor is also zero.

(a) Apply the loop rule to the outer loop and solve for the current through the 4-Ω resistor. The potential difference across the 8-Ω resistor and the capacitor are equal. Set the initial charge on the capacitor equal to zero and solve for the current through the 8-Ω resistor:

$$12 \text{ V} - (4 \ \Omega)I_{4\Omega,0} + 0 = 0, \ I_{4\Omega,0} = \boxed{3 \text{ A}}$$

$$I_{8\Omega,0}(8 \ \Omega) = \frac{Q_0}{C}, \quad I_{8\Omega,0} = \boxed{0}$$

(b) After a long time, the capacitor is fully charged (no more charge flows onto its the plates) and the current through both resistors is the same. Apply the loop rule to the left loop and solve for the current:

$$12 \text{ V} - (4 \ \Omega)I_f - (8 \ \Omega)I_f = 0$$

$$I_f = \boxed{1 \text{ A}}$$

(c) The potential difference across the 8-Ω resistor and the capacitor are equal. Use this to solve for Q_f:

$$I_f(8 \ \Omega) = \frac{Q_f}{C}$$

$$Q_f = (1 \text{ A})(8 \ \Omega)(6 \ \mu\text{F}) = \boxed{48 \ \mu\text{C}}$$

REMARKS The analysis of this circuit at the extreme times when the capacitor is either uncharged or fully charged is simple. When the capacitor is uncharged, it acts like a short circuit between points c and d; that is, the circuit is the same as the one shown in Figure 25-44a, where we have replaced the capacitor by a wire of zero resistance. When the capacitor is fully charged, it acts like an open circuit, as shown in Figure 25-44b.

Energy Conservation in Charging a Capacitor

During the charging process, a total charge $Q_f = \mathcal{E}C$ flows through the battery. The battery therefore does work

$$W = Q_f\mathcal{E} = C\mathcal{E}^2$$

Half of this work is accounted for by the energy stored in the capacitor (see Equation 24-12):

$$U = \tfrac{1}{2}Q_f\mathcal{E}$$

(a)

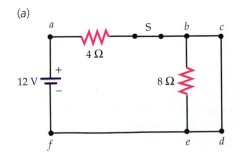

(b)

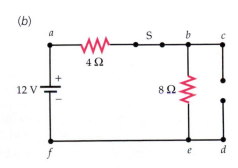

FIGURE 25-44

We now show that the other half of work done by the battery is dissipated as thermal energy by the resistance of the circuit. The rate at which energy is dissipated by the resistance R is

$$\frac{dW_R}{dt} = I^2 R$$

Using Equation 25-37 for the current, we have

$$\frac{dW_R}{dt} = \left(\frac{\mathcal{E}}{R} e^{-t/(RC)}\right)^2 R = \frac{\mathcal{E}^2}{R} e^{-2t/(RC)}$$

We find the total energy dissipated by integrating from $t = 0$ to $t = \infty$:

$$W_R = \int_0^\infty \frac{\mathcal{E}^2}{R} e^{-2t/(RC)}\, dt = \frac{\mathcal{E}^2}{R} \int_0^\infty e^{-at}\, dt$$

where $a = 2/RC$. Thus,

$$W_R = \frac{\mathcal{E}^2}{R} \frac{e^{-at}}{-a}\Bigg|_0^\infty = -\frac{\mathcal{E}^2}{Ra}(0 - 1) = \frac{\mathcal{E}^2}{R}\frac{1}{a} = \frac{\mathcal{E}^2}{R}\frac{RC}{2}$$

The total amount of Joule heating is thus

$$W_R = \frac{1}{2}\mathcal{E}^2 C = \frac{1}{2} Q_f \mathcal{E}$$

where $Q_f = \mathcal{E}C$. This result is independent of the resistance R. Thus, when a capacitor is charged through a resistor by a constant source of emf, half the energy provided by the source of emf is stored in the capacitor and half goes into thermal energy. This thermal energy includes the energy that goes into the internal resistance of the source of emf.

SUMMARY

1. Ohm's law is an empirical law that holds only for certain materials.
2. Current, resistance, and emf are important *defined* quantities.
3. Kirchhoff's rules follow from the conservation of charge and the conservative nature of the electric field.

Topic	Relevant Equations and Remarks
1. Electric Current	Electric current is the rate of flow of electric charge through a cross-sectional area.
	$$I = \frac{\Delta Q}{\Delta t} \qquad \text{25-1}$$
Drift velocity	In a conducting wire, electric current is the result of the slow drift of negatively charged electrons that are accelerated by an electric field in the wire and then collide with the lattice ions. Typical drift velocities of electrons in wires are of the order of a few millimeters per second.
	$$I = qnAv_d \qquad \text{25-3}$$

2. Resistance

Definition of resistance	$R = \dfrac{V}{I}$	25-5
Resistivity, ρ	$R = \rho \dfrac{L}{A}$	25-8
Temperature coefficient of resistivity, α	$\alpha = \dfrac{(\rho - \rho_{20})/\rho_{20}}{t_C - 20°C}$	25-9

3. Ohm's Law

For ohmic materials, the resistance does not depend on the current or the potential drop:

$$V = IR, \quad R \text{ constant} \qquad \text{25-7}$$

4. Power

Supplied to a device or segment	$P = IV$	25-10
Dissipated in a resistor	$P = IV = I^2R = \dfrac{V^2}{R}$	25-11

5. EMF

Source of emf	A device that supplies electrical energy to a circuit.	
Power supplied by an emf source	$P = \mathscr{E}I$	25-12

6. Battery

Ideal	An ideal battery is a source of emf that maintains a constant potential difference between its two terminals, independent of the current through the battery.	
Real	A real battery can be considered as an ideal battery in series with a small resistance called its internal resistance.	
Terminal voltage	$V_a - V_b = \mathscr{E} - Ir$	25-13
Total energy stored	$W = Q\mathscr{E}$	25-15

7. Equivalent Resistance

Resistors in series	$R_{eq} = R_1 + R_2 + R_3 + \ldots$	25-17
Resistors in parallel	$\dfrac{1}{R_{eq}} = \dfrac{1}{R_1} + \dfrac{1}{R_2} + \dfrac{1}{R_3} + \ldots$	25-21

8. Kirchhoff's Rules

1. When any closed-circuit loop is traversed, the algebraic sum of the changes in potential must equal zero.
2. At any junction (branch point) in a circuit where the current can divide, the sum of the currents into the junction must equal the sum of the currents out of the junction.

9. Measuring Devices

Ammeter	An ammeter is a very low resistance device that is placed in series with a circuit element to measure the current in the element.

Voltmeter	A voltmeter is a very high resistance device that is placed in parallel with a circuit element to measure the potential drop across the element.
Ohmmeter	An ohmmeter is a device containing a battery connected in series with a galvanometer and a resistor that is used to measure the resistance of a circuit element placed across its terminals.

10. Discharging a Capacitor

Charge on the capacitor	$Q(t) = Q_0 e^{-t/(RC)} = Q_0 e^{-t/\tau}$	25-31
Current in the circuit	$I = -\dfrac{dQ}{dt} = \dfrac{V_0}{R} e^{-t/(RC)} = I_0 e^{-t/\tau}$	25-33
Time constant	$\tau = RC$	25-32

11. Charging a Capacitor

Charge on the capacitor	$Q = C\mathcal{E}(1 - e^{-t/(RC)}) = Q_f(1 - e^{-t/\tau})$	25-36
Current in the circuit	$I = +\dfrac{dQ}{dt} = \dfrac{\mathcal{E}}{R} e^{-t/(RC)} = I_0 e^{-t/\tau}$	25-37

PROBLEMS

- Single-concept, single-step, relatively easy
- •• Intermediate-level, may require synthesis of concepts
- ••• Challenging
- **SSM** Solution is in the *Student Solutions Manual*
- **iSOLVE** Problems available on iSOLVE online homework service
- **iSOLVE✓** These "Checkpoint" online homework service problems ask students additional questions about their confidence level, and how they arrived at their answer.

In a few problems, you are given more data than you actually need; in a few other problems, you are required to supply data from your general knowledge, outside sources, or informed estimates.

Conceptual Problems

1 • **SSM** In our study of electrostatics, we concluded that there is no electric field within a conductor in electrostatic equilibrium. How is it that we can now discuss electric fields inside a conductor?

2 • Figure 25-8 illustrates a mechanical analog of a simple electric circuit. Devise another mechanical analog in which the current is represented by a flow of water instead of marbles.

3 • Two wires of the same material with the same length have different diameters. Wire A has twice the diameter of wire B. If the resistance of wire B is R, then what is the resistance of wire A? (*a*) R (*b*) $2R$ (*c*) $R/2$ (*d*) $4R$ (*e*) $R/4$

4 •• Discuss the difference between an emf and a potential difference.

5 •• **SSM** A metal bar is to be used as a resistor. Its dimensions are 2 by 4 by 10 units. To get the smallest resistance from this bar, one should attach leads to the opposite sides that have the dimensions of

(*a*) 2 by 4 units.
(*b*) 2 by 10 units.
(*c*) 4 by 10 units.
(*d*) All connections will give the same resistance.
(*e*) None of the above is correct.

6 •• Two cylindrical copper wires have the same mass. Wire A is twice as long as wire B. Their resistances are related by (*a*) $R_A = 8R_B$. (*b*) $R_A = 4R_B$. (*c*) $R_A = 2R_B$. (*d*) $R_A = R_B$.

7 • A resistor carries a current I. The power dissipated in the resistor is P. What is the power dissipated if the same resistor carries current $3I$? (Assume no change in resistance.) (*a*) P (*b*) $3P$ (*c*) $P/3$ (*d*) $9P$ (*e*) $P/9$

8 • The power dissipated in a resistor is P when the potential drop across it is V. If the voltage drop is increased to 2 V (with no change in resistance), what is the power dissipated? (a) P (b) $2P$ (c) $4P$ (d) $P/2$ (e) $P/4$

9 • A heater consists of a variable resistance connected across a constant voltage supply. To increase the heat output, should you decrease the resistance or increase the resistance?

10 • [SSM] Two resistors with resistances R_1 and R_2 are connected in parallel. If $R_1 \gg R_2$, the equivalent resistance of the combination is approximately (a) R_1. (b) R_2. (c) 0. (d) infinity.

11 • Answer Problem 10 with resistors R_1 and R_2 connected in series.

12 • Two resistors are connected in parallel across a potential difference. The resistance of resistor A is twice that of resistor B. If the current carried by resistor A is I, then what is the current carried by resistor B? (a) I (b) $2I$ (c) $I/2$ (d) $4I$ (e) $I/4$

13 • [SSM] Two resistors are connected in series across a potential difference. Resistor A has twice the resistance of resistor B. If the current carried by resistor A is I, then what is the current carried by resistor B? (a) I (b) $2I$ (c) $I/2$ (d) $4I$ (e) $I/4$

14 •• When two identical resistors are connected in series across the terminals of a battery, the power delivered by the battery is 20 W. If these resistors are connected in parallel across the terminals of the same battery, what is the power delivered by the battery? (a) 5 W (b) 10 W (c) 20 W (d) 40 W (e) 80 W

15 • Kirchhoff's loop rule follows from (a) conservation of charge. (b) conservation of energy. (c) Newton's laws. (d) Coulomb's law. (e) quantization of charge.

16 • An ideal voltmeter should have ⎯⎯⎯ internal resistance.

(a) infinite
(b) zero

17 • [SSM] An ideal ammeter should have ⎯⎯⎯ internal resistance.

(a) infinite
(b) zero

18 • An ideal voltage source should have ⎯⎯⎯ internal resistance.

(a) infinite
(b) zero

19 • The capacitor C in Figure 25-45 is initially uncharged. Just after the switch S is closed, (a) the voltage across C equals $\mathcal{E}$. (b) the voltage across R equals $\mathcal{E}$. (c) the current in the circuit is zero. (d) both (a) and (c) are correct.

20 •• During the time it takes to fully charge the capacitor of Figure 25-45, (a) the energy supplied by the battery is $\frac{1}{2}C\mathcal{E}^2$. (b) the energy dissipated in the resistor is $\frac{1}{2}C\mathcal{E}^2$. (c) energy in the resistor is dissipated at a constant rate. (d) the total charge flowing through the resistor is $\frac{1}{2}C\mathcal{E}$.

21 •• [SSM] A battery is connected to a series combination of a switch, a resistor, and an initially uncharged capacitor. The switch is closed at $t = 0$. Which of the following statements is true?

(a) As the charge on the capacitor increases, the current increases.
(b) As the charge on the capacitor increases, the voltage drop across the resistor increases.
(c) As the charge on the capacitor increases, the current remains constant.
(d) As the charge on the capacitor increases, the voltage drop across the capacitor decreases.
(e) As the charge on the capacitor increases, the voltage drop across the resistor decreases.

22 •• A capacitor is discharging through a resistor. If it takes a time T for the charge on a capacitor to drop to half its initial value, how long does it take for the energy to drop to half its initial value?

23 • Which will produce more thermal energy when connected across an ideal battery, a small resistance or a large resistance?

24 • [SSM] All voltage sources have some internal resistance, usually on the order of 100 Ω or less. From this fact, explain the following statement that appears in some electronics textbooks: "A voltage source likes to see a high resistance."

25 • Do Kirchhoff's rules apply to circuits containing capacitors?

26 •• In Figure 25-46, all three resistors are identical. The power dissipated is (a) the same in R_1 as in the parallel combination of R_2 and R_3. (b) the same in R_1 and R_2. (c) greatest in R_1. (d) smallest in R_1.

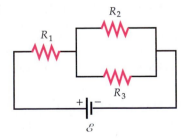

FIGURE 25-46 Problem 26

Estimation and Approximation

27 •• A 16-gauge copper wire insulated with rubber can safely carry a maximum current of 6 A. (a) How great a potential difference can be applied across 40 m of this wire? (b) Find the electric field in the wire when it carries a current of 6 A. (c) Find the power dissipated in the wire when it carries a current of 6 A.

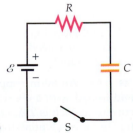

FIGURE 25-45 Problems 19 and 20

28 •• An automobile jumper cable 3 m long is constructed of multiple strands of copper wire that has an equivalent cross-sectional area of 10 mm². (*a*) What is the resistance of the jumper cable? (*b*) When the cable is used to start a car, it carries a current of 90 A. What is the potential drop that occurs across the jumper cable? (*c*) How much power is dissipated in the jumper cable?

29 •• A coil of Nichrome wire is to be used as the heating element in a water boiler that is required to generate 8 g of steam per second. The wire has a diameter of 1.80 mm and is connected to a 120-V power supply. Find the length of wire required.

30 •• SSM Compact fluorescent lightbulbs cost $6 each and have an expected lifetime of 8000 h. These bulbs consume 20 W of power, but produce the illumination equivalent to 75-W incandescent bulbs. Incandescent bulbs cost approximately $1.50 each and have an expected lifetime of 1200 h. If the average household has, on the average, six 75-W incandescent lightbulbs on constantly, and if energy costs 11.5 cents per kilowatt-hour, how much money would a consumer save each year by installing the energy-efficient fluorescent lightbulbs?

31 •• The wires in a house must be large enough in diameter so that they do not get hot enough to start a fire. Suppose a certain wire is to carry a current of 20 A, and it is determined that the joule heating of the wire should not exceed 2 W/m. What diameter must a copper wire have to be safe for this current?

32 •• SSM A laser diode used in making a laser pointer is a highly nonlinear circuit element. For a voltage drop across it less than approximately 2.3 V, it behaves as if it has effectively infinite internal resistance, but for voltages across it higher than this it has a very low internal resistance—effectively zero. (*a*) A laser pointer is made by putting two 1.55 V watch batteries in series across the laser diode. If the batteries each have an internal resistance between 100 Ω and 150 Ω, estimate the current in the laser diode. (*b*) About half of the power delivered to the laser diode goes into radiant energy. Using this fact, estimate the power of the laser diode, and compare this to typical quoted values of about 3 mW. (*c*) If the batteries each have a capacity of 20-mA hours (i.e., they can deliver a constant current of 20 mA for approximately one hour before discharging), estimate how long one can continuously operate the laser pointer before replacing the batteries.

Current and the Motion of Charges

33 • ISOLVE A 10-gauge copper wire carries a current of 20 A. Assuming one free electron per copper atom, calculate the drift velocity of the electrons.

34 • ISOLVE In a fluorescent tube of diameter 3 cm, 2.0×10^{18} electrons and 0.5×10^{18} positive ions (with a charge of $+e$) flow through a cross-sectional area each second. What is the current in the tube?

35 • In a certain electron beam, there are 5.0×10^6 electrons per cubic centimeter. Suppose the kinetic energy of each electron is 10 keV and the beam is cylindrical with a diameter of 1 mm. (*a*) What is the velocity of an electron in the beam? (*b*) Find the beam current.

36 •• A ring of radius a with a linear charge density λ rotates about its axis with angular velocity ω. Find an expression for the current.

37 •• SSM ISOLVE A 10-gauge copper wire and a 14-gauge copper wire are welded together end to end. The wires carry a current of 15 A. If there is one free electron per copper atom in each wire, find the drift velocity of the electrons in each wire.

38 •• In a certain particle accelerator, a proton beam with a diameter of 2 mm constitutes a current of 1 mA. The kinetic energy of each proton is 20 MeV. The beam strikes a metal target and is absorbed by it. (*a*) What is the number n of protons per unit volume in the beam? (*b*) How many protons strike the target in 1 minute? (*c*) If the target is initially uncharged, express the charge of the target as a function of time.

39 •• SSM In a proton supercollider, the protons in a 5-mA beam move with nearly the speed of light. (*a*) How many protons are there per meter of the beam? (*b*) If the cross-sectional area of the beam is 10^{-6} m², what is the number density of protons?

Resistance and Ohm's Law

40 • ISOLVE✔ A 10-m-long wire of resistance 0.2 Ω carries a current of 5 A. (*a*) What is the potential difference across the wire? (*b*) What is the magnitude of the electric field in the wire?

41 • ISOLVE✔ A potential difference of 100 V produces a current of 3 A in a certain resistor. (*a*) What is the resistance of the resistor? (*b*) What is the current when the potential difference is 25 V?

42 • ISOLVE A block of carbon is 3.0 cm long and has a square cross-sectional area with sides of 0.5 cm. A potential difference of 8.4 V is maintained across its length. (*a*) What is the resistance of the block? (*b*) What is the current in this resistor?

43 • ISOLVE A carbon rod with a radius of 0.1 mm is used to make a resistor. The resistivity of this material is 3.5×10^{-5} Ω·m. What length of the carbon rod will make a 10-Ω resistor?

44 • SSM ISOLVE✔ The third (current-carrying) rail of a subway track is made of steel and has a cross-sectional area of about 55 cm². The resistivity of steel is 10^{-7} Ω·m. What is the resistance of 10 km of this track?

45 • ISOLVE✔ What is the potential difference across one wire of a 30-m extension cord made of 16-gauge copper wire carrying a current of 5 A?

46 • How long is a 14-gauge copper wire that has a resistance of 2 Ω?

47 •• A cylinder of glass 1 cm long has a resistivity of 10^{12} Ω·m. How long would a copper wire of the same cross-sectional area need to be to have the same resistance as the glass cylinder?

48 •• An 80-m copper wire 1 mm in diameter is joined end to end with a 49-m iron wire of the same diameter. The current in each is 2 A. (*a*) Find the electric field in each wire. (*b*) Find the potential drop across each wire.

49 •• **SSM** A copper wire and an iron wire with the same length and diameter carry the same current I. (a) Find the ratio of the potential drops across these wires. (b) In which wire is the electric field greater?

50 •• A rubber tube 1 m long with an inside diameter of 4 mm is filled with a salt solution that has a resistivity of 10^{-3} $\Omega \cdot$m. Metal plugs form electrodes at the ends of the tube. (a) What is the resistance of the filled tube? (b) What is the resistance of the filled tube if it is uniformly stretched to a length of 2 m?

51 •• A wire of length 1 m has a resistance of 0.3 Ω. It is uniformly stretched to a length of 2 m. What is its new resistance?

52 •• Currents up to 30 A can be carried by 10-gauge copper wire. (a) What is the resistance of 100 m of 10-gauge copper wire? (b) What is the electric field in the wire when the current is 30 A? (c) How long does it take for an electron to travel 100 m in the wire when the current is 30 A?

53 •• A cube of copper has sides of 2 cm. If it is drawn out to form a 14-gauge wire, what will its resistance be?

54 •• **SSM** A diode is a circuit element with a very nonlinear IV curve. In a diode, $I = I_0(e^{V/(25 \text{ mV})} - 1)$, where $I_0 \sim 2 \times 10^{-9}$ A. Using a spreadsheet program, make a graph of I versus V for a typical diode, for both forward biasing ($V > 0$) and back-biasing ($V < 0$). Show that a plot $\ln(I)$ versus V for forward biasing (using $V > 0.3$ V), is nearly a straight line. What is the slope of the line?

55 •• (a) From the results of Problem 54, show that a diode effectively behaves like a resistor with infinite resistance if the voltage V applied across the diode is less than approximately 0.6 V, and behaves like a resistor with zero resistance if $V > 0.6$ V. (b) Estimate the current flowing through the forward biased diode in the circuit shown in Figure 25-47.

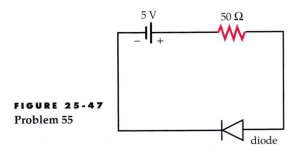

FIGURE 25-47
Problem 55

56 ••• Find the resistance between the ends of the half ring shown in Figure 25-48. The resistivity of the material of the ring is ρ.

FIGURE 25-48 Problem 56

57 ••• The radius of a wire of length L increases linearly along its length according to $r = a + [(b - a)/L]x$, where x is the distance from the small end of radius a. What is the resistance of this wire in terms of its resistivity ρ, length L, radius a, and radius b?

58 ••• **SSM** The space between two concentric spherical-shell conductors is filled with a material that has a resistivity of 10^9 $\Omega \cdot$m. If the inner shell has a radius of 1.5 cm and the outer shell has a radius of 5 cm, what is the resistance between the conductors? (*Hint:* Find the resistance of a spherical-shell element of the material of area $4\pi r^2$ and length dr, and integrate to find the total resistance of the set of shells in series.)

59 ••• **iSOLVE** ✓ The space between two metallic coaxial cylinders of length L and radii a and b is completely filled with a material having a resistivity ρ. (a) What is the resistance between the two cylinders? (See the hint in Problem 58.) (b) Find the current between the two cylinders if $\rho = 30$ $\Omega \cdot$m, $a = 1.5$ cm, $b = 2.5$ cm, $L = 50$ cm, and a potential difference of 10 V is maintained between the two cylinders.

Temperature Dependence of Resistance

60 • **SSM** A tungsten rod is 50 cm long and has a square cross-sectional area with sides of 1 mm. (a) What is its resistance at 20°C? (b) What is its resistance at 40°C?

61 • At what temperature will the resistance of a copper wire be 10 percent greater than it is at 20°C?

62 •• A toaster with a Nichrome heating element has a resistance of 80 Ω at 20°C and an initial current of 1.5 A. When the heating element reaches its final temperature, the current is 1.3 A. What is the final temperature of the heating element?

63 •• An electric space heater has a Nichrome heating element with a resistance of 8 Ω at 20°C. When 120 V are applied, the electric current heats the Nichrome wire to 1000°C. (a) What is the initial current drawn by the cold heating element? (b) What is the resistance of the heating element at 1000°C? (c) What is the operating wattage of this heater?

64 •• A 10-Ω Nichrome resistor is wired into an electronic circuit using copper leads (wires) of diameter 0.6 mm, with a total length of 50 cm. (a) What additional resistance is due to the copper leads? (b) What percentage error in the total added resistance is produced by neglecting the resistance of the copper leads? (c) What change in temperature would produce a change in resistance of the Nichrome wire equal to the resistance of the copper leads?

65 ••• **SSM** A wire of cross-sectional area A, length L_1, resistivity ρ_1, and temperature coefficient α_1 is connected end to end to a second wire of the same cross-sectional area, length L_2, resistivity ρ_2, and temperature coefficient α_2, so that the wires carry the same current. (a) Show that if $\rho_1 L_1 \alpha_1 + \rho_2 L_2 \alpha_2 = 0$, the total resistance R is independent of temperature for small temperature changes. (b) If one wire is made of carbon and the other wire is made of copper, find the ratio of their lengths for which R is approximately independent of temperature.

66 ••• The resistivity of tungsten increases approximately linearly from 56 n$\Omega \cdot$m at 293 K and 1.1 $\mu\Omega \cdot$m at 3500 K. Estimate (a) the resistance and (b) the diameter of a tungsten filament used in a 40-W bulb, assuming that the filament temperature is about 2500 K and that a 100-V dc supply is used to power the lightbulb. Assume that the length of the filament is constant and equal to 0.5 cm.

67 ••• A small light bulb used in an electronics class has a carbon filament in the form of a cylinder with a length of 3 cm and a diameter of $d = 40\mu m$. At temperatures between 500K and 700K, the resistivity of the carbon used in making small light bulb filaments is about 3×10^{-5} $\Omega \cdot m$. (a) Assuming that the bulb is a perfect blackbody radiator, calculate the temperature of the filament when a voltage $V = 5$ V is placed across it. (b) One problem with carbon filament bulbs, unlike tungsten filament bulbs, is that the resistivity of carbon decreases with increasing temperature. Explain why this is a problem.

Energy in Electric Circuits

68 • SSM Find the power dissipated in a resistor connected across a constant potential difference of 120 V if its resistance is (a) 5 Ω and (b) 10 Ω.

69 • ISOLVE A 10,000-Ω carbon resistor used in electronic circuits is rated at 0.25 W. (a) What maximum current can this resistor carry? (b) What maximum voltage can be placed across this resistor?

70 • A 1-kW heater is designed to operate at 240 V. (a) What is the heater's resistance and what current does the heater draw? (b) What is the power dissipated in this resistor if it operates at 120 V? Assume that its resistance is constant.

71 • A battery has an emf of 12 V. How much work does it do in 5 s if it delivers a current of 3 A?

72 • ISOLVE A battery with 12-V emf has a terminal voltage of 11.4 V when it delivers a current of 20 A to the starter of a car. What is the internal resistance r of the battery?

73 • SSM (a) How much power is delivered by the emf of the battery in Problem 72 when it delivers a current of 20 A? (b) How much of this power is delivered to the starter? (c) By how much does the chemical energy of the battery decrease when it delivers a current of 20 A to the starter for 3 min? (d) How much heat is developed in the battery when it delivers a current of 20 A for 3 min?

74 • A battery with an emf of 6 V and an internal resistance of 0.3 Ω is connected to a variable resistance R. Find the current and power delivered by the battery when R is (a) 0, (b) 5 Ω, (c) 10 Ω, and (d) infinite.

75 •• A 12-V automobile battery with negligible internal resistance can deliver a total charge of 160 A·h. (a) What is the total stored energy in the battery? (b) How long could this battery provide 150 W to a pair of headlights?

76 •• A space heater in an old home draws a 12.5-A current. A pair of 12-gauge copper wires carries the current from the fuse box to the wall outlet, a distance of 30 m. The voltage at the fuse box is exactly 120 V. (a) What is the voltage delivered to the space heater? (b) If the fuse will blow at a current of 20 A, how many 60-W bulbs can be supplied by this line when the space heater is on? (Assume that the wires from the wall to the space heater and to the light fixtures have negligible resistance.)

77 •• SSM ISOLVE✔ A lightweight electric car is powered by ten 12-V batteries. At a speed of 80 km/h, the average frictional force is 1200 N. (a) What must be the power of the electric motor if the car is to travel at a speed of 80 km/h? (b) If each battery can deliver a total charge of 160 A·h before recharging, what is the total charge in coulombs that can be delivered by the ten batteries before charging? (c) What is the total electrical energy delivered by the ten batteries before recharging? (d) How far can the car travel at 80 km/h before the batteries must be recharged? (e) What is the cost per kilometer if the cost of recharging the batteries is 9 cents per kilowatt-hour?

78 ••• A 100-W heater is designed to operate with an applied voltage of 120 V. (a) What is the heater's resistance, and what current does the heater draw? (b) Show that if the potential difference V across the heater changes by a small amount ΔV, the power P changes by a small amount ΔP, where $\Delta P/P \approx 2\Delta V/V$. (Hint: Approximate the changes with differentials and assume the resistance is constant.) (c) Find the approximate power dissipated in the heater, if the potential difference is decreased to 115 V.

Combinations of Resistors

79 • SSM (a) Find the equivalent resistance between point a and point b in Figure 25-49. (b) If the potential drop between point a and point b is 12 V, find the current in each resistor.

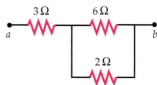

FIGURE 25-49 Problem 79

80 • Repeat Problem 79 for the resistor network shown in Figure 25-50.

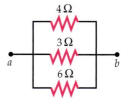

FIGURE 25-50 Problem 80

81 • (a) Show that the equivalent resistance between point a and point b in Figure 25-51 is R. (b) What would be the effect of adding a resistance R between point c and point d?

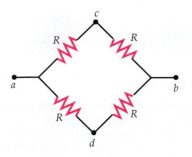

FIGURE 25-51 Problem 81

82 •• The battery in Figure 25-52 has negligible internal resistance. Find (a) the current in each resistor and (b) the power delivered by the battery.

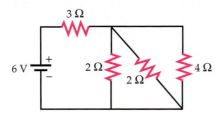

FIGURE 25-52 Problem 82

83 •• SSM A 5-V power supply has an internal resistance of 50 Ω. What is the smallest resistor that we can put in series with the power supply so that the voltage drop across the resistor is larger than 4.5 V?

84 •• A battery has an emf $\mathcal{E}$ and an internal resistance r. When a 5-Ω resistor is connected across the terminals, the current is 0.5 A. When this resistor is replaced by an 11-Ω resistor, the current is 0.25 A. Find (a) the emf $\mathcal{E}$ and (b) the internal resistance r.

85 •• Consider the equivalent resistance of two resistors R_1 and R_2 connected in parallel as a function R_1 and x, where x is the ratio R_2/R_1. (a) Show that $R_{eq} = R_1 x/(1 + x)$. (b) Sketch a plot of R_{eq}/R_1 as a function of x.

86 •• An ideal current source supplies a constant current regardless of the *load* that it is attached to. An almost-ideal current source can be made by putting a large resistor in series with an ideal voltage source. (a) What resistance is needed to turn an ideal 5-V voltage source into an almost-ideal 10-mA current source? (b) If we wish the current to drop by less than 10 percent when we load this current source, what is the largest resistance we can place in series with this current source?

87 •• Repeat Problem 79 for the resistor network shown in Figure 25-53.

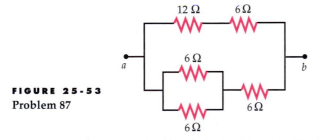

FIGURE 25-53
Problem 87

88 •• Repeat Problem 79 for the resistor network shown in Figure 25-54.

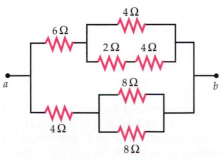

FIGURE 25-54 Problem 88

89 •• SSM A length of wire has a resistance of 120 Ω. The wire is cut into N identical pieces that are then connected in parallel. The resistance of the parallel arrangement is 1.875 Ω. Find N.

90 •• A parallel combination of an 8-Ω resistor and an unknown resistor R is connected in series with a 16-Ω resistor and a battery. This circuit is then disassembled and the three resistors are then connected in series with each other and the same battery. In both arrangements, the current through the 8-Ω resistor is the same. What is the unknown resistance R?

91 •• iSOLVE For the resistance network shown in Figure 25-55, find (a) R_3, so that $R_{ab} = R_1$, (b) R_2, so that $R_{ab} = R_3$; and (c) R_1, so that $R_{ab} = R_1$.

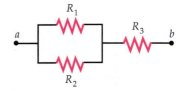

FIGURE 25-55 Problems 91 and 92

92 •• Check your results for Problem 91 using (a) $R_1 = 4\ \Omega$, $R_2 = 6\ \Omega$; (b) $R_1 = 4\ \Omega$, $R_3 = 3\ \Omega$; and (c) $R_2 = 6\ \Omega$, $R_3 = 3\ \Omega$.

Kirchhoff's Rules

93 • SSM In Figure 25-56, the emf is 6 V and $R = 0.5\ \Omega$. The rate of joule heating in R is 8 W. (a) What is the current in the circuit? (b) What is the potential difference across R? (c) What is r?

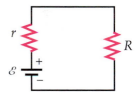

FIGURE 25-56 Problem 93

94 • For the circuit in Figure 25-57, find (a) the current, (b) the power delivered or absorbed by each source of emf, and (c) the rate of joule heating in each resistor. (Assume that the batteries have negligible internal resistance.)

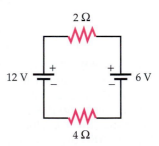

FIGURE 25-57 Problem 94

95 •• A sick car battery with an emf of 11.4 V and an internal resistance of 0.01 Ω is connected to a load of 2 Ω. To help the ailing battery, a second battery with an emf of 12.6 V and an internal resistance of 0.01 Ω is connected by jumper cables to the terminals of the first battery. (a) Draw a diagram of this circuit. (b) Find the current in each part of the circuit. (c) Find the power delivered by the second battery and discuss where this power goes, assuming that the emfs and internal resistances of both batteries remain constant.

96 •• In the circuit in Figure 25-58, the reading of the ammeter is the same with both switches open and both switches closed. Find the resistance R.

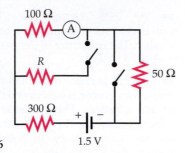

FIGURE 25-58 Problem 96

97 •• **SSM** In the circuit shown in Figure 25-59, the batteries have negligible internal resistance. Find (*a*) the current in each resistor, (*b*) the potential difference between point *a* and point *b*, and (*c*) the power supplied by each battery.

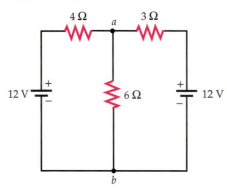

FIGURE 25-59
Problem 97

98 •• Repeat Problem 97 for the circuit in Figure 25-60.

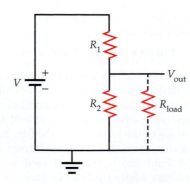

FIGURE 25-60
Problem 98

99 •• Two identical batteries, each with an emf $\mathcal{E}$ and an internal resistance r, can be connected across a resistance R either in series or in parallel. Is the power supplied to R greater when $R < r$ or when $R > r$?

100 •• **SSM** The circuit fragment shown in Figure 25-61 is called a *voltage divider*. (*a*) If R_{load} is not attached, show that $V_{\text{out}} = V(R_2/(R_1 + R_2))$. (*b*) If $R_1 = R_2 = 10$ kΩ, what is the smallest value of R_{load} that can be used so that V_{out} drops by less than 10 percent from its unloaded value? (V_{out} is measured with respect to ground.)

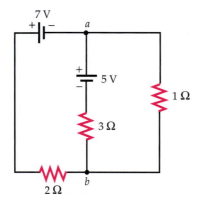

FIGURE 25-61 Problem 100

101 •• Thevenin's theorem states that the voltage divider circuit of Problem 100 can be replaced by a constant voltage source with voltage V' in series with a Thevenin resistance R' in series with the load resistor R_{load}. V' and R' depend only on V, R_1 and R_2. In this arrangement, the voltage drop across R_{load} will be the same as if the load resistor were placed in parallel with R_2 in the voltage divider from Problem 100.

(*a*) Show that $R' = \dfrac{R_1 R_2}{R_1 + R_2}$.

(*b*) Show that $V' = V\dfrac{R_2}{R_1 + R_2}$.

102 •• For the circuit shown in Figure 25-62, find (*a*) the current in each resistor, (*b*) the power supplied by each source of emf, and (*c*) the power dissipated in each resistor.

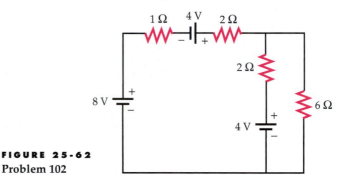

FIGURE 25-62
Problem 102

103 •• For the circuit shown in Figure 25-63, find the potential difference between point *a* and point *b*.

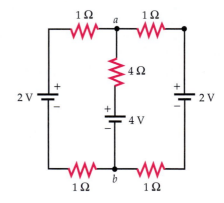

FIGURE 25-63
Problem 103

104 •• You have two batteries, one with $\mathcal{E} = 9$ V and $r = 0.8$ Ω and the other battery with $\mathcal{E} = 3$ V and $r = 0.4$ Ω. (*a*) Show how you would connect the batteries to give the largest current through a resistor R. Find the current for (*b*) $R = 0.2$ Ω, (*c*) $R = 0.6$ Ω, (*d*) $R = 1.0$ Ω, and (*e*) $R = 1.5$ Ω.

Ammeters and Voltmeters

105 •• **SSM** A digital voltmeter can be modeled as an ideal voltmeter with an infinite internal resistance in parallel with a 10 M·Ω resistor. Calculate the voltage measured by the voltmeter in the circuit shown in Figure 25-64 when (*a*) $R = 1$ kΩ, (*b*) $R = 10$ kΩ, (*c*) $R = 1$ MΩ, (*d*) $R = 10$ MΩ, and (*e*) $R = 100$ MΩ. (*f*) What is the largest value of R possible if we wish the measured voltage to be within 10 percent of the *true* voltage (i.e., the voltage drop without the voltmeter in place)?

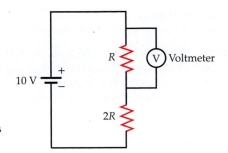

FIGURE 25-64
Problem 105

106 •• You are given a galvanometer meter movement that will deflect full scale if a current of 50 μA runs through the galvanometer. At this current, there is a voltage drop of 0.25 V across the meter. What is the meter's internal resistance?

107 •• We wish to change the meter in Problem 106 into an ammeter that can measure currents up to 100 mA. Show that this can be done by placing a resistor in parallel with the meter, and find the value of its resistance.

108 •• (*a*) If the ammeter from Problem 107 is used to measure the current through a 100-Ω resistor that is hooked up to a 10-V power supply, what current will the meter read? (The question is not as simple as it sounds.) (*b*) What if the ammeter is used to measure the current flowing through a 10-Ω resistor that is hooked up to a 1-V power supply?

109 •• **SSM** Show that the meter movement in Problem 106 can be converted into a voltmeter by placing a large resistance in series with the meter movement, and find the resistance needed for a full-scale deflection when 10 V are placed across it.

110 •• If the voltmeter described in Problem 109 is used to measure the voltage drop across R_1 in the circuit shown in Figure 25-65, what voltage will the voltmeter read?

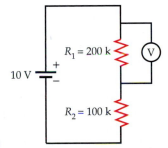

FIGURE 25-65 Problem 110

RC Circuits

111 • **ISOLVE** A 6-μF capacitor is charged to 100 V and is then connected across a 500-Ω resistor. (*a*) What is the initial charge on the capacitor? (*b*) What is the initial current just after the capacitor is connected to the resistor? (*c*) What is the time constant of this circuit? (*d*) How much charge is on the capacitor after 6 ms?

112 • (*a*) Find the initial energy stored in the capacitor of Problem 111. (*b*) Show that the energy stored in the capacitor is given by $U = U_0 e^{-2t/\tau}$, where U_0 is the initial energy and $\tau = RC$ is the time constant. (*c*) Sketch a plot of the energy U in the capacitor versus time t.

113 •• **SSM** **ISOLVE** In the circuit previously shown in Figure 25-40, emf $\mathcal{E} = 50$ V and $C = 2.0$ μF; the capacitor is initially uncharged. At 4 s after switch S is closed, the voltage drop across the resistor is 20 V. Find the resistance of the resistor.

114 •• **SSM** **ISOLVE** A 0.12-μF capacitor is given a charge Q_0. After 4 s, the capacitor's charge is $\frac{1}{2}Q_0$. What is the effective resistance across this capacitor?

115 •• A 1.6-μF capacitor, initially uncharged, is connected in series with a 10-kΩ resistor and a 5-V battery of negligible internal resistance. (*a*) What is the charge on the capacitor after a very long time? (*b*) How long does it take the capacitor to reach 99 percent of its final charge?

116 •• Consider the circuit shown in Figure 25-66. From your knowledge of how capacitors behave in circuits, find (*a*) the initial current through the battery just after the switch is closed, (*b*) the steady-state current through the battery when the switch has been closed for a long time, and (*c*) the maximum voltage across the capacitor.

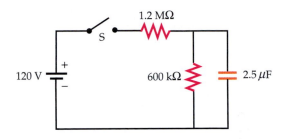

FIGURE 25-66 Problem 116

117 •• A 2-MΩ resistor is connected in series with a 1.5-μF capacitor and a 6.0-V battery of negligible internal resistance. The capacitor is initially uncharged. After a time $t = \tau = RC$, find (*a*) the charge on the capacitor, (*b*) the rate at which the charge is increasing, (*c*) the current, (*d*) the power supplied by the battery, (*e*) the power dissipated in the resistor, and (*f*) the rate at which the energy stored in the capacitor is increasing.

118 •• In the steady state, the charge on the 5-μF capacitor in the circuit shown in Figure 25-67 is 1000 μC. (*a*) Find the battery current. (*b*) Find the resistances R_1, R_2, and R_3.

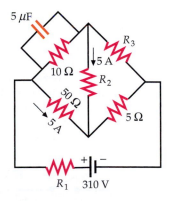

FIGURE 25-67 Problem 118

119 •• Show that Equation 25-35 can be written

$$\frac{dQ}{\mathcal{E}C - Q} = \frac{dt}{RC}$$

Integrate this equation to derive the solution given by Equation 25-36.

120 ••• SSM A photojournalist's flash unit uses a 9-V battery pack to charge a 0.15-μF capacitor, which is then discharged through the flash lamp of 10.5-Ω resistance when a switch is closed. The minimum voltage necessary for the flash discharge is 7 V. The capacitor is charged through an 18-kΩ resistor. (*a*) How much time is required to charge the capacitor to the required 7 V? (*b*) How much energy is released when the lamp flashes? (*c*) How much energy is supplied by the battery during the charging cycle and what fraction of that energy is dissipated in the resistor?

121 ••• For the circuit shown in Figure 25-68, (*a*) what is the initial battery current immediately after switch S is closed? (*b*) What is the battery current a long time after switch S is closed? (*c*) What is the current in the 600-Ω resistor as a function of time?

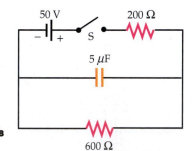

FIGURE 25-68
Problem 121

122 ••• For the circuit shown in Figure 25-69, (*a*) what is the initial battery current immediately after switch S is closed? (*b*) What is the battery current a long time after switch S is closed? (*c*) If the switch has been closed for a long time and is then opened, find the current through the 600-kΩ resistor as a function of time.

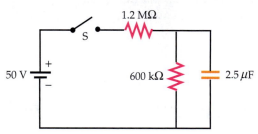

FIGURE 25-69 **Problem 122**

123 ••• In the circuit shown in Figure 25-70, the capacitor has a capacitance of 2.5 μF and the resistor has a resistance of 0.5 MΩ. Before the switch is closed, the potential drop across the capacitor is 12 V, as shown. Switch S is closed at $t = 0$. (*a*) What is the current in R immediately after switch S is closed? (*b*) At what time t is the voltage across the capacitor 24 V?

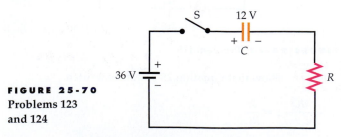

FIGURE 25-70
Problems 123
and 124

124 ••• Repeat Problem 123 if the capacitor is connected with reversed polarity.

General Problems

125 •• SSM In Figure 25-71, $R_1 = 4\ \Omega$, $R_2 = 6\ \Omega$, and $R_3 = 12\ \Omega$. If we denote the currents through these resistors by I_1, I_2, and I_3, respectively, then (*a*) $I_1 > I_2 > I_3$. (*b*) $I_2 = I_3$. (*c*) $I_3 > I_2$. (*d*) none of the above is correct.

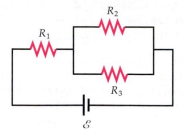

FIGURE 25-71
Problems 125 and 127

126 •• A 25-W lightbulb is connected in series with a 100-W lightbulb and a voltage V is placed across the combination. Which lightbulb is brighter? Explain.

127 • If the battery emf in Figure 25-71 is 24 V and $R_1 = 4\ \Omega$, $R_2 = 6\ \Omega$, and $R_3 = 12\ \Omega$, then (*a*) $I_2 = 4$ A. (*b*) $I_2 = 2$ A. (*c*) $I_2 = 1$ A. (*d*) none of the above is correct.

128 • A 10-Ω resistor is rated as being capable of dissipating 5 W of power. (*a*) What maximum current can this resistor tolerate? (*b*) What voltage across this resistor will produce the maximum current?

129 • A 12-V car battery has an internal resistance of 0.4 Ω. (*a*) What is the current if the battery is shorted momentarily? (*b*) What is the terminal voltage when the battery delivers a current of 20 A to start the car?

130 •• The current drawn from a battery is 1.80 A when a 7-Ω resistor is connected across the battery terminals. If a second 12-Ω resistor is connected in parallel with the 7-Ω resistor, the battery delivers a current of 2.20 A. What are the emf and internal resistance of the battery?

131 •• SSM A closed box has two metal terminals a and b. The inside of the box contains an unknown emf $\mathcal{E}$ in series with a resistance R. When a potential difference of 21 V is maintained between terminal a and terminal b, there is a current of 1 A between the terminals a and b. If this potential difference is reversed, a current of 2 A in the reverse direction is observed. Find $\mathcal{E}$ and R.

132 •• iSOLVE The capacitors in the circuit shown in Figure 25-72 are initially uncharged. (*a*) What is the initial value of the battery current when switch S is closed? (*b*) What is the battery current after a long time? (*c*) What are the final charges on the capacitors?

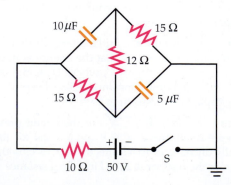

FIGURE 25-72
Probles 132

133 •• **SSM** The circuit shown in Figure 25-73 is a slide-type *Wheatstone bridge*. This bridge is used to determine an unknown resistance R_x, in terms of the known resistances R_1, R_2, and R_0. The resistances R_1 and R_2 comprise a wire 1 m long. Point a is a sliding contact that is moved along the wire to vary these resistances. Resistance R_1 is proportional to the distance from the left end of the wire (labeled 0 cm) to point a, and R_2 is proportional to the distance from point a to the right end of the wire (labeled 100 cm). The sum of R_1 and R_2 remains constant. When points a and b are at the same potential, there is no current in the galvanometer and the bridge is said to be balanced. (Because the galvanometer is used to detect the absence of a current, it is called a *null detector*.) If the fixed resistance $R_0 = 200 \Omega$, find the unknown resistance R_x if (a) the bridge balances at the 18-cm mark, (b) the bridge balances at the 60-cm mark, and (c) the bridge balances at the 95-cm mark.

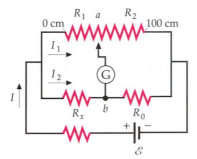

FIGURE 25-73
Problems 133 and 134

134 •• For the Wheatstone bridge presented in Problem 133, the bridge balances at the 98-cm mark when $R_0 = 200 \Omega$. (a) What is the unknown resistance? (b) What effect would an error of 2 mm in the location of the balance point have on the measured value of the unknown resistance? (c) How should R_0 be changed so that the balance point for this unknown resistor will be nearer the 50-cm mark?

135 •• A cyclotron produces a 3.50-μA proton beam of 60-MeV energy. The protons impinge and come to rest inside a 50-g copper target within the vacuum chamber. (a) Determine the number of protons that strike the target per second. (b) Find the energy deposited in the target per second. (c) How much time elapses before the target temperature rises 300°C? (Neglect cooling by radiation.)

136 •• The belt of a Van de Graaff generator carries a surface charge density of 5 mC/m². The belt is 0.5 m wide and moves at 20 m/s. (a) What current does it carry? (b) If this charge is raised to a potential of 100 kV, what is the minimum power of the motor needed to drive the belt?

137 •• Conventional large electromagnets use water cooling to prevent excessive heating of the magnet coils. A large laboratory electromagnet draws 100 A when a voltage of 240 V is applied to the terminals of the energizing coils. To cool the coils, water at an initial temperature of 15°C is circulated through the coils. How many liters per second must pass through the coils if their temperature should not exceed 50°C?

138 •• A parallel-plate capacitor is made from plates of area A separated by a distance d, and are filled with a dielectric with dielectric constant κ and resistivity ρ, show that the product of the resistance R of this dielectric, with the capacitance of the capacitor, is $RC = \epsilon_0 \rho \kappa$.

139 •• Show that the result of Problem 138 is true for a cylindrical capacitor or resistor. Should it be true for a capacitor or resistor of any shape?

140 •• **SSM** (a) Show that a leaky capacitor (one for which the resistance of the dielectric is finite) can be modeled as a capacitor with infinite resistance in parallel with a resistor. (b) Show that the time constant for discharging this capacitor is $\tau = \epsilon_0 \rho \kappa$. ($c$) Mica has a dielectric constant $\kappa = 5$ and a resistivity $\rho = 9 \times 10^{13} \ \Omega \cdot m$. Calculate the time it takes for the charge of a mica-filled capacitor to decrease to 10 percent of its initial value.

141 ••• Figure 25-74 shows the basis of the sweep circuit used in an oscilloscope. Switch S is an electronic switch that closes whenever the potential across the terminals switches reaches a value V_c; switch S opens when the potential has dropped to 0.2 V. The emf $\mathcal{E}$, which is much greater than V_c, charges the capacitor C through a resistor R_1. The resistor R_2 represents the small but finite resistance of the electronic switch. In a typical circuit, $\mathcal{E} = 800$ V, $V_c = 4.2$ V, $R_2 = 0.001 \ \Omega$, $R_1 = 0.5$ MΩ, and $C = 0.02 \ \mu$F. (a) What is the time constant for charging of the capacitor C? (b) Show that in the time required to bring the potential across switch S to the critical potential $V_c = 4.2$ V, the voltage across the capacitor increases almost linearly with time. (*Hint:* Use the expansion of the exponential for small values of exponent.) (c) What should the value of R_1 be so that C charges from 0.2 V to 4.2 V in 0.1 s? (d) How much time elapses during the discharge of C through switch S? (e) At what rate is energy dissipated in the resistor R_1 and in the switch resistance?

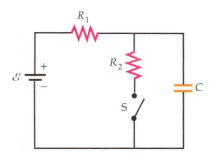

FIGURE 25-74 Problem 141

142 ••• In the circuit shown in Figure 25-75, $R_1 = 2$ MΩ, $R_2 = 5$ MΩ, and $C = 1 \ \mu$F. At $t = 0$, switch S is closed, and at $t = 2.0$ s switch S is opened. (a) Sketch the voltage across C and the current through R_2 between $t = 0$ and $t = 10$ s. (b) Find the voltage across the capacitor at $t = 2$ s and at $t = 8$ s.

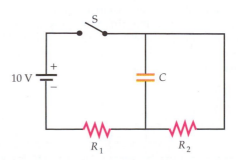

FIGURE 25-75 Problem 142

143 ••• Two batteries with emfs $\mathcal{E}_1$ and $\mathcal{E}_2$ and internal resistances r_1 and r_2 are connected in parallel. Prove that if a resistor is connected in parallel with this combination the optimal load resistance (the resistance at which maximum power is delivered) is $R = r_1 r_2/(r_1 + r_2)$.

144 ••• [SSM] Capacitors C_1 and C_2 are connected in parallel by a resistor and two switches, as shown in Figure 25-76. Capacitor C_1 is initially charged to a voltage V_0, and capacitor C_2 is uncharged. The switches S_1 and S_2 are then closed. (a) What are the final charges on C_1 and C_2? (b) Compare the initial and final stored energies of the system. (c) What caused the decrease in the capacitor-stored energy?

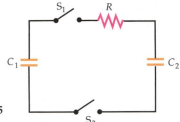

FIGURE 25-76
Problems 144 and 145

145 ••• (a) In Problem 144, find the current through R after the switches S_1 and S_2 are closed as a function of time. (b) Find the energy dissipated in the resistor as a function of time. (c) Find the total energy dissipated in the resistor and compare it with the loss of stored energy found in Part (b) of Problem 144.

146 ••• In the circuit shown in Figure 25-77, the capacitors are initially uncharged. Switch S_2 is closed and then switch S_1 is closed. (a) What is the battery current immediately after S_1 is closed? (b) What is the battery current a long time after both switches are closed? (c) What is the final voltage across C_1? (d) What is the final voltage across C_2? (e) Switch S_2 is opened again after a long time. Give the current in the 150-Ω resistor as a function of time.

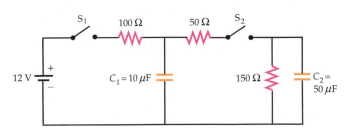

FIGURE 25-77 Problem 146

147 ••• [SSM] The differential resistance[†] of a nonohmic circuit element is defined as $R_d = dV/dI$, where V is the voltage across the element, and I is the current through the element. Show that for $V > 0.6$ V, the differential resistance of a diode (Problem 54) is approximately $R_d = (25 \text{ mV})/I$, and for $V < 0$, R_d increases exponentially with $|V|$. Use this result to justify the approximation given in Problem 55.

148 •• A graph of the voltage as a function of current for an Esaki diode is shown in Figure 25-78. Make a graph of the differential resistance of the diode as a function of voltage. (See Problem 147 for a definition of differential resistance.) At what value of the voltage does the differential resistance become negative?

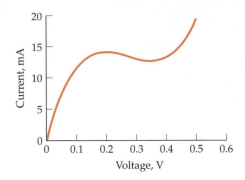

FIGURE 25-78 Problem 148

149 ••• A linear accelerator produces a pulsed beam of electrons. The current is 1.6 A for the 0.1-μs duration of each pulse. (a) How many electrons are in each pulse? (b) What is the average current of the beam if there are 1000 pulses per second? (c) If each electron acquires an energy of 400 MeV, what is the average power output of the accelerator? (d) What is the peak power output? (e) What fraction of the time is the accelerator actually accelerating electrons? (This is called the *duty factor* of the accelerator.)

150 ••• Calculate the equivalent resistance between points a and b for the infinite ladder of resistors shown in Figure 25-79.

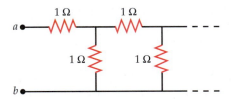

FIGURE 25-79 Problem 150

151 ••• [SSM] Calculate the equivalent resistance between points a and b for the infinite ladder of resistors shown in Figure 25-80, where R_1 and R_2 can take any value.

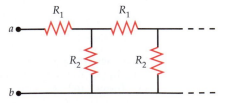

FIGURE 25-80 Problem 151

---

† Differential resistance (dV/dI) is also called dynamic resistance or dynamic impedance.

The Magnetic Field

THE AURORA BOREALIS APPEARS WHEN "SOLAR WIND," CHARGED PARTICLES PRODUCED BY NUCLEAR FUSION REACTIONS IN THE SUN, BECOME TRAPPED IN THE EARTH'S MAGNETIC FIELD.

? **How does the earth's magnetic field act on subatomic particles? (See Example 26-1.)**

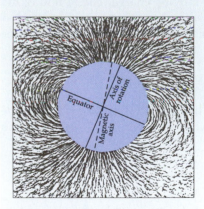

FIGURE 26-1 Magnetic field lines of the earth depicted by iron filings around a uniformly magnetized sphere. The field lines exit from the north magnetic pole, which is near the south geographic pole, and enter the south magnetic pole, which is near the north geographic pole.

More than 2000 years ago, the Greeks were aware that a certain type of stone (now called magnetite) attracts pieces of iron, and there are written references to the use of magnets for navigation dating from the twelfth century.

In 1269, Pierre de Maricourt discovered that a needle laid at various positions on a spherical natural magnet orients itself along lines that pass through points at opposite ends of the sphere. He called these points the poles of the magnet. Subsequently, many experimenters noted that every magnet of any shape has two poles, designated the north and south poles, where the force exerted by the magnet is strongest. It was also noted that the like poles of two magnets repel each other and the unlike poles of two magnets attract each other.

In 1600, William Gilbert discovered that the earth is a natural magnet with magnetic poles near the north and south geographic poles. Since the north pole of a compass needle points toward the south pole of a given magnet, what we call the north pole of the earth is actually a south magnetic pole, as illustrated in Figure 26-1.

Although electric charges and magnetic poles are similar in many respects, there is an important difference: Magnetic poles always occur in pairs. When a magnet is broken in half, equal and opposite poles appear at either side of the break point. The result is two magnets, each with a north and south pole. There has long been speculation about the existence of an isolated magnetic pole, and in recent years considerable experimental effort has been made to find such an object. Thus far, there is no conclusive evidence that an isolated magnetic pole exists.

➤ In this chapter, we consider the effects of a given magnetic field on moving charges and on wires carrying currents. The sources of magnetic fields are discussed in the next chapter.

26-1 The Force Exerted by a Magnetic Field

The existence of a magnetic field $\vec{B}$ at some point in space can be demonstrated with a compass needle. If there is a magnetic field, the needle will align itself in the direction of the field.

Experimentally it is observed that, when a charge q has velocity $\vec{v}$ in a magnetic field, there is a force on the magnetic field that is proportional to q and to v, and to the sine of the angle between the directions of $\vec{v}$ and $\vec{B}$. Surprisingly, the force is perpendicular to both the velocity and the field. These experimental results can be summarized as follows: When a charge q moves with velocity $\vec{v}$ in a magnetic field $\vec{B}$, the magnetic force $\vec{F}$ on the charge is

$$\vec{F} = q\vec{v} \times \vec{B} \qquad \text{26-1}$$

MAGNETIC FORCE ON A MOVING CHARGE

Since $\vec{F}$ is perpendicular to both $\vec{v}$ and $\vec{B}$, $\vec{F}$ is perpendicular to the plane defined by these two vectors. The direction of $\vec{v} \times \vec{B}$ is given by the right-hand rule as $\vec{v}$ is rotated into $\vec{B}$, as illustrated in Figure 26-2. If q is positive, then $\vec{F}$ is in the same direction as $\vec{v} \times \vec{B}$.

Examples of the direction of the forces exerted on moving charges when the magnetic field vector $\vec{B}$ is in the vertical direction are shown in Figure 26-3. Note that the direction of any particular magnetic field $\vec{B}$ can be found experimentally by measuring $\vec{F}$ and $\vec{v}$ for several velocities in different directions and then applying Equation 26-1.

Equation 26-1 defines the **magnetic field** $\vec{B}$ in terms of the force exerted on a moving charge. The SI unit of magnetic field is the **tesla** (T). A charge of one coulomb moving with a velocity of one meter per second perpendicular to a magnetic field of one tesla experiences a force of one newton:

$$1\,\text{T} = 1\frac{\text{N}}{\text{Cm/s}} = 1\,\text{N/(A·m)} \qquad \text{26-2}$$

This unit is rather large. The magnetic field of the earth has a magnitude somewhat less than 10^{-4} T on the earth's surface. The magnetic fields near powerful

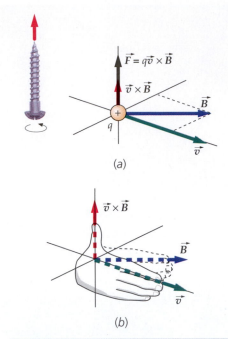

FIGURE 26-2 Right-hand rule for determining the direction of a force exerted on a charge moving in a magnetic field. If q is positive, then $\vec{F}$ is in the same direction as $\vec{v} \times \vec{B}$. (*a*) The cross product $\vec{v} \times \vec{B}$ is perpendicular to both $\vec{v}$ and $\vec{B}$ and is in the direction of the advance of a right-hand-threaded screw if turned in the same direction as to rotate $\vec{v}$ into $\vec{B}$. (*b*) If the fingers of the right hand are in the direction of $\vec{v}$ so that they can be curled toward $\vec{B}$, the thumb points in the direction of $\vec{v} \times \vec{B}$.

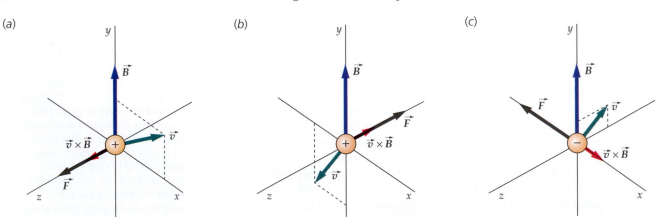

FIGURE 26-3 Direction of the magnetic force on a charged particle moving with velocity $\vec{v}$ in a magnetic field $\vec{B}$.

permanent magnets are about 0.1 T to 0.5 T, and powerful laboratory and industrial electromagnets produce fields of 1 T to 2 T. Fields greater than 10 T are difficult to produce because the resulting magnetic forces will either tear the magnets apart or crush the magnets. A commonly used unit, derived from the cgs system, is the **gauss** (G), which is related to the tesla as follows

$$1 \, G = 10^{-4} \, T \qquad\qquad 26\text{-}3$$

DEFINITION—GAUSS

Since magnetic fields are often given in gauss, which is not an SI unit, remember to convert from gauss to teslas when making calculations.

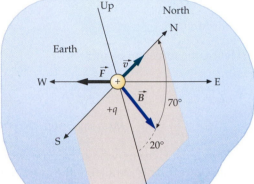

FIGURE 26-4

FORCE ON A PROTON GOING NORTH   **EXAMPLE 26-1**

The magnetic field of the earth is measured at a point on the surface to have a magnitude of 0.6 G and is directed downward and, in the northern hemisphere, northward, making an angle of about 70° with the horizontal, as shown in Figure 26-4. (The earth's magnetic field varies from place to place. These data are approximately correct for the central United States.) A proton ($q = +e$) is moving horizontally in the northward direction with speed $v = 10 \, Mm/s = 10^7 \, m/s$. Calculate the magnetic force on the proton (a) using $F = qvB \sin \theta$ and (b) by expressing $\vec{v}$ and $\vec{B}$ in terms of the unit vectors $\hat{i}, \hat{j}, \hat{k}$, and computing $\vec{F} = q\vec{v} \times \vec{B}$.

PICTURE THE PROBLEM Let the x and y directions be east and north, respectively, and let the z direction be upward (Figure 26-5). The velocity vector is then in the y direction.

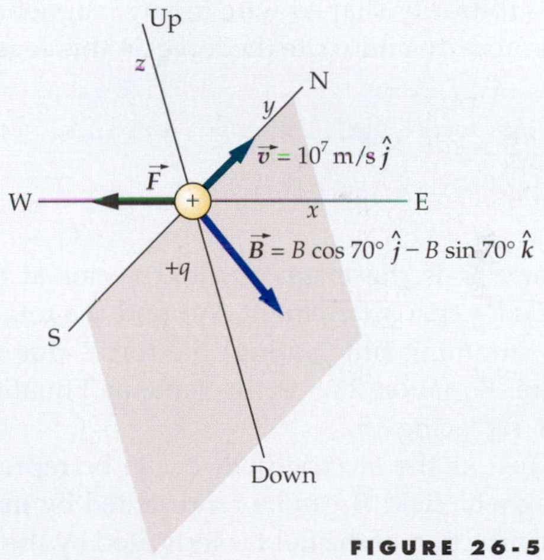

(a) Calculate $F = qvB \sin \theta$ using $\theta = 70°$. From Figure 26-4 we see that the direction of the force is westward:

$$F = qvB \sin 70°$$

$$= (1.6 \times 10^{-19} \, C)(10^7 \, m/s)(0.6 \times 10^{-4} \, T)(0.94)$$

$$= \boxed{9.02 \times 10^{-17} \, N}$$

(b) 1. The magnetic force is the vector product of $q\vec{v}$ and $\vec{B}$:

$$\vec{F} = q\vec{v} \times \vec{B}$$

2. Express $\vec{v}$ and $\vec{B}$ in terms of their components:

$$\vec{v} = v_y \hat{j}$$

$$\vec{B} = B_y \hat{j} + B_z \hat{k}$$

3. Write $\vec{F} = q\vec{v} \times \vec{B}$ in terms of these components:

$$\vec{F} = q\vec{v} \times \vec{B} = q(v_y \hat{j}) \times (B_y \hat{j} + B_z \hat{k})$$

$$= qv_y B_y (\hat{j} \times \hat{j}) + qv_y B_z (\hat{j} \times \hat{k}) = 0 + qv_y B_z \hat{i}$$

4. Evaluate $\vec{F}$:

$$\vec{F} = qv(-B \sin \theta)\hat{i}$$

$$= -(1.6 \times 10^{-19} \, C)(10^7 \, m/s)(0.6 \times 10^{-4} \, T)\sin 70° \, \hat{i}$$

$$= \boxed{-9.02 \times 10^{-17} \, N\hat{i}}$$

FIGURE 26-5

REMARKS Note that the direction of $\hat{i}$ is eastward, so the force is directed westward as shown in Figure 26-5.

EXERCISE Find the force on a proton moving with velocity $\vec{v} = 4 \times 10^6 \, m/s \, \hat{i}$ in a magnetic field $\vec{B} = 2.0 \, T\hat{k}$. (*Answer* $-1.28 \times 10^{-12} \, N\hat{j}$)

When a wire carries a current in a magnetic field, there is a force on the wire that is equal to the sum of the magnetic forces on the charged particles whose motion produces the current. Figure 26-6 shows a short segment of wire of cross-sectional area A and length L carrying a current I. If the wire is in a magnetic field $\vec{B}$, the magnetic force on each charge is $q\vec{v}_d \times \vec{B}$, where $\vec{v}_d$ is the drift velocity of the charge carriers (the drift velocity is the same as the average velocity). The number of charges in the wire segment is the number n per unit volume times the volume AL. Thus, the total force on the wire segment is

$$\vec{F} = (q\vec{v}_d \times \vec{B})nAL$$

From Equation 25-3, the current in the wire is

$$I = nqv_dA$$

Hence, the force can be written

$$\vec{F} = I\vec{L} \times \vec{B} \qquad \text{26-4}$$

MAGNETIC FORCE ON A SEGMENT OF CURRENT-CARRYING WIRE

where $\vec{L}$ is a vector whose magnitude is the length of the wire and whose direction is parallel to the current. For the current in the positive x direction (Figure 26-7) and the magnetic field vector at the segment in the xy plane, the force on the wire is directed along the z axis.

In Equation 26-4 it is assumed that the wire segment is straight and that the magnetic field does not vary over its length. The equation can be generalized for an arbitrarily shaped wire in any magnetic field. If we choose a very small wire segment $d\vec{\ell}$ and write the force on this segment as $d\vec{F}$, we have

$$d\vec{F} = Id\vec{\ell} \times \vec{B} \qquad \text{26-5}$$

MAGNETIC FORCE ON A CURRENT ELEMENT

where $\vec{B}$ is the magnetic field vector at the segment. The quantity $Id\vec{\ell}$ is called a **current element.** We find the total force on a current-carrying wire by summing (integrating) the forces due to all the current elements in the wire. Equation 26-5 is the same as Equation 26-1 with the current element $Id\vec{\ell}$ replacing $q\vec{v}$.

Just as the electric field $\vec{E}$ can be represented by electric field lines, the magnetic field $\vec{B}$ can be represented by **magnetic field lines.** In both cases, the direction of the field is indicated by the direction of the field lines and the magnitude of the field is indicated by their density. There are, however, two important differences between electric field lines and magnetic field lines:

1. Electric field lines are in the direction of the electric force on a positive charge, but the magnetic field lines are perpendicular to the magnetic force on a moving charge.

2. Electric field lines begin on positive charges and end on negative charges; magnetic field lines neither begin nor end.

Figure 26-8 shows the magnetic field lines both inside and outside a bar magnet.

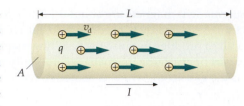

FIGURE 26-6 Wire segment of length L carrying current I. If the wire is in a magnetic field, there will be a force on each charge carrier resulting in a force on the wire.

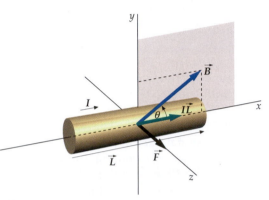

FIGURE 26-7 Magnetic force on a current-carrying segment of wire in a magnetic field. The current is in the x direction, and the magnetic field is in the xy plane and makes an angle θ with the $+x$ direction. The force $\vec{F}$ is in the $+z$ direction, perpendicular to both $\vec{B}$ and $\vec{L}$, and has magnitude $ILB \sin \theta$.

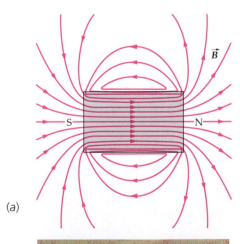

(a)

(b)

FIGURE 26-8 (a) Magnetic field lines inside and outside a bar magnet. The lines emerge from the north pole and enter the south pole, but they have no beginning or end. Instead, they form closed loops. (b) Magnetic field lines outside a bar magnet as indicated by iron filings.

FORCE ON A STRAIGHT WIRE   **EXAMPLE 26-2**

FIGURE 26-9

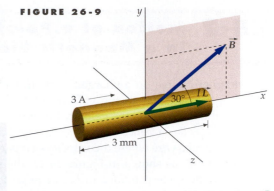

A wire segment 3 mm long carries a current of 3 A in the $+x$ direction. It lies in a magnetic field of magnitude 0.02 T that is in the xy plane and makes an angle of 30° with the $+x$ direction, as shown in Figure 26-9. What is the magnetic force exerted on the wire segment?

PICTURE THE PROBLEM The magnetic force is in the direction of $\vec{L} \times \vec{B}$, which we see from Figure 26-9 is in the positive z direction.

The magnetic force is given by
Equation 26-4:

$$\vec{F} = I\vec{L} \times \vec{B} = ILB \sin 30° \, \hat{k}$$
$$= (3 \text{ A})(0.003 \text{ m})(0.02 \text{ T})(\sin 30°)\hat{k}$$
$$= \boxed{9 \times 10^{-5} \text{ N}\hat{k}}$$

FIGURE 26-10

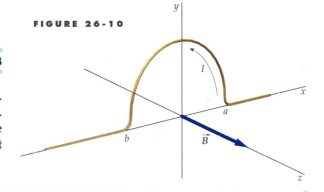

FORCE ON A BENT WIRE   **EXAMPLE 26-3**

A wire bent into a semicircular loop of radius R lies in the xy plane. It carries a current I from point a to point b, as shown in Figure 26-10. There is a uniform magnetic field $\vec{B} = B\hat{k}$ perpendicular to the plane of the loop. Find the force acting on the semicircular loop part of the wire.

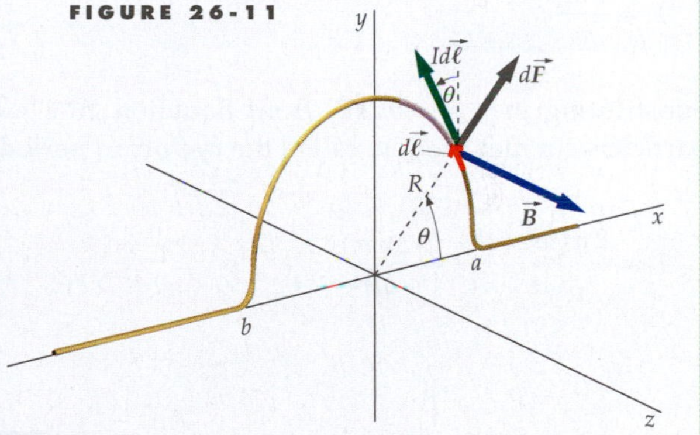

PICTURE THE PROBLEM The force $d\vec{F}$ exerted on a segment of the semicircular wire lies in the xy plane, as shown in Figure 26-11. We find the total force by expressing the x and y components of $d\vec{F}$ in terms of θ and integrating them separately from $\theta = 0$ to $\theta = \pi$.

FIGURE 26-11

1. Write the force $d\vec{F}$ on a current element $d\vec{\ell}$.

$$d\vec{F} = I d\vec{\ell} \times \vec{B}$$

2. Express $d\vec{\ell}$ in terms of the unit vectors $\hat{i}$ and $\hat{j}$:

$$d\vec{\ell} = -d\ell \sin \theta \hat{i} + d\ell \cos \theta \hat{j}$$

3. Compute $I d\vec{\ell}$ using $d\ell = R d\theta$ and $\vec{B} = B\hat{k}$:

$$d\vec{F} = I d\vec{\ell} \times \vec{B}$$
$$= I(-R \sin \theta d\theta \hat{i} + R \cos \theta d\theta \hat{j}) \times B\hat{k}$$
$$= IRB \sin \theta d\theta \hat{j} + IRB \cos \theta d\theta \hat{i}$$

4. Integrate each component of $d\vec{F}$ from $\theta = 0$ to $\theta = \pi$:

$$\vec{F} = \int d\vec{F} = IRB\hat{i} \int_0^\pi \cos \theta d\theta + IRB\hat{j} \int_0^\pi \sin \theta d\theta$$
$$= IRB\hat{i} \, (0) + IRB \, \hat{j} \, (2) = \boxed{2IRB\hat{j}}$$

PLAUSIBILITY CHECK The result that the x component of $\vec{F}$ is zero can be seen from symmetry. For the right half of the loop, $d\vec{F}$ points to the right; for the left half of the loop, $d\vec{F}$ points to the left.

REMARKS The net force on the semicircular wire is the same as if the semicircle were replaced by a straight-line segment of length $2R$ connecting points a and b. (This is a general result, as shown in Problem 30.)

26-2 Motion of a Point Charge in a Magnetic Field

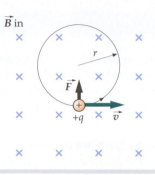

The magnetic force on a charged particle moving through a magnetic field is always perpendicular to the velocity of the particle. The magnetic force thus changes the direction of the velocity but not the velocity's magnitude. Therefore, *magnetic fields do no work on particles and do not change their kinetic energy.*

In the special case where the velocity of a particle is perpendicular to a uniform magnetic field, as shown in Figure 26-12, the particle moves in a circular orbit. The magnetic force provides the centripetal force necessary for the centripetal acceleration v^2/r in circular motion. We can use Newton's second law to relate the radius of the circle to the magnetic field and the speed of the particle. If the velocity is $\vec{v}$, the magnitude of the net force is qvB, since $\vec{v}$ and $\vec{B}$ are perpendicular. Newton's second law gives

$$F = ma$$

$$qvB = m\frac{v^2}{r}$$

or

$$r = \frac{mv}{qB} \qquad \text{26-6}$$

FIGURE 26-12 Charged particle moving in a plane perpendicular to a uniform magnetic field. The magnetic field is into the page as indicated by the crosses. (Each cross represents the tail feathers of an arrow. A field out of the plane of the page would be indicated by dots, each dot representing the point of an arrow.) The magnetic force is perpendicular to the velocity of the particle, causing it to move in a circular orbit.

The period of the circular motion is the time it takes the particle to travel once around the circumference of the circle. The period is related to the speed by

$$T = \frac{2\pi r}{v}$$

Substituting in $r = mv/(qB)$ from Equation 26-6, we obtain the period of the particle's circular motion, called the **cyclotron period:**

$$T = \frac{2\pi(mv/qB)}{v} = \frac{2\pi m}{qB} \qquad \text{26-7}$$

CYCLOTRON PERIOD

(*a*) Circular path of electrons moving in the magnetic field produced by two large coils. The electrons ionize the gas in the tube, causing it to give off a bluish glow that indicates the path of the beam. (*b*) False-color photograph showing tracks of a 1.6-MeV proton (red) and a 7-MeV α particle (yellow) in a cloud chamber. The radius of curvature is proportional to the momentum and inversely proportional to the charge of the particle. For these energies, the momentum of the α particle, which has twice the charge of the proton, is about four times that of the proton and so its radius of curvature is greater.

(*b*)

(*a*)

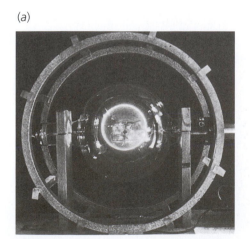

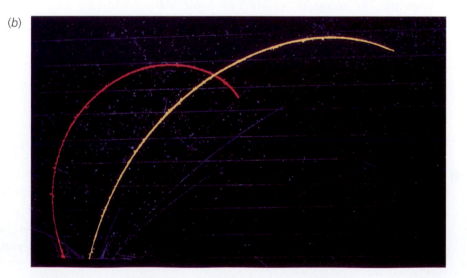

The frequency of the circular motion, called the **cyclotron frequency,** is the reciprocal of the period:

$$f = \frac{1}{T} = \frac{qB}{2\pi m}, \quad \text{so} \quad \omega = 2\pi f = \frac{q}{m}B \qquad\qquad 26\text{-}8$$

CYCLOTRON FREQUENCY

Note that the period and the frequency given by Equations 26-7 and 26-8 depend on the charge-to-mass ratio q/m, but the period and the frequency are independent of the velocity v or the radius r. Two important applications of the circular motion of charged particles in a uniform magnetic field, the mass spectrometer and the cyclotron, are discussed later in this section.

CYCLOTRON PERIOD · **EXAMPLE 26-4**

A proton of mass $m = 1.67 \times 10^{-27}$ kg and charge $q = e = 1.6 \times 10^{-19}$ C moves in a circle of radius $r = 21$ cm perpendicular to a magnetic field $B = 4000$ G. Find (a) the period of the motion and (b) the speed of the proton.

1. Calculate the period T from Equation 26-7 with $B = 4000$ G $= 0.4$ T:

$$T = \frac{2\pi m}{qB} = \frac{2\pi(1.67 \times 10^{-27}\text{ kg})}{(1.6 \times 10^{-19}\text{ C})(0.4\text{ T})}$$

$$= \boxed{1.64 \times 10^{-7}\text{ s} = 164\text{ ns}}$$

2. Calculate the speed v from Equation 26-6:

$$v = \frac{rqB}{m} = \frac{(0.21\text{ m})(1.6 \times 10^{-19}\text{ C})(0.4\text{ T})}{1.67 \times 10^{-27}\text{ kg}}$$

$$= \boxed{8.05 \times 10^6\text{ m/s} = 8.05\text{ m/}\mu\text{s}}$$

REMARKS The radius of the circular motion is proportional to the speed, but the period is independent of both the speed and radius.

PLAUSIBILITY CHECK Note that the product of the speed v and the period T equals the circumference of the circle $2\pi r$ as expected:
$vT = (8.05 \times 10^6\text{ m/s})(1.64 \times 10^{-7}\text{ s}) = 1.32\text{ m}; 2\pi r = 2\pi(0.21\text{ m}) = 1.32\text{ m}.$

Suppose that a charged particle enters a uniform magnetic field with a velocity that is not perpendicular to $\vec{B}$. There is no force component, and thus no acceleration component, parallel to $\vec{B}$, so the component of the velocity parallel to $\vec{B}$ remains constant. The magnetic force on the particle is perpendicular to $\vec{B}$, so the change in motion of the particle due to this force is the same as that just discussed. The path of the particle is thus a helix, as shown in Figure 26-13.

(a)

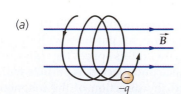

(b)

FIGURE 26-13 (a) When a particle has a velocity component parallel to a magnetic field as well as a velocity component perpendicular to the magnetic field the particle moves in a helical path around the field lines. (b) Cloud-chamber photograph of the helical path of an electron moving in a magnetic field. The path of the electron is made visible by the condensation of water droplets in the cloud chamber.

The motion of charged particles in nonuniform magnetic fields can be quite complex. Figure 26-14 shows a **magnetic bottle,** an interesting magnetic field configuration in which the field is weak at the center and strong at both ends. A detailed analysis of the motion of a charged particle in such a field shows that the particle spirals around the field lines and becomes trapped, oscillating back and forth between points P_1 and P_2 in the figure. Such magnetic field configurations are used to confine dense beams of charged particles, called *plasmas,* in nuclear fusion research. A similar phenomenon is the oscillation of ions back and forth between the earth's magnetic poles in the Van Allen belts (Figure 26-15).

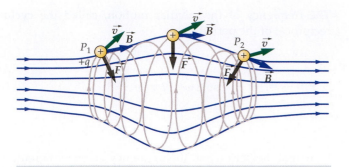

FIGURE 26-14 Magnetic bottle. When a charged particle moves in such a field, which is strong at both ends and weak in the middle, the particle becomes trapped and moves back and forth, spiraling around the field lines.

*The Velocity Selector

The magnetic force on a charged particle moving in a uniform magnetic field can be balanced by an electric force if the magnitudes and directions of the magnetic field and the electric field are properly chosen. Since the electric force is in the direction of the electric field (for positive particles) and the magnetic force is perpendicular to the magnetic field, the electric and magnetic fields in the region through which the particle is moving must be perpendicular to each other if the forces are to balance. Such a region is said to have **crossed fields.**

Figure 26-16 shows a region of space between the plates of a capacitor where there is an electric field and a perpendicular magnetic field (produced by a magnet with poles above and below the paper). Consider a particle of charge q entering this space from the left. The net force on the particle is

$$\vec{F} = q\vec{E} + q\vec{v} \times \vec{B}$$

If q is positive, the electric force of magnitude qE is down and the magnetic force of magnitude qvB is up. If the charge is negative, each of these forces is reversed. The two forces balance if $qE = qvB$ or

$$v = \frac{E}{B} \qquad\qquad 26\text{-}9$$

For given magnitudes of the electric and magnetic fields, the forces balance only for particles with the speed given by Equation 26-9. Any particle with this speed, regardless of its mass or charge, will traverse the space undeflected. A particle with a greater speed will be deflected toward the direction of the magnetic force, and a particle with less speed will be deflected in the direction of the electric force. This arrangement of fields is often used as a **velocity selector,** which is a device that allows only particles with speed, given by Equation 26-9, to pass.

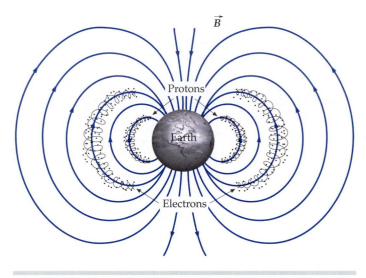

FIGURE 26-15 Van Allen belts. Protons (inner belts) and electrons (outer belts) are trapped in the earth's magnetic field and spiral around the field lines between the north and south poles.

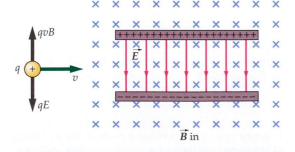

FIGURE 26-16 Crossed electric and magnetic fields. When a positive particle moves to the right, the particle experiences a downward electric force and an upward magnetic force. These forces balance if the speed of the particle is related to the field strengths by $vB = E$.

EXERCISE A proton is moving in the x direction in a region of crossed fields where $\vec{E} = 2 \times 10^5$ N/C $\hat{k}$ and $\vec{B} = -3000$ G $\hat{j}$. (a) What is the speed of the proton if it is not deflected? (b) If the proton moves with twice this speed, in which direction will it be deflected? (*Answer* (a) 667 km/s (b) in the negative z direction)

*Thomson's Measurement of q/m for Electrons

An example of the use of crossed electric and magnetic fields is the famous experiment performed by J. J. Thomson in 1897 where he showed that the rays of a cathode-ray tube can be deflected by electric and magnetic fields, indicating that they must consist of charged particles. By measuring the deflections of these particles Thomson showed that all the particles have the same charge-to-mass ratio q/m. He also showed that particles with this charge-to-mass ratio can be obtained using any material for a source, which means that these particles, now called electrons, are a fundamental constituent of all matter.

Figure 26-17 shows a schematic diagram of the cathode-ray tube Thomson used. Electrons are emitted from the cathode C, which is at a negative potential relative to the slits A and B. An electric field in the direction from A to C accelerates the electrons, and the electrons pass through slits A and B into a field-free region. The electrons then enter the electric field between the capacitor plates D and F that is perpendicular to the velocity of the electrons. This field accelerates the electrons vertically for the short time that they are between the plates. The electrons are deflected and strike the phosphorescent screen S at the far right side of the tube at some deflection Δy from the point at which they strike when there is no field between the plates. The screen glows where the electrons strike the screen, indicating the location of the beam. The initial speed of the electrons v_0 is determined by introducing a magnetic field $\vec{B}$ between the plates in a direction that is perpendicular to both the electric field and the initial velocity of the electrons. The magnitude of $\vec{B}$ is adjusted until the beam is not deflected. The speed is then found from Equation 26-9.

With the magnetic field turned off, the beam is deflected by an amount Δy, which consists of two parts: the deflection Δy_1, which occurs while the electrons are between the plates, and the deflection Δy_2, which occurs after the electrons leave the region between the plates (Figure 26-18).

Let x_1 be the horizontal distance across the deflection plates D and F. If the electron is moving horizontally with speed v_0 when it enters region between the plates, the time spent between the plates is $t_1 = x_1/v_0$, and the vertical velocity when it leaves the plates is

$$v_y = a_y t_1 = \frac{qE_y}{m} t_1 = \frac{qE_y}{m} \frac{x_1}{v_0}$$

where E_y is the upward component of the electric field between the plates. The deflection in this region is

$$\Delta y_1 = \frac{1}{2} a_y t_1^2 = \frac{1}{2} \frac{qE_y}{m} \left(\frac{x_1}{v_0}\right)^2$$

The electron then travels an additional horizontal distance x_2 in the field-free region from the deflection plates to the screen. Since the velocity of the electron is constant in this region, the time to reach the screen is $t_2 = x_2/v_0$, and the additional vertical deflection is

$$\Delta y_2 = v_y t_2 = \frac{qE_y}{m} \frac{x_1}{v_0} \frac{x_2}{v_0}$$

The total deflection at the screen is therefore

$$\Delta y = \Delta y_1 + \Delta y_2 = \frac{1}{2} \frac{qE_y}{mv_0^2} x_1^2 + \frac{qE_y}{mv_0^2} x_1 x_2 \qquad 26\text{-}10$$

The measured deflection Δy can be used to determine the charge-to-mass ratio, q/m, from Equation 26-10.

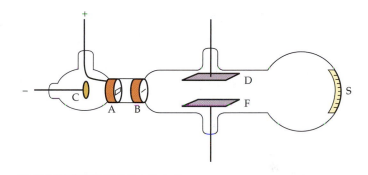

FIGURE 26-17 Thomson's tube for measuring q/m for the particles of cathode rays (electrons). Electrons from the cathode C pass through the slits at A and B and strike a phosphorescent screen S. The beam can be deflected by an electric field between plates D and F or by a magnetic field (not shown).

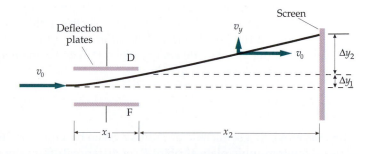

FIGURE 26-18 The total deflection of the beam in the J. J. Thomson experiments consists of the deflection Δy_1 while the electrons are between the plates plus the deflection Δy_2 that occurs in the field-free region between the plates and the screen.

ELECTRON BEAM DEFLECTION

EXAMPLE 26-5

Electrons pass undeflected through the plates of Thomson's apparatus when the electric field is 3000 V/m and there is a crossed magnetic field of 1.40 G. If the plates are 4-cm long and the ends of the plates are 30 cm from the screen, find the deflection on the screen when the magnetic field is turned off.

PICTURE THE PROBLEM The mass and charge of the electron are known: $m = 9.11 \times 10^{-31}$ kg and $q = -e = -1.6 \times 10^{-19}$ C. The speed of the electron can be found from the ratio of the magnetic and electric fields.

1. The total deflection of the electron is given by Equation 26-10:

$$\Delta y = \Delta y_1 + \Delta y_2 = \frac{1}{2} \frac{qE_y}{mv_0^2} x_1^2 + \frac{qE_y}{mv_0^2} x_1 x_2$$

2. The speed v_0 equals E/B:

$$v_0 = \frac{E}{B} = \frac{3000 \text{ V/m}}{1.40 \times 10^{-4} \text{ T}} = 2.14 \times 10^7 \text{ m/s}$$

3. Substitute this value for v_0, the given value of E, and the known values for m and q to find Δy:

$$\Delta y_1 = \frac{1}{2} \frac{(-1.6 \times 10^{-19} \text{ C})(-3000 \text{ V/m})}{(9.11 \times 10^{-31} \text{ kg})(2.14 \times 10^7 \text{ m/s})^2} (0.04 \text{ m})^2$$

$$= 9.20 \times 10^{-4} \text{ m}$$

$$\Delta y_2 = \frac{(-1.6 \times 10^{-19} \text{ C})(-3000 \text{ V/m})}{(9.11 \times 10^{-31} \text{ kg})(2.14 \times 10^7 \text{ m/s})^2} (0.04 \text{ m})(0.30 \text{ m})$$

$$= 1.38 \times 10^{-2} \text{ m}$$

$$\Delta y = \Delta y_1 + \Delta y_2$$

$$= 9.20 \times 10^{-4} \text{ m} + 1.38 \times 10^{-2} \text{ m}$$

$$= 0.92 \text{ mm} + 13.8 \text{ mm} = \boxed{14.7 \text{ mm}}$$

*The Mass Spectrometer

The **mass spectrometer,** first designed by Francis William Aston in 1919, was developed as a means of measuring the masses of isotopes. Such measurements are important in determining both the presence of isotopes and their abundance in nature. For example, natural magnesium has been found to consist of 78.7 percent ^{24}Mg, 10.1 percent ^{25}Mg, and 11.2 percent ^{26}Mg. These isotopes have masses in the approximate ratio 24:25:26.

Figure 26-19 shows a simple schematic drawing of a mass spectrometer. Positive ions are formed by bombarding neutral atoms with X rays or a beam of electrons. (Electrons are knocked out of the atoms by the X rays or bombarding electrons.) These ions are accelerated by an electric field and enter a uniform magnetic field. If the positive ions start from rest and move through a potential difference ΔV, the ions kinetic energy when they enter the magnetic field equals their loss in potential energy, $q|\Delta V|$:

$$\tfrac{1}{2} mv^2 = q|\Delta V| \tag{26-11}$$

The ions move in a semicircle of radius r given by Equation 26-6, $r = mv/qB$, and strike a photographic plate at point P_2, a distance $2r$ from the point P_1 where the ions entered the magnetic field.

The speed v can be eliminated from Equations 26-6 and 26-11 to find m/q in terms of the known quantities ΔV, B, and r. We first solve Equation 26-6 for v and square each term, which gives

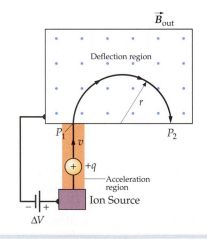

FIGURE 26-19 Schematic drawing of a mass spectrometer. Positive ions from an ion source are accelerated through a potential difference ΔV and enter a uniform magnetic field. The magnetic field is out of the plane of the page as indicated by the dots. The ions are bent into a circular arc and emerge at P_2. The radius of the circle varies with the mass of the ion.

$$v^2 = \frac{r^2 q^2 B^2}{m^2}$$

Substituting this expression for v^2 into Equation 26-11, we obtain

$$\frac{1}{2} m \left(\frac{r^2 q^2 B^2}{m^2} \right) = q |\Delta V|$$

Simplifying this equation and solving for m/q, we obtain

$$\frac{m}{q} = \frac{B^2 r^2}{2|\Delta V|} \qquad\qquad 26\text{-}12$$

In Aston's original mass spectrometer, mass differences could be measured to a precision of about 1 part in 10,000. The precision has been improved by introducing a velocity selector between the ion source and the magnet, which increases the degree of accuracy with which the velocities of the incoming ions can be determined.

Separating Isotopes of Nickel   **EXAMPLE   26-6**

A ^{58}Ni ion of charge $+e$ and mass 9.62×10^{-26} kg is accelerated through a potential drop of 3 kV and deflected in a magnetic field of 0.12 T. (*a*) Find the radius of curvature of the orbit of the ion. (*b*) Find the difference in the radii of curvature of ^{58}Ni ions and ^{60}Ni ions. (Assume that the mass ratio is 58:60.)

PICTURE THE PROBLEM   The radius of curvature r can be found using Equation 26-12. Using the mass dependence of r, we can find the radius for ^{60}Ni ions from the radius for ^{58}Ni ions and then take the difference.

(*a*) Solve Equation 26-12 for r:

$$r = \sqrt{\frac{2m\,|\Delta V|}{qB^2}} = \left[\frac{2(9.62 \times 10^{-26}\ \text{kg})(3000\ \text{V})}{(1.6 \times 10^{-19}\ \text{C})(0.12\ \text{T})^2} \right]^{1/2}$$

$$= \boxed{0.501\ \text{m}}$$

(*b*) 1. Let r_1 and r_2 be the radius of the orbit of the ^{58}Ni ion and the ^{60}Ni ion, respectively. Use the result in Part (*a*) to find the ratio of r_2 to r_1:

$$\frac{r_2}{r_1} = \sqrt{\frac{m_2}{m_1}} = \sqrt{\frac{60}{58}} = 1.017$$

2. Use the result of the previous step to calculate r_2 for ^{60}Ni:

$$r_2 = 1.017\, r_1 = (1.017)(0.501\ \text{m}) = 0.510\ \text{m}$$

3. The difference in orbital radii is $r_2 - r_1$:

$$r_2 - r_1 = 0.510\ \text{m} - 0.501\ \text{m} = \boxed{9\ \text{mm}}$$

The Cyclotron

The cyclotron was invented by E. O. Lawrence and M. S. Livingston in 1934 to accelerate particles, such as protons or deuterons, to high kinetic energies.[†] The high-energy particles are used to bombard atomic nuclei, causing nuclear reactions that are then studied to obtain information about the nucleus. High-energy protons and deuterons are also used to produce radioactive materials and for medical purposes.

---

† A deuteron is the nucleus of heavy hydrogen, ^{2}H, which consists of a proton and neutron tightly bound together.

Figure 26-20 is a schematic drawing of a cyclotron. The particles move in two semicircular metal containers called *dees*, after their shape. The dees are housed in a vacuum chamber that is in a uniform magnetic field provided by an electromagnet. The region in which the particles move must be evacuated so that the particles will not be scattered in collisions with air molecules and lose energy. A potential difference ΔV, which alternates in time with a period T, is maintained between the dees. The period is chosen to be the cyclotron period $T = 2\pi m/(qB)$ (Equation 26-7). The potential difference creates an electric field across the gap between the dees. At the same time, there is no electric field within each dee because the metal dees act as shields.

Positively charged particles are initially injected into dee_1 with a small velocity from an ion source S near the center of the dees. They move in a semicircle in dee_1 and arrive at the gap between dee_1 and dee_2 after a time $\frac{1}{2}T$. The potential is adjusted so that dee_1 is at a higher potential than dee_2 when the particles arrive at the gap between them. Each particle is therefore accelerated across the gap by the electric field and gains kinetic energy equal to $q\,\Delta V$.

Because the particle now has more kinetic energy, the particle moves in a semicircle of larger radius in dee_2. It arrives at the gap again after a time $\frac{1}{2}T$, because the period is independent of the particle's speed. By this time, the potential between the dees has been reversed so that dee_2 is now at the higher potential. Once more the particle is accelerated across the gap and gains additional kinetic energy equal to $q\,\Delta V$. Each time the particle arrives at the gap, it is accelerated and gains kinetic energy equal to $q\,\Delta V$. Thus, the particle moves in larger and larger semicircular orbits until it eventually leaves the magnetic field. In the typical cyclotron, each particle may make 50 to 100 revolutions and exit with energies of up to several hundred mega-electron volts.

The kinetic energy of a particle leaving a cyclotron can be calculated by setting r in Equation 26-6 equal to the maximum radius of the dees and solving the equation for v:

$$r = \frac{mv}{qB}, \quad v = \frac{qBr}{m}$$

Then

$$K = \frac{1}{2}mv^2 = \frac{1}{2}\left(\frac{q^2B^2}{m}\right)r^2 \qquad\qquad 26\text{-}13$$

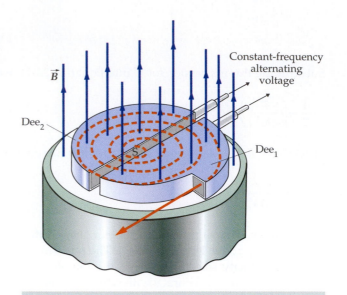

FIGURE 26-20 Schematic drawing of a cyclotron. The upper-pole face of the magnet has been omitted. Charged particles, such as protons, are accelerated from a source at the center by the potential difference across the gap between the dees. When the charged particles arrive at the gap again the potential difference has changed sign so they are again accelerated across the gap and move in a larger circle. The potential difference across the gap alternates with the cyclotron frequency of the particle, which is independent of the radius of the circle.

ENERGY OF ACCELERATED PROTON **EXAMPLE 26-7**

A cyclotron for accelerating protons has a magnetic field of 1.5 T and a maximum radius of 0.5 m. (*a*) What is the cyclotron frequency? (*b*) What is the kinetic energy of the protons when they emerge?

(*a*) The cyclotron frequency is given by Equation 26-8:

$$f = \frac{qB}{2\pi m} = \frac{(1.6 \times 10^{-19}\,\text{C})(1.5\,\text{T})}{2\pi(1.67 \times 10^{-27}\,\text{kg})} = 2.29 \times 10^7\,\text{Hz}$$

$$= \boxed{22.9\,\text{MHz}}$$

(b) 1. The kinetic energy of the emerging protons is given by Equation 26-13:

$$K = \frac{1}{2}\left[\frac{(1.6 \times 10^{-19}\,\text{C})^2(1.5\,\text{T})^2}{1.67 \times 10^{-27}\,\text{kg}}\right](0.5\,\text{m})^2$$

$$= 4.31 \times 10^{-12}\,\text{J}$$

2. The energies of protons and other elementary particles are usually expressed in electron volts. Use $1\,\text{eV} = 1.6 \times 10^{-19}\,\text{J}$ to convert to eV:

$$K = 4.31 \times 10^{-12}\,\text{J} \times \frac{1\,\text{eV}}{1.6 \times 10^{-19}\,\text{J}} = \boxed{26.9\,\text{MeV}}$$

26-3 Torques on Current Loops and Magnets

A current-carrying loop experiences no net force in a uniform magnetic field, but it does experience a torque that tends to twist the current-carrying loop. The orientation of the loop can be described conveniently by a unit vector $\hat{n}$ that is perpendicular to the plane of the loop, as illustrated in Figure 26-21. If the fingers of the right hand curl around the loop in the direction of the current, the thumb points in the direction of $\hat{n}$.

Figure 26-22 shows the forces exerted by a uniform magnetic field on a rectangular loop whose normal unit vector $\hat{n}$ makes an angle θ with the magnetic field $\vec{B}$. The net force on the loop is zero. The forces $\vec{F}_1$ and $\vec{F}_2$ have the magnitude

$$F_1 = F_2 = IaB$$

These forces form a couple, so the torque is the same about any point. Point P in Figure 26-22 is a convenient point about which to compute the torque. The magnitude of the torque is

$$\tau = F_2 b \sin\theta = IaBb \sin\theta = IAB \sin\theta$$

where $A = ab$ is the area of the loop. For a loop with N turns, the torque has the magnitude

$$\tau = NIAB \sin\theta$$

This torque tends to twist the loop so that $\hat{n}$ is in the same direction as $\vec{B}$ (i.e., so that its plane is perpendicular to $\vec{B}$).

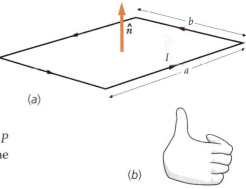

(a)

(b)

FIGURE 26-21 (*a*) The orientation of a current loop is described by the unit vector $\hat{n}$ perpendicular to the plane of the loop. (*b*) Right-hand rule for determining the direction of $\hat{n}$. If the fingers of the right hand curl around the loop in the direction of the current, the thumb points in the direction of $\hat{n}$.

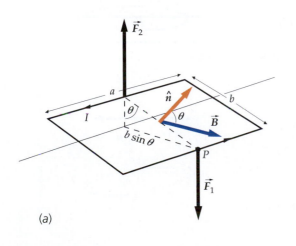

(a)

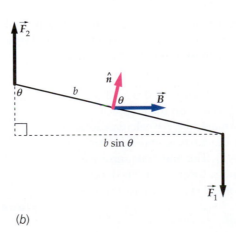

(b)

FIGURE 26-22 (*a*) Rectangular current loop whose unit normal $\hat{n}$ makes an angle θ with a uniform magnetic field $\vec{B}$. (*b*) An edge-on view of the current loop. The torque on the loop has magnitude $IAB \sin\theta$ and is in the direction such that $\hat{n}$ tends to rotate into $\vec{B}$.

The torque can be written conveniently in terms of the **magnetic dipole moment** $\vec{\mu}$ (also referred to simply as the **magnetic moment**) of the current loop, which is defined as

$$\vec{\mu} = NIA\hat{n} \qquad\qquad 26\text{-}14$$

MAGNETIC DIPOLE MOMENT OF A CURRENT LOOP

The SI unit of magnetic moment is the ampere-meter^2 (A·m^2). In terms of the magnetic dipole moment, the torque on the current loop is given by

$$\vec{\tau} = \vec{\mu} \times \vec{B} \qquad\qquad 26\text{-}15$$

TORQUE ON A CURRENT LOOP

Equation 26-15, which we have derived for a rectangular loop, holds in general for a loop of any shape that is in a single plane. The torque on any loop is the cross product of the magnetic moment $\vec{\mu}$ of the loop and the magnetic field $\vec{B}$, where the magnetic moment is defined as a vector that is perpendicular to the plane of the loop (Figure 26-23), has magnitude equal to NIA, and has the same direction as $\hat{n}$. Comparing Equation 26-15 with Equation 21-11 ($\vec{\tau} = \vec{p} \times \vec{E}$) for the torque on an electric dipole, we see that the expression for the torque on a current loop in a magnetic field has the same form as that for the torque on an electric dipole in an electric field.

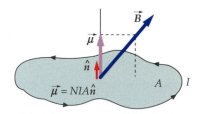

FIGURE 26-23 A flat current loop of arbitrary shape is described by its magnetic moment $\vec{\mu} = NIA\hat{n}$. In a magnetic field $\vec{B}$, the loop experiences a torque $\vec{\mu} \times \vec{B}$.

TORQUE ON A CURRENT LOOP   **EXAMPLE 26-8**

A circular loop of radius 2 cm has 10 turns of wire and carries a current of 3 A. The axis of the loop makes an angle of 30° with a magnetic field of 8000 G. Find the magnitude of the torque on the loop.

The magnitude of the torque is given by Equation 26-15:

$$\tau = |\vec{\mu} \times \vec{B}| = \mu B \sin\theta = NIAB \sin\theta$$

$$= (10)(3\text{ A})\pi(0.02\text{ m})^2(0.8\text{ T})\sin 30°$$

$$= \boxed{1.51 \times 10^{-2}\text{ N·m}^{-2}}$$

TILTING A LOOP   **EXAMPLE 26-9** **Try It Yourself**

A circular wire loop of radius R, mass m, and current I lies on a horizontal surface (Figure 26-24). There is a horizontal magnetic field $\vec{B}$. How large can the current I be before one edge of the loop will lift off the surface?

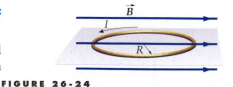

FIGURE 26-24

PICTURE THE PROBLEM The loop (Figure 26-25) will start to rotate when the magnitude of the net torque on the loop is not zero. To eliminate the torque due to the normal force, we calculate torques about the point of contact between the surface and the loop. The magnetic torque is given by $\vec{\tau} = \vec{\mu} \times \vec{B}$. The magnetic torque is the same about any point since the magnetic torque consists of couples. The lever arm for the gravitational torque is the radius of the loop.

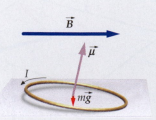

FIGURE 26-25

Cover the column to the right and try these on your own before looking at the answers.

Steps	Answers
1. Find the magnitude of the magnetic torque acting on the loop.	$\tau_m = \mu B = I\pi R^2 B$
2. Find the magnitude of the gravitational torque exerted on the loop.	$\tau_g = mgR$
3. Equate the magnitudes of the torques and solve for the current I.	$I = \boxed{\dfrac{mg}{\pi RB}}$

■ **REMARKS** The torque vectors are equal and opposite.

Potential Energy of a Magnetic Dipole in a Magnetic Field

When a torque is exerted through an angle, work is done. When a dipole is rotated through an angle $d\theta$, the work done is

$$dW = -\tau d\theta = -\mu B \sin\theta d\theta$$

The minus sign arises because the torque tends to decrease θ. Setting this work equal to the decrease in potential energy, we have

$$dU = -dW = +\mu B \sin\theta d\theta$$

Integrating, we obtain

$$U = -\mu B \cos\theta + U_0$$

We choose the potential energy to be zero when $\theta = 90°$. Then $U_0 = 0$ and the potential energy of the dipole is

$$U = -\mu B \cos\theta = -\vec{\boldsymbol{\mu}} \cdot \vec{\boldsymbol{B}} \qquad\qquad 26\text{-}16$$

POTENTIAL ENERGY OF A MAGNETIC DIPOLE

Equation 26-16 gives the potential energy of a magnetic dipole at an angle θ to the direction of a magnetic field.

TORQUE ON A COIL   **EXAMPLE 26-10**

A square 12-turn coil with edge-length 40 cm carries a current of 3 A. It lies in the xy plane, as shown in a uniform magnetic field $\vec{B} = 0.3$ T $\hat{i}$ + 0.4 T $\hat{k}$. Find (a) the magnetic moment of the coil and (b) the torque exerted on the coil. (c) Find the potential energy of the coil.

PICTURE THE PROBLEM From Figure 26-26, we see that the magnetic moment of the loop is in the positive z direction.

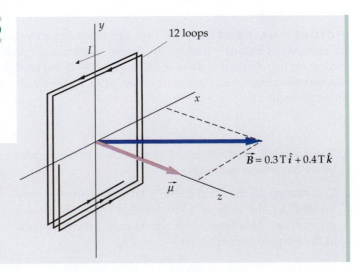

FIGURE 26-26

(a) Calculate the magnetic moment of the loop:

$$\vec{\mu} = NIA\,\hat{k} = (12)(3\text{ A})(0.40\text{ m})^2\,\hat{k}$$

$$= \boxed{5.76\text{ A·m}^2\,\hat{k}}$$

(b) The torque on the current loop is given by Equation 26-15:

$$\vec{\tau} = \vec{\mu} \times \vec{B}$$

$$= (5.76\text{ A·m}^2\,\hat{k}) \times (0.3\text{ T}\,\hat{i} + 0.4\text{ T}\,\hat{k})$$

$$= \boxed{1.73\text{ N·m}\,\hat{j}}$$

(c) The potential energy is the negative dot product of $\vec{\mu}$ and $\vec{B}$:

$$U = -\vec{\mu} \cdot \vec{B}$$

$$= -(5.76\text{ A·m}^2\,\hat{k}) \cdot (0.3\text{ T}\,\hat{i} + 0.4\text{ T}\,\hat{k})$$

$$= \boxed{-2.30\text{ J}}$$

REMARKS We have used $\hat{k} \times \hat{k} = 0$ and $\hat{k} \times \hat{i} = \hat{j}$, $\hat{k} \cdot \hat{i} = 0$ and $\hat{k} \cdot \hat{k} = 1$. The torque is in the y direction.

EXERCISE Calculate U if $\vec{B}$ and the magnetic moment $\vec{\mu}$ are in the same direction. (*Answer* $U = -\mu B = -(5.76\text{ A·m}^2)(0.5\text{ T}) = -2.88\text{ J}$. Note that this potential energy is lower than that found in the example. The potential energy is lowest when $\vec{\mu}$ and $\vec{B}$ are in the same direction.)

When a small permanent magnet, such as a compass needle, is placed in a magnetic field $\vec{B}$, the field exerts a torque on the magnet that tends to rotate the magnet so that it lines up with the field. This effect also occurs with previously unmagnetized iron filings, which become magnetized in the presence of a $\vec{B}$ field. The bar magnet is characterized by a magnetic moment $\vec{\mu}$, a vector that points in the same direction as an arrow drawn from the south pole to the north pole. A small bar magnet thus behaves like a current loop. This is not a coincidence. The origin of the magnetic moment of a bar magnet is, in fact, microscopic current loops that result from the motion of electrons in the atoms of the magnet.

$\vec{\mu}$ OF A ROTATING DISK   **EXAMPLE 26-11**

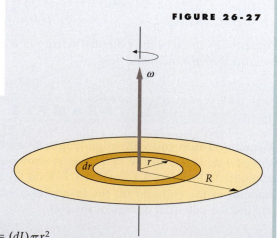

FIGURE 26-27

A thin nonconducting disk of mass m and radius R has a uniform surface charge per unit area σ and rotates with angular velocity $\vec{\omega}$ about its axis. Find the magnetic moment of the rotating disk.

PICTURE THE PROBLEM We find the magnetic moment of a circular element of radius r and width dr and integrate (Figure 26-27). The charge on the element is $dq = \sigma\,dA = \sigma 2\pi r\,dr$. If the charge is positive, the magnetic moment is in the direction of $\vec{\omega}$, so we need only calculate its magnitude.

1. The magnetic moment of the strip shown is the current times the area of the loop:

$$d\mu = (dI)A = (dI)\pi r^2$$

2. The current in the strip is the total charge on the strip divided by the time it takes for this charge to pass a given point. This time is the period that is the reciprocal of the frequency of rotation $f = \omega/(2\pi)$:

$$dI = \frac{dq}{T} = (dq)f = (\sigma\,dA)\frac{\omega}{2\pi}$$

$$= (\sigma 2\pi r\,dr)\frac{\omega}{2\pi} = \sigma\omega r\,dr$$

3. Substitute to obtain the magnetic moment of the strip $d\mu$ in terms of r and dr:

$$d\mu = (dI)\pi r^2 = (\sigma\omega r\, dr)\pi r^2 = \pi\sigma\omega r^3\, dr$$

4. Integrate from $r = 0$ to $r = R$:

$$\mu = \int d\mu = \int_0^R \pi\sigma\omega r^3\, dr = \frac{1}{4}\pi\sigma\omega R^4$$

5. Use the fact that $\vec{\boldsymbol{\mu}}$ is parallel to $\vec{\boldsymbol{\omega}}$ if σ is positive to write the magnetic moment as a vector:

$$\boxed{\vec{\boldsymbol{\mu}} = \tfrac{1}{4}\pi\sigma R^4\,\vec{\boldsymbol{\omega}}}$$

REMARKS In terms of the total charge $Q = \sigma\pi R^2$, the magnetic moment is $\vec{\boldsymbol{\mu}} = \frac{1}{4}QR^2\vec{\boldsymbol{\omega}}$. The angular momentum of the disk is $\vec{L} = (\frac{1}{2}mR^2)\vec{\boldsymbol{\omega}}$, so the magnetic moment can be written $\vec{\boldsymbol{\mu}} = [Q/(2m)]\vec{L}$, which is a more general result. (See Problem 63.)

26-4 The Hall Effect

As we have seen, charges moving in a magnetic field experience a force perpendicular to their motion. When these charges are traveling in a conducting wire, they will be pushed to one side of the wire. This results in a separation of charge in the wire called the **Hall effect.** This phenomenon allows us to determine the sign of the charge on the charge carriers and the number of charge carriers per unit volume n in a conductor. The Hall effect also provides a convenient method for measuring magnetic fields.

Figure 26-28 shows two conducting strips; each conducting strip carries a current I to the right because the left sides of the strips are connected to the positive terminal of a battery and the right sides are connected to the negative terminal. The strips are in a magnetic field that is directed into the paper. Let us assume for the moment that the current in the strip consists of positively charged particles moving to the right, as shown in Figure 26-28a. The magnetic force on these particles is $q\vec{v}_d \times \vec{B}$ (where $\vec{v}_d$ is the drift velocity of the charge carriers). This force is directed upward. The positive particles therefore move up to the top of the strip, leaving the bottom of the strip with an excess negative charge. This separation of charge produces an electric field in the strip that opposes the magnetic force on the charge carriers. When the electric forces and magnetic forces balance, the charge carriers no longer move upward. Since the electric field points in the direction of decreasing potential, the upper part of the strip is at a higher potential than is the lower part of the strip. This potential difference can be measured using a sensitive voltmeter. On the other hand, if the current consists of moving negatively charged particles, as shown in Figure 26-28b, the charge carriers in the strip must move to the left (since the current is still to the right). The magnetic force $q\vec{v}_d \times \vec{B}$ is again up, because the signs of both q and $\vec{v}_d$ have been reversed. Again the carriers are forced to the upper part of the strip, but the upper part of the strip now carries a negative charge (because the charge carriers are negative) and the lower part of the strip now carries a positive charge.

A measurement of the sign of the potential difference between the upper and lower parts of the strip tells us the sign of the charge carriers. In semiconductors, the charge carriers may be negative electrons or positive holes. A measurement of the sign of the potential difference tells us which are dominant for a particular semiconductor. For a normal metallic conductor, we find that the upper part of the strip in Figure 26-28b is at a lower potential than is the lower part of the strip—which means that the upper part must carry a negative charge. Thus, Figure 26-28b is the correct illustration of the current in a normal metallic conductor. It was this type of experiment that led to the discovery that the charge carriers in metallic conductors are negative.

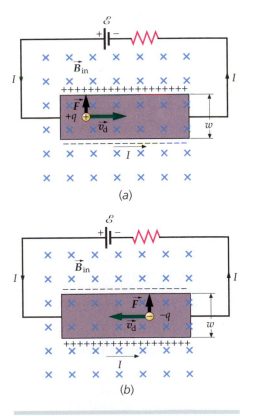

FIGURE 26-28 The Hall effect. The magnetic field is directed into the plane of the page as indicated by the crosses. The magnetic force on a charged particle is upward for a current to the right whether the current is due to (a) positive particles moving to the right or (b) negative particles moving to the left.

The potential difference between the top of the strip and the bottom of the strip is called the **Hall voltage.** We can calculate the magnitude of the Hall voltage in terms of the drift velocity. The magnitude of the magnetic force on the charge carriers in the strip is qv_dB. This magnetic force is balanced by the electrostatic force of magnitude qE_H, where E_H is the electric field due to the charge separation. Thus, we have $E_H = v_dB$. If the width of the strip is w, the potential difference is E_Hw. The Hall voltage is therefore

$$V_H = E_Hw = v_dBw \qquad \text{26-17}$$

EXERCISE A conducting strip of width $w = 2.0$ cm is placed in a magnetic field of 0.8 T. The Hall voltage is measured to be 0.64 μV. Calculate the drift velocity of the electrons. (*Answer* 4.0×10^{-5} m/s)

Since the drift velocity for ordinary currents is very small, we can see from Equation 26-17 that the Hall voltage is very small for ordinary-sized strips and magnetic fields. From measurements of the Hall voltage for a strip of a given size, we can determine the number of charge carriers per unit volume in the strip. The current is given by Equation 25-3:

$$I = nqv_dA$$

where A is the cross-sectional area of the strip. For a strip of width w and thickness t, the cross-sectional area is $A = wt$. Since the charge carriers are electrons, the quantity q is the charge on one electron e. The number density of charge carriers n is thus given by

$$n = \frac{I}{Aqv_d} = \frac{I}{wtev_d} \qquad \text{26-18}$$

Substituting V_H/B for v_dw (Equation 26-17), we have

$$n = \frac{IB}{teV_H} \qquad \text{26-19}$$

CHARGE CARRIER NUMBER DENSITY IN SILVER   **EXAMPLE 26-12**

A silver slab of thickness 1 mm and width 1.5 cm carries a current of 2.5 A in a region in which there is a magnetic field of magnitude 1.25 T perpendicular to the slab. The Hall voltage is measured to be 0.334 μV. (*a*) Calculate the number density of the charge carriers. (*b*) Compare your answer in step 1 to the number density of atoms in silver, which has a mass density of $\rho = 10.5$ g/cm³ and a molar mass of $M = 107.9$ g/mol.

1. Substitute numerical values into Equation 26-19 to find n:

$$n = \frac{IB}{teV_H} = \frac{(2.5 \text{ A})(1.25 \text{ T})}{(0.001 \text{ m})(1.6 \times 10^{-19} \text{ C})(3.34 \times 10^{-7} \text{ V})}$$

$$= \boxed{5.85 \times 10^{28} \text{ electrons/m}^3}$$

2. The number of atoms per unit volume is $\rho N_A/M$:

$$n_a = \rho \frac{N_A}{M} = (10.5 \text{ g/cm}^3) \frac{6.02 \times 10^{23} \text{ atoms/mol}}{107.9 \text{ g/mol}}$$

$$= \boxed{5.86 \times 10^{22} \text{ atoms/cm}^3 = 5.86 \times 10^{28} \text{ atoms/m}^3}$$

REMARKS These results indicate that the number of charge carriers in silver is very nearly one per atom.

The Hall voltage provides a convenient method for measuring magnetic fields. If we rearrange Equation 26-19, we can write for the Hall voltage

$$V_H = \frac{I}{nte} B \qquad\qquad 26\text{-}20$$

A given strip can be calibrated by measuring the Hall voltage for a given current in a known magnetic field. The strip can then be used to measure an unknown magnetic field B by measuring the Hall voltage for a given current.

*The Quantum Hall Effects

According to Equation 26-20, the Hall voltage should increase linearly with magnetic field B for a given current in a given slab. In 1980, while studying the Hall effect in semiconductors at very low temperatures and very large magnetic fields, the German physicist Klaus von Klitzing discovered that a plot of V_H versus B resulted in a series of plateaus, as shown in Figure 26-29, rather than a straight line. That is, the Hall voltage is quantized. For the discovery of the integer quantum Hall effect, von Klitzing won the Nobel Prize in physics in 1985.

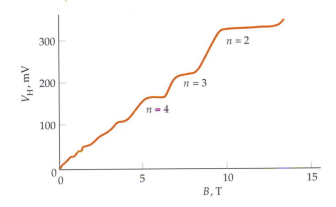

FIGURE 26-29 A plot of the Hall voltage versus applied magnetic field shows plateaus, indicating that the Hall voltage is quantized. These data were taken at a temperature of 1.39 K with the current I held fixed at 25.52 μA.

In the theory of the integer quantum Hall effect, the Hall resistance, defined as $R_H = V_H/I$, can take on only the values

$$R_H = \frac{V_H}{I} = \frac{R_K}{n}, \quad n = 1, 2, 3, \ldots \qquad\qquad 26\text{-}21$$

where n is an integer, and R_K, called the **von Klitzing constant,** is related to the fundamental electronic charge e and Planck's constant h by

$$R_K = \frac{h}{e^2} \qquad\qquad 26\text{-}22$$

Because the von Klitzing constant can be measured to an accuracy of a few parts per billion, the quantum Hall effect is now used to define a standard of resistance. As of January 1990, the **ohm** is defined in terms of the conventional value[†] of the von Klitzing constant R_{K-90}, which has the value

$$R_{K-90} = 25{,}812.807 \ \Omega \ (\text{exact}) \qquad\qquad 26\text{-}23$$

In 1982 it was observed that under certain special conditions the Hall resistance is given by Equation 26-22, but with the integer n replaced by a series of rational fractions. This is called the fractional quantum Hall effect. For the discovery and explanation of the fractional quantum Hall effect, American professors Laughlin, Stormer, and Tsui won the Nobel Prize in physics in 1998.

---

† The value of R_{K-90} differs only slightly from that of R_K. The currently used value of the von Klitzing constant is $R_K = (25\,812.807\,572 \pm 0.000\,095)\ \Omega$

1. The magnetic field describes the condition in space in which moving charges experience a force perpendicular to their velocity.

2. The magnetic force is part of the electromagnetic force, one of the four fundamental forces of nature.

3. The magnitude and direction of a magnetic field $\vec{B}$ are defined by the force $\vec{F} = q\vec{v} \times \vec{B}$ exerted on moving charges.

Topic	Relevant Equations and Remarks	
1. Magnetic Force		
On a moving charge	$\vec{F} = q\vec{v} \times \vec{B}$	26-1
On a current element	$d\vec{F} = I\,d\vec{\ell} \times \vec{B}$	26-5
Unit of the magnetic field	The SI unit of magnetic fields is the tesla (T). A commonly used unit is the gauss (G), which is related to the tesla by $$1\,\text{G} = 10^{-4}\,\text{T}$$	26-3
2. Motion of Point Charges	A particle of mass m and charge q moving with speed v in a plane perpendicular to a uniform magnetic field moves in a circular orbit. The period and frequency of this circular motion are independent of the radius of the orbit and of the speed of the particle.	
Newton's second law	$qvB = m\dfrac{v^2}{r}$	26-6
Cyclotron period	$T = \dfrac{2\pi m}{qB}$	26-7
Cyclotron frequency	$f = \dfrac{1}{T} = \dfrac{qB}{2\pi m}$	26-8
*Velocity selector	A velocity selector consists of crossed electric and magnetic fields so that the electric and magnetic forces balance for a particle moving with speed v. $$E = vB$$	26-9
*Thomson's measurement of q/m	The deflection of a charged particle in an electric field depends on the speed of the particle and is proportional to the charge-to-mass ratio q/m of the particle. J. J. Thomson used crossed electric and magnetic fields to measure the speed of cathode rays and then measured q/m for these particles by deflecting them in an electric field. He showed that all cathode rays consist of particles that all have the same charge-to-mass ratio. These particles are now called electrons.	
*Mass spectrometer	The mass-to-charge ratio of an ion of known speed can be determined by measuring the radius of the circular path taken by the ion in a known magnetic field.	
3. Current Loops		
Magnetic dipole moment	$\vec{\mu} = NIA\hat{n}$	26-14
Torque	$\vec{\tau} = \vec{\mu} \times \vec{B}$	26-15

Potential energy of a magnetic dipole	$U = -\vec{\mu} \cdot \vec{B}$	26-16

Net force	The net force on a current loop in a *uniform* magnetic field is zero.

4. The Hall Effect

When a conducting strip carrying a current is placed in a magnetic field, the magnetic force on the charge carriers causes a separation of charge called the Hall effect. This results in a voltage V_H, called the Hall voltage. The sign of the charge carriers can be determined from a measurement of the sign of the Hall voltage, and the number of carriers per unit volume can be determined from the magnitude of V_H.

Hall voltage	$V_H = E_H w = v_d B w = \dfrac{I}{nte}B$	26-17, 26-20

*Quantum Hall effects	Measurements at very low temperatures in very large magnetic fields indicate that the Hall resistance $R_H = V_H / I$ is quantized and can take on only the values given by	
	$R_H = \dfrac{V_H}{I} = \dfrac{R_K}{n}, \quad n = 1, 2, 3, \ldots$	26-21

*Conventional von Klitzing constant (definition of ohm)	$R_{K-90} = 25{,}812.807 \ \Omega$ (exact)	26-23

PROBLEMS

- Single-concept, single-step, relatively easy
- •• Intermediate-level, may require synthesis of concepts
- ••• Challenging
- SSM Solution is in the *Student Solutions Manual*
- iSOLVE Problems available on iSOLVE online homework service
- iSOLVE✓ These "Checkpoint" online homework service problems ask students additional questions about their confidence level, and how they arrived at their answer.

In a few problems, you are given more data than you actually need; in a few other problems, you are required to supply data from your general knowledge, outside sources, or informed estimates.

Conceptual Problems

1 • SSM When a cathode-ray tube is placed horizontally in a magnetic field that is directed vertically upward, the electrons emitted from the cathode follow one of the dashed paths to the face of the tube in Figure 26-30. The correct path is (a) 1. (b) 2. (c) 3. (d) 4. (e) 5.

FIGURE 26-30
Problem 1

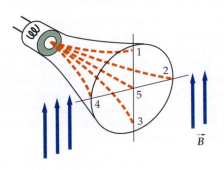

2 • Why not define $\vec{B}$ to be in the direction of $\vec{F}$, as we do for $\vec{E}$?

3 • True or false: The magnetic force does not accelerate a charged particle because the magnetic force is perpendicular to the velocity of the particle.

4 • A beam of positively charged particles passes undeflected from left to right through a velocity selector in which the electric field is up. The beam is then reversed so that it travels from right to left. Will the beam now be deflected in the velocity selector? If so, in which direction?

5 • SSM A *flicker bulb* is a lightbulb with a long, thin filament. When it is plugged in and a magnet is brought near the lightbulb, the filament is seen to oscillate rapidly back and forth. Why does the filament oscillate, and what is the frequency of oscillation?

6 • What orientation of a current loop relative to the direction of the magnetic field gives maximum torque?

7 • True or false:

(a) The magnetic force on a moving charged particle is always perpendicular to the velocity of the particle.

(b) The torque on a magnet by a magnetic field tends to align the magnet's magnetic moment in the direction of the magnetic field.

(c) A current loop in a uniform magnetic field responds to the field in the same manner as a small permanent magnet.

(d) The period of a particle moving in a circle in a magnetic field is proportional to the radius of the circle.

(e) The drift velocity of electrons in a wire can be determined from the Hall effect.

8 • [SSM] The north-seeking pole of a compass needle located on the magnetic equator is the end of the needle that points toward the north, and the direction of any magnetic field $\vec{B}$ is specified as the direction that the north-seeking pole of a compass needle points when the needle is aligned in the field. Suppose that the direction of the magnetic field $\vec{B}$ were instead specified as the direction of a south-seeking pole of a compass needle aligned in the field. Would the right-hand rule shown in Figure 26-2 then give the direction of the magnetic force on the moving positive charge, or would a left-hand rule be required? Explain.

9 • If the magnetic field is directed toward the north and a positively charged particle is moving toward the east, what is the direction of the magnetic force on the particle?

10 • A positively charged particle is moving northward in a magnetic field. The magnetic force on the particle is toward the northeast. What is the direction of the magnetic field? (a) Up (b) West (c) South (d) Down (e) The force cannot be directed toward the northeast.

11 • A ^{7}Li nucleus with a charge of $+3e$ and a mass of 7 u(1 u = 1.66×10^{-27} kg) and a proton with charge $+e$ and mass 1 u are both moving in a plane perpendicular to a magnetic field $\vec{B}$. The magnitude of the momenta of the two particles are equal. The ratio of the radius of curvature of the path of the proton, R_p, to that of the ^{7}Li nucleus, R_{Li}, is (a) $R_p/R_{Li} = 3$. (b) $R_p/R_{Li} = 1/3$. (c) $R_p/R_{Li} = 1/7$. (d) $R_p/R_{Li} = 3/7$. (e) none of these.

12 • [SSM] An electron moving with speed v to the right enters a region of uniform magnetic field directed out of the paper. When the electron enters this region, it will be (a) deflected out of the plane of the paper. (b) deflected into the plane of the paper. (c) deflected upward. (d) deflected downward. (e) undeviated in its motion.

13 ••• The theory of relativity tells us that none of the laws of physics can depend on the absolute velocity of an object, which is in fact impossible to define. Instead, the behavior of physical objects can only depend on the relative velocity between the objects. The development of new physical insights can come from this idea. For example, in Figure 26-31, a magnet moving at high speed flies by an electron that is at rest relative to a physicist observing it in a laboratory. Explain why you are sure that a force must be acting on it. What direction will the force point when the north pole of the magnet passes directly underneath the electron? Explain.

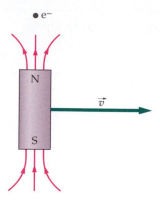

FIGURE 26-31 Problem 13

14 • How are magnetic field lines similar to electric field lines? How are they different?

15 • If a current I in a given wire and a magnetic field $\vec{B}$ are known, the force $\vec{F}$ on the current is uniquely determined. Show that knowing $\vec{F}$ and I does not provide complete knowledge of $\vec{B}$.

Estimation and Approximation

16 •• [SSM] CRT's used in monitors and televisions commonly use magnetic deflection to steer the electron beams. A schematic diagram is shown in Figure 26-32. The electron beam is accelerated through a potential difference and the electron beam is then accelerated through a magnetic field that deflects the electron beam, as shown in the figure. Given the following parameters, estimate the magnitude of the magnetic field needed for maximum deflection: accelerating voltage, $V = 15$ kV; distance over which electron is in magnetic field, $d = 5$ cm; length, $L = 50$ cm; diagonal of CRT, $r = 19$ in.

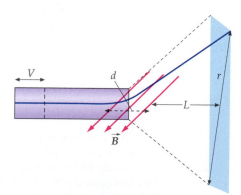

FIGURE 26-32 Problem 16

17 •• (a) Estimate the charge-to-mass ratio of a micrometeorite needed for it to "orbit" the Earth in a low-earth orbit (400 km above the surface of the Earth) under the influence of the Earth's magnetic field alone. Take the magnitude of the Earth's field to be 5×10^{-5} T and assume it perpendicular to the meteorite's velocity. Assume that the speed of the meteorite is about the same as Earth's orbital speed of roughly 30 km/s. (b) If the mass of the micrometeorite is 3×10^{-10} kg, what is its charge?

The Force Exerted by a Magnetic Field

18 • **iSOLVE✓** Find the magnetic force on a proton moving with velocity 4.46 Mm/s in the positive x direction in a magnetic field of 1.75 T in the positive z direction.

19 • A charge $q = -3.64$ nC moves with a velocity of 2.75×10^6 m/s $\hat{\imath}$. Find the force on the charge if the magnetic field is (a) $\vec{B} = 0.38$ T $\hat{\jmath}$, (b) $\vec{B} = 0.75$ T $\hat{\imath} + 0.75$ T $\hat{\jmath}$, (c) $\vec{B} = 0.65$ T $\hat{\imath}$, and (d) $\vec{B} = 0.75$ T $\hat{\imath} + 0.75$ T $\hat{k}$.

20 • **iSOLVE** A uniform magnetic field of magnitude 1.48 T is in the positive z direction. Find the force exerted by the field on a proton if the proton's velocity is (a) $\vec{v} = 2.7$ Mm/s $\hat{\imath}$, (b) $\vec{v} = 3.7$ Mm/s $\hat{\jmath}$, (c) $\vec{v} = 6.8$ Mm/s $\hat{k}$, and (d) $\vec{v} = 4.0$ Mm/s $\hat{\imath} + 3.0$ Mm/s $\hat{\jmath}$.

21 • **iSOLVE** A straight wire segment 2 m long makes an angle of 30° with a uniform magnetic field of 0.37 T. Find the magnitude of the force on the wire if it carries a current of 2.6 A.

22 • **SSM** **iSOLVE✓** A straight wire segment $I\vec{L} = (2.7\ A)(3\ cm\ \hat{\imath} + 4\ cm\ \hat{\jmath})$ is in a uniform magnetic field $\vec{B} = 1.3$ T $\hat{\imath}$. Find the force on the wire.

23 • What is the force (magnitude and direction) on an electron with velocity $\vec{v} = (2\hat{\imath} - 3\hat{\jmath} \times 10^6)$ m/s in a magnetic field $\vec{B} = (0.8\ \hat{\imath} + 0.6\ \hat{\jmath} - 0.4\ \hat{k})$T?

24 •• The wire segment shown in Figure 26-33 carries a current of 1.8 A from a to b. There is a magnetic field $\vec{B} = 1.2$ T $\hat{k}$. Find the total force on the wire and show that the total force is the same as if the wire were a straight segment from a to b.

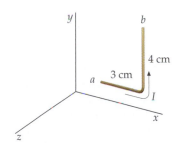

FIGURE 26-33
Problem 24

25 •• **iSOLVE** A straight, stiff, horizontal wire of length 25 cm and mass 50 g is connected to a source of emf by light, flexible leads. A magnetic field of 1.33 T is horizontal and perpendicular to the wire. Find the current necessary to float the wire; that is, find the current so the magnetic force balances the weight of the wire.

26 •• **SSM** **iSOLVE✓** A simple gaussmeter for measuring horizontal magnetic fields consists of a stiff 50-cm wire that hangs vertically from a conducting pivot so that its free end makes contact with a pool of mercury in a dish below. The mercury provides an electrical contact without constraining the movement of the wire. The wire has a mass of 5 g and conducts a current downward. (a) What is the equilibrium angular displacement of the wire from vertical if the horizontal magnetic field is 0.04 T and the current is 0.20 A? (b) If the current is 20 A and a displacement from vertical of 0.5 mm can be detected for the free end, what is the horizontal magnetic field sensitivity of this gaussmeter?

27 •• A current-carrying wire is bent into a semicircular loop of radius R that lies in the xy plane. There is a uniform magnetic field $\vec{B} = B\hat{k}$ perpendicular to the plane of the loop (Figure 26-34). Verify that the force acting on the loop is 0.

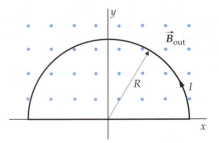

FIGURE 26-34 Problem 27

28 •• A 10-cm length of wire carries a current of 4.0 A in the positive z direction. The force on this wire due to a uniform magnetic field $\vec{B}$ is $\vec{F} = (-0.2\ \hat{\imath} + 0.2\ \hat{\jmath})$N. If this wire is rotated so that the current flows in the positive x direction, the force on the wire is $\vec{F} = 0.2\ \hat{k}$N. Find the magnetic field $\vec{B}$.

29 •• **iSOLVE** A 10-cm length of wire carries a current of 2.0 A in the positive x direction. The force on this wire due to the presence of a magnetic field $\vec{B}$ is $\vec{F} = (3.0\ \hat{\jmath} + 2.0\ \hat{k})$N. If this wire is now rotated so that the current flows in the positive y direction, the force on the wire is $\vec{F} = (-3.0\ \hat{\imath} - 2.0\ \hat{k})$N. Determine the magnetic field $\vec{B}$.

30 ••• A wire bent in some arbitrary shape carries a current I in a uniform magnetic field $\vec{B}$. Show explicitly that the total force on the part of the wire from some point a to some point b is $\vec{F} = I\vec{L} \times \vec{B}$, where $\vec{L}$ is the vector from point a to point b.

Motion of a Point Charge in a Magnetic Field

31 • **SSM** A proton moves in a circular orbit of radius 65 cm perpendicular to a uniform magnetic field of magnitude 0.75 T. (a) What is the period for this motion? (b) Find the speed of the proton. (c) Find the kinetic energy of the proton.

32 • **iSOLVE✓** An electron of kinetic energy 45 keV moves in a circular orbit perpendicular to a magnetic field of 0.325 T. (a) Find the radius of the orbit. (b) Find the frequency and period of the motion.

33 • **iSOLVE** An electron from the sun with a speed of 1×10^7 m/s enters the earth's magnetic field high above the equator where the magnetic field is 4×10^{-7} T. The electron moves nearly in a circle, except for a small drift along the direction of the earth's magnetic field that will take the electron toward the north pole. (a) What is the radius of the circular motion? (b) What is the radius of the circular motion near the north pole where the magnetic field is 2×10^{-5} T?

34 •• Protons and deuterons (each with charge $+e$) and alpha particles (with charge $+2e$) of the same kinetic energy enter a uniform magnetic field $\vec{B}$ that is perpendicular to their velocities. Let R_p, R_d, and R_α be the radii of their circular orbits. Find the ratios R_d/R_p and R_α/R_p. Assume that $m_\alpha = 2m_d = 4m_p$.

35 •• A proton and an alpha particle move in a uniform magnetic field in circles of the same radii. Compare (a) their velocities, (b) their kinetic energies, and (c) their angular momenta. (See Problem 34.)

36 •• A particle of charge q and mass m has momentum $p = mv$ and kinetic energy $K = p^2/2m$. If the particle moves in a circular orbit of radius R perpendicular to a uniform magnetic field $\vec{B}$, show that (a) $p = BqR$ and (b) $K = \frac{1}{2}B^2q^2R^2/m$.

37 •• **SSM** A beam of particles with velocity $\vec{v}$ enters a region of uniform magnetic field $\vec{B}$ that makes a small angle θ with $\vec{v}$. Show that after a particle moves a distance $2\pi(m/qB)v \cos \theta$, measured along the direction of $\vec{B}$, the velocity of the particle is in the same direction as it was when the particle entered the field.

38 •• **iSOLVE** A proton with speed $v = 10^7$ m/s enters a region of uniform magnetic field $B = 0.8$ T, which is into the page, as shown in Figure 26-35. The angle $\theta = 60°$. Find the angle ϕ and the distance d.

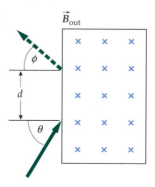

FIGURE 26-35 Problems 38 and 39

39 •• Suppose that in Figure 26-35 $B = 0.6$ T, the distance $d = 0.4$ m, and $\theta = 24°$. Find the speed v and the angle ϕ if the particles are (a) protons and (b) deuterons.

40 •• The galactic magnetic field in some region of interstellar space has a magnitude of 10^{-9} T. A particle of interstellar dust has a mass of 10 μg and a total charge of 0.3 nC. How many years does it take to complete a circular orbit in the magnetic field?

The Velocity Selector

41 • **SSM** **iSOLVE** A velocity selector has a magnetic field of magnitude 0.28 T perpendicular to an electric field of magnitude 0.46 MV/m. (a) What must the speed of a particle be for the particle to pass through undeflected? What energy must (b) protons and (c) electrons have to pass through undeflected?

42 • **iSOLVE**✓ A beam of protons moves along the x axis in the positive x direction with a speed of 12.4 km/s through a region of crossed fields balanced for zero deflection. (a) If there is a magnetic field of magnitude 0.85 T in the positive y direction, find the magnitude and direction of the electric field. (b) Would electrons of the same velocity be deflected by these fields? If so, in what direction?

Thomson's Measurement of q/m for Electrons and the Mass Spectrometer

43 •• **SSM** The plates of a Thomson q/m apparatus are 6.0 cm long and are separated by 1.2 cm. The end of the plates is 30.0 cm from the tube screen. The kinetic energy of the electrons is 2.8 keV. (a) If a potential of 25 V is applied across the deflection plates, by how much will the beam deflect? (b) Find the magnitude of the crossed magnetic field that will allow the beam to pass between the plates undeflected.

44 •• Chlorine has two stable isotopes, ^{35}Cl and ^{37}Cl, whose natural abundances are about 76 percent and 24 percent, respectively. Singly ionized chlorine gas is to be separated into its isotopic components using a mass spectrometer. The magnetic field in the spectrometer is 1.2 T. What is the minimum value of the potential through which these ions must be accelerated so that the separation between them is 1.4 cm?

45 •• **iSOLVE** A singly ionized ^{24}Mg ion (mass 3.983 × 10^{-26} kg) is accelerated through a 2.5-kV potential difference and deflected in a magnetic field of 557 G in a mass spectrometer. (a) Find the radius of curvature of the orbit for the ion. (b) What is the difference in radius for ^{26}Mg ions and for ^{24}Mg ions? (Assume that their mass ratio is 26:24.)

46 •• **SSM** A beam of ^{6}Li and ^{7}Li ions passes through a velocity selector and enters a magnetic spectrometer. If the diameter of the orbit of the ^{6}Li ions is 15 cm, what is the diameter of the orbit for ^{7}Li ions?

The Cyclotron

47 •• In Example 26-6, determine the time required for a ^{58}Ni ion and a ^{60}Ni ion to complete the semicircular path.

48 •• Before entering a mass spectrometer, ions pass through a velocity selector consisting of parallel plates separated by 2.0 mm and having a potential difference of 160 V. The magnetic field between the plates is 0.42 T. The magnetic field in the mass spectrometer is 1.2 T. Find (a) the speed of the ions entering the mass spectrometer and (b) the difference in the diameters of the orbits of singly ionized ^{238}U and ^{235}U. (The mass of a ^{235}U ion is 3.903 × 10^{-25} kg.)

49 •• **SSM** **iSOLVE** A cyclotron for accelerating protons has a magnetic field of 1.4 T and a radius of 0.7 m. (a) What is the cyclotron frequency? (b) Find the maximum energy of the protons when they emerge. (c) How will your answers change if deuterons, which have the same charge but twice the mass, are used instead of protons?

50 •• A certain cyclotron with a magnetic field of 1.8 T is designed to accelerate protons to 25 MeV. (a) What is the cyclotron frequency? (b) What must the minimum radius of the magnet be to achieve a 25-MeV emergence energy? (c) If the alternating potential applied to the dees has a maximum value of 50 kV, how many revolutions must the protons make before emerging with an energy of 25 MeV?

51 •• **iSOLVE** Show that for a certain cyclotron the cyclotron frequencies of deuterons and alpha particles are the same and are half that of a proton in the same magnetic field. (See Problem 34.)

52 •• Show that the radius of the orbit of a charged particle in a cyclotron is proportional to the square root of the number of orbits completed.

Torques on Current Loops and Magnets

53 • **SOLVE** ✓ A small circular coil of 20 turns of wire lies in a uniform magnetic field of 0.5 T, so that the normal to the plane of the coil makes an angle of 60° with the direction of $\vec{B}$. The radius of the coil is 4 cm, and it carries a current of 3 A. (a) What is the magnitude of the magnetic moment of the coil? (b) What is the magnitude of the torque exerted on the coil?

54 • **SOLVE** ✓ What is the maximum torque on a 400-turn circular coil of radius 0.75 cm that carries a current of 1.6 mA and resides in a uniform magnetic field of 0.25 T?

55 • **SSM** **SOLVE** A current-carrying wire is bent into the shape of a square of edge-length $L = 6$ cm and is placed in the xy plane. It carries a current $I = 2.5$ A. What is the magnitude of the torque on the wire if there is a uniform magnetic field of 0.3 T (a) in the z direction and (b) in the x direction?

56 • Repeat Problem 55 if the wire is bent into an equilateral triangle of edge-length 8 cm.

57 •• **SOLVE** ✓ A rigid, circular loop of radius R and mass m carries a current I and lies in the xy plane on a rough, flat table. There is a horizontal magnetic field of magnitude B. What is the minimum value of B so that one edge of the loop will lift off the table?

58 •• A rectangular, 50-turn coil has sides 6-cm long and 8-cm long and carries a current I of 1.75 A. It is oriented and pivoted about the z axis, as shown in Figure 26-36. (a) If the wire in the xy plane makes an angle $\theta = 37°$ with the y axis as shown, what angle does the unit normal $\hat{n}$ make with the x axis? (b) Write an expression for $\hat{n}$ in terms of the unit vectors $\hat{i}$ and $\hat{j}$. (c) What is the magnetic moment of the coil? (d) Find the torque on the coil when there is a uniform magnetic field $\vec{B} = 1.5$ T $\hat{j}$. (e) Find the potential energy of the coil in this field.

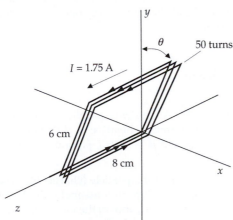

FIGURE 26-36 Problems 58 and 59

59 •• The coil in Problem 58 is pivoted about the z axis and held at various positions in a uniform magnetic field $\vec{B} = 2.0$ T $\hat{j}$. Sketch the position of the coil and find the torque exerted when the unit normal is (a) $\hat{n} = \hat{i}$, (b) $\hat{n} = \hat{j}$, (c) $\hat{n} = -\hat{j}$, and (d) $\hat{n} = (\hat{i} + \hat{j})/\sqrt{2}$.

Magnetic Moments

60 •• **SSM** **SOLVE** ✓ A small magnet of length 6.8 cm is placed at an angle of 60° to the direction of a uniform magnetic field of magnitude 0.04 T. The observed torque has a magnitude of 0.10 N·m. Find the magnetic moment of the magnet.

61 •• **SOLVE** A wire loop consists of two semicircles connected by straight segments (Figure 26-37). The inner and outer radii are 0.3 m and 0.5 m, respectively. A current I of 1.5 A flows in this loop with the current in the outer semicircle in the clockwise direction. What is the magnetic moment of this current loop?

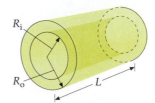

FIGURE 26-37 Problem 61

62 •• A wire of length L is wound into a circular coil of N loops. Show that when this coil carries a current I, its magnetic moment has the magnitude $IL^2/4\pi N$.

63 •• A particle of charge q and mass m moves in a circle of radius R and with angular velocity ω. (a) Show that the average current is $I = q\omega/(2\pi)$ and that the magnetic moment has the magnitude $\mu = \frac{1}{2}q\omega r^2$. (b) Show that the angular momentum of this particle has the magnitude $L = mr^2\omega$ and that the magnetic moment and angular momentum vectors are related by $\vec{\mu} = (\frac{1}{2}q/m)\vec{L}$.

64 ••• **SSM** A hollow cylinder has length L and inner and outer radii R_i and R_o, respectively (Figure 26-38). The cylinder carries a uniform charge density ρ. Derive an expression for the magnetic moment as a function of ω, the angular velocity of rotation of the cylinder about its axis.

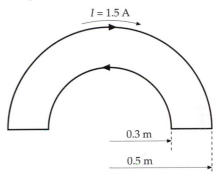

FIGURE 26-38 Problem 64

65 ••• A nonconducting rod of mass m and length L has a uniform charge per unit length λ and rotates with angular velocity ω about an axis through one end and perpendicular to the rod. (a) Consider a small segment of the rod of length dx and charge $dq = \lambda\,dx$ at a distance x from the pivot (Figure 26-39). Show that the magnetic moment of this segment is $\frac{1}{2}\lambda\omega x^2 dx$. (b) Integrate your result to show that the total magnetic moment of the rod is $\mu = \frac{1}{6}\lambda\omega L^3$. (c) Show that the magnetic moment $\vec{\mu}$ and angular momentum $\vec{L}$ are related by $\vec{\mu} = (\frac{1}{2}Q/m)\vec{L}$, where Q is the total charge on the rod.

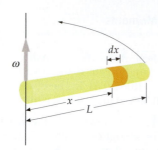

FIGURE 26-39
Problem 65

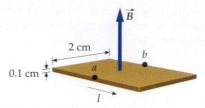

FIGURE 26-41 Problems 70 and 71

66 ••• A nonuniform, nonconducting disk of mass m, radius R, and total charge Q has a surface charge density $\sigma = \sigma_0 r/R$ and a mass per unit area $\sigma_m = (m/Q)\sigma$. The disk rotates with angular velocity ω about its axis. (a) Show that the magnetic moment of the disk has a magnitude $\mu = \frac{1}{5}\pi\omega\sigma_0 R^4 = \frac{3}{10}Q\omega R^2$. (b) Show that the magnetic moment $\vec{\mu}$ and angular momentum $\vec{L}$ are related by $\vec{\mu} = (\frac{1}{2}Q/m)\vec{L}$.

67 ••• A spherical shell of radius R carries a surface charge density σ. The sphere rotates about its diameter with angular velocity ω. Find the magnetic moment of the rotating sphere.

68 ••• A solid sphere of radius R carries a uniform volume charge density ρ. The sphere rotates about its diameter with angular velocity ω. Find the magnetic moment of this rotating sphere.

69 ••• **SSM** A uniform disk of mass m, radius R, and surface charge σ rotates about its center with angular velocity ω in Figure 26-40. A uniform magnetic field of magnitude $\vec{B}$ threads the disk, making an angle θ with respect to the rotation axis of the disk. Calculate (a) the net torque acting on the disk and (b) the precession frequency of the disk in the magnetic field. (See pp. 316–317 for a discussion of precession.)

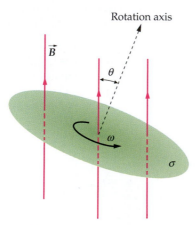

FIGURE 26-40 Problem 69

The Hall Effect

70 • **iSOLVE** ✓ A metal strip 2-cm wide and 0.1-cm thick carries a current of 20 A in a uniform magnetic field of 2 T, as shown in Figure 26-41. The Hall voltage is measured to be 4.27 μV. (a) Calculate the drift velocity of the electrons in the strip. (b) Find the number density of the charge carriers in the strip. (c) Is point a or point b at the higher potential?

71 •• The number density of free electrons in copper is 8.47×10^{22} electrons per cubic centimeter. If the metal strip in Figure 26-41 is copper and the current is 10 A, find (a) the drift velocity v_d and (b) the Hall voltage. (Assume that the magnetic field is 2.0 T.)

72 •• **SSM** **iSOLVE** A copper strip ($n = 8.47 \times 10^{22}$ electrons per cubic centimeter) 2-cm wide and 0.1-cm thick is used to measure the magnitudes of unknown magnetic fields that are perpendicular to the strip. Find the magnitude of B when $I = 20$ A and the Hall voltage is (a) 2.00 μV, (b) 5.25 μV, and (c) 8.00 μV.

73 •• **iSOLVE** Because blood contains charged ions, moving blood develops a Hall voltage across the diameter of an artery. A large artery with a diameter of 0.85 cm has a flow speed of 0.6 m/s. If a section of this artery is in a magnetic field of 0.2 T, what is the maximum possible potential difference across the diameter of the artery?

74 •• **iSOLVE** The Hall coefficient R is defined as $R = E_y/(J_x B_z)$, where J_x is the current per unit area in the x direction in the slab, B_z is the magnetic field in the z direction, and E_y is the resulting Hall field in the y direction. Show that the Hall coefficient is $1/(nq)$, where q is the charge of the charge carriers, -1.6×10^{-19} C if they are electrons. (The Hall coefficients of monovalent metals, such as copper, silver, and sodium are therefore negative.)

75 •• **SSM** Aluminum has a density of 2.7×10^3 kg/m³ and a molar mass of 27 g/mol. The Hall coefficient of aluminum is $R = -0.3 \times 10^{-10}$ m³/C. (See Problem 74 for the definition of R.) Find the number of conduction electrons per aluminum atom.

General Problems

76 • **iSOLVE** ✓ A long wire parallel to the x axis carries a current of 6.5 A in the positive x direction. There is a uniform magnetic field $\vec{B} = 1.35$ T $\hat{j}$. Find the force per unit length on the wire.

77 • **iSOLVE** An alpha particle (charge $+2e$) travels in a circular path of radius 0.5 m in a magnetic field of 1 T. Find (a) the period, (b) the speed, and (c) the kinetic energy (in electron volts) of the alpha particle. Take $m = 6.65 \times 10^{-27}$ kg for the mass of the alpha particle.

78 •• The pole strength q_m of a bar magnet is defined by $q_m = |\vec{\mu}|/L$, where L is the length of the magnet. Show that the torque exerted on a bar magnet in a uniform magnetic field $\vec{B}$ is the same as if a force $+q_m\vec{B}$ is exerted on the north pole and a force $-q_m\vec{B}$ is exerted on the south pole.

79 •• [SSM] A particle of mass m and charge q enters a region where there is a uniform magnetic field $\vec{B}$ along the x axis. The initial velocity of the particle is $\vec{v} = v_{0x}\hat{i} + v_{0y}\hat{j}$, so the particle moves in a helix. (*a*) Show that the radius of the helix is $r = mv_{0y}/qB$. (*b*) Show that the particle takes a time $t = 2\pi m/qB$ to make one orbit around the helix.

80 •• [SSM] [iSOLVE] A metal crossbar of mass m rides on a pair of long, horizontal conducting rails separated by a distance L and connected to a device that supplies constant current I to the circuit, as shown in Figure 26-42. A uniform magnetic field $\vec{B}$ is established, as shown. (*a*) If there is no friction and the bar starts from rest at $t = 0$, show that at time t the bar has velocity $v = (BIL/m)t$. (*b*) In which direction will the bar move? (*c*) If the coefficient of static friction is μ_S, find the minimum field B necessary to start the bar moving.

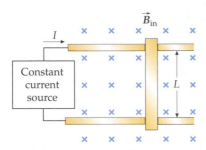

FIGURE 26-42 Problems 80 and 81

81 •• Assume that the rails in Figure 26-42 are frictionless but tilted upward so that they make an angle θ with the horizontal. (*a*) What vertical magnetic field $\vec{B}$ is needed to keep the bar from sliding down the rails? (*b*) What is the acceleration of the bar if B has twice the value found in Part (*a*)?

82 •• A long, narrow bar magnet that has magnetic moment $\vec{\mu}$ parallel to its long axis is suspended at its center as a frictionless compass needle. When placed in a horizontal magnetic field $\vec{B}$, the needle lines up with the field. If it is displaced by a small angle θ, show that the needle will oscillate about its equilibrium position with frequency $f = \frac{1}{2\pi}\sqrt{\mu B/I}$, where I is the moment of inertia about the point of suspension.

83 •• [iSOLVE] A conducting wire is parallel to the y axis. It moves in the positive x direction with a speed of 20 m/s in a magnetic field $\vec{B} = 0.5$ T $\hat{k}$. (*a*) What are the magnitude and direction of the magnetic force on an electron in the conductor? (*b*) Because of this magnetic force, electrons move to one end of the wire leaving the other end positively charged, until the electric field due to this charge separation exerts a force on the electrons that balances the magnetic force. Find the magnitude and direction of this electric field in the steady state. (*c*) Suppose the moving wire is 2-m long. What is the potential difference between its two ends due to this electric field?

84 ••• The rectangular frame shown in Figure 26-43 is free to rotate about the axis A–A on the horizontal shaft. The frame is 10-cm long and 6-cm wide, and the rods that make up the frame have a mass per unit length of 20 g/cm. A uniform magnetic field $B = 0.2$ T is directed, as shown. A current may be sent around the frame by means of the wires attached at the top. (*a*) If no current passes through the frame, what is the period of this physical pendulum for small oscillations? (*b*) If a current of 8 A passes through the frame in the direction indicated by the arrow, what is then the period of this physical

pendulum? (*c*) Suppose the direction of the current is opposite to the direction shown. The frame is displaced from the vertical by some angle θ. What must be the magnitude of the current so that this frame will be in equilibrium?

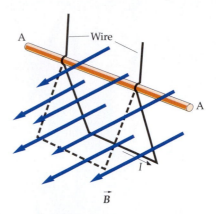

FIGURE 26-43 Problem 84

85 ••• [SSM] A stiff, straight horizontal wire of length 25 cm and mass 20 g is supported by electrical contacts at its ends, but is otherwise free to move vertically upward. The wire is in a uniform, horizontal magnetic field of magnitude 0.4 T perpendicular to the wire. A switch connecting the wire to a battery is closed and the wire flies upward, rising to a maximum height h. The battery delivers a total charge of 2 C during the short time it makes contact with the wire. Find the height h.

86 ••• A circular loop of wire with mass m carries a current I in a uniform magnetic field. It is initially in equilibrium with its magnetic moment vector aligned with the magnetic field. The loop is given a small twist about a diameter and then released. What is the period of the motion? (Assume that the only torque exerted on the loop is due to the magnetic field.)

87 ••• A small bar magnet has a magnetic moment $\vec{\mu}$ that makes an angle θ with the x axis and lies in a nonuniform magnetic field given by $\vec{B} = B_x(x)\hat{i} + B_y(y)\hat{j}$. Use $F_x = -dU/dx$ and $F_y = -dU/dy$ to show that there is a net force on the magnet that is given by

$$\vec{F} = \mu_x \frac{\partial B_x}{\partial x}\hat{i} + \mu_y \frac{\partial B_y}{\partial y}\hat{j}$$

88 ••• [SSM] The special theory of relativity tells us that a particle's mass depends on its speed through the formula:

$$m(v) = \frac{m_0}{\sqrt{1 - \dfrac{v^2}{c^2}}} = \gamma(v)m_0$$

where m_0 is the particle's rest mass and $\gamma(v) = 1/\sqrt{1 - (v^2/c^2)}$ (*a*) *Taking into account the special theory of relativity,* what is the radius and period of a particle's orbit if it has speed v and is moving in a magnetic field with magnitude B that is perpendicular to the direction of the velocity? Assume the force on the particle is given by $\vec{F} = q(\vec{v} \times \vec{B})$. The particle has rest mass m_0 and charge q. (*b*) Using a spreadsheet program, make graphs of the radius and period of the orbit of an electron in a 10-T magnetic field versus $\gamma(v)$ for speeds between $v = 0.1c$ and $v = 0.999c$. Use a logarithmic scale to display $\gamma(v)$.

Sources of the Magnetic Field

? **Have you any idea what the magnetic field of a current-carrying coil looks like? There are illustrations of the magnetic field of a coil in Section 27-2.**

The earliest known sources of magnetism were permanent magnets. One month after Oersted announced his discovery that a compass needle is deflected by an electric current, Jean-Baptiste Biot and Félix Savart announced the results of their measurements of the torque on a magnet near a long, current-carrying wire and they analyzed these results in terms of the magnetic field produced by each element of the current. André-Marie Ampère extended these experiments and showed that current elements also experience a force in the presence of a magnetic field and that two currents exert forces on each other.

➤ **In this chapter, we begin by considering the magnetic field produced by a single moving charge and by the moving charges in a current element. We then calculate the magnetic fields produced by some common current configurations, such as a straight wire segment; a long, straight wire; a current loop; and a solenoid. Next we discuss Ampère's law, which relates the line integral of the magnetic field around a closed loop to the total current that passes through the loop. Finally, we consider the magnetic properties of matter.**

27-1 The Magnetic Field of Moving Point Charges

When a point charge q moves with velocity $\vec{v}$, the moving point charge produces a magnetic field $\vec{B}$ in space, given by[†]

$$\vec{B} = \frac{\mu_0}{4\pi} \frac{q\vec{v} \times \hat{r}}{r^2} \qquad\qquad 27\text{-}1$$

MAGNETIC FIELD OF A MOVING POINT CHARGE

where $\hat{r}$ is a unit vector (see Figure 27-1) that points to the field point P from the charge q moving with velocity $\vec{v}$, and μ_0 is a constant of proportionality called the **permeability of free space**,[‡] which has the exact value

$$\mu_0 = 4\pi \times 10^{-7}\,\text{T} \cdot \text{m/A} = 4\pi \times 10^{-7}\,\text{N/A}^2 \qquad\qquad 27\text{-}2$$

The units of μ_0 are such that B is in teslas when q is in coulombs, v is in meters per second, and r is in meters. The unit N/A^2 comes from the fact that $1\,\text{T} = 1\,\text{N/(A·m)}$. The constant $1/(4\pi)$ is arbitrarily included in Equation 27-1 so that the factor 4π will not appear in Ampère's law (Equation 27-15), which we will study in Section 27-4.

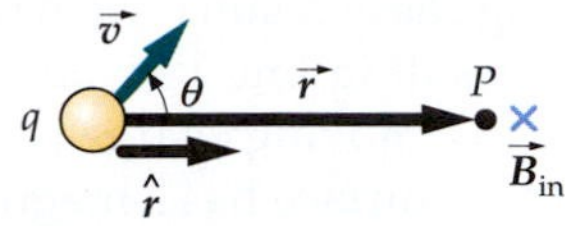

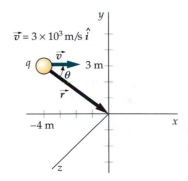

FIGURE 27-1 A positive point charge q moving with velocity $\vec{v}$ produces a magnetic field $\vec{B}$ at a field point P that is in the direction $\vec{v} \times \hat{r}$, where $\hat{r}$ is the unit vector pointing from the charge to the field point. The field varies inversely as the square of the distance from the charge to the field point and is proportional to the sine of the angle between $\vec{v}$ and $\hat{r}$. (The blue × at the field point indicates that the direction of the field is into the page.)

MAGNETIC FIELD OF A MOVING POINT CHARGE   **EXAMPLE 27-1**

A point particle with charge $q = 4.5$ nC is moving with velocity $\vec{v} = 3 \times 10^3$ m/s$\hat{i}$ **parallel to the** x axis along the line $y = 3$ m. Find the magnetic field at the origin produced by this charge when the charge is at the point $x = -4$ m, $y = 3$ m, as shown in Figure 27-2.

FIGURE 27-2

1. The magnetic field is given by Equation 27-1:

$$\vec{B} = \frac{\mu_0}{4\pi} \frac{q\vec{v} \times \hat{r}}{r^2}, \quad \text{with } \vec{v} = v\hat{i}$$

2. Find $\vec{r}$ and r from Figure 27-2 and write $\hat{r}$ in terms of $\hat{i}$ and $\hat{j}$:

$$\vec{r} = 4\,\text{m}\hat{i} - 3\,\text{m}\hat{j}$$

$$r = \sqrt{4^2 + 3^2}\,\text{m} = 5\,\text{m}$$

$$\hat{r} = \frac{\vec{r}}{r} = \frac{4\,\text{m}\hat{i} - 3\,\text{m}\hat{j}}{5\,\text{m}} = 0.8\,\hat{i} - 0.6\,\hat{j}$$

3. Substitute the above results in Equation 27-1 to obtain:

$$\vec{B} = \frac{\mu_0}{4\pi} \frac{q\vec{v} \times \hat{r}}{r^2} = \frac{\mu_0}{4\pi} \frac{q(v\hat{i}) \times (0.8\,\hat{i} - 0.6\,\hat{j})}{r^2} = \frac{\mu_0}{4\pi} \frac{q(-0.6\,v\hat{k})}{r^2}$$

$$= -(10^{-7}\,\text{T·m/A})\frac{(4.5 \times 10^{-9}\,\text{C})(0.6)(3 \times 10^3\,\text{m/s})}{(5\,\text{m})^2}\,\hat{k}$$

$$= \boxed{-3.24 \times 10^{-14}\,\text{T}\hat{k}}$$

[†] This expression is used for speeds much less than the speed of light.
[‡] Some care must be taken not to confuse the constant μ_0 with the magnitude of the magnetic moment vector $\vec{\mu}$.

REMARKS It is also possible to obtain $\vec{B}$ without finding an explicit expression for the unit vector $\hat{r}$. From Figure 27-2 we note that $\vec{v} \times \hat{r}$ is in the negative z direction. In addition, the magnitude of $\vec{v} \times \hat{r}$ is $v \sin \theta$, where $\sin \theta = (3\text{ m})/(5\text{ m}) = 0.6$. Combining these results, we have $\vec{v} \times \hat{r} = v \sin \theta(-\hat{k}) = -v(0.6)\hat{k}$, in agreement with our result in line 1 of step 3. Finally, this example shows that the magnetic field due to a moving charge is quite small. For comparison, the earth's magnetic field near its surface has a magnitude of about 10^{-4} T.

EXERCISE At the same instant, find the magnetic field on the y axis both at $y = 3$ m and at $y = 6$ m. (*Answer* $\vec{B} = 0, \vec{B} = 3.24 \times 10^{-14}$ T$\hat{k}$)

(a)
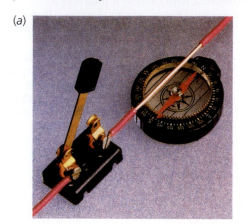

(b)
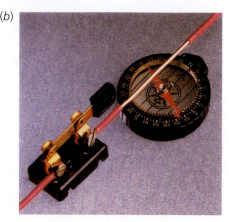

Oersted's experiment. (*a*) With no current in the wire, the compass needle points north. (*b*) When the wire carries a current, the compass needle is deflected in the direction of the resultant magnetic field. The current in the wire is directed upward, from left to right. The insulation has been stripped from the wire to improve the contrast of the photograph.

27-2 The Magnetic Field of Currents: The Biot–Savart Law

In the previous chapter we extended our discussion of forces on point charges to forces on current elements by replacing $q\vec{v}$ with the current element $I\,d\vec{\ell}$. We do the same for the magnetic field produced by a current element. The magnetic field $d\vec{B}$ produced by a current element $I\,d\vec{\ell}$ is given by Equation 27-1, with $q\vec{v}$ replaced by $I\,d\vec{\ell}$:

$$d\vec{B} = \frac{\mu_0}{4\pi} \frac{I\,d\vec{\ell} \times \hat{r}}{r^2}$$

27-3

BIOT–SAVART LAW

Equation 27-3, known as the **Biot–Savart law,** was also deduced by Ampère. The Biot–Savart law and Equation 27-1 are analogous to Coulomb's law for the electric field of a point charge. The source of the magnetic field is a moving charge $q\vec{v}$ or a current element $I\,d\vec{\ell}$, just as the charge q is the source of the electrostatic field. The magnetic field decreases with the square of the distance from the moving charge or current element, just as the electric field decreases with the square of the distance from a point charge. However, the directional aspects of the electric and magnetic fields are quite different. Whereas the electric field points in the radial direction $\hat{r}$ from the point charge to the field point (for a positive charge), the magnetic field is perpendicular to both $\hat{r}$ and to $\vec{v}$, in the case of a point charge, or to $d\vec{\ell}$ in the case of a current element. At a point along the line of a current element, such as point P_2 in Figure 27-3, the magnetic field due to that element is zero. (Equation 27-3 gives $d\vec{B} = 0$ if $d\vec{\ell}$ and $\hat{r}$ are either parallel or antiparallel.)

The magnetic field due to the total current in a circuit can be calculated by using the Biot–Savart law to find the field due to each current element, and then summing (integrating) over all the current elements in the circuit. This calculation is difficult for all but the simplest circuit geometries.

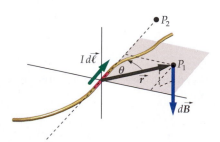

FIGURE 27-3 The current element $I\,d\vec{\ell}$ produces a magnetic field at point P_1 that is perpendicular to both $d\vec{\ell}$ and $\vec{r}$. The current element produces no magnetic field at point P_2, which is along the line of $d\vec{\ell}$.

$\vec{B}$ Due to a Current Loop

Figure 27-4 shows a current element $I\,d\vec{\ell}$ of a current loop of radius R and the unit vector $\hat{r}$ that is directed from the element to the center of the loop. The magnetic field at the center of the loop due to this element is directed along the axis of the loop, and its magnitude is given by

$$dB = \frac{\mu_0}{4\pi}\frac{I\,d\ell\sin\theta}{R^2}$$

where θ is the angle between $d\vec{\ell}$ and $\hat{r}$, which is 90° for each current element, so $\sin\theta = 1$. The magnetic field due to the entire current is found by integrating over all the current elements in the loop. Since R is the same for all elements, we obtain

$$B = \int dB = \frac{\mu_0}{4\pi}\frac{I}{R^2}\oint d\ell$$

The integral of $d\ell$ around the complete loop gives the total length $2\pi R$, the circumference of the loop. The magnetic field due to the entire loop is thus

$$B = \frac{\mu_0}{4\pi}\frac{I}{R^2}\,2\pi R = \frac{\mu_0 I}{2R} \qquad\qquad 27\text{-}4$$

B AT THE CENTER OF A CURRENT LOOP

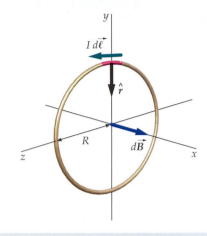

FIGURE 27-4 Current element for calculating the magnetic field at the center of a circular current loop. Each element produces a magnetic field that is directed along the axis of the loop.

EXERCISE Find the current in a circular loop of radius 8 cm that will give a magnetic field of 2 G at the center of the loop. (*Answer* 25.5 A)

Figure 27-5 shows the geometry for calculating the magnetic field at a point on the axis of a circular current loop a distance x from the circular loop's center. We first consider the current element at the top of the loop. Here, as everywhere on the loop, $I\,d\vec{\ell}$ is tangent to the loop and perpendicular to the vector $\vec{r}$ from the current element to the field point P. The magnetic field $d\vec{B}$ due to this element is in the direction shown in the figure, perpendicular to $\hat{r}$ and also perpendicular to $I\,d\vec{\ell}$. The magnitude of $d\vec{B}$ is

$$|d\vec{B}| = \frac{\mu_0}{4\pi}\frac{I|d\vec{\ell}\times\hat{r}|}{r^2} = \frac{\mu_0}{4\pi}\frac{I\,d\ell}{(x^2 + R^2)}$$

where we have used the facts that $r^2 = x^2 + R^2$ and that $d\vec{\ell}$ and $\hat{r}$ are perpendicular, so $|d\vec{\ell}\times\hat{r}| = d\ell$.

When we sum around all the current elements in the loop, the components of $d\vec{B}$ perpendicular to the axis of the loop, such as dB_y in Figure 27-5, sum to zero, which leave only the components dB_x that are parallel to the axis. We thus compute only the x component of the field. From Figure 27-5, we have

$$dB_x = dB\sin\theta = \left(\frac{\mu_0}{4\pi}\frac{I\,d\ell}{(x^2 + R^2)}\right)\left(\frac{R}{\sqrt{x^2 + R^2}}\right) = \frac{\mu_0}{4\pi}\frac{IR\,d\ell}{(x^2 + R^2)^{3/2}}$$

To find the field due to the entire loop of current, we integrate dB_x around the loop:

$$B_x = \oint dB_x = \oint \frac{\mu_0}{4\pi}\frac{IR}{(x^2 + R^2)^{3/2}}\,d\ell$$

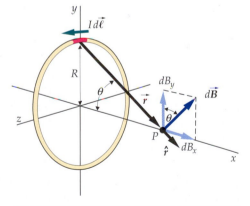

FIGURE 27-5 Geometry for calculating the magnetic field at a point on the axis of a circular current loop.

Since neither x nor R varies as we sum over the elements in the loop, we can remove these quantities from the integral. Then,

$$B_x = \frac{\mu_0}{4\pi} \frac{IR}{(x^2 + R^2)^{3/2}} \oint d\ell$$

The integral of $d\ell$ around the loop gives $2\pi R$. Thus,

$$B_x = \frac{\mu_0}{4\pi} \frac{IR}{(x^2 + R^2)^{3/2}} 2\pi R = \frac{\mu_0}{4\pi} \frac{2\pi R^2 I}{(x^2 + R^2)^{3/2}} \qquad 27\text{-}5$$

B ON THE AXIS OF A CURRENT LOOP

EXERCISE Show that Equation 27-5 reduces to $B_x = \mu_0 I/2R$ (Equation 27-4) at the center of the loop.

At great distances from the loop, $|x|$ is much greater than R, so $(x^2 + R^2)^{3/2} \approx (x^2)^{3/2} = |x|^3$. Then,

$$B_x = \frac{\mu_0}{4\pi} \frac{2I\pi R^2}{|x|^3}$$

or

$$B_x = \frac{\mu_0}{4\pi} \frac{2\mu}{|x|^3} \qquad 27\text{-}6$$

MAGNETIC-DIPOLE FIELD ON THE AXIS OF THE DIPOLE

where $\mu = I\pi R^2$ is the magnitude of the magnetic moment of the loop. Note the similarity of this expression and the electric field on the axis of an electric dipole of moment ρ (Equation 21-10):

$$E_x = \frac{1}{4\pi\epsilon_0} \frac{2\rho}{|x|^3}$$

Although it has not been demonstrated, our result that a current loop produces a magnetic dipole field far away holds in general for any point whether it is on the axis of the loop or off of the axis of the loop. Thus, a current loop behaves as a magnetic dipole because it experiences a torque $\vec{\mu} \times \vec{B}$ when placed in an external magnetic field (as was shown in Chapter 26) and it also produces a magnetic dipole field at a great distance from the current loop. Figure 27-6 shows the magnetic field lines for a current loop.

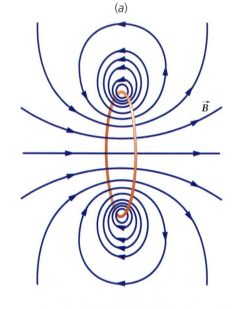

(a)

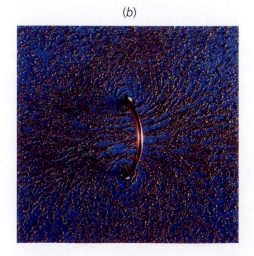

(b)

FIGURE 27-6 (*a*) The magnetic field lines of a circular current loop. (*b*) The magnetic field lines of a circular current loop indicated by iron filings.

FIND $\vec{B}$ *ON AXIS OF COIL*   **EXAMPLE 27-2**

A circular coil of radius 5.0 cm has 12 turns and lies in the $x = 0$ plane and is centered at the origin. It carries a current of 4 A so that the direction of the magnetic moment of the coil is along the x axis. Using Equation 27-5, find the magnetic field on the x axis at (a) $x = 0$, (b) $x = 15$ cm, and (c) $x = 3$ m. (d) Using Equation 27-6, find the magnetic field on the x axis at $x = 3$ m.

PICTURE THE PROBLEM The magnetic field due to a loop with N turns is N times that due to a single turn. (a) At $x = 0$ (center of the loops) $B = \mu_0 N/(2R)$ (from Equation 27-4). Equation 27-5 gives the magnetic field on axis due to the current in a single turn. Far from the loop, as in Part (c), the field can be found using Equation 27-6. In this case, since we have N loops, the magnetic moment is $\mu = NI\pi R^2$.

(a) B_x at the center is N times that given by Equation 27-4 for a single loop:

$$B_x = \frac{\mu_0 NI}{2R}$$

$$= (4\pi \times 10^{-7}\,\text{T·m/A})\,\frac{(12)(4\,\text{A})}{2(0.05\,\text{m})} = \boxed{6.03 \times 10^{-4}\,\text{T}}$$

(b) B_x on the axis is N times that given by Equation 27-5:

$$B_x = \frac{\mu_0}{4\pi}\,\frac{2\pi R^2 NI}{(x^2 + R^2)^{3/2}}$$

$$= (10^{-7}\,\text{T·m/A})\,\frac{2\pi(0.05\,\text{m})^2(12)(4\,\text{A})}{\left[(0.15\,\text{m})^2 + (0.05\,\text{m})^2\right]^{3/2}}$$

$$= \boxed{1.91 \times 10^{-5}\,\text{T}}$$

(c) Use Equation 27-5 again:

$$B_x = \frac{\mu_0}{4\pi}\,\frac{2\pi R^2 NI}{(x^2 + R^2)^{3/2}}$$

$$= (10^{-7}\,\text{T·m/A})\,\frac{2\pi(0.05\,\text{m})^2(12)(4\,\text{A})}{\left[(3\,\text{m})^2 + (0.05\,\text{m})^2\right]^{3/2}}$$

$$= \boxed{2.791 \times 10^{-9}\,\text{T}}$$

(d) 1. Since 3 m is much greater than the radius $R = 0.05$ m, we can use Equation 27-6 for the magnetic field far from the loop:

$$B_x = \frac{\mu_0}{4\pi}\,\frac{2\mu}{|x|^3}$$

2. The magnitude of the magnetic moment of the loop is N/A:

$$\mu = NI\pi R^2 = (12)(4\,\text{A})\pi(0.05\,\text{m})^2 = 0.377\,\text{A·m}^2$$

3. Substitute μ and $x = 3$ m into B_x in step 1:

$$B_x = \frac{\mu_0}{4\pi}\,\frac{2\mu}{|x|^3} = (10^{-7}\,\text{T·m/A})\,\frac{2(0.377\,\text{A·m}^2)}{(3\,\text{m})^3}$$

$$= \boxed{2.793 \times 10^{-9}\,\text{T}}$$

REMARKS In Part (d) $x = 60R$, so we were able to use an approximation that is valid for $x \gg R$. The result differs from the exact value, calculated in Part (c), by less than one tenth of one percent.

EXAMPLE 27-3

In the coil described in Example 27-2 the current is 4 A. Assuming the drift speed is 1.4×10^{-4} m/s, find the number of coulombs of mobile charge in the wire. (The drift speed for a wire carrying a current of 1 A was found to be 3.4×10^{-5} m/s in Example 25-1.)

PICTURE THE PROBLEM The amount of moving charge Q in the wire is the product of the rate at which charge enters one end of the wire and the time it takes the charge to travel the length of the wire. The rate at which charge enters one end of the wire is the current I, and the time for the charge to travel the length L of the wire is L/v_d, which is the drift speed.

1. The amount of moving charge is the product of the current and the time for a charge carrier to travel the length of the wire:

$$Q = I \, \Delta t$$

2. The drift speed is the length of the wire divided by the time:

$$v_d = \frac{L}{\Delta t}$$

3. The length L is the number of turns times the length per turn. Also, we solve the step 2 result for the time:

$$L = N2\pi R = (12)2\pi(.05 \text{ m}) = 3.77 \text{ m}$$

and

$$\Delta t = \frac{L}{v_d} = \frac{3.77 \text{ m}}{1.4 \times 10^{-4} \text{ m/s}} = 2.69 \times 10^4 \text{ s}$$

4. Solve the step 1 result for the amount of moving charge in the wire:

$$Q = I \, \Delta t = (4 \text{ A})(2.69 \times 10^4 \text{ s})$$

$$= \boxed{1.08 \times 10^5 \text{ C}}$$

REMARKS The current consists of more than 10^5 C of moving charges. This is an enormous amount of charge, in comparison to the amount of charge stored in an ordinary capacitor.

EXAMPLE 27-4 **Try It Yourself**

A small bar magnet of magnetic moment $\mu = 0.03$ A·m² is placed at the center of the loop of Example 27-2 so that its magnetic moment vector lies in the xy plane and makes an angle of 30° with the x axis. Neglecting any variation in $\vec{B}$ over the region of the magnet, find the torque on the magnet.

PICTURE THE PROBLEM The torque on a magnetic moment is given by $\vec{\tau} = \vec{\mu} \times \vec{B}$. Since $\vec{B}$ is in the positive x direction, you can see from Figure 27-7 that $\vec{\mu} \times \vec{B}$ is in the negative z direction.

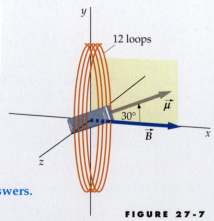

FIGURE 27-7

Cover the column to the right and try these on your own before looking at the answers.

Steps   **Answers**

1. Compute the magnitude of the torque from $\vec{\tau} = \vec{\mu} \times \vec{B}$. $\tau = 9.04 \times 10^{-6}$ N·m

2. Indicate the direction with a unit vector. $\vec{\tau} = \boxed{-(9.04 \times 10^{-6} \text{ N·m})\hat{k}}$

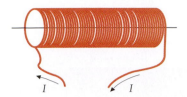

FIGURE 27-8 A tightly wound solenoid can be considered as a set of circular current loops placed side by side that carry the same current. The solenoid produces a uniform magnetic field inside the loops.

$\vec{B}$ Due to a Current in a Solenoid

A **solenoid** is a wire tightly wound into a helix of closely spaced turns, as illustrated in Figure 27-8. A solenoid is used to produce a strong, uniform magnetic field in the region surrounded by its loops. The solenoid's role in magnetism is analogous to that of the parallel-plate capacitor, which produces a strong, uniform electric field between its plates. The magnetic field of a solenoid is essentially that of a set of N identical current loops placed side by side. Figure 27-9 shows the magnetic field lines for two such loops.

Figure 27-10 shows the magnetic field lines for a long, tightly wound solenoid. Inside the solenoid, the field lines are approximately parallel to the axis and are closely and uniformly spaced, indicating a strong, uniform magnetic field. Outside the solenoid, the lines are much less dense. The field lines diverge from one end and converge at the other end. Comparing this figure with Figure 27-8, we see that the field lines of a solenoid, both inside and outside the solenoid, are identical to those of a bar magnet of the same shape as the solenoid.

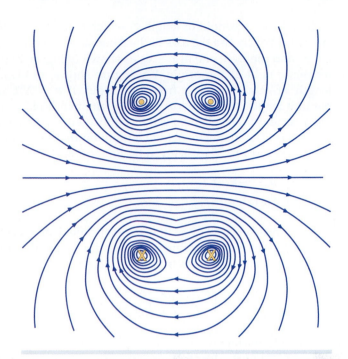

FIGURE 27-9 Magnetic field lines due to two coaxial loops carrying the same current. The points where the loops intersect the plane of the page are each marked by an × where the current enters and by a dot where the current emerges. In the region between the loops near the axis the magnetic fields of the individual loops superpose, so the resultant field is strong and surprisingly uniform. In the regions away from the loops, the resultant field is relatively weak.

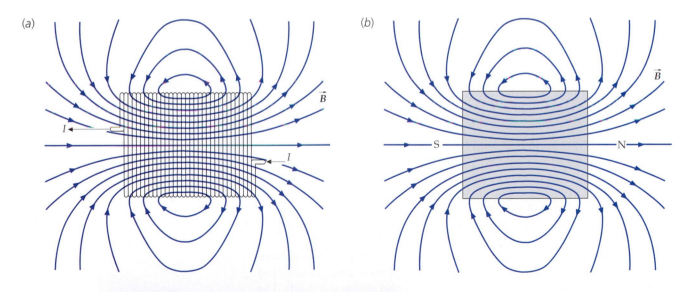

FIGURE 27-10 (*a*) Magnetic field lines of a solenoid. The lines are identical to those of a bar magnet of the same shape, as in Figure 27-10 (*b*). (*c*) Magnetic field lines of a solenoid shown by iron filings.

Consider a solenoid of length L, which consists of N turns of wire carrying a current I. We choose the axis of the solenoid to be the x axis, with the left end at $x = x_1$ and the right end at $x = x_2$, as shown in Figure 27-11. We will calculate the magnetic field at the origin. The figure shows an element of the solenoid of length dx at a distance x from the origin. If $n = N/L$ is the number of turns per unit length, there are $n\,dx$ turns of wire in this element, with each turn carrying a current I. The element is thus equivalent to a single loop carrying a current $di = nI\,dx$. The magnetic field at a point on the x axis due to a loop at the origin carrying a current $nI\,dx$ is given by Equation 27-5 with I replaced by $di = nI\,dx$:

$$dB_x = \frac{\mu_0}{4\pi} \frac{2\pi R^2 nI\,dx}{(x^2 + R^2)^{3/2}}$$

This expression also gives the magnetic field at the origin due to a current loop at x. We find the magnetic field at the origin due to the entire solenoid by integrating this expression from $x = x_1$ to $x = x_2$:

$$B_x = \frac{\mu_0}{4\pi} 2\pi R^2 nI \int_{x_1}^{x_2} \frac{dx}{(x^2 + R^2)^{3/2}} \qquad \text{27-7}$$

The integral in Equation 27-7 can be evaluated using trigonometric substitution with $x = R\tan\theta$. Also, the integral can be looked up in standard tables of integrals. The integral's value is

$$\int_{x_1}^{x_2} \frac{dx}{(x^2 + R^2)^{3/2}} = \frac{x}{R^2\sqrt{x^2 + R^2}}\Bigg|_{x_1}^{x_2} = \frac{1}{R^2}\left(\frac{x_2}{\sqrt{x_2^2 + R^2}} - \frac{x_1}{\sqrt{x_1^2 + R^2}}\right)$$

Substituting this into Equation 27-7, we obtain

$$B_x = \frac{1}{2}\mu_0 nI\left(\frac{x_2}{\sqrt{x_2^2 + R^2}} - \frac{x_1}{\sqrt{x_1^2 + R^2}}\right) \qquad \text{27-8}$$

B_x ON THE AXIS OF A SOLENOID AT $X = 0$

A solenoid is called a long solenoid if its length L is much greater than its radius R. Inside and far from the ends of a long solenoid, the left term in the parentheses tends toward $+1$ and the right term tends toward -1. In the region satisfying these conditions, the magnetic field is

$$B_x = \mu_0 nI \qquad \text{27-9}$$

B_x INSIDE A LONG SOLENOID

If the origin is at the left end of the solenoid, $x_1 = 0$ and $x_2 = L$. Then, if $L \gg R$, the right term in the parentheses of Equation 27-8 is zero and the left term approaches 1, so $B \approx \frac{1}{2}\mu_0 nI$. Thus, the magnitude of $\vec{B}$ at either end of a long solenoid is half the magnitude at points within the solenoid that are distant from either end. Figure 27-12 gives a plot of the magnetic field on the axis of a solenoid versus position x on the axis (with the origin at the center of the solenoid). The approximation that the field is uniform (independent of the position) along the axis is good, except for very near the ends.

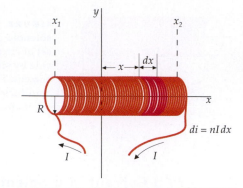

FIGURE 27-11 Geometry for calculating the magnetic field inside a solenoid on its axis. The number of turns in the element dx is $n\,dx$, where $n = N/L$ is the number of turns per unit length. The element dx is treated as a current loop carrying a current $di = nI\,dx$.

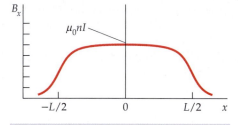

FIGURE 27-12 Graph of the magnetic field on the axis inside a solenoid versus the position x on the axis. The field inside the solenoid is nearly constant except near the ends. The length L of the solenoid is ten times longer than the radius.

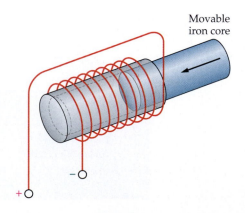

FIGURE 27-13 An automotive starter solenoid. When the solenoid is energized, its magnetic field pulls in the iron core. This engages gears that connect the starter motor to the flywheel of the engine. Once the current to the solenoid is interrupted, a spring disengages the gears and pushes the iron core to the right.

$\vec{B}$ AT CENTER OF A SOLENOID   **EXAMPLE   27-5**

Find the magnetic field at the center of a solenoid of length 20 cm, radius 1.4 cm, and 600 turns that carries a current of 4 A.

PICTURE THE PROBLEM

1. We will calculate the field exactly, using Equation 27-8:

$$B_x = \frac{1}{2}\mu_0 nI \left(\frac{x_2}{\sqrt{x_2^2 + R^2}} - \frac{x_1}{\sqrt{x_1^2 + R^2}} \right)$$

2. For a point at the center of the solenoid, $x_1 = -10$ cm and $x_2 = +10$ cm. Thus, the terms in the parentheses in Equation 27-8 have values of:

$$\frac{x_2}{\sqrt{x_2^2 + R^2}} = \frac{10 \text{ cm}}{\sqrt{(10 \text{ cm})^2 + (1.4 \text{ cm})^2}} = 0.990$$

$$\frac{x_1}{\sqrt{x_1^2 + R^2}} = \frac{-10 \text{ cm}}{\sqrt{(-10 \text{ cm})^2 + (1.4 \text{ cm})^2}} = 0.990$$

3. Substitute these results into B_x in step 1:

$$B_x = \frac{1}{2}(4\pi \times 10^{-7} \text{ T·m/A})[(600 \text{ turns})/(0.2 \text{ m})](4 \text{ A})(0.990 + 0.990)$$

$$= \boxed{1.50 \times 10^{-2} \text{ T}}$$

REMARKS Note that the approximation obtained using Equation 27-9 amounts to replacing 0.99 by 1.00, which differs by only one percent. Note also that the magnitude of the magnetic field inside this solenoid is fairly large—about 250 times the magnetic field of the earth.

EXERCISE Calculate B_x using the long-solenoid approximation. (*Answer* 1.51×10^{-2} T)

A cross section of a doorbell. When the solenoid is energized, its magnetic field pulls on the plunger, causing it to strike the bell (not shown). The spring returns the plunger to its normal position.

$\vec{B}$ Due to a Current in a Straight Wire

Figure 27-14 shows the geometry for calculating the magnetic field $\vec{B}$ at a point P due to the current in the straight wire segment shown. We choose R to be the perpendicular distance from the wire to point P, and we choose the x axis to be along the wire with $x = 0$ at the projection of P onto the x axis.

A typical current element $I\,d\vec{\ell}$ at a distance x from the origin is shown. The vector $\vec{r}$ points from the element to the field point P. The direction of the magnetic field at P due to this element is the direction of $I\,d\vec{\ell} \times \hat{r}$, which is out of the paper. Note that the magnetic fields due to all the current elements of the wire are in this same direction. Thus, we need to compute only the magnitude of the field. The field due to the current element shown has the magnitude (Equation 27-3)

$$dB = \frac{\mu_0}{4\pi}\frac{I\,dx}{r^2}\sin\phi$$

It is more convenient to write this in terms of θ rather than ϕ:

$$dB = \frac{\mu_0}{4\pi}\frac{I\,dx}{r^2}\cos\theta \qquad\qquad 27\text{-}10$$

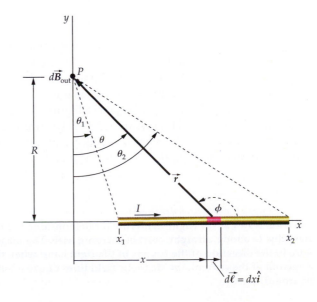

FIGURE 27-14 Geometry for calculating the magnetic field at point P due to a straight current segment. Each element of the segment contributes to the total magnetic field at point P, which is directed out of the paper. The result is expressed in terms of the angles θ_1 and θ_2.

To sum over all the current elements, we need to relate the variables θ, r, and x. It turns out to be easiest to express x and r in terms of θ. We have

$$x = R \tan \theta$$

Then, taking the differential of each side with R as a constant gives

$$dx = R \sec^2 \theta \, d\theta = R \frac{r^2}{R^2} \, d\theta = \frac{r^2}{R} \, d\theta$$

where we have used $\sec \theta = r/R$. Substituting this expression for dx into Equation 27-10, we obtain

$$dB = \frac{\mu_0}{4\pi} \frac{I}{r^2} \frac{r^2 \, d\theta}{R} \cos \theta = \frac{\mu_0}{4\pi} \frac{I}{R} \cos \theta \, d\theta$$

We sum over these elements by integrating from $\theta = \theta_1$ to $\theta = \theta_2$, where θ_1 and θ_2 are shown in Figure 27-14. This gives

$$B = \int_{\theta_1}^{\theta_2} \frac{\mu_0}{4\pi} \frac{I}{R} \cos \theta \, d\theta = \frac{\mu_0}{4\pi} \frac{I}{R} \int_{\theta_1}^{\theta_2} \cos \theta \, d\theta$$

Evaluating the integral, we obtain

$$B = \frac{\mu_0}{4\pi} \frac{I}{R} (\sin \theta_2 - \sin \theta_1) \qquad\qquad 27\text{-}11$$

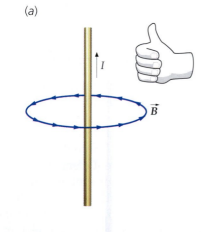

(a)

B DUE TO A STRAIGHT WIRE SEGMENT

This result gives the magnetic field due to any wire segment in terms of the perpendicular distance R and θ_1 and θ_2 are the angles subtended at the field point by the ends of the wire. If the length of the wire approaches infinity in both directions, θ_2 approaches $+90°$ and θ_1 approaches $-90°$. The result for such a very long wire is obtained from Equation 27-11, by setting $\theta_1 = -90°$ and $\theta_2 = +90°$:

$$B = \frac{\mu_0}{4\pi} \frac{2I}{R} \qquad\qquad 27\text{-}12$$

B DUE TO AN INFINITELY LONG, STRAIGHT WIRE

At any point in space, the magnetic field lines of a long, straight, current-carrying wire are tangent to a circle of radius R about the wire, where R is the perpendicular distance from the wire to the field point. The direction of $\vec{B}$ can be determined by applying the right-hand rule, as shown in Figure 27-15a. The magnetic field lines thus encircle the wire, as shown in Figure 27-15b.

The result expressed by Equation 27-12 was found experimentally by Biot and Savart in 1820. From their analysis, Biot and Savart were able to discover the expression given in Equation 27-3 for the magnetic field due to a current element.

(b)

FIGURE 27-15 (a) Right-hand rule for determining the direction of the magnetic field due to a long, straight, current-carrying wire. The magnetic field lines encircle the wire in the direction of the fingers of the right hand when the thumb points in the direction of the current. (b) Magnetic field lines due to a long wire, which is indicated by iron filings.

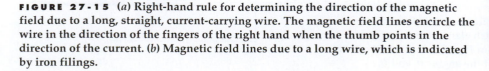

B *At Center of Square Current Loop*   **EXAMPLE   27-6**

Find the magnetic field at the center of a square current loop of edge length $L = 50$ cm, which carries a current of 1.5 A.

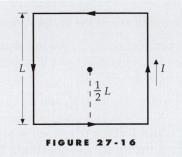

FIGURE 27-16

PICTURE THE PROBLEM The magnetic field at the center of the loop is the sum of the contributions from each of the four sides of the loop. From Figure 27-16, we can see that each side of the loop produces a field of equal magnitude pointing out of the page. Thus, we use Equation 27-11 for a given side, then multiply by 4 for the total field.

1. The total field is 4 times the field B_s due to a side:

$$B = 4B_s$$

2. Calculate the magnetic field B_s due to a given side of the loop. Note from the figure that $R = \frac{1}{2}L$ and $\theta_1 = -45°$ and $\theta_2 = +45°$:

$$B_s = \frac{\mu_0}{4\pi}\frac{I}{R}(\sin\theta_2 - \sin\theta_1) = \frac{\mu_0}{4\pi}\frac{I}{\frac{1}{2}L}[\sin(+45°) - \sin(-45°)]$$

$$= (10^{-7}\text{ T·m/A})\frac{1.5\text{ A}}{0.25\text{ m}}\,2\sin 45° = 8.49 \times 10^{-7}\text{ T}$$

3. Multiply this value by 4 to find the total field:

$$B = 4B_s = 4(8.49 \times 10^{-7}\text{ T}) = \boxed{3.39 \times 19^{-6}\text{ T}}$$

EXERCISE Compare the magnetic field at the center of a circular current loop of radius R with the magnetic field at the center of a square current loop of side $L = 2R$ carrying the same current. Which is larger? (*Answer* B at the center is larger for the circle, by about 10 percent)

EXERCISE Find the distance from a long, straight wire carrying a current of 12 A, where the magnetic field due to the current in the wire is equal in magnitude to 0.6 G (the magnitude of the earth's magnetic field). (*Answer* $R = 4.00$ cm)

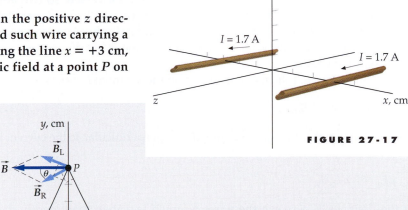

A current gun used to measure electric current. The jaws of the current gun clamp around a current-carrying wire without touching the wire. The magnetic field produced by the wire is measured with a Hall-effect device mounted in the current gun. The Hall-effect device puts out a voltage proportional to the magnetic field, which in turn is proportional to the current in the wire.

B *Due to Two Parallel Wires*   **EXAMPLE   27-7**

A long, straight wire carrying a current of 1.7 A in the positive z direction lies along the line $x = -3$ cm, $y = 0$. A second such wire carrying a current of 1.7 A in the positive z direction lies along the line $x = +3$ cm, $y = 0$, as shown in Figure 27-17. Find the magnetic field at a point P on the y axis at $y = 6$ cm.

FIGURE 27-17

PICTURE THE PROBLEM The magnetic field at point P is the vector sum of the field $\vec{B}_L$ due to the wire on the left in Figure 27-18, and the field $\vec{B}_R$ due to the wire on the right. Since each wire carries the same current, and each wire is the same distance from point P, the magnitudes B_L and B_R are equal. $\vec{B}_L$ is perpendicular to the radius from the left wire to point P, and $\vec{B}_R$ is perpendicular to the radius from the right wire to the point P.

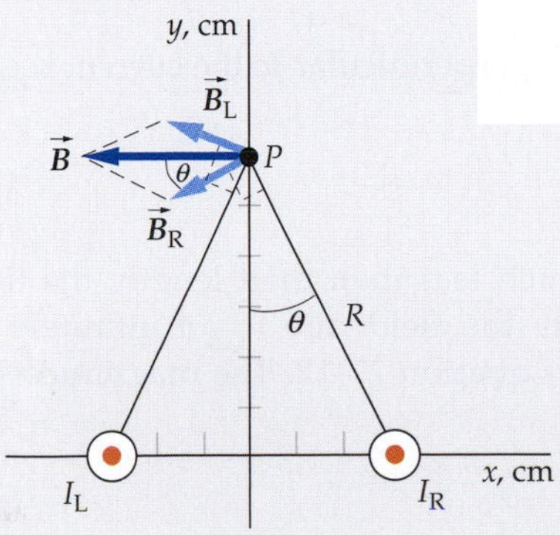

FIGURE 27-18

1. The field at P is the vector sum of the fields $\vec{B}_L$ and $\vec{B}_R$:

$$\vec{B} = \vec{B}_L + \vec{B}_R$$

2. From Figure 27-18, we see that the resultant magnetic field is in the negative x direction and has the magnitude $2B_L \cos \theta$.

$$\vec{B} = -2B_L \cos \theta \, \hat{i}$$

3. The magnitudes of $\vec{B}_L$ and $\vec{B}_R$ are given by Equation 27-12:

$$B_L = B_R = \frac{\mu_0}{4\pi} \frac{2I}{R}$$

4. R is the distance from each wire to the point P. We find R from the figure and substitute R into the expression for B_L and B_R:

$$R = \sqrt{(3 \text{ cm})^2 + (6 \text{ cm})^2} = 6.71 \text{ cm}$$

so

$$B_L = B_R = (10^{-7} \text{ T·m/A}) \frac{2(1.7 \text{ A})}{0.0671 \text{ m}} = 5.07 \times 10^{-6} \text{ T}$$

5. We obtain $\cos \theta$ from the figure:

$$\cos \theta = \frac{6 \text{ cm}}{R} = \frac{6 \text{ cm}}{6.71 \text{ cm}} = 0.894$$

6. Substitute the values of $\cos \theta$ and B_L into the equation in step 2 for $\vec{B}$:

$$\vec{B} = -2(5.07 \times 10^{-6} \text{ T})(0.894)\hat{i} = \boxed{-9.07 \times 10^{-6} \text{ T } \hat{i}}$$

EXERCISE Find $\vec{B}$ at the origin. (*Answer* 0)

EXERCISE Find $\vec{B}$ at the origin assuming that I_R goes into the page. (*Answer* $\vec{B} = 2.27 \times 10^{-5} \text{ T } \hat{j}$)

Magnetic Force Between Parallel Wires

We can use Equation 27-12 for the magnetic field due to a long, straight, current-carrying wire and $d\vec{F} = I \, d\vec{\ell} \times \vec{B}$ (Equation 26-5) for the force exerted by a magnetic field on a segment of a current-carrying wire to find the force exerted by one long straight current on another. Figure 27-19 shows two long parallel wires carrying currents in the same direction. We consider the force on a segment $d\vec{\ell}_2$ carrying current I_2, as shown. The magnetic field $\vec{B}_1$ at this segment due to current I_1 is perpendicular to the segment $I_2 \, d\vec{\ell}_2$, as shown. This is true for all current elements along the wire. The magnetic force $d\vec{F}_2$ on current segment $I_2 \, d\vec{\ell}_2$ is directed toward current I_1, since $d\vec{F}_2 = I_2 \, d\vec{\ell}_2 \times \vec{B}_2$. Similarly, a current segment $I_1 \, d\vec{\ell}_1$ will experience a magnetic force directed toward current I_2 due to a magnetic field arising from current I_2. Thus, two parallel currents attract each other. If one of the currents is reversed the force will be reversed, so two antiparallel currents will repel each other. The attraction or repulsion of parallel or antiparallel currents was discovered experimentally by Ampère one week after he heard of Oersted's discovery of the effect of a current on a compass needle.

The magnitude of the magnetic force on the segment $I_2 \, d\vec{\ell}_2$ is

$$dF_2 = |I_2 \, d\vec{\ell}_2 \times \vec{B}_1|$$

Since the magnetic field at segment $I_2 \, d\vec{\ell}_2$ is perpendicular to the current segment, we have

$$dF_2 = I_2 \, d\ell_2 B_1$$

If the distance R between the wires is much less than their length, the field at $I_2 \, d\vec{\ell}_2$ due to current I_1 will approximate the field due to an infinitely long, current-carrying wire, which is given by Equation 27-12. The magnitude of the force on the segment $I_2 \, d\vec{\ell}_2$ is therefore

$$dF_2 = I_2 \, d\ell_2 \frac{\mu_0 I_1}{2\pi R}$$

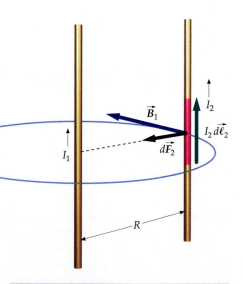

FIGURE 27-19 Two long straight wires carrying parallel currents. The magnetic field $\vec{B}_1$ due to current I_1 is perpendicular to current I_2. The force on current I_2 is toward current I_1. There is an equal and opposite force exerted by current I_2 on I_1. The current-carrying wires thus attract each other.

The force per unit length is

$$\frac{dF_2}{d\ell_2} = I_2 \frac{\mu_0 I_1}{2\pi R} = 2\frac{\mu_0}{4\pi}\frac{I_1 I_2}{R}$$ 27-13

In Chapter 21, the coulomb was defined in terms of the ampere, but the definition of the ampere was deferred. The ampere is defined as follows:

> The ampere is the constant electric current that, when maintained in two straight parallel conductors of infinite length and of negligible circular cross sections placed one meter apart in a vacuum, would produce a force between the conductors equal to 2×10^{-7} newtons per meter of length.

DEFINITION—AMPERE

This definition of the ampere makes the permeability of free space μ_0 equal to exactly $4\pi \times 10^{-7}\,\text{N/A}^2$. It also allows the unit of current (and therefore the unit of electric charge) to be determined by a mechanical measurement. In practice, currents much closer together than 1 m are used so that the force can be measured accurately with long but finite wires.

Figure 27-20 shows a **current balance**, which is a device that can be used to calibrate an ammeter from the definition of the ampere. The upper conductor, directly above the lower conductor, is free to rotate about knife-edge contacts and is balanced so that the wires (or conducting rods) are a small distance apart. The conductors are connected in series to carry the same current but in opposite directions so that the currents will repel each other. Weights are placed on the upper conductor until it balances again at the original separation. The force of repulsion is thus determined by measuring the total weight required to balance the upper conductor.

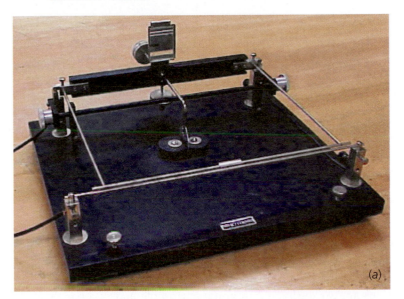

(a)

FIGURE 27-20 (a) A picture of a current balance used in a general physics lab. (b) A schematic diagram of a current balance. The two parallel rods in front carry equal but oppositely directed currents and therefore repel each other. The force of repulsion is balanced by weights placed on the upper rod, which is part of a rectangle that is balanced on knife edges at the back. The mirror on top is used to reflect a beam of laser light to accurately determine the position of the upper rod.

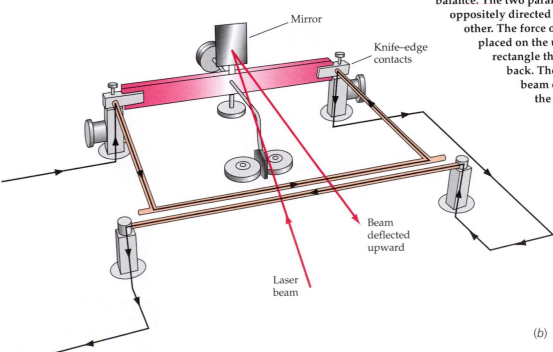

(b)

EXAMPLE 27-8 Try It Yourself

Two straight rods 50-cm long with axes 1.5-mm apart in a current balance carry currents of 15 A each in opposite directions. What mass must be placed on the upper rod to balance the magnetic force of repulsion?

PICTURE THE PROBLEM Equation 27-13 gives the magnitude of the magnetic force per unit length exerted by the lower rod on the upper rod. Find this force for a rod of length L and set it equal to the weight mg.

Cover the column to the right and try these on your own before looking at the answers.

Steps

Answers

1. Set the weight mg equal to the magnetic force of repulsion of the rods.

$$mg = 2\frac{\mu_0}{4\pi}\frac{I_1I_2}{R}L$$

2. Solve for the mass m.

$$m = 1.53 \times 10^{-3}\,\text{kg} = \boxed{1.53\,\text{g}}$$

REMARKS Since only 1.53 g are required to balance the system, we see that the magnetic force between two straight current-carrying wires is relatively small, even for currents as large as 15 A separated by only 1.5 mm.

27-3 Gauss's Law for Magnetism

The magnetic field lines shown in Figure 27-6, Figure 27-9, and Figure 27-10 differ from electric field lines because the lines of $\vec{B}$ form closed curves, whereas lines of $\vec{E}$ begin and end on electric charges. The magnetic equivalent of an electric charge is a magnetic pole, such as appears to be at the ends of a bar magnet. Magnetic field lines appear to diverge from the north-pole end of a bar magnet (Figure 27-10b) and appear to converge on the south-pole end. However, inside the magnet the magnetic field lines neither diverge from a point near the north-pole end, nor do they converge on a point near the south-pole end. Instead, the magnetic field lines pass through the bar magnet from the south-pole end to the north-pole end, as shown in Figure 27-10b. If a Gaussian surface encloses one end of a bar magnet, the number of magnetic field lines that leave through the surface is exactly equal to the number of magnetic field lines that enter through the surface. That is, the net flux $\phi_{\text{m, net}}$ of the magnetic field through any closed surface S is always zero.[†]

$$\phi_{\text{m, net}} = \oint_S B_n\, dA = 0 \qquad\qquad 27\text{-}14$$

GAUSS'S LAW FOR MAGNETISM

where B_n is the component of $\vec{B}$ normal to surface S at area element dA. The definition of the magnetic flux ϕ_m is exactly analogous to the electric flux, with $\vec{B}$ replacing $\vec{E}$. This result is called Gauss's law for magnetism. It is the mathematical statement that there exist no points in space from which magnetic field lines diverge, or to which magnetic field lines converge. That is, isolated magnetic poles do not exist.[‡] The fundamental unit of magnetism is the magnetic

---

[†] Recall that the net flux of the electric field is a measure of the net number of field lines that leave a closed surface and is equal to $Q_{\text{inside}}/\epsilon_0$.

[‡] The existence of magnetic monopoles is a subject of great debate, and the search for magnetic monopoles remains active. To date, however, none have been discovered.

dipole. Figure 27-21 compares the field lines of $\vec{B}$ for a magnetic dipole with the field lines of $\vec{E}$ for an electric dipole. Note that far from the dipoles the field lines are identical. But inside the dipole, the field lines of $\vec{E}$ are directed opposite to the field lines of $\vec{B}$. The field lines of $\vec{E}$ diverge from the positive charge and converge to the negative charge, whereas the field lines of $\vec{B}$ are continuous loops.

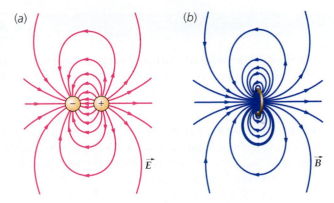

27-4 Ampère's Law

In Chapter 22, we found that for highly symmetric charge distributions we could calculate the electric field more easily using Gauss's law than Coulomb's law. A similar situation exists in magnetism. Ampère's law relates the tangential component B_t of the magnetic field summed (integrated) around a closed curve C to the current I_C that passes through any surface bounded by C. It can be used to obtain an expression for the magnetic field in situations that have a high degree of symmetry. In mathematical form, **Ampère's law** is

$$\oint_C B_t\, d\ell = \oint_C \vec{B} \cdot d\vec{\ell} = \mu_0 I_C, \quad C \text{ is any closed curve} \qquad 27\text{-}15$$

AMPÈRE'S LAW

FIGURE 27-21 (*a*) Electric field lines of an electric dipole. (*b*) Magnetic field lines of a magnetic dipole. Far from the dipoles, the field lines are identical. In the region between the charges in Figure 27-21(*a*), the electric field lines are opposite the direction of the dipole moment, whereas inside the loop in Figure 27-21(*b*), the magnetic field lines are parallel to the direction of the dipole moment.

where I_C is the net current that penetrates any surface S bounded by the curve C. The positive tangential direction for the path integral is related to the choice for the positive direction for the current I_C through S by the right-hand rule shown in Figure 27-22. Ampère's law holds for any curve C, as long as the currents are steady and continuous. This means the current does not change in time and that charge is not accumulating anywhere. Ampère's law is useful in calculating the magnetic field $\vec{B}$ in situations that have a high degree of symmetry so that the line integral $\oint_C \vec{B} \cdot d\vec{\ell}$ can be written as $B \oint_C d\ell$ (the product of B and some distance). The integral $\oint_C \vec{B} \cdot d\vec{\ell}$ is called a **circulation integral.** More specifically, $\oint_C \vec{B} \cdot d\vec{\ell}$ is called the circulation of $\vec{B}$ around curve C. Ampère's law and Gauss's law are both of considerable theoretical importance, and both laws hold whether there is symmetry or there is no symmetry. If there is no symmetry, neither law is very useful in calculating electric or magnetic fields.

The simplest application of Ampère's law is to find the magnetic field of an infinitely long, straight, current-carrying wire. Figure 27-23 shows a circular curve around a long wire with its center at the wire. We know the direction of the magnetic field due to each current element is tangent to this circle from the Biot–Savart law. Assuming that the magnetic field is tangent to this circle, that the magnetic field is in the same direction as $d\vec{\ell}$, and that the magnetic field has the same magnitude B at any point on the circle, Ampère's law ($\oint_C B_t\, d\ell = \mu_0 I_C$) then gives

$$B \oint_C d\ell = \mu_0 I_C$$

where $B = B_t$. We can factor B out of the integral because B has the same value everywhere on the circle. The integral of $d\ell$ around the circle equals $2\pi R$ (the circumference of the circle). The current I_C is the current I in the wire. We thus obtain $B 2\pi R = \mu_0 I$

$$B = \frac{\mu_0 I}{2\pi R}$$

which is Equation 27-12.

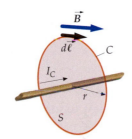

FIGURE 27-22 The positive direction for the path integral for Ampère's law is related to the positive direction for the current passing through the surface by a right-hand rule.

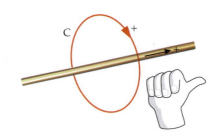

FIGURE 27-23 Geometry for calculating the magnetic field of a long, straight, current-carrying wire using Ampère's law. On a circle around the wire, the magnetic field is constant and tangent to the circle.

$\vec{B}$ *INSIDE AND OUTSIDE A WIRE*   E X A M P L E   2 7 - 9

A long, straight wire of radius R carries a current I that is uniformly distributed over the circular cross section of the wire. Find the magnetic field both outside the wire and inside the wire.

PICTURE THE PROBLEM We can use Ampère's law to calculate $\vec{B}$ because of the high degree of symmetry. At a distance r (Figure 27-24), we know that $\vec{B}$ is tangent to the circle of radius r about the wire and $\vec{B}$ is constant in magnitude everywhere on the circle. The current through the surface S bounded by C depends on whether r is less than or greater than the radius of the wire R.

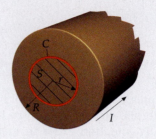

FIGURE 27-24

1. Ampère's law is used to relate the circulation of $\vec{B}$ around curve C to the current passing through the surface S bounded by C:

$$\oint_C \vec{B} \cdot d\vec{\ell} = \mu_0 I_C$$

2. Evaluate the circulation of $\vec{B}$ around a circle of radius r that is coaxial with the wire:

$$\oint_C \vec{B} \cdot d\vec{\ell} = B \oint_C d\ell = B2\pi r$$

3. Substitute into Ampère's law and solve for B:

$$B2\pi r = \mu_0 I_C$$

so

$$B = \frac{\mu_0 I_C}{2\pi r}$$

4. Outside the wire, $r > R$, and the total current passes through the surface bounded by C:

$$I_C = I$$

$$B = \boxed{\frac{\mu_0 I}{2\pi r} \quad (r \geq R)}$$

5. Inside the wire, $r < R$. Assume that the current is distributed uniformly to solve for I_C. Solve for B:

$$\frac{I_C}{\pi r^2} = \frac{I}{\pi R^2}$$

or

$$\left(I_C = \frac{r^2}{R^2} I \right)$$

so

$$B = \frac{\mu_0}{2\pi} \frac{I_C}{r} = \frac{\mu_0}{2\pi} \frac{(r^2/R^2)I}{r} = \boxed{\frac{\mu_0}{2\pi} \frac{I}{R^2} r \quad r \leq R}$$

REMARKS Inside the wire, the field increases with distance from the center of the wire. Figure 27-25 shows the graph of B versus r for this example.

We see from Example 27-9 that the magnetic field due to a current uniformly distributed over a wire of radius R is given by

$$B = \frac{\mu_0 I}{2\pi R^2} r \quad (r \leq R)$$

$$B = \frac{\mu_0}{2\pi} \frac{I}{r} \quad (r \geq R) \qquad \qquad 27\text{-}16$$

For the next application of Ampère's law, we calculate the magnetic field of a tightly wound **toroid,** which consists of loops of wire wound around a doughnut-

FIGURE 27-25

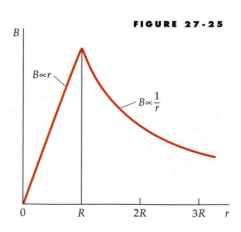

shaped form, as shown in Figure 27-26. There are N turns of wire, each carrying a current I. To calculate B, we evaluate the line integral $\oint_C \vec{B} \cdot d\vec{\ell}$ around a circle of radius r centered in the middle of the toroid. By symmetry, $\vec{B}$ is tangent to this circle and constant in magnitude at every point on the circle. Then,

$$\oint_C \vec{B} \cdot d\vec{\ell} = B2\pi r = \mu_0 I_C$$

Let a and b be the inner and outer radii of the toroid, respectively. The total current through the surface S bounded by a circle of radius r for $a < r < b$ is NI. Ampère's law then gives

$$\oint_C \vec{B} \cdot d\vec{\ell} = \mu_0 I_C, \quad \text{or} \quad (B2\pi r = \mu_0 NI)$$

or

$$B = \frac{\mu_0 NI}{2\pi r}, \quad a < r < b \qquad\qquad 27\text{-}17$$

B INSIDE A TIGHTLY WOUND TOROID

If r is less than a, there is no current through the surface S. If r is greater than b, the total current through S is zero because for each turn of the wire the current penetrates the surface twice (Figure 27-27), once going into the page and once coming out of the page. Thus, the magnetic field is zero for both $r < a$ and $r > b$:

$$B = 0, \quad r < a \quad \text{or} \quad r > b$$

The magnetic field intensity inside the toroid is not uniform but decreases with increasing r. However, if the radius of the loops of the coil, $\frac{1}{2}(b - a)$ is much less than the radius $\frac{1}{2}(b + a)$ of the center of the loops, the variation in r from $r = a$ to $r = b$ is small and B is approximately uniform, as it is in a solenoid.

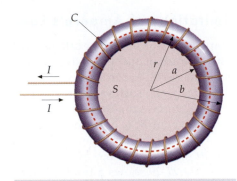

FIGURE 27-26 A toroid consists of loops of wire wound around a doughnut-shaped form. The magnetic field at any distance r can be found by applying Ampère's law to the circle of radius r. The surface S is bounded by curve C. The wire penetrates S once for each turn.

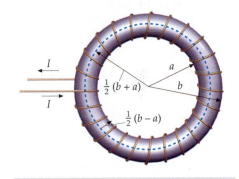

FIGURE 27-27 The toroid has mean radius $r = \frac{1}{2}(b + a)$, where a and b are the inner and outer radii of the toroid. Each turn of the wire is a circle of radius $\frac{1}{2}(b - a)$.

(a)   (b)

(a) The Tokamak fusion-test reactor is a large toroid that produces a magnetic field for confining charged particles. Coils containing over 10 km of water-cooled copper wire carry a pulsed current, which has a peak value of 73,000 A and produces a magnetic field of 5.2 T for about 3 s. (b) Inspection of the assembly of the Tokamak reactor from inside the toroid.

Limitations of Ampère's Law

Ampère's law is useful for calculating the magnetic field only when there is both a steady current and a high degree of symmetry. Consider the current loop shown in Figure 27-28. According to Ampère's law, the line integral $\oint_C \vec{B} \cdot d\vec{\ell} = \oint_C B_t \, d\ell$ around a curve, such as curve C in the figure, equals μ_0 times the current I in the loop. Although Ampère's law is valid for this curve, the tangential component of magnetic field B_t is not constant along any curve encircling the current. Thus, there is not enough symmetry in this situation to allow us to evaluate the integral $\oint_C B_t \, d\ell$ and solve for B_t.

Figure 27-29 shows a finite current segment of length ℓ. We wish to find the magnetic field at point P, which is equidistant from the ends of the segment and at a distance r from the center of the segment. A direct application of Ampère's law gives

$$B = \frac{\mu_0}{2\pi} \frac{I}{r}$$

This result is the same as for an infinitely long wire, since the same symmetry arguments apply. It does not agree with the result obtained from the Biot–Savart law, which depends on the length of the current segment and which agrees with experiment. If the current segment is just one part of a continuous circuit carrying a current, as shown in Figure 27-30, Ampère's law for curve C is valid, but it cannot be used to find the magnetic field at point P because there is insufficient symmetry.

In Figure 27-31, the current in the segment arises from a small spherical conductor with initial charge $+Q$ at the left of the segment and another small spherical conductor at the right with charge $-Q$. When they are connected, a current $I = -dQ/dt$ exists in the segment for a short time, until the spheres are uncharged. For this case, we *do* have the symmetry needed to assume that $\vec{B}$ is tangential to the curve and $\vec{B}$ is constant in magnitude along the curve. For a situation like this, in which the current is discontinuous in space, Ampère's law is not valid. In Chapter 30, we will see how Maxwell was able to modify Ampère's law so that it holds for all currents. When Maxwell's generalized form of Ampère's law is used to calculate the magnetic field for a current segment, such as the current segment shown in Figure 27-31, the result agrees with the result found from the Biot–Savart law.

27-5 Magnetism in Matter

Atoms have magnetic dipole moments due to the motion of their electrons and due to the intrinsic magnetic dipole moment associated with the spin of the electrons. Unlike the situation with electric dipoles, the alignment of magnetic dipoles parallel to an external magnetic field tends to *increase* the field. We can see this difference by comparing the electric field lines of an electric dipole with the magnetic field lines of a magnetic dipole, such as a small current loop, as was shown in Figure 27-21. Far from the dipoles, the field lines are identical. However, between the charges of the electric dipole, the electric field lines are opposite the direction of the dipole moment, whereas inside the current loop, the magnetic field lines are parallel to the magnetic dipole moment. Thus, inside a magnetically polarized material, the magnetic dipoles create a magnetic field that is parallel to the magnetic dipole moment vectors.

Materials fall into three categories—**paramagnetic, diamagnetic,** and **ferromagnetic**—according to the behavior of their magnetic moments in an external magnetic field. Paramagnetism arises from the partial alignment of the electron spins (in metals) or from the atomic or molecular magnetic moments by an

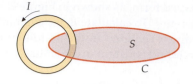

FIGURE 27-28 Ampère's law holds for the curve C encircling the current in the circular loop, but it is not useful for finding B_t, because B_t cannot be factored out of the circulation integral.

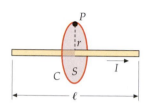

FIGURE 27-29 The application of Ampère's law to find the magnetic field on the bisector of a finite current segment gives an incorrect result.

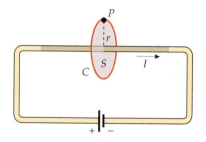

FIGURE 27-30 If the current segment in Figure 27-28 is part of a complete circuit, Ampère's law for the curve C is valid, but there is not enough symmetry to use Ampère's law to find the magnetic field at point P.

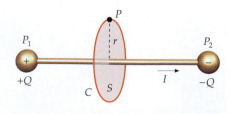

FIGURE 27-31 If the current segment in Figure 27-29 is due to a momentary flow of charge from a small conductor on the left to a small conductor on the right, there is enough symmetry to use Ampère's law to compute the magnetic field at P, but Ampère's law is not valid because the current is not continuous in space.

applied magnetic field in the direction of the field. In paramagnetic materials, the magnetic dipoles do not interact strongly with each other and are normally randomly oriented. In the presence of an applied magnetic field, the dipoles are partially aligned in the direction of the field, thereby increasing the field. However, in external magnetic fields of ordinary strength at ordinary temperatures, only a very small fraction of the molecules are aligned because thermal motion tends to randomize their orientation. The increase in the total magnetic field is therefore very small. Ferromagnetism is much more complicated. Because of a strong interaction between neighboring magnetic dipoles, a high degree of alignment occurs even in weak external magnetic fields, which causes a very large increase in the total field. Even when there is no external magnetic field, a ferromagnetic material may have its magnetic dipoles aligned, as in permanent magnets. Diamagnetism arises from the orbital magnetic dipole moments induced by an applied magnetic field. These magnetic moments are opposite the direction of the applied magnetic field, thereby decreasing the field. This effect actually occurs in all materials; however, because the induced magnetic moments are very small compared to the permanent magnetic moments, diamagnetism is often masked by paramagnetic or ferromagnetic effects. Diamagnetism is thus observed only in materials whose molecules have no permanent magnetic moments.

Magnetization and Magnetic Susceptibility

When some material is placed in a strong magnetic field, such as that of a solenoid, the magnetic field of the solenoid tends to align the magnetic dipole moments (either permanent or induced) inside the material and the material is said to be magnetized. We describe a magnetized material by its **magnetization $\vec{M}$**, which is defined as the net magnetic dipole moment per unit volume of the material:

$$\vec{M} = \frac{d\vec{\mu}}{dV} \qquad\qquad 27\text{-}18$$

Long before we had any understanding of atomic or molecular structure, Ampère proposed a model of magnetism in which the magnetization of materials is due to microscopic current loops inside the magnetized material. We now know that these current loops are a classical model for the orbital motion and spin of the electrons in atoms. Consider a cylinder of magnetized material. Figure 27-32 shows atomic current loops in the cylinder aligned with their magnetic moments along the axis of the cylinder. Because of cancellation of neighboring current loops, the net current at any point inside the material is zero, leaving a net current on the surface of the material (Figure 27-33). This surface current, called an **amperian current,** is similar to the real current in the windings of the solenoid.

Figure 27-34 shows a small disk of cross-sectional area A, length $d\ell$, and volume $dV = A\,d\ell$. Let di be the amperian current on the surface of the disk. The magnitude of the magnetic dipole moment of the disk is the same as that of a current loop of area A carrying a current di:

$$d\mu = A\,di$$

The magnitude of the magnetization of the disk is the magnetic moment per unit volume:

$$M = \frac{d\mu}{dV} = \frac{A\,di}{A\,d\ell} = \frac{di}{d\ell} \qquad\qquad 27\text{-}19$$

Thus, the magnitude of the magnetization vector is the amperian current per unit length along the surface of the magnetized material. We see from this result that the units of M are amperes per meter.

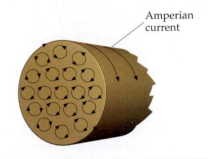

FIGURE 27-32 A model of atomic current loops in which all the atomic dipoles are parallel to the axis of the cylinder. The net current at any point inside the material is zero due to cancellation of neighboring atoms. The result is a surface current similar to that of a solenoid.

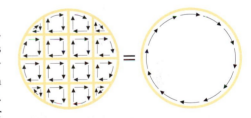

FIGURE 27-33 The currents in the adjacent current loops in the interior of a uniformly magnetized material cancel, leaving only a surface current. Cancellation occurs at every interior point independent of the shape of the loops.

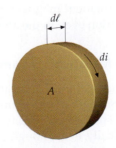

FIGURE 27-34 A disk element for relating the magnetization M to the surface current per unit length.

Consider a cylinder that has a uniform magnetization $\vec{M}$ parallel to its axis. The effect of the magnetization is the same as if the cylinder carried a surface current per unit length of magnitude M. This current is similar to the current carried by a tightly wound solenoid. For a solenoid, the current per unit length is nI, where n is the number of turns per unit length and I is the current in each turn. The magnitude of the magnetic field B_m inside the cylinder and far from its ends is thus given by Equation 27-9 for a solenoid with nI replaced by M:

$$B_m = \mu_0 M \qquad \text{27-20}$$

Suppose we place a cylinder of magnetic material inside a long solenoid with n turns per unit length that carries a current I. The applied field of the solenoid $\vec{B}_{app}$ ($B_{app} = \mu_0 nI$) magnetizes the material so that it has a magnetization $\vec{M}$. The resultant magnetic field at a point inside the solenoid and far from its ends due to the current in the solenoid plus the magnetized material is

$$\vec{B} = \vec{B}_{app} + \mu_0 \vec{M} \qquad \text{27-21}$$

For paramagnetic and ferromagnetic materials, $\vec{M}$ is in the same direction as $\vec{B}_{app}$; for diamagnetic materials, $\vec{M}$ is opposite to $\vec{B}_{app}$. For paramagnetic and diamagnetic materials, the magnetization is found to be proportional to the applied magnetic field that produces the alignment of the magnetic dipoles in the material. We can thus write

$$\vec{M} = \chi_m \frac{\vec{B}_{app}}{\mu_0} \qquad \text{27-22}$$

where χ_m is a dimensionless number called the **magnetic susceptibility**. Equation 27-21 is then

$$\vec{B} = \vec{B}_{app} + \mu_0 \vec{M} = \vec{B}_{app}(1 + \chi_m) = K_m \vec{B}_{app} \qquad \text{27-23}$$

where

$$K_m = 1 + \chi_m \qquad \text{27-24}$$

is called the **relative permeability** of the material. For paramagnetic materials, χ_m is a small positive number that depends on temperature. For diamagnetic materials (other than superconductors), it is a small negative constant independent of temperature. Table 27-1 lists the magnetic susceptibility of various paramagnetic and diamagnetic materials. We see that the magnetic susceptibility for the solids listed is of the order of 10^{-5}, and $K_m \approx 1$.

The magnetization of ferromagnetic materials, which we discuss shortly, is much more complicated. The relative permeability K_m defined as the ratio B/B_{app} is not constant and has maximum values ranging from 5000 to 100,000. In the case of permanent magnets, K_m is not even defined since such materials exhibit magnetization even in the absence of an applied field.

Atomic Magnetic Moments

The magnetization of a paramagnetic or ferromagnetic material can be related to the permanent magnetic moments of the individual atoms or electrons of the material. The orbital magnetic moment of an atomic electron can be derived semiclassically, even though it is quantum mechanical in origin. Consider a particle of mass m and charge q moving with speed v in a circle of radius r, as shown in Figure 27-35. The magnitude of the angular momentum of the particle is

$$L = mvr \qquad \text{27-25}$$

TABLE 27-1

Magnetic Susceptibility of Various Materials at 20°C

Material	χ_m
Aluminum	2.3×10^{-5}
Bismuth	-1.66×10^{-5}
Copper	-0.98×10^{-5}
Diamond	-2.2×10^{-5}
Gold	-3.6×10^{-5}
Magnesium	1.2×10^{-5}
Mercury	-3.2×10^{-5}
Silver	-2.6×10^{-5}
Sodium	-0.24×10^{-5}
Titanium	7.06×10^{-5}
Tungsten	6.8×10^{-5}
Hydrogen (1 atm)	-9.9×10^{-9}
Carbon dioxide (1 atm)	-2.3×10^{-9}
Nitrogen (1 atm)	-5.0×10^{-9}
Oxygen (1 atm)	2090×10^{-9}

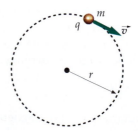

FIGURE 27-35 A particle of charge q and mass m moving with speed v in a circle of radius r. The angular momentum is into the paper and has a magnitude mvr and the magnetic moment is into the paper (if q is positive) and has a magnitude $\frac{1}{2}qvr$.

The magnitude of the magnetic moment is the product of the current and the area of the circle:

$$\mu = IA = I\pi r^2$$

If T is the time for the charge to complete one revolution, the current (charge passing a point per unit time) is q/T. Since the period T is the distance $2\pi r$ divided by the velocity v, the current is

$$I = \frac{q}{T} = \frac{qv}{2\pi r}$$

The magnetic moment is then

$$\mu = IA = \frac{qv}{2\pi r}\pi r^2 = \frac{1}{2}qvr \qquad \text{27-26}$$

Using $vr = L/m$ from Equation 27-25, we have for the magnetic moment

$$\mu = \frac{q}{2m}L$$

If the charge q is positive, the angular momentum and magnetic moment are in the same direction. We can therefore write

$$\vec{\mu} = \frac{q}{2m}\vec{L} \qquad \text{27-27}$$

CLASSICAL RELATION BETWEEN MAGNETIC MOMENT AND ANGULAR MOMENTUM

Equation 27-27 is the general classical relation between magnetic moment and angular momentum. It also holds in the quantum theory of the atom for orbital angular momentum, but the equation does not hold for the intrinsic spin angular momentum of the electron. For electron spin, the magnetic moment is twice that predicted by this equation.[†] The extra factor of 2 is a result from quantum theory that has no analog in classical mechanics.

Since angular momentum is quantized, the magnetic moment of an atom is also quantized. The quantum of angular momentum is $\hbar = h/(2\pi)$, where h is Planck's constant, so we express the magnetic moment in terms of $\vec{L}/\hbar$

$$\vec{\mu} = \frac{q\hbar}{2m}\frac{\vec{L}}{\hbar}$$

For an electron, $m = m_e$ and $q = -e$, so the magnetic moment of the electron due to its orbital motion is

$$\vec{\mu}_\ell = -\frac{e\hbar}{2m_e}\frac{\vec{L}}{\hbar} = -\mu_B\frac{\vec{L}}{\hbar} \qquad \text{27-28}$$

MAGNETIC MOMENT DUE TO THE ORBITAL MOTION OF AN ELECTRON

---

† This result, and the phenomenon of electron spin itself, was predicted in 1927 by Paul Dirac, who combined special relativity and quantum mechanics into a relativistic wave equation called the Dirac equation. Precise measurements indicate that the magnetic moment of the electron due to its spin is 2.00232 times that predicted by Equation 27-27. The fact that the intrinsic magnetic moment of the electron is approximately twice what we would expect makes it clear that the simple model of the electron as a spinning ball is not to be taken literally.

where

$$\mu_B = \frac{e\hbar}{2m_e} = 9.27 \times 10^{-24} \text{ A·m}^2 = 9.27 \times 10^{-24} \text{ J/T}$$

$$= 5.79 \times 10^{-5} \text{ eV/T} \qquad \qquad 27\text{-}29$$

BOHR MAGNETON

is the quantum unit of magnetic moment called a **Bohr magneton.** The magnetic moment of an electron due to its intrinsic spin angular momentum $\vec{S}$ is

$$\vec{\mu}_s = -2 \times \frac{e\hbar}{2m_e} \frac{\vec{S}}{\hbar} = -2\mu_B \frac{\vec{S}}{\hbar} \qquad \qquad 27\text{-}30$$

MAGNETIC MOMENT DUE TO ELECTRON SPIN

Although the calculation of the magnetic moment of any atom is a complicated problem in quantum theory, the result for all electrons, according to both theory and experiment, is that the magnetic moment is of the order of a few Bohr magnetons. For atoms with zero net angular momentum, the net magnetic moment is zero. (The shell structure of atoms is discussed in Chapter 36.)

If all the atoms or molecules in some material have their magnetic moments aligned, the magnetic moment per unit volume of the material is the product of the number of molecules per unit volume n and the magnetic moment μ of each molecule. For this extreme case, the **saturation magnetization** M_s is

$$M_s = n\mu \qquad \qquad 27\text{-}31$$

The number of molecules per unit volume can be found from the molecular mass M, the density ρ of the material, and Avogadro's number N_A:

$$n = \frac{N_A \text{ (atoms/mol)}}{M \text{ (kg/mol)}} \rho(\text{kg/m}^3) \qquad \qquad 27\text{-}32$$

SATURIZATION MAGNETIZATION FOR IRON   **E X A M P L E   2 7 - 1 0**

Find the saturation magnetization and the magnetic field it produces for iron, assuming that each iron atom has a magnetic moment of 1 Bohr magneton.

PICTURE THE PROBLEM We find the number of molecules per unit volume from the density of iron, $\rho = 7.9 \times 10^3 \text{ kg/m}^3$, and its molecular mass $M = 55.8 \times 10^{-3} \text{ kg/mol}$.

1. The saturation magnetization is the product of the number of molecules per unit volume and the magnetic moment of each molecule:

$$M_s = n\mu$$

2. Calculate the number of molecules per unit volume from Avogadro's number, the molecular mass, and the density:

$$n = \frac{N_A}{M}\rho = \frac{6.02 \times 10^{23} \text{ atoms/mol}}{55.8 \times 10^{-3} \text{ kg/mol}} (7.9 \times 10^3 \text{ kg/m}^3)$$

$$= 8.52 \times 10^{28} \text{ atoms/m}^3$$

3. Substitute this result and $\mu = 1$ Bohr magneton to calculate the saturation magnetization:

$$M_s = n\mu$$

$$= (8.52 \times 10^{28} \text{ atoms/m}^3)(9.27 \times 10^{-24} \text{ A·m}^2)$$

$$\boxed{= 7.90 \times 10^5 \text{ A/m}}$$

4. The magnetic field on the axis inside a long iron cylinder resulting from this maximum magnetization is given by $B = \mu_0 M_s$:

$$B = \mu_0 M_s$$

$$= (4\pi \times 10^{-7} \text{ T·A})(7.90 \times 10^5 \text{ A/m})$$

$$\boxed{= 0.993 \text{ T} \approx 1 \text{ T}}$$

REMARKS The measured saturation magnetic field of annealed iron is about 2.16 T, indicating that the magnetic moment of an iron atom is slightly greater than 2 Bohr magnetons. This magnetic moment is due mainly to the spins of two unpaired electrons in the iron atom.

*Paramagnetism

Paramagnetism occurs in materials whose atoms have permanent magnetic moments that interact with each other only very weakly, resulting in a very small, positive magnetic susceptibility χ_m. When there is no external magnetic field, these magnetic moments are randomly oriented. In the presence of an external magnetic field, the magnetic moments tend to line up parallel to the field, but this is counteracted by the tendency for the magnetic moments to be randomly oriented due to thermal motion. The degree to which the moments line up with the field depends on the strength of the field and on the temperature. This degree of alignment usually is small because the energy of a magnetic moment in an external magnetic field is typically much smaller than the thermal energy of an atom of the material, which is of the order of kT, where k is Boltzmann's constant and T is the absolute temperature.

The potential energy of a magnetic dipole of moment $\vec{\mu}$ in an external magnetic field $\vec{B}$ is given by Equation 26-16:

$$U = -\mu B \cos \theta = -\vec{\mu} \cdot \vec{B}$$

The potential energy when the moment is parallel with the field ($\theta = 0$) is thus lower than when the moment is antiparallel ($\theta = 180°$) by the amount $2\mu B$. For a typical atomic magnetic moment of 1 Bohr magneton and a typical strong magnetic field of 1 T, the difference in potential energy is

$$\Delta U = 2\mu_B B = 2(5.79 \times 10^{-5} \text{ eV/T})(1 \text{ T}) = 1.16 \times 10^{-4} \text{ eV}$$

At a normal temperature of $T = 300$ K, the typical thermal energy kT is

$$kT = (8.62 \times 10^{-5} \text{ eV/K})(300 \text{ K}) = 2.59 \times 10^{-2} \text{ eV}$$

which is more than 200 times greater than $2\mu_B B$. Thus, even in a very strong magnetic field of 1 T, most of the magnetic moments will be randomly oriented because of thermal motions (unless the temperature is very low).

Figure 27-36 shows a plot of the magnetization M versus an applied external magnetic field B_{app} at a given temperature. In very strong fields, nearly all the magnetic moments are aligned with the field and $M \approx M_s$. (For magnetic fields attainable in the laboratory, this can occur only for very low temperatures.) When $B_{app} = 0$, $M = 0$, indicating that the orientation of the moments is completely random. In weak fields, the magnetization is approximately proportional to the

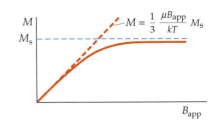

FIGURE 27-36 Plot of magnetization M versus an applied magnetic field B_{app}. In very strong fields, the magnetization approaches the saturation value M_s. This can be achieved only at very low temperatures. In weak fields, the magnetization is approximately proportional to B_{app}, a result known as Curie's law.

applied field, as indicated by the orange dashed line in the figure. In this region, the magnetization is given by

$$M = \frac{1}{3} \frac{\mu B_{app}}{kT} M_s$$

27-33

CURIE'S LAW

Note that $\mu B_{app}/(kT)$ is the ratio of the maximum energy of a dipole in the magnetic field to the characteristic thermal energy. The result that the magnetization varies inversely with the absolute temperature was discovered experimentally by Pierre Curie and is known as **Curie's law.**

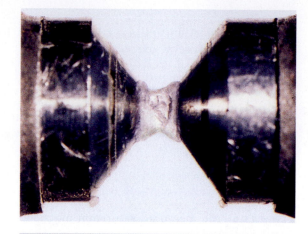

Liquid oxygen, which is paramagnetic, is attracted by the magnetic field of a permanent magnet. A net force is exerted on the magnetic dipoles because the magnetic field is not uniform.

APPLYING CURIE'S LAW   **EXAMPLE 27-11**

If $\mu = \mu_B$, at what temperature will the magnetization be 1 percent of the saturation magnetization in an applied magnetic field of 1 T?

PICTURE THE PROBLEM

1. Curie's law relates M, T, M_s, and B_{app}:

$$M = \frac{1}{3} \frac{\mu B_{app}}{kT} M_s$$

2. Solve for T using $\mu = \mu_B$ and $M/M_s = 0.01$:

$$T = \frac{\mu_B B_{app}}{3k} \frac{M_s}{M} = \frac{(5.79 \times 10^{-5} \text{ eV/T})(1 \text{ T})}{3(8.62 \times 10^{-5} \text{ eV/K})} 100$$

$$= \boxed{22.4 \text{ K}}$$

REMARKS From this example, we see that even in a strong applied magnetic field of 1 T, the magnetization is less than 1 percent of saturation at temperatures above 22.4 K.

EXERCISE If $\mu = \mu_B$, what fraction of the saturation magnetization is M at 300 K for an external magnetic field of 1.5 T? (*Answer* $M/M_s = 1.12 \times 10^{-3}$)

*Ferromagnetism

Ferromagnetism occurs in pure iron, cobalt, and nickel as well as in alloys of these metals with each other. It also occurs in gadolinium, dysprosium, and a few compounds. Ferromagnetism arises from a strong interaction between the electrons in a partially full band in a metal or between the localized electrons that form magnetic moments on neighboring atoms or molecules. This interaction, called the **exchange interaction,** lowers the energy of a pair of electrons with parallel spins.

Ferromagnetic materials have very large positive values of magnetic susceptibility χ_m (as measured under conditions described, which follow). In these substances, a small external magnetic field can produce a very large degree of alignment of the atomic magnetic dipole moments. In some cases, the alignment can persist even when the external magnetizing field is removed. This alignment persists because the magnetic dipole moments exert strong forces on their neighbors so that over a small region of space the moments are aligned with each other even when there is no external field. The region of space over which the magnetic dipole moments are aligned is called a **magnetic domain.** The size of a domain is usually microscopic. Within the domain, all the permanent atomic magnetic moments are aligned, but the direction of alignment varies from domain to domain so that the net magnetic moment of a macroscopic piece of ferromagnetic

A Canadian quarter that is attracted by a magnet. Canadian coins often contain significant amounts of nickel, which is ferromagnetic.

(a)

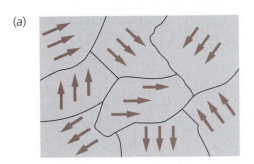

(b)

FIGURE 27-37 (a) Schematic illustration of ferromagnetic domains. Within a domain, the magnetic dipoles are aligned, but the direction of alignment varies from domain to domain so that the net magnetic moment is zero. A small external magnetic field may cause the enlargement of those domains that are aligned parallel to the field, or it may cause the alignment within a domain to rotate. In either case, the result is a net magnetic moment parallel to the field. (b) Magnetic domains on the surface of an FE−3 percent Si crystal observed using a scanning electron microscope with polarization analysis. The four colors indicate four possible domain orientations.

material is zero in the normal state. Figure 27-37 illustrates this situation. The dipole forces that produce this alignment are predicted by quantum theory but cannot be explained with classical physics. At temperatures above a critical temperature, called the **Curie temperature,** thermal agitation is great enough to break up this alignment and ferromagnetic materials become paramagnetic.

When an external magnetic field is applied, the boundaries of the domains may shift or the direction of alignment within a domain may change so that there is a net macroscopic magnetic moment in the direction of the applied field. Since the degree of alignment is large for even a small external field, the magnetic field produced in the material by the dipoles is often much greater than the external field.

Let us consider what happens when we magnetize a long iron rod by placing it inside a solenoid and gradually increase the current in the solenoid windings. We assume that the rod and the solenoid are long enough to permit us to neglect end effects. Since the induced magnetic moments are in the same direction as the applied field, $\vec{B}_{app}$ and $\vec{M}$ are in the same direction. Then,

$$B = B_{app} + \mu_0 M = \mu_0 n I + \mu_0 M \qquad \text{27-34}$$

In ferromagnetic materials, the magnetic field $\mu_0 M$ due to the magnetic moments is often greater than the magnetizing field B_{app} by a factor of several thousand.

A chunk of magnetite (lodestone) attracts the needle of a compass.

(a)

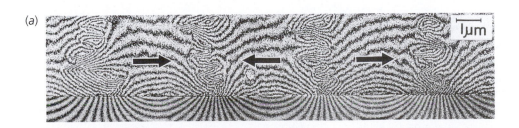

(b)

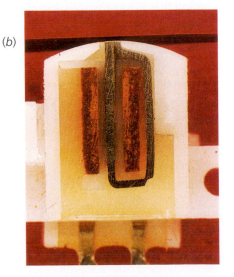

(a) Magnetic field lines on a cobalt magnetic recording tape. The solid arrows indicate the encoded magnetic bits. (b) Cross section of a magnetic tape recording head. Current from an audio amplifier is sent to wires around a magnetic core in the recording head where it produces a magnetic field. When the tape passes over a gap in the core of the recording head, the fringing magnetic field encodes information on the tape.

Figure 27-38 shows a plot of B versus the magnetizing field B_{app}. As the current is gradually increased from zero, B increases from zero along the part of the curve from the origin O to point P_1. The flattening of this curve near point P_1 indicates that the magnetization M is approaching its saturation value M_s, at which all the atomic magnetic moments are aligned. Above saturation, B increases only because the magnetizing field $B_{app} = \mu_0 nI$ increases. When B_{app} is gradually decreased from point P_1, there is not a corresponding decrease in the magnetization. The shift of the domains in a ferromagnetic material is not completely reversible, and some magnetization remains even when B_{app} is reduced to zero, as indicated in the figure. This effect is called **hysteresis,** from the Greek word *hysteros* meaning later or behind, and the curve in Figure 27-38 is called a **hysteresis curve.** The value of the magnetic field at point r when B_{app} is zero is called the **remnant field** B_{rem}. At this point, the iron rod is a permanent magnet. If the current in the solenoid is now reversed so that B_{app} is in the opposite direction, the magnetic field B is gradually brought to zero at point c. The remaining part of the hysteresis curve is obtained by further increasing the current in the opposite direction until point P_2 is reached, which corresponds to saturation in the opposite direction, and then decreasing the current to zero at point P_3 and increasing it again in its original direction.

Since the magnetization M depends on the previous history of the material, and since it can have a large value even when the applied field is zero, it is not simply related to the applied field B_{app}. However, if we confined ourselves to that part of the magnetization curve from the origin to point P_1 in Figure 27-38, $\vec{B}_{app}$ and $\vec{M}$ are parallel and M is zero when B_{app} is zero. We can then define the magnetic susceptibility as in Equation 27-22,

$$M = \chi_m \frac{B_{app}}{\mu_0}$$

and

$$B = B_{app} + \mu_0 M = B_{app}(1 + \chi_m) = K_m \mu_0 nI = \mu nI \qquad \text{27-35}$$

where

$$\mu = (1 + \chi_m)\mu_0 = K_m \mu_0 \qquad \text{27-36}$$

is called the **permeability** of the material. (For paramagnetic and diamagnetic materials, χ_m is much less than 1 so the permeability μ and the permeability of free space μ_0 are very nearly equal.)

Since B does not vary linearly with B_{app}, as can be seen from Figure 27-38, the relative permeability is not constant. The maximum value of K_m occurs at a magnetization that is considerably less than the saturation magnetization. Table 27-2 lists the saturation magnetic field $\mu_0 M_s$ and the maximum values of K_m for some ferromagnetic materials. Note that the maximum values of K_m are much greater than 1.

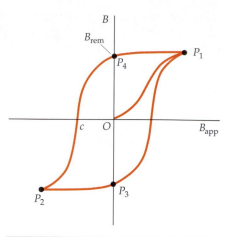

FIGURE 27-38 Plot of B versus the applied magnetizing field B_{app}. The outer curve is called a hysteresis curve. The field B_{rem} is called the remnant field. It remains when the applied field returns to zero.

TABLE 27-2

Maximum Values of $\mu_0 M$ and K_m for Some Ferromagnetic Materials

Material	$\mu_0 M_s$, T	K_m
Iron (annealed)	2.16	5,500
Iron-silicon (96 percent Fe, 4 percent Si)	1.95	7,000
Permalloy (55 percent Fe, 45 percent Ni)	1.60	25,000
Mu-metal (77 percent Ni, 16 percent Fe, 5 percent Cu, 2 percent Cr)	0.65	100,000

The area enclosed by the hysteresis curve is proportional to the energy dissipated as heat in the irreversible process of magnetizing and demagnetizing. If the hysteresis effect is small, so that the area inside the curve is small, indicating a small energy loss, the material is called **magnetically soft.** Soft iron is an example. The hysteresis curve for a magnetically soft material is shown in Figure 27-39. Here the remnant field B_{rem} is nearly zero, and the energy loss per cycle is small. Magnetically soft materials are used for transformer cores to allow the magnetic field B to change without incurring large energy losses as the field alternates. On the other hand, a large remnant field is desirable in a permanent magnet. **Magnetically hard** materials, such as carbon steel and the alloy Alnico 5, are used for permanent magnets.

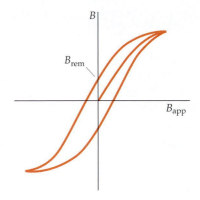

FIGURE 27-39 Hysteresis curve for a magnetically soft material. The remnant field is very small compared with the remnant field for a magnetically hard material such as that shown in Figure 27-38.

(a)

(a) An extremely high-capacity, hard-disk drive for magnetic storage of information, capable of storing over 250 gigabytes of information. (b) A magnetic test pattern on a hard disk, magnified 2400 times. The light and dark regions correspond to oppositely directed magnetic fields. The smooth region just outside the pattern is a region of the disk that has been erased just prior to writing.

(b)

SOLENOID WITH IRON CORE   **EXAMPLE 27-12**

A long solenoid with 12 turns per centimeter has a core of annealed iron. When the current is 0.50 A, the magnetic field inside the iron core is 1.36 T. Find (a) the applied field B_{app}, (b) the relative permeability K_m, and (c) the magnetization M.

PICTURE THE PROBLEM The applied field is just that of a long solenoid given by $B_{app} = \mu_0 nI$. Since the total magnetic field is given, we can find the relative permeability from its definition ($K_m = B/B_{app}$) and we can find M from $B = B_{app} + \mu_0 M$.

1. The applied field is given by Equation 27-9:

$$B_{app} = \mu_0 nI$$
$$= (4\pi \times 10^{-7}\,\text{T·m/A})(1200\,\text{turns/m})(0.5\,\text{A})$$
$$= 7.54 \times 10^{-4}\,\text{T}$$

2. The relative permeability is the ratio of B to B_{app}:

$$K_m = \frac{B}{B_{app}} = \frac{1.36\,\text{T}}{7.54 \times 10^{-4}\,\text{T}} = 1.80 \times 10^3$$

3. The magnetization M is found from Equation 27-34:

$$\mu_0 M = B - B_{app}$$
$$= 1.36\,\text{T} - 7.54 \times 10^{-4}\,\text{T} \approx B = 1.36\,\text{T}$$

$$M = \frac{B}{\mu_0} = \frac{1.36\,\text{T}}{4\pi \times 10^{-7}\,\text{T·m/A}} = \boxed{1.08 \times 10^6\,\text{A/m}}$$

REMARKS The applied magnetic field of 7.54×10^{-4} T is a negligible fraction of the total field of 1.36 T. Note that the value for K_m of 1800 is considerably smaller than the maximum value of 5500 in Table 27-2. Note also that the susceptibility $\chi_m = K_m - 1 \approx K_m$ to the three-place accuracy with which we calculated K_m.

*Diamagnetism

Diamagnetic materials are those materials that have very small negative values of magnetic susceptibility χ_m. Diamagnetism was discovered by Michael Faraday in 1845 when Faraday found that a piece of bismuth is repelled by either pole of a magnet, indicating that the external field of the magnet induces a magnetic moment in bismuth in the direction opposite the field.

We can understand this effect qualitatively from Figure 27-40, which shows two positive charges moving in circular orbits with the same speed but in opposite directions. Their magnetic moments are in opposite directions and therefore cancel.[†] In the presence of an external magnetic field $\vec{B}$ directed into the paper, the charges experience an extra force $q\vec{v} \times \vec{B}$, which is along the radial direction. For the charge on the left, this extra force is inward, increasing the centripetal force. If the charge is to remain in the same circular orbit, it must speed up so that mv^2/r equals the total centripetal force.[‡] Its magnetic moment, which is outward, is thus increased. For the charge on the right, the additional force is outward, so the particle must slow down to maintain its circular orbit. Its magnetic moment, which is

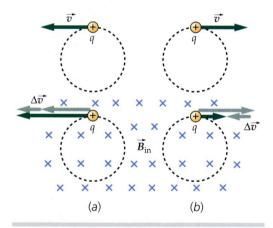

(a)   (b)

FIGURE 27-40 (a) A positive charge moving counterclockwise in a circle has its magnetic moment directed out of the page. When an external, magnetic field directed into the page is turned on, the magnetic force increases the centripetal force so the speed of the particle must increase. The change in the magnetic moment is out of the page. (b) A positive charge moving clockwise in a circle has its magnetic moment directed into the page. When an external, magnetic field directed into the page is turned on, the magnetic force decreases the centripetal force so the speed of the particle must decrease. As in (a), the change in the magnetic moment is directed out of the page.

[†] It is simpler to consider positive charges even though it is the negatively charged electrons that provide the magnetic moments in matter.

[‡] The electron speeds up because of an electric field induced by the changing magnetic field, an effect called induction, which we discuss in Chapter 28.

inward, is decreased. In each case, the *change* in the magnetic moment of the charges is in the direction out of the page, opposite that of the external applied field. Since the permanent magnetic moments of the two charges are equal and oppositely directed they add to zero, leaving only the induced magnetic moments which are both opposite the direction of the applied magnetic field.

A material will be diamagnetic if its atoms have zero net angular momentum and therefore no permanent magnetic moment. (The net angular momentum of an atom depends on the electronic structure of the atom, which is a subject that we will study in Chapter 35.) The induced magnetic moments that cause diamagnetism have magnitudes of the order of 10^{-5} Bohr magnetons. Since this is much smaller than the permanent magnetic moments of the atoms of paramagnetic or ferromagnetic materials, the diamagnetic effect in these atoms is masked by the alignment of their permanent magnetic moments. However, since this alignment decreases with temperature, all materials are theoretically diamagnetic at sufficiently high temperatures.

When a superconductor is placed in an external magnetic field, electric currents are induced on the superconductor's surface so that the net magnetic field in the superconductor is zero. Consider a superconducting rod inside a solenoid of n turns per unit length. When the solenoid is connected to a source of emf so that it carries a current I, the magnetic field due to the solenoid is $\mu_0 nI$. A surface current of $-nI$ per unit length is induced on the superconducting rod that cancels out the field due to the solenoid so that the net field inside the superconductor is zero. From Equation 27-23,

$$\vec{B} = \vec{B}_{app}(1 + \chi_m) = 0$$

so

$$\chi_m = -1$$

A superconductor is thus a perfect diamagnet with a magnetic susceptibility of -1.

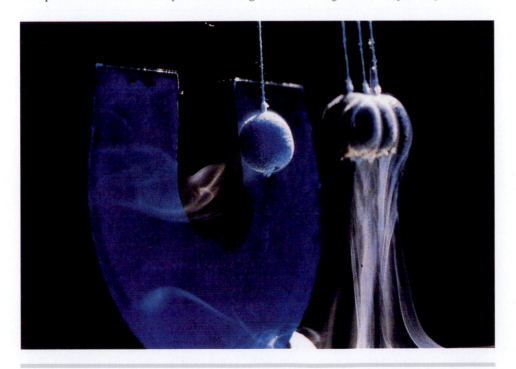

A superconductor is a perfect diamagnet. Here, the superconducting pendulum bob is repelled by the permanent magnet.

SUMMARY

1. Magnetic fields arise from moving charges, and therefore from currents.
2. The Biot–Savart law describes the magnetic field produced by a current element.
3. Ampère's law relates the line integral of the magnetic field along some closed curve to the current that passes through any surface bounded by the curve.
4. The magnetization vector $\vec{M}$ describes the magnetic moment per unit volume of matter.
5. The classical relation $\vec{\mu} = [q/(2m)]\,\vec{L}$ is derived from the definitions of angular momentum and magnetic moment.
6. The Bohr magneton is a convenient unit for atomic and nuclear magnetic moments.

Topic	Relevant Equations and Remarks

1. Magnetic Field $\vec{B}$

Due to a moving point charge

$$\vec{B} = \frac{\mu_0}{4\pi}\frac{q\vec{v}\times\hat{r}}{r^2}$$ 27-1

where $\hat{r}$ is a unit vector that points to the field point P from the charge q moving with velocity $\vec{v}$, and μ_0 is a constant of proportionality called the permeability of free space:

$$\mu_0 = 4\pi\times 10^{-7}\,\text{T·m/A} = 4\pi\times 10^{-7}\,\text{N/A}^2$$ 27-2

Due to a current element (Biot–Savart law)

$$d\vec{B} = \frac{\mu_0}{4\pi}\frac{I\,d\vec{\ell}\times\hat{r}}{r^2}$$ 27-3

On the axis of a current loop

$$B_x = \frac{\mu_0}{4\pi}\frac{2\pi R^2 I}{(x^2 + R^2)^{3/2}}$$ 27-5

On the axis of a current loop, at great distances from the loop

$$B_x = \frac{\mu_0}{4\pi}\frac{2\mu}{|x|^3}$$

where μ is the magnitude of the magnetic moment of the loop. 27-6

Inside a long solenoid, far from the ends

$$B_x = \mu_0 n I$$

where n is the number of turns per unit length. 27-9

Due to a straight wire segment

$$B = \frac{\mu_0}{4\pi}\frac{I}{R}(\sin\theta_2 - \sin\theta_1)$$ 27-11

where R is the perpendicular distance to the wire and θ_1 and θ_2 are the angles subtended at the field point by the ends of the wire.

Due to a long, straight wire

$$B = \frac{\mu_0}{4\pi}\frac{2I}{R}$$ 27-12

The direction of $\vec{B}$ is such that the magnetic field lines of $\vec{B}$ encircle the wire in the direction of the fingers of the right hand if the thumb points in the direction of the current.

Inside a tightly wound toroid

$$B = \frac{\mu_0}{2\pi}\frac{NI}{r},\quad a < r < b$$ 27-17

2. Magnetic Field Lines

The magnetic field is indicated by lines parallel to $\vec{B}$ at any point whose density is proportional to the magnitude of $\vec{B}$. Magnetic lines do not begin or end at any point in space. Instead, they form continuous loops.

3. Gauss's Law for Magnetism

$$\phi_{m,\,net} = \oint_S B_n \, dA = 0$$

27-14

4. Magnetic Poles

Magnetic poles always occur in pairs. Isolated magnetic poles have not been found.

5. Ampère's Law

$$\oint_C \vec{B} \cdot d\vec{\ell} = \mu_0 I_C$$

where C is any closed curve.

27-15

Validity of Ampère's law

Ampère's law is valid only if the currents are steady and continuous. It can be used to derive expressions for the magnetic field for situations with a high degree of symmetry, such as a long, straight, current-carrying wire or a long, tightly wound solenoid.

6. Magnetism in Matter

Matter can be classified as paramagnetic, ferromagnetic, or diamagnetic.

Magnetization

A magnetized material is described by its magnetization vector $\vec{M}$, which is defined as the net magnetic dipole moment per unit volume of the material:

$$\vec{M} = \frac{d\vec{\mu}}{dV}$$

27-18

The magnetic field due to a uniformly magnetized cylinder is the same as if the cylinder carried a current per unit length of magnitude M on its surface. This current, which is due to the intrinsic motion of the atomic charges in the cylinder, is called an amperian current.

7. $\vec{B}$ in Magnetic Materials

$$\vec{B} = \vec{B}_{app} + \mu_0 \vec{M}$$

27-21

Magnetic susceptibility χ_m

$$\vec{M} = \chi_m \frac{\vec{B}_{app}}{\mu_0}$$

27-22

For paramagnetic materials, χ_m is a small positive number that depends on temperature. For diamagnetic materials (other than superconductors), it is a small negative constant independent of temperature. For superconductors, $\chi_m = -1$. For ferromagnetic materials, the magnetization depends not only on the magnetizing current but also on the past history of the material.

Relative permeability

$$\vec{B} = K_m \vec{B}_{app}$$

27-23

where

$$K_m = 1 + \chi_m$$

27-24

8. Atomic Magnetic Moments

$$\vec{\mu} = \frac{q}{2m} \vec{L}$$

27-27

where $\vec{L}$ is the orbital angular momentum of the particle.

Due to the orbital motion of an electron

$$\vec{\mu}_\ell = -\frac{e\hbar}{2m_e} \frac{\vec{L}}{\hbar} = -\mu_B \frac{\vec{L}}{\hbar}$$

27-28

Due to electron spin

$$\vec{\mu}_s = -2 \times \frac{e\hbar}{2m_e} \frac{\vec{S}}{\hbar} = -2\mu_B \frac{\vec{S}}{\hbar}$$

27-30

Bohr magneton

$$\mu_B = \frac{e\hbar}{2m_e} = 9.27 \times 10^{-24} \text{ A·m}^2$$

$$= 9.27 \times 10^{-24} \text{ J/T} = 5.79 \times 10^{-5} \text{ eV/T}$$

27-29

where

$$\hbar = \frac{h}{2\pi} = 1.05 \times 10^{-34} \text{ J·s}$$

and $h = 6.626 \times 10^{-34}$ J·s is Planck's constant.

*9. Paramagnetism

Paramagnetic materials have permanent atomic magnetic moments that have random directions in the absence of an applied magnetic field. In an applied field these dipoles are aligned with the field to some degree, producing a small contribution to the total field that adds to the applied field. The degree of alignment is small except in very strong fields and at very low temperatures. At ordinary temperatures, thermal motion tends to maintain the random directions of the magnetic moments.

Curie's law

In weak fields, the magnetization is approximately proportional to the applied field and inversely proportional to the absolute temperature.

$$M = \frac{1}{3} \frac{\mu B_{\text{app}}}{kT} M_s \qquad \qquad 27\text{-}33$$

*10. Ferromagnetism

Ferromagnetic materials have small regions of space called magnetic domains in which all the permanent atomic magnetic moments are aligned. When the material is unmagnetized, the direction of alignment in one domain is independent of that in another domain so that no net magnetic field is produced. When the material is magnetized, the domains of a ferromagnetic material are aligned, producing a very strong contribution to the magnetic field. This alignment can persist even when the external field is removed, thus leading to permanent magnetism.

*11. Diamagnetism

Diamagnetic materials are those materials in which the magnetic moments of all electrons in each atom cancel, leaving each atom with zero magnetic moment in the absence of an external field. In an external field, a very small magnetic moment is induced that tends to weaken the field. This effect is independent of temperature. Superconductors are diamagnetic with a magnetic susceptibility equal to -1.

PROBLEMS

- • Single-concept, single-step, relatively easy
- •• Intermediate-level, may require synthesis of concepts
- ••• Challenging
- **SSM** Solution is in the *Student Solutions Manual*
- **iSOLVE** Problems available on iSOLVE online homework service
- **iSOLVE✓** These "Checkpoint" online homework service problems ask students additional questions about their confidence level, and how they arrived at their answer.

In a few problems, you are given more data than you actually need; in a few other problems, you are required to supply data from your general knowledge, outside sources, or informed estimates.

Conceptual Problems

1 • **SSM** Compare the directions of the electric force and the magnetic force between two positive charges, which move along parallel paths (*a*) in the same direction and (*b*) in opposite directions.

2 • Is $\vec{B}$ uniform everywhere within a current loop? Explain.

3 • Sketch the field lines for the electric dipole and the magnetic dipole shown in Figure 27-41. How do they differ in appearance close to the center of each dipole?

Electric dipole   Magnetic dipole

FIGURE 27-41 Problem 3

4 • Two wires lie in the plane of the paper and carry equal currents in opposite directions, as shown in Figure 27-42. At a point midway between the wires, the magnetic field is (*a*) zero. (*b*) into the page. (*c*) out of the page. (*d*) toward the top or bottom of the page. (*e*) toward one of the two wires.

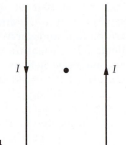

FIGURE 27-42 Problem 4

5 • Two parallel wires carry currents I_1 and $I_2 = 2I_1$ in the same direction. The forces F_1 and F_2 on the wires are related by (*a*) $F_1 = F_2$. (*b*) $F_1 = 2F_2$. (*c*) $2F_1 = F_2$. (*d*) $F_1 = 4F_2$. (*e*) $4F_1 = F_2$.

6 • **SSM** A wire carries an electrical current straight up. What is the direction of the magnetic field due to the wire a distance of 2 m north of the wire? (*a*) North (*b*) East (*c*) West (*d*) South (*e*) Upward

7 • Two current-carrying wires are perpendicular to each other. The current in one wire flows vertically upward and the current in the other wire flows horizontally toward the east. The horizontal wire is 1 m south of the vertical wire. What is the direction of the net magnetic force on the horizontal wire? (*a*) North (*b*) East (*c*) West (*d*) South (*e*) There is no net magnetic force on the horizontal wire.

8 • Make a field-line sketch of the magnetic field due to the currents in the pair of coaxial coils (Figure 27-43). The currents in the coils have the same magnitude and are in the same direction in each coil.

FIGURE 27-43 Problems 8 and 9

9 • **SSM** Make a field-line sketch of the magnetic field due to the currents in the pair of coaxial coils (Figure 27-43). The currents in the coils have the same magnitude but are opposite in direction in each coil.

10 • Ampère's law is valid (*a*) when there is a high degree of symmetry. (*b*) when there is no symmetry. (*c*) when the current is constant. (*d*) when the magnetic field is constant. (*e*) in all of these situations if the current is continuous.

11 • True or false:
(*a*) Diamagnetism is the result of induced magnetic dipole moments.
(*b*) Paramagnetism is the result of the partial alignment of permanent magnetic dipole moments.

12 • **SSM** If the magnetic susceptibility is positive, (*a*) paramagnetic effects or ferromagnetic effects must be greater than diamagnetic effects. (*b*) diamagnetic effects must be greater than paramagnetic effects. (*c*) diamagnetic effects must be greater than ferromagnetic effects. (*d*) ferromagnetic effects must be greater than paramagnetic effects. (*e*) paramagnetic effects must be greater than ferromagnetic effects.

13 • True or false:
(*a*) The magnetic field due to a current element is parallel to the current element.
(*b*) The magnetic field due to a current element varies inversely with the square of the distance from the element.

(*c*) The magnetic field due to a long wire varies inversely with the square of the distance from the wire.
(*d*) Ampère's law is valid only if there is a high degree of symmetry.
(*e*) Ampère's law is valid only for continuous currents.

14 • Can a particle have angular momentum and not have a magnetic moment?

15 • Can a particle have a magnetic moment and not have angular momentum?

16 • A circular loop of wire carries a current I. Is there angular momentum associated with the magnetic moment of the loop? If so, why is it not noticed?

17 • A hollow tube carries a current. Inside the tube, $\vec{B} = 0$. Why is this the case, because $\vec{B}$ is strong inside a solenoid?

18 • **SSM** When a current is passed through the wire in Figure 27-44, will the wire tend to bunch up or form a circle?

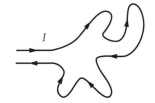

FIGURE 27-44 Problem 18

19 • Which of the four gases listed in Table 27-1 are diamagnetic and which of the four gases are paramagnetic?

Estimation and Approximation

20 •• The magnetic moment of the earth is about 9×10^{22} A·m^2. (*a*) If the magnetization of the earth's core were 1.5×10^9 A/m, what is the core volume? (*b*) What is the radius of such a core if it were spherical and centered with the earth?

21 •• **SSM** Estimate the transient magnetic field 100 m away from a lightning bolt if a charge of about 30 C is transferred from cloud to ground and the average speed of the charges is 10^6 m/s.

22 •• **SSM** The rotating disk of Problem 125 (page 896) can be used as a model for the magnetic field due to a sunspot. If the sunspot radius is approximately 10^7 m rotating at an angular velocity of about 10^{-2} rad/s, calculate the total charge Q on the sunspot needed to create a magnetic field of order 0.1 T at the center of the sunspot. What is the electrical field magnitude just above the center of the sunspot due to this charge?

The Magnetic Field of Moving Point Charges

23 • At time $t = 0$, a particle with charge $q = 12 \ \mu C$ is located at $x = 0$, $y = 2$ m; the particle's velocity at that time is $\vec{v} = 30$ m/s $\hat{i}$. Find the magnetic field at (*a*) the origin; (*b*) $x = 0$, $y = 1$ m; (*c*) $x = 0$, $y = 3$ m; and (*d*) $x = 0$, $y = 4$ m.

24 • For the particle in Problem 23, find the magnetic field at (*a*) $x = 1$ m, $y = 3$ m; (*b*) $x = 2$ m, $y = 2$ m; and (*c*) $x = 2$ m, $y = 3$ m.

25 • A proton (charge $+e$) traveling with a velocity of $\vec{v} = 1 \times 10^4$ m/s $\hat{i} + 2 \times 10^4$ m/s $\hat{j}$ is located at $x = 3$ m, $y = 4$ m at some time t. Find the magnetic field at (*a*) $x = 2$ m, $y = 2$ m; (*b*) $x = 6$ m, $y = 4$ m; and (*c*) $x = 3$ m, $y = 6$ m.

26 • **SOLVE** An electron orbits a proton at a radius of 5.29×10^{-11} m. What is the magnetic field at the proton due to the orbital motion of the electron?

27 •• **SSM** Two equal charges q located at $(0, 0, 0)$ and at $(0, b, 0)$ at time zero are moving with speed v in the positive x direction ($v \ll c$). Find the ratio of the magnitudes of the magnetic force and electrostatic force on each charge.

The Magnetic Field of Currents: The Biot–Savart Law

28 • A small current element $I\, d\vec{\ell}$ with $d\vec{\ell} = 2$ mm$\hat{k}$ and $I = 2$ A, is centered at the origin. Find the magnetic field $d\vec{B}$ at the following points: (a) on the x axis at $x = 3$ m, (b) on the x axis at $x = -6$ m, (c) on the z axis at $z = 3$ m, and (d) on the y axis at $y = 3$ m.

29 • For the current element in Problem 28, find the magnitude and direction of $d\vec{B}$ at $x = 0, y = 3$ m, $z = 4$ m.

30 • **SSM** For the current element in Problem 28, find the magnitude of $d\vec{B}$ and indicate its direction on a diagram at (a) $x = 2$ m, $y = 4$ m, $z = 0$ and (b) $x = 2$ m, $y = 0$, $z = 4$ m.

$\vec{B}$ Due to a Current Loop

31 • A single loop of wire with radius 3 cm carries a current of 2.6 A. What is the magnitude of B on the axis of the loop at (a) the center of the loop, (b) 1 cm from the center, (c) 2 cm from the center, and (d) 35 cm from the center?

32 • **SSM** **SOLVE** A single-turn circular loop of radius 10.0 cm is to produce a field at its center that will just cancel the earth's magnetic field at the equator, which is 0.7 G directed north. Find the current in the loop and make a sketch that shows the orientation of the loop and the current.

33 •• For the loop of wire in Problem 32, at what point along the axis of the loop is the magnetic field (a) 10 percent of the field at the center, (b) 1 percent of the field at the center, and (c) 0.1 percent of the field at the center?

34 •• **SOLVE** A single-turn circular loop of radius 8.5 cm is to produce a field at its center that will just cancel the earth's field of magnitude 0.7 G directed at 70° below the horizontal north direction. Find the current in the loop and make a sketch that shows the orientation of the loop and the current in the loop.

35 •• A circular current loop of radius R carrying a current $I = 10$ A is centered at the origin with its axis along the x axis. Its current is such that it produces a magnetic field in the positive x direction. (a) Using a spreadsheet program or graphing calculator, construct a graph of B_x versus x/R for points on the x/R axis $-5 < x/R < +5$. Compare this graph with that for E_x due to a charged ring of the same size. (b) A second, identical current loop, carrying an equal current in the same sense, is in a plane parallel to the yz plane with its center at $x = R$. Make separate graphs of B_x on the x axis due to each loop and also graph the resultant field due to the two loops. Show from your sketch that dB_x/dx is zero midway between the two loops.

36 •• A pair of identical coils, each of radius r, are separated by a distance r. Called *Helmholtz coils*, the coils are coaxial and carry equal currents such that their axial fields add. A feature of Helmholtz coils is that the resultant magnetic field in the region between the coils is very uniform. Let $r = 30$ cm, $I = 15$ A and $N = 250$ turns for each coil. Using a spreadsheet program calculate and graph the magnetic field as a function of x, the distance from the center of the coils along the common axis, for $-r < x < r$. Over what range of x does the field vary by less than 20%?

37 ••• Two Helmholtz coils with radii R have their axes along the x axis (see Problem 36). One coil is in the yz plane and the second coil is in a parallel plane at $x = R$. Show that at the midpoint of the coils $dB_x/dx = 0$, $d^2B_x/dx^2 = 0$, and $d^3B_x/dx^3 = 0$. (*Note:* This shows that the magnetic field at points near the midpoint is approximately equal to that at the midpoint.)

38 ••• **SSM** *Anti-Helmholtz* coils are used in many physics applications, such as laser cooling and trapping, where a spatially inhomogeneous field with a uniform gradient is desired. These coils have the same construction as a Helmholtz coil, except that the currents flow in opposite directions, so that the axial fields subtract, and the coil separation is $r\sqrt{3}$ rather than r. Graph the magnetic field as a function of x, the axial distance from the center of the coils, for an anti-Helmholtz coil using the same parameters as in Problem 36.

39 •• Two concentric coplaner conducting circular loops have radii $r_1 = 10$ cm and $r_2 > r_1$ are in a horizontal plane. A current $I = 1$ A flows in each coil, but in opposite directions, with the current in the inner coil being counterclockwise as viewed from above. Using a spreadsheet program, calculate and graph the magnetic field as a function of the height x above the center of the coils for $r_2 = $ (a) 10.1 cm, (b) 11 cm, (c) 15 cm and (d) 20 cm.

40 ••• Two concentric circular loops of wire in the same plane have radii $r_1 = 10$ cm and $r_2 > r_1$. A current $I = 1$ A flows in each loop but in the opposite direction. Using a spreadsheet program, calculate and graph the magnetic field component B_x on the axis of the loops as a function of the distance x from the center of the coils. Construct a separate curve for $r_2 = $ (a) 10.1 cm, (b) 11 cm, (c) 15 cm, and (d) 20 cm.

41 ••• For the coils considered in Problem 40, show that if $r_2 = r_1 + \Delta r$, where $\Delta r \ll r_1$, then

$$B(x) \approx \left(\frac{\mu_0 I \Delta r}{2}\right)\left(\frac{2rx^2 - r^3}{(x^2 + r_1^2)^{5/2}}\right).$$

Straight-Line Current Segments

42 •• For the coils considered in Problem 41, show that when $x \gg r_1$, then

$$B(x) \approx -\left(\frac{\mu_0 I \Delta r}{2}\right)\left(\frac{2r_1}{x^3}\right).$$

Compare this to the results of Problem 39 (a).

Problems 43 to 48 refer to Figure 27-45, which shows two long straight wires in the xy plane and parallel to the x axis. One wire is at $y = -6$ cm and the other wire is at $y = +6$ cm. The current in each wire is 20 A.

FIGURE 27-45
Problems 43–48

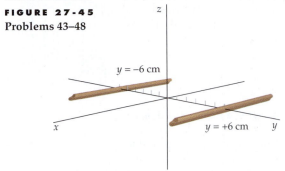

43 • **SSM** If the currents in Figure 27-45 are in the negative x direction, find $\vec{B}$ at the points on the y axis at (a) $y = -3$ cm, (b) $y = 0$, (c) $y = +3$ cm, and (d) $y = +9$ cm.

44 •• Using a spreadsheet program or graphing calculator, graph B_z versus y for points on the y axis when both currents are in the negative x direction.

45 • Find $\vec{B}$ at points on the y axis, as in Problem 43, when the current in the wire at $y = -6$ cm is in the negative x direction and the current in the wire at $y = +6$ cm is in the positive x direction.

46 •• Using a spreadsheet program or graphing calculator, graph B_z versus y for points on the y axis when the directions of the currents are opposite to those in Problem 45.

47 • Find $\vec{B}$ on the z axis at $z = +8$ cm if (a) the currents are parallel, as in Problem 43 and (b) the currents are antiparallel, as in Problem 45.

48 • **iSOLVE** Find the magnitude of the force per unit length exerted by one wire on the other.

49 • **iSOLVE** Two long, straight parallel wires 8.6 cm apart carry currents of equal magnitude I. The parallel wires repel each other with a force per unit length of 3.6 nN/m. (a) Are the currents parallel or antiparallel? (b) Find I.

50 •• **iSOLVE** The current in the wire shown in Figure 27-46 is 8 A. Find B at point P due to each wire segment and sum to find the resultant B.

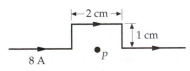

FIGURE 27-46 Problem 50

51 •• **iSOLVE** A wire of length 16 cm is suspended by flexible leads above a long straight wire. Equal but opposite currents are established in the wires so that the 16-cm wire floats 1.5 mm above the long wire with no tension in its suspension leads. If the mass of the 16-cm wire is 14 g, what is the current?

52 •• **SSM** Three long, parallel straight wires pass through the corners of an equilateral triangle of sides 10 cm, as shown in Figure 27-47, where a dot means that the current is out of the paper and a cross means that the current is into the paper. If each current is 15 A, find

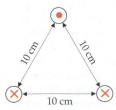

FIGURE 27-47 Problems 52 and 53

(a) the force per unit length on the upper wire and (b) the magnetic field B at the upper wire due to the two lower wires.

53 •• Rework Problem 52, with the current in the lower right corner of Figure 27-47 reversed.

54 •• An infinitely long insulated wire lies along the x axis and carries current I in the positive x direction. A second infinitely long insulated wire lies along the y axis and carries current I in the positive y direction. Where in the xy plane is the resultant magnetic field zero?

55 •• An infinitely long wire lies along the z axis and carries a current of 20 A in the positive z direction. A second infinitely long wire is parallel to the z axis at $x = 10$ cm. (a) Find the current in the second wire if the magnetic field at $x = 2$ cm is zero. (b) What is the magnetic field at $x = 5$ cm?

56 •• Three very long parallel wires are at the corners of a square, as shown in Figure 27-48. The wires each carry a current of magnitude I. Find the magnetic field B at the unoccupied corner of the square when (a) all the currents are into the paper, (b) I_1 and I_3 are into the paper and I_2 is out, and (c) I_1 and I_2 are into the paper and I_3 is out.

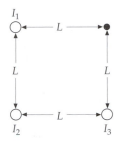

FIGURE 27-48 Problem 56

57 •• **SSM** Four long, straight parallel wires each carry current I. In a plane perpendicular to the wires, the wires are at the corners of a square of side a. Find the force per unit length on one of the wires if (a) all the currents are in the same direction and (b) the currents in the wires at adjacent corners are oppositely directed.

58 •• An infinitely long nonconducting cylinder of radius R lies along the z axis. Five long conducting wires are parallel to the cylinder and are spaced equally on the upper half of the cylinder's surface. Each wire carries a current I in the positive z direction. Find the magnetic field on the z axis.

$\vec{B}$ Due to a Current in a Solenoid

59 • A solenoid with length 30 cm, radius 1.2 cm, and 300 turns carries a current of 2.6 A. Find B on the axis of the solenoid (a) at the center, (b) inside the solenoid at a point 10 cm from one end, and (c) at one end.

60 • **SSM** **iSOLVE** A solenoid 2.7-m long has a radius of 0.85 cm and 600 turns. It carries a current I of 2.5 A. What is the approximate magnetic field B on the axis of the solenoid?

61 ••• A solenoid has n turns per unit length and radius R and carries a current I. Its axis is along the x axis with one end at $x = -\frac{1}{2}\ell$ and the other end at $x = +\frac{1}{2}\ell$, where ℓ is the total length of the solenoid. Show that the magnetic field B at a point on the axis outside the solenoid is given by

$$B = \frac{1}{2}\mu_0 nI(\cos\theta_1 - \cos\theta_2)$$

where
$$\cos\theta_1 = \frac{x + \frac{1}{2}\ell}{[R^2 + (x + \frac{1}{2}\ell)^2]^{1/2}}$$

and
$$\cos\theta_2 = \frac{x - \frac{1}{2}\ell}{[R^2 + (x - \frac{1}{2}\ell)^2]^{1/2}}$$

62 ••• In Problem 61, a formula for the magnetic field along the axis of a solenoid is given. For $x \gg \ell$ and $\ell > R$, the angles θ_1 and θ_2 are very small, so the small-angle approximation $\cos\theta \approx 1 - \theta^2/2$ is valid. (a) Draw a diagram and show that

$$\theta_1 \approx \frac{R}{x + \frac{1}{2}\ell}$$

and

$$\theta_2 \approx \frac{R}{x - \frac{1}{2}\ell}$$

(b) Show that the magnetic field at a point far from either end of the solenoid can be written

$$B = \frac{\mu_0}{4\pi}\left(\frac{q_m}{r_1^2} - \frac{q_m}{r_2^2}\right)$$

where $r_1 = x - \frac{1}{2}\ell$ is the distance to the near end of the solenoid, $r_1 = x + \frac{1}{2}\ell$ is the distance to the far end, and $q_m = nI\pi R^2 = \mu/\ell$, where $\mu = NI\pi R^2$ is the magnetic moment of the solenoid.

Ampère's Law

63 • **SSM** **SOLVE** A long, straight, thin-walled cylindrical shell of radius R carries a current I. Find B inside the cylinder and outside the cylinder.

64 • In Figure 27-49, one current is 8 A into the paper, the other current is 8 A out of the paper, and each curve is a circular path. (a) Find $\oint_C \vec{B} \cdot d\vec{\ell}$ for each path indicated, where each integral is taken with $d\vec{\ell}$ counterclockwise. (b) Which path, if any, can be used to find B at some point due to these currents?

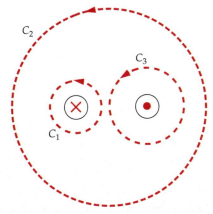

FIGURE 27-49
Problem 64

65 • A very long coaxial cable consists of an inner wire and a concentric outer cylindrical conducting shell of radius R. At one end, the wire is connected to the shell. At the other end, the wire and shell are connected to opposite terminals of a battery, so there is a current down the wire and back up the shell. Assume that the cable is straight. Find B (a) at points between the wire and the shell far from the ends and (b) outside the cable.

66 •• **SOLVE** A wire of radius 0.5 cm carries a current of 100 A that is uniformly distributed over its cross-sectional area. Find B (a) 0.1 cm from the center of the wire, (b) at the surface of the wire, and (c) at a point outside the wire 0.2 cm from the surface of the wire. (d) Sketch a graph of B versus the distance from the center of the wire.

67 •• **SSM** Show that a uniform magnetic field with no fringing field, such as that shown in Figure 27-50, is impossible because it violates Ampère's law. Do this by applying Ampère's law to the rectangular curve shown by the dashed lines.

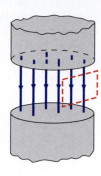

FIGURE 27-50 Problem 67

68 •• A coaxial cable consists of a solid inner cylindrical conductor of radius 1.00 mm and an outer cylindrical shell conductor of inner radius 2.00 mm with an outer radius of 3.00 mm. There is a current of 15 A down the inner wire and an equal return current in the outer wire. The currents are uniform over the cross section of each conductor. Using a spreadsheet program or graphing calculator, graph the magnitude of the magnetic field B as a function of the distance r from the cable axis for 0 mm $< r <$ 3.00 mm. What is the field outside the wire?

69 •• An infinitely long, thick cylindrical shell of inner radius a and outer radius b carries a current I uniformly distributed across a cross section of the shell. Find the magnetic field for (a) $r < a$, (b) $a < r < b$, and (c) $r > b$.

70 •• Figure 27-51 shows a solenoid carrying a current I with n turns per unit length. Apply Ampère's law to the rectangular curve shown in the figure to derive an expression for B, assuming that B is uniform inside the solenoid and that B is zero outside the solenoid.

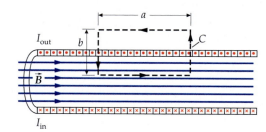

FIGURE 27-51 Problem 70

71 •• **SOLVE** A tightly wound toroid of inner radius 1 cm and outer radius 2 cm has 1000 turns of wire and carries a current of 1.5 A. (a) What is the magnetic field at a distance of 1.1 cm from the center? (b) What is the magnetic field at a distance of 1.5 cm from the center?

72 •• **SSM** The xz plane contains an infinite sheet of current in the positive z direction. The current per unit length (along the x direction) is λ. Figure 27-52a shows a point P above the sheet ($y > 0$) and two portions of the current sheet labeled I_1 and I_2. (a) What is the direction of the magnetic field $\vec{B}$ at point P due to the two portions of the current shown? (b) What is the direction of the magnetic field $\vec{B}$ at point P due to the entire sheet? (c) What is the direction of $\vec{B}$ at a point below the sheet ($y < 0$)? (d) Apply Ampère's law to the rectangular curve shown in Figure 27-52b to show that the magnetic field at any point above the sheet is given by $\vec{B} = -\frac{1}{2}\mu_0\lambda\hat{i}$

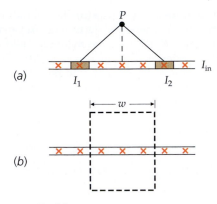

FIGURE 27-52 Problem 72

Magnetization and Magnetic Susceptibility

73 • A tightly wound solenoid 20-cm long has 400 turns and carries a current of 4 A so that its axial field is in the z direction. Neglecting end effects, find B and B_{app} at the center when (a) there is no core in the solenoid and (b) there is an iron core with a magnetization $M = 1.2 \times 10^6$ A/m.

74 • If the solenoid of Problem 73 has an aluminum core, find B_{app}, M, and B at the center, neglecting end effects.

75 • Repeat Problem 74 for a tungsten core.

76 • A long solenoid is wound around a tungsten core and carries a current. (a) If the core is removed while the current is held constant, does the magnetic field inside the solenoid decrease or increase? (b) By what percentage does the magnetic field inside the solenoid decrease or increase?

77 • **ISOLVE** When a sample of liquid is inserted into a solenoid carrying a constant current, the magnetic field inside the solenoid decreases by 0.004 percent. What is the magnetic susceptibility of the liquid?

78 • **ISOLVE** A long solenoid carrying a current of 10 A has 50 turns/cm. What is the magnetic field in the interior of the solenoid when the interior is (a) a vacuum, (b) filled with aluminum, and (c) filled with silver?

79 •• **SSM** A cylinder of magnetic material is placed in a long solenoid of n turns per unit length and current I. The values for magnetic field B within the material versus nI is given below. Use these values to plot B versus B_{app} and K_m versus nI.

nI, A/m	0	50	100	150	200	500	1000	10,000
B, T	0	0.04	0.67	1.00	1.2	1.4	1.6	1.7

80 •• A small magnetic sample is in the form of a disk that has a radius of 1.4 cm, a thickness of 0.3 cm, and a uniform magnetization along its axis throughout its volume. The magnetic moment of the sample is 1.5×10^{-2} A·m². (a) What is the magnetization $\vec{M}$ of the sample? (b) If this magnetization is due to the alignment of N electrons, each with a magnetic moment of 1 μ_B, what is N? (c) If the magnetization is along the axis of the disk, what is the magnitude of the amperian surface current?

81 •• A cylindrical shell in the shape of a flat washer has inner radius r, outer radius R, and thickness (length) t, where $t \ll R$. The material of the shell has a uniform magnetization of magnitude M parallel to its axis. Show that the magnetic field due to the cylinder can be modeled using the concentric conducting loops model of Problem 39. What is the amperian current I which we must use to model this field?

Atomic Magnetic Moments

82 •• **SSM** **ISOLVE** Nickel has a density of 8.7 g/cm³ and a molecular mass of 58.7 g/mol. Nickel's saturation magnetization is given by $\mu_0 M_s = 0.61$ T. Calculate the magnetic moment of a nickel atom in Bohr magnetons.

83 •• **ISOLVE** Repeat Problem 82 for cobalt, which has a density of 8.9 g/cm³, a molecular mass of 58.9 g/mol, and a saturation magnetization given by $\mu_0 M_s = 1.79$ T.

*Paramagnetism

84 • Show that Curie's law predicts that the magnetic susceptibility of a paramagnetic substance is $\chi_m = \mu_0 M_s/3kT$.

85 •• In a simple model of paramagnetism, we can consider that some fraction f of the molecules have their magnetic moments aligned with the external magnetic field and that the rest of the molecules are randomly oriented and therefore do not contribute to the magnetic field. (a) Use this model and Curie's law to show that at temperature T and external magnetic field B the fraction of aligned molecules is $f = \mu B/3kT$. (b) Calculate this fraction for $T = 300$ K, $B = 1$ T, assuming μ to be 1 Bohr magneton.

86 •• **SSM** Assume that the magnetic moment of an aluminum atom is 1 Bohr magneton. The density of aluminum is 2.7 g/cm³, and its molecular mass is 27 g/mol. (a) Calculate M_s and $\mu_0 M_s$ for aluminum. (b) Use the results of Problem 84 to calculate χ_m at $T = 300$ K. (c) Explain why the result for Part (b) is larger than the value listed in Table 27-1.

87 •• A toroid with N turns carrying a current I has a mean radius R and a cross-sectional radius r, where $r \ll R$ (Figure 27-53). When the toroid is filled with material, it is called a *Rowland ring*. Find B_{app} and B in such a ring, assuming a magnetization $\vec{M}$ everywhere parallel to $\vec{B}_{app}$.

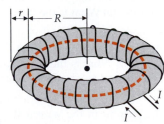

FIGURE 27-53 Problem 87

88 •• A toroid is filled with liquid oxygen that has a susceptibility of 4×10^{-3}. The toroid has 2000 turns and carries a current of 15 A. Its mean radius is 20 cm, and the radius of its cross section is 0.8 cm. (a) What is the magnetization M? (b) What is the magnetic field B? (c) What is the percentage increase in B produced by the liquid oxygen?

89 •• A toroid has an average radius of 14 cm and a cross-sectional area of 3 cm². It is wound with fine wire, 60 turns/cm measured along its mean circumference, and the wire carries a current of 4 A. The core is filled with a paramagnetic material of magnetic susceptibility 2.9×10^{-4}. (a) What is the magnitude of the magnetic field within the substance? (b) What is the magnitude of the magnetization? (c) What would the magnitude of the magnetic field be if there were no paramagnetic core present?

*Ferromagnetism

90 • [SSM] For annealed iron, the relative permeability K_m has its maximum value of approximately 5500 at $B_{app} = 1.57 \times 10^{-4}$ T. Find M and B when K_m is maximum.

91 •• The saturation magnetization for annealed iron occurs when $B_{app} = 0.201$ T. Find the permeability μ and the relative permeability K_m of annealed iron at saturation. (See Table 27-2.)

92 •• [SOLVE] The coercive force is defined to be the applied magnetic field needed to bring B back to zero along the hysteresis curve (which is point c in Figure 27-38). For a certain permanent bar magnet, the coercive force $B_{app} = 5.53 \times 10^{-2}$ T. The bar magnet is to be demagnetized by placing it inside a 15-cm-long solenoid with 600 turns. What minimum current is needed in the solenoid to demagnetize the magnet?

93 •• A long solenoid with 50 turns/cm carries a current of 2 A. The solenoid is filled with iron and B is measured to be 1.72 T. (a) Neglecting end effects, what is B_{app}? (b) What is M? (c) What is the relative permeability K_m?

94 •• When the current in Problem 93 is 0.2 A, the magnetic field is measured to be 1.58 T. (a) Neglecting end effects, what is B_{app}? (b) What is M? (c) What is the relative permeability K_m?

95 •• A long, iron-core solenoid with 2000 turns/m carries a current of 20 mA. At this current, the relative permeability of the iron core is 1200. (a) What is the magnetic field within the solenoid? (b) With the iron core removed, what current will produce the same field within the solenoid?

96 •• [SSM] Two long straight wires 4-cm apart are embedded in a uniform insulator that has a relative permeability of $K_m = 120$. The wires carry 40 A in opposite directions. (a) What is the magnetic field at the midpoint of the plane of the wires? (b) What is the force per unit length on the wires?

97 •• The toroid of Problem 88 has its core filled with iron. When the current is 10 A, the magnetic field in the toroid is 1.8 T. (a) What is the magnetization M? (b) Find the values for K_m, μ, and χ_m for the iron sample.

98 •• Find the magnetic field in the toroid of Problem 89 if the current in the wire is 0.2 A and soft iron, which has a relative permeability of 500, is substituted for the paramagnetic core?

99 •• A long straight wire with a radius of 1.0 mm is coated with an insulating ferromagnetic material that has a thickness of 3.0 mm and a relative magnetic permeability of $K_m = 400$. The coated wire is in air and the wire itself is nonmagnetic. The wire carries a current of 40 A. (a) Find the magnetic field inside the wire as a function of radius R. (b) Find the magnetic field inside the ferromagnetic material as a function of radius R. (c) Find the magnetic field outside the ferromagnetic material as a function of R. (d) What must the magnitudes and directions of the amperian currents be on the surfaces of the ferromagnetic material to account for the magnetic fields observed?

General Problems

100 • Find the magnetic field at point P in Figure 27-54.

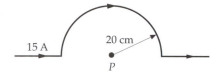

FIGURE 27-54 Problem 100

101 • [SSM] In Figure 27-55, find the magnetic field at point P, which is at the common center of the two semicircular arcs.

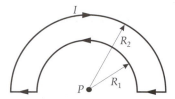

FIGURE 27-55 Problem 101

102 •• A wire of length ℓ is wound into a circular coil of N loops and carries a current I. Show that the magnetic field at the center of the coil is given by $B = \mu_0 \pi N^2 I / \ell$.

103 •• A very long wire carrying a current I is bent into the shape shown in Figure 27-56. Find the magnetic field at point P.

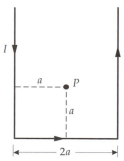

FIGURE 27-56 Problem 103

104 •• [SSM] A power cable carrying 50 A is 2 m below the earth's surface, but the cable's direction and precise position are unknown. Show how you could locate the cable using a compass. Assume that you are at the equator, where the earth's magnetic field is 0.7 G north.

105 •• A long straight wire carries a current of 20 A, as shown in Figure 27-57. A rectangular coil with two sides parallel to the straight wire has sides 5 cm and 10 cm with the near side a distance 2 cm from the wire. The coil carries a current of 5 A. (a) Find the force on each segment of the rectangular coil due to the current in the long straight wire.

(b) What is the net force on the coil?

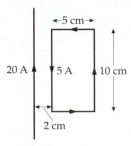

FIGURE 27-57 Problem 105

106 •• **iSOLVE** The closed loop shown in Figure 27-58 carries a current of 8 A in the counterclockwise direction. The radius of the outer arc is 60 cm, that of the inner arc is 40 cm. Find the magnetic field at point P.

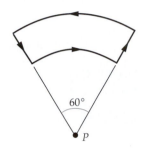

FIGURE 27-58 Problem 106

107 •• **iSOLVE** A closed circuit consists of two semicircles of radii 40 cm and 20 cm that are connected by straight segments, as shown in Figure 27-59. A current of 3 A flows around this circuit in the clockwise direction. Find the magnetic field at point P.

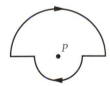

FIGURE 27-59 Problem 107

108 •• **SSM** **iSOLVE** A very long straight wire carries a current of 20 A. An electron 1 cm from the center of the wire is moving with a speed of 5.0×10^6 m/s. Find the force on the electron when it moves (a) directly away from the wire, (b) parallel to the wire in the direction of the current, and (c) perpendicular to the wire and tangent to a circle around the wire.

109 •• A current I of 5 A is uniformly distributed over the cross section of a long straight wire of radius $r_0 = 2.55$ mm. Using a spreadsheet program, graph the magnitude of the magnetic field as a function of r and the distance from the center of the wire for $0 \le r \le 10r_0$.

110 •• A large, 50-turn circular coil of radius 10 cm carries a current of 4 A. At the center of the large coil is a small 20-turn coil of radius 0.5 cm carrying a current of 1 A. The planes of the two coils are perpendicular. Find the torque exerted by the large coil on the small coil. (Neglect any variation in B due to the large coil over the region occupied by the small coil.)

111 •• **SSM** Figure 27-60 shows a bar magnet suspended by a thin wire that provides a restoring torque $-\kappa\theta$. The magnet is 16-cm long, has a mass of 0.8 kg, a dipole moment of $\mu = 0.12$ A·m², and it is located in a region where a uniform magnetic field B can be established. When the external magnetic field is 0.2 T and the magnet is given a small angular displacement $\Delta\theta$, the bar magnet oscillates about its equilibrium position with a period of 0.500 s. Determine the constant κ and the period of this torsional pendulum when $B = 0$.

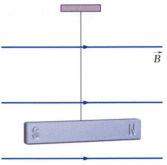

FIGURE 27-60 Problem 111

112 •• A long, narrow bar magnet that has magnetic moment μ parallel to its long axis is suspended at its center as a frictionless compass needle. When placed in a magnetic field $\vec{B}$, the needle lines up with the field. If it is displaced by a small angle θ, show that the needle will oscillate about its equilibrium position with frequency $f = (\frac{1}{2})\pi\sqrt{\mu B/I}$, where I is the moment of inertia about the point of suspension.

113 •• A small bar magnet of mass 0.1 kg, length 1 cm, and magnetic moment $\mu = 0.04$ A·m² is located at the center of a 100-turn loop of 0.2 m diameter. The loop carries a current of 5.0 A. At equilibrium, the bar magnet is aligned with the field due to the current loop. The bar magnet is given a displacement along the axis of the loop and released. Show that if the displacement is small, the bar magnet executes simple harmonic motion, and find the period of this motion.

114 •• Suppose the needle in Problem 112 is a uniformly magnetized iron rod that is 8 cm long and has a cross-sectional area of 3 mm². Assume that the magnetic dipole moment for each iron atom is 2.2 μ_B and that all the iron atoms have their dipole moments aligned. Calculate the frequency of small oscillations about the equilibrium position when the magnetic field is 0.5 G.

115 •• The needle of a magnetic compass has a length of 3 cm, a radius of 0.85 mm, and a density of 7.96×10^3 kg/m³. The needle is free to rotate in a horizontal plane, where the horizontal component of the earth's magnetic field is 0.6 G. When disturbed slightly, the compass executes simple harmonic motion about its midpoint with a frequency of 1.4 Hz. (a) What is the magnetic dipole moment of the needle? (b) What is the magnetization M? (c) What is the amperian current on the surface of the needle? (See Problem 112.)

116 •• **SSM** An iron bar of length 1.4 m has a diameter of 2 cm and a uniform magnetization of 1.72×10^6 A/m directed along the bar's length. The bar is stationary in space and is suddenly demagnetized so that its magnetization disappears. What is the rotational angular velocity of the bar if its angular momentum is conserved? (Assume that Equation 27-27 holds where m is the mass of an electron and $q = -e$.)

117 •• The magnetic dipole moment of an iron atom is 2.219 μ_B. (*a*) If all the atoms in an iron bar of length 20 cm and cross-sectional area 2 cm² have their dipole moments aligned, what is the dipole moment of the bar? (*b*) What torque must be supplied to hold the iron bar perpendicular to a magnetic field of 0.25 T?

118 •• **SSM** A relatively inexpensive ammeter, called a *tangent galvanometer*, can be made using the earth's field. A plane circular coil of N turns and radius R is oriented so the field B_c it produces in the center of the coil is either east or west. A compass is placed at the center of the coil. When there is no current in the coil, the compass needle points north. When there is a current I, the compass needle points in the direction of the resultant magnetic field $\vec{B}$ at an angle θ to the north. Show that the current I is related to θ and to the horizontal component of the earth's field B_e by

$$I = \frac{2RB_c}{\mu_0 N} \tan\theta$$

119 •• **SOLVE** An infinitely long straight wire is bent, as shown in Figure 27-61. The circular portion has a radius of 10 cm with its center a distance r from the straight part. Find r so that the magnetic field at the center of the circular portion is zero.

FIGURE 27-61
Problem 119

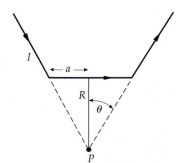

120 •• (*a*) Find the magnetic field at point P for the wire carrying current I, as shown in Figure 27-62. (*b*) Use your result from Part (*a*) to find the field at the center of a polygon of N sides. Show that when N is very large, your result approaches that for the magnetic field at the center of a circle.

FIGURE 27-62
Problem 120

121 •• The current in a long cylindrical conductor of radius $R = 10$ cm varies with distance from the axis of the cylinder according to the relation $I(r) = (50\ \text{A/m})r$. Find the magnetic field at (*a*) $r = 5$ cm, (*b*) $r = 10$ cm, and (*c*) $r = 20$ cm.

122 •• Figure 27-63 shows a square loop, 20 cm per side, in the xy plane with its center at the origin. The loop carries a current of 5 A. Above it at $y = 0$, $z = 10$ cm is an infinitely long wire parallel to the x axis carrying a current of 10 A. (*a*) Find the torque on the loop. (*b*) Find the net force on the loop.

FIGURE 27-63
Problem 122

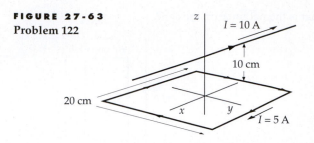

123 •• A current balance is constructed in the following way: A 10-cm-long section of wire is placed on top of the pan of an electronic balance used in a chemistry lab. Leads are clipped to it running into a power supply and through the supply to another segment of wire that is suspended directly above it, parallel with it. (See figure below.) The distance between the two wires is $L = 2.0$ cm. The power supply provides a current I running through the wires. When the power supply is switched on, the reading on the balance increases by 5.0 mg (1 mg = 10^{-6} kg). What is the current running through the wire?

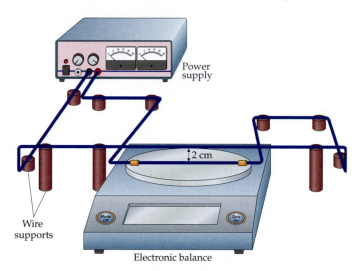

FIGURE 27-64 Problem 123

124 •• Consider the current balance of Problem 123. If the sensitivity of the balance is 0.1 mg, what is the minimum current detectable using this current balance? Discuss any advantages or disadvantages of this type of current balance versus the "standard" current balance discussed in the chapter.

125 ••• **SSM** A disk of radius R carries a fixed charge density σ and rotates with angular velocity ω. (*a*) Consider a circular strip of radius r and width dr with charge dq. Show that the current produced by this strip $dI = (\omega/2\pi)\,dq = \omega\sigma r\,dr$. (*b*) Use your result from Part (*a*) to show that the magnetic field at the center of the disk is $B = \frac{1}{2}\mu_0\sigma\omega R$. (*c*) Use your result from Part (*a*) to find the magnetic field at a point on the axis of the disk a distance x from the center.

126 ••• A square loop of side ℓ lies in the yz plane with its center at the origin. It carries a current I. Find the magnetic field B at any point on the x axis and show from your expression that for x much larger than ℓ,

$$B \approx \frac{\mu_0\,2\mu}{4\pi\,x^3}$$

where $\mu = I\ell^2$ is the magnetic moment of the loop.

Magnetic Induction

DEMONSTRATION OF INDUCED EMF.
WHEN THE MAGNET IS MOVING
TOWARD OR AWAY FROM THE COIL,
AN EMF IS INDUCED IN THE COIL, AS
SHOWN BY THE GALVANOMETER'S
DEFLECTION. NO DEFLECTION IS
OBSERVED WHEN THE MAGNET
IS STATIONARY.

? **What causes the current
when the magnet moves? This
is discussed in Section 28-2.**

In the early 1830s, Michael Faraday in England and Joseph Henry in America independently discovered that in a *changing* magnetic field a changing magnetic flux through a surface bounded by a closed stationary loop of wire induces a current in the wire. The emfs and currents caused by such changing magnetic fluxes are called **induced emfs** and **induced currents.** The process itself is referred to as **induction.** Faraday and Henry also discovered that in a *static* magnetic field a changing magnetic flux through a surface bounded by a moving loop of wire induces an emf in the wire. An emf caused by the motion of a conductor in a region with a magnetic field is called a **motional emf.**

When you pull the plug of an electric cord from its socket, you sometimes observe a small spark. Before the cord is disconnected, the cord carries a current that produces a magnetic field encircling the current. When the cord is disconnected, the current abruptly ceases and the magnetic field around the cord collapses. This changing magnetic field induces an emf that tends to maintain the original current, resulting in a spark at the points of the disconnect. Once the magnetic field collapses to zero it is no longer changing, and the induced emf is zero.

Changing magnetic fields can result from changing currents or from moving magnets. The chapter-opening photo illustrates a simple classroom demonstration of emf induced by a changing magnetic field. The ends of a coil are attached to a galvanometer and a strong magnet is moved toward or away from the coil. The momentary deflection shown by the galvanometer *during* the motion indicates that there is an induced electric current in the coil–galvanometer circuit. A current is also induced if the coil is moved toward a stationary magnet, away from a stationary magnet, or if the coil is rotated in a region with a static magnetic field. A coil rotating in a static magnetic field is the basic element of a generator, which converts mechanical energy into electrical energy.

> This chapter will explore the various methods of magnetic induction, all of which can be summarized by a single relation known as Faraday's law. Faraday's law relates the induced emf in a circuit to the rate of change in magnetic flux through the circuit. (The *magnetic flux through the circuit* refers to the flux of the magnetic field through any surface bounded by the circuit.)

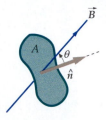

FIGURE 28-1 When $\vec{B}$ makes an angle θ with the normal to the area of a loop, the flux through the loop is $\vec{B} \cdot \hat{n} A = BA \cos \theta$.

28-1 Magnetic Flux

The flux of any vector field through a surface is calculated in the same way as the flux of an electric field through a surface (Section 22-2). Let dA be an element of area on the surface S, and let $\hat{n}$ be a unit normal, a unit vector normal to the area element (Figure 28-1). There are two directions normal to any area element, and which of the two directions is selected for the direction of $\hat{n}$ is a matter of choice. However, the sign of the flux does depend on this choice. The magnetic flux ϕ_m through S is

$$\phi_m = \int_S \vec{B} \cdot \hat{n} \, dA = \int_S B_n \, dA \qquad \text{28-1}$$

MAGNETIC FLUX

The unit of magnetic flux is that of magnetic field intensity times area, teslameter squared, which is called a **weber** (Wb):

$$1 \text{ Wb} = 1 \text{ T·m}^2 \qquad \text{28-2}$$

Since B is proportional to the number of field lines per unit area, the magnetic flux is proportional to the number of lines through an element of area.

EXERCISE Show that a weber per second is a volt.

If the surface is flat with area A, and if $\vec{B}$ is uniform (has the same magnitude and direction) over the surface, the magnetic flux through the surface is

$$\phi_m = \vec{B} \cdot \hat{n} A = BA \cos \theta = B_n A$$

where θ is the angle between the direction of $\vec{B}$ and the positive normal direction. We are often interested in the flux through a surface bounded by a coil that contains several turns of wire. If the coil contains N turns, the flux through the surface is N times the flux through each turn (Figure 28-2). That is,

$$\phi_m = NBA \cos \theta \qquad \text{28-3}$$

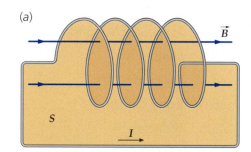

(a)

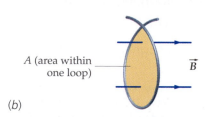

A (area within one loop)

(b)

FIGURE 28-2 (*a*) The flux through the surface *S* bounded by a coil with *N* turns is proportional to the number of field lines penetrating the surface. The coil shown has 4 turns. For the two field lines shown, each line penetrates the surface four times, once for each turn, so the flux through *S* is four times greater than the flux through the surface "bounded" by a single turn of the coil. The coil shown is not tightly wound so the surface *S* can better be observed. (*b*) The area *A* of the flat surface that is (almost) bounded by a single turn.

where A is the area of the flat surface bounded by a single turn. (*Note:* Only a closed curve can actually bound a surface.) A single turn of a multiturn coil is not closed, so a single turn can not actually bound a surface. However, if a coil is tightly wound a single turn is almost closed, and A is the area of the flat surface that it (almost) bounds.

FLUX THROUGH A SOLENOID   **EXAMPLE 28-1**

Find the magnetic flux through a solenoid that is 40-cm long, has a radius of 2.5 cm, has 600 turns, and carries a current of 7.5 A.

PICTURE THE PROBLEM The magnetic field $\vec{B}$ inside the solenoid is uniform and parallel with the axis of the solenoid. It is therefore perpendicular to the plane of each coil. Therefore, we need to find B inside the solenoid and then multiply B by NA.

1. The magnetic flux is the product of the number of turns, the magnetic field strength, and the area bounded by one turn:

$$\phi_m = NBA$$

2. The magnetic field inside the solenoid is given by $B = \mu_0 nI$, where $n = N/\ell$ is the number of turns per unit length:

$$\phi_m = N\mu_0 nIA = N\mu_0 \frac{N}{\ell} IA = \frac{\mu_0 N^2 IA}{\ell}$$

3. Express the area A in terms of its radius:

$$A = \pi r^2$$

4. Substitute the given values to calculate the flux:

$$\phi_m = \frac{\mu_0 N^2 I \pi r^2}{\ell}$$

$$= \frac{(4\pi \times 10^{-7}\ \text{T·m/A})(600\ \text{turns})^2(7.5\ \text{A})\pi(0.025\ \text{m})^2}{0.40\ \text{m}}$$

$$= \boxed{1.66 \times 10^{-2}\ \text{Wb}}$$

REMARKS Note that since $\phi_m = NBA$ and B is proportional to the number of turns N, ϕ_m is proportional to N^2.

28-2 Induced EMF and Faraday's Law

Experiments by Faraday, Henry, and others showed that if the magnetic flux through a surface bounded by a circuit is changed by any means, an emf equal in magnitude to the rate of change of the flux is induced in the circuit. We usually detect the emf by observing a current in the circuit, but the emf is present even if the circuit is nonexistent or incomplete (not closed) and there is no current. Previously we considered emfs that were localized in a specific part of the circuit, such as between the terminals of the battery. However, induced emfs can be distributed throughout the circuit.

The magnetic flux through a surface bounded by a circuit can be changed in several ways. The current producing the magnetic field may be increased or decreased, permanent magnets may be moved toward the surface or away from the surface, the circuit itself may be rotated in a region with a static magnetic field or translated in a region with a nonuniform static magnetic field $\vec{B}$, the orientation of the circuit may be changed, or the area of the surface in a region with a uniform static magnetic field may be increased or decreased. In every case, an emf $\mathcal{E}$ is induced in the circuit that is equal in magnitude to the rate of change of the magnetic flux through (a surface bounded by) the circuit. That is

$$\mathcal{E} = -\frac{d\phi_m}{dt} \qquad\qquad 28\text{-}4$$

FARADAY'S LAW

This result is known as **Faraday's law.** The negative sign in Faraday's law has to do with the direction of the induced emf, which is addressed shortly.

Figure 28-3 shows a single stationary loop of wire in a magnetic field. The flux through the loop is changing because the magnetic field strength is increasing, so an emf is induced in the loop. Since emf is the work done per unit charge, we know there must be forces exerted on the mobile charges doing work on them. Magnetic forces can do no work, therefore, we cannot attribute the emf to the work done by magnetic forces. It is electric forces associated with a nonconservative electric field $\vec{E}_{nc}$ doing the work on the mobile charges. The line integral of this electric field around a complete circuit equals the work done per unit charge, which is the induced emf in the circuit.

The electric fields that we studied in earlier chapters resulted from static electric charges. Such electric fields are conservative, meaning that their circulation about any curve C is zero. (The circulation of a vector field $\vec{A}$ about a closed curve C is defined as $\oint_C \vec{A} \cdot d\vec{\ell}$.) However, the electric field associated with a changing magnetic field is nonconservative. Its circulation about C is an induced emf, equal to the negative of the rate of change of the magnetic flux through any surface S bounded by C:

$$\mathcal{E} = \oint_C \vec{E}_{nc} \cdot d\vec{\ell} = -\frac{d}{dt}\int_S \vec{B} \cdot \hat{n}\, dA = -\frac{d\phi_m}{dt} \qquad\qquad 28\text{-}5$$

INDUCED EMF IN A STATIONARY CIRCUIT IN A CHANGING MAGNETIC FIELD

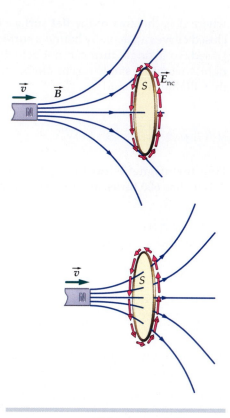

FIGURE 28-3 If the magnetic flux through the stationary wire loop is changing, an emf is induced in the loop. The emf is distributed throughout the loop, which is due to a nonconservative electric field $\vec{E}_{nc}$ tangent to the wire.

INDUCED EMF IN A CIRCULAR COIL I **EXAMPLE 28-2**

A uniform magnetic field makes an angle of 30° with the axis of a circular coil of 300 turns and a radius of 4 cm. The magnitude of the magnetic field increases at a rate of 85 T/s while its direction remains fixed. Find the magnitude of the induced emf in the coil.

PICTURE THE PROBLEM The induced emf equals N times the rate of change of the flux through a single turn. Since $\vec{B}$ is uniform, the flux through each turn is simply $\phi_m = BA\cos\theta$, where $A = \pi r^2$ is the area of the coil.

1. The magnitude of the induced emf is given by Faraday's law: $\qquad \mathcal{E} = -\dfrac{d\phi_m}{dt}$

2. For a uniform field, the flux is: $\qquad \phi_m = N\vec{B} \cdot \hat{n}A = NBA\cos\theta$

3. Substitute this expression for ϕ_m and calculate $\mathcal{E}$:

$$\mathcal{E} = -\frac{d\phi_m}{dt} = -\frac{d}{dt}(NBA\cos\theta) = -N\pi r^2 \cos\theta\,\frac{dB}{dt}$$

$$= -(300)\pi(0.04\text{ m})^2 \cos 30°(85\text{ T/s}) = -111\text{ V}$$

$$|\mathcal{E}| = \boxed{111\text{ V}}$$

EXERCISE If the resistance of the coil is 200 Ω, what is the induced current?
(*Answer* 0.555 A)

INDUCED EMF IN A CIRCULAR COIL II **EXAMPLE 28-3** **Try It Yourself**

An 80-turn coil of radius 5 cm and resistance of 30 Ω sits in a region with a uniform magnetic field normal to the plane of the coil. At what rate must the magnitude of the magnetic field change to produce a current of 4 A in the coil?

PICTURE THE PROBLEM The rate of change of the magnetic field is related to the rate of change of the flux, which is related to the induced emf by Faraday's law. The emf in the coil equals IR.

Cover the column to the right and try these on your own before looking at the answers.

Steps	Answers
1. Write the magnetic flux in terms of B, N, and the radius r, and solve for B.	$\phi_m = NBA = NB\pi r^2$ $$B = \frac{\phi_m}{N\pi r^2}$$
2. Take the time derivative of B.	$$\frac{dB}{dt} = \frac{1}{N\pi r^2}\frac{d\phi_m}{dt}$$
3. Use Faraday's law to relate the rate of change of the flux to the emf.	$$\mathcal{E} = -\frac{d\phi_m}{dt}$$
4. Calculate the magnitude of the emf in the coil from the current and resistance of the coil.	$|\mathcal{E}| = IR = 120\ \text{V}$
5. Substitute numerical values of E, N, and r to calculate dB/dt.	$$\left|\frac{dB}{dt}\right| = \frac{1}{N\pi r^2}|\mathcal{E}| = \boxed{191\ \text{T/s}}$$

A sign convention allows us to use Equation 28-5 to find the direction of both the induced electric field and the induced emf. According to this convention, the positive tangential direction along the integration path C is related to the direction of the unit normal $\hat{n}$ on the surface S bounded by C by a right-hand rule (Figure 28-4). By placing your right thumb in the direction of $\hat{n}$, the fingers of your hand curl in the positive tangential direction on C. If $d\phi_m/dt$ is positive, then in accord with Faraday's law (Equation 28-5), both $\vec{E}_{nc}$ and $\mathcal{E}$ are in the negative tangential direction. (The direction of both $\vec{E}_{nc}$ and $\mathcal{E}$ can be determined via Lenz's law, which is discussed in Section 28-3.)

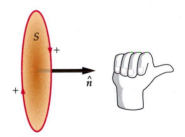

FIGURE 28-4 By placing your right thumb in the direction of $\hat{n}$ on the surface S, the fingers of your hand curl in the positive tangential direction on C.

INDUCED NONCONSERVATIVE ELECTRIC FIELD **EXAMPLE 28-4**

A magnetic field $\vec{B}$ is perpendicular to the plane of the page. $\vec{B}$ is uniform throughout a circular region of radius R, as shown in Figure 28-5. Outside this region, B equals zero. The direction of $\vec{B}$ remains fixed and rate of change of B is dB/dt. What are the magnitude and direction of the induced electric field in the plane of the page (a) a distance $r < R$ from the center of the circular region and (b) a distance $r > R$ from the center, where $B = 0$.

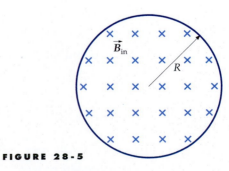

FIGURE 28-5

PICTURE THE PROBLEM The magnetic field $\vec{B}$ is into the page and uniform over a circular region of radius R, as shown in Figure 28-6. As B increases or decreases, the magnetic flux through a surface bounded by closed curve C also changes, and an emf $\mathcal{E} = \oint_C \vec{E} \cdot d\vec{\ell}$ is induced around C. The induced electric field is found by applying $\oint_C \vec{E} \cdot d\vec{\ell} = -d\phi_m/dt$ (Equation 28-5). To take advantage of the system's symmetry, we choose C to be a circular curve of radius r and then evaluate the line integral. By symmetry, $\vec{E}$ is tangent to circle C and has the same magnitude at any point on the circle. We will assign into the page as the direction of $\hat{n}$. The sign convention then tells us that the positive tangential direction is clockwise. We then calculate the magnetic flux ϕ_m, take its time derivative, and solve for E_t.

FIGURE 28-6

(a) 1. The $\vec{E}$ and $\vec{B}$ fields are related by Equation 28-5:

$$\oint_C \vec{E} \cdot d\vec{\ell} = -\frac{d\phi_m}{dt}$$

where

$$\phi_m = \int_S \vec{B} \cdot \hat{n}\, dA$$

2. E_t (the tangential component of $\vec{E}$) is found from the line integral for a circle of radius $r < R$. $\vec{E}$ is tangent to the circle and has a constant magnitude:

$$\oint_C \vec{E} \cdot d\vec{\ell} = \oint_C E_t\, d\ell = E_t \oint_C d\ell = E_t\, 2\pi r$$

3. For $r < R$, $\vec{B}$ is uniform on the flat surface S bounded by the circle C. We choose into the page as the direction of $\hat{n}$. Because $\vec{B}$ is also into the page, the flux through S is simply BA:

$$\phi_m = \int_S \vec{B} \cdot \hat{n}\, dA = \int_S B_n\, dA = B_n \int_S dA$$

$$= BA = B\pi r^2$$

4. Calculate the time derivative of ϕ_m:

$$\frac{d\phi_m}{dt} = \frac{d}{dt}(B\pi r^2) = \frac{dB}{dt}\pi r^2$$

5. Substitute the step 2 and step 4 results into the step 1 result and solve for E_t. The positive tangential direction is clockwise.

$$E_t\, 2\pi r = -\frac{dB}{dt}\pi r^2$$

so

$$\boxed{E_t = -\frac{r}{2}\frac{dB}{dt}, \quad r < R}$$

6. For the choice for the direction of $\hat{n}$ in step 3, the positive tangential direction is clockwise:

E_t is negative, so $\vec{E}$ is $\boxed{\text{counterclockwise}}$.

(b) 1. For a circle of radius $r > R$ (the region where the magnetic field is zero), the line integral is the same as before:

$$\oint_C \vec{E} \cdot d\vec{\ell} = E_t\, 2\pi r$$

2. Since $B = 0$ for $r > R$, the magnetic flux through S is $B\pi R^2$:

$$\phi_m = B\pi R^2$$

3. Apply Faraday's law to find E_t:

$$E_t\, 2\pi r = -\frac{dB}{dt}\pi R^2$$

$$E_t = -\frac{R^2}{2r}\frac{dB}{dt}, \quad r > R$$

E_t is negative, so $\vec{E}$ is $\boxed{\text{counterclockwise}}$.

REMARKS The positive tangential direction is clockwise. When $d\phi_m/dt$ is positive, E_t is negative and the electric field direction is counterclockwise, as shown in Figure 28-7. Note that the electric field in this example is produced by a changing magnetic field rather than by electric charges. Note also that $\vec{E}$, and thus the emf, exists along any closed curve bounding the area through which the magnetic flux is changing, whether there is a wire or circuit along the curve or there is not.

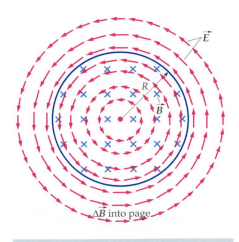

FIGURE 28-7 The magnetic field is into the page and increasing in magnitude. The induced electric field is counterclockwise.

28-3 Lenz's Law

The negative sign in Faraday's law has to do with the direction of the induced emf. This can be obtained by the sign convention described in the previous section, or by a general physical principle known as **Lenz's law:**

> The induced emf is in such a direction as to oppose, or tend to oppose, the change that produces it.

LENZ'S LAW

Note that Lenz's law does not specify just what kind of change causes the induced emf and current. The statement of Lenz's law is purposely left vague to cover a variety of conditions, which we will now illustrate.

Figure 28-8 shows a bar magnet moving toward a loop that has a resistance R. It is the motion of the bar magnetic to the right that induces an emf and current in the loop. Lenz's law tells us that this induced emf and current must be in a direction to oppose the motion of the bar magnet. That is, the current induced in the loop produces a magnetic field of its own, and this magnetic field must exert a force to the left on the approaching bar magnet. Figure 28-9 shows the induced magnetic moment of the current loop when the magnet is moving toward it. The loop acts like a small magnet with its north pole to the left and its south pole to the right. Since like poles repel, the induced magnetic moment of the loop repels the bar magnet; that is, it opposes its motion toward the loop. This means the direction of the induced current in the loop must be as shown in Figure 28-9.

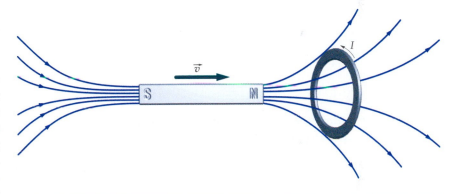

FIGURE 28-8 When the bar magnet is moving to the right, toward the loop, the emf induced in the loop produces an induced current in the direction shown. The magnetic field due to this induced current in the loop produces a magnetic field that exerts a force on the bar magnet opposing its motion to the right.

Suppose the induced current in the loop shown in Figure 28-9 was opposite to the direction shown. Then there would be a magnetic force on the approaching bar magnet to the right, causing it to gain speed. This gain in speed would cause an increase in the induced current, which in turn would cause the force on the bar magnet to increase, and so forth. This is too good to be true. Any time we nudge a bar magnetic toward a conducting loop it would move toward the loop with ever increasing speed and with no significant effort on our part. Were this to occur, it would be a violation of energy conservation. However, the reality is that energy is conserved, and the statement called Lenz's law is consistent with this reality.

FIGURE 28-9 The magnetic moment of the loop $\vec{\mu}$ (shown in outline as if it were a bar magnet) due to the induced current is such as to oppose the motion of the bar magnet. The bar magnet is moving toward the loop, so the induced magnetic moment repels the bar magnet.

An alternative statement of Lenz's law in terms of magnetic flux is frequently of use. This statement is:

> When a magnetic flux through a surface changes, the magnetic field due to any induced current produces a flux of its own—through the same surface and in opposition to the change.

ALTERNATIVE STATEMENT OF LENZ'S LAW

For an example of how this alternative statement is applied, see Example 28-5.

LENZ'S LAW AND INDUCED CURRENT   **E X A M P L E   2 8 - 5**

Using the alternative statement of Lenz's law, find the direction of the induced current in the loop shown in Figure 28-8.

PICTURE THE PROBLEM Use the alternative statement of Lenz's law to determine the direction of the magnetic field due to the current induced in the loop. Then use a right-hand rule to determine the direction of the induced current.

1. Draw a sketch of the loop bounding the flat surface S (Figure 28-10). On surface S draw the vector $\Delta \vec{B}_1$, which is the change in the magnetic field $\vec{B}_1$ of the approaching bar magnet:

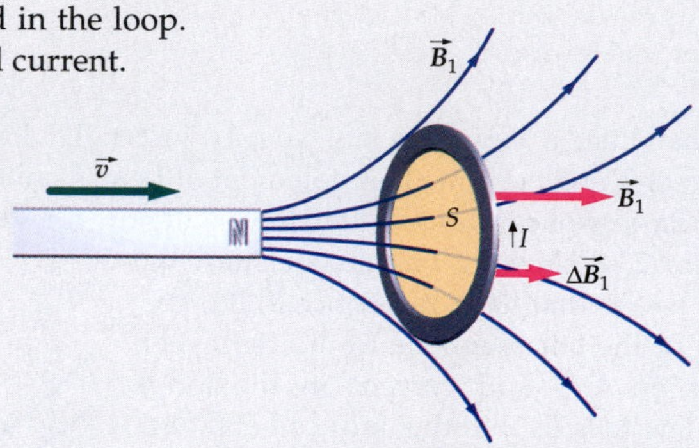

FIGURE 28-10

2. On the sketch draw the vector $\vec{B}_2$, which is the magnetic field of the current induced in the loop (Figure 28-11). Use the alternative statement of Lenz's law to determine the direction of $\vec{B}_2$:

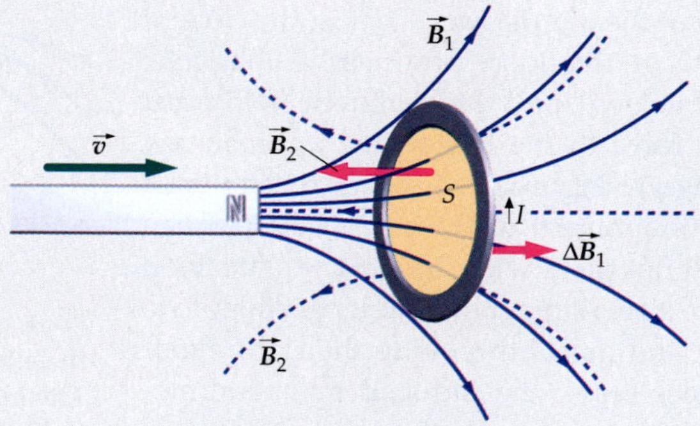

FIGURE 28-11

3. Using the right-hand rule and the direction of $\vec{B}_2$, determine the direction of the current induced in the loop (Figure 28-12):

FIGURE 28-12

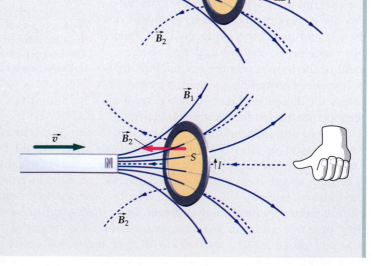

EXERCISE Using the alternative statement of Lenz's law, find the direction of the induced current in the loop shown in Figure 28-8 if the magnet is moving to the left (away from the loop). (*Answer* Opposite to the direction shown in Figure 28-12)

In Figure 28-13, the bar magnet is at rest and the loop is moving away from the magnet. The induced current and magnetic moment are shown in the figure. In this case, the bar magnet attracts the loop, thus opposing the motion of the loop as required by Lenz's law.

In Figure 28-14, when the current in circuit 1 is changing, there is a changing flux through circuit 2. Suppose that the switch S in circuit 1 is initially open so that there is no current in the circuit (Figure 28-14a). When we close the switch (Figure 28-14b), the current in circuit 1 does not reach its steady value $\mathcal{E}_1/R_1$ instantaneously but takes some time to change from zero to this value. During the time the current is increasing, the flux through circuit 2 is changing and a current is induced in circuit 2 in the direction shown. When the current in circuit 1 reaches its steady value, the flux through circuit 2 is no longer changing, so there is no longer an induced current in circuit 2. An induced current in circuit 2 in the opposite direction appears momentarily when the switch in circuit 1 is opened (Figure 28-14c) and the current in circuit 1 is decreasing to zero. It is important to understand that there is an induced emf *only while the flux is changing*. The emf does not depend on the magnitude of the flux itself, but only on its rate of change. If there is a large steady flux through a circuit, there is no induced emf.

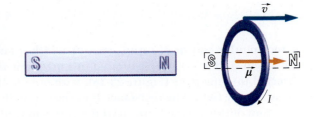

FIGURE 28-13 When the loop is moving away from the stationary bar magnet, the bar magnet attracts the magnetic moment of the loop, again opposing the relative motion.

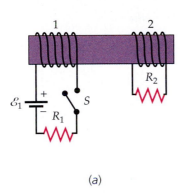

(a)

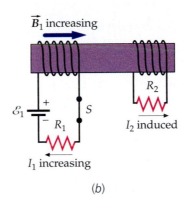

(b)

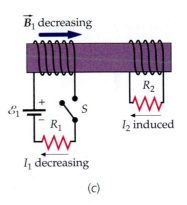

(c)

FIGURE 28-14 (*a*) Two adjacent circuits. (*b*) Just after the switch is closed, I_1 is increasing in the direction shown. The changing flux through circuit 2 induces the current I_2. The flux through circuit 2 due to I_2 opposes the change in flux due to I_1. (*c*) As the switch is opened, I_1 decreases and the flux through circuit 2 changes. The induced current I_2 then tends to maintain the flux through circuit 2.

For our next example, we consider the single isolated circuit shown in Figure 28-15. If there is a current in the circuit, there is a magnetic flux through the coil due to its own current. If the current is changing, the flux in the coil is changing and there is an induced emf in the circuit while the flux is changing. This *self-induced emf* opposes the change in the current. It is therefore called a **back emf**. Because of this self-induced emf, the current in a circuit cannot jump instantaneously from zero to some finite value or from some finite value to zero. Henry first noticed this effect when he was experimenting with a circuit consisting of many turns of a wire like that in Figure 28-15. This arrangement gives a large flux through the circuit for even a small current. Joseph Henry noticed a spark across the switch when he tried to break the circuit. Such a spark is due to the large induced emf that occurs when the current varies rapidly, as during the opening of the switch. In this case, the induced emf is directed so as to maintain the original current. The large induced emf produces a large potential difference across the switch as it is opened. The electric field between the contacts of the switch is large enough to produce dielectric breakdown in the surrounding air. When dielectric breakdown occurs, the air conducts electric current in the form of a spark.

FIGURE 28-15 The coil with many turns of wire gives a large flux for a given current in the circuit. Thus, when the current changes, there is a large emf induced in the coil opposing the change.

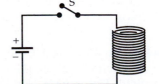

LENZ'S LAW AND A MOVING COIL   **E X A M P L E   2 8 - 6**

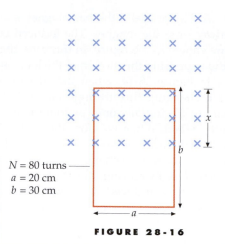

A rectangular coil of N turns, each of width a and length b; where $N = 80$, $a = 20$ cm, and $b = 30$ cm; is located in a magnetic field of magnitude $B = 0.8$ T directed into the page (Figure 28-16), with only half of the coil in the region of the magnetic field. The resistance R of the coil is 30 Ω. Find the magnitude and direction of the induced current if the coil is moved with a speed of 2 m/s (*a*) to the right, (*b*) up, and (*c*) down.

$N = 80$ turns
$a = 20$ cm
$b = 30$ cm

FIGURE 28-16

PICTURE THE PROBLEM   The induced current equals the induced emf divided by the resistance. We can calculate the emf induced in the circuit as the coil moves by calculating the rate of change of the flux through the coil. The flux is proportional to the distance x. The direction of the induced current is found from Lenz's law.

(*a*) 1. The induced current equals the emf divided by the resistance:

$$I = \frac{\mathcal{E}}{R}$$

2. The induced emf and the magnetic flux are related by Faraday's law:

$$\mathcal{E} = -\frac{d\phi_m}{dt}$$

3. The flux through the coil is N times the flux through each turn of the coil. We choose into the page as the direction of $\hat{n}$. The flux through the surface S bounded by a single turn is Bax:

$$\phi_m = N\vec{B} \cdot \hat{n}A = NBax$$

4. When the coil is moving to the right (or to the left), the flux does not change (until the coil leaves the region of magnetic field). The current is therefore zero:

$$\mathcal{E} = -\frac{d\phi_m}{dt} = 0$$

so

$$I = \boxed{0}$$

(*b*) 1. Compute the rate of change of the flux when the coil is moving up. In this case x is increasing, so dx/dt is positive:

$$\frac{d\phi_m}{dt} = \frac{d}{dt}(NBax) = NBa\frac{dx}{dt}$$

2. Calculate the magnitude of the current:

$$I = \frac{\mathcal{E}}{R} = \frac{NBa(dx/dt)}{R}$$

$$= \frac{(80)(0.8 \text{ T})(0.20 \text{ m})(2 \text{ m/s})}{30 \text{ }\Omega} = 0.853 \text{ A}$$

3. As the coil moves upward, the flux of $\vec{B}$ through S is increasing. The induced current must produce a magnetic field whose flux through S decreases as x increases. That would be a magnetic field whose dot product with $\hat{n}$ is negative. Such a magnetic field is directed out of the page on S. To produce a magnetic field in this direction the induced current must be counterclockwise:

$$\boxed{I = 0.853 \text{ A, counterclockwise}}$$

(*c*) As the coil moves downward, the flux of $\vec{B}$ through S is decreasing. The induced current must produce a magnetic field whose flux through S increases as x decreases. That would be a magnetic field whose dot product with $\hat{n}$ is positive. Such a magnetic field is directed into the page on S. To produce a magnetic field in this direction the induced current must be clockwise:

$$\boxed{I = 0.853 \text{ A, clockwise}}$$

REMARKS   In this example the magnetic field is static, so there is no nonconservative electric field. Thus, the emf is not the work done by a nonconservative electric field. This issue is examined in the next section.

28-4 Motional EMF

The emf induced in a conductor moving through a magnetic field is called **motional emf.** More generally,

> Motional emf is any emf induced by the motion of a conductor in a magnetic field.

DEFINITION—MOTIONAL EMF

TOTAL CHARGE THROUGH A FLIPPED COIL   **EXAMPLE 28-7**

A small coil of N turns has its plane perpendicular to a uniform static magnetic field $\vec{B}$, as shown in Figure 28-17. The coil is connected to a current integrator (C.I.), which is a device used to measure the total charge passing through the coil. Find the charge passing through the coil if the coil is rotated through 180° about the axis shown.

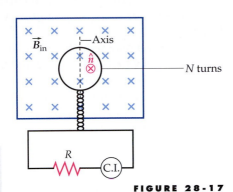

FIGURE 28-17

PICTURE THE PROBLEM When the coil in Figure 28-17 is rotated, the magnetic flux through the coil changes, causing an induced emf $\mathcal{E}$. The emf in turn causes a current $I = \mathcal{E}/R$, where R is the total resistance of the circuit. Since $I = dq/dt$, we can find the charge Q passing through the integrator by integrating I; that is, $Q = \int dq = \int I\,dt$.

1. The increment of charge dq equals the current I times the increment of time dt:

$$dq = I\,dt$$

2. The emf $\mathcal{E}$ is related to I by Ohm's law:

$$\mathcal{E} = RI$$

so

$$\mathcal{E}\,dt = RI\,dt$$

3. The emf is related to the flux ϕ_m by Faraday's law:

$$\mathcal{E} = -\frac{d\phi_m}{dt}$$

or

$$\mathcal{E}\,dt = -d\phi_m$$

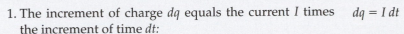

4. Substitute $-d\phi_m$ for $\mathcal{E}\,dt$ and dq for $I\,dt$ in the step 2 result and solve for dq:

$$-d\phi_m = R\,dq$$

so

Before rotation   After rotation

FIGURE 28-18

$$dq = -\frac{1}{R}d\phi_m$$

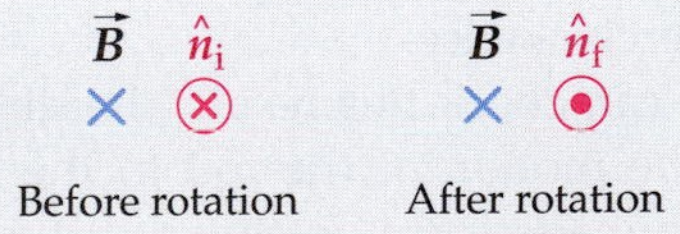

5. Integrate to find the total charge Q:

$$Q = \int_0^Q dq = -\frac{1}{R}\int_{\phi_{m,i}}^{\phi_{m,f}} d\phi_m = -\frac{1}{R}(\phi_{m,f} - \phi_{m,i}) = -\frac{\Delta\phi_m}{R}$$

6. The flux through the coil is $\phi_m = N\vec{B}\cdot\hat{n}A$, where $\hat{n}$ is the normal to the flat surface bounded by the coil (Figure 28-18). Initially, the normal is directed into the page. When the coil rotates, so does the surface and its normal. Find the change in ϕ_m when the coil rotates 180°:

$$\Delta\phi_m = \phi_{m,f} - \phi_{m,i} = N\vec{B}\cdot\hat{n}_f A - N\vec{B}\cdot\hat{n}_i A$$
$$= NA(\vec{B}\cdot\hat{n}_f - \vec{B}\cdot\hat{n}_i) = NA[(-B) - (+B)] = -2NBA$$

7. Combining the previous two results yields Q:

$$Q = \boxed{\frac{2NBA}{R}}$$

REMARKS Note that the charge Q does not depend on whether or not the coil is rotated slowly or quickly—all that matters is the change in magnetic flux through the coil. A coil used in this way is called a *flip coil*. It is used to measure magnetic fields. For example, if the current integrator (C.I.) measures a total charge Q passing through the coil when it is flipped, the magnetic field strength can be found from $B = RQ/(2NA)$.

EXERCISE A flip coil of 40 turns has a radius of 3 cm, a resistance of 16 Ω, and the plane of the coil is initially perpendicular to a static, uniform 0.50-T magnetic field. If the coil is flipped through 90°, how much charge passes through the coil? (*Answer* 3.53 mC)

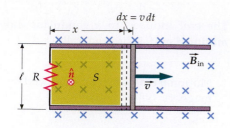

FIGURE 28-19 A conducting rod sliding on conducting rails in a magnetic field. As the rod moves to the right, the area of the surface S increases, so the magnetic flux through S into the paper increases. An emf of magnitude $B\ell v$ is induced in the circuit, inducing a counterclockwise current that produces flux through the surface S directed out of the paper opposing the change in flux due to the motion of the rod.

Figure 28-19 shows a thin conducting rod sliding to the right along conducting rails that are connected by a resistor. A uniform magnetic field $\vec{B}$ is directed into the page.

Consider the magnetic flux through the flat surface S bounded by the circuit. Let the normal $\hat{n}$ to the surface be into the page. As the rod moves to the right the surface S increases, as does the magnetic flux through the surface S. Thus, an emf is induced in the circuit. Let ℓ be the separation of the rails and x be the distance from the left end of the rails to the rod. The area of surface S is then ℓx, and the magnetic flux through S is

$$\phi_m = \vec{B} \cdot \hat{n} A = B_n A = B\ell x$$

When x increases by dx, the area of surface S increases by $dA = \ell\,dx$ and the flux ϕ_m increases by $d\phi_m = B\ell\,dx$. The rate of change of the flux is

$$\frac{d\phi_m}{dt} = B\ell\frac{dx}{dt} = B\ell v$$

where $v = dx/dt$ is the speed of the rod. The emf induced in this circuit is therefore

$$\mathcal{E} = -\frac{d\phi_m}{dt} = -B\ell v$$

where the negative sign tells us that the emf is in the negative tangential direction. Put your right thumb in the direction of $\hat{n}$ (into the page) and your fingers will curl in the positive tangential direction (clockwise). Thus, the induced emf is counterclockwise.

We can check this result (the direction of the induced emf) using Lenz's law. It is the motion of the rod to the right that produces the induced current, so the magnetic force on this rod due to the induced current must be to the left. The magnetic force on a current-carrying conductor is given by $I\vec{L} \times \vec{B}$ (Equation 26-4), where $\vec{L}$ is in the direction of the current. If $\vec{L}$ is upward the force is to the left, which affirms our previous result (that the induced emf is counterclockwise). If the rod is given some initial velocity $\vec{v}$ to the right and is then released, the force due to the induced current slows the rod until it stops. To maintain the motion of the rod, an external force pushing the rod to the right must be maintained.

A second check on the direction of the induced emf and current is implemented by considering the direction of the magnetic force on the charge carriers moving to the right with the rod. The charge carriers move rightward with the same velocity $\vec{v}$ as the rod, so the charge carriers experience a magnetic force $\vec{F} = q\vec{v} \times \vec{B}$. If q is positive this force is upward, which means the induced emf is counterclockwise.

Emf is the work per unit charge on the charge carriers, but what is the force that is doing this work in the circuit shown in Figure 28-19? It turns out this work is done by the superposition of a magnetic force and an electric force (Figure 28-20). To see how this comes about, consider that the current in the rod is upward, so the drift velocity $\vec{v}_d$ of the assumed positive charge carriers is upward. Thus, a magnetic force ($\vec{F}_L = q\vec{v}_d \times \vec{B}$) toward the left acts on the charge carriers and, as a result, the rod becomes polarized—its left side positively charged and its right side negatively charged. These surface charges produce an electric field $\vec{E}_\perp$ inside the rod toward the right, and this field exerts an electric force ($\vec{F}_R = q\vec{E}_\perp$) toward the right on the charge carriers. The sum $\vec{F}_L + \vec{E}_\perp = 0$, since the net horizontal force on the charge carriers is zero. In addition, an upward magnetic force $\vec{F}_U = q\vec{v} \times \vec{B}$, where $\vec{v}$ is the velocity of both the charge carriers and the rod to the right. The total work done by all three of these forces on a charge carrier traversing the rod is just the work done by $\vec{F}_U$, and this work is $F_U\ell = qvB\ell$. Thus, the work per unit charge is $vB\ell$, which is obtained by dividing the total work by the charge q. The magnitude of the emf equals this work.

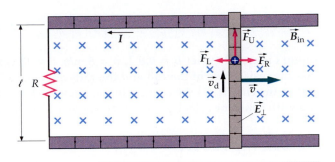

FIGURE 28-20 As a positive charge carrier moves along the moving rod, electric forces and magnetic forces act on the charge carrier. The net electromagnetic force on the charge carrier is directed upward, in the direction of the drift velocity. The work per unit charge done by this force on the charge carrier as it transverses the rod is the motional emf.

$$\mathcal{E} = vB\ell \qquad\qquad 28\text{-}6$$

MAGNITUDE OF EMF FOR A ROD MOVING PERPENDICULAR TO BOTH THE ROD AND $\vec{B}$

The magnitude of the emf is the total work per unit charge done by all three forces $\vec{F}_L$, $\vec{F}_R$, and $\vec{F}_U$. Taken together, $\vec{F}_L$ and $\vec{F}_U$ constitute the total magnetic force. The total magnetic force, however, is perpendicular to the velocity of the charge carriers and thus does no work. Therefore, the total work done by all three forces is done solely by the electric force $\vec{F}_R$.

Figure 28-21 shows a positive charge carrier in a conducting rod that is moving at constant speed through a uniform magnetic field directed into the paper. Because the charge carrier is moving horizontally with the rod, there is an upward magnetic force on the charge carrier of magnitude qvB. Responding to this force, the charge carriers in the rod move upward, producing a net positive charge at the top of the rod and leaving a net negative charge at the bottom of the rod. The charge carriers continue to move upward until the electric field $\vec{E}_\parallel$ produced by the separated charges exerts a downward force of magnitude $qE_\parallel$ on the separated charges, which balances the upward magnetic force qvB. In equilibrium, the magnitude of this electric field in the rod is

$$E_\parallel = vB$$

The direction of this electric field is parallel to the rod, directed downward. The associated potential difference across the length ℓ of the rod is

$$\Delta V = E_\parallel \ell = vB\ell$$

with the potential being higher at the top. That is, when there is no current through the rod, the potential difference across the rod equals $vB\ell$ (the motional emf). When there is a current I through the rod, the potential difference is

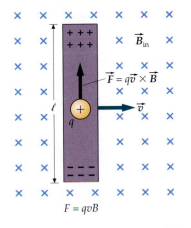

FIGURE 28-21 A positive charge carrier in a conducting rod that is moving through a magnetic field experiences a magnetic force that has an upward component. Some of these charge carriers move to the top of the rod, leaving the bottom of the rod negative. The charge separation produces a downward electric field of magnitude $E_\parallel = vB$ in the rod. Thus, the potential at the top of the rod is greater than the potential at the bottom of the rod by $E_\parallel \ell = vB\ell$.

$$\Delta V = vB\ell - Ir \qquad\qquad 28\text{-}7$$

POTENTIAL DIFFERENCE ACROSS A MOVING ROD

where r is the resistance of the rod.

EXERCISE A rod 40-cm long moves at 12 m/s in a plane perpendicular to a magnetic field of 0.30 T. The rod's velocity is perpendicular to its length. Find the emf induced in the rod. (*Answer* 1.44 V)

A U-Shaped Conductor and a Sliding Rod   **E X A M P L E   2 8 - 8**   **Try It Yourself**

Using Figure 28-19, let $B = 0.6$ T, $v = 8$ m/s, $\ell = 15$ cm, and $R = 25\ \Omega$; assume that the resistances of the rod and the rails are negligible. Find (*a*) the induced emf in the circuit, (*b*) the current in the circuit, (*c*) the force needed to move the rod with constant velocity, and (*d*) the power dissipated in the resistor.

PICTURE THE PROBLEM

Cover the column to the right and try these on your own before looking at the answers.

Steps	Answers
1. Calculate the induced emf from Equation 28-6.	$\mathcal{E} = Bv\ell =$ $\boxed{0.720\text{ V}}$
2. Find the current from Ohm's law.	$I = \dfrac{\mathcal{E}}{R} =$ $\boxed{28.8\text{ mA}}$
3. The force needed to move the rod with constant velocity is equal and opposite to the force exerted by the magnetic field on the rod, which has the magnitude $I\ell B$ (Equation 26-4). Calculate the magnitude of this force.	$F = IB\ell =$ $\boxed{2.59\text{ mN}}$
4. Find the power dissipated in the resistor.	$P = I^2R =$ $\boxed{20.7\text{ mW}}$

🔋 **PLAUSIBILITY CHECK** Using $P = Fv$, we confirm that the power is 20.7 mW.

REMARKS The potential at the top of the rod is greater than the potential at the bottom of the rod by the emf.

Magnetic Drag   **E X A M P L E   2 8 - 9**

A rod of mass m slides on frictionless conducting rails in a region of static uniform magnetic field $\vec{B}$ directed into the page (Figure 28-22). An external agent is pushing the rod, maintaining its motion to the right at constant speed v_0. At time $t = 0$, the agent abruptly stops pushing and the rod continues forward, being slowed by the magnetic force. Find the speed v of the rod as a function of time.

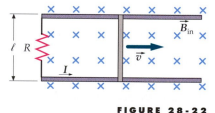

FIGURE 28-22

PICTURE THE PROBLEM The speed of the rod changes because a magnetic force acts on the induced current. The motion of the rod through a magnetic field induces an emf $\mathcal{E} = B\ell v$ and, therefore, a current in the rod, $I = \mathcal{E}/R$. This causes a magnetic force to act on the rod, $F = IB\ell$. With the force known, we apply Newton's second law to find the speed as a function of time. Take the positive x direction as being to the right.

1. Apply Newton's second law to the rod:	$F_x = ma_x = m\dfrac{dv}{dt}$
2. The force exerted on the rod is the magnetic force, which is proportional to the current and in the negative x direction, as shown in Figure 28-22:	$F_x = -IB\ell$
3. The current equals the motional emf divided by the resistance of the rod:	$I = \dfrac{\mathcal{E}}{R} = \dfrac{B\ell v}{R}$
4. Combining these results, we find the magnitude of the magnetic force exerted on the rod:	$F_x = -IB\ell = -\dfrac{B\ell v}{R}B\ell = -\dfrac{B^2\ell^2 v}{R}$
5. Newton's second law then gives:	$-\dfrac{B^2\ell^2 v}{R} = m\dfrac{dv}{dt}$

6. Separate the variables, then integrate the velocity from v_0 to v_f and integrate the time from 0 to t_f:

$$\frac{dv}{v} = -\frac{B^2\ell^2}{mR} dt$$

$$\int_{v_0}^{v_f} \frac{dv}{v} = -\frac{B^2\ell^2}{mR} \int_0^{t_f} dt$$

$$\ln\frac{v_f}{v_0} = -\frac{B^2\ell^2}{mR} t_f$$

7. Let $v = v_f$ and $t = t_f$, then solve for v:

$$v = \boxed{v_0 e^{-t/\tau}, \text{ where } \tau = \frac{mR}{B^2\ell^2}}$$

REMARKS If the force were constant, the rod's speed would decrease linearly with time. However, because the force is proportional to the rod's speed, as found in step 4, the force is large initially but the force decreases as the speed decreases. In principle, the rod never stops moving. Even so, the rod travels only a finite distance. (See Problem 37.)

The general equation for motional emf is

$$\mathcal{E} = \oint_C (\vec{v} \times \vec{B}) \cdot d\vec{\ell} = -\frac{d\phi_m}{dt} \qquad \text{28-8}$$

GENERAL EQUATION FOR MOTIONAL EMF

where $\vec{v}$ is the velocity of the wire at the element $d\vec{\ell}$. The integral is taken at an instant in time.

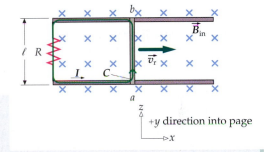

FIGURE 28-23 The positive x, y, and z directions are to the right, into the page, and up the page respectively. The rod moves to the right with the velocity $\vec{v}_r$, and there is a uniform static magnetic field directed into the page.

VERIFYING $\mathcal{E} = vB\ell$ **EXAMPLE 28-10**

Integrate $\mathcal{E} = \oint_C(\vec{v} \times \vec{B}) \cdot d\vec{\ell}$ to show that the emf in the circuit in Figure 28-23 is given by Equation 28-6.

PICTURE THE PROBLEM The circuit C can be divided into two parts: part C_1, which is moving, and part C_2, which is stationary.

1. Divide the circuit into two parts, C_1 and C_2.
   On C_1, $\vec{v} = \vec{v}_r$ and on C_2, $\vec{v} = 0$:

$$\oint_C (\vec{v} \times \vec{B}) \cdot d\vec{\ell} = \int_{a\,C_1}^{b} (\vec{v} \times \vec{B}) \cdot d\vec{\ell} + \int_{b\,C_2}^{a} (\vec{v} \times \vec{B}) \cdot d\vec{\ell}$$

$$= \int_{a\,C_1}^{b} (\vec{v}_r \times \vec{B}) \cdot d\vec{\ell} + 0$$

2. Evaluate $(\vec{v}_r \times \vec{B}) \cdot d\vec{\ell}$ on C_1:

$$\vec{v}_r \times \vec{B} = v_r\hat{i} \times B\hat{j} = v_r B\,\hat{k}$$

and

$$d\vec{\ell} = d\ell\,\hat{k}$$

so

$$(\vec{v}_r \times \vec{B}) \cdot d\vec{\ell} = v_r B\hat{k} \cdot d\ell\,\hat{k} = v_r B\,d\ell$$

3. Evaluate the integral and find the emf:

$$\mathcal{E} = \int_{a\,C_1}^{b} (\vec{v}_r \times \vec{B}) \cdot d\vec{\ell} = \int_{a\,C_1}^{b} v_r B\,d\ell = v_r B \int_{a\,C_1}^{b} d\ell$$

$$= \boxed{v_r B\ell}$$

28-5 Eddy Currents

In the examples we have discussed, currents were induced in thin wires or rods. Often a changing flux sets up circulating currents, which are called *eddy currents,* in a piece of bulk metal like the core of a transformer. The heat produced by such current constitutes a power loss in the transformer. Consider a conducting slab between the pole faces of an electromagnet (Figure 28-24). If the magnetic field $\vec{B}$ between the pole faces is changing with time (as it will if the current in the magnet windings is alternating current), the flux through any closed loop in the slab, such as through the curve C indicated in the figure, will be changing. Since path C is in a conductor, there will be an induced emf around C.

The existence of eddy currents can be demonstrated by pulling a copper or aluminum sheet between the poles of a strong permanent magnet (Figure 28-25). Part of the area enclosed by curve C in this figure is in the magnetic field, and part of the area enclosed by curve C is outside the magnetic field. As the sheet is pulled to the right, the flux through this curve decreases (assuming that into the paper is the positive normal direction). A clockwise emf is induced around this curve. This emf drives a current that is directed upward in the region between the pole faces, and the magnetic field exerts a force on this current to the left opposing motion of the sheet. You can feel this drag force on the sheet if you pull a conducting sheet rapidly through a strong magnetic field.

Eddy currents are usually unwanted because power is lost due to joule heating by the current, and this dissipated energy must be transferred to the environment. The power loss can be reduced by increasing the resistance of the possible paths for the eddy currents, as shown in Figure 28-26*a.* Here the conducting slab is laminated; that is, the conducting slab is made up of small strips glued together. Because insulating glue separates the strips, the eddy currents are essentially confined to the strips. The large eddy-current loops are broken up, and the power loss is greatly reduced. Similarly, if there are cuts in the sheet, as shown in Figure 28-26*b,* the eddy currents are lessened and the magnetic force is greatly reduced.

Eddy currents are not always undesirable. For example, eddy currents are often used to damp unwanted oscillations. With no damping present, sensitive mechanical balance scales that are used to weigh small masses might oscillate back and forth around their equilibrium reading many times. Such scales are usually designed so that a small sheet of aluminum (or some other metal) moves between the poles of a magnet as the scales oscillate. The resulting eddy currents dampen the oscillations so that equilibrium is quickly reached. Eddy currents also play a role in the magnetic braking systems of some rapid transit cars. A large electromagnet is positioned in the vehicle over the rails. If the magnet is energized by a current in its windings, eddy currents are induced in the rails by the motion of the magnet and the magnetic forces provide a drag force on the magnet that slows the car.

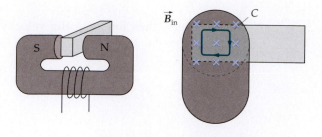

FIGURE 28-24 Eddy currents. When the magnetic field through a metal slab is changing, and emf is induced in any closed loop in the metal, such as loop C. The induced emfs drive currents, which are called eddy currents.

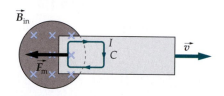

FIGURE 28-25 Demonstration of eddy currents. When the metal sheet is pulled to the right, there is a magnetic force to the left on the induced current opposing the motion.

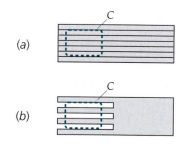

(a)

(b)

FIGURE 28-26 Disrupting the conduction paths in the metal slab can reduce the eddy current. (*a*) If the slab is constructed from strips of metal glued together, the insulating glue between the slabs increases the resistance of the closed loop C. (*b*) Slots cut into the metal slab also reduce the eddy current.

28-6 Inductance

Self-Inductance

The magnetic flux through a circuit is related to the current in that circuit and the currents in other nearby circuits.[†] Consider a coil carrying a current I. The current in the coil produces a magnetic field $\vec{B}$ that varies from point to point, but the value of $\vec{B}$ at each point is proportional to I. The magnetic flux through the coil is therefore also proportional to I:

---

† We are assuming that there are no permanent magnets around.

$$\phi_m = LI \qquad\qquad 28\text{-}9$$

DEFINITION—SELF-INDUCTANCE

where L, the proportionality constant, is called the self-inductance of the coil. The self-inductance depends on the geometric shape of the coil. The SI unit of inductance is the **henry** (H). From Equation 28-9, we can see that the unit of inductance equals the unit of flux divided by the unit of current:

$$1\,\text{H} = 1\,\frac{\text{Wb}}{\text{A}} = 1\,\frac{\text{T·m}^2}{\text{A}}$$

In principle, the self-inductance of any coil or circuit can be calculated by assuming a current I, calculating $\vec{B}$ at every point on a surface bounded by the coil, calculating the flux ϕ_m, and using $L = \phi_m/I$. In actual practice, the calculation is often very challenging. However, the self-inductance of a long, tightly wound solenoid can be calculated directly. The magnetic flux through a solenoid of length ℓ and N turns carrying a current I was calculated in Example 28-1:

$$\phi_m = \frac{\mu_0 N^2 I A}{\ell} = \mu_0 n^2 I A \ell \qquad\qquad 28\text{-}10$$

where $n = N/\ell$ is the number of turns per unit length. As expected, the flux is proportional to the current I. The proportionality constant is the self-inductance:

$$L = \frac{\phi_m}{I} = \mu_0 n^2 A \ell \qquad\qquad 28\text{-}11$$

SELF-INDUCTANCE OF A SOLENOID

The self-inductance of a solenoid is proportional to the square of the number of turns per unit length n and to the volume $A\ell$. Thus, like capacitance, self-inductance depends only on geometric factors.[†] From the dimensions of Equation 28-11, we can see that μ_0 can be expressed in henrys per meter:

$$\mu_0 = 4\pi \times 10^{-7}\,\text{H/m}$$

† If the inductor has an iron core, the self-inductance also depends on properties of the core.

SELF-INDUCTANCE OF A SOLENOID **EXAMPLE 28-11**

Find the self-inductance of a solenoid of length 10 cm, area 5 cm², and 100 turns.

PICTURE THE PROBLEM We can calculate the self-inductance in henrys from Equation 28-11.

1. L is given by Equation 28-11:

$$L = \mu_0 n^2 A \ell$$

2. Convert the given quantities to SI units:

$$\ell = 10\,\text{cm} = 0.1\,\text{m}$$

$$A = 5\,\text{cm}^2 = 5 \times 10^{-4}\,\text{m}^2$$

$$n = N/\ell = (100\,\text{turns})/(0.1\,\text{m}) = 1000\,\text{turns/m}$$

$$\mu_0 = 4\pi \times 10^{-7}\,\text{H/m}$$

3. Substitute the given quantities:

$$L = \mu_0 n^2 A \ell$$

$$= (4\pi \times 10^{-7}\,\text{H/m})(10^3\,\text{turns/m})^2(5 \times 10^{-4}\,\text{m}^2)(0.1\,\text{m})$$

$$= \boxed{6.28 \times 10^{-5}\,\text{H}}$$

When the current in a circuit is changing, the magnetic flux due to the current is also changing, so an emf is induced in the circuit. Because the self-inductance of a circuit is constant, the change in flux is related to the change in current by

$$\frac{d\phi_m}{dt} = \frac{d(LI)}{dt} = L\frac{dI}{dt}$$

According to Faraday's law, we have

$$\mathcal{E} = -\frac{d\phi_m}{dt} = -L\frac{dI}{dt} \qquad\qquad 28\text{-}12$$

Thus, the self-induced emf is proportional to the rate of change of the current. A coil or solenoid with many turns has a large self-inductance and is called an **inductor.** In circuits, it is denoted by the symbol ⌒⌒⌒. Typically, we can neglect the self-inductance of the rest of the circuit compared with that of an inductor. The potential difference across an inductor is given by

$$\Delta V = \mathcal{E} - Ir = -L\frac{dI}{dt} - Ir \qquad\qquad 28\text{-}13$$

POTENTIAL DIFFERENCE ACROSS AN INDUCTOR

where r is the internal resistance of the inductor.[†] For an ideal inductor, $r = 0$.

EXERCISE At what rate must the current in the solenoid of Example 28-11 change to induce a back emf of 20 V? (*Answer* 3.18×10^5 A/s)

Mutual Inductance

When two or more circuits are close to each other, as in Figure 28-27, the magnetic flux through one circuit depends not only on the current in that circuit but also on the current in the nearby circuits. Let I_1 be the current in circuit 1, on the left in Figure 28-27, and let I_2 be the current in circuit 2, on the right in Figure 28-27. The magnetic field $\vec{B}$ at surface S_2 is the superposition of $\vec{B}_1$ due to I_1, and $\vec{B}_2$ due to I_2, where $\vec{B}_1$ is proportional to I_1 (and $\vec{B}_2$ is proportional to I_2). We can therefore write the flux of $\vec{B}_1$ through circuit 2, $\phi_{m2,1}$ as:

$$\phi_{m2,1} = M_{2,1}I_1 \qquad\qquad 28\text{-}14a$$

DEFINITION—MUTUAL INDUCTANCE

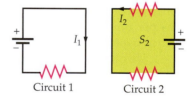

FIGURE 28-27 Two adjacent circuits. The magnetic field on S_2 is partly due to current I_1 and partly due to current I_2. The flux through the magnetic field is the sum of two terms, one proportional to I_1 and the other to I_2.

where $M_{2,1}$ is called the **mutual inductance** of the two circuits. The mutual inductance depends on the geometrical arrangement of the two circuits. For instance, if the circuits are far apart, the flux of $\vec{B}_1$ through circuit 2 will be small and the mutual inductance will be small. (The net flux ϕ_{m2} of $\vec{B} = \vec{B}_1 + \vec{B}_2$ through circuit 2 is given by $\phi_{m2} = \phi_{m2,2} + \phi_{m2,1}$.) An equation similar to Equation 28-14a can be written for the flux of $\vec{B}_2$ through circuit 1:

$$\phi_{m1,2} = M_{1,2}I_2 \qquad\qquad 28\text{-}14b$$

We can calculate the mutual inductance for two tightly wound concentric solenoids like the solenoids shown in Figure 28-28. Let ℓ be the length of both solenoids, and let the inner solenoid have N_1 turns and radius r_1 and the outer

---

† If the inductor has an iron core, the internal resistance includes properties of the core.

solenoid have N_2 turns and radius r_2. We will first calculate the mutual inductance $M_{2,1}$ by assuming that the inner solenoid carries a current I_1 and finding the magnetic flux ϕ_{m2} due to this current through the outer solenoid.

The magnetic field $\vec{B}_1$ due to the current in the inner solenoid is constant in the space within the inner solenoid and has magnitude

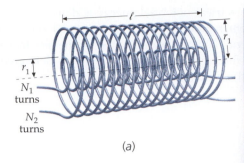

$$B_1 = \mu_0(N_1/\ell)I_1 = \mu_0 n_1 I_1, \quad r < r_1 \qquad 28\text{-}15$$

and outside the inner solenoid this magnetic field B_1 is negligible. The flux of $\vec{B}_1$ through the outer solenoid is therefore

$$\phi_{m2} = N_2 B_1(\pi r_1^2) = n_2 \ell B_1(\pi r_1^2) = \mu_0 n_2 n_1 \ell(\pi r_1^2)I_1$$

Note that the area used to compute the flux through the outer solenoid is not the area of that solenoid, πr_2^2, but rather is the area of the inner solenoid, πr_1^2, because the magnetic field due to the inner solenoid is zero outside the inner solenoid. The mutual inductance $M_{1,2}$ is thus

$$M_{2,1} = \frac{\phi_{m2,1}}{I_1} = \mu_0 n_2 n_1 \ell \pi r_1^2 \qquad 28\text{-}16$$

EXERCISE Calculate the mutual inductance $M_{1,2}$ of the concentric solenoids of Figure 28-28 by finding the flux through the inner solenoid due to a current I_2 in the outer solenoid. (*Answer* $M_{1,2} = M_{2,1} = \mu_0 n_2 n_1 \ell \pi r_1^2$)

Note from the exercise that $M_{1,2} = M_{2,1}$. It can be shown that this is a general result. We will therefore drop the subscripts for mutual inductance and simply write M.

(b)

FIGURE 28-28 (*a*) A long narrow solenoid inside a second solenoid of the same length. A current in either solenoid produces magnetic flux in the other. (*b*) A tesla coil illustrating the geometry of the wires in Figure 28-28*a*. Such a device functions as a transformer.[†] Here, low-voltage alternating current in the outer winding is transformed into a higher-voltage alternating current in the inner winding. The emf induced in the inner coil by the field of the charging current in the outer coil is high enough to light the bulb above the coils.

28-7 Magnetic Energy

An inductor stores magnetic energy, just as a capacitor stores electrical energy. Consider the circuit shown in Figure 28-29, which consists of an inductance L and a resistance R in series with a battery of emf $\mathcal{E}_0$ and a switch S. We assume that R and L are the resistance and inductance of the entire circuit. The switch is initially open, so there is no current in the circuit. A short time after the switch is closed there is a current I in the circuit, a potential difference $-IR$ across the resistor, and a potential difference $-L\, dI/dt$ across the inductor. (For an inductor with negligible resistance, the difference in potential across the inductor equals the back emf, which was given in Equation 28-12.) Applying Kirchhoff's loop rule to this circuit gives

$$\mathcal{E}_0 - IR - L\frac{dI}{dt} = 0 \qquad 28\text{-}17$$

If we multiply each term by the current I and rearrange, we obtain

$$\mathcal{E}_0 I = I^2 R + LI\frac{dI}{dt} \qquad 28\text{-}18$$

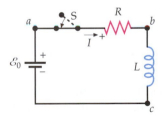

FIGURE 28-29 Just after the switch S is closed in this circuit, the current begins to increase and a back emf of magnitude $L\, dI/dt$ is induced in the inductor. The potential drop across the resistor IR plus the potential drop across the inductor $L\, dI/dt$ equals the emf of the battery.

---

† The transformer is discussed in Chapter 29.

The term $\mathcal{E}_0 I$ is the rate at which electrical potential energy is delivered by the battery. The term $I^2 R$ is the rate at which potential energy is delivered to the resistor. (It is also the rate at which potential energy is dissipated by the resistance in the circuit.) The term $LI \, dI/dt$ is the rate at which potential energy is delivered to the inductor. If U_m is the energy in the inductor, then

$$\frac{dU_m}{dt} = LI \frac{dI}{dt}$$

which implies

$$dU_m = LI \, dI$$

Integrating this equation from time $t = 0$, when the current is zero, to $t = \infty$, when the current has reached its final value I_f, we obtain

$$U_m = \int dU_m = \int_0^{I_f} LI \, dI = \frac{1}{2} LI_f^2$$

The energy stored in an inductor carrying a current I is thus given by

$$U_m = \frac{1}{2} LI^2 \qquad\qquad 28\text{-}19$$

ENERGY STORED IN AN INDUCTOR

When a current is produced in an inductor, a magnetic field is created in the space within the inductor coil. We can think of the energy stored in an inductor as energy stored in the magnetic field. For the special case of a long solenoid, the magnetic field is related to the current I and the number of turns per unit length n by

$$B = \mu_0 nI$$

and the self-inductance is given by Equation 28-11:

$$L = \mu_0 n^2 A\ell$$

where A is the cross-sectional area and ℓ is the length. Substituting $B/(\mu_0 n)$ for I and $\mu_0 n^2 A\ell$ for L in Equation 28-19, we obtain

$$U_m = \frac{1}{2} LI^2 = \frac{1}{2} \mu_0 n^2 A\ell \left(\frac{B}{\mu_0 n} \right)^2 = \frac{B^2}{2\mu_0} A\ell$$

The quantity $A\ell$ is the volume of the space within the solenoid containing the magnetic field. The energy per unit volume is the **magnetic energy density** u_m:

$$u_m = \frac{B^2}{2\mu_0} \qquad\qquad 28\text{-}20$$

MAGNETIC ENERGY DENSITY

Although we derived this by considering the special case of the magnetic field in a long solenoid, it is a general result. Whenever there is a magnetic field in space, the magnetic energy per unit volume is given by Equation 28-20. Note the similarity to the energy density in an electric field (Equation 24-13):

$$u_e = \frac{1}{2} \epsilon_0 E^2$$

ELECTROMAGNETIC ENERGY DENSITY   **EXAMPLE 28-12**

A certain region of space contains a uniform magnetic field of 0.020 T and a uniform electric field of 2.5×10^6 N/C. Find (a) the total electromagnetic energy density and (b) the energy in a cubical box of edge length $\ell = 12$ cm.

PICTURE THE PROBLEM   The total energy density u is the sum of the electrical and magnetic energy densities, $u = u_e + u_m$. The energy in a volume $\mathcal{V}$ is given by $U = u\mathcal{V}$.

(a) 1. Calculate the electrical energy density:

$$u_e = \frac{1}{2}\epsilon_0 E^2$$

$$= \frac{1}{2}(8.85 \times 10^{-12}\,\text{C}^2/\text{N·m}^2)(2.5 \times 10^6\,\text{N/C})^2$$

$$= 27.7\,\text{J/m}^3$$

2. Calculate the magnetic energy density:

$$u_m = \frac{B^2}{2\mu_0} = \frac{(0.02\,\text{T})^2}{2(4\pi \times 10^{-7}\,\text{N/A}^2)} = 159\,\text{J/m}^3$$

3. The total energy density is the sum of the above two contributions:

$$u = u_e + u_m = 27.7\,\text{J/m}^3 + 159\,\text{J/m}^3 = \boxed{187\,\text{J/m}^3}$$

(b) The total energy in the box is $U = u\mathcal{V}$, where $\mathcal{V} = \ell^3$ is the volume of the box:

$$U = u\mathcal{V} = u\ell^3 = (187\,\text{J/m}^3)(0.12\,\text{m})^3 = \boxed{0.323\,\text{J}}$$

*28-8 RL Circuits

A circuit containing a resistor and an inductor, such as that shown in Figure 28-29, is called an **RL circuit.** Because all circuits have resistance and self-inductance at room temperature, the analysis of an RL circuit can be applied to some extent to all circuits.[†]

For the circuit shown in Figure 28-29, application of Kirchhoff's loop rule (Equation 28-17) gave us

$$\mathcal{E}_0 - IR - L\frac{dI}{dt} = 0$$

Let us look at some general features of the current before we solve this equation. Just after we close the switch in the circuit the current is still zero, so IR is zero, and $L\,dI/dt$ equals the emf of the battery, $\mathcal{E}_0$. Setting $I = 0$ in Equation 28-17, we get

$$\left.\frac{dI}{dt}\right|_{I=0} = \frac{\mathcal{E}_0}{L} \qquad\qquad 28\text{-}21$$

As the current increases IR increases, and dI/dt decreases. Note that the current cannot abruptly jump from zero to some finite value as it would if there were no inductance. When the inductance L is not negligible dI/dt is finite, and therefore the current must be continuous in time. After a short time, the current has reached a positive value I, and the rate of change of the current is

$$\frac{dI}{dt} = \frac{\mathcal{E}_0}{L} - \frac{IR}{L}$$

[†] All circuits also have some capacitance between parts of the circuits at different potentials. We will consider the effects of capacitance in Chapter 29 when we study ac circuits. Here we will neglect capacitance to simplify the analysis and to focus on the effects of inductance.

At this time the current is still increasing, but its rate of increase is less than at $t = 0$. The final value of the current can be obtained by setting dI/dt equal to zero:

$$I_f = \frac{\mathcal{E}_0}{R} \qquad\qquad 28\text{-}22$$

Figure 28-30 shows the current in this circuit as a function of time. This figure is the same as that for the charge on a capacitor as a function of time when the capacitor is charged in an *RC* circuit (Figure 25-41).

Equation 28-17 is of the same form as Equation 25-36 for the charging of a capacitor and can be solved in the same way—by separating variables and integrating. The result is

$$I = \frac{\mathcal{E}_0}{R}(1 - e^{-(R/L)t}) = I_f(1 - e^{-t/\tau}) \qquad\qquad 28\text{-}23$$

where $I_f = \mathcal{E}_0/R$ is the current as $t \rightarrow \infty$, and

$$\tau = \frac{L}{R} \qquad\qquad 28\text{-}24$$

is the **time constant** of the circuit. The larger the self-inductance L or the smaller the resistance R, the longer it takes for the current to reach any specified fraction of its final current I_f.

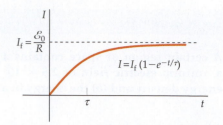

FIGURE 28-30 Current versus time in an *RL* circuit. At a time $t = \tau = L/R$, the current is at 63 percent of its maximum value $\mathcal{E}_0/R$.

ENERGIZING A COIL   **EXAMPLE 28-13**

A coil of self-inductance 5 mH and a resistance of 15 Ω is placed across the terminals of a 12-V battery of negligible internal resistance. (*a*) What is the final current? (*b*) What is the time constant? (*c*) How many time constants does it take for the current to reach 99 percent of its final value?

PICTURE THE PROBLEM The final current is the current when $dI/dt = 0$, as given in Equation 28-22. The current as a function of time is given by Equation 28-23, $I = I_f(1 - e^{-t/\tau})$, where $\tau = L/R$.

1. Use Equation 28-22 to find the final current, I_f:

$$I_f = \frac{\mathcal{E}_0}{R} = \frac{12 \text{ V}}{15 \ \Omega} = \boxed{0.800 \text{ A}}$$

2. Calculate the time constant τ.

$$\tau = \frac{L}{R} = \frac{5 \times 10^{-3} \text{ H}}{15 \ \Omega} = \boxed{333 \ \mu s}$$

3. Use Equation 28-23 and calculate the time t for $I = 0.99 I_f$:

$$I = I_f(1 - e^{-t/\tau})$$

so

$$e^{-t/\tau} = \left(1 - \frac{I}{I_f}\right)$$

and

$$-\frac{t}{\tau} = \ln\left(1 - \frac{I}{I_f}\right)$$

Thus,

$$t = -\tau \ln\left(1 - \frac{I}{I_f}\right) = -\tau \ln(1 - 0.99)$$

$$= -\tau \ln(0.01) = +\tau \ln 100 = \boxed{4.61 \ \tau}$$

REMARKS In five time constants the current is within one percent of its final value.

EXERCISE How much energy is stored in this inductor when the final current has been attained? (*Answer* $U_m = \frac{1}{2}LI_f^2 = 1.6 \times 10^{-3}\,\text{J}$)

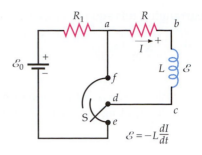

In Figure 28-31, the circuit has a make-before-break switch (shown in Figure 28-32) that allows us to remove the battery from the circuit without interrupting the current through the inductor. The resistor R_1 protects the battery so that the battery is not shorted when the switch is thrown. If the switch pole is in position e, the battery, the inductor, and the two resistors are connected in series and the current builds up in the circuit as just discussed, except that the total resistance is now $R_1 + R$ and the final current is $\mathcal{E}_0/(R + R_1)$. Suppose that the pole has been in position e for a long time, so that the current remains at its final value, which we will call I_0. At time $t = 0$ we rapidly move the pole to position f (to remove the battery from consideration completely). We now have a circuit (loop $abcda$) with just a resistor and an inductor carrying an initial current I_0. Applying Kirchoff's loop rule to this circuit gives

$$-IR - L\frac{dI}{dt} = 0$$

FIGURE 28-31 An *RL* circuit with a make-before-break switch so that the battery can be removed from the circuit without interrupting the current through the inductor. The current in the inductor reaches its maximum value with the switch pole in position e. The pole is then rapidly moved to position f.

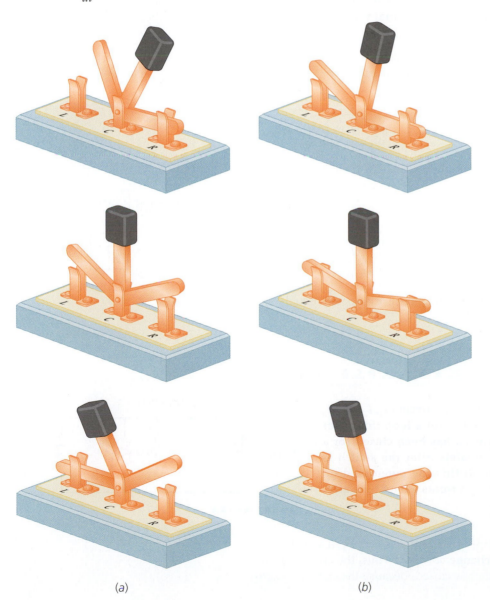

(a)   (b)

FIGURE 28-32 (*a*) The standard single-pole, double-throw switch is a break-before-make switch. That is, it breaks the first contact before making the second contact. (*b*) In a make-before-break, single-pole, double-throw switch the throw makes the second contact before breaking the first contact. With the throw in the middle position, the throw is in electrical contact with contact *L* and contact *R*.

Rearranging this equation to separate the variables I and t gives

$$\frac{dI}{I} = -\frac{R}{L} dt \qquad\qquad\qquad 28\text{-}25$$

Equation 28-25 is of the same form as Equation 25-31 for the discharge of a capacitor. Integrating and then solving for I gives

$$I = I_0 e^{-t/\tau} \qquad\qquad\qquad 28\text{-}26$$

where $\tau = L/R$ is the time constant. Figure 28-33 shows the current as a function of time.

EXERCISE What is the time constant of a circuit of resistance 85 Ω and inductance 6 mH? (*Answer* 70.6 μs)

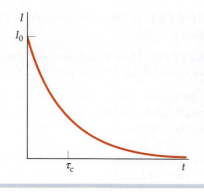

FIGURE 28-33 Current versus time for the circuit in Figure 28-31. The current decreases exponentially with time.

ENERGY DISSIPATED **EXAMPLE 28-14**

Find the total energy dissipated in the resistor R, as shown in Figure 28-31, when the current in the inductor decreases from its initial value of I_0 to 0.

PICTURE THE PROBLEM The rate of energy dissipation I^2R varies with time so to calculate the total energy dissipated requires that we integrate.

1. The rate of heat production is I^2R: $\qquad\qquad P = I^2R$

2. The total energy U dissipated in the resistor is the integral of $P\,dt$ from $t = 0$ to $t = \infty$: $\qquad\qquad U = \int_0^\infty I^2R\,dt$

3. The current I is given by Equation 28-26: $\qquad\qquad I = I_0 e^{-(R/L)t}$

4. Substitute this current into the integral: $\qquad\qquad U = \int_0^\infty I^2R\,dt = \int_0^\infty I_0^2 e^{-2(R/L)t}R\,dt = I_0^2R\int_0^\infty e^{-2(R/L)t}\,dt$

5. The integration can be done by substituting $x = 2Rt/L$: $\qquad\qquad U = I_0^2R\left.\frac{e^{-2(R/L)t}}{-2(R/L)}\right|_0^\infty = I_0^2R\frac{-L}{2R}(0 - 1) = \boxed{\frac{1}{2}LI_0^2}$

PLAUSIBILITY CHECK The total amount of energy dissipated equals the energy $\frac{1}{2}LI_0^2$ originally stored in the inductor.

INITIAL CURRENTS AND FINAL CURRENTS **EXAMPLE 28-15**

For the circuit shown in Figure 28-34, find the currents I_1, I_2, and I_3 (*a*) immediately after switch S is closed and (*b*) a long time after switch S has been closed. After the switch has been closed for a long time the switch is opened. Immediately after the switch is opened (*c*) find the three currents and (*d*) find the potential drop across the 20-Ω resistor. (*e*) Find all three currents a long time after switch S was opened.

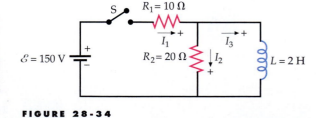

FIGURE 28-34

PICTURE THE PROBLEM (*a*) We simplify our calculations by using the fact that the current in an inductor cannot change abruptly. Thus, the current in the inductor must be zero just after the switch is closed, because the current is zero before the switch is closed. (*b*) When the current reaches its final value dI/dt equals zero, so there is no potential drop across the inductor. The inductor thus acts like a short circuit; that is, the inductor acts like a wire with zero resistance. (*c*) Immediately after the switch is opened, the current in the inductor is the same as it was before. (*d*) A long time after the switch is opened, all the currents must be zero.

(*a*) 1. The switch is just opened. The current through the inductor is zero, just as it was before the switch was closed. Apply the junction rule to relate I_1 and I_2:

$$I_3 = \boxed{0}$$

$$I_1 = I_2 + I_3$$

so

$$I_1 = I_2$$

2. The current in the left loop is obtained by applying the loop rule to the loop on the left:

$$\mathcal{E} - I_1 R_1 - I_2 R_2 = 0$$

so

$$I_1 = \frac{\mathcal{E}}{R_1 + R_2} = \frac{150\ \text{V}}{10\ \Omega + 20\ \Omega} = \boxed{5\ \text{A}} = I_2$$

(*b*) 1. After a long time, the currents are steady and the inductor acts like a short circuit, so the potential drop across R_2 is zero. Apply the loop rule to the right loop and solve for I_2:

$$-L\frac{dI_3}{dt} + I_2 R_2 = 0$$

$$0 + I_2 R_2 = 0 \quad \Rightarrow \quad I_2 = \boxed{0}$$

2. Apply the loop rule to the left loop and solve for I_1:

$$\mathcal{E} - I_1 R_1 - I_2 R_2 = 0$$

$$\mathcal{E} - I_1 R_1 - 0 = 0$$

so

$$I_1 = \frac{\mathcal{E}}{R_1} = \frac{150\ \text{V}}{10\ \Omega} = \boxed{15\ \text{A}}$$

3. Apply the junction rule and solve for I_3:

$$I_1 = I_2 + I_3$$

$$15\ \text{A} = 0 + I_3$$

so

$$I_3 = \boxed{15\ \text{A}}$$

(*c*) When the switch is reopened, I_1 *instantly* becomes zero. The current I_3 in the inductor changes continuously, so at that instant $I_3 = 15$ A. Apply the junction rule and solve for I_2:

$$I_3 = \boxed{15\ \text{A}}$$

$$I_1 = I_2 + I_3$$

so

$$I_2 = I_1 - I_3 = 0 - 15\ \text{A} = \boxed{-15\ \text{A}}$$

(*d*) Apply Ohm's law to find the potential drop across R_2:

$$V = I_2 R_2 = (15\ \text{A})(20\ \Omega) = \boxed{300\ \text{V}}$$

(*e*) A long time after the switch is opened, all the currents must equal zero.

$$I_1 = I_2 = I_3 = \boxed{0}$$

REMARKS Were you surprised to find the potential drop across R_2 in Part (*d*) to be larger than the emf of the battery? This potential drop is equal to the emf of the inductor.

EXERCISE Suppose $R_2 = 200\ \Omega$ and the switch has been closed for a long time. What is the potential drop across it immediately after the switch is then opened? (*Answer* 3000 V)

*28-9 Magnetic Properties of Superconductors

Superconductors have resistivities of zero below a critical temperature T_c, which varies from material to material. In the presence of a magnetic field $\vec{B}$, the critical temperature is lower than the critical temperature is when there is no field. As the magnetic field increases, the critical temperature decreases. If the magnetic field magnitude is greater than some critical field B_c, superconductivity does not exist at any temperature.

*Meissner Effect

As a superconductor is cooled below the critical temperature in an applied magnetic field, the magnetic field inside the superconductor becomes zero (Figure 28-35). This effect was discovered by Walter Meissner and Robert Ochsenfeld in 1933 and is now known as the **Meissner effect.** The magnetic field becomes zero because superconducting currents induced on the surface of the superconductor produce a second magnetic field that cancels out the applied one. The magnetic levitation (see the following photo) results from the repulsion between the permanent magnet producing the applied field and the magnetic field produced by the currents induced in the superconductor. Only certain superconductors, called **type I superconductors,** exhibit the complete Meissner effect. Figure 28-36a shows a plot of the magnetization M times μ_0 versus the applied magnetic field B_{app} for a type I superconductor. For a magnetic field less than the critical field B_c, the magnetic field $\mu_0 M$ induced in the superconductor is equal and opposite to the applied magnetic field. The values of B_c for type I superconductors are always too small for such materials to be useful in the coils of a superconducting magnet.

Other materials, known as **type II superconductors,** have a magnetization curve similar to that in Figure 28-36b. Such materials are usually alloys or metals that have large resistivities in the normal state. Type II superconductors exhibit the electrical properties of superconductors except for the Meissner effect up to the critical field B_{c2}, which may be several hundred times the typical values of critical fields for type I superconductors. For example, the alloy Nb_3Ge has a critical field $B_{c2} = 34$ T. Such materials can be used for high-field superconducting magnets. Below the critical field B_{c1}, the behavior of a type II superconductor is the same as that of a type I superconductor. In the region between fields B_{c1} and B_{c2}, the superconductor is said to be in a vortex state.

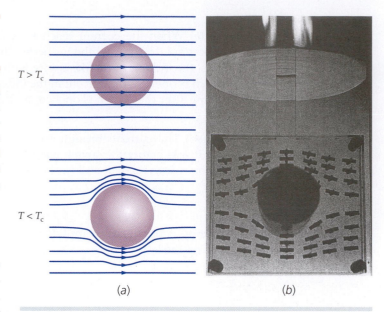

(a)   (b)

FIGURE 28-35 (a) The Meissner effect in a superconducting solid sphere cooled in a constant applied magnetic field. As the temperature drops below the critical temperature T_c, the magnetic field inside the sphere becomes zero. (b) Demonstration of the Meissner effect. A superconducting tin cylinder is situated with its axis perpendicular to a horizontal magnetic field. The directions of the field lines are indicated by weakly magnetized compass needles mounted in a Lucite sandwich so that they are free to turn.

The cube is a superconductor. The magnetic levitation results from the repulsion between the permanent magnet producing the applied field and the magnetic field produced by the currents induced in the superconductor.

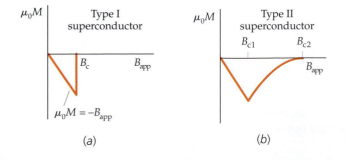

(a)   (b)

FIGURE 28-36 Plots of μ_0 times the magnetization M versus applied magnetic field for type I and type II superconductors. (a) In a type I superconductor, the resultant magnetic field is zero below a critical applied field B_c because the field due to induced currents on the surface of the superconductor exactly cancels the applied field. Above the critical field, the material is a normal conductor and the magnetization is too small to be seen on this scale. (b) In a type II superconductor, the magnetic field starts to penetrate the superconductor at a field B_{c1}, but the material remains superconducting up to a field B_{c2}, after which the material becomes a normal conductor.

*Flux Quantization

Consider a superconducting ring of area A carrying a current. There can be a magnetic flux $\phi_m = B_n A$ through the flat surface S bounded by the ring due to the current in the ring and due also perhaps to other currents external to the ring. According to Equation 28-5, if the flux through S changes, an electric field will be induced in the ring whose circulation is proportional to the rate of change of the flux. But there can be no electric field in a superconducting ring because it has no resistance, so a finite electric field would drive an infinite current. The flux through the ring is thus frozen and cannot change.

Another effect, which results from the quantum-mechanical treatment of superconductivity, is that the total flux through surface S is quantized and is given by

$$\phi_m = n\frac{h}{2e}, \quad n = 1, 2, 3, \ldots \qquad 28\text{-}27$$

The smallest unit of flux, called a **fluxon,** is

$$\phi_0 = \frac{h}{2e} = 2.0678 \times 10^{-15}\ \text{T·m}^2 \qquad 28\text{-}28$$

SUMMARY

1. Faraday's law and Lenz's law are fundamental laws of physics.
2. Self-inductance is a property of a circuit element that relates the flux through the element to the current.

Topic	Relevant Equations and Remarks	
1. Magnetic Flux ϕ_m		
General definition	$\phi_m = \int_S \vec{B} \cdot \hat{n}\, dA$	28-1
Uniform field, flat surface bounded by coil of N turns	$\phi_m = NBA \cos\theta$ where A is the area of the flat surface bounded by a single turn.	28-3
Units	$1\ \text{Wb} = 1\ \text{T·m}^2$	28-2
Due to current in a circuit	$\phi_m = LI$	28-9
Due to current in two circuits	$\phi_{m1} = L_1 I_1 + M I_2$ $\phi_{m2} = L_2 I_2 + M I_1$	28-14
*Quantization	$\phi_m = n\dfrac{h}{2e}, \quad n = 1, 2, 3, \cdots$	28-27
*Fluxon	$\phi_0 = \dfrac{h}{2e} = 2.0678 \times 10^{-15}\ \text{T·m}^2$	28-28

2. EMF

Faraday's law (includes both induction and motional emf)	$\mathcal{E} = -\dfrac{d\phi_m}{dt}$	28-4
Induction (time varying magnetic field, C stationary)	$\mathcal{E} = \displaystyle\oint_C \vec{E}\cdot d\vec{\ell}$	28-5
Motional (static magnetic field, C not stationary)	$\mathcal{E} = \displaystyle\oint_C (\vec{v} \times \vec{B})\cdot d\vec{\ell}$ where $\vec{v}$ is the velocity of the conducting path.	28-8
Rod moving perpendicular to both itself and $\vec{B}$	$\mathcal{E} = vB\ell$	28-6
Self-induced (back emf)	$\mathcal{E} = -L\dfrac{dI}{dt}$	28-12

3. Faraday's Law

$$\mathcal{E} = -\dfrac{d\phi_m}{dt} \qquad\qquad \text{28-4}$$

4. Lenz's Law

The induced emf and induced current are in such a direction as to oppose, or tend to oppose, the change that produces them.

Alternative statement	When a magnetic flux through a surface changes, the magnetic field due to any induced current produces a flux of its own—through the same surface and in opposition to the change.

5. Inductance

Self-inductance	$L = \dfrac{\phi_m}{I}$	28-9
Self-inductance of a solenoid	$L = \mu_0 n^2 A\ell$	28-11
Mutual inductance	$M = \dfrac{\phi_{m2,1}}{I_1} = \dfrac{\phi_{m1,2}}{I_2}$	28-16
Units	$1\,\text{H} = 1\dfrac{\text{Wb}}{\text{A}} = 1\dfrac{\text{T}\cdot\text{m}^2}{\text{A}}$ $\mu_0 = 4\pi \times 10^{-7}\,\text{H/m}$	

6. Magnetic Energy

Energy stored in an inductor	$U_m = \dfrac{1}{2}LI^2$	28-19
Energy density in a magnetic field	$u_m = \dfrac{B^2}{2\mu_0} = \dfrac{1}{2}\mu_0^{-1}B^2$	28-20

*7. *RL* Circuits

Potential difference across an inductor	$\Delta V = \mathcal{E} - Ir = -L\dfrac{dI}{dt} - Ir$ where r is the internal resistance of the inductor. For an ideal inductor $r = 0$.	28-13

Energizing an inductor with a battery	In a circuit consisting of a resistance R, an inductance L, and a battery of emf $\mathcal{E}_0$ in series, the current does not reach its maximum value I_f instantaneously, but rather takes some time to build up. If the current is initially zero, its value at some later time t is given by $$I = \frac{\mathcal{E}_0}{R}(1 - e^{-t/\tau}) = I_f(1 - e^{-t/\tau})$$	28-23
Time constant τ	$$\tau = \frac{L}{R}$$	28-24
De-energizing an inductor through a resistor	In a circuit consisting of a resistance R and an inductance L, the current does not drop to zero instantaneously, but rather takes some time to decrease. If the current is initially I_0, its value at some later time t is given by $$I = I_0 e^{-t/\tau}$$	28-26

PROBLEMS

- • Single-concept, single-step, relatively easy
- •• Intermediate-level, may require synthesis of concepts
- ••• Challenging
- **SSM** Solution is in the *Student Solutions Manual*
- **iSOLVE** Problems available on iSOLVE online homework service
- **iSOLVE✓** These "Checkpoint" online homework service problems ask students additional questions about their confidence level, and how they arrived at their answer.

In a few problems, you are given more data than you actually need; in a few other problems, you are required to supply data from your general knowledge, outside sources, or informed estimates.

Conceptual Problems

1 • **SSM** **iSOLVE** A conducting loop lies in the plane of this page and carries a clockwise induced current. Which of the following statements could be true? (a) A constant magnetic field is directed into the page. (b) A constant magnetic field is directed out of the page. (c) An increasing magnetic field is directed into the page. (d) A decreasing magnetic field is directed into the page. (e) A decreasing magnetic field is directed out of the page.

2 • Give the direction of the induced current in the circuit, shown on the right in Figure 28-37, when the resistance in the circuit on the left is suddenly (a) increased and (b) decreased.

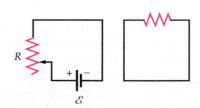

FIGURE 28-37 Problem 2

3 •• The two circular loops in Figure 28-38 have their planes parallel to each other. As viewed from the left, there is a counterclockwise current in loop A. Give the direction of the current in loop B and state whether the loops attract or repel each other if the current in loop A is (a) increasing and (b) decreasing.

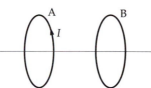

FIGURE 28-38 Problem 3

4 •• A bar magnet moves with constant velocity along the axis of a loop, as shown in Figure 28-39. (a) Make a qualitative graph of the flux ϕ_m through the loop as a function of time. Indicate the time t_1 when the magnet is halfway through the loop. (b) Sketch a graph of the current I in the loop versus time, choosing I to be positive when it is clockwise as viewed from the left.

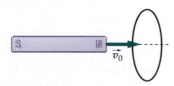

FIGURE 28-39 Problem 4

5 •• A bar magnet is mounted on the end of a coiled spring in such a way that it moves with simple harmonic motion along the axis of a loop, as shown in Figure 28-40. (a) Make a qualitative graph of the flux ϕ_m through the loop as a function of time. Indicate the time t_1 when the magnet is halfway through the loop. (b) Sketch the current I in the loop versus time, choosing I to be positive when it is clockwise as viewed from above.

FIGURE 28-40 Problem 5

6 • **SSM** If the current through an inductor were doubled, the energy stored in the inductor would be (a) the same. (b) doubled. (c) quadrupled. (d) halved. (e) quartered.

7 • Inductors in circuits that are switched on and off are often protected by being placed in parallel with a diode, as shown in Figure 28-41. (A diode is a one-way valve for current; current can flow in the direction of the arrow, but not opposite to the direction of the arrow.) Why is such protection needed? Explain how the voltage across the inductor changes with no diode protection if a switch is opened suddenly while current is flowing through the inductor.

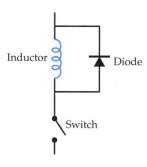

FIGURE 28-41 Problem 7

8 • Two inductors are made from identical lengths of wire wrapped around identical circular cores of the same radius. However, one inductor has three times the number of coils per unit length as the other. Which coil has the higher self-inductance? What is the ratio of the self-inductance of the two coils?

9 • True or false:

(a) The induced emf in a circuit is proportional to the magnetic flux through the circuit.
(b) There can be an induced emf at an instant when the flux through the circuit is zero.
(c) Lenz's law is related to the conservation of energy.
(d) The inductance of a solenoid is proportional to the rate of change of the current in the solenoid.
(e) The magnetic energy density at some point in space is proportional to the square of the magnetic field at that point.

10 • **SSM** A pendulum is fabricated from a thin, flat piece of aluminum. At the bottom of its arc, it passes between the poles of a strong permanent magnet. In Figure 28-42a, the metal sheet is continuous, whereas in Figure 28-42b, there are slots in it. The pendulum with slots swings back and forth many times, but the pendulum without slots comes to a stop in no more than one complete oscillation. Explain why.

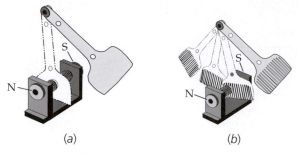

(a)   (b)

FIGURE 28-42 Problem 10

11 • A bar magnet is dropped inside a long vertical tube. If the tube is made of metal, the magnet quickly approaches a terminal speed, but if the tube is made of cardboard, the magnet does not. Explain.

12 •• An experimental setup for a jumping ring demonstration is shown in Figure 28-43. A metal ring is placed on top of a large coil, with an iron rod threading the center of the ring and the coil. When a current is suddenly started in the coil, the ring jumps several feet into the air. Explain how the demonstration works. Will the demonstration work if a slot is cut into the ring?

FIGURE 28-43 Problem 12

Estimation and Approximation

13 •• **SSM** A physics teacher attempts the following emf demonstration. She has two of her students hold a long wire connected to a voltmeter. The wire is held slack, so that there is a large arc in it. When she says "start," the students begin rotating the wire in a large vertical arc, as if they were playing jump rope. The students stand 3.0 m apart, and the sag in the wire is about 1.5 m. (You may idealize the shape of the wire as a perfect semicircular arc of diameter $d = 1.5$ m.) The induced emf from the jump rope is then measured on the voltmeter. (a) Estimate a reasonable value for the maximum angular velocity that the students can rotate the wire. (b) From this, estimate the maximum emf induced in the wire. The magnitude of the Earth's magnetic field is approximately 0.7 G. (c) Can the students rotate the jump rope fast enough to generate an emf of 1 V? (d) Suggest modifications to the demonstration that would allow higher emfs to be generated.

14 • Compare the energy density stored in the earth's electric field, which has a value $E \sim 100$ V/m at the surface of the earth, to that of the earth's magnetic field, where $B \sim 5 \times 10^{-5}$ T.

15 •• A lightning strike transfers roughly 30 C of charge from the sky to the ground in approximately 1 μs. Estimate the maximum emf induced by the lightning strike in an antenna consisting of a single loop of wire with cross-sectional area 0.1 m² a distance 300 m away from the lightning strike.

Magnetic Flux

16 • A uniform magnetic field of magnitude 2000 G is parallel to the x axis. A square coil of side 5 cm has a single turn and makes an angle θ with the z axis, as shown in Figure 28-44. Find the magnetic flux through the coil when (a) $\theta = 0°$, (b) $\theta = 30°$, (c) $\theta = 60°$, and (d) $\theta = 90°$.

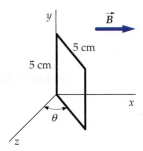

FIGURE 28-44 Problem 16

17 • **SSM** A circular coil has 25 turns and a radius of 5 cm. It is at the equator, where the earth's magnetic field is 0.7 G north. Find the magnetic flux through the coil when its plane is (a) horizontal, (b) vertical with its axis pointing north, (c) vertical with its axis pointing east, and (d) vertical with its axis making an angle of 30° with north.

18 • **ISOLVE✔** A magnetic field of 1.2 T is perpendicular to a square coil of 14 turns. The length of each side of the coil is 5 cm. (a) Find the magnetic flux through the coil. (b) Find the magnetic flux through the coil if the magnetic field makes an angle of 60° with the normal to the plane of the coil.

19 • **ISOLVE✔** A uniform magnetic field $\vec{B}$ is perpendicular to the base of a hemisphere of radius R. Calculate the magnetic flux through the spherical surface of the hemisphere.

20 •• Find the magnetic flux through a 400-turn solenoid of length 25 cm and radius 1 cm that carries a current of 3 A.

21 •• **ISOLVE** Rework Problem 20 for an 800-turn solenoid of length 30 cm and radius 2 cm that carries a current of 2 A.

22 •• A circular coil of 15 turns of radius 4 cm is in a uniform magnetic field of 4000 G in the positive x direction. Find the flux through the coil when the unit vector perpendicular to the plane of the coil is (a) $\hat{n} = \hat{i}$, (b) $\hat{n} = \hat{j}$, (c) $\hat{n} = (\hat{i} + \hat{j})/\sqrt{2}$, (d) $\hat{n} = \hat{k}$, and (e) $\hat{n} = 0.6\hat{i} + 0.8\hat{j}$.

23 •• A solenoid has n turns per unit length, radius R_1, and carries a current I. (a) A large circular loop of radius $R_2 > R_1$ and N turns encircles the solenoid at a point far away from the ends of the solenoid. Find the magnetic flux through the loop. (b) A small circular loop of N turns and radius $R_3 < R_1$ is completely inside the solenoid, far from its ends, with its axis parallel to that of the solenoid. Find the magnetic flux through this small loop.

24 •• **SSM** **ISOLVE✔** A long straight wire carries a current I. A rectangular loop with two sides parallel to the straight wire has sides a and b, with its near side a distance d from the straight wire, as shown in Figure 28-45. (a) Compute the magnetic flux through the rectangular loop. (Hint: Calculate the flux through a strip of area $dA = b\,dx$ and integrate from $x = d$ to $x = d + a$.) (b) Evaluate your answer for $a = 5$ cm, $b = 10$ cm, $d = 2$ cm, and $I = 20$ A.

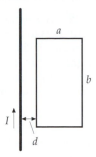

FIGURE 28-45 Problem 24

25 ••• A long cylindrical conductor of radius R carries a current I that is uniformly distributed over its cross-sectional area. Find the magnetic flux per unit length through the area indicated in Figure 28-46.

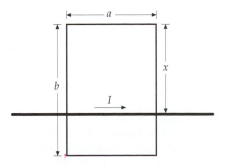

FIGURE 28-46 Problem 25

26 ••• A rectangular coil in the plane of the page has dimensions a and b. A long wire that carries a current I is placed directly above the coil (Figure 28-47). (a) Obtain an expression for the magnetic flux through the coil as a function of x for $0 \le x \le 2b$. (b) For what value of x is the flux through the coil a maximum? For what value of x is the flux a minimum?

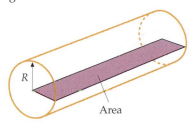

FIGURE 28-47 Problem 26

Induced EMF and Faraday's Law

27 • **SSM** A uniform magnetic field $\vec{B}$ is established perpendicular to the plane of a loop of radius 5 cm, resistance 0.4 Ω, and negligible self-inductance. The magnitude of $\vec{B}$ is increasing at a rate of 40 mT/s. Find (a) the induced emf $\mathcal{E}$ in the loop, (b) the induced current in the loop, and (c) the rate of joule heating in the loop.

28 • The flux through a loop is given by $\phi_m = (t^2 - 4t) \times 10^{-1}$ Wb, where t is in seconds. (a) Find the induced emf $\mathcal{E}$ as a function of time. (b) Find both ϕ_m and $\mathcal{E}$ at $t = 0$, $t = 2$ s, $t = 4$ s, and $t = 6$ s.

29 • (a) For the flux given in Problem 28, sketch graphs of ϕ_m and $\mathcal{E}$ versus t. (b) At what time is the flux minimum? What is the emf at this time? (c) At what times is the flux zero? What is the emf at these times?

30 • **SOLVE✓** A solenoid of length 25 cm and radius 0.8 cm with 400 turns is in an external magnetic field of 600 G that makes an angle of 50° with the axis of the solenoid. (a) Find the magnetic flux through the solenoid. (b) Find the magnitude of the emf induced in the solenoid if the external magnetic field is reduced to zero in 1.4 s.

31 •• **SSM** A 100-turn circular coil has a diameter of 2 cm and resistance of 50 Ω. The plane of the coil is perpendicular to a uniform magnetic field of magnitude 1 T. The direction of the field is suddenly reversed. (a) Find the total charge that passes through the coil. If the reversal takes 0.1 s, find (b) the average current in the coil and (c) the average emf in the coil.

32 •• **SOLVE✓** At the equator, a 1000-turn coil with a cross-sectional area of 300 cm² and a resistance of 15 Ω is aligned with its plane perpendicular to the earth's magnetic field of 0.7 G. If the coil is flipped over, how much charge flows through the coil?

33 •• **SOLVE** A circular coil of 300 turns and radius 5 cm is connected to a current integrator. The total resistance of the circuit is 20 Ω. The plane of the coil is originally aligned perpendicular to the earth's magnetic field at some point. When the coil is rotated through 90°, the charge that passes through the current integrator is measured to be 9.4 μC. Calculate the magnitude of the earth's magnetic field at that point.

34 •• The wire in Problem 26 is placed at $x = b/4$. (a) Obtain an expression for the emf induced in the coil if the current varies with time according to $I = 2t$. (b) If $a = 1.5$ m and $b = 2.5$ m, what should be the resistance of the coil so that the induced current is 0.1 A? What is the direction of this current?

35 •• Repeat Problem 34 if the wire is placed at $x = b/3$.

Motional EMF

36 • **SSM** **SOLVE** A rod 30 cm long moves at 8 m/s in a plane perpendicular to a magnetic field of 500 G. The velocity of the rod is perpendicular to its length. Find (a) the magnetic force on an electron in the rod, (b) the electrostatic field $\vec{E}$ in the rod, and (c) the potential difference V between the ends of the rod.

37 • **SOLVE** Find the speed of the rod in Problem 36 if the potential difference between the ends is 6 V.

38 • In Figure 28-22, let B be 0.8 T, $v = 10$ m/s, $\ell = 20$ cm, and $R = 2$ Ω. Find (a) the induced emf in the circuit, (b) the current in the circuit, and (c) the force needed to move the rod with constant velocity assuming negligible friction. Find (d) the power input by the force found in Part (c), and (e) the rate of joule heat production I^2R.

FIGURE 28-22
Problem 38

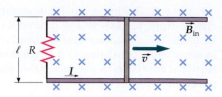

39 •• A 10-cm by 5-cm rectangular loop with resistance 2.5 Ω is pulled through a region of uniform magnetic field $B = 1.7$ T (Figure 28-48) with constant speed $v = 2.4$ cm/s. The front of the loop enters the region of the magnetic field at time $t = 0$. (a) Find and graph the flux through the loop as a function of time. (b) Find and graph the induced emf and the current in the loop as functions of time. Neglect any self-inductance of the loop and extend your graphs from $t = 0$ to $t = 16$ s.

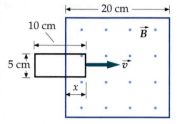

FIGURE 28-48 Problem 39

40 •• **SOLVE✓** A uniform magnetic field of magnitude 1.2 T is in the z direction. A conducting rod of length 15 cm lies parallel to the y axis and oscillates in the x direction with displacement given by $x = (2$ cm$)$ cos $120 \pi t$. What is the emf induced in the rod?

41 •• In Figure 28-49, the rod has a resistance R and the rails are horizontal and have negligible resistance. A battery of emf $\mathcal{E}$ and negligible internal resistance is connected between points a and b so that the current in the rod is downward. The rod is placed at rest at $t = 0$. (a) Find the force on the rod as a function of the speed v and write Newton's second law for the rod when it has speed v. (b) Show that the rod moves at a terminal speed and find an expression for it. (c) What is the current when the rod will approach its terminal speed?

FIGURE 28-49
Problems 41 and 44

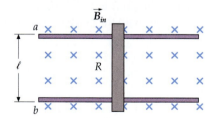

42 •• **SSM** In Example 28-9, find the total energy dissipated in the resistance and show that it is equal to mv_0^2.

43 •• Find the total distance traveled by the rod in Example 28-9.

44 •• In Figure 28-49, the rod has a resistance R and the rails have negligible resistance. A capacitor with charge Q_0 and capacitance C is connected between points a and b so that the current in the rod is downward. The rod is placed at rest at $t = 0$. (a) Write the equation of motion for the rod on the rails. (b) Show that the terminal speed of the rod down the rails is related to the final charge on the capacitor.

45 •• **SSM** In Figure 28-50, a conducting rod of mass m and negligible resistance is free to slide without friction along two parallel rails of negligible resistance separated by a distance ℓ and connected by a resistance R. The rails are attached to a long inclined plane that makes an angle θ with the horizontal. There is a magnetic field B directed upward. (a) Show that there is a retarding force directed up the incline given by $F = (B^2\ell^2 v \cos^2 \theta)/R$. (b) Show that the terminal speed of the rod is $v_t = (mgR \sin \theta)/(B^2\ell^2 \cos^2 \theta)$.

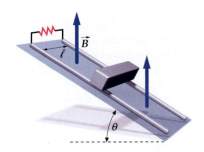

FIGURE 28-50 Problems 45 and 49

46 •• A square loop of a conducting wire (area A) is pulled out of a region of constant, very high magnetic field B that is directed perpendicular to the plane of the wire. Half of the wire is in the field and half of the wire is out of the field when the wire is pulled out. A constant force F is exerted on the wire to pull the wire out. The wire is pulled out in time t. All else being equal, if the force were doubled, approximately how long would it take to pull the wire out? (a) t (b) $t/\sqrt{2}$ (c) $t/2$ (d) $t/4$

47 •• If instead of doubling the force the resistance of the wire in Problem 46 were halved (all else being equal), what would the new time be? (a) t (b) $2t$ (c) $t/2$ (d) $t\sqrt{2}$

48 •• A wire lies along the z axis and carries current $I = 20$ A in the positive z direction. A small conducting sphere of radius $R = 2$ cm is initially at rest on the y axis at a distance $h = 45$ m above the wire. The sphere is dropped at time $t = 0$. (a) What is the electric field at the center of the sphere at $t = 3$ s? Assume that the only magnetic field is the magnetic field produced by the wire. (b) What is the voltage across the sphere at $t = 3$ s?

49 •• **SOLVE** In Figure 28-50, let $\theta = 30°$; $m = 0.4$ kg, $\ell = 15$ m, and $R = 2\ \Omega$. The rod starts from rest at the top of the inclined plane at $t = 0$. The rails have negligible resistance. There is a constant, vertically directed magnetic field of magnitude $B = 1.2$ T. (a) Find the emf induced in the rod as a function of its velocity down the rails. (b) Write Newton's law of motion for the rod; show that the rod will approach a terminal speed and determine its value.

50 ••• A solid conducting cylinder of radius 0.1 m and mass 4 kg rests on horizontal conducting rails (Figure 28-51). The rails, separated by a distance $a = 0.4$ m, have a rough surface, so the cylinder rolls rather than slides. A 12-V battery is connected to the rails as shown. The only significant resistance in the circuit is the contact resistance of 6 Ω between the cylinder and rails. The system is in a uniform vertical magnetic field. The cylinder is initially at rest next to the battery. (a) What must be the magnitude and the direction of $\vec{B}$ so that the cylinder has an initial acceleration of 0.1 m/s² to the right? (b) Find the force on the cylinder as a function of its speed v. (c) Find the terminal velocity of the cylinder. (d) What is the kinetic energy of the cylinder when it has reached its terminal velocity? (Neglect the magnetic field due to the current in

the battery–rails–cylinder loop, and assume that the current density in the cylinder is uniform.)

FIGURE 28-51 Problem 50

51 ••• **SSM** The loop in Problem 24 moves away from the wire with a constant speed v. At time $t = 0$, the left side of the loop is a distance d from the long straight wire. (a) Compute the emf in the loop by computing the motional emf in each segment of the loop that is parallel to the long wire. Explain why you can neglect the emf in the segments that are perpendicular to the wire. (b) Compute the emf in the loop by first computing the flux through the loop as a function of time and then using $\mathcal{E} = -d\phi_m/dt$. Compare your answer with that obtained in Part (a).

52 ••• A conducting rod of length ℓ rotates at constant angular velocity about one end, in a plane perpendicular to a uniform magnetic field B (Figure 28-52). (a) Show that the magnetic force on a body whose charge is q at a distance r from the pivot is $Bqr\omega$. (b) Show that the potential difference between the ends of the rod is $V = \frac{1}{2}B\omega\ell^2$. (c) Draw any radial line in the plane from which to measure $\theta = \omega t$. Show that the area of the pie-shaped region between the reference line and the rod is $A = \frac{1}{2}\ell^2\theta$. Compute the flux through this area, and show that $\mathcal{E} = \frac{1}{2}B\omega\ell^2$ follows when Faraday's law is applied to this area.

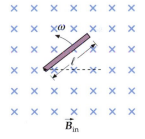

FIGURE 28-52 Problem 52

Inductance

53 • A coil with a self-inductance of 8 H carries a current of 3 A that is changing at a rate of 200 A/s. Find (a) the magnetic flux through the coil and (b) the induced emf in the coil.

54 • **SSM** A coil with self-inductance L carries a current I, given by $I = I_0 \sin 2\pi ft$. Find and graph the flux ϕ_m and the self-induced emf as functions of time.

55 •• **SOLVE** A solenoid has a length of 25 cm, a radius of 1 cm, 400 turns, and carries a 3-A current. Find (a) B on the axis at the center of the solenoid; (b) the flux through the solenoid, assuming B to be uniform; (c) the self-inductance of the solenoid; and (d) the induced emf in the solenoid when the current changes at 150 A/s.

56 •• **SOLVE** Two solenoids of radii 2 cm and 5 cm are coaxial. They are each 25 cm long and have 300 turns and 1000 turns, respectively. Find their mutual inductance.

57 •• SSM ISOLVE ✓ A long insulated wire with a resistance of 18 Ω/m is to be used to construct a resistor. First, the wire is bent in half and then the doubled wire is wound in a cylindrical form, as shown in Figure 28-53. The diameter of the cylindrical form is 2 cm, its length is 25 cm, and the total length of wire is 9 m. Find the resistance and inductance of this wire-wound resistor.

FIGURE 28-53 Problem 57

58 ••• In Figure 28-54, circuit 2 has a total resistance of 300 Ω. A total charge of 2×10^{-4} C flows through the galvanometer in circuit 2 when switch S in circuit 1 is closed. After a long time, the current in circuit 1 is 5 A. What is the mutual inductance between the two coils?

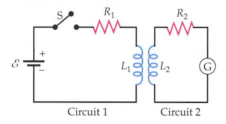

FIGURE 28-54 Problem 58

59 ••• Show that the inductance of a toroid of rectangular cross section, as shown in Figure 28-55, is given by

$$L = \frac{\mu_0 N^2 H \ln(b/a)}{2\pi}$$

where N is the total number of turns, a is the inside radius, b is the outside radius, and H is the height of the toroid.

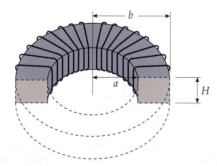

FIGURE 28-55 Problem 59

Magnetic Energy

60 • A coil with a self-inductance of 2 H and a resistance of 12 Ω is connected across a 24-V battery of negligible internal resistance. (a) What is the final current? (b) How much energy is stored in the inductor when the final current is attained?

61 •• SSM In a plane electromagnetic wave, such as a light wave, the magnitudes of the electric fields and magnetic fields are related by $E = cB$, where $c = 1/\sqrt{\epsilon_0\mu_0}$ is the speed of light. Show that in this case the electric energy and the magnetic energy densities are equal.

62 •• A solenoid of 2000 turns, area 4 cm², and length 30 cm carries a current of 4 A. (a) Calculate the magnetic energy stored in the solenoid from $\frac{1}{2}LI^2$. (b) Divide your answer in Part (a) by the volume of the solenoid to find the magnetic energy per unit volume in the solenoid. (c) Find B in the solenoid. (d) Compute the magnetic energy density from $u_m = B^2/2\mu_0$, and compare your answer with your result for Part (b).

63 •• ISOLVE A long cylindrical wire of radius $a = 2$ cm carries a current $I = 80$ A uniformly distributed over its cross-sectional area. Find the magnetic energy per unit length within the wire.

64 •• SSM You are given a length d of wire that has radius a and are told to wind it into an inductor in the shape of a cylinder with a circular cross section of radius r. The windings are to be as close together as possible without overlapping. Show that the self-inductance of this inductor is

$$L = \mu_0 \left(\frac{rd}{4a}\right)$$

65 • Using the result of Problem 64, calculate the self-inductance of an inductor wound from 10 cm of wire with a diameter of 1 mm into a coil with radius $R = 0.25$ cm.

66 •• A toroid of mean radius 25 cm and circular cross section of radius 2 cm is wound with a superconducting wire of length 1000 m that carries a current of 400 A. (a) What is the number of turns on the coil? (b) What is the magnetic field at the mean radius? (c) Assuming that B is constant over the area of the coil, calculate the magnetic energy density and the total energy stored in the toroid.

*RL Circuits

67 • A coil of resistance 8 Ω and self-inductance 4 H is suddenly connected across a constant potential difference of 100 V. Let $t = 0$ be the time of connection, at which the current is zero. Find the current I and its rate of change dI/dt at times (a) $t = 0$, (b) $t = 0.1$ s, (c) $t = 0.5$ s, and (d) $t = 1.0$ s.

68 • The current in a coil with a self-inductance of 1 mH is 2 A at $t = 0$, when the coil is shorted through a resistor. The total resistance of the coil plus the resistor is 10 Ω. Find the current after (a) 0.5 ms and (b) 10 ms.

69 •• SSM In the circuit shown Figure 28-29, let $\mathcal{E}_0 = 12$ V, $R = 3$ Ω, and $L = 0.6$ H. The switch is closed at time $t = 0$. At time $t = 0.5$ s, find (a) the rate at which the battery supplies power, (b) the rate of joule heating, and (c) the rate at which energy is being stored in the inductor.

70 •• Rework Problem 69 for the times $t = 1$ s and $t = 100$ s.

71 •• The current in an RL circuit is zero at time $t = 0$ and increases to half its final value in 4 s. (a) What is the time constant of this circuit? (b) If the total resistance is 5 Ω, what is the self-inductance?

72 •• How many time constants must elapse before the current in an *RL* circuit that is initially zero reaches (*a*) 90 percent, (*b*) 99 percent, and (*c*) 99.9 percent of its final value?

73 •• A coil with inductance 4 mH and resistance 150 Ω is connected across a battery of emf 12 V and negligible internal resistance. (*a*) What is the initial rate of increase of the current? (*b*) What is the rate of increase when the current is half its final value? (*c*) What is the final current? (*d*) How long does it take for the current to reach 99 percent of its final value?

74 •• **SOLVE** A large electromagnet has an inductance of 50 H and a resistance of 8 Ω. It is connected to a dc power source of 250 V. Find the time for the current to reach (*a*) 10 A and (*b*) 30 A.

75 ••• **SSM** Given the circuit shown in Figure 28-56, assume that the switch S has been closed for a long time so that steady currents exist in the inductor, and that the inductor *L* has negligible resistance. (*a*) Find the battery current, the current in the 100 Ω resistor, and the current through the inductor. (*b*) Find the initial voltage across the inductor when the switch S is opened. (*c*) Using a spreadsheet program, make graphs of the current and voltage across the inductor as a function of time.

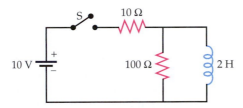

FIGURE 28-56 Problem 75

76 •• Compute the initial slope *dI/dt* at *t* = 0 from Equation 28-26, and show that if the current decreased steadily at this rate the current would be zero after one time constant.

77 •• An inductance *L* and resistance *R* are connected in series with a battery, as shown in Figure 28-31. A long time after switch S_1 is closed, the current is 2.5 A. When the battery is switched out of the circuit by opening switch S_1 and closing S_2, the current drops to 1.5 A in 45 ms. (*a*) What is the time constant for this circuit? (*b*) If *R* = 0.4 Ω, what is *L*?

78 • **SOLVE** When the current in a certain coil is 5 A and the current is increasing at the rate of 10 A/s, the potential difference across the coil is 140 V. When the current is 5 A and the current is decreasing at the rate of 10 A/s, the potential difference is 60 V. Find the resistance and self-inductance of the coil.

79 •• For the circuit shown in Figure 28-57, (*a*) find the rate of change of the current in each inductor and in the resistor just after the switch is closed. (*b*) What is the final current? (Use the result from Problem 88.)

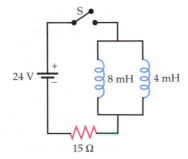

FIGURE 28-57 Problem 79

80 •• **SSM** For the circuit of Example 28-11, find the time at which the power dissipation in the resistor equals the rate at which magnetic energy is stored in the inductor.

81 ••• In the circuit shown in Figure 28-29, let $\mathcal{E}_0$ = 12 V, *R* = 3 Ω, and *L* = 0.6 H. The switch is closed at time *t* = 0. From time *t* = 0 to *t* = τ, find (*a*) the total energy that has been supplied by the battery, (*b*) the total energy that has been dissipated in the resistor, and (*c*) the energy that has been stored in the inductor. (*Hint:* Find the rates as functions of time and integrate from *t* = 0 to *t* = τ = *L/R*.)

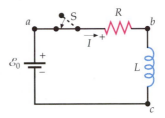

FIGURE 28-29 Problem 81

General Problems

82 • A circular coil of radius 3 cm has 6 turns. A magnetic field *B* = 5000 G is perpendicular to the coil. (*a*) Find the magnetic flux through the coil. (*b*) Find the magnetic flux through the coil if the coil makes an angle of 20° with the magnetic field.

83 • The magnetic field in Problem 82 is steadily reduced to zero in 1.2 s. Find the emf induced in the coil when (*a*) the magnetic field is perpendicular to the coil and (*b*) the magnetic field makes an angle of 20° with the normal to the coil.

84 • **SOLVE** A 100-turn coil has a radius of 4 cm and a resistance of 25 Ω. At what rate must a perpendicular magnetic field change to produce a current of 4 A in the coil?

85 •• **SSM** Figure 28-58 shows an ac generator. The generator consists of a rectangular loop of dimensions *a* and *b* with *N* turns connected to slip rings. The loop rotates with an angular velocity ω in a uniform magnetic field $\vec{B}$. (*a*) Show that the potential difference between the two slip rings is $\mathcal{E} = NBab\omega \sin \omega t$. (*b*) If *a* = 1 cm, *b* = 2 cm, *N* = 1000, and *B* = 2 T, at what angular frequency ω must the coil rotate to generate an emf whose maximum value is 110 V?

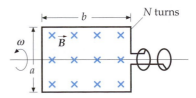

FIGURE 28-58 Problems 85 and 86

86 •• **SOLVE** Prior to 1960, magnetic field strength was measured by means of a rotating coil gaussmeter. This device used a small loop of many turns rotating on an axis perpendicular to the magnetic field at fairly high speed, which was connected to an ac voltmeter by means of slip rings, like those shown in Figure 28-58. The sensing coil for a rotating coil gaussmeter has 400 turns and an area of 1.4 cm².

The coil rotates at 180 rpm. If the magnetic field strength is 0.45 T, find the maximum induced emf in the coil and the orientation of the coil relative to the field for which this maximum induced emf occurs.

87 •• Show that the effective inductance for two inductors L_1 and L_2 connected in series, so that none of the flux from either passes through the other, is given by $L_{eff} = L_1 + L_2$.

88 •• SSM Show that the effective inductance for two inductors L_1 and L_2 connected in parallel, so that none of the flux from either passes through the other, is given by

$$\frac{1}{L_{eff}} = \frac{1}{L_1} + \frac{1}{L_2}$$

89 •• SSM Figure 28-59(a) shows an experiment designed to measure the acceleration of gravity. A large plastic tube is encircled by a wire, which is arranged in single loops separated by a distance of 10 cm. A strong magnet is dropped through the top of the loop. As the magnet falls through each loop the voltage rises and then the voltage rapidly falls through 0 to a large negative value as the magnet passes through the loop and then returns to 0. The shape of the voltage signal is shown in Figure 28-59(b). (a) Explain how this experiment works. (b) Explain why the tube cannot be made of a conductive material. (c) Qualitatively explain the shape of the voltage signal in Figure 28-59(b). (d) The times at which the voltage crosses 0 as the magnet falls through each loop in succession are given in the table in the next column. Use these data to calculate a value for g.

Loop Number	Zero Crossing Time(s)
1	0.011189
2	0.063133
3	0.10874
4	0.14703
5	0.18052
6	0.21025
7	0.23851
8	0.26363
9	0.28853
10	0.31144
11	0.33494
12	0.35476
13	0.37592
14	0.39107

90 •• The rectangular coil shown in Figure 28-60 has 80 turns, is 25 cm wide, is 30 cm long, and is located in a magnetic field $B = 1.4$ T directed out of the page, as shown, with only half of the coil in the region of the magnetic field. The resistance of the coil is 24 Ω. Find the magnitude and the direction of the induced current if the coil is moved with a speed of 2 m/s (a) to the right, (b) up, (c) to the left, and (d) down.

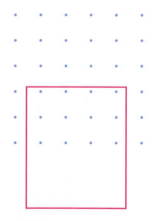

FIGURE 28-60 Problem 90

91 •• SSM Suppose the coil of Problem 90 is rotated about its vertical centerline at constant angular velocity of 2 rad/s. Find the induced current as a function of time.

92 •• Show that if the flux through each turn of an N-turn coil of resistance R changes from ϕ_{m1} to ϕ_{m2}, the total charge passing through the coil is given by $Q = N(\phi_{m1} - \phi_{m2})/R$.

93 •• A long solenoid has n turns per unit length and carries a current given by $I = I_0 \sin \omega t$. The solenoid has a circular cross section of radius R. Find the induced electric field at a radius r from the axis of the solenoid for (a) $r < R$ and (b) $r > R$.

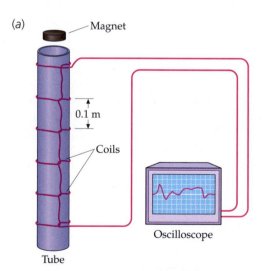

(a) Magnet

0.1 m

Coils

Oscilloscope

Tube

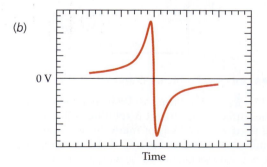

(b)

0 V

Time

FIGURE 28-59 Problem 89

94 ••• A coaxial cable consists of two very thin-walled conducting cylinders of radii r_1 and r_2 (Figure 28-61). Current I goes in one direction down the inner cylinder and in the opposite direction in the outer cylinder. (a) Use Ampère's law to find B. Show that $B = 0$, except in the region between the conductors. (b) Show that the magnetic energy density in the region between the cylinders is

$$u_m = \frac{\mu_0 I^2}{8\pi^2 r^2}$$

(c) Find the magnetic energy in a cylindrical shell volume element of length ℓ and volume $dV = \ell 2\pi r\, dr$, and integrate your result to show that the total magnetic energy in the volume of length ℓ is

$$U_m = \frac{\mu_0}{4\pi} I^2 \ell \ln\frac{r_2}{r_1}$$

(d) Use the result in Part (c) and $U_m = \frac{1}{2}LI^2$ to show that the self-inductance per unit length is

$$\frac{L}{\ell} = \frac{\mu_0}{2\pi}\ln\frac{r_2}{r_1}$$

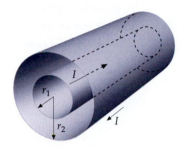

FIGURE 28-61 Problems 94 and 95

95 ••• Using Figure 28-61, compute the flux through a rectangular area of sides ℓ and $r_2 - r_1$ between the conductors. Show that the self-inductance per unit length can be found from $\phi_m = LI$ (see Part (d) of Problem 94).

96 ••• **SSM** Figure 28-62 shows a rectangular loop of wire, 0.30 m wide and 1.50 m long, in the vertical plane and perpendicular to a uniform magnetic field $B = 0.40$ T, directed inward as shown. The portion of the loop not in the magnetic field is 0.10 m long. The resistance of the loop is 0.20 Ω and its mass is 0.05 kg. The loop is released from rest at $t = 0$. (a) What is the magnitude and direction of the induced current when the loop has a downward velocity v? (b) What is the force that acts on the loop as a result of this current? (c) What is the net force acting on the loop? (d) Write the equation of motion of the loop. (e) Obtain an expression for the velocity of the loop as a function of time. (f) Integrate the

expression obtained in Part (e) to find the displacement y as a function of time. (g) Using a spreadsheet program, make a graph of the position y of the loop as a function of time for values of y between 0 m and 1.4 m (i.e., when the loop leaves the magnetic field). At what time t does $y = 1.4$ m? Compare this to the time it would have taken if $B = 0$.

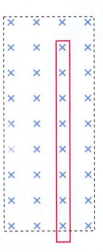

FIGURE 28-62 Problems 96 and 97

97 ••• The loop of Problem 96 is attached to a plastic spring of spring constant κ (see Figure 28-62). (a) When $B = 0$, the period of small-amplitude vertical oscillations of the mass–spring system is 0.8 s. Find the spring constant κ. (b) When $B \neq 0$, a current is induced in the loop as a result of its up and down motion. Obtain an expression for the induced current as a function of time when $B = 0.40$ T, and the displacement of the center of the loop is $y = 0.05$ m downward. (c) Show that the force on the loop is of the form $-\beta v$, where $v = dy/dt$, and find an expression for β in terms of B, w, and R, where w is the width of the wire loop and R is its resistance. (d) Using a spreadsheet program, make graphs of the position y and the velocity v of the center of the loop as a function of time, use the parameters given.

98 ••• A coil of N turns and area A hangs from a wire that provides a linear restoring torque with torsion constant κ. The two ends of the coil are connected to each other, the coil has resistance R, and the moment of inertia of the coils is I. The plane of the coil is vertical, and parallel to a uniform horizontal magnetic field B when the wire is not twisted (i.e., $\theta = 0$). The coil is twisted and released from a small angle $\theta = \theta_0$. Show that the orientation of the coil will undergo damped harmonic oscillation according to $\theta(t) = \theta_0 e^{-\beta t} \cos \omega t$, where

$$\omega = \sqrt{\kappa/I} \quad \text{and} \quad \beta = \frac{N^2 B^2 A^2}{RI}.$$

Alternating Current Circuits

THIS HIP-LOOKING LISTENER DIALS IN HER FAVORITE RADIO STATION. THIS CHANGES THE RESONANT FREQUENCY OF AN OSCILLATING ELECTRIC CIRCUIT WITHIN THE TUNER, SO ONLY THE STATION SHE SELECTS IS AMPLIFIED.

? **What component of the circuit is modified as she turns the dial? To find out more about the workings of a radio turner, see Example 29-9.**

More than 99 percent of the electrical energy used today is produced by electrical generators in the form of alternating current, which has a great advantage over direct current, because electrical energy can be transported over long distances at very high voltages and low currents to reduce energy losses due to Joule heating. Electrical energy can then be transformed, with almost no energy loss, to lower and safer voltages and correspondingly higher currents for everyday use. The transformer that accomplishes these changes in potential difference and current works on the basis of magnetic induction. In North America, power is delivered by a sinusoidal current of frequency 60 Hz. Devices such as radios, television sets, and microwave ovens detect or generate alternating currents of much higher frequencies.

Alternating current is produced by motional emf or magnetic induction in an ac generator, which is designed to provide a sinusoidal emf.

➤ In this chapter, we will see that when the generator output is sinusoidal, the current in an inductor, a capacitor, or a resistor is also sinusoidal, although it is generally not in phase with the generator's emf. When the emf and current are both sinusoidal, their maximum values are related. The study of sinusoidal currents is particularly important because even currents that are not sinusoidal can be analyzed in terms of sinusoidal components using Fourier analysis.

29-1 Alternating Current Generators

Figure 29-1 shows a simple **ac generator** that consists of a coil of area A and N turns rotating in a uniform magnetic field. The ends of the coil are connected to rings, called slip rings, that rotate with the coil. They make electrical contact through stationary conducting brushes that are in contact with the rings.

When the normal to the plane of the coil makes an angle θ with a uniform magnetic field $\vec{B}$, as shown in the figure, the magnetic flux through the coil is

$$\phi_m = NBA \cos \theta \qquad 29\text{-}1$$

where A is the area of the flat surface bounded by a single turn of the coil and N is the number of turns. When the coil is mechanically rotated, the flux through the coil will change, and an emf will be induced. If ω is the angular velocity of rotation and the initial angle is δ, the angle at some later time t is given by

$$\theta = \omega t + \delta$$

Then

$$\phi_m = NBA \cos(\omega t + \delta) = NBA \cos(2\pi f t + \delta)$$

The emf in the coil will then be

$$\mathcal{E} = -\frac{d\phi_m}{dt} = -NBA\frac{d}{dt}\cos(\omega t + \delta) = +NBA\omega \sin(\omega t + \delta) \qquad 29\text{-}2$$

where $NBA\omega$ is the peak (maximum) emf. Thus,

$$\mathcal{E} = \mathcal{E}_{peak} \sin(\omega t + \delta) \qquad 29\text{-}3$$

where the emf amplitude is given by

$$\mathcal{E}_{peak} = NBA\omega \qquad 29\text{-}4$$

We can thus produce a sinusoidal emf in a coil by rotating the coil with constant angular velocity in a magnetic field. Although practical generators are considerably more complicated, they produce a sinusoidal emf either via induction or via motional emf. In circuit diagrams, an ac generator is represented by the symbol ⊖.

The same coil in a static magnetic field that can be used to generate an alternating emf can also be used as an **ac motor**. Instead of mechanically rotating the coil to generate an emf, we apply an ac potential difference generated by another ac generator to the coil. This produces an ac current in the coil, and the magnetic field exerts forces on the wires producing a torque that rotates the coil. As the coil rotates in the magnetic field, a back emf is generated that tends to counter the

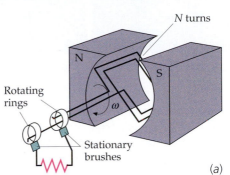
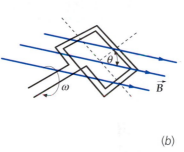
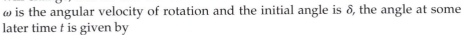

(a)   (b)

FIGURE 29-1 (*a*) An ac generator. A coil rotating with constant angular frequency ω in a static magnetic field $\vec{B}$ generates a sinusoidal emf. Energy from a waterfall or a steam turbine is used to rotate the coil to produce electrical energy. The emf is supplied to an external circuit by the brushes that are in contact with the rings. (*b*) At this instant, the normal to the plane of the coil makes an angle θ with the magnetic field, and the flux through the flat surface bounded by the coil is $BA \cos \theta$.

(a)

(b)

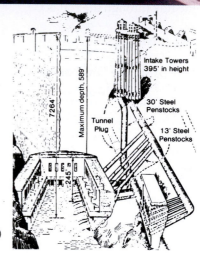

(c)

applied potential difference that produces the current. When the motor is first turned on, there is no back emf and the current is very large, being limited only by the resistance in the circuit. As the motor begins to rotate, the back emf increases and the current decreases.

EXERCISE A 250-turn coil has an area per turn of 3 cm². If it rotates in a magnetic field of 0.4 T at 60 Hz, what is $\mathcal{E}_{peak}$? (*Answer* $\mathcal{E}_{peak} = 11.3$ V)

29-2 Alternating Current in a Resistor

Figure 29-2 shows a simple ac circuit that consists of an ideal generator and a resistor. (A generator is ideal if its internal resistance, self-inductance, and capacitance are negligible.) The voltage drop across the resistor V_R is equal to the emf $\mathcal{E}$ of the generator. If the generator produces an emf given by Equation 29-3, we have

$$V_R = \mathcal{E} = \mathcal{E}_{peak} \sin(\omega t + \delta) = V_{R,peak} \sin(\omega t + \delta)$$

where $V_{R,peak} = \mathcal{E}_{peak}$. In this equation, the phase constant δ is arbitrary. It is convenient to choose $\delta = \pi/2$ so that

$$V_R = V_{R,peak} \sin\left(\omega t + \frac{\pi}{2}\right) = V_{R,peak} \cos \omega t$$

Applying Ohm's law, we have

$$V_R = IR \qquad\qquad 29\text{-}5$$

Thus,

$$V_{R,peak} \cos \omega t = IR \qquad\qquad 29\text{-}6$$

so the current in the resistor is

$$I = \frac{V_{R,peak}}{R} \cos \omega t = I_{peak} \cos \omega t \qquad\qquad 29\text{-}7$$

(*a*) The mechanical energy of falling water drives turbines (*b*) for the generation of electricity. (*c*) Schematic drawing of the Hoover Dam showing the intake towers and pipes (penstocks) that carry the water to the generators below.

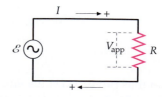

FIGURE 29-2 An ac generator in series with a resistor R.

where

$$I_{peak} = \frac{V_{R,peak}}{R} \qquad 29\text{-}8$$

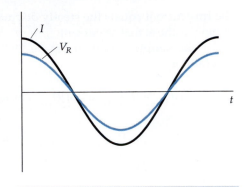

Note that the current through the resistor is in phase with the potential drop across the resistor, as shown in Figure 29-3.

The power dissipated in the resistor varies with time. Its instantaneous value is

$$P = I^2R = (I_{peak}\cos\omega t)^2 R = I_{peak}^2 R \cos^2\omega t \qquad 29\text{-}9$$

FIGURE 29-3 The voltage drop across a resistor is in phase with the current.

Figure 29-4 shows the power as a function of time. The power varies from zero to its peak value $I_{peak}^2 R$, as shown. We are usually interested in the average power over one or more complete cycles:

$$P_{av} = (I^2R)_{av} = I_{peak}^2 R(\cos^2\omega t)_{av}$$

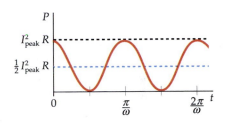

The average value of $\cos^2\omega t$ over one or more periods is $\frac{1}{2}$. This can be seen from the identity $\cos^2\omega t + \sin^2\omega t = 1$. A plot of $\sin^2\omega t$ looks the same as a plot of $\cos^2\omega t$ except that the plot is shifted by 90°. Both have the same average value over one or more periods, and since their sum is 1, the average value of each must be $\frac{1}{2}$. The average power dissipated in the resistor is thus

$$P_{av} = (I^2R)_{av} = \frac{1}{2}I_{peak}^2 R \qquad 29\text{-}10$$

FIGURE 29-4 Plot of the power dissipated in the resistor shown in Figure 29-2 versus time. The power varies from zero to a peak value $I_{peak}^2 R$. The average power is half the peak power.

Root-Mean-Square Values

Most ac ammeters and voltmeters are designed to measure the **root-mean-square (rms) values** of current and potential difference rather than the peak values. The rms value of a current I_{rms} is defined by

$$I_{rms} = \sqrt{(I^2)_{av}} \qquad 29\text{-}11$$

DEFINITION—RMS CURRENT

For a sinusoidal current, the average value of I^2 is

$$(I^2)_{av} = \left[(I_{peak}\cos\omega t)^2\right]_{av} = \frac{1}{2}I_{peak}^2$$

Substituting $\frac{1}{2}I_{peak}^2$ for $(I^2)_{av}$ in Equation 29-11, we obtain

$$I_{rms} = \frac{1}{\sqrt{2}}I_{peak} \approx 0.707 I_{peak} \qquad 29\text{-}12$$

RMS VALUE RELATED TO PEAK VALUE

The rms value of any quantity that varies sinusoidally equals the peak value of that quantity divided by $\sqrt{2}$.

Substituting I_{rms}^2 for $\frac{1}{2}I_{peak}^2$ in Equation 29-10, we obtain for the average power dissipated in the resistor

$$P_{av} = I_{rms}^2 R \qquad 29\text{-}13$$

The rms current equals the steady dc current that would produce the same Joule heating as the actual ac current.

For the simple circuit in Figure 29-2, the average power delivered by the generator is:

$$P_{av} = (\mathcal{E}I)_{av} = [(\mathcal{E}_{peak} \cos \omega t)(I_{peak} \cos \omega t)]_{av} = \mathcal{E}_{peak} I_{peak} (\cos^2 \omega t)_{av}$$

or

$$P_{av} = \tfrac{1}{2}\mathcal{E}_{peak} I_{peak}$$

Using $I_{rms} = I_{peak}/\sqrt{2}$ and $\mathcal{E}_{rms} = \mathcal{E}_{peak}/\sqrt{2}$, this can be written

$$P_{av} = \mathcal{E}_{rms} I_{rms} \qquad\qquad 29\text{-}14$$

AVERAGE POWER DELIVERED BY A GENERATOR

The rms current is related to the rms potential drop in the same way that the peak current is related to the peak potential drop. We can see this by dividing each side of Equation 29-8 by $\sqrt{2}$ and using $I_{rms} = I_{peak}/\sqrt{2}$ and $V_{R,rms} = V_{R,peak}/\sqrt{2}$.

$$I_{rms} = \frac{V_{R,rms}}{R} \qquad\qquad 29\text{-}15$$

Equations 29-13, 29-14, and 29-15 are of the same form as the corresponding equations for direct-current circuits with I replaced by I_{rms} and V_R replaced by $V_{R,rms}$. We can therefore calculate the power input and the heat generated using the same equations that we used for direct current, if we use rms values for the current and potential drop.

EXERCISE The sinusoidal potential drop across a 12-Ω resistor has a peak value of 48 V. Find (*a*) the rms current, (*b*) the average power, and (*c*) the maximum power. (*Answer* (*a*) 2.83 A, (*b*) 96 W, (*c*) 192 W)

The ac power supplied to domestic wall outlets and light fixtures in the United States has an rms potential difference of 120 V at a frequency of 60 Hz. This potential difference is maintained, independent of the current. If you plug a 1600-W space heater into a wall outlet it will draw a current of

$$I_{rms} = \frac{P_{av}}{V_{rms}} = \frac{1600 \text{ W}}{120 \text{ V}} = 13.3 \text{ A}$$

All appliances plugged into the outlets of a single 120-V circuit are connected in parallel. If you plug a 500-W toaster into another outlet of the same circuit, it will draw a current of 500 W/120 V = 4.17 A, and the total current through the parallel combination will be 17.5 A. Typical household wall outlets are rated at 15 A and are part of a circuit using wires rated at either 15 A or 20 A, with each circuit having several outlets. The wire in each circuit is rated at 15 A or 20 A, correspondingly. A total current greater than the rated current for the wiring is likely to overheat the wiring and is a fire hazard. Each circuit is therefore equipped with a circuit breaker (or a fuse in older houses) that trips (or blows) when the total current exceeds the 15-A or 20-A rating.

High-power domestic appliances, such as electric clothes dryers, kitchen ranges, and hot water heaters, typically require power delivered at 240-V rms. For a given power requirement, only half as much current is required at 240 V as at 120 V, but 240 V is more likely to deliver a fatal shock or to start a fire than 120 V.

SAWTOOTH WAVEFORM   **EXAMPLE   29-1**

Find (*a*) the average current and (*b*) the rms current for the sawtooth waveform shown in Figure 29-5. In the region $0 < t < T$, the current is given by $I = (I_0/T)t$.

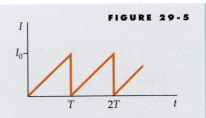

FIGURE 29-5

PICTURE THE PROBLEM The average of any quantity over a time interval T is the integral of the quantity over the interval divided by T. We use this to find both the average current, I_{av}, and the average of the current squared, $(I^2)_{av}$.

(*a*) Calculate I_{av} by integrating I from $t = 0$ to $t = T$ and dividing by T:

$$I_{av} = \frac{1}{T}\int_0^T I\, dt = \frac{1}{T}\int_0^T \frac{I_0}{T} t\, dt = \frac{I_0}{T^2}\frac{T^2}{2} = \frac{1}{2}I_0$$

(*b*) 1. Find $(I^2)_{av}$ by integrating I^2:

$$(I^2)_{av} = \frac{1}{T}\int_0^T I^2\, dt = \frac{1}{T}\left(\frac{I_0}{T}\right)^2\int_0^T t^2\, dt = \frac{I_0^2}{T^2}\frac{T^3}{3} = \frac{1}{3}I_0^2$$

2. The rms current is the square root of $(I^2)_{av}$:

$$I_{rms} = \sqrt{(I^2)_{av}} = \boxed{\frac{I_0}{\sqrt{3}}}$$

29-3 Alternating Current Circuits

Alternating current behaves differently than direct current in inductors and capacitors. When a capacitor becomes fully charged in a dc circuit, the capacitor blocks the current; that is, the capacitor acts like an open circuit. However, if the current alternates, charge continually flows onto the plates or off the plates of the capacitor. We will see that at high frequencies, a capacitor hardly impedes the current at all. That is, the capacitor acts like a short circuit. Conversely, an induction coil usually has a low resistance and is essentially a short circuit for direct current; however, when the current is changing, a back emf is generated in an inductor that is proportional to dI/dt. At high frequencies, the back emf is large and the inductor acts like an open circuit.

Inductors in Alternating Current Circuits

Figure 29-6 shows an inductor coil in series with an ac generator. When the current changes in the inductor, a back emf of magnitude $L\, dI/dt$ is generated due to the changing flux. Usually this back emf is much greater than the IR drop due to the resistance of the coil, so we normally neglect the resistance of the coil. The potential drop across the inductor V_L is then given by

$$V_L = L\frac{dI}{dt} \qquad\qquad 29\text{-}16$$

In this circuit, the potential drop V_L across the inductor equals the emf $\mathcal{E}$ of the generator. That is,

$$V_L = \mathcal{E} = \mathcal{E}_{max}\cos \omega t = V_{L,peak}\cos \omega t$$

where $V_{L,peak} = \mathcal{E}_{peak}$. Substituting for V_L in Equation 29-16 gives

$$V_{L,peak}\cos \omega t = L\frac{dI}{dt} \qquad\qquad 29\text{-}17$$

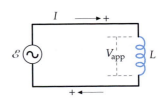

FIGURE 29-6 An ac generator in series with an inductor L. The arrow indicates the positive direction along the wire. Note that for a positive value of dI/dt, the voltage drop V_L across the inductor is positive. That is, if you traverse the inductor in the direction of the direction arrow you go in the direction of decreasing potential.

Rearranging, we obtain

$$dI = \frac{V_{L,\text{peak}}}{L} \cos \omega t \, dt \qquad\qquad 29\text{-}18$$

We solve for the current I by integrating both sides of the equation:

$$I = \frac{V_{L,\text{peak}}}{L} \int \cos \omega t \, dt = \frac{V_{L,\text{peak}}}{\omega L} \sin \omega t + C \qquad\qquad 29\text{-}19$$

where the constant of integration C is the dc component of the current. Setting the dc component of the current to be zero, we have

$$I = \frac{V_{L,\text{peak}}}{\omega L} \sin \omega t = I_{\text{peak}} \sin \omega t \qquad\qquad 29\text{-}20$$

where

$$I_{\text{peak}} = \frac{V_{L,\text{peak}}}{\omega L} \qquad\qquad 29\text{-}21$$

The potential drop $V_L = V_{L,\text{peak}} \cos \omega t$ across the inductor is 90° out of phase with the current $I = I_{\text{peak}} \sin \omega t$. From Figure 29-7, which shows I and V_L as functions of time, we can see that the peak value of the potential drop occurs 90° or one-fourth period prior to the corresponding peak value of the current. The potential drop across an inductor is said to *lead the current by 90°*. We can understand this physically. When I is zero but increasing, dI/dt is maximum, so the back emf induced in the inductor is at its maximum. One-quarter cycle later, I is maximum. At this time, dI/dt is zero, so V_L is zero. Using the trigonometric identity $\sin \theta = \cos\left(\theta - \frac{\pi}{2}\right)$, where $\theta = \omega t$, Equation 29-20 for the current can be written

$$I = I_{\text{peak}} \cos\left(\omega t - \frac{\pi}{2}\right) \qquad\qquad 29\text{-}22$$

The relation between the peak current and the peak potential drop (or between the rms current and rms potential drop) for an inductor can be written in a form similar to Equation 29-15 for a resistor. From Equation 29-21, we have

$$I_{\text{peak}} = \frac{V_{L,\text{peak}}}{\omega L} = \frac{V_{L,\text{peak}}}{X_L} \qquad\qquad 29\text{-}23$$

where

$$X_L = \omega L \qquad\qquad 29\text{-}24$$

DEFINITION—INDUCTIVE REACTANCE

is called the **inductive reactance**. Since $I_{\text{rms}} = I_{\text{peak}}/\sqrt{2}$ and $V_{L,\text{rms}} = V_{L,\text{peak}}/\sqrt{2}$ the rms current is given by

$$I_{\text{rms}} = \frac{V_{L,\text{rms}}}{X_L} \qquad\qquad 29\text{-}25$$

Like resistance, inductive reactance has units of ohms. As we can see from Equation 29-25, the larger the reactance for a given potential drop, the smaller

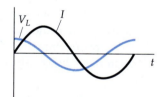

FIGURE 29-7 Current and potential drop across the inductor shown in Figure 29-6 as functions of time. The maximum potential drop occurs one-fourth period before the maximum current. Thus, the potential drop is said to lead the current by one-fourth period or 90°.

the peak current. Unlike resistance, the inductive reactance depends on the frequency of the current—the greater the frequency, the greater the reactance.

The *instantaneous* power delivered to the inductor from the generator is

$$P = V_L I = (V_{L,\text{peak}} \cos \omega t)(I_{\text{peak}} \sin \omega t) = V_{L,\text{peak}} I_{\text{peak}} \cos \omega t \sin \omega t$$

The *average* power delivered to the inductor is zero. We can see this by using the trigonometric identity

$$2 \cos \omega t \sin \omega t = \sin 2\omega t$$

The value of $\sin 2\omega t$ oscillates twice during each cycle and is negative as often as it is positive. Thus, on the average, no energy is dissipated in an inductor. (This is the case only if the resistance of the inductor is negligible.)

INDUCTIVE REACTANCE   **E X A M P L E   29 - 2**

The potential drop across a 40-mH inductor is sinusoidal with a peak potential drop of 120 V. Find the inductive reactance and the peak current when the frequency is (a) 60 Hz and (b) 2000 Hz.

PICTURE THE PROBLEM We calculate the inductive reactance at each frequency and use Equation 29-23 to find the peak current.

(a) 1. The peak current equals the peak potential drop divided by the inductive reactance. The peak potential drop equals the emf:

$$I_{\text{peak}} = \frac{V_{L,\text{peak}}}{X_L}$$

2. Compute the inductive reactance at 60 Hz:

$$X_{L1} = \omega_1 L = 2\pi f_1 L = (2\pi)(60 \text{ Hz})(40 \times 10^{-3} \text{ H})$$

$$= \boxed{15.1 \ \Omega}$$

3. Use this value of X_L to compute the peak current at 60 Hz:

$$I_{1,\text{peak}} = \frac{120 \text{ V}}{15.1 \ \Omega} = \boxed{7.95 \text{ A}}$$

(b) 1. Compute the inductive reactance at 2000 Hz:

$$X_{L2} = \omega_2 L = 2\pi f_2 L$$

$$= (2\pi)(2000 \text{ Hz})(40 \times 10^{-3} \text{ H}) = \boxed{503 \ \Omega}$$

2. Use this value of X_L to compute the peak current at 2000 Hz:

$$I_{2,\text{peak}} = \frac{120 \text{ V}}{503 \ \Omega} = \boxed{0.239 \text{ A}}$$

Capacitors in Alternating Current Circuits

When a capacitor is connected across the terminals of an ac generator (Figure 29-8), the voltage drop across the capacitor is

$$V_C = \frac{Q}{C} \qquad\qquad 29\text{-}26$$

where Q is the charge on the upper plate of the capacitor.

In this circuit, the potential drop V_C across the capacitor equals the emf $\mathcal{E}$ of the generator. That is,

$$V_C = \mathcal{E} = \mathcal{E}_{\text{peak}} \cos \omega t = V_{C,\text{peak}} \cos \omega t$$

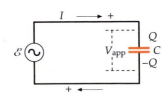

FIGURE 29-8 An ac generator in series with a capacitor C. The positive direction along the circuit is such that when the current is positive the charge Q on the upper capacitor plate is increasing, so the current is related to the charge by $I = +dQ/dt$.

where $V_{C,peak} = \mathcal{E}_{peak}$. Substituting for V_C in Equation 29-26 and solving for Q gives

$$Q = V_C C = V_{C,peak} C \cos \omega t$$

The current is

$$I = \frac{dQ}{dt} = -\omega V_{C,peak} C \sin \omega t = -I_{peak} \sin \omega t$$

where

$$I_{peak} = \omega V_{C,peak} C \qquad \qquad 29\text{-}27$$

Using the trigonometric identity $\sin \theta = -\cos\left(\theta + \frac{\pi}{2}\right)$, where $\theta = \omega t$, we obtain

$$I = -\omega C V_{C,peak} \sin \omega t = I_{peak} \cos\left(\omega t + \frac{\pi}{2}\right) \qquad \qquad 29\text{-}28$$

As with the inductor, the voltage drop $V_C = V_{C,peak} \cos \omega t$ across the capacitor is out of phase with the current

$$I = I_{peak} \cos\left(\omega t + \frac{\pi}{2}\right)$$

in the circuit. From Figure 29-9, we see that the maximum value of the potential drop occurs 90° or one-fourth period *after* the maximum value of the current. Thus, *the potential drop across a capacitor lags the current by 90°*. Again, we can understand this physically. The charge Q is proportional to the potential drop V_C. The maximum value of $dQ/dt = I$ occurs when the charge Q, and therefore when V_C, is zero. As the charge on the capacitor plate increases the current decreases until, one-fourth period later, the charge Q, and therefore V_C, is a maximum and the current is zero. The current then becomes negative as the charge Q decreases.

Again, we can relate the current to the potential drop in a form similar to Equation 29-8 for a resistor. From Equation 29-27, we have

$$I_{peak} = \omega C V_{C,peak} = \frac{V_{C,peak}}{1/(\omega C)} = \frac{V_{C,peak}}{X_C}$$

and, similarly,

$$I_{rms} = \frac{V_{C,rms}}{X_C} \qquad \qquad 29\text{-}29$$

where

$$X_C = \frac{1}{\omega C} \qquad \qquad 29\text{-}30$$

DEFINITION—CAPACITIVE REACTANCE

is called the **capacitive reactance** of the circuit. Like resistance and inductive reactance, capacitive reactance has units of ohms and, like inductive reactance, capacitive reactance depends on the frequency of the current. In this case, the

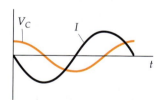

FIGURE 29-9 Current and potential drop across the capacitor shown in Figure 29-8 versus time. The maximum potential drop occurs one-fourth period after the maximum current. Thus, the potential drop is said to lag the current by 90°.

greater the frequency, the smaller the reactance. The average power delivered to a capacitor from an ac generator is zero, as it is for an inductor. This occurs because the potential drop is proportional to $\cos \omega t$ and the current is proportional to $\sin \omega t$ and $(\cos \omega t \sin \omega t)_{av} = 0$. Thus, like inductors with no resistance, capacitors dissipate no energy.

Since charge cannot pass across the space between the plates of a capacitor, it may seem strange that there is a continuing alternating current in the circuit shown in Figure 29-8. Suppose we choose the time to be zero at the instant that the voltage drop V_C across the capacitor is both zero and increasing. (At this same instant, the charge Q on the upper plate of the capacitor is also both zero and increasing.) As V_C then increases, positive charge flows off the lower plate and onto the upper plate, and Q reaches its maximum value Q_{peak} a quarter period later. After Q reaches its maximum value Q continues to change, reaching zero at the half-period point, $-Q_{peak}$ at the three-quarter-period point, and zero (again) at the completion of the cycle at the full-period point. The charge Q_{peak} flows through the generator each quarter period. If we double the frequency, we halve the period. Thus, if we double the frequency we halve the time for the charge Q_{peak} to flow through the generator, so we have doubled the current amplitude I_{peak}. Hence, the greater the frequency, the less the capacitor impedes the flow of charge.

CAPACITIVE REACTANCE   **EXAMPLE 29-3**

A 20-μF capacitor is placed across an ac generator that applies a potential drop with an amplitude (peak value) of 100 V. Find the capacitive reactance and the current amplitude when the frequency is 60 Hz and when the frequency is 6000 Hz.

PICTURE THE PROBLEM The capacitive reactance is $X_C = 1/(\omega C)$ and the peak current is $I_{peak} = V_{C,peak}/X_C$.

1. Calculate the capacitive reactance at 60 Hz and at 6000 Hz:

$$X_{C1} = \frac{1}{\omega_1 C} = \frac{1}{2\pi f_1 C} = \frac{1}{2\pi (60 \text{ Hz})(20 \times 10^{-6} \text{ F})} = \boxed{133 \ \Omega}$$

$$X_{C2} = \frac{1}{\omega_2 C} = \frac{1}{2\pi f_2 C} = \frac{1}{2\pi (6000 \text{ Hz})(20 \times 10^{-6} \text{ F})} = \boxed{1.33 \ \Omega}$$

2. Use these values of X_C to find the peak currents:

$$I_{1,peak} = \frac{V_{C,peak}}{X_{C1}} = \frac{100 \text{ V}}{133 \ \Omega} = \boxed{0.752 \text{ A}}$$

$$I_{2,peak} = \frac{V_{C,peak}}{X_{C2}} = \frac{100 \text{ V}}{1.33 \ \Omega} = \boxed{75.2 \text{ A}}$$

REMARKS Note that the capacitive reactance is inversely proportional to the frequency, so increasing the frequency by two orders of magnitude decreases the reactance by two orders of magnitude. The current is directly proportional to the frequency, as expected.

---

*29-4 Phasors

Until this point, the circuits considered contained an ideal ac generator and only a single passive element (i.e., resistor, inductor, or capacitor). In these circuits, the potential drop across the passive element equaled the emf of the generator. In circuits that contain an ideal ac generator and two or more additional elements connected in series, the sum of the potential drops across the elements is equal to the generator emf; which is the same as with dc circuits. However, in ac circuits these potential drops typically are not in phase, so the sum of their rms values does not equal the rms value of the generator emf.

Two-dimensional vectors, which are called phasors, can represent the phase relations between the current and the potential drops across resistors, capacitors, or inductors. In Figure 29-10, the potential drop across a resistor V_R is represented by a vector $\vec{V}_R$ that has magnitude $I_{peak}R$ and makes an angle θ with the x axis. This potential drop is in phase with the current. In general, the current in a steady-state ac circuit varies with time, as

$$I = I_{peak} \cos \theta = I_{peak} \cos (\omega t - \delta) \qquad 29\text{-}31$$

where ω is the angular frequency and δ is some phase constant. The potential drop across a resistor is then given by

$$V_R = IR = I_{peak}R \cos (\omega t - \delta) \qquad 29\text{-}32$$

The potential drop across a resistor is thus equal to the x component of the phasor vector $\vec{V}_R$, which rotates counterclockwise with an angular frequency ω. The current I may be written as the x component of a phasor $\vec{I}$ having the same direction as $\vec{V}_R$.

When several components are connected together in a series combination, their potential drops add. When several components are connected in parallel, their currents add. Unfortunately, adding sines or cosines of different amplitudes and phases algebraically is awkward. It is much easier to do this by vector addition.[†]

Let us look at how phasors are used. Any ac current or potential drop is written in the form $A \cos(\omega t - \delta)$, which in turn is treated as A_x, the x component of a phasor that makes an angle $(\omega t - \delta)$ with the positive x direction. Instead of adding two potential drops or currents algebraically, as $A \cos(\omega t - \delta_1) + B \cos(\omega t - \delta_2)$, we represent these quantities as phasors $\vec{A}$ and $\vec{B}$ and find the phasor sum $\vec{C} = \vec{A} + \vec{B}$ geometrically. The resultant potential drop or current is then the x component of the resultant phasor, $C_x = A_x + B_x$. The geometric representation conveniently shows the relative amplitudes and phases of the phasors.

Consider a circuit that contains an inductor L, a capacitor C, and a resistor R connected in series. They all carry the same current, which is represented as the x component of the current phasor $\vec{I}$. The potential drop across the inductor V_L is represented by a phasor $\vec{V}_L$ that has magnitude $I_{peak}X_L$ and leads the current phasor $\vec{I}$ by 90°. Similarly, the potential drop across the capacitor V_C is represented by a phasor $\vec{V}_C$ that has magnitude $I_{peak}X_C$ and lags the current by 90°. Figure 29-11 shows the phasors $\vec{V}_R$, $\vec{V}_L$, and $\vec{V}_C$. As time passes, the three phasors rotate counterclockwise with an angular frequency ω, so the relative positions of the vectors do not change. At any time, the instantaneous value of the potential drop across any of these elements equals the x component of the corresponding phasor.

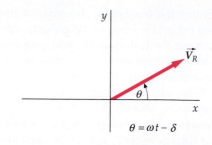

FIGURE 29-10 The potential drop across a resistor can be represented by a vector $\vec{V}_R$, which is called a phasor, that has magnitude $I_{peak}R$ and makes an angle $\theta = \omega t - \delta$ with the x axis. The phasor rotates with an angular frequency ω. The potential drop $V_R = IR$ is the x component of $\vec{V}_R$.

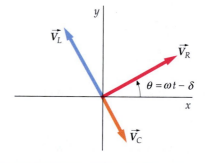

FIGURE 29-11 Phasor representations of the potential drops V_R, V_L, and V_C. Each vector rotates in the counterclockwise direction with an angular frequency ω. At any instant, the potential drop across an element equals the x component of the corresponding phasor, and the potential drop across the RLC-series combination, which equals the sum of the potential drops, equals the x component of the vector sum $\vec{V}_R + \vec{V}_L + \vec{V}_C$.

*29-5 *LC* and *RLC* Circuits Without a Generator

Figure 29-12 shows a simple circuit with inductance and capacitance but with no resistance. Such a circuit is called an *LC* circuit. We assume that the upper capacitor plate carries an initial positive charge Q_0 and that the switch is initially open.

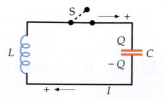

FIGURE 29-12 An *LC* circuit. When the switch is closed, the initially charged capacitor discharges through the inductor, producing a back emf.

---

† It is also easier to do using complex numbers.

After the switch is closed at $t = 0$, the charge begins to flow through the inductor. Let Q be the charge on the upper plate of the capacitor and let the positive direction around the circuit be counterclockwise, as shown. Then,

$$I = +\frac{dQ}{dt}$$

Applying Kirchhoff's loop rule to the circuit, we have

$$L\frac{dI}{dt} + \frac{Q}{C} = 0 \qquad\qquad 29\text{-}33$$

Substituting dQ/dt for I gives

$$L\frac{d^2Q}{dt^2} + \frac{Q}{C} = 0 \qquad\qquad 29\text{-}34$$

This equation is of the same form as Equation 14-2 for the acceleration of a mass on a spring:

$$m\frac{d^2x}{dt^2} + kx = 0$$

The behavior of an *LC* circuit is thus analogous to that of a mass on a spring, with L analogous to the mass m, Q analogous to the position x, and $1/C$ analogous to the spring constant k. Also, the current I is analogous to the velocity v, since $v = dx/dt$ and $I = dQ/dt$. In mechanics, the mass of an object describes the inertia of the object. The greater the mass, the more difficult it is to change the velocity of the object. Similarly, the inductance L can be thought of as the inertia of an ac circuit. The greater the inductance, the more opposition there is to changes in the current I.

If we divide each term in Equation 29-34 by L and rearrange, we obtain

$$\frac{d^2Q}{dt^2} = -\frac{1}{LC}Q \qquad\qquad 29\text{-}35$$

which is analogous to

$$\frac{d^2x}{dt^2} = -\frac{k}{m}x \qquad\qquad 29\text{-}36$$

In Chapter 14, we found that we could write the solution of Equation 29-36 for simple harmonic motion in the form

$$x = A\cos(\omega t - \delta)$$

where $\omega = \sqrt{k/m}$ is the angular frequency, A is the displacement amplitude, and δ is the phase constant, which depends on the initial conditions. The solution to Equation 29-35 is thus

$$Q = A\cos(\omega t - \delta)$$

with

$$\omega = \frac{1}{\sqrt{LC}} \qquad\qquad 29\text{-}37$$

The current I is found by differentiating:

$$I = \frac{dQ}{dt} = -\omega A \sin(\omega t - \delta)$$

If we choose our initial conditions to be $Q = Q_{peak}$ and $I = 0$ at $t = 0$, the phase constant δ is zero and $A = Q_{peak}$. Our solutions are then

$$Q = Q_{peak} \cos \omega t \qquad\qquad\qquad 29\text{-}38$$

and

$$I = -\omega Q_{peak} \sin \omega t = -I_{peak} \sin \omega t \qquad\qquad 29\text{-}39$$

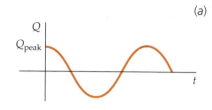

(a)

where $I_{peak} = \omega Q_{peak}$.

Figure 29-13 shows graphs of Q and I versus time. The charge oscillates between the values $+Q_{peak}$ and $-Q_{peak}$ with angular frequency $\omega = 1/\sqrt{(LC)}$. The current oscillates between $+\omega Q_{peak}$ and $-\omega Q_{peak}$ with the same frequency. Also, the current leads the charge by 90° (see Problem 29-37). The current is maximum when the charge is zero and the current is zero when the charge is maximum.

In our study of the oscillations of a mass on a spring, we found that the total energy is constant, and that the total energy oscillates between potential energy and kinetic energy. We also have two kinds of energy in the LC circuit, electric energy and magnetic energy. The electric energy stored in the capacitor is

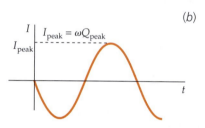

(b)

$$U_e = \frac{1}{2} Q V_C = \frac{1}{2} \frac{Q^2}{C}$$

FIGURE 29-13 Graphs of (a) Q versus t and (b) I versus t for the LC circuit shown in Figure 29-12.

Substituting $Q_{peak} \cos \omega t$ for Q, we have for the electric energy

$$U_e = \frac{1}{2} \frac{Q_{peak}^2}{C} \cos^2 \omega t \qquad\qquad 29\text{-}40$$

The electric energy oscillates between its maximum value $Q_0^2/(2C)$ and zero at an angular frequency of 2ω (see Problem 29-37). The magnetic energy stored in the inductor is

$$U_m = \frac{1}{2} L I^2 \qquad\qquad\qquad 29\text{-}41$$

Substituting $I = -\omega Q_{peak} \sin \omega t$ (Equation 29-39), we get

$$U_m = \frac{1}{2} L \omega^2 Q_{peak}^2 \sin^2 \omega t = \frac{1}{2} \frac{Q_{peak}^2}{C} \sin^2 \omega t \qquad 29\text{-}42$$

where we have used $\omega^2 = 1/LC$. The magnetic energy also oscillates between its maximum value of $Q_{peak}^2/2C$ and zero at an angular frequency of 2ω. The sum of the electrostatic energy and the magnetic energy is the total energy, which is constant in time:

$$U_{total} = U_e + U_m = \frac{1}{2} \frac{Q_{peak}^2}{C} \cos^2 \omega t + \frac{1}{2} \frac{Q_{peak}^2}{C} \sin^2 \omega t = \frac{1}{2} \frac{Q_{peak}^2}{C}$$

This sum equals the energy initially stored on the capacitor.

LC OSCILLATOR

EXAMPLE 29-4

A 2-μF capacitor is charged to 20 V and the capacitor is then connected across a 6-μH inductor. (*a*) What is the frequency of oscillation? (*b*) What is the peak value of the current?

PICTURE THE PROBLEM In (b), the current is maximum when dQ/dt is maximum, so the current amplitude is ωQ_{peak}. $Q = Q_{peak}$ when $V = V_{peak}$, where V is the voltage across the capacitor.

(*a*) The frequency of oscillation depends only on the values of the capacitance and the inductance:

$$f = \frac{\omega}{2\pi} = \frac{1}{2\pi\sqrt{LC}} = \frac{1}{2\pi\sqrt{(6 \times 10^{-6}\,\text{H})(2 \times 10^{-6}\,\text{F})}}$$

$$= \boxed{4.59 \times 10^4\,\text{Hz}}$$

(*b*) 1. The peak value of the current is related to the peak value of the charge:

$$I_{peak} = \omega Q_{peak} = \frac{Q_{peak}}{\sqrt{LC}}$$

2. The peak charge on the capacitor is related to the peak potential drop across the capacitor:

$$Q_{peak} = CV_{peak}$$

3. Substitute CV_{peak} for Q_{peak} and calculate I_{peak}:

$$I_{peak} = \frac{CV_{peak}}{\sqrt{LC}} = \frac{(2\,\mu\text{F})(20\,\text{V})}{\sqrt{(6\,\mu\text{H})(2\,\mu\text{F})}} = \boxed{11.5\,\text{A}}$$

EXERCISE A 5-μF capacitor is charged and is then discharged through an inductor. What should the value of the inductance be so that the current oscillates with frequency 8 kHz? (*Answer* 79.2 μH)

If we include a resistor in series with the capacitor and the inductor, as in Figure 29-14, we have an **RLC circuit.** Kirchhoff's loop rule gives

$$L\frac{dI}{dt} + IR + \frac{Q}{C} = 0 \qquad\qquad 29\text{-}43a$$

or

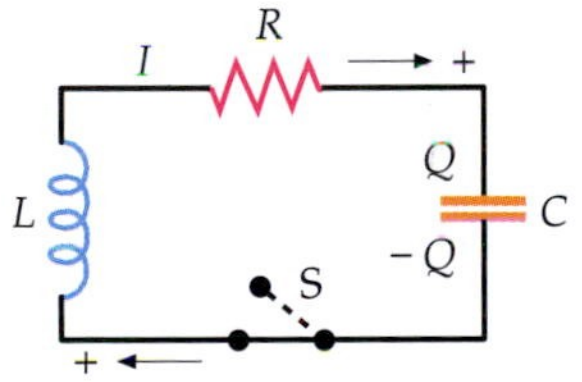

FIGURE 29-14 An *RLC* circuit.

$$L\frac{d^2Q}{dt^2} + R\frac{dQ}{dt} + \frac{1}{C}Q = 0 \qquad\qquad 29\text{-}43b$$

where we have used $I = dQ/dt$ as before. Equations 29-43*a* and 29-43*b* are analogous to the equation for a damped harmonic oscillator (see Equation 14-35):

$$m\frac{d^2x}{dt^2} + b\frac{dx}{dt} + kx = 0$$

The first term, $L\,dI/dt = L\,d^2Q/dt^2$, is analogous to the mass times the acceleration, $m\,dv/dt = m\,d^2x/dt^2$; the second term, $IR = R\,dQ/dt$, is analogous to the damping term, $bv = b\,dx/dt$; and the third term, Q/C, is analogous to the restoring force kx. In the oscillation of a mass on a spring, the damping constant b leads to a dissipation of mechanical energy. In an *RLC* circuit, the resistance R is analogous to the damping constant b and leads to a dissipation of electrical energy.

If the resistance is small, the charge and the current oscillate with (angular) frequency[†] that is very nearly equal to $\omega_0 = 1/\sqrt{LC}$, which is called the

---

† As in Chapter 14 when we discussed mechanical oscillations, we usually omit the word *angular* when the omission will not cause confusion.

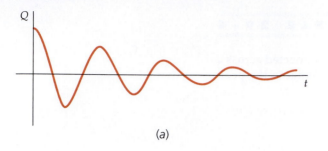

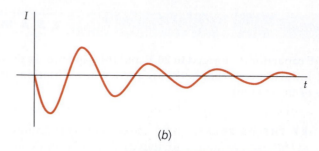

(a)   (b)

natural frequency of the circuit, but the oscillations are damped. We can understand this qualitatively from energy considerations. If we multiply each term in Equation 29-43a by the current I, we obtain

$$LI\frac{dI}{dt} + I^2R + I\frac{Q}{C} = 0 \qquad 29\text{-}44$$

The magnetic energy in the inductor is given by $\frac{1}{2}LI^2$ (see Equation 28-20). Note that

$$\frac{d(\frac{1}{2}LI^2)}{dt} = LI\frac{dI}{dt}$$

where $LI\,dI/dt$ is the first term in Equation 29-44. If $LI\,dI/dt$ is positive, it equals the rate at which electrical potential energy is transformed into magnetic energy. If $LI\,dI/dt$ is negative, it equals the rate at which magnetic energy is transformed back into electrical potential energy. Note that $LI\,dI/dt$ is positive or negative depending on whether I and dI/dt have the same sign or different signs. The second term in Equation 29-44 is I^2R, the rate at which electrical potential energy is dissipated in the resistor. I^2R is never negative. Note that

$$\frac{d(\frac{1}{2}Q^2/C)}{dt} = \frac{Q}{C}\frac{dQ}{dt} = I\frac{Q}{C}$$

where IQ/C is the third term in Equation 29-44. This is the rate of change of the electric potential energy of the capacitor, which may be positive or negative. The sum of the electric and magnetic energies is not constant for this circuit because energy is continually dissipated in the resistor. Figure 29-15 shows graphs of Q versus t and I versus t for a small resistance R in an RLC circuit. If we increase R, the oscillations become more heavily damped until a critical value of R is reached for which there is not even one oscillation. Figure 29-16 shows a graph of Q versus t in an RLC circuit when the value of R is greater than the critical damping value.

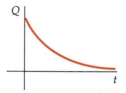

*29-6 Driven RLC Circuits

Series RLC Circuit

Figure 29-17 shows a series RLC circuit being sinusoidally driven by an ac generator. If the potential drop applied by the generator to the series RLC combination is $V_{app} = V_{app,peak} \cos \omega t$, applying Kirchhoff's loop rule gives

$$V_{app,peak} \cos \omega t - L\frac{dI}{dt} - IR - \frac{Q}{C} = 0$$

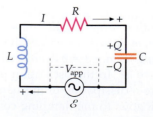

Using $I = dQ/dt$ and rearranging, we obtain

$$L\frac{d^2Q}{dt^2} + R\frac{dQ}{dt} + \frac{1}{C}Q = V_{\text{app,peak}}\cos\omega t \qquad\qquad 29\text{-}45$$

This equation is analogous to Equation 14-51 for the forced oscillation of a mass on a spring:

$$m\frac{d^2x}{dt^2} + b\frac{dx}{dt} + m\omega_0^2 x = F_0\cos\omega t$$

(In Equation 14-51, the force constant k was written in terms of the mass m and the natural angular frequency ω_0 using $k = m\omega_0^2$. The capacitance in Equation 29-45 could be similarly written in terms of L and the natural angular frequency using $1/C = L\omega_0^2$.)

We will discuss the solution of Equation 29-45 qualitatively as we did with Equation 14-51 for the forced oscillator. The current in the circuit consists of a transient current that depends on the initial conditions (e.g., the initial phase of the generator and the initial charge on the capacitor) and a steady-state current that does not depend on the initial conditions. We will ignore the transient current, which decreases exponentially with time and is eventually negligible, and concentrate on the steady-state current. The steady-state current obtained by solving Equation 29-45 is

$$I = I_{\text{peak}}\cos(\omega t - \delta) \qquad\qquad 29\text{-}46$$

where the phase angle δ is given by

$$\tan\delta = \frac{X_L - X_C}{R} \qquad\qquad 29\text{-}47$$

PHASE CONSTANT FOR A SERIES *RLC* CIRCUIT

The peak current is

$$I_{\text{peak}} = \frac{V_{\text{app,peak}}}{\sqrt{R^2 + (X_L - X_C)^2}} = \frac{V_{\text{app,peak}}}{Z} \qquad\qquad 29\text{-}48$$

PEAK CURRENT IN A SERIES *RLC* CIRCUIT

where

$$Z = \sqrt{R^2 + (X_L - X_C)^2} \qquad\qquad 29\text{-}49$$

IMPEDANCE OF A SERIES *RLC* CIRCUIT

The quantity $X_L - X_C$ is called the **total reactance,** and Z is called the **impedance.** Combining these results, we have

$$I = \frac{V_{\text{app,peak}}}{Z}\cos(\omega t - \delta) \qquad\qquad 29\text{-}50$$

Equation 29-50 can also be obtained from a simple diagram using the phasor representations. Figure 29-18 shows the phasors representing the potential drops across the resistance, the inductance, and the capacitance. The x component of each of these vectors equals the instantaneous potential drop across the corresponding element. Since the sum of the x components equals the x component of the sum, the sum of the x components equals the sum of the potential drops across these elements, which by Kirchhoff's loop rule equals the instantaneous applied potential drop.

If we represent the potential drop applied across the series combination $V_{app} = V_{app,peak} \cos \omega t$ as a phasor $\vec{V}_{app}$ that has the magnitude $V_{app,peak}$, we have

$$\vec{V}_{app} = \vec{V}_R + \vec{V}_L + \vec{V}_C \qquad 29\text{-}51$$

In terms of the magnitudes,

$$V_{app,peak} = |\vec{V}_R + \vec{V}_L + \vec{V}_C| = \sqrt{V_{R,peak}^2 + (V_{L,peak} - V_{C,peak})^2}$$

But $V_R = I_{peak}R$, $V_L = I_{peak}X_L$, and $V_C = I_{peak}X_C$. Thus,

$$V_{app,peak} = I_{peak}\sqrt{R^2 + (X_L - X_C)^2} = I_{peak}Z$$

The phasor $\vec{V}_{app}$ makes an angle δ with $\vec{V}_R$, as shown in Figure 29-18. From the figure, we can see that

$$\tan \delta = \frac{|\vec{V}_L + \vec{V}_C|}{|\vec{V}_R|} = \frac{I_{peak}X_L - I_{peak}X_C}{I_{peak}R} = \frac{X_L - X_C}{R}$$

in agreement with Equation 29-47. Since $\vec{V}_{app}$ makes an angle ωt with the x axis, $\vec{V}_R$ makes an angle $\omega t - \delta$ with the x axis. This applied potential drop is in phase with the current, which is therefore given by

$$I = I_{peak}\cos(\omega t - \delta) = \frac{V_{app,peak}}{Z}\cos(\omega t - \delta)$$

This is Equation 29-50. The relation between the impedance Z, the resistance R, and the total reactance $X_L - X_C$ is best remembered by using the right triangle shown in Figure 29-19.

Resonance

When X_L and X_C are equal, the total reactance is zero, and the impedance Z has its smallest value R. Then I_{peak} has its greatest value and the phase angle δ is zero, which means that the current is in phase with the applied potential drop. Let ω_{res} be the value of ω for which X_L and X_C are equal. It is obtained from

$$X_L = X_C$$

$$\omega_{res}L = \frac{1}{\omega_{res}C}$$

or

$$\omega_{res} = \frac{1}{\sqrt{LC}}$$

which equals the natural frequency ω_0. When the frequency of the applied potential drop equals the natural frequency ω_0, the impedance is smallest, I_{peak} is

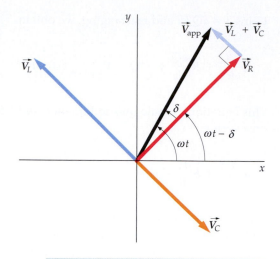

FIGURE 29-18 Phase relations among potential drops in a series RLC circuit. The potential drop across the resistor is in phase with the current. The potential drop across the inductor V_L leads the current by 90°. The potential drop across the capacitor lags the current by 90°. The sum of the vectors representing these potential drops gives a vector at an angle δ with the current representing the applied emf. For the case shown here, V_L is greater than V_C, and the current lags the applied potential drop by δ.

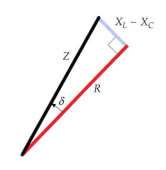

FIGURE 29-19 A right triangle relating capacitive and inductive reactance, resistance, impedance, and the phase angle in an RLC circuit.

greatest, and the circuit is said to be at **resonance.** The natural frequency ω_0 is therefore also called the **resonance frequency.** This resonance condition in a driven *RLC* circuit is similar to that in a driven simple harmonic oscillator.

Since neither an inductor nor a capacitor dissipates energy, the average power delivered to a series *RLC* circuit is the average power supplied to the resistor. The instantaneous power supplied to the resistor is

$$P = I^2 R = [I_{peak} \cos(\omega t - \delta)]^2 R$$

Averaging over one or more cycles and using $(\cos^2 \theta)_{av} = \frac{1}{2}$, we obtain for the average power

$$P_{av} = \frac{1}{2} I_{peak}^2 R = I_{rms}^2 R \qquad\qquad 29\text{-}52$$

Using $R/Z = \cos \delta$ from Figure 29-19 and $I_{peak} = V_{app,peak}/Z$, this can be written

$$P_{av} = \frac{1}{2} V_{app,peak} I_{peak} \cos \delta = V_{app,rms} I_{rms} \cos \delta \qquad\qquad 29\text{-}53$$

The quantity $\cos \delta$ is called the **power factor** of the *RLC* circuit. At resonance, δ is zero, and the power factor is 1.

The power can also be expressed as a function of the angular frequency ω. Using $I_{rms} = V_{app,rms}/Z$ Equation 29-52 becomes

$$P_{av} = I_{rms}^2 R = V_{app,rms}^2 \frac{R}{Z^2}$$

From the definition of impedance Z, we have

$$Z^2 = (X_L - X_C)^2 + R^2 = \left(\omega L - \frac{1}{\omega C} \right)^2 + R^2$$

$$= \frac{L^2}{\omega^2} \left(\omega^2 - \frac{1}{LC} \right)^2 + R^2$$

$$= \frac{L^2}{\omega^2} (\omega^2 - \omega_0^2)^2 + R^2$$

where we have used $\omega_0 = 1/\sqrt{LC}$. Using this expression for Z^2, we obtain the average power as a function of ω:

$$P_{av} = \frac{V_{app,rms}^2 R \omega^2}{L^2(\omega^2 - \omega_0^2)^2 + \omega^2 R^2} \qquad\qquad 29\text{-}54$$

Figure 29-20 shows the average power supplied by the generator to the series combination as a function of generator frequency for two different values of the resistance R. These curves, called **resonance curves,** are the same as the power-versus-frequency curves for a driven damped oscillator (see Section 14-5). The average power is greatest when the generator frequency equals the resonance frequency. When the resistance is small, the resonance curve is narrow; when the resistance is large, the resonance curve is broad. A resonance curve can be characterized by the **resonance width** $\Delta \omega$. As shown in Figure 29-20, the resonance width is the frequency difference between the two points on the curve where the power is half its maximum value. When the width is small compared with the resonance frequency, the resonance is sharp; that is, the resonance curve is narrow.

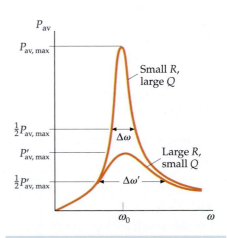

FIGURE 29-20 Plot of average power versus frequency for a series *RLC* circuit. The power is maximum when the frequency of the generator ω equals the natural frequency of the circuit $\omega_0 = 1/\sqrt{LC}$. If the resistance is small, the Q factor is large and the resonance is sharp. The resonance width $\Delta \omega$ of the curves is measured between points where the power is half its maximum value.

In Chapter 14, the Q factor for a mechanical oscillator is defined as $Q = \omega_0 m/b$ where m is the mass and b is the damping constant. We then saw that for an underdamped oscillator $Q = 2\pi E/|\Delta E|$, where E is the total energy of the system at the beginning of a cycle and ΔE is the energy dissipated during the cycle. The **Q factor** for an RLC circuit can be defined in a similar way. Since L is analogous to the mass m and R is analogous to the damping constant b, the Q factor for an RLC circuit is given by

$$Q = 2\pi \frac{E}{|\Delta E|} = \frac{\omega_0 L}{R} \qquad\qquad 29\text{-}55$$

When the resonance curve is reasonably narrow (that is, when Q is greater than about 2 or 3), the Q factor can be approximated by

$$Q = \frac{\omega_0}{\Delta\omega} = \frac{f_0}{\Delta f} \qquad\qquad 29\text{-}56$$

Q FACTOR FOR AN RLC CIRCUIT

Resonance circuits are used in radio receivers, where the resonance frequency of the circuit is varied either by varying the capacitance or the inductance. Resonance occurs when the natural frequency of the circuit equals one of the frequencies of the radio waves picked up at the antenna. At resonance, there is a relatively large current in the antenna circuit. If the Q factor of the circuit is sufficiently high, currents due to other station frequencies off resonance will be negligible compared with those currents due to the station frequency to which the circuit is tuned.

DRIVEN SERIES RLC CIRCUIT   **EXAMPLE 29-5**

A series RLC combination with $L = 2$ H, $C = 2$ μF, and $R = 20$ Ω is driven by an ideal generator with a peak emf of 100 V and a frequency that can be varied. Find (a) the resonance frequency f_0, (b) the Q value, (c) the width of the resonance Δf, and (d) the current amplitude at resonance.

PICTURE THE PROBLEM The resonance frequency is found from $\omega_0 = 1/\sqrt{LC}$ and the Q value is found from $Q = \omega_0 L/R$.

1. The resonance frequency is $f_0 = \omega_0/2\pi$:

$$f_0 = \frac{\omega_0}{2\pi} = \frac{1}{2\pi\sqrt{LC}}$$

$$= \frac{1}{2\pi\sqrt{(2\text{ H})(2\times10^{-6}\text{ F})}} = \boxed{79.6\text{ Hz}}$$

2. Use this result to calculate Q:

$$Q = \frac{\omega_0 L}{R} = \frac{2\pi(79.6\text{ Hz})(2\text{ H})}{20\ \Omega} = \boxed{50}$$

3. Use the value of Q to find the width of the resonance Δf:

$$\Delta f = \frac{f_0}{Q} = \frac{79.6\text{ Hz}}{50} = \boxed{1.59\text{ Hz}}$$

4. At resonance, the impedance is R and I_{peak} is $V_{app,peak}/R$:

$$I_{max} = \frac{V_{app,peak}}{R} = \frac{\mathcal{E}_{peak}}{R} = \frac{100\text{ V}}{20\ \Omega} = \boxed{5\text{ A}}$$

REMARKS The width of 1.59 Hz is less than 2 percent of the resonance frequency of 79.6 Hz, so the resonance peak is quite sharp.

DRIVEN SERIES RLC CIRCUIT CURRENT, PHASE, AND POWER   **EXAMPLE 29-6**   Try It Yourself

If the generator in Example 29-5 has a frequency of 60 Hz, find (a) the current amplitude, (b) the phase constant δ, (c) the power factor, and (d) the average power delivered.

PICTURE THE PROBLEM The current amplitude is the amplitude of the applied potential drop divided by the total impedance of the series combination. The phase angle δ is found from $\tan \delta = (X_L - X_C)/R$. You can use either Equation 29-52 or Equation 29-53 to find the average power delivered.

Cover the column to the right and try these on your own before looking at the answers.

Steps

Answers

(a) 1. Write the peak current in terms of $V_{app,peak}$ and the impedance.

$$I_{peak} = \frac{V_{app,peak}}{Z} = \frac{\mathcal{E}_{peak}}{Z}$$

2. Calculate the capacitive and inductive reactances and the total reactance.

$X_C = 1326 \ \Omega, X_L = 754 \ \Omega$

so

$X_L - X_C = -572 \ \Omega$

3. Calculate the total impedance Z.

$Z = 573 \ \Omega$

4. Use the results of steps 2 and 3 to calculate I_{peak}.

$I_{peak} = \boxed{0.175 \text{ A}}$

(b) Use the results of Part (a) steps 2 and 3 to calculate δ.

$\delta = \tan^{-1} \dfrac{X_L - X_C}{R} = \boxed{-88.0°}$

(c) Use your value of δ to compute the power factor.

$\cos \delta = 0.0349$

(d) Calculate the average power delivered from Equation 29-52.

$P_{av} = \frac{1}{2}I_{peak}^2 R = \boxed{0.305 \text{ W}}$

⬤ PLAUSIBILITY CHECK To check our result for the average power using the power factor found in Part (c), we have $P_{av} = \frac{1}{2}V_{app,peak}I_{peak} \cos \delta = \frac{1}{2}\mathcal{E}_{peak}I_{peak} \cos \delta = 0.305$ W. This is in agreement with our result for Part (d).

REMARKS The frequency of 60 Hz is well below the resonance frequency of 79.6 Hz. (Recall that the width as calculated in Example 29-5 is only 1.59 Hz.) As a result, the total reactance is much greater in magnitude than the resistance. This is always the case far from resonance. Similarly, an I_{peak} of 0.175 A is much less than an I_{peak} at resonance, which was found to be 5 A. Finally, we see from Figure 29-18 that a negative phase angle δ means that the current leads the applied potential drop.

DRIVEN SERIES RLC CIRCUIT AT RESONANCE   **EXAMPLE 29-7**   Try It Yourself

Find the peak potential drop across the resistor, the inductor, and the capacitor at resonance for the circuit in Example 29-5.

PICTURE THE PROBLEM The peak potential drop across the resistor is I_{peak} times R. Similarly, the peak potential drop across the inductor or capacitor is I_{peak} times the appropriate reactance. We found that at resonance $I_{peak} = 5$ A and $f_0 = 79.6$ Hz in Example 29-5.

Cover the column to the right and try these on your own before looking at the answers.

Steps	Answers
1. Calculate $V_{R,\text{peak}} = I_{\text{peak}}R$.	$V_{R,\text{peak}} = I_{\text{peak}}R = \boxed{100 \text{ V}}$
2. Express $V_{L,\text{peak}}$ in terms of I_{peak} and X_L.	$V_{L,\text{peak}} = I_{\text{peak}}X_L = I_{\text{peak}}\omega_0 L = \boxed{5000 \text{ V}}$
3. Express $V_{C,\text{peak}}$ in terms of I_{peak} and X_C.	$V_{C,\text{peak}} = I_{\text{peak}}X_C = \dfrac{I_{\text{peak}}}{\omega_0 C} = \boxed{5000 \text{ V}}$

REMARKS The inductive and capacitive reactances are equal, as we would expect, since we found the resonance frequency by setting them equal. The phasor diagram for the potential drops across the resistor, capacitor, and inductor is shown in Figure 29-21. The peak potential drop across the resistor is a relatively safe 100 V, equal to the peak emf of the generator. However, the peak potential drops across the inductor and the capacitor are a dangerously high 5000 V. These potential drops are 180° out of phase. At resonance, the potential drop across the inductor at any instant is the negative of that across the capacitor, so they always sum to zero, leaving the potential drop across the resistor equal to the emf in the circuit.

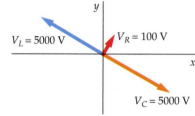

FIGURE 29-21

$V_L = 5000 \text{ V}$ $V_R = 100 \text{ V}$

$V_C = 5000 \text{ V}$

RC LOW-PASS FILTER   **EXAMPLE 29-8**

A resistor R and capacitor C are in series with a generator, as shown in Figure 29-22. The generator applies a potential drop across the RC combination given by $V_{\text{app}} = \sqrt{2}V_{\text{app,rms}} \cos \omega t$. Find the rms potential drop across the capacitor $V_{\text{out,rms}}$ as a function of frequency ω.

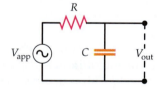

FIGURE 29-22 The peak output voltage decreases as frequency increases.

PICTURE THE PROBLEM The rms potential drop across the capacitor is the product of the rms current and the capacitive reactance. The rms current is found from the potential drop applied by the generator and the impedance of the series RC combination.

1. The potential drop across the capacitor is I_{rms} times X_C:

$$V_{\text{out,rms}} = I_{\text{rms}}X_C$$

2. The rms current depends on the applied rms potential drop and the impedance:

$$I_{\text{rms}} = \frac{V_{\text{app,rms}}}{Z}$$

3. In this circuit, only R and X_C contribute to the total impedance:

$$Z = \sqrt{R^2 + X_C^2}$$

4. Substitute these values and $X_C = 1/(\omega C)$ to find the output rms potential drop:

$$V_{\text{out,rms}} = I_{\text{rms}}X_C = \frac{V_{\text{app,rms}}}{Z}X_C = \frac{V_{\text{app,rms}}X_C}{\sqrt{R^2 + X_C^2}}$$

$$= \frac{V_{\text{app,rms}}\left(\dfrac{1}{\omega C}\right)}{\sqrt{R^2 + \left(\dfrac{1}{\omega C}\right)^2}} = \boxed{\frac{V_{\text{app,rms}}}{\sqrt{1 + \omega^2(RC)^2}}}$$

REMARKS This circuit is called an *RC low-pass filter,* since it transmits low frequencies with greater amplitude than high frequencies. In fact, the output potential drop equals the potential drop applied by the generator in the limit that $\omega \to 0$, but approaches zero for $\omega \to \infty$, as shown in the graph of the ratio of output potential drop to applied potential drop in Figure 29-23.

EXERCISE Find the output potential drop for this circuit if the capacitor is replaced by an inductor L. (*Answer* $V_{out,rms} = V_{in,rms}/\sqrt{1 + (R/L)^2/\omega^2}$. This circuit is a *high-pass filter.*)

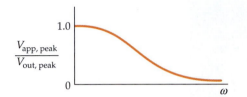

$\dfrac{V_{app,\,peak}}{V_{out,\,peak}}$

FIGURE 29-23

AN FM TUNER   **EXAMPLE 29-9**   **Put It in Context**

You have been tinkering with building a radio tuner using your new knowledge of physics. You know that the FM dial gives its frequencies in megahertz, and you would like to determine what percentage of change in an inductor would allow you to tune for the whole FM range. You decide to start at midrange and determine a percent increase and decrease needed for inductance. A variable inductor is usually an iron-core solenoid, and the inductance is increased by further inserting the core. The FM dial goes from 88 MHz to 108 MHz.

PICTURE THE PROBLEM We can relate inductance to the resonant frequency with $\omega = 2\pi f$ and $\omega = 1/\sqrt{LC}$. Then, if we find the percent change in frequency, we can determine the percent change in inductance. The capacitance C does not vary.

1. The resonant angular frequency ω is related to the inductance L:

$$\omega = 1/\sqrt{LC}$$

and

$$\omega = 2\pi f$$

so

$$f = \frac{1}{2\pi\sqrt{LC}}$$

2. L is inversely proportional to f^2:

$$L = af^{-2}$$

where

$$a = (4\pi^2 C)^{-1}$$

3. Express the fractional change in L in terms of the frequencies: When L is maximum, f is minimum and vice versa. The middle frequency f_{mid} is halfway between the maximum and minimum frequency, and L_{mid} is the inductance when $f = f_{mid}$:

$$\frac{\Delta L}{L} = \frac{L_{max} - L_{min}}{L_{mid}} = \frac{af_{max}^{-2} - af_{min}^{-2}}{af_{mid}^{-2}}$$

$$= f_{mid}^2\left(\frac{1}{f_{max}^2} - \frac{1}{f_{min}^2}\right) = 98^2\left(\frac{1}{108^2} - \frac{1}{88^2}\right)$$

$$= -0.417$$

4. The negative sign is not relevant, except as an indication that when the inductance increases the resonant frequency dcreases. Express the step 3 result as a percentage:

The inductance varies by about $\boxed{42 \text{ percent}}$

A shipboard radio, circa 1920. Exposed at the operator's left are the inductance coils and capacitor plates of the tuning circuit.

Parallel *RLC* Circuit

Figure 29-24 shows a resistor *R*, a capacitor *C*, and an inductor *L* connected in parallel across an ac generator. The total current *I* from the generator divides into three currents. The current I_R in the resistor, the current I_C in the capacitor, and the current I_L in the inductor. The instantaneous potential drop V_{app} is the same across each element. The current in the resistor is in phase with the potential drop and the phasor $\vec{I}_R$ has magnitude V_{peak}/R. Since the potential drop across an inductor *leads* the current in the inductor by 90°, I_L *lags* the potential drop by 90°, and the phasor $\vec{I}_L$ has magnitude V_{peak}/X_L. Similarly, the I_C leads the potential drop by 90° and the phasor $\vec{I}_C$ has magnitude V_{peak}/X_C. These currents are represented by phasors in Figure 29-25. The total current *I* is the *x* component of the vector sum of the individual currents as shown in the figure. The magnitude of the total current is

$$I = \sqrt{I_R^2 + (I_L - I_C)^2} = \sqrt{\left(\frac{V_{peak}}{R}\right)^2 + \left(\frac{V_{peak}}{X_L} - \frac{V_{peak}}{X_C}\right)^2} = \frac{V_{peak}}{Z} \qquad 29\text{-}57$$

where the total impedance *Z* is related to the resistance and the capacitive and inductive reactances by

$$\frac{1}{Z} = \sqrt{\left(\frac{1}{R}\right)^2 + \left(\frac{1}{X_L} - \frac{1}{X_C}\right)^2} \qquad 29\text{-}58$$

At resonance, the currents in the inductor and capacitor are 180° out of phase, so the total current is a minimum and is just the current in the resistor. We see from Equation 29-57 that this occurs if *Z* is maximum, so $1/Z$ is minimum. Then, we see from Equation 29-58 that if $X_L = X_C$, $1/Z$ has its minimum value $1/R$. Equating X_L with X_C and solving for ω obtains the resonant frequency, which equals the natural frequency $\omega_0 = 1/\sqrt{LC}$.

*29-7 The Transformer

A transformer is a device used to raise or lower the voltage in a circuit without an appreciable loss of power. Figure 29-26 shows a simple transformer consisting of two wire coils around a common iron core. The coil carrying the input power is

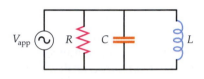

FIGURE 29-24 A parallel *RLC* circuit.

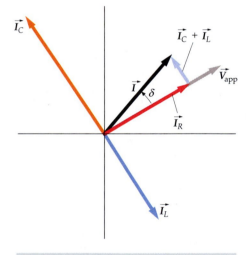

FIGURE 29-25 A phasor diagram for the currents in the parallel *RLC* circuit shown in Figure 29-24. The potential drop is the same across each element. The current in the resistor is in phase with the potential drop. The current in the capacitor leads the potential drop by 90° and the current in the inductor lags the potential drop by 90°. The phase difference δ between the total current and the potential drop depends on the relative magnitudes of the currents, which depend on the values of the resistance and of the capacitive and inductive reactances.

called the primary, and the other coil is called the secondary. Either coil of a transformer can be used for the primary or secondary. The transformer operates on the principle that an alternating current in one circuit induces an alternating emf in a nearby circuit due to the mutual inductance of the two circuits. The iron core increases the magnetic field for a given current and guides it so that nearly all the magnetic flux through one coil goes through the other coil. If no power were lost, the product of the potential difference across and the current in the secondary windings would equal the product of the potential drop across and the current in the primary windings. Thus, if the potential difference across the secondary coil is higher than the potential drop across the primary circuit, the current in the secondary coil is lower than the current in the primary coil, and vice versa. Power losses arise because of the Joule heating in the small resistances in both coils, or in current loops within the core,[†] and from hysteresis in the iron cores. We will neglect these losses and consider an ideal transformer of 100 percent efficiency, for which all of the power supplied to the primary coil appears in the secondary coil. Actual transformers are often 90 percent to 95 percent efficient.

Consider a transformer with a potential drop V_1 across the primary coil of N_1 turns; the secondary coil of N_2 turns is an open circuit. Because of the iron core, there is a large flux through each coil even when the magnetizing current I_m in the primary circuit is very small. We can ignore the resistances of the coils, which are negligible in comparison with their inductive reactances. The primary circuit is then a simple circuit consisting of an ac generator and a pure inductance, like that discussed in Section 29-3. The current magnetizing in the primary coil and the voltage drop across the primary coil are out of phase by 90°, and the average power dissipated in the primary coil is zero. If ϕ_{turn} is the magnetic flux through a single turn of the primary coil, the potential drop across the primary coil is equal to the back emf, so

$$V_1 = N_1 \frac{d\phi_{turn}}{dt} \qquad\qquad 29\text{-}59$$

If there is no flux leakage out of the iron core, the flux through each turn is the same for both coils. Thus, the total flux through the secondary coil is $N_2 \phi_{turn}$, and the potential difference across the secondary coil is

$$V_2 = N_2 \frac{d\phi_{turn}}{dt} \qquad\qquad 29\text{-}60$$

Comparing Equations 29-59 and 29-60, we can see that

$$V_2 = \frac{N_2}{N_1} V_1 \qquad\qquad 29\text{-}61$$

If N_2 is greater than N_1, the potential difference across the secondary coil is greater than the potential drop across the primary coil, and the transformer is called a step-up transformer. If N_2 is less than N_1, the potential difference across the secondary coil is less than the potential drop across the primary coil, and the transformer is called a step-down transformer.

When we put a resistance R, called a load resistance, across the secondary coil, there will then be a current I_2 in the secondary circuit that is in phase with the potential drop V_2 across the resistance. This current sets up an additional

---

[†] The induced currents, called eddy currents, can be greatly reduced by using a core of laminated metal to break up current paths.

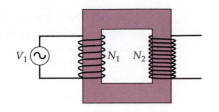

FIGURE 29-26 A transformer with N_1 turns in the primary and N_2 turns in the secondary.

(a)

(b)

(a) A power box with a transformer for stepping down voltage for distribution to homes. (b) A suburban power substation where transformers step down voltage from high-voltage transmission lines.

flux ϕ'_{turn} through each turn that is proportional to N_2I_2. This flux opposes the original flux set up by the original magnetizing current I_m in the primary. However, the potential drop across the primary coil is determined by the generator emf, which is unaffected by the secondary circuit. According to Equation 29-60, the flux in the iron core must change at the original rate; that is, the total flux in the iron core must be the same as when there is no load across the secondary. The primary coil thus draws an additional current I_1 to maintain the original flux ϕ_{turn}. The flux through each turn produced by this additional current is proportional to N_1I_1. Since this flux equals $-\phi'_{turn}$, the additional current I_1 in the primary is related to the current I_2 in the secondary by

$$N_1I_1 = -N_2I_2 \qquad\qquad 29\text{-}62$$

These currents are 180° out of phase and produce counteracting fluxes. Since I_2 is in phase with V_2, the additional current I_1 is in phase with the potential drop across the primary circuit. The power input from the generator is $V_{1,rms}I_{1,rms}$, and the power output is $V_{2,rms}I_{2,rms}$. (The magnetizing current does not contribute to the power input because it is 90° out of phase with the generator voltage.) If there are no losses,

$$V_{1,rms}I_{1,rms} = V_{2,rms}I_{2,rms} \qquad\qquad 29\text{-}63$$

In most cases, the additional current in the primary I_1 is much greater than the original magnetizing current I_m that is drawn from the generator when there is no load. This can be demonstrated by putting a lightbulb in series with the primary coil. The lightbulb is much brighter when there is a load across the secondary circuit than when the secondary circuit is open. If I_m can be neglected, Equation 29-63 relates the total currents in the primary and secondary circuits.

DOORBELL TRANSFORMER   **E X A M P L E   2 9 - 1 0**

A doorbell requires 0.4 A at 6 V. It is connected to a transformer whose primary, containing 2000 turns, is connected to a 120-V ac line. (a) How many turns should there be in the secondary? (b) What is the current in the primary?

PICTURE THE PROBLEM We can find the number of turns from the turns ratio, which equals the voltage ratio. The primary current can be found by equating the power out to the power in.

1. The turns ratio can be obtained from Equation 29-61. Solve for the number of turns in the secondary, N_2:

$$\frac{N_2}{N_1} = \frac{V_2}{V_1}$$

so

$$N_2 = \frac{V_{2,rms}}{V_{1,rms}}N_1 = \frac{6\text{ V}}{120\text{ V}}\,2000\text{ turns} = \boxed{100\text{ turns}}$$

2. Since we are assuming 100 percent efficiency in power transmission, the input and output currents are related by Equation 29-62. Solve for the current in the primary, I_1:

$$V_2I_2 = V_1I_1$$

so

$$I_1 = \frac{V_2}{V_1}I_2 = \frac{6\text{ V}}{120\text{ V}}(0.4\text{ A}) = \boxed{0.02\text{ A}}$$

An important use of transformers is in the transport of electrical power. To minimize the I^2R heat loss (Joule heating) in transmission lines, it is economical to use a high voltage and a low current. On the other hand, safety and other considerations require that power be delivered to consumers at lower voltages and therefore with higher currents. Suppose, for example, that each person in a city with a population of 50,000 uses 1.2 kW of electric power. (The per capita consumption of power in the United States is actually somewhat higher than this.) At 120 V, the current required for each person would be

$$I = \frac{1200 \text{ W}}{120 \text{ V}} = 10 \text{ A}$$

The total current for 50,000 people would then be 500,000 A. The transport of such a current from a power-plant generator to a city many kilometers away would require conductors of enormous thickness, and the I^2R power loss would be substantial. Rather than transmit the power at 120 V, step-up transformers are used at the power plant to step up the voltage to some very large value, such as 600,000 V. For this voltage, the current needed is only

$$I = \frac{120 \text{ V}}{600,000 \text{ V}} (500,000 \text{ A}) = 100 \text{ A}$$

To reduce the voltage to a safer level for transport within a city, power substations are located just outside the city to step down the voltage to a safer value, such as 10,000 V. Transformers in boxes attached to the power poles outside each house again step down the voltage to 120 V (or 240 V) for distribution to the house. Because of the ease of stepping the voltage up or down with transformers, alternating current rather than direct current is in common use.

TRANSMISSION LOSSES **E X A M P L E 2 9 - 1 1**

A transmission line has a resistance of 0.02 Ω/km. Calculate the I^2R power loss if 200 kW of power is transmitted from a power generator to a city 10 km away at (a) 240 V and (b) 4.4 kV.

PICTURE THE PROBLEM First, note that the total resistance of 10 km of wire is $R = (0.02 \ \Omega/\text{km})(10 \text{ km}) = 0.2 \ \Omega$. In each case, begin by finding the current needed to transmit 200 kW using $P = IV$, then find the power loss using I^2R.

(a) 1. Find the current needed to transmit 200 kW of power at 240 V: $\quad I = \dfrac{P}{V} = \dfrac{200 \text{ kW}}{240 \text{ V}} = 833 \text{ A}$

2. Calculate the power loss: $\quad I^2R = (833 \text{ A})^2(0.2 \ \Omega) = \boxed{139,000 \text{ W}}$

(b) 1. Now, find the current needed to transmit 200 kW of power at 4.4 kV: $\quad I = \dfrac{P}{V} = \dfrac{200 \text{ kW}}{4.4 \text{ kV}} = 45.5 \text{ A}$

2. Calculate the power loss: $\quad I^2R = (45.5 \text{ A})^2(0.2 \ \Omega) = \boxed{414 \text{ W}}$

REMARKS Note that with a transmission voltage of 240 V almost 70 percent of the power is wasted through heat loss, and there is an IR (voltage) drop across the transmission line of 167 V, so the power is delivered at only 73 V. However, with transmission at 4.4 kV only about 0.2 percent of the power is lost in transmission, and there is an IR drop across the transmission line of only 9 V, so the power is delivered with only a 0.2 percent voltage drop. This illustrates the advantages of high-voltage power transmission.

1. Reactance is a frequency-dependent property of capacitors and inductors that is analogous to the resistance of a resistor.
2. Impedance is a frequency-dependent property of an ac circuit or circuit loop that is analogous to the resistance in a dc circuit.
3. Phasors are two-dimensional vectors that allow us to picture the phase relations in a circuit.
4. Resonance occurs when the frequency of the generator equals the natural frequency of the oscillating circuit.

Topic	Relevant Equations and Remarks	
1. Alternating Current Generators	An ac generator is a device that transforms mechanical energy into electrical energy. This transformation can be accomplished by using the mechanical energy to either rotate a conducting coil in a magnetic field or rotating a magnet in a conducting coil.	
EMF generated	$\mathcal{E} = \mathcal{E}_{peak} \sin(\omega t + \delta) = NBA\omega \sin(\omega t + \delta)$	**29-3, 29-4**
2. Current		
RMS current	$I_{rms} = \sqrt{(I^2)_{av}}$	**29-11**
RMS current and peak current	$I_{rms} = \dfrac{1}{\sqrt{2}} I_{peak}$	**29-12**
For a resistor	$I_{rms} = \dfrac{V_{R,rms}}{R}$	**29-15**
	potential drop and current in phase	
For an inductor	$I_{rms} = \dfrac{V_{L,rms}}{\omega L} = \dfrac{V_{L,rms}}{X_L}$	**29-25**
	potential drop leads current by 90°	
For a capacitor	$I_{rms} = \dfrac{V_{C,rms}}{1/\omega L} = \dfrac{V_{C,rms}}{X_C}$	**29-29**
	potential drop lags current by 90°	
3. Reactance		
Inductive reactance	$X_L = \omega L$	**29-24**
Capacitive reactance	$X_C = \dfrac{1}{\omega C}$	**29-30**
4. Average Power Dissipation		
By a resistor	$P_{av} = V_{R,rms} I_{rms} = I^2_{rms} R$	**29-13, 29-15**
By an inductor or by a capacitor	$P_{av} = 0$	

5. ***Phasors**

Phasors are two-dimensional vectors that represent the current $\vec{I}$, the potential drop across a resistor $\vec{V}_R$, the potential drop across a capacitor $\vec{V}_C$, and the potential drop across an inductor $\vec{V}_L$ in an ac circuit. These phasors rotate in the counterclockwise direction with an angular velocity that is equal to the angular frequency ω of the current. $\vec{V}_R$ is in phase with the current, $\vec{V}_L$ leads the current by 90°, and $\vec{V}_C$ lags the current by 90°. The x component of each phasor equals the magnitude of the current or the corresponding potential drop at any instant.

6. ***LC and RLC Series Circuits**

If a capacitor is discharged through an inductor, the charge and the voltage on the capacitor oscillate with angular frequency

$$\omega = \frac{1}{\sqrt{LC}} \qquad\qquad 29\text{-}37$$

The current in the inductor oscillates with the same frequency, but it is out of phase with the charge by 90°. The energy oscillates between electric energy in the capacitor and magnetic energy in the inductor. If the circuit also has resistance, the oscillations are damped because energy is dissipated in the resistor.

7. **Series RLC Circuit Driven by an Applied Potential Drop of Frequency ω**

Applied potential drop	$V_{app} = V_{app,peak} \cos \omega t$	
Current	$I = \dfrac{V_{app,peak}}{Z} \cos(\omega t - \delta)$	29-50
Impedance Z	$Z = \sqrt{R^2 + (X_L - X_C)^2}$	29-49
Phase angle δ	$\tan \delta = \dfrac{X_L - X_C}{R}$	29-47
Average power	$P_{av} = I_{rms}^2 R = V_{app,rms} I_{rms} \cos \delta = \dfrac{V_{app,rms}^2 R \omega^2}{L^2(\omega^2 - \omega_0^2)^2 + \omega^2 R^2}$	29-52, 29-53, 29-54

Power factor

The quantity $\cos \delta$ in Equation 29-53 is called the power factor of the RLC circuit. At resonance, δ is zero, the power factor is 1, and

$$P_{av} = V_{app,rms} I_{rms}$$

Resonance

When the rms current is maximum, the circuit is said to be at resonance. The conditions for resonance are

$$X_L = X_C, \quad \text{so} \quad Z = \sqrt{R^2 + (X_L - X_C)^2} = R$$

$$\omega = \omega_0 = \frac{1}{\sqrt{LC}} \quad \text{and} \quad \delta = 0$$

8. **Q Factor**

The sharpness of the resonance curve is described by the Q factor

$$Q = \frac{\omega_0 L}{R} \qquad\qquad 29\text{-}55$$

When the resonance curve is reasonably narrow, the Q factor can be approximated by

$$Q = \frac{\omega_0}{\Delta\omega} = \frac{f_0}{\Delta f} \qquad\qquad 29\text{-}56$$

9. Transformers

A transformer is a device used to raise or lower the voltage in a circuit without an appreciable loss in power. For a transformer with N_1 turns in the primary and N_2 turns in the secondary, the potential difference across the secondary coil is related to the potential drop across the primary coil by

$$V_2 = \frac{N_2}{N_1} V_1$$

29-61

If there are no power losses,

$$V_{1,\text{rms}} I_{1,\text{rms}} = V_{2,\text{rms}} I_{2,\text{rms}}$$

29-63

PROBLEMS

- Single-concept, single-step, relatively easy
- •• Intermediate-level, may require synthesis of concepts
- ••• Challenging
- SSM Solution is in the *Student Solutions Manual*
- ISOLVE Problems available on iSOLVE online homework service
- ISOLVE✓ These "Checkpoint" online homework service problems ask students additional questions about their confidence level, and how they arrived at their answer.

Conceptual Problems

1 • SSM As the frequency in the simple ac circuit in Figure 29-27 increases, the rms current through the resistor (a) increases. (b) does not change. (c) may increase or decrease depending on the magnitude of the original frequency. (d) may increase or decrease depending on the magnitude of the resistance. (e) decreases.

FIGURE 29-27 Problem 1

2 • If the rms voltage in an ac circuit is doubled, the peak voltage is (a) increased by a factor of 2. (b) decreased by a factor of 2. (c) increased by a factor of $\sqrt{2}$. (d) decreased by a factor of $\sqrt{2}$. (e) not changed.

3 • If the frequency in the circuit shown in Figure 29-28 is doubled, the inductance of the inductor will (a) increase by a factor of 2. (b) not change. (c) decrease by a factor of 2. (d) increase by a factor of 4. (e) decrease by a factor of 4.

FIGURE 29-28 Problems 3 and 4

4 • If the frequency in the circuit shown in Figure 29-28 is doubled, the inductive reactance of the inductor will (a) increase by a factor of 2. (b) not change. (c) decrease by a factor of 2. (d) increase by a factor of 4. (e) decrease by a factor of 4.

5 • SSM If the frequency in the circuit in Figure 29-29 is doubled, the capacitive reactance of the circuit will (a) increase by a factor of 2. (b) not change. (c) decrease by a factor of 2. (d) increase by a factor of 4. (e) decrease by a factor of 4.

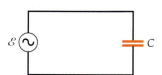

FIGURE 29-29 Problem 5

6 • In a circuit consisting of a generator and an inductor, are there any times when the inductor absorbs power from the generator? Are there any times when the inductor supplies power to the generator?

7 • In a circuit consisting of a generator and a capacitor, are there any times when the capacitor absorbs power from the generator? Are there any times when the capacitor supplies power to the generator?

8 • The SI units of inductance times capacitance are (a) seconds squared. (b) hertz. (c) volts. (d) amperes. (e) ohms.

9 •• SSM Making *LC* circuits with oscillation frequencies of thousands of hertz or more is easy, but making *LC* circuits that have small frequencies is difficult. Why?

10 • True or false:

(a) An *RLC* circuit with a high *Q* factor has a narrow resonance curve.

(b) At resonance, the impedance of an *RLC* circuit equals the resistance *R*.

(c) At resonance, the current and generator voltage are in phase.

11 • Does the power factor depend on the frequency?

12 • SSM Are there any disadvantages to having a radio tuning circuit with an extremely large *Q* factor?

13 • What is the power factor for a circuit that has inductance and capacitance but no resistance?

14 • A transformer is used to change (a) capacitance, (b) frequency, (c) voltage, (d) power, (e) none of these.

15 • True or false: If a transformer increases the current, it must decrease the voltage.

16 •• An ideal transformer has N_1 turns on its primary and N_2 turns on its secondary. The power dissipated in a load resistance *R* connected across the secondary is P_2 when the primary voltage is V_1. The current in the primary windings is then (a) P_2/V_1. (b) $(N_1/N_2)(P_2/V_1)$. (c) $(N_2/N_1)(P_2/V_1)$. (d) $(N_2/N_1)^2(P_2/V_1)$.

17 • True or false:

(a) Alternating current in a resistance dissipates no power because the current is negative as often as the current is positive.

(b) At very high frequencies, a capacitor acts like a short circuit.

Estimation and Approximation

18 •• SSM The impedances of motors, transformers, and electromagnets have inductive reactance. Suppose that the phase angle of the total impedance of a large industrial plant is 25° when the plant is under full operation and using 2.3 MW of power. The power is supplied to the plant from a substation 4.5 km from the plant; the 60 Hz rms line voltage at the plant is 40,000 V. The resistance of the transmission line from the substation to the plant is 5.2 Ω. The cost per kilowatt-hour is 0.07 dollars. The plant pays only for the actual energy used. (a) What are the resistance and inductive reactance of the plant's total load? (b) What is the current in the power lines and what must be the rms voltage at the substation to maintain the voltage at the plant at 40,000 V? (c) How much power is lost in transmission? (d) Suppose that the phase angle of the plant's impedance were reduced to 18° by adding a bank of capacitors in series with the load. How much money would be saved by the electric utility during one month of operation, assuming the plant operates at full capacity for 16 h each day? (e) What must be the capacitance of this bank of capacitors?

Alternating Current Generators

19 • A 200-turn coil has an area of 4 cm² and rotates in a magnetic field of 0.5 T. (a) What frequency will generate a maximum emf of 10 V? (b) If the coil rotates at 60 Hz, what is the maximum emf?

20 • In what magnetic field must the coil of Problem 19 be rotating to generate a maximum emf of 10 V at 60 Hz?

21 • SSM A 2-cm by 1.5-cm rectangular coil has 300 turns and rotates in a magnetic field of 4000 G. (a) What is the maximum emf generated when the coil rotates at 60 Hz? (b) What must its frequency be to generate a maximum emf of 110 V?

22 • The coil of Problem 21 rotates at 60 Hz in a magnetic field *B*. What value of *B* will generate a maximum emf of 24 V?

Alternating Current in a Resistor

23 • SSM A 100-W lightbulb is plugged into a standard 120-V (rms) outlet. Find (a) I_{rms}, (b) I_{max}, and (c) the maximum power.

24 • ISOLVE A circuit breaker is rated for a current of 15 A rms at a voltage of 120 V rms. (a) What is the largest value of I_{max} that the breaker can carry? (b) What average power can be supplied by this circuit?

Alternating Current in Inductors and Capacitors

25 • What is the reactance of a 1-mH inductor at (a) 60 Hz, (b) 600 Hz, and (c) 6 kHz?

26 • ISOLVE An inductor has a reactance of 100 Ω at 80 Hz. (a) What is its inductance? (b) What is its reactance at 160 Hz?

27 • ISOLVE At what frequency would the reactance of a 10-μF capacitor equal that of a 1-mH inductor?

28 • What is the reactance of a 1-nF capacitor at (a) 60 Hz, (b) 6 kHz, and (c) 6 MHz?

29 • SSM An emf of 10 V maximum and frequency 20 Hz is applied to a 20-μF capacitor. Find (a) I_{max} and (b) I_{rms}.

30 • ISOLVE At what frequency is the reactance of a 10-μF capacitor (a) 1 Ω, (b) 100 Ω, and (c) 0.01 Ω?

31 •• ISOLVE Two ac voltage sources are connected in series with a resistor $R = 25$ Ω. One source is given by

$$V_1 = (5 \text{ V}) \cos(\omega t - \alpha),$$

and the other source is

$$V_2 = (5 \text{ V}) \cos(\omega t + \alpha),$$

with $\alpha = \pi/6$. (a) Find the current in *R* using a trigonometric identity for the sum of two cosines. (b) Use phasor diagrams to find the current in *R*. (c) Find the current in *R* if $\alpha = \pi/4$ and the amplitude of V_2 is increased from 5 V to 7 V.

LC and *RLC* Circuits Without a Generator

32 • SSM Show from the definitions of the henry and the farad that $1/\sqrt{LC}$ has the unit s^{-1}.

33 • (a) What is the period of oscillation of an *LC* circuit consisting of a 2-mH coil and a 20-μF capacitor? (b) What inductance is needed with an 80-μF capacitor to construct an *LC* circuit that oscillates with a frequency of 60 Hz?

34 •• An LC circuit has capacitance C_1 and inductance L_1. A second circuit has capacitance $C_2 = \frac{1}{2}C_1$ and $L_2 = 2L_1$, and a third circuit has capacitance $C_3 = 2C_1$ and $L_3 = \frac{1}{2}L_1$. (a) Show that each circuit oscillates with the same frequency. (b) In which circuit would the maximum current be greatest if the capacitor in each were charged to the same potential V?

35 •• **SOLVE** A 5-μF capacitor is charged to 30 V and is then connected across a 10-mH inductor. (a) How much energy is stored in the system? (b) What is the frequency of oscillation of the circuit? (c) What is the maximum current in the circuit?

36 • **SOLVE** A coil can be considered to be a resistance and an inductance in series. Assume that $R = 100\ \Omega$ and $L = 0.4$ H. The coil is connected across a 120-V rms, 60-Hz line. Find (a) the power factor, (b) the rms current, and (c) the average power supplied.

37 •• **SSM** An inductor and a capacitor are connected, as shown in Figure 29-30. With the switch open, the left plate of the capacitor has charge Q_0. The switch is closed and the charge and current vary sinusoidally with time. (a) Plot both Q versus t and I versus t and explain how to interpret these two plots to illustrate that the current leads the charge by 90°. (b) Using a trig identity, show the expression for the current (Equation 29-38) leads the expression for the charge (Equation 29-39) by 90°. That is, show

$$I = -I_{peak} \sin \omega t = I_{peak} \cos\left(\omega t + \frac{\pi}{2}\right).$$

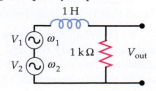

FIGURE 29-30 Problem 37

RL Circuits With a Generator

38 •• **SOLVE** A resistance R and a 1.4-H inductance are in series across a 60-Hz ac voltage. The voltage across the resistor is 30 V and the voltage across the inductor is 40 V. (a) What is the resistance R? (b) What is the ac input voltage?

39 •• **SOLVE** A coil has a dc resistance of 80 Ω and an impedance of 200 Ω at a frequency of 1 kHz. Neglect the wiring capacitance of the coil at this frequency. What is the inductance of the coil?

40 •• A single transmission line carries two voltage signals given by $V_1 = 10$ V cos $100t$ and $V_2 = 10$ V cos $10,000\ t$, where t is in seconds. A series inductor of 1 H and a shunting resistor of 1 kΩ are inserted into the transmission line, as indicated in Figure 29-31. (a) What is the voltage signal observed at the output side of the transmission line? (b) What is the ratio of the low-frequency amplitude to the high-frequency amplitude?

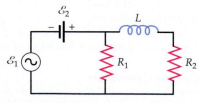

FIGURE 29-31 Problem 40

41 •• **SOLVE** A coil with resistance and inductance is connected to a 120-V rms, 60-Hz line. The average power supplied to the coil is 60 W, and the rms current is 1.5 A. Find (a) the power factor, (b) the resistance of the coil, and (c) the inductance of the coil. (d) Does the current lag or lead the voltage? What is the phase angle δ?

42 •• **SOLVE** A 36-mH inductor with a resistance of 40 Ω is connected to a source whose voltage is $\mathcal{E} = 345$ V cos $150\pi t$, where t is in seconds. Determine the maximum current in the circuit, the maximum and rms voltages across the inductor, the average power dissipation, and the maximum and average energy stored in the magnetic field of the inductor.

43 •• A coil of resistance R, inductance L, and negligible capacitance has a power factor of 0.866 at a frequency of 60 Hz. What is the power factor for a frequency of 240 Hz?

44 •• **SSM** A resistor and an inductor are connected in parallel across an emf $\mathcal{E} = \mathcal{E}_{max}$ as shown in Figure 29-32. Show that (a) the current in the resistor is $I_R = \mathcal{E}_{max}/R \cos \omega t$, (b) the current in the inductor is $I_L = \mathcal{E}_{max}/X_L \cos(\omega t - 90°)$, and (c) $I = I_R + I_L = I_{max} \cos(\omega t - \delta)$, where tan $\delta = R/X_L$ and $I_{max} = \mathcal{E}_{max}/Z$ with $Z^{-2} = R^{-2} + X_L^{-2}$.

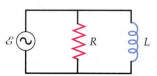

FIGURE 29-32 Problem 44

45 •• Figure 29-33 shows a load resistor $R_L = 20\ \Omega$ connected to a high-pass filter consisting of an inductor $L = 3.2$ mH and a resistor $R = 4\ \Omega$. The input voltage is $\mathcal{E} = 100$ V cos $2\pi ft$. Find the rms currents in R, L, and R_L if (a) $f = 500$ Hz and (b) $f = 2000$ Hz. (c) What fraction of the total power delivered by the voltage source is dissipated in the load resistor if the frequency is 500 Hz and if the frequency is 2000 Hz?

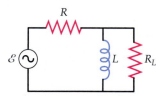

FIGURE 29-33 Problem 45

46 •• An ac source $\mathcal{E}_1 = 20$ V cos $2\pi ft$ in series with a battery whose emf is $\mathcal{E}_2 = 16$ V is connected to a circuit consisting of resistors $R_1 = 10\ \Omega$ and $R_2 = 8\ \Omega$ and an inductor $L = 6$ mH (Figure 29-34). Find the power dissipated in R_1 and R_2 if (a) $f = 100$ Hz, (b) $f = 200$ Hz, and (c) $f = 800$ Hz.

FIGURE 29-34 Problem 46

47 •• A 100-V rms voltage is applied to a series RC circuit. The rms voltage across the capacitor is 80 V. What is the voltage across the resistor?

Filters and Rectifiers

48 •• [SSM] The circuit shown in Figure 29-35 is called an *RC* high-pass filter because it transmits signals with a high-input frequency with greater amplitude than low-frequency signals. If the input voltage is $V_{in} = V_{peak} \cos \omega t$, show that the output voltage is $V_{out} = V_H \cos(\omega t - \delta)$ where

$$V_H = \frac{V_{peak}}{\sqrt{1 + \left(\dfrac{1}{\omega RC}\right)^2}}$$

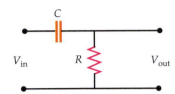

FIGURE 29-35 Problem 48

49 •• (a) Show that the phase constant δ in Problem 48 is given by

$$\tan \delta = -\left(\frac{1}{\omega RC}\right)$$

(b) What is the value of δ in the limit as $\omega \to 0$? (c) What is the value of δ in the limit $\omega \to \infty$?

50 •• Assume that the resistor of Problem 48 has value $R = 20\ k\Omega$ and the capacitor has value $C = 15\ nF$. (a) At what frequency f is $V_{out} = V_{in}/\sqrt{2}$? (This is known as the 3 dB frequency, or f_{3dB} for the circuit.) (b) Using a spreadsheet program, make a graph of V_{out} versus f. Use a logarithmic scale for each variable. Make sure that the scale extends from at least $0.1\ f_{3dB}$ to $10\ f_{3dB}$ (c) Make a graph of δ versus f and graph f on a logarithmic scale. What value does δ have at $f = f_{3dB}$?

51 ••• Show that if an arbitrary voltage signal is fed into the high-pass filter of Problem 48, in which the time variance of the signal is much slower than $1/(RC)$, the output of the circuit will be proportional to the time derivative of the input.

52 •• We define the output from the high-pass filter from Problem 48 in the decibel scale as

$$\beta = 20 \log_{10} \frac{V_H}{V_{peak}}$$

Show that for $f \ll f_{3dB}$, where f_{3dB} is defined in Problem 50, the output drops at a rate of 6 dB per octave. That is, every time the frequency is halved, the output drops by 6 dB.

53 •• [SSM] Show that the average power dissipated in the resistor of the high-pass filter of Problem 48 is given by

$$P_{ave} = \frac{V_{peak}^2}{2R} \left(\frac{(\omega RC)^2}{1 + (\omega RC)^2}\right)$$

54 •• One application of the high-pass filter of Problem 48 is that of a noise filter for electronic circuits (i.e., one that blocks out low-frequency noise). Using $R = 20\ k\Omega$, pick a value for C for a high-pass filter that attenuates an input voltage signal at $f = 60\ Hz$ by a factor of 10.

55 •• The circuit shown in Figure 29-36 is a low-pass filter. If the input voltage is

$$V_{in} = V_{peak} \cos \omega t \text{ show that the output voltage is}$$

$$V_{out} = V_L \cos(\omega t - \delta) \text{ where}$$

$$V_L = \frac{V_{peak}}{\sqrt{1 + (\omega RC)^2}}$$

Discuss the behavior of the output voltage in the limiting cases $\omega \to 0$ and $\omega \to \infty$.

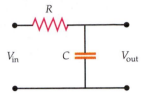

FIGURE 29-36 Problem 55

56 •• Show that δ for the low-pass filter of Problem 55 is given by the expression $\tan \delta = \omega RC$. Find the value of δ in the limit $\omega \to 0$ and $\omega \to \infty$.

57 •• [SSM] Using a spreadsheet program, make a graph of V_L versus $f = \omega/2\pi$ and δ versus f for the low-pass filter of Problem 55. Use $R = 10\ k\Omega$ and $C = 5\ nF$.

58 ••• Show that if an arbitrary voltage signal is fed into the low-pass filter of Problem 55, in which the time variance of the signal is much faster than $1/(RC)$, the output of the circuit will be proportional to the integral of the input.

59 ••• [SSM] Show the *trap* filter, shown in Figure 29-37, acts to reject signals at a frequency $\omega = 1/\sqrt{LC}$. How does the width of the frequency band rejected depend on the resistance R?

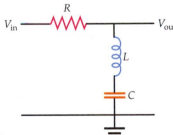

FIGURE 29-37 Problem 59

60 •• A half-wave rectifier for transforming an ac voltage into a dc voltage is shown in Figure 29-38. The diode in the figure can be thought of as a one-way valve for current, allowing current to pass in the forward (upward) direction when the voltage between points A and B is greater than $+0.6\ V$. The resistance of the diode is effectively infinite when the voltage is less than $+0.6\ V$. Using the same axes plot two cycles of both V_{in} and V_{out} versus t when $V_{in} = V_{peak} \cos \omega t$.

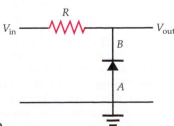

FIGURE 29-38 Problem 60

61 •• (a) The output of rectifier of Problem 60, Figure 29-39a, can be smoothed by putting its output through a low-pass filter. The resulting output is a dc voltage with a small amount of ripple on it, as shown in Figure 29-39b. If the input frequency $f = \omega/2\pi = 60$ Hz and the resistance is $R = 1$ kΩ, find an approximate value for C, so that the output voltage varies by less than 50 percent of the mean value over one cycle.

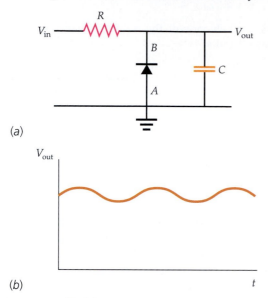

(a)

(b)

t

FIGURE 29-39 Problem 61

LC Circuits With a Generator

62 •• The generator voltage in Figure 29-40 is given by $\mathcal{E} = (100$ V$) \cos 2\pi ft$. (a) For each branch, what is the amplitude of the current and what is its phase relative to the applied voltage? (b) What is the angular frequency ω so that the current in the generator vanishes? (c) At this resonance, what is the current in the inductor? What is the current in the capacitor? (d) Draw a phasor diagram showing the general relationships between the applied voltage, the generator current, the capacitor current, and the inductor current for the case where the inductive reactance is larger than the capacitive reactance.

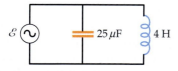

25 μF   4 H

FIGURE 29-40 Problem 62

63 •• **SOLVE** The charge on the capacitor of a series LC circuit is given by $Q = (15\ \mu C) \cos(1250t + \frac{\pi}{4})$, where t is in seconds. (a) Find the current as a function of time. (b) Find C if $L = 28$ mH. (c) Write expressions for the electrical energy U_e, the magnetic energy U_m, and the total energy U.

64 ••• **SSM** One method for measuring the compressibility of a dielectric material uses an LC circuit with a parallel-plate capacitor. The dielectric is inserted between the plates and the change in resonance frequency is determined as the capacitor plates are subjected to a compressive stress. In such an arrangement, the resonance frequency is 120 MHz when a dielectric of thickness 0.1 cm and dielectric constant

$\kappa = 6.8$ is placed between the capacitor plates. Under a compressive stress of 800 atm, the resonance frequency decreases to 116 MHz. Find Young's modulus of the dielectric material.

65 ••• Figure 29-41 shows an inductance L and a parallel plate capacitor of width $w = 20$ cm and thickness 0.2 cm. A dielectric with dielectric constant $\kappa = 4.8$ that can completely fill the space between the capacitor plates can be slid between the plates. The inductor has an inductance $L = 2$ mH. When half the dielectric is between the capacitor plates (i.e., when $x = \frac{1}{2} w$), the resonant frequency of this LC combination is 90 MHz. (a) What is the capacitance of the capacitor without the dielectric? (b) Find the resonance frequency as a function of x.

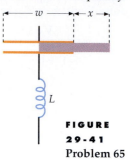

FIGURE 29-41
Problem 65

RLC Circuits With a Generator

66 • A series RLC circuit in a radio receiver is tuned by a variable capacitor, so that it can resonate at frequencies from 500 to 1600 kHz. If $L = 1\ \mu$H, find the range of capacitances necessary to cover this range of frequencies.

67 • (a) Find the power factor for the circuit in Example 29-5 when $\omega = 400$ rad/s. (b) At what angular frequency is the power factor 0.5?

68 • **SOLVE** An ac generator with a maximum emf of 20 V is connected in series with a 20-μF capacitor and an 80-Ω resistor. There is no inductance in the circuit. Find (a) the power factor, (b) the rms current, and (c) the average power if the angular frequency of the generator is 400 rad/s.

69 •• **SSM** Show that the formula $P_{av} = R\mathcal{E}_{rms}^2/Z^2$ gives the correct result for a circuit containing only a generator and (a) a resistor, (b) a capacitor, and (c) an inductor.

70 •• **SOLVE** A series RLC circuit with $L = 10$ mH, $C = 2\ \mu$F, and $R = 5\ \Omega$ is driven by a generator with a maximum emf of 100 V and a variable angular frequency ω. Find (a) the resonant frequency ω_0 and (b) I_{rms} at resonance. When $\omega = 8000$ rad/s, find (c) X_C and X_L, (d) Z and I_{rms}, and (e) the phase angle δ.

71 •• For the circuit in problem 70, let the generator frequency be $f = \omega/2\pi = 1$ kHz. Find (a) the resonance frequency $f_0 = \omega_0/2\pi$, (b) X_C and X_L, (c) the total impedance Z and I_{rms}, and (d) the phase angle δ.

72 •• Find the power factor and the phase angle δ for the circuit in Problem 70 when the generator frequency is (a) 900 Hz, (b) 1.1 kHz, and (c) 1.3 kHz.

73 •• Find (a) the Q factor and (b) the resonance width for the circuit in Problem 70. (c) What is the power factor when $\omega = 8000$ rad/s?

74 •• **SSM** **SOLVE** FM radio stations have carrier frequencies that are separated by 0.20 MHz. When the radio is tuned to a station, such as 100.1 MHz, the resonance width of the receiver circuit should be much smaller than 0.2 MHz, so that adjacent stations are not received. If $f_0 = 100.1$ MHz and $\Delta f = 0.05$ MHz, what is the Q factor for the circuit?

75 •• **SOLVE** A coil is connected to a 60-Hz, 100-V ac generator. At this frequency, the coil has an impedance of 10 Ω and a reactance of 8 Ω. (*a*) What is the current in the coil? (*b*) What is the phase angle between the current and the applied voltage? (*c*) What series capacitance is required so that the current and voltage are in phase? (*d*) What is the voltage measured across the capacitor?

76 •• An 0.25-H inductor and a capacitor *C* are connected in series with a 60-Hz ac generator. An ac voltmeter is used to measure the rms voltages across the inductor and capacitor separately. The rms voltage across the capacitor is 75 V and that across the inductor is 50 V. (*a*) Find the capacitance *C* and the rms current in the circuit. (*b*) What would be the measured rms voltage across both the capacitor and inductor together?

77 •• (*a*) Show that Equation 29-47 can be written as

$$\tan \delta = \frac{L(\omega^2 - \omega_0^2)}{\omega R}$$

Find δ approximately at (*b*) very low frequencies and (*c*) very high frequencies.

78 •• (*a*) Show that in a series *RC* circuit with no inductance, the power factor is given by

$$\cos \delta = \frac{RC\omega}{\sqrt{1 + (RC\omega)^2}}$$

(*b*) Using a spreadsheet program, graph the power factor versus ω.

79 •• **SSM** In the circuit shown in Figure 29-42, the ac generator produces an rms voltage of 115 V when operated at 60 Hz. What is the rms voltage across points (*a*) *AB*, (*b*) *BC*, (*c*) *CD*, (*d*) *AC*, and (*e*) *BD*?

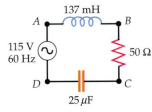

FIGURE 29-42 Problem 79

80 •• When an *RLC* series circuit is connected to a 120-V rms, 60-Hz line, the current is $I_{rms} = 11$ A and the current leads the voltage by 45°. (*a*) Find the power supplied to the circuit. (*b*) What is the resistance? (*c*) If the inductance $L = 0.05$ H, find the capacitance *C*. (*d*) What capacitance or inductance should you add to make the power factor 1?

81 •• **SOLVE** A series *RLC* circuit is driven at a frequency of 500 Hz. The phase angle between the applied voltage and current is determined from an oscilloscope measurement to be δ = 75°. If the total resistance is known to be 35 Ω and the inductance is 0.15 H, what is the capacitance of the circuit?

82 •• A series *RLC* circuit with $R = 400$ Ω, $L = 0.35$ H, and $C = 5$ μF is driven by a generator of variable frequency *f*. (*a*) What is the resonance frequency f_0? Find *f* and f/f_0 when the phase angle δ is (*b*) 60° and (*c*) −60°.

83 •• Sketch the impedance *Z* versus ω for (*a*) a series *LR* circuit, (*b*) a series *RC* circuit, and (*c*) a series *RLC* circuit.

84 •• **SSM** Show that Equation 29-48 can be written as

$$I_{max} = \frac{\omega \mathcal{E}_{max}}{\sqrt{L^2(\omega^2 - \omega_0^2)^2 + \omega^2 R^2}}$$

85 •• In a series *RLC* circuit, $X_C = 16$ Ω and $X_L = 4$ Ω at some frequency. The resonance frequency is $\omega_0 = 10^4$ rad/s. (*a*) Find *L* and *C*. If $R = 5$ Ω and $\mathcal{E}_{max} = 26$ V, find (*b*) the *Q* factor, and (*c*) the maximum current.

86 •• In a series *RLC* circuit connected to an ac generator whose maximum emf is 200 V, the resistance is 60 Ω and the capacitance is 8 μF. The inductance can be varied from 8 mH to 40 mH, by the insertion of an iron core in the solenoid. The angular frequency of the generator is 2500 rad/s. If the capacitor voltage is not to exceed 150 V, find (*a*) the maximum current and (*b*) the range of inductance that is safe to use.

87 •• A certain electrical device draws 10 A rms and has an average power of 720 W when connected to a 120-V rms, 60-Hz power line. (*a*) What is the impedance of the device? (*b*) What series combination of resistance and reactance is this device equivalent to? (*c*) If the current leads the emf, is the reactance inductive or capacitive?

88 •• **SSM** A method for measuring inductance is to connect the inductor in series with a known capacitance, a known resistance, an ac ammeter, and a variable-frequency signal generator. The frequency of the signal generator is varied and the emf is kept constant until the current is maximum. (*a*) If $C = 10$ μF, $\mathcal{E}_{max} = 10$ V, $R = 100$ Ω, and *I* is maximum at ω = 5000 rad/s, what is *L*? (*b*) What is I_{max}?

89 •• A resistor and a capacitor are connected in parallel across a sinusoidal emf $\mathcal{E} = \mathcal{E}_{max} \cos \omega t$, as shown in Figure 29-43. (*a*) Show that the current in the resistor is $I_R = (\mathcal{E}_{max}/R) \cos \omega t$. (*b*) Show that the current in the capacitor branch is $I_C = (\mathcal{E}_{max}/X_C) \cos(\omega t + 90°)$. (*c*) Show that the total current is given by $I = I_R + I_C = I_{max} \cos(\omega t + \delta)$, where $\tan \delta = R/X_C$ and $I_{max} = \mathcal{E}_{max}/Z$ with $Z^{-2} = R^{-2} + X_C^{-2}$.

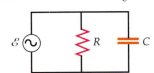

FIGURE 29-43 Problem 89

90 •• **SSM** In the circuit shown in Figure 29-44, $R = 10$ Ω, $R_L = 30$ Ω, $L = 150$ mH, and $C = 8$ μF; the frequency of the ac source is 10 Hz and its amplitude is 100 V. (*a*) Using phasor diagrams, determine the impedance of the circuit when switch S is closed. (*b*) Determine the impedance of the circuit when switch S is open. (*c*) What are the voltages across the load resistor R_L when switch S is closed and when it is open? (*d*) Repeat Parts (*a*), (*b*), and (*c*) with the frequency of the source changed to 1000 Hz. (*e*) Which arrangement is a better low-pass filter, S open or S closed?

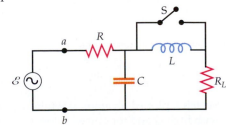

FIGURE 29-44
Problem 90

91 •• In the circuit shown in Figure 29-45, $R_1 = 2\ \Omega$, $R_2 = 4\ \Omega$, $L = 12$ mH, $C = 30\ \mu$F, and $\mathcal{E} = (40$ V$) \cos \omega t$. (a) Find the resonance frequency. (b) At the resonance frequency, what are the rms currents in each resistor and the rms current supplied by the source emf?

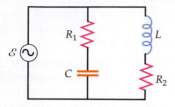

FIGURE 29-45 Problems 91, 103, and 104

92 •• For the circuit in Figure 29-24, derive an expression for the Q of the circuit, assuming the resonance is sharp.

93 •• **ISOLVE** For the circuit in Figure 29-24, $L = 4$ mH. (a) What capacitance C will result in a resonance frequency of 4 kHz? (b) When C has the value found in Part (a), what should be the resistance R, so that the Q of the circuit is 8?

94 •• If the capacitance of C in Problem 93 is reduced to half the value found in Problem 93, what then are the resonance frequency and the Q of the circuit? What should be the resistance R to give $Q = 8$?

95 •• **ISOLVE** A series circuit consists of a 4.0-nF capacitor, a 36-mH inductor, and a 100-Ω resistor. The circuit is connected to a 20-V ac source whose frequency can be varied over a wide range. (a) Find the resonance frequency f_0 of the circuit. (b) At resonance, what is the rms current in the circuit and what are the rms voltages across the inductor and capacitor? (c) What is the rms current and what are the rms voltages across the inductor and capacitor at $f = f_0 + \frac{1}{2}\Delta f$, where Δf is the width of the resonance?

96 ••• In the parallel circuit shown in Figure 29-46, $V_{max} = 110$ V. (a) What is the impedance of each branch? (b) For each branch, what is the current amplitude and its phase relative to the applied voltage? (c) Give the current phasor diagram, and use it to find the total current and its phase relative to the applied voltage.

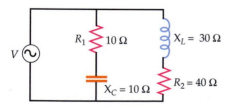

FIGURE 29-46 Problem 96

97 ••• **SSM** (a) Show that Equation 29-47 can be written as

$$\tan \delta = \frac{Q(\omega^2 - \omega_0^2)}{\omega \omega_0}$$

(b) Show that near resonance

$$\tan \delta \approx \frac{2Q(\omega - \omega)}{\omega}$$

(c) Sketch a plot of δ versus x, where $x = \omega/\omega_0$, for a circuit with high Q and for one with low Q.

98 ••• Show by direct substitution that the current given by Equation 29-46 with δ and I_{max} given by Equations 29-47 and 29-48, respectively, satisfies Equation 29-45. (*Hint:* Use trigonometric identities for the sine and cosine of the sum of two angles, and write the equation in the form $A \sin \omega t + B \cos \omega t = 0$. Because this equation must hold for all times, $A = 0$ and $B = 0$.)

99 ••• An ac generator is in series with a capacitor and an inductor in a circuit with negligible resistance. (a) Show that the charge on the capacitor obeys the equation

$$L\frac{d^2Q}{dt^2} + \frac{Q}{C} = \mathcal{E}_{max} \cos \omega t$$

(b) Show by direct substitution that this equation is satisfied by $Q = Q_{max} \cos \omega t$ if,

$$Q_{max} = -\frac{\mathcal{E}_{max}}{L(\omega^2 - \omega_0^2)}$$

(c) Show that the current can be written as $I = I_{max} \cos(\omega t - \delta)$, where

$$I_{max} = \frac{\omega \mathcal{E}_{max}}{L|\omega^2 - \omega_0^2|} = \frac{\mathcal{E}_{max}}{|X_L - X_C|}$$

and $\delta = -90°$ for $\omega < \omega_0$ and $\delta = 90°$ for $\omega > \omega_0$.

100 ••• Figure 29-20 shows a plot of average power P_{av} versus generator frequency ω for an RLC circuit with a generator. The average power P_{av} is given by Equation 29-54. The full width at half-maximum, $\Delta \omega$, is the width of the resonance curve between the two points, where P_{av} is one-half its maximum value. Show that for a sharply peaked resonance, $\Delta \omega \approx R/L$ and, hence, that $Q \approx \omega_0/\Delta \omega$ in this case (Equation 29-56). (*Hint:* At resonance, the denominator of the expression on the right of Equation 29-54 is $\omega^2 R^2$. The half-power points will occur when the denominator is twice the value near resonance; that is, when $L^2(\omega^2 - \omega_0^2)^2 = \omega^2 R^2 \approx \omega_0^2 R^2$. Let ω_1 and ω_2 be the solutions of this equation. For a sharply peaked resonance, $\omega_1 \approx \omega_0$ and $\omega_2 \approx \omega_0$. Then, using the fact that $\omega + \omega_0 \approx 2\omega_0$, one finds that $\Delta \omega = \omega_2 - \omega_1 \approx R/L$.)

101 • Show by direct substitution that

$$L\frac{d^2Q}{dt^2} + R\frac{dQ}{dt} + \frac{1}{C}Q = 0$$

(Equation 29-43b) is satisfied by

$$Q = Q_0 e^{-Rt/2L} \cos \omega' t$$

where

$$\omega' = \sqrt{(1/LC - (R/2L)^2)}$$

and Q_0 is the charge on the capacitor at $t = 0$.

102 ••• **SSM** One method for measuring the magnetic susceptibility of a sample uses an LC circuit consisting of an air-core solenoid and a capacitor. The resonant frequency of the circuit without the sample is determined and then measured again with the sample inserted in the solenoid. Suppose the solenoid is 4 cm long, 0.3 cm in diameter, and has 400 turns of fine wire. Assume that the sample that is inserted in the solenoid is also 4 cm long and fills the air space. Neglect end effects. (In practice, a test sample of known susceptibility of the same shape as the unknown is used to calibrate the instrument.) (a) What is the inductance of the empty solenoid? (b) What should be the capacitance of the capacitor so that the resonance frequency of the circuit without a sample is 6.0000 MHz? (c) When a sample is inserted in the solenoid, the resonance frequency drops to 5.9989 MHz. Determine the sample's susceptibility.

103 ••• (a) Find the angular frequency ω for the circuit in Problem 91 so that the magnitude of the reactance of the two parallel branches are equal. (b) At that frequency, what is the power dissipation in each of the two resistors?

104 ••• (a) For the circuit of Problem 91, find the angular frequency ω for which the power dissipation in the two resistors is the same. (b) At that angular frequency, what is the reactance of each of the two parallel branches? (c) Draw a phasor diagram showing the current through each of the two parallel branches. (d) What is the impedance of the circuit?

*The Transformer

105 • **SSM** An ac voltage of 24 V is required for a device whose impedance is 12 Ω. (a) What should the turn ratio of a transformer be, so that the device can be operated from a 120-V line? (b) Suppose the transformer is accidentally connected reversed (i.e., with the secondary winding across the 120-V line and the 12-Ω load across the primary). How much current will then flow in the primary winding?

106 • A transformer has 400 turns in the primary and 8 turns in the secondary. (a) Is this a step-up or a step-down transformer? (b) If the primary is connected across 120 V rms, what is the open-circuit voltage across the secondary? (c) If the primary current is 0.1 A, what is the secondary current, assuming negligible magnetization current and no power loss?

107 • **iSOLVE** The primary of a step-down transformer has 250 turns and is connected to a 120-V rms line. The secondary is to supply 20 A at 9 V. Find (a) the current in the primary and (b) the number of turns in the secondary, assuming 100 percent efficiency.

108 • A transformer has 500 turns in its primary, which is connected to 120 V rms. Its secondary coil is tapped at three places to give outputs of 2.5 V, 7.5 V, and 9 V. How many turns are needed for each part of the secondary coil?

109 • The distribution circuit of a residential power line is operated at 2000 V rms. This voltage must be reduced to 240 V rms for use within residences. If the secondary side of the transformer has 400 turns, how many turns are in the primary?

110 •• **SSM** An audio oscillator (ac source) with an internal resistance of 2000 Ω and an open-circuit rms output voltage of 12 V is to be used to drive a loudspeaker with a resistance of 8 Ω. What should be the ratio of primary to secondary turns of a transformer, so that maximum power is transferred to the speaker? Suppose a second identical speaker is connected in parallel with the first speaker. How much power is then supplied to the two speakers combined?

111 •• One use of a transformer is for *impedance matching*. For example, the output impedance of a stereo amplifier is matched to the impedance of a speaker by a transformer. In Equation 29-63, the currents I_1 and I_2 can be related to the impedance Z in the secondary because $I_2 = V_2/Z$. Using Equations 29-61 and 29-62, show that

$$I_1 = \mathcal{E}/[(N_1/N_2)^2 Z]$$

and, therefore, $Z_{eff} = (N_1/N_2)^2 Z$.

General Problems

112 • A 5-kW electric clothes dryer runs on 240 V rms. Find (a) I_{rms} and (b) I_{max}. (c) Find the same quantities for a dryer of the same power that operates at 120 V rms.

113 • Find the reactance of a 10.0-μF capacitor at (a) 60 Hz, (b) 6 kHz, and (c) 6 MHz.

114 •• **iSOLVE** A resistance R carries a current $I = 5$ A sin $120\pi t + 7$ A sin $240\pi t$. (a) What is the rms current? (b) If the resistance R is 12 Ω, what is the power dissipated in the resistor? (c) What is the rms voltage across the resistor?

115 •• **SSM** Figure 29-47 shows the voltage V versus time t for a *square-wave* voltage. If $V_0 = 12$ V, (a) what is the rms voltage of this waveform? (b) If this alternating waveform is rectified by eliminating the negative voltages, so that only the positive voltages remain, what now is the rms voltage of the rectified waveform?

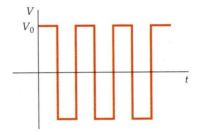

FIGURE 29-47 Problem 115

116 •• **iSOLVE** A pulsed current has a constant value of 15 A for the first 0.1 s of each second and is then 0 for the next 0.9 s of each second. (a) What is the rms value for this current waveform? (b) Each current pulse is generated by a voltage pulse of maximum value 100 V. What is the average power delivered by the pulse generator?

117 •• A circuit consists of two capacitors, a 24-V battery, and an ac voltage connected, as shown in Figure 29-48. The ac voltage is given by $\mathcal{E} = 20$ V cos $120\pi t$, where t is in seconds. (a) Find the charge on each capacitor as a function of time. Assume transient effects have had sufficient time to decay. (b) What is the steady-state current? (c) What is the maximum energy stored in the capacitors? (d) What is the minimum energy stored in the capacitors?

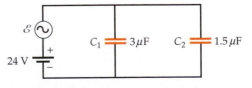

FIGURE 29-48 Problem 117

118 •• What are the average values and rms values of current for the two current waveforms shown in Figure 29-49?

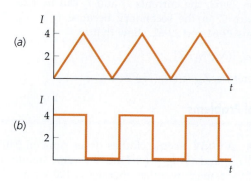

FIGURE 29-49 Problem 118

119 •• **iSOLVE** In the circuit shown in Figure 29-50, $\mathscr{E}_1 = (20\ V) \cos 2\pi ft$, $f = 180$ Hz; $\mathscr{E}_2 = 18$ V, and $R = 36\ \Omega$. Find the maximum, minimum, average, and rms values of the current through the resistor.

FIGURE 29-50
Problems 119 through 121

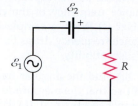

120 •• **SSM** Repeat Problem 119 if the resistor R is replaced by a 2-μF capacitor.

121 •• Repeat Problem 119 if the resistor R is replaced by a 12-mH inductor.

Maxwell's Equations and Electromagnetic Waves

THE 70-M ANTENNA AT GOLDSTONE, CALIFORNIA. THE GOLDSTONE DEEP SPACE COMMUNICATIONS COMPLEX, LOCATED IN THE MOJAVE DESERT IN CALIFORNIA, IS ONE OF THREE COMPLEXES THAT COMPRISE NASA'S DEEP SPACE NETWORK. THIS NETWORK PROVIDES RADIO COMMUNICATIONS FOR ALL OF NASA'S INTERPLANETARY SPACECRAFT AND IS ALSO UTILIZED FOR RADIO ASTRONOMY AND RADAR OBSERVATIONS OF THE SOLAR SYSTEM AND THE UNIVERSE.

? **Did you ever wonder** whether a radio antenna generates a wave equally in all directions? This topic is discussed in Section 30-3.

Maxwell's equations, first proposed by the great Scottish physicist James Clerk Maxwell, relate the electric and magnetic field vectors $\vec{E}$ and $\vec{B}$ and their sources, which are electric charges and currents. These equations summarize the experimental laws of electricity and magnetism—the laws of Coulomb, Gauss, Biot–Savart, Ampère, and Faraday. These experimental laws hold in general except for Ampère's law, which applies only to steady continuous currents.

➤ **In this chapter, we will see how Maxwell was able to generalize Ampère's law with the invention of the displacement current (Section 30-1). Maxwell was then able to show that the generalized laws of electricity and magnetism imply the existence of electromagnetic waves.**

Maxwell's equations play a role in classical electromagnetism analogous to that of Newton's laws in classical mechanics. In principle, all problems in classical electricity and magnetism can be solved using Maxwell's equations, just as all problems in classical mechanics can be solved using Newton's laws. Maxwell's equations are considerably more complicated than Newton's laws, however, and their application to most problems involves mathematics beyond the scope of this book. Nevertheless, Maxwell's equations are of great theoretical importance. For example, Maxwell showed that these equations can be combined to yield a wave equation for the electric and magnetic field vectors $\vec{E}$ and $\vec{B}$. Such **electromagnetic waves** are caused by accelerating charges, (e.g., the charges in an alternating current in an antenna). These electromagnetic waves were first produced in the laboratory by Heinrich Hertz in 1887. Maxwell showed that his equations predicted the speed of electromagnetic waves in free space to be

$$c = \frac{1}{\sqrt{\mu_0 \epsilon_0}}$$ 30-1

THE SPEED OF ELECTROMAGNETIC WAVES

where ϵ_0, the permittivity of free space, is the constant appearing in Coulomb's and Gauss's laws and μ_0, the permeability of free space, is the constant appearing in the Biot–Savart law and Ampère's law. Maxwell noticed with great excitement the coincidence that the measure for the speed of light equaled $1/\sqrt{\mu_0 \epsilon_0}$, and Maxwell correctly surmised that light itself is an electromagnetic wave. Today, the value of c is defined as 2.99792458×10^8 m/s, the value of μ_0 is defined as $4\pi \times 10^7$ N/A^2, and the value of ϵ_0 is defined by Equation 30-1.

30-1 Maxwell's Displacement Current

Ampère's law (Equation 27-15) relates the line integral of the magnetic field around some closed curve C to the current that passes through any surface bounded by that curve:

$$\oint_C \vec{B} \cdot d\vec{\ell} = \mu_0 I_s, \text{ for any closed curve } C$$ 30-2

Maxwell recognized a flaw in Ampère's law. Figure 30-1 shows two different surfaces, S_1 and S_2, bounded by the same curve C, which encircles a wire carrying current to a capacitor plate. The current through surface S_1 is I, but there is no current through surface S_2 because the charge stops on the capacitor plate. Thus, there is ambiguity in the phrase "the current through any surface bounded by the curve." Such a problem arises when the current is not continuous.

Maxwell showed that the law can be generalized to include all situations if the current I in the equation is replaced by the sum of the conduction current I and another term I_d, called **Maxwell's displacement current,** defined as

$$I_d = \epsilon_0 \frac{d\phi_e}{dt}$$ 30-3

DEFINITION—DISPLACEMENT CURRENT

where ϕ_e is the flux of the electric field through the same surface bounded by the curve C. The generalized form of Ampère's law is then

$$\oint_C \vec{B} \cdot d\vec{\ell} = \mu_0(I + I_d) = \mu_0 I + \mu_0 \epsilon_0 \frac{d\phi_e}{dt}$$ 30-4

GENERALIZED FORM OF AMPÈRE'S LAW

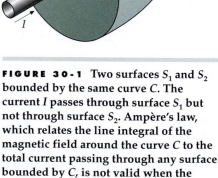
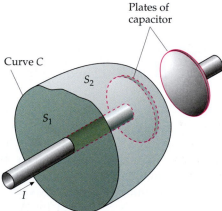

FIGURE 30-1 Two surfaces S_1 and S_2 bounded by the same curve C. The current I passes through surface S_1 but not through surface S_2. Ampère's law, which relates the line integral of the magnetic field around the curve C to the total current passing through any surface bounded by C, is not valid when the current is not continuous, as when it stops at the capacitor plate here.

We can understand this generalization by considering Figure 30-1 again. Let us call the sum $I + I_d$ the generalized current. According to the argument just stated, the same generalized current must cross any surface bounded by the curve C. Thus, there can be no net generalized current into or out of the volume bounded by the two surfaces S_1 and S_2, which together form a closed surface. If there is a net conduction current I into the volume, there must be an equal net displacement current I_d out of the volume. In the volume in the figure, there is a net conduction current I into the volume that increases the charge Q_{inside} within the volume:

$$I = \frac{dQ_{inside}}{dt}$$

The flux of the electric field out of the volume is related to the charge by Gauss's law:

$$\phi_{e,\,net} = \oint_S E_n \, dA = \frac{1}{\epsilon_0} Q_{inside}$$

Solving for the charge gives

$$Q_{inside} = \epsilon_0 \, \phi_{e,\,net}$$

and taking the derivative of each side gives

$$\frac{dQ_{inside}}{dt} = \epsilon_0 \frac{d\phi_{e,\,net}}{dt}$$

The rate of increase of the charge is thus proportional to the rate of increase of the net flux out of the volume:

$$\frac{dQ_{inside}}{dt} = \epsilon_0 \frac{d\phi_{e,\,net}}{dt} = I_d$$

Thus, the net conduction current into the volume equals the net displacement current out of the volume. The generalized current is thus continuous, and this is *always* the case.

It is interesting to compare Equation 30-4 to Equation 28-5:

$$\mathcal{E} = \oint_C \vec{E} \cdot d\vec{\ell} = -\frac{d\phi_m}{dt} = -\int_S \frac{\partial B_n}{\partial t} \, dA \qquad\qquad 30\text{-}5$$

which in this chapter will be referred to as Faraday's law. (Equation 30-5 is a restricted form of Faraday's law, a form that does not include motional emfs. Equation 30-5 does include emfs associated with a time varying magnetic field.) According to Faraday's law, a changing magnetic flux produces an electric field whose line integral around a closed curve is proportional to the rate of change of magnetic flux through any surface bounded by the curve. Maxwell's modification of Ampère's law shows that a changing electric flux produces a magnetic field whose line integral around a curve is proportional to the rate of change of the electric flux. We thus have the interesting reciprocal result that a changing magnetic field produces an electric field (Faraday's law) and a changing electric field produces a magnetic field (generalized form of Ampère's law). Note, there is no magnetic analog of a conduction current I. This is because the magnetic monopole, the magnetic analog of an electric charge, does not exist.[†]

[†] The question of the existence of magnetic monopoles has theoretical importance. There have been numerous attempts to observe magnetic monopoles but to date no one has been successful at doing so.

EXAMPLE 30-1

A parallel-plate capacitor has closely spaced circular plates of radius R. Charge is flowing onto the positive plate and off the negative plate at the rate $I = dQ/dt = 2.5$ A. Compute the displacement current through surface S passing between the plates (Figure 30-2) by directly computing the rate of change of the flux of $\vec{E}$ through surface S.

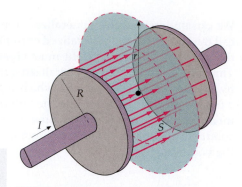

PICTURE THE PROBLEM The displacement current is $I_d = \epsilon_0 \, d\phi_e/dt$, where ϕ_e is the electric flux through the surface between the plates. Since the parallel plates are closely spaced, in the region between the plates the electric field is uniform and perpendicular to the plates. Outside the capacitor the electric field is negligible. Thus, the electric flux is simply $\phi_e = EA$, where E is the electric field between the plates and A is the plate area.

FIGURE 30-2 The surface S passes between the capacitor plates. The charge Q is increasing at 2.5 C/s = 2.5 A. The distance between the plates is not drawn to scale. The plates are much closer together than the plates shown in the figure.

1. The displacement current is found by taking the time derivative of the electric flux:

$$I_d = \epsilon_0 \frac{d\phi_e}{dt}$$

2. The flux equals the electric field magnitude times the plate area:

$$\phi_e = EA$$

3. The electric field is proportional to the charge density on the plates, which we treat as uniformly distributed:

$$E = \frac{\sigma}{\epsilon_0} = \frac{Q/A}{\epsilon_0}$$

4. Substitute these results to calculate I_d:

$$I_d = \epsilon_0 \frac{d(EA)}{dt} = \epsilon_0 A \frac{dE}{dt} = \epsilon_0 A \frac{d}{dt}\left(\frac{Q}{A\epsilon_0}\right)$$

$$= \frac{dQ}{dt} = \boxed{2.5 \text{ A}}$$

REMARKS Note that the displacement current through the surface passing between the plates of the capacitor is equal to the conduction current in the wires carrying current to and from the capacitor.

EXAMPLE 30-2

The circular plates in Example 30-1 have a radius of $R = 3.0$ cm. Find the magnetic field strength B at a point between the plates a distance $r = 2.0$ cm from the axis of the plates when the current into the positive plate is 2.5 A.

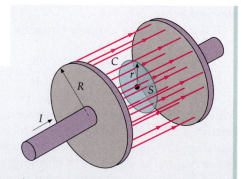

PICTURE THE PROBLEM We find B from the generalized form of Ampère's law (Equation 30-4). We chose a circular path C of radius $r = 2.0$ cm about the centerline joining the plates, as shown in Figure 30-3. We then calculate the displacement current through the surface S bounded by C. By symmetry, $\vec{B}$ is tangent to C and has the same magnitude everywhere on C.

FIGURE 30-3 The space distance between the plates is not drawn to scale. The plates are much closer together than they appear.

1. We find B from the generalized form of Ampère's law:

$$\oint_C \vec{B} \cdot d\vec{\ell} = \mu_0(I + I_d)$$

where

$$I_d = \epsilon_0 \frac{d\phi_e}{dt}$$

2. The line integral is B times the circumference of the circle:

$$\oint_C \vec{B} \cdot d\vec{\ell} = B(2\pi r)$$

3. Since there is no conduction current between the plates of the capacitor, $I = 0$. The generalized current is just the displacement current:

$$\oint_C \vec{B} \cdot d\vec{\ell} = \mu_0 I + \mu_0 \epsilon_0 \frac{d\phi_e}{dt}$$

$$B(2\pi r) = 0 + \mu_0 \epsilon_0 \frac{d\phi_e}{dt}$$

4. The electric flux equals the product of the uniform field E and the area of the flat surface bounded by the curve:

$$\phi_e = \pi r^2 E = \pi r^2 \frac{\sigma}{\epsilon_0} = \pi r^2 \frac{Q}{\epsilon_0 \pi R^2}$$

$$= \frac{Q r^2}{\epsilon_0 R^2}$$

5. Substitute these results into step 3 and solve for B:

$$B(2\pi r) = \mu_0 \epsilon_0 \frac{d}{dt}\left(\frac{Q r^2}{\epsilon_0 R^2}\right) = \mu_0 \frac{r^2}{R^2} \frac{dQ}{dt}$$

$$B = \frac{\mu_0}{2\pi} \frac{r}{R^2} \frac{dQ}{dt} = \frac{\mu_0}{2\pi} \frac{r}{R^2} I$$

$$= (2 \times 10^{-7} \text{ T·m/A}) \frac{0.02 \text{ m}}{(0.03 \text{ m})^2} (2.5 \text{ A})$$

$$= \boxed{1.11 \times 10^{-5} \text{ T}}$$

30-2 Maxwell's Equations

Maxwell's equations are

$$\oint_S E_n \, dA = \frac{1}{\epsilon_0} Q_{\text{inside}} \qquad\qquad 30\text{-}6a$$

$$\oint_S B_n \, dA = 0 \qquad\qquad 30\text{-}6b$$

$$\oint_C \vec{E} \cdot d\vec{\ell} = -\frac{d}{dt}\int_S B_n \, dA = -\int_S \frac{\partial B_n}{\partial t} \, dA \qquad\qquad 30\text{-}6c$$

$$\oint_C \vec{B} \cdot d\vec{\ell} = \mu_0(I + I_d)$$

$$= \mu_0 I + \mu_0 \epsilon_0 \frac{d}{dt}\int_S E_n \, dA = \mu_0 I + \mu_0 \epsilon_0 \int_S \frac{\partial E_n}{\partial t} \, dA \qquad\qquad 30\text{-}6d$$

MAXWELL'S EQUATIONS[†]

Equation 30-6a is Gauss's law; it states that the flux of the electric field through any closed surface equals $1/\epsilon_0$ times the net charge inside the surface. As discussed in Chapter 22, Gauss's law implies that the electric field due to a point charge varies inversely as the square of the distance from the charge. This law describes how electric field lines diverge from a positive charge and converge on a negative charge. Its experimental basis is Coulomb's law.

[†] In all four equations, the integration paths C and the integration surfaces S are at rest and the integrations take place at an instant in time.

Equation 30-6b, sometimes called Gauss's law for magnetism, states that the flux of the magnetic field vector $\vec{B}$ is zero through *any* closed surface. This equation describes the experimental observation that magnetic field lines do not diverge from any point in space or converge on any point; that is, it implies that isolated magnetic poles do not exist.

Equation 30-6c is Faraday's law; it states that the integral of the electric field around any closed curve C, which is the emf, equals the (negative) rate of change of the magnetic flux through any surface S bounded by the curve. (S is not a closed surface, so the magnetic flux through S is not necessarily zero.) Faraday's law describes how electric field lines encircle any area through which the magnetic flux is changing, and it relates the electric field vector $\vec{E}$ to the rate of change of the magnetic field vector $\vec{B}$.

Equation 30-6d, which is Ampère's law modified to include Maxwell's displacement current, states that the line integral of the magnetic field $\vec{B}$ around any closed curve C equals μ_0 times the current through any surface S bounded by the curve plus $\mu_0 \epsilon_0$ times the rate of change of the electric flux through the same surface S. This law describes how the magnetic field lines encircle an area through which a current is passing or through which the electric flux is changing.

In Section 30-4, we show how wave equations for both the electric field $\vec{E}$ and the magnetic field $\vec{B}$ can be derived from Maxwell's equations.

30-3 Electromagnetic Waves

Figure 30-4 shows the electric and magnetic field vectors of an electromagnetic wave. The electric and magnetic fields are perpendicular to each other and perpendicular to the direction of propagation of the wave. Electromagnetic waves are thus transverse waves. The electric and magnetic fields are in phase and, at each point in space and at each instant in time, their magnitudes are related by

$$E = cB \qquad\qquad\qquad 30\text{-}7$$

where $c = 1/\sqrt{\mu_0 \epsilon_0}$ is the speed of the wave. The direction of propagation of an electromagnetic wave is the direction of the cross product $\vec{E} \times \vec{B}$.

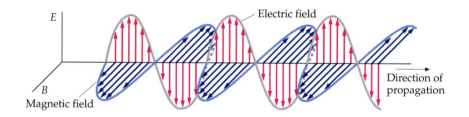

FIGURE 30-4 The electric and magnetic field vectors in an electromagnetic wave. The fields are in phase, perpendicular to each other, and perpendicular to the direction of propagation of the wave.

The Electromagnetic Spectrum

The various types of electromagnetic waves—light, radio waves, X rays, gamma rays, microwaves, and others—differ only in wavelength and frequency, which are related to the speed c in the usual way, $f\lambda = c$. Table 30-1 gives the **electromagnetic spectrum** and the names usually associated with the various frequency and wavelength ranges. These ranges are often not well defined and sometimes overlap. For example, electromagnetic waves with wavelengths of approximately 0.1 nm are usually called X rays, but if the electromagnetic waves originate from nuclear radioactivity, they are called gamma rays.

The human eye is sensitive to electromagnetic radiation with wavelengths from approximately 400 nm to 700 nm, which is the range called **visible light.** The shortest wavelengths in the visible spectrum correspond to violet light and

TABLE 30-1

The Electromagnetic Spectrum

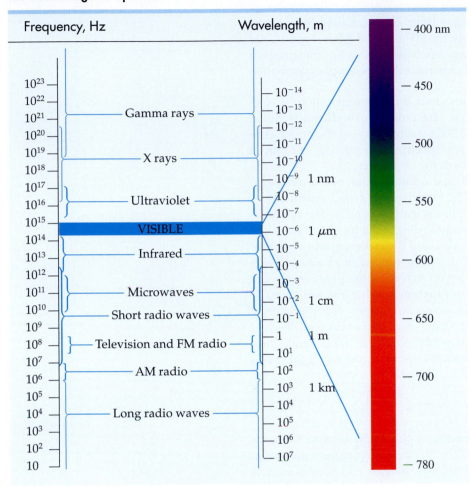

the longest wavelengths to red light, with all the colors of the rainbow falling between these extremes. Electromagnetic waves with wavelengths just beyond the visible spectrum on the short-wavelength side are called **ultraviolet rays,** and those with wavelengths just beyond the visible spectrum on the long-wavelength side are called **infrared waves.** Heat radiation given off by bodies at ordinary temperatures is in the infrared region of the electromagnetic spectrum. There are no limits on the wavelengths of electromagnetic radiation; that is, all wavelengths (or frequencies) are theoretically possible.

The differences in wavelengths of the various kinds of electromagnetic waves have important physical consequences. As we know, the behavior of waves depends strongly on the relative sizes of the wavelengths and the physical objects or apertures the waves encounter. Since the wavelengths of light are in the rather narrow range from approximately 400 nm to 700 nm, they are much smaller than most obstacles, so the ray approximation (introduced in Section 15-4) is often valid. The wavelength and frequency are also important in determining the kinds of interactions between electromagnetic waves and matter. X rays, for example, have very short wavelengths and high frequencies. They easily penetrate many materials that are opaque to lower-frequency light waves, which are absorbed by the materials. Microwaves have wavelengths of the order of a few centimeters and frequencies that are close to the natural resonance frequencies of water molecules in solids and liquids. Microwaves are therefore readily absorbed by the water molecules in foods, which is the mechanism by which food is heated in microwave ovens.

Production of Electromagnetic Waves

Electromagnetic waves are produced when free electric charges accelerate or when electrons bound to atoms and molecules make transitions to lower energy states. Radio waves, which have frequencies from approximately 550 kHz to 1600 kHz for AM and from approximately 88 MHz to 108 MHz for FM, are produced by macroscopic electric currents oscillating in radio transmission antennas. The frequency of the emitted waves equals the frequency of oscillation of the charges.

A continuous spectrum of X rays is produced by the deceleration of electrons when they crash into a metal target. The radiation produced is called **bremsstrahlung** (German for braking radiation). Accompanying the broad, continuous bremsstrahlung spectrum is a discrete spectrum of X-ray lines produced by transitions of inner electrons in the atoms of the target material.

Synchrotron radiation arises from the circular orbital motion of charged particles (usually electrons or positrons) in nuclear accelerators called synchrotrons. Originally considered a nuisance by accelerator scientists, synchrotron radiation X rays are now produced and used as a medical diagnostic tool because of the ease of manipulating the beams with reflection and diffraction optics. Synchrotron radiation is also emitted by charged particles trapped in magnetic fields associated with stars and galaxies. It is believed that most low-frequency radio waves reaching the earth from outer space originate as synchrotron radiation.

Heat is radiated by the thermally excited molecular charges. The spectrum of heat radiation is the blackbody radiation spectrum discussed in Section 20-4.

Light waves, which have frequencies of the order of 10^{14} Hz, are generally produced by transitions of bound atomic charges. We discuss sources of light waves in Chapter 31.

Electric Dipole Radiation

Figure 30-5 is a schematic drawing of an electric-dipole radio antenna that consists of two conducting rods along a line fed by an alternating current generator. At time $t = 0$ (Figure 30-5a), the ends of the rods are charged, and there is an electric field near the rod parallel to the rod. There is also a magnetic field, which is not shown, encircling the rods due to the current in the rods. The fluctuations in these fields move out away from the rods with the speed of light. After one-fourth period, at $t = T/4$ (Figure 30-5b), the rods are uncharged, and the electric field near the rod is zero. At $t = T/2$ (Figure 30-5c), the rods are again charged, but the charges are opposite those at $t = 0$. The electric and magnetic fields at a great distance from the antenna are quite different from the fields near the antenna. Far from the antenna, the electric and magnetic fields oscillate in phase with simple harmonic motion, perpendicular to each other and to the direction of propagation of the wave. Figure 30-6 shows the electric and magnetic fields far from an electric dipole antenna.

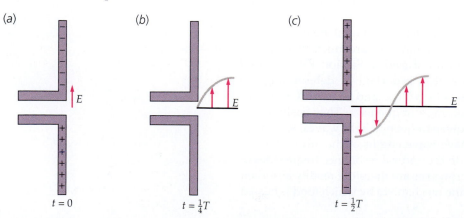

(a) E $t = 0$

(b) E $t = \frac{1}{4}T$

(c) E $t = \frac{1}{2}T$

FIGURE 30-5 An electric dipole radio antenna for radiating electromagnetic waves. Alternating current is supplied to the antenna by a generator (not shown). The fluctuations in the electric field due to the fluctuations in the charges in the antenna propagates outward at the speed of light. There is also a fluctuating magnetic field (not shown) perpendicular to the paper due to the current in the antenna.

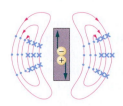

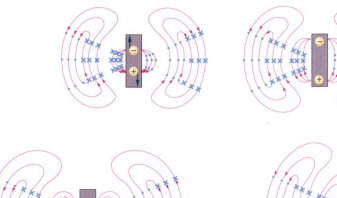

FIGURE 30-6 Electric field lines (in red) and magnetic field lines (in blue) produced by an oscillating electric dipole. Each magnetic field line is a circle with the dipole along its axis. The cross product $\vec{E} \times \vec{B}$ is directed away from the dipole at all points.

Electromagnetic waves of radio or television frequencies can be detected by an electric dipole antenna placed parallel to the electric field of the incoming wave, so that it induces an alternating current in the antenna (Figure 30-7). These electromagnetic waves can also be detected by a loop antenna placed perpendicular to the magnetic field, so that the changing magnetic flux through the loop induces a current in the loop (Figure 30-8). Electromagnetic waves of frequency in the visible light range are detected by the eye or by photographic film, both of which are mainly sensitive to the electric field.

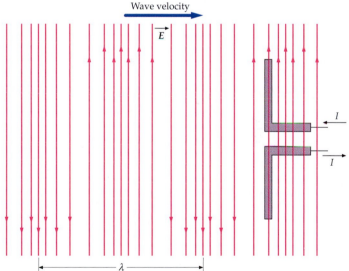

FIGURE 30-7 An electric dipole antenna for detecting electromagnetic waves. The alternating electric field of the incoming wave produces an alternating current in the antenna.

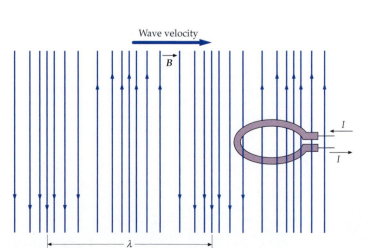

FIGURE 30-8 Loop antenna for detecting electromagnetic radiation. The alternating magnetic flux through the loop due to the magnetic field of the radiation induces an alternating current in the loop.

The radiation from a dipole antenna, such as that shown in Figure 30-5, is called electric dipole radiation. Many electromagnetic waves exhibit the characteristics of electric dipole radiation. An important feature of this type of radiation is that the intensity of the electromagnetic waves radiated by a dipole antenna is zero along the axis of the antenna and maximum in the radial direction (away from the axis). If the dipole is in the y direction with its center at the origin, as in Figure 30-9, the intensity is zero along the y axis and maximum in the xz plane. In the direction of a line making an angle θ with the y axis, the intensity is proportional to $\sin^2 \theta$.

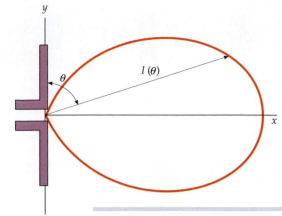

FIGURE 30-9 Polar plot of the intensity of electromagnetic radiation from an electric dipole antenna versus angle. The intensity $I(\theta)$ is proportional to the length of the arrow. The intensity is maximum perpendicular to the antenna at $\theta = 90°$ and minimum along the antenna at $\theta = 0°$ or $\theta = 180°$.

EMF INDUCED IN A LOOP ANTENNA   **EXAMPLE 30-3**

A loop antenna consisting of a single 10-cm radius loop of wire is used to detect electromagnetic waves for which $E_{rms} = 0.15$ V/m. Find the rms emf induced in the loop if the wave frequency is (*a*) 600 kHz and (*b*) 60 MHz.

PICTURE THE PROBLEM The induced emf in the wire is related to the rate of change of the magnetic flux through the loop by Faraday's law (Equation 30-5). Using Equation 30-7, we can obtain the rms value of the magnetic field from the given rms value of the electric field.

(*a*) 1. Faraday's law relates the magnitude of the emf to the rate of change of the magnetic flux through the flat stationary surface bounded by the loop:

$$|\mathcal{E}| = \frac{d\phi_m}{dt}$$

2. The wavelength of a 600 kHz wave traveling at speed c is $\lambda = c/f = 500$ m. Over the flat surface bounded by the 10-cm radius loop, $\vec{B}$ is quite uniform.

$$\phi_m = BA = \pi r^2 B, \text{ so } |\mathcal{E}| = \frac{d\phi_m}{dt} = \pi r^2 \frac{\partial B}{\partial t}$$

and

$$\mathcal{E}_{rms} = \pi r^2 \left(\frac{\partial B}{\partial t}\right)_{rms}$$

3. Compute dB_{rms}/dt from a sinusoidal B:

$$B = B_0 \sin(kx - \omega t)$$

$$\frac{\partial B}{\partial t} = -\omega B_0 \cos(kx - \omega t)$$

4. Calculate the rms value of $\partial B/\partial t$. The rms value of any sinusoidal function of time equals $1/\sqrt{2}$, and the peak value divided by $\sqrt{2}$ equals the rms value:

$$\left(\frac{\partial B}{\partial t}\right)_{rms} = \omega B_0 [-\cos(kx - \omega t)]_{rms} = \omega B_0/\sqrt{2} = \omega B_{rms}$$

5. Using Equation 30-7 ($E = cB$), relate the rms value of $\partial B/\partial t$ to E_{rms}:

$$E = cB$$

so

$$B_{rms} = \frac{E_{rms}}{c}$$

6. Substituting into the step 3 result gives:

$$\left(\frac{\partial B}{\partial t}\right)_{rms} = \omega B_{rms} = \omega \frac{E_{rms}}{c} = \frac{2\pi f}{c} E_{rms}$$

7. Substituting the step 6 result into the step 2 result, calculate $\mathcal{E}_{rms}$ at $f = 600$ kHz:

$$\mathcal{E}_{rms} = \pi r^2 \left(\frac{\partial B}{\partial t}\right)_{rms} = \pi r^2 \frac{2\pi f}{c} E_{rms}$$

$$= \pi (0.1 \text{ m})^2 \frac{2\pi (6 \times 10^5 \text{ Hz})}{3 \times 10^8 \text{ m/s}} (0.15 \text{ V/m})$$

$$\boxed{= 5.92 \times 10^{-5} \text{ V} = 59.2 \ \mu\text{V}}$$

(b) The induced emf is proportional to the frequency (step 4), so at 60 MHz it will be 100 times greater than at 600 kHz:

$$\mathcal{E}_{rms} = (100)(5.92 \times 10^{-5} \text{ V}) = 0.00592 \text{ V}$$

$$\boxed{= 5.92 \text{ mV}}$$

REMARKS For part (b) the frequency is 60 MHz, so $\lambda = c/f = 5$ m. $\vec{B}$ is not as uniform over the surface bounded by the 10-cm radius loop when $\lambda = 5$ m as it is when $\lambda = 500$ m, as in part (a). Thus, $\vec{B}$ on the surface when $\lambda = 5$ m is uniform enough that the part (b) result is sufficiently accurate for most purposes.

Energy and Momentum in an Electromagnetic Wave

Like other waves, electromagnetic waves carry energy and momentum. The energy carried is described by the intensity, which is the average power per unit area incident on a surface perpendicular to the direction of propagation. The momentum per unit time per unit area carried by an electromagnetic wave is called the **radiation pressure.**

Intensity Consider an electromagnetic wave traveling toward the right and a cylindrical region of length L and cross-sectional area A with its axis oriented from left to right. The average amount of electromagnetic energy U_{av} within this region equals $u_{av}V$, where u_{av} is the average energy density and $V = LA$ is the volume of the region. In the time it takes the electromagnetic wave to travel the distance L, all of this energy passes through the right end of the region. The time Δt for the wave to travel the distance L is L/c, so the power P_{av} (the energy per unit time) passing out the right end of the region is

$$P_{av} = U_{av}/\Delta t = u_{av}LA/(L/c) = u_{av}Ac$$

and the intensity I (the average power per unit area) is

$$I = P_{av}/A = u_{av}c$$

The total energy density in the wave u is the sum of the electric and magnetic energy densities. The electric energy density u_e (Equation 24-13) and magnetic energy density u_m (Equation 28-20) are given by

$$u_e = \frac{1}{2}\epsilon_0 E^2 \quad \text{and} \quad u_m = \frac{B^2}{2\mu_0}$$

In an electromagnetic wave in free space, E equals cB, so we can express the magnetic energy density in terms of the electric field:

$$u_m = \frac{B^2}{2\mu_0} = \frac{(E/c)^2}{2\mu_0} = \frac{E^2}{2\mu_0 c^2} = \frac{1}{2}\epsilon_0 E^2$$

where we have used $c^2 = 1/(\epsilon_0\mu_0)$. Thus, the electric and magnetic energy densities are equal. Using $E = cB$, we may express the total energy density in several useful ways:

$$u = u_e + u_m = \epsilon_0 E^2 = \frac{B^2}{\mu_0} = \frac{EB}{\mu_0 c} \qquad 30\text{-}8$$

ENERGY DENSITY IN AN ELECTROMAGNETIC WAVE

To compute the average energy density, we replace the instantaneous fields E and B by their rms values $E_{rms} = E_0/\sqrt{2}$ and $B_{rms} = B_0/\sqrt{2}$, where E_0 and B_0 are the maximum values of the fields. The intensity is then

$$I = u_{av}c = \frac{E_{rms}B_{rms}}{\mu_0} = \frac{1}{2}\frac{E_0 B_0}{\mu_0} = |\vec{S}|_{av} \qquad 30\text{-}9$$

INTENSITY OF AN ELECTROMAGNETIC WAVE

where the vector

$$\vec{S} = \frac{\vec{E} \times \vec{B}}{\mu_0} \qquad 30\text{-}10$$

DEFINITION—POYNTING VECTOR

is called the **Poynting vector** after its discoverer, John Poynting. The average magnitude of $\vec{S}$ is the intensity of the wave, and the direction of $\vec{S}$ is the direction of propagation of the wave.

Radiation Pressure   We now show by a simple example that an electromagnetic wave carries momentum. Consider a wave moving along the x axis that is incident on a stationary charge, as shown in Figure 30-10. For simplicity, we assume that $\vec{E}$ is in the y direction and $\vec{B}$ is in the z direction, and we neglect the time dependence of the fields. The particle experiences a force $q\vec{E}$ in the y direction and is thus accelerated by the electric field. At any time t, the velocity in the y direction is

$$v_y = at = \frac{qE}{m}t$$

After a short time t_1, the charge has acquired kinetic energy equal to

$$K = \frac{1}{2}mv_y^2 = \frac{1}{2}\frac{mq^2E^2t_1^2}{m^2} = \frac{1}{2}\frac{q^2E^2}{m}t_1^2 \qquad 30\text{-}11$$

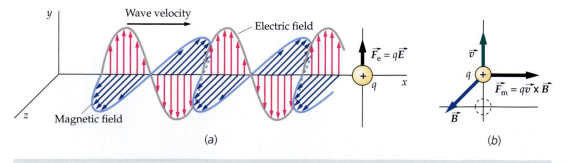

(a)   (b)

FIGURE 30-10 An electromagnetic wave incident on a point charge that is initially at rest on the x axis. (*a*) The electric force $q\vec{E}$ accelerates the charge in the upward direction. (*b*) When the velocity $\vec{v}$ of the charge is upward, the magnetic force $q\vec{v} \times \vec{B}$ accelerates the charge in the direction of the wave.

When the charge is moving in the y direction, it experiences a magnetic force

$$\vec{F}_m = q\vec{v} \times \vec{B} = qv_y\hat{j} \times B\hat{k} = qv_yB\hat{i} = \frac{q^2EB}{m}t\hat{i}$$

Note that this force is in the direction of propagation of the wave. Using $dp_x = F_x\,dt$, we find for the momentum p_x transferred by the wave to the particle in time t_1:

$$p_x = \int_0^{t_1} F_x\,dt = \int_0^{t_1} \frac{q^2EB}{m}t\,dt = \frac{1}{2}\frac{q^2EB}{m}t_1^2$$

If we use $B = E/c$, this becomes

$$p_x = \frac{1}{c}\left(\frac{1}{2}\frac{q^2E^2}{m}t_1^2\right) \tag{30-12}$$

Comparing Equations 30-11 and 30-12, we see that the momentum acquired by the charge in the direction of the wave is $1/c$ times the energy. Although our simple calculation was not rigorous, the results are correct. The magnitude of the momentum carried by an electromagnetic wave is $1/c$ times the energy carried by the wave:

$$p = \frac{U}{c} \tag{30-13}$$

MOMENTUM AND ENERGY IN AN ELECTROMAGNETIC WAVE

Since the intensity is the energy per unit area per unit time, the intensity divided by c is the momentum carried by the wave per unit area per unit time. The momentum carried per unit time is a force. The intensity divided by c is thus a force per unit area, which is a pressure. This pressure is the radiation pressure P_r:

$$P_r = \frac{I}{c} \tag{30-14}$$

RADIATION PRESSURE AND INTENSITY

We can relate the radiation pressure to the electric or magnetic fields by using Equation 30-9 to relate I to E and B, and Equation 30-7 to eliminate either E or B:

$$P_r = \frac{I}{c} = \frac{E_0B_0}{2\mu_0 c} = \frac{E_{rms}B_{rms}}{\mu_0 c} = \frac{E_0^2}{2\mu_0 c^2} = \frac{B_0^2}{2\mu_0} \tag{30-15}$$

RADIATION PRESSURE IN TERMS OF E AND B

Consider an electromagnetic wave incident normally on some surface. If the surface absorbs energy U from the electromagnetic wave, it also absorbs momentum p given by Equation 30-13, and the pressure exerted on the surface equals the radiation pressure. If the wave is reflected, the momentum transferred is $2p$ because the wave now carries momentum in the opposite direction. The pressure exerted on the surface by the wave is then twice that given by Equation 30-15.

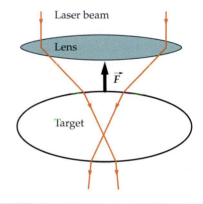

Laser beam
Lens
$\vec{F}$
Target

"Laser tweezers" make use of the momentum carried by electromagnetic waves to manipulate targets on a molecular scale. The two rays shown are refracted as they pass through a transparent target, such as a biological cell, or on an even smaller scale, as a tiny transparent bead attached to a large molecule within a cell. At each refraction, the rays are bent downward, which increases the downward component of momentum of the rays. The target thus exerts a downward force on the laser beams, and the laser beams exert an upward force on the target, which pulls the target toward the laser source. The force is typically of the order of piconewtons. Laser tweezers have been used to accomplish such astonishing feats as stretching out coiled DNA.

RADIATION PRESSURE 3 M FROM A LIGHTBULB   **EXAMPLE 30-4**

A lightbulb emits spherical electromagnetic waves uniformly in all directions. Find (*a*) the intensity, (*b*) the radiation pressure, and (*c*) the electric and magnetic field magnitudes at a distance of 3 m from the lightbulb, assuming that 50 W of electromagnetic radiation is emitted.

PICTURE THE PROBLEM At a distance r from the lightbulb, the energy is spread uniformly over an area $4\pi r^2$. The intensity is the power divided by the area. The radiation pressure can then be found from $P_r = I/c$.

(*a*) 1. Divide the power output by the area to find the intensity:

$$I = \frac{50\ \text{W}}{4\pi r^2}$$

2. Substitute $r = 3$ m:

$$I = \frac{50\ \text{W}}{4\pi (3\ \text{m})^2} = \boxed{0.442\ \text{W/m}^2}$$

(*b*) The radiation pressure is the intensity divided by the speed of light:

$$P_r = \frac{I}{c} = \frac{0.442\ \text{W/m}^2}{3 \times 10^8\ \text{m/s}} = \boxed{1.47 \times 10^{-9}\ \text{Pa}}$$

(*c*) 1. B_0 is related to P_r by Equation 30-15:

$$B_0 = \sqrt{2\mu_0 P_r}$$
$$= [2(4\pi \times 10^{-7}\ \text{T·m/A})(1.47 \times 10^{-9}\ \text{Pa})]^{1/2}$$
$$= 6.08 \times 10^{-8}\ \text{T}$$

2. The maximum value of the electric field E_0 is c times B_0:

$$E_0 = cB_0 = (3 \times 10^8\ \text{m/s})(6.08 \times 10^{-8}\ \text{T})$$
$$= 18.2\ \text{V/m}$$

3. The electric and magnetic field magnitudes at that point are of the form:

$$\boxed{\begin{array}{l} E = E_0 \sin \omega t \quad \text{and} \quad B = B_0 \sin \omega t \\ \text{with } E_0 = 18.2\ \text{V/m} \quad \text{and} \quad B_0 = 6.08 \times 10^{-8}\ \text{T} \end{array}}$$

 REMARKS Only about 2 percent of the power consumed by incandescent bulbs is transformed into visible light. Note that the radiation pressure calculated in Part (*b*) is very small compared with the atmospheric pressure, which is of the order of 10^5 Pa.

A LASER ROCKET   **EXAMPLE 30-5**

You are stranded in space a distance of 20 m from your spaceship. You carry a 1-kW laser. If your total mass, including your space suit and laser, is 95 kg, how long will it take you to reach the spaceship if you point the laser directly away from it?

PICTURE THE PROBLEM The laser emits light, which carries with it momentum. By momentum conservation, you are given an equal and opposite momentum toward the spaceship. The momentum carried by light is $p = U/c$, where U is the energy of the light. If the power of the laser is $P = dU/dt$, then the rate of change of momentum produced by the laser is $dp/dt = (dU/dt)/c = P/c$. This is the force exerted on you, which is constant.

1. The time taken is related to the distance and the acceleration. We assume that you are initially at rest relative to the spaceship:

$$x = \frac{1}{2}at^2; \quad t = \sqrt{\frac{2x}{a}}$$

2. Your acceleration is the force divided by your mass, and the force is the power divided by c:

$$a = \frac{F}{m} = \frac{P/c}{m} = \frac{P}{mc}$$

3. Use this acceleration to calculate the time t:

$$t = \sqrt{\frac{2x}{a}} = \sqrt{\frac{2xmc}{P}}$$

$$= \sqrt{\frac{2(20 \text{ m})(95 \text{ kg})(3 \times 10^8 \text{ m/s})}{1000 \text{ W}}}$$

$$= 3.38 \times 10^4 \text{ s} = \boxed{9.38 \text{ h}}$$

REMARKS Note that the acceleration is extremely small—only about 10^{-9} g. Your speed when you reach the spaceship would be $v = at = 1.19$ mm/s, which is practically imperceptible.

EXERCISE How long would it take you to reach the spaceship if you took off one of your shoelaces and threw it as fast as you could in the direction opposite the ship? (To answer this, you must first estimate the mass of the shoelace and the maximum speed that you can throw the shoelace.) (*Answer* About 5 h for a 10-g shoelace thrown at 10 m/s)

*30-4 The Wave Equation for Electromagnetic Waves

In Section 15-1, we saw that waves on a string obey a partial differential equation called the **wave equation:**

$$\frac{\partial^2 y(x, t)}{\partial x^2} = \frac{1}{v^2} \frac{\partial^2 y(x, t)}{\partial t^2} \qquad \text{30-16}$$

where $y(x, t)$ is the wave function, which for string waves is the displacement of the string. The velocity of the wave is given by $v = \sqrt{F/\mu}$, where F is the tension and μ is the linear mass density. The general solution to this equation is

$$y(x, t) = f_1(x - vt) + f_2(x + vt)$$

The general solution functions can be expressed as a superposition of harmonic wave functions of the form

$$y(x, t) = y_0 \sin(kx - \omega t) \quad \text{and} \quad y(x, t) = y_0 \sin(kx + \omega t)$$

where $k = 2\pi/\lambda$ is the wave number and $\omega = 2\pi f$ is the angular frequency.

Maxwell's equations imply that $\vec{E}$ and $\vec{B}$ obey wave equations similar to Equation 30-16. We consider only free space, in which there are no charges or currents, and we assume that the electric and magnetic fields $\vec{E}$ and $\vec{B}$ are functions of time and one space coordinate only, which we will take to be the x coordinate. Such a wave is called a **plane wave,** because $\vec{E}$ and $\vec{B}$ are uniform throughout any plane perpendicular to the x axis. For a plane electromagnetic wave traveling parallel to the x axis, the x components of the fields are zero, so the vectors $\vec{E}$ and $\vec{B}$ are perpendicular to the x axis and each obeys the wave equation:

$$\frac{\partial^2 \vec{E}}{\partial x^2} = \frac{1}{c^2} \frac{\partial^2 \vec{E}}{\partial t^2} \qquad \text{30-17a}$$

WAVE EQUATION FOR $\vec{E}$

$$\frac{\partial^2 \vec{B}}{\partial x^2} = \frac{1}{c^2} \frac{\partial^2 \vec{B}}{\partial t^2}$$

30-17b

WAVE EQUATION FOR $\vec{B}$

where $c = 1/\sqrt{\mu_0 \epsilon_0}$ is the speed of the waves. (*Note:* Dimensional reasoning helps in remembering these equations. For each equation, the numerators on both sides are the same and the denominators on both sides have the dimension of length squared.)

*Derivation of the Wave Equation

We can relate the space derivative of one of the field vectors to the time derivative of the other field vector by applying Faraday's law (Equation 30-6c) and the modified version of Ampère's law (Equation 30-6d) to appropriately chosen curves in space. We first relate the space derivative of E_y to the time derivative of B_z by applying Equation 30-6c (Faraday's law) to the rectangular curve of sides Δx and Δy lying in the xy plane (Figure 30-11). The circulation of $\vec{E}$ around C (the line integral of $\vec{E}$ around curve C) is

$$\oint_C \vec{E} \cdot d\vec{\ell} = E_y(x_2)\Delta y - E_y(x_1)\Delta y$$

where $E_y(x_1)$ is the value of E_y at the point x_1 and $E_y(x_2)$ is the value of E_y at the point x_2. The contributions of the type $E_x \Delta x$ from the top and bottom of this curve are zero because $E_x = 0$. Since Δx is very small (compared to the wavelength), we can approximate the difference in E_y on the left and right sides of this curve (at x_1 and at x_2) by

$$E_y(x_2) - E_y(x_1) = \Delta E_y \approx \frac{\partial E_y}{\partial x}\Delta x$$

Then

$$\oint_C \vec{E} \cdot d\vec{\ell} \approx \frac{\partial E_y}{\partial x}\Delta x \, \Delta y$$

Faraday's law is

$$\oint_C \vec{E} \cdot d\vec{\ell} = -\int_S \frac{\partial B_n}{\partial t} dA$$

The flux of $\partial B_n/\partial t$ through the rectangular surface bounded by this curve is approximately

$$\int_S B_n \, dA \approx \frac{\partial B_z}{\partial t} \Delta x \, \Delta y$$

Faraday's law then gives

$$\frac{\partial E_y}{\partial x}\Delta x \, \Delta y = -\frac{\partial B_z}{\partial t}\Delta x \, \Delta y$$

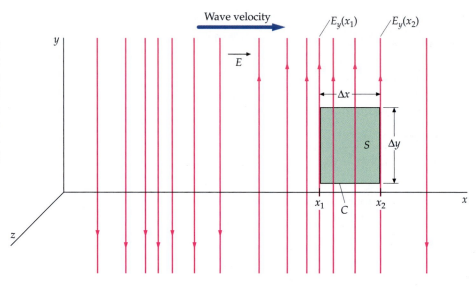

FIGURE 30-11 A rectangular curve in the xy plane for the derivation of Equation 30-18.

or

$$\frac{\partial E_y}{\partial x} = -\frac{\partial B_z}{\partial t}$$

30-18

Equation 30-18 implies that if there is a component of the electric field E_y that depends on x, there must be a component of the magnetic field B_z that depends on time or, conversely, if there is a component of the magnetic field B_z that depends on time, there must be a component of the electric field E_y that depends on x. We can get a similar equation relating the space derivative of the magnetic field B_z to the time derivative of the electric field E_y by applying Ampère's law (Equation 30-6d) to the curve of sides Δx and Δz in the xz plane shown in Figure 30-12.

For the case of no conduction currents ($I = 0$), Equation 30-6d is

$$\oint_C \vec{B} \cdot d\vec{\ell} = \mu_0 \epsilon_0 \int_S \frac{\partial E_n}{\partial t} dA$$

The details of this calculation are similar to those for Equation 30-18. The result is

$$\frac{\partial B_z}{\partial x} = -\mu_0 \epsilon_0 \frac{\partial E_y}{\partial t}$$

30-19

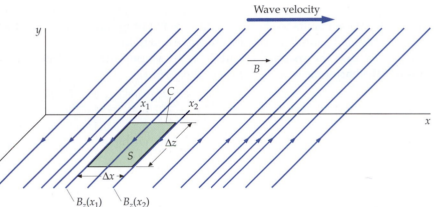

FIGURE 30-12 A rectangular curve in the xz plane for the derivation of Equation 30-19.

We can eliminate either B_z or E_y from Equations 30-18 and 30-19 by differentiating both sides of either equation with respect to either x or t. If we differentiate both sides of Equation 30-18 with respect to x, we obtain

$$\frac{\partial}{\partial x}\left(\frac{\partial E_y}{\partial x}\right) = -\frac{\partial}{\partial x}\left(\frac{\partial B_z}{\partial t}\right)$$

Interchanging the order of the time and space derivatives on the term to the right of the equal sign gives

$$\frac{\partial^2 E_y}{\partial x^2} = -\frac{\partial}{\partial t}\left(\frac{\partial B_z}{\partial x}\right)$$

Using Equation 30-19, we substitute for $\partial B_z / \partial x$ to obtain

$$\frac{\partial^2 E_y}{\partial x^2} = -\frac{\partial}{\partial t}\left(-\mu_0 \epsilon_0 \frac{\partial E_y}{\partial t}\right)$$

which yields the wave equation

$$\frac{\partial^2 E_y}{\partial x^2} = \mu_0 \epsilon_0 \frac{\partial^2 E_y}{\partial t^2}$$

30-20

Comparing Equation 30-20 with Equation 30-16, we see that E_y obeys a wave equation for waves with speed $c = 1/\sqrt{\mu_0 \epsilon_0}$, which is Equation 30-1.

If we had instead chosen to eliminate E_y from Equations 30-18 and 30-19 (by differentiating Equation 30-18 with respect to t, for example), we would have obtained an equation identical to Equation 30-20 except with B_z replacing E_y. We can thus see that both the electric field E_y and the magnetic field B_z obey a wave equation for waves traveling with the velocity $1/\sqrt{\mu_0 \epsilon_0}$, which is the velocity of light.

By following the same line of reasoning as used above, and applying Equation 30-6c (Faraday's law) to the curve in the xz plane (Figure 30-12), we would obtain

$$\frac{\partial E_z}{\partial x} = \frac{\partial B_y}{\partial t} \tag{30-21}$$

Similarly, the application of Equation 30-6d to the curve in the xy plane (Figure 30-11) gives

$$\frac{\partial B_y}{\partial x} = \mu_0 \epsilon_0 \frac{\partial E_z}{\partial t} \tag{30-22}$$

We can use these results to show that, for a wave propagating in the x direction, the components E_z and B_y also obey the wave equation.

To show that the magnetic field B_z is in phase with the electric field E_y, consider the harmonic wave function of the form

$$E_y = E_{y0} \sin(kx - \omega t) \tag{30-23}$$

If we substitute this solution into Equation 30-18, we have

$$\frac{\partial B_z}{\partial t} = -\frac{\partial E_y}{\partial x} = -k E_{y0} \cos(kx - \omega t)$$

To solve for B_z, we take the integral of $\partial B_z / \partial t$ with respect to time. Doing so yields

$$B_z = \int \frac{\partial B_z}{\partial t} \, dt = \frac{k}{\omega} E_{y0} \sin(kx - \omega t) + f(x) \tag{30-24}$$

where $f(x)$ is an arbitrary function of x.

EXERCISE Verify Equation 30-24 by taking $\partial B_z / \partial t$, where $B_z = (k/\omega) E_{y0} \sin(kx - \omega t) + f(x)$.

The result should be $-k E_{y0} \cos(kx - \omega t)$, which is the right hand side of the previous equation.

We next substitute the solution (Equation 30-23) into Equation 30-19 and obtain

$$\frac{\partial B_z}{\partial x} = -\mu_0 \epsilon_0 \frac{\partial E_y}{\partial t} = \omega \mu_0 \epsilon_0 E_{y0} \cos(kx - \omega t)$$

Solving for B_z gives

$$B_z = \int \frac{\partial B_z}{\partial x} \, dx = \frac{\omega \mu_0 \epsilon_0}{k} E_{y0} \sin(kx - \omega t) + g(t) \tag{30-25}$$

where $g(t)$ is an arbitrary function of time. Equating the right sides of Equations 30-24 and 30-25 gives

$$\frac{k}{\omega} E_{y0} \sin(kx - \omega t) + f(x) = \frac{\omega \mu_0 \epsilon_0}{k} E_{y0} \sin(kx - \omega t) + g(t)$$

Substituting c for ω / k and $1/c^2$ for $\mu_0 \epsilon_0$ gives

$$\frac{1}{c}E_{y0}\sin(kx - \omega t) + f(x) = \frac{1}{c}E_{y0}\sin(kx - \omega t) + g(t)$$

which implies $f(x) = g(t)$ for all values of x and t. These remain equal only if $f(x) = g(t) = $ constant (independent of both x and t). Thus, Equation 30-24 becomes

$$B_z = \frac{k}{\omega}E_{y0}\sin(kx - \omega t) + \text{constant} = B_{z0}\sin(kx - \omega t) \qquad 30\text{-}26$$

where $B_{z0} = (k/\omega)E_{y0} = (1/c)E_{y0}$. The integration constant was dropped because it plays no part in the wave. It merely allows for the presence of a static uniform magnetic field. Since the electric and magnetic fields oscillate in phase with the same frequency, we have the general result that the magnitude of the electric field is c times the magnitude of the magnetic field for an electromagnetic wave:

$$E = cB$$

which is Equation 30-7.

We see that Maxwell's equations imply wave equations 30-17a and 30-17b for the electric and magnetic fields; and that if E_y varies harmonically, as in Equation 30-23, the magnetic field B_z is in phase with E_y and has an amplitude related to the amplitude of E_y by $B_z = E_y/c$. The electric and magnetic fields are perpendicular to each other and to the direction of the wave propagation, as shown in Figure 30-4.

$\vec{B}(x, t)$ FOR A LINEARLY POLARIZED PLANE WAVE   **EXAMPLE 30-6**

The electric field of an electromagnetic wave is given by $\vec{E}(x, t) = E_0 \cos(kx - \omega t)\hat{k}$. (*a*) What is the direction of propagation of the wave? (*b*) What is the direction of the magnetic field in the $x = 0$ plane at time $t = 0$? (*c*) Find the magnetic field of the same wave. (*d*) Compute $\vec{E} \times \vec{B}$.

PICTURE THE PROBLEM The argument of the cosine gives the direction of propagation. $\vec{B}$ is perpendicular to both $\vec{E}$ and to the direction of propagation. $\vec{B}$ and $\vec{E}$ are in phase.

(*a*) The argument of the cosine function $(kx - \omega t)$ tells us the direction of propagation:

> The direction of propagation is the direction of increasing x, which is the direction of $\hat{i}$.

(*b*) 1. $\vec{B}$ is in phase with $\vec{E}$ and is perpendicular to both $\vec{E}$ and the direction of propagation $\hat{k}$. (That is, $\vec{B}$ is perpendicular to both $\hat{i}$ and $\hat{k}$.) That means:

$$\vec{B}(x, t) = \pm B_0 \cos(kx - \omega t)\hat{j}$$

2. $\vec{E} \times \vec{B}$ is in the direction of propagation $\hat{i}$. Use the expressions for $\vec{E}$ and $\vec{B}$ and take the cross product:

$$\vec{E} \times \vec{B} = E_0 \cos(kx - \omega t)\hat{k} \times (\pm B_0 \cos(kx - \omega t)\hat{j})$$
$$= E_0(\pm B_0)\cos^2(kx - \omega t)(\hat{k} \times \hat{j})$$
$$= E_0(\pm B_0)\cos^2(kx - \omega t)(-\hat{i})$$

3. Choose the sign so that $\vec{E} \times \vec{B}$ is in the $\hat{i}$ direction:

$$\vec{E} \times \vec{B} = E_0(-B_0)\cos^2(kx - \omega t)(-\hat{i})$$

so

$$\vec{B}(x, t) = -B_0 \cos(kx - \omega t)\hat{j}$$

4. Evaluate $\vec{B}$ when both x and t equal zero.

$$\vec{B}(0, 0) = -B_0 \cos[k(0) - \omega(0)]\hat{j} = -B_0\hat{j}$$

$$\therefore \boxed{\vec{B}(0, 0) \text{ is in the negative } y \text{ direction.}}$$

(c) In an electromagnetic wave, $E_0 = cB_0$ and $\vec{B}$ and $\vec{E}$ are in phase. Thus:

$$\vec{B}(x, t) = \boxed{-B_0 \cos(kx - \omega t)\,\hat{j}, \text{ where } B_0 = E_0/c}$$

(d) Calculate $\vec{E} \times \vec{B}$. Let $\theta = kx - \omega t$ and do the calculation:

$$\vec{E} \times \vec{B} = (E_0 \cos\theta\,\hat{k}) \times (-B_0 \cos\theta\,\hat{j})$$

$$= -E_0 B_0 \cos^2\theta\,(\hat{k} \times \hat{j})$$

$$= \boxed{E_0 B_0 \cos^2\theta\,\hat{i}, \text{ where } \theta = kx - \omega t}$$

REMARKS The Part (d) result confirms the Part (a) result, because for an electromagnetic wave $\vec{E} \times \vec{B}$ is always in the direction of propagation.

$\vec{B}(x, t)$ FOR A CIRCULAR POLARIZED PLANE WAVE   **EXAMPLE 30-7**

The electric field of an electromagnetic wave is given by $\vec{E}(x, t) = E_0 \sin(kx - \omega t)\,\hat{j} + E_0 \cos(kx - \omega t)\,\hat{k}$. (a) Find the magnetic field of the same wave. (b) Compute $\vec{E} \cdot \vec{B}$ and $\vec{E} \times \vec{B}$.

PICTURE THE PROBLEM We can solve this using the principle of superposition. The given electric field is the superposition of two fields, the one given in Equation 30-23 and the one given in the problem statement of Example 30-6.

(a) 1. From the phase (the argument of the trig functions) we can see that the direction of propagation is the positive x direction:

The phase is for a wave traveling in the positive x direction.

2. The given electric field can be considered as the superposition of $\vec{E}_1 = E_0 \sin(kx - \omega t)\,\hat{j}$ and $\vec{E}_2 = E_0 \cos(kx - \omega t)\,\hat{k}$. Find the magnetic fields $\vec{B}_1$ and $\vec{B}_2$ associated with these electric fields, respectively. Use the procedure followed in Example 30-6:

For $\vec{E}_1 = E_0 \sin(kx - \omega t)\,\hat{j}$, $\vec{B}_1 = B_0 \sin(kx - \omega t)\,\hat{k}$

and

For $\vec{E}_2 = E_0 \cos(kx - \omega t)\,\hat{k}$, $\vec{B}_2 = -B_0 \cos(kx - \omega t)\,\hat{j}$

where

$$E_0 = cB_0$$

3. The superposition of magnetic fields gives the resultant magnetic fields:

$$\boxed{\begin{aligned}\vec{B}(x, t) &= \vec{B}_1 + \vec{B}_2 \\ &= B_0 \sin(kx - \omega t)\,\hat{k} - B_0 \cos(kx - \omega t)\,\hat{j} \\ \text{where} \\ B_0 &= E_0/c\end{aligned}}$$

(b) 1. Let $\theta = kx - \omega t$ to simplify the notation and calculate $\vec{E} \cdot \vec{B}$:

$$\vec{E} \cdot \vec{B} = (E_0 \sin\theta\,\hat{j} + E_0 \cos\theta\,\hat{k}) \cdot (B_0 \sin\theta\,\hat{k} - B_0 \cos\theta\,\hat{j})$$

$$= E_0 B_0 \sin^2\theta\,\hat{j} \cdot \hat{k} - E_0 B_0 \sin\theta \cos\theta\,\hat{j} \cdot \hat{j}$$

$$+ E_0 B_0 \cos\theta \sin\theta\,\hat{k} \cdot \hat{k} - E_0 B_0 \cos^2\theta\,\hat{k} \cdot \hat{j}$$

$$= 0 - E_0 B_0 \sin\theta \cos\theta + E_0 B_0 \cos\theta \sin\theta - 0 = \boxed{0}$$

2. Calculate $\vec{E} \times \vec{B}$:

$$\vec{E} \times \vec{B} = (E_0 \sin\theta\,\hat{j} + E_0 \cos\theta\,\hat{k}) \times (-B_0 \cos\theta\,\hat{j} + B_0 \sin\theta\,\hat{k})$$

$$= -E_0 B_0 \sin\theta \cos\theta\,(\hat{j} \times \hat{j}) + E_0 B_0 \sin^2\theta\,(\hat{j} \times \hat{k})$$

$$- E_0 B_0 \cos^2\theta\,(\hat{k} \times \hat{j}) + E_0 B_0 \cos\theta \sin\theta\,(\hat{k} \times \hat{k})$$

$$= 0 + E_0 B_0 \sin^2\theta\,\hat{i} + E_0 B_0 \cos^2\theta\,\hat{i} + 0 = \boxed{E_0 B_0\,\hat{i}}$$

REMARKS We see that $\vec{E}$ and $\vec{B}$ are perpendicular to one another, and that $\vec{E} \times \vec{B}$ is in the direction of propagation of the wave. This type of electromagnetic wave is said to be *circularly polarized*. At a fixed value of x, both $\vec{E}$ and $\vec{B}$ rotate in a circle in a plane perpendicular to $\hat{i}$ with angular frequency ω.

EXERCISE Calculate $\vec{E} \cdot \vec{E}$ and $\vec{B} \cdot \vec{B}$. [*Answer* $\vec{E} \cdot \vec{E} = E_y^2 + E_z^2 = E_0^2 \sin^2(kx - \omega t) + E_0^2 \cos^2(kx - \omega t) = E_0^2$ and $\vec{B} \cdot \vec{B} = B_y^2 + B_2^2 = B_0^2 \cos^2(kx - \omega t) + B_0^2 \sin^2(kx - \omega t) = B_0^2$]

■ **REMARKS** The fields $\vec{E}$ and $\vec{B}$ are constant in magnitude.

SUMMARY

1. Maxwell's equations summarize the fundamental laws of physics that govern electricity and magnetism.

2. Electromagnetic waves include light, radio and television waves, X rays, gamma rays, microwaves, and others.

Topic	Relevant Equations and Remarks	
1. Maxwell's Displacement Current	Ampère's law can be generalized to apply to currents that are not steady (and not continuous) if the conduction current I is replaced by $I + I_d$, where I_d is Maxwell's displacement current:	
	$$I_d = \epsilon_0 \frac{d\phi_e}{dt}$$	30-3
Generalized form of Ampère's law	$$\oint_C \vec{B} \cdot d\vec{\ell} = \mu_0(I + I_d) = \mu_0 I + \mu_0 \epsilon_0 \frac{d\phi_e}{dt}$$	30-4
2. Maxwell's Equations	The laws of electricity and magnetism are summarized by Maxwell's equations.	
Gauss's law	$$\oint_S E_n \, dA = \frac{1}{\epsilon_0} Q_{inside}$$	30-6a
Gauss's law for magnetism (isolated magnetic poles do not exist)	$$\oint_S B_n \, dA = 0$$	30-6b
Faraday's law	$$\oint_C \vec{E} \cdot d\vec{\ell} = -\frac{d}{dt} \int_S B_n \, dA = -\int_S \frac{\partial B_n}{\partial t} \, dA$$	30-6c
Ampère's law modified	$$\oint_C \vec{B} \cdot d\vec{\ell} \, \mu_0(I + I_d)$$	
	$$= \mu_0 I + \mu_0 \epsilon_0 \frac{d}{dt} \int_S E_n \, dA = \mu_0 I + \mu_0 \epsilon_0 \int_S \frac{\partial E_n}{\partial t} \, dA$$	30-6d
3. Electromagnetic Waves	In an electromagnetic wave, the electric and magnetic field vectors are perpendicular to each other and to the direction of propagation. Their magnitudes are related by	
	$$E = cB$$	30-7
Wave speed	$$c = \frac{1}{\sqrt{\mu_0 \epsilon_0}} \approx 3 \times 10^8 \text{ m/s}$$	30-1

Electromagnetic spectrum	The various types of electromagnetic waves—light, radio waves, X rays, gamma rays, microwaves, and others—differ only in wavelength and frequency. The human eye is sensitive to the range from about 400 nm to 700 nm.
Electric dipole radiation	Electromagnetic waves are produced when free electric charges accelerate. Oscillating charges in an electric dipole antenna radiate electromagnetic waves with an intensity that is greatest in directions perpendicular to the antenna. There is no radiated intensity along the axis of the antenna. Perpendicular to the antenna and far away from it, the electric field of the electromagnetic wave is parallel to the antenna.

Energy density in an electromagnetic wave

$$u = u_e + u_m = \epsilon_0 E^2 = \frac{B^2}{\mu_0} = \frac{EB}{\mu_0 c}$$

30-8

Intensity of an electromagnetic wave

$$I = u_{av} c = \frac{E_{rms} B_{rms}}{\mu_0} = \frac{1}{2} \frac{E_0 B_0}{\mu_0} = |\vec{S}|_{av}$$

30-9

Poynting vector

$$\vec{S} = \frac{\vec{E} \times \vec{B}}{\mu_0}$$

30-10

Momentum and energy in an electromagnetic wave

$$p = \frac{U}{c}$$

30-13

Radiation pressure and intensity

$$P_r = \frac{I}{c}$$

30-14

4. ***The Wave Equation for Electromagnetic Waves**

Maxwell's equations imply that the electric and magnetic field vectors in free space obey a wave equation.

$$\frac{\partial^2 \vec{E}}{\partial x^2} = \frac{1}{c^2} \frac{\partial^2 \vec{E}}{\partial t^2}$$

30-17a

$$\frac{\partial^2 \vec{B}}{\partial x^2} = \frac{1}{c^2} \frac{\partial^2 \vec{B}}{\partial t^2}$$

30-17b

PROBLEMS

- • Single-concept, single-step, relatively easy
- •• Intermediate-level, may require synthesis of concepts
- ••• Challenging
- **SSM** Solution is in the *Student Solutions Manual*
- **iSOLVE** Problems available on iSOLVE online homework service
- **iSOLVE✓** These "Checkpoint" online homework service problems ask students additional questions about their confidence level, and how they arrived at their answer.

In a few problems, you are given more data than you actually need; in a few other problems, you are required to supply data from your general knowledge, outside sources, or informed estimates.

Conceptual Problems

1 • **SSM** True or false:

(a) Maxwell's equations apply only to fields that are constant over time.

(b) The wave equation can be derived from Maxwell's equations.

(c) Electromagnetic waves are transverse waves.

(d) In an electromagnetic wave in free space, the electric and magnetic fields are in phase.

(e) In an electromagnetic wave in free space, the electric and magnetic field vectors $\vec{E}$ and $\vec{B}$ are equal in magnitude.

(f) In an electromagnetic wave in free space, the electric and magnetic energy densities are equal.

2 •• Theorists have speculated about the possible existence of magnetic monopoles, and there have been several, as yet unsuccessful, experimental searches for such monopoles. Suppose magnetic monopoles were found and that the magnetic field at a distance r from a monopole of strength q_m is given by $B = (\mu_0/4\pi)q_m/r^2$. How would Maxwell's equations have to be modified to be consistent with such a discovery?

3 • Which waves have greater frequencies, light waves or X rays?

4 • **SSM** Are the frequencies of ultraviolet radiation greater or less than those of infrared radiation?

5 • What kind of waves have wavelengths of the order of a few meters?

6 • The detection of radio waves can be accomplished with either a dipole antenna or a loop antenna. The dipole antenna detects the (pick one) *electric/magnetic* field of the wave, and the loop antenna detects the *electric/magnetic* field of the wave.

7 • A transmitter uses a loop antenna with the loop in the horizontal plane. What should be the orientation of a dipole antenna at the receiver for optimum signal reception?

8 • **SSM** A helium-neon laser has a red beam. It is shone in turn on a red plastic filter (of the kind used for theater lighting) and a green plastic filter. (A red theater-lighting filter transmitts only red light.) On which filter will the laser exert a larger force?

Estimation and Approximation

9 •• Estimate the intensity and total power needed in a laser beam to lift a 15-μm diameter plastic bead against the force of gravity. Make any assumptions you think reasonable.

10 ••• Some science fiction writers have used solar sails to propel interstellar spaceships. Imagine a giant sail erected on a spacecraft subjected to the solar radiation pressure. (a) Show that the spacecraft's acceleration is given by

$$a = \frac{P_s A}{4\pi r^2 cm}$$

where P_s is the power output of the sun and is equal to 3.8×10^{26} W, A is the surface area of the sail, m is the total mass of the spacecraft, r is the distance from the sun, and c is the speed of light. (b) Show that the velocity of the spacecraft at a distance r from the sun is found from

$$v^2 = v_0^2 + \left(\frac{P_s A}{2\pi mc}\right)\left(\frac{1}{r_0} - \frac{1}{r}\right)$$

where v_0 is the initial velocity at r_0. (c) Compare the relative accelerations due to the radiation pressure and the gravitational force. Use reasonable values for A and m. Will such a system work?

11 •• The intensity of sunlight striking the earth's upper atmosphere (called the solar constant) is 1.37 kW/m². (a) Find E_{rms} and B_{rms} due to the sun at the upper atmosphere of the earth. (b) Find the average power output of the sun. (c) Find the intensity and the radiation pressure at the surface of the sun.

12 •• **SSM** Estimate the radiation pressure force exerted on the earth by the sun, and compare the radiation pressure force to the gravitational attraction of the sun. At the earth's orbit the intensity of sunlight is 1.37 kW/m²

13 •• **SSM** Repeat Problem 12 for the planet Mars. Which planet has the larger ratio of radiation pressure to gravitational attraction. Why?

14 •• **SSM** In the new field of laser cooling and trapping, the forces associated with radiation pressure are used to slow down atoms from thermal speeds of hundreds of meters per second at room temperature to speeds of just a few meters per second or slower. An isolated atom will absorb radiation only at specific resonant frequencies. If the frequency of the laser-beam radiation is one of the resonant frequencies of the target atom, then the radiation is absorbed via a process called resonant absorption. The effective cross-sectional area of the atom for resonant absorption is approximately equal to λ^2, where λ is the wavelength of the laser beam. (a) Estimate the acceleration of a rubidium atom (atomic mass 85 g/mol) in a laser beam whose wavelength is 780 nm and intensity is 10 W/m². (b) About how long would it take such a light beam to slow a rubidium atom in a gas at room temperature (300 K) down to near-zero velocity?

Maxwell's Displacement Current

15 • **ISOLVE** A parallel-plate capacitor in air has circular plates of radius 2.3 cm separated by 1.1 mm. Charge is flowing onto the upper plate and off the lower plate at a rate of 5 A. (a) Find the time rate of change of the electric field between the plates. (b) Compute the displacement current between the plates and show that the displacement current equals 5 A.

16 • **ISOLVE** In a region of space, the electric field varies according to $E = (0.05$ N/C$)$ sin $2000t$, where t is in seconds. Find the maximum displacement current through a 1-m² area perpendicular to $\vec{E}$.

17 •• For Problem 15, show that at a distance r from the axis of the plates the magnetic field between the plates is given by $B = (1.89 \times 10^{-3}$ T/m$)r$, if r is less than the radius of the plates.

18 •• (a) Show that for a parallel-plate capacitor the displacement current is given by $I_d = C\,dV/dt$, where C is the capacitance and V is the voltage across the capacitor. (b) A 5-nF parallel-plate capacitor is connected to an emf $\mathcal{E} = \mathcal{E}_0 \cos \omega t$, where $\mathcal{E}_0 = 3$ V and $\omega = 500\pi$. Find the displacement current between the plates as a function of time. Neglect any resistance in the circuit.

19 •• **SSM** **ISOLVE** Current of 10 A flows into a capacitor having plates with areas of 0.5 m². (a) What is the displacement current between the plates? (b) What is dE/dt between the plates for this current? (c) What is the line integral of $\vec{B} \cdot d\vec{\ell}$ around a circle of radius 10 cm that lies within the plates and parallel to the plates?

20 •• A parallel-plate capacitor with circular plates is given a charge Q_0. Between the plates is a leaky dielectric having a dielectric constant of κ and a resistivity ρ. (a) Find the conduction current between the plates as a function of time. (b) Find the displacement current between the plates as a function of time. What is the total (conduction plus displacement) current? (c) Find the magnetic field produced between the plates by the leakage discharge current as a function of time. (d) Find the magnetic field between the plates produced by the displacement current as a function of time. (e) What is the total magnetic field between the plates during discharge of the capacitor?

21 •• The leaky capacitor of Problem 20 has plate separation d. It is being charged such that the voltage across the capacitor is given by $V(t) = (0.01 \text{ V/s})t$. (a) Find the conduction current as a function of time. (b) Find the displacement current. (c) Find the time for which the displacement current is equal to the conduction current.

22 •• The space between the plates of a capacitor is filled with a material of resistivity $\rho = 10^4 \ \Omega \cdot \text{m}$ and dielectric constant $\kappa = 2.5$. The parallel plates are circular with a radius of 20 cm and are separated by 1 mm. The voltage across the plates is given by $V_0 \cos \omega t$, with $V_0 = 40$ V and $\omega = 120\pi$ rad/s. (a) What is the displacement current density? (b) What is the conduction current between the plates? (c) At what angular frequency is the total current $45°$ out of phase with the applied voltage?

23 ••• [SSM] Show that the generalized form of Ampère's law (Equation 30-4) and the Biot–Savart law give the same result in a situation in which they both can be used. Figure 30-13 shows two charges $+Q$ and $-Q$ on the x axis at $x = -a$ and $x = +a$, with a current $I = -dQ/dt$ along the line between them. Point P is on the y axis at $y = R$. (a) Use the Biot–Savart law to show that the magnitude of B at point P is

$$B = \frac{\mu_0 I a}{2\pi R} \frac{1}{\sqrt{R^2 + a^2}}$$

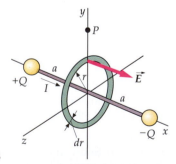

FIGURE 30-13 Problem 23

(b) Consider a circular strip of radius r and width dr in the yz plane with its center at the origin. Show that the flux of the electric field through this strip is

$$E_x \, dA = \frac{Q}{\epsilon_0} a(r^2 + a^2)^{-3/2} r \, dr$$

(c) Use your result from Part (b) to find the total flux ϕ_e through a circular area of radius R. Show that

$$\epsilon_0 \, \phi_e = Q\left(1 - \frac{a}{\sqrt{a^2 + R^2}}\right)$$

(d) Find the displacement current I_d, and show that

$$I + I_d = I\frac{a}{\sqrt{a^2 + R^2}}$$

(e) Finally, show that Equation 30-4 gives the same result for B as the result found in Part (a).

Maxwell's Equations and the Electromagnetic Spectrum

24 •• Show that the normal component of the magnetic field $\vec{B}$ is continuous across a surface, by applying Gauss's law for $\vec{B}$ ($\int B_n \, dA = 0$) to a pillbox Gaussian surface that has a face on each side of the surface.

25 • [SSM] [iSOLVE✓] Find the wavelength for (a) a typical AM radio wave with a frequency of 1000 kHz and (b) a typical FM radio wave with a frequency of 100 MHz.

26 • [SSM] [iSOLVE] What is the frequency of a 3-cm microwave?

27 • [iSOLVE] What is the frequency of an X ray with a wavelength of 0.1 nm?

Electric Dipole Radiation

28 •• [iSOLVE] The intensity of radiation from an electric dipole is proportional to $\sin^2 \theta / r^2$, where θ is the angle between the electric dipole moment and the position vector $\vec{r}$. A radiating electric dipole lies along the z axis (its dipole moment is in the z direction). Let I_1 be the intensity of the radiation at a distance $r = 10$ m and at angle $\theta = 90°$. Find the intensity (in terms of I_1) at (a) $r = 30$ m, $\theta = 90°$; (b) $r = 10$ m, $\theta = 45°$; and (c) $r = 20$ m, $\theta = 30°$.

29 •• (a) For the situation described in Problem 28, at what angle is the intensity at $r = 5$ m equal to I_1? (b) At what distance is the intensity equal to I_1 at $\theta = 45°$?

30 •• [iSOLVE] The transmitting antenna of a station is a dipole located atop a mountain 2000 m above sea level. The intensity of the signal on a nearby mountain 4 km distant and also 2000 m above sea level is 4×10^{-12} W/m^2. What is the intensity of the signal at sea level and 1.5 km from the transmitter? (See Problem 28.)

31 ••• A radio station that uses a vertical dipole antenna broadcasts at a frequency of 1.20 MHz with total power output of 500 kW. The radiation pattern is as shown in Figure 30-8 (i.e., the intensity of the signal varies as $\sin^2 \theta$, where θ is the angle between the direction of propagation and the vertical and is independent of azimuthal angle). Calculate the intensity of the signal at a horizontal distance of 120 km from the station. What is the intensity at that point as measured in photons per square centimeter per second?

32 ••• [SSM] At a distance of 30 km from a radio station broadcasting at a frequency of 0.8 MHz, the intensity of the electromagnetic wave is 2×10^{-13} W/m^2. The transmitting antenna is a vertical dipole. What is the total power radiated by the station?

33 ••• [iSOLVE✓] A small private plane approaching an airport is flying at an altitude of 2500 m above ground. The airport's flight control system transmits 100 W at 24 MHz, using a vertical dipole antenna. What is the intensity of the signal at the plane's receiving antenna when the plane's position on a map is 4 km from the airport?

Energy and Momentum in an Electromagnetic Wave

34 • **ISOLVE✓** An electromagnetic wave has an intensity of 100 W/m². Find (a) the radiation pressure P_r, (b) E_{rms}, and (c) B_{rms}.

35 • **ISOLVE✓** The amplitude of an electromagnetic wave is $E_0 = 400$ V/m. Find (a) E_{rms}, (b) B_{rms}, (c) the intensity I, and (d) the radiation pressure P_r.

36 • **ISOLVE** The rms value of the electric field in an electromagnetic wave is $E_{rms} = 400$ V/m. (a) Find B_{rms}, (b) the average energy density, and (c) the intensity.

37 • Show that the units of $E = cB$ are consistent; that is, show that when B is in teslas and c is in meters per second, the units of cB are volts per meter or newtons per coulomb.

38 • **SSM** **ISOLVE** The rms value of the magnitude of the magnetic field in an electromagnetic wave is $B_{rms} = 0.245$ μT. Find (a) E_{rms}, (b) the average energy density, and (c) the intensity.

39 •• **ISOLVE** (a) An electromagnetic wave of intensity 200 W/m² is incident normally on a rectangular black card with sides of 20 cm and 30 cm that absorbs all the radiation. Find the force exerted on the card by the radiation. (b) Find the force exerted by the same wave if the card reflects all the radiation incident on it.

40 •• Find the force exerted by the electromagnetic wave on the reflecting card in Part (b) of Problem 39 if the radiation is incident at an angle of 30° to the normal.

41 •• **SSM** An AM radio station radiates an isotropic sinusoidal wave with an average power of 50 kW. What are the amplitudes of E_{max} and B_{max} at a distance of (a) 500 m, (b) 5 km, and (c) 50 km?

42 •• **ISOLVE✓** A laser beam has a diameter of 1.0 mm and average power of 1.5 mW. Find (a) the intensity of the beam, (b) E_{rms}, (c) B_{rms}, and (d) the radiation pressure.

43 •• **SSM** **ISOLVE** Instead of sending power by a 750-kV, 1000-A transmission line, one desires to beam this energy via an electromagnetic wave. The beam has a uniform intensity within a cross-sectional area of 50 m². What are the rms values of the electric and the magnetic fields?

44 •• **ISOLVE✓** A laser pulse has an energy of 20 J and a beam radius of 2 mm. The pulse duration is 10 ns and the energy density is constant within the pulse. (a) What is the spatial length of the pulse? (b) What is the energy density within the pulse? (c) Find the electric and magnetic amplitudes of the laser pulse.

45 •• **SSM** The electric field of an electromagnetic wave oscillates in the y direction and the Poynting vector is given by

$$\vec{S}(x, t) = (100 \text{ W/m}^2) \cos^2[10x - (3 \times 10^9)t]\hat{i}$$

where x is in meters and t is in seconds. (a) What is the direction of propagation of the wave? (b) Find the wavelength and the frequency. (c) Find the electric and magnetic fields.

46 •• A parallel-plate capacitor is being charged. The capacitor consists of two circular parallel plates of area A and separation d. (a) Show that the displacement current in the capacitor gap has the same value as the conduction current in the capacitor leads. (b) What is the direction of the Poynting vector in the region of space between the capacitor plates? (c) Calculate the Poynting vector S in this region and show that the flux of S into this region is equal to the rate of change of the energy stored in the capacitor.

47 •• **ISOLVE✓** A pulsed laser fires a 1000-MW pulse of 200-ns duration at a small object of mass 10 mg suspended by a fine fiber 4 cm long. If the radiation is completely absorbed without other effects, what is the maximum angle of deflection of this pendulum?

48 •• The mirrors used in a particular type of laser are 99.99% reflecting. (a) If the laser has an average output power of 15 W, what is the average power of the radiation incident on one of the mirrors? (b) What is the force due to radiation pressure on one of the mirrors?

49 •• A 10-cm by 15-cm card has a mass of 2 g and is perfectly reflecting. The card hangs in a vertical plane and is free to rotate about a horizontal axis through the top edge. The card is illuminated uniformly by an intense light that causes the card to make an angle of 1° with the vertical. Find the intensity of the light.

*The Wave Equation for Electromagnetic Waves

50 • Show by direct substitution that Equation 30-17a is satisfied by the wave function

$$E_y = E_0 \sin(kx - \omega t) = E_0 \sin k(x - ct)$$

where $c = \omega/k$.

51 • Use the known values of μ_0 and ϵ_0 in SI units to compute $c = 1/\sqrt{\epsilon_0\mu_0}$, and show that it is approximately 3×10^8 m/s.

52 ••• **SSM** (a) Using arguments similar to those given in the text, show that for a plane wave, in which E and B are independent of y and z,

$$\frac{\partial E_z}{\partial x} = \frac{\partial B_y}{\partial t} \quad \text{and} \quad \frac{\partial B_y}{\partial x} = \mu_0\epsilon_0\frac{\partial E_z}{\partial t}$$

(b) Show that E_z and B_y also satisfy the wave equation.

53 ••• Show that any function of the form $y(x, t) = f(x - vt)$ or $y(x, t) = g(x + vt)$ satisfies the wave Equation 30-16.

General Problems

54 • (a) Show that if E is in volts per meter and B is in teslas, the units of the Poynting vector $\vec{S} = (\vec{E} \times \vec{B})/\mu_0$ are watts per square meter. (b) Show that if the intensity I is in watts per square meter, the units of radiation pressure $P_r = I/c$ are newtons per square meter.

55 •• A loop antenna that may be rotated about a vertical axis is used to locate an unlicensed amateur radio transmitter. If the output of the receiver is proportional to the intensity of the received signal, how does the output of the receiver vary with the orientation of the loop antenna?

56 •• An electromagnetic wave has a frequency of 100 MHz and is traveling in a vacuum. The magnetic field is given by $\vec{B}(z, t) = (10^{-8} \text{ T}) \cos(kz - \omega t)\hat{i}$. (a) Find the wavelength and the direction of propagation of this wave. (b) Find the electric vector $\vec{E}(z, t)$. (c) Give Poynting's vector, and find the intensity of this wave.

57 •• SSM A circular loop of wire can be used to detect electromagnetic waves. Suppose a 100-MHz FM radio station radiates 50 kW uniformly in all directions. What is the maximum rms voltage induced in a loop of radius 30 cm at a distance of 10^5 m from the station?

58 •• ISOLVE The electric field from a radio station some distance from the transmitter is given by $E = (10^{-4} \text{ N/C}) \cos 10^6 t$, where t is in seconds. (a) What voltage is picked up on a 50-cm wire oriented along the electric field direction? (b) What voltage can be induced in a loop of radius 20 cm?

59 •• A circular capacitor of radius a has a thin wire of resistance R connecting the centers of the two plates. A voltage $V_0 \sin \omega t$ is applied between the plates. (a) What is the current drawn by this capacitor? (b) What is the magnetic field as a function of radial distance r from the centerline within the plates of this capacitor? (c) What is the phase angle between the current and the applied voltage?

60 •• ISOLVE A 20-kW beam of radiation is incident normally on a surface that reflects half of the radiation. What is the force on this surface?

61 •• SSM The electric fields of two harmonic waves of angular frequency ω_1 and ω_2 are given by $\vec{E}_1 = E_{1,0} \cos(k_1 x - \omega_1 t)\hat{j}$ and by $\vec{E}_2 = E_{2,0} \cos(k_2 x - \omega_2 t + \delta)\hat{j}$. Find (a) the instantaneous Poynting vector for the resultant wave motion and (b) the time-average Poynting vector. If the direction of propagation of the second wave is reversed so $\vec{E}_2 = E_{2,0} \cos(k_2 x + \omega_2 t + \delta)\hat{j}$, find (c) the instantaneous Poynting vector for the resultant wave motion and (d) the time-average Poynting vector.

62 •• SSM ISOLVE At the surface of the earth, there is an approximate average solar flux of 0.75 kW/m². A family wishes to construct a solar energy conversion system to power their home. If the conversion system is 30 percent efficient and the family needs a maximum of 25 kW, what effective surface area is needed for perfectly absorbing collectors?

63 •• ISOLVE✓ Suppose one has an excellent radio capable of detecting a signal as weak as 10^{-14} W/m². This radio has a 2000-turn coil antenna that has a radius of 1 cm wound on an iron core that increases the magnetic field by a factor of 200. The radio frequency is 140 KHz. (a) What is the amplitude of the magnetic field in this wave? (b) What is the emf induced in the antenna? (c) What would be the emf induced in a 2-m wire oriented in the direction of the electric field?

64 •• Show that

$$\frac{\partial B_z}{\partial x} = -\mu_0 \epsilon_0 \frac{\partial E_y}{\partial t}$$

(Equation 30-19) follows from

$$\oint_C \vec{B} \cdot d\vec{\ell} = \mu_0 \epsilon_0 \int_S \frac{\partial E_n}{\partial t} dA$$

(Equation 30-6d with $I = 0$) by integrating along a suitable curve C and over a suitable surface S in a manner that parallels the derivation of Equation 30-18.

65 ••• SSM A long cylindrical conductor of length L, radius a, and resistivity ρ carries a steady current I that is uniformly distributed over its cross-sectional area. (a) Use Ohm's law to relate the electric field E in the conductor to I, ρ, and a. (b) Find the magnetic field B just outside the conductor. (c) Use the results from Part (a) and Part (b) to compute the Poynting vector $\vec{S} = (\vec{E} \times \vec{B})/\mu_0$ at $r = a$ (the edge of the conductor). In what direction is $\vec{S}$? (d) Find the flux $\oint S_n \, dA$ through the surface of the conductor into the conductor, and show that the rate of energy flow into the conductor equals $I^2 R$, where R is the resistance of the cylinder. (Here, S_n is the *inward* component of $\vec{S}$ perpendicular to the surface of the conductor.)

66 ••• A long solenoid of n turns per unit length has a current that slowly increases with time. The solenoid has radius R, and the current in the windings has the form $I(t) = at$. (a) Find the induced electric field at a distance $r < R$ from the solenoid axis. (b) Find the magnitude and direction of the Poynting vector $\vec{S}$ at the cylindrical surface $r = R$ just inside the solenoid windings. (c) Calculate the flux $\oint S_n \, dA$ into the solenoid, and show that the flux equals the rate of increase of the magnetic energy inside the solenoid. (Here, S_n is the *inward* component of $\vec{S}$ perpendicular to the surface of the solenoid.)

67 ••• SSM ISOLVE Small particles might be blown out of solar systems by the radiation pressure of sunlight. Assume that the particles are spherical with a radius r and a density of 1 g/cm³ and that they absorb all the radiation in a cross-sectional area of πr^2. The particles are a distance R from the sun, which has a power output of 3.83×10^{26} W. What is the radius r for which the radiation force of repulsion just balances the gravitational force of attraction to the sun?

68 ••• When an electromagnetic wave is reflected at normal incidence on a perfectly conducting surface, the electric field vector of the reflected wave at the reflecting surface is the negative of that of the incident wave. (a) Explain why this should be. (b) Show that the superposition of incident and reflected waves results in a standing wave. (c) What is the relationship between the magnetic field vector of the incident waves and reflected waves at the reflecting surface?

69 ••• SSM An intense point source of light radiates 1 MW isotropically. The source is located 1 m above an infinite, perfectly reflecting plane. Determine the force that acts on the plane.

Properties of Light

LIGHT IS TRANSMITTED BY TOTAL INTERNAL REFLECTION THROUGH TINY GLASS FIBERS.

?   **How large must the angle of incidence of the light on the wall of the tube be so that no light escapes? (See Example 31-5.)**

The human eye is sensitive to electromagnetic radiation with wavelengths from approximately 400 nm to 700 nm. The shortest wavelengths in the visible spectrum correspond to violet light and the longest to red light. The perceived colors of light are the result of the physiological and psychological response of the eye–brain sensing system to the different frequencies of visible light. Although the correspondence between perceived color and frequency is quite good, there are many interesting deviations. For example, a mixture of red light and green light is perceived by the eye–brain sensing system as yellow even in the absence of light in the yellow region of the spectrum.

➤ In this chapter, we study how light is produced; how its speed is measured; and how light is scattered, reflected, refracted, and polarized.

31-1 Wave-Particle Duality

The wave nature of light was first demonstrated by Thomas Young, who observed the interference pattern of two coherent light sources produced by illuminating a pair of narrow, parallel slits with a single source. The wave theory of light culminated in 1860 with Maxwell's prediction of electromagnetic waves. The particle nature of light was first proposed by Albert Einstein in 1905 in his explanation of the photoelectric effect.[†] A particle of light called a **photon** has energy E that is related to the frequency f and wavelength λ of the light wave by the Einstein equation

$$E = hf = \frac{hc}{\lambda} \qquad \text{31-1}$$

EINSTEIN'S EQUATION FOR PHOTON ENERGY

where c is the speed of light and h is Planck's constant:

$$h = 6.626 \times 10^{-34} \, \text{J·s} = 4.136 \times 10^{-15} \, \text{eV·s}$$

Since energies are often given in electron volts and wavelengths are given in nanometers, it is convenient to express the combination hc in eV·nm. We have

$$hc = (4.136 \times 10^{-15} \, \text{eV·s})(2.998 \times 10^8 \, \text{m/s}) = 1.240 \times 10^{-6} \, \text{eV·m}$$

or

$$hc = 1240 \, \text{eV·nm} \qquad \text{31-2}$$

The propagation of light is governed by its wave properties, whereas the exchange of energy between light and matter is governed by its particle properties. This wave–particle duality is a general property of nature. For example, the propagation of electrons (and other so-called particles) is also governed by wave properties, whereas the exchange of energy between the electrons and other particles is governed by particle properties.

31-2 Light Spectra

Newton was the first to recognize that white light is a mixture of light of all colors of approximately equal intensity. He demonstrated this by letting sunlight fall on a glass prism and observing the spectrum of the refracted light (Figure 31-1). Because the angle of refraction produced by a glass prism depends slightly on wavelength, the refracted beam is spread out in space into its component colors or wavelengths, like a rainbow.

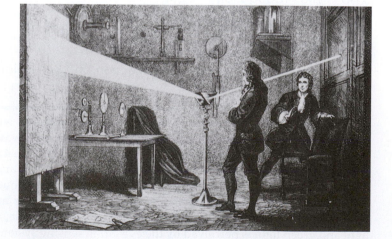

FIGURE 31-1 Newton demonstrating the spectrum of sunlight with a glass prism.

---

† The photoelectric effect is discussed in Chapter 34.

Figure 31-2 shows a spectroscope, which is a device for analyzing the spectra of a light source. Light from the source passes through a narrow slit, traverses a lens to make the beam parallel, and falls on a glass prism. The refracted beam is viewed with a telescope, which is mounted on a rotating platform so that the angle of the refracted beam, which depends on the wavelength, can be measured. The spectrum of the light source can thus be analyzed in terms of its component wavelengths. The spectrum of sunlight contains a continuous range of wavelengths and is therefore called a **continuous spectrum.** The light emitted by the atoms in low-pressure gases, such as mercury atoms in a fluorescent lamp, contains only a discrete set of wavelengths. Each wavelength emitted by the source produces a separate image of the collimating slit in the spectroscope. Such a spectrum is called a **line spectrum.** The continuous visible spectrum and the line spectra from several elements are shown in the photograph.

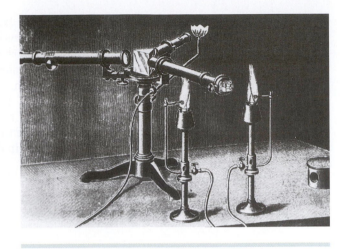

FIGURE 31-2 A late nineteenth-century spectroscope belonging to Gustav Kirchhoff. Modern student spectroscopes usually share the same general design.

31-3 Sources of Light

Line Spectra

The most common sources of visible light are transitions of the outer electrons in atoms. Normally an atom is in its ground state with its electrons at their lowest allowed energy levels consistent with the exclusion principle. (The exclusion principle, which was first enunciated by Wolfgang Pauli in 1925 to explain the electronic structure of atoms, states that no two electrons in an atom can be in the same quantum state.) The lowest energy electrons are closest to the nucleus and are tightly bound, forming a stable inner core. The one or two electrons in the highest energy states are much farther from the nucleus and are relatively easily excited to vacant higher energy states. These outer electrons are responsible for the energy changes in the atom that result in the emission or absorption of visible light.

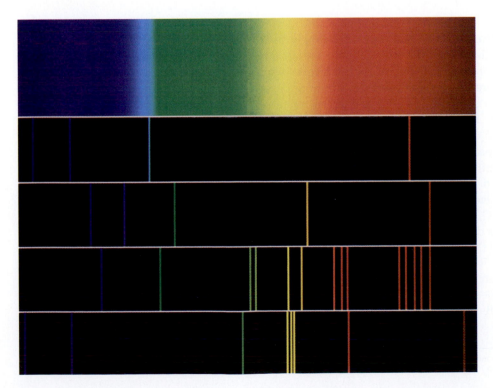

The continuous visible spectrum (top) and the line spectra of (from top to bottom) hydrogen, helium, barium, and mercury.

When an atom collides with another atom or with a free electron, or when the atom absorbs electromagnetic energy, the outer electrons can be excited to higher energy states. After a time of approximately 10 ns (1 ns = 10^{-9} s), these outer electrons spontaneously make transitions to lower energy states with the emission of a photon. This process, called **spontaneous emission,** is random; the photons emitted from two different atoms are not correlated. The emitted light is thus incoherent. By conservation of energy, the energy of an emitted photon is the energy difference $|\Delta E|$ between the initial state and the final state. The frequency of the light wave is related to the energy by the Einstein equation, $|\Delta E| = hf$. The wavelength of the emitted light is then

$$\lambda = \frac{c}{f} = \frac{hc}{hf} = \frac{hc}{|\Delta E|}$$

31-3

The photon energies corresponding to shortest wavelengths (400 nm) and longest (700 nm) wavelengths in the visible spectrum are

$$E_{400\,nm} = \frac{hc}{\lambda} = \frac{1240\ eV\cdot nm}{400\ nm} = 3.10\ eV$$

31-4a

and

$$E_{700\,nm} = \frac{hc}{\lambda} = \frac{1240\ eV\cdot nm}{700\ nm} = 1.77\ eV$$

31-4b

Because the energy levels in atoms form a discrete set, the emission spectrum of light from single atoms or from atoms in low-pressure gases consists of a set of sharp discrete lines that are characteristic of the element. These narrow lines are broadened somewhat by Doppler shifts, due to the motion of the atom relative to the observer and by collisions with other atoms; but, generally, if the gas density is low enough, the lines are narrow and well separated from one another. The study of the line spectra of hydrogen and other atoms led to the first understanding of the energy levels of atoms.

Continuous Spectra   When atoms are close together and interact strongly, as in liquids and solids, the energy levels of the individual atoms are spread out into energy bands, resulting in essentially continuous bands of energy levels. When the bands overlap, as they often do, the result is a continuous spectrum of possible energies and a continuous emission spectrum. In an incandescent material such as a hot metal filament, electrons are randomly accelerated by frequent collisions, resulting in a broad spectrum of thermal radiation. The rate at which an object radiates thermal energy is proportional to the fourth power of its absolute temperature.[†] The radiation emitted by an object at temperatures below approximately 600°C is concentrated in the infrared and is not visible. As an object is heated, the energy radiated extends to shorter and shorter wavelengths. Between approximately 600°C and 700°C, enough of the radiated energy is in the visible spectrum for the object to glow a dull red. At higher and higher temperatures, the object becomes bright red and then white. For a given temperature, the wavelength λ_{peak} at which the emitted power is a maximum varies inversely with the temperature, a result known as Wien's displacement law. The surface of the sun at $T = 6000$ K emits a continuous spectrum of approximately constant intensity over the visible range of wavelengths.

[†] This is known as the Stefan–Boltzmann law. This and other properties of thermal radiation, such as Wien's displacement law, are discussed more fully in Section 20-4.

Absorption, Scattering, Spontaneous Emission, and Stimulated Emission

When radiation is emitted, an atom makes a transition from an excited state to a state of lower energy; when radiation is absorbed, an atom makes a transition from a lower state to a higher state. When atoms are irradiated with a continuous spectrum of radiation, the transmitted spectrum shows dark lines corresponding to the absorption of light at discrete wavelengths. The absorption spectra of atoms were the first line spectra observed. Since atoms and molecules at normal temperatures are in either their ground states or low-lying excited states, only transitions from a ground state (or a near ground state) to a more highly excited state are observed. Thus, absorption spectra usually have far fewer lines than emission spectra have.

Figure 31-3 illustrates several interesting phenomena that can occur when a photon is incident on an atom. In Figure 31-3a, the energy of the incoming photon is too small to excite the atom to an excited state, so the atom remains in its ground state and the photon is said to be scattered. Since the incoming and outgoing or scattered photons have the same energy, the scattering is said to be elastic. If the wavelength of the incident light is large compared with the size of the atom, the scattering can be described in terms of classical electromagnetic theory and is called **Rayleigh scattering** after Lord Rayleigh, who worked out the theory in 1871. The probability of Rayleigh scattering varies as $1/\lambda^4$. This means that blue light is scattered much more readily than red light, which accounts for the bluish color of the sky. The removal of blue light by Rayleigh scattering also accounts for some of the reddish color of the transmitted light seen in sunsets.

Inelastic scattering, also called **Raman scattering,** occurs when an incident photon with just the right amount of energy is absorbed and the molecule undergoes a transition to a more energetic state. Then the molecule emits a photon as it undergoes a transition to a less energetic state, whose energy differs from that of the initial state. If the energy of the scattered photon hf' is less than that of the incident photon hf (Figure 31-3b), it is called **Stokes Raman scattering.** If the energy of the scattered photon is greater than that of the incident photon (Figure 31-3c), it is called **anti-Stokes Raman scattering.**

In Figure 31-3d, the energy of the incident photon is just equal to the difference in energy between the initial state and a more energetic state. The atom absorbs the photon and makes a transition to the more excited state in a process called **resonance absorption.**

In Figure 31-3e, an atom in an excited state spontaneously undergoes a transition to a less energetic state, in a process called **spontaneous emission.** Often an atom in an excited state undergoes transitions to one or more intermediate states as it returns to the ground state. A common example occurs when an atom is excited by ultraviolet light and emits visible light as it returns via multiple transitions to its ground state. This process, often called **fluorescence,** occurs in a thin film lining the inside of the glass tubes of fluorescent light bulbs. Since the lifetime of a typical excited atomic energy state is of the order of 10 ns, this process appears to occur instantaneously. However, some excited states have much longer lifetimes—of the order of milliseconds or occasionally seconds or even minutes. Such a state is called a **metastable state. Phosphorescent materials** have very long-lived metastable states and emit light long after the original excitation.

Figure 31-3f illustrates the photoelectric effect, in which the absorption of the photon ionizes the atom by causing the emission of an electron. Figure 31-3g illustrates **stimulated emission.** This process occurs if the atom or molecule is initially in an excited state of energy E_H, and the energy of the incident photon is equal to $E_H - E_L$, where E_L is the energy of a lower state. In this case, the oscillating electromagnetic field associated with the incident photon can stimulate the excited atom or molecule, which then emits a photon in the same direction as the incident photon and in phase with it. The photons from the stimulated atoms or molecules can stimulate the emission of additional photons propagating in the same direction

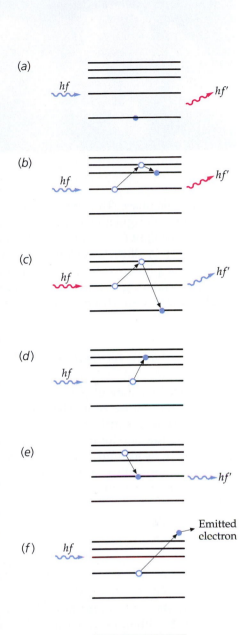

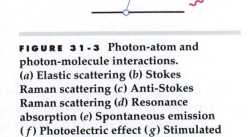

FIGURE 31-3 Photon-atom and photon-molecule interactions. (*a*) Elastic scattering (*b*) Stokes Raman scattering (*c*) Anti-Stokes Raman scattering (*d*) Resonance absorption (*e*) Spontaneous emission (*f*) Photoelectric effect (*g*) Stimulated emission (*h*) Compton scattering.

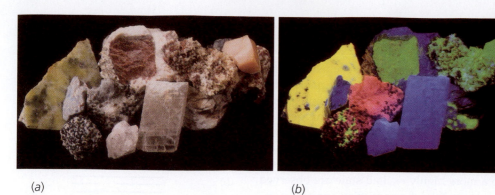

(a)

(b)

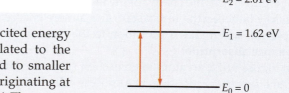

(c)

with the same phase. This process amplifies the initially emitted photon, yielding a beam of light originating from different atoms that is coherent. As a result, interference of the light from a large number of atoms can easily be observed.

Figure 31-3*h* illustrates **Compton scattering,** which occurs if the energy of the incident photon is much greater than the ionization energy. Note that in Compton scattering, a photon is absorbed and a photon is emitted, whereas in the photoelectric effect, a photon is absorbed with none emitted.

A collection of minerals in (*a*) daylight and in (*b*) ultraviolet light (sometimes called *black light*). Identified by number in the schematic (*c*), they are 1, powerllite; 2, willemite; 3, scheelite; 4, calcite; 5, calcite and willemite composite; 6, optical calcite; 7, willemite; and 8, opal. The change in color is due to the minerals fluorescing under the ultraviolet light. In optical calcite, both fluorescence and phosphorescence occur.

RESONANT ABSORPTION AND EMISSION **EXAMPLE 31-1**

The first excited state of potassium is $E_1 = 1.62$ eV above the ground state E_0, which we take to be zero. The second and third excited states of potassium have energy levels at $E_2 = 2.61$ eV and $E_3 = 3.07$ eV above the ground state. (*a*) What is the maximum wavelength of radiation that can be absorbed by potassium in its ground state? Calculate the wavelength of the emitted photon when the atom makes a transition from (*b*) the second excited state (E_2) to the ground state and from (*c*) the third excited state (E_3) to the second excited state (E_2).

PICTURE THE PROBLEM The ground state and the first three excited energy levels are shown in Figure 31-4. (*a*) Since the wavelength is related to the energy of a photon by $\lambda = hc/\Delta E$, longer wavelengths correspond to smaller energy differences. The smallest energy difference for a transition originating at the ground state is from the ground state to the first excited state. (*b*) The wavelengths of the photons given off when the atom de-excites are related to the energy differences by $\lambda = hc/|\Delta E|$.

$E_3 = 3.07$ eV
$E_2 = 2.61$ eV

$E_1 = 1.62$ eV

$E_0 = 0$

FIGURE 31-4

(*a*) Calculate the wavelength of radiation absorbed in a transition from the ground state to the first excited state:

$$\lambda = \frac{hc}{\Delta E} = \frac{hc}{E_1 - E_0} = \frac{1240 \text{ eV·nm}}{1.62 \text{ eV} - 0} = \boxed{765 \text{ nm}}$$

(*b*) For the transition from E_3 to the ground state, the photon energy is $E_3 - E_0 = E_3$. Calculate the wavelength of radiation emitted in this transition:

$$\lambda = \frac{hc}{|\Delta E|} = \frac{hc}{E_3 - E_0} = \frac{1240 \text{ eV·nm}}{3.07 \text{ eV} - 0} = \boxed{404 \text{ nm}}$$

(*c*) For the transition from E_3 to E_2, the photon energy is $E_3 - E_2$. Calculate the wavelength of radiation emitted in this transition:

$$\lambda = \frac{hc}{|\Delta E|} = \frac{hc}{E_3 - E_2} = \frac{1240 \text{ eV·nm}}{3.07 \text{ eV} - 2.61 \text{ eV}}$$

$$= \boxed{2700 \text{ nm}}$$

REMARKS The wavelength of radiation emitted in the transition from E_1 to the ground state E_0 is 765 nm, the same as that for radiation absorbed in the transition from the ground state to E_1. This transition and the transmission from E_3 to the ground state both result in photons in the visible spectrum.

Lasers

The *laser* (light *a*mplification by *s*timulated *e*mission of *r*adiation) is a device that produces a strong beam of coherent photons by stimulated emission. Consider a system consisting of atoms that have a ground state of energy E_1 and an excited metastable state of energy E_2. If these atoms are irradiated by photons of energy $E_2 - E_1$, those atoms in the ground state can absorb a photon and make the transition to state E_2, whereas those atoms already in the excited state may be stimulated to decay back to the ground state. The relative probabilities of absorption and stimulated emission, first worked out by Einstein, are equal. Ordinarily, nearly all the atoms of the system at normal temperature will initially be in the ground state, so absorption will be the main effect. To produce more stimulated-emission transitions than absorption transitions, we must arrange to have more atoms in the excited state than in the ground state. This condition, called population inversion, can be achieved by a method called optical pumping in which atoms are *pumped* up to levels of energy greater than E_2 by the absorption of an intense auxiliary radiation. The atoms then decay down to state E_2 either by spontaneous emission or by nonradiative transitions, such as those due to collisions.

Figure 31-5 shows a schematic diagram of the first laser, a ruby laser built by Theodore Maiman in 1960. The laser consists of a ruby rod a few centimeters long surrounded by a helical gaseous flashtube that emits a broad spectrum of light. The ends of the ruby rod are flat and perpendicular to the axis of the rod. Ruby is a transparent crystal of Al_2O_3 with a small amount (about 0.05 percent) of chromium. It appears red because the chromium ions (Cr^{3+}) have strong absorption bands in the blue and green regions of the visible spectrum, as shown in Figure 31-6. The energy levels of chromium—important for the operation of a ruby laser—are shown in Figure 31-7. When the flashtube is fired, there is an intense burst of light that lasts several milliseconds. Photon absorption excites many of the chromium ions to the bands of energy levels indicated by the shading in Figure 31-7. The excited chromium ions then rapidly drop down to a closely spaced pair of metastable states labeled E_2 in the figure. These metastable states are approximately 1.79 eV above the ground state. The expected lifetime for a chromium ion to remain in one of these metastable states is about 5 ms, after which the chromium ion spontaneously emits a photon and decays to the ground state. A millisecond is a long time for an atomic process. Consequently, if the flash is intense enough, the number of chromium ions populating the two metastable states will exceed the population of chromium ions in the ground state. It follows that during the time that the flashtube is firing, the populations of ions in the ground state and the metastable states are inverted. When the chromium ions in the state E_2 decay to the ground state by spontaneous emission, they emit photons of energy 1.79 eV and wavelength 694.3 nm. These photons have just the right energy to stimulate chromium ions in the metastable states to emit photons of the same energy (and wavelength) as they undergo the transition to the ground state. The photons also have just the right energy to stimulate chromium ions in the ground state to absorb a photon as they undergo the transition to one of the metastable states. These are competing processes, and the stimulated emission process dominates as long as the population of chromium ions in the metastable states exceeds the population in the ground state.

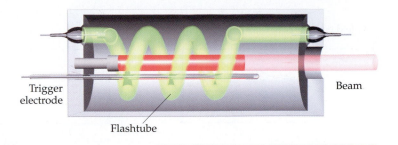

Trigger electrode

Beam

Flashtube

FIGURE 31-5 Schematic diagram of the first ruby laser.

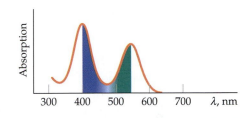

FIGURE 31-6 Absorption versus wavelength for Cr^{3+} in ruby. Ruby appears red because of the strong absorption of green and blue light by the chromium ions.

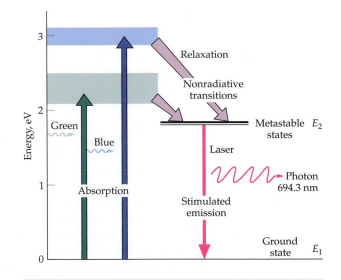

FIGURE 31-7 Energy levels in a ruby laser. To make the population of the metastable states greater than that of the ground state, the ruby crystal is subjected to intense radiation that contains energy in the green and blue wavelengths. This excites atoms from the ground state to the bands of energy levels indicated by the shading, from which the atoms decay to the metastable states by nonradiative transitions. Then, by stimulated emission, the atoms undergo the transition from the metastable states to the ground state.

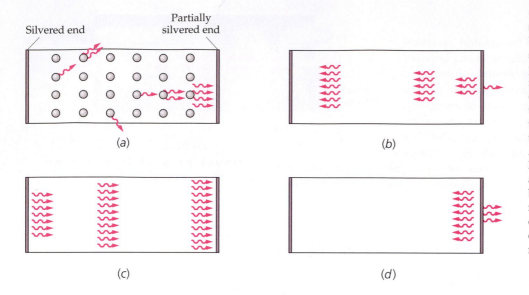

Silvered end

Partially
silvered end

(a)

(b)

(c)

(d)

FIGURE 31-8 Buildup of photon beam in a laser. (*a*) When irradiated, some atoms spontaneously emit photons, some of which travel to the right and stimulate other atoms to emit photons parallel to the axis of the crystal. (*b*) Of the four photons that strike the right face, one is transmitted and three are reflected. As the reflected photons traverse the laser crystal, they stimulate other atoms to emit photons, and the beam builds up. By the time the beam reaches the right face again (*c*), it comprises many photons. (*d*) Some of these photons are transmitted, the rest of the photons are reflected.

In the ruby laser, one end of the crystal is fully silvered, so it is 100 percent reflecting; the other end of the crystal, called the output coupler, is partially silvered, leaving it about 85 percent reflecting. When photons traveling parallel to the axis of the crystal strike the silvered ends, all are reflected from the back face and 85 percent are reflected from the front face, with 15 percent of the photons escaping through the partially silvered front face. During each pass through the crystal, the photons stimulate more and more atoms so that an intense beam is emitted from the partially silvered end (Figure 31-8). Because the duration of each flash of the flashtube is between two and three seconds, the laser beam is produced in pulses lasting a few milliseconds. Modern ruby lasers generate intense light beams with energies ranging from 50 J to 100 J. The beam can have a diameter as small as 1 mm and an angular divergence as small as 0.25 milliradian to about 7 milliradians.

Population inversion is achieved somewhat differently in the continuous helium–neon laser. The energy levels of helium and neon that are important for the operation of the laser are shown in Figure 31-9. Helium has an excited energy state $E_{2,\text{He}}$ that is 20.61 eV above its ground state. Helium atoms are excited to state $E_{2,\text{He}}$ by an electric discharge. Neon has an excited state $E_{3,\text{Ne}}$ that is 20.66 eV above its ground state. This is just 0.05 eV above the first excited state of helium. The neon atoms are excited to state $E_{3,\text{Ne}}$ by collisions with excited helium atoms. The kinetic energy of the helium atoms provides the extra 0.05 eV of energy needed to excite the neon atoms. There is another excited state of neon $E_{2,\text{Ne}}$ that is 18.70 eV above its ground state and 1.96 eV below state $E_{3,\text{Ne}}$. Since state $E_{2,\text{Ne}}$ is normally unoccupied, population inversion between states $E_{3,\text{Ne}}$ and $E_{2,\text{Ne}}$ is obtained immediately. The stimulated emission that occurs between these states results in photons of energy 1.96 eV and wavelength 632.8 nm, which produces a bright red light. After stimulated emission, the atoms in state $E_{2,\text{Ne}}$ decay to the ground state by spontaneous emission.

Note that there are four energy levels involved in the helium–neon laser, whereas the ruby laser involved only three levels. In a three-level laser, population inversion is

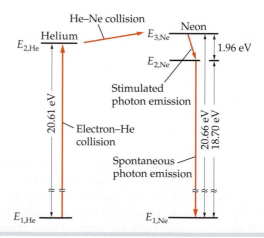

FIGURE 31-9 Energy levels of helium and neon that are important for the helium–neon laser. The helium atoms are excited by electrical discharge to an energy state 20.61 eV above the ground state. They collide with neon atoms, exciting some neon atoms to an energy state 20.66 eV above the ground state. Population inversion is thus achieved between this level and one 1.96 eV below it. The spontaneous emission of photons of energy 1.96 eV stimulates other atoms in the upper state to emit photons of energy 1.96 eV.

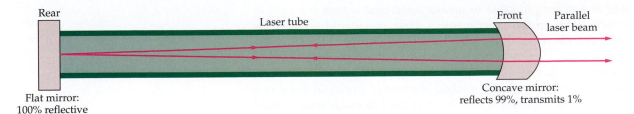

Rear
Laser tube
Front   Parallel
laser beam
Flat mirror:
100% reflective
Concave mirror:
reflects 99%, transmits 1%

FIGURE 31-10 Schematic drawing of a helium–neon laser. The use of a concave mirror rather than a second plane mirror makes the alignment of the mirrors less critical than it is for the ruby laser. The concave mirror on the right also serves as a lens that focuses the emitted light into a parallel beam.

difficult to achieve because more than half the atoms in the ground state must be excited. In a four-level laser, population inversion is easily achieved because the state after stimulated emission is not the ground state but an excited state that is normally unpopulated.

Figure 31-10 shows a schematic diagram of a helium–neon laser commonly used for physics demonstrations. The helium–neon laser consists of a gas tube that contains 15 percent helium gas and 85 percent neon gas. A totally reflecting flat mirror is mounted at one end of the gas tube and a 99 percent reflecting concave mirror is placed at the other end of the gas tube. The concave mirror focuses parallel light at the flat mirror and also acts as a lens that transmits part of the light, so that the light emerges as a parallel beam.

A laser beam is coherent, very narrow, and intense. Its coherence makes the laser beam useful in the production of holograms, which we discuss in Chapter 33. The precise direction and small angular spread of the laser beam make it useful as a surgical tool for destroying cancer cells or reattaching a detached retina. Lasers are also used by surveyors for precise alignment over large distances. Distances can be accurately measured by reflecting a laser pulse from a mirror and measuring the time the pulse takes to travel to the mirror and back. The distance to the moon has been measured to within a few centimeters using a mirror placed on the moon for that purpose. Laser beams are also used in fusion research. An intense laser pulse is focused on tiny pellets of deuterium–tritium in a combustion chamber. The beam heats the pellets to temperatures of the order of 10^8 K in a very short time, causing the deuterium and tritium to fuse and release energy.

Laser technology is advancing so quickly that it is possible to mention only a few of the recent developments. In addition to the ruby laser, there are many other solid-state lasers with output wavelengths that range from approximately 170 nm to 3900 nm. Lasers that generate more than 1 kW of continuous power have been constructed. Pulsed lasers can now deliver nanosecond pulses of power exceeding 10^{14} W. Various gas lasers can now produce beams of wavelengths that range from the far infrared to the ultraviolet. Semiconductor lasers (also known as diode lasers or junction lasers) have shrunk in just 10 years from the size of a pinhead to mere billionths of a meter. Liquid lasers that use chemical dyes can be tuned over a range of wavelengths (approximately 70 nm for continuous lasers and more than 170 nm for pulsed lasers). A relatively new laser, the free-electron laser, extracts light energy from a beam of free electrons moving through a spatially varying magnetic field. The free-electron laser has the potential for very high power and high efficiency and can be tuned over a large range of wavelengths. There appears to be no limit to the variety and uses of modern lasers.

31-4 The Speed of Light

Prior to the seventeenth century the speed of light was thought by many to be infinite, and an effort to measure the speed of light was made by Galileo. He and a partner stood on hilltops about three kilometers apart, each with a lantern and a shutter to cover it. Galileo proposed to measure the time it took for light to travel back and forth between the experimenters. First, one would uncover his lantern, and when the other saw the light, he would uncover his. The time between the first partner's uncovering his lantern and his seeing the light from the other lantern would be the time it took for light to travel back and forth between the experimenters. Though this method is sound in principle, the speed of light is so great that the time interval to be measured is much smaller than fluctuations in human response time, so Galileo was unable to obtain a value for the speed of light.

The first indication of the true magnitude of the speed of light came from astronomical observations of the period of Io, one of the moons of Jupiter. This period is determined by measuring the time between eclipses of Io behind Jupiter. The eclipse period is about 42.5 h, but measurements made when the earth is moving away from Jupiter along path *ABC* in Figure 31-11 give a greater time for this period than do measurements made when the earth is moving toward Jupiter along path *CDA* in the figure. Since these measurements differ from the average value by only about 15 s, the discrepancies were difficult to measure accurately. In 1675, the astronomer Ole Römer attributed these discrepancies to the fact that the speed of light is finite, and that during the 42.5 h between eclipses of Jupiter's moon, the distance between the earth and Jupiter changes, making the path for the light longer or shorter. Römer devised the following method for measuring the cumulative effect of these discrepancies. Jupiter is moving much more slowly than the earth, so we can neglect its motion. When the earth is at point *A*, nearest to Jupiter, the distance between the earth and Jupiter is changing negligibly. The period of Io's eclipse is measured, providing the time between the beginnings of successive eclipses. Based on this measurement, the number of occultations during 6 months is computed, and the time when an eclipse should begin a half-year later when the earth is at point *C* is predicted. When the earth is actually at point *C*, the observed beginning of the eclipse is about 16.6 min later than predicted. This is the time it takes light to travel a distance equal to the diameter of the earth's orbit. This calculation neglects the distance traveled by Jupiter toward the earth. However, because the orbital speed of Jupiter is so much slower than that of the earth, the distance Jupiter moves toward (or away from) the earth during the 6 months is much less than the diameter of the earth's orbit.

EXERCISE Calculate (*a*) the distance traveled by the earth between successive eclipses of Io and (*b*) the speed of light, given that the time between successive eclipses is 15 s longer than average when the earth is moving directly away from Jupiter. (*Answer* (*a*) 4.59×10^6 km (*b*) 3.06×10^8 m/s)

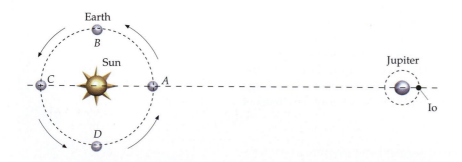

FIGURE 31-11 Römer's method of measuring the speed of light. The time between eclipses of Jupiter's moon Io appears to be greater when the earth is moving along path *ABC* than when the earth is moving along path *CDA*. The difference is due to the time it takes light to travel the distance traveled by the earth along the line of sight during one period of Io. (The distance traveled by Jupiter in one earth year is negligible.)

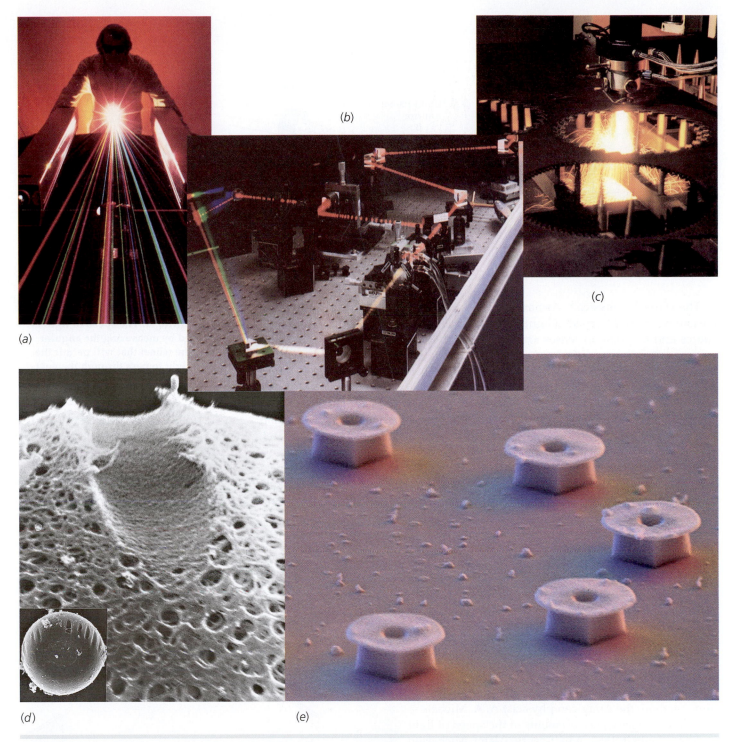

(*a*) Beams from a krypton laser and an argon laser, split into their component wavelengths. In these gas lasers, krypton and argon atoms have been stripped of multiple electrons, forming positive ions. The light-emitting energy transitions occur when excited electrons in the ions decay from one upper energy level to another. Here, several energy transitions are occurring at once, each corresponding to emitted light of a different wavelength. (*b*) A femtosecond pulsed laser. By a technique known as *modelocking*, different excited modes within a laser's cavity can be made to interfere with one another and create a series of ultrashort pulses, which are picoseconds long, that correspond to the time it takes light to bounce back and forth once within the cavity. Ultrashort pulses have been used as probes to study the behavior of molecules during chemical reactions. (*c*) A carbon dioxide laser takes just 2 minutes to cut out a steel saw blade. (*d*) A groove etched in the zona pellucida (protective outer covering) of a mouse egg by a *laser scissor* facilitates implantation. This technique has already been applied in human fertility therapies. Several effects contribute to the ability of the finely focused laser to cut on such a delicate scale—photon absorption may heat the target, break molecular bonds, or drive chemical reactions. (*e*) The so-called nanolasers shown are semiconductor disks mere microns in diameter and fractions of a micron in width. These tiny lasers work like their larger counterparts. Exploiting quantum effects that prevail on this microscopic scale, nanolasers promise great efficiency and they are being explored as ultrafast, low-energy switching devices.

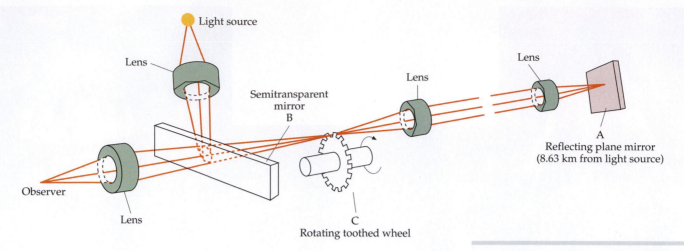

Light source

Lens

Semitransparent mirror B

Lens

Lens

A
Reflecting plane mirror (8.63 km from light source)

Observer

Lens

C
Rotating toothed wheel

FIGURE 31-12 Fizeau's method of measuring the speed of light. Light from the source is reflected by mirror B and is transmitted through a gap in the toothed wheel to mirror A. The speed of light is determined by measuring the angular speed of the wheel that will permit the reflected light to pass through the next gap in the toothed wheel so that an image of the source is observed.

The French physicist Armand Fizeau made the first nonastronomical measurement of the speed of light in 1849. On a hill in Paris, Fizeau placed a light source and a system of lenses arranged so that the light reflected from a semitransparent mirror was focused on a gap in a toothed wheel, as shown in Figure 31-12. On a distant hill (about 8.63 km away) Fizeau placed a mirror to reflect the light back, to be viewed by an observer as shown. The toothed wheel was rotated, and the speed of rotation was varied. At low speeds of rotation, no light was visible because the light that passed through a gap in the rotating wheel and was reflected back by the mirror was obstructed by the next tooth of the wheel. The speed of rotation was then increased. The light suddenly became visible when the rotation speed was such that the reflected light passed through the next gap in the wheel. The time for the wheel to rotate through the angle between successive gaps equals the time for the light to make the round trip to the distant mirror and back.

Fizeau's method was improved upon by Jean Foucault, who replaced the toothed wheel with a rotating mirror, as shown in Figure 31-13. Light strikes the rotating mirror, the light is reflected to a distant fixed mirror, the light is then reflected back to the rotating mirror, and then to an observing telescope. During the time taken for the light to travel from the rotating mirror to the distant fixed mirror and back, the mirror rotates through a small angle. By measuring the angle θ, the time for the light to travel to the distant mirror and back is determined. In approximately 1850, Foucault measured the speed of light in air and in water, and he showed that the speed of light is less in water than the speed of light in air. Using essentially the same method, the American physicist A. A. Michelson made more precise measurements of the speed of light in approximately 1880. A half-century later, Michelson made even more precise measurements of the speed of light, using an octagonal rotating mirror (Figure 31-14). In these measurements, the mirror rotates through one-eighth of a turn during the time it takes for the light to travel to the fixed mirror and back. The rotation rate is varied until another face of the mirror is in the right position for the reflected light to enter the telescope.

Another method of determining the speed of light involves the measurement of the electrical constants ϵ_0 and μ_0 to determine c from $c = 1/\sqrt{\epsilon_0\mu_0}$.

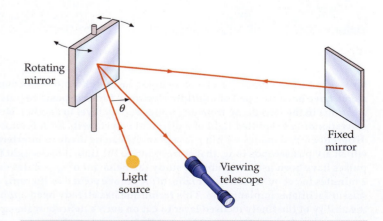

Rotating mirror

θ

Fixed mirror

Light source

Viewing telescope

FIGURE 31-13 Simplified drawing of Foucault's method of measuring the speed of light.

The various methods we have discussed for measuring the speed of light are all in general agreement. Today, the speed of light is defined to be exactly

$$c = 299{,}792{,}458 \text{ m/s} \qquad\qquad 31\text{-}5$$

DEFINITION—SPEED OF LIGHT

and the standard unit of length, the meter, is defined in terms of this speed and the standard unit of time. The meter is the distance light travels (in a vacuum) in $1/299{,}792{,}458$ s. The value 3×10^8 m/s for the speed of light is accurate enough for nearly all calculations. The speed of radio waves and all other electromagnetic waves (in a vacuum) is the same as the speed of light.

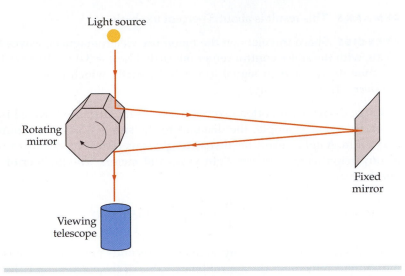

FIGURE 31-14 Simplified drawing of Michelson's method of measuring the speed of light at Mt. Wilson in the late 1920s.

THE SPEED OF LIGHT   **EXAMPLE   31-2**

What is the speed of light in feet per nanosecond?

PICTURE THE PROBLEM This is an exercise in unit conversions. There are $\sim$30 cm = 0.3 m in 1 ft.

1. Convert m/s to ft/ns:
$$c = 3 \times 10^8 \text{ m/s} \times \left(\frac{1 \text{ ft}}{0.3 \text{ m}}\right) \times \left(\frac{1 \text{ s}}{10^9 \text{ ns}}\right) = \boxed{1 \text{ ft/ns}}$$

FIZEAU'S DETERMINATION OF C   **EXAMPLE   31-3**

In Fizeau's experiment, his wheel had 720 teeth, and light was observed when the wheel rotated at 25.2 revolutions per second. If the distance from the wheel to the distant mirror was 8.63 km, what was Fizeau's value for the speed of light?

PICTURE THE PROBLEM The time taken for the light to travel from the wheel to the mirror and back is the time for the wheel to rotate one Nth of a revolution, where $N = 720$ is the total number of teeth.

1. The speed is the distance divided by the time. The distance from the wheel to the mirror is L:
$$c = \frac{2L}{\Delta t}$$

2. The angular displacement equals the angular speed times the time:
$$\Delta\theta = \omega \Delta t$$

3. Solve for the time:
$$\Delta t = \frac{\Delta\theta}{\omega}$$

4. Substitute for Δt and solve for c:
$$c = \frac{2L\omega}{\Delta\theta} = \frac{2(8.63 \times 10^3 \text{ m})(25.2 \text{ rev/s})}{\dfrac{1}{720} \text{ rev}}$$

$$= \boxed{3.14 \times 10^8 \text{ m/s}}$$

REMARKS This result is about 5 percent too high.

EXERCISE Space travelers on the moon use electromagnetic waves to communicate with the space control center on earth. Use $c = 3.00 \times 10^8$ m/s to calculate the time delay for their signal to reach the earth, which is 3.84×10^8 m away. (*Answer* 1.28 s each way)

Large distances are often given in terms of the distance traveled by light in a given time. For example, the distance to the sun is 8.33 light-minutes, written 8.33 c-min. A light-year is the distance light travels in one year. We can easily find a conversion factor between light-years and meters. The number of seconds in one year is

$$1\,y = 1\,y \times \frac{365.24\,d}{1\,y} \times \frac{24\,h}{1\,d} \times \frac{3600\,s}{1\,h} = 3.156 \times 10^7\,s$$

(*Note:* There are approximately π times 10^7 seconds per year, which is how some individuals remember the approximate value of the conversion.) The number of meters in one light-year is thus

$$1c\text{-year} = (2.998 \times 10^8\,\text{m/s})(3.156 \times 10^7\,\text{s}) = 9.46 \times 10^{15}\,\text{m} \qquad 31\text{-}6$$

31-5 The Propagation of Light

The propagation of light is governed by the wave equation discussed in Chapter 30. But long before Maxwell's theory of electromagnetic waves, the propagation of light and other waves was described empirically by two interesting and very different principles attributed to the Dutch physicist Christian Huygens (1629–1695) and the French mathematician Pierre de Fermat (1601–1665).

Huygens's Principle

Figure 31-15 shows a portion of a spherical wavefront emanating from a point source. The wavefront is the locus of points of constant phase. If the radius of the wavefront is r at time t, its radius at time $t + \Delta t$ is $r + c\,\Delta t$, where c is the speed of the wave. However, if a part of the wave is blocked by some obstacle or if the wave passes through a different medium, as in Figure 31-16, the determination of the new wavefront position at time $t + \Delta t$ is much more difficult. The propagation of any wavefront through space can be described using a geometric method discovered by Huygens in approximately 1678, which is now known as **Huygens's principle** or **Huygens's construction:**

> Each point on a primary wavefront serves as the source of spherical secondary wavelets that advance with a speed and frequency equal to those of the primary wave. The primary wavefront at some later time is the envelope of these wavelets.

HUYGENS'S PRINCIPLE

Figure 31-17 shows the application of Huygens's principle to the propagation of a plane wave and the propagation of a spherical wave. Of course, if each point on a wavefront were really a point source, there would be waves in the backward direction as well. Huygens ignored these back waves.

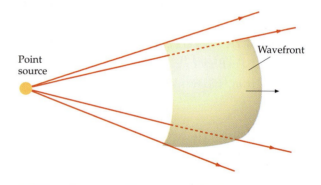

FIGURE 31-15 Spherical wavefront from a point source.

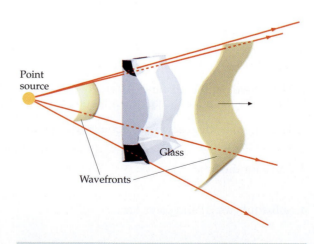

FIGURE 31-16 Wavefront from a point source before and after passing through a piece of glass of varied thickness.

Huygens's principle was later modified by Augustin Fresnel, so that the new wavefront was calculated from the old wavefront by superposition of the wavelets considering their relative amplitudes and phases. Kirchhoff later showed that the Huygens–Fresnel principle was a consequence of the wave equation (Equation 30-17), thus putting it on a firm mathematical basis. Kirchhoff showed that the intensity of each wavelet depends on the angle and is zero at 180° (the backward direction).

We will use Huygens's principle to derive the laws of reflection and refraction in Section 31-8. In Chapter 33, we apply Huygens's principle with Fresnel's modification to calculate the diffraction pattern of a single slit. Because the wavelength of light is so small, we can often use the ray approximation to describe its propagation.

Fermat's Principle

The propagation of light can also be described by Fermat's principle:

> The path taken by light traveling from one point to another is such that the time of travel is a minimum.[†]

FERMAT'S PRINCIPLE

In Section 31-8, we will use Fermat's principle to derive the laws of reflection and refraction.

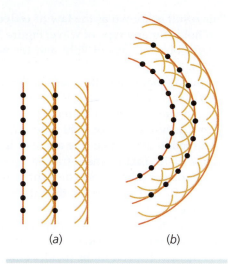

(a)　　　　　(b)

FIGURE 31-17 Huygens's construction for the propagation to the right of (a) a plane wave and (b) an outgoing spherical, or circular, wave.

31-6 Reflection and Refraction

The speed of light in a transparent medium such as air, water, or glass is less than the speed $c = 3 \times 10^8$ m/s in vacuum. A transparent medium is characterized by the **index of refraction,** n, which is defined as the ratio of the speed of light in a vacuum, c, to the speed in the medium, v:

$$n = \frac{c}{v} \qquad 31\text{-}7$$

DEFINITION—INDEX OF REFRACTION

For water, $n = 1.33$, whereas for glass n ranges from approximately 1.50 to 1.66, depending on the type of glass. Diamond has a very high index of refraction—approximately 2.4. The index of refraction of air is approximately 1.0003, so for most purposes we can assume that the speed of light in air is the same as the speed of light in vacuum.

When a beam of light strikes a boundary surface separating two different media, such as an air–glass interface, part of the light energy is reflected and part of the light energy enters the second medium. If the incident light is not perpendicular to the surface, then the transmitted beam is not parallel to the incident beam. The change in direction of the transmitted ray is called **refraction.** Figure 31-18 shows a light ray striking a smooth air–glass interface. The angle θ_1 between the incident ray and the normal (the line perpendicular to the surface) is called the **angle of incidence,** and the plane defined by these two lines is called the **plane of incidence.** The reflected ray lies in the plane of incidence and makes an angle θ_1' with the normal that is equal to the angle of incidence as shown in the figure:

$$\theta_1' = \theta_1 \qquad 31\text{-}8$$

LAW OF REFLECTION

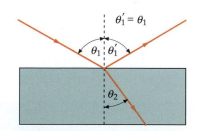

FIGURE 31-18 The angle of reflection θ_1' equals the angle of incidence θ_1. The angle of refraction θ_2 is less than the angle of incidence if the light speed in the second medium is less than that in the incident medium.

[†] A more complete and general statement is that the time of travel is stationary with respect to variations in path; that is, if t is expressed in terms of some parameter x, the path taken will be such that $dt/dx = 0$. The important characteristic of a stationary path is that the time taken along nearby paths will be approximately the same as that along the true path.

This result is known as the **law of reflection.** The law of reflection holds for any type of wave. Figure 31-19 illustrates the law of reflection for rays of light and for wavefronts of ultrasonic waves.

The ray that enters the glass in Figure 31-18 is called the refracted ray, and the angle θ_2 is called the angle of refraction. When a wave crosses a boundary where the wave speed is reduced, as in the case of light entering glass from air, the angle of refraction is less than the angle of incidence θ_1, as shown in Figure 31-18; that is, the refracted ray is bent toward the normal. If, on the other hand, the light beam originates in the glass and is refracted into the air, then the refracted ray is bent away from the normal.

The angle of refraction θ_2 depends on the angle of incidence and on the relative speed of light waves in the two mediums. If v_1 is the wave speed in the incident medium and v_2 is the wave speed in the transmission medium, the angles of incidence and refraction are related by

$$\frac{1}{v_1}\sin\theta_1 = \frac{1}{v_2}\sin\theta_2 \qquad \text{31-9}a$$

Equation 31-9a holds for the refraction of any kind of wave incident on a boundary separating two media.

In terms of the indexes of refraction of the two media n_1 and n_2, Equation 31-9a is

$$n_1\sin\theta_1 = n_2\sin\theta_2 \qquad \text{31-9}b$$

SNELL'S LAW OF REFRACTION

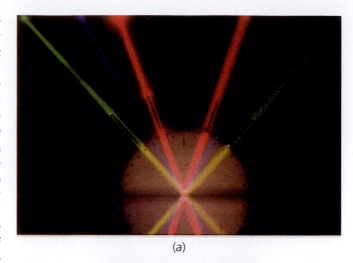

(a)

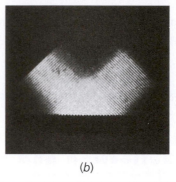

(b)

FIGURE 31-19 (a) Light rays reflecting from an air–glass interface showing equal angles of incidence and reflection. (b) Ultrasonic plane waves in water reflecting from a steel plate.

This result was discovered experimentally in 1621 by the Dutch scientist Willebrord Snell and is known as **Snell's law** or the **law of refraction.** It was independently discovered a few years later by the French mathematician and philosopher René Descartes.

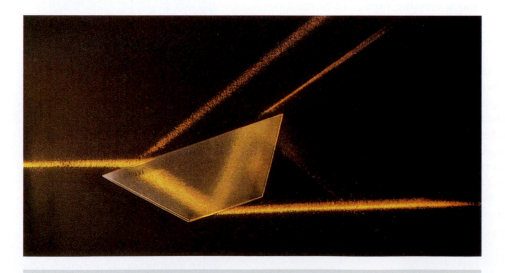

Reflection and refraction of a beam of light incident on a glass slab.

Physical Mechanisms for Reflection and Refraction

The physical mechanism of the reflection and refraction of light can be understood in terms of the absorption and reradiation of the light by the atoms in the reflecting or refracting medium. When light traveling in air strikes a glass surface, the atoms in the glass absorb the light and reradiate it at the same frequency in all directions. The waves radiated backward by the glass atoms interfere constructively at an angle equal to the angle of incidence to produce the reflected wave.

The transmitted wave is the result of the interference of the incident wave and the wave produced by the absorption and reradiation of light energy by the atoms in the medium. For light entering glass from air, there is a phase lag between the reradiated wave and the incident wave. There is, therefore, also a phase lag between the resultant wave and the incident wave. This phase lag means that the position of a wave crest of the transmitted wave is retarded relative to the position of a wave crest of the incident wave in the medium. As a result, a transmitted wave crest does not travel as far in a given time as the original incident wave crest; that is, the velocity of the transmitted wave is less than that of the incident wave. The index of refraction is therefore greater than 1. The frequency of the light in the second medium is the same as the frequency of the incident light—the atoms absorb and reradiate the light at the same frequency—but the wave speed is different, so the wavelength of the transmitted light is different from that of the incident light. If λ is the wavelength of light in a vacuum, then $\lambda f = c$, and if λ' is the wavelength in a medium in which it has speed v, then $\lambda'f = v$. Combining these two relations gives $\lambda/\lambda' = c/v$, or

$$\lambda' = \frac{\lambda}{c/v} = \frac{\lambda}{n} \qquad\qquad 31\text{-}10$$

where $n = c/v$ is the index of refraction of the medium.

Specular Reflection and Diffuse Reflection

Figure 31-20*a* shows a bundle of light rays from a point source P that are reflected from a flat surface. After reflection, the rays diverge exactly as if they came from a point P' behind the surface. (This point is called the *image point*. We will study the formation of images by reflecting and refracting surfaces in the next chapter.) When these rays enter the eye, they cannot be distinguished from rays actually diverging from a source at P'.

FIGURE 31-20 (*a*) Specular reflection from a smooth surface. (*b*) Specular reflection of trees from water.

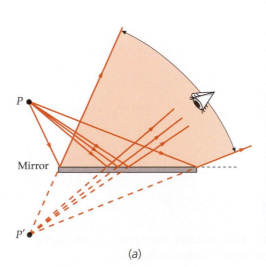

(a)

(b)

Reflection from a smooth surface is called **specular reflection.** It differs from **diffuse reflection,** which is illustrated in Figure 31-21. Here, because the surface is rough, the rays from a point reflect in random directions and do not diverge from any point, so there is no image. The reflection of light from the page of this book is diffuse reflection. The glass used in picture frames is sometimes ground slightly to give diffuse reflection and thereby cut down on glare from the light used to illuminate the picture. Diffuse reflection from the surface of the road allows you to see the road when you are driving at night because some of the light from your headlights reflects back toward you. In wet weather the reflection is mostly specular; therefore, little light is reflected back toward you, which makes the road difficult to see.

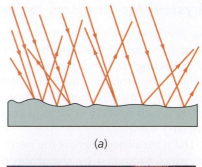

(a)

Relative Intensity of Reflected and Transmitted Light

The fraction of light energy reflected at a boundary, such as an air–glass interface, depends in a complicated way on the angle of incidence, the orientation of the electric field vector associated with the wave, and the indexes of refraction of the two media. For the special case of normal incidence ($\theta_1 = \theta_1' = 0$), the reflected intensity can be shown to be

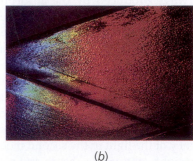

(b)

$$I = \left(\frac{n_1 - n_2}{n_1 + n_2}\right)^2 I_0$$

31-11

FIGURE 31-21 (a) Diffuse reflection from a rough surface. (b) Diffuse reflection of colored lights from a sidewalk.

where I_0 is the incident intensity and n_1 and n_2 are the indexes of refraction of the two media. For a typical case of reflection from an air–glass interface for which $n_1 = 1$ and $n_2 = 1.5$, Equation 31-11 gives $I = I_0/25$. Only about 4 percent of the energy is reflected; the remainder of the energy is transmitted.

REFRACTION FROM AIR TO WATER   **EXAMPLE 31-4**

Light traveling in air enters water with an angle of incidence of 45°. If the index of refraction of water is 1.33, what is the angle of refraction?

PICTURE THE PROBLEM The angle of refraction is found using Snell's law. Let subscripts 1 and 2 refer to the air and water, respectively. Then $n_1 = 1$, $\theta_1 = 45°$, $n_2 = 1.33$, and θ_2 is the angle of refraction (Figure 31-22).

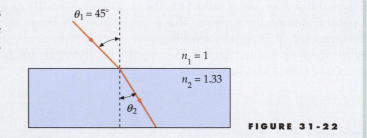

$\theta_1 = 45°$

$n_1 = 1$

$n_2 = 1.33$

θ_2

FIGURE 31-22

1. Use Snell's law to solve for $\sin \theta_2$, the sine of the angle of refraction:

$$n_1 \sin \theta_1 = n_2 \sin \theta_2$$

so

$$\sin \theta_2 = \frac{n_1}{n_2} \sin \theta_1$$

2. Find the angle whose sine is 0.532:

$$\theta_2 = \sin^{-1}\left(\frac{n_1}{n_2} \sin \theta_1\right) = \sin^{-1}\left(\frac{1.00}{1.33} \sin 45°\right)$$

$$= \sin^{-1}(0.532) = \boxed{32.1°}$$

MASTER the CONCEPT

WEB

REMARKS Note that the light is bent toward the normal as the light travels into the medium with the larger index of refraction.

(a)

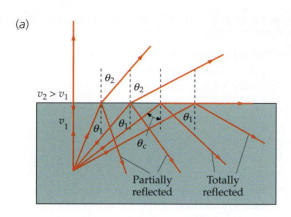

(b)

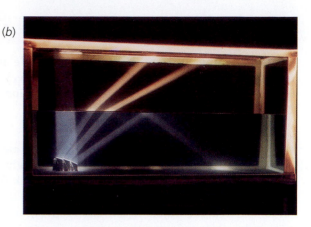

Total Internal Reflection

Figure 31-23 shows a point source in glass with rays striking the glass–air interface at various angles. All the rays not perpendicular to the interface are bent away from the normal. As the angle of incidence is increased, the angle of refraction increases until a critical angle of incidence θ_c is reached for which the angle of refraction is 90°. For incident angles greater than this critical angle, there is no refracted ray. All the energy is reflected. This phenomenon is called **total internal reflection**. The critical angle can be found in terms of the indexes of refraction of the two media by solving Equation 31-9b ($n_1 \sin \theta_1 = n_2 \sin \theta_2$) for $\sin \theta_1$ and setting θ_2 equal to 90°. That is,

FIGURE 31-23 (a) Total internal reflection. As the angle of incidence is increased, the angle of refraction is increased until, at a critical angle of incidence θ_c, the angle of refraction is 90°. For angles of incidence greater than the critical angle, there is no refracted ray. (b) A photograph of refraction and total internal reflection from a water–air interface.

$$\sin \theta_c = \frac{n_2}{n_1} \sin 90° = \frac{n_2}{n_1} \qquad \text{31-12}$$

CRITICAL ANGLE FOR TOTAL INTERNAL REFLECTION

Note that total internal reflection occurs only when the incident light is in the medium with the higher index of refraction. Mathematically, if n_2 is greater than n_1, Snell's law of refraction cannot be satisfied because there is no real-valued angle whose sine is greater than 1.

TOTAL INTERNAL REFLECTION   **EXAMPLE 31-5   Try It Yourself**

A particular glass has an index of refraction of $n = 1.50$. What is the critical angle for total internal reflection for light leaving this glass and entering air, for which $n = 1.00$ (Figure 31-24)?

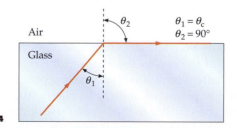

$\theta_1 = \theta_c$
$\theta_2 = 90°$

FIGURE 31-24

Cover the column to the right and try these on your own before looking at the answers.

Steps   **Answers**

1. Make a diagram showing the incident and refracted rays. For the critical angle the angle of refraction is 90°.

2. Apply the law of refraction Equation 31-9b. The critical angle is the angle of incidence. $\theta_c = \boxed{41.8°}$

You are enjoying a nice break at the pool. While under the water, you look up and notice that you see objects above water level in a circle of light of radius approximately 2.0 m, and the rest of your vision is the color of the sides of the pool. How deep are you in the pool?

PICTURE THE PROBLEM We can determine the depth of the pool from the radius of the light and the angle at which the light is entering our eye from the edge of the circle. At the edge of the circle the light is entering the water at 90°, so the angle of refraction at the air–water surface is the critical angle for total internal refraction at the water–air surface. From Figure 31-25, we see that the depth y is related to this angle and the radius of the circle R by $\tan \theta_c = R/y$. The critical angle is found from Equation 31-12 with $n_2 = 1$ and $n_1 = 1.33$.

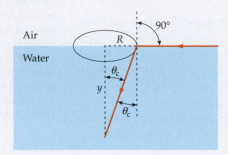

FIGURE 31-25

1. The depth y is related to the radius of the circle R and the critical angle θ_c:

$$\tan \theta_c = R/y$$

2. Solve for the depth y:

$$y = \frac{R}{\tan \theta_c}$$

3. Find the critical angle for total internal refraction at a water–air surface:

$$\sin \theta_c = \frac{n_2}{n_1} = \frac{1}{1.33} = 0.752$$

$$\theta_c = 48.8°$$

4. Solve for the depth y:

$$y = \frac{R}{\tan \theta_c} = \frac{2.0 \text{ m}}{\tan 48.8°} = \boxed{1.75 \text{ m}}$$

Figure 31-26*a* shows light incident normally on one of the short sides of a 45–45–90° glass prism. If the index of refraction of the prism is 1.5, the critical angle for total internal reflection is 41.8°, as you found in Example 31-5. Since the angle of incidence of the ray on the glass–air interface is 45°, the light will be totally reflected and will exit perpendicular to the other face of the prism, as shown. In Figure 31-26*b*, the light is incident perpendicular to the hypotenuse of the prism and is totally reflected twice so that it emerges at 180° to its original direction. Prisms are used to change the directions of light rays. In binoculars, two prisms are used on each side. These prisms reflect the light, thus shortening the required length, and reinvert the image (first inverted by a lens).[†] Diamonds have a very high index of refraction ($n \approx 2.4$), so nearly all the light that enters a diamond is eventually reflected back out, giving the diamond its sparkle.

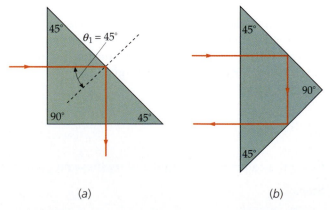

(a)   (b)

FIGURE 31-26 (*a*) Light entering through one of the short sides of a 45–45–90° glass prism is totally reflected. (*b*) Light entering through the long side of the prism is totally reflected twice.

---

† The image produced by the objective lens of a telescope is discussed in Section 32-4.

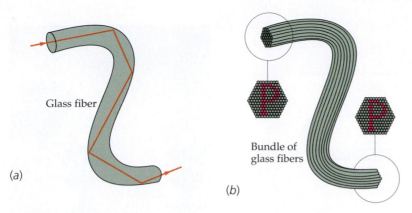

(a)

(b)

(c)

Fiber Optics An interesting application of total internal reflection is the transmission of a beam of light down a long, narrow, transparent glass fiber (Figure 31-27a). If the beam begins approximately parallel to the axis of the fiber, it will strike the walls of the fiber at angles greater than the critical angle (if the bends in the fiber are not too sharp) and no light energy will be lost through the walls of the fiber. A bundle of such fibers can be used for imaging, as illustrated in Figure 31-27b. Fiber optics has many applications in medicine and in communications. In medicine, light is transmitted along tiny fibers to visually probe various internal organs without surgery. In communications, the rate at which information can be transmitted is related to the signal frequency. A transmission system using light of frequencies of the order of 10^{14} Hz can transmit information at a much greater rate than one using radio waves, which have frequencies of the order of 10^6 Hz. In telecommunication systems, a single glass fiber the thickness of a human hair can transmit audio or video information equivalent to 32,000 voices speaking simultaneously.

FIGURE 31-27 (a) A light pipe. Light inside the pipe is always incident at an angle greater than the critical angle, so no light escapes the pipe by refraction. (b) Light from the object is transported by a bundle of glass fibers to form an image of the object at the other end of the pipe. (c) Light emerging from a bundle of glass fibers.

Mirages

When the index of refraction of a medium changes gradually, the refraction is continuous, leading to a gradual bending of the light. An interesting example of this is the formation of a mirage. On a hot and sunny day, the surface of exposed rocks, pavement, and sand often gets very hot. In this case there is often a layer of air near the ground that is warmer, and therefore less dense, than the air just above it. The speed

(a)

(b)

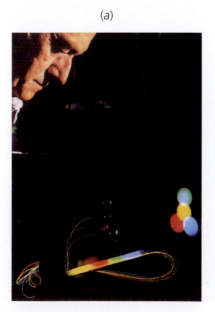

(a) In this demonstration at the Naval Research Laboratory, a combination of laser sources generates different colors that excite adjacent fiber sensor elements, leading to a separation of the information as indicated by the separation of the colors. (b) The tip of a light guide preform is softened by heat and drawn into a long, tiny fiber. The colors in the preform indicate a layered structure of differing compositions, which is retained in the fiber.

(a)

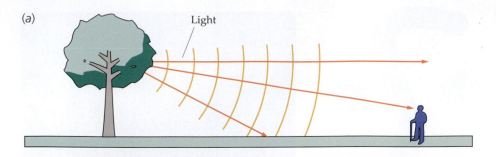

(c)

(b)

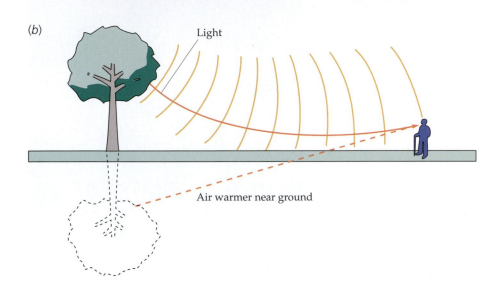

Air warmer near ground

FIGURE 31-28 A mirage. (*a*) When the air is at a uniform temperature, the wavefronts of the light from the tree are spherical. (*b*) When the air near the ground is warmer, the wavefronts are not spherical and the light from the tree is continuously refracted into a curved path. (*c*) Apparent reflections of motorcycles on a hot road.

of any light waves is slightly greater in this less dense layer, so a light beam passing from the cooler layer into the warmer layer is bent. Figure 31-28a shows the light from a tree when all the surrounding air is at the same temperature. The wavefronts are spherical, and the rays are straight lines. In Figure 31-28b, the air near the ground is warmer, resulting in the wavefronts traveling faster there. The portions of the wavefronts near the hot ground get ahead of the higher portions, creating a nonspherical wavefront and causing a curving of the rays. Thus, the two rays shown initially heading for the ground are bent upward. As a result, the viewer sees an image of the tree looking as if it were reflected off a water surface on the ground. When driving on a hot sunny day, you may have noticed apparent wet spots on the highway ahead that disappear as you approach them. These mirages are due to the refraction of light from the sky by a layer of air that has been heated due to its proximity to the hot pavement.

Dispersion

The index of refraction of a material has a slight dependence on wavelength. For many materials, n decreases slightly as the wavelength increases, as shown in Figure 31-29. The dependence of the index of refraction on wavelength (and therefore on frequency) is called **dispersion**. When a beam of white light is incident at some angle on the surface of a glass prism, the angle of refraction (which is measured relative to the normal) for the shorter wavelengths is slightly smaller than the angle of refraction for the longer wavelengths. The light of shorter wavelength (toward the violet end of the spectrum) is therefore bent more toward the normal than that of longer wavelength. The beam of white light is thus spread out or dispersed into its component colors or wavelengths (Figure 31-30).

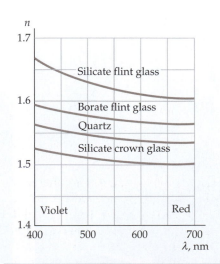

FIGURE 31-29 Index of refraction versus wavelength for various materials.

FIGURE 31-30 A beam of white light incident on a glass prism is dispersed into its component colors. The index of refraction decreases as the wavelength increases so that the longer wavelengths (red) are bent less than the shorter wavelengths (blue).

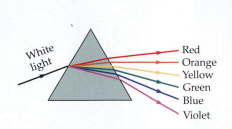

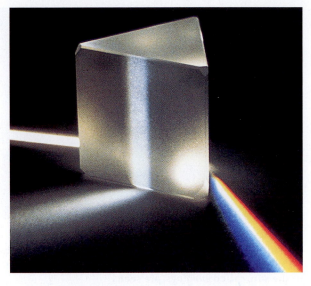

Rainbows The rainbow is a familiar example of dispersion, in this case the dispersion of sunlight. Figure 31-31 is a diagram originally drawn by Descartes, showing parallel rays of light from the sun entering a spherical water drop. First, the rays are refracted as they enter the drop. The rays are then reflected from the water–air interface on the other side of the drop and finally are refracted again as they leave the drop.

From Figure 31-31, we can see that the angle made by the emerging rays and the diameter (along ray 1) reaches a maximum around ray 7 and then decreases. The concentration of rays emerging at approximately the maximum angle gives rise to the rainbow. By construction, using the law of refraction, Descartes showed that the maximum angle is about 42°. To observe a rainbow, we must therefore look at the water drops at an angle of 42° relative to the line back to the sun, as shown in Figure 31-32. The angular radius of the rainbow is therefore 42°.

The separation of the colors in the rainbow results from the fact that the index of refraction of water depends slightly on the wavelength of light. The angular radius of the bow will therefore depend slightly on the wavelength of the light. The observed rainbow is made up of light rays from many different droplets of water (Figure 31-33). The color seen at a particular angular radius corresponds to the wavelength of light that allows the light to reach the eye from the droplets at that angular radius. Because n_{water} is smaller for red light than for blue light, the red part of the rainbow is at a slightly greater angular radius than the blue part of the rainbow, so red is at the outer side of the rainbow.

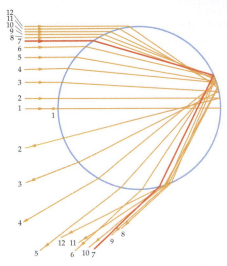

FIGURE 31-31 Descartes's construction of parallel rays of light entering a spherical water drop. Ray 1 enters the drop along a diameter and is reflected back along its incident path. Ray 2 enters slightly above the diameter and emerges below the diameter at a small angle with the diameter. The rays entering farther and farther away from the diameter emerge at greater and greater angles up to ray 7, shown as the heavy line. Rays entering above ray 7 emerge at smaller and smaller angles with the diameter.

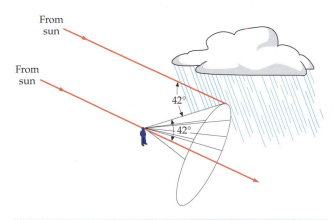

FIGURE 31-32 A rainbow is viewed at an angle of 42° from the line to the sun, as predicted by Descartes's construction, as shown in Figure 31-31.

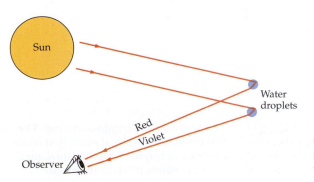

FIGURE 31-33 The rainbow results from light rays from many different water droplets.

(a)

(b)

(a) **This 22° halo around the sun results from refraction by hexagonal ice crystals that are randomly oriented in the upper atmosphere.**
(b) **When the ice crystals are not randomly oriented but are falling with their flat bases horizontal, only parts of the halo on each side of the sun, called *sun dogs*, are seen.**

When a light ray strikes a surface separating water and air, part of the light is reflected and part of the light is refracted. A secondary rainbow results from the light rays that are reflected twice within a droplet (Figure 31-34). The secondary bow has an angular radius of 51°, and its color sequence is the reverse of that of the primary bow; that is, the violet is on the outside in the secondary bow. Because of the small fraction of light reflected from a water–air interface, the secondary bow is considerably fainter than the primary bow.

*Calculating the Angular Radius of the Rainbow   We can calculate the angular radius of the rainbow from the laws of reflection and refraction. Figure 31-35 shows a ray of light incident on a spherical water droplet at point A. The angle of refraction θ_2 is related to the angle of incidence θ_1 by Snell's law of refraction:

$$n_{\text{air}} \sin \theta_1 = n_{\text{water}} \sin \theta_2 \qquad\qquad 31\text{-}13$$

Point P in Figure 31-35 is the intersection of the line of the incident ray and the line of the emerging ray. The angle ϕ_{d} is called the angle of deviation of the ray, and ϕ_{d} and 2β form a straight angle. Thus,

$$\phi_{\text{d}} + 2\beta = \pi \qquad\qquad 31\text{-}14$$

We wish to relate the angle of deviation ϕ_{d} to the angle of incidence θ_1. From the triangle AOB, we have

$$2\theta_2 + \alpha = \pi \qquad\qquad 31\text{-}15$$

Similarly, from the triangle AOP, we have

$$\theta_1 + \beta + \alpha = \pi \qquad\qquad 31\text{-}16$$

Eliminating α from Equations 31-15 and 31-16 and solving for β gives

$$\beta = \pi - \theta_1 - \alpha = \pi - \theta_1 - (\pi - 2\theta_2) = 2\theta_2 - \theta_1$$

Substituting this value for β into Equation 31-14 gives the angle of deviation:

$$\phi_{\text{d}} = \pi - 2\beta = \pi - 4\theta_2 + 2\theta_1 \qquad\qquad 31\text{-}17$$

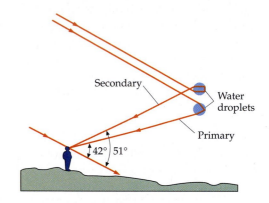

FIGURE 31-34 The secondary rainbow results from light rays that are reflected twice within a water droplet.

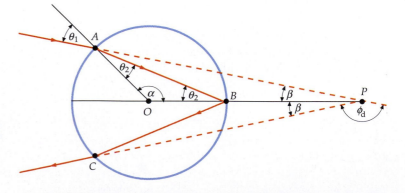

FIGURE 31-35 Light ray incident on a spherical water drop. The refracted ray strikes the back of the water droplet at point B. It makes an angle θ_2 with the radial line OB and is reflected at an equal angle. The ray is refracted again at point C, where it leaves the droplet.

Equation 31-17 can be combined with Equation 31-13 to eliminate θ_2 and give the angle of deviation ϕ_d in terms of the angle of incidence θ_1:

$$\phi_d = \pi + 2\theta_1 - 4\sin^{-1}\left(\frac{n_{air}}{n_{water}}\sin\theta_1\right) \qquad \text{31-18}$$

Figure 31-36 shows a plot of ϕ_d versus θ_1. The angle of deviation ϕ_d has its minimum value when $\theta_1 \approx 60°$. At an angle of incidence of 60°, the angle of deviation is $\phi_{d,min} = 138°$. This angle is called the **angle of minimum deviation.** At incident angles that are slightly greater or slightly smaller than 60°, the angle of deviation is approximately the same. Therefore, the intensity of the light reflected by the water droplet will be a maximum at the angle of minimum deviation. We can see from Figure 31-35 that the maximum value of β corresponds to the minimum value of ϕ_d. Thus, angular radius of the intensity maximum, given by $2\beta_{max}$, is

$$2\beta_{max} = \pi - \phi_{d,min} = 180° - 138° = 42° \qquad \text{31-19}$$

The index of refraction of water varies slightly with wavelength. Thus, for each wavelength (color), the intensity maxima occurs at an angular radius slightly different than that of neighboring wavelengths.

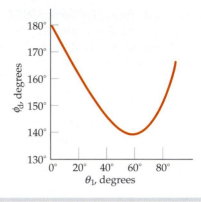

FIGURE 31-36 Plot of the angle of deviation ϕ_d as a function of incident angle θ_1. The angle of deviation has its minimum value of 138° when the angle of incidence is 60°. Since $d\phi_d/d\theta_1 = 0$ at minimum deviation, the deviation of rays with incident angles slightly less or slightly greater than 60° will be approximately the same.

31-7 Polarization

In a transverse mechanical wave, the vibration is perpendicular to the direction of propagation of the wave. If the vibration remains parallel to a plane, the wave is said to be **plane polarized** or **linearly polarized.** We can visualize polarization most easily by considering mechanical waves on a taut horizontal string. Let the x axis be along the string, let the z axis be vertical, and let the y axis be horizontal, perpendicular to both the x and z axes. If one end of the string is shaken up and down, the resulting waves on the string are linearly polarized with each element of the string vibrating up and down. Similarly, if one end is shaken back and forth along the y axis, the displacements of the string are linearly polarized with each element vibrating parallel with the y axis. If one end of the string is moved with constant speed in an ellipse in the $x = 0$ plane, the resulting wave is said to be **elliptically polarized.** In this case, each element of the string moves in an ellipse in a plane of constant x. Unpolarized waves can be produced by moving the end of the string in the $x = 0$ plane in a random way. Then the vibrations will have both y and z components that vary randomly. A linearly polarized electromagnetic wave is one in which the electric field remains parallel to a line. A wave produced by an electric dipole antenna is polarized with the electric field vector at any field point remaining parallel with the plane containing the field point and the antenna axis. Waves produced by numerous sources are usually unpolarized. A typical light source, for example, contains millions of atoms acting independently. The electric field for such a wave can be resolved into x and y components that vary randomly, because there is no correlation between the individual atoms producing the light.

The polarization of electromagnetic waves can be demonstrated with microwaves, which have wavelengths on the order of centimeters. In a typical microwave generator, polarized waves are radiated by an electric dipole antenna. In Figure 31-37, the electric dipole antenna is vertical, so the electric field vector $\vec{E}$ of the horizontally radiated waves is also vertical. An absorber can be made of a screen of parallel straight wires. When the wires are vertical, as in Figure 31-37a, the electric field parallel to the wires sets up currents in the wires and energy is absorbed. When the wires are horizontal and therefore perpendicular to $\vec{E}$, as in Figure 31-37b, no currents are set up and the waves are transmitted.

(a)

(b)

FIGURE 31-37 Demonstration showing the polarization of microwaves. The electric field of the microwaves is vertical, parallel to the vertical dipole antenna. (a) When the metal wires of the absorber are vertical, electric currents are set up in the wires and energy is absorbed, as indicated by the low reading on the microwave detector. (b) When the wires are horizontal, no currents are set up and the microwaves are transmitted, as indicated by the high reading on the detector.

There are four phenomena that produce polarized electromagnetic waves from unpolarized waves: (1) absorption, (2) reflection, (3) scattering, and (4) birefringence (also called double refraction), each of which is examined in the upcoming sections.

Polarization by Absorption

Several naturally occurring crystals, when cut into appropriate shapes, absorb and transmit light differently depending on the polarization of the light. These crystals can be used to produce linearly polarized light. In 1938, E. H. Land invented a simple commercial polarizing sheet called Polaroid. This material contains long-chain hydrocarbon molecules that are aligned when the sheet is stretched in one direction during the manufacturing process. These chains become conducting at optical frequencies when the sheet is dipped in a solution containing iodine. When light is incident with its electric field vector parallel to the chains, electric currents are set up along the chains, and the light energy is absorbed, just as the microwaves are absorbed by the wires in Figure 31-37. If the electric field is perpendicular to the chains, the light is transmitted. The direction perpendicular to the chains is called the **transmission axis.** We will make the simplifying assumption that all the light is transmitted when the electric field is parallel to the transmission axis and all the light is absorbed when it is perpendicular to the transmission axis. In reality, Polaroid absorbs some of the light, even when the electric field is parallel to the transmission axis.

Consider an unpolarized light beam incident on a polarizing sheet with its transmission axis along the x direction, as shown in Figure 31-38. The beam is incident on a second polarizing sheet, the analyzer, whose transmission axis makes an angle θ with the x axis. If E is the electric field amplitude of the incident beam, the component parallel with the transmission axis is $E_\parallel = E \cos \theta$, and the component perpendicular to the transmission axis is $E_\perp = E \sin \theta$. The sheet absorbs $E_\perp$ and transmits $E_\parallel$, so the transmitted beam has an electric field amplitude of $E_\parallel = E \cos \theta$ and is linearly polarized in the direction of the transmission axis. Because the intensity of light is proportional to the square of the magnitude of the electric field amplitude, the intensity I of light transmitted by the sheet is given by

$$I = I_0 \cos^2 \theta \qquad \text{31-20}$$

LAW OF MALUS

where I_0 is the intensity of the incident beam. If we have an incident beam of unpolarized light of intensity I_0 incident on a polarizing sheet, the direction of the incident electric field varies from location to location on the sheet, and at each location it fluctuates in time. At each location the angle between the electric field and the transmission axis is, on average, 45°, so applying Equation 31-20 gives $I = I_0 \cos^2 45° = \frac{1}{2}I_0$, where I is the intensity of the transmitted beam.

When two polarizing elements are placed in succession in a beam of unpolarized light, the first polarizing element is called the **polarizer** and the second polarizing element is called the **analyzer.** If the polarizer and the analyzer are crossed, that is, if their transmission axes are perpendicular to each other, no light gets through. Equation 31-20 is known as the **law of Malus** after its discoverer, E. L. Malus (1775–1812). It applies to any two polarizing elements whose transmission axes make an angle θ with each other.

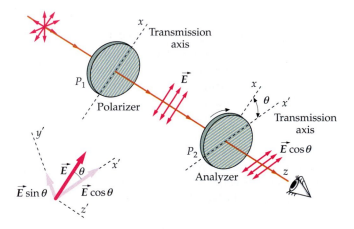

FIGURE 31-38 A vertically polarized beam is incident on a polarizing sheet with its transmission axes making an angle θ with the vertical. Only the component $E \cos \theta$ is transmitted through the second sheet, and the transmitted beam is linearly polarized in the direction of the transmission axis. If the intensity between the sheets is I_0, the intensity transmitted by both sheets is $I_0 \cos^2 \theta$.

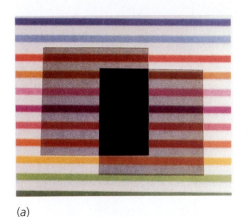

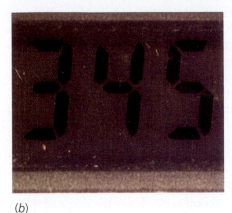

(a)   (b)

(a) Cross polarizers block out all of the light. (b) In a liquid crystal display, the crystal is between crossed polarizers. Light incident on the crystal is transmitted because the crystal rotates the direction of polarization of the light 90°. The light is reflected back out through the crystal by a mirror behind the crystal, and a uniform background is seen. When a voltage is applied across a small segment of the crystal, the polarization is not rotated, so no light is transmitted and the segment appears black.

INTENSITY TRANSMITTED   **E X A M P L E   3 1 - 7**

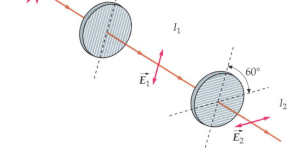

FIGURE 31-39

Unpolarized light of intensity 3.0 W/m² is incident on two polarizing sheets whose transmission axes make an angle of 60° (Figure 31-39). What is the intensity of light transmitted by the second sheet?

PICTURE THE PROBLEM   The incident light is unpolarized, so the intensity transmitted by the first polarizing sheet is half the incident intensity. The second sheet further reduces the intensity by a factor of $\cos^2 \theta$, with $\theta = 60°$.

1. The intensity I_1 transmitted by the first sheet is half the intensity I_0 of unpolarized light incident on the first sheet:

$$I_1 = \tfrac{1}{2} I_0$$

2. The intensity I_2 transmitted by the second sheet is related to the intensity I_1 of the light incident on the second sheet by Equation 31-20:

$$I_2 = I_1 \cos^2 \theta$$

3. Combine these results and substitute the given data:

$$I_2 = \tfrac{1}{2} I_0 \cos^2 60° = \tfrac{1}{2} (3.0 \text{ W/m}^2)(0.500)^2$$

$$= \boxed{0.375 \text{ W/m}^2}$$

REMARKS   Half the intensity passes through the first sheet no matter what the orientation of that sheet's transmission axis. Note that the second sheet rotates the plane of polarization by 60°.

Polarization by Reflection

When unpolarized light is reflected from a plane surface boundary between two transparent media, such as air and glass or air and water, the reflected light is partially polarized. The degree of polarization depends on the angle of incidence and on the ratio of the wave speeds in the two media. For a certain angle of incidence called the polarizing angle θ_p, the reflected light is completely polarized. At the polarizing angle, the reflected and refracted rays are perpendicular to each other. David Brewster (1781–1868), a Scottish scientist and an inventor of numerous instruments (including the kaleidoscope), discovered this experimentally in 1812. The polarizing angle is also referred to as Brewster's angle.

Figure 31-40 shows light incident at the polarizing angle θ_p for which the reflected light is completely polarized. The electric field of the incident light can be resolved into components parallel and perpendicular to the plane of incidence. The reflected light is linearly polarized with its electric field perpendicular to the plane of incidence. We can relate the polarizing angle to the indexes of refraction of the media using Snell's law (the law of refraction). If n_1 is the index of refraction of the first medium and n_2 is the index of refraction of the second medium, the law of refraction gives

$$n_1 \sin \theta_p = n_2 \sin \theta_2$$

where θ_2 is the angle of refraction. From Figure 31-40, we can see that the sum of the angle of reflection and the angle of refraction is 90°. Since the angle of reflection equals the angle of incidence, we have

$$\theta_2 = 90° - \theta_p$$

Then

$$n_1 \sin \theta_p = n_2 \sin(90° - \theta_p) = n_2 \cos \theta_p$$

or

$$\tan \theta_p = \frac{n_2}{n_1} \qquad \text{31-21}$$

POLARIZING ANGLE

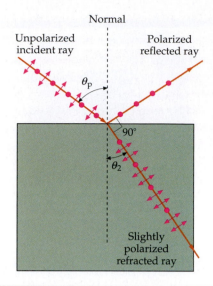

FIGURE 31-40 Polarization by reflection. The incident wave is unpolarized and has components of the electric field parallel to the plane of incidence (arrows) and components perpendicular to this plane (dots). For incidence at the polarizing angle, the reflected wave is completely polarized, with its electric field perpendicular to the plane of incidence.

Although the reflected light is completely polarized for this angle of incidence, the transmitted light is only partially polarized because only a small fraction of the incident light is reflected. If the incident light itself is polarized with the electric field in the plane of incidence, there is no reflected light when the angle of incidence is θ_p. We can understand this qualitatively from Figure 31-41. If we consider the molecules next to the surface of the second medium to be oscillating parallel to the electric field of the refracted ray, there can be no reflected ray because for an electric dipole antenna no energy is radiated along the line of oscillation. (Each of the oscillating molecules are a small electric dipole antenna.)

Because of the polarization of reflected light, sunglasses that contain a polarizing sheet can be very effective in cutting out glare. If light is reflected from a horizontal surface, such as a lake surface or snow on the ground, the electric field of the reflected light will be predominantly horizontal and the plane of incidence on the glasses will be predominantly vertical. Polarized sunglasses with a vertical transmission axis will then reduce glare by absorbing much of the reflected light. If you have polarized sunglasses, you can observe this effect by looking through the glasses at reflected light and then rotating the glasses 90°; much more of the light will be transmitted.

Polarization by Scattering

The phenomenon of absorption and reradiation is called **scattering.** Scattering can be demonstrated by passing a light beam through a container of water to which a small amount of powdered milk has been added. The milk particles absorb light and reradiate it, making the light beam visible. Similarly, laser beams can be made visible by introducing chalk or smoke particles into the air to scatter the light. A familiar example of light scattering is that from air molecules, which tend to scatter short wavelengths more than long wavelengths, thereby giving the sky its blue color.

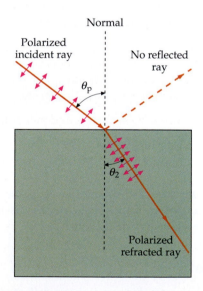

FIGURE 31-41 Polarized light incident at the polarizing angle. When the incident light is polarized with $\vec{E}$ in the plane of incidence, there is no reflected ray.

We can understand polarization by scattering if we think of a scattering molecule as an electric dipole antenna that radiates waves with a maximum intensity in directions perpendicular to the antenna axis and zero intensity in the direction along the antenna axis. The electric field vector of the scattered light perpendicular to the direction of propagation is in the plane of the antenna axis and the field point. Figure 31-42 shows a beam of unpolarized light that initially travels along the z axis, striking a molecule at the origin. The electric field in the light beam has components in both the x and y directions perpendicular to the direction of motion of the light beam. These fields set up oscillations of the charges within the molecule in the $z = 0$ plane, and there is no oscillation along the z direction. These oscillations can be thought of as a superposition of an oscillation along the x axis and another along the y axis, with each of these oscillations producing dipole radiation. Thus, the oscillation along the x axis produces no radiation along the x axis, which means the light radiated along the x axis is produced only by the oscillation along the y axis. It follows that the light radiated along the x axis is polarized with its electric field parallel with the y axis. There is nothing special about the choice of axes for this discussion, so the result can be generalized. That is, the light scattered in a direction perpendicular to the incident light beam is polarized with its electric field perpendicular to both the incident beam and the direction of propagation of the scattered light. This can be seen easily by examining the scattered light with a piece of polarizing sheet.

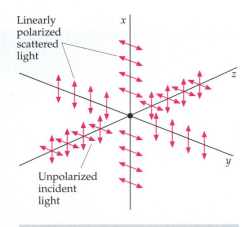

FIGURE 31-42 Polarization by scattering. Unpolarized light propagating in the z direction is incident on a scattering center at the origin. The light scattered in the $z = 0$ plane along the x direction is polarized parallel with the y axis (and the light scattered in the y direction is polarized parallel with the x axis).

Polarization by Birefringence

Birefringence is a complicated phenomenon that occurs in calcite and other noncubic crystals and in some stressed plastics, such as cellophane. Most materials are **isotropic**, that is, the speed of light passing through the material is the same in all directions. Because of their atomic structure, birefringent materials are **anisotropic**. The speed of light depends on the plane of polarization and on the direction of propagation of the light. When a light ray is incident on such materials, it may be separated into two rays called the *ordinary ray* and the *extraordinary ray*. These rays are polarized in mutually perpendicular directions, and they travel with different speeds. Depending on the relative orientation of the material and the incident light beam, the rays may also travel in different directions.

There is one particular direction in a birefringent material in which both rays propagate with the same speed. This direction is called the **optic axis** of the material. (The optic axis is actually a *direction* rather than a line in the material.) Nothing unusual happens when light travels in the direction of the optic axis. However, when light is incident at an angle to the optic axis, as shown in Figure 31-43, the rays travel in different directions and emerge separated in space. If the material is rotated, the extraordinary ray (the e ray in the figure) revolves in space around the ordinary ray (o ray).

If light is incident on a birefringent plate perpendicular to its crystal face and perpendicular to the optic axis, the two rays travel in the same direction but at different speeds. The number of wavelengths in the two rays in the plate is different because the wavelengths ($\lambda = v/f$) of the rays differ. The rays emerge with a phase difference that depends on the thickness of the plate and on the wavelength of the incident light. In a **quarter-wave plate**, the thickness is such that there is a 90° phase difference between the waves of a particular wavelength when they emerge. In a **half-wave plate**, the rays emerge with a phase difference of 180°.

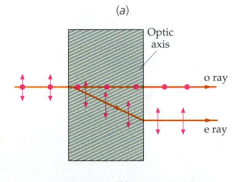

(a)

(b)

FIGURE 31-43 (*a*) A narrow beam of light incident on a birefringent crystal such as calcite is split into two beams, called the ordinary ray (o ray) and the extraordinary ray (e ray), that have mutually perpendicular polarizations. If the crystal is rotated, the extraordinary ray rotates in space. (*b*) A double image of the cross hatching is produced by this birefringent crystal of calcium carbonate.

(a)

(b)

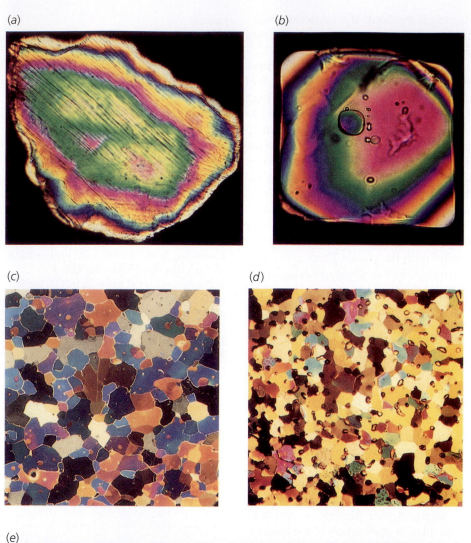

(c)

(d)

When the transmission axes of two polarizing sheets are perpendicular, the polarizers are said to be crossed and no light is transmitted. However, many materials are birefringent or become so under stress. Such materials rotate the direction of polarization of the light so that light of a particular wavelength is transmitted through both polarizers. When a birefringent material is viewed between crossed polarizers, information about its internal structure is revealed. (*a*) A shocked quartz grain from the site of a meteorite crater. The layered structure, evidenced by the parallel lines, arises from the shock of the impact of the meteor. (*b*) A grain of quartz typically found in silicic volcanic rocks. No shock lines are seen. (*c*) Thin sections of ice core from the antarctic ice sheet reveal bubbles of trapped CO_2, which appear amber-colored. This sample was taken from a depth of 194 m, corresponding to air trapped 1600 years ago, whereas the sample in (*d*) is from a depth of 56 m, corresponding to air trapped 450 years ago. Ice core measurements have replaced the less reliable technique of analyzing carbon in tree rings to compare current atmospheric CO_2 levels with those of the recent past. (*e*) Robert Mark of the Princeton School of Architecture examines the stress patterns in a plastic model of the nave structure of Chartres Cathedral.

(e)

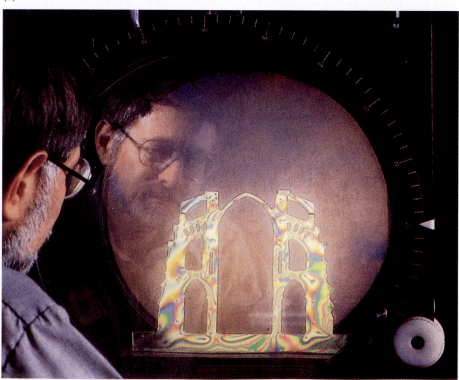

Suppose that the incident light is linearly polarized so that the electric field vector is at 45° to the optic axis, as illustrated in Figure 31-44. The ordinary and extraordinary rays start out in phase and have equal amplitudes. With a quarter-wave plate, the waves emerge with a phase difference of 90°, so the resultant electric field has components $E_x = E_0 \sin \omega t$ and $E_y = E_0 \sin(\omega t + 90°) = E_0 \cos \omega t$. The electric field vector thus rotates in a circle and the wave is circularly polarized.

With a half-wave plate, the waves emerge with a phase difference of 180°, so the resultant electric field is linearly polarized with components $E_x = E_0 \sin \omega t$ and $E_y = E_0 \sin(\omega t + 180°) = -E_0 \sin \omega t$. The net effect is that the direction of polarization of the wave is rotated by 90° relative to that of the incident light, as shown in Figure 31-45.

Interesting and beautiful patterns can be observed by placing birefringent materials, such as cellophane or stressed plastic, between two polarizing sheets with their transmission axes perpendicular to each other. Ordinarily, no light is transmitted through crossed polarizing sheets. However, if we place a birefringent material between the crossed polarizing sheets, the material acts as a half-wave plate for light of a certain color depending on the material's thickness. The direction of polarization is rotated and some light gets through both sheets. Various glasses and plastics become birefringent when under stress. The stress patterns can be observed when the material is placed between crossed polarizing sheets.

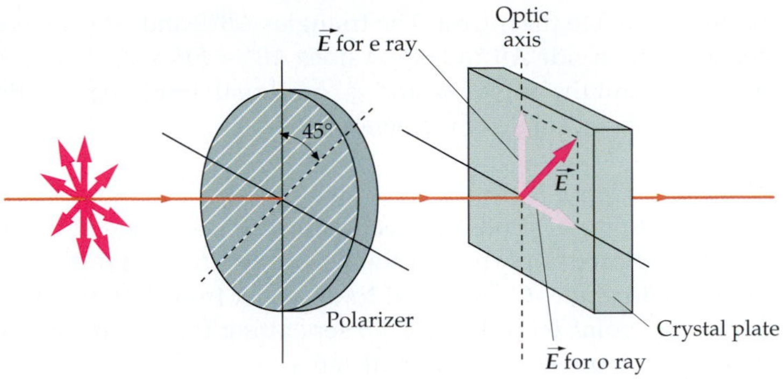

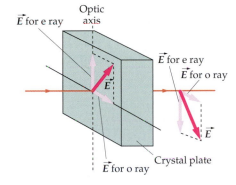

FIGURE 31-44 Polarized light emerging from the polarizer is incident on a birefringent crystal so that the electric field vector makes a 45° angle with the optic axis, which is perpendicular to the light beam. The ordinary and extraordinary rays travel in the same direction but at different speeds. The polarization of the emerging light depends on the thickness of the crystal and the wavelength of the light.

31-8 Derivation of the Laws of Reflection and Refraction

The laws of reflection and refraction can be derived from either Huygens's principle or Fermat's principle.

Huygens's Principle

Reflection Figure 31-46 shows a plane wavefront AA' striking a mirror at point A. As can be seen from the figure, the angle ϕ_1 between the wavefront and the mirror is the same as the angle of incidence θ_1, which is the angle between the normal to the mirror and the rays (which are perpendicular to the wavefronts). According to Huygens's principle, each point on a given wavefront can be considered a point source of secondary wavelets. The position of the wavefront after a time t is found by constructing wavelets of radius ct with their centers on the wavefront AA'. Wavelets that have not yet reached the mirror form the portion of the new wavefront BB'. Wavelets that have already reached the mirror are reflected and form the portion of the new wavefront $B''B$. By a similar construction, the wavefront $C''C$ is obtained from the Huygens's wavelets originating on the wavefront $B''B$. Figure 31-47 is an enlargement of a portion of Figure 31-46 showing AP, which is part of the initial position of the wavefront. During the time t, the wavelet from point P reaches the mirror at point B, and the wavelet from point A reaches point B''. The reflected wavefront $B''B$ makes an angle ϕ_1' with the mirror that is equal to the angle of reflection θ_1' between the reflected ray

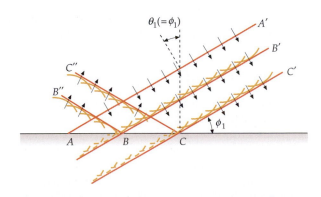

FIGURE 31-45 If the birefringent crystal in Figure 31-44 is a half-wave plate, and if the electric field vector of the incident light makes an angle of 45° with the optic axis, then the direction of polarization of the emerging light is rotated by 90°.

FIGURE 31-46 Plane wave reflected at a plane mirror. The angle θ_1 between the incident ray and the normal to the mirror is the angle of incidence. It is equal to the angle ϕ_1 between the incident wavefront and the mirror.

and the normal to the mirror. The triangles $AB''B$ and APB are both right triangles with a common side AB and equal sides $AB'' = BP = ct$. Hence, these triangles are congruent, and the angles ϕ_1 and ϕ_1' are equal, implying that the angle of reflection θ_1' equals the angle of incidence θ_1.

Refraction Figure 31-48 shows a plane wave incident on an air–glass interface. We apply Huygens's construction to find the wavefront in the transmitted wave. Line AP indicates a portion of the wavefront in medium 1 that strikes the glass surface at an angle ϕ_1. In time t, the wavelet from P travels the distance v_1t and reaches the point B on the line AB separating the two media, while the wavelet from point A travels a shorter distance v_2t into the second medium. The new wavefront BB' is not parallel to the original wavefront AP because the speeds v_1 and v_2 are different. From the triangle APB,

$$\sin \phi_1 = \frac{v_1t}{AB}$$

or

$$AB = \frac{v_1t}{\sin \phi_1} = \frac{v_1t}{\sin \theta_1}$$

using the fact that the angle ϕ_1 equals the angle of incidence θ_1. Similarly, from triangle $AB'B$,

$$\sin \phi_2 = \frac{v_2t}{AB}$$

or

$$AB = \frac{v_2t}{\sin \phi_2} = \frac{v_2t}{\sin \theta_2}$$

where $\theta_2 = \phi_2$ is the angle of refraction. Equating the two values for AB, we obtain

$$\frac{\sin \theta_1}{v_1} = \frac{\sin \theta_2}{v_2} \qquad\qquad 31\text{-}22$$

Substituting $v_1 = c/n_1$ and $v_2 = c/n_2$ in this equation and multiplying by c, we obtain $n_1 \sin \theta_1 = n_2 \sin \theta_2$, which is Snell's law.

Fermat's Principle

Reflection Figure 31-49 shows two paths in which light leaves point A, strikes the plane surface, which we can consider to be a mirror, and travels to point B. The problem for the application of Fermat's principle to reflection can be stated as follows: At what point P in the figure must the light strike the mirror so that it will travel from point A to point B in the least time? Since the light is traveling in the same medium for this problem, the time will be minimum when the distance is minimum. In Figure 31-49 the distance APB is the same as the distance $A'PB$, where point A' lies along the perpendicular from A to the mirror and is equidistant behind the mirror. As we vary point P, the distance $A'PB$ is least when the points A', P, and B lie on a straight line. We can see from the figure that this occurs when the angle of incidence equals the angle of reflection.

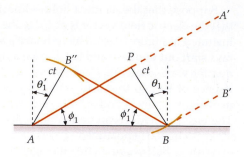

FIGURE 31-47 Geometry of Huygens's construction for the calculation of the law of reflection. The wavefront AP initially strikes the mirror at point A. After a time t, the Huygens wavelet from P strikes the mirror at point B, and the Huygens wavelet from point A reaches point B.

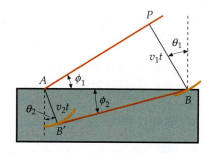

FIGURE 31-48 Application of Huygens's principle to the refraction of plane waves at the surface separating a medium in which the wave speed is v_1 from a medium in which the wave speed v_2 is less than v_1. The angle of refraction in this case is less than the angle of incidence.

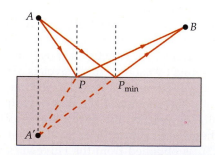

FIGURE 31-49 Geometry for deriving the law of reflection from Fermat's principle. The time it takes for the light to travel from point A to point B is a minimum for light striking the surface at point P.

Refraction   The derivation of Snell's law of refraction from Fermat's principle is slightly more complicated. Figure 31-50 shows the possible paths for light traveling from point A in air to point B in glass. Point P_1 is on the straight line between A and B, but this path is not the one for the shortest travel time because light travels with a smaller speed in the glass. If we move slightly to the right of P_1, the total path length is greater, but the distance traveled in the slower medium is less than for the path through P_1. It is not apparent from the figure which path is the path of least time, but it is not surprising that a path slightly to the right of the straight-line path takes less time because the time gained by traveling a shorter distance in the glass more than compensates for the time lost traveling a longer distance in the air. As we move the point of intersection of the possible path to the right of point P_1, the total time of travel from point A to point B decreases until we reach a minimum at point P_{min}. Beyond this point, the time saved by traveling a shorter distance in the glass does not compensate for the greater time required for the greater distance traveled in the air.

Figure 31-51 shows the geometry for finding the path of least time. If L_1 is the distance traveled in medium 1 with index of refraction n_1, and L_2 is the distance traveled in medium 2 with index of refraction n_2, the time for light to travel the total path AB is

$$t = \frac{L_1}{v_1} + \frac{L_2}{v_2} = \frac{L_1}{c/n_1} + \frac{L_2}{c/n_2} = \frac{n_1 L_1}{c} + \frac{n_2 L_2}{c} \qquad 31\text{-}23$$

We wish to find the point P_{min} for which this time is a minimum. We do this by expressing the time in terms of a single parameter x, as shown in the figure, indicating the position of point P_{min}. In terms of the distance x,

$$L_1^2 = a^2 + x^2 \quad \text{and} \quad L_2^2 = b^2 + (d - x)^2 \qquad 31\text{-}24$$

Figure 31-52 shows the time t as a function of x. At the value of x for which the time is a minimum, the slope of the graph of t versus x is zero:

$$\frac{dt}{dx} = 0$$

Differentiating each term in Equation 31-23 with respect to x and setting the result equal to zero, we obtain

$$\frac{dt}{dx} = \frac{1}{c}\left(n_1 \frac{dL_1}{dx} + n_2 \frac{dL_2}{dx} \right) = 0 \qquad 31\text{-}25$$

We can compute these derivatives from Equations 31-24. We have

$$2L_1 \frac{dL_1}{dx} = 2x \quad \text{or} \quad \frac{dL_1}{dx} = \frac{x}{L_1}$$

where x/L_1 is just $\sin\theta_1$ and θ_1 is the angle of incidence. Thus,

$$\frac{dL_1}{dx} = \sin\theta_1 \qquad 31\text{-}26$$

Similarly,

$$2L_2 \frac{dL_2}{dx} = 2(d - x)(-1)$$

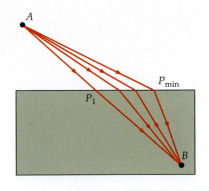

FIGURE 31-50 Geometry for deriving Snell's law from Fermat's principle. The point P_{min} is the point at which light must strike the glass in order that the travel time from point A to point B is a minimum.

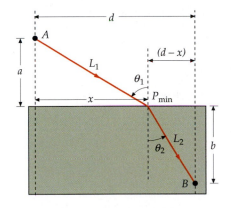

FIGURE 31-51 Geometry for calculating the minimum time in the derivation of Snell's law from Fermat's principle.

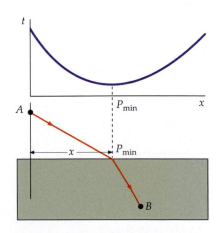

FIGURE 31-52 Graph of the time it takes for light to travel from point A to point B versus x, measured along the refracting surface. The time is a minimum at the point at which the angles of incidence and refraction obey Snell's law.

or

$$\frac{dL_2}{dx} = -\frac{d-x}{L_2} = -\sin \theta_2 \qquad\qquad 31\text{-}27$$

where θ_2 is the angle of refraction. From Equation 31-25,

$$n_1 \frac{dL_1}{dx} + n_2 \frac{dL_2}{dx} = 0 \qquad\qquad 31\text{-}28$$

Substituting the results of Equations 31-26 and 31-27 for dL_1/dx and dL_2/dx gives

$$n_1 \sin \theta_1 + n_2(-\sin \theta_2) = 0$$

or

$$n_1 \sin \theta_1 = n_2 \sin \theta_2$$

which is Snell's law.

SUMMARY

Topic	Relevant Equations and Remarks
1. **Visible Light**	The human eye is sensitive to electromagnetic radiation with wavelengths from approximately 400 nm (violet) to 700 nm (red). The photon energies range from approximately 1.8 eV to 3.1 eV. A uniform mixture of wavelengths, such as the wavelengths emitted by the sun, appears white to our eyes.
2. **Wave–Particle Duality**	Light propagates like a wave, but interacts with matter like a particle.
Photon energy	$E = hf = \dfrac{hc}{\lambda}$ **31-1**
Planck's constant	$h = 6.626 \times 10^{-34}\ \text{J·s} = 4.136 \times 10^{-15}\ \text{eV·s}$
hc	$hc = 1240\ \text{eV·nm}$ **31-2**
3. **Emission of Light**	Light is emitted when an outer atomic electron makes a transition from an excited state to a state of lower energy.
Line spectra	Atoms in dilute gases emit a discrete set of wavelengths called a line spectra. The photon energy $E = hf = hc/\lambda$ equals the difference in energy of the initial and final states of the atom.
Continuous spectra	Atoms in high-density gases, liquids, or solids have continuous bands of energy levels, so they emit a continuous spectrum of light. Thermal radiation is visible if the temperature of the emitting object is above approximately 600°C.
Spontaneous emission	An atom in an excited state will spontaneously make a transition to a lower state with the emission of a photon. This process is random, with a characteristic lifetime of about 10^{-8} s. The photons from two or more atoms are not correlated, so the light is incoherent.

Stimulated emission	Stimulated emission occurs if an atom is initially in an excited state and a photon of energy equal to the energy difference between that state and a lower state is incident on the atom. The oscillating electromagnetic field of the incident photon stimulates the excited atom to emit another photon in the same direction and in phase with the incident photon. The emitted light is coherent.
4. Lasers	A laser produces an intense, coherent, and narrow beam of photons as the result of stimulated emission. The operation of a laser depends on population inversion, in which there are more atoms in an excited state than in the ground state or a lower state.
5. Speed of Light	The SI unit of length, the meter, is defined so that the speed of light in vacuum is exactly $$c = 299{,}792{,}458 \text{ m/s} \qquad \text{31-5}$$
v in a transparent medium	$$v = \frac{c}{n} \qquad \text{31-7}$$ where n is the index of refraction.
6. Huygens's Principle	Each point on a primary wavefront serves as the source of spherical secondary wavelets that advance with a speed and frequency equal to that of the primary wave. The primary wavefront at some later time is the envelope of these wavelets.
7. Reflection and Refraction	When light is incident on a surface separating two media in which the speed of light differs, part of the light energy is transmitted and part of the light energy is reflected.
Law of reflection	The reflected ray lies in the plane of incidence and makes an angle θ_1' with the normal that is equal to the angle of incidence. $$\theta_1' = \theta_1 \qquad \text{31-8}$$
Reflected intensity, normal incidence	$$I = \left(\frac{n_1 - n_2}{n_1 + n_2}\right)^2 I_0 \qquad \text{31-11}$$
Index of refraction	$$n = \frac{c}{v} \qquad \text{31-7}$$
Law of refraction (Snell's law)	$$n_1 \sin\theta_1 = n_2 \sin\theta_2 \qquad \text{31-9}b$$
Total internal reflection	When light is traveling in a medium with an index of refraction n_1 and is incident on the boundary of a second medium with a lower index of refraction $n_2 < n_1$, the light is totally reflected if the angle of incidence is greater than the critical angle θ_c given by
Critical angle	$$\sin\theta_c = \frac{n_2}{n_1} \qquad \text{31-12}$$
Dispersion	The speed of light in a medium, and therefore the index of refraction of that medium, depends on the wavelength of light. Because of dispersion, a beam of white light incident on a refracting prism is dispersed into its component colors. Similarly, the reflection and refraction of sunlight by raindrops produces a rainbow.
8. Polarization	Transverse waves can be polarized. The four phenomena that produce polarized electromagnetic waves from unpolarized waves are: (1) absorption, (2) scattering, (3) reflection, and (4) birefringence.
Malus's law	When two polarizers have their transmission axes at an angle θ, the intensity transmitted by the second polarizer is reduced by the factor $\cos^2\theta$. $$I = I_0 \cos^2\theta \qquad \text{31-20}$$

PROBLEMS

- Single-concept, single-step, relatively easy
- •• Intermediate-level, may require synthesis of concepts
- ••• Challenging
- **SSM** Solution is in the *Student Solutions Manual*
- **iSOLVE** Problems available on iSOLVE online homework service
- **iSOLVE✔** These "Checkpoint" online homework service problems ask students additional questions about their confidence level, and how they arrived at their answer.

In a few problems, you are given more data than you actually need; in a few other problems, you are required to supply data from your general knowledge, outside sources, or informed estimates.

Conceptual Problems

1 • Why is helium needed in a helium–neon laser? Why not just use neon?

2 •• When a beam of visible white light passes through a gas of atomic hydrogen and is viewed with a spectroscope, dark lines are observed at the wavelengths of the emission series. The atoms that participate in the resonance absorption then emit this same wavelength light as they return to the ground state. Explain why the observed spectrum nevertheless exhibits pronounced dark lines.

3 • How does a thin layer of water on the road affect the light you see reflected off the road from your own headlights? How does it affect the light you see reflected from the headlights of an oncoming car?

4 • A ray of light passes from air into water, striking the surface of the water with an angle of incidence of 45°. Which of the following four quantities change as the light enters the water: (1) wavelength, (2) frequency, (3) speed of propagation, (4) direction of propagation? (*a*) 1 and 2 only; (*b*) 2, 3, and 4 only; (*c*) 1, 3, and 4 only; (*d*) 3 and 4 only; or (*e*) 1, 2, 3, and 4.

5 •• **SSM** The density of the atmosphere decreases with height, as does the index of refraction. Explain how one can see the sun after it has set. Why does the setting sun appear flattened?

6 • A physics student playing pocket billiards wants to strike her cue ball so that it hits a cushion and then hits the eight ball squarely. She chooses several points on the cushion and for each point measures the distance from it to the cue ball and to the eight ball. She aims at the point for which the sum of these distances is least. (*a*) Will her cue ball hit the eight ball? (*b*) How is her method related to Fermat's principle?

7 • A swimmer at point *S* in Figure 31-53 develops a leg cramp while swimming near the shore of a calm lake and calls for help. A lifeguard at point *L* hears the call. The lifeguard can run 9 m/s and swim 3 m/s. She knows physics and chooses a path that will take the least time to reach the swimmer. Which of the paths shown in Figure 31-53 does the lifeguard take?

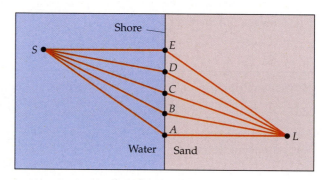

FIGURE 31-53 Problem 7

8 • Two polarizers have their transmission axes at an angle *θ*. Unpolarized light of intensity *I* is incident upon the first polarizer. What is the intensity of the light transmitted by the second polarizer? (*a*) $I \cos^2 \theta$, (*b*) $(I \cos^2 \theta)/2$, (*c*) $(I \cos^2 \theta)/4$, (*d*) $I \cos \theta$, (*e*) $(I \cos \theta)/4$, or (*f*) none of the answers are correct.

9 • Which of the following is *not* a phenomenon whereby polarized light can be produced from unpolarized light? (*a*) absorption (*b*) reflection (*c*) birefringence (*d*) diffraction (*e*) scattering

10 •• **SSM** We learned in Chapter 30, Section 30-3, that an oscillating electric dipole produces electromagnetic radiation (see Figure 30-8). Assuming that the light reflected off and refracted into the surface of a piece of transparent material is caused by such dipoles, show that the condition for Brewster's angle (Equation 31-21) is exactly the same as saying that the refracted ray is perpendicular to the axis of the radiating dipoles for light polarized in the plane of incidence.

11 •• Draw a diagram to explain how Polaroid sunglasses reduce glare from sunlight reflected from a smooth horizontal surface, such as the surface found on a pool of water. Your diagram should clearly indicate the direction of polarization of the light as it propagates from the sun to the reflecting surface and then through the sunglasses into the eye.

12 • **iSOLVE** True or false:

(*a*) Light and radio waves travel with the same speed through a vacuum.
(*b*) Most of the light incident normally on an air–glass interface is reflected.

(c) The angle of refraction of light is always less than the angle of incidence.

(d) The index of refraction of water is the same for all wavelengths in the visible spectrum.

(e) Longitudinal waves cannot be polarized.

13 •• **iSOLVE** ✓ Of the following statements about the speeds of the various colors of light in glass, which are true?

(a) All colors of light have the same speed in glass.
(b) Violet has the highest speed, red the lowest.
(c) Red has the highest speed, violet the lowest.
(d) Green has the highest speed, red and violet the lowest.
(e) Red and violet have the highest speed, green the lowest.

14 •• **SSM** It is a common experience that on a calm, sunny day one can hear voices of persons in a boat over great distances. Explain this phenomenon, keeping in mind that sound is reflected from the surface of the water and that the temperature of the air just above the water's surface is usually less than that at a height of 10 m or 20 m above the water.

15 • The human eye perceives color using a structure called a cone, located on the retina. The molecules of the cones come in three types that respond in a process similar to resonance absorption to red, green, and blue light, respectively. Use this fact to explain why the color of a blue object (450 nm in air) does not appear to change when immersed in clear colorless water, in spite of the fact that the wavelength of the light is shortened in accordance with Equation 31-10.

Estimation and Approximation

16 • Estimate the time required for light to make the round trip in Galileo's experiment to determine the speed of light.

17 • Ole Römer's method for measuring the speed of light requires the precise prediction of the time of occurrence for the eclipse of Jupiter's moon Io. Assuming an eclipse took place on June 1 at midnight when the earth was in location A, as shown in Figure 31-11, predict the expected time of the eclipse one-quarter year later at location B, assuming (a) the speed of light is infinite and (b) the speed of light is the presently defined value of 2.998×10^8 m/s.

18 •• If the angle of incidence is small enough, the approximation $\sin \theta \approx \theta$ may be used to simplify Snell's law. Calculate the angle of incidence that would make the error in calculating the angle of refraction using this small angle approximation no worse than 1 percent when compared to the exact formula. This approximation will be used in connection with image formation by spherical surfaces in Chapter 32.

Sources of Light

19 • **iSOLVE** A pulse from a ruby laser has an average power of 10 MW and lasts 1.5 ns. (a) What is the total energy of the pulse? (b) How many photons are emitted in this pulse?

20 • **iSOLVE** A helium–neon laser emits light of wavelength 632.8 nm and has a power output of 4 mW. How many photons are emitted per second by this laser?

21 • **iSOLVE** ✓ The first excited state of an atom of a gas is 2.85 eV above the ground state. (a) What is the wavelength of radiation for resonance absorption? (b) If the gas is irradiated with monochromatic light of 320 nm wavelength, what is the wavelength of the Raman scattered light?

22 •• A gas is irradiated with monochromatic ultraviolet light of 368 nm wavelength. Scattered light of the same wavelength and of 658 nm wavelength is observed. Assuming that the gas atoms were in their ground state prior to irradiation, find the energy difference between the ground state and the atomic state excited by the irradiation.

23 •• Sodium has excited states 2.11 eV, 3.2 eV, and 4.35 eV above the ground state. (a) What is the maximum wavelength of radiation that will result in resonance fluorescence? What is the wavelength of the fluorescent radiation? (b) What wavelength will result in excitation of the state 4.35 eV above the ground state? If that state is excited, what are the possible wavelengths of resonance fluorescence that might be observed?

24 •• **SSM** Singly ionized helium is a hydrogen-like atom with a nuclear charge of $2e$. Its energy levels are given by $E_n = -4E_0/n^2$, where $E_0 = 13.6$ eV. If a beam of visible white light is sent through a gas of singly ionized helium, at what wavelengths will dark lines be found in the spectrum of the transmitted radiation?

The Speed of Light

25 • Mission Control sends a brief wake-up call to astronauts in a far away spaceship. Five seconds after the call is sent, Mission Control can hear the groans of the astronauts. How far away (at most) from the earth is the spaceship? (a) 7.5×10^8 m. (b) 15×10^8 m. (c) 30×10^8 m. (d) 45×10^8 m. (e) The spaceship is on the moon.

26 • **iSOLVE** The spiral galaxy in the Andromeda constellation is about 2×10^{19} km away from us. How many light-years is this?

27 • **iSOLVE** On a spacecraft sent to Mars to take pictures, the camera is triggered by radio waves, which like all electromagnetic waves, travel with the speed of light. What is the time delay between sending the signal from the earth and receiving the signal on Mars? (Take the distance to Mars to be 9.7×10^{10} m.)

28 • The distance from a point on the surface of the earth to one on the surface of the moon is measured by aiming a laser light beam at a reflector on the surface of the moon and measuring the time required for the light to make a round trip. The uncertainty in the measured distance Δx is related to the uncertainty in the time Δt by $\Delta x = c \, \Delta t$. If the time intervals can be measured to ± 1 ns, find the uncertainty of the distance in meters.

29 •• **SSM** **iSOLVE** In Galileo's attempt to determine the speed of light, he and his assistant were located on hilltops about 3 km apart. Galileo flashed a light and received a return flash from his assistant. (a) If his assistant had an instant reaction, what time difference would Galileo need to be able to measure for this method to be successful? (b) How does this time compare with human reaction time, which is about 0.2 s?

Reflection and Refraction

30 • **iSOLVE** ✓ Calculate the fraction of light energy reflected from an air–water interface at normal incidence.

31 •• **SSM** A ray of light is incident on one of a pair of mirrors set at right angles to each other. The plane of incidence is perpendicular to both mirrors. Show that after reflecting off each mirror the ray will emerge in the opposite direction, regardless of the angle of incidence.

32 •• (a) A beam of light in air is incident on an air–water interface. Using a spreadsheet or graphing program, plot the angle of refraction as a function of the angle of incidence from 0° to 90°. (b) Repeat Part (a), but for a beam of light initially in water, incident on a water–air interface. For Part (b), what is the meaning of your graph for angles of incidence that are greater than the critical angle?

33 • **iSOLVE** ✓ Find the speed of light in water and in glass.

34 • **iSOLVE** The index of refraction for silicate flint glass is 1.66 for light with a wavelength of 400 nm and 1.61 for light with a wavelength of 700 nm. Find the angles of refraction for light of these wavelengths that is incident on this glass at an angle of 45°.

35 •• **iSOLVE** ✓ A slab of glass with an index of refraction of 1.5 is submerged in water with an index of refraction of 1.33. Light in the water is incident on the glass. Find the angle of refraction if the angle of incidence is (a) 60°, (b) 45°, and (c) 30°.

36 •• Repeat Problem 35 for a beam of light initially in the glass that is incident on the glass–water interface at the same angles.

37 •• **SSM** **iSOLVE** ✓ Light is incident normally on a slab of glass with an index of refraction $n = 1.5$. Reflection occurs at both surfaces of the slab. Approximately what percentage of the incident light energy is transmitted by the slab?

38 •• This problem is a refraction analogy. A band is marching down a football field with a constant speed v_1. About midfield, the band comes to a section of muddy ground that has a sharp boundary making an angle of 30° with the 50-yd line, as shown in Figure 31-54. In the mud, the marchers move with speed $v_2 = v_1/2$. Diagram how each line of marchers is bent as it encounters the muddy section of the field so that the band is eventually marching in a different direction. Indicate the original direction by a ray, the final direction by a second ray, and find the angles between the rays and the line perpendicular to the boundary. Is their direction of motion bent toward the perpendicular to the boundary or away from it?

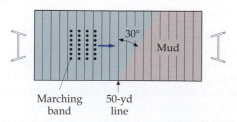

FIGURE 31-54 Problem 38

39 •• In Figure 31-55, light is initially in a medium (e.g., air) of index of refraction n_1. It is incident at angle θ_1 on the surface of a liquid (e.g., water) of index of refraction n_2. The light passes through the layer of water and enters glass of index of refraction n_3. If θ_3 is the angle of refraction in the glass, show that $n_1 \sin \theta_1 = n_3 \sin \theta_3$. That is, show that the second medium can be neglected when finding the angle of refraction in the third medium.

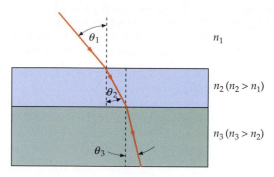

FIGURE 31-55 Problem 39

40 ••• **SSM** Figure 31-56 shows a beam of light incident on a glass plate of thickness d and index of refraction n. (a) Find the angle of incidence so that the perpendicular separation between the ray reflected from the top surface and the ray reflected from the bottom surface and exiting the top surface is a maximum. (b) What is this angle of incidence if the index of refraction of the glass is 1.60? What is the separation of the two beams if the thickness of the glass plate is 4.0 cm?

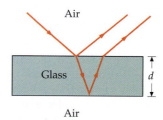

FIGURE 31-56 Problem 40

Total Internal Reflection

41 • What is the critical angle for total internal reflection for light traveling initially in water that is incident on a water–air interface?

42 •• **iSOLVE** A glass surface ($n = 1.50$) has a layer of water ($n = 1.33$) on it. Light in the glass is incident on the glass–water interface. Find the critical angle for total internal reflection.

43 •• **iSOLVE** ✓ A point source of light is located 5 m below the surface of a large pool of water. Find the area of the largest circle on the pool's surface through which light coming directly from the source can emerge.

44 •• Light is incident normally on the largest face of an isosceles-right-triangle prism. What is the speed of light in this prism if the prism is just barely able to produce total internal reflection?

45 •• **ISOLVE✓** A point source of light is located at the bottom of a steel tank, and an opaque circular card of radius 6 cm is placed over it. A transparent fluid is gently added to the tank so that the card floats on the surface with its center directly above the light source. No light is seen by an observer above the surface until the fluid is 5 cm deep. What is the index of refraction of the fluid?

46 •• **SSM** An optical fiber allows rays of light to propagate long distances through total internal reflection. As shown in Figure 31-57, the fiber consists of a core material with index of refraction n_2 and radius b, surrounded by a cladding material of index $n_3 < n_2$. The numerical aperture of the fiber is defined as $\sin\theta_1$, where θ_1 is the angle of incidence of a ray of light impinging the end of the fiber that reflects off the core-cladding interface at the critical angle. Using the figure as a guide, show that the numerical aperture is given by

$$\sqrt{n_2^2 - n_3^2}$$

assuming the ray is incident from air. (*Hint: Use of the Pythagorean theorem may be required.*)

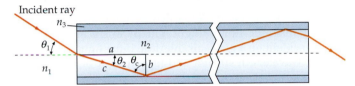

FIGURE 31-57 Problems 46, 47, and 48

47 • Find the maximum angle of incidence of a ray that would propagate through an optical fiber with a core index of refraction of 1.492, a core radius of 50 μm, and a cladding index of 1.489. See Problem 46 and Figure 31-57.

48 •• Calculate the difference in time needed for two pulses of light to travel down 15 km of the fiber of Problem 47. Assume that one pulse enters the fiber at normal incidence, and the second pulse enters the fiber at the maximum angle of incidence calculated in Problem 47 (see Figure 31-57). In fiber optics, this effect is known as modal dispersion.

49 ••• Investigate how a thin film of water on a glass surface affects the critical angle for total reflection. Take $n = 1.5$ for glass and $n = 1.33$ for water. (*a*) What is the critical angle for total internal reflection at the glass–water interface? (*b*) Is there any range of incident angles that are greater than θ_c for glass-to-air refraction and for which light rays will leave the glass and the water and pass into the air?

50 ••• A laser beam is incident on a plate of glass of thickness 3 cm. The glass has an index of refraction of 1.5 and the angle of incidence is 40°. The top and bottom surfaces of the glass are parallel and both produce reflected beams of nearly the same intensity. What is the perpendicular distance d between the two adjacent reflected beams?

Dispersion

51 •• **SSM** **ISOLVE** A beam of light strikes the plane surface of silicate flint glass at an angle of incidence of 45°. The index of refraction of the glass varies with wavelength, as shown in the graph in Figure 31-26. How much smaller is the angle of refraction for violet light of wavelength 400 nm than that for red light of wavelength 700 nm?

52 •• Different colors (frequencies) of light travel at different speeds (a phenomena referred to as dispersion). This can cause problems in fiber-optic communications systems where pulses of light must travel very long distances in glass. Assuming a fiber is made of silicate crown glass, calculate the difference in time needed for two short pulses of light to travel 15 km of fiber if the first pulse has a wavelength of 700 nm and the second pulse has a wavelength of 500 nm.

Polarization

53 • **ISOLVE✓** What is the polarizing angle for (*a*) water with $n = 1.33$ and (*b*) glass with $n = 1.5$?

54 • Light known to be polarized in the horizontal direction is incident on a polarizing sheet. It is observed that only 15 percent of the intensity of the incident light is transmitted through the sheet. What angle does the transmission axis of the sheet make with the horizontal? (*a*) 8.6°, (*b*) 21°, (*c*) 23°, (*d*) 67°, or (*e*) 81°.

55 • Two polarizing sheets have their transmission axes crossed so that no light gets through. A third sheet is inserted between the first two so that its transmission axis makes an angle θ with that of the first sheet. Unpolarized light of intensity I_0 is incident on the first sheet. Find the intensity of the light transmitted through all three sheets if (*a*) $\theta = 45°$ and (*b*) $\theta = 30°$.

56 •• A horizontal 5 mW laser beam polarized in the vertical direction is incident on a pair of polarizers. The first is oriented so that its transmission axis is vertical and the second is oriented with its transmission axis at an angle of 27° with respect to the first. What is the power of the transmitted beam?

57 •• The polarizing angle for a certain substance is 60°. (*a*) What is the angle of refraction of light incident at this angle? (*b*) What is the index of refraction of this substance?

58 •• Two polarizing sheets have their transmission axes crossed and a third sheet is inserted so that its transmission axis makes an angle θ with that of the first sheet, as in Problem 55. Find the intensity of the transmitted light as a function of θ. Show that the intensity transmitted through all three sheets is maximum when $\theta = 45°$.

59 •• If the middle polarizing sheet in Problem 58 is rotating at an angular velocity ω about an axis parallel to the light beam, find the intensity transmitted through all three sheets as a function of time. Assume that $\theta = 0$ at time $t = 0$.

60 •• **SSM** A stack of $N + 1$ ideal polarizing sheets is arranged with each sheet rotated by an angle of $\pi/(2N)$ rad with respect to the preceding sheet. A plane, linearly polarized light wave of intensity I_0 is incident normally on the stack. The incident light is polarized along the transmission axis of the first sheet and is therefore perpendicular to the transmission axis of the last sheet in the stack. (*a*) Show that the transmitted intensity through the stack is given by the expression

$$I_0 \cos^{2N}\left(\frac{\pi}{2N}\right)$$

(b) Using a spreadsheet or graphing program, plot the transmitted intensity as a function of N for values of N from 2 to 100. *(c)* What is the direction of polarization of the transmitted beam in each case?

61 •• The device described in Problem 60 could serve as a polarization *rotator,* one that takes the linear plane of polarization from one direction to another. The efficiency of such a device is measured by taking the ratio of the output intensity at the desired polarization to the input intensity. The result of the previous problem suggests that the best way to do this would be to use a large number N. But in a real polarizer, a small amount of intensity is lost regardless of the input polarization. For each polarizer, assume the transmitted intensity is 98 percent of the amount predicted by the law of Malus and use a spreadsheet or graphing program to determine the optimum number of sheets you should use to rotate the polarization 90°.

62 •• **SSM** Show that a linearly polarized wave can be thought of as a superposition of a right and a left circularly polarized wave.

63 •• Suppose that the middle sheet in Problem 55 is replaced by two polarizing sheets. If the angles between the directions of polarization of adjacent sheets is 30°, what is the intensity of the transmitted light? How does this compare with the intensity obtained in Problem 55, Part *(a)*?

64 •• **SSM** Show that the electric field of a circularly polarized wave propagating in the x direction can be expressed by

$$\vec{E} = E_0 \sin(kx - \omega t)\hat{j} + E_0 \cos(kx - \omega t)\hat{k}$$

65 •• A circularly polarized wave is said to be *right circularly polarized* if the electric and magnetic fields rotate clockwise when viewed along the direction of propagation and *left circularly polarized* if the fields rotate counterclockwise. What is the sense of the circular polarization for the wave described by the expression in Problem 64? What would be the corresponding expression for a circularly polarized wave of the opposite sense?

General Problems

66 • **iSOLVE** A beam of monochromatic red light with a wavelength of 700 nm in air travels in water. *(a)* What is the wavelength in water? *(b)* Does a swimmer underwater observe the same color or a different color for this light?

67 •• The critical angle for total internal reflection for a substance is 45°. What is the polarizing angle for this substance?

68 •• **SSM** Figure 31-58 shows two plane mirrors that make an angle θ with each other. Show that the angle between the incident and reflected rays is 2θ.

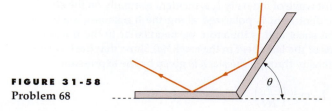

FIGURE 31-58
Problem 68

69 •• A silver coin sits on the bottom of a swimming pool that is 4 m deep. A beam of light reflected from the coin emerges from the pool making an angle of 20° with respect to the water's surface and enters the eye of an observer. Draw a ray from the coin to the eye of the observer. Extend this ray, which goes from the water–air interface to the eye, straight back until it intersects with the vertical line drawn through the coin. What is the apparent depth of the swimming pool to this observer?

70 •• Fishermen always insist on silence because noise on shore will scare fish away. Suppose a fisherman cast a baited hook 20 m from the shore of a calm lake to a point where the depth is 15 m. Show that noise on shore cannot possibly be sensed by fish at that point. (*Note:* The speed of sound in air is 330 m/s; the speed of sound in water is 1450 m/s.)

71 •• **SSM** **iSOLVE** A swimmer at the bottom of a pool 3 m deep looks up and sees a circle of light. If the index of refraction of the water in the pool is 1.33, find the radius of the circle.

72 •• Show that when a mirror is rotated through an angle θ, the reflected beam of light is rotated through 2θ.

73 •• Use Figure 31-25 to calculate the critical angles for total internal reflection for light initially in silicate flint glass that is incident on a glass–air interface if the light is *(a)* violet light of wavelength 400 nm and *(b)* red light of wavelength 700 nm.

74 •• Show that for normally incident light, the intensity transmitted through a glass slab with an index of refraction of n is approximately given by

$$I_T = I_0 \left[\frac{4n}{(n+1)^2} \right]^2$$

75 •• A ray of light begins at the point $x = -2$ m, $y = 2$ m, strikes a mirror in the xz plane at some point x, and reflects through the point $x = 2$ m, $y = 6$ m. *(a)* Find the value of x that makes the total distance traveled by the ray a minimum. *(b)* What is the angle of incidence on the reflecting plane? What is the angle of reflection?

76 •• **SSM** A Brewster window is used in lasers to preferentially transmit light of one polarization, as shown in Figure 31-59. Show that if θ_{P1} is the polarizing angle for the n_1/n_2 interface, then θ_{P2} is the polarizing angle for the n_2/n_1 interface.

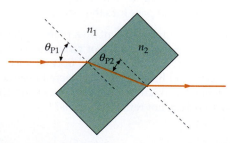

FIGURE 31-59 Problem 76

77 •• From the data provided in Figure 31-29, calculate the polarization angle for an air–glass interface, using light of wavelength 550 nm in each of the four types of glass shown in the figure.

78 ••• Light passes symmetrically through a prism with an apex angle of α, as shown in Figure 31-60. (*a*) Show that the angle of deviation δ is given by

$$\sin \frac{\alpha + \delta}{2} = n \sin \frac{\alpha}{2}$$

(*b*) If the refractive index for red light is 1.48 and the refractive index for violet light is 1.52, what is the angular separation of visible light for a prism with an apex angle of 60°?

FIGURE 31-60
Problems 78 and 89

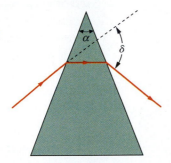

79 •• **SSM** (*a*) For a light ray inside a transparent medium that has a planar interface with a vacuum, show that the polarizing angle and the critical angle for internal reflection satisfy $\tan \theta_p = \sin \theta_c$. (*b*) Which angle is larger?

80 •• **ISOLVE** Light is incident from air on a transparent substance at an angle of 58° with the normal. The reflected and refracted rays are observed to be mutually perpendicular. (*a*) What is the index of refraction of the transparent substance? (*b*) What is the critical angle for total internal reflection in this substance?

81 •• **ISOLVE** A light ray in dense flint glass with an index of refraction of 1.655 is incident on the glass surface. An unknown liquid condenses on the surface of the glass. Total internal reflection on the glass–liquid interface occurs for an angle of incidence on the glass–liquid interface of 53.7°. (*a*) What is the refractive index of the unknown liquid? (*b*) If the liquid is removed, what is the angle of incidence for total internal reflection? (*c*) For the angle of incidence found in Part (*b*), what is the angle of refraction of the ray into the liquid film? Does a ray emerge from the liquid film into the air above? Assume the glass and liquid have perfect planar surfaces.

82 •• Given that the index of refraction for red light in water is 1.3318 and that the index of refraction for blue light in water is 1.3435, find the angular separation of these colors in the primary rainbow. (Use the equation given in Problem 86.)

83 •• (*a*) Use the result from Problem 74 to find the ratio of the transmitted intensity to the incident intensity through N parallel slabs of glass for light of normal incidence. (*b*) Find this ratio for three slabs of glass with $n = 1.5$. (*c*) How many slabs of glass with $n = 1.5$ will reduce the intensity to 10 percent of the incident intensity?

84 •• Light is incident on a slab of transparent material at an angle θ_1, as shown in Figure 31-61. The slab has a thickness t and an index of refraction n. Show that

$$n = \frac{\sin \theta_1}{\sin[\arctan(d/t)]}$$

where d is the distance shown in the figure and $\arctan(d/t)$ is the angle whose tangent is d/t.

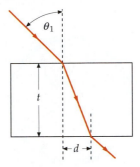

FIGURE 31-61 Problem 84

85 •• **SSM** Suppose rain falls vertically from a stationary cloud 10,000 m above a confused marathoner running in a circle with constant speed of 4 m/s. The rain has a terminal speed of 9 m/s. (*a*) What is the angle that the rain appears to make with the vertical to the marathoner? (*b*) What is the apparent motion of the cloud as observed by the marathoner? (*c*) A star on the axis of the earth's orbit appears to have a circular orbit of angular diameter of 41.2 seconds of arc. How is this angle related to the earth's speed in its orbit and the velocity of photons received from this distant star? (*d*) What is the speed of light as determined from the data in Part (*c*)?

86 ••• Equation 31-18 gives the relation between the angle of deviation ϕ_d of a light ray incident on a spherical drop of water in terms of the incident angle θ_1 and the index of refraction of water. (*a*) Assume that $n_{air} = 1$, and differentiate ϕ_d with respect to θ_1. [*Hint:* If $y = \arcsin x$, then $dy/dx = (1 - x^2)^{-1/2}$.] (*b*) Set $d\phi_d/d\theta_1 = 0$ and show that the angle of incidence $\theta_{1\,m}$ for minimum deviation is given by

$$\cos \theta_{1m} = \sqrt{\frac{n^2 - 1}{3}}$$

and find θ_{1m} for water, where the index of refraction for water is 1.33.

87 ••• **ISOLVE** Show that a light ray incident on a rectangular glass slab of thickness t at angle of incidence θ_1 emerges parallel to the incident ray but displaced from it by an amount s given by

$$s = \frac{t \sin(\theta_1 - \theta_2)}{\cos(\theta_2)}$$

where θ_2 is the angle of refraction.

88 •• Use the result of Problem 87 to find the lateral displacement of a laser beam incident at an angle of 30° on a 15 mm thick rectangular slab of glass with index of refraction 1.5.

89 ••• Show that the angle of deviation δ is a minimum if the angle of incidence is such that the ray passes through the prism symmetrically, as shown in Figure 31-60.

Optical Images

NOTE THAT THE PHOTOGRAPH SHOWS EVIDENCE THAT THE HANDS ON THE MIRROR IMAGE OF AN ORDINARY CLOCK ROTATE NOT CLOCKWISE, BUT COUNTERCLOCKWISE. IT IS ALSO TRUE THAT IF YOU COULD LOOK AT A CLOCK FACE FROM BEHIND THE CLOCK IT WOULD LOOK JUST LIKE THE MIRROR IMAGE OF THE CLOCK FACE.

? **When looked at from behind, do the hands of a clock rotate clockwise or counterclockwise? A number of observations concerning mirror images are discussed in Section 32-1.**

Because the wavelength of light is very small compared with most obstacles and openings, diffraction—the bending of waves around corners—is often negligible, and the ray approximation, in which waves are considered to propagate in straight lines, accurately describes observations.
➤ In this chapter, we apply the laws of reflection and refraction to the formation of images by mirrors and lenses.

32-1 Mirrors

Plane Mirrors

Figure 32-1 shows a bundle of light rays emanating from a point source P and reflected from a plane mirror. After reflection, the rays diverge exactly as if they came from a point P' behind the plane of the mirror. The point P' is called the **image** of the **object** P. When these reflected rays enter the eye, they cannot be distinguished from rays diverging from a source at P' with no mirror present. This image at P' is called a **virtual image** because the light does not actually emanate from it. The plane of the mirror is the perpendicular bisector of the line from the object point P to the image point P' as shown. The image can be seen by an eye located anywhere in the shaded region indicated, in which a straight line from

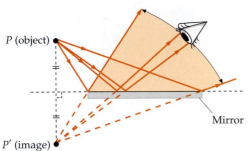

FIGURE 32-1 Image formed by a plane mirror. The rays from point P that strike the mirror and enter the eye appear to come from the image point P' behind the mirror. The image can be seen by an eye located anywhere in the shaded region.

the image to the eye passes through the mirror. The object need not be directly in front of the mirror. As long as the object is not behind the plane of the mirror, there is some location at which the eye can be located to view the image.

If you hold up your right hand and look in the mirror, the image you see is neither magnified nor reduced, but it looks like a left hand (Figure 32-2). This right-to-left reversal is a result of **depth inversion**—the hand is transformed from a right hand to a left hand because the front and the back of the hand are reversed by the mirror. Depth inversion is also illustrated in Figure 32-3. Figure 32-4 shows the image of a simple rectangular coordinate system. The mirror transforms a right-handed coordinate system, for which $\hat{i} \times \hat{j} = \hat{k}$, into a left-handed coordinate system, for which $\hat{i} \times \hat{j} = -\hat{k}$.

Figure 32-5 shows an arrow of height y standing parallel to a plane mirror a distance s from the mirror. We can locate the image of the arrowhead (and of any other point on the arrow) by drawing two rays. One ray, drawn perpendicular to the mirror, hits the mirror at point A and is reflected back onto itself. The other ray, making an angle θ with the normal to the mirror, is reflected, making an equal angle θ with the x axis. The extension of these two rays back behind the mirror locates the image of the arrowhead, as shown by the dashed lines in the figure. We can see from this figure that the image is the same distance behind the mirror as the object is in front of the mirror, and that the image is upright (points in the same direction as the object) and the image is the same size as the object.

FIGURE 32-2 The image of a right hand in a plane mirror is transformed to a left hand. This right-to-left reversal is a result of depth inversion.

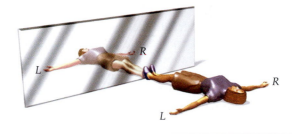

FIGURE 32-3 A person lying down with her feet against the mirror. The image is depth inverted.

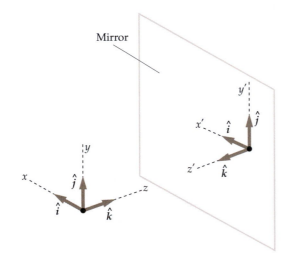

FIGURE 32-5 Ray diagram for locating the image of an arrow in a plane mirror.

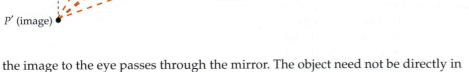

FIGURE 32-4 Image of a rectangular coordinate system in a plane mirror. The arrow along the z axis is reversed in the image. The image of the original right-handed coordinate system, for which $\hat{i} \times \hat{j} = \hat{k}$, is a left-handed coordinate system, for which $\hat{i} \times \hat{j} = -\hat{k}$.

FIGURE 32-6 Images formed by two plane mirrors. P_1' is the image of the object P in mirror 1, and P_2' is the image of the object in mirror 2. Point $P_{1,2}''$ is the image of P_1' in mirror 2, which is seen when light rays from the object reflect first from mirror 1 and then from mirror 2. The image P_2' does not have an image in mirror 1 because it is behind that mirror.

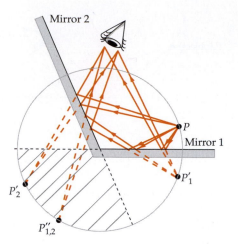

The formation of multiple images by two plane mirrors that make an angle with each other is illustrated in Figure 32-6. We frequently see this phenomenon in clothing stores that provide adjacent mirrors. The light from source point P that is reflected from mirror 1 strikes mirror 2 just as if it came from the image point P_1'. The image P_1' is the object for mirror 2. Its image is behind mirror 2 at point $P_{1,2}''$. This image will be formed whenever the image point P_1' is in front of the plane of mirror 2. The image at point P_2' is due to rays from P that reflect directly from mirror 2. Since P_2' is behind the plane of mirror 1, it cannot serve as an object point for a further image in mirror 1. The image at point P_2' cannot serve as an object for mirror 1 because the geometry dictates that none of the rays from P that reflect directly from mirror 2 can then strike mirror 1. An alternative way of stating this is that since P_2' is behind the plane of mirror 1, the image at P_2' cannot serve as an object for mirror one. The number of multiple images formed by two mirrors depends on the angle between the mirrors and the position of the object.

EXERCISE Show that a source point and all consequent image points formed by two plane mirrors are equidistant from the intersection of the two planes.

Suppose your friend Ben is standing at point P, is wearing a sweatshirt with BEN printed on it, and is waving his right hand. Also, suppose that you are standing at the location of the eye. You can see an image of Ben at all three image locations. For the images at P_1' and P_2', Ben is waving his left hand and the printing on his sweatshirt appears as **NƎB**. However, for the image at $P_{1,2}''$, Ben is waving his right hand and the printing appears as BEN. For the image at $P_{1,2}''$ depth inversion occurs twice, once for each reflection, so the result is as if no depth inversion occurs.

EXERCISE Which of the images of himself can Ben see? (*Answer* Ben can see only the image at P_1'.)

Figure 32-7 illustrates the fact that a horizontal ray reflected from two perpendicular vertical mirrors is reflected back along a parallel path no matter what angle the ray makes with the mirrors. If three mirrors are placed perpendicular to each other, like the sides of an inside corner of a box, any ray incident on any of the mirrors from any direction is reflected back on a path parallel to that of the incident

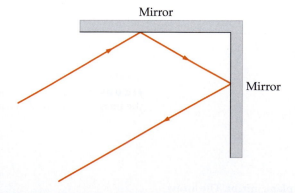

FIGURE 32-7 A horizontal ray striking one of two perpendicular plane mirrors is reflected from the second mirror in the direction opposite the original direction for any angle of incidence.

ray. A set of three mirrors arranged in this manner is called a corner-cube reflector. An array of corner-cube reflectors was placed on the moon in the Sea of Tranquility by the Apollo 11 astronauts in 1969. A laser beam from the earth that is directed at the mirrors is reflected back to the same place on the earth. Such a beam has been used to measure the distance to the mirrors to within a few centimeters by measuring the time it takes for the light to reach the mirrors and return.

Spherical Mirrors

Figure 32-8 shows a bundle of rays from a point source P on the axis of a concave spherical mirror reflecting from the mirror and converging at point P'. (A concave mirror is shaped like a cave when you look into it.) The rays then diverge from this point, just as if there were an object at the point. This image is called a **real image,** because light actually does emanate from the image point. The image can be seen by an eye at the left of the image looking into the mirror. It could also be observed on a small viewing screen[†] or on a small piece of photographic film placed at the image point. A virtual image, such as that formed by a plane mirror as discussed in the previous section, cannot be observed on a screen at the image point because there is no light at the image point. Despite this distinction between real and virtual images, the eye makes no distinction between them. The light rays diverging from a real image and those appearing to diverge from a virtual image are the same to the eye.

From Figure 32-9, we can see that only rays that strike the spherical mirror at points near the axis AV are reflected through the image point. Rays almost parallel with the axis and near to it are called **paraxial rays.** Rays that strike the mirror at points far from the axis upon reflection pass near the image point, but not through it. Such rays cause the image to appear blurred, an effect called **spherical aberration.** The image can be sharpened by blocking off all but the central part of the mirror, so that rays far from the axis do not strike it. The image is then sharper, but its brightness is reduced because less light is reflected to the image point.

We wish to obtain an equation relating the position of the image point to the position of the object point. To do this, we draw two rays (Figure 32-10a) from an arbitrarily positioned object point P. One ray passes through point C, the center of curvature of the mirror, and the other ray strikes point A, an arbitrarily positioned point on the mirror. The image point P' is where these two rays intersect after reflecting off the mirror. Using the law of reflection, we obtain the location of P'. The ray passing through point C strikes the mirror at normal incidence, so the ray reflects back upon itself. The ray striking the mirror at A makes angle θ with the normal, so, as shown, the reflected ray also makes angle θ with the normal. (Any line normal to a spherical surface passes through the center of curvature.) The image distance s' and object distance s are measured from the plane tangent to the mirror at its vertex V. The angle β is an exterior angle to the triangle PAC, therefore, $\beta = \alpha + \theta$. Similarly, from the triangle PAP', $\gamma = \alpha + 2\theta$. Eliminating θ from these equations gives $2\beta = \alpha + \gamma$. By assuming all rays are paraxial, we can substitute using the small-angle approximations: $\alpha \approx \ell/s$, $\beta \approx \ell/r$, and $\gamma \approx \ell/s'$. Equation 32-1 follows directly:

$$\frac{1}{s} + \frac{1}{s'} = \frac{2}{r}$$

32-1

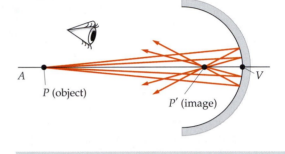

FIGURE 32-8 Rays from a point object P on the axis AV of a concave spherical mirror form an image at P'. The image is sharp if the rays strike the mirror near the axis and if the rays are almost parallel with the axis.

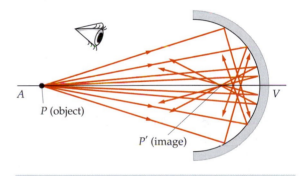

FIGURE 32-9 Spherical aberration of a mirror. Nonparaxial rays that strike the mirror at points far from the axis AV are not reflected through the image point P' formed by the paraxial rays. The nonparaxial rays blur the image.

---

† A viewing screen must produce either diffuse reflection or diffuse transmission of the light. Ground glass is commonly used for this purpose. The screen must be small, so that it does not block all of the light from the source from reaching the mirror.

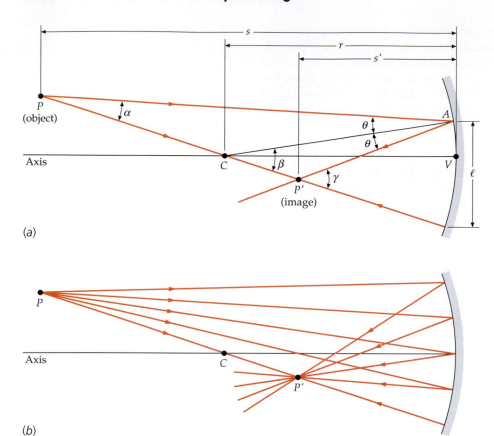

(a)

(b)

FIGURE 32-10 (a) Geometry for calculating the image distance s' from the object distance s and the radius of curvature r. The angle β is an exterior angle to the triangle PAC; therefore, $\beta = \alpha + \theta$. Similarly, from the triangle PAP', $\gamma = \alpha + 2\theta$. Eliminating θ from these equations gives $2\beta = \alpha + \gamma$. Equation 32-1 follows directly, if we assume the following small-angle approximations: $\alpha \approx \ell/s$, $\beta \approx \ell/r$, and $\gamma \approx \ell/s'$. (b) All paraxial rays from object point P pass through image point P' after reflecting off the mirror.

This equation relates the object and image distances with the radius of curvature. The striking thing about this equation is that it contains absolutely nothing about the location of point A. Therefore, the equation is valid for *any* choice for the location of point A, as long as point A is on the surface of the mirror and all rays are paraxial. That is, as shown in Figure 32-10b, *all* paraxial rays emanating from an object point will, upon reflection, pass through a *single* image point.

Equation 32-1 specifies the image position in terms of its distance from the mirror. We now specify the image position in terms of its distance from the axis. We first draw a single ray (Figure 32-11) that reflects off the mirror at its vertex. The two right triangles formed are similar. Corresponding sides of similar triangles are equal, so

$$\frac{y'}{y} = -\frac{s'}{s}$$

32-2

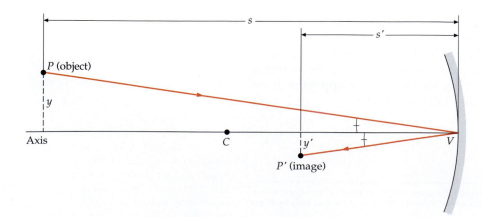

FIGURE 32-11 Geometry for calculating the position y' of the image point with respect to its distance from the axis.

The negative sign takes into account that y'/y is negative as P and P' are on opposite sides of the axis. Thus, if y is positive, y' is negative and if y is negative, y' is positive.

EXERCISE   For the image point and object point shown in Figure 32-11, show that

$$\frac{y'}{y} = -\frac{r/2}{s - (r/2)}$$

(*Hint:* Solve Equation 32-1 for s' and substitute your result into Equation 32-2.)

When the object distance is large compared with the radius of curvature of the mirror, the term $1/s$ in Equation 32-1 is much smaller than $2/r$ and can be neglected. That is, as $s \rightarrow \infty$, $s' \rightarrow \frac{1}{2}r$, where s' is the image distance. This distance is called the **focal length** f of the mirror, and the plane on which parallel rays incident on the mirror are focused is called the **focal plane**. The intersection of the axis with the focal plane is called the **focal point** F, as illustrated in Figure 32-12a. (Again, only paraxial rays are focused at a single point.)

$$f = \tfrac{1}{2}r \qquad\qquad 32\text{-}3$$

FOCAL LENGTH FOR A MIRROR

EXERCISE   Show that solving Equation 32-1 for s' gives

$$s' = \frac{r}{2 - \dfrac{r}{s}}$$

Then show that as $s \rightarrow \infty$, $s' \rightarrow \frac{1}{2}r$.

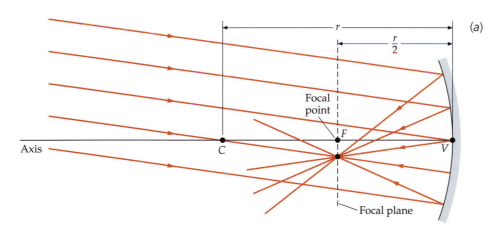

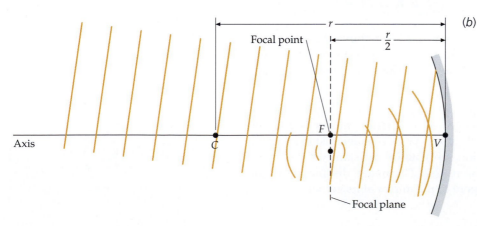

FIGURE 32-12 (*a*) Parallel rays strike a concave mirror and are reflected to a point on the focal plane a distance *r*/2 to the left of the mirror. (*b*) The incoming wavefronts are plane waves; upon reflection, they become spherical wavefronts that converge to, and then diverge from, the focal point.

The focal length of a spherical mirror is half the radius of curvature. In terms of the focal length f, Equation 32-1 is

$$\frac{1}{s} + \frac{1}{s'} = \frac{1}{f}$$

32-4

MIRROR EQUATION

Equation 32-4 is called the **mirror equation.**

When an object point is very far from the mirror, the rays are parallel, and the wavefronts are approximately planes (Figure 32-12b). In Figure 32-12b, note that the last part of each wavefront to reflect off the concave mirror surface is the part just below the vertex *V*. This results in a spherical wavefront upon reflection. Figure 32-13 shows both the wavefronts and the rays for plane waves striking a convex mirror. In this case, the central part of the wavefront strikes the mirror first, and the reflected waves appear to come from the focal point behind the mirror.

Figure 32-14 illustrates a property of waves called **reversibility.** If we reverse the direction of a reflected ray, the law of reflection assures us that the reflected ray will be along the original incoming ray, but in the opposite direction. (Reversibility holds also for refracted rays, which are discussed in later sections.) Thus, if we have a real image of an object formed by a reflecting (or refracting) surface, we can place an object at the image point and a new, real image will be formed at the position of the original object.

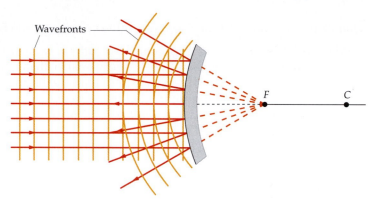

FIGURE 32-13 Reflection of plane waves from a convex mirror. The outgoing wavefronts are spherical, as if emanating from the focal point *F* behind the mirror. The rays are normal to the wavefronts and appear to diverge from *F*.

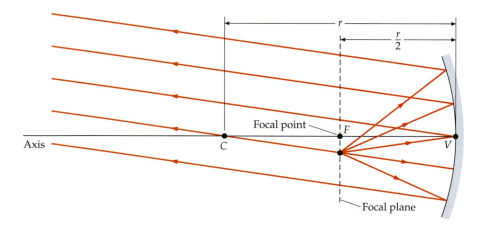

FIGURE 32-14 Reversibility. Rays diverging from a point source on the focal plane of a concave mirror are reflected from the mirror as parallel rays. The rays follow the same paths as in Figure 32-12a but in the reverse direction.

IMAGE IN A CONCAVE MIRROR **EXAMPLE 32-1**

A point object is 12 cm from a concave mirror and 3 cm above the axis of the mirror. The radius of curvature of the mirror is 6 cm. Find (*a*) the focal length of the mirror and (*b*) the image distance. (*c*) Find the position of the image relative to the axis.

PICTURE THE PROBLEM The focal length of a spherical mirror is half the radius of curvature. Once the focal length is known, the image distance can be found using the mirror equation (Equation 32-4), and the distance of the image from the axis can be found using Equation 32-2. The image distance from the mirror is the distance from the plane tangent to the mirror at its vertex.

(*a*) The focal length is half the radius of curvature:

$$f = \tfrac{1}{2}r = \tfrac{1}{2}(6 \text{ cm}) = \boxed{3 \text{ cm}}$$

(*b*) 1. Use the mirror equation to find a relation for the image distance s':

$$\frac{1}{s} + \frac{1}{s'} = \frac{1}{f}$$

or

$$\frac{1}{12 \text{ cm}} + \frac{1}{s'} = \frac{1}{3 \text{ cm}}$$

2. Solve for s':

$$\frac{1}{s'} = \frac{4}{12 \text{ cm}} - \frac{1}{12 \text{ cm}} = \frac{3}{12 \text{ cm}}$$

$$s' = \boxed{4 \text{ cm}}$$

(*c*) 1. Use Equation 32-2 to find the distance y' of image from the axis:

$$\frac{y'}{y} = -\frac{s'}{s}$$

2. Solve for y':

$$y' = -\frac{s'}{s}y = -\frac{4 \text{ cm}}{12 \text{ cm}}(3 \text{ cm}) = \boxed{-1 \text{ cm}}$$

EXERCISE A concave mirror has a focal length of 4 cm. (*a*) What is the mirror's radius of curvature? (*b*) Find the image distance for an object 2 cm from the mirror. (*Answer* (*a*) 8 cm (*b*) $s' = -4$ cm)

REMARKS A negative image distance means the image is on the opposite side of the mirror from the reflected light, as is the case with a plane mirror.

■ **EXERCISE** What is the radius of curvature of a plane mirror? (*Answer* Infinity)

Ray Diagrams for Mirrors

A useful method to locate images is by geometric construction of a **ray diagram,** as illustrated in Figure 32-15, where the object is a human figure perpendicular to the axis a distance s from the mirror. By the judicious choice of rays from the head of the figure, we can quickly locate the image. There are three **principal rays** that are convenient to use:

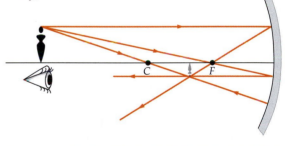

1. The **parallel ray,** drawn parallel to the axis. This ray is reflected through the focal point.

2. The **focal ray,** drawn through the focal point. This ray is reflected parallel to the axis.

3. The **radial ray,** drawn through the center of curvature. This ray strikes the mirror perpendicular to its surface and is thus reflected back on itself.

FIGURE 32-15 Ray diagram for the location of the image by geometric construction.

PRINCIPAL RAYS FOR A MIRROR

These rays are shown in Figure 32-15. The intersection of any two paraxial rays locates the image point of the head. The three principal rays are easier to draw than any of the other rays. Typically, you draw two of the principal rays to locate the image, and then draw the third principal ray as a check to verify the result. Ray diagrams are best drawn with the mirror replaced by a straight line that extends as far as necessary to intercept the rays, as shown in Figure 32-16. Note that the image in this case is inverted and smaller than the object.

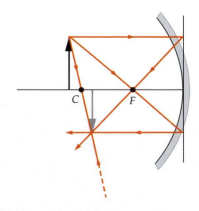

FIGURE 32-16 Ray diagrams are easier to construct if the curved surface is replaced by a plane tangent to the surface at the vertex.

When the object is between the mirror and its focal point, the rays reflected from the mirror do not converge but appear to diverge from a point behind the mirror, as illustrated in Figure 32-17. In this case, the image is virtual and upright (*upright* meaning not inverted relative to the object). For an object between the mirror and the focal point, s is less than $r/2$, so the image distance s' calculated from Equation 32-1 turns out to be negative. We can apply Equations 32-1, 32-2, 32-3, and 32-4 to this case and to convex mirrors if we adopt a convenient sign convention. Whether the mirror is convex or concave, real images can be formed only in front of the mirror, that is, on the same side of the mirror as the reflected light (and the object). Virtual images are formed behind the mirror where there is no actual light from the object. Our sign convention is as follows:

1. s is positive if the object is on the incident-light side of the mirror.
2. s' is positive if the image is on the reflected-light side of the mirror.
3. r (and f) is positive if the mirror is concave so the center of curvature is on the reflected-light side of the mirror.

SIGN CONVENTIONS FOR REFLECTION

The incident-light side and the reflected-light side are, of course, the same. The parameters s, s', r, and f are all positive if a real object[†] is in front of a concave mirror that forms a real image. A parameter is negative if it does not meet the stated condition for being positive.

The ratio of the image size to the object size is defined as the **lateral magnification** m of the image. From Figure 32-18 and Equation 32-2, we see that the lateral magnification is

$$m = \frac{y'}{y} = -\frac{s'}{s}$$ (32-5)

LATERAL MAGNIFICATION

A negative magnification, which occurs when both s and s' are positive, indicates that the image is inverted.

For plane mirrors, the radius of curvature is infinite. The focal length given by Equation 32-3 is then also infinite. Equation 32-4 then gives $s' = -s$, indicating that the image is behind the mirror at a distance equal to the object distance. The magnification given by Equation 32-5 is then $+1$, indicating that the image is upright and the same size as the object.

Although the preceding equations, coupled with our sign conventions, are relatively easy to use, we often need to know only the approximate location and magnification of the image and whether it is real or virtual and upright or inverted. This knowledge is usually easiest to obtain by constructing a ray diagram. It is always a good idea to use both the graphical method and the algebraic method to locate an image, so that one method serves as a check on the results of the other.

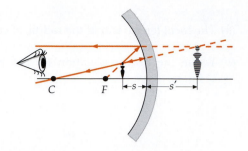

FIGURE 32-17 A virtual image is formed by a concave mirror when the object is inside the focal point. Here the image is located by the radial ray, which is reflected back on itself, and the focal ray, which is reflected parallel to the axis. The two reflected rays appear to diverge from an image point behind the mirror. This image point is found by constructing extensions to the reflected rays.

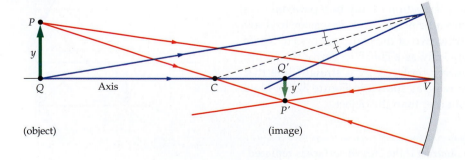

(object)   (image)

FIGURE 32-18 Geometry for showing the lateral magnification. Rays from the top of the object at P, upon reflection, intersect at P'; and rays from the bottom of the object at Q intersect at Q', where points P and P' have vertical positions y and y', respectively. The lateral magnification m is given by the ratio y'/y. In accord with Equation 32-2, $y'/y = -s'/s$. The minus sign results from the fact that y'/y is negative when s and s' are both positive. A negative m means the image is inverted.

---

† An object is real if it is on the same side of the mirror as the incident light.

Convex Mirrors   Figure 32-19 shows a ray diagram for an object in front of a convex mirror. The central ray heading toward the center of curvature C is perpendicular to the mirror and is reflected back on itself. The parallel ray is reflected as if it came from the focal point F behind the mirror. The focal ray (not shown) would be drawn toward the focal point and would be reflected parallel to the axis. We can see from the figure that the image is behind the mirror and is therefore virtual. The image is also upright and smaller than the object.

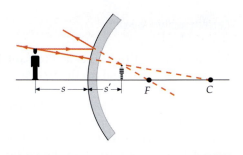

FIGURE 32-19 Ray diagram for an object in front of a convex mirror.

IMAGE IN A CONVEX MIRROR	**EXAMPLE 32-2**

An object 2 cm high is 10 cm from a convex mirror with a radius of curvature of 10 cm. (a) Locate the image and (b) find the height of the image.

PICTURE THE PROBLEM The ray diagram for this problem is the same as shown in Figure 32-19. From this figure, we see that the image is upright, virtual, and smaller than the object. To find the exact location and height of the image, we use the mirror equation, with $s = 10$ cm and $r = -10$ cm.

(a) 1. The image distance s' is related to the object distance s and the focal length f by the mirror equation:

$$\frac{1}{s} + \frac{1}{s'} = \frac{1}{f}$$

2. Calculate the focal length of the mirror:

$$f = \tfrac{1}{2}r = \tfrac{1}{2}(-10 \text{ cm}) = -5 \text{ cm}$$

3. Substitute $s = 10$ cm and $f = -5$ cm into the mirror equation to find the image distance:

$$\frac{1}{10 \text{ cm}} + \frac{1}{s'} = \frac{1}{-5 \text{ cm}}$$

4. Solve for s':

$$s' = \boxed{-3.33 \text{ cm}}$$

(b) 1. The height of the image is m times the height of the object:

$$y' = my$$

2. Calculate the magnification m:

$$m = -\frac{s'}{s} = -\frac{-3.33 \text{ cm}}{10 \text{ cm}} = +0.333$$

3. Use m to find the height of the image:

$$y' = my = (0.333)(2 \text{ cm}) = \boxed{0.666 \text{ cm}}$$

REMARKS The image distance is negative, indicating a virtual image behind the mirror. The magnification is positive and less than one, indicating that the image is upright and smaller than the object.

EXERCISE Find the image distance and magnification for an object 5 cm away from the mirror in Example 32-2, and draw a ray diagram. (*Answer* $s' = -2.5$ cm, $m = +0.5$; the image is upright, virtual, and reduced in size. The ray diagram is shown in Figure 32-20.)

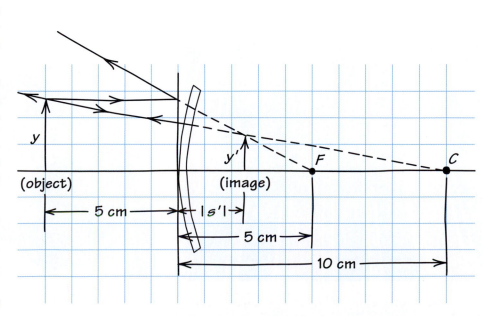

FIGURE 32-20

You have a part-time job at Pleasant Hills Golf Course. The fairway of the 16th hole is horizontal for the first 50 yd and then goes down a not-too-steep hill (Figure 32-21), so the people on the tee cannot see the party in front. To prevent people from driving off the tee into the party in front, a convex mirror is mounted on a pole, enabling golfers on the tee to see if the party in front is out of range. Your boss says that a range finder that works by triangulation could be placed facing the mirror, so the golfers could measure how the image of the party in front is behind the mirror. Then the golfers could be given a chart telling them how far the next party is from the tee. Your boss knows you are taking a physics course, so he asks you to calculate the distance of the image behind the mirror if the next party is 250 yd from the tee. The radius of curvature of the mirror has a magnitude of 20 yd.

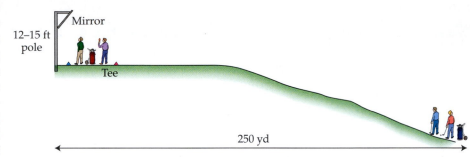

FIGURE 32-21

PICTURE THE PROBLEM The image distance is related to the object distance by the mirror formula, and the focal length of the lens is half the radius of curvature.

1. Use the mirror equation. For a convex mirror, the radius of curvature is negative:

$$\frac{1}{s} + \frac{1}{s'} = \frac{1}{f}$$

and

$$f = \frac{2}{r}$$

so

$$\frac{1}{250 \text{ yd}} + \frac{1}{s'} = \frac{2}{-20 \text{ yd}}$$

2. The image is 9.62 yd behind the mirror:

$$s' = \boxed{-9.62 \text{ yd}}$$

EXERCISE What is the distance to the party in front if the image is 9.75 yd behind the mirror? (*Answer* 390 yd)

(a)

(b)

(*a*) A convex mirror resting on paper with equally spaced parallel stripes. Note the large number of lines imaged in a small space and the reduction in size and distortion in shape of the image. (*b*) A convex mirror is used for security in a store.

32-2 Lenses

Images Formed by Refraction

One end of a long transparent cylinder is machined and pol-
ished to form a convex spherical surface. Figure 32-22 illustrates the formation of an image by refraction at such a surface.
Suppose the cylinder is submerged in a transparent liquid
with index of refraction n_1, and suppose the cylinder is made
of a plastic material with index of refraction n_2, where n_2 is
greater than n_1. Again, only paraxial rays converge to one
point. An equation relating the image distance to the object
distance, the radius of curvature, and the indexes of refraction
can be derived by applying the law of refraction (Snell's law)
to these rays and using small-angle approximations. The
geometry is shown in Figure 32-23. The result is

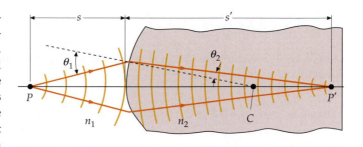

FIGURE 32-22 Image formed by refraction at a spherical
surface between two media where the waves move slower in
the second medium.

$$\frac{n_1}{s} + \frac{n_2}{s'} = \frac{n_2 - n_1}{r}$$

32-6

REFRACTION AT A SINGLE SURFACE

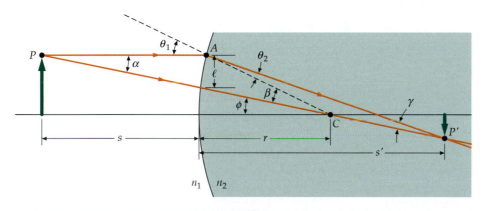

FIGURE 32-23 Geometry for relating the image position to the object position for
refraction at a single spherical surface. The angles θ_1 and θ_2 are related by Snell's law
of refraction: $n_1 \sin \theta_1 = n_2 \sin \theta_2$. The small-angle approximation $\sin \theta = \theta$ gives
$n_1 \theta_1 = n_2 \theta_2$. From triangle ACP', we have $\beta = \theta_2 + \gamma = (n_1/n_2)\theta_1 + \gamma$. We can obtain
another relation for θ_1 from triangle PAC: $\theta_1 = \alpha + \beta$. Eliminate θ_1 from these two
equations: $n_1\alpha + n_1\beta + n_2\gamma = n_2\beta$. Simplify: $n_1\alpha + n_2\gamma = (n_2 - n_1)\beta$. Using the small-
angle approximations $\alpha \approx \ell/s$, $\beta \approx \ell/r$, and $\gamma \approx \ell/s'$ gives Equation 32-6.

In refraction, real images are formed in back of the surface, which we will call
the refracted-light side, whereas virtual images occur on the incident-light side,
in front of the surface. The sign conventions we use for refraction are similar to
those for reflection:

1. s is positive for objects on the incident-light side of the surface.

2. s' is positive for images on the refracted-light side of the surface.

3. r is positive if the center of curvature is on the refracted-light side of
   the surface.

SIGN CONVENTIONS FOR REFRACTION[†]

[†] The sign convention of choice for advanced work on optical design is the Cartesian sign convention. It can
readily be found on the Internet.

We see that parameters s, s', and r are all positive if a real object is in front of a convex refracting surface that forms a real image. A parameter is negative if it does not meet the stated condition for being positive.

MAGNIFICATION BY A REFRACTING SURFACE **E X A M P L E 3 2 - 4** **Try It Yourself**

Derive an expression for the magnification $m = y'/y$ of an image formed by a spherical refracting surface.

PICTURE THE PROBLEM The magnification is the ratio of y' to y. Using Figure 32-18 and Figure 32-23 as guides, draw a ray diagram suitable for this derivation. These heights are related to the tangents of the angles θ_1 and θ_2, as shown in Figure 32-24. The angles are related by Snell's law. For paraxial rays, you can use the approximations $\tan\theta \approx \sin\theta \approx \theta$, and $\cos\theta \approx 1$.

Cover the column to the right and try these on your own before looking at the answers.

Steps

Answers

1. Using Figure 32-18 and Figure 32-23 as guides, draw a ray diagram suitable for this derivation. This drawing should include an object, a real image, a refracting surface, and an axis. Then draw an incident ray from the top of the object to the intersection of the axis with the refracting surface, and draw the refracted ray to the corresponding image point.

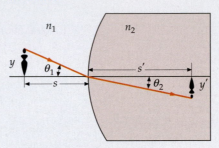

FIGURE 32-24

2. Write expressions for $\tan\theta_1$ and $\tan\theta_2$ in terms of the heights y and $-y'$ and the object and image distances s and s'. (Since y' is negative, use $-y'$, so that $\tan\theta_2$ is positive.)

$$\tan\theta_1 = \frac{y}{s}; \quad \tan\theta_2 = \frac{-y'}{s'}$$

3. Apply the small-angle approximation $\tan\theta \approx \theta$ to your expressions.

$$\theta_1 = \frac{y}{s}; \quad \theta_2 = \frac{-y'}{s'}$$

4. Write Snell's law of refraction relating the angles θ_1 and θ_2 using the small-angle approximation $\sin\theta \approx \theta$.

$$n_1 \sin\theta_1 = n_2 \sin\theta_2$$
$$n_1\theta_1 = n_2\theta_2$$

5. Substitute the expressions for θ_1 and θ_2 found in step 3.

$$n_1\left(\frac{y}{s}\right) = n_2\left(\frac{-y'}{s'}\right)$$

6. Solve for the magnification $m = y'/y$.

$$m = \frac{y'}{y} = -\frac{n_1 s'}{n_2 s}$$

We see from Example 32-4 that the magnification due to refraction at a spherical surface is

$$m = \frac{y'}{y} = -\frac{n_1 s'}{n_2 s} \qquad \qquad 32\text{-}7$$

MAGNIFICATION FOR A REFRACTING BOUNDARY

EXAMPLE   3 2 - 5

Goldie the goldfish is in a 15-cm-radius spherical bowl of water with an index of refraction of 1.33. Fluffy the cat is sitting on the table with her nose 10 cm from the surface of the bowl (Figure 32-25). The light from Fluffy's nose is refracted by the air–water boundary to form an image. Find (*a*) the image distance and (*b*) the magnification of the image of Fluffy's nose. Neglect any effect of the bowl's thin glass wall.

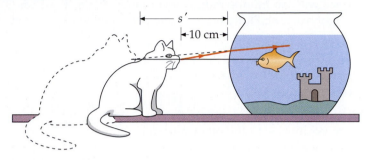

FIGURE 32-25

PICTURE THE PROBLEM We find the image distance s' using Equation 32-6 and the magnification using Equation 32-7. Since we are interested in light that goes from Fluffy's nose to the bowl, it follows that the air–water boundary is convex, and that air is the incident-light side of boundary and water is the refracted-light side of boundary. With these identifications, we have $n_1 = 1$, $n_2 = 1.33$, $s = +10$ cm, and $r = +15$ cm.

(*a*) 1. The equation relating the object distance to the image distance is Equation 32-6:

$$\frac{n_1}{s} + \frac{n_2}{s'} = \frac{n_2 - n_1}{r}$$

2. Identify and assign signs to the parameters in the previous step:

$n_1 = 1, n_2 = 1.33, s = +10$ cm, and $r = +15$ cm

3. Substitute numerical values and solve for s':

$$\frac{1}{10 \text{ cm}} + \frac{1.33}{s'} = \frac{1.33 - 1}{15 \text{ cm}}$$

so

$$s' = \boxed{-17.1 \text{ cm}}$$

(*b*) Substitute numerical values into Equation 32-7 to find the magnification *m*:

$$m = -\frac{n_1 s'}{n_2 s} = -\frac{(1)(-17.1 \text{ cm})}{(1.33)(10 \text{ cm})} = \boxed{1.29}$$

REMARKS Since s' is negative, the image is virtual; that is, the image is on the opposite side of the refracting surface as the refracted light, as shown in Figure 32-25. The fish, Goldie, would see Fluffy to be slightly farther away ($|s'| > s$) than she actually is, and larger ($|m| > 1$) than she actually is. That *m* is positive indicates the image is upright.

EXERCISE If Goldie is 7.5 cm from the side of the bowl nearest Fluffy, find (*a*) the location and (*b*) the magnification of Goldie's image, as seen by Fluffy. (*Answer* $n_1 = 1.33$, $n_2 = 1$, $s = 7.5$ cm, $r = -15$ cm; thus, (*a*) $s' = -6.44$ cm and (*b*) $m = 1.14$. Fluffy sees Goldie to be slightly closer and slightly larger than she actually is.)

EXAMPLE   3 2 - 6

During the summer months, Goldie the fish spends much of her time in a small pond in her owner's backyard. While enjoying a rest at the bottom of the 1-m deep pond, Goldie is being watched by Fluffy the cat, who is perched on a tree limb 3 m above the surface of the pond. How far below the surface is the image of the fish that Fluffy sees? (The index of refraction of water is 1.33.)

PICTURE THE PROBLEM The surface of the pond is a spherical refracting surface with an infinite radius of curvature. Thus, Equation 32-6 applies. Since the light reaching Fluffy originates in the water, use $n_1 = \frac{4}{3}$ and $n_2 = 1$.

1. Draw a picture of the situation. Label the object distance and the indexes of refraction of the media. Goldie is the object (Figure 32-26):

2. Using Equation 32-6, relate the image position s' to the other relevant parameters:

$$\frac{n_1}{s} + \frac{n_2}{s'} = \frac{n_2 - n_1}{r}$$

3. The refracting surface is flat. Using $r = \infty$, solve for s':

$$s' = -\frac{n_2}{n_1} s$$

4. Using the given values $n_1 = 1.33$, $n_2 = 1$, and $s = 1$ m, substitute to obtain s':

$$s' = -\frac{1}{1.33}(1 \text{ m}) = -0.75 \text{ m}$$

That the image is negative means that the image is on the side of the surface opposite the refracted light. That is, it is 0.75 m below the surface.

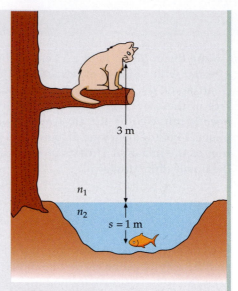

FIGURE 32-26

REMARKS (1) This image can be seen at the calculated position only when the object is viewed from directly overhead, or nearly so. From that observation point the rays are paraxial, a condition necessary for Equation 32-6 to be valid. If Fluffy is standing on the edge of the pond, the rays will not satisfy the paraxial approximation and Equation 32-6 will not correctly predict the location of the image. (2) The distance $(n_2/n_1)s$ is called the apparent depth of the submerged object. If $n_2 = 1$, the apparent depth equals s/n_1.

EXERCISE Draw a ray diagram for the image of Goldie, as described in Example 32-6. That is, draw several rays diverging from an object point P on Goldie, and show that after refracting the rays appear to diverge from an image point P' somewhat above the object point (Figure 32-27).

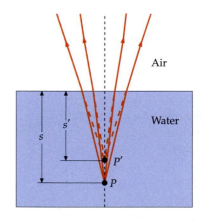

FIGURE 32-27 Ray diagram for the image of an object in water as viewed from directly overhead. The depth of the image is less than the depth of the object.

Thin Lenses

The most important application of Equation 32-6 for refraction at a single surface is finding the position of the image formed by a lens. This is done by considering the refraction at each surface of the lens separately to derive an equation relating the image distance to the object distance, the radius of curvature of each surface of the lens, and the index of refraction of the lens.

We will consider a thin lens of index of refraction n with air on both sides. Let the radii of curvature of the surfaces of the lens be r_1 and r_2. If an object is at a distance s from the first surface (and therefore from the lens), the distance s'_1 of the image due to refraction at the first surface can be found using Equation 32-6:

$$\frac{1}{s} + \frac{n}{s'_1} = \frac{n-1}{r_1}$$

32-8

The light refracted at the first surface is again refracted at the second surface. Figure 32-28 shows the case when the image distance s_1' for the first surface is negative, indicating a virtual image to the left of the surface. Rays in the glass refracted from the first surface diverge as if they came from the image point P_1'. The rays strike the second surface at the same angles as if there were an object at image point P_1'. The image for the first surface therefore becomes the object for the second surface. Since the lens is of negligible thickness, the object distance is equal in magnitude to s_1'. Object distances for objects on the incident-light side of a surface are positive, whereas image distances for images located there are negative. Thus, the object distance for the second surface is $s_2 = -s_1'$.[†] We now write Equation 32-6 for the second surface with $n_1 = n$, $n_2 = 1$, and $s = -s_1'$. The image distance for the second surface is the final image distance s' for the lens:

$$\frac{n}{-s_1'} + \frac{1}{s'} = \frac{1 - n}{r_2}$$

32-9

We can eliminate the image distance for the first surface s_1' by adding Equations 32-8 and 32-9. We obtain

$$\frac{1}{s} + \frac{1}{s'} = (n - 1)\left(\frac{1}{r_1} - \frac{1}{r_2}\right)$$

32-10

Equation 32-10 gives the image distance s' in terms of the object distance s and the properties of the thin lens—r_1, r_2, and n. As with mirrors, the focal length f of a thin lens is defined as the image distance when the object distance is infinite. Setting s equal to infinity and writing f for the image distance s', we obtain

$$\frac{1}{f} = (n - 1)\left(\frac{1}{r_1} - \frac{1}{r_2}\right)$$

32-11

LENS-MAKER'S EQUATION

Equation 32-11 is called the **lens-maker's equation**; it gives the focal length of a thin lens in terms of the properties of the lens. Substituting $1/f$ for the right side of Equation 32-10, we obtain

$$\frac{1}{s} + \frac{1}{s'} = \frac{1}{f}$$

32-12

THIN-LENS EQUATION

This **thin-lens equation** is the same as the mirror equation (Equation 32-4). Recall, however, that the sign conventions for refraction are somewhat different from those for reflection. For lenses, the image distance s' is positive when the image is on the refracted-light side of the lens, that is, when it is on the side opposite the incident-light side. The sign of the focal length (see Equation 32-11) is determined by the sign convention for a single refracting boundary. That is, r is positive if the center of curvature is on the same side of the surface as the refracted light. For a lens like that shown in Figure 32-28, r_1 is positive and r_2 is negative, so f is positive.

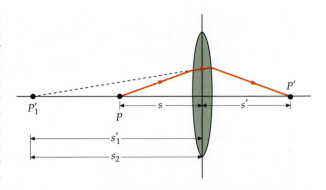

FIGURE 32-28 Refraction occurs at both surfaces of a lens. Here, the refraction at the first surface leads to a virtual image at P_1'. The rays strike the second surface as if they came from P_1'. Image distances are negative when the image is on the incident-light side of the surface, whereas object distances are positive for objects located there. Thus, $s_2 = -s_1'$ is the object distance for the second surface of the lens.

[†] If s_1' were positive, the rays would be converging as they strike the second surface. The object for the second surface would then be a virtual object located to the right of the second surface. This object would be a virtual object. Again, $s_2 = -s_1'$.

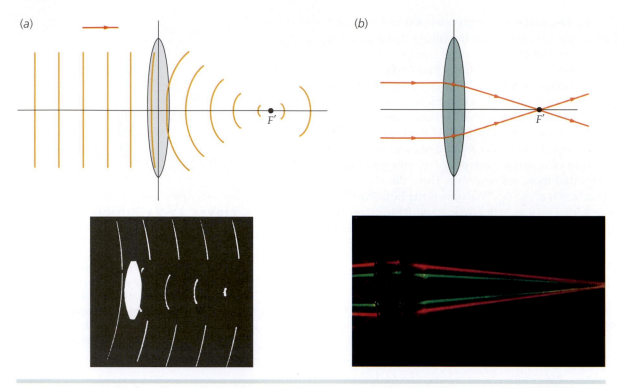

FIGURE 32-29 (*a*) *Top:* Wavefronts for plane waves striking a converging lens. The central part of the wavefront is retarded more by the lens than the outer part, resulting in a spherical wave that converges at the focal point F'. *Bottom:* Wavefronts passing through a lens, shown by a photographic technique called *light-in-flight recording* that uses a pulsed laser to make a hologram of the wavefronts of light. (*b*) *Top:* Rays for plane waves striking a converging lens. The rays are bent at each surface and converge at the focal point. *Bottom:* A photograph of rays focused by a converging lens.

Figure 32-29*a* shows the wavefronts of plane waves incident on a double convex lens. The central part of the wavefront strikes the lens first. Since the wave speed in the lens is less than that in air (assuming $n > 1$), the central part of the wavefront lags behind the outer parts, resulting in a spherical wavefront that converges at the focal point F'. The rays for this situation are shown in Figure 32-29*b*. Such a lens is called a **converging lens.** Since its focal length as calculated from Equation 32-11 is positive, it is also called a **positive lens.** Any lens that is thicker in the middle than at the edges is a converging lens (providing that the index of refraction of the lens is greater than that of the surrounding medium). Figure 32-30 shows the wavefronts and rays for plane waves incident on a double concave lens. In this case, the outer part of the wavefronts lag behind the central parts, resulting in outgoing spherical waves that diverge from a focal point on the incident-light side of the lens. The focal length of this lens is negative. Any lens that is thinner in the middle than at the edges is a **diverging,** or **negative,** lens (providing that the index of refraction of the lens is greater than that of the surrounding medium).

FIGURE 32-30 (*a*) Wavefronts for plane waves striking a diverging lens. Here, the outer part of the wavefront is retarded more than the central part, resulting in a spherical wave that diverges as it moves out, as if it came from the focal point F' to the left of the lens. (*b*) Rays for plane waves striking the same diverging lens. The rays are bent outward and diverge, as if they came from the focal point F'. (*c*) A photograph of rays passing through a diverging lens.

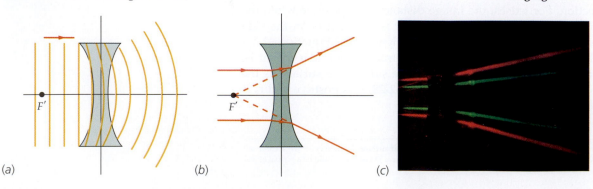

(*a*) (*b*) (*c*)

THE LENS-MAKER'S FORMULA   **EXAMPLE 32-7**

A double convex, thin glass lens with index of refraction $n = 1.5$ has radii of curvature of magnitude 10 cm and 15 cm, as shown in Figure 32-31. Find its focal length.

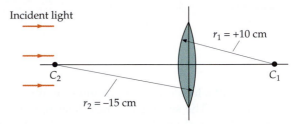

FIGURE 32-31

PICTURE THE PROBLEM We can find the focal length using the lens-maker's equation (Equation 32-11). Here, light is incident on the surface with the smaller radius of curvature. The center of curvature of this surface, C_1, is on the refracted-light side of the lens; thus, $r_1 = +10$ cm. For the second surface, the center of curvature, C_2, is on the incident-light side; therefore, $r_2 = -15$ cm.

Numerical substitution in Equation 32-11 yields the focal length f:

$$\frac{1}{f} = (n - 1)\left(\frac{1}{r_1} - \frac{1}{r_2}\right)$$

$$= (1.5 - 1)\left(\frac{1}{10 \text{ cm}} - \frac{1}{-15 \text{ cm}}\right) = 0.5\left(\frac{1}{6 \text{ cm}}\right)$$

$$f = \boxed{12 \text{ cm}}$$

REMARKS Note that both surfaces tend to converge the light rays; therefore, they both make a positive contribution to the focal length of the lens.

EXERCISE A double convex thin lens has an index of refraction $n = 1.6$ and radii of curvature of equal magnitude. If its focal length is 15 cm, what is the magnitude of the radius of curvature of each surface? (*Answer* 18 cm)

EXERCISE Show that if you reverse the direction of the incoming light for the lens shown in Example 32-7, so that it is incident on the surface with the greater radius of curvature, you get the same result for the focal length.

If parallel light strikes the lens of Example 32-7 from the left, it is focused at a point 12 cm to the right of the lens; whereas if parallel light strikes the lens from the right, it is focused at 12 cm to the left of the lens. Both of these points are focal points of the lens. Using the reversibility property of light rays, we can see that light diverging from a focal point and striking a lens will leave the lens as a parallel beam, as shown in Figure 32-32. In a particular lens problem in which the direction of the incident light is specified, the object point for which light emerges as a parallel beam is called the **first focal point** F, and the point at which parallel light is focused is called the **second focal point** F'. For a positive lens, the first focal point is on the incident-light side and the second focal point is on the refracted-light side. If parallel light is incident on the lens at a small angle with the axis, as in Figure 32-33, it is focused at a point in the **focal plane** a distance f from the lens.

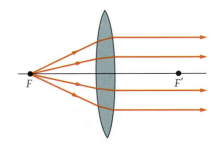

FIGURE 32-32 Light rays diverging from the focal point of a positive lens emerge parallel to the axis.

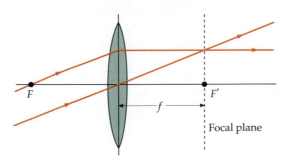

FIGURE 32-33 Parallel rays incident on the lens at an angle to its axis are focused at a point in the focal plane of the lens.

The reciprocal of the focal length is called the **power of a lens.** When the focal length is expressed in meters, the power is given in reciprocal meters, called **diopters** (D):

$$P = \frac{1}{f} \qquad\qquad\qquad\qquad \text{32-13}$$

The power of a lens measures its ability to focus parallel light at a short distance from the lens. The shorter the focal length, the greater the power. For example, a lens with a focal length of 25 cm = 0.25 m has a power of 4 D. A lens with a focal length of 10 cm = 0.10 m has a power of 10 D. Since the focal length of a diverging lens is negative, its power is negative.

POWER OF A LENS

E X A M P L E   3 2 - 8

The lens shown in Figure 32-34 has an index of refraction of 1.5 and radii of curvature of magnitude 10 cm and 13 cm. Find (*a*) its focal length and (*b*) its power.

PICTURE THE PROBLEM For the orientation of the lens relative to the incident light shown in Figure 32-34, the radius of curvature of the first surface is $r_1 = +10$ cm and that of the second surface is $r_2 = +13$ cm.

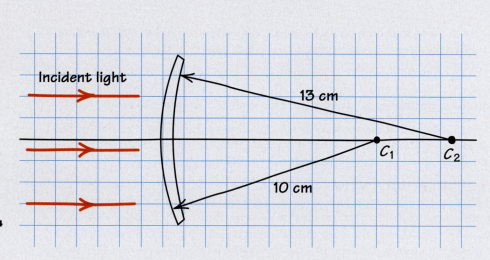

FIGURE 32-34

1. Calculate f from the lens-maker's equation using the given value of n and the values of r_1 and r_2 for the orientation shown:

$$\frac{1}{f} = (n - 1)\left(\frac{1}{r_1} - \frac{1}{r_2}\right)$$

$$= (1.5 - 1)\left(\frac{1}{10 \text{ cm}} - \frac{1}{13 \text{ cm}}\right)$$

$$f = \boxed{86.7 \text{ cm}}$$

2. The power is the reciprocal of the focal length expressed in meters:

$$P = \frac{1}{f} = \frac{1}{0.867 \text{ m}} = \boxed{1.15 \text{ D}}$$

REMARKS We obtain the same result no matter which surface the light strikes first.

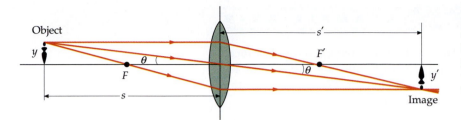

FIGURE 32-35 Ray diagram for a thin converging lens. We assume that all the bending of light takes place at the central plane. The ray through the center is undeflected because the lens surfaces there are parallel and close together.

In laboratory experiments involving lenses, it is usually much easier to measure the focal length than to calculate the focal length from the radii of curvature of the surfaces.

Ray Diagrams for Lenses

As with images formed by mirrors, it is convenient to locate the images of lenses by graphical methods. Figure 32-35 illustrates the graphical method for a thin converging lens. We consider the rays to bend at the plane through the center of the lens. The three principal rays are as follows:

1. The **parallel ray,** drawn parallel to the axis. The emerging ray is directed toward (or away from) the second focal point of the lens.

2. The **central ray,** drawn through the center (the vertex) of the lens. This ray is undeflected. (The faces of the lens are parallel at this point, so the ray emerges in the same direction but displaced slightly. Since the lens is thin, the displacement is negligible.)

3. The **focal ray,** drawn through the first focal point.[†] This ray emerges parallel to the axis.

PRINCIPAL RAYS FOR A THIN LENS

The weight and bulk of a large-diameter lens can be reduced by constructing the lens from annular segments at different angles so that light from a point is refracted by the segments into a parallel beam. Such an arrangement is called a Fresnel lens. Several Fresnel lenses are used in this lighthouse to produce intense parallel beams of light from a source at the focal point of the lenses. The illuminated surface of an overhead projector is a Fresnel lens.

These three rays converge to the image point, as shown in Figure 32-35. In this case, the image is real and inverted. From the figure, we have $\tan \theta = y/s = -y'/s'$. The lateral magnification is then

$$m = \frac{y'}{y} = -\frac{s'}{s}$$

32-14

This expression is the same as the expression for mirrors. Again, a negative magnification indicates that the image is inverted. The ray diagram for a diverging lens is shown in Figure 32-36.

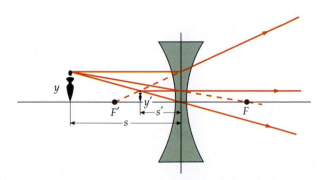

FIGURE 32-36 Ray diagram for a diverging lens. The parallel ray is bent away from the axis, as if it came from the second focal point F'. The ray toward the first focal point F emerges parallel to the axis.

[†] The focal ray is drawn toward the first focal point for a diverging lens.

EXAMPLE 3 2 - 9

An object 1.2 cm high is placed 4 cm from a double convex lens with a focal length of 12 cm. Locate the image both graphically and algebraically, state whether the image is real or virtual, and find its height. Place an eye on the figure positioned and oriented so as to view the image.

1. Draw the parallel ray. This ray leaves the object parallel to the axis, then is bent by the lens to pass through the second focal point, F' (Figure 32-37):

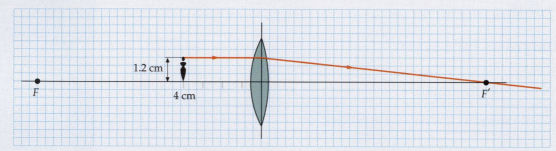

FIGURE 32-37

2. Draw the central ray, which passes undeflected through the center of the lens. Since the two rays are diverging on the refracted-light side, we extend them back to the incident-light side to find the image (Figure 32-38):

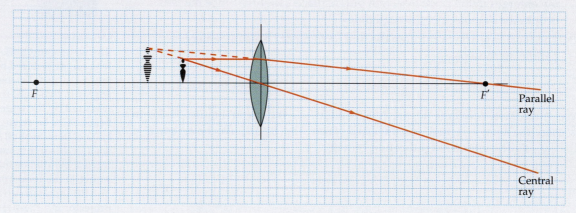

FIGURE 32-38

3. As a check, we also draw the focal ray. This ray leaves the object on a line passing through the first focal point, then emerges parallel to the axis. Note that the image is virtual, upright, and enlarged (Figure 32-39):

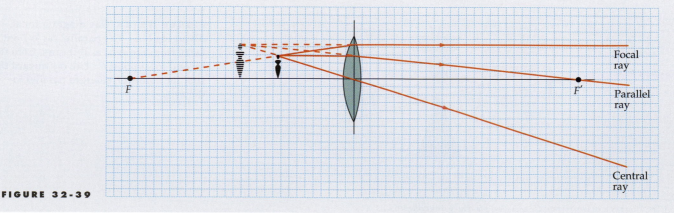

FIGURE 32-39

4. The eye must be positioned so the light from the image enters the eye.

5. We now verify the results of the ray diagram algebraically. First, find the image distance using Equation 32-12:

$$\frac{1}{4 \text{ cm}} + \frac{1}{s'} = \frac{1}{12 \text{ cm}}$$

$$\frac{1}{s'} = \frac{1}{12 \text{ cm}} - \frac{1}{4 \text{ cm}} = -\frac{1}{6 \text{ cm}}$$

$$s' = -6 \text{ cm}$$

6. The height of the image is found from the height of the object and the magnification:

$$h' = mh$$

7. The magnification m is given by Equation 32-14:

$$m = -\frac{s'}{s} = -\frac{-6 \text{ cm}}{4 \text{ cm}} = \boxed{+1.5}$$

8. Using this result we find the height of the image, h':

$$h' = mh = (1.5)(1.2 \text{ cm}) = \boxed{1.8 \text{ cm}}$$

REMARKS Note the agreement between the algebraic and ray diagram results. Algebraically, we find that the image is 6 cm from the lens on the incident-light side (since $s' < 0$); that is, the image is 2 cm to the left of the object. Since $m > 0$, it follows that the image is upright, and because $m > 1$, the image is enlarged. It is good practice to process lens problems both graphically and algebraically and to compare the results.

EXERCISE An object is placed 15 cm from a double convex lens of focal length 10 cm. Find the image distance and the magnification. Draw a ray diagram. Is the image real or virtual? Is the image upright or inverted? (*Answer* $s' = 30$ cm, $m = -2$; real, inverted)

EXERCISE Work the previous exercise for an object placed 5 cm from a lens with a focal length of 10 cm. (*Answer* $s' = -10$ cm, $m = 2$; virtual, upright)

Combinations of Lenses

If we have two or more thin lenses, we can find the final image produced by the system by finding the image distance for the first lens and then using it, along with the distance between the lenses, to find the object distance for the second lens. That is, we consider each image, whether it is real or virtual—and whether it is actually formed or not—as the object for the next lens.

IMAGE FORMED BY A SECOND LENS   **EXAMPLE   32-10**

A second lens of focal length +6 cm is placed 12 cm to the right of the lens in Example 32-9. Locate the final image.

PICTURE THE PROBLEM The principal rays used to locate the image of the first lens will not necessarily be principal rays for the second lens. In this example, however, we have chosen the position of the second lens (Figure 32-40a) so that the parallel ray for the first lens turns out to be the central ray for the second lens. Also, the focal ray for the first lens emerges parallel to the axis and is therefore the parallel ray for the second lens. If additional principal rays for the second lens are needed, we simply draw them from the image formed by the first lens. For example, in Figure 32-40*b* we added such a ray, drawn from the first image through the first focal point F_2 of the second lens.

FIGURE 32-40

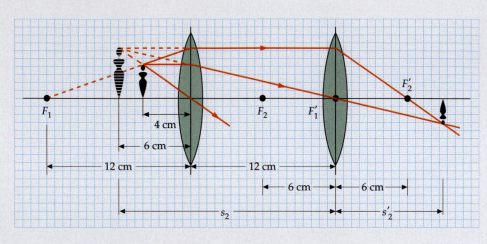

(a)

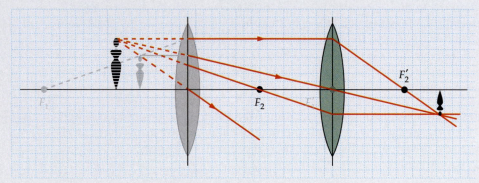

(b)

Algebraically we use $s_2 = 18$ cm, because the first image is 6 cm to the left of the first lens and therefore 18 cm to the left of the second lens.

Use $s_2 = 18$ cm and $f = 6$ cm to calculate s_2':

$$\frac{1}{s_2} + \frac{1}{s_2'} = \frac{1}{f_2}$$

$$\frac{1}{18 \text{ cm}} + \frac{1}{s_2'} = \frac{1}{6 \text{ cm}}$$

$$s_2' = \boxed{9 \text{ cm}}$$

A COMBINATION OF TWO LENSES   **EXAMPLE 32-11**   **Try It Yourself**

Two lenses, each of focal length 10 cm, are 15 cm apart. Find the final image of an object 15 cm from one of the lenses.

PICTURE THE PROBLEM Use a ray diagram to find the location of the image formed by lens 1. When these rays strike lens 2 they are further refracted, leading to the final image. More accurate results are obtained algebraically using the thin-lens equation for both lens 1 and lens 2.

Cover the column to the right and try these on your own before looking at the answers.

| Steps | Answers |

1. Draw the (*a*) parallel, (*b*) central, and (*c*) focal rays for lens 1 (Figure 32-41). If lens 2 did not alter these rays, they would form an image at I_1.

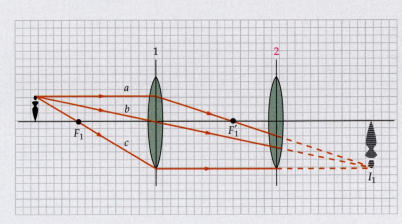

FIGURE 32-41

2. To locate the final image, add three principal rays (*d*, *e*, and *f*) for lens 2. The intersection of these rays gives the image location (Figure 32-42).

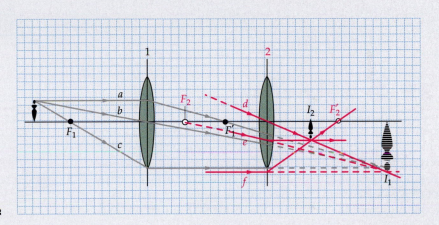

FIGURE 32-42

3. To proceed algebraically, use the thin-lens equation to find the image distance s_1' produced by lens 1.

$s_1' = 30 \text{ cm}$

4. For lens 2, the image, I_1 is 15 cm from the lens on the refracted-light side; hence, $s_2 = -15$ cm. Use this to find the final image distance s_2'.

$s_2' = \boxed{6 \text{ cm}}$

REMARKS From the ray diagram we see that the final image is real, inverted, and slightly reduced.

Compound Lenses

When two thin lenses of focal lengths f_1 and f_2 are placed together, the effective focal length of the combination f_{eff} is given by

$$\frac{1}{f_{\text{eff}}} = \frac{1}{f_1} + \frac{1}{f_2}$$

32-15

as is shown in the following Example 32-12. The power of two lenses in contact is given by

$$P_{\text{eff}} = P_1 + P_2$$

32-16

For two lenses very close together, derive the relation $\dfrac{1}{f_{eff}} = \dfrac{1}{f_1} + \dfrac{1}{f_2}$.

PICTURE THE PROBLEM Apply the thin-lens equation to each lens using the fact that the distance between the lenses is zero, so the object distance for the second lens is the negative of the image distance for the first lens.

Cover the column to the right and try these on your own before looking at the answers.

Steps	Answers
1. Write the thin-lens equation for lens 1.	$\dfrac{1}{s} + \dfrac{1}{s_1'} = \dfrac{1}{f_1}$
2. Using $s_2 = -s_1'$, write the thin-lens equation for lens 2.	$\dfrac{1}{-s_1'} + \dfrac{1}{s'} = \dfrac{1}{f_2}$
3. Add your two resulting equations to eliminate s_1'.	$\boxed{\dfrac{1}{s} + \dfrac{1}{s'} = \dfrac{1}{f_1} + \dfrac{1}{f_2} = \dfrac{1}{f_{eff}}}$

*32-3   Aberrations

When all the rays from a point object are not focused at a single image point, the resultant blurring of the image is called **aberration.** Figure 32-43 shows rays from a point source on the axis traversing a thin lens with spherical surfaces. Rays that strike the lens far from the axis are bent much more than are the rays near the axis, with the result that not all the rays are focused at a single point. Instead, the image appears as a circular disk. The **circle of least confusion** is at point C, where the diameter is minimum. This type of aberration in a lens is called **spherical aberration;** it is the same as the spherical aberration of mirrors discussed in Section 32-1. Similar but more complicated aberrations called *coma* (for the comet-shaped image) and *astigmatism* occur when objects are off axis. The aberration in the shape of the image of an extended object that occurs, because the magnification depends on the distance of the object point from the axis, is called **distortion.** We will not discuss these aberrations further, except to point out that they do not arise from any defect in the lens or mirror but instead result from the application of the laws of refraction and reflection to spherical surfaces. These aberrations are not evident in our simple equations, because we used small-angle approximations in the derivation of these equations.

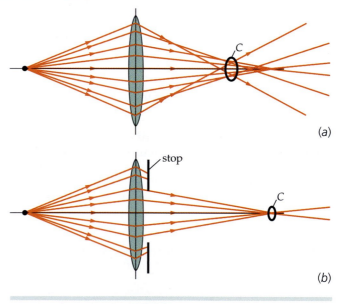

(a)

(b)

FIGURE 32-43 Spherical aberration in a lens. (*a*) Rays from a point object on the axis are not focused at a point. (*b*) Spherical aberration can be reduced by using a stop to block off the outer parts of the lens, but this also reduces the amount of light reaching the image.

Some aberrations can be eliminated or partially corrected by using nonspherical surfaces for mirrors or lenses, but nonspherical surfaces are usually much more difficult and costly to produce than spherical surfaces. One example of a nonspherical reflecting surface is the parabolic mirror illustrated in Figure 32-44. Rays that are parallel to the axis of a parabolic surface are reflected and focused at a common point, no matter how far the rays are from the axis. Parabolic reflecting surfaces are sometimes used in large astronomical telescopes, which need a large reflecting surface to gather as much light as possible to make the image as intense as possible (reflecting telescopes are described in the upcoming optional Section 32-4). Satellite dishes use parabolic surfaces to focus microwaves from communications satellites. A parabolic surface can also be used in a searchlight to produce a parallel beam of light from a small source placed at the focal point of the surface.

An important aberration found with lenses but not found with mirrors is **chromatic aberration,** which is due to variations in the index of refraction with wavelength. From Equation 32-11, we can see that the focal length of a lens depends on its index of refraction and is therefore different for different wavelengths. Since n is slightly greater for blue light than for red light, the focal length for blue light will be shorter than the focal length for red light. Because chromatic aberration does not occur for mirrors, many large telescopes use a large mirror instead of the large, light-gathering (objective) lens.

Chromatic aberration and other aberrations can be partially corrected by using combinations of lenses instead of a single lens. For example, a positive lens and a negative lens of greater focal length can be used together to produce a converging lens system that has much less chromatic aberration than a single lens of the same focal length. The lens of a good camera typically contains six elements to correct the various aberrations that are present.

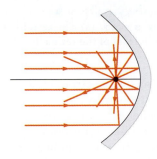

FIGURE 32-44 A parabolic mirror focuses all rays parallel to the axis to a single point with no spherical aberration.

*32-4 Optical Instruments

*The Eye

The optical system of prime importance is the eye, which is shown in Figure 32-45. Light enters the eye through a variable aperture, the pupil. The light is focused by the cornea, with assistance from the lens, on the retina, which has a film of nerve fibers covering the back surface. The retina contains tiny sensing structures called *rods* and *cones,* which detect the light and transmit the information along the optic nerve to the brain. The shape of the crystalline lens can be altered slightly by the action of the ciliary muscle. When the eye is focused on an object far away, the muscle is relaxed and the cornea–lens system has its maximum focal length, about 2.5 cm, which is the distance from the cornea to the retina. When the object is brought closer to the eye, the ciliary muscle increases the curvature of the lens slightly, thereby decreasing its focal length, so that the image is again focused on the retina. This process is called *accommodation.* If the object is too close to the eye, the lens cannot focus the light on the retina and the image is blurred. The closest point for which the lens can focus the image on the retina is called the **near point.** The distance from the eye to the near point varies greatly from one person to another and changes with age. At 10 years, the near point may be as close as 7 cm, whereas at 60 years it may recede to 200 cm because of the loss of flexibility of the lens. The standard value taken for the near point is 25 cm.

If the eye underconverges, resulting in the images being focused behind the retina, the person is said to be farsighted. A farsighted person can see distant objects where little convergence is required, but has trouble seeing close objects. Farsightedness is corrected with a converging (positive) lens (Figure 32-46).

On the other hand, the eye of a nearsighted person overconverges and focuses light from distant objects in front of the retina. A nearsighted person can see nearby objects for which the widely diverging incident rays can be focused on

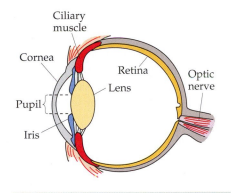

FIGURE 32-45 The human eye. The amount of light entering the eye is controlled by the iris, which regulates the size of the pupil. The lens thickness is controlled by the ciliary muscle. The cornea and lens together focus the image on the retina, which contains approximately 125 million receptors, called rods and cones, and approximately 1 million optic-nerve fibers.

(a)

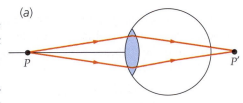

(b)

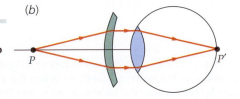

FIGURE 32-46 (*a*) A farsighted eye focuses rays from a nearby object to a point behind the retina. (*b*) A converging lens corrects this defect by bringing the image onto the retina. These diagrams, and those following, are drawn as if all the focusing of the eye is done at the lens; whereas, in fact, the lens and cornea system act more like a spherical refracting surface than a thin lens.

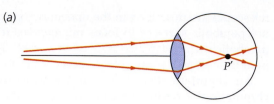
(a)

FIGURE 32-47 (*a*) A nearsighted eye focuses rays from a distant object to a point in front of the retina. (*b*) A diverging lens corrects this defect.

the retina, but has trouble seeing distant objects clearly. Nearsightedness is corrected with a diverging (negative) lens (Figure 32-47).

Another common defect of vision is astigmatism, which is caused by the cornea being not quite spherical but having a different curvature in one plane than in another. This results in a blurring of the image of a point object into a short line. Astigmatism is corrected by glasses using lenses of cylindrical rather than spherical shape.

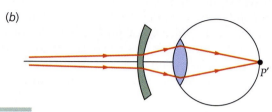
(b)

FOCAL LENGTH OF THE CORNEA–LENS SYSTEM **EXAMPLE 32-13**

Both a thin lens and a spherical mirror have a focal length given by the formula $\frac{1}{s} + \frac{1}{s'} = \frac{1}{f}$, where f is a constant. Using the same formula, we define the focal length of a spherical refracting surface. However, in this case, the focal length is not constant but depends upon s. By how much does the focal length of the cornea–lens system of the eye change if the object is moved from infinity to the near point at 25 cm? Assume that all the focusing is done at the cornea, and that the distance from the cornea to the retina is 2.5 cm.

PICTURE THE PROBLEM With the object at infinity, the focal length is 2.5 cm. We use the thin-lens equation to calculate the focal length when $s = 25$ cm and $s' = 2.5$ cm.

1. Use the thin-lens equation to calculate f:

$$\frac{1}{f} = \frac{1}{s} + \frac{1}{s'} = \frac{1}{25 \text{ cm}} + \frac{1}{2.5 \text{ cm}}$$

$$= \frac{1}{25 \text{ cm}} + \frac{10}{25 \text{ cm}} = \frac{11}{25 \text{ cm}}$$

so

$$f = 2.27 \text{ cm}$$

2. Subtract the original focal length of 2.5 cm to find the change: $\Delta f = 2.27 \text{ cm} - 2.5 \text{ cm} = \boxed{-0.23 \text{ cm}}$

REMARKS In terms of the power of the cornea–lens system, when the focal length is 2.5 cm = 0.025 m for distant objects, the power is $P = 1/f = 40$ D. When the focal length is 2.27 cm, the power is 44 D.

EXERCISE Find the change in the focal length of the eye when an object originally at 4 m is brought to 40 cm from the eye. (Assume that the distance from the cornea to the retina is 2.5 cm.) (*Answer* -0.13 cm)

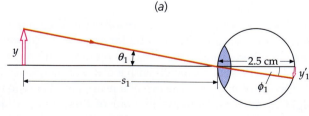
(a)

The apparent size of an object is determined by the actual size of the image on the retina. The larger the image on the retina, the greater the number of rods and cones activated. From Figure 32-48, we see that the size of the image on the retina is greater when the

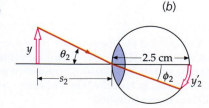
(b)

FIGURE 32-48 (*a*) A distant object of height *y* looks small because the image on the retina is small. (*b*) When the same object is closer, it looks larger because the image on the retina is larger.

object is close than it is when the object is far away. The apparent size of an object is thus greater when it is closer to the eye. The image size is proportional to the angle θ subtended by the object at the eye. For Figure 32-48,

$$\phi \approx \frac{y'}{2.5 \text{ cm}} \quad \text{and} \quad \theta \approx \frac{y}{s}$$

32-17

for small angles. Applying the law of refraction gives $n_{\text{Air}} \sin \theta = n \sin \phi$, where $n_{\text{Air}} = 1.00$ and n is the refractive index inside the eye. For small angles this becomes

$$\theta \approx n\phi$$

32-18

Combining Equations 32-17 and 32-18 gives

$$\frac{y}{s} \approx n\frac{y'}{2.5 \text{ cm}}, \quad \text{or} \quad y' \approx \frac{2.5 \text{ cm}}{n}\frac{y}{s}$$

32-19

The size of the image on the retina is proportional to the size of the object and inversely proportional to its distance from the eye. Since the near point is the closest point to the eye for which a sharp image can be formed on the retina, the distance to the near point is called the *distance of most distinct vision.*

READING GLASSES   **E X A M P L E   3 2 - 1 4**

The near-point distance of a person's eye is 75 cm. With a reading glasses lens placed a negligible distance from the eye, the near-point distance of the lens–eye system is 25 cm. That is, if an object is placed 25 cm in front of the lens, then the lens forms a virtual image of the object a distance 75 cm in front of the lens. (*a*) What power is the reading glasses lens and (*b*) what is the lateral magnification of the image formed by the lens? (*c*) Which produces the bigger image on the retina, (1) the object at the near point of, and viewed by, the unaided eye or (2) the object at the near point of the lens–eye system and viewed through the lens that is immediately in front of the eye?

PICTURE THE PROBLEM A near-point distance of the lens–eye system of 25 cm means the lens forms a virtual image 75 cm in front of the lens if an object is placed 25 cm in front of the lens. Figure 32-49a shows a diagram of an object 25 cm from a converging lens that produces a virtual, upright image at $s' = -75$ cm. Figure 32-49b shows the image on the retina formed by the focusing power of the eye.

FIGURE 32-49

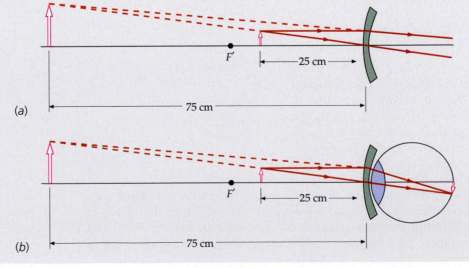

(a)

(b)

1. Use the thin-lens equation with $s = 25$ cm and $s' = -75$ cm to calculate the power, $1/f$:

$$\frac{1}{f} = \frac{1}{s} + \frac{1}{s'} = \frac{1}{25 \text{ cm}} + \frac{1}{-75 \text{ cm}}$$

$$= \frac{2}{75 \text{ cm}} = \frac{2}{0.75 \text{ m}} = \boxed{2.67 \text{ D}}$$

2. Using $m = -s'/s$, find m:

$$m = -\frac{s'}{s} = -\frac{-75 \text{ cm}}{25 \text{ cm}} = \boxed{3}$$

3. In both cases, the rays entering the eye appear to diverge from an image 75 cm in front of the eye. However, with the lens in place, the image there is larger by a factor of 3:

$\boxed{\text{Option 2}}$

REMARKS (1) If your near point is 75 cm, you are farsighted. To read a book you must hold it at least 75 cm from your eye to be able to focus on the print. The image of the print on your retina is then very small. The reading glasses lens produces an image also 75 cm from your eye, and this image is three times larger than the actual print. Thus, looking through the lens, the image of the print on the retina is larger by a factor of 3. (2) In this example, the distance from the lens to the eye was negligible. The results are slightly different if this distance is not negligible and is factored into the calculations.

EXERCISE Calculate the power of the eye for which the near point is 75 cm and the cornea–retina distance is 2.5 cm, and calculate the combined power of the lens and eye when they are in contact. Compare this with the power of a lens for which $s' = 2.5$ cm, when $s = 25$ cm. (*Answer* $P_{eye} = 41.33$ D; $P_c = 41.33$ D + 2.67 D = 44 D; $P = 44$ D)

*The Simple Magnifier

We saw in Example 32-14 that the *apparent* size of an object can be increased by using a converging lens placed next to the eye. A converging lens is called a **simple magnifier** if it is placed next to the eye and if the object is placed closer to the lens than its focal length, as was the case for the lens in Example 32-14. In that example, the lens formed a virtual image at the near point of the eye, the same location that the object must be placed for best viewing by the unaided eye. So, with the lens in place, the magnitude of the image distance $|s'|$ was greater than the object distance s, so the image seen by the eye is magnified by $m = |s'|/s$. If the actual height of the object was y, then the height y' of the image formed by the lens would have been my. To the eye, this image subtended an angle θ (Figure 32-50) given approximately by

$$\theta = \frac{my}{|s'|} = m\frac{y}{|s'|} = \frac{|s'|}{s}\frac{y}{|s'|} = \frac{y}{s}$$

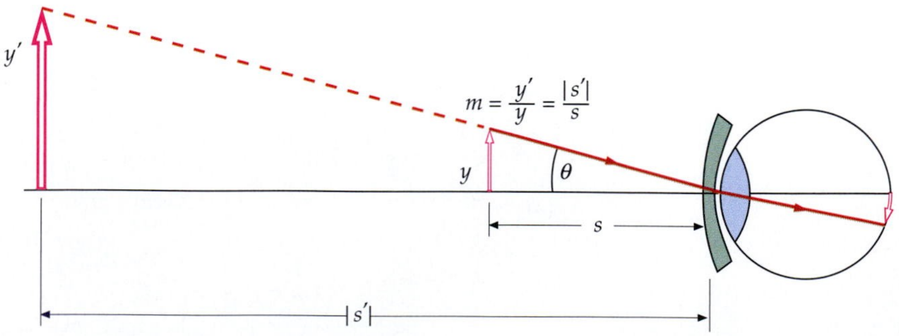

FIGURE 32-50

$$m = \frac{y'}{y} = \frac{|s'|}{s}$$

which is the *very same angle* the object would subtend were the lens removed while the object and the eye were left in place. That is, the apparent size of the image seen by the eye through the lens is the same as the apparent size of the object that would be seen by the eye were the lens removed (assuming the eye could focus at that distance). Thus, the apparent size of the object seen through the lens is inversely proportional to the distance from the object to the eye to the distance from the object to the eye with the lens in place. The smaller s is, the larger the subtended angle θ and the larger the apparent size of the object.

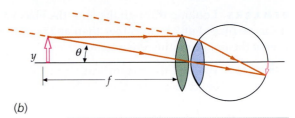

(a)   (b)

In Figure 32-51a, a small object of height y is at the near point of the eye at a distance x_{np}. The angle subtended, θ_0, is given approximately by

$$\theta_0 = \frac{y}{x_{np}}$$

In Figure 32-51b, a converging lens of focal length f that is smaller than x_{np} is placed a negligible distance in front of the eye, and the object is placed in the focal plane of the lens. The rays emerge from the lens parallel, indicating that the image is located an infinite distance in front of the lens. The parallel rays are focused by the relaxed eye on the retina. The angle subtended by this image is equal to the angle subtended by the object (assuming that the lens is a negligible distance from the eye). The angle subtended by the object is approximately

$$\theta = \frac{y}{f}$$

The ratio θ / θ_0 is called the *angular magnification* or *magnifying power M* of the lens:

$$M = \frac{\theta}{\theta_0} = \frac{x_{np}}{f} \qquad\qquad \text{32-20}$$

Simple magnifiers are used as eyepieces (called oculars) in microscopes and telescopes to view the image formed by another lens or lens system. To correct aberrations, combinations of lenses that result in a short positive focal length may be used in place of a single lens, but the principle of the simple magnifier is the same.

FIGURE 32-51 (*a*) An object at the near point subtends an angle θ_0 at the naked eye. (*b*) When the object is at the focal point of the converging lens, the rays emerge from the lens parallel and enter the eye as if they came from an object a very large distance away. The image can thus be viewed at infinity by the relaxed eye. When f is less than the near-point distance, the converging lens allows the object to be brought closer to the eye. This increases the angle subtended by the object to θ, thereby increasing the size of the image on the retina.

Angular Magnification of a Simple Magnifier   **E X A M P L E   3 2 - 1 5**   **Try It Yourself**

A person with a near point of 25 cm uses a 40-D lens as a simple magnifier. What angular magnification is obtained?

PICTURE THE PROBLEM The angular magnification is found from the focal length f (Equation 32-20), which is the reciprocal of the power.

Cover the column to the right and try these on your own before looking at the answers.

Steps   **Answers**

1. Calculate the focal length of the lens. $f = 2.5$ cm

2. Use your result from step 1 and incorporate the result into $M = \boxed{10}$
   Equation 32-20 to calculate the angular magnification.

REMARKS Looking through the lens, the object appears 10 times larger because it can be placed at 2.5 cm rather than at 25 cm from the eye, thus increasing the size of the image on the retina tenfold.

EXERCISE What is the magnification in this example if the near point of the person is 30 cm rather than 25 cm? (*Answer* $M = 12$)

*The Compound Microscope

The compound microscope (Figure 32-52) is used to look at very small objects at short distances. In its simplest form, it consists of two converging lenses. The lens nearest the object, called the **objective,** forms a real image of the object. This image is enlarged and inverted. The lens nearest the eye, called the **eyepiece** or **ocular,** is used as a simple magnifier to view the image formed by the objective.

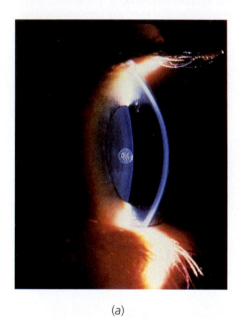

(a)

(b)

(*a*) The human eye in profile. (*b*) The lens of the eye is kept in place by the ciliary muscle (shown here in the upper left), which rings the lens. When the ciliary muscle contracts, the lens tends to bulge. The greater lens curvature enables the eye to focus on nearby objects. (*c*) Some of the 120 million rods and 7 million cones in the eye, magnified approximately 5000 times. The rods (the more slender of the two) are more sensitive in dim light, whereas the cones are more sensitive to color. The rods and cones form the bottom layer of the retina and are covered by nerve cells, blood vessels, and supporting cells. Most of the light entering the eye is reflected or absorbed before reaching the rods and cones. The light that does reach them triggers electrical impulses along nerve fibers that ultimately reach the brain. (*d*) A neural net used in the vision system of certain robots. Loosely modeled on the human eye, it contains 1920 sensors.

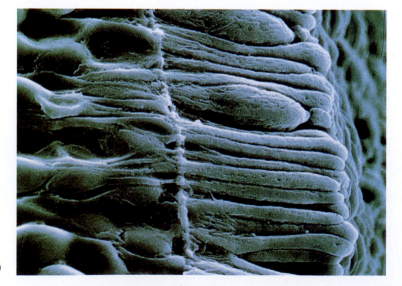

(c)

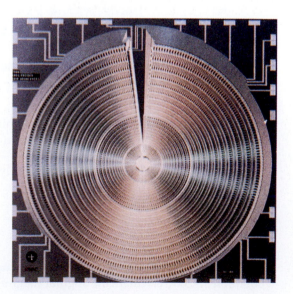

(d)

The eyepiece is placed so that the image formed by the objective falls at the first focal point of the eyepiece. The light from each point on the object thus emerges from the eyepiece as a parallel beam, as if it were coming from a point a great distance in front of the eye. (This is commonly called *viewing the image at infinity.*)

The distance between the second focal point of the objective and the first focal point of the eyepiece is called the **tube length** L. It is typically fixed at approximately 16 cm. The object is placed just outside the first focal point of the objective so that an enlarged image is formed at the first focal point of the eyepiece a distance $L + f_o$ from the objective, where f_o is the focal length of the objective. From Figure 32-52, $\tan \beta = y/f_o = -y'/L$. The lateral magnification of the objective is therefore

$$m_o = \frac{y'}{y} = -\frac{L}{f_o} \qquad \text{32-21}$$

The angular magnification of the eyepiece (from Equation 32-20) is

$$M_e = \frac{x_{np}}{f_e}$$

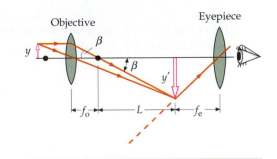

FIGURE 32-52 Schematic diagram of a compound microscope consisting of two positive lenses, the objective of focal length f_o and the ocular, or eyepiece, of focal length f_e. The real image of the object formed by the objective is viewed by the eyepiece, which acts as a simple magnifier. The final image is at infinity.

where x_{np} is the near-point distance of the viewer, and f_e is the focal length of the eyepiece. The magnifying power of the compound microscope is the product of the lateral magnification of the objective and the angular magnification of the eyepiece:

$$M = m_o M_e = -\frac{L}{f_o}\frac{x_{np}}{f_e} \qquad \text{32-22}$$

MAGNIFYING POWER OF A MICROSCOPE

THE COMPOUND MICROSCOPE   **EXAMPLE 32-16**

A microscope has an objective lens of focal length 1.2 cm and an eyepiece of focal length 2 cm. These lenses are separated by 20 cm. (a) Find the magnifying power if the near point of the viewer is 25 cm. (b) Where should the object be placed if the final image is to be viewed at infinity?

(a) 1. The magnifying power is given by Equation 32-22:

$$M = -\frac{L}{f_o}\frac{x_{np}}{f_e}$$

2. The tube length L is the distance between the lenses minus the focal distances:

$$L = 20 \text{ cm} - 2 \text{ cm} - 1.2 \text{ cm} = 16.8 \text{ cm}$$

3. Substitute this value for L and the given values of $x_{np}, f_o,$ and f_e to calculate M:

$$M = -\frac{L}{f_o}\frac{x_{np}}{f_e} = -\frac{16.8 \text{ cm}}{1.2 \text{ cm}}\frac{25 \text{ cm}}{2 \text{ cm}}$$

$$= \boxed{-175}$$

(b) 1. Calculate the object distance s in terms of the image distance for the objective s' and the focal length f_o:

$$\frac{1}{s} + \frac{1}{s'} = \frac{1}{f_o}$$

2. From Figure 32-52, the image distance for the image of the objective is $f_o + L$:

$$s' = f_o + L = 1.2 \text{ cm} + 16.8 \text{ cm}$$
$$= 18 \text{ cm}$$

3. Substitute to calculate s:

$$\frac{1}{s} + \frac{1}{18 \text{ cm}} = \frac{1}{1.2 \text{ cm}}$$

$$s = \boxed{1.29 \text{ cm}}$$

REMARKS The object should thus be placed at 1.29 cm from the objective or 0.09 cm outside its first focal point.

*The Telescope

A telescope is used to view objects that are far away and are often large. The telescope works by creating a real image of the object that is much closer than the object. The astronomical telescope, illustrated schematically in Figure 32-53, consists of two positive lenses—an objective lens that forms a real, inverted image and an eyepiece that is used as a simple magnifier to view that image. Because the object is very far away, the image of the objective lies in the focal plane of the objective, and the image distance equals the focal length f_o. The image formed by the objective is much smaller than the object because the object distance is much larger than the focal length of the objective. For example, if we are looking at the moon, the image of the moon formed by the objective is much smaller than the moon itself. The purpose of the objective is not to magnify the object, but to produce an image that is close to us so it can be viewed by the eyepiece. The eyepiece is placed a distance f_e from the image, where f_e is the focal length of the eyepiece, so the final image can be viewed at infinity. Since this image is at the second focal plane of the objective and at the first focal plane of the eyepiece, the objective and eyepiece must be separated by the sum of the focal lengths of the objective and eyepiece, $f_o + f_e$.

The magnifying power of the telescope is the angular magnification θ_e/θ_o, where θ_e is the angle subtended by the final image as viewed through the eyepiece and θ_o is the angle subtended by the object when it is viewed directly by the unaided eye. The angle θ_o is the same as that subtended by the object at the objective shown in Figure 32-53. (The distance from a distant object, such as the moon, to the objective is essentially the same as the distance to the eye.) From this figure, we can see that

$$\tan \theta_o = \frac{y}{s} = -\frac{y'}{f_o} \approx \theta_o$$

where we have used the small-angle approximation $\tan \theta \approx \theta$. The angle θ_e in the figure is that subtended by the image at infinity formed by the eyepiece:

$$\tan \theta_e = \frac{y'}{f_e} \approx \theta_e$$

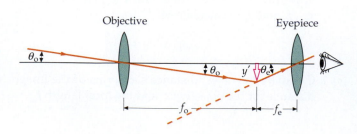

FIGURE 32-53 Schematic diagram of an astronomical telescope. The objective lens forms a real, inverted image of a distant object near its second focal point, which coincides with the first focal point of the eyepiece. The eyepiece serves as a simple magnifier to view the image.

Since y' is negative, θ_e is negative, indicating that the image is inverted. The magnifying power of the telescope is then

$$M = \frac{\theta_e}{\theta_o} = -\frac{f_o}{f_e} \qquad \text{32-23}$$

MAGNIFYING POWER OF A TELESCOPE

From Equation 32-23, we can see that a large magnifying power is obtained with an objective of large focal length and an eyepiece of short focal length.

EXERCISE The world's largest refracting telescope is at the Yerkes Observatory of the University of Chicago at Williams Bay, Wisconsin. The telescope's objective has a diameter of 102 cm and a focal length of 19.5 m. The focal length of the eyepiece is 10 cm. What is its magnifying power? (*Answer* -195)

The main consideration with an astronomical telescope is not its magnifying power but its light-gathering power, which depends on the size of the objective. The larger the objective, the brighter the image. Very large lenses without aberrations are difficult to produce. In addition, there are mechanical problems in supporting very large, heavy lenses by their edges. A reflecting telescope (Figure 32-54 and Figure 32-55) uses a concave mirror instead of a lens for its objective. This offers several advantages. For one, a mirror does not produce chromatic aberration. In addition, mechanical support is much simpler, since the mirror weighs far less than a lens of equivalent optical quality, and the mirror can be supported over its entire back surface. In modern earth-based telescopes, the objective mirror consists of several dozen adaptive mirror segments that can be adjusted individually to correct for minute variations in gravitational stress when the telescope is tilted, and to compensate for thermal expansions and contractions and other changes caused by climatic conditions. In addition, they can adjust to nullify the distortions produced by atmospheric fluctuations.

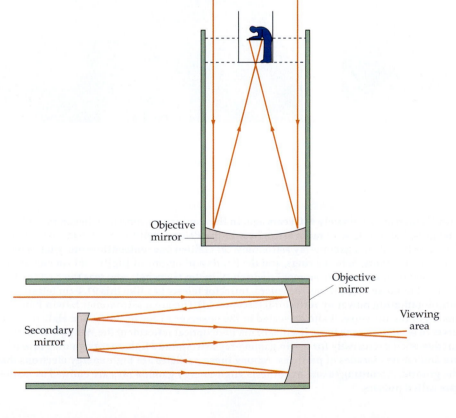

Objective mirror

Objective mirror

Secondary mirror

Viewing area

FIGURE 32-54 A reflecting telescope uses a concave mirror instead of a lens for its objective. Because the viewer compartment blocks off some of the incoming light, the arrangement shown here is used only in telescopes with very large objective mirrors.

FIGURE 32-55 This reflecting telescope has a secondary mirror to redirect the light through a small hole in the objective mirror, thus providing more room for auxiliary instruments in the viewing area.

(a)

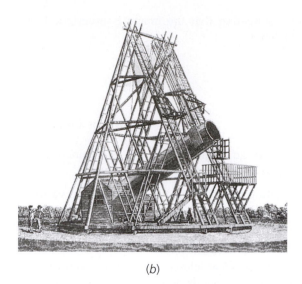

(b)

(c)

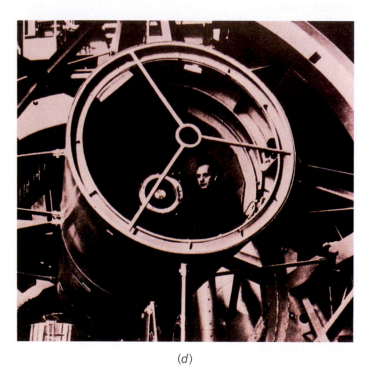

(d)

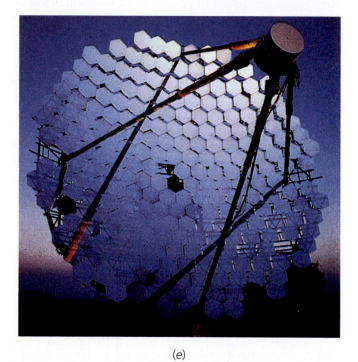

(e)

Astronomy at optical wavelengths began with Galileo approximately 400 years ago. In this century, astronomers began to explore the electromagnetic spectrum at other wavelengths; beginning with radio astronomy in the 1940s, satellite-based X-ray astronomy in the early 1960s, and more recently, ultraviolet, infrared, and gamma-ray astronomy. (*a*) Galileo's seventeenth-century telescope, with which he discovered mountains on the moon, sunspots, Saturn's rings, and the bands and moons of Jupiter. (*b*) An engraving of the reflector telescope built in the 1780s and used by the great astronomer Friedrich Wilhelm Herschel, who was the first to observe galaxies outside our own. (*c*) Because it is difficult to make large, flaw-free lenses, refractor telescopes like this 91.4-cm telescope at Lick Observatory have been superseded in light-gathering power by reflector telescopes. (*d*) The great astronomer Edwin Powell Hubble, who discovered the apparent expansion of the universe, is shown seated in the observer's cage of the 5.08-m Hale reflecting telescope, which is large enough for the observer to sit at the prime focus itself. (*e*) This 10-m optical reflector at the Whipple Observatory in southern Arizona is the largest instrument designed exclusively for use in gamma-ray astronomy. High-energy gamma rays of unknown origin strike the upper atmosphere and create cascades of particles. Among these particles are high-energy electrons that emit Cerenkov radiation observable from the ground. According to one hypothesis, high-energy gamma rays are emitted as matter is accelerated toward ultradense rotating stars called pulsars.

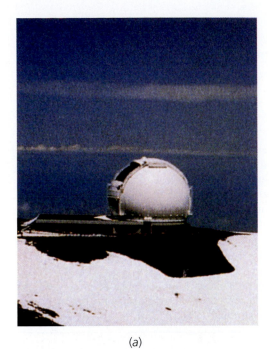

(a)

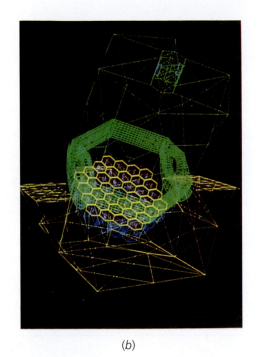

(b)

(c)

(a) The Keck Observatory, atop the inactive volcano of Mauna Kea, Hawaii, houses the world's largest optical telescope. The clear, dry air and lack of light pollution make the remote heights of Mauna Kea an ideal site for astronomical observations. (b) The Keck telescope is composed of 36 hexagonal mirror segments performing together as if they were a single mirror 10-m wide—roughly twice as large as the largest single-mirror telescope presently in operation. (c) Beneath each Keck mirror is a system of computer-controlled sensors and motor-driven actuators that can continuously vary the mirror's shape. These variations, which are sensitive to within 100 nm, enable the system to correct for variations in the alignments of the segments due to minute variations in gravitational stress when the telescope is tilted and to compensate for thermal expansions and contractions and fluctuations caused by gusts of wind on the mountaintop.

The Hubble Space Telescope is high above the atmospheric turbulence that limits the ability of ground-based telescopes to resolve images at optical wavelengths.

Topic	Relevant Equations and Remarks
1. Virtual and Real Images and Objects	
Images	An image is real if actual light rays converge to each image point. This can occur in front of a mirror, or on the refracted-light side of a thin lens or refracting surface. An image is virtual if only extensions of the actual light rays converge to each image point. This can occur behind a mirror or on the incident-light side of a lens or refracting surface.
Virtual object	An object is real if actual light rays diverge from each object point. This can occur only on the incident-light side of a mirror, lens, or refracting surface. A real object is either an actual object or a real image. An object is virtual if only extensions of actual light rays diverge from each object point. This can occur only behind a mirror or on the refracted-light side of a lens or refracting surface.
2. Spherical Mirrors	
Focal length	The focal length is the image distance when the object is at infinity, so the incident light is parallel to the axis:
Mirror equation (for locating an image)	$$\frac{1}{s} + \frac{1}{s'} = \frac{1}{f} \qquad \text{32-4}$$ where $$f = \frac{r}{2} \qquad \text{32-3}$$
Lateral magnification	$$m = \frac{y'}{y} = -\frac{s'}{s} \qquad \text{32-5}$$
Ray diagrams	Images can be located by a ray diagram using any two paraxial rays. The parallel, focal, and radial rays are the easiest to draw: 1. The parallel ray, drawn parallel to the axis, is reflected through the focal point. 2. The focal ray, drawn through the focal point, is reflected parallel to the axis. 3. The radial ray, drawn through the center of curvature, strikes the mirror perpendicular to its surface and is thus reflected back on itself.
Sign conventions for reflection	1. s is positive if the object is on the incident-light side of the mirror. 2. s' is positive if the image is on the reflected-light side of the mirror. 3. r (and f) is positive if the mirror is concave so the center of curvature is on the reflected-light side of the mirror.
3. Images Formed by Refraction	
Refraction at a single surface	$$\frac{n_1}{s} + \frac{n_2}{s'} = \frac{n_2 - n_1}{r} \qquad \text{32-6}$$ where n_1 is the index of refraction of the medium on the incident-light side of the surface.
Magnification	$$m = \frac{y'}{y} = -\frac{n_1 s'}{n_2 s} \qquad \text{32-7}$$

Sign conventions for refraction	1. s is positive for objects on the incident-light side of the surface.
	2. s' is positive for images on the refracted-light side of the surface.
	3. r is positive if the center of curvature is on the refracted-light side of the surface.

4. Thin Lenses

Focal length (lens-maker's equation)	$$\frac{1}{f} = (n - 1)\left(\frac{1}{r_1} - \frac{1}{r_2}\right)$$	32-11

A positive lens ($f > 0$) is a converging lens (like a double convex lens). A negative lens ($f < 0$) is a diverging lens (like a double concave lens).

First and second focal points	Incident rays parallel to the axis emerge directed either toward or away from the *first focal point F'*. Incident rays directed either toward or away from the *second focal point F* emerge parallel with the axis.

Power	$$P = \frac{1}{f}$$	32-13

Thin-lens equation (for locating image)	$$\frac{1}{s} + \frac{1}{s'} = \frac{1}{f}$$	32-12

Magnification	$$m = \frac{y'}{y} = -\frac{s'}{s}$$	32-14

Ray diagrams	Images can be located by a ray diagram using any two paraxial rays. The parallel, central, and focal rays are the easiest to draw:
	1. The parallel ray, drawn parallel to the axis, emerges directed toward (or away from) the second focal point of the lens.
	2. The central ray, drawn through the center of the lens, is not deflected.
	3. The focal ray, drawn through (or toward) the first focal point, emerges parallel to the axis.

Sign conventions for lenses	The sign conventions are the same as for refraction at a spherical surface.

5. *Aberrations

Blurring of the image of a single object point is called aberration. Spherical aberration results from the fact that a spherical surface focuses only paraxial rays (those that travel close to the axis) at a single point. Nonparaxial rays are focused at nearby points depending on the angle made with the axis. Spherical aberration can be reduced by blocking the rays farthest from the axis. This, of course, reduces the amount of light reaching the image.

Chromatic aberration, which occurs with lenses but not mirrors, results from the variation in the index of refraction with wavelength. Lens aberrations are most commonly reduced by using a series of lens elements.

6. *The Eye

The cornea–lens system of the eye focuses light on the retina, where it is sensed by the rods and cones that send information along the optic nerve to the brain. When the eye is relaxed, the focal length of the cornea–lens system is about 2.5 cm, which is the distance to the retina. When objects are brought near the eye, the lens changes shape to decrease the overall focal length so that the image remains focused on the retina. The closest distance for which the image can be focused on the retina is called the near point, typically about 25 cm. The apparent size of an object depends on the size of the image on the retina. The closer the object, the larger the image on the retina and therefore the larger the apparent size of the object.

7. *The Simple Magnifier

A simple magnifier consists of a lens with a positive focal length that is smaller than the near point.

Magnifying power (angular magnification)	$M = \dfrac{\theta}{\theta_0} = \dfrac{x_{np}}{f}$	32-20

8. ***The Compound Microscope**

The compound microscope is used to look at very small objects that are nearby. It consists of two converging lenses (or lens systems), an objective, and an ocular or eyepiece. The object to be viewed is placed just outside the focal point of the objective, which forms an enlarged image of the object at the focal plane of the eyepiece. The eyepiece acts as a simple magnifier to view the final image.

Magnifying power (angular magnification)	$M = m_o M_e = -\dfrac{L}{f_o}\dfrac{x_{np}}{f_e}$	32-22

where L is the tube length, the distance between the second focal point of the objective and the first focal point of the eyepiece.

9. ***The Telescope**

The telescope is used to view objects that are far away. The objective of the telescope forms a real image of the object that is much smaller than the object but much closer. The eyepiece is then used as a simple magnifier to view the image. A reflecting telescope uses a mirror for its objective.

Magnifying power (angular magnification)	$M = \dfrac{\theta_e}{\theta_o} = -\dfrac{f_o}{f_e}$	32-23

PROBLEMS

- • Single-concept, single-step, relatively easy
- •• Intermediate-level, may require synthesis of concepts
- ••• Challenging
- **SSM** Solution is in the *Student Solutions Manual*
- **iSOLVE** Problems available on iSOLVE online homework service
- **iSOLVE✓** These "Checkpoint" online homework service problems ask students additional questions about their confidence level, and how they arrived at their answer.

In a few problems, you are given more data than you actually need; in a few other problems, you are required to supply data from your general knowledge, outside sources, or informed estimates.

Conceptual Problems

1 • Can a virtual image be photographed?

2 • Suppose each axis of a coordinate system, like the one shown in Figure 32-4, is painted a different color. One photograph is taken of the coordinate system and another is taken of its image in a plane mirror. Is it possible to tell that one of the photographs is of a mirror image, rather than both being photographs of the real coordinate system from different angles?

3 •• **iSOLVE** True or False

(a) The virtual image formed by a concave mirror is always smaller than the object.

(b) A concave mirror always forms a virtual image.

(c) A convex mirror never forms a real image of a real object.

(d) A concave mirror never forms an enlarged real image of an object.

4 •• **SSM** Under what condition will a concave mirror produce (a) an upright image, (b) a virtual image, (c) an image smaller than the object, and (d) an image larger than the object?

5 •• Answer Problem 4 for a convex mirror.

6 •• Convex mirrors are often used for rearview mirrors on cars and trucks to give a wide-angle view. Below the mirror is written, "Warning, objects are closer than they appear." Yet, according to a ray diagram, such as the diagram shown in Figure 32-19, the image distance for distant objects is much smaller than the object distance. Why then do they appear more distant?

7 •• As an object is moved from a great distance toward the focal point of a concave mirror, the image moves from (a) a great distance toward the focal point and is always real. (b) the focal point to a great distance from the mirror and is always real. (c) the focal point toward the center of curvature of the mirror and is always real. (d) the focal point to a great distance from the mirror and changes from a real image to a virtual image.

8 • A bird above the water is viewed by a scuba diver submerged beneath the water's surface directly below the bird. Does the bird appear to the diver to be closer to or farther from the surface than it actually is?

9 • **SSM** Under what conditions will the focal length of a thin lens be (a) positive and (b) negative? Consider both the case where the index of refraction of the lens is greater than and less than the surrounding medium.

10 • The focal length of a simple lens is different for different colors of light. Why?

11 •• An object is placed 40 cm from a lens of focal length −10 cm. The image is (a) real, inverted, and diminished. (b) real, inverted, and enlarged. (c) virtual, inverted, and diminished. (d) virtual, upright, and diminished. (e) virtual, upright, and enlarged.

12 •• If a real object is placed just inside the focal point of a converging lens, the image is (a) real, inverted, and enlarged. (b) virtual, upright, and diminished. (c) virtual, upright, and enlarged. (d) real, inverted, and diminished.

13 • Both the eye and the camera work by forming real images, the eye's image forming on the retina and the camera's image forming on the film. Explain the difference between the ways in which these two systems accommodate objects located at different object distances and still keep a focused image.

14 • **SSM** If an object is placed 25 cm from the eye of a farsighted person who does not wear corrective lenses, a sharp image is formed (a) behind the retina, and the corrective lens should be convex. (b) behind the retina, and the corrective lens should be concave. (c) in front of the retina, and the corrective lens should be convex. (d) in front of the retina, and the corrective lens should be concave.

15 •• Myopic (nearsighted) persons sometimes claim to see better under water without corrective lenses. Why? (a) The accommodation of the eye's lens is better under water. (b) Refraction at the water–cornea interface is less than at the air–cornea interface. (c) Refraction at the water–cornea interface is greater than at the air–cornea interface. (d) No reason; the effect is only an illusion and not really true.

16 •• A nearsighted person who wears corrective lenses would like to examine an object at close distance. Identify the correct statement. (a) The corrective lenses give an enlarged image and should be worn while examining the object. (b) The corrective lenses give a reduced image of the object and should be removed. (c) The corrective lenses result in a magnification of unity; it does not matter whether they are worn or removed.

17 • **SSM** The image of a real object formed by a convex mirror (a) is always real and inverted. (b) is always virtual and enlarged. (c) may be real. (d) is always virtual and diminished.

18 • The image of a real object formed by a converging lens (a) is always real and inverted. (b) is always virtual and enlarged. (c) may be real. (d) is always virtual and diminished.

19 • The glass of a converging lens has an index of refraction of 1.6. When the lens is in air, its focal length is 30 cm. If the lens is immersed in water, its focal length will be (a) greater than 30 cm. (b) less than 30 cm. (c) the same as before, 30 cm. (d) negative.

20 •• True or false:
(a) A virtual image cannot be displayed on a screen.
(b) A negative image distance implies that the image is virtual.
(c) All rays parallel to the axis of a spherical mirror are reflected through a single point.
(d) A diverging lens cannot form a real image from a real object.
(e) The image distance for a positive lens is always positive.

21 • **SSM** Explain the following statement: A microscope is an object magnifier, but a telescope is an angle magnifier.

Estimation and Approximation

22 •• The lens-maker's equation contains three design parameters, the index of refraction of the lens and the radius of curvature of its two surfaces. Thus, there are many ways to design a lens with a particular focal length. Use the lens-maker's equation to design three different thin converging lenses, each with a focal length of 27 cm and each made from glass with an index of refraction of 1.6. Draw a sketch of each of your designs.

23 •• Repeat Problem 22, but for a diverging lens of focal length −27 cm.

24 •• **SSM** Estimate the maximum value that could be usefully obtained for the magnifying power of a simple magnifier, using Equation 32-20. (*Hint: Think about the smallest focal length lens that could be made from glass and still be used as a magnifier.*)

Plane Mirrors

25 • The image of the point object *P* in Figure 32-56 is viewed by an eye, as shown. Draw a bundle of rays from the object that reflect from the mirror and enter the eye. For this object position and mirror, indicate the region of space in which the eye can be positioned and still see the image.

FIGURE 32-56 Problem 25

26 • A person 1.62 m tall wants to be able to see her full image in a plane mirror. (a) What must be the minimum height of the mirror? (b) How far above the floor should the mirror be placed, assuming that the top of the person's head is 15 cm above her eye level? Draw a ray diagram.

27 • **SSM** Two plane mirrors make an angle of 90°. The light from an object point that is arbitrarily positioned in front of the mirrors produces images at three locations. For each image location, draw two rays from the object that, upon one or two reflections, appear to come from the image location.

28 • (a) Two plane mirrors make an angle of 60° with each other. Draw a sketch to show the location of all the images formed of a point object on the bisector of the angle between the mirrors. (b) Repeat for an angle of 120°.

29 •• When two plane mirrors are parallel, such as on opposite walls in a barber shop, multiple images arise because each image in one mirror serves as an object for the other mirror. A point object is placed between parallel mirrors separated by 30 cm. The object is 10 cm in front of the left mirror and 20 cm in front of the right mirror. (*a*) Find the distance from the left mirror to the first four images in that mirror. (*b*) Find the distance from the right mirror to the first four images in that mirror.

Spherical Mirrors

30 •• [SSM] A concave spherical mirror has a radius of curvature of 24 cm. Draw ray diagrams to locate the image (if one is formed) for an object at a distance of (*a*) 55 cm, (*b*) 24 cm, (*c*) 12 cm, and (*d*) 8 cm from the mirror. For each case, state whether the image is real or virtual; upright or inverted; and enlarged, reduced, or the same size as the object.

31 • Use the mirror equation (Equation 32-4) to locate and describe the images for the object distances and mirror of Problem 30.

32 •• Repeat Problem 30 for a convex mirror with the same radius of curvature.

33 • Use the mirror equation (Equation 32-4) to locate and describe the images for the object distances and convex mirror of Problem 32.

34 • Show that a convex mirror cannot form a real image of a real object, no matter where the object is placed, by showing that s' is always negative for a positive s.

35 • [SSM] [iSOLVE]✓ A dentist wants a small mirror that will produce an upright image with a magnification of 5.5 when the mirror is located 2.1 cm from a tooth. (*a*) What should the radius of curvature of the mirror be? (*b*) Should the mirror be concave or convex?

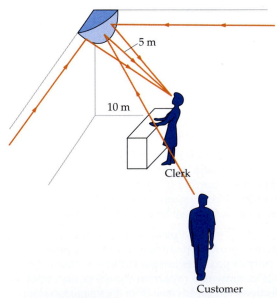

FIGURE 32-57 Problem 36

36 •• [iSOLVE]✓ Convex mirrors are used in stores to provide a wide angle of surveillance for a reasonable mirror size. The mirror shown in Figure 32-57 allows a clerk 5 m away

from the mirror to survey the entire store. It has a radius of curvature of 1.2 m. (*a*) If a customer is 10 m from the mirror, how far from the mirror surface is his image? (*b*) Is the image in front of or behind the mirror? (*c*) If the customer is 2 m tall, how tall is his image?

37 •• [iSOLVE] A certain telescope uses a concave spherical mirror of radius 8 m. Find the location and diameter of the image of the moon formed by this mirror. The moon has a diameter of 3.5×10^6 m and is 3.8×10^8 m from the earth.

38 •• A concave spherical mirror has a radius of curvature of 6 cm. A point object is on the axis 9 cm from the mirror. Construct a precise ray diagram showing rays from the object that make angles of 5°, 10°, 30°, and 60° with the axis, which strike the mirror and are reflected back across the axis. (Use a compass to draw the mirror, and use a protractor to measure the angles needed to find the reflected rays.) What is the spread δx of the points where these rays cross the axis?

39 •• [SSM] A concave mirror has a radius of curvature 6 cm. Draw rays parallel to the axis at 0.5 cm, 1 cm, 2 cm, and 4 cm above the axis, and find the points at which the reflected rays cross the axis. (Use a compass to draw the mirror and a protractor to find the angle of reflection for each ray.) (*a*) What is the spread δx of the points where these rays cross the axis? (*b*) By what percentage could this spread be reduced if the edge of the mirror were blocked off so that parallel rays more than 2 cm from the axis could not strike the mirror?

40 •• [iSOLVE] An object located 100 cm from a concave mirror forms a real image 75 cm from the mirror. The mirror is then turned around so that its convex side faces the object. The mirror is moved so that the image is now 35 cm behind the mirror. How far was the mirror moved? Was it moved toward the object or away from the object?

41 •• Parallel light from a distant object strikes the large mirror shown in Figure 32-58 at $r = 5$ m and is reflected by the small mirror that is 2 m from the large mirror. The small mirror is actually spherical, not planar as shown. The light is focused at the vertex of the large mirror. (*a*) What is the radius of curvature of the small mirror? (*b*) Is the mirror convex or concave?

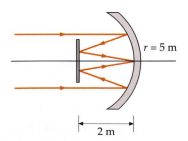

FIGURE 32-58 Problem 41

Images Formed by Refraction

42 • A sheet of paper with writing on it is protected by a thick glass plate having an index of refraction of 1.5. If the plate is 2 cm thick, at what distance beneath the top of the plate does the writing appear when it is viewed from directly overhead?

43 • **ISOLVE ✓** A fish is 10 cm from the front surface of a fish bowl of radius 20 cm. (*a*) Where does the fish appear to be to someone viewing the fish from in front of the bowl? (*b*) Where does the fish appear to be when it is 30 cm from the front surface of the bowl?

44 •• **SSM** A very long glass rod of 3.5-cm diameter has one end ground to a convex spherical surface of radius 7.2 cm. Its index of refraction is 1.5. (*a*) A point object in air is on the axis of the rod 35 cm from the surface. Find the image and state whether the image is real or virtual. Repeat (*b*) for an object 6.5 cm from the surface and (*c*) for an object very far from the surface. Draw a ray diagram for each case.

45 •• At what distance from the glass rod of Problem 44 should the object be placed, so that the light rays in the rod are parallel? Draw a ray diagram for this situation.

46 •• Repeat Problem 44 for a glass rod with a concave hemispherical surface of radius −7.2 cm.

47 •• Repeat Problem 44 when the glass rod and the objects are immersed in water.

48 •• Repeat Problem 44 for a glass rod with a concave hemispherical surface of radius −7.5 cm when the glass rod and the objects are immersed in water.

49 •• **SSM** **ISOLVE** A glass rod 96 cm long with an index of refraction of 1.6 has its ends ground to convex spherical surfaces of radii 8 cm and 16 cm. A point object is in air on the axis of the rod 20 cm from the end with the 8-cm radius. (*a*) Find the image distance due to refraction at the first surface. (*b*) Find the final image due to refraction at both surfaces. (*c*) Is the final image real or virtual?

50 •• Repeat Problem 49 for a point object in air on the axis of the glass rod 20 cm from the end with the 16-cm radius.

Thin Lenses

51 • The following thin lenses are made of glass with an index of refraction of 1.5. Make a sketch of each lens, and find its focal length in air: (*a*) double convex, $r_1 = 15$ cm and $r_2 = -26$ cm; (*b*) plano-convex, $r_1 = \infty$ and $r_2 = -15$ cm; (*c*) double concave, $r_1 = -15$ cm and $r_2 = +15$ cm; and (*d*) plano-concave, $r_1 = \infty$ and $r_2 = +26$ cm.

52 • Find the focal length of a glass lens of index of refraction 1.62 that has a concave surface with radius of magnitude 100 cm and a convex surface with a radius of magnitude 40 cm.

53 • **SSM** A double concave lens of index of refraction 1.45 has radii of magnitudes 30 cm and 25 cm. An object is located 80 cm to the left of the lens. Find (*a*) the focal length of the lens, (*b*) the location of the image, and (*c*) the magnification of the image. (*d*) Is the image real or virtual? Is the image upright or inverted?

54 • **ISOLVE** The following thin lenses are made of glass of index of refraction 1.6. Make a sketch of each lens, and find its focal length in air: (*a*) $r_1 = 20$ cm and $r_2 = 10$ cm, (*b*) $r_1 = 10$ cm and $r_2 = 20$ cm, and (*c*) $r_1 = -10$ cm and $r_2 = -20$ cm.

55 • **SSM** **ISOLVE ✓** An object 3 cm high is placed 25 cm in front of a thin lens of power 10 D. Draw a precise ray diagram to find the position and the size of the image, and check your results using the thin-lens equation.

56 • Repeat Problem 55 for an object 1.5 cm high that is placed 20 cm in front of a thin lens of power 10 D.

57 • Repeat Problem 55 for an object 1.5 cm high that is placed 20 cm in front of a thin lens of power −10 D.

58 •• (*a*) What is meant by a negative object distance? How can a negative object distance occur? Find the image distance and the magnification and state whether the image is virtual or real and upright or inverted for a thin lens in air when (*b*) $s = -20$ cm, $f = +20$ cm and (*c*) $s = -10$ cm, $f = -30$ cm. Draw a ray diagram for each of these cases.

59 •• **SSM** **ISOLVE ✓** Two converging lenses, each of focal length 10 cm, are separated by 35 cm. An object is 20 cm to the left of the first lens. (*a*) Find the position of the final image using both a ray diagram and the thin-lens equation. (*b*) Is the image real or virtual? Is the image upright or inverted? (*c*) What is the overall lateral magnification of the image?

60 •• Rework Problem 59 for a second lens that is a diverging lens of focal length −15 cm.

61 •• (*a*) Show that to obtain a magnification of magnitude *m* with a converging thin lens of focal length *f*, the object distance must be given by $s = (m - 1)f/m$. (*b*) A camera lens with a 50-mm focal length is used to take a picture of a person 1.75 m tall. How far from the camera should the person stand so that the image size is 24 mm?

62 •• A converging lens has a focal length of $f = 12$ cm. (*a*) Using a spreadsheet program or graphing calculator, plot the image distance *s'* as a function of the object distance *s*, for values of *s* ranging from $s = 1.1f$ to $s = 10f$. (*b*) On the same graph used in Part (*a*), but using a different *y* axis, plot the magnification of the lens as a function of the object distance *s*. (*c*) What type of image is produced for this range of object distances, real or virtual, upright or inverted? (*d*) Discuss the significance of any asymptotic limits your graph has.

63 •• A converging lens has a focal length of $f = 12$ cm. (*a*) Using a spreadsheet program or graphing calculator, plot the image distance *s'* as a function of the object distance *s*, for values of *s* ranging from $s = 0.01f$ to $s = 0.9f$. (*b*) On the same graph used in Part (*a*), but using a different *y* axis, plot the magnification of the lens as a function of the object distance *s*. (*c*) What type of image is produced for this range of object distances, real or virtual, upright or inverted? (*d*) Discuss the significance of any asymptotic limits your graph has.

64 •• **SSM** An object is 15 cm in front of a positive lens of focal length 15 cm. A second positive lens of focal length 15 cm is 20 cm from the first lens. Find the final image and draw a ray diagram.

65 •• Rework Problem 64 for a second lens with a focal length of −15 cm.

66 ••• In a convenient form of the thin-lens equation used by Newton, the object and image distances are measured from the focal points. Show that if $x = s - f$ and $x' = s' - f$, the thin-lens equation can be written as $xx' = f^2$, and the lateral magnification is given by $m = -x'/f = -f/x$. Indicate x and x' on a sketch of a lens.

67 ••• In *Bessel's method* for finding the focal length f of a lens, an object and a screen are separated by distance D, where $D > 4f$. It is then possible to place the lens at either of two locations, both between the object and the screen, so that there is an image of the object on the screen, in one case magnified and in the other case reduced. Show that if the distance between the two lens locations is given by L, that

$$f = \frac{D^2 - L^2}{4D}$$

(*Hint: Refer to Figure 32-59.*) The two lens locations are such that the object distance with the lens in the one setting is equal to the image distance with the lens in the other setting and vice versa.

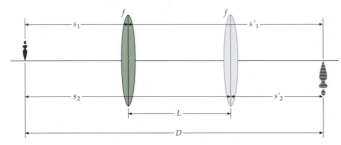

FIGURE 32-59 Problems 67 and 68

68 •• An optician uses *Bessel's method* to find the focal length of a lens, as described in Problem 67. The object-to-image distance was set at 1.7 m. The position of the lens was then adjusted to get a sharp image on the screen. A second image was found when the lens was moved a distance of 72 cm. (*a*) Using the result from Problem 67, find the focal length of the lens. (*b*) What were the two locations of the lens with respect to the object?

69 ••• An object is 17.5 cm to the left of a lens of focal length 8.5 cm. A second lens of focal length -30 cm is 5 cm to the right of the first lens. (*a*) Find the distance between the object and the final image formed by the second lens. (*b*) What is the overall magnification? (*c*) Is the final image real or virtual? Is the final image upright or inverted?

*Aberrations

70 • **SSM** Chromatic aberration is a common defect of (*a*) concave and convex lenses. (*b*) concave lenses only. (*c*) concave and convex mirrors. (*d*) all lenses and mirrors.

71 • True or false:

(*a*) Aberrations occur only for real images.
(*b*) Chromatic aberration does not occur with mirrors.

72 • A double convex lens of radii $r_1 = +10$ cm and $r_2 = -10$ cm is made from glass with indexes of refraction of 1.53 for blue light and 1.47 for red light. Find the focal length of this lens for (*a*) red light and (*b*) blue light.

*The Eye

73 •• **SSM** The Model Eye I: A simple model for the eye is a lens with variable power P located a fixed distance d in front of a screen, with the space between the lens and the screen filled by air. Refer to Figure 32-60. The "eye" can focus for all values of s such that $x_{np} \leq s \leq x_{fp}$. This "eye" is said to be normal if it can focus on very distant objects. (*a*) Show that for a normal "eye," the minimum value of P is

$$P_{min} = \frac{1}{d}$$

(*b*) Show that the maximum value of P is

$$P_{max} = \frac{1}{x_{np}} + \frac{1}{d}$$

(*c*) The difference $A = P_{max} - P_{min}$ is called the accommodation. Find the minimum power and accommodation for a model eye with $d = 2.5$ cm and $x_{np} = 25$ cm.

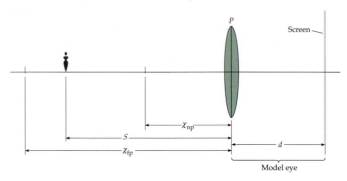

FIGURE 32-60 Problems 73, 74, and 75

74 •• The Model Eye II: In an eye that exhibits nearsightedness, the eye cannot focus on distant objects. Refer to Figure 32-60 and to Problem 73. (*a*) Show that for a nearsighted model eye capable of focusing out to a maximum distance x_{fp}, the minimum value of P is greater than that of a normal eye and is given by

$$P_{min} = \frac{1}{x_{fp}} + \frac{1}{d}$$

(*b*) To correct for nearsightedness, a contact lens may be placed directly in front of the model-eye's lens. What power contact lens would be needed to correct the vision of a nearsighted model eye with $x_{fp} = 50$ cm?

75 •• The Model Eye III: In an eye that exhibits farsightedness, the eye may be able to focus on distant objects but cannot focus on close objects. Refer to Figure 32-60 and to Problem 73. (*a*) Show that for a farsighted model eye capable of focusing only as close as a distance x'_{np}, the maximum value of P is given by

$$P_{max} = \frac{1}{x'_{np}} + \frac{1}{d}$$

(*b*) Show that compared to a model eye capable of focusing as close as a distance x_{np} (where $x_{np} < x'_{np}$), the maximum power of the farsighted lens is too small by

$$\frac{1}{x_{np}} - \frac{1}{x'_{np}}$$

(c) What power contact lens would be needed to correct the vision of a farsighted model eye, with x_{np} = 150 cm, so that the eye may focus on objects as close as 15 cm?

76 • **ISOLVE✓** Suppose the eye were designed like a camera with a lens of fixed focal length f = 2.5 cm that could move toward or away from the retina. Approximately how far would the lens have to move to focus the image of an object 25 cm from the eye onto the retina? (*Hint:* Find the distance from the retina to the image behind it for an object at 25 cm.)

77 • **ISOLVE✓** Find the change in the focal length of the eye when an object originally at 3 m is brought to 30 cm from the eye.

78 • A farsighted person requires lenses with a power of 1.75 D to read comfortably from a book that is 25 cm from the eye. What is that person's near point without the lenses?

79 • **SSM** If two point objects close together are to be seen as two distinct objects, the images must fall on the retina on two different cones that are not adjacent. That is, there must be an unactivated cone between them. The separation of the cones is about 1 μm. Model the eye as a uniform 2.5-cm-diameter sphere with a refractive index of 1.34. (a) What is the smallest angle the two points can subtend? (See Figure 32-61.) (b) How close together can two points be if they are 20 m from the eye?

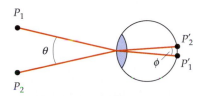

FIGURE 32-61 Problem 79

80 •• A person with a near point of 80 cm needs to read from a computer screen that is 45 cm from her eye. (a) Find the focal length of the lenses in reading glasses that will produce an image of the screen at 80 cm from her eye. (b) What is the power of the lenses?

81 •• A nearsighted person cannot focus clearly on objects that are more distant than 225 cm from her eye. What power lenses are required for her to see distant objects clearly?

82 •• Since the index of refraction of the lens of the eye is not very different from that of the surrounding material, most of the refraction takes place at the cornea, where n changes abruptly from 1.0 in air to approximately 1.4. Assuming the cornea to be a homogeneous sphere with an index of refraction of 1.4, calculate the cornea's radius if it focuses parallel light on the retina a distance 2.5 cm away. Do you expect your result to be larger or smaller than the actual radius of the cornea?

83 •• The near point of a certain person is 80 cm. Reading glasses are prescribed so that he can read a book at 25 cm from his eye. The glasses are 2 cm from the eye. What diopter lens should be used in the glasses?

84 ••• At age 45, a person is fitted for reading glasses of power 2.1 D in order to read at 25 cm. By the time she

reaches 55, she discovers herself holding her newspaper at a distance of 40 cm in order to see it clearly with her glasses on. (a) Where was her near point at age 45? (b) Where is her near point at age 55? (c) What power is now required for the lenses of her reading glasses so that she can again read at 25 cm? (Assume the glasses are 2.2 cm from her eyes.)

*The Simple Magnifier

85 • **SSM** **ISOLVE** A person with a near-point distance of 30 cm uses a simple magnifier of power 20 D. What is the magnification obtained if the final image is at infinity?

86 • A person with a near-point distance of 25 cm wishes to obtain a magnifying power of 5 with a simple magnifier. What should be the focal length of the lens used?

87 • What is the magnifying power of a lens of focal length 7 cm when the image is viewed at infinity by a person whose near point is at 35 cm?

88 •• A lens of focal length 6 cm is used as a simple magnifier with the image at infinity by one person whose near point is 25 cm and by another person whose near point is 40 cm. What is the effective magnifying power of the lens for each person? Compare the size of the image on the retina when each person looks at the same object with the magnifier.

89 •• A botanist examines a leaf using a convex lens of power 12 D as a simple magnifier. What is the expected angular magnification if (a) the final image is at infinity and (b) the final image is at 25 cm?

90 •• **SSM** (a) Show that if the final image of a simple magnifier is to be at the near point of the eye rather than at infinity, the angular magnification is given by

$$M = \frac{x_{np}}{f} + 1$$

(b) Find the magnification of a 20-D lens for a person with a near point of 30 cm if the final image is at the near point. Draw a ray diagram for this situation.

91 •• Show that when the image of a simple magnifier is viewed at the near point, the lateral and angular magnification of the magnifier are equal.

*The Microscope

92 •• **ISOLVE✓** A microscope objective has a focal length of 17 mm. It forms an image at 16 cm from its second focal point. (a) How far from the objective is the object located? (b) What is the magnifying power for a person whose near point is at 25 cm if the focal length of the eyepiece is 51 mm?

93 •• **SSM** A microscope has an objective of focal length 8.5 mm and an eyepiece that gives an angular magnification of 10 for a person whose near point is 25 cm. The tube length is 16 cm. (a) What is the lateral magnification of the objective? (b) What is the magnifying power of the microscope?

94 •• [ISOLVE]✓ A crude, symmetric handheld microscope consists of two converging 20-D lenses fastened in the ends of a tube 30 cm long. (a) What is the tube length of this microscope? (b) What is the lateral magnification of the objective? (c) What is the magnifying power of the microscope? (d) How far from the objective should the object be placed?

95 •• [SSM] A compound microscope has an objective lens with a power of 45 D and an eyepiece with a power of 80 D. The lenses are separated by 28 cm. Assuming that the final image is formed 25 cm from the eye, what is the magnifying power?

96 ••• A microscope has a magnifying power of 600, and an eyepiece of angular magnification of 15. The objective lens is 22 cm from the eyepiece. Without making any approximations, calculate (a) the focal length of the eyepiece, (b) the location of the object so that it is in focus for a normal relaxed eye, and (c) the focal length of the objective lens.

*The Telescope

97 • [ISOLVE]✓ A simple telescope has an objective with a focal length of 100 cm and an eyepiece of focal length 5 cm. It is used to look at the moon, which subtends an angle of about 0.009 rad. (a) What is the diameter of the image formed by the objective? (b) What angle is subtended by the final image at infinity? (c) What is the magnifying power of the telescope?

98 • The objective lens of the refracting telescope at the Yerkes Observatory has a focal length of 19.5 m. When it is used to look at the moon, which subtends an angle of about 0.009 rad, what is the diameter of the image of the moon formed by the objective?

99 •• [SSM] The 200-in (5.1-m) mirror of the reflecting telescope at Mt. Palomar has a focal length of 1.68 m. (a) By what factor is the light-gathering power increased over the 40-in (1.016-m) diameter refracting lens of the Yerkes Observatory telescope? (b) If the focal length of the eyepiece is 1.25 cm, what is the magnifying power of this telescope?

100 •• [ISOLVE] An astronomical telescope has a magnifying power of 7. The two lenses are 32 cm apart. Find the focal length of each lens.

101 •• A disadvantage of the astronomical telescope for terrestrial use (e.g., at a football game) is that the image is inverted. A Galilean telescope uses a converging lens as its objective, but a diverging lens as its eyepiece. The image formed by the objective is behind the eyepiece at its focal point so that the final image is virtual, upright, and at infinity. (a) Show that the magnifying power is $M = -f_o/f_e$, where f_o is the focal length of the objective and f_e is that of the eyepiece (which is negative). (b) Draw a ray diagram to show that the final image is indeed virtual, upright, and at infinity.

102 •• A Galilean telescope (see Problem 101) is designed so that the final image is at the near point, which is 25 cm (rather than at infinity). The focal length of the objective is 100 cm and that of the eyepiece is −5 cm. (a) If the object distance is 30 m, where is the image of the objective? (b) What is the object distance for the eyepiece so that the final image is at the near point? (c) How far apart are the lenses? (d) If the object height is 1.5 m, what is the height of the final image? What is the angular magnification?

103 ••• If you look into the wrong end of a telescope, that is, into the objective, you will see distant objects reduced in size. For a refracting telescope with an objective of focal length 2.25 m and an eyepiece of focal length 1.5 cm, by what factor is the angular size of the object reduced?

General Problems

104 • Show that a diverging lens can never form a real image from a real object. (Hint: Show that s′ is always negative.)

105 • [SSM] A camera uses a positive lens to focus light from an object onto film. Unlike the eye, the camera lens has a fixed focal length, but the lens itself can be moved slightly to vary the image distance to the image on the film. A telephoto lens has a focal length of 200 mm. By how much must the lens move to change from focusing on an object at infinity to an object at a distance of 30 m?

106 • A wide-angle lens of a camera has a focal length of 28 mm. By how much must the lens move to change from focusing on an object at infinity to an object at a distance of 5 m? (See Problem 105.)

107 • A thin converging lens of focal length 10 cm is used to obtain an image that is twice as large as a small object. Find the object and image distances if (a) the image is to be upright and (b) the image is to be inverted. Draw a ray diagram for each case.

108 •• You are given two converging lenses with focal lengths of 75 mm and 25 mm. (a) Show how the lenses should be arranged to form an astronomical telescope. State which lens to use as the objective, which lens to use as the eyepiece, how far apart to place the lenses and, what angular magnification you expect. (b) Draw a ray diagram to show how rays from a distant object are magnified by the telescope.

109 •• (a) Show how the same two lenses in Problem 108 should be arranged as a compound microscope with a tube length of 160 mm. State which lens to use as the objective, which lens to use as the eyepiece, how far apart to place the lenses, and what overall magnification you expect to get, assuming the user has a near point of 25 cm. (b) Draw a ray diagram to show how rays from a close object are magnified into a larger image.

110 •• [SSM] A scuba diver wears a diving mask with a face plate that bulges outward with a radius of curvature of 0.5 m. There is thus a convex spherical surface between the water and the air in the mask. A fish is 2.5 m in front of the diving mask. (a) Where does the fish appear to be? (b) What is the magnification of the image of the fish?

111 •• [ISOLVE] A 35-mm camera has a picture size of 24 mm by 36 mm. It is used to take a picture of a person 175-cm tall so that the image just fills the height (24 mm) of the film. How far should the person stand from the camera if the focal length of the lens is 50 mm?

112 •• A 35-mm camera with interchangeable lenses is used to take a picture of a hawk that has a wing span of 2 m. The hawk is 30 m away. What would be the ideal focal length of the lens used so that the image of the wings just fills the width of the film, which is 36 mm?

113 •• An object is placed 12 cm to the left of a lens of focal length 10 cm. A second lens of focal length 12.5 cm is placed 20 cm to the right of the first lens. (*a*) Find the position of the final image. (*b*) What is the magnification of the image? (*c*) Sketch a ray diagram showing the final image.

114 •• (*a*) Show that if *f* is the focal length of a thin lens in air, its focal length in water is

$$f' = \frac{n_w(n-1)}{n-n_w}$$

where n_w is the index of refraction of water and n is that of the lens. (*b*) Calculate the focal length in air and in water of a double concave lens of index of refraction $n = 1.5$ that has radii of magnitude 30 cm and 35 cm.

115 •• [SSM] (*a*) Find the focal length of a *thick*, double convex lens with an index of refraction of 1.5, a thickness of 4 cm, and radii of +20 cm and −20 cm. (*b*) Find the focal length of this lens in water.

116 •• A 2-cm-thick layer of water ($n = 1.33$) floats on top of a 4-cm-thick layer of carbon tetrachloride ($n = 1.46$) in a tank. How far below the top surface of the water does the bottom of the tank appear to be to an observer looking from above at normal incidence?

117 •• While sitting in your car, you see a jogger in your side mirror, which is convex with a radius of curvature of magnitude 2 m. The jogger is 5 m from the mirror and is approaching at 3.5 m/s. How fast does the jogger appear to be running when viewed in the mirror?

118 •• In the seventeenth century, Antonie van Leeuwenhoek, the first great microscopist, used simple spherical lenses made first of water droplets and then of glass for his first instruments. He made staggering discoveries with these simple lenses. Consider a glass sphere of radius 2.0 mm with an index of refraction of 1.50. Find the focal length of this lens. (*Hint:* Use the equation for refraction at a single spherical surface to find the image distance for an infinite object distance for the first surface. Then use this image point as the object point for the second surface.)

119 ••• An object is 15 cm to the left of a thin convex lens of focal length 10 cm. A concave mirror of radius 10 cm is 25 cm to the right of the lens. (*a*) Find the position of the final image formed by the mirror and lens. (*b*) Is the image real or virtual?

Is the image upright or inverted? (*c*) On a diagram, show where your eye must be to see this image.

120 ••• [SSM] When a bright light source is placed 30 cm in front of a lens, there is an upright image 7.5 cm from the lens. There is also a faint inverted image 6 cm in front of the lens due to reflection from the front surface of the lens. When the lens is turned around, this weaker, inverted image is 10 cm in front of the lens. Find the index of refraction of the lens.

121 ••• A horizontal concave mirror with radius of curvature of 50 cm holds a layer of water with an index of refraction of 1.33 and a maximum depth of 1 cm. At what height above the mirror must an object be placed so that its image is at the same position as the object?

122 ••• A lens with one concave side with a radius of magnitude 17 cm and one convex side with a radius of magnitude 8 cm has a focal length in air of 27.5 cm. When placed in a liquid with an unknown index of refraction, the focal length increases to 109 cm. What is the index of refraction of the liquid?

123 ••• A glass ball of radius 10 cm has an index of refraction of 1.5. The back half of the ball is silvered so that it acts as a concave mirror (Figure 32-62). Find the position of the final image seen by an eye positioned to the left of the object and ball, for an object at (*a*) 30 cm and (*b*) 20 cm to the left of the front surface of the ball.

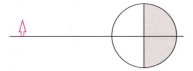

FIGURE 32-62 Problem 123

124 ••• (*a*) Show that a small change dn in the index of refraction of a lens material produces a small change in the focal length df given approximately by $df/f = -dn/(n-1)$. (*b*) Use this result to find the focal length of a thin lens for blue light, for which $n = 1.53$, if the focal length for red light, for which $n = 1.47$, is 20 cm.

125 ••• [SSM] The lateral magnification of a spherical mirror or a thin lens is given by $m = -s'/s$. Show that for objects of small horizontal extent, the longitudinal magnification is approximately $-m^2$. (*Hint:* Show that $ds'/ds = -s'^2/s^2$.)

Interference and Diffraction

WHITE LIGHT IS REFLECTED OFF A SOAP BUBBLE. WHEN LIGHT OF ONE WAVELENGTH IS INCIDENT ON A THIN SOAP-AND-WATER FILM, LIGHT IS REFLECTED BOTH OFF THE FRONT SURFACE AND OFF THE BACK SURFACE OF THE FILM. IF THE ORDER OF MAGNITUDE OF THE THICKNESS OF THE FILM IS THAT OF THE WAVELENGTH OF THE LIGHT, THE TWO REFLECTED LIGHT WAVES INTERFERE. IF THE TWO REFLECTED WAVES ARE 180° OUT OF PHASE, THE REFLECTED WAVE INTERFERES DESTRUCTIVELY, SO THE NET RESULT IS THAT NO LIGHT IS REFLECTED. IF WHITE LIGHT, WHICH CONTAINS MANY WAVELENGTHS, IS INCIDENT ON THE THIN FILM, THEN THE REFLECTED WAVES WILL INTERFERE DESTRUCTIVELY ONLY FOR CERTAIN WAVELENGTHS, AND FOR OTHER WAVELENGTHS THEY WILL INTERFERE CONSTRUCTIVELY. THIS PROCESS PRODUCES THE COLORED FRINGES THAT YOU SEE IN THE SOAP BUBBLE.

 **Have you ever wondered if the phenomena that produces the bands that you see in the light reflected off a soap bubble has any practical applications? Example 33-2 and Problem 21 reveal how the density of the bands relates to the difference in the thickness of the film for a given distance along the film.**

nterference and diffraction are the important phenomena that distinguish waves from particles.[†] Interference is the combining by superposition of two or more waves that meet at one point in space. Diffraction is the bending of waves around corners that occurs when a portion of a wavefront is cut off by a barrier or obstacle.

➤ **In this chapter, we will see how the pattern of the resulting wave can be calculated by treating each point on the original wavefront as a point source, according to Huygens's principle, and calculating the interference pattern resulting from these sources.**

33-1 Phase Difference and Coherence

When two harmonic waves of the same frequency and wavelength but differing in phase combine, the resultant wave is a harmonic wave whose amplitude depends on the phase difference. If the phase difference is zero or an integer times 360°,

---

[†] Before you study this chapter, you may wish to review Chapter 15 and Chapter 16, where the general topics of interference and diffraction of waves are first discussed.

the waves are in phase and interfere constructively. The resultant amplitude equals the sum of the individual amplitudes, and the intensity (which is proportional to the square of the amplitude) is maximum. If the phase difference is 180° or any odd integer times 180°, the waves are out of phase and interfere destructively. The resultant amplitude is then the difference between the individual amplitudes, and the intensity is a minimum. If the amplitudes are equal, the maximum intensity is four times that of either source and the minimum intensity is zero.

A phase difference between two waves is often the result of a difference in path length. A path difference of one wavelength produces a phase difference of 360°, which is equivalent to no phase difference at all. A path difference of one-half wavelength produces a 180° phase difference. In general, a path difference of Δr contributes a phase difference δ given by

$$\delta = \frac{\Delta r}{\lambda} 2\pi = \frac{\Delta r}{\lambda} 360° \qquad \text{33-1}$$

PHASE DIFFERENCE DUE TO A PATH DIFFERENCE

PHASE DIFFERENCE **EXAMPLE 33-1**

(*a*) **What is the minimum path difference that will produce a phase difference of 180° for light of wavelength 800 nm? (*b*) What phase difference will that path difference produce in light of wavelength 700 nm?**

PICTURE THE PROBLEM The phase difference is to 360° as the path length difference is to the wavelength.

(a) The phase difference δ is to 360° as the path length difference Δr is to the wavelength λ. We know that $\lambda = 800$ nm and $\delta = 180°$:

$$\frac{\delta}{360°} = \frac{\Delta r}{\lambda}$$

$$\Delta r = \frac{\delta}{360°} \lambda = \frac{180°}{360°}(800 \text{ nm}) = \boxed{400 \text{ nm}}$$

(b) Set $\lambda = 700$ nm, $\Delta r = 400$ nm, and solve for δ:

$$\delta = \frac{\Delta r}{\lambda} 360° = \frac{400 \text{ nm}}{700 \text{ nm}} 360°$$

$$= \boxed{206° = 3.59 \text{ rad}}$$

Another cause of phase difference is the 180° phase change a wave sometimes undergoes upon reflection from a boundary surface. This phase change is analogous to the inversion of a pulse on a string when it reflects from a point where the density suddenly increases, such as when a light string is attached to a heavier string or rope. The inversion of the reflected pulse is equivalent to a phase change of 180° for a sinusoidal wave (which can be thought of as a series of pulses). When light traveling in air strikes the surface of a medium in which light travels more slowly, such as glass or water, there is a 180° phase change in the reflected light. When light is originally traveling in glass or water, there is no phase change in the light reflected from the glass–air or water–air interface. This is analogous to the reflection without inversion of a pulse on a heavy string at a point where the heavy string is attached to a lighter string.

If light traveling in one medium strikes the surface of a medium in which light travels more slowly, there is a 180° phase change in the reflected light.

PHASE DIFFERENCE DUE TO REFLECTION

As we saw in Chapter 16, interference of waves is observed when two or more coherent waves overlap. Interference of overlapping waves from two sources is not observed unless the sources are coherent. Because the light from each source is usually the result of millions of atoms radiating independently, the phase difference between the waves from such sources fluctuates randomly many times per second, so two light sources are usually not coherent. Coherence in optics is often achieved by splitting the light beam from a single source into two or more beams that can then be combined to produce an interference pattern. The light beam can be split by reflecting the light from the two closely spaced surfaces of a thin film (Section 33-2), by diffracting the beam through two small openings or slits in an opaque barrier (Section 33-3), or by using a single point source and its image in a plane mirror for the two sources (Section 33-3). Today, lasers are the most important sources of coherent light in the laboratory.

Light from an ideal monochromatic source is an infinitely long sinusoidal wave, and light from certain lasers approach this ideal. However, light from conventional *monochromatic* sources, such as gas discharge tubes designed for this purpose, consists of packets of sinusoidal light that are only a few million wavelengths long. The light from such a source consists of many such packets, each approximately the same length. The packets have essentially the same wavelength, but the packets differ in phase in a random manner. The length of one of these packets is called the **coherence length** of the light, and the time it takes one of the packets to pass a point in space is the **coherence time.** The light emitted by a gas discharge tube designed to produce monochromatic light has a coherence length of only a few millimeters. By comparison, some highly stable lasers produce light with a coherence length many kilometers long.

33-2 Interference in Thin Films

You have probably noticed the colored bands in a soap bubble or in the film on the surface of oily water. These bands are due to the interference of light reflected from the top and bottom surfaces of the film. The different colors arise because of variations in the thickness of the film, causing interference for different wavelengths at different points.

Consider a thin film of water (such as a small section of a soap bubble) of uniform thickness viewed at small angles with the normal, as shown in Figure 33-1. Part of the light is reflected from the upper air–water interface where it undergoes a 180° phase change. Most of the light enters the film and part of it is reflected by the bottom water–air interface. There is no phase change in this reflected light. If the light is nearly perpendicular to the surfaces, both the ray reflected from the top surface and the ray reflected from the bottom surface can enter the eye at point P in the figure. The path difference between these two rays is 2t, where t is the thickness of the film. This path difference produces a phase difference of $(2t/\lambda')360°$, where $\lambda' = \lambda/n$ is the wavelength of the light in the film, and n is the index of refraction of the film. The total phase difference between these two rays is thus 180° plus the phase difference due to the path difference. Destructive interference occurs when the path difference 2t is zero or a whole number of wavelengths λ' (in the film). Constructive interference occurs when the path difference is an odd number of half-wavelengths.

When a thin film of water lies on a glass surface, as in Figure 33-2, the ray that reflects from the lower water–glass interface also undergoes a 180° phase change, because the index of refraction of glass (approximately 1.50) is greater than that of water (approximately 1.33). Thus, both the rays shown in the figure have undergone a 180° phase change upon reflection. The phase difference between these rays is due solely to the path difference and is given by $\delta = (2t/\lambda')360°$.

When a thin film of varying thickness is viewed with monochromatic light, such as the yellow light from a sodium lamp, alternating bright and dark bands

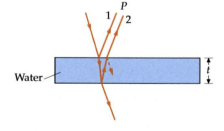

FIGURE 33-1 Light rays reflected from the top and bottom surfaces of a thin film are coherent because both rays come from the same source. If the light is incident almost normally, the two reflected rays will be very close to each other and will produce interference.

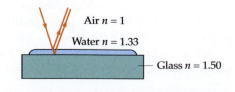

FIGURE 33-2 The interference of light reflected from a thin film of water resting on a glass surface. In this case, both rays undergo a change in phase of 180° upon reflection.

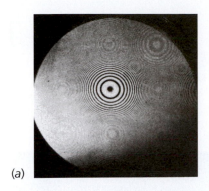

(a)

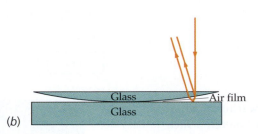

(b)

FIGURE 33-3 (a) Newton's rings observed with light reflected from a thin film of air between a plane glass surface and a spherical glass surface. At the center, the thickness of the air film is negligible and the interference is destructive because of the 180° phase change of one of the rays upon reflection. (b) Glass surfaces for the observation of Newton's rings shown in Figure 33-3a. The thin film in this case is the film of air between the glass surfaces.

or lines called **fringes** are observed. The distance between a bright fringe and a dark fringe is that distance over which the film's thickness changes so that the path difference $2t$ is $\lambda'/2$. Figure 33-3a shows the interference pattern observed when light is reflected from an air film between a spherical glass surface and a plane glass surface in contact. These circular interference fringes are known as **Newton's rings.** Typical rays reflected at the top and bottom of the air film are shown in Figure 33-3b. Near the point of contact of the surfaces, where the path difference between the ray reflected from the upper glass–air interface and the ray reflected from the lower air–glass interface is essentially zero or is at least small compared with the wavelength of light, the interference is perfectly destructive because of the 180° phase shift of the ray reflected from the lower air–glass interface. This central region in Figure 33-3a is therefore dark. The first bright fringe occurs at the radius at which the path difference is $\lambda/2$, which contributes a phase difference of 180°. This adds to the phase shift due to reflection to produce a total phase difference of 360°, which is equivalent to a zero phase difference. The second dark region occurs at the radius at which the path difference is λ, and so on.

FIGURE 33-4 The angle θ, which is less than 0.02°, is exaggerated. The incoming and outgoing rays are essentially perpendicular to all air–glass interfaces.

A WEDGE OF AIR	**EXAMPLE** **33-2**

A wedge-shaped film of air is made by placing a small slip of paper between the edges of two flat pieces of glass, as shown in Figure 33-4. Light of wavelength 500 nm is incident normally on the glass, and interference fringes are observed by reflection. If the angle θ made by the plates is 3×10^{-4} rad, how many dark interference fringes per centimeter are observed?

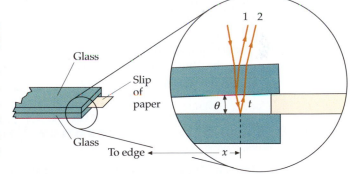

PICTURE THE PROBLEM We find the number of fringes per centimeter by finding the horizontal distance x to the mth fringe and solving for m/x. Because the ray reflected from the bottom plate undergoes a 180° phase shift, the point of contact (where the path difference is zero) will be dark. The first dark fringe after this point occurs when $2t = \lambda'$, where $\lambda' = \lambda$ is the wavelength in the air film, and t is the plate separation at x, as shown in Figure 33-4. Since the angle θ is small, we can use the small-angle approximation $\theta \approx t/x$.

1. The mth dark fringe occurs when the path difference $2t$ equals m wavelengths:

$$2t = m\lambda' = m\lambda$$

$$m = \frac{2t}{\lambda}$$

2. The thickness t is related to the angle θ:

$$\theta = \frac{t}{x}$$

3. Substitute $t = x\theta$ into the equation for m:

$$m = \frac{2x\theta}{\lambda}$$

4. Calculate m/x:

$$\frac{m}{x} = \frac{2\theta}{\lambda} = \frac{2(3 \times 10^{-4})}{5 \times 10^{-7}\,\text{m}} = 1200\,\text{m}^{-1}$$

$$= \boxed{12\,\text{cm}^{-1}}$$

REMARKS We therefore observe 12 dark fringes per centimeter. In practice, the number of fringes per centimeter, which is easy to count, can be used to determine the angle. Note that if the angle of the wedge is increased, the fringes become more closely spaced.

EXERCISE How many dark fringes per centimeter are observed if light of wavelength 650 nm is used? (*Answer* 9.2 cm^{-1})

Figure 33-5*a* shows interference fringes produced by a wedge-shaped air film between two flat glass plates, as in Example 33-2. Plates that produce straight fringes, such as those in Figure 33-5*a*, are said to be **optically flat**. To be optically flat, a surface must be flat to within a small fraction of a wavelength. A similar wedge-shaped air film formed by two ordinary glass plates yields the irregular fringe pattern in Figure 33-5*b*, which indicates that these plates are not optically flat.

One application of interference effects in thin films is in nonreflecting lenses, which are made by covering a lens with a thin film of a material that has an index of refraction of approximately 1.38, which is between that of glass and air. Then the intensities of the light reflected from the top and bottom surfaces of the film are approximately equal, and since both rays undergo a 180° phase change, there is no phase difference between the rays due to reflection. The thickness of the film is chosen to be $\lambda'/4 = \lambda/4n$, where λ is in the middle of the visible spectrum, so that there is a phase change of 180° due to the path difference of $\lambda'/2$. Reflection from the coated surface is thus minimized, whereas transmission through the surface is maximized.

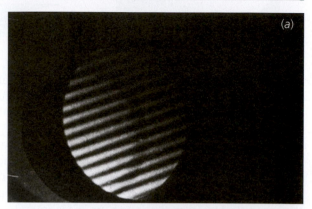

(a)

(b)

FIGURE 33-5 (*a*) Straight-line fringes from a wedge-shaped film of air, like that shown in Figure 33-4. The straightness of the fringes indicates that the glass plates are optically flat. (*b*) Fringes from a wedge-shaped film of air between glass plates that are not optically flat.

33-3 Two-Slit Interference Pattern

Interference patterns of light from two or more sources can be observed only if the sources are coherent. The interference in thin films discussed previously can be observed because the two beams come from the same light source but are separated by reflection. In Thomas Young's famous experiment, in which he demonstrated the wave nature of light, two coherent light sources are produced by illuminating two very narrow parallel slits with a single light source. We saw in Chapter 15 that when a wave encounters a barrier with a very small opening, the opening acts as a point source of waves (Figure 33-6). In Young's experiment, diffraction causes each slit to act as a line source (which is equivalent to a point source in two dimensions). The interference pattern is observed on a screen far from the slits (Figure 33-7*a*). At very large distances from the slits, the lines from

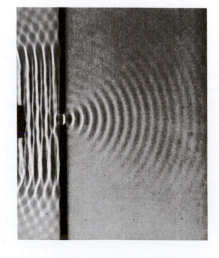

FIGURE 33-6 Plane water waves in a ripple tank encountering a barrier with a small opening. The waves to the right of the barrier are circular waves that are concentric about the opening, just as if there were a point source at the opening.

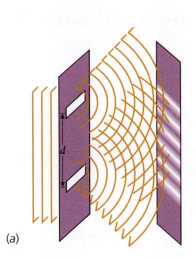

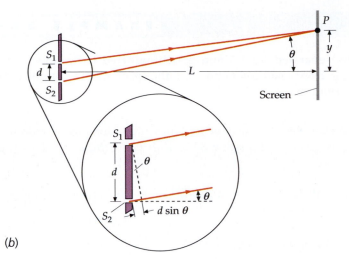

(a)   (b)

FIGURE 33-7 (a) Two slits act as coherent sources of light for the observation of interference in Young's experiment. Cylindrical waves from the slits overlap and produce an interference pattern on a screen. (b) Geometry for relating the distance y measured along the screen to L and θ. When the screen is very far away compared with the slit separation, the rays from the slits to a point on the screen are approximately parallel, and the path difference between the two rays is d sin θ.

the two slits to some point P on the screen are approximately parallel, and the path difference is approximately $d \sin \theta$, where d is the separation of the slits, as shown in Figure 33-7b. When the path difference is equal to an integral number of wavelengths, the interference is constructive. We thus have interference maxima at an angle θ_m given by

$$d \sin \theta_m = m\lambda, \qquad m = 0, 1, 2, \ldots \qquad \text{33-2}$$

TWO-SLIT INTERFERENCE MAXIMA

where m is called the **order number.** The interference minima occur at

$$d \sin \theta_m = (m - \tfrac{1}{2})\lambda, \qquad m = 1, 2, 3, \ldots \qquad \text{33-3}$$

TWO-SLIT INTERFERENCE MINIMA

The phase difference δ at a point P is related to the path difference $d \sin \theta$ by

$$\frac{\delta}{2\pi} = \frac{d \sin \theta}{\lambda} \qquad \text{33-4}$$

We can relate the distance y_m measured along the screen from the central point to the mth bright fringe (see Figure 33-7b) to the distance L from the slits to the screen:

$$\tan \theta_m = \frac{y_m}{L}$$

For small angles, $\tan \theta \approx \sin \theta$. Substituting y_m/L for $\sin \theta_m$ in Equation 33-2 and solving for y_m gives

$$y_m = m\frac{\lambda L}{d} \qquad \text{33-5}$$

DISTANCE ON SCREEN TO THE MTH BRIGHT FRINGE

From this result, we see that for small angles the fringes are equally spaced on the screen.

FRINGE SPACING FROM SLIT SPACING   **EXAMPLE 33-3** **Try It Yourself**

Two narrow slits separated by 1.5 mm are illuminated by yellow light of wavelength 589 nm from a sodium lamp. Find the spacing of the bright fringes observed on a screen 3 m away.

PICTURE THE PROBLEM The distance y_m measured along the screen to the mth bright fringe is given by Equation 33-2, with $L = 3$ m, $d = 1.5$ mm, and $\lambda = 589$ nm.

Cover the column to the right and try these on your own before looking at the answers.

Steps	Answers
1. Make a sketch of the situation (Figure 33-8).	
2. Using the sketch, obtain an expression for the spacing between fringes.	fringe spacing $= \dfrac{y_3}{3}$
3. Apply Equation 33-2 to the $m = 3$ fringe.	$d \sin \theta_3 = 3\lambda$
4. Using trig, relate y_3 and θ_3.	$\sin \theta_3 \approx \tan \theta_3 = \dfrac{y_3}{L}$
5. Substitute into the step 3 result and solve for the fringe spacing.	$\dfrac{y_3}{3} = \boxed{1.18 \text{ mm}}$

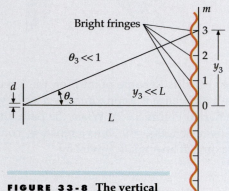

FIGURE 33-8 The vertical scale of the figure is expanded.

REMARKS The fringes are uniformly spaced only to the degree that the small-angle approximation is valid. That is, to the degree that $\lambda/d \ll 1$. In this example, $\lambda/d = (589 \text{ nm})/(1.5 \text{ mm}) = 0.0004$.

EXERCISE A point source of light ($\lambda = 589$ nm) is placed 0.4 mm above the surface of a glass mirror. Interference fringes are observed on a screen 6 m away, and the interference is between the light reflected off the front surface of the glass and the light traveling from the source directly to the screen. Find the spacing of the fringes. (*Answer* 4.42 mm)

Calculation of Intensity

To calculate the intensity of the light on the screen at a general point P, we need to add two harmonic wave functions that differ in phase.[†] The wave functions for electromagnetic waves are the electric field vectors. Let E_1 be the electric field at some point P on the screen due to the waves from slit 1, and let E_2 be the electric field at that point due to waves from slit 2. Since the angles of interest are small, we can assume that these fields are parallel. Both electric fields oscillate with the same frequency (they result from a single source that illuminates both slits) and they have the same amplitude. (The path difference is only of the order of a few wavelengths of light at most.) They have a phase difference δ given by Equation 33-4. If we represent these wave functions by

$$E_1 = A_0 \sin \omega t$$

and

$$E_2 = A_0 \sin(\omega t + \delta)$$

the resultant wave function is

[†] We did this in Chapter 16 where we first discussed the superposition of two waves.

$$E = E_1 + E_2 = A_0 \sin \omega t + A_0 \sin(\omega t + \delta)$$
$$= 2A_0 \cos \tfrac{1}{2}\delta \sin(\omega t + \tfrac{1}{2}\delta)$$
$$\text{33-6}$$

where we used the identity

$$\sin \alpha + \sin \beta = 2 \cos \tfrac{1}{2}(\alpha - \beta) \sin \tfrac{1}{2}(\alpha + \beta) \qquad \text{33-7}$$

The amplitude of the resultant wave is thus $2A_0 \cos \tfrac{1}{2}\delta$. It has its maximum value of $2A_0$ when the waves are in phase and is zero when they are 180° out of phase. Since the intensity is proportional to the square of the amplitude, the intensity at any point P is

$$I = 4I_0 \cos^2 \tfrac{1}{2}\delta \qquad \text{33-8}$$

INTENSITY IN TERMS OF PHASE DIFFERENCE

where I_0 is the intensity of the light on the screen from either slit separately. The phase angle δ is related to the position on the screen by Equation 33-4.

Figure 33-9a shows the intensity pattern as seen on a screen. A graph of the intensity as a function of $\sin \theta$ is shown in Figure 33-9b. For small θ, this is equivalent to a plot of intensity versus y (since $y = L \tan \theta \approx L \sin \theta$). The intensity I_0 is that from each slit separately. The dashed line in Figure 33-9b shows the average intensity $2I_0$, which is the result of averaging over a distance containing many interference maxima and minima. This is the intensity that would arise from the two sources if they acted independently without interference, that is, if they were not coherent. Then the phase difference between the two sources would fluctuate randomly, so that only the average intensity would be observed.

Figure 33-10 shows another method of producing the two-slit interference pattern, an arrangement known as **Lloyd's mirror.** A single slit is placed at a distance $\tfrac{1}{2}d$ above the plane of a mirror. Light striking the screen directly from the source interferes with the light that is reflected from the mirror. The reflected light can be considered to come from the virtual image of the slit formed by the mirror. Because of the 180° change in phase upon reflection at the mirror, the interference pattern is that of two coherent line sources that differ in phase by 180°. The pattern is the same as that shown in Figure 33-9 for two slits, except that the maxima and minima are interchanged. Constructive interference occurs at points for which the path difference is a half-wavelength or any odd number of half-wavelengths. At these points, the 180° phase difference due to the path difference combines with the 180° phase difference of the sources to produce constructive interference.

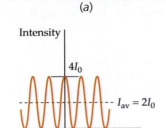

(a)

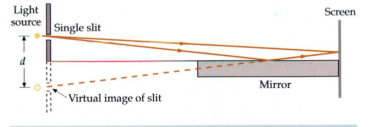

(b)

FIGURE 33-9 (a) The interference pattern observed on a screen far away from the two slits shown in Figure 33-7. (b) Plot of intensity versus $\sin \theta$. The maximum intensity is $4I_0$, where I_0 is the intensity due to each slit separately. The average intensity (dashed line) is $2I_0$.

FIGURE 33-10 Lloyd's mirror for producing a two-slit interference pattern. The two sources (the slit and its image) are coherent and are 180° out of phase. The central interference band at the point equidistant from the sources is dark.

33-4 Diffraction Pattern of a Single Slit

In our discussion of the interference patterns produced by two or more slits, we assumed that the slits were very narrow so that we could consider the slits to be line sources of cylindrical waves, which in our two-dimensional diagrams are point sources of circular waves. We could therefore assume that the intensity due to one slit acting alone was the same (I_0) at any point P on the screen, independent of the angle θ made between the ray to point P and the normal line between the slit and the screen. When the slit is not narrow, the intensity on a screen far away is not independent of angle but decreases as the angle increases. Consider a slit of width a. Figure 33-11 shows the intensity pattern on a screen far away from the slit of width a as a function of sin θ. We can see that the intensity is maximum in the forward direction (sin $\theta = 0$) and decreases to zero at an angle that depends on the slit width a and the wavelength λ.

Most of the light intensity is concentrated in the broad **central diffraction maximum,** although there are minor secondary maxima bands on either side of the central maximum. The first zeroes in the intensity occur at angles specified by

$$\sin \theta_1 = \lambda/a \qquad \text{33-9}$$

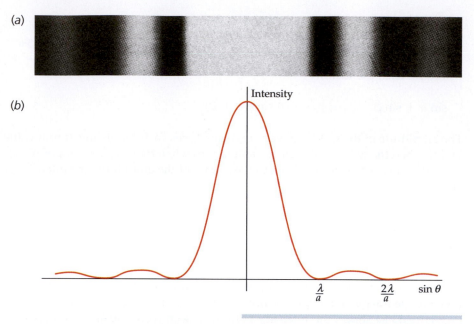

FIGURE 33-11 (a) Diffraction pattern of a single slit as observed on a screen far away. (b) Plot of intensity versus sin θ for the pattern in Figure 33-11a.

Note that for a given wavelength λ, Equation 33-9 describes how variations in the slit width result in variations in the angular width of the central maximum. If we *increase* the slit width a, the angle θ_1 at which the intensity first becomes zero *decreases*, giving a more narrow central diffraction maximum. Conversely, if we *decrease* the slit width, the angle of the first zero *increases*, giving a wider central diffraction maximum. When a is smaller than λ, then sin θ_1 would have to exceed 1 to satisfy Equation 33-9. Thus, for a less than λ, there are no points of zero intensity in the pattern, and the slit acts as a line source (a point source in two dimensions) radiating light energy essentially equal in all directions.

Multiplying both sides of Equation 33-9 by $a/2$ gives

$$\tfrac{1}{2}a \sin \theta_1 = \tfrac{1}{2}\lambda \qquad \text{33-10}$$

The quantity $\tfrac{1}{2}a \sin \theta_1$ is the path difference between a light ray leaving the middle of the upper half of the slit and one leaving the middle of the lower half of the slit. We see that the first diffraction *minimum* occurs when these two rays are 180° out of phase, that is, when their path difference equals a half-wavelength. We can understand this result by considering each point on a wavefront to be a point source of light in accordance with Huygens's principle. In Figure 33-12, we have placed a line of dots on the wavefront at the slit to represent these point sources schematically. Suppose, for example, that we have 100 such dots and that we look at an angle θ_1 for which $a \sin \theta_1 = \lambda$. Let us consider the slit to be divided into two halves, with the first 50 sources in the upper half and sources 51 through 100 in the lower half. When the path difference between the middle of the upper half and the middle of the lower half of the slit equals a half-wavelength, the path difference between source 1 (the first source in the upper half) and source 51 (the first source in the lower half) is $\tfrac{1}{2}\lambda$. The waves from these two sources will be out of phase by 180° and will thus cancel. Similarly, waves from the second source in each region (source 2 and source 52) will cancel. Continuing this argument, we can see that the waves from each pair of sources separated by $a/2$ will cancel. Thus, there will be no light energy at this angle. We can extend this argument to the second and third minima in the diffraction pattern of Figure 33-11. At an angle θ_2 where $a \sin \theta_2 = 2\lambda$, we can divide the slit into four regions, two regions for the top half and two regions for the bottom half. Using this same argument, the light intensity from the top half is zero because of the cancellation of pairs of sources, and, similarly, the light intensity from the bottom half is zero. The general expression for the points of zero intensity in the diffraction pattern of a single slit is thus

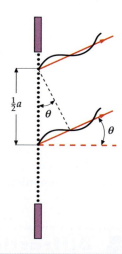

FIGURE 33-12 A single slit is represented by a large number of point sources of equal amplitude. At the first diffraction minimum of a single slit, the waves from each point source in the upper half of the slit 180° out of phase with the wave from the point source a distance $a/2$ lower in the slit. Thus, the interference from each such pair of point sources is destructive.

$$a \sin \theta_m = m\lambda, \qquad m = 1, 2, 3 \ldots \qquad \text{33-11}$$

POINTS OF ZERO INTENSITY FOR A SINGLE-SLIT DIFFRACTION PATTERN

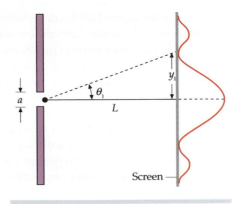

Usually, we are just interested in the first occurrence of a minimum in the light intensity because nearly all of the light energy is contained in the central diffraction maximum.

In Figure 33-13, the distance y_1 from the central maximum to the first diffraction minimum is related to the angle θ_1 and the distance L from the slit to the screen by

$$\tan \theta_1 = \frac{y_1}{L}$$

FIGURE 33-13 The distance y_1 measured along the screen from the central maximum to the first diffraction minimum is related to the angle θ_1 by $\tan \theta_1 = y_1/L$, where L is the distance to the screen.

WIDTH OF THE CENTRAL DIFFRACTION MAXIMUM **EXAMPLE 33-4**

In a lecture demonstration of single-slit diffraction, a laser beam of wavelength 700 nm passes through a vertical slit 0.2 mm wide and hits a screen 6 m away. Find the width of the central diffraction maximum on the screen; that is, find the distance between the first minimum on the left and the first minimum on the right of the central maximum.

PICTURE THE PROBLEM Referring to Figure 33-13, the width of the central diffraction maximum is $2y_1$.

1. The half-width of the central maxima y_1 is related to the angle θ_1 by:

$$\tan \theta_1 = \frac{y_1}{L}$$

2. The angle θ_1 is related to the slit width a by Equation 33-11:

$$\sin \theta_1 = \lambda/a$$

3. Solve the step 2 result for θ_1, substitute into the step 1 result, and solve for $2y_1$:

$$2y_1 = 2L \tan \theta_1 = 2L \tan\left(\sin^{-1} \frac{\lambda}{a}\right)$$

$$= 2(6 \text{ m}) \tan\left(\sin^{-1} \frac{700 \times 10^{-9} \text{ m}}{0.0002 \text{ m}}\right)$$

$$= 4.2 \times 10^{-2} \text{ m} = \boxed{4.20 \text{ cm}}$$

REMARKS Since $\sin \theta_1 = \lambda/a = (700 \text{ nm})/(0.2 \text{ mm}) = 0.0035$, we can use the small-angle approximation to evaluate $2y_1$. In this approximation, $\sin \theta_1 = \tan \theta_1$, so $\lambda/a = y_1/L$ and $2y_1 = 2L\lambda/a = 2(6 \text{ m})(700 \text{ nm})/(0.2 \text{ mm}) = 4.20 \text{ cm}$. (This approximate value is in agreement with the exact value to within 0.0006 percent.)

Interference–Diffraction Pattern of Two Slits

When there are two or more slits, the intensity pattern on a screen far away is a combination of the single-slit diffraction pattern and the multiple-slit interference pattern we have studied. Figure 33-14 shows the intensity pattern on a screen far from two slits whose separation d is 10 times the width a of each slit. The pattern is the same as the two-slit pattern with very narrow slits (Figure 33-11) except that it is modulated by the single-slit diffraction pattern; that is, the intensity due to each slit separately is now not constant but decreases with angle, as shown in Figure 33-14b.

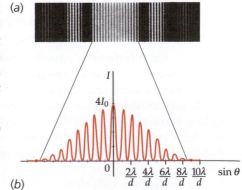

FIGURE 33-14 (a) Interference–diffraction pattern for two slits whose separation d is equal to 10 times their width a. The tenth interference maximum on either side of the central interference maximum is missing because it falls at the first diffraction minimum. (b) Plot of intensity versus $\sin \theta$ for the central band of the pattern in Figure 33-14a.

Note that the central diffraction maximum in Figure 33-14 contains 19 interference maxima—the central interference maximum and 9 maxima on either side. The tenth interference maximum on either side of the central one is at the angle θ_{10}, given by $\sin \theta_{10} = 10\lambda/d = \lambda/a$, since $d = 10a$. This coincides with the position of the first diffraction minimum, so this interference maximum is not seen. At these points, the light from the two slits would be in phase and would interfere constructively, but there is no light from either slit because the points are diffraction minima. In general, we can see that if $m = d/a$, the mth interference maximum will fall at the first diffraction minimum. Since the mth fringe is not seen, there will be $m - 1$ fringes on each side of the central fringe for a total of N fringes in the central maximum, where N is given by

$$N = 2(m - 1) + 1 = 2m - 1 \qquad\qquad 33\text{-}12$$

INTERFERENCE AND DIFFRACTION **EXAMPLE 3 3 - 5**

Two slits of width $a = 0.015$ mm are separated by a distance $d = 0.06$ mm and are illuminated by light of wavelength $\lambda = 650$ nm. How many bright fringes are seen in the central diffraction maximum?

PICTURE THE PROBLEM We need to find the value of m for which the mth interference maximum coincides with the first diffraction minimum. Then there will be $N = 2m - 1$ fringes in the central maximum.

1. Find the angle θ_1 of the first diffraction minimum:

$$\sin \theta_1 = \frac{\lambda}{a} \text{ (first diffraction minimum)}$$

2. Find the angle θ_m of the mth interference maxima:

$$\sin \theta_m = \frac{m\lambda}{d} \text{ (} m\text{th interference maxima)}$$

3. Set these angles equal and solve for m:

$$\frac{m\lambda}{d} = \frac{\lambda}{a}$$

$$m = \frac{d}{a} = \frac{0.06 \text{ mm}}{0.015 \text{ mm}} = 4$$

4. The first diffraction minimum coincides with the fourth bright fringe. Therefore, there are 3 bright fringes visible on either side of the central diffraction maximum. These 6 maxima, plus the central interference maximum, combine for a total of 7 bright fringes in the central diffraction maximum:

$$N = \boxed{7 \text{ bright fringes}}$$

---

*33-5 Using Phasors to Add Harmonic Waves

To calculate the interference pattern produced by three, four, or more coherent light sources and to calculate the diffraction pattern of a single slit, we need to combine several harmonic waves of the same frequency that differ in phase. A simple geometric interpretation of harmonic wave functions leads to a method of adding harmonic waves of the same frequency by geometric construction.

Let the wave functions for two waves at some point be $E_1 = A_1 \sin \alpha$ and $E_2 = A_2 \sin(\alpha + \delta)$, where $\alpha = \omega t$. Our problem is then to find the sum:

$$E_1 + E_2 = A_1 \sin \alpha + A_2 \sin(\alpha + \delta)$$

We can represent each wave function by a two-dimensional vector, as shown in Figure 33-15. The geometric method of addition is based on the fact that the y (or x) component of the resultant of two vectors equals the sum of the y (or x) components of the vectors, as illustrated in the figure. The wave function E_1 is represented by the vector $\vec{A}_1$. As the time varies, this vector rotates in the xy plane with angular frequency ω. Such a vector is called a **phasor**. (We encountered phasors in our study of ac circuits in Section 29-4.) The wave function E_2 is the y component of a phasor of magnitude A_2 that makes an angle $\alpha + \delta$ with the x axis. By the laws of vector addition, the sum of these components equals the y component of the resultant phasor $\vec{A}$, as shown in Figure 33-15. The y component of the resultant phasor, $A \sin(\alpha + \delta')$, is a harmonic wave function that is the sum of the two original wave functions:

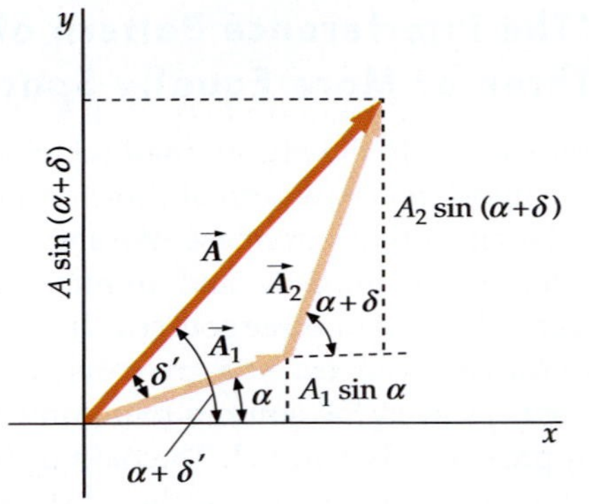

$$A_1 \sin \alpha + A_2 \sin(\alpha + \delta) = A \sin(\alpha + \delta') \qquad \text{33-13}$$

where A (the amplitude of the resultant wave) and δ' (the phase of the resultant wave relative to the first wave) are found by adding the phasors representing the waves. As time varies, α varies. The phasors representing the two wave functions and the resultant phasor representing the resultant wave function rotate in space, but their relative positions do not change because they all rotate with the same angular velocity ω.

FIGURE 33-15 Phasor representation of wave functions.

WAVE SUPERPOSITION USING PHASORS **EXAMPLE 33-6 Try It Yourself**

Use the phasor method of addition to derive Equation 33-16 for the superposition of two waves of the same amplitude.

PICTURE THE PROBLEM Represent the waves $y_1 = A_0 \sin \alpha$ and $y_2 = A_0 \sin(\alpha + \delta)$ by vectors (phasors) of length A_0 making an angle δ with one another. The resultant wave $y_r = A \sin(\alpha + \delta')$ is represented by the sum of these vectors, which form an isosceles triangle, as shown in Figure 33-16.

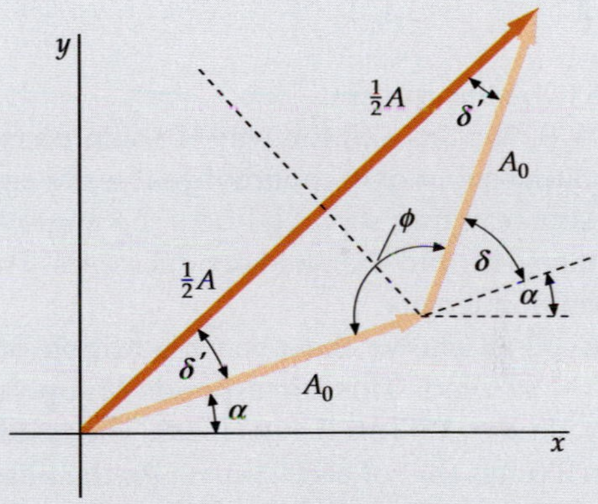

FIGURE 33-16

Cover the column to the right and try these on your own before looking at the answers.

Steps	Answers
1. Find the phase angle δ' in terms of ϕ from the fact that the three angles in the triangle must sum to $180°$.	$\delta' + \delta' + \phi = 180°$
2. Relate ϕ to δ.	$\delta + \phi = 180°$
3. Eliminate ϕ from the step 1 and step 2 results and solve for δ'.	$\delta' = \frac{1}{2}\delta$
4. Write $\cos \delta'$ in terms of A and A_0.	$\cos \delta' = \dfrac{\frac{1}{2}A}{A_0}$
5. Solve for A in terms of δ.	$A = 2A_0 \cos \delta' = 2A_0 \cos \frac{1}{2}\delta$
6. Use your results for A and δ' to write the resultant wave function.	$y_r = A \sin(\alpha + \delta')$
	$= \boxed{2A_0 \cos(\tfrac{1}{2}\delta) \sin(\alpha + \tfrac{1}{2}\delta)}$

EXERCISE Find the resultant wave function of the two waves $E_1 = 4 \sin(\omega t)$ and $E_2 = 3 \sin(\omega t + 90°)$ [*Answer* $E_1 + E_2 = 5 \sin(\omega t + 37°)$]

*The Interference Pattern of Three or More Equally Spaced Sources

We can apply the phasor method of addition to calculate the interference pattern of three or more equally spaced, coherent sources in phase. We are most interested in the interference maxima and minima. Figure 33-17 illustrates the case of three sources. The geometry is the same as for two sources. At a great distance from the sources, the rays from the sources to a point P on the screen are approximately parallel. The path difference between the first and second source is then $d \sin \theta$, as before, and the path difference between the first and third source is $2d \sin \theta$. The wave at point P is the sum of three waves. Let $\alpha = \omega t$ be the phase of the first wave at point P. We thus have the problem of adding three waves of the form

$$E_1 = A_0 \sin \alpha$$

$$E_2 = A_0 \sin(\alpha + \delta)$$

$$E_3 = A_0 \sin(\alpha + 2\delta) \qquad \text{33-14}$$

where

$$\delta = \frac{2\pi}{\lambda} d \sin \theta \approx \frac{2\pi yd}{\lambda L} \qquad \text{33-15}$$

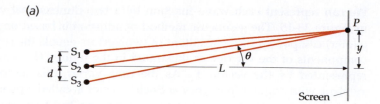

(a)

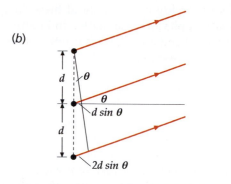

(b)

FIGURE 33-17 Geometry for calculating the intensity pattern far away from three equally spaced, coherent sources that are in phase.

as in the two-slit problem.

At $\theta = 0$, $\delta = 0$, so all the waves are in phase. The amplitude of the resultant wave is 3 times that of each individual wave and the intensity is 9 times that due to each source acting separately. As the angle δ increases from $\theta = 0$, the phase angle δ increases and the intensity decreases. The position $\theta = 0$ is thus a position of maximum intensity.

Figure 33-18 shows the phasor addition of three waves for a phase angle $\delta = 30° = \pi/6$ rad. This corresponds to a point P on the screen for which θ is given by $\sin \theta = \lambda\delta/(2\pi d) = \lambda/(12d)$. The resultant amplitude A is considerably less than 3 times that of each source. As the phase angle δ increases, the resultant amplitude decreases until the amplitude is zero at $\delta = 120°$. For this phase difference, the three phasors form an equilateral triangle (Figure 33-19). This first interference minimum for three sources occurs at a smaller phase angle δ (and therefore at a smaller space angle θ) than it does for only two sources (for which the first minimum occurs at $\delta = 180°$). As δ increases from 120°, the resultant amplitude increases, reaching a secondary maximum near $\delta = 180°$. At the phase angle $\delta = 180°$, the amplitude is the same as that from a single source, since the waves from the first two sources cancel each other, leaving only the third. The intensity of the secondary maximum is one-ninth that of the maximum at $\theta = 0$. As δ increases beyond 180°, the amplitude again decreases and is zero at $\delta = 180° + 60° = 240°$. For δ greater than 240°, the amplitude increases and is again 3 times that of each source when $\delta = 360°$. This phase angle corresponds to a path difference of 1 wavelength for the waves from the first two sources and 2 wavelengths for the waves from the first and third sources. Hence, the

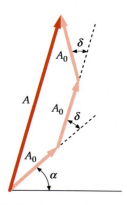

FIGURE 33-18 Phasor diagram for determining the resultant amplitude A due to three waves, each of amplitude A_0, that have phase differences of δ and 2δ due to path differences of $d \sin \theta$ and $2d \sin \theta$. The angle $\alpha = \omega t$ varies with time, but this does not affect the calculation of A.

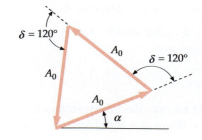

FIGURE 33-19 The resultant amplitude for the waves from three sources is zero when δ is 120°. This interference minimum occurs at a smaller angle θ than does the first minimum for two sources, which occurs when δ is 180°.

three waves are in phase at this point. The largest maxima, called the principal maxima, are at the same positions as for just two sources, which are those points corresponding to the angles θ given by

$$d \sin \theta_m = m\lambda, \quad m = 0, 1, 2, \ldots \quad \text{33-16}$$

These maxima are stronger and narrower than those for two sources. They occur at points for which the path difference between adjacent sources is zero or an integral number of wavelengths.

These results can be generalized to more than three sources. For four equally spaced sources that are in phase, the principal interference maxima are again given by Equation 33-16, but these maxima are even more intense, they are narrower, and there are two small secondary maxima between each pair of principal maxima. At $\theta = 0$, the intensity is 16 times that due to a single source. The first interference minimum occurs when δ is 90°, as can be seen from the phasor diagram of Figure 33-20. The first secondary maximum is near $\delta = 132°$, leaving only the wave from the fourth source. The intensity of the secondary maximum is approximately one-sixteenth that of the central maximum. There is another minimum at $\delta = 180°$, another secondary maximum near $\delta = 228°$, and another minimum at $\delta = 270°$ before the next principal maximum at $\delta = 360°$.

Figure 33-21 shows the intensity patterns for two, three, and four equally spaced coherent sources. Figure 33-22 shows a graph of I/I_0, where I_0 is the intensity due to each source acting separately. For three sources, there is a very small secondary maximum between each pair of principal maxima, and the principal maxima are sharper and more intense than those due to just two sources. For four sources, there are two small secondary maxima between each pair of principal maxima, and the principal maxima are even more narrow and intense.

From this discussion, we can see that as we increase the number of sources, the intensity becomes more and more concentrated in the principal maxima given by Equation 33-16, and these maxima become narrower. For N sources, the intensity of the principal maxima is N^2 times that due to a single source. The first minimum occurs at a phase angle of $\delta = 360°/N$, for which the N phasors form a closed polygon of N sides. There are $N - 2$ secondary maxima between each pair of principal maxima. These secondary maxima are very weak compared with the principal maxima. As the number of sources is increased, the principal maxima become sharper and more intense, and the intensities of the secondary maxima become negligible compared to those of the principal maxima.

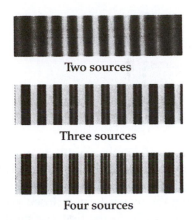

FIGURE 33-20 Phasor diagram for the first minimum for four equally spaced in-phase sources. The amplitude is zero when the phase difference of the waves from adjacent sources is 90°.

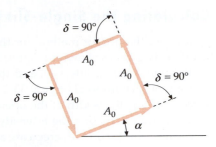

FIGURE 33-21 Intensity patterns for two, three, and four equally spaced coherent sources. There is a secondary maximum between each pair of principal maxima for three sources, and two secondary maxima for four sources.

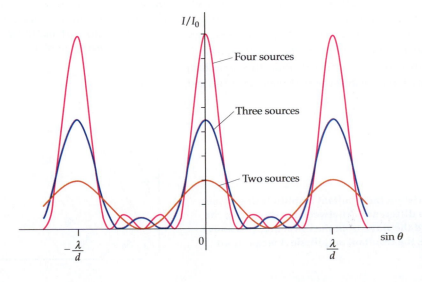

FIGURE 33-22 Plot of relative intensity versus $\sin \theta$ for two, three, and four equally spaced coherent sources.

*Calculating the Single-Slit Diffraction Pattern

We now use the phasor method for the addition of harmonic waves to calculate the intensity pattern shown in Figure 33-11. We assume that the slit of width a is divided into N equal intervals and that there is a point source of waves at the midpoint of each interval (Figure 33-23). If d is the distance between two adjacent sources and a is the width of the opening, we have $d = a/N$. Since the screen on which we are calculating the intensity is far from the sources, the rays from the sources to a point P on the screen are approximately parallel. The path difference between any two adjacent sources is $d \sin \theta$, and the phase difference δ is related to the path difference by

$$\frac{\delta}{2\pi} = \frac{d \sin \theta}{\lambda}$$

If A_0 is the amplitude due to a single source, the amplitude at the central maximum, where $\theta = 0$ and all the waves are in phase, is $A_{max} = NA_0$ (Figure 33-24).

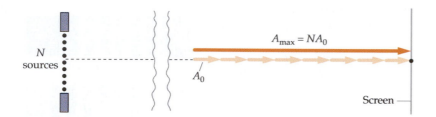

$$A_{max} = NA_0$$

N sources

A_0

Screen

We can find the amplitude at some other point at an angle θ by using the phasor method for the addition of harmonic waves. As in the addition of two, three, or four waves, the intensity is zero at any point where the phasors representing the waves form a closed polygon. In this case, the polygon has N sides (Figure 33-25). At the first minimum, the wave from the first source just below the top of the opening and the wave from the source just below the middle of the opening are 180° out of phase. In this case, the waves from the source near the top of the opening differ from those from the bottom of the opening by nearly 360°. [The phase difference is, in fact, $360° - (360°/N)$.] Thus, if the number of sources is very large, $360°/N$ is negligible and we get complete cancellation if the waves from the first and last sources are out of phase by 360°, corresponding to a path difference of 1 wavelength, in agreement with Equation 33-11.

We will now calculate the amplitude at a general point at which the waves from two adjacent sources differ in phase by δ. Figure 33-26 shows the phasor diagram for the addition of N waves, where the subsequent waves differ in phase from the first wave by δ, 2δ, ..., $(N - 1)\delta$. When N is very large and δ is very small, the phasor diagram approximates the arc of a circle. The resultant amplitude A is the length of the chord of this arc. We will calculate this resultant amplitude in terms of the phase difference ϕ between the first wave and the last wave. From Figure 33-26, we have

$$\sin \tfrac{1}{2}\phi = \frac{A/2}{r}$$

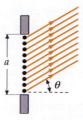

FIGURE 33-23 Diagram for calculating the diffraction pattern far away from a narrow slit. The slit width a is assumed to contain a large number of in-phase point sources separated by a distance d. The rays from these sources to a point far away are approximately parallel. The path difference for the waves from adjacent sources is $d \sin \theta$.

FIGURE 33-24 A single slit is represented by N sources, each of amplitude A_0. At the central maximum point, where $\theta = 0$, the waves from the sources add in phase, giving a resultant amplitude $A_{max} = NA_0$.

$$\delta = \frac{360°}{N}$$

FIGURE 33-25 Phasor diagram for calculating the first minimum in the single-slit diffraction pattern. When the waves from the N sources completely cancel, the N phasors form a closed polygon. The phase difference between the waves from adjacent sources is then $\delta = 360°/N$. When N is very large, the waves from the first and last sources are approximately in phase.

FIGURE 33-26 Phasor diagram for calculating the resultant amplitude due to the waves from N sources in terms of the phase difference ϕ between the wave from the first source just below the top of the slit and the wave from the last source just above the bottom of the slit. When N is very large, the resultant amplitude A is the chord of a circular arc of length $NA_0 = A_{max}$.

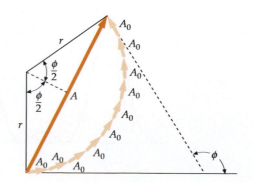

or

$$A = 2r \sin \tfrac{1}{2}\phi \qquad\qquad\qquad 33\text{-}17$$

where r is the radius of the arc. Since the length of the arc is $A_{max} = NA_0$ and the angle subtended is ϕ, we have

$$\phi = \frac{A_{max}}{r} \qquad\qquad\qquad 33\text{-}18$$

or

$$r = \frac{A_{max}}{\phi}$$

Substituting this into Equation 33-17 gives

$$A = \frac{2A_{max}}{\phi} \sin\frac{1}{2}\phi = A_{max}\frac{\sin\frac{1}{2}\phi}{\frac{1}{2}\phi}$$

Since the amplitude at the center of the central maximum ($\theta = 0$) is A_{max}, the ratio of the intensity at any other point to that at the center of the central maximum is

$$\frac{I}{I_0} = \frac{A^2}{A_{max}^2} = \left(\frac{\sin\frac{1}{2}\phi}{\frac{1}{2}\phi}\right)^2$$

or

$$I = I_0\left(\frac{\sin\frac{1}{2}\phi}{\frac{1}{2}\phi}\right)^2 \qquad\qquad\qquad 33\text{-}19$$

INTENSITY FOR A SINGLE-SLIT DIFFRACTION PATTERN

The phase difference ϕ between the first and last waves is related to the path difference $a \sin\theta$ between the top and bottom of the opening by:

$$\frac{\phi}{2\pi} = \frac{a \sin\theta}{\lambda} \qquad\qquad\qquad 33\text{-}20$$

Equation 33-19 and Equation 33-20 describe the intensity pattern shown in Figure 33-11. The first minimum occurs at $a \sin\theta = \lambda$, which is the point where the waves from the middle of the upper half and the middle of the lower half of the slit have a path difference of $\lambda/2$ and are 180° out of phase. The second minimum occurs at $a \sin\theta = 2\lambda$, where the waves from the upper half of the upper half of the slit and those from the lower half of the upper half of the slit have a path difference of $\lambda/2$ and are 180° out of phase.

There is a secondary maximum approximately midway between the first and second minima at $a \sin\theta \approx \tfrac{3}{2}\lambda$. Figure 33-27 shows the phasor diagram for determining the approximate intensity of this secondary maximum. The phase difference between the first and last waves is approximately 360° + 180°. The phasors thus complete $1\tfrac{1}{2}$ circles. The resultant amplitude is the diameter of a circle with a circumference that is two-thirds the total length A_{max}. If $C = \tfrac{2}{3}A_{max}$ is the circumference, the diameter A is

$$A = \frac{C}{\pi} = \frac{\tfrac{2}{3}A_{max}}{\pi} = \frac{2}{3\pi}A_{max}$$

and

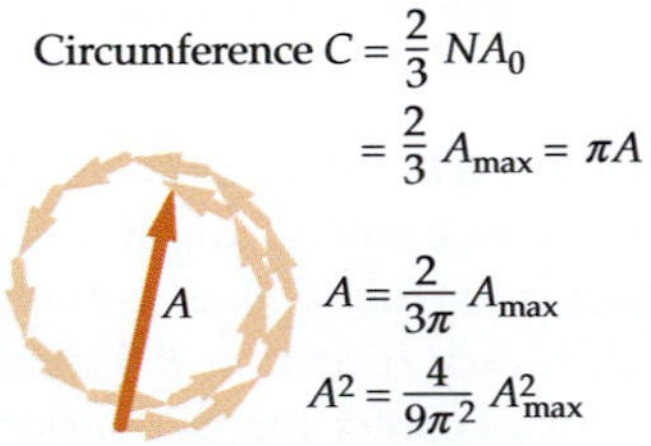

Circumference $C = \dfrac{2}{3} NA_0$

$\qquad\qquad = \dfrac{2}{3} A_{max} = \pi A$

$A = \dfrac{2}{3\pi} A_{max}$

$A^2 = \dfrac{4}{9\pi^2} A_{max}^2$

FIGURE 33-27 Phasor diagram for calculating the approximate amplitude of the first secondary maximum of the single-slit diffraction pattern. This secondary maximum occurs near the midpoint between the first and second minima when the N phasors complete $1\tfrac{1}{2}$ circles.

$$A^2 = \frac{4}{9\pi^2} A^2_{max}$$

The intensity at this point is

$$I = \frac{4}{9\pi^2} I_0 \approx \frac{1}{22.2} I_0 \qquad\qquad 33\text{-}21$$

*Calculating the Interference–Diffraction Pattern of Multiple Slits

The intensity of the two-slit interference–diffraction pattern can be calculated from the two-slit pattern (Equation 33-8) with the intensity of each slit (I_0 in that equation) replaced by the diffraction pattern intensity due to each slit, I, given by Equation 33-19. The intensity for the two-slit interference–diffraction pattern is thus

$$I = 4I_0\left(\frac{\sin\frac{1}{2}\phi}{\frac{1}{2}\phi}\right)^2 \cos^2\frac{1}{2}\delta \qquad\qquad 33\text{-}22$$

INTERFERENCE–DIFFRACTION INTENSITY FOR TWO SLITS

where ϕ is the difference in phase between rays from the top and bottom of each slit, which is related to the width of each slit by

$$\phi = \frac{2\pi}{\lambda} a \sin\theta$$

and δ is the difference in phase between rays from the centers of two adjacent slits, which is related to the slit separation by

$$\delta = \frac{2\pi}{\lambda} d \sin\theta$$

In Equation 33-22, the intensity I_0 is the intensity at $\theta = 0$ due to one slit alone.

FIVE-SLIT INTERFERENCE–DIFFRACTION PATTERN   **EXAMPLE 33 - 7**

Find the interference–diffraction intensity pattern for five equally spaced slits, where a is the width of each slit and d is the distance between adjacent slits.

PICTURE THE PROBLEM First, find the interference intensity pattern for the five slits, assuming no angular variations in the intensity due to diffraction. To do this, first construct a phasor diagram to find the amplitude of the resultant wave in an arbitrary direction θ. Intensity is proportional to the square of the amplitude. Next, correct for the variation of intensity with θ by using the single-slit diffraction pattern intensity relation (Equation 33-19 and Equation 33-20).

1. The diffraction pattern intensity I' due to a slit of width a is given by Equation 33-19 and Equation 33-20:

$$I' = I_0\left(\frac{\sin\frac{1}{2}\phi}{\frac{1}{2}\phi}\right)^2$$

where

$$\phi = \frac{2\pi}{\lambda} a \sin\theta$$

2. The interference pattern intensity I is proportional to the square of the amplitude A of the superposition of the wave functions for the light from the five slits:

$I \propto A^2$

where

$$A \sin(\alpha + \delta') = A_0 \sin \alpha + A_0 \sin(\alpha + \delta) + A_0 \sin(\alpha + 2\delta)$$
$$+ A_0 \sin(\alpha + 3\delta) + A_0 \sin(\alpha + 4\delta)$$

with $\alpha = \omega t$ and $\delta = \dfrac{2\pi}{\lambda} d \sin \theta$

3. To solve for A, we construct a phasor diagram (Figure 33-28). The amplitude A equals the sum of the projections of the individual phasors onto the resultant phasor:

$\delta' = \beta + \delta$

so

$\beta = \delta' - \delta = 2\delta - \delta = \delta$

FIGURE 33-28

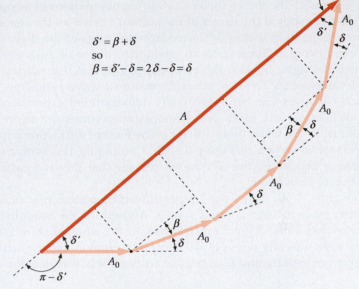

4. To find δ', we add the exterior angles. The sum of the exterior angles equals 2π. (If you walk the perimeter of a polygon you rotate through the sum of the exterior angles, and you rotate through 2π radians):

$2(\pi - \delta') + 4\delta = 2\pi$

so

$\delta' = 2\delta$

5. Solve for A from the figure:

$A = 2A_0 \cos \delta' + 2A_0 \cos \beta + A_0$

6. Substitute for δ' using the step 4 result, and substitute for β using the relation $\beta = \delta$:

$A = A_0(2 \cos 2\delta + 2 \cos \delta + 1)$

7. Square both sides to relate the intensities. Recall, I' is the intensity from a single slit, and A_0 is the amplitude from a single slit:

$A^2 = A_0^2(2 \cos 2\delta + 2 \cos \delta + 1)^2$

so

$I = I'(2 \cos 2\delta + 2 \cos \delta + 1)^2$

8. Substitute for I' using the step 1 result:

$$\boxed{I = I_0 \left(\frac{\sin \frac{1}{2}\phi}{\frac{1}{2}\phi} \right)^2 (2 \cos 2\delta + 2 \cos \delta + 1)^2}$$

$$\text{where } \phi = \frac{2\pi}{\lambda} a \sin \theta \text{ and } \delta = \frac{2\pi}{\lambda} d \sin \theta$$

⦿ PLAUSIBILITY CHECK If $\theta = 0$, both $\phi = 0$ and $\delta = 0$. So, for $\theta = 0$, step 5 becomes $A = 5A_0$ and step 8 becomes $I = 5^2 I_0 = 25 I_0$ as expected.

33-6 Fraunhofer and Fresnel Diffraction

Diffraction patterns, like the single-slit pattern shown in Figure 33-11, that are observed at points for which the rays from an aperture or an obstacle are nearly parallel are called **Fraunhofer diffraction patterns**. Fraunhofer patterns can be observed at great distances from the obstacle or the aperture so that the rays reaching any point are approximately parallel, or they can be observed using a lens to focus parallel rays on a viewing screen placed in the focal plane of the lens.

The diffraction pattern observed near an aperture or an obstacle is called a **Fresnel diffraction pattern.** Because the rays from an aperture or an obstacle close to a screen cannot be considered parallel, Fresnel diffraction is much more difficult to analyze. Figure 33-29 illustrates the difference between the Fresnel and the Fraunhofer patterns for a single slit.[†]

Figure 33-30a shows the Fresnel diffraction pattern of an opaque disk. Note the bright spot at the center of the pattern caused by the constructive interference of the light waves diffracted from the edge of the disk. This pattern is of some historical interest. In an attempt to discredit Augustin Fresnel's wave theory of light, Siméon Poisson pointed out that it predicted a bright spot at the center of the shadow, which he assumed was a ridiculous contradiction of fact. However, Fresnel immediately demonstrated experimentally that such a spot does, in fact, exist. This demonstration convinced many doubters of the validity of the wave theory of light. The Fresnel diffraction pattern of a circular aperture is shown in Figure 33-30b. Comparing this with the pattern of the opaque disk in Figure 33-30a, we can see that the two patterns are complements of each other.

Figure 33-31a shows the Fresnel diffraction pattern of a straight edge illuminated by light from a point source. A graph of the intensity versus distance (measured along a line perpendicular to the edge) is shown in Figure 33-31b. The light intensity does not fall abruptly to zero in the geometric shadow, but it decreases rapidly and is negligible within a few wavelengths of the edge. The

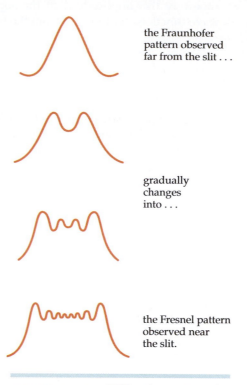

As the screen is moved closer,

the Fraunhofer pattern observed far from the slit . . .

gradually changes into . . .

the Fresnel pattern observed near the slit.

FIGURE 33-29 Diffraction patterns for a single slit at various screen distances.

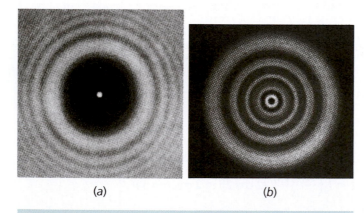

(a)   (b)

FIGURE 33-30 (a) The Fresnel diffraction pattern of an opaque disk. At the center of the shadow, the light waves diffracted from the edge of the disk are in phase and produce a bright spot called the *Poisson spot.* (b) The Fresnel diffraction pattern of a circular aperture. Compare this with Figure 33-30a.

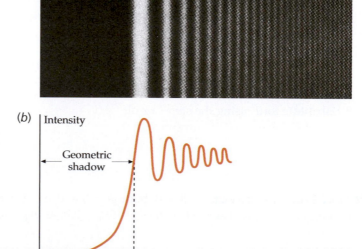

(a)

(b) Intensity

Geometric shadow

Edge   Distance

FIGURE 33-31 (a) The Fresnel diffraction of a straightedge. (b) A graph of intensity versus distance along a line perpendicular to the edge.

[†] See Richard E. Haskel, "A Simple Experiment on Fresnel Diffraction," *American Journal of Physics* 38 (1970): 1039.

Fresnel diffraction pattern of a rectangular aperture is shown in Figure 33-32. These patterns cannot be seen with extended light sources like an ordinary light-bulb, because the dark fringes of the pattern produced by light from one point on the source overlap the bright fringes of the pattern produced by light from another point.

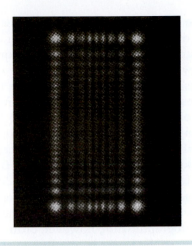

FIGURE 33-32 The Fresnel diffraction pattern of a rectangular aperture.

33-7 Diffraction and Resolution

Diffraction due to a circular aperture has important implications for the resolution of many optical instruments. Figure 33-33 shows the Fraunhofer diffraction pattern of a circular aperture. The angle θ subtended by the first diffraction minimum is related to the wavelength and the diameter of the opening D by

$$\sin \theta = 1.22 \frac{\lambda}{D} \qquad 33\text{-}23$$

Equation 33-23 is similar to Equation 33-9 except for the factor 1.22, which arises from the mathematical analysis, and is similar to the equation for a single slit but more complicated because of the circular geometry. In many applications, the angle θ is small, so $\sin \theta$ can be replaced by θ. The first diffraction minimum is then at an angle θ given by

$$\theta \approx 1.22 \frac{\lambda}{D} \qquad 33\text{-}24$$

FIGURE 33-33 The Fraunhofer diffraction pattern of a circular aperture.

Figure 33-34 shows two point sources that subtend an angle α at a circular aperture far from the sources. The intensities of the Fraunhofer diffraction pattern are also indicated in this figure. If α is much greater than $1.22\lambda/D$, the sources will be seen as two sources. However, as α is decreased, the overlap of the diffraction patterns increases, and it becomes difficult to distinguish the two sources from one source. At the critical angular separation, α_c, given by

$$\alpha_c = 1.22 \frac{\lambda}{D} \qquad 33\text{-}25$$

(a)

the first minimum of the diffraction pattern of one source falls on the central maximum of the other source. These objects are said to be just resolved by **Rayleigh's criterion for resolution.** Figure 33-35 shows the diffraction patterns for two sources when α is greater than the critical angle for resolution and when α is just equal to the critical angle for resolution.

Equation 33-25 has many applications. The *resolving power* of an optical instrument, such as a microscope or telescope, is the ability of the instrument to resolve

(b)

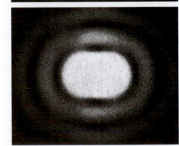

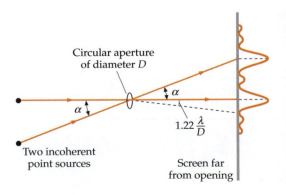

FIGURE 33-34 Two distant sources that subtend an angle α. If α is much greater than $1.22\lambda/D$, where λ is the wavelength of light and D is the diameter of the aperture, the diffraction patterns have little overlap and the sources are easily seen as two sources. If α is not much greater than $1.22\lambda/D$, the overlap of the diffraction patterns makes it difficult to distinguish two sources from one.

Circular aperture of diameter D

$1.22 \frac{\lambda}{D}$

Two incoherent point sources

Screen far from opening

FIGURE 33-35 The diffraction patterns for a circular aperture and two incoherent point sources when (a) α is much greater than $1.22\lambda/D$ and (b) when α is at the limit of resolution, $\alpha_c = 1.22\lambda/D$.

two objects that are close together. The images of the objects tend to overlap because of diffraction at the entrance aperture of the instrument. We can see from Equation 33-25 that the resolving power can be increased either by increasing the diameter D of the lens (or mirror) or by decreasing the wavelength λ. Astronomical telescopes use large objective lenses or mirrors to increase their resolution as well as to increase their light-gathering power. An array of 27 radio antennas (Figure 33-36) mounted on rails can be configured to form a single telescope with a resolution distance of 36 km (22 mi). In a microscope, a film of transparent oil with index of refraction of approximately 1.55 is sometimes used under the objective to decrease the wavelength of the light ($\lambda' = \lambda/n$). The wavelength can be reduced further by using ultraviolet light and photographic film; however, ordinary glass is opaque to ultraviolet light, so the lenses in an ultraviolet microscope must be made from quartz or fluorite. To obtain very high resolutions, electron microscopes are used—microscopes that use electrons rather than light. The wavelengths of electrons vary inversely with the square root of their kinetic energy and can be made as small as desired.[†]

FIGURE 33-36 The very large array (VLA) of radio antennas is located near Socorro, New Mexico. The 25-m-diameter antennas are mounted on rails, which can be arranged in several configurations, and can be extended over a diameter of 36 km. The data from the antennas are combined electronically, so the array is really a single high-resolution telescope.

[†] The wave properties of electrons are discussed in Chapter 34.

PHYSICS IN THE LIBRARY   **EXAMPLE 33-8**   **Put It in Context**

While studying in the library, you lean back in your chair and ponder the small holes you notice in the ceiling tiles. You notice that the holes are approximately 5 mm apart. You can clearly see the holes directly above you, about 2 m up, but the tiles far away do not appear to have these holes. You wonder if the reason you cannot see the distant holes is because they are not within the criteria for resolution established by Rayleigh. Is this a feasible explanation for the disappearance of the holes? You notice the holes disappear about 20 m from you.

PICTURE THE PROBLEM We will need to make assumptions about the situation. If we use Equation 33-25, we will need to know the wavelength of light and the aperture diameter. Assuming our pupil is the aperture, we can assume approximately 5 mm for the diameter. (This is the number used in our physics textbook.) The light is probably centered around 500 nm or so.

1. The angular limit for resolution by the eye depends on the ratio of the wavelength and the diameter of the pupil:

$$\theta_c \approx 1.22 \frac{\lambda}{D}$$

2. The angle subtended by two holes depends on their separation distance d and their distance L from your eye:

$$\theta \approx \frac{d}{L}$$

3. Equating the two angles and putting in the numbers gives:

$$\frac{d}{L} \approx 1.22 \frac{\lambda}{D}$$

$$\frac{5 \text{ mm}}{L} \approx 1.22 \frac{500 \text{ nm}}{5 \text{ mm}}$$

4. Solving for L gives:

$$L = 40 \text{ m}$$

5. By a factor of 2, 40 m is too large. However, you are suspect of the value given for the pupil diameter in your physics textbook. You know the pupil is smaller when the light is bright, and the library ceiling is very bright and colored white. An online search for eye pupil diameter soon turns up the information you need. The pupil diameter ranges from 2 to 3 mm up to 7 mm:

| Success. If the pupil diameter is 2.5 mm, the value of L is 20 m. |

It is instructive to compare the limitation on resolution of the eye due to diffraction, as seen in Example 33-8, with the limitation on resolution due to the separation of the receptors (cones) on the retina. To be seen as two distinct objects, the images of the objects must fall on the retina on two nonadjacent cones. (See Problem 79 in Chapter 32.) Because the retina is about 2.5 cm from the eye lens, the distance y on the retina corresponding to an angular separation of 1.5×10^{-4} rad is found from

$$\alpha_c = 1.5 \times 10^{-4} \text{ rad} = \frac{y}{2.5 \text{ cm}}$$

or

$$y = 3.75 \times 10^{-4} \text{ cm} = 3.75 \times 10^{-6} \text{ m} = 3.75 \ \mu\text{m}$$

The actual separation of the cones in the fovea centralis, where the cones are the most tightly packed, is about 1 μm. Outside this region, they are about 3 μm to 5 μm apart.

*33-8 Diffraction Gratings

A useful tool for measuring the wavelength of light is the **diffraction grating,** which consists of a large number of equally spaced lines or slits on a flat surface. Such a grating can be made by cutting parallel, equally spaced grooves on a glass or metal plate with a precision ruling machine. With a reflection grating, light is reflected from the ridges between the lines or grooves. Phonograph records and compact disks exhibit some of the properties of reflection gratings. In a transmission grating, the light passes through the clear gaps between the rulings. Inexpensive, optically produced plastic gratings with 10,000 or more slits per centimeter are common items. The spacing of the slits in a grating with 10,000 slits per centimeter is $d = (1 \text{ cm})/10,000 = 10^{-4}$ cm.

Consider a plane light wave incident normally on a transmission grating (Figure 33-37). Assume that the width of each slit is very small so that it produces a widely diffracted beam. The interference pattern produced on a screen a large distance from the grating is due to a large number of equally spaced light sources. Suppose we have N slits with separation d between adjacent slits. At $\theta = 0$, the light from each slit is in phase with that from all the other slits, so the

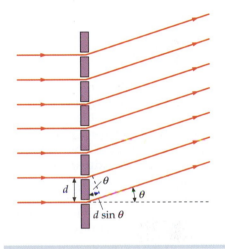

FIGURE 33-37 Light incident normally on a diffraction grating. At an angle θ, the path difference between rays from adjacent slits is $d \sin \theta$.

Compact disks act as reflection gratings.

amplitude of the wave is NA_0, where A_0 is the amplitude from each slit, and the intensity is $N^2 I_0$, where I_0 is the intensity due to a single slit alone. At an angle θ such that $d \sin \theta = \lambda$, the path difference between any two successive slits is λ, so again the light from each slit is in phase with that from all the other slits and the intensity is $N^2 I_0$. The interference maxima are thus at angles θ given by

$$d \sin \theta_m = m\lambda, \qquad m = 0, 1, 2, \ldots \qquad \text{33-26}$$

The position of an interference maximum does not depend on the number of sources, but the more sources there are, the sharper and more intense the maximum will be.

To see that the interference maxima will be sharper when there are many slits, consider the case of N illuminated slits, where N is large ($N \gg 1$). The distance from the first slit to the Nth slit is $(N - 1)d \approx Nd$. When the path difference for the light from the first slit and that from the Nth slit is λ, the resulting intensity will be zero. (We saw this in our discussion of single-slit diffraction.) Since the first and Nth slits are separated by approximately Nd, the intensity will be zero at angle θ_{min} given by

$$Nd \sin \theta_{min} = \lambda$$

so

$$\theta_{min} \approx \sin \theta_{min} = \frac{\lambda}{Nd}$$

The width of the interference maximum $2\theta_{min}$ is thus inversely proportional to N. Therefore, the greater the number of illuminated slits N, the sharper the maximum. Since the intensity in the maximum is proportional to $N^2 I_0$, the intensity in the maximum times the width of the maximum is proportional to $N I_0$. The intensity times the width is a measure of power per unit length in the maximum.

Figure 33-38a shows a student spectroscope that uses a diffraction grating to analyze light. In student laboratories, the light source is typically a glass tube containing atoms of a gas (e.g., helium or sodium vapor) that are excited by a bombardment of electrons accelerated by high voltage across the tube. The light emitted by such a source contains only certain wavelengths that are characteristic of the atoms in the source. Light from the source passes through a narrow collimating slit and is made parallel by a lens. Parallel light from the lens is incident on the grating. Instead of falling on a screen a large distance away, the parallel light from the grating is focused by a telescope and viewed by the eye. The telescope is mounted on a rotating platform that has been calibrated so that

FIGURE 33-38 (a) A typical student spectroscope. Light from a collimating slit near the source is made parallel by a lens and falls on a grating. The diffracted light is viewed with a telescope at an angle that can be accurately measured. (b) Aerial view of the very large array (VLA) radio telescope in New Mexico. Radio signals from distant galaxies add constructively when Equation 33-26 is satisfied, where d is the distance between two adjacent telescopes.

(a)

(b)

the angle θ can be measured. In the forward direction ($\theta = 0$), the central maximum for all wavelengths is seen. If light of a particular wavelength λ is emitted by the source, the first interference maximum is seen at the angle θ given by Equation 33-26 with $m = 1$. Each wavelength emitted by the source produces a separate image of the collimating slit in the spectroscope called a **spectral line.** The set of lines corresponding to $m = 1$ is called the **first-order spectrum.** The **second-order spectrum** corresponds to $m = 2$ for each wavelength. Higher orders may be seen, providing the angle θ given by Equation 33-26 is less than 90°. Depending on the wavelengths, the orders may be mixed; that is, the third-order line for one wavelength may occur before the second-order line for another wavelength. If the spacing of the slits in the grating is known, the wavelengths emitted by the source can be determined by measuring the angle θ.

RESOLVING THE SODIUM D LINES   **EXAMPLE 33-9**

Sodium light is incident on a diffraction grating with 12,000 lines per centimeter. At what angles will the two yellow lines (called the sodium D lines) of wavelengths 589.00 nm and 589.59 nm be seen in the first order?

PICTURE THE PROBLEM Apply $d \sin \theta_m = m\lambda$ to each wavelength, with $m = 1$ and $d = 1 \text{ cm}/12{,}000$.

1. The angle θ_m is given by $d \sin \theta_m = m\lambda$ with $m = 1$:

$$\sin \theta_1 = \frac{\lambda}{d}$$

2. Calculate θ_1 for $\lambda = 589.00$ nm:

$$\theta_1 = \sin^{-1}\left[\frac{589.00 \times 10^{-9} \text{ m}}{(1 \text{ cm}/12{,}000)} \times \left(\frac{100 \text{ cm}}{1 \text{ m}}\right)\right]$$

$$= \boxed{44.98°}$$

3. Repeat the calculation for $\lambda = 589.59$ nm:

$$\theta_1 = \sin^{-1}\left[\frac{589.59 \times 10^{-9} \text{ m}}{(1 \text{ cm}/12{,}000)} \times \left(\frac{100 \text{ cm}}{1 \text{ m}}\right)\right]$$

$$= \boxed{45.03°}$$

REMARKS Note that light of longer wavelength is diffracted through larger angles.

EXERCISE Find the angles for the two yellow lines if the grating has 15,000 lines per centimeter. (*Answer* 62.07° and 62.18°)

An important feature of a spectroscope is its ability to resolve spectral lines of two nearly equal wavelengths λ_1 and λ_2. For example, the two prominent yellow lines in the spectrum of sodium have wavelengths 589.00 and 589.59 nm. These can be seen as two separate wavelengths if their interference maxima do not overlap. According to Rayleigh's criterion for resolution, these wavelengths are resolved if the angular separation of their interference maxima is greater than the angular separation between an interference maximum and the first interference minimum on either side of it. The **resolving power** of a diffraction grating is defined to be $\lambda/|\Delta\lambda|$, where $|\Delta\lambda|$ is the smallest difference between two nearby wavelengths, each approximately equal to λ, that may be resolved. The resolving power is proportional to the number of slits illuminated because the more slits illuminated, the sharper the interference maxima. The resolving power R can be shown to be

$$R = \frac{\lambda}{|\Delta\lambda|} = mN$$

33-27

where N is the number of slits and m is the order number (see Problem 76). We can see from Equation 33-27 that to resolve the two yellow lines in the sodium spectrum the resolving power must be

$$R = \frac{589.00 \text{ nm}}{589.59 \text{ nm} - 589.00 \text{ nm}} = 998$$

Thus, to resolve the two yellow sodium lines in the first order ($m = 1$), we need a grating containing 998 or more slits in the area illuminated by the light.

*Holograms

An interesting application of diffraction gratings is the production of a three-dimensional photograph called a **hologram** (Figure 33-39). In an ordinary photograph, the intensity of reflected light from an object is recorded on a film. When the film is viewed by transmitted light, a two-dimensional image is produced. In a hologram, a beam from a laser is split into two beams, a reference beam and an object beam. The object beam reflects from the object to be photographed and the interference pattern between it, and the reference beam is recorded on a photographic film. This can be done because the laser beam is coherent so that the relative phase difference between the reference beam and the object beam can be kept constant during the exposure. The interference fringes on the film act as a diffraction grating. When the film is illuminated with a laser, a three-dimensional image of the object is produced.

Holograms that you see on credit cards or postage stamps, called rainbow holograms, are more complicated. A horizontal strip of the original hologram is used to make a second hologram. The three-dimensional image can be seen as the viewer moves from side to side, but if viewed with laser light, the image disappears when the viewer's eyes move above or below the slit image. When viewed with white light, the image is seen in different colors as the viewer moves in the vertical direction.

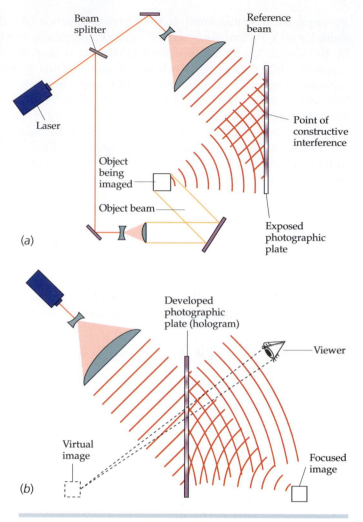

FIGURE 33-39 (*a*) The production of a hologram. The interference pattern produced by the reference beam and object beam is recorded on a photographic film. (*b*) When the film is developed and illuminated by coherent laser light, a three-dimensional image is seen.

(a)

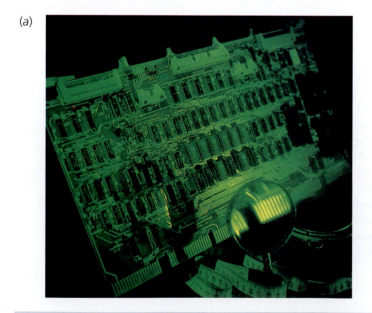

(b)

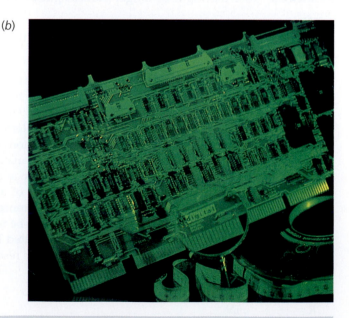

A hologram viewed from two different angles. Note that different parts of the circuit board appear behind the front magnifying lens.

Topic	Relevant Equations and Remarks	
1. Interference	Two superposing light waves interfere if their phase difference remains constant for a time long enough to observe. They interfere constructively if their phase difference is zero or an integer times 360°. They interfere destructively if their phase difference is 180° or an odd integer times 180°.	
Phase difference due to a path difference	$$\frac{\delta}{2\pi} = \frac{\Delta r}{\lambda}$$	33-1
Phase difference due to reflection	A phase difference of 180° is introduced when a light wave is reflected from a boundary between two media for which the wave speed is greater on the incident-wave side of the boundary.	
Thin films	The interference of light waves reflected from the front and back surfaces of a thin film produces interference fringes, commonly observed in soap films or oil films. The difference in phase between the two reflected waves results from the path difference of twice the thickness of the film plus any phase change due to reflection of one or both of the rays.	
Two slits	The path difference at an angle θ on a screen far away from two narrow slits separated by a distance d is $d \sin \theta$. If the intensity due to each slit separately is I_0, the intensity at points of constructive interference is $4I_0$, and the intensity at points of destructive interference is zero.	
Interference maxima (sources in phase)	$d \sin \theta_m = m\lambda, \qquad m = 0, 1, 2, \ldots$	33-2
Interference minima (sources 180° out of phase)	$d \sin \theta_m = (m - \tfrac{1}{2})\lambda, \qquad m = 1, 2, 3, \ldots$	33-3
2. Diffraction	Diffraction occurs whenever a portion of a wavefront is limited by an obstacle or aperture. The intensity of light at any point in space can be computed using Huygens's principle by taking each point on the wavefront to be a point source and computing the resulting interference pattern.	
Fraunhofer patterns	Fraunhofer patterns are observed at great distances from the obstacle or aperture so that the rays reaching any point are approximately parallel, or they can be observed using a lens to focus parallel rays on a viewing screen placed in the focal plane of the lens.	
Fresnel patterns	Fresnel patterns are observed at points close to the source.	
Single slit	When light is incident on a single slit of width a, the intensity pattern on a screen far away shows a broad central diffraction maximum that decreases to zero at an angle θ_1 given by	
	$$\sin \theta_1 = \frac{\lambda}{a}$$	33-9
	The width of the central maximum is inversely proportional to the width of the slit. The zeros in the single-slit diffraction pattern occur at angles given by	
	$a \sin \theta_m = m\lambda, \qquad m = 1, 2, 3, \ldots$	33-11
	The maxima on either side of the central maxima have intensities that are much smaller than the intensity of the central maxima.	

Two slits	The interference–diffraction pattern of two slits is the two-slit interference pattern modulated by the single-slit diffraction pattern.
Resolution of two sources	When light from two point sources that are close together passes through an aperture, the diffraction patterns of the sources may overlap. If the overlap is too great, the two sources cannot be resolved as two separate sources. When the central diffraction maximum of one source falls at the diffraction minimum of the other source, the two sources are said to be just resolved by Rayleigh's criterion for resolution. For a circular aperture of diameter D, the critical angular separation of two sources for resolution by Rayleigh's criterion is
Rayleigh's criterion	$$\alpha_c = 1.22 \frac{\lambda}{D} \qquad \qquad \text{33-25}$$
*Gratings	A diffraction grating consisting of a large number of equally spaced lines or slits is used to measure the wavelength of light emitted by a source. The positions of the mth order interference maxima from a grating are at angles given by $$d \sin \theta_m = m\lambda, \quad m = 0, 1, 2 \ldots \qquad \text{33-26}$$ The resolving power of a grating is $$R = \frac{\lambda}{|\Delta\lambda|} = mN \qquad \qquad \text{33-27}$$ where N is the number of slits of the grating that are illuminated and m is the order number.
3. *Phasors	Two or more harmonic waves can be added by representing each wave as a two-dimensional vector called a phasor. The phase difference between two harmonic waves is represented as the angle between the phasors.

PROBLEMS

- Single-concept, single-step, relatively easy
- •• Intermediate-level, may require synthesis of concepts
- ••• Challenging
- **SSM** Solution is in the *Student Solutions Manual*
- **iSOLVE** Problems available on iSOLVE online homework service
- **iSOLVE✓** These "Checkpoint" online homework service problems ask students additional questions about their confidence level, and how they arrived at their answer.

In a few problems, you are given more data than you actually need; in a few other problems, you are required to supply data from your general knowledge, outside sources, or informed estimates.

Conceptual Problems

1 • **SSM** When destructive interference occurs, what happens to the energy in the light waves?

2 • Which of the following pairs of light sources are coherent: (*a*) two candles, (*b*) one point source and its image in a plane mirror, (*c*) two pinholes uniformly illuminated by the same point source, (*d*) two headlights of a car, (*e*) two images of a point source due to reflection from the front and back surfaces of a soap film.

3 • The spacing between Newton's rings decreases rapidly as the diameter of the rings increases. Explain qualitatively why this occurs.

4 •• If the angle of a wedge-shaped air film, such as that in Example 33-2, is too large, fringes are not observed. Why?

5 •• Why must a film that is used to observe interference colors be thin?

6 • **SSM** A loop of wire is dipped in soapy water and held so that the soap film is vertical. (*a*) Viewed by reflection with white light, the top of the film appears black. Explain why. (*b*) Below the black region are colored bands. Is the first band red or violet? (*c*) Describe the appearance of the film when it is viewed by *transmitted* light.

7 • As the width of a slit producing a single-slit diffraction pattern is slowly and steadily reduced, how will the diffraction pattern change?

8 • Equation 33-2, which is $d \sin \theta_m = m\lambda$, and Equation 33-11, which is $a \sin \theta_m = m\lambda$, are sometimes confused. For each equation, define the symbols and explain the equation's application.

9 • When a diffraction grating is illuminated with white light, the first-order maximum of green light (*a*) is closer to the central maximum than that of red light. (*b*) is closer to the central maximum than that of blue light. (*c*) overlaps the second-order maximum of red light. (*d*) overlaps the second-order maximum of blue light.

10 • **SSM** A double-slit interference experiment is set up in a chamber that can be evacuated. Using monochromatic light, an interference pattern is observed when the chamber is open to air. As the chamber is evacuated, one will note that (*a*) the interference fringes remain fixed. (*b*) the interference fringes move closer together. (*c*) the interference fringes move farther apart. (*d*) the interference fringes disappear completely.

11 • True or false:

(*a*) When waves interfere destructively, the energy is converted into heat energy.

(*b*) Interference is observed only for waves from coherent sources.

(*c*) In the Fraunhofer diffraction pattern for a single slit, the narrower the slit, the wider the central maximum of the diffraction pattern.

(*d*) A circular aperture can produce both a Fraunhofer diffraction pattern and a Fresnel diffraction pattern.

(*e*) The ability to resolve two point sources depends on the wavelength of the light.

Estimation and Approximation

12 • **SSM** It is claimed that the Great Wall of China is the only human object that can be seen from space with the naked eye. Make an argument in support of this claim based on the resolving power of the human eye. Evaluate the validity of your argument for observers both in low-earth orbit (~ 400 km altitude) and on the moon.

13 • Naturally occuring coronas (brightly colored rings) are sometimes seen around the moon or the sun when viewed through a thin cloud. (Warning: When viewing a sun corona, be sure that the entire sun is blocked by the edge of a building, a tree, or a traffic pole to safeguard your eyes.) These coronas are due to diffraction of light by small water droplets in the cloud. A typical angular diameter for a coronal ring is about $10°$. From this, estimate the size of the water droplets in the cloud. Assume that the water droplets can be modeled as opaque disks with the same radius as the droplet, and that the Fraunhofer diffraction pattern from an opaque disk is the same as the pattern from an aperture of the same diameter. (This statement is known as *Babinet's principle*.)

14 • An artificial corona (see Problem 13) can be made by placing a suspension of polystyrene microspheres in water. Polystyrene microspheres are small, uniform spheres made of plastic with an index of refraction of 1.59. Assuming that the water has a refractive index of 1.33, what is the angular diameter of such an artificial corona if 5 μm diameter particles are illuminated by a helium–neon laser with wavelength in air $\lambda = 632.8$ nm?

15 • Coronas (see Problem 13) can be caused by pollen grains, typically of birch or pine. Such grains are irregular in shape, but they can be treated as if they had an average diameter of approximately 25 μm. What is the coronal radius (in degrees) for blue light? What is the coronal radius (in degrees) for red light?

16 •• **SSM** Human hair has a diameter of approximately 70 μm. If we illuminate a hair using a helium–neon laser with wavelength $\lambda = 632.8$ nm and intercept the light scattered from the hair on a screen 10 m away, what will be the separation of the first diffraction peak from the center? (The diffraction pattern of a hair with diameter d is the same as the diffraction pattern of a single slit with width $a = d$.)

Phase Difference and Coherence

17 • **ISOLVE** (*a*) What minimum path difference is needed to introduce a phase shift of $180°$ in light of wavelength 600 nm? (*b*) What phase shift will that path difference introduce in light of wavelength 800 nm?

18 • **ISOLVE** Light of wavelength 500 nm is incident normally on a film of water 10^{-4} cm thick. The index of refraction of water is 1.33. (*a*) What is the wavelength of the light in the water? (*b*) How many wavelengths are contained in the distance $2t$, where t is the thickness of the film? (*c*) What is the phase difference between the wave reflected from the top of the air–water interface and the wave reflected from the bottom of the water–air interface after it has traveled this distance?

19 •• **SSM** **ISOLVE✓** Two coherent microwave sources that produce waves of wavelength 1.5 cm are in the xy plane, one on the y axis at $y = 15$ cm and the other at $x = 3$ cm, $y = 14$ cm. If the sources are in phase, find the difference in phase between the two waves from these sources at the origin.

Interference in Thin Films

20 • A wedge-shaped film of air is made by placing a small slip of paper between the edges of two flat plates of glass. Light of wavelength 700 nm is incident normally on the glass plates, and interference bands are observed by reflection. (*a*) Is the first band near the point of contact of the plates dark or bright? Why? (*b*) If there are five dark bands per centimeter, what is the angle of the wedge?

21 •• **SSM** The diameters of fine fibers can be accurately measured using interference patterns. Two optically flat pieces of glass of length L are arranged with the wire between them, as shown in Figure 33-40. The setup is illuminated by monochromatic light, and the resulting interference fringes are detected. Suppose that $L = 20$ cm and that yellow sodium light ($\lambda \approx 590$ nm) is used for illumination. If 19 bright fringes are seen along this 20-cm distance, what are the limits on the diameter of the wire? (*Hint:* The nineteenth fringe might not be right at the end, but you do not see a twentieth fringe at all.)

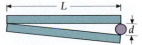

FIGURE 33-40 Problem 21

22 •• **SOLVE**✔ Light of wavelength 600 nm is used to illuminate two glass plates at normal incidence. The plates are 22 cm in length, touch at one end, and are separated at the other end by a wire of radius 0.025 mm. How many bright fringes appear along the total length of the plates?

23 •• A thin film having an index of refraction of 1.5 is surrounded by air. It is illuminated normally by white light and is viewed by reflection. Analysis of the resulting reflected light shows that the wavelengths 360, 450, and 602 nm are the only missing wavelengths in or near the visible portion of the spectrum. That is, for these wavelengths, there is destructive interference. (a) What is the thickness of the film? (b) What visible wavelengths are brightest in the reflected interference pattern? (c) If this film were resting on glass with an index of refraction of 1.6, what wavelengths in the visible spectrum would be missing from the reflected light?

24 •• **SOLVE** A drop of oil ($n = 1.22$) floats on water ($n = 1.33$). When reflected light is observed from above, as shown in Figure 33-41, what is the thickness of the drop at the point where the second red fringe, counting from the edge of the drop, is observed? Assume red light has a wavelength of 650 nm.

FIGURE 33-41
Problem 24

25 •• A film of oil of index of refraction $n = 1.45$ rests on an optically flat piece of glass of index of refraction $n = 1.6$. When illuminated with white light at normal incidence, light of wavelengths 690 nm and 460 nm is predominant in the reflected light. Determine the thickness of the oil film.

26 •• **SSM** **SOLVE**✔ A film of oil of index of refraction $n = 1.45$ floats on water ($n = 1.33$). When illuminated with white light at normal incidence, light of wavelengths 700 nm and 500 nm is predominant in the reflected light. Determine the thickness of the oil film.

Newton's Rings

27 •• **SSM** A Newton's ring apparatus consists of a plano-convex glass lens with radius of curvature R that rests on a flat glass plate, as shown in Figure 33-42. The thin film is air of variable thickness. The pattern is viewed by reflected light. (a) Show that for a thickness t the condition for a bright (constructive) interference ring is

$$t = \left(m + \frac{1}{2}\right)\frac{\lambda}{2}, \qquad m = 0, 1, 2, \ldots$$

(b) Apply the Pythagorean theorem to the triangle of sides r, $R - t$, and hypotenuse R to show that for $t \ll R$, the radius of a fringe is related to t by

$$r = \sqrt{2tR}$$

(c) How would the transmitted pattern look in comparison with the reflected one? (d) Use $R = 10$ m and a lens diameter of 4 cm. How many bright fringes would you see if the apparatus was illuminated by yellow sodium light ($\lambda \approx 590$ nm) and viewed by reflection? (e) What would be the diameter of the sixth bright fringe? (f) If the glass used in the apparatus has an index of refraction $n = 1.5$ and water ($n_w = 1.33$) is placed between the two pieces of glass, what change will take place in the bright-fringe pattern?

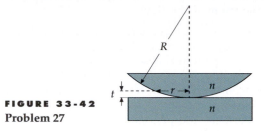

FIGURE 33-42
Problem 27

28 •• **SOLVE**✔ A plano-convex glass lens of radius of curvature 2.0 m rests on an optically flat glass plate. The arrangement is illuminated from above with monochromatic light of 520-nm wavelength. The indexes of refraction of the lens and plate are 1.6. Determine the radii of the first and second bright fringe in the reflected light. (Use the results from Problem 27b to relate r to t.)

29 •• Suppose that before the lens of Problem 28 is placed on the plate, a film of oil of refractive index 1.82 is deposited on the plate. What will then be the radii of the first and second bright fringes? (Use the results from Problem 27b to relate r to t.)

Two-Slit Interference Pattern

30 • **SSM** Two narrow slits separated by 1 mm are illuminated by light of wavelength 600 nm, and the interference pattern is viewed on a screen 2 m away. Calculate the number of bright fringes per centimeter on the screen.

31 • **SOLVE**✔ Using a conventional two-slit apparatus with light of wavelength 589 nm, 28 bright fringes per centimeter are observed on a screen 3 m away. What is the slit separation?

32 • Light of wavelength 633 nm from a helium–neon laser is shone normally on a plane containing two slits. The first interference maximum is 82 cm from the central maximum on a screen 12 m away. (a) Find the separation of the slits. (b) How many interference maxima can be observed?

33 •• **SOLVE** Two narrow slits are separated by a distance d. Their interference pattern is to be observed on a screen a large distance L away. (a) Calculate the spacing Δy of the maxima on the screen for light of wavelength 500 nm, when $L = 1$ m and $d = 1$ cm. (b) Would you expect to be able to observe the interference of light on the screen for this situation? (c) How close together should the slits be placed for the maxima to be separated by 1 mm for this wavelength and screen distance?

34 •• Light is incident at an angle ϕ with the normal to a vertical plane containing two slits of separation d (Figure 33-43). Show that the interference maxima are located at angles θ_m given by $\sin \theta_m + \sin \phi = m\lambda/d$.

FIGURE 33-43 Problems 34 and 35

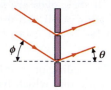

35 •• SSM White light falls at an angle of 30° to the normal of a plane containing a pair of slits separated by 2.5 μm. What visible wavelengths give a bright interference maximum in the transmitted light in the direction normal to the plane? (See Problem 34.)

36 •• Two small speakers are separated by a distance of 5 cm, as shown in Figure 33-44. The speakers are driven in phase with a sine wave signal of frequency 10 kHz. A small microphone is placed a distance 1 m away from the speakers on the axis running through the middle of the two speakers, and the microphone is then moved perpendicular to the axis. Where does the microphone record the first minimum and the first maximum of the interference pattern from the speakers? The speed of sound in air is 343 m/s.

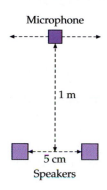

FIGURE 33-44 Problem 36

Diffraction Pattern of a Single Slit

37 • SOLVE Light of wavelength 600 nm is incident on a long narrow slit. Find the angle of the first diffraction minimum if the width of the slit is (a) 1 mm, (b) 0.1 mm, and (c) 0.01 mm.

38 • SOLVE✓ Plane microwaves are incident on a thin metal sheet with a long, narrow slit of width 5 cm in it. The first diffraction minimum is observed at $\theta = 37°$. What is the wavelength of the microwaves?

39 •• SSM Measuring the distance to the moon (lunar ranging) is routinely done by firing short-pulse lasers and measuring the time it takes for the pulses to reflect back from the moon. A pulse is fired from the earth; to send it out, the pulse is expanded so that it fills the aperture of a 6-in-diameter telescope. (a) Assuming the only thing spreading the beam out to be diffraction, how large will the beam be when it reaches the moon, 382,000 km away? (b) The pulse is reflected off a retroreflecting mirror left by the Apollo 11 astronauts. If the diameter of the mirror is 20 in, how large will the beam be when it gets back to the earth? (c) What fraction of the power of the beam is reflected back to the earth? (d) If the beam is refocused

on return by the same 6 in telescope, what fraction of the original beam energy is recaptured? Ignore any atmospheric losses.

Interference–Diffraction Pattern of Two Slits

40 • How many interference maxima will be contained in the central diffraction maximum in the interference–diffraction pattern of two slits if the separation d of the slits is 5 times their width a? How many will there be if $d = na$ for any value of n?

41 •• A two-slit Fraunhofer interference–diffraction pattern is observed with light of wavelength 500 nm. The slits have a separation of 0.1 mm and a width of a. (a) Find the width a if the fifth interference maximum is at the same angle as the first diffraction minimum. (b) For this case, how many bright interference fringes will be seen in the central diffraction maximum?

42 •• SOLVE✓ A two-slit Fraunhofer interference–diffraction pattern is observed with light of wavelength 700 nm. The slits have widths of 0.01 mm and are separated by 0.2 mm. How many bright fringes will be seen in the central diffraction maximum?

43 •• SSM Suppose that the *central* diffraction maximum for two slits contains 17 interference fringes for some wavelength of light. How many interference fringes would you expect in the first *secondary* diffraction maximum?

44 •• SOLVE Light of wavelength 550 nm illuminates two slits of width 0.03 mm and separation 0.15 mm. (a) How many interference maxima fall within the full width of the central diffraction maximum? (b) What is the ratio of the intensity of the third interference maximum to the side of the centerline (not counting the center interference maximum) to the intensity of the center interference maximum?

*Using Phasors to Add Harmonic Waves

45 • Find the resultant of the two waves $E_1 = 2 \sin \omega t$ and $E_2 = 3 \sin(\omega t + 270°)$.

46 • SSM Find the resultant of the two waves $E_1 = 4 \sin \omega t$ and $E_2 = 3 \sin(\omega t + 60°)$.

47 •• At the second secondary maximum of the diffraction pattern of a single slit, the phase difference between the waves from the top and bottom of the slit is approximately 5π. The phasors used to calculate the amplitude at this point complete 2.5 circles. If I_0 is the intensity at the central maximum, find the intensity I at this second secondary maximum.

48 •• (a) Show that the positions of the interference minima on a screen a large distance L away from three equally spaced sources (spacing d, with $d \gg \lambda$) are given approximately by

$$y = \frac{n\lambda L}{3d}, \text{ where } n = 1, 2, 4, 5, 7, 8, 10, \ldots$$

that is, n is not a multiple of 3. (b) For $L = 1$ m, $\lambda = 5 \times 10^{-7}$ m, and $d = 0.1$ mm, calculate the width of the principal interference maxima (the distance between successive minima) for three sources.

49 •• (a) Show that the positions of the interference minima on a screen a large distance L away from four equally spaced sources (spacing d, with $d \gg \lambda$) are given approximately by

$$y = \frac{n\lambda L}{4d}, \text{ where } n = 1, 2, 3, 5, 6, 7, 9, 10, \dots$$

that is, n is not a multiple of 4. (b) For $L = 2$ m, $\lambda = 6 \times 10^{-7}$ m, and $d = 0.1$ mm, calculate the width of the principal interference maxima (the distance between successive minima) for four sources. Compare this width with that for two sources with the same spacing.

50 •• Light of wavelength 480 nm falls normally on four slits. Each slit is 2 μm wide and is separated from the next slit by 6 μm. (a) Find the angle from the center to the first point of zero intensity of the single-slit diffraction pattern on a distant screen. (b) Find the angles of any bright interference maxima that lie inside the central diffraction maximum. (c) Find the angular spread between the central interference maximum and the first interference minimum on either side of it. (d) Sketch the intensity as a function of angle.

51 ••• Three slits, each separated from its neighbor by 0.06 mm, are illuminated by a coherent light source of wavelength 550 nm. The slits are extremely narrow. A screen is located 2.5 m from the slits. The intensity on the centerline is 0.05 W/m². Consider a location 1.72 cm from the centerline. (a) Draw the phasors, according to the phasor model for the addition of harmonic waves, appropriate for this location. (b) From the phasor diagram, calculate the intensity of light at this location.

52 ••• SSM For single-slit diffraction, calculate the first three values of ϕ (the total phase difference between rays from each edge of the slit) that produce subsidiary maxima by (a) using the phasor model and (b) setting $dI/d\phi = 0$, where I is given by Equation 33-19.

Diffraction and Resolution

53 • SOLVE✓ Light of wavelength 700 nm is incident on a pinhole of diameter 0.1 mm. (a) What is the angle between the central maximum and the first diffraction minimum for a Fraunhofer diffraction pattern? (b) What is the distance between the central maximum and the first diffraction minimum on a screen 8 m away?

54 • Two sources of light of wavelength 700 nm are 10 m away from the pinhole of Problem 53. How far apart must the sources be for their diffraction patterns to be resolved by Rayleigh's criterion?

55 • SSM Two sources of light of wavelength 700 nm are separated by a horizontal distance x. They are 5 m from a vertical slit of width 0.5 mm. What is the least value of x for which the diffraction pattern of the sources can be resolved by Rayleigh's criterion?

56 • The headlights on a small car are separated by 112 cm. At what maximum distance could you resolve the headlights if the diameter of your pupil is 5 mm and the effective wavelength of the light is 550 nm?

57 • You are told not to shoot until you see the whites of their eyes. If their eyes are separated by 6.5 cm and the diameter of your pupil is 5 mm, at what distance can you resolve the two eyes using light of wavelength 550 nm?

58 •• SOLVE✓ The ceiling of your lecture hall is probably covered with acoustic tile, which has small holes separated by about 6 mm. (a) Using light with a wavelength of 500 nm, how far could you be from this tile and still resolve these holes? The diameter of the pupil of your eye is about 5 mm. (b) Could you resolve these holes better with red light or with violet light?

59 •• SOLVE The telescope on Mount Palomar has a diameter of 200 in. Suppose a double star were 4 light-years away. Under ideal conditions, what must be the minimum separation of the two stars for their images to be resolved using light of wavelength 550 nm?

60 •• SSM The star Mizar in Ursa Major is a binary system of stars of nearly equal magnitudes. The angular separation between the two stars is 14 seconds of arc. What is the minimum diameter of the pupil that allows resolution of the two stars using light of wavelength 550 nm?

*Diffraction Gratings

61 • SOLVE A diffraction grating with 2000 slits per centimeter is used to measure the wavelengths emitted by hydrogen gas. At what angles θ in the first-order spectrum would you expect to find the two violet lines of wavelengths 434 nm and 410 nm?

62 • SSM With the diffraction grating used in Problem 61, two other lines in the first-order hydrogen spectrum are found at angles $\theta_1 = 9.72 \times 10^{-2}$ rad and $\theta_2 = 1.32 \times 10^{-1}$ rad. Find the wavelengths of these lines.

63 • Repeat Problem 61 for a diffraction grating with 15,000 slits per centimeter.

64 • SOLVE What is the longest wavelength that can be observed in the fifth-order spectrum using a diffraction grating with 4000 slits per centimeter?

65 • The colors of many butterfly wings and beetle carapaces are due to effects of diffraction. The *Morpho* butterfly has structural elements on its wings that effectively act as a diffraction grating with spacing 880 nm. At what angle θ_1 will normally incident blue light of wavelength $\lambda = 440$ nm be diffracted by the *Morpho's* wings?

66 •• A diffraction grating of 2000 slits per centimeter is used to analyze the spectrum of mercury. (a) Find the angular separation in the first-order spectrum of the two lines of wavelength 579 nm and 577 nm. (b) How wide must the beam on the grating be for these lines to be resolved?

67 •• SSM A diffraction grating with 4800 lines per centimeter is illuminated at normal incidence with white light (wavelength range of 400 nm to 700 nm). For how many orders can one observe the complete spectrum in the transmitted light? Do any of these orders overlap? If so, describe the overlapping regions.

68 •• **SOLVE** A square diffraction grating with an area of 25 cm² has a resolution of 22,000 in the fourth order. At what angle should you look to see a wavelength of 510 nm in the fourth order?

69 •• Sodium light of wavelength 589 nm falls normally on a 2-cm-square diffraction grating ruled with 4000 lines per centimeter. The Fraunhofer diffraction pattern is projected onto a screen at 1.5 m by a lens of focal length 1.5 m placed immediately in front of the grating. Find (a) the positions of the first two intensity maxima on one side of the central maximum, (b) the width of the central maximum, and (c) the resolution in the first order.

70 •• The spectrum of neon is exceptionally rich in the visible region. Among the many lines are two lines at wavelengths of 519.313 nm and 519.322 nm. If light from a neon discharge tube is normally incident on a transmission grating with 8400 lines per centimeter and the spectrum is observed in second order, what must be the width of the grating that is illuminated, so that these two lines can be resolved?

71 •• **SSM** Mercury has several stable isotopes, among them ¹⁹⁸Hg and ²⁰²Hg. The strong spectral line of mercury, at about 546.07 nm, is a composite of spectral lines from the various mercury isotopes. The wavelengths of this line for ¹⁹⁸Hg and ²⁰²Hg are 546.07532 nm and 546.07355 nm, respectively. What must be the resolving power of a grating capable of resolving these two isotopic lines in the third-order spectrum? If the grating is illuminated over a 2-cm-wide region, what must be the number of lines per centimeter of the grating?

72 •• **SOLVE** A transmission grating is used to study the spectral region extending from 480 nm to 500 nm. The angular spread of this region is 12° in the third order. (a) Find the number of lines per centimeter. (b) How many orders are visible?

73 •• **SOLVE** White light is incident normally on a transmission grating and the spectrum is observed on a screen 8.0 m from the grating. In the second-order spectrum, the separation between light of 520-nm wavelength and 590-nm wavelength is 8.4 cm. (a) Determine the number of lines per centimeter of the grating. (b) What is the separation between these two wavelengths in the first-order spectrum and the third-order spectrum?

74 ••• A diffraction grating has n lines per unit length. Show that the angular separation of two lines of wavelengths λ and $\lambda + \Delta\lambda$ is approximately

$$\Delta\theta = \Delta\lambda \Big/ \sqrt{\frac{1}{n^2 m^2} - \lambda^2}$$

where m is the order number.

75 ••• For a diffraction grating in which all the surfaces are normal to the incident radiation, most of the energy goes into the zeroth order, which is useless from a spectroscopic point of view, since in zeroth order all the wavelengths are at 0°. Therefore, modern reflection gratings have shaped, or *blazed*, grooves, as shown in Figure 33-45. This shifts the specular reflection, which contains most of the energy, from the zeroth order to some higher order. (a) Calculate the blaze angle ϕ_m in terms of the groove separation d, the wavelength λ, and the order number m in which specular reflection is to occur for $m = 1, 2, \ldots$. (b) Calculate the proper blaze angle for the specular reflection to occur in the second order for light of wavelength 450 nm incident on a grating with 10,000 lines per centimeter.

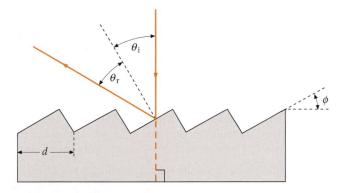

FIGURE 33-45 Problem 75

76 ••• In this problem, you will derive Equation 33-27 for the resolving power of a diffraction grating containing N slits separated by a distance d. To do this, you will calculate the angular separation between the maximum and minimum for some wavelength λ and set it equal to the angular separation of the mth-order maximum for two nearby wavelengths. (a) Show that the phase difference ϕ between the light from two adjacent slits is given by

$$\phi = \frac{2\pi d}{\lambda} \sin\theta$$

(b) Differentiate this expression to show that a small change in angle $d\theta$ results in a change in phase of $d\phi$ given by

$$d\phi = \frac{2\pi d}{\lambda} \cos\theta \, d\theta$$

(c) For N slits, the angular separation between an interference maximum and an interference minimum corresponds to a phase change of $d\phi = 2\pi/N$. Use this to show that the angular separation $d\theta$ between the maximum and minimum for some wavelength λ is given by

$$d\theta = \frac{\lambda}{Nd \cos\theta}$$

(d) The angle of the mth-order interference maximum for wavelength λ is given by Equation 33-26. Compute the differential of each side of this equation to show that angular separation of the mth-order maximum for two nearly equal wavelengths differing by $d\lambda$ is given by

$$d\theta \approx \frac{m \, d\lambda}{d \cos\theta}$$

(e) According to Rayleigh's criterion, two wavelengths will be resolved in the mth order if the angular separation of the wavelengths, given by the Part (d) result, equals the angular separation of the interference maximum and the interference minimum given by the Part (c) result. Use this to derive Equation 33-27 for the resolving power of a grating.

General Problems

77 • SSM A long, narrow horizontal slit lies 1 mm above a plane mirror, which is in the horizontal plane. The interference pattern produced by the slit and its image is viewed on a screen 1 m from the slit. The wavelength of the light is 600 nm. (*a*) Find the distance from the mirror to the first maximum. (*b*) How many dark bands per centimeter are seen on the screen?

78 •• A radio telescope is situated at the edge of a lake. The telescope is looking at light from a radio galaxy that is just rising over the horizon. If the height of the antenna is 20 m above the surface of the lake, at what angle above the horizon will the light from the radio galaxy go through its first interference maximum? The wavelength of the radio waves received by the telescope is $\lambda = 20$ cm. Remember that the light has a 180° phase shift on reflection from the water.

79 • ISOLVE In a lecture demonstration, a laser beam of wavelength 700 nm passes through a vertical slit 0.5 mm wide and hits a screen 6 m away. Find the horizontal width of the principal diffraction maximum on the screen; that is, find the distance between the first minimum on the left and the first minimum on the right of the central maximum.

80 • What minimum aperture, in millimeters, is required for opera glasses (binoculars) if an observer is to be able to distinguish the soprano's individual eyelashes (separated by 0.5 mm) at an observation distance of 25 m? Assume the effective wavelength of the light to be 550 nm.

81 • The diameter of the aperture of the radio telescope at Arecibo, Puerto Rico, is 300 m. What is the resolving power of the telescope when tuned to detect microwaves of 3.2-cm wavelength?

82 •• SSM ISOLVE A thin layer of a transparent material with an index of refraction of 1.30 is used as a nonreflective coating on the surface of glass with an index of refraction of 1.50. What should the thickness of the material be for the material to be nonreflecting for light of wavelength 600 nm?

83 •• A *Fabry–Perot interferometer* consists of two parallel, half-silvered mirrors separated by a small distance *a*. Show that when light is incident on the interferometer with an angle of incidence θ, the transmitted light will have maximum intensity when $a = (m\lambda/2) \cos \theta$.

84 •• A mica sheet 1.2 μm thick is suspended in air. In reflected light, there are gaps in the visible spectrum at 421, 474, 542, and 633 nm. Find the index of refraction of the mica sheet.

85 •• A camera lens is made of glass with an index of refraction of 1.6. This lens is coated with a magnesium fluoride film ($n = 1.38$) to enhance its light transmission. This film is to produce zero reflection for light of wavelength 540 nm.

Treat the lens surface as a flat plane and the film as a uniformly thick flat film. (*a*) How thick must the film be to accomplish its objective? (*b*) Would there be destructive interference for any other visible wavelengths? (*c*) By what factor would the reflection for light of wavelengths 400 nm and 700 nm be reduced by this film? Neglect the variation in the reflected light amplitudes from the two surfaces.

86 •• In a pinhole camera, the image is fuzzy because of geometry (rays arrive at the film after passing through different parts of the pinhole) and because of diffraction. As the pinhole is made smaller, the fuzziness due to geometry is reduced, but the fuzziness due to diffraction is increased. The optimum size of the pinhole for the sharpest possible image occurs when the spread due to diffraction equals the spread due to the geometric effects of the pinhole. Estimate the optimum size of the pinhole if the distance from the pinhole to the film is 10 cm and the wavelength of the light is 550 nm.

87 •• SSM ISOLVE ✔ The Impressionist painter Georges Seurat used a technique called *pointillism*, in which his paintings are composed of small, closely spaced dots of pure color, each about 2 mm in diameter. The illusion of the colors blending together smoothly is produced in the eye of the viewer by diffraction effects. Calculate the minimum viewing distance for this effect to work properly. Use the wavelength of visible light that requires the *greatest* distance, so that you are sure the effect will work for *all* visible wavelengths. Assume the pupil of the eye has a diameter of 3 mm.

88 ••• SSM A *Jamin refractometer* is a device for measuring or for comparing the indexes of refraction of gases. A beam of monochromatic light is split into two parts, each of which is directed along the axis of a separate cylindrical tube before being recombined into a single beam that is viewed through a telescope. Suppose that each tube is 0.4 m long and that sodium light of wavelength 589 nm is used. Both tubes are initially evacuated, and constructive interference is observed in the center of the field of view. As air is slowly allowed to enter one of the tubes, the central field of view changes to dark and back to bright a total of 198 times. (*a*) What is the index of refraction of air? (*b*) If the fringes can be counted to ± 0.25 fringe, where one fringe is equivalent to one complete cycle of intensity variation at the center of the field of view, to what accuracy can the index of refraction of air be determined by this experiment?

89 ••• Light of wavelength λ is diffracted through a single slit of width *a*, and the resulting pattern is viewed on a screen a long distance *L* away from the slit. (*a*) Show that the width of the central maximum on the screen is approximately $2L\lambda/a$. (*b*) If a slit of width $2L\lambda/a$ is cut in the screen and is illuminated by light of the same wavelength, show that the width of its central diffraction maximum at the same distance *L* is *a* to the same approximation.

PART VI MODERN PHYSICS: Quantum Mechanics, Relativity, and the Structure of Matter

Wave–Particle Duality and Quantum Physics

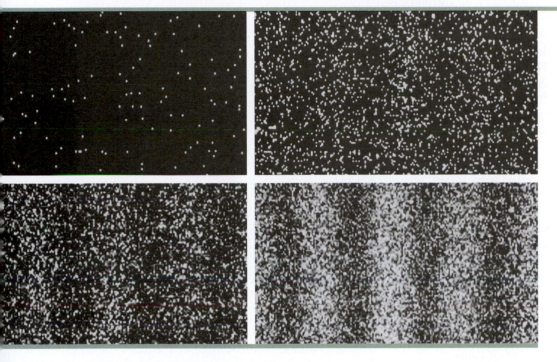

ELECTRON INTERFERENCE PATTERN PRODUCED BY ELECTRONS INCIDENT ON A BARRIER CONTAINING TWO SLITS: (A) 10 ELECTRONS, (B) 100 ELECTRONS, (C) 3000 ELECTRONS, AND (D) 70,000 ELECTRONS. THE MAXIMA AND MINIMA DEMONSTRATE THE WAVE NATURE OF THE ELECTRON AS IT TRAVERSES THE SLITS. INDIVIDUAL DOTS ON THE SCREEN INDICATE THE PARTICLE NATURE OF THE ELECTRON AS IT EXCHANGES ENERGY WITH THE DETECTOR. THE PATTERN IS THE SAME WHETHER ELECTRONS OR PHOTONS (PARTICLES OF LIGHT) ARE USED.

? **How do you calculate the wavelength of an electron? This is revealed in Example 34-4.**

We have seen that the propagation of waves through space is quite different from the propagation of particles. Waves bend around corners (diffraction) and interfere with one another, producing interference patterns. If a wave encounters a small aperture, the wave spreads out on the other side as if the aperture were a point source. If two coherent waves of equal intensity I_0 meet in space, the result can be a wave of intensity $4I_0$ (constructive interference), an intensity of zero (destructive interference), or a wave of intensity between zero and $4I_0$, depending on the phase difference between the waves at their meeting point.

The propagation of particles is quite unlike the propagation of waves. Particles travel in straight lines until they collide with something, after which the particles again travel in straight lines. If two particle beams meet in space, they never produce an interference pattern.

Particles and waves also exchange energy differently. Particles exchange energy in collisions that occur at specific points in space and in time. The energy of waves, on the other hand, is spread out in space and deposited continuously as the wave fronts interact with matter.

Sometimes the propagation of a wave cannot be distinguished from the propagation of a beam of particles. If the wavelength λ is very small compared to distances from the edges of objects, diffraction effects are negligible and the wave travels in straight lines. Also, interference maxima and minima are so close together in space as to be unobservable. The result is that the wave interacts with a detector, like a beam of numerous small particles each exchanging a small amount of energy; the exchange cannot distinguish particles from waves. For example, you do not observe the individual air molecules bouncing off your face if the wind blows on it. Instead, the interaction of billions of particles appears to be continuous, like that of a wave.

At the beginning of the twentieth century, it was thought that sound, light, and other electromagnetic radiation (e.g., radio) were waves; whereas electrons, protons, atoms, and similar constituents of nature were understood to be particles. The first 30 years of the century revealed startling developments in theoretical and experimental physics, such as the finding that light, thought to be a wave, actually exchanges energy in discrete lumps or quanta, just like particles; and the finding that an electron, thought to be a particle, exhibits diffraction and interference as it propagates through space, just like a wave.

The fact that light exchanges energy like a particle implies that light energy is not continuous but is *quantized*. Similarly, the wave nature of the electron, along with the fact that the standing wave condition requires a discrete set of frequencies, implies that the energy of an electron in a confined region of space is not continuous, but is quantized to a discrete set of values.

➤ **In this chapter, we begin by discussing some basic properties of light and electrons, examining their wave and particle characteristics. We then consider some of the detailed properties of matter waves, showing, in particular, how standing waves imply the quantization of energy. Finally, we discuss some of the important features of the theory of quantum physics, which was developed in the 1920s and which has been extremely successful in describing nature. Quantum physics is now the basis of our understanding of both the microscopic and very low temperature worlds.**

34-1 Light

The question of whether light consists of a beam of particles or waves in motion is one of the most interesting in the history of science. Isaac Newton used a particle theory of light to explain the laws of reflection and refraction; however, for refraction, Newton needed to assume that light travels faster in water or glass than in air, an assumption later shown to be false. The chief early proponents of the wave theory were Robert Hooke and Christian Huygens, who explained refraction by assuming that light travels more slowly in glass or water than it does in air. Newton rejected the wave theory because, in his time, light was

8 • Equation 33-2, which is $d \sin \theta_m = m\lambda$, and Equation 33-11, which is $a \sin \theta_m = m\lambda$, are sometimes confused. For each equation, define the symbols and explain the equation's application.

9 • When a diffraction grating is illuminated with white light, the first-order maximum of green light (a) is closer to the central maximum than that of red light. (b) is closer to the central maximum than that of blue light. (c) overlaps the second-order maximum of red light. (d) overlaps the second-order maximum of blue light.

10 • **SSM** A double-slit interference experiment is set up in a chamber that can be evacuated. Using monochromatic light, an interference pattern is observed when the chamber is open to air. As the chamber is evacuated, one will note that (a) the interference fringes remain fixed. (b) the interference fringes move closer together. (c) the interference fringes move farther apart. (d) the interference fringes disappear completely.

11 • True or false:

(a) When waves interfere destructively, the energy is converted into heat energy.
(b) Interference is observed only for waves from coherent sources.
(c) In the Fraunhofer diffraction pattern for a single slit, the narrower the slit, the wider the central maximum of the diffraction pattern.
(d) A circular aperture can produce both a Fraunhofer diffraction pattern and a Fresnel diffraction pattern.
(e) The ability to resolve two point sources depends on the wavelength of the light.

Estimation and Approximation

12 • **SSM** It is claimed that the Great Wall of China is the only human object that can be seen from space with the naked eye. Make an argument in support of this claim based on the resolving power of the human eye. Evaluate the validity of your argument for observers both in low-earth orbit (~ 400 km altitude) and on the moon.

13 • Naturally occuring coronas (brightly colored rings) are sometimes seen around the moon or the sun when viewed through a thin cloud. (Warning: When viewing a sun corona, be sure that the entire sun is blocked by the edge of a building, a tree, or a traffic pole to safeguard your eyes.) These coronas are due to diffraction of light by small water droplets in the cloud. A typical angular diameter for a coronal ring is about 10°. From this, estimate the size of the water droplets in the cloud. Assume that the water droplets can be modeled as opaque disks with the same radius as the droplet, and that the Fraunhofer diffraction pattern from an opaque disk is the same as the pattern from an aperture of the same diameter. (This statement is known as *Babinet's principle*.)

14 • An artificial corona (see Problem 13) can be made by placing a suspension of polystyrene microspheres in water. Polystyrene microspheres are small, uniform spheres made of plastic with an index of refraction of 1.59. Assuming that the water has a refractive index of 1.33, what is the angular diameter of such an artificial corona if 5 μm diameter particles are illuminated by a helium–neon laser with wavelength in air $\lambda = 632.8$ nm?

15 • Coronas (see Problem 13) can be caused by pollen grains, typically of birch or pine. Such grains are irregular in shape, but they can be treated as if they had an average diameter of approximately 25 μm. What is the coronal radius (in degrees) for blue light? What is the coronal radius (in degrees) for red light?

16 •• **SSM** Human hair has a diameter of approximately 70 μm. If we illuminate a hair using a helium–neon laser with wavelength $\lambda = 632.8$ nm and intercept the light scattered from the hair on a screen 10 m away, what will be the separation of the first diffraction peak from the center? (The diffraction pattern of a hair with diameter d is the same as the diffraction pattern of a single slit with width $a = d$.)

Phase Difference and Coherence

17 • **SOLVE** (a) What minimum path difference is needed to introduce a phase shift of 180° in light of wavelength 600 nm? (b) What phase shift will that path difference introduce in light of wavelength 800 nm?

18 • **SOLVE** Light of wavelength 500 nm is incident normally on a film of water 10^{-4} cm thick. The index of refraction of water is 1.33. (a) What is the wavelength of the light in the water? (b) How many wavelengths are contained in the distance $2t$, where t is the thickness of the film? (c) What is the phase difference between the wave reflected from the top of the air–water interface and the wave reflected from the bottom of the water–air interface after it has traveled this distance?

19 •• **SSM** **SOLVE** ✓ Two coherent microwave sources that produce waves of wavelength 1.5 cm are in the xy plane, one on the y axis at $y = 15$ cm and the other at $x = 3$ cm, $y = 14$ cm. If the sources are in phase, find the difference in phase between the two waves from these sources at the origin.

Interference in Thin Films

20 • A wedge-shaped film of air is made by placing a small slip of paper between the edges of two flat plates of glass. Light of wavelength 700 nm is incident normally on the glass plates, and interference bands are observed by reflection. (a) Is the first band near the point of contact of the plates dark or bright? Why? (b) If there are five dark bands per centimeter, what is the angle of the wedge?

21 •• **SSM** The diameters of fine fibers can be accurately measured using interference patterns. Two optically flat pieces of glass of length L are arranged with the wire between them, as shown in Figure 33-40. The setup is illuminated by monochromatic light, and the resulting interference fringes are detected. Suppose that $L = 20$ cm and that yellow sodium light ($\lambda \approx 590$ nm) is used for illumination. If 19 bright fringes are seen along this 20-cm distance, what are the limits on the diameter of the wire? (*Hint:* The nineteenth fringe might not be right at the end, but you do not see a twentieth fringe at all.)

FIGURE 33-40 Problem 21

22 •• $\boxed{\text{SOLVE}}$✓ Light of wavelength 600 nm is used to illuminate two glass plates at normal incidence. The plates are 22 cm in length, touch at one end, and are separated at the other end by a wire of radius 0.025 mm. How many bright fringes appear along the total length of the plates?

23 •• A thin film having an index of refraction of 1.5 is surrounded by air. It is illuminated normally by white light and is viewed by reflection. Analysis of the resulting reflected light shows that the wavelengths 360, 450, and 602 nm are the only missing wavelengths in or near the visible portion of the spectrum. That is, for these wavelengths, there is destructive interference. (a) What is the thickness of the film? (b) What visible wavelengths are brightest in the reflected interference pattern? (c) If this film were resting on glass with an index of refraction of 1.6, what wavelengths in the visible spectrum would be missing from the reflected light?

24 •• $\boxed{\text{SOLVE}}$ A drop of oil ($n = 1.22$) floats on water ($n = 1.33$). When reflected light is observed from above, as shown in Figure 33-41, what is the thickness of the drop at the point where the second red fringe, counting from the edge of the drop, is observed? Assume red light has a wavelength of 650 nm.

FIGURE 33-41
Problem 24

25 •• A film of oil of index of refraction $n = 1.45$ rests on an optically flat piece of glass of index of refraction $n = 1.6$. When illuminated with white light at normal incidence, light of wavelengths 690 nm and 460 nm is predominant in the reflected light. Determine the thickness of the oil film.

26 •• $\boxed{\text{SSM}}$ $\boxed{\text{SOLVE}}$✓ A film of oil of index of refraction $n = 1.45$ floats on water ($n = 1.33$). When illuminated with white light at normal incidence, light of wavelengths 700 nm and 500 nm is predominant in the reflected light. Determine the thickness of the oil film.

Newton's Rings

27 •• $\boxed{\text{SSM}}$ A Newton's ring apparatus consists of a plano-convex glass lens with radius of curvature R that rests on a flat glass plate, as shown in Figure 33-42. The thin film is air of variable thickness. The pattern is viewed by reflected light. (a) Show that for a thickness t the condition for a bright (constructive) interference ring is

$$t = \left(m + \frac{1}{2}\right)\frac{\lambda}{2}, \qquad m = 0, 1, 2, \ldots$$

(b) Apply the Pythagorean theorem to the triangle of sides r, $R - t$, and hypotenuse R to show that for $t \ll R$, the radius of a fringe is related to t by

$$r = \sqrt{2tR}$$

(c) How would the transmitted pattern look in comparison with the reflected one? (d) Use $R = 10$ m and a lens diameter of 4 cm. How many bright fringes would you see if the apparatus was illuminated by yellow sodium light ($\lambda \approx 590$ nm) and viewed by reflection? (e) What would be the diameter of the sixth bright fringe? (f) If the glass used in the apparatus has an index of refraction $n = 1.5$ and water ($n_w = 1.33$) is placed between the two pieces of glass, what change will take place in the bright-fringe pattern?

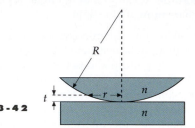

FIGURE 33-42
Problem 27

28 •• $\boxed{\text{SOLVE}}$✓ A plano-convex glass lens of radius of curvature 2.0 m rests on an optically flat glass plate. The arrangement is illuminated from above with monochromatic light of 520-nm wavelength. The indexes of refraction of the lens and plate are 1.6. Determine the radii of the first and second bright fringe in the reflected light. (Use the results from Problem 27b to relate r to t.)

29 •• Suppose that before the lens of Problem 28 is placed on the plate, a film of oil of refractive index 1.82 is deposited on the plate. What will then be the radii of the first and second bright fringes? (Use the results from Problem 27b to relate r to t.)

Two-Slit Interference Pattern

30 • $\boxed{\text{SSM}}$ Two narrow slits separated by 1 mm are illuminated by light of wavelength 600 nm, and the interference pattern is viewed on a screen 2 m away. Calculate the number of bright fringes per centimeter on the screen.

31 • $\boxed{\text{SOLVE}}$✓ Using a conventional two-slit apparatus with light of wavelength 589 nm, 28 bright fringes per centimeter are observed on a screen 3 m away. What is the slit separation?

32 • Light of wavelength 633 nm from a helium–neon laser is shone normally on a plane containing two slits. The first interference maximum is 82 cm from the central maximum on a screen 12 m away. (a) Find the separation of the slits. (b) How many interference maxima can be observed?

33 •• $\boxed{\text{SOLVE}}$ Two narrow slits are separated by a distance d. Their interference pattern is to be observed on a screen a large distance L away. (a) Calculate the spacing Δy of the maxima on the screen for light of wavelength 500 nm, when $L = 1$ m and $d = 1$ cm. (b) Would you expect to be able to observe the interference of light on the screen for this situation? (c) How close together should the slits be placed for the maxima to be separated by 1 mm for this wavelength and screen distance?

34 •• Light is incident at an angle ϕ with the normal to a vertical plane containing two slits of separation d (Figure 33-43). Show that the interference maxima are located at angles θ_m given by $\sin \theta_m + \sin \phi = m\lambda/d$.

believed to travel through a medium only in straight lines—diffraction had not yet been observed.

Because of Newton's great reputation and authority, his particle theory of light was accepted for more than a century. Then, in 1801, Thomas Young demonstrated the wave nature of light in a famous experiment in which two coherent light sources are produced by illuminating a pair of narrow, parallel slits with a single source (Figure 34-1). In Chapter 33, we saw that when light encounters a small opening, the opening acts as a point source of waves (Figure 33-7). In Young's experiment, each slit acts as a line source, which is equivalent to a point source in two dimensions. The interference pattern is observed on a screen placed behind the slits. Interference maxima occur at angles so that the path difference is an integral number of wavelengths. Similarly, interference minima occur if the path difference is one-half wavelength or any odd number of half wavelengths. Figure 34-1b shows the intensity pattern as seen on the screen. Young's experiment and many other experiments demonstrate that light propagates like a wave.

In the early nineteenth century, the French physicist Augustin Fresnel (1788–1827) performed extensive experiments on interference and diffraction and put the wave theory on a rigorous mathematical basis. Fresnel showed that the observed straight-line propagation of light is a result of the very short wavelengths of visible light.

The classical wave theory of light culminated in 1860 when James Clerk Maxwell published his mathematical theory of electromagnetism. This theory yielded a wave equation that predicted the existence of electromagnetic waves that propagate with a speed that can be calculated from the laws of electricity and magnetism. The fact that the result of this calculation was $c \approx 3 \times 10^8$ m/s, the same as the speed of light, suggested to Maxwell that light is an electromagnetic wave. The eye is sensitive to electromagnetic waves with wavelengths in the range from approximately 400 nm (1 nm = 10^{-9} m) to approximately 700 nm. This range is called *visible light*. Other electromagnetic waves (e.g., microwaves, radio, television, and X rays) differ from light only in wavelength and in frequency.

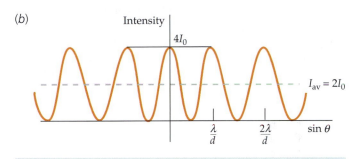

FIGURE 34-1 (a) Two slits act as coherent sources of light for the observation of interference in Young's experiment. Cylindrical waves from the slits overlap and produce an interference pattern on a screen far away. (b) The intensity pattern produced in Figure 34-1a. The intensity is maximum at points where the path difference is an integral number of wavelengths, and the intensity is zero where the path difference is an odd number of half wavelengths.

34-2 The Particle Nature of Light: Photons

The diffraction of light and the existence of an interference pattern in the two-slit experiment give clear evidence that light has wave properties. However, early in the twentieth century, it was found that light energy comes in discrete amounts.

The Photoelectric Effect

The quantum nature of light and the quantization of energy were suggested by Albert Einstein in 1905 in his explanation of the photoelectric effect. Einstein's work marked the beginning of quantum theory, and for his work, Einstein received the Nobel Prize for physics. Figure 34-2 shows a schematic diagram of the basic apparatus for studying the photoelectric effect. Light of a single frequency enters an evacuated chamber and falls on a clean metal surface C (C for cathode), causing electrons to be emitted. Some of these electrons strike the second metal plate A (A for anode), constituting an electric current between the plates. Plate A is negatively charged, so the electrons are repelled by it, with only the most energetic electrons reaching the plate. The maximum kinetic energy of the emitted electrons is measured by slowly increasing the voltage until the current becomes zero. Experiments give the surprising result that the maximum kinetic energy of the emitted electrons is *independent of the intensity* of the incident light. Classically, we would expect that increasing the rate at which light energy falls on the metal surface would increase the energy absorbed by individual electrons and, therefore, would increase the maximum kinetic energy of the electrons emitted. Experimentally, this is not what happens. The maximum kinetic energy of the emitted electrons is the same for a given wavelength of incident light, no matter how intense the light. Einstein demonstrated that this experimental result can be explained if light energy is quantized in small bundles called **photons.** The energy E of each photon is given by

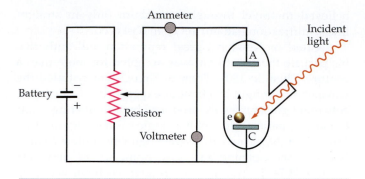

FIGURE 34-2 A schematic drawing of the apparatus for studying the photoelectric effect. Light of a single frequency enters an evacuated chamber and strikes the cathode C, which then ejects electrons. The current in the ammeter measures the number of these electrons that reach the anode A per unit time. The anode is made electrically negative with respect to the cathode to repel the electrons. Only those electrons with enough initial kinetic energy to overcome the repulsion can reach the anode. The voltage between the two plates is slowly increased until the current becomes zero, which happens when even the most energetic electrons do not make it to plate A.

$$E = hf = \frac{hc}{\lambda} \qquad\qquad \text{34-1}$$

EINSTEIN EQUATION FOR PHOTON ENERGY

where f is the frequency, and h is a constant now known as **Planck's constant.**[†] The measured value of this constant is

$$h = 6.626 \times 10^{-34}\,\text{J·s} = 4.136 \times 10^{-15}\,\text{eV·s} \qquad\qquad \text{34-2}$$

PLANCK'S CONSTANT

Equation 34-1 is sometimes called the **Einstein equation.**

At the fundamental level, a light beam consists of a beam of particles—photons—each with energy hf. The intensity (power per unit area) of a monochromatic light beam is the number of photons per unit area per unit of time, times the energy per photon. The interaction of the light beam with the metal surface consists of collisions between photons and electrons. In these collisions, the photons disappear, with each photon giving all its energy to an electron, and the electron emitted from the surface thus receives its energy from a single photon. If the intensity of light is increased, more photons fall on the surface per unit time, and more electrons are emitted. However, each photon still has the same energy hf, so the energy absorbed by each electron is unchanged.

If ϕ is the minimum energy necessary to remove an electron from a metal surface, the maximum kinetic energy of the electrons emitted is given by

$$K_{max} = \left(\tfrac{1}{2}mv^2\right)_{max} = hf - \phi \qquad\qquad \text{34-3}$$

EINSTEIN'S PHOTOELECTRIC EQUATION

The quantity ϕ, called the **work function,** is a characteristic of the particular metal. (Some electrons will have kinetic energies less than $hf - \phi$, because of the loss of energy from traveling through the metal.)

[†] In 1900, the German physicist Max Planck introduced this constant to explain discrepancies between the theoretical curves and experimental data on the spectrum of blackbody radiation. Planck also assumed that the radiation was emitted and absorbed by a blackbody in quanta of energy hf, but he considered his assumption to be just a calculational device rather than a fundamental property of electromagnetic radiation. Blackbody radiation was discussed in Chapter 20.

According to Einstein's photoelectric equation, a plot of K_{max} versus frequency f should be a straight line with the slope h. This was a bold prediction, because, at the time, there was no evidence that Planck's constant had any application outside of blackbody radiation. In addition, there was no experimental data on K_{max} versus frequency f, because no one before had even suspected that the frequency of the light was related to K_{max}. This prediction was difficult to verify experimentally, but careful experiments by R. A. Millikan approximately 10 years later showed that Einstein's equation was correct. Figure 34-3 shows a plot of Millikan's data.

Photons with frequencies less than a **threshold frequency** f_t, and therefore with wavelengths greater than a **threshold wavelength** $\lambda_t = c/f_t$, do not have enough energy to eject an electron from a particular metal. The threshold frequency and the corresponding threshold wavelength can be related to the work function ϕ by setting the maximum kinetic energy of the electrons equal to zero in Equation 34-3. Then

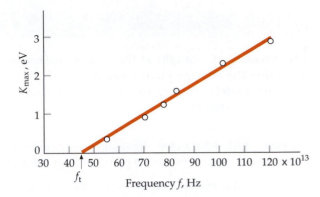

FIGURE 34-3 Millikan's data for the maximum kinetic energy K_{max} versus frequency f for the photoelectric effect. The data fall on a straight line that has a slope h, as predicted by Einstein approximately a decade before the experiment was performed.

$$\phi = hf_t = \frac{hc}{\lambda_t} \tag{34-4}$$

Work functions for metals are typically a few electron volts. Since wavelengths are usually given in nanometers and energies in electron volts, it is useful to have the value of hc in electron volt–nanometers:

$$hc = (4.1357 \times 10^{-15} \text{ eV·s})(2.9979 \times 10^8 \text{ m/s}) = 1.240 \times 10^{-6} \text{ eV·m}$$

or

$$hc = 1240 \text{ eV·nm} \tag{34-5}$$

PHOTON ENERGIES FOR VISIBLE LIGHT   **EXAMPLE 34-1**

Calculate the photon energies for light of wavelengths 400 nm (violet) and 700 nm (red). (These are the approximate wavelengths at the two extremes of the visible spectrum.)

1. The energy is related to the wavelength by Equation 34-1: $E = hf = \dfrac{hc}{\lambda}$

2. For $\lambda = 400$ nm, the energy is: $E = \dfrac{hc}{\lambda} = \dfrac{1240 \text{ eV·nm}}{400 \text{ nm}} = \boxed{3.10 \text{ eV}}$

3. For $\lambda = 700$ nm, the energy is: $E = \dfrac{hc}{\lambda} = \dfrac{1240 \text{ eV·nm}}{700 \text{ nm}} = \boxed{1.77 \text{ eV}}$

REMARKS We can see from these calculations that visible light contains photons with energies that range from approximately 1.8 eV to 3.1 eV. X rays, which have much shorter wavelengths, contain photons with energies of the order of keV. Gamma rays emitted by nuclei have even shorter wavelengths and photons of energy of the order of MeV.

EXERCISE Find the energy of a photon corresponding to electromagnetic radiation in the FM radio band of wavelength 3 m. (*Answer* 4.13×10^{-7} eV)

EXERCISE Find the wavelength of a photon whose energy is (*a*) 0.1 eV, (*b*) 1 keV, and (*c*) 1 MeV. (*Answer* (*a*) 12.4 μm, (*b*) 1.24 nm, (*c*) 1.24 pm)

THE NUMBER OF PHOTONS PER SECOND IN SUNLIGHT   **E X A M P L E 3 4 - 2**   **Try It Yourself**

The intensity of sunlight at the earth's surface is approximately 1400 W/m². Assuming the average photon energy is 2 eV (corresponding to a wavelength of approximately 600 nm), calculate the number of photons that strike an area of 1 cm² each second.

PICTURE THE PROBLEM Intensity (power per unit area) is given as is the area. From this, we can calculate the power, which is the energy per unit time.

Cover the column to the right and try these on your own before looking at the answers.

Steps	Answers
1. The energy ΔE is related to the number N of photons and the energy per photon $hf = 2$ eV.	$\Delta E = Nhf$
2. The intensity I (power per unit area) and the area A are given, so we can find the power.	$I = \dfrac{P}{A}$
3. Knowing the power (energy per unit time) and the time, we can find the energy.	$\Delta E = P\Delta t$
4. Substitute the results from steps 1–3 and solve for N. Take care to get the units correct.	$N = \dfrac{IA\Delta t}{hf} = \boxed{4.38 \times 10^{17}}$

REMARKS This is an enormous number of photons. In most everyday situations, the number of photons is so great that the quantization of light is not noticeable.

EXERCISE Calculate the photon density (in photons per cubic centimeter) of the sunlight in Example 34-2. The number arriving on an area of 1 cm² in one second is the number in a column whose cross section is 1 cm² and whose height is the distance light travels in one second. (*Answer* 1.46×10^7 cm^{-3})

Compton Scattering

The first use of the photon concept was to explain the results of photoelectric-effect experiments. The photon concept was used by Arthur H. Compton to explain the results of his measurements of the scattering of X rays by free electrons in 1923. According to classical theory, if an electromagnetic wave of frequency f_1 is incident on material containing free charges, the charges will oscillate with this frequency and reradiate electromagnetic waves of the same frequency. Compton considered these reradiated waves as scattered photons, and he pointed out that if the scattering process were a collision between a photon and an electron (Figure 34-4), the electron would recoil and thus absorb energy. The scattered photon would then have less energy, and therefore a lower frequency and longer wavelength, than the incident photon.

According to classical electromagnetic theory (see Section 30-3), the energy and momentum of an electromagnetic wave are related by

$$E = pc \qquad\qquad 34\text{-}6$$

The momentum of a photon is thus related to its wavelength λ by $p = E/c = hf/c = h/\lambda$.

$$p = \dfrac{h}{\lambda} \qquad\qquad 34\text{-}7$$

MOMENTUM OF A PHOTON

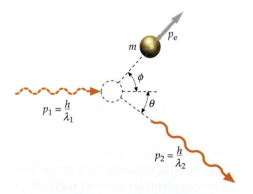

FIGURE 34-4 The scattering of light by an electron is considered as a collision of a photon of momentum h/λ_1 and a stationary electron. The scattered photon has less energy and therefore has a greater wavelength than does the incident electron.

Compton applied the laws of conservation of momentum and energy to the collision of a photon and an electron to calculate the momentum p_2 and thus the wavelength $\lambda_2 = h/p_2$ of the scattered photon (see Figure 34-4). Applying conservation of momentum to the collision gives

$$\vec{p}_1 = \vec{p}_2 + \vec{p}_e \qquad\qquad 34\text{-}8$$

where $\vec{p}_1$ is the momentum of the incident photon and $\vec{p}_e$ is the momentum of the electron after the collision. The initial momentum of the electron is zero. Rearranging Equation 34-8, we have $\vec{p}_e = \vec{p}_1 - \vec{p}_2$. Taking the dot product of each side with itself gives

$$p_e^2 = p_1^2 + p_2^2 - 2p_1 p_2 \cos\theta \qquad\qquad 34\text{-}9$$

where θ is the angle the scattered photon makes with the direction of the incident photon. Because the kinetic energy of the electron after the collision can be a significant fraction of the rest energy of an electron, the relativistic expression relating the energy E of the electron to its momentum is used. This expression (Equation R-17) is

$$E = \sqrt{p_e^2 c^2 + (m_e c^2)^2}$$

where m_e is the rest mass of the electron. Applying conservation of energy to the collision gives

$$p_1 c + m_e c^2 = p_2 c + \sqrt{p_e^2 c^2 + (m_e c^2)^2} \qquad\qquad 34\text{-}10$$

where pc (Equation 34-6) has been used to express the energies of the photons. Eliminating p_e^2 from Equation 34-9 and Equation 34-10 gives

$$\frac{1}{p_2} - \frac{1}{p_1} = \frac{1}{m_e c}(1 - \cos\theta)$$

and substituting for p_1 and p_2, using Equation 34-7 gives

$$\lambda_2 - \lambda_1 = \frac{h}{m_e c}(1 - \cos\theta) \qquad\qquad 34\text{-}11$$

COMPTON EQUATION

The increase in wavelengths is independent of the wavelength λ_1 of the incident photon. The quantity $h/m_e c$ has dimensions of length and is called the Compton wavelength. Its value is

$$\lambda_C = \frac{h}{m_e c} = \frac{hc}{m_e c^2} = \frac{1240 \text{ eV·nm}}{5.11 \times 10^5 \text{ eV}} = 2.43 \times 10^{-12} \text{ m} = 2.43 \text{ pm} \qquad 34\text{-}12$$

Because $\lambda_2 - \lambda_1$ is small, it is difficult to observe unless λ_1 is so small that the fractional change $(\lambda_2 - \lambda_1)/\lambda_1$ is appreciable. Compton used X rays of wavelength 71.1 pm (1 pm $= 10^{-12}$ m $= 10^{-3}$ nm). The energy of a photon of this wavelength is $E = hc/\lambda = (1240 \text{ eV·nm})/(0.0711 \text{ nm}) = 17.4$ keV. Since this is much greater than the binding energy of the valence electrons in most atoms (which is of the order of a few eV), these electrons can be considered to be essentially free. Compton's measurements of $\lambda_2 - \lambda_1$ as a function of scattering angle θ agreed with Equation 34-11, thereby confirming the correctness of the photon concept (i.e., of the particle nature of light).

EXAMPLE 34-3

An X-ray photon of wavelength 6 pm makes a head-on collision with an electron, so that the scattered photon goes in a direction opposite to that of the incident photon. The electron is initially at rest. (*a*) How much longer is the wavelength of the scattered photon than that of the incident photon? (*b*) What is the kinetic energy of the recoiling electron?

PICTURE THE PROBLEM We can calculate the increase in wavelength, and thus the new wavelength, from Equation 34-11. We then use the new wavelength to find the energy of the scattered photon and then to find the kinetic energy of the recoiling electron from conservation of energy (Figure 34-5).

FIGURE 34-5

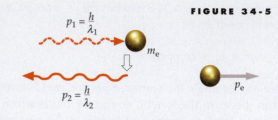

(*a*) Use Equation 34-11 to calculate the increase in wavelength:

$$\Delta\lambda = \lambda_2 - \lambda_1$$

$$= \frac{h}{m_e c}(1 - \cos\theta)$$

$$= (2.43 \text{ pm})(1 - \cos 180°) = \boxed{4.86 \text{ pm}}$$

(*b*) 1. The kinetic energy of the recoiling electron equals the energy of the incident photon E_1 minus the energy of the scattered photon E_2:

$$K_e = E_1 - E_2 = \frac{hc}{\lambda_1} - \frac{hc}{\lambda_2}$$

2. Calculate λ_2 from the given wavelength of the incident photon and the change found in Part (*a*):

$$\lambda_2 = \lambda_1 + \Delta\lambda = 6 \text{ pm} + 4.86 \text{ pm}$$

$$= 10.86 \text{ pm}$$

3. Substitute the calculated values of E_1 and E_2 to find the energy of the recoiling electron:

$$K_e = \frac{hc}{\lambda_1} - \frac{hc}{\lambda_2}$$

$$= \frac{1240 \text{ eV·nm}}{6.0 \text{ pm}} - \frac{1240 \text{ eV·nm}}{10.86 \text{ pm}}$$

$$= \frac{1.24 \text{ keV·nm}}{6.0 \times 10^{-3} \text{ nm}} - \frac{1.24 \text{ keV·nm}}{10.86 \times 10^{-3} \text{ nm}}$$

$$= 207 \text{ keV} - 114 \text{ keV}$$

$$= \boxed{93 \text{ keV}}$$

REMARKS The kinetic energy of the scattered electron is 93 keV and the rest energy of an electron is 511 keV, so the kinetic energy is 18 percent of the rest energy. Thus, the nonrelativistic formula for the kinetic energy ($\frac{1}{2}m_e v^2$) is not valid.

EXERCISE What is the speed of the scattered electron given by the nonrelativistic formula for the kinetic energy ($\frac{1}{2}m_e v^2$)? (*Answer* 0.6c)

34-3 Energy Quantization in Atoms

Ordinary white light has a continuous spectrum; that is, it contains *all* the wavelengths in the visible spectrum. But if atoms in a gas at low pressure are excited by an electric discharge, they emit light of specific wavelengths that are characteristic of the element or the compound. Since the energy of a photon is related to its wavelength by $E = hf = hc/\lambda$, a discrete set of wavelengths implies a discrete set of energies. Conservation of energy then implies that if an atom absorbs or emits a photon, its internal energy changes by a discrete amount, which is $\pm$ the energy of the photon. In 1913, this led Niels Bohr to postulate that the internal energy of an atom can have only a discrete set of values. That is, the internal energy of an atom is **quantized.** If an atom radiates light of frequency f, the atom makes a transition from one allowed level to another level that is lower in energy by $\Delta E = hf$. Bohr was able to construct a semiclassical model of the hydrogen atom that had a discrete set of energy levels consistent with the observed spectrum of emitted light.[†] However, the *reason* for the quantization of energy levels in atoms and other systems remained a mystery until the wave nature of electrons was discovered a decade later.

34-4   Electrons and Matter Waves

In 1897, J. J. Thomson showed that the rays of a cathode-ray tube (Figure 34-6) can be deflected by electric and magnetic fields and therefore must consist of electrically charged particles. By measuring the deflections of these particles, Thomson showed that all the particles have the same charge-to-mass ratio q/m. He also showed that particles with this charge-to-mass ratio can be obtained using any material for the cathode, which means that these particles, now called **electrons,** are a fundamental constituent of all matter.

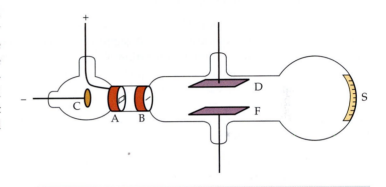

FIGURE 34-6 Schematic diagram of the cathode-ray tube Thomson used to measure q/m for the particles that comprise cathode rays (electrons). Electrons from the cathode C pass through the slits at A and B and strike a phosphorescent screen S. The beam can be deflected by an electric field between plates D and F or by a magnetic field (not shown).

The de Broglie Hypothesis

Since light seems to have both wave and particle properties, it is natural to ask whether matter (e.g., electrons, protons) might also have both wave and particle characteristics. In 1924, a French physics student, Louis de Broglie, suggested this idea in his doctoral dissertation. de Broglie's work was highly speculative, because there was no evidence at that time of any wave aspects of matter.

For the wavelength of electron waves, de Broglie chose

$$\lambda = \frac{h}{p} \qquad \qquad 34\text{-}13$$

DE BROGLIE RELATION FOR THE WAVELENGTH OF ELECTRON WAVES

where p is the momentum of the electron. Note that this is the same as Equation 34-7 for a photon. For the frequency of electron waves, de Broglie chose the Einstein equation relating the frequency and energy of a photon.

$$f = \frac{E}{h} \qquad \qquad 34\text{-}14$$

DE BROGLIE RELATION FOR THE FREQUENCY OF ELECTRON WAVES

These equations are thought to apply to all matter. However, for macroscopic objects, the wavelengths calculated from Equation 34-13 are so small that it is impossible to observe the usual wave properties of interference or diffraction. Even a dust particle with a mass as small as 1 μg is much too massive for any wave characteristics to be noticed, as we see in the following example.

[†] We will study the Bohr model in Chapter 36.

THE DE BROGLIE WAVELENGTH   **E X A M P L E   3 4 - 4**   **Try It Yourself**

Find the de Broglie wavelength of a 10^{-6} g particle moving with a speed of 10^{-6} m/s.

Cover the column to the right and try this on your own before looking at the answers.

Steps	Answers
Write the definition of the de Broglie wavelength and substitute the given data.	$\lambda = \dfrac{h}{p} = \dfrac{h}{mv} = \dfrac{6.63 \times 10^{-34} \text{ J·s}}{(10^{-9} \text{ kg})(10^{-6} \text{ m/s})}$
	$= \boxed{6.63 \times 10^{-19} \text{ m}}$

REMARKS This wavelength is several orders of magnitude smaller than the diameter of an atomic nucleus, which is about 10^{-15} m.

Since the wavelength found in Example 34-4 is so small, much smaller than any possible apertures or obstacles, diffraction or interference of such waves cannot be observed. In fact, the propagation of waves of very small wavelengths is indistinguishable from the propagation of particles. The momentum of the particle in Example 34-4 was only 10^{-15} kg·m/s. A macroscopic particle with a greater momentum would have an even smaller de Broglie wavelength. We therefore do not observe the wave properties of such macroscopic objects as baseballs and billiard balls.

EXERCISE Find the de Broglie wavelength of a baseball of mass 0.17 kg moving at 100 km/h. (*Answer* 1.4×10^{-34} m)

The situation is different for low-energy electrons and other microscopic particles. Consider a particle with kinetic energy K. Its momentum is found from

$$K = \frac{p^2}{2m}$$

or

$$p = \sqrt{2mK}$$

Its wavelength is then

$$\lambda = \frac{h}{p} = \frac{h}{\sqrt{2mK}}$$

If we multiply the numerator and the denominator by c, we obtain

$$\lambda = \frac{hc}{\sqrt{2mc^2K}} = \frac{1240 \text{ eV·nm}}{\sqrt{2mc^2K}} \qquad\qquad 34\text{-}15$$

WAVELENGTH ASSOCIATED WITH A PARTICLE OF MASS M

where we have used $hc = 1240$ eV·nm. For electrons, $mc^2 = 0.511$ MeV. Then,

$$\lambda = \frac{1240 \text{ eV·nm}}{\sqrt{2mc^2K}} = \frac{1240 \text{ eV·nm}}{\sqrt{2(0.511 \times 10^6 \text{ eV})K}}$$

or

$$\lambda = \frac{1.226}{\sqrt{K}}\text{ nm}, \quad K \text{ in electron volts} \qquad 34\text{-}16$$

ELECTRON WAVELENGTH

Equation 34-15 and Equation 34-16 do not hold for relativistic particles whose kinetic energies are a significant fraction of their rest energies mc^2. (Rest energies were discussed in Chapter 7 and in Chapter R.)

EXERCISE Find the wavelength of an electron whose kinetic energy is 10 eV. (*Answer* 0.388 nm. From this result, we see that a 10-eV electron has a de Broglie wavelength of about 0.4 nm. This is on the same order of magnitude as the size of the atom and the spacing of atoms in a crystal.)

Electron Interference and Diffraction

The observation of diffraction and interference of electron waves would provide the crucial test of the existence of wave properties of electrons. This was first discovered serendipitously in 1927 by C. J. Davisson and L. H. Germer as they were studying electron scattering from a nickel target at the Bell Telephone Laboratories. After heating the target to remove an oxide coating that had accumulated during an accidental break in the vacuum system, they found that the scattered electron intensity as a function of the scattering angle showed maxima and minima. Their target had crystallized, and by accident they had observed electron diffraction. Davisson and Germer then prepared a target consisting of a single crystal of nickel and investigated this phenomenon extensively. Figure 34-7a illustrates their experiment. Electrons from an electron gun are directed at a crystal and detected at some angle ϕ that can be varied. Figure 34-7b shows a typical pattern observed. There is a strong scattering maximum at an angle of 50°. The angle for maximum scattering of waves from a crystal depends on the wavelength of the waves and the spacing of the atoms in the crystal. Using the known spacing of atoms in their crystal, Davisson and Germer calculated the wavelength that could produce such a maximum and found that it agreed with the de Broglie equation (Equation 34-16) for the electron energy they were using. By varying the energy of the incident electrons, they could vary the electron wavelengths and produce maxima and minima at different locations in the diffraction patterns. In all cases, the measured wavelengths agreed with de Broglie's hypothesis.

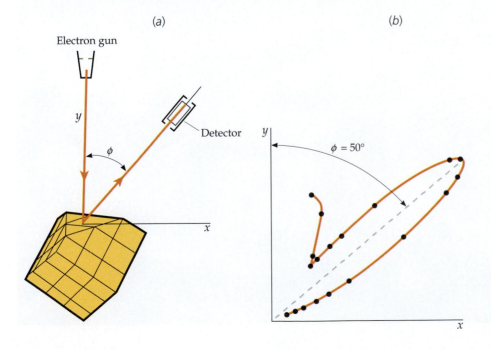

(a)   (b)

Electron gun

Detector

$\phi = 50°$

FIGURE 34-7 The Davisson–Germer experiment. (*a*) Electrons are scattered from a nickel crystal into a detector. (*b*) Intensity of scattered electrons versus scattering angle. The maximum is at the angle predicted by the diffraction of waves of wavelength λ given by the de Broglie formula.

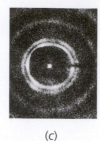
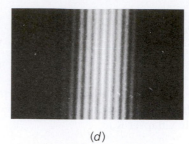

(a) (b) (c) (d)

FIGURE 34-8 (*a*) The diffraction pattern produced by X rays of wavelength 0.071 nm on an aluminum foil target. (*b*) The diffraction pattern produced by 600-eV electrons ($\lambda = 0.050$ nm) on an aluminum foil target. (*c*) The diffraction of 0.0568 eV neutrons ($\lambda = 0.12$ nm) incident on a copper foil. (*d*) A two-slit electron diffraction-interference pattern.

Another demonstration of the wave nature of electrons was provided in the same year by G. P. Thomson (son of J. J. Thomson) who observed electron diffraction in the transmission of electrons through thin metal foils. A metal foil consists of tiny, randomly oriented crystals. The diffraction pattern resulting from such a foil is a set of concentric circles. Figure 34-8*a* and Figure 34-8*b* show the diffraction pattern observed using X rays and electrons on an aluminum foil target. Figure 34-8*c* shows the diffraction patterns of neutrons on a copper foil target. Note the similarity of the patterns. The diffraction of hydrogen and helium atoms was observed in 1930. In all cases, the measured wavelengths agree with the de Broglie predictions. Figure 34-8*d* shows a diffraction pattern produced by electrons incident on two narrow slits. This experiment is equivalent to Young's famous double-slit experiment with light. The pattern is identical to the pattern observed with photons of the same wavelength. (Compare with Figure 34-1*b*.)

Shortly after the wave properties of the electron were demonstrated, it was suggested that electrons rather than light might be used to *see* small objects. As discussed in Chapter 33, reflected waves or transmitted waves can resolve details of objects only if the details are larger than the wavelength of the reflected wave. Beams of electrons, which can be focused electrically, can have very small wavelengths—much shorter than visible light. Today, the electron microscope (Figure 34-9) is an important research tool used to visualize specimens at scales far smaller than those possible with a light microscope.

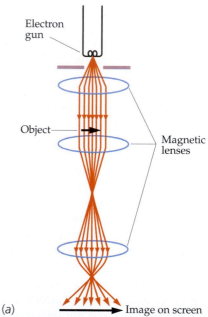

Electron gun

Object

Magnetic lenses

Image on screen

(a)

(b)

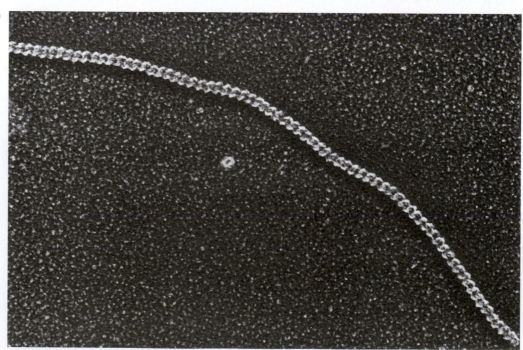

FIGURE 34-9
(*a*) An electron microscope. Electrons from a heated filament (the electron gun) are accelerated by a large potential difference. The electron beam is made parallel by a magnetic focusing lens. The electrons strike a thin target and are then focused by a second magnetic lens. The third magnetic lens projects the electron beam onto a fluorescent screen to produce the image.
(*b*) An electron micrograph of a DNA molecule.

Standing Waves and Energy Quantization

Given that electrons have wave-like properties, it should be possible to produce standing electron waves. If energy is associated with the frequency of a standing wave, as in $E = hf$ (Equation 34-14), then standing waves imply quantized energies.

The idea that the discrete energy states in atoms could be explained by standing waves led to the development by Erwin Schrödinger and others in 1926 of a detailed mathematical theory known as quantum theory, quantum mechanics, or wave mechanics. In this theory, the electron is described by a wave function ψ that obeys a wave equation called the Schrödinger equation. The form of the Schrödinger equation of a particular system depends on the forces acting on the particle, which are described by the potential energy functions associated with those forces. In Chapter 35 we discuss this equation, which is somewhat similar to the classical wave equations for sound or for light. Schrödinger solved the standing wave problem for the hydrogen atom, the simple harmonic oscillator, and other systems of interest. He found that the allowed frequencies, combined with $E = hf$, resulted in the set of energy levels found experimentally for the hydrogen atom, thereby demonstrating that quantum theory provides a general method of finding the quantized energy levels for a given system. Quantum theory is the basis for our understanding of the modern world, from the inner workings of the atomic nucleus to the radiation spectra of distant galaxies.

34-5 The Interpretation of the Wave Function

The wave function for waves on a string is the string displacement y. The wave function for sound waves can be either the displacement of the air molecules s, or the pressure P. The wave function for electromagnetic waves is the electric field $\vec{E}$ and the magnetic field $\vec{B}$. What is the wave function for electron waves? The symbol we use for this wave function is ψ (the Greek letter psi). When Schrödinger published his wave equation, neither he nor anyone else knew just how to interpret the wave function ψ. We can get a hint about how to interpret ψ by considering the quantization of light waves. For classical waves, such as sound or light, the energy per unit volume in the wave is proportional to the square of the wave function. Since the energy of a light wave is quantized, the energy per unit volume is proportional to the number of photons per unit volume. We might therefore expect the square of the photon's wave function to be proportional to the number of photons per unit volume in a light wave. But suppose we have a very low-energy source of light that emits just one photon at a time. In any unit volume, there is either one photon or none. The square of the wave function must then describe the *probability* of finding a photon in some unit volume.

The Schrödinger equation describes a single particle. The square of the wave function for a particle must then describe the *probability density*, which is the probability per unit volume, of finding the particle at a location. The probability of finding the particle in some volume element must also be proportional to the size of the volume element dV. Thus, in one dimension, the probability of finding a particle in a region dx at the position x is $\psi^2(x)\, dx$. If we call this probability $P(x)\, dx$, where $P(x)$ is the **probability density,** we have

$$P(x) = \psi^2(x)$$

34-17

PROBABILITY DENSITY

Generally the wave function depends on time as well as position, and is written $\psi(x,t)$. However, for standing waves, the probability density is independent of time. Since we will be concerned mostly with standing waves in this chapter, we omit the time dependence of the wave function and write it $\psi(x)$ or just ψ.

The probability of finding the particle in dx at point x_1 or at point x_2 is the sum of the separate probabilities $P(x_1)\, dx + P(x_2)\, dx$. If we have a particle at all, the probability of finding the particle somewhere must be 1. Then the sum of the probabilities over all the possible values of x must equal 1. That is,

$$\int_{-\infty}^{\infty} \psi^2 \, dx = 1 \qquad\qquad \text{34-18}$$

NORMALIZATION CONDITION

Equation 34-18 is called the **normalization condition**. If ψ is to satisfy the normalization condition, it must approach zero as x approaches infinity. This places a restriction on the possible solutions of the Schrödinger equation. There are solutions to the Schrödinger equation that do not approach zero as x approaches infinity. However, these are not acceptable as wave functions.

PROBABILITY CALCULATION FOR A CLASSICAL PARTICLE **E X A M P L E 3 4 - 5**

It is known that a classical point particle moves back and forth with constant speed between two walls at $x = 0$ and $x = 8$ cm (Figure 34-10). No additional information about the location of the particle is known. (*a*) What is the probability density $P(x)$? (*b*) What is the probability of finding the particle at $x = 2$ cm? (*c*) What is the probability of finding the particle between $x = 3.0$ cm and $x = 3.4$ cm?

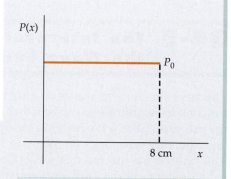

FIGURE 34-10 The probability function $P(x)$.

PICTURE THE PROBLEM We do not know the initial position of the particle. Since the particle moves with constant speed, it is equally likely to be anywhere in the region $0 < x < 8$ cm. The probability density $P(x)$ is therefore independent of x, for $0 < x < 8$ cm, and zero outside of this range. We can find $P(x)$, for $0 < x < 8$ cm, by normalization, that is, by requiring that the probability that the particle is somewhere between $x = 0$ and $x = 8$ cm is 1.

(*a*) 1. The probability density $P(x)$ is uniform between the walls and zero elsewhere:

$$P(x) = P_0, \quad 0 < x < 8 \text{ cm}$$
$$P(x) = 0, \quad x < 0 \quad \text{or} \quad x > 8 \text{ cm}$$

2. Apply the normalization condition:

$$\int_{-\infty}^{+\infty} P(x)\, dx = \int_0^{8 \text{ cm}} P_0\, dx = P_0 \,(8 \text{ cm}) = 1$$

3. Solve for P_0:

$$P_0 = \boxed{\dfrac{1}{8 \text{ cm}}}$$

(*b*) The probability of finding the particle in some range Δx is proportional to $P_0 \Delta x = \Delta x/(8 \text{ cm})$. Since it is given that $\Delta x = 0$, the probability of finding the particle at the point $x = 2$ cm is 0. Alternatively, since there is an infinite number of points between $x = 0$ and $x = 8$ cm, and the particle is equally likely to be at any point, the chance that the particle will be at any one particular point must be zero.

The probability of finding the particle at the point $x = 2$ cm is 0.

(c) Since the probability density is constant, the probability of a particle being in some range Δx in the region $0 < x < 8$ cm is $P_0 \Delta x$. The probability of the particle being in the region 3.0 cm $< x <$ 3.4 cm is thus:

$$P_0 \Delta x = \left(\frac{1}{8 \text{ cm}}\right) 0.4 \text{ cm} = \boxed{0.05}$$

REMARKS Note in step 2 of Part (a) that we need only integrate from 0 to 8 cm, because $P(x)$ is zero outside this range.

34-6 Wave–Particle Duality

We have seen that light, which we ordinarily think of as a wave, exhibits particle properties when it interacts with matter, as in the photoelectric effect or in Compton scattering. Electrons, which we usually think of as particles, exhibit the wave properties of interference and diffraction when they pass near the edges of obstacles. All carriers of momentum and energy (e.g., electrons, atoms, light, or sound), exhibit both wave and particle characteristics. It might be tempting to say that an electron, for example, is both a wave and a particle, but what does this mean? In classical physics, the concepts of waves and particles are mutually exclusive. A **classical particle** behaves like a piece of shot; it can be localized and scattered, it exchanges energy suddenly at a point in space, and it obeys the laws of conservation of energy and momentum in collisions. It does *not* exhibit interference or diffraction. A **classical wave,** on the other hand, behaves like a water wave; it exhibits diffraction and interference, and its energy is spread out continuously in space and time. They are mutually exclusive. Nothing can be both a classical particle and a classical wave at the same time.

After Thomas Young observed the two-slit interference pattern with light in 1801, light was thought to be a classical wave. On the other hand, the electrons discovered by J. J. Thomson were thought to be classical particles. We now know that these classical concepts of waves and particles do not adequately describe the complete behavior of any phenomenon.

Everything propagates like a wave and exchanges energy like a particle.

Often the concepts of the classical particle and the classical wave give the same results. If the wavelength is very small, diffraction effects are negligible, so the waves travel in straight lines like classical particles. Also, interference is not seen for waves of very short wavelength, because the interference fringes are too closely spaced to be observed. It then makes no difference which concept we use. If diffraction is negligible, we can think of light as a wave propagating along rays, as in geometrical optics, or as a beam of photon particles. Similarly, we can think of an electron as a wave propagating in straight lines along rays or, more commonly, as a particle.

We can also use either the wave or particle concept to describe exchanges of energy if we have a large number of particles and we are interested only in the average values of energy and momentum exchanges.

The Two-Slit Experiment Revisited

The wave–particle duality of nature is illustrated by the analysis of the experiment in which a single electron is incident on a barrier with two slits. The analysis is the same whether we use an electron or a photon (light). To describe the propagation of an electron, we must use wave theory. Let us assume the source is a point source, such as a needlepoint, so we have spherical waves spreading out from the source. After passing through the two slits, the wave-fronts spread out—as if each slit was a source of wavefronts. The wave function ψ at a point on a screen or film far from the slits depends on the difference in path lengths from the source to the point, one path through one slit, and the other path through the other slit. At points on the screen for which the difference in path lengths is either zero or an integral number of wavelengths, the amplitude of the wave function is a maximum. Since the probability of detecting the electron is proportional to ψ^2, the electron is very likely to arrive near these points. At points for which the path difference is an odd number of half wave-lengths, the wave function ψ is zero, so the electron is very unlikely to arrive near these points. The chapter opening photos show the interference pattern produced by 10 electrons, 100 electrons, 3000 electrons, and 70,000 electrons. Note that, although the electron propagates through the slits like a wave, the electron interacts with the screen at a single point like a particle.

The Uncertainty Principle

An important principle consistent with the wave–particle duality of nature is the uncertainty principle. It states that, in principle, it is impossible to simultaneously measure both the position and the momentum of a particle with unlimited precision. A common way to measure the position of an object is to look at the object with light. If we do this, we scatter light from the object and determine the position by the direction of the scattered light. If we use light of wavelength λ, we can measure the position x only to an uncertainty Δx of the order of λ because of diffraction effects.

$$\Delta x \sim \lambda$$

To reduce the uncertainty in position, we therefore use light of very short wavelength, perhaps even X rays. In principle, there is no limit to the accuracy of such a position measurement, because there is no limit on how small the wavelength λ can be.

We can determine the momentum p_x of the object if we know the mass and can determine its velocity. The momentum of the object can be found by measuring the object's position at two nearby times and computing its velocity. If we use light of wavelength λ, the photons carry momentum h/p_x. If these photons are scattered by the object we are looking at, the scattering changes the momentum of the object in an uncontrollable way. Each photon carries momentum h/λ, so the uncertainty in the momentum Δp_x of the object, introduced by looking at it, is of the order of h/λ:

$$\Delta p_x \sim \frac{h}{\lambda}$$

If the wavelength of the radiation is small, the momentum of each photon will be large and the momentum measurement will have a large uncertainty. Reducing the intensity of light cannot eliminate this uncertainty; such a reduction merely reduces the number of photons in the beam. To *see* the object, we must scatter at least one photon. Therefore, the uncertainty in the momentum measurement of the object will be large if λ is small, and the uncertainty in the position measurement of the object will be large if λ is large.

Of course we could always *look at* the objects by scattering electrons instead of photons, but the same difficulty remains. If we use low-momentum electrons to reduce the uncertainty in the momentum measurement, we have a large uncertainty in the position measurement because of diffraction of the electrons. The relation between the wavelength and momentum $\lambda = h/p_x$ is the same for electrons as it is for photons.

The product of the intrinsic uncertainties in position and momentum is

$$\Delta x \, \Delta p_x \sim \lambda \times \frac{h}{\lambda} = h$$

If we define precisely what we mean by uncertainties in measurement, we can give a precise statement of the uncertainty principle. If Δx and Δp are defined to be the standard deviations in the measurements of position and momentum, it can be shown that their product must be greater than or equal to $\hbar/2$.

$$\Delta x \, \Delta p_x \geq \tfrac{1}{2}\hbar \qquad\qquad 34\text{-}19$$

where $\hbar = h/2\pi$.[†]

Equation 34-19 provides a statement of the uncertainty principle first enunciated by Werner Heisenberg in 1927. In practice, the experimental uncertainties are usually much greater than the intrinsic lower limit that results from wave–particle duality.

34-7 A Particle in a Box

We can illustrate many of the important features of quantum physics without solving the Schrödinger equation by considering a simple problem of a particle of mass m confined to a one-dimensional box of length L, like the particle in Example 34-5. This can be considered a crude description of an electron confined within an atom or a proton confined within a nucleus. If a classical particle bounces back and forth between the walls of the box, the particle's energy and momentum can have any values. However, according to quantum theory, the particle is described by a wave function ψ, whose square describes the probability of finding the particle in some region. Since we are assuming that the particle is indeed inside the box, the wave function must be zero everywhere outside the box. If the box is between $x = 0$ and $x = L$, we have

$$\psi = 0, \text{ for } x \leq 0 \text{ and for } x \geq L$$

In particular, if we assume the wave function to be continuous everywhere, it must be zero at the end points of the box $x = 0$ and $x = L$. This is the same condition as the condition for standing waves on a string fixed at $x = 0$ and $x = L$, and the results are the same. The allowed wavelengths for a particle in the box are those where the length L equals an integral number of half wavelengths (Figure 34-11).

$$L = n\frac{\lambda_n}{2}, \qquad n = 1, 2, 3, \ldots \qquad\qquad 34\text{-}20$$

STANDING-WAVE CONDITION FOR A PARTICLE IN A BOX OF LENGTH L

The total energy of the particle is its kinetic energy

$$E = \frac{1}{2}mv^2 = \frac{p^2}{2m}$$

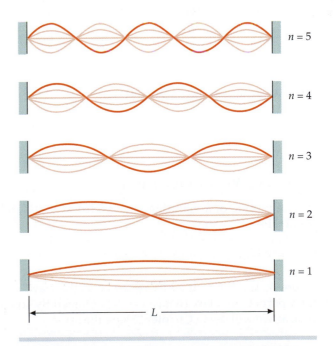

FIGURE 34-11 Standing waves on a string fixed at both ends. The standing-wave condition is the same as for standing electron waves in a box.

[†] The combination $h/2\pi$ occurs so often it is given a special symbol, somewhat analogous to giving the special symbol ω for $2\pi f$, which occurs often in oscillations.

Substituting the de Broglie relation $p_n = h/\lambda_n$,

$$E_n = \frac{p_n^2}{2m} = \frac{(h/\lambda_n)^2}{2m} = \frac{h^2}{2m\lambda_n^2}$$

Then the standing-wave condition $\lambda_n = 2L/n$ gives the allowed energies.

$$E_n = n^2 \frac{h^2}{8mL^2} = n^2 E_1 \qquad\qquad 34\text{-}21$$

ALLOWED ENERGIES FOR A PARTICLE IN A BOX

where

$$E_1 = \frac{h^2}{8mL^2} \qquad\qquad 34\text{-}22$$

GROUND-STATE ENERGY FOR A PARTICLE IN A BOX

is the energy of the lowest state, which is the ground state.

The condition $\psi = 0$ at $x = 0$ and $x = L$ is called a **boundary condition.** Boundary conditions in quantum theory lead to energy quantization. Figure 34-12 shows the energy-level diagram for a particle in a box. Note that the lowest energy is not zero. This result is a general feature of quantum theory. If a particle is confined to some region of space, the particle has a minimum kinetic energy, which is called the **zero-point energy.** The smaller the region of space the particle is confined to, the greater its zero-point energy. In Equation 34-22, this is indicated by the fact that E_1 varies as $1/L^2$.

If an electron is confined (i.e., bound to an atom) in some energy state E_i, the electron can make a transition to another energy state E_f with the emission of a photon (if $E_f < E_i$; if E_f is greater than E_i, the system absorbs a photon). The transition from state 3 to the ground state is indicated in Figure 34-12 by the vertical arrow. The frequency of the emitted photon is found from the conservation of energy[†]

$$hf = E_i - E_f \qquad\qquad 34\text{-}23$$

The wavelength of the photon is then

$$\lambda = \frac{c}{f} = \frac{hc}{E_i - E_f} \qquad\qquad 34\text{-}24$$

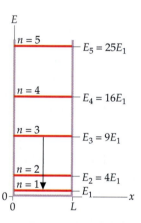

FIGURE 34-12 Energy-level diagram for a particle in a box. Classically, a particle can have any energy value. Quantum mechanically, only those energy values given by Equation 34-22 are allowed. A transition between the state $n = 3$ and the ground state $n = 1$ is indicated by the vertical arrow.

Standing-Wave Functions

The amplitude of a vibrating string fixed at $x = 0$ and $x = L$ is given by Equation 16-15:

$$y_n = A_n \sin k_n x$$

where A_n is a constant and $k_n = 2\pi/\lambda_n$ is the wave number. The wave functions for a particle in a box (which can be obtained by solving the Schrödinger equation, as we will see in Chapter 35) are the same

---

[†] This equation was first proposed by Niels Bohr in his semiclassical model of the hydrogen atom in 1913, about 10 years before de Broglie's suggestion that electrons have wave properties. We will study the Bohr model in Chapter 36.

$$\psi_n(x) = A_n \sin k_n x$$

where $k_n = 2\pi/\lambda_n$. Using $\lambda_n = 2L/n$, we have

$$k_n = \frac{2\pi}{\lambda_n} = \frac{2\pi}{2L/n} = \frac{n\pi}{L}$$

The wave functions can thus be written

$$\psi_n(x) = A_n \sin\left(n\pi\frac{x}{L}\right)$$

The constant A_n is determined by the normalization condition (Equation 34-18):

$$\int_{-\infty}^{\infty} \psi^2\, dx = \int_0^L A_n^2 \sin^2\left(n\pi\frac{x}{L}\right) dx = 1$$

Note that we need integrate only from $x = 0$ to $x = L$ because $\psi(x)$ is zero everywhere else. The result of evaluating the integral and solving for A_n is

$$A_n = \sqrt{\frac{2}{L}}$$

independent of n. The normalized wave functions for a particle in a box are thus

$$\psi_n(x) = \sqrt{\frac{2}{L}} \sin\left(n\pi\frac{x}{L}\right) \qquad \text{34-25}$$

WAVE FUNCTIONS FOR A PARTICLE IN A BOX

These standing-wave functions for $n = 1$, $n = 2$, and $n = 3$ are shown in Figure 34-13.

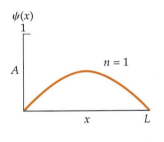
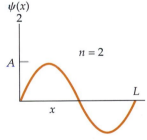
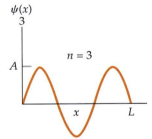

FIGURE 34-13 Standing-wave functions for $n = 1$, $n = 2$, and $n = 3$.

The number n is called a **quantum number.** It characterizes the wave function for a particular state and for the energy of that state. In our one-dimensional problem, a quantum number arises from the boundary condition on the wave function that it must be zero at $x = 0$ and $x = L$. In three-dimensional problems, three quantum numbers arise, one associated with a boundary condition in each dimension.

Figure 34-14 shows plots of ψ^2 for the ground state $n = 1$, the first excited state $n = 2$, the second excited state $n = 3$, and the state $n = 10$. In the ground state, the particle is most likely to be found near the center of the box, as indicated by the maximum value of ψ^2 at $x = L/2$. In the first excited state, the particle is least likely to be found near the center of the box because ψ^2 is small near $x = L/2$.

For very large values of n, the maxima and minima of ψ^2 are very close together, as illustrated for $n = 10$. The average value of ψ^2 is indicated in this figure by the dashed line. For very large values of n, the maxima are so closely spaced that ψ^2 cannot be distinguished from its average value. The fact that $(\psi^2)_{av}$ is constant across the whole box means that the particle is equally likely to be found anywhere in the box—the same as in the classical result. This is an example of **Bohr's correspondence principle:**

> In the limit of very large quantum numbers, the classical calculation and the quantum calculation must yield the same results.

BOHR'S CORRESPONDENCE PRINCIPLE

The region of very large quantum numbers is also the region of very large energies. For large energies, the percentage change in energy between adjacent quantum states is very small, so energy quantization is not important (see Problem 83).

We are so accustomed to thinking of the electron as a classical particle that we tend to think of an electron in a box as a particle bouncing back and forth between the walls. But the probability distributions shown in Figure 34-14 are stationary; that is, they do not depend on time. A better picture for an electron in a bound state is a cloud of charge with the charge density proportional to ψ^2. Figure 34-14 can then be thought of as plots of the charge density versus x for the various states. In the ground state, $n = 1$, the electron cloud is centered in the middle of the box and is spread out over most of the box, as indicated in Figure 34-14a. In the first excited state, $n = 2$, the charge density of the electron cloud has two maxima, as indicated in Figure 34-14b. For very large values of n, there are many closely spaced maxima and minima in the charge density resulting in an average charge density that is approximately uniform throughout the box. This electron-cloud picture of an electron is very useful in understanding the structure of atoms and molecules. However, it should be noted that whenever an electron is observed to interact with matter or radiation, it is always observed as a whole unit charge.

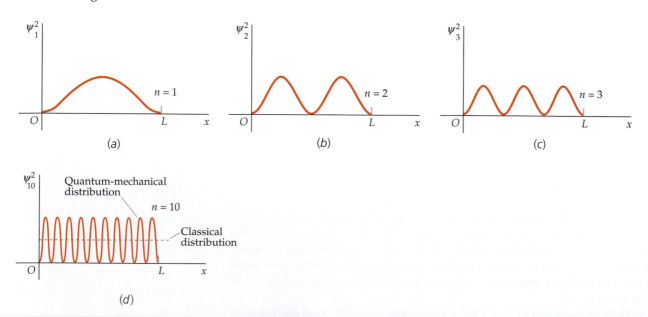

FIGURE 34-14 ψ^2 versus x for a particle in a box of length L for (a) the ground state, $n = 1$; (b) the first excited state, $n = 2$; (c) the second excited state, $n = 3$; and (d) the state $n = 10$. For $n = 10$, the maxima and minima of ψ^2 are so close together that individual maxima may be hard to distinguish. The dashed line indicates the average value of ψ^2. It gives the classical prediction that the particle is equally likely to be found near any point in the box.

EXAMPLE 34-6

An electron is in a one-dimensional box of length 0.1 nm. (*a*) Find the ground-state energy. (*b*) Find the energy in electron volts of the five lowest states, and then sketch an energy-level diagram. (*c*) Find the wavelength of the photon emitted for each transition from the state $n = 3$ to a lower-energy state.

PICTURE THE PROBLEM For Part (*a*) and Part (*b*), the energies are given by $E_n = n^2 E_1$ (Equation 34-21), where the ground-state energy $E_1 = h^2/8mL^2$ (Equation 34-22). For Part (*c*), the photon wavelengths are given by $\lambda = hc/(E_i - E_f)$ (Equation 34-24).

(*a*) Use $hc = 1240$ eV·nm and $mc^2 = 5.11 \times 10^5$ eV to calculate E_1:

$$E_1 = \frac{(hc)^2}{8(mc^2)L^2}$$

$$= \frac{(1240 \text{ eV·nm})^2}{8(5.11 \times 10^5 \text{ eV})(0.1 \text{ nm})^2} = \boxed{37.6 \text{ eV}}$$

(*b*) Calculate $E_n = n^2 E_1$ for $n = 2, 3, 4,$ and 5:

$$E_2 = (2)^2(37.6 \text{ eV}) = \boxed{150 \text{ eV}}$$

$$E_3 = (3)^2(37.6 \text{ eV}) = \boxed{338 \text{ eV}}$$

$$E_4 = (4)^2(37.6 \text{ eV}) = \boxed{602 \text{ eV}}$$

$$E_5 = (5)^2(37.6 \text{ eV}) = \boxed{940 \text{ eV}}$$

(*c*) 1. Use the energies found in Part (*b*) to calculate the wavelength for a transition from state 3 to state 2:

$$\lambda = \frac{hc}{E_3 - E_2}$$

$$= \frac{1240 \text{ eV·nm}}{338 \text{ eV} - 150 \text{ eV}} = \boxed{6.60 \text{ nm}}$$

2. Then use the energies in Part (*a*) and Part (*b*) to calculate the wavelength for a transition from state 3 to state 1:

$$\lambda = \frac{hc}{E_3 - E_1}$$

$$= \frac{1240 \text{ eV·nm}}{338 \text{ eV} - 37.6 \text{ eV}} = \boxed{4.13 \text{ nm}}$$

REMARKS The energy-level diagram is shown in Figure 34-15. The transitions from $n = 3$ to $n = 2$ and from $n = 3$ to $n = 1$ are indicated by the vertical arrows. The ground-state energy of 37.6 eV is on the same order of magnitude as the kinetic energy of the electron in the ground state of the hydrogen atom, which is 13.6 eV. In the hydrogen atom, the electron also has potential energy of -27.2 eV in the ground state, giving it a total ground-state energy of -13.6 eV.

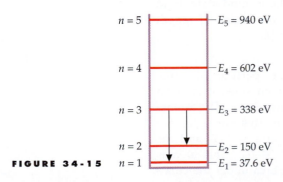

$n = 5$ $E_5 = 940$ eV

$n = 4$ $E_4 = 602$ eV

$n = 3$ $E_3 = 338$ eV

$n = 2$ $E_2 = 150$ eV

FIGURE 34-15 $n = 1$ $E_1 = 37.6$ eV

EXERCISE Calculate the wavelength of the photon emitted if the electron in the box makes a transition from $n = 4$ to $n = 3$. (*Answer* 4.69 nm)

34-8 Expectation Values

The solution of a classical mechanics problem is typically specified by giving the position of a particle as a function of time. But the wave nature of matter prevents us from doing this for microscopic systems. The most that we can know is the probability of measuring a certain value of the position x. If we measure the position for a large number of identical systems, we get a range of values corresponding to the probability distribution. The average value of x obtained from such measurements is called the **expectation value** and is written $<x>$. The expectation value of x is the same as the average value of x that we would expect to obtain from a measurement of the positions of a large number of particles with the same wave function $\psi(x)$.

Since $\psi^2(x)\, dx$ is the probability of finding a particle in the region dx, the expectation value of x is

$$<x> = \int_{-\infty}^{+\infty} x\psi^2(x)\, dx \qquad\qquad 34\text{-}26$$

EXPECTATION VALUE OF X DEFINED

The expectation value of any function $f(x)$ is given by

$$<f(x)> = \int_{-\infty}^{+\infty} f(x)\psi^2(x)\, dx \qquad\qquad 34\text{-}27$$

EXPECTATION VALUE OF $F(X)$ DEFINED

*Calculating Probabilities and Expectation Values

The problem of a particle in a box allows us to illustrate the calculation of the probability of finding the particle in various regions of the box and the expectation values for various energy states. We give two examples, using the wave functions given by Equation 34-25.

THE PROBABILITY OF THE PARTICLE BEING FOUND IN A SPECIFIED REGION OF A BOX
                                                    EXAMPLE 34-7

A particle in a one-dimensional box of length L is in the ground state. Find the probability of finding the particle (*a*) anywhere in the region of length $\Delta x = 0.01L$, centered at $x = \frac{1}{2}L$ and (*b*) in the region $0 < x < \frac{1}{4}L$.

PICTURE THE PROBLEM The probability P of finding the particle in some infinitesimal range dx is $\psi^2\, dx$. For a particle in the ground state, ψ is given by Equation 34-25, with $n = 1$; ψ^2 is illustrated in Figure 34-14. The probability of finding x in some region is just the area under this curve for the region. For Part (*a*), the region is $\Delta x = 0.01L$, centered at $x = L/2$, and the area under the ψ^2 versus x curve is shown in Figure 34-16a. This area is $\approx \psi^2\, \Delta x$. For Part (*b*), the region is $0 < x < L/4$, and the area under the curve is shown in Figure 34-16b. To calculate this area, we must integrate ψ^2 from $x = 0$ to $x = L/4$.

(a)

(b)

FIGURE 34-16

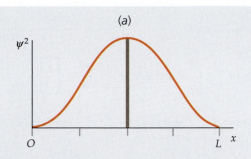

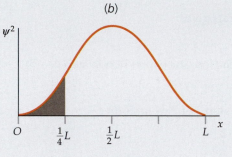

(a) 1. The probability of finding the particle is the area under the curve shown in Figure 34-16a. To calculate this area, we need to calculate the height of curve at $x = L/2$:

$$\psi(x) = \sqrt{\frac{2}{L}} \sin\left(\pi \frac{x}{L}\right)$$

so

$$\psi^2(L/2) = \frac{2}{L} \sin^2 \frac{\pi}{2} = \frac{2}{L}$$

2. The area is the height times the width, and the width is $\Delta x = 0.01L$:

$$P = \psi^2(L/2)\Delta x = \frac{2}{L} \times 0.01L = \boxed{0.02}$$

(b) 1. The probability of finding the particle is the area under the curve shown in Figure 34-16b. To calculate this area, we need to integrate from $x = 0$ to $x = L/4$:

$$P = \int_0^{L/4} \psi^2(x)\, dx = \int_0^{L/4} \frac{2}{L} \sin^2 \frac{\pi x}{L}\, dx$$

2. The integral can be evaluated a number of ways. If a table of integrals is used, a change in the integration variable in required. Changing the integration variable to $\theta = \pi x/L$ gives:

$$P = \frac{2}{\pi} \int_0^{\pi/4} \sin^2 \theta\, d\theta$$

3. The integral can be found in tables:

$$\int_0^{\pi/4} \sin^2 \theta\, d\theta = \left(\frac{\theta}{2} - \frac{\sin^2 2\theta}{4}\right)\Bigg|_0^{\pi/4} = \left(\frac{\pi}{8} - \frac{1}{4}\right)$$

4. Use the result from Part (b), step 3 to calculate the probability:

$$P = \frac{2}{\pi}\left(\frac{\pi}{8} - \frac{1}{4}\right) = \boxed{0.091}$$

REMARKS An integral was not necessary for Part (a) because the area of interest could be well approximated by a rectangle of height ψ^2 and width x. The chance of finding the particle in the region $\Delta x = 0.01L$ at $x = \frac{1}{2}L$ is approximately 2 percent. The chance of finding the particle in the region $0 < x < L/4$ is about 9.1 percent.

CALCULATING EXPECTATION VALUES **EXAMPLE 34-8**

Find (a) $<x>$ and (b) $<x^2>$ for a particle in its ground state in a box of length L.

PICTURE THE PROBLEM We use $<f(x)> = \int f(x)\psi^2(x)\, dx$, with

$$\psi_n(x) = \sqrt{\frac{2}{L}} \sin \frac{n\pi x}{L}.$$

(a) 1. Write $<x>$ using the ground-state wave function given by Equation 34-25, with $n = 1$:

$$<x> = \int_{-\infty}^{+\infty} x\psi^2(x)\, dx = \frac{2}{L} \int_0^L x \sin^2\left(\frac{\pi x}{L}\right) dx$$

2. To evaluate this integral by using a table of integrals, first change the integration variable to $\theta = \pi x/L$:

$$<x> = \frac{2}{L}\left(\frac{L}{\pi}\right)^2 \int_0^\pi \theta \sin^2 \theta\, d\theta$$

$$= \frac{2L}{\pi^2} \int_0^\pi \theta \sin^2 \theta\, d\theta$$

3. The table of integrals gives:

$$\int_0^\pi \theta \sin^2 \theta \, d\theta = \left[\frac{\theta^2}{4} - \frac{\theta \sin 2\theta}{4} - \frac{\cos 2\theta}{8} \right]_0^\pi = \frac{\pi^2}{4}$$

4. Substitute this value into the expression in step 2:

$$<x> = \frac{2L}{\pi^2} \int_0^\pi \theta \sin^2 \theta \, d\theta = \frac{2L}{\pi^2} \frac{\pi^2}{4} = \boxed{\frac{L}{2}}$$

(b) 1. Repeat step 1 and step 2 of Part (a) for $<x^2>$:

$$<x^2> = \int_{-\infty}^{+\infty} x^2 \psi^2(x) \, dx = \int_0^L x^2 \frac{2}{L} \sin^2(\pi x/L) \, dx$$

$$= \frac{2}{L} \left(\frac{L}{\pi}\right)^3 \int_0^\pi \theta^2 \sin^2 \theta \, d\theta = \frac{2L^2}{\pi^3} \int_0^\pi \theta^2 \sin^2 \theta \, d\theta$$

2. Evaluating the integral using a table of integrals gives:

$$\int_0^\pi \theta^2 \sin^2 \theta \, d\theta = \left[\frac{\theta^3}{6} - \left(\frac{\theta^2}{4} - \frac{1}{8}\right) \sin 2\theta - \frac{\theta \cos 2\theta}{4} \right]\Big|_0^\pi$$

$$= \frac{\pi^3}{6} - \frac{\pi}{4}$$

3. Substitute this value into the expression in step 1 of Part (b):

$$<x^2> = \frac{2L^2}{\pi^3} \left(\frac{\pi^3}{6} - \frac{\pi}{4}\right) = \left(\frac{1}{3} - \frac{1}{2\pi^2}\right) L^2 = \boxed{0.283L^2}$$

REMARKS The expectation value of x is $L/2$, as we would expect, because the probability distribution is symmetric about the midpoint of the box. Note that $<x^2>$ is not equal to $<x>^2$.

34-9 Energy Quantization in Other Systems

The quantized energies of a system are generally determined by solving the Schrödinger equation for that system. The form of the Schrödinger equation depends on the potential energy of the particle. The potential energy for a one-dimensional box from $x = 0$ to $x = L$ is shown in Figure 34-17. This potential energy function is called an **infinite square-well potential,** and it is described mathematically by

$$U(x) = 0, \quad 0 < x < L$$

$$U(x) = \infty, \quad x < 0 \quad \text{or} \quad x > L \tag{34-28}$$

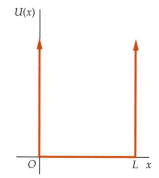

FIGURE 34-17 The infinite square-well potential energy. For $x < 0$ and $x > L$, the potential energy $U(x)$ is infinite. The particle is confined to the region in the well $0 < x < L$.

Inside the box the particle moves freely, so the potential energy is uniform. For convenience, we choose the value of this potential energy to be zero. Outside the box the potential energy is infinite, so the particle cannot exist outside the box no matter what its energy. We did not need to solve the Schrödinger equation for this potential because the wave functions and quantized frequencies are the same as for a string fixed at both ends, which we studied in Chapter 16. Although this problem seems artificial, actually it is useful for some physical problems, such as a neutron inside a nucleus.

The Harmonic Oscillator

More realistic than the particle in a box is the harmonic oscillator, which applies to an object of mass m on a spring of force constant k or to any system undergoing small oscillations about a stable equilibrium. Figure 34-18 shows the potential energy function

$$U(x) = \tfrac{1}{2}kx^2 = \tfrac{1}{2}m\omega_0^2 x^2$$

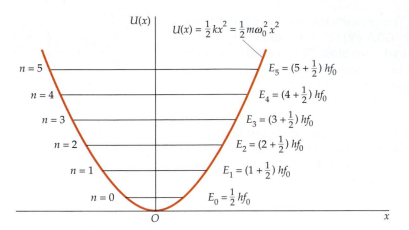

FIGURE 34-18 Harmonic oscillator potential energy function. The allowed energy levels are indicated by the equally spaced horizontal lines. Also, $\omega_0 = 2\pi f_0$.

where $\omega_0 = \sqrt{k/m}$ is the natural frequency of the oscillator. Classically, the object oscillates between $x = +A$ and $x = -A$. Its total energy is $E = \frac{1}{2}m\omega_0^2 A^2$, which can have any nonnegative value, including zero.

In quantum theory, the particle is represented by the wave function $\psi(x)$, which is determined by solving the Schrödinger equation for this potential. Normalizable wave functions $\psi_n(x)$ occur only for discrete values of the energy E_n given by

$$E_n = (n + \tfrac{1}{2})hf_0, \quad n = 0, 1, 2, 3, \dots \qquad \qquad 34\text{-}29$$

where $f_0 = \omega_0/2\pi$ is the classical frequency of the oscillator. Note that the energy levels of a harmonic oscillator are evenly spaced with separation hf, as indicated in Figure 34-18. Compare this with the uneven spacing of the energy levels for the particle in a box, as shown in Figure 34-12. If a harmonic oscillator makes a transition from energy level n to the next lowest energy level $n - 1$, the frequency f of the photon emitted is given by $hf = E_i - E_f$ (Equation 34-23). Applying this equation gives

$$hf = E_n - E_{n-1} = (n + \tfrac{1}{2})hf_0 - (n - 1 + \tfrac{1}{2})hf_0 = hf_0$$

The frequency f of the emitted photon is therefore equal to the classical frequency f_0 of the oscillator.

The Hydrogen Atom

In the hydrogen atom, an electron is bound to a proton by the electrostatic force of attraction (discussed in Chapter 21). This force varies inversely as the square of the separation distance (exactly like the gravitational attraction of the earth and sun). The potential energy of the electron–proton system therefore varies inversely with separation distance (Equation 23-9). As in the case of gravitational potential energy, the potential energy of the electron–proton system is chosen to be zero if the electron is an infinite distance from the proton. Then for all finite distances, the potential energy is negative. Like the case of an object orbiting the earth, the electron–proton system is a bound system if its total energy is negative. Like the energies of a particle in a box and of a harmonic oscillator, the energies are described by a quantum number n. As we will see in Chapter 36, the allowed energies of the hydrogen atom are given by

$$E_n = -\frac{13.6 \text{ eV}}{n^2}, \quad n = 1, 2, 3, \dots \qquad \qquad 34\text{-}30$$

The lowest energy corresponds to $n = 1$. The ground-state energy is thus -13.6 eV. The energy of the first excited state is $-(13.6 \text{ eV})/2^2 = -3.40$ eV. Figure 34-19 shows the energy-level diagram for the hydrogen atom. The vertical arrows indicate transitions from a higher state to a lower state with the emission of electromagnetic radiation. Only those transitions ending at the first excited state ($n = 2$) involve energy differences in the range of visible light of 1.77 eV to 3.10 eV, as calculated in Example 34-1.

Other atoms are more complicated than the hydrogen atom, but their energy levels are similar in many ways to those of hydrogen. Their ground-state energies are of the order of -1 eV to -10 eV, and many transitions involve energies corresponding to photons in the visible range.

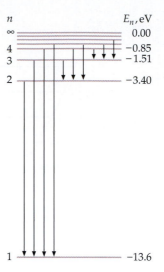

FIGURE 34-19 Energy-level diagram for the hydrogen atom. The energy of the ground state is -13.6 eV. As n approaches ∞ the energy approaches 0, which is the highest energy state for an electron bound to the atom.

SUMMARY

1. All phenomenon propagate like waves and interact like particles.

2. The quantum of light is called a photon and has energy $E = hf$, where h is Planck's constant.

3. The relation between wavelength and momentum of electrons, photons, and other particles is given by the de Broglie relation $\lambda = h/p$.

4. Energy quantization in bound systems arises from standing-wave conditions, which are equivalent to boundary conditions on the wave function.

5. The uncertainty principle is a fundamental law of nature that places theoretical restrictions on the precision of a simultaneous measurement of the position and momentum of a particle. It follows from the general properties of waves.

Topic	Relevant Equations and Remarks	
1. Constants and Values		
Planck's constant	$h = 6.626 \times 10^{-34}$ J·s $= 4.136 \times 10^{-15}$ eV·s	**34-2**
hc	$hc = 1240$ eV·nm	**34-5**
Compton Wavelength	$\lambda_C = \dfrac{h}{m_e c} = 2.43$ pm	**34-12**
2. The Particle Nature of Light: Photons	Energy is quantized.	
Photon energy	$E = hf$	**34-1**
Momentum of a photon	$p = \dfrac{h}{\lambda}$	**34-7**

3. **Frequency–Wavelength (energy–momentum) Relations**

Photons	$E = pc,$ so $\lambda f = c$	R-17

Nonrelativistic particles	$K = \dfrac{p^2}{2m}$, so $\lambda = \dfrac{hc}{\sqrt{2mc^2 K}}$	34-15

Electrons	$\lambda = \dfrac{1.226}{\sqrt{K}}$ nm, K in electron volts	34-16

Photoelectric effect	$K_{\text{max}} = (\tfrac{1}{2}mv^2)_{\text{max}} = hf - \phi$	34-3

where ϕ is the work function of the cathode.

Compton scattering	$\lambda_2 - \lambda_1 = \dfrac{h}{m_e c}(1 - \cos\theta) = \lambda_C(1 - \cos\theta)$	34-11

4. **Quantum Mechanics**

The state of a particle, such as an electron, is described by its wave function ψ, which is the solution of the Schrödinger wave equation.

Probability density

The probability of finding the particle in some region of space dx is given by

$$P(x) = \psi^2(x)\,dx \qquad \text{34-17}$$

Normalization condition

$$\int_{-\infty}^{\infty} \psi^2\,dx = 1 \qquad \text{34-18}$$

Quantum number

The wave function for a particular energy state is characterized by a quantum number n. In three dimensions there are three quantum numbers, one associated with a boundary condition in each dimension.

Expectation value

The expectation value of x is the same as the average value of x that we would expect to obtain from a measurement of the positions of a large number of particles with the same wave function $\psi(x)$.

$$<x> = \int_{-\infty}^{+\infty} x\psi^2(x)\,dx \qquad \text{34-26}$$

$$<f(x)> = \int_{-\infty}^{+\infty} f(x)\psi^2(x)\,dx \qquad \text{34-27}$$

5. **Wave–Particle Duality**

Light, electrons, neutrons, and all carriers of momentum and energy exhibit both wave and particle properties. Everything propagates like a classical wave exhibiting diffraction and interference, but exchanges energy in discrete lumps like a classical particle. Because the wavelength of macroscopic objects is so small, diffraction and interference are not observed. Also, if a macroscopic amount of energy is exchanged, so many quanta are involved that the particle nature of the energy is not evident.

6. **Uncertainty Principle**

The wave–particle duality of nature leads to the uncertainty principle, which states that the product of the uncertainty in a measurement of position and the uncertainty in a measurement of momentum must be greater than or equal to $\tfrac{1}{2}\hbar$, where $\hbar$ is Planck's constant divided by 2π.

$$\Delta x\,\Delta p \geq \tfrac{1}{2}\hbar \qquad \text{34-19}$$

PROBLEMS

- Single-concept, single-step, relatively easy
- •• Intermediate-level, may require synthesis of concepts
- ••• Challenging
- **SSM** Solution is in the *Student Solutions Manual*
- **iSOLVE** Problems available on iSOLVE online homework service
- **iSOLVE✓** These "Checkpoint" online homework service problems ask students additional questions about their confidence level, and how they arrived at their answer.

In a few problems, you are given more data than you actually need; in a few other problems, you are required to supply data from your general knowledge, outside sources, or informed estimates.

Conceptual Problems

1 • **SSM** The quantized character of electromagnetic radiation is revealed by (a) the Young double-slit experiment. (b) diffraction of light by a small aperture. (c) the photoelectric effect. (d) the J. J. Thomson cathode-ray experiment.

2 •• Two monochromatic light sources, A and B, emit the same number of photons per second. The wavelength of A is $\lambda_A = 400$ nm, and the wavelength of B is $\lambda_B = 600$ nm. The power radiated by source B is (a) equal to the power of source A. (b) less than the power of source A. (c) greater than the power of source A. (d) cannot be compared to the power from source A using the available data.

3 • True or false:
(a) In the photoelectric effect, the current is proportional to the intensity of the incident light.
(b) In the photoelectric effect, the work function of a metal depends on the frequency of the incident light.
(c) In the photoelectric effect, the maximum kinetic energy of electrons emitted varies linearly with the frequency of the incident light.
(d) In the photoelectric effect, the energy of a photon is proportional to its frequency.

4 • In the photoelectric effect, the number of electrons emitted per second is (a) independent of the light intensity. (b) proportional to the light intensity. (c) proportional to the work function of the emitting surface. (d) proportional to the frequency of the light.

5 • **SSM** The work function of a surface is ϕ. The threshold wavelength for emission of photoelectrons from the surface is (a) hc/ϕ. (b) ϕ/hf. (c) hf/ϕ. (d) none of the answers are correct.

6 •• When light of wavelength λ_1 is incident on a certain photoelectric cathode, no electrons are emitted, no matter how intense the incident light is. Yet, when light of wavelength $\lambda_2 < \lambda_1$ is incident, electrons are emitted, even when the incident light has low intensity. Explain.

7 • True or false:
(a) The de Broglie wavelength of an electron varies inversely with its momentum.
(b) Electrons can be diffracted.
(c) Neutrons can be diffracted.
(d) An electron microscope is used to look at electrons.

8 • If the de Broglie wavelength of an electron and a proton are equal, then (a) the velocity of the proton is greater than the velocity of the electron. (b) the velocity of the proton and the electron are equal. (c) the velocity of the proton is less than the velocity of the electron. (d) the energy of the proton is greater than the energy of the electron. (e) both (a) and (d) are correct.

9 • A proton and an electron have equal kinetic energies. It follows that the de Broglie wavelength of the proton is (a) greater than that of the electron. (b) equal to that of the electron. (c) less than that of the electron.

10 • Can the expectation value of x ever equal a value for which the probability function $P(x)$ is zero? Give a specific example.

11 • **SSM** Explain why the maximum kinetic energy of electrons emitted in the photoelectric effect does not depend on the intensity of the incident light, but the total number of electrons emitted does depend on the intensity of the incident light.

12 •• A six-sided die has the numeral 1 painted on three sides and the numeral 2 painted on the other three sides. (a) What is the probability of a 1 coming up when the die is thrown? (b) What is the expectation value of the numeral that comes up when the die is thrown?

13 •• It was once believed that if two identical experiments are done on identical systems under the same conditions, the results must be identical. Explain why this is not true and how it can be modified, so that it is consistent with quantum physics.

Estimation and Approximation

14 •• Students in a physics lab are trying to determine the value of Planck's constant h, using a photoelectric apparatus similar to the one shown in Figure 34-2. For a light source, the students use a helium–neon laser with tunable wavelength. The data that the students obtain for the maximum electron kinetic energy are

λ	544 nm	594 nm	604 nm	612 nm	633 nm
K_{max}	0.360 eV	0.199 eV	0.156 eV	0.117 eV	0.062 eV

(*a*) Using a spreadsheet program or graphing calculator, convert the wavelengths to frequencies and plot K_{max} versus frequency. (*b*) Use the graph to estimate the value of Planck's constant implied by the students' data. (*Note:* You may wish to use the linear regression function of your spreadsheet program or graphing calculator.) (*c*) Compare your result with the accepted value for Planck's constant.

15 •• The cathode that was used by the students in the experiment described in Problem 14 is known to be constructed from one of the following metals, with the given work function

Metals	Tungsten	Silver	Potassium	Cesium
Work function	4.58 eV	2.4 eV	2.1 eV	1.9 eV

Solve this problem using the same data as given in Problem 14. (*a*) Using a spreadsheet program or graphing calculator, convert the wavelengths to frequencies, and plot K_{max} versus frequency. (*b*) Use the graph to estimate the value of the work function implied by the students' data. (*Note:* You may wish to use the linear regression function of your spreadsheet program or graphing calculator.) (*c*) Which metal was most likely used for the cathode in their experiment?

16 •• SSM Students in an advanced physics lab use X rays to measure the Compton wavelength, λ_C. The students obtain the following wavelength shifts $\lambda_2 - \lambda_1$ as a function of scattering angle θ

θ	45°	75°	90°	135°	180°
$\lambda_2 - \lambda_1$	0.647 pm	1.67 pm	2.45 pm	3.98 pm	4.95 pm

Use their data to estimate the value for the Compton wavelength. Compare this number with the accepted value.

17 •• SSM Baseball, tennis, golf, and soccer are sports that involve placing a ball in play with a certain speed. Estimate which of these sports has a ball with the smallest de Broglie wavelength when the ball is moving with the highest speed typically created by a professional athlete.

The Particle Nature of Light: Photons

18 • iSOLVE✓ Find the photon energy in joules and in electron volts for an electromagnetic wave of frequency (*a*) 100 MHz in the FM radio band and (*b*) 900 kHz in the AM radio band.

19 • What are the frequencies of photons that have the following energies? (*a*) 1 eV, (*b*) 1 keV, and (*c*) 1 MeV.

20 • SSM Find the photon energy for light of wavelength (*a*) 450 nm, (*b*) 550 nm, and (*c*) 650 nm.

21 • Find the photon energy if the wavelength is (*a*) 0.1 nm (about 1 atomic diameter) and (*b*) 1 fm (1 fm = 10^{-15} m, about 1 nuclear diameter).

22 •• iSOLVE The wavelength of light emitted by a 3-mW helium–neon laser is 632 nm. If the diameter of the laser beam is 1 mm, what is the density of photons in the beam?

23 • SSM Lasers used in the telecommunications network typically have a wavelength near 1.55 μm. How many photons per second are being transmitted if such a laser has an output power of 2.5 mW?

The Photoelectric Effect

24 • The work function for tungsten is 4.58 eV. (*a*) Find the threshold frequency and wavelength for the photoelectric effect. Find the maximum kinetic energy of the electrons if the wavelength of the incident light is (*b*) 200 nm and (*c*) 250 nm.

25 • iSOLVE✓ When light of wavelength 300 nm is incident on potassium, the emitted electrons have maximum kinetic energy of 2.03 eV. (*a*) What is the energy of an incident photon? (*b*) What is the work function for potassium? (*c*) What would be the maximum kinetic energy of the electrons if the incident light had a wavelength of 430 nm? (*d*) What is the threshold wavelength for the photoelectric effect with potassium?

26 • The threshold wavelength for the photoelectric effect for silver is 262 nm. (*a*) Find the work function for silver. (*b*) Find the maximum kinetic energy of the electrons if the incident radiation has a wavelength of 175 nm.

27 • The work function for cesium is 1.9 eV. (*a*) Find the threshold frequency and threshold wavelength for the photoelectric effect. Find the maximum kinetic energy of the electrons if the wavelength of the incident light is (*b*) 250 nm and (*c*) 350 nm.

28 •• SSM iSOLVE When a surface is illuminated with light of wavelength 780 nm, the maximum kinetic energy of the emitted electrons is 0.37 eV. What is the maximum kinetic energy if the surface is illuminated with light of wavelength 410 nm?

Compton Scattering

29 • Find the shift in wavelength of photons scattered by electrons at $\theta = 60°$.

30 • When photons are scattered by electrons in carbon, the shift in wavelength is 0.33 pm. Find the scattering angle.

31 • The wavelength of Compton-scattered photons is measured at $\theta = 135°$. If $\Delta\lambda/\lambda$ is to be 2.3 percent, what should the wavelength of the incident photons be?

32 • SSM iSOLVE Compton used photons of wavelength 0.0711 nm. (*a*) What is the energy of these photons? (*b*) What is the wavelength of the photon scattered at $\theta = 180°$? (*c*) What is the energy of the photon scattered at this angle?

33 • For the photons used by Compton (see Problem 32), find the momentum of the incident photon and the momentum of the photon scattered at 180°; use the conservation of momentum to find the momentum of the recoil electron in this case.

34 •• An X-ray photon of wavelength 6 pm that collides with an electron is scattered by an angle of 90°. (*a*) What is the change in wavelength of the photon? (*b*) What is the kinetic energy of the scattered electron?

35 •• How many head-on, Compton-scattering events are necessary to double the wavelength of a photon with an initial wavelength of 200 pm?

Electrons and Matter Waves

36 • Use Equation 34-16 to calculate the de Broglie wavelength for an electron of kinetic energy (*a*) 2.5 eV, (*b*) 250 eV, (*c*) 2.5 keV, and (*d*) 25 keV.

37 • An electron is moving at $v = 2.5 \times 10^5$ m/s. Find the electron's de Broglie wavelength.

38 • **SOLVE** An electron has a wavelength of 200 nm. Find (*a*) its momentum and (*b*) its kinetic energy.

39 •• **SSM** An electron, a proton, and an alpha particle (the nucleus of a helium atom) each have a kinetic energy of 150 keV. Find (*a*) their momenta and (*b*) their de Broglie wavelengths.

40 • **SOLVE** A neutron in a reactor has kinetic energy of approximately 0.02 eV. Calculate the de Broglie wavelength of this neutron from Equation 34-15, where $mc^2 = 940$ MeV is the rest energy of the neutron.

41 • Use Equation 34-15 to find the de Broglie wavelength of a proton (rest energy $mc^2 = 938$ MeV) that has a kinetic energy of 2 MeV.

42 • **SSM** A proton is moving at $v = 0.003c$, where c is the speed of light. Find the electron's de Broglie wavelength.

43 • What is the kinetic energy of a proton whose de Broglie wavelength is (*a*) 1 nm and (*b*) 1 fm?

44 • Which sport has a ball with the longest de Broglie wavelength; baseball, with a ball weighing 5 oz and moving at 95 mi/h, or tennis, with a ball weighing 2 oz and moving at 130 mi/h?

45 • The energy of the electron beam in Davisson and Germer's experiment was 54 eV. Calculate the wavelength for these electrons.

46 • The distance between Li^+ and Cl^+ ions in a LiCl crystal is 0.257 nm. Find the energy of electrons that have a wavelength equal to this spacing.

47 • **SSM** An electron microscope uses electrons of energy 70 keV. Find the wavelength of these electrons.

48 • What is the de Broglie wavelength of a neutron with speed 10^6 m/s?

Wave–Particle Duality

49 • Suppose you have a spherical object of mass 4 g moving at 100 m/s. What size aperture is necessary for the object to show diffraction? Show that no common objects would be small enough to squeeze through such an aperture.

50 • A neutron has a kinetic energy of 10 MeV. What size object is necessary to observe neutron diffraction effects? Is there anything in nature of this size that could serve as a target to demonstrate the wave nature of 10-MeV neutrons?

51 • What is the de Broglie wavelength of an electron of kinetic energy 200 eV? What are some common targets that could demonstrate the wave nature of such an electron?

A Particle in a Box

52 •• **SSM** Use a spreadsheet program or graphing calculator to plot $\psi(x)$ and the probability distribution $\psi^2(x)$ of a particle in a box for the states $n = 1$, $n = 2$, and $n = 3$.

53 •• (*a*) Find the energy of the ground state ($n = 1$) and the first two excited states of a proton in a one-dimensional box of length $L = 10^{-15}$ m = 1 fm. (These are the order of magnitude of nuclear energies.) Make an energy-level diagram for this system, and calculate the wavelength of electromagnetic radiation emitted when the proton makes a transition from (*b*) $n = 2$ to $n = 1$, (*c*) $n = 3$ to $n = 2$, and (*d*) $n = 3$ to $n = 1$.

54 •• **SOLVE** ✔ (*a*) Find the energy of the ground state ($n = 1$) and the first two excited states of a proton in a one-dimensional box of length 0.2 nm (about the diameter of a H_2 molecule). Calculate the wavelength of electromagnetic radiation emitted when the proton makes a transition from (*b*) $n = 2$ to $n = 1$, (*c*) $n = 3$ to $n = 2$, and (*d*) $n = 3$ to $n = 1$.

*Calculating Probabilities and Expectation Values

55 •• **SOLVE** A particle is in the ground state of a box of length L. Find the probability of finding the particle in the interval $\Delta x = 0.002L$ at (*a*) $x = L/2$, (*b*) $x = 2L/3$, and (*c*) $x = L$. (Since Δx is very small, you need not do any integration because the wave function is slowly varying.)

56 •• **SSM** **SOLVE** ✔ Repeat Problem 55 for a particle in the first excited state ($n = 2$).

57 •• (*a*) Find $<x>$ for the second excited state ($n = 2$) for a particle in a box of length L and (*b*) find $<x^2>$.

58 •• A particle in a one-dimensional box is in the first excited state ($n = 2$). (*a*) Sketch $\psi^2(x)$ versus x for this state. (*b*) What is the expectation value $<x>$ for this state? (*c*) What is the probability of finding the particle in some small region dx centered at $x = L/2$? (*d*) Are your answers for Part (*b*) and Part (*c*) contradictory? If not, explain.

59 •• A particle of mass m has a wave function given by $\psi(x) = Ae^{-|x|/a}$, where A and a are constants. (*a*) Find the normalization constant A. (*b*) Calculate the probability of finding the particle in the region $-a \le x \le a$.

60 •• A one-dimensional box is on the x-axis in the region of $0 \le x \le L$. A particle in this box is in its ground state. Calculate the probability that the particle will be found in the region (*a*) $0 < x < \frac{1}{2}L$, (*b*) $0 < x < L/3$, and (*c*) $0 < x < 3L/4$.

61 •• Repeat Problem 60 for a particle in the first excited state of the box.

62 •• The classical probability distribution function for a particle in a box of length L is given by $P(x) = 1/L$. Use this to show that $<x> = L/2$ and $<x^2> = L^2/3$ for a classical particle in the box described in Problem 60.

63 •• (*a*) For the wave functions

$$\psi_n(x) = \sqrt{\frac{2}{L}} \sin \frac{n\pi x}{L}, \quad n = 1, 2, 3, \ldots$$

corresponding to a particle in the nth state of the box described in Problem 60, show that $<x> = L/2$ and $<x^2> = L^2/3 - L^2/2n^2\pi^2$. (b) Compare this result for $n >> 1$ with your answer for the classical distribution of Problem 62.

64 •• **SSM** (a) Use a spreadsheet program or graphing calculator to plot the expectation value for position $<x>$ and the square of the position $<x^2>$ as a function of the quantum number n for the particle in the box described in Problem 60, for values of n from 1 to 100. Assume $L = 1$ m for your graph. Refer to Problem 63. (b) Comment on the significance of any asymptotic limits that your graph shows.

65 •• The wave functions for a particle of mass m in a one-dimensional box of length L *centered at the origin* (so that the ends are at $x = \pm L/2$) are given by

$$\psi(x) = \sqrt{\frac{2}{L}} \cos \frac{n\pi x}{L}, \qquad n = 1, 3, 5, 7, \ldots$$

and

$$\psi(x) = \sqrt{\frac{2}{L}} \sin \frac{n\pi x}{L}, \qquad n = 2, 4, 6, 8, \ldots$$

Calculate $<x>$ and $<x^2>$ for the ground state.

66 •• Calculate $<x>$ and $<x^2>$ for the first excited state of the box described in Problem 65.

General Problems

67 • **SSM** **iSOLVE** A light beam of wavelength 400 nm has an intensity of 100 W/m². (a) What is the energy of each photon in the beam? (b) How much energy strikes an area of 1 cm² perpendicular to the beam in 1 s? (c) How many photons strike this area in 1 s?

68 • **iSOLVE** A 1-μg particle is moving with a speed of approximately 10^{-1} cm/s in a box of length 1 cm. Treating this as a one-dimensional particle in a box, calculate the approximate value of the quantum number n.

69 • (a) For the classical particle of Problem 68, find Δx and Δp, assuming that these uncertainties are given by $\Delta x/L = 0.01$ percent and $\Delta p/p = 0.01$ percent. (b) What is $(\Delta x \Delta p)/\hbar$?

70 • **iSOLVE** In 1987, a laser at Los Alamos National Laboratory produced a flash that lasted 1×10^{-12} s and that had a power of 5×10^{15} W. Estimate the number of emitted photons if their wavelength was 400 nm.

71 • **iSOLVE** You cannot "see" anything smaller than the wavelength λ used. What is the minimum energy of an electron needed in an electron microscope to "see" an atom, which has a diameter of about 0.1 nm?

72 • **iSOLVE** A common flea that has a mass of 0.008 g can jump vertically as high as 20 cm. Estimate the de Broglie wavelength for the flea immediately after takeoff.

73 •• **SSM** **iSOLVE**✓ Suppose that a 100-W source radiates light of wavelength 600 nm uniformly in all directions and that the eye can detect this light if only 20 photons per second enter a dark-adapted eye with a pupil 7 mm in diameter. How far from the source can the light be detected under these rather extreme conditions?

74 •• **iSOLVE** The diameter of the pupil of the eye under room-light conditions is approximately 5 mm. (It can vary from approximately 1 mm to 8 mm.) Find the intensity of light of wavelength 600 nm so that 1 photon per second passes through the pupil.

75 •• **iSOLVE** A lightbulb radiates 90 W uniformly in all directions. (a) Find the intensity at a distance of 1.5 m. (b) If the wavelength is 650 nm, find the number of photons per second that strike a surface of area 1 cm², which is oriented so that the line to the bulb is perpendicular to the surface.

76 •• When light of wavelength λ_1 is incident on the cathode of a photoelectric tube, the maximum kinetic energy of the emitted electrons is 1.8 eV. If the wavelength is reduced to $\lambda_1/2$, the maximum kinetic energy of the emitted electrons is 5.5 eV. Find the work function ϕ of the cathode material.

77 •• A photon of energy E undergoes Compton scattering at an angle of θ. Show that the energy E' of the scattered photon is given by

$$E' = \frac{E}{(E/m_e c^2)(1 - \cos \theta) + 1}$$

78 •• **iSOLVE**✓ A particle is confined to a one-dimensional box. In making a transition from the state n to the state $n - 1$, radiation of 114.8 nm is emitted; in the transition from the state $n - 1$ to the state $n - 2$, radiation of wavelength 147 nm is emitted. The ground-state energy of the particle is 1.2 eV. Determine n.

79 •• **SSM** The Pauli exclusion principle states that no more than one electron may occupy a particular quantum state at a time. Therefore, if we wish to model an atom as a collection of electrons trapped in a one-dimensional box, each electron in the box must have a unique value of the quantum number n. Calculate the energy that the most energetic electron would have for the uranium atom with atomic number 92, assuming the box has a width of 0.05 nm. How does this energy compare to the rest-mass energy of the electron?

80 •• **iSOLVE** A beam of electrons, each with the same kinetic energy, illuminates a pair of slits separated by a distance $d = 54$ nm. The beam forms bright and dark fringes on a screen located a distance $L = 1.5$ m beyond the two slits. The arrangement is otherwise identical to that used in the optical two-slit interference experiment (described in Chapter 33 and in Figure 33-7) and the fringes have the appearance shown in Figure 34-18(d). The bright fringes are found to be separated by a distance of 0.68 mm. What is the kinetic energy of the electrons in the beam?

81 •• **iSOLVE** When a surface is illuminated with light of wavelength λ, the maximum kinetic energy of the emitted electrons is 1.2 eV. If the wavelength $\lambda' = 0.8\lambda$ is used, the maximum kinetic energy increases to 1.76 eV. For wavelength $\lambda' = 0.6\lambda$, the maximum kinetic energy of the emitted electrons is 2.676 eV. Determine the work function of the surface and the wavelength λ.

82 •• A simple pendulum of length 1 m has a bob of mass 0.3 kg. The energy of this oscillator is quantized to the values $E_n = (n + \frac{1}{2})hf_0$, where n is an integer and f_0 is the frequency of the pendulum. (a) Find n if the angular amplitude is 10°. (b) Find n if the energy changes by 0.01 percent.

83 •• SSM (a) Show that for large n, the fractional difference in energy between state n and state $n + 1$ for a particle in a one-dimensional box is given approximately by

$$(E_{n+1} - E_n)/E_n \approx 2/n$$

(b) What is the approximate percentage energy difference between the states $n_1 = 1000$ and $n_2 = 1001$? (c) Comment on how this result is related to Bohr's correspondence principle.

84 •• A mode-locked, titanium–sapphire laser has a wavelength of 850 nm and produces 100 million pulses of light each second. Each pulse has a duration of 125 femtoseconds (1 fs = 10^{-15} s) and contains 5×10^9 photons. What is the average power produced by the laser?

85 •• ISOLVE This problem is one of estimating the time lag (expected classically but not observed) in the photoelectric effect. Let the intensity of the incident radiation be 0.01 W/m². (a) If the area of the atom is 0.01 nm², find the energy per second falling on an atom. (b) If the work function is 2 eV, how long would it take classically for this much energy to fall on one atom?

Applications of the Schrödinger Equation

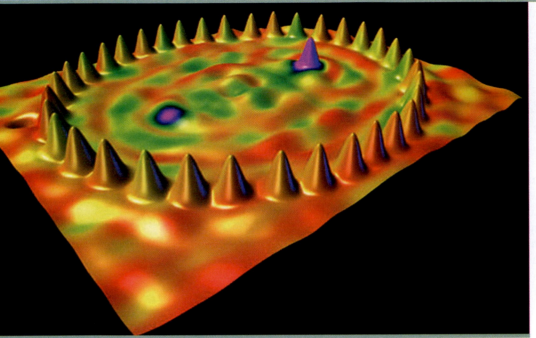

A QUANTUM MIRAGE. THE SCANNING TUNNELING MICROSCOPE (STM) ALLOWS ONE TO PUSH INDIVIDUAL ATOMS AROUND ON A SURFACE AND TO IMAGE THEM. ESPECIALLY INTRIGUING ARE IMAGES OF *QUANTUM CORRALS*, WHICH ARE CIRCULAR OR ELLIPTICAL ARRANGEMENTS ON A SURFACE INSIDE OF WHICH THE WAVES CORRESPONDING TO ELECTRONS NEAR THE SUBSTRATE SURFACE CAN BE REVEALED. THIS IMAGE COMES FROM IBM, WHERE PHYSICISTS PLACED THIRTY-SIX COBALT ATOMS IN AN ELLIPTICAL "STONEHENGE" PATTERN ON A COPPER SURFACE. AN EXTRA MAGNETIC COBALT ATOM WAS PLACED AT ONE OF THE TWO FOCI OF THE ELLIPSE, CAUSING VISIBLE INTERACTIONS WITH THE SURFACE ELECTRON WAVES. BUT THE WAVES ALSO SEEM TO BE INTERACTING WITH A PHANTOM COBALT ATOM AT THE OTHER FOCUS, AN ATOM THAT IS NOT REALLY THERE.

? **Could the phantom cobalt atom described above be caused by reflections of waves from the corral of cobalt atoms? The reflection and transmissions of one-dimensional waves is discussed in Section 35-4.**

In Chapter 34, we found that electrons and other particles have wave properties and are described by a wave function $\Psi(x, t)$. The probability of finding the particle in some region of space is proportional to the square of the wave function. We mentioned that the wave function is a solution of the Schrödinger equation, and we discussed some solutions qualitatively without reference to the equation itself. In particular, we showed how the standing-wave conditions lead to quantization of energy for a particle confined to a one-dimensional box.

➤ **This chapter is a continuation of the material introduced in Chapter 34. We discuss the Schrödinger equation and apply the equation to the particle in the box problem and to several other situations in which a particle is confined to a region of space to illustrate how boundary conditions lead to energy quantization. We then show how the Schrödinger equation leads to barrier penetration and discuss the extension of the Schrödinger equation to more than one dimension and to more than one particle.**

35-1 The Schrödinger Equation

Like the classical wave equation (Equation 15-9b), the Schrödinger equation is a partial differential equation in space and time. Like Newton's laws of motion, the Schrödinger equation cannot be derived. Its validity, like that of Newton's laws, lies in its agreement with experiment. In one dimension, the Schrödinger equation is[†]

$$-\frac{\hbar^2}{2m}\frac{\partial^2\Psi(x,t)}{\partial x^2} + U\Psi(x,t) = i\hbar\frac{\partial\Psi(x,t)}{\partial t} \qquad 35\text{-}1$$

TIME-DEPENDENT SCHRÖDINGER EQUATION

where U is the potential energy function. Equation 35-1 is called the **time-dependent Schrödinger equation.** Unlike the classical wave equation, it relates the second space derivative of the wave function to the *first* time derivative of the wave function, and it contains the imaginary number $i = \sqrt{-1}$. The wave functions that are solutions of this equation are not necessarily real. $\Psi(x,t)$ is not a measurable function like the classical wave functions for sound or electromagnetic waves. The probability of finding a particle in some region of space dx is certainly real though, so we must modify slightly the equation for probability density given in Chapter 34 (Equation 34-17). We take for the probability of finding a particle in some region dx

$$P(x,t)\,dx = |\Psi(x,t)|^2\,dx = \Psi^*\Psi\,dx \qquad 35\text{-}2$$

where Ψ^*, the complex conjugate of Ψ, is obtained from Ψ by replacing i by $-i$ wherever it appears.[‡]

In classical mechanics, the standing-wave solutions to the wave equation (Equation 16-16) are of great interest and value. Not surprisingly, standing-wave solutions to the Schrödinger wave equation are also of great interest and value. The wave function for the standing-wave motion of a uniform taut string is $A\sin(kx)\cos(\omega t + \delta)$, and this is representative of all standing waves. A standing wave function can always be expressed as a function of position multiplied by a function of time, where the function of time is one that varies sinusoidally with time. Standing-wave solutions to the one-dimensional Schrödinger wave equation are thus expressed

$$\Psi(x,t) = \psi(x)e^{-i\omega t} \qquad 35\text{-}3$$

where $e^{-i\omega t} = \cos(\omega t) - i\sin(\omega t)$. [In Appendix D, it is shown that $e^{-i\omega t} = \cos(\omega t) - i\sin(\omega t)$.] The right side of Equation 35-1 is then

$$i\hbar\frac{\partial\Psi(x,t)}{\partial t} = i\hbar(-i\omega)\psi(x)e^{-i\omega t} = \hbar\omega\psi(x)e^{-i\omega t} = E\psi(x)e^{-i\omega t}$$

where $E = \hbar\omega$ is the energy of the particle.

The Schrödinger wave equation has standing-wave solutions only if the potential energy function depends upon position alone. Substituting $\psi(x)e^{-i\omega t}$ into Equation 35-1 and canceling the common factor $e^{-i\omega t}$, we obtain an equation for $\psi(x)$, called the **time-independent Schrödinger equation:**

[†] Although we simply state the Schrödinger equation, Schrödinger himself had a vast knowledge of classical wave theory that led him to this equation.

[‡] Every complex number can be written in the form $z = a + bi$, where a and b are real numbers and $i = \sqrt{-1}$. The complex conjugate of z is $z^* = a - bi$, so $z^*z = (a + bi)(a - bi) = a^2 + b^2 = |z|^2$. Complex numbers are discussed more fully in Appendix D.

$$-\frac{\hbar^2}{2m}\frac{d^2\psi(x)}{dx^2} + U(x)\psi(x) = E\psi(x) \qquad\qquad 35\text{-}4$$

TIME-INDEPENDENT SCHRÖDINGER EQUATION

where we have written U as $U(x)$ to emphasize that while U may depend on position, U does not depend on time. The function $U(x)$ represents the environment of the particle being described. It is this potential energy function in the Schrödinger equation that establishes the difference between different problems, just as the expressions for forces acting on a particle play in classical physics.

The calculation of the allowed energy levels in a system involves only the time-independent Schrödinger equation, whereas finding the probabilities of transition between these levels requires the solution of the time-dependent equation. In this book, we will be concerned only with the time-independent Schrödinger equation.

The solution of Equation 35-4 depends on the form of the potential energy function $U(x)$. When $U(x)$ is such that the particle is confined to some region of space, only certain discrete energies E_n give solutions ψ_n that can satisfy the normalization condition (Equation 34-18):

$$\int_{-\infty}^{\infty} |\psi|^2\, dx = 1$$

The complete time-dependent wave functions are then given, from Equation 35-3, by

$$\Psi_n(x, t) = \psi_n(x)e^{-i\omega_n t} = \psi_n(x)e^{-i(E_n/\hbar)t} \qquad\qquad 35\text{-}5$$

A Particle in an Infinite Square-Well Potential

We will illustrate the use of the time-independent Schrödinger equation by solving it for the problem of a particle in a box. The potential energy for a one-dimensional box from $x = 0$ to $x = L$ is shown in Figure 35-1. It is called an **infinite square-well potential** and is described mathematically by

$$U(x) = 0, \quad 0 < x < L$$

$$U(x) = \infty, \quad x < 0 \text{ or } x > L \qquad\qquad 35\text{-}6$$

Inside the box, the potential energy is zero, whereas outside the box it is infinite. Since we require the particle to be in the box, we have $\psi(x) = 0$ everywhere outside the box. We then need to solve the Schrödinger equation inside the box subject to the condition that $\psi(x)$ must be zero at $x = 0$ and at $x = L$.

Inside the box $U(x) = 0$, so the Schrödinger equation is written

$$-\frac{\hbar^2}{2m}\frac{d^2\psi(x)}{dx^2} = E\psi(x)$$

or

$$\frac{d^2\psi(x)}{dx^2} + k^2\psi(x) = 0 \qquad\qquad 35\text{-}7$$

where

$$k^2 = \frac{2mE}{\hbar^2} \qquad\qquad 35\text{-}8$$

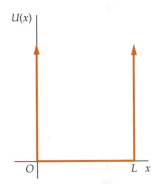

FIGURE 35-1 The infinite square-well potential energy. For $x < 0$ and $x > L$, the potential energy $U(x)$ is infinite. The particle is confined to the region in the well $0 < x < L$.

The general solution of Equation 35-7 can be written as

$$\psi(x) = A \sin kx + B \cos kx \qquad \text{35-9}$$

where A and B are constants. At $x = 0$, we have

$$\psi(0) = A \sin(k0) + B \cos(k0) = 0 + B$$

The boundary condition $\psi(x) = 0$ at $x = 0$ thus gives $B = 0$, and Equation 35-9 becomes

$$\psi(x) = A \sin kx \qquad \text{35-10}$$

The wave function is thus a sine wave with the wavelength λ related to the wave number k in the usual way, $\lambda = 2\pi/k$. The boundary condition $\psi(x) = 0$ at $x = L$ restricts the possible values of k and therefore the values of the wavelength λ, and (from Equation 35-8) the energy $E = \hbar^2 k^2/2m$. We have

$$\psi(L) = A \sin kL = 0 \qquad \text{35-11}$$

This condition is satisfied if kL is π or any integer times π, that is, if k is restricted to the values k_n given by

$$k_n = n\frac{\pi}{L}, \quad n = 1, 2, 3, \dots \qquad \text{35-12}$$

The condition (Equation 35-11) is also satisfied for $n = 0$. The function $\psi(x) = A \sin 0 = 0$ for all values of x is a solution to the wave equation. However, this solution is rejected as a wave function on physical grounds. It cannot be normalized and cannot be a wave function for a particle. Substituting this result into Equation 35-8 and solving for E gives us the allowed energy values:

$$E_n = \frac{\hbar^2 k_n^2}{2m} = \frac{\hbar^2}{2m}\left(n\frac{\pi}{L}\right)^2 = n^2\left(\frac{h^2}{8mL^2}\right) = n^2 E_1 \qquad \text{35-13}$$

where

$$E_1 = \frac{h^2}{8mL^2} \qquad \text{35-14}$$

Equation 35-14 is the same as Equation 34-22, which we obtained by fitting an integral number of half-wavelengths into the box.

For each value of n, there is wave function $\psi_n(x)$ given by

$$\psi_n(x) = A_n \sin \frac{n\pi x}{L} \qquad \text{35-15}$$

which is the same as Equation 34-25 with the constant $A_n = \sqrt{2/L}$ determined by normalization.[†]

35-2 A Particle in a Finite Square Well

The quantization of energy that we found for a particle in an infinite square well is a result that follows from the general solution of the Schrödinger equation for any particle confined to some region of space. We will illustrate this by considering the qualitative behavior of the wave function for a slightly more general potential energy function, the finite square well, which is shown in Figure 35-2.

---

† See Equation 34-18.

This potential energy function is described mathematically by

$$U(x) = U_0, \quad x < 0$$

$$U(x) = 0, \quad 0 < x < L \tag{35-16}$$

$$U(x) = U_0, \quad x > L$$

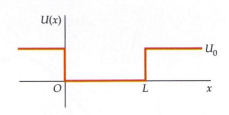

FIGURE 35-2 The finite square-well potential energy.

This potential energy function is discontinuous at $x = 0$ and $x = L$, but it is finite everywhere. The solutions of the Schrödinger equation for this type of potential energy function depend on whether the total energy E is greater or less than U_0. We will not discuss the case of $E > U_0$, except to remark that in that case the particle is not confined and any value of the energy is allowed; that is, there is no energy quantization. Here we assume that $0 \le E < U_0$.

Inside the well, $U(x) = 0$, and the time-independent Schrödinger equation is the same as for the infinite well (Equation 35-7):

$$-\frac{\hbar^2}{2m}\frac{d^2\psi(x)}{dx^2} = E\psi(x)$$

or

$$\frac{d^2\psi(x)}{dx^2} + k^2\psi(x) = 0$$

where $k^2 = 2mE/\hbar^2$. The general solution is of the form

$$\psi(x) = A \sin kx + B \cos kx$$

In this case, $\psi(x)$ is not required to be zero at $x = 0$, so B is not zero. Outside the well, the time-independent Schrödinger equation is

$$-\frac{\hbar^2}{2m}\frac{d^2\psi(x)}{dx^2} + U_0\psi(x) = E\psi(x)$$

or

$$\frac{d^2\psi(x)}{dx^2} - \alpha^2\psi(x) = 0 \tag{35-17}$$

where

$$\alpha^2 = \frac{2m}{\hbar^2}(U_0 - E) > 0 \tag{35-18}$$

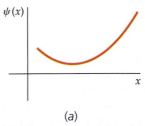

(a)

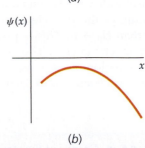

(b)

The wave functions and allowed energies for the particle can be found by solving Equation 35-17 for $\psi(x)$ outside the well and then requiring that $\psi(x)$ and $d\psi(x)/dx$ be continuous at the boundaries $x = 0$ and $x = L$. The solution of Equation 35-17 is not difficult (for positive values of x, it is of the form $\psi(x) = \psi(x) = Ce^{-\alpha x}$), but applying the boundary conditions involves much tedious algebra and is not important for our purpose. The important feature of Equation 35-17 is that the second derivative of $\psi(x)$, which is related to the concavity of the wave function, has the same sign as the wave function ψ. If ψ is positive, $d^2\psi/dx^2$ is also positive and the wave function curves away from the axis, as shown in Figure 35-3a. Similarly, if ψ is negative, $d^2\psi/dx^2$ is negative and ψ again curves away from the axis, as shown in Figure 35-3b. This behavior is

FIGURE 35-3 (a) A positive function with positive concavity. (b) A negative function with negative concavity.

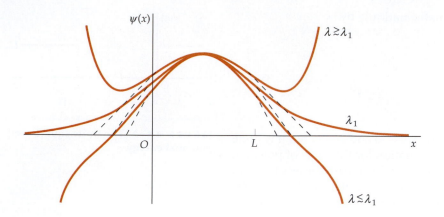

FIGURE 35-4 Functions satisfying the Schrödinger equation with wavelengths near the wavelength λ_1 corresponding to the ground-state energy $E_1 = \hbar^2/2m\lambda_1^2$ in the finite well. If λ is slightly greater than λ_1, the function approaches infinity, like the function in Figure 35-3a. At the critical wavelength λ_1, the function and its slope approach zero together. If λ is slightly less than λ_1, the function crosses the x axis while the slope is still negative. The slope then becomes more negative because its rate of change $d^2\psi/dx^2$ is now negative. This function approaches negative infinity as x approaches infinity.

very different from the behavior inside the well, where ψ and $d^2\psi/dx^2$ have opposite signs so that ψ always curves toward the axis like a sine or cosine function. Because of this behavior outside the well, for most values of the energy E in Equation 35-17, $\psi(x)$ becomes infinite as x approaches $\pm\infty$; that is, most wave functions $\psi(x)$ are not well behaved outside the well. Though they satisfy the Schrödinger equation, such functions are not proper wave functions because they cannot be normalized. The solutions of the Schrödinger equation are well behaved (i.e., they approach 0 as $|x|$ becomes very large) only for certain values of the energy. These energy values are the allowed energies for the finite square well.

Figure 35-4 shows a well-behaved wave function, with a wavelength λ_1 inside the well corresponding to the ground-state energy. The behavior of the wave functions corresponding to nearby wavelengths and energies is also shown. Figure 35-5 shows the wave functions and probability distributions for the ground state and first two excited states. From this figure, we can see that the wavelengths inside the well are slightly longer than the corresponding wavelengths for the infinite well (Figure 34-14), so the corresponding energies are slightly less than those for the infinite well. Another feature of the finite-well problem is that there are only a finite number of allowed energies. For very small values of U_0, there is only one allowed energy.

Note that the wave function penetrates beyond the edges of the well at $x = L$ and $x = 0$, indicating that there is some small probability of finding the particle in the region in which its total energy E is less than its potential energy U_0. This region is called the *classically forbidden region* because the kinetic energy, $E - U_0$, would be negative when $U_0 > E$. Since negative kinetic energy has no meaning in classical physics, it is interesting to speculate on the result of an attempt to observe the particle in the classically forbidden region. It can be shown from the uncertainty principle that if an attempt is made to localize the particle in the classically forbidden region, such a measurement introduces an uncertainty in the momentum of the particle corresponding to a minimum kinetic energy that is greater than $U_0 - E$. This is just great enough to prevent us from measuring a negative kinetic energy. The penetration of the wave function into a classically forbidden region does have important consequences in barrier penetration, which will be discussed in Section 35-4.

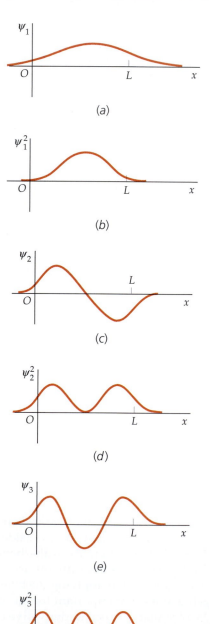

FIGURE 35-5 Graphs of the wave functions $\psi_n(x)$ and probability distributions $\psi_n^2(x)$ for $n = 1$, $n = 2$, and $n = 3$ for the finite square well. Compare these graphs with those of Figure 34-14 for the infinite square well, where the wave functions are zero at $x = 0$ and $x = L$. The wavelengths here are slightly longer than the corresponding wavelengths for the infinite well, so the allowed energies are somewhat smaller.

Much of our discussion of the finite-well problem applies to any problem in which $E > U(x)$ in some region and $E < U(x)$ outside that region, as we see in the next section.

35-3 The Harmonic Oscillator

The potential energy for a particle of mass m attached to a spring of force constant k is

$$U(x) = \tfrac{1}{2}kx^2 = \tfrac{1}{2}m\omega_0^2 x^2 \qquad\qquad 35\text{-}19$$

where $\omega_0 = \sqrt{k/m}$ is the natural frequency of the oscillator. Classically, the object oscillates between $x = +A$ and $x = -A$. The object's total energy is $E = \tfrac{1}{2}m\omega_0^2 A^2$, which can have any positive value or zero.

This potential energy function, shown in Figure 35-6, applies to virtually any system undergoing small oscillations about a position of stable equilibrium. For example, it could apply to the oscillations of the atoms of a diatomic molecule, such as H_2 or HCl, oscillating about their equilibrium separation. Between the classical turning points ($|x| < A$), the total energy is greater than the potential energy, and the Schrödinger equation can be written

$$\frac{d^2\psi(x)}{dx^2} = -k^2\psi(x) \qquad\qquad 35\text{-}20$$

where $k^2 = (2m/\hbar^2)[E - U(x)]$ now depends on x. The solutions of this equation are no longer simple sine or cosine functions because the wave number $k = 2\pi/\lambda$ now varies with x; but since $d^2\psi/dx^2$ and ψ have opposite signs, ψ will always curve toward the axis and the solutions will oscillate.

Outside the classical turning points ($|x| > A$), the potential energy is greater than the total energy and the Schrödinger equation is similar to Equation 35-17:

$$\frac{d^2\psi(x)}{dx^2} = +\alpha^2\psi(x) \qquad\qquad 35\text{-}21$$

except that here $\alpha^2 = (2m/\hbar^2)[U(x) - E] > 0$ depends on x. For $|x| > A$, $d^2\psi/dx^2$ and ψ have the same sign, so ψ will curve away from the axis and there will be only certain values of E for which solutions exist that approach zero as x approaches infinity.

For the harmonic oscillator potential energy function, the Schrödinger equation is

$$-\frac{\hbar^2}{2m}\frac{d^2\psi(x)}{dx^2} + \tfrac{1}{2}m\omega_0^2 x^2\psi(x) = E\psi(x) \qquad\qquad 35\text{-}22$$

Wave Functions and Energy Levels

Rather than pursue a general solution to the Schrödinger equation for this system, we simply present the solution for the ground state and the first excited state.

The ground-state wave function $\psi_0(x)$ is found to be a Gaussian function centered at the origin:

$$\psi_0(x) = A_0 e^{-ax^2} \qquad\qquad 35\text{-}23$$

where A_0 and a are constants. This function and the wave function for the first excited state are shown in Figure 35-7.

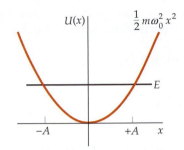

FIGURE 35-6 Harmonic oscillator potential.

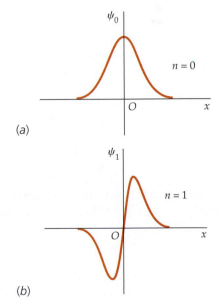

FIGURE 35-7 (a) The ground-state wave function for the harmonic oscillator potential. (b) The wave function for the first excited state of the harmonic oscillator potential.

VERIFYING THE GROUND-STATE WAVE FUNCTION   **E X A M P L E   3 5 - 1**

Verify that $\psi_0(x) = A_0 e^{-ax^2}$, **where** a **is a positive constant, is a solution of the Schrödinger equation for the harmonic oscillator.**

PICTURE THE PROBLEM We take the first and second derivative of ψ with respect to x and substitute into Equation 35-22. Since this is the ground-state wave function, we write E_0 for the energy E.

1. Compute $d\psi_0/dx$:

$$\frac{d\psi_0(x)}{dx} = \frac{d}{dx}(A_0 e^{-ax^2}) = -2axA_0 e^{-ax^2}$$

2. Compute $d^2\psi_0/dx^2$:

$$\frac{d^2\psi_0(x)}{dx^2} = -2aA_0 e^{-ax^2} + 4a^2x^2 A_0 e^{-ax^2}$$

3. Substitute these derivatives into the Schrödinger equation:

$$-\frac{\hbar^2}{2m}\frac{d^2\psi(x)}{dx^2} + \frac{1}{2}m\omega_0^2 x^2\psi(x) = E\psi(x)$$

$$-\frac{\hbar^2}{2m}(-2aA_0 e^{-ax^2} + 4a^2x^2 A_0 e^{-ax^2}) + \frac{1}{2}m\omega_0^2 x^2 A_0 e^{-ax^2} = E_0 A_0 e^{-ax^2}$$

4. Cancel the common factor $A_0 e^{-ax^2}$ and show the result in standard polynomial form:

$$-\frac{\hbar^2}{2m}(-2a + 4a^2x^2) + \frac{1}{2}m\omega_0^2 x^2 = E_0$$

so

$$\left(-\frac{2\hbar^2 a^2}{m} + \frac{1}{2}m\omega_0^2\right)x^2 + \left(\frac{\hbar^2 a}{m} - E_0\right) = 0$$

5. The equation in step 4 must hold for all x. Set $x = 0$ and solve for E_0:

$$0 + \left(\frac{\hbar^2 a}{m} - E_0\right) = 0$$

so

$$E_0 = \frac{\hbar^2 a}{m}$$

6. Substitute this result for E_0 into the equation in step 4 and simplify:

$$\left(-\frac{2\hbar^2 a^2}{m} + \frac{1}{2}m\omega_0^2\right)x^2 + 0 = 0$$

7. It follows that the coefficient of x^2 must equal zero:

$$-\frac{2\hbar^2 a^2}{m} + \frac{1}{2}m\omega_0^2 = 0$$

8. Solve for a:

$$a = \frac{m\omega_0}{2\hbar}$$

9. Substitute this result into the equation for E_0 in step 5:

$$E_0 = \frac{\hbar^2 a}{m} = \frac{1}{2}\hbar\omega_0$$

> We have shown that the given function satisfies the Schrödinger equation for any value of A_0, as long as the energy is given by $E_0 = \frac{1}{2}\hbar\omega_0$.

REMARKS The step 4 equation is a polynomial that is equal to zero. A theorem that would have simplified the solution is: If a polynomial is equal to zero over a continuous range of values of x, then each of the polynomial coefficients is equal to zero. For example, if $Ax^3 + Bx^2 + Cx + D = 0$ on the interval $1 < x < 2$, then $A = B = C = D = 0$. The proof of this result is the topic of Problem 43.

We see from this example that the ground-state energy is given by

$$E_0 = \frac{\hbar^2 a}{m} = \frac{1}{2}\hbar\omega_0 \qquad\qquad 35\text{-}24$$

The first excited state has a node in the center of the potential well, just as with the particle in a box.[†] The wave function $\psi_1(x)$ is

$$\psi_1(x) = A_1 x e^{-ax^2} \qquad\qquad 35\text{-}25$$

where $a = m\omega_0/2\hbar$, as in Example 35-1. This function is also shown in Figure 35-7. Substituting $\psi_1(x)$ into the Schrödinger equation, as was done for $\psi_0(x)$ in Example 35-1, yields the energy of the first excited state,

$$E_1 = \tfrac{3}{2}\hbar\omega_0$$

In general, the energy of the nth excited state of the harmonic oscillator is

$$E_n = (n + \tfrac{1}{2})\hbar\omega_0, \quad n = 0, 1, 2, \dots \qquad 35\text{-}26$$

as indicated in Figure 35-8. The fact that the energy levels are evenly spaced by the amount $\hbar\omega_0$ is a peculiarity of the harmonic oscillator potential. As we saw in Chapter 34, the energy levels for a particle in a box, or for the hydrogen atom, are not evenly spaced. The precise spacing of energy levels is closely tied to the particular form of the potential energy function.

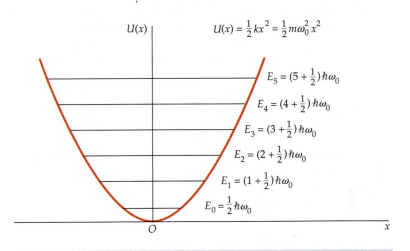

FIGURE 35-8 Energy levels in the harmonic oscillator potential.

35-4 Reflection and Transmission of Electron Waves: Barrier Penetration

In Sections 35-2 and 35-3, we were concerned with bound-state problems in which the potential energy is larger than the total energy for large values of $|x|$. In this section, we consider some simple examples of unbound states for which E is greater than $U(x)$. For these problems, $d^2\psi/dx^2$ and ψ have opposite signs, so $\psi(x)$ curves toward the axis and does not become infinite at large values of $|x|$.

Step Potential

Consider a particle of energy E moving in a region in which the potential energy is the step function

$$U(x) = 0, \quad x < 0$$

$$U(x) = U_0, \quad x > 0$$

as shown in Figure 35-9. We are interested in what happens when a particle moving from left to right encounters the step.

The classical answer is simple. To the left of the step, the particle moves with a speed $v = \sqrt{2E/m}$. At $x = 0$, an impulsive force acts on the particle. If the initial energy E is less than U_0, the particle will be turned around and will then move to the left at its original speed; that is, the particle will be reflected by the step. If E is greater than U_0, the particle will continue to move to the right but with reduced speed given by $v = \sqrt{2(E - U_0)/m}$. We can picture this classical problem as a ball rolling along a level surface and coming to a steep hill of height h given by $mgh = U_0$. If the initial kinetic energy of the ball is less than

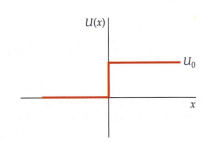

FIGURE 35-9 Step potential. A classical particle incident from the left, with total energy $E > U_0$, is always transmitted. The change in potential energy at $x = 0$ merely provides an impulsive force that reduces the speed of the particle. A wave incident from the left is partially transmitted and partially reflected because the wavelength changes abruptly at $x = 0$.

[†] Each higher-energy state has one additional node in the wave function.

mgh, the ball will roll part way up the hill and then back down and to the left along the lower surface at its original speed. If *E* is greater than *mgh,* the ball will roll up the hill and proceed to the right at a lesser speed.

The quantum-mechanical result is similar when *E* is less than U_0. Figure 35-10 shows the wave function for the case $E < U_0$. The wave function does not go to zero at $x = 0$ but rather decays exponentially, like the wave function for the bound state in a finite square-well problem. The wave penetrates slightly into the classically forbidden region $x > 0$, but it is eventually completely reflected. This problem is somewhat similar to that of total internal reflection in optics.

For $E > U_0$, the quantum-mechanical result differs markedly from the classical result. At $x = 0$, the wavelength changes abruptly from $\lambda_1 = h/p_1 = h/\sqrt{2mE}$ to $\lambda_2 = h/p_2 = h/\sqrt{2m(E - U_0)}$. We know from our study of waves that when the wavelength changes suddenly, part of the wave is reflected and part of the wave is transmitted. Since the motion of an electron (or other particle) is governed by a wave equation, the electron sometimes will be transmitted and sometimes will be reflected. The probabilities of reflection and transmission can be calculated by solving the Schrödinger equation in each region of space and comparing the amplitudes of the transmitted waves and reflected waves with that of the incident wave. This calculation and its result are similar to finding the fraction of light reflected from an air–glass interface. If *R* is the probability of reflection, called the reflection coefficient, this calculation gives

$$R = \frac{(k_1 - k_2)^2}{(k_1 + k_2)^2} \qquad\qquad 35\text{-}27$$

where k_1 is the wave number for the incident wave and k_2 is the wave number for the transmitted wave. This result is the same as the result in optics for the reflection of light at normal incidence from the boundary between two media having different indexes of refraction *n* (Equation 31-11). The probability of transmission *T*, called the **transmission coefficient,** can be calculated from the reflection coefficient, since the probability of transmission plus the probability of reflection must equal 1:

$$T + R = 1 \qquad\qquad 35\text{-}28$$

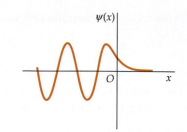

FIGURE 35-10 When the total energy *E* is less than U_0, the wave function penetrates slightly into the region $x > 0$. However, the probability of reflection for this case is 1, so no energy is transmitted.

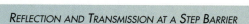

REFLECTION AND TRANSMISSION AT A STEP BARRIER **EXAMPLE 35-2**

A particle of energy E_0 traveling in a region in which the potential energy is zero is incident on a potential barrier of height $U_0 = 0.2E_0$. Find the probability that the particle will be reflected.

PICTURE THE PROBLEM We need to calculate the wave numbers k_1 and k_2 and use them to calculate the reflection coefficient *R* from Equation 35-27. The wave numbers are related to the kinetic energy *K* by $K = p^2/2m = \hbar^2 k^2/2m$.

1. The probability of reflection is the reflection coefficient:
$$R = \frac{(k_1 - k_2)^2}{(k_1 + k_2)^2}$$

2. Calculate k_1 from the initial kinetic energy E_0:
$$E_0 = \frac{\hbar^2 k_1^2}{2m}$$
$$k_1 = \sqrt{2mE_0/\hbar^2} = 1.41\sqrt{mE_0}/\hbar^2$$

3. Relate k_2 to the final kinetic energy K_2:
$$\frac{\hbar k_2^2}{2m} = K_2 = E_0 - U_0 = E_0 - 0.2E_0 = 0.8E_0$$

4. Solve for k_2:

$$k_2 = \sqrt{2m(0.8E_0)/\hbar^2} = 1.26\sqrt{mE_0}/\hbar^2$$

5. Substitute these values into Equation 35-27 to calculate R:

$$R = \frac{(k_1 - k_2)^2}{(k_1 + k_2)^2} = \frac{(1.41 - 1.26)^2}{(1.41 + 1.26)^2} = \boxed{0.00316}$$

REMARKS   The probability of reflection is only 0.3 percent. This probability is small because the barrier height reduces the kinetic energy by only 20 percent. Since k is proportional to the square root of the kinetic energy, the wave number and therefore the wavelength is changed by only 10 percent.

 EXERCISE   Express the index of refraction n of light in terms of the wave number k, and show that Equation 31-11 for the reflection of light at normal incidence is the same as Equation 35-27.

In quantum mechanics, a localized particle is represented by a wave packet, which has a maximum at the most probable position of the particle. Figure 35-11 shows a wave packet representing a particle of energy E incident on a step potential of height U_0, which is less than E. After the encounter, there are two wave packets. The relative heights of the transmitted packet and reflected packet indicate the relative probabilities of transmission and reflection. For the situation shown here, E is much greater than U_0, and the probability of transmission is much greater than that of reflection.

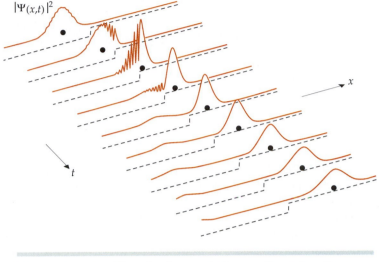

FIGURE 35-11   Time development of a one-dimensional wave packet representing a particle incident on a step potential for $E > U_0$. The position of a classical particle is indicated by the dot. Note that part of the packet is transmitted and part is reflected.

Barrier Penetration

Figure 35-12a shows a rectangular potential barrier of height U_0 and width a given by

$$U(x) = 0, \quad x < 0$$

$$U(x) = U_0, \quad 0 < x < a$$

$$U(x) = 0, \quad x > a$$

We consider a particle of energy E, which is slightly less than U_0, that is incident on the barrier from the left. Classically, the particle would always be reflected. However, a wave incident from the left does not decrease immediately to zero at the barrier, but it will instead decay exponentially in the classically forbidden region $0 < x < a$. Upon reaching the far wall of the barrier ($x = a$), the wave function must join smoothly to a sinusoidal wave function to the right of the barrier, as shown in Figure 35-12b. This implies that there is some probability of the particle (which is represented by the wave function) being found on the far side of the barrier even though, classically, it should never pass through the barrier. For the case in which the quantity $\alpha a = \sqrt{2ma^2(U_0 - E)/\hbar^2}$ is much greater than 1, the transmission coefficient is proportional to $e^{-2\alpha a}$:

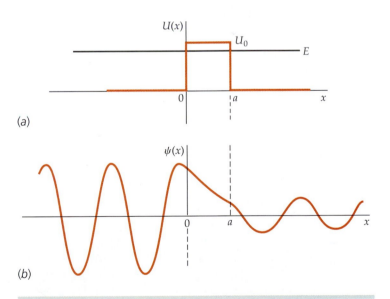

FIGURE 35-12   (a) A rectangular potential barrier. (b) The penetration of the barrier by a wave with total energy less than the barrier energy. Part of the wave is transmitted by the barrier even though, classically, the particle cannot enter the region $0 < x < a$ in which the potential energy is greater than the total energy. To the left of the barrier, there is both an incident and a reflected wave. These waves form a resultant wave so that ψ is a superposition of a standing wave and a traveling wave (traveling toward the barrier). To the right of the barrier is only the transmitted wave that is traveling away from the barrier.

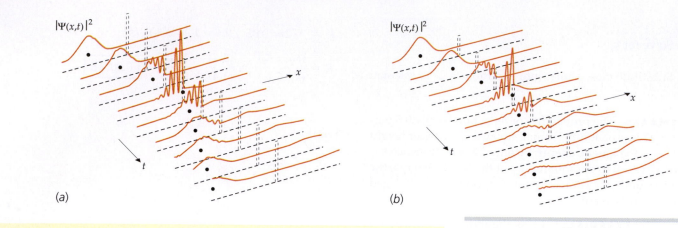

(a)   (b)

$$T \propto e^{-2\alpha a}$$

35-29

TRANSMISSION THROUGH A BARRIER

FIGURE 35-13 Barrier penetration. (a) The same particle incident on a barrier of height much greater than the energy of the particle. A very small part of the packet tunnels through the barrier. In both drawings, the position of a classical particle is indicated by a dot. (b) A wave packet representing a particle incident on a barrier of height just slightly greater than the energy of the particle. For this particular choice of energies, the probability of transmission is approximately equal to the probability of reflection, as indicated by the relative sizes of the transmitted and reflected packets.

with $\alpha = \sqrt{2m(U_0 - E)/\hbar^2}$. The probability of penetration of the barrier thus decreases exponentially with the barrier thickness a and with the square root of the relative barrier height $(U_0 - E)$.

Figure 35-13a shows a wave packet incident on a potential barrier of height U_0 that is considerably greater than the energy of the particle. The probability of penetration is very small, as indicated by the relative sizes of the reflected and transmitted packets. In Figure 35-13b, the barrier is just slightly greater than the energy of the particle. In this case, the probability of penetration is about the same as the probability of reflection. Figure 35-14 shows a particle incident on two potential barriers of height just slightly greater than the energy of the particle.

As we have mentioned, the penetration of a barrier is not unique to quantum mechanics. When light is totally reflected from a glass–air interface, the light wave can penetrate the air barrier if a second piece of glass is brought within a few wavelengths of the first. This effect can be demonstrated with a laser beam and two 45° prisms (Figure 35-15). Similarly, water waves in a ripple tank can penetrate a gap of deep water (Figure 35-16).

The theory of barrier penetration was used by George Gamow in 1928 to explain the enormous variation in the half-lives for α decay of radioactive nuclei.

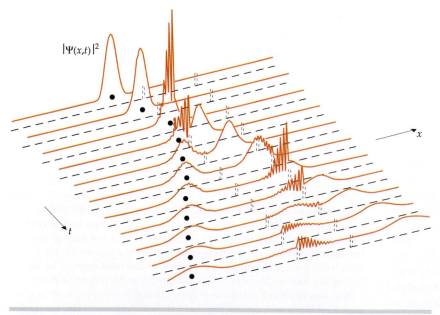

FIGURE 35-14 A wave packet representing a particle incident on two barriers. At each encounter, part of the packet is transmitted and part reflected, resulting in part of the packet being trapped between the barriers for some time.

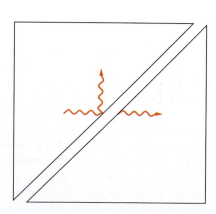

FIGURE 35-15 The penetration of an optical barrier. If the second prism is close enough to the first, part of the wave penetrates the air barrier even when the angle of incidence in the first prism is greater than the critical angle.

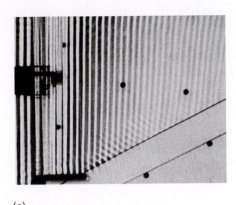
(a)

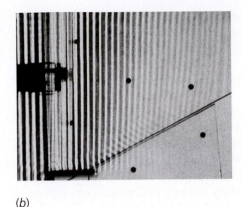
(b)

FIGURE 35-16 The penetration of a barrier by water waves in a ripple tank. In Figure 35-16a, the waves are totally reflected from a gap of deeper water. When the gap is very narrow, as in Figure 35-16b, a transmitted wave appears. The dark circles are spacers that are used to support the prisms from below.

(Alpha particles are helium nuclei emitted from larger atoms in radioactive decay; they consist of two protons and two neutrons tightly bound together.) In general, the smaller the energy of the emitted α particle, the longer the half-life. The energies of α particles from natural radioactive sources range from approximately 4 MeV to 7 MeV, whereas the half-lives range from approximately 10^{-5} second to 10^{10} years. Gamow represented a radioactive nucleus by a potential well containing an α particle, as shown in Figure 35-17. Without knowing very much about the nuclear force that is exerted on the α particle within the nucleus, Gamow represented it by a square well. Just outside the well, the α particle with its charge of $+2e$ is repelled by the nucleus with its charge $+Ze$, where Ze is the remaining nuclear charge. This force is represented by the Coulomb potential energy $+k(2e)(Ze)/r$. The energy E is the measured kinetic energy of the emitted α particle, because when it is far from the nucleus its potential energy is zero. After the α particle is formed inside the radioactive nucleus, it bounces back and forth inside the nucleus, hitting the barrier at the nuclear radius R. Each time the α particle strikes the barrier, there is some small probability of the particle penetrating the barrier and appearing outside the nucleus. We can see from Figure 35-17 that a small increase in E reduces the relative height of the barrier $U - E$ and also the barrier's thickness. Because the probability of penetration is so sensitive to the barrier thickness and relative height, a small increase in E leads to a large increase in the probability of transmission and therefore to a shorter lifetime. Gamow was able to derive an expression for the half-life as a function of E that is in excellent agreement with experimental results.

In the **scanning tunneling electron microscope** developed in the 1980s, a thin space between a material specimen and a tiny probe acts as a barrier to electrons bound in the specimen. A small voltage applied between the probe and the specimen causes the electrons to *tunnel* through the vacuum separating the two surfaces if the surfaces are close enough together. The tunneling current is extremely sensitive to the size of the gap between the probe and specimen. If a constant tunneling current is maintained as the probe scans the specimen, the surface of the specimen can be mapped out by the motions of the probe. In this way, the surface features of a specimen can be measured with a resolution of the order of the size of an atom.

FIGURE 35-17 Model of a potential energy function for an α particle in a radioactive nucleus. The strong attractive nuclear force when r is less than the nuclear radius R can be approximately described by the potential well shown. Outside the nucleus the nuclear force is negligible, and the potential is given by Coulomb's law, $U(r) = +k(2e)(Ze)/r$, where Ze is the nuclear charge and $2e$ is the charge of the α particle. The wave function of the alpha particle, shown in red, is placed on the graph.

35-5 The Schrödinger Equation in Three Dimensions

The one-dimensional time-independent Schrödinger equation is easily extended to three dimensions. In rectangular coordinates, it is

$$-\frac{\hbar^2}{2m}\left(\frac{\partial^2 \psi}{\partial x^2} + \frac{\partial^2 \psi}{\partial y^2} + \frac{\partial^2 \psi}{\partial z^2}\right) + U\psi = E\psi \qquad 35\text{-}30$$

where the wave function ψ and the potential energy U are generally functions of all three coordinates, x, y, and z. To illustrate some of the features of problems in three dimensions, we consider a particle in a three-dimensional infinite square well given by $U(x, y, z) = 0$ for $0 < x < L$, $0 < y < L$, and $0 < z < L$. Outside this cubical region, $U(x, y, z) = \infty$. For this problem, the wave function must be zero at the edges of the well.

There are standard methods in partial differential equations for solving Equation 35-30. We can guess the form of the solution from our knowledge of probability. For a one-dimensional box along the x axis, we have found the probability that a particle is in the region dx at x to be $A_1^2 \sin^2(k_1 x)\, dx$ (from Equation 35-10), where A_1 is a normalization constant and $k_1 = n\pi/L$ is the wave number. Similarly, for a box along the y axis, the probability of a particle being in a region dy at y is $A_2^2 \sin^2(k_2 y)\, dy$. The probability of two independent events occurring is the product of the probabilities of each event occurring.[†] So the probability of a particle being in region dx at x *and* in region dy at y is $A_1^2 \sin^2(k_1 x)\, dx\, A_2^2 \sin^2(k_2 y)\, dy = A_1^2 \sin^2(k_1 x) A_2^2 \sin^2(k_2 y)\, dx\, dy$. The probability of a particle being in the region dx, dy, and dz is $\psi(x, y, z)\, dx\, dy\, dz$, where $\psi(x, y, z)$ is the solution of Equation 35-30. This solution is of the form

$$\psi(x, y, z) = A \sin^2(k_1 x) \sin^2(k_2 y) \sin^2(k_3 z)\, dx\, dy\, dz \qquad \text{35-31}$$

where the constant A is determined by normalization. Inserting this solution into Equation 35-30, we obtain for the energy

$$E = \frac{\hbar^2}{2m}(k_1^2 + k_2^2 + k_3^2)$$

which is equivalent to $E = (p_x^2 + p_y^2 + p_z^2)/(2m)$, with $p_x = \hbar k_1$, and so on. The wave function will be zero at $x = L$ if $k_1 = n_1 \pi/L$, where n_1 is an integer. Similarly, the wave function will be zero at $y = L$ if $k_2 = n_2 \pi/L$, and the wave function will be zero at $z = L$ if $k_3 = n_3 \pi/L$. (It is also zero at $x = 0$, $y = 0$, and $z = 0$.) The energy is thus quantized to the values

$$E_{n_1, n_2, n_3} = \frac{\hbar^2 \pi^2}{2mL^2}(n_1^2 + n_2^2 + n_3^2) = E_1 (n_1^2 + n_2^2 + n_3^2) \qquad \text{35-32}$$

where n_1, n_2, and n_3 are integers and E_1 is the ground-state energy of the one-dimensional well. Note that the energy and wave function are characterized by three quantum numbers, each arising from the boundary conditions for one of the coordinates.

The lowest energy state (the ground state) for the cubical well occurs when $n_1 = n_2 = n_3 = 1$ and has the value

$$E_{1,1,1} = \frac{3\hbar^2 \pi^2}{2mL^2} = 3E_1$$

The first excited energy level can be obtained in three different ways: $n_1 = 2$, $n_2 = n_3 = 1$; $n_2 = 2$, $n_1 = n_3 = 1$; or $n_3 = 2$, $n_1 = n_2 = 1$. Each has a different wave function. For example, the wave function for $n_1 = 2$ and $n_2 = n_3 = 1$ is

$$\psi_{2,1,1} = A \sin \frac{2\pi x}{L} \sin \frac{\pi y}{L} \sin \frac{\pi z}{L} \qquad \text{35-33}$$

(where the value of the normalization constant A is different than the value of the normalization constant in Equation 35-31). There are thus three different quantum states as described by the three different wave functions corresponding to the same energy level. An energy level with which more than one wave function is associated is said to be **degenerate.** In this case, there is threefold degeneracy.

[†] For example, if you throw two dice, the probability of the first die coming up 6 is 1/6 and the probability of the second die coming up an odd number is 1/2. The probability of the first die coming up 6 *and* the second die coming up an odd number is $(1/6)(1/2) = 1/12$.

Degeneracy is related to the spatial symmetry of the system. If, for example, we consider a noncubic well, where $U = 0$ for $0 < x < L_1$, $0 < y < L_2$, and $0 < z < L_3$, the boundary conditions at the edges would lead to the quantum conditions $k_1 L_1 = n_1 \pi$, $k_2 L_2 = n_2 \pi$, and $k_3 L_3 = n_3 \pi$, and the total energy would be

$$E_{n_1,n_2,n_3} = \frac{\hbar^2 \pi^2}{2m}\left(\frac{n_1^2}{L_1^2} + \frac{n_2^2}{L_2^2} + \frac{n_3^2}{L_3^2}\right) \qquad \text{35-34}$$

These energy levels are not degenerate if L_1, L_2, and L_3 are all different. Figure 35-18 shows the energy levels for the ground state and first two excited states for an infinite cubic well in which the excited states are degenerate and for a noncubic infinite well in which L_1, L_2, and L_3 are all slightly different so that the excited levels are slightly split apart and the degeneracy is removed. The ground state is the state where the quantum numbers n_1, n_2, and n_3 all equal 1. None of the three quantum numbers can be zero. If any one of n_1, n_2, and n_3 were zero, the corresponding wave number k would also equal zero and the corresponding wave function (Equation 35-31) would equal zero for all values of x, y, and z.

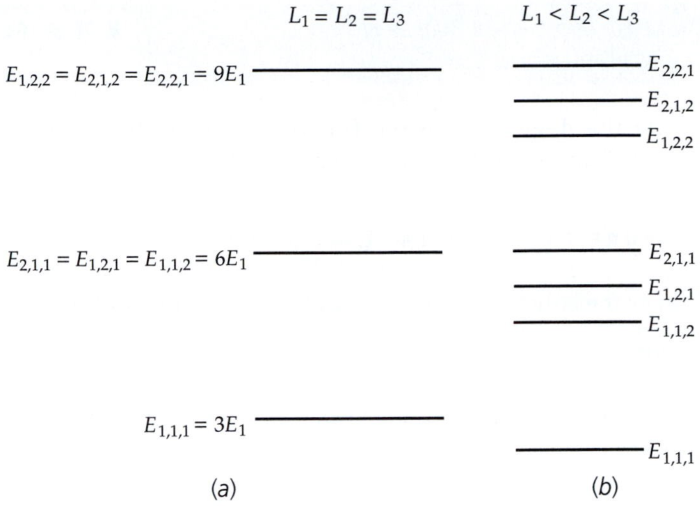

Energy-level diagram (a):
$L_1 = L_2 = L_3$

$E_{1,2,2} = E_{2,1,2} = E_{2,2,1} = 9E_1$ ——————

$E_{2,1,1} = E_{1,2,1} = E_{1,1,2} = 6E_1$ ——————

$E_{1,1,1} = 3E_1$ ——————

(a)

Energy-level diagram (b):
$L_1 < L_2 < L_3$

—————— $E_{2,2,1}$
—————— $E_{2,1,2}$
—————— $E_{1,2,2}$

—————— $E_{2,1,1}$
—————— $E_{1,2,1}$
—————— $E_{1,1,2}$

—————— $E_{1,1,1}$

(b)

FIGURE 35-18 Energy-level diagrams for (a) a cubic infinite well and (b) a noncubic infinite well. In Figure 35-18a the energy levels are degenerate; that is, there are two or more wave functions having the same energy. The degeneracy is removed when the symmetry of the potential is removed, as in Figure 35-18b.

ENERGY LEVELS FOR A PARTICLE IN A THREE-DIMENSIONAL BOX

EXAMPLE 35-3

A particle is in a three-dimensional box with $L_3 = L_2 = 2L_1$. Give the quantum numbers n_1, n_2, and n_3 that correspond to the thirteen quantum states of this box that have the lowest energies.

PICTURE THE PROBLEM We can use Equation 35-34 to write the energies in terms of the ratios $L_2/L_1 = 2$ and $L_3/L_1 = 2$, then find by inspection the values of the quantum numbers that give the lowest energies.

1. The energy of a level is given by Equation 35-34:

$$E_{n_1,n_2,n_3} = \frac{\hbar^2 \pi^2}{2m}\left(\frac{n_1^2}{L_1^2} + \frac{n_2^2}{L_2^2} + \frac{n_3^2}{L_3^2}\right)$$

2. Factor out $1/L_1^2$:

$$E_{n_1,n_2,n_3} = \frac{\hbar^2 \pi^2}{2m L_1^2}\left(n_1^2 + n_2^2\frac{L_1^2}{L_2^2} + n_3^2\frac{L_1^2}{L_3^2}\right)$$

$$= E_1(n_1^2 + n_2^2/4 + n_3^2/4)$$

3. The lowest energy is $E_{1,1,1}$:

$$E_{1,1,1} = E_1(1^2 + 1^2\tfrac{1}{4} + 1^2\tfrac{1}{4}) = 1.5E_1 \qquad \text{(1st)}$$

4. The energy increases the least when we increase n_2 or n_3. Try various values of the quantum numbers:

$$E_{1,2,1} = E_{1,1,2} = E_1(1^2 + 2^2\tfrac{1}{4} + 1\tfrac{1}{4}) = 2.25E_1 \qquad \text{(2nd and 3rd)}$$

$$E_{1,2,2} = E_1(1^2 + 2^2\tfrac{1}{4} + 2^2\tfrac{1}{4}) = 3.0E_1 \qquad \text{(4th)}$$

$$E_{1,3,1} = E_{1,1,3} = E_1(1^2 + 3^2\tfrac{1}{4} + 1^2\tfrac{1}{4}) = 3.50E_1 \qquad \text{(5th and 6th)}$$

$$E_{1,3,2} = E_{1,2,3} = E_1(1^2 + 3^2\tfrac{1}{4} + 2^2\tfrac{1}{4}) = 4.25E_1 \qquad \text{(7th and 8th)}$$

$$E_{2,1,1} = E_1(2^2 + 1^2\tfrac{1}{4} + 1^2\tfrac{1}{4}) = 4.5E_1 \qquad \text{(9th)}$$

$$E_{2,2,1} = E_{2,1,2} = E_1(2^2 + 2^2\tfrac{1}{4} + 1^2\tfrac{1}{4}) = 5.25E_1 \left.\begin{array}{c} \\ \\ \end{array}\right\}$$

$$E_{1,4,1} = E_{1,1,4} = E_1(1^2 + 4^2\tfrac{1}{4} + 1^2\tfrac{1}{4}) = 5.25E_1 \quad \text{(10th, 11th, 12th, and 13th)}$$

REMARKS Note the degeneracy of the levels.

EXERCISE Find the quantum numbers and energies of the next two energy levels in step 4. (*Answer* $E_{1,3,3} = 5.5E_1$, $E_{1,4,2} = E_{1,2,4} = E_{2,2,2} = 6.0E_1$)

Write the degenerate wave functions for the fourth and fifth excited states
(levels 5 and 6) of the results in step 4 of Example 35-3.

PICTURE THE PROBLEM Use Equation 35-33 with $k_i = n_i \pi / L$.

Cover the column to the right and try these on your own before looking at the answers.

Steps

Write the wave functions corresponding to the energies
$E_{1,3,1}$ and $E_{1,1,3}$.

Answers

$$\psi_{1,3,1} = A \sin \frac{\pi x}{L} \sin \frac{3\pi y}{L} \sin \frac{\pi z}{L}$$

$$\psi_{1,1,3} = A \sin \frac{\pi x}{L} \sin \frac{\pi y}{L} \sin \frac{3\pi z}{L}$$

35-6 The Schrödinger Equation for Two Identical Particles

Our discussion of quantum mechanics has thus far been limited to situations
in which a single particle moves in some force field characterized by a potential
energy function U. The most important physical problem of this type is the
hydrogen atom, in which a single electron moves in the Coulomb potential of the
proton nucleus. This problem is actually a two-body problem, since the proton
also moves in the field of the electron. However, the motion of the much more
massive proton requires only a very small correction to the energy of the atom
that is easily made in both classical and quantum mechanics. When we consider
more complicated problems, such as the helium atom, we must apply quantum
mechanics to two or more electrons moving in an external field. Such problems
are complicated by the interaction of the electrons with each other and also by
the fact that the electrons are identical.

The interaction of two electrons with each other is electromagnetic and is
essentially the same as the classical interaction of two charged particles. The
Schrödinger equation for an atom with two or more electrons cannot be solved
exactly, so approximation methods must be used. This is not very different from
the situation in classical problems with three or more particles. However, the
complications arising from the identity of electrons are purely quantum mechanical and have no classical counterpart. They are due to the fact that it is impossible to keep track of which electron is which. Classically, identical particles can be
identified by their positions, which in principle can be determined with unlimited
accuracy. This is impossible quantum mechanically because of the uncertainty
principle. Figure 35-19 offers a schematic illustration of the problem.

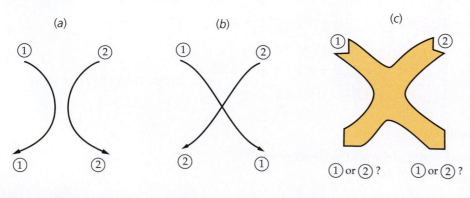

FIGURE 35-19 Two possible
classical electron paths (Figure 35-19*a* and
Figure 35-19*b*). If electrons were classical
particles, they could be distinguished by
the paths followed. However, because of
the quantum-mechanical wave properties
of electrons, the paths are spread out,
as indicated by the shaded region in
Figure 35-19*c*. It is impossible to
distinguish which electron is
which after they separate.

The indistinguishability of identical particles has important consequences. For instance, consider the very simple case of two identical, noninteracting particles in a one-dimensional infinite square well. The time-independent Schrödinger equation for two particles, each of mass m, is

$$-\frac{\hbar^2}{2m}\frac{\partial^2 \psi(x_1, x_2)}{\partial x_1^2} - \frac{\hbar^2}{2m}\frac{\partial^2 \psi(x_1, x_2)}{\partial x_2^2} + U\psi(x_1, x_2) = E\psi(x_1, x_2) \qquad 35\text{-}35$$

where x_1 and x_2 are the coordinates of the two particles. If the particles interact, the potential energy U contains terms with both x_1 and x_2 that cannot be separated into separate terms containing only x_1 or x_2. For example, the electrostatic repulsion of two electrons in one dimension is represented by the potential energy $ke^2/|x_2 - x_1|$. However, if the particles do not interact (as we are assuming here), we can write $U = U_1(x_1) + U_2(x_2)$. For the infinite square well, we need only solve the Schrödinger equation inside the well where $U = 0$, and require that the wave function be zero at the walls of the well. With $U = 0$, Equation 35-35 looks just like the expression for a particle in a two-dimensional well (Equation 35-30, with no z and with y replaced by x_2).

Solutions of this equation can be written in the form[†]

$$\psi_{n,m} = \psi_n(x_1)\psi_m(x_2) \qquad 35\text{-}36$$

where ψ_n and ψ_m are the single-particle wave functions for a particle in an infinite well and n and m are the quantum numbers of particles 1 and 2, respectively. For example, for $n = 1$ and $m = 2$, the wave function is

$$\psi_{1,2} = A\sin\frac{\pi x_1}{L}\sin\frac{2\pi x_2}{L} \qquad 35\text{-}37$$

The probability of finding particle 1 in dx_1 and particle 2 in dx_2 is $\psi_{n,m}^2(x_1, x_2)\,dx_1\,dx_2$, which is just the product of the separate probabilities $\psi_n^2(x_1)\,dx_1$ and $\psi_m^2(x_2)\,dx_2$. However, even though we have labeled the particles 1 and 2, we cannot distinguish which is in dx_1 and which is in dx_2 if they are identical. The mathematical descriptions of identical particles must be the same if we interchange the labels. The probability density $\psi^2(x_1, x_2)$ must therefore be the same as $\psi^2(x_2, x_1)$:

$$\psi^2(x_2, x_1) = \psi^2(x_1, x_2) \qquad 35\text{-}38$$

Equation 35-38 is satisfied if $\psi(x_2, x_1)$ is either **symmetric** or **antisymmetric** on the exchange of particles—that is, if either

$$\psi(x_2, x_1) = \psi(x_1, x_2), \quad \text{symmetric} \qquad 35\text{-}39$$

or

$$\psi(x_2, x_1) = -\psi(x_1, x_2), \quad \text{antisymmetric} \qquad 35\text{-}40$$

Note that the wave functions given by Equations 35-36 and 35-37 are neither symmetric nor antisymmetric. If we interchange x_1 and x_2 in these wave functions, we get a different wave function, which implies that the particles can be distinguished.

We can find symmetric and antisymmetric wave functions that are solutions of the Schrödinger equation by adding or subtracting $\psi_{n,m}$ and $\psi_{m,n}$. Adding them, we obtain

---

[†] Again, this result can be obtained by solving Equation 35-35, but it also can be understood in terms of our knowledge of probability. The probability of electron 1 being in region dx_1 and electron 2 being in region dx_2 is the product of the individual probabilities.

$$\psi_S = A'[\psi_n(x_1)\psi_m(x_2) + \psi_n(x_2)\psi_m(x_1)], \quad \text{symmetric} \qquad 35\text{-}41$$

and subtracting them, we obtain

$$\psi_A = A'[\psi_n(x_1)\psi_m(x_2) - \psi_n(x_2)\psi_m(x_1)], \quad \text{antisymmetric} \qquad 35\text{-}42$$

For example, the symmetric and antisymmetric wave functions for the first excited state of two identical particles in an infinite square well would be

$$\psi_S = A'\left(\sin\frac{\pi x_1}{L}\sin\frac{2\pi x_2}{L} + \sin\frac{\pi x_2}{L}\sin\frac{2\pi x_1}{L}\right) \qquad 35\text{-}43$$

and

$$\psi_A = A'\left(\sin\frac{\pi x_1}{L}\sin\frac{2\pi x_2}{L} - \sin\frac{\pi x_2}{L}\sin\frac{2\pi x_1}{L}\right) \qquad 35\text{-}44$$

There is an important difference between antisymmetric and symmetric wave functions. If $n = m$, the antisymmetric wave function is identically zero for all values of x_1 and x_2, whereas the symmetric wave function is not. Thus, if the wave function describing two identical particles is antisymmetric, the quantum numbers n and m of two particles cannot be the same. This is an example of the **Pauli exclusion principle,** which was first stated by Wolfgang Pauli for electrons in an atom:

No two electrons in an atom can have the same quantum numbers.

PAULI EXCLUSION PRINCIPLE

It is found that electrons, protons, neutrons, and some other particles have antisymmetric wave functions and obey the Pauli exclusion principle. These particles are called **fermions.** Other particles (e.g., α particles, deuterons, photons, and mesons) have symmetric wave functions and do not obey the Pauli exclusion principle. These particles are called **bosons.**

SUMMARY

1. The Schrödinger equation is a differential equation that relates the second spatial derivative of a wave function to its first time derivative. Wave functions that describe physical situations are solutions of this differential equation.

2. Because a wave function must be normalizable, it must be well behaved; that is, it must approach zero as x approaches infinity. For bound systems such as a particle in a box, a simple harmonic oscillator, or an electron in an atom, this requirement leads to energy quantization.

3. The well-behaved wave functions for bound systems describe standing waves.

Topic	Relevant Equations and Remarks	
1. **Time-Independent Schrödinger Equation**	$-\dfrac{\hbar^2}{2m}\dfrac{d^2\psi(x)}{dx^2} + U(x)\psi(x) = E\psi(x)$	35-4

Allowable solutions	In addition to satisfying the Schrödinger equation, a wave function $\psi(x)$ must be continuous and (if U is not infinite) must have a continuous first derivative $d\psi/dx$. Because the probability of finding an electron somewhere must be 1, the wave function must obey the normalization condition $$\int_{-\infty}^{\infty} |\psi|^2 \, dx = 1$$ This condition implies the boundary condition that ψ must approach 0 as x approaches $\pm\infty$. Such boundary conditions lead to the quantization of energy.
2. Confined Particles	When the total energy E is greater than the potential energy $U(x)$ in some region (the classically allowed region) and less than $U(x)$ outside that region, the wave function oscillates within the classically allowed region and increases or decreases exponentially outside that region. The wave function approaches zero as x approaches ∞ only for certain values of the total energy E. The energy is thus quantized.
In a finite square well	In a finite well of height U_0, there are only a finite number of allowed energies, and these are slightly less than the corresponding energies in an infinite well.
In the simple harmonic oscillator	In the oscillator potential energy function $U(x) = \frac{1}{2}m\omega_0^2 x^2$, the allowed energies are equally spaced and given by $$E_n = (n + \tfrac{1}{2})\hbar\omega_0 \qquad\qquad \textbf{35-26}$$ The ground-state wave function is given by $$\psi_0(x) = A_0 e^{-ax^2} \qquad\qquad \textbf{35-23}$$ where A_0 is the normalization constant and $a = m\omega_0/(2\hbar)$.
3. Reflection and Barrier Penetration	When the potential changes abruptly over a small distance, a particle may be reflected even though $E > U(x)$. A particle may penetrate a region in which $E < U(x)$. Reflection and penetration of electron waves are similar to those for other kinds of waves.
4. The Schrödinger Equation in Three Dimensions	The wave function for a particle in a three-dimensional box can be written $$\psi(x, y, z) = \psi_1(x)\psi_2(y)\psi_3(z)$$ where ψ_1, ψ_2, and ψ_3 are wave functions for a one-dimensional box.
Degeneracy	When more than one wave function is associated with the same energy level, the energy level is said to be degenerate. Degeneracy arises because of spatial symmetry.
5. The Schrödinger Equation for Two Identical Particles	A wave function that describes two identical particles must be either symmetric or antisymmetric when the coordinates of the particles are exchanged. Fermions (which include electrons, protons, and neutrons) are described by antisymmetric wave functions and obey the Pauli exclusion principle, which states that no two particles can have the same values for their quantum number. Bosons (which include α particles, deuterons, photons, and mesons) have symmetric wave functions and do not obey the Pauli exclusion principle.

- Single-concept, single-step, relatively easy
- •• Intermediate-level, may require synthesis of concepts
- ••• Challenging
- SSM Solution is in the *Student Solutions Manual*
- iSOLVE Problems available on iSOLVE online homework service
- iSOLVE✓ These "Checkpoint" online homework service problems ask students additional questions about their confidence level, and how they arrived at their answer.

In a few problems, you are given more data than you actually need; in a few other problems, you are required to supply data from your general knowledge, outside sources, or informed estimates.

Conceptual Problems

1 • True or false: Boundary conditions on the wave function lead to energy quantization.

2 • Sketch (a) the wave function and (b) the probability distribution for the $n = 4$ state for the finite square-well potential.

3 • Sketch (a) the wave function and (b) the probability distribution for the $n = 5$ state for the finite square-well potential.

Estimation and Approximation

4 • SSM The Schrödinger equation could be applied equally well to baseballs as to electrons; yet, we would never analyze the motion of a baseball with a wave function. Explain why this is the case by estimating the quantum mechanically predicted lowest energy level of a baseball trapped inside a locker. You can treat the locker as if it were a one-dimensional infinite potential well. What value of the quantum number n would you need for a ball rolling around in a locker, after you toss the ball in, so that the kinetic energy is approximately equal to the quantum mechanically calculated energy?

The Schrödinger Equation

5 •• Show that if $\psi_1(x)$ and $\psi_2(x)$ are each solutions to the time-independent Schrödinger equation (Equation 35-4), then $\psi_3(x) = \psi_1(x) + \psi_2(x)$ is also a solution. This is known as the superposition principle and it applies to the solutions of all linear differential equations.

The Harmonic Oscillator

6 •• The harmonic oscillator problem may be used to describe molecules. For example, the hydrogen molecule H_2 is found to have equally spaced energy levels separated by 8.7×10^{-20} J. What value of spring constant would be needed to get this energy spacing, assuming that the molecule can be modeled as a single hydrogen atom attached to a spring?

7 •• Show that the expectation value $<x> = \int x|\psi|^2 \, dx$ is zero for both the ground state and the first excited state of the harmonic oscillator.

8 •• SSM Use the procedure of Example 35-1 to verify that the energy of the first excited state of the harmonic oscillator is $E_1 = \frac{3}{2}\hbar\omega_0$. (Note: *Rather than solve for a again, use the result $a = m\omega_0/(2\hbar)$ obtained in Example 35-1.*)

9 •• Show that the normalization constant A_0 of Equation 35-23 is $A_0 = (2m\omega_0/h)^{1/4}$.

10 •• Show that for the ground state of the harmonic oscillator $<x^2> = \int x^2|\psi|^2 \, dx = \hbar/(2m\omega_0) = 1/(4a)$. Use this result to show that the average potential energy equals half the total energy.

11 •• The quantity $\sqrt{<x^2> - <x>^2}$ is a measure of the average spread in the location of a particle. (a) Consider an electron trapped in a harmonic oscillator potential. Its lowest energy level is found to be 2.1×10^{-4} eV. Calculate $\sqrt{<x^2> - <x>^2}$ for this electron. (See Problems 7 and 10.) (b) Now consider an electron trapped in an infinite square-well potential. If the width of the well is equal to $\sqrt{<x^2> - <x>^2}$, what would be the lowest energy level for this electron?

12 ••• Classically, the average kinetic energy of the harmonic oscillator equals the average potential energy. We may assume that this is also true for the quantum mechanical harmonic oscillator. Use this condition to determine the expectation value of p^2 for the ground state of the harmonic oscillator.

13 ••• We know that for the classical harmonic oscillator, $p_{av} = 0$. It can be shown that for the quantum mechanical harmonic oscillator, $<p> = 0$. Use the results of Problem 7, Problem 10, and Problem 12 to determine the uncertainty product $\Delta x \, \Delta p$ for the ground state of the harmonic oscillator.

Reflection and Transmission of Electron Waves: Barrier Penetration

14 •• SSM A particle of mass m with wave number k_1 is traveling to the right along the negative x axis. The potential energy of the particle is equal to zero everywhere on the negative x axis and is equal to U_0 everywhere on the positive x axis, $U_0 > 0$. (a) Show that if the total energy is $E = \alpha U_0$, where $\alpha \geq 1$, wave number k_2 in the region $x > 0$ is given by

$$k_2 = \sqrt{\frac{\alpha - 1}{\alpha}} k_1$$

(b) Using a spreadsheet program or graphing calculator, graph the reflection coefficient R and the transmission coefficient T for $1 \leq \alpha \leq 5$.

15 •• [I SOLVE ✓]   Suppose that the potential in Problem 14 drops from zero to $-U_0$ at $x = 0$ so that the particle speeds up instead of slowing down. The wave number for the incident particle is again k_1, and the total energy is $2U_0$. (a) What is the wave number for the particle in the region of positive x? (b) Calculate the reflection coefficient R at $x = 0$. (c) What is the transmission coefficient T? (d) If one million particles with wave number k_1 are incident upon the potential drop, how many particles are expected to continue along in the positive x direction? How does this compare with the classical prediction?

16 •• [I SOLVE ✓]   A particle of energy E approaches a step barrier of height U_0. What should be the ratio E/U_0 so that the reflection coefficient is $1/2$?

17 •• [I SOLVE]   Use Equation 35-29 to calculate the order of magnitude of the probability that a proton will tunnel out of a nucleus in one collision with the nuclear barrier if it has energy 6 MeV below the top of the potential barrier and the barrier thickness is 10^{-15} m.

18 •• [SSM]   A 10-eV electron is incident on a potential barrier of height 25 eV and width of 1 nm. (a) Use Equation 35-29 to calculate the order of magnitude of the probability that the electron will tunnel through the barrier. (b) Repeat your calculation for a width of 0.1 nm.

19 •••   To understand how a small change in α-particle energy can dramatically change the tunneling probability from a nucleus, consider an α particle emitted by a uranium nucleus ($Z = 92$). (a) Referring to Figure 35-17, calculate the distance of closest approach r_1 that α particles with kinetic energies of 4 MeV and 7 MeV could make to the uranium nucleus. (b) Use the result from Part (a) to calculate the relative transmission coefficient $e^{-2\alpha a}$ for the same α particles. (Note: *The actual half-lives of uranium nuclei vary over nine orders of magnitude. Your calculation will show a smaller range than this; however, to find half-life, you must also include the frequency with which the α particle strikes the barrier.*)

The Schrödinger Equation in Three Dimensions

20 •   A particle is confined to a three-dimensional box that has sides L_1, $L_2 = 2L_1$, and $L_3 = 3L_1$. Give the quantum numbers n_1, n_2, n_3 that correspond to the lowest ten quantum states of this box.

21 •   Give the wave functions for the lowest ten quantum states of the particle in Problem 20.

22 •   (a) Repeat Problem 20 for the case $L_2 = 2L_1$ and $L_3 = 4L_1$. (b) What quantum numbers correspond to degenerate energy levels?

23 • [SSM]   Give the wave functions for the lowest ten quantum states of the particle in Problem 22.

24 •   A particle moves in a potential well given by $U(x, y, z) = 0$ for $-L/2 < x < L/2$, $0 < y < L$, and $0 < z < L$; and $U = \infty$ outside these ranges. (a) Write an expression for the ground-state wave function for this particle. (b) How do the allowed energies compare with those for a well having $U = 0$ for $0 < x < L$, rather than for $-L/2 < x < L/2$?

25 ••   A particle moves freely in the two-dimensional region defined by $0 \le x \le L$ and $0 \le y \le L$. (a) Find the wave functions satisfying the Schrödinger equation. (b) Find the

corresponding energies. (c) Find the lowest two states that are degenerate. Give the quantum numbers for this case. (d) Find the lowest three states that have the same energy. Give the quantum numbers for the three states having the same energy.

The Schrödinger Equation for Two Identical Particles

26 •   Show that Equation 35-37 satisfies Equation 35-35 with $U = 0$, and find the energy of this state.

27 • [I SOLVE]   What is the ground-state energy of ten noninteracting bosons in a one-dimensional box of length L?

28 • [SSM] [I SOLVE]   What is the ground-state energy of ten noninteracting fermions, such as neutrons, in a one-dimensional box of length L? (Because the quantum number associated with spin can have two values, each spatial state can hold two neutrons.)

Orthogonality of Wave Functions

The integral of two functions over some space interval is somewhat analogous to the dot product of two vectors. If this integral is zero, the functions are said to be orthogonal, which is analogous to two vectors being perpendicular. The following problems illustrate the general principle that any two wave functions corresponding to different energy levels in the same potential are orthogonal. A general hint for all these problems is that the integral of an antisymmetric integrand over symmetric limits is equal to zero.

29 ••   Show that the ground-state wave function and the wave function of the first excited state of the harmonic oscillator are orthogonal; that is, show that $\int \psi_0(x)\, \psi_1(x)\, dx = 0$.

30 ••   The wave function for the state $n = 2$ of the harmonic oscillator is $\psi_2(x) = A_2(2ax^2 - \frac{1}{2})e^{-ax^2}$, where A_2 is the normalization constant for this wave function. Show that the wave functions for the states $n = 1$ and $n = 2$ of the harmonic oscillator are orthogonal.

31 ••   For the wave functions $\psi_n(x) = \sqrt{2/L}\, \sin(n\pi x/L)$ corresponding to a particle in an infinite square-well potential from 0 to L, show that $\int \psi_n(x)\, \psi_m(x)\, dx = 0$; that is, show that ψ_n and ψ_m are orthogonal.

General Problems

32 ••   Consider a particle in a one-dimensional box of length L that is centered at the origin (see Problem 65 in Chapter 34). (a) What are the values of $\psi_1(0)$ and $\psi_2(0)$? (b) What are the values of $<x>$ for the states $n = 1$ and $n = 2$? (c) Evaluate $<x^2>$ for the states $n = 1$ and $n = 2$.

33 •• [SSM]   Eight identical noninteracting fermions (e.g., neutrons) are confined to a two-dimensional square box of side length L. Determine the energies of the three lowest states. (See Problem 26.)

34 •• [I SOLVE]   A particle is confined to a two-dimensional box defined by the following boundary conditions: $U(x, y) = 0$, for $-L/2 \le x \le L/2$ and $-3L/2 \le y \le 3L/2$; and $U(x, y) = \infty$ elsewhere. (a) Determine the energies of the lowest three bound states. Are any of these states degenerate? (b) Identify the quantum numbers of the lowest doubly degenerate bound state and determine its energy.

35 ••• The classical probability distribution function for a particle in a one-dimensional box of length L is $P = 1/L$. (See Example 34-5.) (a) Show that the classical expectation value of x^2 for a particle in a one-dimensional box of length L centered at the origin (Problem 32) is $L^2/12$. (b) Find the quantum expectation value of x^2 for the nth state of a particle in the one-dimensional box of Problem 32, and show that it approaches the classical limit $L^2/12$ for $n \gg 1$.

36 •• Show that Equation 35-27 and Equation 35-28 imply that the transmission coefficient for particles of energy E incident on a step barrier $U_0 < E$ is given by

$$T = \frac{4k_1 k_2}{(k_1 + k_2)^2} = \frac{4r}{(1 + r)^2}$$

where $r = k_2/k_1$.

37 •• (a) Show that for the case of a particle of energy E incident on a step barrier $U_0 < E$, the wave numbers k_1 and k_2 are related by

$$\frac{k_2}{k_1} = r = \sqrt{1 - \frac{U_0}{E}}$$

(b) Use this result to show that $R = (1 - r)^2/(1 + r)^2$

38 •• (a) Using a spreadsheet program or graphing calculator and the results of Problem 36 and Problem 37, graph the transmission coefficient T and reflection coefficient R as a function of incident energy E for values of E ranging from $E = U_0$ to $E = 10.0U_0$. (b) What limiting values do your graphs indicate?

39 ••• Determine the normalization constant A_2 in Problem 30.

40 ••• Consider the time-independent, one-dimensional Schrödinger equation when the potential function is symmetric about the origin, that is, when $U(x)$ is even.[†] (a) Show that if $\psi(x)$ is a solution of the Schrödinger equation with energy E, then $\psi(-x)$ is also a solution with the same energy E, and that, therefore, $\psi(x)$ and $\psi(-x)$ can differ by only a multiplicative constant. (b) Write $\psi(x) = C\psi(-x)$, and show that $C = \pm 1$. Note that $C = +1$ means that $\psi(x)$ is an even function of x, and $C = -1$ means that $\psi(x)$ is an odd function of x.

41 ••• **SSM** In this problem you will derive the ground-state energy of the harmonic oscillator using the precise form of the uncertainty principle, $\Delta x \, \Delta p \geq \hbar/2$, where Δx and Δp are defined to be the standard deviations $(\Delta x)^2 = [(x - x_{av})^2]_{av}$ and $(\Delta p)^2 = [(p - p_{av})^2]_{av}$ (see Equation 17-35a). Proceed as follows:

1. Write the total classical energy in terms of the position x and momentum p using $U(x) = m\omega_0^2 x^2$ and $K = p^2/2m$.

2. Use the result of Equation 17-35 to write $(\Delta x)^2 = [(x - x_{av})^2]_{av} = (x^2)_{av} - x_{av}^2$ and $(\Delta p)^2 = [(p - p_{av})^2]_{av} = (p^2)_{av} - p_{av}^2$.

3. Use the symmetry of the potential energy function to argue that x_{av} and p_{av} must be zero, so that $(\Delta x)^2 = (x^2)_{av}$ and $(\Delta p)^2 = (p^2)_{av}$.

4. Assume that $\Delta p = \hbar/(2\Delta x)$ to eliminate $(p^2)_{av}$ from the average energy $E_{av} = (p^2)_{av}/(2m) + \frac{1}{2}m\omega^2(x^2)_{av}$ and write E_{av} as $E_{av} = \hbar^2/(8mZ) + \frac{1}{2}m\omega^2 Z$, where $Z = (x^2)_{av}$.

5. Set $dE/dZ = 0$ to find the value of Z for which E is a minimum.

6. Show that the minimum energy is given by $(E_{av})_{min} = +\frac{1}{2}\hbar\omega_0$.

42 ••• A particle of mass m near the earth's surface at $z = 0$ can be described by the potential energy

$$U = mgz, \quad z > 0$$

$$U = \infty, \quad z < 0$$

For some positive value of total energy E, indicate the classically allowed region on a sketch of $U(z)$ versus z. Sketch also the classical kinetic energy versus z. The Schrödinger equation for this problem is quite difficult to solve. Using arguments similar to those in Section 35-2 about the curvature of the wave function as given by the Schrödinger equation, sketch your guesses for the shape of the wave function for the ground state and for the first two excited states.

[†] A function $f(x)$ is even if $f(x) = f(-x)$ for all x.

Atoms

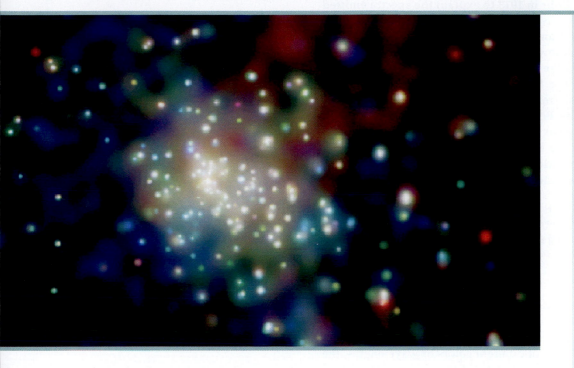

AT A DISTANCE OF 6,000 LIGHT YEARS FROM EARTH, THE STAR CLUSTER RCW 38 IS A RELATIVELY CLOSE STAR-FORMING REGION. THIS IMAGE COVERS AN AREA ABOUT 5 LIGHT YEARS ACROSS AND CONTAINS THOUSANDS OF HOT, VERY YOUNG STARS FORMED LESS THAN A MILLION YEARS AGO. X RAYS FROM THE HOT UPPER ATMOSPHERES OF 190 OF THESE STARS WERE DETECTED BY CHANDRA, AN X-RAY OBSERVATORY ORBITING EARTH. THE MECHANISMS GENERATING THESE X RAYS IS NOT KNOWN.

ON EARTH, X-RAY MACHINES PRODUCE X RAYS BY BOMBARDING A TARGET WITH HIGH-ENERGY ELECTRONS. THE ATOMIC NUMBER OF THE ATOMS THAT MAKE UP THE TARGET CAN BE DETERMINED BY ANALYZING THE RESULTING X-RAY SPECTRA.

? **How is the atomic number obtained from the spectral analysis? Example 36-8 shows one way to accomplish this task.**

There are 113 chemical elements that have been discovered, and there are a couple of additional chemical elements that recently have been reported. Each element is characterized by an atom that contains a number of protons Z, an equal number of electrons, and a number of neutrons N. The number of protons Z is called the **atomic number**. The lightest atom, hydrogen (H), has $Z = 1$; the next lightest atom, helium (He), has $Z = 2$; the third lightest, lithium (Li), has $Z = 3$; and so forth. Nearly all the mass of the atom is concentrated in a tiny nucleus, which contains the protons and neutrons. Typically, the nuclear radius is approximately 1 fm to 10 fm (1 fm = 10^{-15} m). The distance between the nucleus and the electrons is approximately 0.1 nm = 100,000 fm. This distance determines the *size* of the atom.

The chemical properties and physical properties of an element are determined by the number and arrangement of the electrons in the atom. Because each proton has a positive charge $+e$, the nucleus has a total positive charge $+Ze$. The electrons are negatively charged ($-e$), so they are attracted to the nucleus and repelled by each other. Since electrons and protons have equal but opposite charges, and there are an equal number of electrons and protons in an atom,

1171

atoms are electrically neutral. Atoms that lose or gain one or more electrons are then electrically charged and are called *ions*.

➤ We will begin our study of atoms by discussing the Bohr model, a semi-classical model developed by Niels Bohr in 1913 to explain the spectra emitted by hydrogen atoms. Although this *prequantum mechanics* model has many shortcomings, it provides a useful framework for the discussion of atomic phenomena. For example, we now know that the electron does not circle the nucleus in well-defined orbits, as in the Bohr model, but instead is described by a wave function that satisfies the Schrödinger equation. However, the probability distributions that follow from the full quantum theory do in fact have maxima at the positions of the Bohr orbits. After discussing the Bohr model, we will apply our knowledge of quantum mechanics from Chapter 35 to give a qualitative description of the hydrogen atom. We will then discuss the structure of other atoms and the periodic table of the elements. Finally, we will discuss both optical and X-ray spectra.

36-1 The Nuclear Atom

Atomic Spectra

By the beginning of the twentieth century, a large body of data had been collected on the emission of light by atoms in a gas when the atoms are excited by an electric discharge. Viewed through a spectroscope with a narrow-slit aperture, this light appears as a discrete set of lines of different colors or wavelengths; the spacing and intensities of the lines are characteristic of the element. The wavelengths of these spectral lines could be accurately determined, and much effort went into finding regularities in the spectra. Figure 36-1 shows line spectra for hydrogen and for mercury.

In 1885 a Swiss schoolteacher, Johann Balmer, found that the wavelengths of the lines in the visible spectrum of hydrogen can be represented by the formula

$$\lambda = (364.6 \text{ nm}) \frac{m^2}{m^2 - 4}, \quad m = 3, 4, 5, \dots \qquad \text{36-1}$$

Balmer suggested that this might be a special case of a more general expression that would be applicable to the spectra of other elements. Such an expression, found by Johannes R. Rydberg and Walter Ritz and known as the **Rydberg–Ritz formula,** gives the reciprocal wavelength as

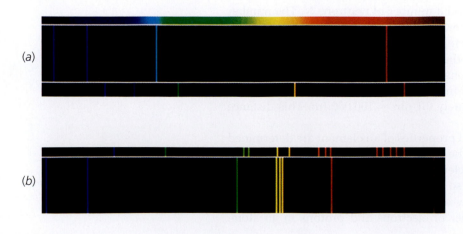

(a)

(b)

FIGURE 36-1 (*a*) Line spectrum of hydrogen and (*b*) line spectrum of mercury.

$$\frac{1}{\lambda} = R\left(\frac{1}{n_2^2} - \frac{1}{n_1^2}\right) \qquad\qquad 36\text{-}2$$

where n_1 and n_2 are integers, with $n_1 > n_2$, and R is the Rydberg constant. The Rydberg constant is the same for all spectral series of the same element and varies only slightly in a regular way from element to element. For hydrogen, R has the value

$$R_{\text{H}} = 1.097776 \times 10^7 \text{ m}^{-1}$$

The Rydberg–Ritz formula gives the wavelengths for all the lines in the spectra of hydrogen, as well as alkali elements such as lithium and sodium. The hydrogen Balmer series given by Equation 36-1 is also given by Equation 36-2, with $R = R_{\text{H}}$, $n_2 = 2$, and $n_1 = m$.

Many attempts were made to construct a model of the atom that would yield these formulas for its radiation spectrum. The most popular model, due to J. J. Thomson, considered various arrangements of electrons embedded in some kind of fluid that contained most of the mass of the atom and had enough positive charge to make the atom electrically neutral. Thomson's model, called the "plum pudding" model, is illustrated in Figure 36-2. Since classical electromagnetic theory predicted that a charge oscillating with frequency f would radiate electromagnetic energy of that frequency, Thomson searched for configurations that were stable and that had normal modes of vibration of frequencies equal to those of the spectrum of the atom. A difficulty of this model and all other models was that, according to classical physics, electric forces alone cannot produce stable equilibrium. Thomson was unsuccessful in finding a model that predicted the observed frequencies for any atom.

The Thomson model was essentially ruled out by a set of experiments by H. W. Geiger and E. Marsden, under the supervision of E. Rutherford in approximately 1911, in which alpha particles from radioactive radium were scattered by atoms in a gold foil. Rutherford showed that the number of alpha particles scattered at large angles could not be accounted for by an atom in which the positive charge was distributed throughout the atomic size (known to be about 0.1 nm in diameter) but required that the positive charge and most of the mass of the atom be concentrated in a very small region, now called the nucleus, of diameter of the order of 10^{-6} nm = 1 fm.

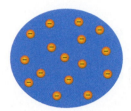

FIGURE 36-2 J. J. Thomson's plum pudding model of the atom. In this model, the negative electrons are embedded in a fluid of positive charge. For a given configuration in such a system, the resonance frequencies of oscillations of the electrons can be calculated. According to classical theory, the atom should radiate light of frequency equal to the frequency of oscillation of the electrons. Thomson could not find any configuration that would give frequencies in agreement with the measured frequencies of the spectrum of any atom.

36-2 The Bohr Model of the Hydrogen Atom

Niels Bohr, working in the Rutherford laboratory in 1912, proposed a model of the hydrogen atom that extended the work of Planck, Einstein, and Rutherford and successfully predicted the observed spectra. According to Bohr's model, the electron of the hydrogen atom moves under the influence of the Coulomb attraction to the positive nucleus according to classical mechanics, which predicts circular or elliptical orbits with the force center at one focus, as in the motion of the planets around the sun. For simplicity, Bohr chose a circular orbit, as shown in Figure 36-3.

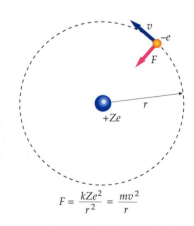

$$F = \frac{kZe^2}{r^2} = \frac{mv^2}{r}$$

FIGURE 36-3 Electron of charge $-e$ traveling in a circular orbit of radius r around the nuclear charge $+Ze$. The attractive electrical force kZe^2/r^2 keeps the electron in its orbit.

Energy for a Circular Orbit

Consider an electron of charge $-e$ moving in a circular orbit of radius r about a positive charge Ze such as the nucleus of a hydrogen atom ($Z = 1$) or of a singly ionized helium atom ($Z = 2$). The total energy of the electron can be related to the radius of the orbit. The potential energy of the electron of charge $-e$ at a distance r from a positive charge Ze is

$$U = \frac{kq_1q_2}{r} = \frac{k(Ze)(-e)}{r} = -\frac{kZe^2}{r}$$

36-3

where k is the Coulomb constant. The kinetic energy K can be obtained as a function of r by using Newton's second law, $F_{net} = ma$. Setting the Coulomb attractive force equal to the mass times the centripetal acceleration gives

$$\frac{kZe^2}{r^2} = m\frac{v^2}{r}$$

36-4a

Multiplying both sides by $r/2$ gives

$$K = \frac{1}{2}mv^2 = \frac{1}{2}\frac{kZe^2}{r}$$

36-4b

Thus, the kinetic energy and the potential energy vary inversely with r. Note that the magnitude of the potential energy is twice that of the kinetic energy:

$$U = -2K$$

36-5

This is a general result in $1/r^2$ force fields. It also holds for circular orbits in a gravitational field (see Example 11-6 in Section 11-3). The total energy is the sum of the kinetic energy and the potential energy:

$$E = K + U = \frac{1}{2}\frac{kZe^2}{r} - \frac{kZe^2}{r}$$

or

$$E = -\frac{1}{2}\frac{kZe^2}{r}$$

36-6

ENERGY IN A CIRCULAR ORBIT FOR A $1/r^2$ FORCE

Although mechanical stability is achieved because the Coulomb attractive force provides the centripetal force necessary for the electron to remain in orbit, classical *electromagnetic* theory says that such an atom would be unstable electrically. The atom would be unstable because the electron must accelerate when moving in a circle and therefore radiate electromagnetic energy of frequency equal to that of its motion. According to the classical theory, such an atom would quickly collapse, with the electron spiraling into the nucleus as it radiates away its energy.

Bohr's Postulates

Bohr circumvented the difficulty of the collapsing atom by *postulating* that only certain orbits, called stationary states, are allowed, and that in these orbits the electron does not radiate. An atom radiates only when the electron makes a transition from one allowed orbit (stationary state) to another.

> The electron in the hydrogen atom can move only in certain nonradiating, circular orbits called stationary states.
>
> BOHR'S FIRST POSTULATE—NONRADIATING ORBITS

The second postulate relates the frequency of radiation to the energies of the stationary states. If E_i and E_f are the initial and final energies of the atom, the frequency of the emitted radiation during a transition is given by

$$f = \frac{E_i - E_f}{h} \qquad \text{36-7}$$

BOHR'S SECOND POSTULATE—PHOTON FREQUENCY FROM ENERGY CONSERVATION

where h is Planck's constant. This postulate is equivalent to the assumption of conservation of energy with the emission of a photon of energy hf. Combining Equation 36-6 and Equation 36-7, we obtain for the frequency

$$f = \frac{E_1 - E_2}{h} = \frac{1}{2}\frac{kZe^2}{h}\left(\frac{1}{r_2} - \frac{1}{r_1}\right) \qquad \text{36-8}$$

where r_1 and r_2 are the radii of the initial and final orbits.

To obtain the frequencies implied by the Rydberg–Ritz formula, $f = c/\lambda = cR(1/n_2^2 - 1/n_1^2)$, it is evident that the radii of stable orbits must be proportional to the squares of integers. Bohr searched for a quantum condition for the radii of the stable orbits that would yield this result. After much trial and error, Bohr found that he could obtain it if he postulated that the angular momentum of the electron in a stable orbit equals an integer times $\hbar$ ("bar," Planck's constant divided by 2π). Since the angular momentum of a circular orbit is just mvr, this postulate is

$$mvr = \frac{nh}{2\pi} = n\hbar, \quad n = 1, 2, 3, \ldots \qquad \text{36-9}$$

BOHR'S THIRD POSTULATE—QUANTIZED ANGULAR MOMENTUM

where $\hbar = h/2\pi = 1.055 \times 10^{-34}$ J·s $= 6.582 \times 10^{-16}$ eV·s.

Equation 36-9 relates the speed v to the radius r. Equation 36-4a, from Newton's second law, gives us another equation relating the speed to the radius:

$$\frac{kZe^2}{r^2} = m\frac{v^2}{r}$$

or

$$v^2 = \frac{kZe^2}{mr} \qquad \text{36-10}$$

We can determine r by eliminating v between Equations 36-9 and 36-10. Solving Equation 36-9 for v and squaring gives

$$v^2 = n^2\frac{\hbar^2}{m^2r^2}$$

Equating this expression for v^2 with the expression given by Equation 36-10, we get

$$n^2 \frac{\hbar^2}{m^2 r^2} = \frac{kZe^2}{mr}$$

Solving for r, we obtain

$$r = n^2 \frac{\hbar^2}{mkZe^2} = n^2 \frac{a_0}{Z} \qquad 36\text{-}11$$

RADIUS OF THE BOHR ORBITS

where a_0 is called the **first Bohr radius**.

$$a_0 = \frac{\hbar^2}{mke^2} \approx 0.0529 \text{ nm} \qquad 36\text{-}12$$

FIRST BOHR RADIUS

Substituting the expressions for r in Equation 36-11 into Equation 36-8 for the frequency gives

$$f = \frac{1}{2} \frac{kZe^2}{h} \left(\frac{1}{r_2} - \frac{1}{r_1} \right) = Z^2 \frac{mk^2 e^4}{4\pi\hbar^3} \left(\frac{1}{n_2^2} - \frac{1}{n_1^2} \right) \qquad 36\text{-}13$$

If we compare this expression with $Z = 1$ for $f = c/\lambda$ with the empirical Rydberg–Ritz formula (Equation 36-2), we obtain for the Rydberg constant

$$R = \frac{mk^2 e^4}{4\pi c\hbar^3} \qquad 36\text{-}14$$

Using the values of m, e, and $\hbar$ known in 1913, Bohr calculated R and found his result to agree (within the limits of the uncertainties of the constants) with the value obtained from spectroscopy.

STANDING-WAVE CONDITION IMPLIES
QUANTIZATION OF ANGULAR MOMENTUM

EXAMPLE 36-1

For waves in a circle, the standing-wave condition is that there is an integral number of wavelengths in the circumference. That is, $n\lambda = 2\pi r$, where $n = 1, 2, 3$, and so on. Show that this condition for electron waves implies quantization of angular momentum.

1. Write the standing-wave condition:

$$n\lambda = 2\pi r$$

2. Use the de Broglie relation (Equation 34-10) to relate the momentum p to λ:

$$p = \frac{h}{\lambda} = \frac{nh}{2\pi r} = n\frac{\hbar}{r}$$

3. The angular momentum of an electron in a circular orbit is $mvr = pr$, where $p = mv$:

$$L = mvr = pr = \boxed{n\hbar}$$

Energy Levels

The total mechanical energy of the electron in the hydrogen atom is related to the radius of the circular orbit by Equation 36-6. If we substitute the quantized values of r as given by Equation 36-11, we obtain

$$E_n = -\frac{1}{2}\frac{kZe^2}{r} = -\frac{1}{2}\frac{kZ^2e^2}{n^2a_0} = -\frac{1}{2}\frac{mk^2Z^2e^4}{n^2\hbar^2}$$

or

$$E_n = -Z^2\frac{E_0}{n^2} \qquad\qquad 36\text{-}15$$

ENERGY LEVELS IN THE HYDROGEN ATOM

where

$$E_0 = \frac{mk^2e^4}{2\hbar^2} = \frac{1}{2}\frac{ke^2}{a_0} \approx 13.6 \text{ eV} \qquad\qquad 36\text{-}16$$

The energies E_n with $Z = 1$ are the quantized allowed energies for the hydrogen atom.

Transitions between these allowed energies result in the emission or absorption of a photon whose frequency is given by $f = (E_i - E_f)/h$, and whose wavelength is

$$\lambda = \frac{c}{f} = \frac{hc}{E_i - E_f} \qquad\qquad 36\text{-}17$$

As we found in Chapter 34, it is convenient to have the value of hc in electron-volt nanometers:

$$hc = 1240 \text{ eV·nm} \qquad\qquad 36\text{-}18$$

Since the energies are quantized, the frequencies and wavelengths of the radiation emitted by the hydrogen atom are quantized in agreement with the observed line spectrum.

Figure 36-4 shows the energy-level diagram for hydrogen. The energy of the hydrogen atom in the ground state is $E_1 = -13.6$ eV. As n approaches infinity the energy approaches zero. The process of removing an electron from an atom is called ionization, and the energy required to remove the electron is the **ionization energy.** The ionization energy of the ground-state hydrogen atom, which is also its binding energy, is 13.6 eV. A few transitions from a higher state to a lower state are indicated in Figure 36-4. When Bohr published his model of the hydrogen atom, the Balmer series, corresponding to $n_2 = 2$ and $n_1 = 3, 4, 5$, and so on; and the Paschen series, corresponding to $n_2 = 3$ and $n_1 = 4, 5, 6$, and so on, were known. In 1916, T. Lyman found the series corresponding to $n_2 = 1$, and in 1922 and 1924, F. Brackett and H. A. Pfund, respectively, found the series corresponding to $n_2 = 4$ and $n_2 = 5$. Only the Balmer series lies in the visible portion of the electromagnetic spectrum.

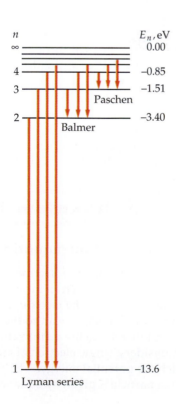

FIGURE 36-4 Energy-level diagram for hydrogen showing the first few transitions in each of the Lyman, Balmer, and Paschen series. The energies of the levels are given by Equation 36-15.

E X A M P L E   3 6 - 2

Find (*a*) the energy and (*b*) the wavelength of the line with the longest wavelength in the Lyman series.

PICTURE THE PROBLEM From Figure 36-4, we can see that the Lyman series corresponds to transitions ending at the ground-state energy, $E_f = E_1 = -13.6$ eV. Since λ varies inversely with energy, the transition with the longest wavelength is the transition with the lowest energy, which is that from the first excited state $n = 2$ to the ground state $n = 1$.

1. The energy of the photon is the difference in the energies of the initial and final atomic state:

$$E_{photon} = \Delta E_{atom} = E_i - E_f$$

$$= E_2 - E_1 = \frac{-13.6 \text{ eV}}{2^2} - \frac{-13.6 \text{ eV}}{1^2}$$

$$= -3.40 \text{ eV} + 13.6 \text{ eV} = \boxed{10.2 \text{ eV}}$$

2. The wavelength of the photon is:

$$\lambda = \frac{hc}{E_2 - E_1} = \frac{1240 \text{ eV·nm}}{10.2 \text{ eV}} = \boxed{121.6 \text{ nm}}$$

REMARKS This photon is outside the visible spectrum, in the ultraviolet region. Since all the other lines in the Lyman series have even greater energies and shorter wavelengths, the Lyman series is completely in the ultraviolet region.

EXERCISE Find the shortest wavelength for a line in the Lyman series. (*Answer* 91.2 nm)

Despite its spectacular successes, the Bohr model of the hydrogen atom had many shortcomings. There was no justification for the postulates of stationary states or for the quantization of angular momentum other than the fact that these postulates led to energy levels that agreed with spectroscopic data. Furthermore, attempts to apply the model to more complicated atoms had little success. The quantum-mechanical theory resolves these difficulties. The stationary states of the Bohr model correspond to the standing-wave solutions of the Schrödinger equation analogous to the standing electron waves for a particle in a box discussed in Chapter 34 and Chapter 35. Energy quantization is a direct consequence of the standing-wave solutions of the Schrödinger equation. For hydrogen, these quantized energies agree with those obtained from the Bohr model and with experiment. The quantization of angular momentum that had to be postulated in the Bohr model is predicted by the quantum theory.

36-3 Quantum Theory of Atoms

The Schrödinger Equation in Spherical Coordinates

In quantum theory, the electron is described by its wave function ψ. The probability of finding the electron in some volume dV of space equals the product of the absolute square of the electron wave function $|\psi|^2$ and dV. Boundary conditions on the wave function lead to the quantization of the wavelengths and frequencies and thereby to the quantization of the electron energy.

Consider a single electron of mass m moving in three dimensions in a region in which the potential energy is U. The time-independent Schrödinger equation for such a particle is given by Equation 35-30:

$$-\frac{\hbar^2}{2m}\left(\frac{\partial^2\psi}{\partial x^2} + \frac{\partial^2\psi}{\partial y^2} + \frac{\partial^2\psi}{\partial z^2}\right) + U\psi = E\psi \qquad\qquad 36\text{-}19$$

For a single isolated atom, the potential energy U depends only on the radial distance $r = \sqrt{x^2 + y^2 + z^2}$. The problem is then most conveniently treated using the spherical coordinates r, θ, and ϕ, which are related to the rectangular coordinates x, y, and z by

$$z = r\cos\theta$$

$$x = r\sin\theta\cos\phi$$

$$y = r\sin\theta\sin\phi \qquad\qquad 36\text{-}20$$

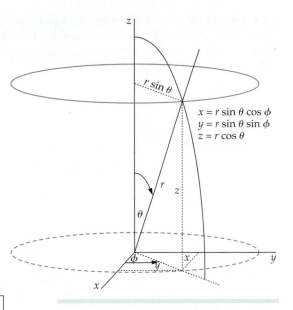

$x = r\sin\theta\cos\phi$
$y = r\sin\theta\sin\phi$
$z = r\cos\theta$

FIGURE 36-5 Geometric relations between spherical coordinates and rectangular coordinates.

These relations are shown in Figure 36-5. The transformation of the bracketed term in Equation 36-19 is straightforward but involves much tedious calculation, which we will omit. The result is

$$\frac{\partial^2\psi}{\partial x^2} + \frac{\partial^2\psi}{\partial y^2} + \frac{\partial^2\psi}{\partial z^2} = \frac{1}{r^2}\frac{\partial}{\partial r}\left(r^2\frac{\partial\psi}{\partial r}\right) + \frac{1}{r^2}\left[\frac{1}{\sin\theta}\frac{\partial}{\partial\theta}\left(\sin\theta\frac{\partial\psi}{\partial\theta}\right) + \frac{1}{\sin^2\theta}\frac{\partial^2\psi}{\partial\phi^2}\right]$$

Substituting into Equation 36-19 gives

$$-\frac{\hbar^2}{2mr^2}\frac{\partial}{\partial r}\left(r^2\frac{\partial\psi}{\partial r}\right) - \frac{\hbar^2}{2mr^2}\left[\frac{1}{\sin\theta}\frac{\partial}{\partial\theta}\left(\sin\theta\frac{\partial\psi}{\partial\theta}\right) + \frac{1}{\sin^2\theta}\frac{\partial^2\psi}{\partial\phi^2}\right] + U(r)\psi = E\psi$$

$$36\text{-}21$$

Despite the formidable appearance of this equation, it was not difficult for Schrödinger to solve because it is similar to other partial differential equations in classical physics that had been thoroughly studied. We will not solve this equation but merely discuss qualitatively some of the interesting features of the wave functions that satisfy it.

The first step in the solution of a partial differential equation, such as Equation 36-21, is to separate the variables by writing the wave function $\psi(r, \theta, \phi)$ as a product of functions of each single variable:

$$\psi(r, \theta, \phi) = R(r)f(\theta)g(\phi) \qquad\qquad 36\text{-}22$$

where R depends only on the radial coordinate r, f depends only on θ, and g depends only on ϕ. When this form of $\psi(r, \theta, \phi)$ is substituted into Equation 36-21, the partial differential equation can be transformed into three ordinary differential equations, one for $R(r)$, one for $f(\theta)$, and one for $g(\phi)$. The potential energy $U(r)$ appears only in the equation for $R(r)$, which is called the **radial equation.** The particular form of $U(r)$ given in Equation 36-19 therefore has no effect on the solutions of the equations for $f(\theta)$ and $g(\phi)$, and therefore has no effect on the angular dependence of the wave function $\psi(r, \theta, \phi)$. These solutions are applicable to any problem in which the potential energy depends only on r.

Quantum Numbers in Spherical Coordinates

In three dimensions, the requirement that the wave function be continuous and normalizable introduces three quantum numbers, one associated with each spatial dimension. In spherical coordinates the quantum number associated with r is labeled n, that associated with θ is labeled ℓ, and that associated with ϕ is labeled m_ℓ.[†] The quantum numbers n_1, n_2, and n_3 that we found in Chapter 35

[†] For simplicity, m_ℓ is sometimes written as m.

for a particle in a three-dimensional square well in rectangular coordinates x, y, and z were independent of one another, but the quantum numbers associated with wave functions in spherical coordinates are interdependent. The possible values of these quantum numbers are

$$n = 1, 2, 3, \ldots$$

$$\ell = 0, 1, 2, 3, \ldots, n - 1$$

$$m_\ell = -\ell, (-\ell + 1), \ldots, -2, -1, 0, 1, 2, \ldots, (\ell + 1), \ell \qquad 36\text{-}23$$

QUANTUM NUMBERS IN SPHERICAL COORDINATES

That is, n can be any positive integer; ℓ can be 0 or any positive integer up to $n - 1$; and m_ℓ can have $2\ell + 1$ possible values, ranging from $-\ell$ to $+\ell$ in integral steps.

The number n is called the **principal quantum number.** It is associated with the dependence of the wave function on the distance r and therefore with the probability of finding the electron at various distances from the nucleus. The quantum numbers ℓ and m_ℓ are associated with the angular momentum of the electron and with the angular dependence of the electron wave function. The quantum number ℓ is called the **orbital quantum number.** The magnitude L of the orbital angular momentum $\vec{L}$ is related to the orbital quantum number ℓ by

$$L = \sqrt{\ell(\ell + 1)}\hbar \qquad 36\text{-}24$$

The quantum number m_ℓ is called the **magnetic quantum number.** It is related to the component of the angular momentum along some direction in space. All spatial directions are equivalent for an isolated atom, but placing the atom in a magnetic field results in the direction of the magnetic field being separated out from the other directions. The convention is that the z direction is chosen for the magnetic-field direction. Then the z component of the angular momentum of the electron is given by the quantum condition

$$L_z = m_\ell \hbar \qquad 36\text{-}25$$

This quantum condition arises from the boundary condition on the azimuth coordinate ϕ that the probability of finding the electron at some arbitrary angle ϕ_1 must be the same as that of finding the electron at angle $\phi_1 + 2\pi$ because these are the same points in space.

If we measure the angular momentum of the electron in units of $\hbar$, we see that the angular-momentum magnitude is quantized to the value $\sqrt{\ell(\ell + 1)}$ units and that its component along any direction can have only the $2\ell + 1$ values ranging from $-\ell$ to $+\ell$ units. Figure 36-6 shows a vector-model diagram illustrating the possible orientations of the angular-momentum vector for $\ell = 2$. Note that only specific values of θ are allowed; that is, the directions in space are quantized.

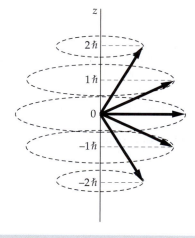

FIGURE 36-6 Vector-model diagram illustrating the possible values of the z component of the angular-momentum vector for the case $\ell = 2$. The magnitude of the angular momentum is
$$L = \hbar\sqrt{\ell(\ell + 1)} = \hbar\sqrt{2(2 + 1)} = \hbar\sqrt{6}.$$

THE DIRECTIONS OF THE ANGULAR MOMENTUM **EXAMPLE 36-3**

If the angular momentum is characterized by the quantum number $\ell = 2$, what are the possible values of L_z, and what is the smallest possible angle between $\vec{L}$ and the z axis?

PICTURE THE PROBLEM The possible orientations of $\vec{L}$ and the z axis are shown in Figure 36-6. The z-axis direction is parallel with that of the external magnetic field in the vicinity of the atom.

1. Write the possible values of L_z:

$$\boxed{L_z = m_\ell \hbar, \text{ where } m_\ell = -2, -1, 0, 1, 2}$$

2. Express the angle θ between $\vec{L}$ and the z axis in terms of L and L_z:

$$\cos \theta = \frac{L_z}{L} = \frac{m\hbar}{\sqrt{\ell(\ell+1)}\hbar} = \frac{m}{\sqrt{\ell(\ell+1)}}$$

3. The smallest angle occurs when $m_\ell = \ell = 2$:

$$\cos \theta_{min} = \frac{2}{\sqrt{2(2+1)}} = \frac{2}{\sqrt{6}} = 0.816$$

$$\theta_{min} = \boxed{35.3°}$$

REMARKS We note the somewhat strange result that the angular-momentum vector cannot lie along the z axis.

EXERCISE An atom in a region with a magnetic field has an angular momentum characterized by the quantum number $\ell = 4$. What are the possible values of m_ℓ? (*Answer* $-4, -3, -2, -1, 0, 1, 2, 3, 4$)

36-4 Quantum Theory of the Hydrogen Atom

We can treat the simplest atom, the hydrogen atom, as a stationary nucleus, a proton, that has a single moving particle, an electron, with kinetic energy $p^2/2m$. The potential energy $U(r)$ due to the electrostatic attraction between the electron and the proton[†] is

$$U(r) = -\frac{kZe^2}{r} \qquad\qquad 36\text{-}26$$

For this potential-energy function, the Schrödinger equation can be solved exactly. In the lowest energy state, which is the ground state, the principal quantum number n has the value 1, ℓ is 0, and m_ℓ is 0.

Energy Levels

The allowed energies of the hydrogen atom that result from the solution of the Schrödinger equation are

$$E_n = -\frac{mk^2e^4}{2\hbar^2 n^2} = -Z^2 \frac{E_0}{n^2}, \quad n = 1, 2, 3, \ldots \qquad\qquad 36\text{-}27$$

ENERGY LEVELS FOR HYDROGEN

where

$$E_0 = -\frac{mk^2e^4}{2\hbar^2} \approx 13.6 \text{ eV} \qquad\qquad 36\text{-}28$$

---

† We include the factor Z, which is 1 for hydrogen, so that we can apply our results to other one-electron atoms, such as singly ionized helium He$^+$, for which Z = 2.

These energies are the same as in the Bohr model. Note that the energy is negative, indicating that the electron is bound to the nucleus (thus the term *bound state*), and that the energy depends only on the principal quantum number n. The fact that the energy does not depend on the orbital quantum number ℓ is a peculiarity of the inverse-square force and holds only for an inverse r potential such as Equation 36-26. For more complicated atoms having several electrons, the interaction of the electrons leads to a dependence of the energy on ℓ. In general, the lower the value of ℓ, the lower the energy for such atoms. Since there is usually no preferred direction in space, the energy for any atom does not ordinarily depend on the magnetic quantum number m_ℓ, which is related to the z component of the angular momentum. The energy does depend on m_ℓ if the atom is in a magnetic field.

Figure 36-7 shows an energy-level diagram for hydrogen. This diagram is similar to Figure 36-4, except that the states with the same value of n but with different values of ℓ are shown separately. These states (called *terms*) are referred to by giving the value of n along with a code letter: s for $\ell = 0$, p for $\ell = 1$, d for $\ell = 2$, and f for $\ell = 3$.[†] (Lowercase letters s, p, d, f, and so on, are used to identify the orbital angular momentum of an individual electron; uppercase letters S, P, D, F, and so on, are used to identify the orbital angular momentum for the entire multielectron atom. For hydrogen, either uppercase or lowercase letters will do, but most people use lowercase as we have done.) When an atom makes a transition from one allowed energy state to another, electromagnetic radiation in the form of a photon is emitted or absorbed. Such transitions result in spectral lines that are characteristic of the atom. The transitions obey the **selection rules:**

$$\Delta m_\ell = 0 \text{ or } \pm 1$$

$$\Delta \ell = \pm 1 \qquad\qquad 36\text{-}29$$

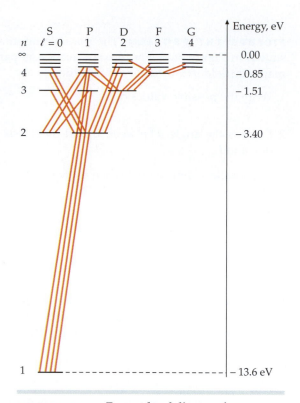

FIGURE 36-7 Energy-level diagram for hydrogen. The diagonal lines show transitions that involve emission or absorption of radiation and obey the selection rule $\Delta\ell = \pm1$. States with the same value of n but with different values of ℓ have the same energy $-E_0/n^2$, where $E_0 = 13.6$ eV as in the Bohr model.

These selection rules are related to the conservation of angular momentum and to the fact that the photon itself has an intrinsic angular momentum that has a maximum component along any axis of $1\hbar$. The wavelengths of the light emitted by hydrogen (and by other atoms) are related to the energy levels by

$$hf = \frac{hc}{\lambda} = E_i - E_f \qquad\qquad 36\text{-}30$$

where E_i and E_f are the energies of the initial and final states.

Wave Functions and Probability Densities

The solutions of the Schrödinger equation in spherical coordinates are characterized by the quantum numbers n, ℓ, and m_ℓ, and are written $\psi_{n\ell m_\ell}$. For any given value of n, there are n possible values of ℓ ($\ell = 0, 1, \ldots, n - 1$), and for each value of ℓ, there are $2\ell + 1$ possible values of m_ℓ. For hydrogen, the energy depends only on n, so there are generally many different wave functions that correspond to the same energy (except at the lowest energy level, for which $n = 1$ and therefore ℓ and m_ℓ must be 0). These energy levels are therefore degenerate (see Section 35-5). The origins of this degeneracy are the $1/r$ dependence of the potential energy and the fact that, in the absence of any external fields, there is no preferred direction in space.[‡]

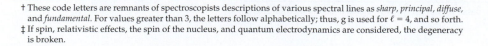

---

† These code letters are remnants of spectroscopists descriptions of various spectral lines as *sharp, principal, diffuse,* and *fundamental.* For values greater than 3, the letters follow alphabetically; thus, g is used for $\ell = 4$, and so forth.

‡ If spin, relativistic effects, the spin of the nucleus, and quantum electrodynamics are considered, the degeneracy is broken.

The Ground State In the lowest energy state, the ground state of hydrogen, the principal quantum number n has the value 1, ℓ is 0, and m_ℓ is 0. The energy is -13.6 eV, and the angular momentum is zero. (In the Bohr model of the atom the angular momentum in the ground state is equal to $\hbar$, not zero.) The wave function for the ground state is

$$\psi_{1,0,0} = C_{1,0,0}e^{-Zr/a_0} \qquad\qquad 36\text{-}31$$

where

$$a_0 = \frac{\hbar^2}{mke^2} = 0.0529 \text{ nm}$$

is the first Bohr radius and $C_{1,0,0}$ is a constant that is determined by normalization. In three dimensions, the normalization condition is

$$\int |\psi|^2\, dV = 1$$

where dV is a volume element and the integration is performed over all space. In spherical coordinates, the volume element (Figure 36-8) is

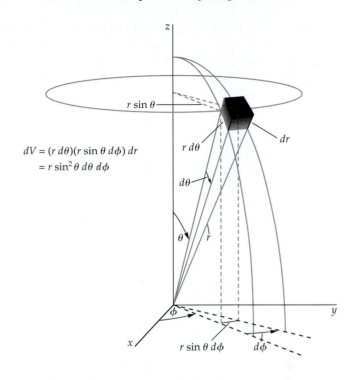

$$dV = (r\, d\theta)(r\sin\theta\, d\phi)\, dr$$
$$= r\sin^2\theta\, d\theta\, d\phi$$

FIGURE 36-8 Volume element in spherical coordinates.

$$dV = (r\sin\theta\, d\phi)(r\, d\theta)\, dr = r^2\sin\theta\, d\theta\, d\phi\, dr$$

We integrate over all space by integrating over ϕ, from $\phi = 0$ to $\phi = 2\pi$, over θ, from $\theta = 0$ to $\theta = \pi$; and over r, from $r = 0$ to $r = \infty$. The normalization condition is thus

$$\int |\psi|^2\, dV = \int_0^\infty \left[\int_0^\pi \left(\int_0^{2\pi} |\psi|^2\, r^2 \sin\theta\, d\phi \right) d\theta \right] dr$$

$$= \int_0^\infty \left[\int_0^\pi \left(\int_0^{2\pi} C_{1,0,0}^2 e^{-2Zr/a_0} r^2 \sin\theta\, d\phi \right) d\theta \right] dr = 1$$

Since there is no θ or ϕ dependence in $\psi_{1,0,0}$, the triple integral can be factored into the product of three integrals. This gives

$$\int |\psi|^2\, dV = \left(\int_0^{2\pi} d\phi \right)\left(\int_0^\pi \sin\theta\, d\theta \right)\left(\int_0^\infty C_{1,0,0}^2 e^{-2Zr/a_0} r^2\, dr \right)$$

$$= 2\pi \cdot 2 \cdot C_{1,0,0}^2 \left(\int_0^\infty r^2 e^{-2Zr/a_0}\, dr \right) = 1$$

The remaining integral is of the form $\int_0^\infty x^n e^{-ax}\, dx$, with n a positive integer and with $a > 0$. Using successive integration-by-parts operations[†] yields the result

$$\int_0^\infty x^n e^{-ax}\, dx = \frac{n!}{a^{n+1}}$$

---

† This integral can also be looked up in a table of integrals.

so

$$\int_0^\infty r^2 e^{-2Zr/a_0}\, dr = \frac{a_0^3}{4Z^3}$$

Then

$$4\pi C_{1,0,0}^2 \left(\frac{a_0^3}{4Z^3}\right) = 1$$

so

$$C_{1,0,0} = \frac{1}{\sqrt{\pi}}\left(\frac{Z}{a_0}\right)^{3/2} \qquad\qquad 36\text{-}32$$

The normalized ground-state wave function is thus

$$\psi_{1,0,0} = \frac{1}{\sqrt{\pi}}\left(\frac{Z}{a_0}\right)^{3/2} e^{-Zr/a_0} \qquad\qquad 36\text{-}33$$

The probability of finding the electron in a volume dV is $|\psi|^2\, dV$. The probability density $|\psi|^2$ is illustrated in Figure 36-9. Note that this probability density is spherically symmetric; that is, the probability density depends only on r, and is independent of θ or ϕ. The probability density is maximum at the origin.

We are more often interested in the probability of finding the electron at some radial distance r between r and $r + dr$. This radial probability $P(r)\, dr$ is the probability density $|\psi|^2$ times the volume of the spherical shell of thickness dr, which is $dV = 4\pi r^2\, dr$. The probability of finding the electron in the range from r to $r + dr$ is thus $P(r)\, dr = |\psi|^2 4\pi r^2\, dr$, and the **radial probability density** is

$$P(r) = 4\pi r^2 |\psi|^2 \qquad\qquad 36\text{-}34$$

RADIAL PROBABILITY DENSITY

For the hydrogen atom in the ground state, the radial probability density is

$$P(r) = 4\pi r^2 |\psi|^2 = 4\pi C_{1,0,0}^2 r^2 e^{-2Zr/a_0} = 4\left(\frac{Z}{a_0}\right)^3 r^2 e^{-2Zr/a_0} \qquad\qquad 36\text{-}35$$

Figure 36-10 shows the radial probability density $P(r)$ as a function of r. The maximum value of $P(r)$ occurs at $r = a_0/Z$, which for $Z = 1$ is the first Bohr radius. In contrast to the Bohr model, in which the electron stays in a well-defined orbit at $r = a_0$, we see that it is possible for the electron to be found at any distance from the nucleus. However, the most probable distance is a_0 (assuming $Z = 1$), and the chance of finding the electron at a much different distance is small. It is often useful to think of the electron in an atom as a charged cloud of charge density $\rho = -e|\psi|^2$, but we should remember that when it interacts with matter, an electron is always observed as a single charge.

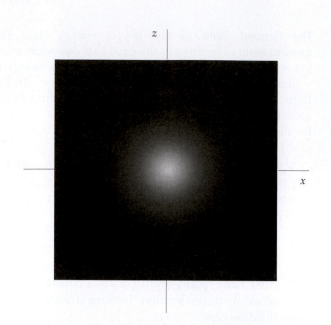

FIGURE 36-9 Computer-generated picture of the probability density $|\psi|^2$ for the ground state of hydrogen. The quantity $-e|\psi|^2$ can be thought of as the electron charge density in the atom. The density is spherically symmetric, is greatest at the origin, and decreases exponentially with r.

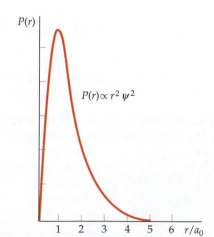

$P(r) \propto r^2 \psi^2$

FIGURE 36-10 Radial probability density $P(r)$ versus r/a_0 for the ground state of the hydrogen atom. $P(r)$ is proportional to $r^2 \psi^2$. The value of r for which $P(r)$ is maximum is the most probable distance $r = a_0$.

PROBABILITY THAT THE ELECTRON IS IN A THIN SPHERICAL SHELL **EXAMPLE 36-4**

Find the probability of finding the electron in a thin spherical shell of radius r and thickness $\Delta r = 0.06a_0$ at (a) $r = a_0$ and (b) $r = 2a_0$ for the ground state of the hydrogen atom.

PICTURE THE PROBLEM Because the range Δr is so small compared to r, the variation in the radial probability density $P(r)$ in the shell can be neglected. The probability of finding the electron in some small range Δr is then $P(r)\,\Delta r$.

1. Use Equation 36-35 with $Z = 1$ and $r = a_0$:

$$P(r)\,\Delta r = \left[4\left(\frac{1}{a_0}\right)^3 r^2 e^{-2r/a_0}\right]\Delta r = \left[4\left(\frac{1}{a_0}\right)^3 a_0^2 e^{-2}\right](0.06a_0) = \boxed{0.0325}$$

2. Use Equation 36-35 with $Z = 1$ and $r = 2a_0$:

$$P(r)\,\Delta r = \left[4\left(\frac{1}{a_0}\right)^3 r^2 e^{-2r/a_0}\right]\Delta r = \left[4\left(\frac{1}{a_0}\right)^3 4a_0^2 e^{-4}\right](0.06a_0) = \boxed{0.0176}$$

REMARKS There is approximately a 3 percent chance of finding the electron in this range at $r = a_0$, but at $r = 2a_0$ the chance is slightly less than 2 percent.

The First Excited State In the first excited state, $n = 2$ and ℓ can be either 0 or 1. For $\ell = 0$, $m_\ell = 0$, and we again have a spherically symmetric wave function, this time given by

$$\psi_{2,0,0} = C_{2,0,0}\left(2 - \frac{Zr}{a_0}\right)e^{-Zr/(2a_0)} \qquad\qquad 36\text{-}36$$

For $\ell = 1$, m_ℓ can be $+1$, 0, or -1. The corresponding wave functions are

$$\psi_{2,1,0} = C_{2,1,0}\frac{Zr}{a_0}e^{-Zr/(2a_0)}\cos\theta \qquad\qquad 36\text{-}37$$

$$\psi_{2,1,\pm1} = C_{2,1,1}\frac{Zr}{a_0}e^{-Zr/(2a_0)}\sin\theta\, e^{\pm i\phi} \qquad\qquad 36\text{-}38$$

where $C_{2,0,0}$, $C_{2,1,0}$, and $C_{2,1,1}$ are normalization constants. The probability densities are given by

$$\psi_{2,0,0}^2 = C_{2,0,0}^2\left(2 - \frac{Zr}{a_0}\right)^2 e^{-Zr/a_0} \qquad\qquad 36\text{-}39$$

$$\psi_{2,1,0}^2 = C_{2,1,0}^2\left(\frac{Zr}{a_0}\right)^2 e^{-Zr/a_0}\cos^2\theta \qquad\qquad 36\text{-}40$$

$$|\psi_{2,1,\pm1}|^2 = C_{2,1,1}^2\left(\frac{Zr}{a_0}\right)^2 e^{-Zr/a_0}\sin^2\theta \qquad\qquad 36\text{-}41$$

The wave functions and probability densities for $\ell \neq 0$ are not spherically symmetric, but instead depend on the angle θ. The probability densities do not depend on ϕ. Figure 36-11 shows the probability density $|\psi|^2$ for $n = 2$, $\ell = 0$, and $m_\ell = 0$ (Figure 36-11a); for $n = 2$, $\ell = 1$, and $m_\ell = 0$ (Figure 36-11b); and for $n = 2$, $\ell = 1$, and $m_\ell = \pm1$ (Figure 36-11c). An important feature of these plots is that the electron cloud is spherically symmetric for $\ell = 0$ and is not spherically symmetric for $\ell \neq 0$. These angular distributions of the electron charge density depend only on the values of ℓ and m_ℓ and not on the radial part of the wave function.

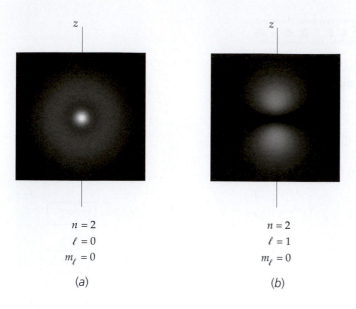

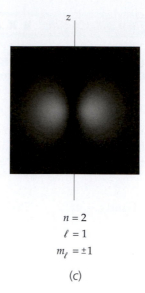

$n = 2$
$\ell = 0$
$m_\ell = 0$

(a)

$n = 2$
$\ell = 1$
$m_\ell = 0$

(b)

$n = 2$
$\ell = 1$
$m_\ell = \pm 1$

(c)

FIGURE 36-11
Computer-generated picture of the probability densities $|\psi|^2$ for the electron in the $n = 2$ states of hydrogen. All three images represent figures of revolution about the z axis. (a) For $\ell = 0$, $|\psi|^2$ is spherically symmetric. (b) For $\ell = 1$ and $m_\ell = 0$, $|\psi|^2$ is proportional to $\cos^2 \theta$. (c) For $\ell = 1$ and $m_\ell = +1$ or -1, $|\psi|^2$ is proportional to $\sin^2 \theta$.

Similar charge distributions for the valence electrons of more complicated atoms play an important role in the chemistry of molecular bonding.

Figure 36-12 shows the probability of finding the electron at a distance r as a function of r for $n = 2$, when $\ell = 1$ and when $\ell = 0$. We can see from the figure that the probability distribution depends on ℓ as well as on n.

For $n = 1$, we found that the most likely distance between the electron and the nucleus is a_0, which is the first Bohr radius, whereas for $n = 2$ and $\ell = 1$, the most likely distance between the electron and the nucleus is $4a_0$. These are the orbital radii for the first and second Bohr orbits (Equation 36-11). For $n = 3$ (and $\ell = 2$),[†] the most likely distance between the electron and nucleus is $9a_0$, which is the radius of the third Bohr orbit.

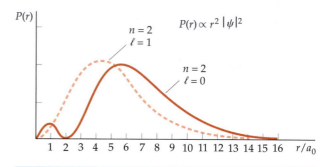

FIGURE 36-12 Radial probability density $P(r)$ versus r/a_0 for the $n = 2$ states of hydrogen. For $\ell = 1$, $P(r)$ is maximum at the Bohr value $r = 2^2 a_0$. For $\ell = 0$, there is a maximum near this value and a much smaller maximum near the origin.

36-5 The Spin-Orbit Effect and Fine Structure

The orbital magnetic moment of an atomic electron can be derived semiclassically, even though it is quantum mechanical in origin.[‡] Consider a particle of mass m and charge q moving with speed v in a circle of radius r. The magnitude of the angular momentum of the particle is $L = mvr$, and the magnitude of the magnetic moment is the product of the current and the area of the circle $\mu = IA = I\pi r^2$. If T is the time for the charge to complete one revolution, the current (charge passing a point per unit time) is q/T. Since the period T is the distance $2\pi r$ divided by the velocity v, the current is $I = q/T = qv/(2\pi r)$. The magnetic moment is then

$$\mu = IA = \frac{qv}{2\pi r} \pi r^2 = \frac{1}{2} qvr = \frac{q}{2m} L$$

where we have substituted L/m for vr. If the charge q is positive, the angular momentum and magnetic moment are in the same direction. We can therefore write

$$\vec{\mu} = \frac{q}{2m} \vec{L}$$

36-42

[†] The correspondence with the Bohr model is closest for the maximum value of ℓ, which is $n - 1$.

[‡] This topic was first presented in Section 27-5.

Equation 36-42 is the general classical relation between magnetic moment and angular momentum. It also holds in the quantum theory of the atom for orbital angular momentum, but not for the intrinsic spin angular momentum of the electron. For electron spin, the magnetic moment is twice that predicted by Equation 36-42.[†] The extra factor of 2 is a result from quantum theory that has no analog in classical mechanics.

The quantum of angular momentum is $\hbar$, so we express the magnetic moment in terms of $\vec{L}/\hbar$:

$$\vec{\mu} = \frac{q\hbar}{2m}\frac{\vec{L}}{\hbar}$$

For an electron, $m = m_e$ and $q = -e$, so the magnetic moment of the electron due to its orbital motion is

$$\vec{\mu}_\ell = -\frac{e\hbar}{2m_e}\frac{\vec{L}}{\hbar} = -\mu_B\frac{\vec{L}}{\hbar}$$

where $\mu_B = e\hbar/(2m_e) = 5.79 \times 10^{-5}$ eV/T is the quantum unit of magnetic moment called a Bohr magneton. The magnetic moment of an electron due to its intrinsic spin angular momentum $\vec{S}$ is

$$\vec{\mu}_S = -2 \times \frac{e\hbar}{2m_e}\frac{\vec{S}}{\hbar} = -2\mu_B\frac{\vec{S}}{\hbar}$$

In general, an electron in an atom has both orbital angular momentum characterized by the quantum number ℓ and spin angular momentum characterized by the quantum number s. Analogous classical systems that have two kinds of angular momentum are the earth, which is spinning about its axis of rotation in addition to revolving about the sun, and a precessing gyroscope that has angular momentum of precession in addition to its spin. The total angular momentum $\vec{J}$ is the sum of the orbital angular momentum $\vec{L}$ and the spin angular momentum $\vec{S}$, where

$$\vec{J} = \vec{L} + \vec{S} \qquad\qquad 36\text{-}43$$

Classically $\vec{J}$ is an important quantity because the resultant torque on a system equals the rate of change of the total angular momentum, and in the case of only central forces, the total angular momentum is conserved. For a classical system, the direction of the total angular momentum $\vec{J}$ is without restrictions and the magnitude of $\vec{J}$ can take on any value between $J_{max} = L + S$ and $J_{min} = |L - S|$. However, in quantum mechanics, the directions of both $\vec{L}$ and $\vec{S}$ are more restricted and the magnitudes L and S are both quantized. Furthermore, like $\vec{L}$ and $\vec{S}$, the direction of the total angular momentum $\vec{J}$ is restricted and the magnitude of $\vec{J}$ is quantized. For an electron with orbital angular momentum characterized by the quantum number ℓ and spin $s = \frac{1}{2}$, the total angular-momentum magnitude J is equal to $\sqrt{j(j + 1)}\hbar$, where the quantum number j is given by

$$j = +\tfrac{1}{2}, \quad \ell = 0$$

[†] This result, and the phenomenon of electron spin itself, was predicted in 1927 by Paul Dirac, who combined special relativity and quantum mechanics into a relativistic wave equation called the Dirac equation. Precise measurements indicate that the magnetic moment of the electron due to its spin is 2.00232 times that predicted by Equation 36-42. The fact that the intrinsic magnetic moment of the electron is approximately twice what we would expect makes it clear that the simple model of the electron as a spinning ball is not to be taken literally.

and either

$$j = \ell + \tfrac{1}{2} \quad \text{or} \quad j = \ell - \tfrac{1}{2}, \quad \ell > 0 \qquad \qquad 36\text{-}44$$

Figure 36-13 is a vector model illustrating the two possible combinations $j = \tfrac{3}{2}$ and $j = \tfrac{1}{2}$ for the case of $\ell = 1$. The lengths of the vectors are proportional to $\sqrt{\ell(\ell + 1)}\hbar$, $\sqrt{s(s + 1)}\hbar$, and $\sqrt{j(j + 1)}\hbar$. The spin angular momentum and the orbital angular momentum are said to be *parallel* when $j = \ell + s$ and *antiparallel* when $j = \ell - s$.

Atomic states with the same n and ℓ values but with different j values have slightly different energies because of the interaction of the spin of the electron with its orbital motion. This effect is called the **spin–orbit effect**. The resulting splitting of spectral lines is called **fine-structure splitting**.

In the notation $n\ell_j$, the ground state of the hydrogen atom is written $1s_{1/2}$, where the 1 indicates that $n = 1$, the s indicates that $\ell = 0$, and the 1/2 indicates that $j = \tfrac{1}{2}$. The $n = 2$ states can have either $\ell = 0$ or $\ell = 1$, and the $\ell = 1$ state can have either $j = \tfrac{3}{2}$ or $j = \tfrac{1}{2}$. These states are thus denoted by $2s_{1/2}$, $2p_{3/2}$, and $2p_{1/2}$. Because of the spin–orbit effect, the $2p_{3/2}$ and $2p_{1/2}$ states have slightly different energies resulting in the fine-structure splitting of the transitions $2p_{3/2} \to 2p_{1/2}$ and $2p_{1/2} \to 2s_{1/2}$.

We can understand the spin–orbit effect qualitatively from a simple Bohr-model picture, as shown in Figure 36-14. In this figure, the electron moves in a circular orbit around a fixed proton. In Figure 36-14a, the orbital angular momentum $\vec{L}$ is up. In an inertial reference frame in which the electron is momentarily at rest (see Figure 36-14b), the proton is moving at right angles to the line connecting the proton and the electron. The moving proton produces a magnetic field $\vec{B}$ at the position of the electron. The direction of $\vec{B}$ is up, parallel to $\vec{L}$. The energy of the electron depends on its spin because of the magnetic moment $\vec{\mu}_s$ associated with the electron's spin. The energy is lowest when $\vec{\mu}_s$ is parallel to $\vec{B}$ and the energy is highest when it is antiparallel. This energy is given by (Equation 36-16)

$$U = -\vec{\mu}_s \cdot \vec{B} = -\mu_{s_z}B \approx -\mu_B B \qquad \qquad 36\text{-}45^\dagger$$

Since $\vec{\mu}_s$ is directed opposite to its spin (because the electron has a negative charge), the energy is lowest when the spin $\vec{S}$ is antiparallel to $\vec{B}$ and thus to $\vec{L}$. The energy of the $2p_{1/2}$ state in hydrogen, in which $\vec{L}$ and $\vec{S}$ are antiparallel (Figure 36-15), is therefore slightly lower than that of the $2p_{3/2}$ state, in which $\vec{L}$ and $\vec{S}$ are parallel.

---

† Transferring the energy of the dipole to the frame of the proton gives a factor of 2, which is included in this result.

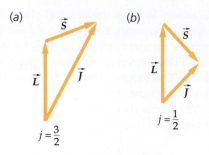

FIGURE 36-13 Vector diagrams illustrating the addition of orbital angular momentum and spin angular momentum for the case $\ell = 1$ and $s = \tfrac{1}{2}$. There are two possible values of the quantum number for the total angular momentum: $j = \ell + s = \tfrac{3}{2}$ and $j = \ell - s = \tfrac{1}{2}$.

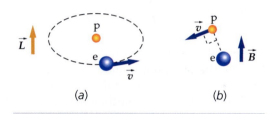

(a) (b)

FIGURE 36-14 (a) An electron moving about a proton in a circular orbit in the horizontal plane with angular momentum $\vec{L}$ up. (b) In an inertial reference frame in which the electron is momentarily at rest there is, at the location of the electron, a magnetic field $\vec{B}$ due to the motion of the proton that is also directed up. When the electron spin $\vec{S}$ is parallel to $\vec{L}$, its magnetic moment $\vec{\mu}_s$ is antiparallel to $\vec{L}$ and $\vec{B}$, so the spin–orbit energy is at its greatest.

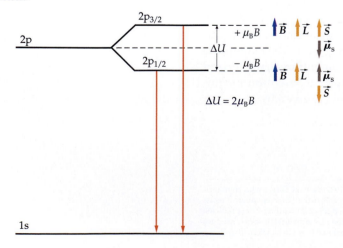

FIGURE 36-15 Fine-structure energy-level diagram. On the left, the levels in the absence of a magnetic field are shown. The effect of the field is shown on the right. Because of the spin–orbit interaction, the magnetic field splits the 2p level into two energy levels, with the $j = \tfrac{3}{2}$ level having slightly greater energy than the $j = \tfrac{1}{2}$ level. The spectral line due to the transition 2p → 1s is therefore split into two lines of slightly different wavelengths.

DETERMINING B BY FINE-STRUCTURE SPLITTING

EXAMPLE 36-5

As a consequence of fine-structure splitting, the energies of the $2p_{3/2}$ and $2p_{1/2}$ levels in hydrogen differ by 4.5×10^{-5} eV. If the 2p electron sees an internal magnetic field of magnitude B, the spin–orbit energy splitting will be of the order of $\Delta E = 2\mu_B B$, where μ_B is the Bohr magneton. From this, estimate the magnetic field that the 2p electron in hydrogen experiences.

1. Write the spin–orbit energy splitting in terms of the magnetic moment:

$$\Delta E = 2\mu_B B = 4.5 \times 10^{-5} \text{ eV}$$

2. Solve for the magnetic field B:

$$B = \frac{4.5 \times 10^{-5} \text{ eV}}{2\mu_B} = \frac{4.5 \times 10^{-5} \text{ eV}}{2(5.79 \times 10^{-5} \text{ eV/T})}$$

$$= \boxed{0.389 \text{ T}}$$

36-6 The Periodic Table

For atoms with more than one electron, the Schrödinger equation cannot be solved exactly. However, powerful approximation methods allow us to determine the energy levels of the atoms and wave functions of the electrons to a high degree of accuracy. As a first approximation, the Z electrons in an atom are assumed to be noninteracting. The Schrödinger equation can then be solved, and the resulting wave functions used to calculate the interaction of the electrons, which in turn can be used to better approximate the wave functions. Because the spin of an electron can have two possible components along an axis, there is an additional quantum number m_s, which can have the possible values $+\frac{1}{2}$ or $-\frac{1}{2}$. The state of each electron is thus described by the four quantum numbers n, ℓ, m, and m_s. The energy of the electron is determined mainly by the principal quantum number n (which is related to the radial dependence of the wave function) and by the orbital angular-momentum quantum number ℓ. Generally, the lower the values of n, the lower the energy; and for a given value of n, the lower the value of ℓ, the lower the energy. The dependence of the energy on ℓ is due to the interaction of the electrons in the atom with each other. In hydrogen, of course, there is only one electron, and the energy is independent of ℓ. The specification of n and ℓ for each electron in an atom is called the **electron configuration.** Customarily, ℓ is specified according to the same code used to label the states of the hydrogen atom rather than by its numerical value. The code is

	s	p	d	f	g	h
ℓ value	0	1	2	3	4	5

The n values are sometimes referred to as shells, which are identified by another letter code: $n = 1$ denotes the K shell;[†] $n = 2$, the L shell; and so on.

The electron configuration of atoms is constrained by the Pauli exclusion principle, which states that no two electrons in an atom can be in the same quantum state; that is, no two electrons can have the same set of values for the quantum numbers n, ℓ, m_ℓ, and m_s. Using the exclusion principle and the restrictions on the quantum numbers discussed in the previous sections (n is a positive integer, ℓ is an integer that ranges from 0 to $n - 1$, m_ℓ can have $2\ell + 1$ values from $-\ell$ to ℓ in integral steps, and m_s can be either $+\frac{1}{2}$ or $-\frac{1}{2}$), we can understand much of the structure of the periodic table.

---

† The designation of the $n = 1$ shell as K is usually found when dealing with X-ray levels where the final shell in an inner electron transition is labeled as K, L, M, and so on.

We have already discussed the lightest element, hydrogen, which has just one electron. In the ground (lowest energy) state, the electron has $n = 1$ and $\ell = 0$, with $m_\ell = 0$ and $m_s = +\frac{1}{2}$ or $-\frac{1}{2}$. We call this a 1s electron. The 1 signifies that $n = 1$, and the s signifies that $\ell = 0$.

As electrons are added to make the heavier atoms, the electrons go into those states that will give the lowest total energy consistent with the Pauli exclusion principle.

Helium ($Z = 2$)

The next element after hydrogen is helium ($Z = 2$), which has two electrons. In the ground state, both electrons are in the K shell with $n = 1$, $\ell = 0$, and $m_\ell = 0$; one electron has $m_s = +\frac{1}{2}$ and the other has $m_s = -\frac{1}{2}$. This configuration is lower in energy than any other two-electron configuration. The resultant spin of the two electrons is zero. Since the orbital angular momentum is also zero, the total angular momentum is zero. The electron configuration for helium is written $1s^2$. The 1 signifies that $n = 1$, the s signifies that $\ell = 0$, and the superscript 2 signifies that there are two electrons in this state. Since ℓ can be only 0 for $n = 1$, these two electrons fill the K ($n = 1$) shell. The energy required to remove the most loosely bound electron from an atom in the ground state is called the **ionization energy.** This energy is the binding energy of the last electron placed in the atom. For helium, the ionization energy is 24.6 eV, which is relatively large. Helium is therefore basically inert.

ELECTRON INTERACTION ENERGY IN HELIUM   **EXAMPLE 36-6**

(*a*) **Use the measured ionization energy to calculate the energy of interaction of the two electrons in the ground state of the helium atom. (*b*) Use your result to estimate the average separation of the two electrons.**

PICTURE THE PROBLEM The energy of one electron in the ground state of helium is E_1 (which is negative) given by Equation 36-27, with $n = 1$ and $Z = 2$. If the electrons did not interact, the energy of the second electron would also be E_1, the same as that of the first electron. Thus, for an atom with noninteracting electrons, the ionization energy would be $|E_1|$ and the ground-state energy would be $E_{non} = 2E_1$. This is represented by the lowest level in Figure 36-16. Because of the interaction energy, the ground-state energy is greater than $2E_1$. This is represented by the higher level labeled E_g in the figure. When we add $E_{ion} = 24.6$ eV to ionize He, we obtain ionized helium, written He$^+$, which has just one electron and therefore energy E_1.

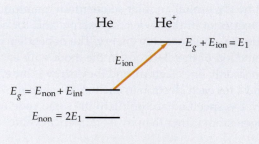

FIGURE 36-16

(*a*) 1. The energy of interaction plus the energy of two noninteracting electrons equals the ground-state energy of helium:

$$E_{int} + E_{non} = E_g$$

2. Solve for E_{int} and substitute $E_{non} = 2E_1$:

$$E_{int} = E_g - E_{non} = E_g - 2E_1$$

3. Use Equation 36-27 to calculate the energy E_1 of one electron in the ground state:

$$E_n = -Z^2 \frac{E_0}{n^2}$$

so

$$E_1 = -(2)^2 \frac{13.6 \text{ eV}}{1^2} = -54.4 \text{ eV}$$

4. Substitute this value for E_1:

$$E_{int} = E_g - 2E_1 = E_g - 2(-54.4 \text{ eV})$$
$$= E_g + 108.8 \text{ eV}$$

5. The ground-state energy of He, E_g, plus the ionization energy equals the ground-state energy of He$^+$, which is E_1:

$$E_g + E_{ion} = E_1 = -54.4 \text{ eV}$$

6. Substitute $E_{ion} = 24.6 \text{ eV}$ to calculate E_g:

$$E_g = E_1 - E_{ion} = -54.4 \text{ eV} - 24.6 \text{ eV}$$
$$= -79 \text{ eV}$$

7. Substitute this result for E_g to obtain E_{int}:

$$E_{int} = E_g + 108.8 \text{ eV} = -79 \text{ eV} + 108.8 \text{ eV}$$
$$= \boxed{29.8 \text{ eV}}$$

(b) 1. The energy of interaction of two electrons separated by distance r_s apart is the potential energy:

$$U = +\frac{ke^2}{r_s}$$

2. Set U equal to 29.8 eV, and solve for r. It is convenient to express r in terms of a_0, the radius of the first Bohr orbit in hydrogen, and to use Equation 36-16:

$$r_s = \frac{ke^2}{U} = \frac{ke^2}{a_0}\frac{a_0}{U} = 2\frac{ke^2}{2a_0}\frac{a_0}{U} = 2\frac{E_0}{U}a_0$$
$$= 2\frac{13.6 \text{ eV}}{29.8 \text{ eV}}a_0 = \boxed{0.913 a_0}$$

PLAUSIBILITY CHECK This separation is approximately the size of the diameter d_1 of the first Bohr orbit for an electron in helium, which is $d_1 = 2r_1 = 2a_0/Z = a_0$.

Lithium (Z = 3)

The next element, lithium, has three electrons. Since the K shell ($n = 1$) is completely filled with two electrons, the third electron must go into a higher energy shell. The next lowest energy shell after $n = 1$ is the $n = 2$ or L shell. The outer electron is much farther from the nucleus than are the two inner $n = 1$ electrons. It is most likely to be found at a radius near that of the second Bohr orbit, which is four times the radius of the first Bohr orbit.

The nuclear charge is partially screened from the outer electron by the two inner electrons. Recall that the electric field outside a spherically symmetric charge density is the same as if all the charge were at the center of the sphere. If the outer electron were completely outside the charge cloud of the two inner electrons, the electric field the outer electron would see would be that of a single charge $+e$ at the center due to the nuclear charge of $+3e$ and the charge $-2e$ of the inner electron cloud. However, the outer electron does not have a well-defined orbit; instead, it is itself a charge cloud that penetrates the charge cloud of the inner electrons to some extent. Because of this penetration, the effective nuclear charge $Z'e$ is somewhat greater than $+1e$. The energy of the outer electron at a distance r from a point charge $+Z'e$ is given by Equation 36-6, with the nuclear charge $+Z$ replaced by $+Z'$.

$$E = -\frac{1}{2}\frac{kZ'e^2}{r} \qquad\qquad 36\text{-}46$$

The greater the penetration of the inner electron cloud, the greater the effective nuclear charge $Z'e$ and the lower the energy. Because the penetration is greater for ℓ values closer to zero (see Figure 36-12), the energy of the outer electron in lithium is lower for the s state ($\ell = 0$) than for the p state ($\ell = 1$). The electron configuration of lithium in the ground state is therefore $1s^2 2s$. The ionization energy of lithium is only 5.39 eV. Because its outer electron is so loosely bound to the atom, lithium is very active chemically. It behaves like a one-electron atom, similar to hydrogen.

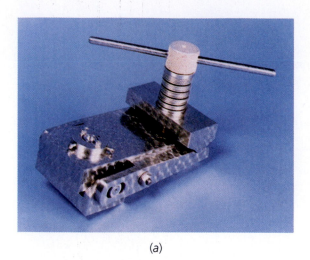

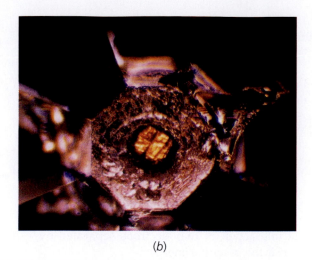

(a)

(b)

(a) A diamond anvil cell, in which the facets of two diamonds (approximately 1 mm² each) are used to compress a sample substance, subjecting it to very high pressure. (b) Samarium monosulfide (SmS) is normally a black, dull-looking semiconductor. When it is subjected to pressure above 7000 atm, an electron from the 4f state is dislocated into the 5d state. The resulting compound glitters like gold and behaves like a metal.

EFFECTIVE NUCLEAR CHARGE FOR AN OUTER ELECTRON   **E X A M P L E   3 6 - 7**

Suppose the electron cloud of the outer electron in the lithium atom in the ground state were completely outside the electron clouds of the two inner electrons, the nuclear charge would be shielded by the two inner electrons and the effective nuclear charge would be $Z'e = 1e$. Then the energy of the outer electron would be $-(13.6 \text{ eV})/2^2 = -3.4 \text{ eV}$. However, the ionization energy of lithium is 5.39 eV, not 3.4 eV. Use this fact to calculate the effective nuclear charge Z' seen by the outer electron in lithium.

PICTURE THE PROBLEM Because the outer electron is in the $n = 2$ shell, we will take $r = 4a_0$ for its average distance from the nucleus. We can then calculate Z' from Equation 36-46. Since r is given in terms of a_0, it will be convenient to use the fact that $E_0 = ke^2/(2a_0) = 13.6 \text{ eV}$ (Equation 36-16).

1. Equation 36-46 relates the energy of the outer electron to its average distance r and the effective nuclear charge Z':

$$E = \frac{1}{2}\frac{kZ'e^2}{r}$$

2. Substitute the given values $r = 4a_0$ and $E = -5.39 \text{ eV}$:

$$-5.39 \text{ eV} = -\frac{1}{2}\frac{kZ'e^2}{4a_0}$$

3. Use $ke^2/(2a_0) = E_0 = 13.6 \text{ eV}$ and solve for Z':

$$-5.39 \text{ eV} = -\frac{Z'}{4}\frac{ke^2}{2a_0} = -\frac{Z'}{4}(13.6 \text{ eV})$$

so

$$Z' = 4\frac{5.39 \text{ eV}}{13.6 \text{ eV}} = \boxed{1.59}$$

REMARKS This calculation is interesting but not very rigorous. We essentially used the radius ($r = 4a_0$) for the circular orbit from the semiclassical Bohr model and the measured ionization energy to calculate the effective inner charge seen by the outer electron. We know, of course, that this outer electron does not move in a circular orbit of constant radius, but is better represented by a stationary charged cloud of charge density $|\psi|^2$ that penetrates the charged clouds of the inner electrons.

Beryllium (Z = 4)

The energy of the beryllium atom is a minimum if both outer electrons are in the 2s state. There can be two electrons with $n = 2$, $\ell = 0$, and $m_\ell = 0$ because of the two possible values for the spin quantum number m_s. The configuration of beryllium is thus $1s^2 2s^2$.

Boron to Neon (Z = 5 to Z = 10)

Since the 2s subshell is filled, the fifth electron must go into the next available (lowest energy) subshell, which is the 2p subshell, with $n = 2$ and $\ell = 1$. Since there are three possible values of m ($+1, 0,$ and -1) and two values of m_s for each value of m_ℓ, there can be six electrons in this subshell. The electron configuration for boron is $1s^2 2s^2 2p$. The electron configurations for the elements carbon ($Z = 6$) to neon ($Z = 10$) differ from that for boron only in the number of electrons in the 2p subshell. The ionization energy increases with Z for these elements, reaching the value of 21.6 eV for the last element in the group, neon. Neon has the maximum number of electrons allowed in the $n = 2$ shell. The electron configuration for neon is $1s^2 2s^2 2p^6$. Because of its very high ionization energy, neon, like helium, basically is chemically inert. The element just before neon, fluorine, has a hole in the 2p subshell; that is, it has room for one more electron. It readily combines with elements such as lithium that have one outer electron. Lithium, for example, will donate its single outer electron to the fluorine atom to make an F^- ion and a Li^+ ion. These ions then bond together to form a molecule of lithium fluoride.

Sodium to Argon (Z = 11 to Z = 18)

The eleventh electron must go into the $n = 3$ shell. Since this electron is very far from the nucleus and from the inner electrons, it is weakly bound in the sodium ($Z = 11$) atom. The ionization energy of sodium is only 5.14 eV. Sodium therefore combines readily with atoms such as fluorine. With $n = 3$, the value of ℓ can be 0, 1, or 2. Because of the lowering of the energy due to penetration of the electron shield formed by the other ten electrons (similar to that discussed for lithium) the 3s state is lower than the 3p or 3d states. This energy difference between subshells of the same n value becomes greater as the number of electrons increases. The electron configuration of sodium is $1s^2 2s^2 2p^6 3s^1$. As we move to elements with higher values of Z, the 3s subshell and then the 3p subshell fill. These two subshells can accommodate $2 + 6 = 8$ electrons. The configuration of argon ($Z = 18$) is $1s^2 2s^2 2p^6 3s^2 3p^6$. One might expect the nineteenth electron to go into the third subshell (the d subshell with $\ell = 2$), but the penetration effect is now so strong that the energy of the next electron is lower in the 4s subshell than in the 3d subshell. There is thus another large energy difference between the eighteenth and nineteenth electrons, and so argon, with its full 3p subshell, is basically stable and inert.

Elements With Z > 18

The nineteenth electron in potassium ($Z = 19$) and the twentieth electron in calcium ($Z = 20$) go into the 4s subshell rather than the 3d subshell. The electron configurations of the next ten elements, scandium ($Z = 21$) through zinc ($Z = 30$),

Hydrogen

Carbon

Silicon

Iron

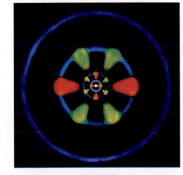

Silver

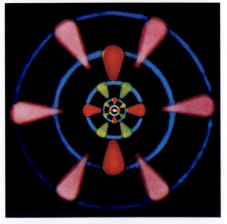

Europium

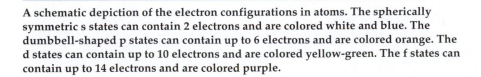

A schematic depiction of the electron configurations in atoms. The spherically symmetric s states can contain 2 electrons and are colored white and blue. The dumbbell-shaped p states can contain up to 6 electrons and are colored orange. The d states can contain up to 10 electrons and are colored yellow-green. The f states can contain up to 14 electrons and are colored purple.

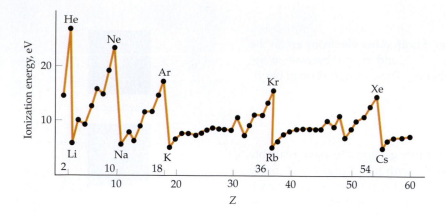

FIGURE 36-17 Ionization energy versus Z for $Z = 1$ to $Z = 60$. This energy is the binding energy of the last electron in the atom. The binding energy increases with Z until a shell is closed at $Z = 2, 10, 18, 36$, and 54. Elements with a closed shell plus one outer electron, such as sodium ($Z = 11$), have very low binding energies because the outer electron is very far from the nucleus and is shielded by the inner core electrons.

differ only in the number of electrons in the 3d shell, except for chromium ($Z = 24$) and copper ($Z = 29$), each of which has only one 4s electron. These ten elements are called **transition elements.**

Figure 36-17 shows a plot of the ionization energy versus Z for $Z = 1$ to $Z = 60$. The peaks in ionization energy at $Z = 2, 10, 18, 36$, and 54 mark the closing of a shell or subshell. Table 36-1 gives the ground-state electron configurations of the elements up to atomic number 109.

TABLE 36-1

Electron Configurations of the Atoms in Their Ground States

For some of the rare-earth elements ($Z = 57$ to 71) and the heavy elements ($Z > 89$) the configurations are not firmly established.

		Shell (n):	K(1)	L(2)		M(3)			N(4)				O(5)				P(6)			Q(7)
			s	s	p	s	p	d	s	p	d	f	s	p	d	f	s	p	d	s
Z	Element	Subshell (ℓ):	(0)	(0)	(1)	(0)	(1)	(2)	(0)	(1)	(2)	(3)	(0)	(1)	(2)	(3)	(0)	(1)	(2)	(1)
1	H hydrogen		1																	
2	He helium		2																	
3	Li lithium		2	1																
4	Be beryllium		2	2																
5	B boron		2	2	1															
6	C carbon		2	2	2															
7	N nitrogen		2	2	3															
8	O oxygen		2	2	4															
9	F fluorine		2	2	5															
10	Ne neon		2	2	6															
11	Na sodium		2	2	6	1														
12	Mg magnesium		2	2	6	2														
13	Al aluminum		2	2	6	2	1													
14	Si silicon		2	2	6	2	2													
15	P phosphorus		2	2	6	2	3													
16	S sulfur		2	2	6	2	4													
17	Cl chlorine		2	2	6	2	5													
18	Ar argon		2	2	6	2	6													
19	K potassium		2	2	6	2	6	.	1											

TABLE 36-1 (continued)

Electron Configurations of the Atoms in Their Ground States

For some of the rare-earth elements ($Z = 57$ to 71) and the heavy elements ($Z > 89$) the configurations are not firmly established.

		Shell (n):	K (1)	L (2)		M (3)			N (4)				O (5)				P (6)			Q (7)
		Subshell (ℓ):	s (0)	s (0)	p (1)	s (0)	p (1)	d (2)	s (0)	p (1)	d (2)	f (3)	s (0)	p (1)	d (2)	f (3)	s (0)	p (1)	d (2)	s (1)
Z	Element																			
20	Ca calcium		2	2	6	2	6	.	2											
21	Sc scandium		2	2	6	2	6	1	2											
22	Ti titanium		2	2	6	2	6	2	2											
23	V vanadium		2	2	6	2	6	3	2											
24	Cr chromium		2	2	6	2	6	5	1											
25	Mn manganese		2	2	6	2	6	5	2											
26	Fe iron		2	2	6	2	6	6	2											
27	Co cobalt		2	2	6	2	6	7	2											
28	Ni nickel		2	2	6	2	6	8	2											
29	Cu copper		2	2	6	2	6	10	1											
30	Zn zinc		2	2	6	2	6	10	2											
31	Ga gallium		2	2	6	2	6	10	2	1										
32	Ge germanium		2	2	6	2	6	10	2	2										
33	As arsenic		2	2	6	2	6	10	2	3										
34	Se selenium		2	2	6	2	6	10	2	4										
35	Br bromine		2	2	6	2	6	10	2	5										
36	Kr krypton		2	2	6	2	6	10	2	6										
37	Rb rubidium		2	2	6	2	6	10	2	6	.	.	1							
38	Sr strontium		2	2	6	2	6	10	2	6	.	.	2							
39	Y yttrium		2	2	6	2	6	10	2	6	1	.	2							
40	Zr zirconium		2	2	6	2	6	10	2	6	2	.	2							
41	Nb niobium		2	2	6	2	6	10	2	6	4	.	1							
42	Mo molybdenum		2	2	6	2	6	10	2	6	5	.	1							
43	Tc technetium		2	2	6	2	6	10	2	6	6	.	1							
44	Ru ruthenium		2	2	6	2	6	10	2	6	7	.	1							
45	Rh rhodium		2	2	6	2	6	10	2	6	8	.	1							
46	Pd palladium		2	2	6	2	6	10	2	6	10	.	.							
47	Ag silver		2	2	6	2	6	10	2	6	10	.	1							
48	Cd cadmium		2	2	6	2	6	10	2	6	10	.	2							
49	In indium		2	2	6	2	6	10	2	6	10	.	2	1						
50	Sn tin		2	2	6	2	6	10	2	6	10	.	2	2						
51	Sb antimony		2	2	6	2	6	10	2	6	10	.	2	3						
52	Te tellurium		2	2	6	2	6	10	2	6	10	.	2	4						
53	I iodine		2	2	6	2	6	10	2	6	10	.	2	5						
54	Xe xenon		2	2	6	2	6	10	2	6	10	.	2	6						
55	Cs cesium		2	2	6	2	6	10	2	6	10	.	2	6	.	.	1			
56	Ba barium		2	2	6	2	6	10	2	6	10	.	2	6	.	.	2			
57	La lanthanum		2	2	6	2	6	10	2	6	10	.	2	6	1	.	2			

TABLE 36-1 (continued)

Electron Configurations of the Atoms in Their Ground States

For some of the rare-earth elements ($Z = 57$ to 71) and the heavy elements ($Z > 89$) the configurations are not firmly established.

			Shell (n):	K (1)	L (2)		M (3)			N (4)				O (5)				P (6)			Q (7)
				s	s	p	s	p	d	s	p	d	f	s	p	d	f	s	p	d	s
Z	Element		Subshell (ℓ):	(0)	(0)	(1)	(0)	(1)	(2)	(0)	(1)	(2)	(3)	(0)	(1)	(2)	(3)	(0)	(1)	(2)	(1)
58	Ce	cerium		2	2	6	2	6	10	2	6	10	1	2	6	1	.	2			
59	Pr	praseodymium		2	2	6	2	6	10	2	6	10	3	2	6	.	.	2			
60	Nd	neodymium		2	2	6	2	6	10	2	6	10	4	2	6	.	.	2			
61	Pm	promethium		2	2	6	2	6	10	2	6	10	5	2	6	.	.	2			
62	Sm	samarium		2	2	6	2	6	10	2	6	10	6	2	6	.	.	2			
63	Eu	europium		2	2	6	2	6	10	2	6	10	7	2	6	.	.	2			
64	Gd	gadolinium		2	2	6	2	6	10	2	6	10	7	2	6	1	.	2			
65	Tb	terbium		2	2	6	2	6	10	2	6	10	9	2	6	.	.	2			
66	Dy	dysprosium		2	2	6	2	6	10	2	6	10	10	2	6	.	.	2			
67	Ho	holmium		2	2	6	2	6	10	2	6	10	11	2	6	.	.	2			
68	Er	erbium		2	2	6	2	6	10	2	6	10	12	2	6	.	.	2			
69	Tm	thulium		2	2	6	2	6	10	2	6	10	13	2	6	.	.	2			
70	Yb	ytterbium		2	2	6	2	6	10	2	6	10	14	2	6	.	.	2			
71	Lu	lutetium		2	2	6	2	6	10	2	6	10	14	2	6	1	.	2			
72	Hf	hafnium		2	2	6	2	6	10	2	6	10	14	2	6	2	.	2			
73	Ta	tantalum		2	2	6	2	6	10	2	6	10	14	2	6	3	.	2			
74	W	tungsten (wolfram)		2	2	6	2	6	10	2	6	10	14	2	6	4	.	2			
75	Re	rhenium		2	2	6	2	6	10	2	6	10	14	2	6	5	.	2			
76	Os	osmium		2	2	6	2	6	10	2	6	10	14	2	6	6	.	2			
77	Ir	iridium		2	2	6	2	6	10	2	6	10	14	2	6	7	.	2			
78	Pt	platinum		2	2	6	2	6	10	2	6	10	14	2	6	9	.	1			
79	Au	gold		2	2	6	2	6	10	2	6	10	14	2	6	10	.	1			
80	Hg	mercury		2	2	6	2	6	10	2	6	10	14	2	6	10	.	2			
81	Tl	thallium		2	2	6	2	6	10	2	6	10	14	2	6	10	.	2	1		
82	Pb	lead		2	2	6	2	6	10	2	6	10	14	2	6	10	.	2	2		
83	Bi	bismuth		2	2	6	2	6	10	2	6	10	14	2	6	10	.	2	3		
84	Po	polonium		2	2	6	2	6	10	2	6	10	14	2	6	10	.	2	4		
85	At	astatine		2	2	6	2	6	10	2	6	10	14	2	6	10	.	2	5		
86	Rn	radon		2	2	6	2	6	10	2	6	10	14	2	6	10	.	2	6		
87	Fr	francium		2	2	6	2	6	10	2	6	10	14	2	6	10	.	2	6	.	1
88	Ra	radium		2	2	6	2	6	10	2	6	10	14	2	6	10	.	2	6	.	2
89	Ac	actinium		2	2	6	2	6	10	2	6	10	14	2	6	10	.	2	6	1	2
90	Th	thorium		2	2	6	2	6	10	2	6	10	14	2	6	10	.	2	6	2	2
91	Pa	protactinium		2	2	6	2	6	10	2	6	10	14	2	6	10	1	2	6	2	2
92	U	uranium		2	2	6	2	6	10	2	6	10	14	2	6	10	3	2	6	1	2
93	Np	neptunium		2	2	6	2	6	10	2	6	10	14	2	6	10	4	2	6	1	2
94	Pu	plutonium		2	2	6	2	6	10	2	6	10	14	2	6	10	6	2	6	.	2
95	Am	americium		2	2	6	2	6	10	2	6	10	14	2	6	10	7	2	6	.	2

TABLE 36-1 (continued)

Electron Configurations of the Atoms in Their Ground States
For some of the rare-earth elements (Z = 57 to 71) and the heavy elements (Z > 89) the configurations are not firmly established.

Shell (n):		K (1)	L (2)		M (3)			N (4)				O (5)				P (6)			Q (7)
		s	s	p	s	p	d	s	p	d	f	s	p	d	f	s	p	d	s
Z	Element	Subshell (ℓ): (0)	(0)	(1)	(0)	(1)	(2)	(0)	(1)	(2)	(3)	(0)	(1)	(2)	(3)	(0)	(1)	(2)	(1)
96	Cm curium	2	2	6	2	6	10	2	6	10	14	2	6	10	7	2	6	1	2
97	Bk berkelium	2	2	6	2	6	10	2	6	10	14	2	6	10	8	2	6	1	2
98	Cf californium	2	2	6	2	6	10	2	6	10	14	2	6	10	10	2	6	.	2
99	Es einsteinium	2	2	6	2	6	10	2	6	10	14	2	6	10	11	2	6	.	2
100	Fm fermium	2	2	6	2	6	10	2	6	10	14	2	6	10	12	2	6	.	2
101	Md mendelevium	2	2	6	2	6	10	2	6	10	14	2	6	10	13	2	6	.	2
102	No nobelium	2	2	6	2	6	10	2	6	10	14	2	6	10	14	2	6	.	2
103	Lr lawrencium	2	2	6	2	6	10	2	6	10	14	2	6	10	14	2	6	1	2
104	Rf rutherfordium	2	2	6	2	6	10	2	6	10	14	2	6	10	14	2	6	2	2
105	Db dubnium	2	2	6	2	6	10	2	6	10	14	2	6	10	14	2	6	3	2
106	Sg seaborgium	2	2	6	2	6	10	2	6	10	14	2	6	10	14	2	6	4	2
107	Bh bohrium	2	2	6	2	6	10	2	6	10	14	2	6	10	14	2	6	5	2
108	Hs hassium	2	2	6	2	6	10	2	6	10	14	2	6	10	14	2	6	6	2
109	Mt meitnerium	2	2	6	2	6	10	2	6	10	14	2	6	10	14	2	6	7	2

36-7 Optical Spectra and X-Ray Spectra

When an atom is in an excited state (i.e., when it is in an energy state above the ground state), it makes transitions to lower energy states, and in doing so emits electromagnetic radiation. The wavelength of the electromagnetic radiation emitted is related to the initial and final states by the Bohr formula (Equation 36-17), $\lambda = hc/(E_i - E_f)$, where E_i and E_f are the initial and final energies and h is Planck's constant. The atom can be excited to a higher energy state by bombarding the atom with a beam of electrons, as in a spectral tube with a high voltage across it. Since the excited energy states of an atom form a discrete (rather than continuous) set, only certain wavelengths are emitted. These wavelengths of the emitted radiation constitute the emission spectrum of the atom.

Optical Spectra

To understand atomic spectra we need to understand the excited states of the atom. The situation for an atom with many electrons is, in general, much more complicated than that of hydrogen with just one electron. An excited state of the atom may involve a change in the state of any one of the electrons, or even two or more electrons. Fortunately, in most cases, an excited state of an atom involves the excitation of just one of the electrons in the atom. The energies of excitation of the outer, valence electrons of an atom are of the order of a few electron volts. Transitions involving these electrons result in photons in or near the visible or **optical spectrum.** (Recall that the energies of visible photons range from approximately 1.5 eV to 3 eV.) The excitation energies can often be calculated from a simple model in which the atom is pictured as a single electron plus a stable core consisting of the nucleus plus the other inner electrons. This model works particularly well for the alkali metals: Li, Na, K, Rb, and Cs. These elements are in the first column of the periodic table. The optical spectra of these elements are similar to the optical spectra of hydrogen.

Figure 36-18 shows an energy-level diagram for the optical transitions in sodium, whose electrons form a neon core plus one outer electron. Since the total spin angular momentum of the core adds up to zero, the spin of each state of sodium is $\frac{1}{2}$. Because of the spin–orbit effect, the states with $J = L - \frac{1}{2}$ have a slightly different energy than those with $J = L + \frac{1}{2}$ (except for states with $L = 0$). Each state (except for the $L = 0$ states) is therefore split into two states, called a doublet. The doublet splitting is very small and not evident on the energy scale of this diagram. The usual spectroscopic notation is that these states are labeled with a superscript given by $2S + 1$, followed by a letter denoting the orbital angular momentum, followed by a subscript denoting the total angular momentum J. For states with a total spin angular momentum $S = \frac{1}{2}$ the superscript is 2, indicating the state is a doublet. Thus $^2P_{3/2}$, read as "doublet P three halves," denotes a state in which $L = 1$ and $J = \frac{3}{2}$. (The $L = 0$, or S, states are customarily labeled as if they were doublets even though they are not.) In the first excited state, the outer electron is excited from the 3s level to the 3p level, which is approximately 2.1 eV above the ground state. The energy difference between the $P_{3/2}$ and $P_{1/2}$

A neon sign outside a Chinatown restaurant in Paris. Neon atoms in the tube are excited by an electron current passing through the tube. The excited neon atoms emit light in the visible range as they decay toward their ground states. The colors of neon signs result from the characteristic red-orange spectrum of neon plus the color of the glass tube itself.

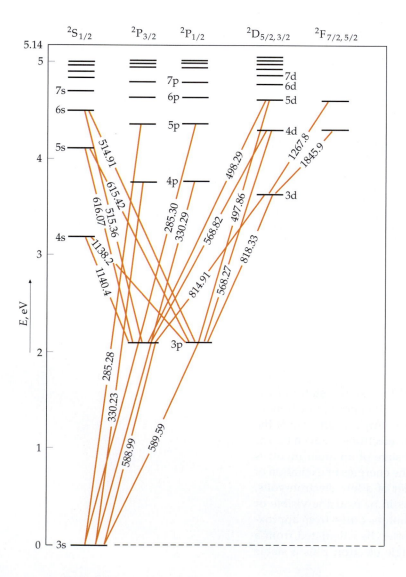

FIGURE 36-18 Energy-level diagram for sodium. The diagonal lines show observed optical transitions, with wavelengths given in nanometers. The energy of the ground state has been chosen as the zero point for the scale on the left.

states due to the spin–orbit effect is about 0.002 eV. Transitions from these states to the ground state give the familiar sodium yellow doublet:

$$3p(^2P_{1/2}) \rightarrow 3s(^2S_{1/2}), \quad \lambda = 589.6 \text{ nm}$$

$$3p(^2P_{3/2}) \rightarrow 3s(^2S_{1/2}), \quad \lambda = 589.0 \text{ nm}$$

The energy levels and spectra of other alkali metal atoms are similar to those for sodium. The optical spectrum for atoms such as helium, beryllium, and magnesium that have two outer electrons is considerably more complex because of the interaction of the two outer electrons.

X-Ray Spectra

X rays are usually produced in the laboratory by bombarding a target element with a high-energy beam of electrons in an X-ray tube. The result (Figure 36-19) consists of a continuous spectrum that depends only on the energy of the bombarding electrons and a line spectrum that is characteristic of the target element. The characteristic spectrum results from excitation of the inner core electrons in the target element.

The energy needed to excite an inner core electron—for example, an electron in the $n = 1$ state (K shell)—is much greater than the energy required to excite an outer, valence electron. An inner electron cannot be excited to any of the filled states (e.g., the $n = 2$ states in sodium) because of the exclusion principle. The energy required to excite an inner core electron to an unoccupied state is typically of the order of several kilo-electron volts. If an electron is knocked out of the $n = 1$ state (K shell), there is a vacancy left in this shell. This vacancy can be filled if an electron in the L shell (or in a higher shell) makes a transition into the K shell. The photons emitted by electrons making such transitions also have energies of the order of kilo-electron volts and produce the sharp peaks in the X-ray spectrum, as shown in Figure 36-19. The K_α line arises from transitions from the $n = 2$ (L) shell to the $n = 1$ (K) shell. The K_β line arises from transitions from the $n = 3$ shell to the $n = 1$ shell. These and other lines arising from transitions ending at the $n = 1$ shell make up the K series of the characteristic X-ray spectrum of the target element. Similarly, a second series, the L series, is produced by transitions from higher energy states to a vacated place in the $n = 2$ (L) shell. The letters K, L, M, and so on, designate the final shell of the electron making the transition and the series α, β, and so on, designates the number of shells above the final shell for the initial state of the electron.

In 1913, the English physicist Henry Moseley measured the wavelengths of the characteristic K_α X-ray spectra for approximately forty elements. Using this data, Moseley showed that a plot of $\lambda^{-1/2}$ versus the order in which the elements appeared in the periodic table resulted in a straight line (with a few gaps and a few outliers). From his data, Moseley was able to accurately determine the atomic number Z for each known element, and to predict the existence of some elements that were later discovered. The equation of the straight line of his plot is given by

$$\frac{1}{\sqrt{\lambda_{K_\alpha}}} = a(Z - 1)$$

The work of Bohr and Moseley can be combined to obtain an equation relating the wavelength of the emitted photon and the atomic number. According to the Bohr model of a single-electron atom (see Equation 36-13), the wavelength of the emitted photon when the electron makes the transition from $n = 2$ to $n = 1$ is given by

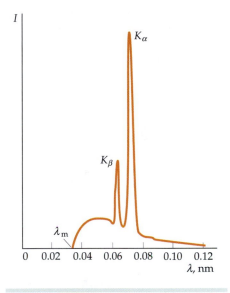

FIGURE 36-19 X-ray spectrum of molybdenum. The sharp peaks labeled K_α and K_β are characteristic of the element. The cutoff wavelength λ_{m} is independent of the target element and is related to the voltage V of the X-ray tube by $\lambda_{\mathrm{m}} = hc/eV$.

$$\frac{1}{\lambda} = Z^2 \frac{E_0}{hc}\left(1 - \frac{1}{2^2}\right)$$

where $E_0 = 13.6$ eV is the binding energy of the ground-state hydrogen atom. Taking the square root of both sides gives

$$\frac{1}{\sqrt{\lambda_{K_\alpha}}} = \left[\frac{E_0}{hc}\left(1 - \frac{1}{2^2}\right)\right]^{1/2} Z$$

Moseley's equation and this equation are in agreement if $Z - 1$ is substituted for Z in Moseley's equation and if $a = 3E_0/(4hc)$. This raises the question, why a factor of $Z - 1$ instead of a factor of Z? Part of the explanation is that the formula from the Bohr theory ignores the shielding of the nuclear charge. In a multi-electron atom, electrons in the $n = 2$ states are electrically shielded from the nuclear charge by the two electrons in the $n = 1$ state, so the $n = 2$ state electrons are attracted by an effective nuclear charge of about $(Z - 2)e$. However, when there is only one electron in the K shell, the $n = 2$ electrons are attracted by an effective nuclear charge of about $(Z - 1)e$. When an electron from state n drops into the vacated state in the $n = 1$ shell, a photon of energy $E_2 - E_1$ is emitted. The wavelength of this photon is

$$\lambda_{K_\alpha} = \frac{hc}{(Z - 1)^2 E_0\left(1 - \frac{1}{2^2}\right)} \qquad\qquad 36\text{-}47$$

which is obtained from the previous equation with $Z - 1$ substituted for Z.

IDENTIFYING THE ELEMENT FROM THE K_α X-RAY LINE   **E X A M P L E   3 6 - 8**

The wavelength of the K_α X-ray line for a certain element is $\lambda = 0.0721$ nm. What is the element?

PICTURE THE PROBLEM The K_α line corresponds to a transition from $n = 2$ to $n = 1$. The wavelength is related to the atomic number Z by Equation 36-47.

1. Solve Equation 36-47 for $(Z - 1)^2$:

$$\lambda_{K_\alpha} = \frac{hc}{(Z - 1)^2 E_0\left(1 - \frac{1}{2^2}\right)}$$

so

$$(Z - 1)^2 = \frac{4hc}{3\lambda_{K_\alpha} E_0}$$

2. Substitute the given data and solve for Z:

$$(Z - 1)^2 = \frac{4(1240 \text{ eV·nm})}{3(0.0721 \text{ nm})(13.6 \text{ eV})} = 1686$$

so

$$Z = 1 + \sqrt{1686} = 42.06$$

3. Since Z is an integer, we round to the nearest integer: $Z = 42$

> The element is molybdenum.

1. The Bohr model is important historically because it was the first model to succeed at explaining the discrete optical spectrum of atoms in terms of the quantization of energy. It has been superceded by the quantum-mechanical model.

2. The quantum theory of atoms results from the application of the Schrödinger equation to a bound system consisting of nucleus of charge $+Ze$ and Z electrons of charge $-e$.

3. For the simplest atom, hydrogen, consisting of one proton and one electron, the time independent Schrödinger equation can be solved exactly to obtain the wave functions ψ, which depend on the quantum numbers n, ℓ, m_ℓ, and m_s.

4. The electron configuration of atoms is governed by the Pauli exclusion principle, which states that no two electrons in an atom can have the same set of values for the quantum numbers n, ℓ, m_ℓ, and m_s. Using the exclusion principle and the restrictions on the quantum numbers, we can understand much of the structure of the periodic table.

Topic	Relevant Equations and Remarks	
1. The Bohr Model of the Hydrogen Atom		
Postulates for the hydrogen atom		
Nonradiating orbits	The electron moves in a circular nonradiating orbit around the proton.	
Photon frequency from energy conservation	$f = \dfrac{E_i - E_f}{h}$	36-7
Quantized angular momentum	$mvr = n\hbar, \quad n = 1, 2, 3, \ldots$	36-9
First Bohr radius	$a_0 = \dfrac{\hbar^2}{mke^2} \approx 0.0529 \text{ nm}$	36-12
Radius of the Bohr orbits	$r = n^2 \dfrac{a_0}{Z}$	36-11
Energy levels in the hydrogen atom	$E_n = -\dfrac{mk^2e^4}{2\hbar^2} \dfrac{Z^2}{n^2} = -Z^2 \dfrac{E_0}{n^2}$	36-15
	where	
	$E_0 = \dfrac{mk^2e^4}{2\hbar^2} = \dfrac{1}{2} \dfrac{ke^2}{a_0} \approx 13.6 \text{ eV}$	36-16
Wavelengths emitted by the hydrogen atom	$\lambda = \dfrac{c}{f} = \dfrac{hc}{E_i - E_f} = \dfrac{1240 \text{ eV·nm}}{E_i - E_f}$	36-17, 36-18
2. Quantum Theory of Atoms	The electron is described by a wave function ψ that is a solution of the Schrödinger equation. Energy quantization arises from standing-wave conditions. ψ is described by the quantum numbers n, ℓ, and m_ℓ, and the spin quantum number $m_s = \pm\frac{1}{2}$.	
Schrödinger equation in spherical coordinates	$-\dfrac{\hbar^2}{2mr^2} \dfrac{\partial}{\partial r}\left(r^2 \dfrac{\partial \psi}{\partial r}\right) - \dfrac{\hbar^2}{2mr^2}\left[\dfrac{1}{\sin\theta}\dfrac{\partial}{\partial\theta}\left(\sin\theta \dfrac{\partial\psi}{\partial\theta}\right) + \dfrac{1}{\sin^2\theta}\dfrac{\partial^2\psi}{\partial\phi^2}\right] + U(r)\psi = E\psi$	36-21

The solutions can be written as products of functions of r, θ, and ϕ separately.	$\psi(r, \theta, \phi) = R(r)f(\theta)g(\phi)$	36-22

Quantum numbers in spherical coordinates

Principal quantum number	$n = 1, 2, 3, \ldots$	
Orbital quantum number	$\ell = 0, 1, 2, 3, \ldots, n - 1$	
Magnetic quantum number	$m_\ell = -\ell, (-\ell + 1), \ldots, -2, -1, 0, 1, 2, \ldots, (\ell + 1), \ell$	36-23
Orbital angular momentum	$L = \sqrt{\ell(\ell + 1)}\hbar$	36-24
z component of angular momentum	$L_z = m_\ell \hbar$	36-25

3. Quantum Theory of the Hydrogen Atom

Energy levels for hydrogen (same as for the Bohr model)	$E_n = -\dfrac{mk^2e^4}{2\hbar^2} = -Z^2\dfrac{E_0}{n^2}, \quad n = 1, 2, 3, \ldots$	36-27

where

$$E_0 = \frac{mk^2e^4}{2\hbar^2} \approx 13.6 \text{ eV}$$

36-28

Wavelengths emitted by the hydrogen atom (same as for Bohr model)	$\lambda = \dfrac{c}{f} = \dfrac{hc}{E_i - E_f} = \dfrac{1240 \text{ eV·nm}}{E_i - E_f}$	36-17, 36-18

Wave functions

The ground state	$\psi_{1,0,0} = C_{1,0,0}e^{-Zr/a_0} = \dfrac{1}{\sqrt{\pi}}\left(\dfrac{Z}{a_0}\right)^{3/2}e^{-Zr/a_0}$	36-31, 36-33
The first excited state	$\psi_{2,0,0} = C_{2,0,0}\left(2 - \dfrac{Zr}{a_0}\right)e^{-Zr/(2a_0)}$	36-36
	$\psi_{2,1,0} = C_{2,1,0}\dfrac{Zr}{a_0}e^{-Zr/(2a_0)}\cos\theta$	36-37
	$\psi_{2,1,\pm1} = C_{2,1,1}\dfrac{Zr}{a_0}e^{-Zr/(2a_0)}\sin\theta\, e^{\pm i\phi}$	36-38

Probability densities	For $\ell = 0$, $|\psi|^2$ is spherically symmetric. For $\ell \neq 0$, $|\psi|^2$ depends on the angle θ.	
Radial probability density	$P(r) = 4\pi r^2|\psi|^2$	36-34

The radial probability density is maximum at the distances corresponding roughly to the Bohr orbits.

4. The Spin–Orbit Effect and Fine Structure

The total angular momentum of an electron in an atom is a combination of the orbital angular momentum and spin angular momentum. It is characterized by the quantum number j, which can be either $|\ell - \frac{1}{2}|$ or $\ell + \frac{1}{2}$. Because of the interaction of the orbital and spin magnetic moments, the state $j = |\ell - \frac{1}{2}|$ has lower energy than the state $j = \ell + \frac{1}{2}$, for $\ell \geq 1$. This small splitting of the energy states gives rise to a small splitting of the spectral lines called fine structure.

5. The Periodic Table

Beginning with hydrogen, each larger neutral atom adds one electron. The electrons go into those states that will give the lowest energy consistent with the Pauli exclusion principle.

The state of an atom is described by its electron configuration, which gives the values of n and ℓ for each electron. The ℓ values are specified by a code:

	s	p	d	f	g	h
ℓ values	0	1	2	3	4	5

Pauli exclusion principle

No two electrons in an atom can have the same set of values for the quantum numbers n, ℓ, m_ℓ, and m_s.

6. Atomic Spectra

Atomic spectra include optical spectra and X-ray spectra. Optical spectra result from transitions between energy levels of a single outer electron moving in the field of the nucleus and core electrons of the atom. Characteristic X-ray spectra result from the excitation of an inner core electron and the subsequent filling of the vacancy by other electrons in the atom.

Selection rules

Transitions between energy states with the emission of a photon are governed by the following selection rules

$$\Delta m_\ell = 0 \text{ or } \pm 1 \qquad\qquad\qquad 36\text{-}29$$

$$\Delta \ell = \pm 1$$

PROBLEMS

- Single-concept, single-step, relatively easy
- •• Intermediate-level, may require synthesis of concepts
- ••• Challenging
- SSM Solution is in the *Student Solutions Manual*
- ISOLVE Problems available on iSOLVE online homework service
- ISOLVE✓ These "Checkpoint" online homework service problems ask students additional questions about their confidence level, and how they arrived at their answer.

In a few problems, you are given more data than you actually need; in a few other problems, you are required to supply data from your general knowledge, outside sources, or informed estimates.

Conceptual Problems

1 • SSM As n increases, does the spacing of adjacent energy levels increase or decrease?

2 • ISOLVE The energy of the ground state of doubly ionized lithium ($Z = 3$) is _____, where $E_0 = 13.6$ eV. (a) $-9E_0$, (b) $-3E_0$, (c) $-E_0/3$, (d) $-E_0/9$.

3 • Bohr's quantum condition on electron orbits requires (a) that the angular momentum of the electron about the hydrogen nucleus equals $n\hbar$. (b) that no more than one electron occupy a given stationary state. (c) that the electrons spiral into the nucleus while radiating electromagnetic waves. (d) that the energies of an electron in a hydrogen atom be equal to nE_0, where E_0 is a constant and n is an integer. (e) none of the above.

4 • According to the Bohr model, if an electron moves to a larger orbit, does the electron's total energy increase or decrease? Does the electron's kinetic energy increase or decrease?

5 • According to the Bohr model, the kinetic energy of the electron in the ground state of hydrogen is 13.6 eV $= E_0$. The kinetic energy of the electron in the state $n = 2$ is (a) $4E_0$. (b) $2E_0$. (c) $E_0/2$. (d) $E_0/4$.

6 • ISOLVE✓ According to the Bohr model, the radius of the $n = 1$ orbit in the hydrogen atom is $a_0 = 0.053$ nm. What is the radius of the $n = 5$ orbit? (a) $5a_0$, (b) $25a_0$, (c) a_0, (d) $a_0/5$, or (e) $a_0/25$.

7 • SSM For the principal quantum number $n = 4$, how many different values can the orbital quantum number ℓ have? (a) 4, (b) 3, (c) 7, (d) 16, or (e) 6.

8 • For the principal quantum number $n = 4$, how many different values can the magnetic quantum number m_ℓ have? (a) 4, (b) 3, (c) 7, (d) 16, or (e) 6.

9 • The p state of an electron configuration corresponds to (a) $n = 2$, (b) $\ell = 2$, (c) $\ell = 1$, (d) $n = 0$, or (e) $\ell = 0$.

10 •• [SSM] Why is the energy of the 3s state considerably lower than the energy of the 3p state for sodium, whereas in hydrogen these states have essentially the same energy?

11 •• Discuss the evidence from the periodic table of the need for a fourth quantum number. How would the properties of helium differ if there were only three quantum numbers, n, ℓ, and m_ℓ?

12 •• Separate the following six elements—potassium, calcium, titanium, chromium, manganese, and copper—into two groups of three each, so that those in a group have similar properties.

13 • What element has the electron configuration (a) $1s^2 2s^2 2p^6 3s^2 3p^3$ and (b) $1s^2 2s^2 2p^6 3s^2 3p^6 3d^5 4s^1$?

14 • [SSM] For the principal quantum number $n = 3$, what are the possible values of the quantum numbers ℓ and m_ℓ?

15 • An electron in the L shell means that (a) $\ell = 0$, (b) $\ell = 1$, (c) $n = 1$, (d) $n = 2$, or (e) $m_\ell = 2$.

16 •• The Bohr theory and the Schrödinger theory of the hydrogen atom give the same results for the energy levels. Discuss the advantages and disadvantages of each model.

17 •• The Sommerfeld–Hosser displacement theorem states that the optical spectrum of any neutral atom is very similar to the spectrum of the singly charged positive ion of the element immediately following it in the periodic table. Discuss why this is true.

18 •• [SSM] The Ritz combination principle states that for any atom, one can find different spectral lines λ_1, λ_2, λ_3, and λ_4, so that $1/\lambda_1 + 1/\lambda_2 = 1/\lambda_3 + 1/\lambda_4$. Show why this is true using an energy-level diagram.

19 • Using the triplet of numbers (n, ℓ, m_ℓ) to represent an electron with principal quantum number n, orbital quantum number ℓ, and magnetic quantum number m_ℓ, which of the following transitions is allowed? (a) $(5, 2, 2) \to (3, 1, 2)$; (b) $(2, 0, 0) \to (3, 0, 1)$; (c) $(4, 3, -2) \to (3, 2, 0)$; (d) $(1, 0, 0) \to (2, 1, -1)$; or (e) $(2, 1, 0) \to (3, 0, 0)$.

Estimation and Approximation

20 •• [SSM] In laser cooling and trapping, atoms in a beam traveling in one direction are slowed by interaction with an intense laser beam in the opposite direction. The photons scatter off the atoms via resonance absorption, a process by which the incident photon is absorbed by the atom, and a short time later a photon of equal energy is emitted in a random direction. The net result of a single such scattering event is a transfer of momentum to the atom in a direction opposite to the motion of the atom, followed by a second transfer of momentum to the atom in a random direction. Thus, during photon absorption the atom loses speed, but during photon emission the change in speed of the atom is, on average, zero (because the directions of the emitted photons are random). An analogy often made to this process is that of slowing down a bowling ball by bouncing ping-pong balls off of it. (a) Given a typical photon energy used in these experiments of about 1 eV and a momentum typical for an atom with a thermal speed appropriate to a temperature of about 500 K (a typical temperature for an atomic beam), estimate the number of photon-atom collisions that are required to bring an atom to rest. (The average kinetic energy of an atom is equal to $\frac{3}{2}kT$, where k is the Boltzmann constant. Use this to estimate the speed of the atoms.) (b) Compare this with the number of ping-pong ball–bowling ball collisions that are required to bring the bowling ball to rest. (Assume the speeds of the incident ping-pong balls are all equal to the initial speed of the bowling ball.) (c) ^{85}Rb is a type of atom often used in cooling experiments. The wavelength of the light resonant with the cooling transition is $\lambda = 780.24$ nm. Estimate the number of photons needed to slow down an ^{85}Rb atom from a typical thermal velocity of 300 m/s to a stop.

21 •• (a) We can define a thermal de Broglie wavelength λ_T for an atom in a gas at temperature T as being the de Broglie wavelength for an atom moving at the rms speed appropriate to that temperature. (The average kinetic energy of an atom is equal to $\frac{3}{2}kT$, where k is the Boltzman constant. Use this to calculate the rms speed of the atoms.) Show that

$$\lambda_T = \sqrt{\frac{h^2}{3mkT}},$$ where m is the mass of the atom. (b) Cooled neutral atoms can form a *Bose condensate* (a new state of matter) when their thermal de Broglie wavelength becomes larger than the average interatomic spacing. From this criterion, estimate the temperature needed to create a Bose condensate in a gas of ^{85}Rb atoms whose number density is 10^{12} atoms/cm^3.

The Bohr Model of the Hydrogen Atom

22 • Use the known values of the constants in Equation 36-12 to show that a_0 is approximately 0.0529 nm.

23 • [ISOLVE✓] The longest wavelength in the Lyman series was calculated in Example 36-2. Find the wavelengths for the transitions (a) $n_1 = 3$ to $n_2 = 1$ and (b) $n_1 = 4$ to $n_2 = 1$.

24 • Find the photon energy for the three longest wavelengths in the Balmer series and calculate the wavelengths.

25 •• (a) Find the photon energy and wavelength for the series limit (shortest wavelength) in the Paschen series ($n_2 = 3$). (b) Calculate the wavelength for the three longest wavelengths in this series and indicate their positions on a horizontal linear scale.

26 •• [SSM] Repeat Problem 25 for the Brackett series, $n_2 = 4$.

27 •• The hydrogen spectrum is found by collimating the light from a hydrogen discharge tube and shining it on a grating to disperse the light into its various colors. The grating spacing is $d = 3.377$ μm. A bright red line ($m = 1$) is seen at an angle of 11.233° from the center of the spectroscope (see Figure 36-1). (a) What is the wavelength of this spectral line? (b) Assuming this line is from a transition from level $n_1 = 3$ to level $n_2 = 2$ (i.e., the longest wavelength Balmer series transition), what do you calculate for the value of the Rydberg constant?

28 ••• In this problem, you will estimate the radius and the energy of the lowest stationary state of the hydrogen atom using the uncertainty principle. The total energy of the electron with momentum p and mass m a distance r from the proton in the hydrogen atom is given by

$$E = p^2/2m - ke^2/r$$

where k is the Coulomb constant. Assume that the minimum value of $p^2 \approx (\Delta p)^2 = \hbar^2/r^2$, where Δp is the uncertainty in p and we have taken $\Delta r \sim r$ for the order of magnitude of the uncertainty in position; the energy is then

$$E = \hbar^2/2mr^2 - ke^2/r$$

Find the radius r for which this energy is a minimum, and calculate the minimum value of E in electron volts.

29 ••• **SSM** In the center-of-mass reference frame of a hydrogen atom, the electron and nucleus have equal and opposite momenta of magnitude p. (a) Show that the total kinetic energy of the electron and nucleus can be written $K = p^2/(2\mu)$ where $\mu = m_e M/(M + m_e)$ is called the reduced mass, m_e is the mass of the electron, and M is the mass of the nucleus. (b) For the equations for the Bohr model of the atom, the motion of the nucleus can be taken into account by replacing the mass of the electron with the reduced mass. Use Equation 36-14 to calculate the Rydberg constant for a hydrogen atom with a nucleus of mass $M = m_p$. Find the approximate value of the Rydberg constant by letting M go to infinity in the reduced mass formula. To how many figures does this approximate value agree with the actual value? (c) Find the percentage correction for the ground-state energy of the hydrogen atom by using the reduced mass in Equation 36-16. *Remark: In general, the reduced mass for a two-body problem with masses m_1 and m_2 is given by*

$$\mu = \frac{m_1 m_2}{m_1 + m_2}$$

30 •• **SSM** The Pickering series of the spectrum of He^+ (singly-ionized helium) consists of spectral lines due to transitions to the $n = 4$ state of He^+. Experimentally, every other line of the Pickering series is very close to a spectral line in the Balmer series for hydrogen transitions to $n = 2$. (a) Show that this is true. (b) Calculate the wavelength of a transition from the $n = 6$ level to the $n = 4$ level of He^+, and show that it corresponds to one of the Balmer lines.

Quantum Numbers in Spherical Coordinates

31 • For $\ell = 1$, find (a) the magnitude of the angular momentum L and (b) the possible values of m_ℓ. (c) Draw a vector diagram to scale showing the possible orientations of $\vec{L}$ with the z axis.

32 • Repeat Problem 31 for $\ell = 3$.

33 • If $n = 3$, (a) what are the possible values of ℓ? (b) For each value of ℓ in Part (a), list the possible values of m_ℓ. (c) Using the fact that there are two quantum states for each value of ℓ and m_ℓ because of electron spin, find the total number of electron states with $n = 3$.

34 • Find the total number of electron states with (a) $n = 4$ and (b) $n = 2$. (See Problem 33.)

35 •• **SSM** **ISOLVE** Find the minimum value of the angle θ between $\vec{L}$ and the z axis for (a) $\ell = 1$, (b) $\ell = 4$, and (c) $\ell = 50$.

36 •• What are the possible values of n and m_ℓ if (a) $\ell = 3$, (b) $\ell = 4$, and (c) $\ell = 0$.

37 •• For an $\ell = 2$ state, find (a) the square magnitude of the angular momentum $\vec{L}^2$, (b) the maximum value of L_z^2, and (c) the smallest value of $L_x^2 + L_y^2$.

Quantum Theory of the Hydrogen Atom

38 • For the ground state of the hydrogen atom, find the values of (a) ψ, (b) ψ^2, and (c) the radial probability density $P(r)$ at $r = a_0$. Give your answers in terms of a_0.

39 • **SSM** **ISOLVE** (a) If spin is not included, how many different wave functions are there corresponding to the first excited energy level $n = 2$ for hydrogen? (b) List these functions by giving the quantum numbers for each state.

40 •• For the ground state of the hydrogen atom, calculate the probability of finding the electron in the range $\Delta r = 0.03a_0$ at (a) $r = a_0$ and (b) $r = 2a_0$.

41 •• The value of the constant $C_{2,0,0}$ in Equation 36-36 is given by

$$C_{2,0,0} = \frac{1}{4\sqrt{2\pi}}\left(\frac{Z}{a_0}\right)^{3/2}$$

Find the values of (a) ψ, (b) ψ^2, and (c) the radial probability density $P(r)$ at $r = a_0$ for the state $n = 2$, $\ell = 0$, and $m_\ell = 0$ in hydrogen. Give your answers in terms of a_0.

42 ••• Show that the radial probability density for the $n = 2$, $\ell = 1$, and $m_\ell = 0$ state of a one-electron atom can be written as $P(r) = A\cos^2\theta\, r^4 e^{-Zr/a_0}$, where A is a constant.

43 ••• Calculate the probability of finding the electron in the range $\Delta r = 0.02a_0$ for (a) $r = a_0$ and (b) $r = 2a_0$ for the state $n = 2$, $\ell = 0$, and $m_\ell = 0$ in hydrogen. (See Problem 41 for the value of $C_{2,0,0}$.)

44 •• **SSM** Show that the ground-state hydrogen wave function (Equation 36-33) is a solution to Schödinger's equation (Equation 36-21) and the potential energy function equation (Equation 36-26).

45 •• Show by unit cancellation that the expression for the hydrogen ground-state energy given by Equation 36-28 has the dimensions of energy.

46 •• By dimensional analysis, show that the expression for the first Bohr radius given by Equation 36-12 has the dimensions of length.

47 •• The radial probability distribution function for a one-electron atom in its ground state can be written $P(r) = Cr^2 e^{-2Zr/a_0}$, where C is a constant. Show that $P(r)$ has its maximum value at $r = a_0/Z$.

48 ••• Show that the number of states in the hydrogen atom for a given n is $2n^2$.

49 ••• **ISOLVE** Calculate the probability that the electron in the ground state of a hydrogen atom is in the region $0 < r < a_0$.

The Spin–Orbit Effect and Fine Structure

50 • **SSM** **iSOLVE** The potential energy of a magnetic moment in an external magnetic field is given by $U = -\vec{\mu} \cdot \vec{B}$. (a) Calculate the difference in energy between the two possible orientations of an electron in a magnetic field $\vec{B} = 1.50\ \text{T}\hat{k}$. (b) If these electrons are bombarded with photons of energy equal to this energy difference, "spin flip" transitions can be induced. Find the wavelength of the photons needed for such transitions. This phenomenon is called *electron spin resonance*.

51 • The total angular momentum of a hydrogen atom in a certain excited state has the quantum number $j = \frac{1}{2}$. What can you say about the orbital angular-momentum quantum number ℓ?

52 • A hydrogen atom is in the state $n = 3$, $\ell = 2$. What are the possible values of j?

53 • Using a scaled vector diagram, show how the orbital angular momentum $\vec{L}$ combines with the spin orbital angular momentum $\vec{S}$ to produce the two possible values of total angular momentum $\vec{J}$ for the $\ell = 3$ state of the hydrogen atom.

The Periodic Table

54 • The total number of quantum states of hydrogen with quantum number $n = 4$ is (a) 4, (b) 16, (c) 32, (d) 36, or (e) 48.

55 • How many of oxygen's eight electrons are found in the p state? (a) 0, (b) 2, (c) 4, (d) 6, or (e) 8.

56 • **SSM** Write the electron configuration of (a) carbon and (b) oxygen.

57 • Give the possible values of the z component of the orbital angular momentum of (a) a d electron, and (b) an f electron.

Optical Spectra and X-Ray Spectra

58 • The optical spectra of atoms with two electrons in the same outer shell are similar, but they are quite different from the spectra of atoms with just one outer electron because of the interaction of the two electrons. Separate the following elements into two groups so that those in each group have similar spectra: lithium, beryllium, sodium, magnesium, potassium, calcium, chromium, nickel, cesium, barium.

59 • Write down the possible electron configurations for the first excited state of (a) hydrogen, (b) sodium, and (c) helium.

60 • **iSOLVE** Indicate which of the following elements should have optical spectra similar to hydrogen and which should be similar to helium: Li, Ca, Ti, Rb, Hg, Ag, Cd, Ba, Fr, and Ra.

61 • **SSM** (a) Calculate the next two longest wavelengths in the K series (after the K_α line) of molybdenum. (b) What is the wavelength of the shortest wavelength in this series?

62 • The wavelength of the K_α line for a certain element is 0.3368 nm. What is the element?

63 • Calculate the wavelength of the K_α line in (a) magnesium ($Z = 12$) and (b) copper ($Z = 29$).

General Problems

64 • What is the energy of the shortest wavelength photon emitted by the hydrogen atom?

65 • The wavelength of a spectral line of hydrogen is 97.254 nm. Identify the transition that results in this line under the assumption that the transition is to the ground state.

66 • The wavelength of a spectral line of hydrogen is 1093.8 nm. Identify the transition that results in this line.

67 • Spectral lines of the following wavelengths are emitted by singly ionized helium: 164 nm, 230.6 nm, and 541 nm. Identify the transitions that result in these spectral lines.

68 •• **SSM** We are often interested in finding the quantity ke^2/r in electron volts when r is given in nanometers. Show that $ke^2 = 1.44\ \text{eV·nm}$.

69 •• The wavelengths of the photons emitted by potassium corresponding to transitions from the $4P_{3/2}$ and $4P_{1/2}$ states to the ground state are 766.41 nm and 769.90 nm. (a) Calculate the energies of these photons in electron volts. (b) The difference in the energies of these photons equals the difference in energy ΔE between the $4P_{3/2}$ and $4P_{1/2}$ states in potassium. Calculate ΔE. (c) Estimate the magnetic field that the 4p electron in potassium experiences.

70 •• To observe the characteristic K lines of the X-ray spectrum, one of the $n = 1$ electrons must be ejected from the atom. This is generally accomplished by bombarding the target material with electrons of sufficient energy to eject this tightly bound electron. What is the minimum energy required to observe the K lines of (a) tungsten, (b) molybdenum, and (c) copper?

71 •• **SSM** The combination of physical constants $\alpha = e^2k/\hbar c$, where k is the Coulomb constant, is known as the *fine-structure constant*. It appears in numerous relations in atomic physics. (a) Show that α is dimensionless. (b) Show that in the Bohr model of hydrogen $v_n = c\alpha/n$, where v_n is the speed of the electron in the stationary state of quantum number n.

72 •• The *positron* is a particle that is identical to the electron except that it carries a positive charge of e. *Positronium* is the bound state of an electron and positron. (a) Calculate the energies of the five lowest energy states of positronium using the reduced mass, as given in Problem 29. (b) Do transitions between any of the levels found in Part (a) fall in the visible range of wavelengths? If so, which transitions are these?

73 • In 1947, Lamb and Retherford showed that there was a very small energy difference between the $2S_{1/2}$ and the $2P_{1/2}$ states of the hydrogen atom. They measured this difference essentially by causing transitions between the two states using very long wavelength electromagnetic radiation. The energy difference (the Lamb shift) is 4.372×10^{-6} eV and is explained by quantum electrodynamics as being due to fluctuations in the energy level of the vacuum. (a) What is the frequency of a photon whose energy is equal to the Lamb shift energy? (b) What is the wavelength of this photon? In what spectral region does it belong?

74 • **SSM** A Rydberg atom is one in which an outer shell electron is placed into a *very* high excited state (n ≈ 40 or higher). Such atoms are useful for experiments that probe the transition from quantum-mechanical behavior to classical. Furthermore, these excited states have extremely long lifetimes (i.e., the electron will stay in this high excited state for a very long time). A hydrogen atom is in the $n = 45$ state. (a) What is the ionization energy of the atom when it is in this state? (b) What is the energy level separation (in eV) between this state and the $n = 44$ state? (c) What is the wavelength of a photon resonant with this transition? (d) How large is the atom in the $n = 45$ state?

75 •• The deuteron, the nucleus of deuterium (heavy hydrogen), was first recognized from the spectrum of hydrogen. The deuteron has a mass twice that of the proton. (a) Calculate the Rydberg constant for hydrogen and for deuterium using the reduced mass as given in Problem 29. (b) Using the result

obtained in Part (a), determine the wavelength difference between the longest wavelength Balmer lines of hydrogen and deuterium.

76 •• The muonium atom is a hydrogen atom with the electron replaced by a μ^- particle. The μ^- is identical to an electron but has a mass 207 times as great as the electron. (a) Calculate the energies of the five lowest energy levels of muonium using the reduced mass as given in Problem 29. (b) Do transitions between any of the levels found in Part (a) fall in the visible range of wavelengths, (i.e., between $\lambda = 700$ nm and 400 nm)? If so, which transitions are these?

77 •• The triton, a nucleus consisting of a proton and two neutrons, is unstable with a fairly long half-life of approximately 12 years. Tritium is the bound state of an electron and a triton. (a) Calculate the Rydberg constant of tritium using the reduced mass as given in Problem 29. (b) Using the result obtained in Part (a) and the result obtained in Part (b) of Problem 76, determine the wavelength difference between the longest wavelength Balmer lines of tritium and deuterium and between tritium and hydrogen.

Molecules

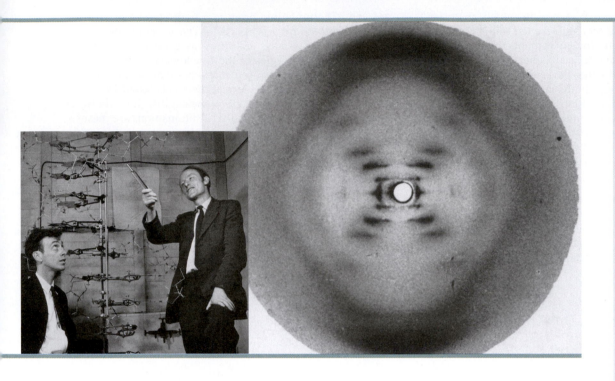

Most atoms bond together to form molecules or solids. Molecules may exist as separate entities, as in gaseous O_2 or N_2, or they may bond together to form liquids or solids. A molecule is the smallest constituent of a substance that retains its chemical properties.

➤ **In this chapter, we use our understanding of quantum mechanics to discuss molecular bonding and the energy levels and spectra of diatomic molecules. Much of our discussion will be qualitative because, as in atomic physics, the quantum-mechanical calculations are very difficult.**

37-1 Molecular Bonding

There are two extreme views that we can take of a molecule. Consider, for example, H_2. We can think of H_2 either as two H atoms joined together or as a quantum-mechanical system of two protons and two electrons. The latter picture is more fruitful in this case because neither of the electrons in the H_2 molecule can be identified as belonging to either proton. Instead, the wave function for each electron is spread out in space throughout the whole molecule. For more complicated molecules, however, an intermediate picture is useful. For example, the nitrogen molecule N_2 consists of 14 protons and 14 electrons, but only two of the electrons take part in the bonding. We therefore can consider this molecule as two N^+ ions and two electrons that belong to the molecule as a whole. The molecular wave functions for these bonding electrons are called **molecular**

THE DISCOVERERS OF THE STRUCTURE OF DNA JAMES WATSON AT LEFT AND FRANCIS CRICK ARE SHOWN WITH THEIR MODEL OF PART OF A DNA MOLECULE IN 1953. CRICK AND WATSON MET AT THE CAVENDISH LABORATORY, CAMBRIDGE, IN 1951. THEIR WORK ON THE STRUCTURE OF DNA WAS PERFORMED WITH A KNOWLEDGE OF CHARGAFF'S RATIOS OF THE BASES IN DNA AND SOME ACCESS TO THE X-RAY CRYSTALLOGRAPHY OF MAURICE WILKINS AND ROSALIND FRANKLIN AT KING'S COLLEGE LONDON. COMBINING ALL OF THIS WORK LED TO THE DEDUCTION THAT DNA EXISTS AS A DOUBLE HELIX, THUS TO ITS STRUCTURE. CRICK, WATSON, AND WILKINS SHARED THE 1962 NOBEL PRIZE FOR PHYSIOLOGY OR MEDICINE; FRANKLIN DIED FROM CANCER IN 1958.

X-RAY DIFFRACTION PATTERN OF THE B FORM OF DNA ROSALIND FRANKLIN'S COLLEAGUE MAURICE WILKINS, WITHOUT OBTAINING HER PERMISSION, MADE AVAILABLE TO WATSON AND CRICK HER THEN UNPUBLISHED X-RAY DIFFRACTION PATTERN OF THE B FORM OF DNA, WHICH WAS CRUCIAL EVIDENCE FOR THE HELICAL STRUCTURE. IN HIS ACCOUNT OF THIS DISCOVERY, WATSON WROTE: "THE INSTANT I SAW THE PICTURE, MY MOUTH FELL OPEN AND PULSE BEGAN TO RACE. . . . THE BLACK CROSS OF REFLECTIONS WHICH DOMINATED THE PICTURE COULD ARISE ONLY FROM A HELICAL STRUCTURE. . . . MERE INSPECTION OF THE X-RAY PICTURE GAVE SEVERAL OF THE VITAL HELICAL PERAMETERS" (FROM STENT, GUNTHER, *THE DOUBLE HELIX*, NEW YORK: NORTON, 1980).

orbitals. In many cases, these molecular wave functions can be constructed from combinations of the atomic wave functions with which we are familiar.

The two principal types of bonds responsible for the formation of molecules are the ionic bond and the covalent bond. Other types of bonds that are important in the bonding of liquids and solids are van der Waals bonds, metallic bonds, and hydrogen bonds. In many cases, bonding is a mixture of these mechanisms.

The Ionic Bond

The simplest type of bond is the **ionic bond,** which is found in salts such as sodium chloride (NaCl). The sodium atom has one 3s electron outside a stable core. The ionization energy of sodium is the energy needed to remove this electron from an isolated sodium atom. This energy is just 5.14 eV (see Figure 36-18). The removal of this electron leaves an isolated positive ion with a spherically symmetric, closed-shell electron core. Chlorine, on the other hand, is one electron short of having a closed shell. The energy released by an isolated atom's acquisition of one electron is called its **electron affinity,** which in the case of chlorine is 3.62 eV. The acquisition of one electron by chlorine results in a negative ion with a spherically symmetric, closed-shell electron core. Thus, the formation of a Na^+ ion and a Cl^- ion by the donation of one electron of sodium to chlorine requires only 5.14 eV − 3.62 eV = 1.52 eV at infinite separation. The electrostatic potential energy U_e of the two ions when they are a distance r apart is $-ke^2/r$. When the separation of the ions is less than approximately 0.95 nm, the negative potential energy of attraction is of greater magnitude than the 1.52 eV of energy needed to create the ions. Thus, at separation distances less than 0.95 nm, it is energetically favorable (i.e., the total energy of the system is reduced) for the sodium atom to donate an electron to the chlorine atom to form NaCl.

Since the electrostatic attraction increases as the ions get closer together, it might seem that equilibrium could not exist. However, when the separation of the ions is very small, there is a strong repulsion that is quantum mechanical in nature and is related to the exclusion principle. This **exclusion-principle repulsion** is responsible for the repulsion of the atoms in all molecules (except $H_2)^\dagger$ for all bonding mechanisms. We can understand it qualitatively as follows. When the ions are very far apart, the wave function for a core electron in one of the ions does not overlap that of any electron in the other ion. We can distinguish the electrons by the ion to which they belong. This means that electrons in the two ions can have the same quantum numbers because they occupy different regions of space. However, as the distance between the ions decreases, the wave functions of the core electrons begin to overlap; that is, the electrons in the two ions begin to occupy the same region of space. Because of the exclusion principle, some of these electrons must go into higher energy quantum states.‡ But energy is required to shift the electrons into higher energy quantum states. This increase in energy when the ions are pushed closer together is equivalent to a repulsion of the ions. It is not a sudden process. The energy states of the electrons change gradually as the ions are brought together. A sketch of the potential energy $U(r)$ of the Na^+ and Cl^- ions versus separation distance r is shown in Figure 37-1. The energy is lowest at an equilibrium separation r_0 of approximately 0.236 nm. At smaller separations, the energy rises steeply

FIGURE 37-1 Potential energy for Na^+ and Cl^- ions as a function of separation distance r. The energy at infinite separation is chosen to be 1.52 eV, corresponding to the energy $-\Delta E$ needed to form the ions from neutral atoms. The minimum energy is at the equilibrium separation $r_0 = 0.236$ nm for the ions in the molecule.

---

† In H_2, the repulsion is simply that of the two positively charged protons.

‡ Recall from our discussion in Chapter 35 that the exclusion principle is related to the fact that the wave function for two identical electrons is antisymmetric on the exchange of the electrons and that an antisymmetric wave function for two electrons with the same quantum numbers is zero if the space coordinates of the electrons are the same.

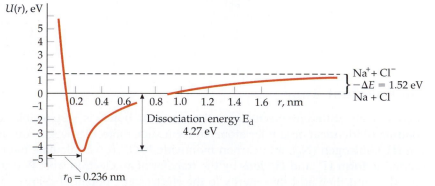

as a result of the exclusion principle. The energy required to separate the ions and form neutral sodium and chlorine atoms is called the **dissociation energy** E_d, which is approximately 4.27 eV for NaCl.

The equilibrium separation distance of 0.236 nm is for gaseous diatomic NaCl, which can be obtained by evaporating solid NaCl. Normally, NaCl exists in a cubic crystal structure, with the Na^+ and Cl^- ions at the alternate corners of a cube. The separation of the ions in a crystal is somewhat larger, approximately 0.28 nm. Because of the presence of neighboring ions of opposite charge, the electrostatic energy per ion pair is lower when the ions are in a crystal.

THE ENERGY OF A SODIUM-FLUORIDE MOLECULE　　　**EXAMPLE 37-1**

The electron affinity of fluorine is 3.40 eV, and the equilibrium separation of sodium fluoride (NaF) is 0.193 nm. (a) How much energy is needed to form Na^+ and F^- ions from neutral sodium and fluorine atoms? (b) What is the electrostatic potential energy of the Na^+ and F^- ions at their equilibrium separation? (c) The dissociation energy of NaF is 5.38 eV. What is the energy due to repulsion of the ions at the equilibrium separation?

PICTURE THE PROBLEM (a) The energy ΔE needed to form Na^+ and F^- ions from the neutral sodium and fluorine atoms is the difference between the ionization energy of sodium (5.14 eV) and the electron affinity of fluorine. (b) The electrostatic potential energy with $U = 0$ at infinity is $U_e = -ke^2/r$. (c) If we choose the potential energy at infinity to be ΔE, the total potential energy is $U_{tot} = U_e + \Delta E + U_{rep}$, where U_{rep} is the energy of repulsion, which is found by setting the dissociation energy equal to $-U_{tot}$.

(a) Calculate the energy needed to form Na^+ and F^- ions from the neutral sodium and fluorine atoms (see Picture the Problem):

$$\Delta E = 5.14 \text{ eV} - 3.40 \text{ eV} = \boxed{1.74 \text{ eV}}$$

(b) 1. Calculate the electrostatic potential energy at the equilibrium separation of $r = 0.193$ nm:

$$U_e = -\frac{ke^2}{r}$$

$$= -\frac{(8.99 \times 10^9 \text{ N·m}^2/\text{C}^2)(1.60 \times 10^{-19} \text{ C})^2}{1.93 \times 10^{-10} \text{ m}}$$

$$= -1.19 \times 10^{-18} \text{ J}$$

2. Convert from joules to electron volts:

$$U_e = -1.19 \times 10^{-18} \text{ J}\left(\frac{1 \text{ eV}}{1.60 \times 10^{-19} \text{ J}}\right) = \boxed{-7.45 \text{ eV}}$$

(c) The dissociation energy equals the negative of the total potential energy:

$$E_d = -U_{tot} = -(U_e + \Delta E + U_{rep})$$

so

$$U_{rep} = -(E_d + \Delta E + U_e)$$

$$= -(5.38 \text{ eV} + 1.74 \text{ eV} - 7.45 \text{ eV})$$

$$= \boxed{0.33 \text{ eV}}$$

The Covalent Bond

A completely different mechanism, the **covalent bond,** is responsible for the bonding of identical or similar atoms to form such molecules as gaseous hydrogen (H_2), nitrogen (N_2), and carbon monoxide (CO). If we calculate the energy needed to form H^+ and H^- ions by the transfer of an electron from one atom to the other and then add this energy to the electrostatic potential energy, we find that there is no separation distance for which the total energy is negative. The bond thus cannot be ionic. Instead, the attraction of two hydrogen atoms is an entirely quantum-mechanical effect. The decrease in energy when two hydrogen atoms approach each other is due to the sharing of the two electrons by both atoms. It is intimately connected with the symmetry properties of the wave functions of electrons.

We can gain some insight into covalent bonding by considering a simple, one-dimensional quantum-mechanics problem of two finite square wells. We first consider a single electron that is equally likely to be in either well. Because the wells are identical, the probability distribution, which is proportional to $|\psi^2|$, must be symmetric about the midpoint between the wells. Then ψ must be either symmetric or antisymmetric with respect to the two wells. The two possibilities for the ground state are shown in Figure 37-2a for the case in which the wells are far apart and in Figure 37-2b for the case in which the wells are close together. An important feature of Figure 37-2b is that in the region between the wells the symmetric wave function is large and the antisymmetric wave function is small.

Now consider adding a second electron to the two wells. We saw in Chapter 35 that the wave functions for particles that obey the exclusion principle are antisymmetric on exchange of the particles. Thus the total wave function for the two electrons must be antisymmetric on exchange of the electrons. Note that exchanging the electrons in the wells here is the same as exchanging the wells. The total wave function for two electrons can be written as a product of a space part and a spin part. So, an antisymmetric wave function can be the product of a symmetric space part and an antisymmetric spin part or of a symmetric spin part and an antisymmetric space part.

To understand the symmetry of the total wave function, we must therefore understand the symmetry of the spin part of the wave function. The spin of a single electron can have two possible values for its quantum number m_S: $m_S = +\frac{1}{2}$, which we call spin up, or $m_S = -\frac{1}{2}$, which we call spin down. We will use arrows to designate the spin wave function for a single electron: $\uparrow_1$ or $\uparrow_2$ for electron 1 or electron 2 with spin up and $\downarrow_1$ or $\downarrow_2$ for electron 1 or electron 2 with spin down. The total spin quantum number for two electrons can be $S = 1$, with $m_S = +1, 0$, or -1; or $S = 0$, with $m_S = 0$. We use ϕ_{S, m_S} to denote the spin wave function for two electrons. The spin state $\phi_{1, +1}$, corresponding to $S = 1$ and $m_S = +1$, can be written

$$\phi_{1,+1} = \uparrow_1\uparrow_2, \quad S = 1, m_S = +1 \qquad 37\text{-}1$$

Similarly, the spin state for $S = 1$, $m_S = -1$ is

$$\phi_{1,-1} = \downarrow_1\downarrow_2, \quad S = 1, m_S = -1 \qquad 37\text{-}2$$

Note that both of these states are symmetric upon exchange of the electrons. The spin state corresponding to $S = 1$ and $m_S = 0$ is not quite so obvious. It turns out to be proportional to

$$\phi_{1,0} = \uparrow_1\downarrow_2 + \uparrow_2\downarrow_1, \quad S = 1, m_S = 0 \qquad 37\text{-}3$$

This spin state is also symmetric upon exchange of the electrons. The spin state for two electrons with antiparallel spins ($S = 0$) is

$$\phi_{0,0} = \uparrow_1\downarrow_2 - \uparrow_2\downarrow_1, \quad S = 0, m_S = 0 \qquad 37\text{-}4$$

This spin state is antisymmetric upon exchange of electrons.

We thus have the important result that the *spin* part of the wave function is symmetric for parallel spins ($S = 1$) and antisymmetric for antiparallel spins

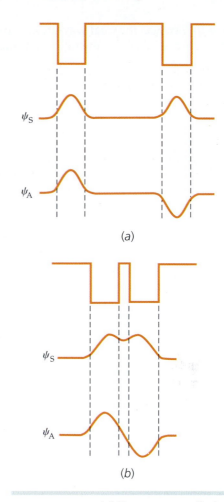

(a)

(b)

FIGURE 37-2 (*a*) Two square wells far apart. The electron wave function can be either symmetric (ψ_S) or antisymmetric (ψ_A) in space. The probability distributions and energies are the same for the two wave functions when the wells are far apart. (*b*) Two square wells that are close together. Between the wells, the antisymmetric space wave function is approximately zero, whereas the symmetric space wave function is quite large.

($S = 0$). Because the total wave function is the product of the space function and spin function, we have the following important result:

> For the total wave function of two electrons to be antisymmetric, the space part of the wave function must be antisymmetric for parallel spins ($S = 1$) and symmetric for antiparallel spins ($S = 0$).

SPIN ALIGNMENT AND WAVE-FUNCTION SYMMETRY

We can now consider the problem of two hydrogen atoms. Figure 37-3a shows a spatially symmetric wave function ψ_S and a spatially antisymmetric wave function ψ_A for two hydrogen atoms that are far apart, and Figure 37-3b shows the same two wave functions for two hydrogen atoms that are close together. The squares of these two wave functions are shown in Figure 37-3c. Note that the probability distribution $|\psi|^2$ in the region between the protons is large for the symmetric wave function and small for the antisymmetric wave function. Thus, when the space part of the wave function is symmetric ($S = 0$), the electrons are often found in the region between the protons. The negatively charged electron cloud representing these electrons is concentrated in the space between the protons, as shown in the upper part of Figure 37-3c, and the protons are bound together by this negatively charged cloud. Conversely, when the space part of the wave function is antisymmetric ($S = 1$), the electrons spend little time between the protons, and the atoms do not bind together to form a molecule. In this case, the electron cloud is not concentrated in the space between the protons, as shown in the lower part of Figure 37-3c.

The total electrostatic potential energy for the H_2 molecule consists of the positive energy of repulsion of the two electrons and the negative potential energy of attraction of each electron for each proton. Figure 37-4 shows the electrostatic potential energy for two hydrogen atoms versus separation for the case in which the space part of the electron wave function is symmetric (U_S) and for the case in which it is antisymmetric (U_A). We can see that the potential energy for the symmetric state is the lower of the two and that the shape of this potential energy curve is similar to that for ionic bonding. The equilibrium separation for H_2 is $r_0 = 0.074$ nm, and the binding energy is 4.52 eV. For the antisymmetric state, the potential energy is never negative and there is no bonding.

We can now see why three hydrogen atoms do not bond to form H_3. If a third hydrogen atom is brought near an H_2 molecule, the third electron cannot be in a 1s state and have its spin antiparallel to the spin of both of the other electrons. If this electron is in an antisymmetric space state with respect to exchange with one

FIGURE 37-3 One-dimensional symmetric and antisymmetric wave functions for two hydrogen atoms (*a*) far apart and (*b*) close together. (*c*) Electron probability distributions ($|\psi|^2$) for the wave functions in Figure 37-3*b*. For the symmetric wave function, the electron charge density is large between the protons. This negative charge density holds the protons together in the hydrogen molecule H_2. For the antisymmetric wave function, the electron charge density is not large between the protons.

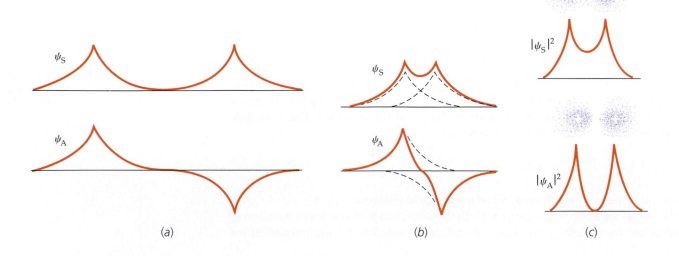

(a) (b) (c)

of the electrons, the repulsion of this atom is greater than the attraction of the other. As the three atoms are pushed together, the third electron is, in effect, forced into a higher quantum-energy state by the exclusion principle. The bond between two hydrogen atoms is called a **saturated bond** because there is no room for another electron. The two shared electrons essentially fill the 1s states of both atoms.

We can also see why two helium atoms do not normally bond together to form the He_2 molecule. There are no valence electrons that can be shared. The electrons in the closed shells are forced into higher energy states when the two atoms are brought together. At low temperatures or high pressures, helium atoms do bond together due to van der Waals forces, which we will discuss next. This bonding is so weak that at atmospheric pressure helium boils at 4 K, and it does not form a solid at any temperature unless the pressure is greater than about 20 atm.

When two identical atoms bond, as in O_2 or N_2, the bonding is purely covalent. However, the bonding of two dissimilar atoms is often a mixture of covalent and ionic bonding. Even in NaCl, the electron donated by sodium to chlorine has some probability of being at the sodium atom because its wave function in the vicinity of the sodium atom, while small, is not zero. Thus, this electron is partially shared in a covalent bond, although this bonding is only a small part of the total bond, which is mainly ionic.

A measure of the degree to which a bond is ionic or covalent can be obtained from the electric dipole moment of the molecule. For example, if the bonding in NaCl were purely ionic, the center of positive charge would be at the Na^+ ion and the center of negative charge would be at the Cl^- ion. The electric dipole moment would have the magnitude

$$p_{ionic} = er_0 \qquad\qquad 37\text{-}5$$

where $r_0 = 2.36 \times 10^{-10}$ m is the equilibrium separation of the ions. Thus, the dipole moment of NaCl would be (from Figure 37-1)

$$p_{ionic} = er_0$$
$$= (1.60 \times 10^{-19}\,\text{C})(2.36 \times 10^{-10}\,\text{m}) = 3.78 \times 10^{-29}\,\text{C·m}$$

The actual measured electric dipole moment of NaCl is

$$p_{measured} = 3.00 \times 10^{-29}\,\text{C·m}$$

We can define the ratio of $p_{measured}$ to p_{ionic} as the fractional amount of ionic bonding. For NaCl, this ratio is $3.00/3.78 = 0.79$. Thus, the bonding in NaCl is about 79 percent ionic.

EXERCISE The equilibrium separation of HCl is 0.128 nm and its measured electric dipole moment is 3.60×10^{-30} C·m. What is the percentage of ionic bonding in HCl? (*Answer* 18 percent)

Other Bonding Types

The van der Waals Bond Any two separated molecules will be attracted to one another by electrostatic forces called van der Waals forces. So will any two atoms that do not form ionic or covalent bonds. The **van der Waals bonds** due to these

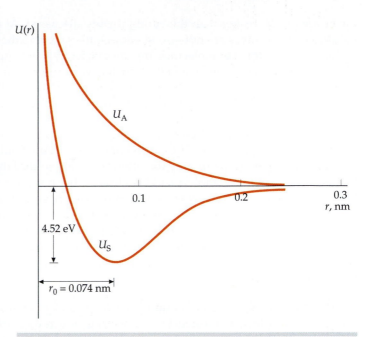

FIGURE 37-4 Potential energy versus separation for two hydrogen atoms. The curve labeled U_S is for a wave function with a symmetric space part, and the curve labeled U_A is for a wave function with an antisymmetric space part.

forces are much weaker than the bonds already discussed. At high enough temperatures, these forces are not strong enough to overcome the ordinary thermal agitation of the atoms or molecules, but at sufficiently low temperatures, thermal agitation becomes negligible, and the van der Waals forces will cause virtually all substances to condense into a liquid and then a solid form.[†] The van der Waals forces arise from the interaction of the instantaneous electric dipole moments of the molecules.

Figure 37-5 shows how two polar molecules—molecules with *permanent* electric dipole moments, such as H_2O—can bond. The electric field due to the dipole moment of one molecule orients the other molecule so that the two dipole moments attract. Nonpolar molecules also attract other nonpolar molecules via the van der Waals forces. Although nonpolar molecules have zero electric dipole moments on the average, they have instantaneous dipole moments that are generally not zero because of fluctuations in the positions of the charges. When two nonpolar molecules are near each other, the fluctuations in the instantaneous dipole moments tend to become correlated so as to produce attraction. This is illustrated in Figure 37-6.

The Hydrogen Bond Another bonding mechanism of great importance is the hydrogen bond, which is formed by the sharing of a proton (the nucleus of the hydrogen atom) between two atoms, frequently two oxygen atoms. This sharing of a proton is similar to the sharing of electrons responsible for the covalent bond already discussed. It is facilitated by the small mass of the proton and by the absence of inner core electrons in hydrogen. The hydrogen bond often holds groups of molecules together and is responsible for the cross-linking that allows giant biological molecules and polymers to hold their fixed shapes. The well-known helical structure of DNA is due to hydrogen-bond linkages across turns of the helix (Figure 37-7).

The Metallic Bond In a metal, two atoms do not bond together by exchanging or sharing an electron to form a molecule. Instead, each valence electron is shared by many atoms. The bonding is thus distributed throughout the entire metal. A metal can be thought of as a lattice of positive ions held together by a *gas* of essentially free electrons that roam throughout the solid. In the quantum-mechanical picture, these free electrons form a cloud of negative charge density between the positively charged lattice ions that holds the ions together. In this

---

[†] Helium is the only element that does not solidify at any temperature at atmospheric pressure.

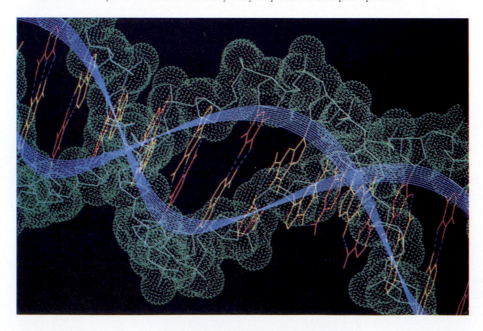

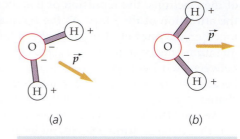

(a)　　　　(b)

FIGURE 37-5 Bonding of H_2O molecules because of the attraction of the electric dipoles. The dipole moment of each molecule is indicated by $\vec{p}$. The field of one dipole orients the other dipole so the moments tend to be parallel. When the dipole moments are approximately parallel, the center of negative charge of one molecule is close to the center of positive charge of the other molecule, and the molecules attract.

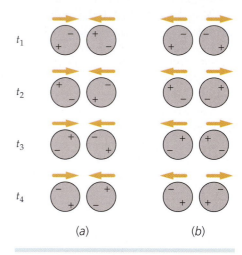

(a)　　　　(b)

FIGURE 37-6 van der Waals attraction of molecules with zero average dipole moments. (*a*) Possible orientations of instantaneous dipole moments at different times leading to attraction. (*b*) Possible orientations leading to repulsion. The electric field of the instantaneous dipole moment of one molecule tends to polarize the other molecule; thus the orientations leading to attraction (Figure 37-6*a*) are much more likely than those leading to repulsion (Figure 37-6*b*).

FIGURE 37-7 The DNA molecule.

respect, the metallic bond is somewhat similar to the covalent bond. However, with the metallic bond, there are far more than just two atoms involved, and the negative charge is distributed uniformly throughout the volume of the metal. The number of free electrons varies from metal to metal but is of the order of one per atom.

*37-2 Polyatomic Molecules

Molecules with more than two atoms range from such relatively simple molecules as water, which has a molecular mass number of 18, to such giants as proteins and DNA, which can have molecular masses of hundreds of thousands up to many millions. As with diatomic molecules, the structure of polyatomic molecules can be understood by applying basic quantum mechanics to the bonding of individual atoms. The bonding mechanisms for most polyatomic molecules are the covalent bond and the hydrogen bond. We will discuss only some of the simplest polyatomic molecules—H_2O, NH_3, and CH_4—to illustrate both the simplicity and complexity of the application of quantum mechanics to molecular bonding.

The basic requirement for the sharing of electrons in a covalent bond is that the wave functions of the valence electrons in the individual atoms must overlap as much as possible. As our first example, we will consider the water molecule. The ground-state configuration of the oxygen atom is $1s^2 2s^2 2p^4$. The 1s and 2s electrons are in closed-shell states and do not contribute to the bonding. The 2p shell has room for six electrons, two in each of the three space states (orbitals) corresponding to $\ell = 1$. In an isolated atom, we describe these space states by the hydrogen-like wave functions corresponding to $\ell = 1$ and $m = +1$, 0, and -1. Since the energy is the same for these three space states, we could equally well use any linear combination of these wave functions. When an atom participates in molecular bonding, certain combinations of these atomic wave functions are important. These combinations are called the p_x, p_y, and p_z **atomic orbitals.** The angular dependence of these orbitals is

$$p_x \propto \sin \theta \cos \phi \qquad\qquad 37\text{-}6$$

$$p_y \propto \sin \theta \sin \phi \qquad\qquad 37\text{-}7$$

$$p_z \propto \cos \phi \qquad\qquad 37\text{-}8$$

The electron charge distribution for these orbitals is maximum along the x, y, or z axis, respectively, as shown in Figure 37-8.

FIGURE 37-8 Computer-generated dot plot illustrating the spatial dependence of the electron charge distribution in the p_x, p_y, and p_z atomic orbitals.

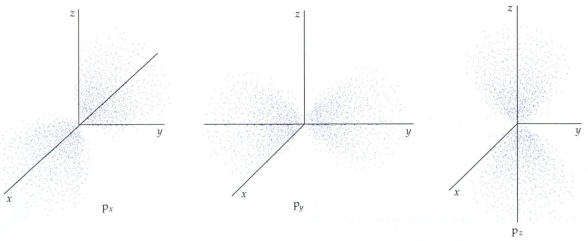

For the oxygen in an H_2O molecule, maximum overlap of the electron wave functions occurs when two of the four 2p electrons are paired with their spins antiparallel in one of the atomic orbitals (for this example, assume the p_z orbital), one of the other electrons is in a second orbital (the p_x orbital), and the other electron is in the third orbital (the p_y orbital). Each of the unpaired electrons (in the p_x and p_y orbitals, in this illustration) forms a bond with the electron of a hydrogen atom, as shown in Figure 37-9. Because of the repulsion of the two hydrogen atoms, the angle between the O–H bonds is actually greater than 90°. The effect of this repulsion can be calculated, and the result is in agreement with the measured angle of 104.5°.

Similar reasoning leads to an understanding of the bonding in NH_3. In the ground state, nitrogen has three electrons in the 2p state. When these three electrons are in the p_x, p_y, and p_z atomic orbitals, they bond to the electrons of hydrogen atoms. Again, because of the repulsion of the hydrogen atoms, the angles between the bonds are somewhat larger than 90°.

The bonding of carbon atoms is somewhat more complicated. Carbon forms a wide variety of different types of molecular bonds, leading to a great diversity in the kinds of organic molecules. The ground-state configuration of carbon is $1s^2 2s^2 2p^2$. From our previous discussion, we might expect carbon to be divalent—that is, bonding only through its two 2p electrons—with the two bonds forming at approximately 90°. However, one of the most important features of the chemistry of carbon is that tetravalent carbon compounds, such as CH_4, are overwhelmingly favored.

The observed valence of 4 for carbon comes about in an interesting way. One of the first excited states of carbon occurs when a 2s electron is excited to a 2p state, giving a configuration of $1s^2 2s^1 2p^3$. In this excited state, we can have four unpaired electrons, one each in the 2s, $2p_x$, $2p_y$, and $2p_z$ atomic orbitals. We might expect there to be three similar bonds corresponding to the three p orbitals and one different bond corresponding to the s orbital. However, when carbon forms tetravalent bonds, these four atomic orbitals become mixed and form four new *equivalent* molecular orbitals called **hybrid orbitals**. This mixing of atomic orbitals, called **hybridization,** is among the most important features involved in the physics of complex molecular bonds. Figure 37-10 shows the tetrahedral structure of the methane molecule (CH_4), and Figure 37-11 shows the structure of the ethane molecule (CH_3–CH_3), which is similar to two joined methane molecules in which one of the C–H bonds is replaced with a C–C bond.

Carbon orbitals can also hybridize, with the s, p_x, and p_y orbitals combining to form three hybrid orbitals in the *xy* plane with 120° bonds and the p_z orbital remaining unmixed. An example of this configuration is graphite, in which the bonds in the *xy* plane provide the strongly layered structure characteristic of the material.

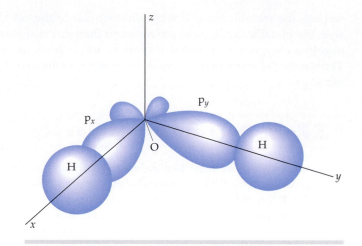

FIGURE 37-9 Electron charge distribution in the H_2O molecule.

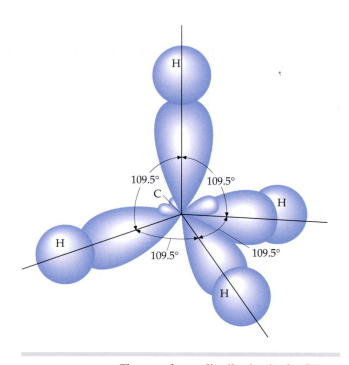

FIGURE 37-10 Electron charge distribution in the CH_4 (methane) molecule.

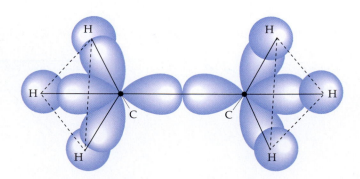

FIGURE 37-11 Electron charge distribution in the CH_3–CH_3 (ethane) molecule.

37-3 Energy Levels and Spectra of Diatomic Molecules

As is the case with an atom, a molecule often emits electromagnetic radiation when it makes a transition from an excited energy state to a state of lower energy. Conversely, a molecule can absorb radiation and make a transition from a lower energy state to a higher energy state. The study of molecular emission and absorption spectra thus provides us with information about the energy states of molecules. For simplicity, we will consider only diatomic molecules here.

The energy of a molecule can be conveniently separated into three parts: electronic, due to the excitation of the electrons of the molecule; vibrational, due to the oscillations of the atoms of the molecule; and rotational, due to the rotation of the molecule about its center of mass. The magnitudes of these energies are sufficiently different that they can be treated separately. The energies due to the electronic excitations of a molecule are of the order of magnitude of 1 eV, the same as for the excitation of an atom. The energies of vibration and rotation are much smaller than this.

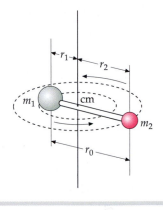

FIGURE 37-12 Diatomic molecule rotating about an axis through its center of mass.

Rotational Energy Levels

Figure 37-12 shows a simple schematic model of a diatomic molecule consisting of a mass m_1 and a mass m_2 separated by a distance r and rotating about its center of mass. Classically, the kinetic energy of rotation (see Equation 9-11) is

$$E = \tfrac{1}{2} I \omega^2 \qquad\qquad 37\text{-}9$$

where I is the moment of inertia and ω is the angular frequency of rotation. If we write this in terms of the angular momentum $L = I\omega$, we have

$$E = \frac{(I\omega)^2}{2I} = \frac{L^2}{2I} \qquad\qquad 37\text{-}10$$

The solution of the Schrödinger equation for rotation leads to quantization of the angular momentum with values given by

$$L^2 = \ell(\ell + 1)\hbar^2, \quad \ell = 0, 1, 2, \dots \qquad\qquad 37\text{-}11$$

where ℓ is the **rotational quantum number.** This is the same quantum condition on angular momentum that holds for the orbital angular momentum of an electron in an atom. Note, however, that L in Equation 37-10 refers to the angular momentum of the entire molecule rotating about its center of mass. The energy levels of a rotating molecule are therefore given by

$$E_\ell = \frac{\ell(\ell + 1)\hbar^2}{2I} = \ell(\ell + 1)E_{0r}, \quad \ell = 0, 1, 2, \dots \qquad\qquad 37\text{-}12$$

ROTATIONAL ENERGY LEVELS

where E_{0r} is the characteristic rotational energy of a particular molecule, which is inversely proportional to its moment of inertia:

$$E_{0r} = \frac{\hbar^2}{2I} \qquad\qquad 37\text{-}13$$

CHARACTERISTIC ROTATIONAL ENERGY

A measurement of the rotational energy of a molecule from its rotational spectrum can be used to determine the moment of inertia of the molecule, which can then be used to find the separation of the atoms in the molecule. The moment of inertia about an axis through the center of mass of a diatomic molecule (see Figure 37-12) is

$$I = m_1 r_1^2 + m_2 r_2^2$$

Using $m_1 r_1 = m_2 r_2$, where r_1 is the distance of atom 1 from the center of mass, r_2 is the distance of atom 2 from the center of mass, and $r_0 = r_1 + r_2$, we can write the moment of inertia (see Problem 31) as

$$I = \mu r_0^2 \qquad\qquad 37\text{-}14$$

where μ, called the **reduced mass**, is

$$\mu = \frac{m_1 m_2}{m_1 + m_2} \qquad\qquad 37\text{-}15$$

DEFINITION—REDUCED MASS

If the masses are equal ($m_1 = m_2 = m$), as in H_2 and O_2, the reduced mass is $\mu = \frac{1}{2}m$ and

$$I = \frac{1}{2} m r_0^2 \qquad\qquad 37\text{-}16$$

A unit of mass convenient for discussing atomic and molecular masses is the **unified mass unit**, u, which is defined as one-twelfth the mass of the neutral carbon-12 (^{12}C) atom. The mass of one ^{12}C atom is thus 12 u. The mass of an atom in unified mass units is therefore numerically equal to the molar mass of the atom in grams. The unified mass unit is related to the gram and kilogram by

$$1\,u = \frac{1\,g}{N_A} = \frac{10^{-3}\,kg}{6.0221 \times 10^{23}} = 1.6606 \times 10^{-27}\,kg \qquad\qquad 37\text{-}17$$

where N_A is Avogadro's number.

THE REDUCED MASS OF A DIATOMIC MOLECULE **EXAMPLE 37-2**

Find the reduced mass of the HCl molecule.

PICTURE THE PROBLEM We find the masses of the hydrogen and chlorine atoms in the periodic table[†] in Appendix C and use the definition in Equation 37-15.

1. The reduced mass μ is related to the individual masses m_H and m_{Cl}:

$$\mu = \frac{m_H m_{Cl}}{m_H + m_{Cl}}$$

2. Find the masses in the periodic table:

$$m_H = 1.01\,u, \quad m_{Cl} = 35.5\,u$$

3. Substitute to calculate the reduced mass:

$$\mu = \frac{m_H m_{Cl}}{m_H + m_{Cl}} = \frac{(1.01\,u)(35.5\,u)}{1.01\,u + 35.5\,u}$$

$$= \boxed{0.982\,u}$$

---

† The masses in these tables are weighted according to the natural isotopic distribution. Thus, the mass of carbon is given as 12.011 rather than 12.000 because natural carbon consists of about 98.9 percent ^{12}C and 1.1 percent ^{13}C. Similarly, natural chlorine consists of about 76 percent ^{35}Cl and 24 percent ^{37}Cl.

REMARKS Note that the reduced mass is less than the mass of either atom in the molecule, and that it is approximately equal to the mass of the hydrogen atom. When one atom of a diatomic molecule is much more massive than the other, the center of mass of the molecule is approximately at the center of the more massive atom, and the reduced mass is approximately equal to the mass of the lighter atom.

ROTATIONAL KINETIC ENERGY OF A MOLECULE   **EXAMPLE   37-3**

Estimate the characteristic rotational energy of an O_2 molecule, assuming that the separation of the atoms is 0.1 nm.

1. The characteristic rotational energy is inversely propor- $E_{0r} = \dfrac{\hbar^2}{2I}$
   tional to the moment of inertia:

2. Calculate the moment of inertia: $I = \mu r_0^2 = \tfrac{1}{2} m r_0^2$

3. Substitute this expression for I into the expression for $E_{0r} = \dfrac{\hbar^2}{m r_0^2}$
   E_{0r}:

4. Use $m = 16$ u for the mass of oxygen and the given val- $E_{0r} = \dfrac{\hbar^2}{m r_0^2}$
   ues of the constants to calculate E_{0r}:

$$= \frac{(1.055 \times 10^{-34}\ \text{J·s})^2}{(16\ \text{u})(10^{-10}\ \text{m})^2} \times \left(\frac{1\ \text{u}}{1.66 \times 10^{-27}\ \text{kg}}\right)$$

$$= 4.19 \times 10^{-23}\ \text{J} = \boxed{2.62 \times 10^{-4}\ \text{eV}}$$

We can see from Example 37-3 that the rotational energy levels are several orders of magnitude smaller than energy levels due to electron excitation, which have energies of the order of 1 eV or higher. Transitions within a given set of rotational energy levels yield photons in the microwave region of the electromagnetic spectrum. The rotational energies are also small compared with the typical thermal energy kT at normal temperatures. For $T = 300$ K, for example, kT is about 2.6×10^{-2} eV, which is approximately 100 times the characteristic rotational energy as calculated in Example 37-3, and approximately 1 percent of the typical electronic energy. Thus, at ordinary temperatures, a molecule can be easily excited to the lower rotational energy levels by collisions with other molecules. But such collisions cannot excite the molecule to its electronic energy levels above the ground state.

Vibrational Energy Levels

The quantization of energy in a simple harmonic oscillator was one of the first problems solved by Schrödinger in his paper proposing his wave equation. Solving the Schrödinger equation for a simple harmonic oscillator gives

$$E_\nu = (\nu + \tfrac{1}{2})hf, \quad \nu = 0, 1, 2, \ldots \tag{37-18}$$

VIBRATIONAL ENERGY LEVELS

where f is the frequency of the oscillator and ν (lowercase Greek nu) is the **vibrational quantum number.**[†] An interesting feature of this result is that the energy levels are equally spaced with intervals equal to hf. The frequency of vibration

---

[†] We use ν here rather than n so as not to confuse the vibrational quantum number with the principal quantum number n for electronic energy levels.

of a diatomic molecule can be related to the force exerted by one atom on the other. Consider two objects of mass m_1 and m_2 connected by a spring of force constant k_F. The frequency of oscillation of this system (see Problem 36) can be shown to be

$$f = \frac{1}{2\pi}\sqrt{\frac{k_F}{\mu}} \qquad\qquad 37\text{-}19$$

where μ is the reduced mass given by Equation 37-15. The effective force constant k_F of a diatomic molecule can thus be determined from a measurement of the frequency of oscillation of the molecule.

A selection rule on transitions between vibrational states (of the same electronic state) requires that the vibrational quantum number ν can change only by ± 1, so the energy of a photon emitted by such a transition is hf and the frequency is f, the same as the frequency of vibration. There is a similar selection rule that ℓ must change by ± 1 for transitions between rotational states.

A typical measured frequency of a transition between vibrational states is 5×10^{13} Hz, which gives

$$E \approx hf = (4.14 \times 10^{-15}\text{ eV·s})(5 \times 10^{13}\text{ s}^{-1}) = 0.2\text{ eV}$$

for the order of magnitude of vibrational energies. This typical vibrational energy is approximately 1000 times greater than the typical rotational energy E_{0r} of the O_2 molecule we found in Example 37-3 and about 8 times greater than the typical thermal energy $kT = 0.026$ eV at $T = 300$ K. Thus, the vibrational levels are almost never excited by molecular collisions at ordinary temperatures.

DETERMINING THE FORCE CONSTANT **EXAMPLE 37-4**

The frequency of vibration of the CO molecule is 6.42×10^{13} Hz. What is the effective force constant for this molecule?

PICTURE THE PROBLEM We use Equation 37-19 to relate k_F to the frequency and the reduced mass, and calculate μ from its definition.

1. The effective force constant is related to the frequency and reduced mass by Equation 37-19:

$$f = \frac{1}{2\pi}\sqrt{\frac{k_F}{\mu}}$$

$$k_F = (2\pi f)^2 \mu$$

2. Calculate the reduced mass using 12 u for the mass of the carbon atom and 16 u for the mass of the oxygen atom:

$$\mu = \frac{m_1 m_2}{m_1 + m_2} = \frac{(12\text{ u})(16\text{ u})}{12\text{ u} + 16\text{ u}} = 6.86\text{ u}$$

3. Substitute this value of μ into the equation for k_F in Step 1 and convert to SI units:

$$k_F = (2\pi f)^2 \mu$$

$$= 4\pi^2 (6.42 \times 10^{13}\text{ Hz})^2 (6.86\text{ u})$$

$$= 1.12 \times 10^{30}\text{ u/s}^2 \times \left(\frac{1.66 \times 10^{-27}\text{ kg}}{1\text{ u}}\right)$$

$$= \boxed{1.85 \times 10^3\text{ N/m}}$$

MASTER the CONCEPT
WEB

Emission Spectra

Figure 37-13 shows schematically some electronic, vibrational, and rotational energy levels of a diatomic molecule. The vibrational levels are labeled with the quantum number ν and the rotational levels are labeled with ℓ. The lower vibrational levels are evenly spaced, with $\Delta E = hf$. For higher vibrational levels, the approximation that the vibration is simple harmonic is not valid and the levels are not quite evenly spaced. Note that the potential energy curves representing the force between the two atoms in the molecule do not have exactly the same shape for the electronic ground and excited states. This implies that the fundamental frequency of vibration f is different for different electronic states. For transitions between vibrational states of different electronic states, the selection rule $\Delta\nu = \pm1$ does not hold. Such transitions result in the emission of photons of wavelengths in or near the visible spectrum, so the emission spectrum of a molecule for electronic transitions is also sometimes called the optical spectrum.

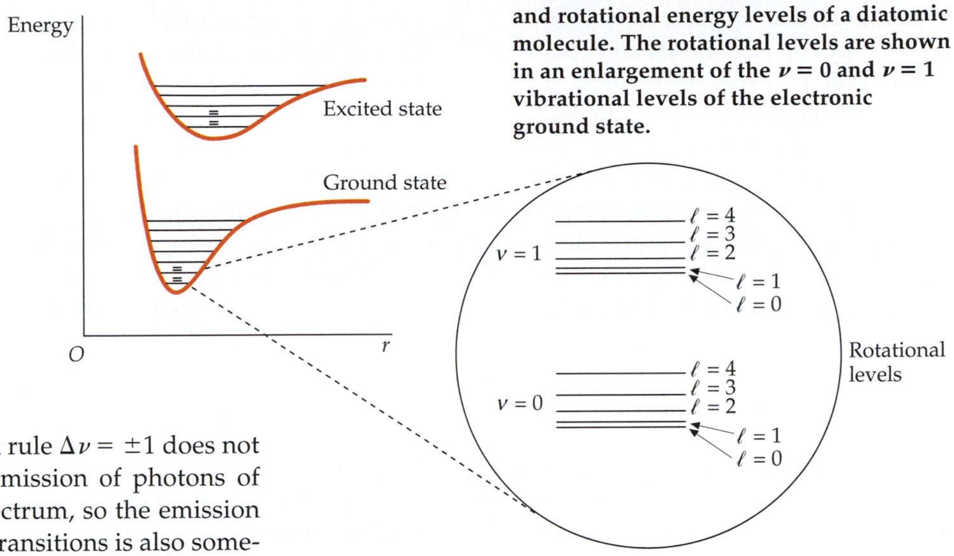

FIGURE 37-13 Electronic, vibrational, and rotational energy levels of a diatomic molecule. The rotational levels are shown in an enlargement of the $\nu = 0$ and $\nu = 1$ vibrational levels of the electronic ground state.

The spacing of the rotational levels increases with increasing values of ℓ. Since the energies of rotation are so much smaller than those of vibrational excitation or electronic excitation of a molecule, molecular rotation shows up in optical spectra as a fine splitting of the spectral lines. When the fine structure is not resolved, the spectrum appears as bands, as shown in Figure 37-14a. Close inspection of these bands reveals that they have a fine structure due to the rotational energy levels, as shown in the enlargement in Figure 37-14c.

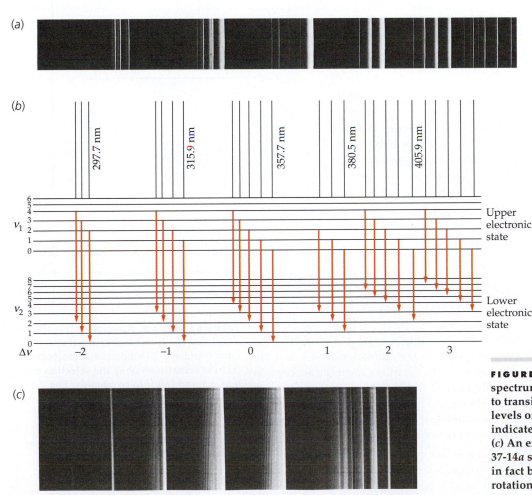

FIGURE 37-14 (a) Part of the emission spectrum of N_2. The spectral lines are due to transitions between the vibrational levels of two electronic states, as indicated in the energy level diagram (b). (c) An enlargement of part of Figure 37-14a shows that the apparent lines are in fact bands with structure caused by rotational levels.

Absorption Spectra

Much molecular spectroscopy is done using infrared absorption techniques in which only the vibrational and rotational energy levels of the ground-state electronic level are excited. For ordinary temperatures, the vibrational energies are sufficiently large in comparison with the thermal energy kT that most of the molecules are in the lowest vibrational state $\nu = 0$, for which the energy is $E_0 = \frac{1}{2}hf$. The transition from $\nu = 0$ to $\nu = 1$ is the predominant transition in absorption. The rotational energies, however, are sufficiently less than the thermal energy kT that the molecules are distributed among several rotational energy states. If the molecule is originally in a vibrational state characterized by $\nu = 0$ and a rotational state characterized by the quantum number ℓ, the molecule's initial energy is

$$E_\ell = \tfrac{1}{2}hf + \ell(\ell + 1)E_{0r} \qquad\qquad 37\text{-}20$$

where E_{0r} is given by Equation 37-13. From this state, two transitions are permitted by the selection rules. For a transition to the next higher vibrational state $\nu = 1$ and a rotational state characterized by $\ell + 1$, the final energy is

$$E_{\ell+1} = \tfrac{3}{2}hf + (\ell + 1)(\ell + 2)E_{0r} \qquad\qquad 37\text{-}21$$

For a transition to the next higher vibrational state and to a rotational state characterized by $\ell - 1$, the final energy is

$$E_{\ell-1} = \tfrac{3}{2}hf + (\ell - 1)\ell E_{0r} \qquad\qquad 37\text{-}22$$

The energy differences are

$$\Delta E_{\ell \to \ell+1} = E_{\ell+1} - E_\ell = hf + 2(\ell + 1)E_{0r} \qquad\qquad 37\text{-}23$$

where $\ell = 0, 1, 2$, and so on, and

$$\Delta E_{\ell \to \ell-1} = E_{\ell-1} - E_\ell = hf - 2\ell E_{0r} \qquad\qquad 37\text{-}24$$

where $\ell = 1, 2, 3$, and so on. (In Equation 37-24, ℓ begins at $\ell = 1$ because from $\ell = 0$ only the transition $\ell \to \ell + 1$ is possible.) Figure 37-15 illustrates these transitions. The frequencies of these transitions are given by

$$f_{\ell \to \ell+1} = \frac{\Delta E_{\ell \to \ell+1}}{h} = f + \frac{2(\ell + 1)E_{0r}}{h}, \quad \ell = 0, 1, 2, \ldots \quad 37\text{-}25$$

and

$$f_{\ell \to \ell-1} = \frac{\Delta E_{\ell \to \ell-1}}{h} = f - \frac{2\ell E_{0r}}{h}, \quad \ell = 1, 2, 3, \ldots \quad 37\text{-}26$$

The frequencies for the transitions $\ell \to \ell + 1$ are thus $f + 2(E_{0r}/h), f + 4(E_{0r}/h)$, $f + 6(E_{0r}/h)$, and so forth; those corresponding to the transition $\ell \to \ell - 1$ are $f - 2(E_{0r}/h), f - 4(E_{0r}/h), f - 6(E_{0r}/h)$, and so forth. We thus expect the absorption spectrum to contain frequencies equally spaced by $2E_{0r}/h$ except for a gap of $4E_{0r}/h$ at the vibrational frequency f, as shown in Figure 37-16. A measurement of the position of the gap gives f and a measurement of the spacing of the absorption peaks gives E_{0r}, which is inversely proportional to the moment of inertia of the molecule.

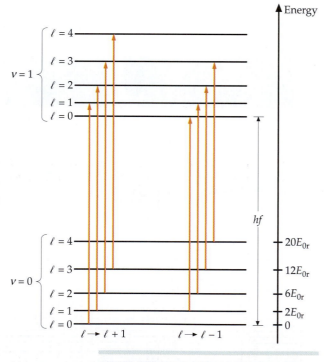

FIGURE 37-15 Absorptive transitions between the lowest vibrational states $\nu = 0$ and $\nu = 1$ in a diatomic molecule. These transitions obey the selection rule $\Delta \ell \pm 1$ and fall into two bands. The energies of the $\ell \to \ell + 1$ band are $hf + 2E_{0r}, hf + 4E_{0r}, hf + 6E_{0r}$, and so forth; whereas the energies of the $\ell \to \ell - 1$ band are $hf - 2\,E_{0r}, hf - 4\,E_{0r}$, $hf - 6E_{0r}$, and so forth.

Figure 37-17 shows the absorption spectrum of HCl. The double-peak structure results from the fact that chlorine occurs naturally in two isotopes, ^{35}Cl and ^{37}Cl, which gives HCl with two different moments of inertia. If all the rotational levels were equally populated initially, we would expect the intensities of each absorption line to be equal. However, the population of a rotational level is proportional to the degeneracy of the level, that is, to the number of states with the same value of ℓ, which is $2\ell + 1$, and to the Boltzmann factor $e^{-E/kT}$, where E is the energy of the state. For low values of ℓ, the population increases slightly because of the degeneracy factor, whereas for higher values of ℓ, the population decreases because of the Boltzmann factor. The intensities of the absorption lines therefore increase with ℓ for low values of ℓ and then decrease with ℓ for high values of ℓ, as can be seen from the figure.

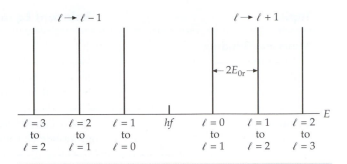

FIGURE 37-16 Expected absorption spectrum of a diatomic molecule. The right branch corresponds to transitions $\ell \to \ell + 1$ and the left branch corresponds to the transitions $\ell \to \ell - 1$. The lines are equally spaced by $2E_{0r}$. The energy midway between the branches is hf, where f is the frequency of vibration of the molecule.

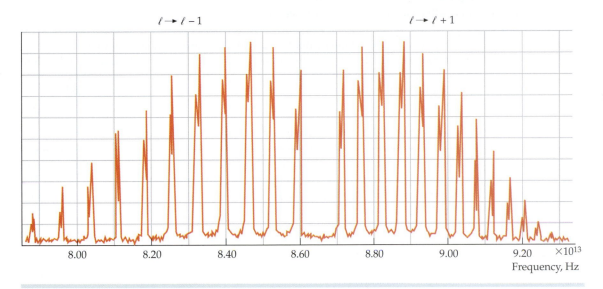

FIGURE 37-17 Absorption spectrum of the diatomic molecule HCl. The double-peak structure results from the two isotopes of chlorine, ^{35}Cl (abundance 75.5 percent) and ^{37}Cl (abundance 24.5 percent). The intensities of the peaks vary because the population of the initial state depends on ℓ.

SUMMARY

1. Atoms are usually found in nature bonded to form molecules or in the lattices of crystalline solids.

2. Ionic bonds and covalent bonds are the principal mechanisms responsible for forming molecules. van der Waals bonds and metallic bonds are important in the formation of liquids and solids. Hydrogen bonds enable large biological molecules to maintain their shape.

3. Like atoms, molecules emit electromagnetic radiation when making a transition from a higher energy state to a lower energy state. The internal energy of a molecule can be separated into three parts: electronic, vibrational, and rotational energy.

4. The molecules in liquids are characterized by a temporary short-range order. The molecules or ions in solids have a more lasting order. Amorphous solids maintain a short-range order similar to the short-range order of a liquid. Crystalline solids display a long-range order determined by their minimum potential energy state.

Topic	Relevant Equations and Remarks

1. Molecular Bonding

Ionic
Ionic bonds result when an electron is transferred from one atom to another, resulting in a positive ion and a negative ion that bond together.

Covalent
The covalent bond is a quantum-mechanical effect that arises from the sharing of one or more electrons by atoms.

van der Waals
The van der Waals bonds are weak bonds that result from the interaction of the instantaneous electric dipole moments of molecules.

Hydrogen
The hydrogen bond results from the sharing of a proton of the hydrogen atom by other atoms.

Metallic
In the metallic bond, the positive lattice ions of the metal are held together by a cloud of negative charge comprised of free electrons.

Mixed
A diatomic molecule formed from two identical atoms, such as O_2, must bond by covalent bonding. The bonding of two nonidentical atoms is often a mixture of covalent and ionic bonding. The percentage of ionic bonding can be found from the ratio of the measured electric dipole moment to the ionic electric dipole moment defined by

$$p_{ionic} = er_0 \qquad \text{37-5}$$

where r_0 is the equilibrium separation of the ions.

2. *Polyatomic Molecules
The shapes of such polyatomic molecules as H_2O and NH_3 can be understood from the spatial distribution of the atomic-orbital or molecular-orbital wave functions. The tetravalent nature of the carbon atom is a result of the hybridization of the 2s and 2p atomic orbitals.

3. Diatomic Molecules

Moment of inertia
$$I = \mu r_0^2 \qquad \text{37-14}$$

where r_0 is the equilibrium separation, and μ is the reduced mass.

Reduced mass
$$\mu = \frac{m_1 m_2}{m_1 + m_2} \qquad \text{37-15}$$

Rotational energy levels
$$E_\ell = \frac{\ell(\ell + 1)\hbar^2}{2I} = \ell(\ell + 1)E_{0r}, \quad \ell = 0, 1, 2, \ldots \qquad \text{37-12}$$

where

$$E_{0r} = \frac{\hbar^2}{2I} \qquad \text{37-13}$$

Vibrational energy levels
$$E_\nu = (\nu + \tfrac{1}{2})hf, \quad \nu = 0, 1, 2, \ldots \qquad \text{37-18}$$

Effective force constant k_F
$$f = \frac{1}{2\pi}\sqrt{\frac{k_F}{\mu}} \qquad \text{37-19}$$

4. Molecular Spectra
The optical spectra of molecules have a band structure due to transitions between rotational levels. Information about the structure and bonding of a molecule can be found from its rotational and vibrational absorption spectrum involving transitions from one vibrational–rotational level to another. These transitions obey the selection rules

$$\Delta\nu = \pm 1, \quad \Delta\ell = \pm 1$$

PROBLEMS

- • Single-concept, single-step, relatively easy
- •• Intermediate-level, may require synthesis of concepts
- ••• Challenging
- SSM Solution is in the *Student Solutions Manual*
- iSOLVE Problems available on iSOLVE online homework service
- iSOLVE✓ These "Checkpoint" online homework service problems ask students additional questions about their confidence level, and how they arrived at their answer.

In a few problems, you are given more data than you actually need; in a few other problems, you are required to supply data from your general knowledge, outside sources, or informed estimates.

Conceptual Problems

1 • SSM Would you expect the NaCl molecule to be polar or nonpolar?

2 • Would you expect the N_2 molecule to be polar or nonpolar?

3 • Does neon occur naturally as Ne or Ne_2? Why?

4 • What type of bonding mechanism would you expect for (*a*) the HF molecule, (*b*) the KBr molecule, (*c*) the N_2 molecule? (*d*) Ag atoms in a solid?

5 •• SSM The elements on the far right column of the periodic table are sometimes called noble gases because they virtually never react with other atoms to form molecules. However, this behavior is sometimes modified if the resulting molecule is formed in an electronic excited state. An example is ArF. When it is formed in the excited state, it is written ArF* and is called an excimer (for excited dimer). Refer to Figure 37-13 and discuss how this diagram would look for ArF in which the ArF ground state is unstable but the ArF* excited state is stable. *Remark: Excimers are used in certain kinds of lasers.*

6 • Find other elements with the same subshell electron configuration in the two outermost orbitals as carbon. Would you expect the same type of hybridization for these elements as for carbon?

7 • How does the value of the effective force constant calculated for the CO molecule in Example 37-4 compare with the value of the force constant of the suspension springs on a typical automobile, which is about 1.5 kN/m?

8 • Explain why the moment of inertia of a diatomic molecule increases slightly with increasing angular momentum.

9 • Why would you expect the separation distance between the two protons to be larger in the H_2^+ ion than in the H_2 molecule?

10 • Why does an atom usually absorb radiation only from the ground state, whereas a diatomic molecule can absorb radiation from many different rotational states?

11 •• The vibrational energy levels of diatomic molecules are described by a single vibrational frequency f that is the frequency of vibration between the two atoms of the molecule. What would you expect to see in the case of polyatomic molecules? Consider in particular the water molecule H_2O (Figure 37-9).

Estimation and Approximation

12 •• The Anharmonic Oscillator: The potential energy between the atoms in a diatomic molecule has a minimum as shown in Figure 37-13. Near this minimum the graph for the energy as a function of distance between the atoms may be approximated as a parabola, leading to the harmonic oscillator model for the vibrating molecule. An improved approximation is called the anharmonic oscillator and leads to a modification of the energy formula (Equation 37-18). The improved formula for energy is

$$E_\nu = (\nu + \tfrac{1}{2})hf - (\nu + \tfrac{1}{2})^2 hf\alpha$$

For the O_2 molecule, the constants have the values $f = 4.74 \times 10^{13}\ \text{s}^{-1}$ and $\alpha = 7.6 \times 10^{-3}$. Use this formula to estimate the value of the quantum number ν for which the improved formula corrects the original formula by 10 percent.

13 •• To understand why quantum mechanics is not needed to describe many macroscopic systems, estimate the quantum number ℓ and spacing between adjacent energy levels for a baseball ($m \sim 300$ g, $r \sim 3$ cm) spinning about its own axis at 20 rev/min. *Hint: Pick ℓ so the quantum energy formula (Equation 37-12) gives the correct energy for the given system. Then find the energy increase for the next highest energy level.*

14 •• SSM Repeat Problem 13, finding the quantum number ν and spacing between adjacent energy levels for a 5-kg mass attached to 1500-N/m spring vibrating with an amplitude of 2 cm. *Hint: Pick ν so that the quantum energy formula (Equation 37-18) gives the correct energy for the given system. Then find the energy increase for the next highest energy level.*

Molecular Bonding

15 • iSOLVE✓ Calculate the separation of Na^+ and Cl^- ions, for which the potential energy is -1.52 eV.

16 • The dissociation energy of Cl_2 is 2.48 eV. Consider the formation of NaCl according to the reaction $2Na + Cl \rightarrow 2NaCl$. Does this reaction absorb energy or release energy? How much energy per molecule is absorbed or released?

17 • The dissociation energy is sometimes expressed in kilocalories per mole (kcal/mol). (*a*) Find the relation between the units eV/molecule and kcal/mol. (*b*) Find the dissociation energy of molecular NaCl in kcal/mol.

18 • [SSM] [SOLVE] The equilibrium separation of the HF molecule is 0.0917 nm, and its measured electric dipole moment is 6.40×10^{-30} C·m. What percentage of the bonding is ionic?

19 •• The dissociation energy of RbF is 5.12 eV, and the equilibrium separation is 0.227 nm. The electron affinity of fluorine is 3.40 eV, and the ionization energy of rubidium is 4.18 eV. Determine the core-repulsion energy of RbF.

20 •• [SOLVE] The equilibrium separation of the K^+ and Cl^- ions in KCl is about 0.267 nm. (a) Calculate the potential energy of attraction of the ions, assuming them to be point charges at this separation. (b) The ionization energy of potassium is 4.34 eV, and the electron affinity of chlorine is 3.62 eV. Find the dissociation energy neglecting any energy of repulsion. (See Figure 37-1.) The measured dissociation energy is 4.49 eV. What is the energy due to repulsion of the ions at the equilibrium separation?

21 •• Indicate the mean value of r for two vibration levels in the potential energy curve for a diatomic molecule. Show that because of the asymmetry in the curve, r_{av} increases with increasing vibration energy, and therefore solids expand when heated.

22 •• [SOLVE] Calculate the potential energy of attraction between the Na^+ and Cl^- ions at the equilibrium separation $r_0 = 0.236$ nm. Compare this result with the dissociation energy given in Figure 37-1. What is the energy due to repulsion of the ions at the equilibrium separation?

23 •• [SOLVE]✓ The equilibrium separation of the K^+ and F^- ions in KF is about 0.217 nm. (a) Calculate the potential energy of attraction of the ions, assuming them to be point charges at this separation. (b) The ionization energy of potassium is 4.34 eV, and the electron affinity of fluorine is 3.40 eV. Find the dissociation energy neglecting any energy of repulsion. (c) The measured dissociation energy is 5.07 eV. Calculate the energy due to repulsion of the ions at the equilibrium separation.

24 ••• [SSM] Assume that the potential energy associated with the core repulsion of the two ions of a diatomic molecule with ionic bonding can be represented by a potential energy of the form $U_{rep} = C/r^n$, so the total potential energy is $U = U_e + U_{rep} + \Delta E$, where $U_e = -ke^2/r$. ΔE is the energy of the two ions at infinite separation less the energy of the two neutral atoms at infinite separation (see Figure 37-1). Use $\frac{dU}{dr} = 0$ at $r = r_0$ to show that $n = \frac{|U_e(r_0)|}{U_{rep}(r_0)}$

25 ••• (a) Find U_{rep} at $r = r_0$ for NaCl. (b) Assume $U_{rep} = C/r^n$ and find C and n for NaCl. (See Problem 24.)

Energy Levels of Spectra of Diatomic Molecules

26 • [SOLVE] The characteristic rotational energy E_{0r} for the rotation of the N_2 molecule is 2.48×10^{-4} eV. From this, find the separation distance of the 2 nitrogen atoms.

27 • [SSM] [SOLVE]✓ The separation of the two oxygen atoms in a molecule of O_2 is actually slightly greater than the 0.1 nm used in Example 37-3, and the characteristic energy of rotation E_{0r} is 1.78×10^{-4} eV rather than the result obtained in that example. Use this value to calculate the separation distance of the two oxygen atoms.

28 •• Show that the reduced mass is smaller than either mass in a diatomic molecule and calculate it for (a) H_2, (b) N_2, (c) CO, and (d) HCl. Express your answers in unified mass units.

29 •• The CO molecule has a binding energy of approximately 11 eV. Find the vibrational quantum number v that would cause the molecule to have this much energy, and thus cause it to "shake" apart.

30 •• [SSM] [SOLVE]✓ The equilibrium separation between the nuclei of the LiH molecule is 0.16 nm. Determine the energy separation between the $\ell = 3$ and $\ell = 2$ rotational levels of this diatomic molecule.

31 •• [SSM] Derive Equations 37-14 and 37-15 for the moment of inertia in terms of the reduced mass of a diatomic molecule.

32 •• [SOLVE]✓ Use the separation of the K^+ and Cl^- ions given in Problem 20 and the reduced mass of KCl to calculate the characteristic rotational energy E_{0r}.

33 •• [SOLVE] The central frequency for the absorption band of HCl shown in Figure 37-17 is at $f = 8.66 \times 10^{13}$ Hz, and the absorption peaks are separated by about $f = 6 \times 10^{11}$ Hz. Use this information to find (a) the lowest (zero-point) vibrational energy for HCl, (b) the moment of inertia of HCl, and (c) the equilibrium separation of the atoms.

34 •• [SOLVE] Calculate the effective force constant for HCl from its reduced mass and the fundamental vibrational frequency obtained from Figure 37-17.

35 •• To see how the population of rotational states of the oxygen molecule depends on the angular momentum quantum number ℓ, use a spreadsheet program or graphing calculator to graph the function $(2\ell + 1)e^{-E_\ell/kT}$, where $E_\ell = \ell(\ell + 1)E_{0r}$ for values of $0 \le \ell \le 10$ at $T = 100K$, $200K$, $300K$, and $500K$.

36 •• [SSM] Two objects of mass m_1 and m_2 are attached to a spring of force constant k and equilibrium length r_0. (a) Show that when m_1 is moved a distance Δr_1 from the center of mass, the force exerted by the spring is

$$F = -k\left(\frac{m_1 + m_2}{m_2}\right)\Delta r_1$$

(b) Show that the frequency of oscillation is $f = \left(\frac{1}{2\pi}\right)\sqrt{k/\mu}$, where μ is the reduced mass.

37 ••• Calculate the reduced mass for the $H^{35}Cl$ and $H^{37}Cl$ molecules and the fractional difference $\Delta\mu/\mu$. Show that the mixture of isotopes in HCl leads to a fractional difference in the frequency of a transition from one rotational state to another given by $\Delta f/f = -\Delta\mu/\mu$. Compute $\Delta f/f$ and compare your result with Figure 37-17.

General Problems

38 • Show that when one atom in a diatomic molecule is much more massive than the other the reduced mass is approximately equal to the mass of the lighter atom.

39 •• The equilibrium separation between the nuclei of the CO molecule is 0.113 nm. Determine the energy difference between the $\ell = 2$ and $\ell = 1$ rotational energy levels of this molecule.

40 •• [SSM] The effective force constant for the HF molecule is 970 N/m. Find the frequency of vibration for this molecule.

41 •• [iSOLVE] The frequency of vibration of the NO molecule is 5.63×10^{13} Hz. Find the effective force constant for NO.

42 •• The effective force constant of the hydrogen bond in the H_2 molecule is 580 N/m. Obtain the energies of the four lowest vibrational levels of the H_2, HD, and D_2 molecules and the wavelengths of photons resulting from transitions between adjacent vibrational levels of these molecules.

43 •• The potential energy between two atoms in a molecule can often be described rather well by the Lenard–Jones potential, which can be written as

$$U = U_0 \left[\left(\frac{a}{r} \right)^{12} - 2 \left(\frac{a}{r} \right)^6 \right],$$

where U_0 and a are constants. Find the interatomic separation r_0 in terms of a for which the potential energy is a minimum. Find the corresponding value of U_{min}. Use Figure 37-4 to obtain numerical values of r_0 and U_0 for the H_2 molecule, and express your answers in nanometers and electron volts.

44 •• In this problem, you are to find how the van der Waals force between a polar molecule and a nonpolar molecule depends on the distance between the molecules. Let the dipole moment of the polar molecule be in the x direction and the nonpolar molecule be a distance x away. (*a*) How does the electric field due to an electric dipole depend on distance x? (*b*) Use the facts that (1) the potential energy of an electric dipole of moment $\vec{p}$ in an electric field $\vec{E}$ is $U = -\vec{p} \cdot \vec{E}$, and (2) the induced dipole moment of the nonpolar molecule is

proportional to E, to find how the potential energy of interaction of the two molecules depends on separation distance. (*c*) Using $F_x = -dU/dx$, find the x dependence of the force between the two molecules.

45 •• Find the dependence of the force on separation distance between two polar molecules. (See Problem 44.)

46 •• Use the infrared absorption spectrum of HCl in Figure 37-17 to obtain (*a*) the characteristic rotational energy E_{0r} (in eV) and (*b*) the vibrational frequency f and the vibrational energy hf (in eV).

47 •• [SSM] For a molecule such as CO, which has a permanent electric dipole moment, radiative transitions obeying the selection rule $\Delta \ell = \pm 1$ between two rotational energy levels of the same vibrational level are allowed. (That is, the selection rule $\Delta v = \pm 1$ does not hold.) (*a*) Find the moment of inertia of CO and calculate the characteristic rotational energy E_{0r} (in eV). (*b*) Make an energy-level diagram for the rotational levels for $\ell = 0$ to $\ell = 5$ for some vibrational level. Label the energies in electron volts, starting with $E = 0$ for $\ell = 0$. Indicate on your diagram the transitions that obey $\Delta \ell = -1$, and calculate the energy of the photon emitted. (*c*) Find the wavelength of the photons emitted for each transition in (*b*). In what region of the electromagnetic spectrum are these photons?

48 ••• [SSM] Use the results of Problem 24 to calculate the vibrational frequency of the LiCl molecule. The dissociation energy of LiCl is 4.86 eV, and the equilibrium separation is 0.202 nm. The electron affinity of chlorine is 3.62 eV, and the ionization energy of lithium is 5.39 eV. To do this, expand the potential about $r = r_0$, where r_0 is the equilibrium separation, in a Taylor series. Retain only the term proportional to $(r - r_0)^2$. Recall that the potential energy of a simple harmonic oscillator is given by $U_{SHO} = \frac{1}{2} m \omega^2 x^2$. What is the wavelength resulting from transitions between adjacent harmonic oscillator levels of this molecule?

Solids

SILICON INGOT SILICON IS A SEMICONDUCTOR, AND SLICES OF INGOTS LIKE THIS ARE USED TO PRODUCE TRANSISTORS AND OTHER ELECTRONIC DEVICES. TO MAKE A TRANSISTOR, ATOMS OF ARSENIC AND GALLIUM ARE INJECTED INTO THE SILICON.

? **Do you know how many atoms of arsenic it takes to increase the charge-carrier density by a factor of 5 million? For more on this topic, see Example 38-7.**

The first microscopic model of electric conduction in metals was proposed by Paul K. Drude in 1900 and developed by Hendrik A. Lorentz about 1909. This model successfully predicts that the current is proportional to the potential drop (Ohm's law) and relates the resistivity ρ of conductors to the mean speed v_{av} and the mean free path[†] λ of the free electrons within the conductor. However, when v_{av} and λ are interpreted classically, there is a disagreement between the calculated values and the measured values of the resistivity, and a similar disagreement between the predicted temperature dependence and the observed temperature dependence. Thus, the classical theory fails to adequately describe the resistivity of metals. Furthermore, the classical theory says nothing about the most striking property of solids, namely that some materials are conductors, others are insulators, and still others are semiconductors, which are materials whose resistivity falls between that of conductors and insulators.

---

† The mean free path is the average distance traveled between collisions.

When v_{av} and λ are interpreted using quantum theory, both the magnitude and the temperature dependence of the resistivity are correctly predicted. In addition, quantum theory allows us to determine if a material will be a conductor, an insulator, or a semiconductor.

➤ **In this chapter, we use our understanding of quantum mechanics to discuss the structure of solids and solid-state semiconducting devices. Much of our discussion will be qualitative because, as in atomic physics, the quantum-mechanical calculations are very difficult.**

38-1 The Structure of Solids

The three phases of matter we observe—gas, liquid, and solid—result from the relative strengths of the attractive forces between molecules and the thermal energy of the molecules. Molecules in the gas phase have a relatively high thermal kinetic energy, and such molecules have little influence on one another except during their frequent but brief collisions. At sufficiently low temperatures (depending on the type of molecule), van der Waals forces will cause practically every substance to condense into a liquid and then into a solid. In liquids, the molecules are close enough—and their kinetic energy is low enough—that they can develop a temporary **short-range order.** As thermal kinetic energy is further reduced, the molecules form solids, which are characterized by a lasting order.

If a liquid is cooled slowly so that the kinetic energy of its molecules is reduced slowly, the molecules (or atoms or ions) may arrange themselves in a regular crystalline array, producing the maximum number of bonds and leading to a minimum potential energy. However, if the liquid is cooled rapidly so that its internal energy is removed before the molecules have a chance to arrange themselves, the solid formed is often not crystalline but instead resembles a snapshot of the liquid. Such a solid is called an **amorphous solid.** It displays short-range order but not the **long-range order** (over many molecular diameters) that is characteristic of a crystal. Glass is a typical amorphous solid. A characteristic result of the long-range ordering of a crystal is that it has a well-defined melting point, whereas an amorphous solid merely softens as its temperature is increased. Many materials may solidify into either an amorphous state or a crystalline state depending on how the materials are prepared; others exist only in one form or the other.

Most common solids are polycrystalline; that is, they consist of many single crystals that meet at grain boundaries. The size of a single crystal is typically a fraction of a millimeter. However, large single crystals do occur naturally and can be produced artificially. The most important property of a single crystal is the symmetry and regularity of its structure. It can be thought of as having a single unit structure that is repeated throughout the crystal. This smallest unit of a crystal is called the **unit cell;** its structure depends on the type of bonding—ionic, covalent, metallic, hydrogen, van der Waals—between the atoms, ions, or molecules. If more than one kind of atom is present, the structure will also depend on the relative sizes of the atoms.

Figure 38-1 shows the structure of the ionic crystal sodium chloride (NaCl). The Na^+ and Cl^- ions are spherically symmetric, and the Cl^- ion is approximately twice as large as the Na^+ ion. The minimum potential energy for this crystal occurs when an ion of either kind has six nearest neighbors of the other kind. This structure is called *face-centered-cubic* (fcc). Note that the Na^+ and Cl^- ions in solid NaCl are *not* paired into NaCl molecules.

The net attractive part of the potential energy of an ion in a crystal can be written

$$U_{att} = -\alpha \frac{ke^2}{r} \qquad 38\text{-}1$$

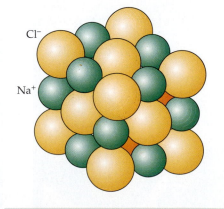

FIGURE 38-1 Face-centered-cubic structure of the NaCl crystal.

Cl^-

Na^+

where r is the separation distance between neighboring ions (0.281 nm for the Na^+ and Cl^- ions in crystalline NaCl), and α, called the **Madelung constant,** depends on the geometry of the crystal. If only the six nearest neighbors of each ion were important, α would be six. However, in addition to the six neighbors of the opposite charge at a distance r, there are twelve ions of the same charge at a distance $\sqrt{2}r$, eight ions of opposite charge at a distance $\sqrt{3}r$, and so on. The Madelung constant is thus an infinite sum:

$$\alpha = 6 - \frac{12}{\sqrt{2}} + \frac{8}{\sqrt{3}} - \ldots \qquad\qquad 38\text{-}2$$

The result for face-centered-cubic structures is $\alpha = 1.7476$.[†]

[†] A large number of terms are needed to calculate the Madelung constant accurately because the sum converges very slowly.

Crystal structure. (*a*) The hexagonal symmetry of a snowflake arises from a hexagonal symmetry in its lattice of hydrogen atoms and oxygen atoms. (*b*) NaCl (salt) crystals, magnified approximately thirty times. The crystals are built up from a cubic lattice of sodium and chloride ions. In the absence of impurities, an exact cubic crystal is formed. This (false-color) scanning electron micrograph shows that in practice the basic cube is often disrupted by dislocations, giving rise to crystals with a wide variety of shapes. The underlying cubic symmetry, though, remains evident. (*c*) A crystal of quartz (SiO_2, silicon dioxide), the most abundant and widespread mineral on the earth. If molten quartz solidifies without crystallizing, glass is formed. (*d*) A soldering iron tip, ground down to reveal the copper core within its iron sheath. Visible in the iron is its underlying microcrystalline structure.

(*a*)

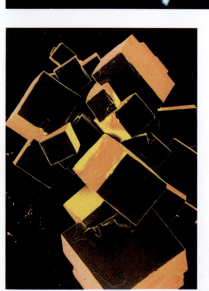

(*b*)

(*c*)

(*d*)

When Na$^+$ and Cl$^-$ ions are very close together, they repel each other because of the overlap of their electrons and the exclusion-principle repulsion discussed in Section 37-1. A simple empirical expression for the potential energy associated with this repulsion that works fairly well is

$$U_{\text{rep}} = \frac{A}{r^n}$$

where A and n are constants. The total potential energy of an ion is then

$$U = -\alpha \frac{ke^2}{r} + \frac{A}{r^n} \tag{38-3}$$

The equilibrium separation $r = r_0$ is that at which the force $F = -dU/dr$ is zero. Differentiating and setting $dU/dr = 0$ at $r = r_0$, we obtain

$$A = \frac{\alpha ke^2 r_0^{n-1}}{n} \tag{38-4}$$

The total potential energy can thus be written

$$U = -\alpha \frac{ke^2}{r_0} \left[\frac{r_0}{r} - \frac{1}{n} \left(\frac{r_0}{r} \right)^n \right] \tag{38-5}$$

At $r = r_0$, we have

$$U(r_0) = -\alpha \frac{ke^2}{r_0} \left(1 - \frac{1}{n} \right) \tag{38-6}$$

If we know the equilibrium separation r_0, the value of n can be found approximately from the *dissociation energy* of the crystal, which is the energy needed to break up the crystal into atoms.

SEPARATION DISTANCE BETWEEN Na$^+$ AND Cl$^-$ IN NaCl   **EXAMPLE 38-1**

Calculate the equilibrium spacing r_0 for NaCl from the measured density of NaCl, which is $\rho = 2.16$ g/cm^3.

PICTURE THE PROBLEM We consider each ion to occupy a cubic volume of side r_0. The mass of 1 mol of NaCl is 58.4 g, which is the sum of the atomic masses of sodium and chlorine. There are $2N_A$ ions in 1 mol of NaCl, where $N_A = 6.02 \times 10^{23}$ is Avogadro's number.

1. The volume v per mol of NaCl equals the number of ions times the volume per ion:

$$v = 2N_A r_0^3$$

2. Relate r_0 to the density ρ:

$$\rho = \frac{M}{v} = \frac{M}{2N_A r_0^3}$$

3. Solve for r_0^3 and substitute the known values:

$$r_0^3 = \frac{M}{2N_A \rho} = \frac{58.4 \text{ g}}{2(6.02 \times 10^{23})(2.16 \text{ g/cm}^3)}$$

$$= 2.25 \times 10^{-23} \text{ cm}^3$$

so

$$r_0 = 2.82 \times 10^{-8} \text{ cm} = \boxed{0.282 \text{ nm}}$$

The measured dissociation energy of NaCl is 770 kJ/mol. Using 1 eV = 1.602×10^{-19} J, and the fact that 1 mol of NaCl contains N_A pairs of ions, we can express the dissociation energy in electron volts per ion pair. The conversion between electron volts per ion pair and kilojoules per mole is

$$1\frac{\text{eV}}{\text{ion pair}} \times \frac{6.022 \times 10^{23} \text{ ion pairs}}{1 \text{ mol}} \times \frac{1.602 \times 10^{-19} \text{ J}}{1 \text{ eV}}$$

The result is

$$1\frac{\text{eV}}{\text{ion pair}} = 96.47 \frac{\text{kJ}}{\text{mol}} \qquad\qquad 38\text{-}7$$

Thus, 770 kJ/mol = 7.98 eV per ion pair. Substituting -7.98 eV for $U(r_0)$, 0.282 nm for r_0, and 1.75 for α in Equation 38-6, we can solve for n. The result is $n = 9.35 \approx 9$.

Most ionic crystals, such as LiF, KF, KCl, KI, and AgCl, have a face-centered-cubic structure. Some elemental solids that have fcc structure are silver, aluminum, gold, calcium, copper, nickel, and lead.

Figure 38-2 shows the structure of CsCl, which is called the *body-centered-cubic* (bcc) structure. In this structure, each ion has eight nearest neighbor ions of the opposite charge. The Madelung constant for these crystals is 1.7627. Elemental solids with bcc structure include barium, cesium, iron, potassium, lithium, molybdenum, and sodium.

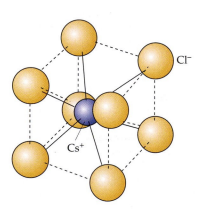

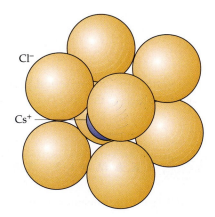

FIGURE 38-2 Body-centered-cubic structure of the CsCl crystal.

Figure 38-3 shows another important crystal structure: the *hexagonal close-packed* (hcp) structure. This structure is obtained by stacking identical spheres, such as bowling balls. In the first layer, each ball touches six others; thus, the name *hexagonal*. In the next layer, each ball fits into a triangular depression of the first layer. In the third layer, each ball fits into a triangular depression of the second layer, so it lies directly over a ball in the first layer. Elemental solids with hcp structure include beryllium, cadmium, cerium, magnesium, osmium, and zinc.

In some solids with covalent bonding, the crystal structure is determined by the directional nature of the bonds. Figure 38-4 illustrates the diamond structure of carbon, in which each atom is bonded to four other atoms as a result of hybridization, which is discussed in Section 38-2. This is also the structure of germanium and silicon.

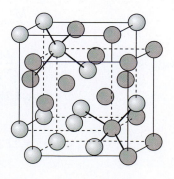

FIGURE 38-3 Hexagonal close-packed crystal structure.

FIGURE 38-4 Diamond crystal structure. This structure can be considered to be a combination of two interpenetrating face-centered-cubic structures.

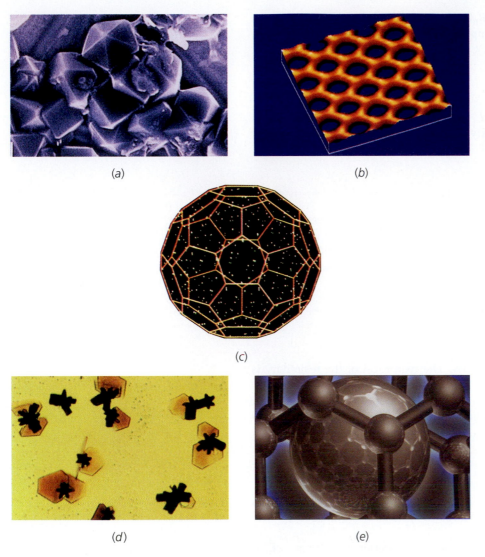

(a)

(b)

(c)

(d)

(e)

Carbon exists in three well-defined crystalline forms: diamond, graphite, and fullerenes (short for "buckminsterfullerenes"), the third of which was predicted and discovered less than two decades ago. The forms differ in how the carbon atoms are packed together in a lattice. A fourth form of carbon, in which no well-defined crystalline form exists, is common charcoal. (*a*) Synthetic diamonds, magnified approximately 75,000 times. In diamond, each carbon atom is centered in a tetrahedron of four other carbon atoms. The strength of these bonds accounts for the hardness of a diamond. (*b*) An atomic-force micrograph of graphite. In graphite, carbon atoms are arranged in sheets, with each sheet made up of atoms in hexagonal rings. The sheets slide easily across one another, a property that allows graphite to function as a lubricant. (*c*) A single sheet of carbon rings can be closed on itself if certain rings are allowed to be pentagonal, instead of hexagonal. A computer-generated image of the smallest such structure, C_{60}, is shown here. Each of the sixty vertices corresponds to a carbon atom; twenty of the faces are hexagons and twelve of the faces are pentagons. The same geometric pattern is encountered in a soccer ball. (*d*) Fullerene crystals, in which C_{60} molecules are close-packed. The smaller crystals tend to form thin brownish platelets; larger crystals are usually rod-like in shape. Fullerenes exist in which more than sixty carbon atoms appear. In the crystals shown here, about one-sixth of the molecules are C_{70}. (*e*) Carbon nanotubes have very interesting electrical properties. A single graphite sheet is a semimetal, which means that it has properties intermediate between those of semiconductors and those of metals. When a graphite sheet is rolled into a nanotube, not only do the carbon atoms have to line up around the circumference of the tube, but the wavefunctions of the electrons must also match up. This boundary-matching requirement places restrictions on these wavefunctions, which affects the motion of the electrons. Depending on exactly how the tube is rolled up, the nanotube can be either a semiconductor or a metal.

38-2 A Microscopic Picture of Conduction

We consider a metal as a regular three-dimensional lattice of ions filling some volume V and containing a large number N of electrons that are free to move throughout the whole metal. Experimentally, the number of free electrons in a metal is approximately one electron to four electrons per atom. In the absence of an electric field, the free electrons move about the metal randomly, much the way gas molecules move about in a container. We will often refer to these free electrons in a metal as an electron gas.

The current in a conducting wire segment is proportional to the voltage drop across the segment:

$$I = \frac{V}{R}, \quad \text{or } (V = IR)$$

The resistance R is proportional to the length L of the wire segment and inversely proportional to the cross-sectional area A:

$$R = \rho \frac{L}{A}$$

where ρ is the resistivity. Substituting $\rho L/A$ for R, and EL for V, we can write the current in terms of the electric field strength E and the resistivity. We have

$$I = \frac{V}{R} = \frac{EL}{\rho L/A} = \frac{1}{\rho}EA$$

Dividing both sides by the area A gives $I/A = (1/\rho)E$, or $J = (1/\rho)E$, where $J = I/A$ is the magnitude of the **current density** vector $\vec{J}$. The current density vector is defined as

$$\vec{J} = qn\vec{v}_d \qquad\qquad 38\text{-}8$$

DEFINITION—CURRENT DENSITY

where q, n, and $\vec{v}_d$ are the charge, the number density, and the drift velocity of the charge carrier. (This follows from Equation 25-3). In vector form, the relation between the current density and the electric field is

$$\vec{J} = \frac{1}{\rho}\vec{E} \qquad\qquad 38\text{-}9$$

This relation is the point form of Ohm's law. The reciprocal of the resistivity is called the **conductivity.**

According to Ohm's law, the resistivity is independent of both the current density and the electric field $\vec{E}$. Combining Equation 38-8 and Equation 38-9 gives

$$-en_e\vec{v}_d = \frac{1}{\rho}\vec{E} \qquad\qquad 38\text{-}10$$

where $-e$ has been substituted for q. According to Equation 38-10, the drift velocity $\vec{v}_d$ is proportional to $\vec{E}$.

In the presence of an electric field, a free electron experiences a force $-e\vec{E}$. If this were the only force acting, the electron would have a constant acceleration $-e\vec{E}/m_e$. However, Equation 38-10 implies a steady-state situation with a constant drift velocity that is proportional to the field $\vec{E}$. In the microscopic model, it is assumed that a free electron is accelerated for a short time and then makes a collision with a lattice ion. The velocity of the electron immediately after the collision is completely unrelated to the drift velocity. The justification for this assumption is that the magnitude of the drift velocity is extremely small compared with the random thermal speeds of the electrons.

For a typical electron, its velocity a time t after its last collision is $\vec{v}_0 - (e\vec{E}/m_e)t$, where $\vec{v}_0$ is its velocity immediately after that collision. Since the direction of $\vec{v}_0$ is random, it does not contribute to the average velocity of the electrons. Thus, the average velocity or drift velocity of the electrons is

$$\vec{v}_d = -\frac{e\vec{E}}{m_e}\tau \qquad\qquad 38\text{-}11$$

where τ is the average time since the last collision. Substituting for $\vec{v}_d$ in Equation 38-10, we obtain

$$-n_e e\left(-\frac{e\vec{E}}{m_e}\tau\right) = \frac{1}{\rho}\vec{E}$$

so

$$\rho = \frac{m_e}{n_e e^2 \tau} \qquad\qquad 38\text{-}12$$

The time τ, called the **collision time,** is also the average time between collisions.[†] The average distance an electron travels between collisions is $v_{av}\tau$, which is called the mean free path λ:

$$\lambda = v_{av}\tau \qquad\qquad 38\text{-}13$$

where v_{av} is the mean speed of the electrons. (The mean speed is many orders of magnitude greater than the drift speed.) In terms of the mean free path and the mean speed, the resistivity is

$$\rho = \frac{m_e v_{av}}{n_e e^2 \lambda} \qquad\qquad 38\text{-}14$$

RESISTIVITY IN TERMS OF v_{AV} AND λ

According to Ohm's law, the resistivity ρ is independent of the electric field $\vec{E}$. Since m_e, n_e, and e are constants, the only quantities that could possibly depend on $\vec{E}$ are the mean speed v_{av} and the mean free path λ. Let us examine these quantities to see if they can possibly depend on the applied field $\vec{E}$.

Classical Interpretation of v_{av} and λ

Classically, at $T = 0$ all the free electrons in a conductor should have zero kinetic energy. As the conductor is heated, the lattice ions acquire an average kinetic energy of $\frac{3}{2}kT$, which is imparted to the electron gas by the collisions between the electrons and the ions. (This is a result of the equipartition theorem studied in Chapters 17 and 18.) The electron gas would then have a Maxwell–Boltzmann distribution just like a gas of molecules. In equilibrium, the electrons would be expected to have a mean kinetic energy of $\frac{3}{2}kT$, which at ordinary temperatures ($\sim$300 K) is approximately 0.04 eV. At $T = 300$ K, their root-mean-square (rms) speed,[‡] which is slightly greater than the mean speed, is

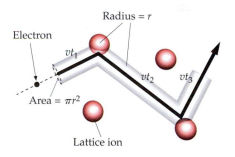

$$v_{av} \approx v_{rms} = \sqrt{\frac{3kT}{m_e}} = \sqrt{\frac{3(1.38 \times 10^{-23}\,\text{J/K})(300\,\text{K})}{9.11 \times 10^{-31}\,\text{kg}}} \qquad 38\text{-}15$$

$$= 1.17 \times 10^5\,\text{m/s}$$

Note that this is about nine orders of magnitude greater than the typical drift speed of 3.5×10^{-5} m/s, which was calculated in Example 25-1. The very small drift speed caused by the electric field therefore has essentially no effect on the very large mean speed of the electrons, so v_{av} in Equation 38-14 cannot depend on the electric field $\vec{E}$.

The mean free path is related classically to the size of the lattice ions in the conductor and to the number of ions per unit volume. Consider one electron moving with speed v through a region of stationary ions that are assumed to be hard spheres (Figure 38-5). The size of the electron is assumed to be negligible.

FIGURE 38-5 Model of an electron moving through the lattice ions of a conductor. The electron, which is considered to be a point particle, collides with an ion if it comes within a distance r of the center of the ion, where r is the radius of the ion. If the electron speed is v, it collides in time t with all the ions whose centers are in the volume $\pi r^2 vt$. While this picture is in accord with the classical Drude model for conduction in metals, it is in conflict with the current quantum-mechanical model presented later in this chapter.

---

† It is tempting but incorrect to think that if τ is the average time between collisions, the average time since its last collision is $\frac{1}{2}\tau$ rather than τ. (If you find this confusing, you may take comfort in the fact that Drude used the incorrect result $\frac{1}{2}\tau$ in his original work.)

‡ See Equation 17-23.

The electron will collide with an ion if it comes within a distance r from the center of the ion, where r is the radius of the ion. In some time t_1, the electron moves a distance vt_1. If there is an ion whose center is in the cylindrical volume $\pi r^2 vt_1$, the electron will collide with the ion. The electron will then change directions and collide with another ion in time t_2 if the center of the ion is in the volume $\pi r^2 vt_2$. Thus, in the total time $t = t_1 + t_2 + \ldots$, the electron will collide with the number of ions whose centers are in the volume $\pi r^2 vt$. The number of ions in this volume is $n_{\text{ion}} \pi r^2 vt$, where n_{ion} is the number of ions per unit volume. The total path length divided by the number of collisions is the mean free path:

$$\lambda = \frac{vt}{n_{\text{ion}} \pi r^2 vt} = \frac{1}{n_{\text{ion}} \pi r^2} = \frac{1}{n_{\text{ion}} A} \qquad \text{38-16}$$

where $A = \pi r^2$ is the cross-sectional area of a lattice ion.

Successes and Failures of the Classical Model

Neither n_{ion} nor r depends on the electric field $\vec{E}$, so λ also does not depend on $\vec{E}$. Thus, according to the classical interpretation of v_{av} and λ, neither depend on $\vec{E}$, so the resistivity ρ does not depend on $\vec{E}$, in accordance with Ohm's law. However, the classical theory gives an incorrect temperature dependence for the resistivity. Because λ depends only on the radius and the number density of the lattice ions, the only quantity in Equation 38-14 that depends on temperature in the classical theory is v_{av}, which is proportional to $\sqrt{T}$. But experimentally, ρ varies linearly with temperature. Furthermore, when ρ is calculated at $T = 300$ K using the Maxwell–Boltzmann distribution for v_{av} and Equation 38-16 for λ, the numerical result is about six times greater than the measured value.

The classical theory of conduction fails because electrons are not classical particles. The wave nature of the electrons must be considered. Because of the wave properties of electrons and the exclusion principle (to be discussed in the following section), the energy distribution of the free electrons in a metal is not even approximately given by the Maxwell–Boltzmann distribution. Furthermore, the collision of an electron with a lattice ion is not similar to the collision of a baseball with a tree. Instead, it involves the scattering of electron waves by the lattice. To understand the quantum theory of conduction, we need a qualitative understanding of the energy distribution of free electrons in a metal. This will also help us understand the origin of contact potentials between two dissimilar metals in contact and the contribution of free electrons to the heat capacity of metals.

38-3 The Fermi Electron Gas

We have used the term *electron gas* to describe the free electrons in a metal. Whereas the molecules in an ordinary gas, such as air, obey the classical Maxwell–Boltzmann energy distribution, the free electrons in a metal do not. Instead, they obey a quantum energy distribution called the Fermi–Dirac distribution. Because the behavior of this electron gas is so different from a gas of molecules, the electron gas is often called a **Fermi electron gas.** The main features of a Fermi electron gas can be understood by considering an electron in a metal to be a particle in a box, a problem whose one-dimensional version we studied extensively in Chapter 34. We discuss the main features of a Fermi electron gas semiquantitatively in this section and leave the details of the Fermi–Dirac distribution to Section 38-9.

Energy Quantization in a Box

In Chapter 34, we found that the wavelength associated with an electron of momentum p is given by the de Broglie relation:

$$\lambda = \frac{h}{p}$$

38-17

where h is Planck's constant. When a particle is confined to a finite region of space, such as a box, only certain wavelengths λ_n given by standing-wave conditions are allowed. For a one-dimensional box of length L, the standing-wave condition is

$$n\frac{\lambda_n}{2} = L$$

38-18

This results in the quantization of energy:

$$E_n = \frac{p_n^2}{2m} = \frac{(h/\lambda_n)^2}{2m} = \frac{h^2}{2m}\frac{1}{\lambda_n^2} = \frac{h^2}{2m}\frac{1}{(2L/n)^2}$$

or

$$E_n = n^2\frac{h^2}{8mL^2}$$

38-19

The wave function for the nth state is given by

$$\psi_n(x) = \sqrt{\frac{2}{L}}\sin\frac{n\pi x}{L}$$

38-20

The quantum number n characterizes the wave function for a particular state and the energy of that state. In three-dimensional problems, three quantum numbers arise, one associated with each dimension.

The Exclusion Principle

The distribution of electrons among the possible energy states is dominated by the exclusion principle, which states that no two electrons in an atom can be in the same quantum state; that is, they cannot have the same set of values for their quantum numbers. The exclusion principle applies to all "spin one-half" particles, which include electrons, protons, and neutrons. These particles have a *spin* quantum number m_s which has two possible values, $+\frac{1}{2}$ and $-\frac{1}{2}$. The quantum state of a particle is characterized by the spin quantum number m_s plus the quantum numbers associated with the spatial part of the wave function. Because the spin quantum numbers have just two possible values, the exclusion principle can be stated in terms of the spatial states:

There can be at most two electrons with the same set of values for their *spatial* quantum numbers.

EXCLUSION PRINCIPLE IN TERMS OF SPATIAL STATES

When there are more than two electrons in a system, such as an atom or metal, only two can be in the lowest energy state. The third and fourth electrons must go into the second-lowest state, and so on.

BOSON-SYSTEM ENERGY VERSUS FERMION-SYSTEM ENERGY   **EXAMPLE 38-2**

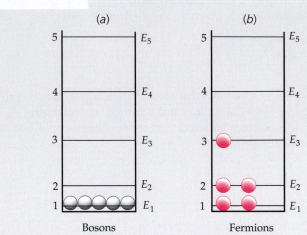

FIGURE 38-6

Compare the total energy of the ground state of five identical bosons of mass m in a one-dimensional box with that of five identical fermions of mass m in the same box.

PICTURE THE PROBLEM The ground state is the lowest possible energy state. The energy levels in a one-dimensional box are given by $E_n = n^2 E_1$, where $E_1 = h^2/(8\,mL^2)$. The lowest energy for five bosons occurs when all the bosons are in the state $n = 1$, as shown in Figure 38-6a. For fermions, the lowest state occurs with two fermions in the state $n = 1$, two fermions in the state $n = 2$, and one fermion in the state $n = 3$, as shown in Figure 38-6b.

1. The energy of five bosons in the state $n = 1$ is:

$$E = 5E_1$$

2. The energy of two fermions in the state $n = 1$, two fermions in the state $n = 2$, and one fermion in the state $n = 3$ is:

$$E = 2E_1 + 2E_2 + 1E_3 = 2E_1 + 2(2)^2 E_1 + 1(3)^2 E_1$$
$$= 2E_1 + 8E_1 + 9E_1 = 19E_1$$

3. Compare the total energies:

> The five identical fermions have 3.8 times the total energy of the five identical bosons.

REMARKS We see that the exclusion principle has a large effect on the total energy of a multiple-particle system.

The Fermi Energy

When there are many electrons in a box, at $T = 0$ the electrons will occupy the lowest energy states consistent with the exclusion principle. If we have N electrons, we can put two electrons in the lowest energy level, two electrons in the next lowest energy level, and so on. The N electrons thus fill the lowest $N/2$ energy levels (Figure 38-7). The energy of the last filled (or half-filled) level at $T = 0$ is called the Fermi energy E_F. If the electrons moved in a one-dimensional box, the Fermi energy would be given by Equation 38-19, with $n = N/2$:

$$E_F = \left(\frac{N}{2}\right)^2 \frac{h^2}{8m_e L^2} = \frac{h^2}{32m_e}\left(\frac{N}{L}\right)^2 \qquad 38\text{-}21$$

FERMI ENERGY AT $T = 0$ IN ONE DIMENSION

In a one-dimensional box, the Fermi energy depends on the number of free electrons per unit length of the box.

EXERCISE Suppose there is an ion, and therefore a free electron, every 0.1 nm in a one-dimensional box. Calculate the Fermi energy. *Hint:* Write Equation 38-21 as

$$E_F = \frac{(hc)^2}{32m_e c^2}\left(\frac{N}{L}\right)^2 = \frac{(1240 \text{ eV·nm})^2}{32(0.511 \text{ MeV})}\left(\frac{N}{L}\right)^2$$

(*Answer* $E_F = 9.4$ eV)

In our model of conduction, the free electrons move in a *three-dimensional* box of volume V. The derivation of the Fermi energy in three dimensions is somewhat difficult, so we will just give the result. In three dimensions, the Fermi energy at $T = 0$ is

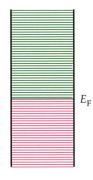

FIGURE 38-7 At $T = 0$ the electrons fill up the allowed energy states to the Fermi energy E_F. The levels are so closely spaced that they can be assumed to be continuous.

$$E_F = \frac{h^2}{8m_e}\left(\frac{3N}{\pi V}\right)^{2/3}$$

38-22a

FERMI ENERGY AT $T = 0$ IN THREE DIMENSIONS

The Fermi energy depends on the number density of electrons N/V. Substituting numerical values for the constants gives

$$E_F = (0.365 \text{ eV·nm}^2)\left(\frac{N}{V}\right)^{2/3}$$

38-22b

FERMI ENERGY AT $T = 0$ IN THREE DIMENSIONS

THE FERMI ENERGY FOR COPPER

EXAMPLE 38-3

The number density for electrons in copper was calculated in Example 25-1 and found to be 84.7/nm³. Calculate the Fermi energy at $T = 0$ for copper.

1. The Fermi energy is given by Equation 38-22:

$$E_F = (0.365 \text{ eV·nm}^2)\left(\frac{N}{V}\right)^{2/3}$$

2. Substitute the given number density for copper:

$$E_F = (0.365 \text{ eV·nm}^2)(84.7/\text{nm}^3)^{2/3}$$

$$= \boxed{7.04 \text{ eV}}$$

REMARKS Note that the Fermi energy is much greater than kT at ordinary temperatures. For example, at $T = 300$ K, kT is only about 0.026 eV.

EXERCISE Use Equation 38-22b to calculate the Fermi energy at $T = 0$ for gold, which has a free-electron number density of 59.0/nm³. (*Answer* 5.53 eV)

Table 38-1 lists the free-electron number densities and Fermi energies at $T = 0$ for several metals.

The average energy of a free electron can be calculated from the complete energy distribution of the electrons, which is discussed in Section 38-9. At $T = 0$, the average energy turns out to be

$$E_{av} = \tfrac{3}{5}E_F$$

38-23

AVERAGE ENERGY OF ELECTRONS IN A FERMI GAS AT $T = 0$

For copper, E_{av} is approximately 4 eV. This average energy is huge compared with typical thermal energies of about $kT \approx 0.026$ eV at a normal temperature of $T = 300$ K. This result is very different from the classical Maxwell–Boltzmann distribution result that at $T = 0$, $E = 0$, and that at some temperature T, E is of the order of kT.

TABLE 38-1

Free-Electron Number Densities† and Fermi Energies at $T = 0$ for Selected Elements

	Element	N/V, electrons/nm³	E_F, eV
Al	Aluminum	181	11.7
Ag	Silver	58.6	5.50
Au	Gold	59.0	5.53
Cu	Copper	84.7	7.04
Fe	Iron	170	11.2
K	Potassium	14.0	2.11
Li	Lithium	47.0	4.75
Mg	Magnesium	86.0	7.11
Mn	Manganese	165	11.0
Na	Sodium	26.5	3.24
Sn	Tin	148	10.2
Zn	Zinc	132	9.46

† Number densities are measured using the Hall effect, discussed in Section 26-4.

The Fermi Factor at $T = 0$

The probability of an energy state being occupied is called the **Fermi factor,** $f(E)$. At $T = 0$ all the states below E_F are filled, whereas all those above this energy are empty, as shown in Figure 38-8. Thus, at $T = 0$ the Fermi factor is simply

$$f(E) = 1, \qquad E < E_F$$
$$f(E) = 0, \qquad E > E_F \qquad\qquad 38\text{-}24$$

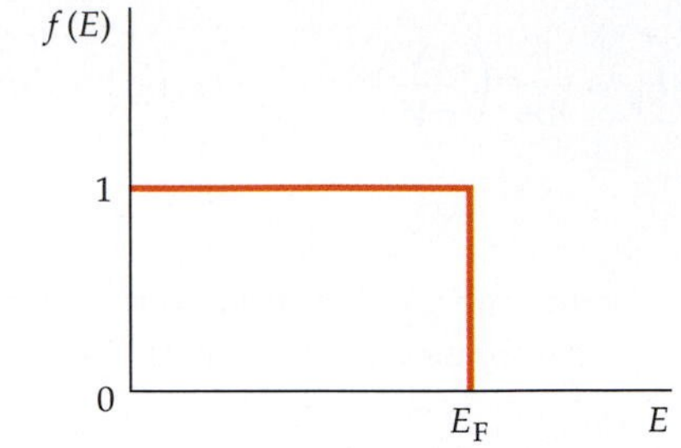

The Fermi Factor for $T > 0$

At temperatures greater than $T = 0$, some electrons will occupy higher energy states because of thermal energy gained during collisions with the lattice. However, an electron cannot move to a higher or lower state unless it is unoccupied. Since the kinetic energy of the lattice ions is of the order of kT, electrons cannot gain much more energy than kT in collisions with the lattice ions. Therefore, only those electrons with energies within about kT of the Fermi energy can gain energy as the temperature is increased. At 300 K, kT is only 0.026 eV, so the exclusion principle prevents all but a very few electrons near the top of the energy distribution from gaining energy through random collisions with the lattice ions. Figure 38-9 shows the Fermi factor for some temperature T. Since for $T > 0$ there is no distinct energy that separates filled levels from unfilled levels, the definition of the Fermi energy must be slightly modified. At temperature T, the Fermi energy is defined to be that energy for which the probability of being occupied is $\frac{1}{2}$. For all but extremely high temperatures, the difference between the Fermi energy at temperature T and the Fermi energy at temperature $T = 0$ is very small.

The **Fermi temperature** T_F is defined by

$$kT_F = E_F \qquad\qquad 38\text{-}25$$

For temperatures much lower than the Fermi temperature, the average energy of the lattice ions will be much less than the Fermi energy, and the electron energy distribution will not differ greatly from that at $T = 0$.

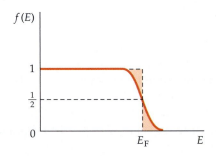

FIGURE 38-8 Fermi factor versus energy at $T = 0$.

FIGURE 38-9 The Fermi factor for some temperature T. Some electrons with energies near the Fermi energy are excited, as indicated by the shaded regions. The Fermi energy is that value of E for which $f(E) = \frac{1}{2}$.

THE FERMI TEMPERATURE FOR COPPER   **EXAMPLE 38-4**

Find the Fermi temperature for copper.

Use $E_F = 7.04$ eV and $k = 8.62 \times 10^{-5}$ eV/K in Equation 38-25: $T_F = \dfrac{E_F}{k} = \dfrac{7.04 \text{ eV}}{8.62 \times 10^{-5} \text{ eV/K}} = \boxed{81{,}700 \text{ K}}$

REMARKS We can see from this example that the Fermi temperature of copper is much greater than any temperature T for which copper remains a solid.

Because an electric field in a conductor accelerates all of the conduction electrons together, the exclusion principle does not prevent the free electrons in filled states from participating in conduction. Figure 38-10 shows the Fermi factor in one dimension versus *velocity* for an ordinary temperature. The factor is

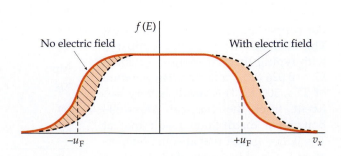

FIGURE 38-10 Fermi factor versus velocity in one dimension with no electric field (solid) and with an electric field in the $-x$ direction (dashed). The difference is greatly exaggerated.

approximately 1 for speeds v_x in the range $-u_F < v_x < u_F$, where the Fermi speed u_F is related to the Fermi energy by $E_F = \frac{1}{2}mu_F^2$. Then

$$u_F = \sqrt{\frac{2E_F}{m_e}} \qquad\qquad\qquad 38\text{-}26$$

THE FERMI SPEED FOR COPPER **EXAMPLE 38-5**

Calculate the Fermi speed for copper.

Use Equation 38-26 with $E_F = 7.04$ eV:

$$u_F = \sqrt{\frac{2(7.04 \text{ eV})}{9.11 \times 10^{-31} \text{ kg}}\left(\frac{1.60 \times 10^{-19} \text{ J}}{1 \text{ eV}}\right)} = \boxed{1.57 \times 10^6 \text{ m/s}}$$

The dashed curve in Figure 38-10 shows the Fermi factor after the electric field has been acting for some time t. Although all of the free electrons have their velocities shifted in the direction opposite to the electric field, the net effect is equivalent to shifting only the electrons near the Fermi energy.

Contact Potential

When two different metals are placed in contact, a potential difference V_{contact} called the **contact potential** develops between them. The contact potential depends on the work functions of the two metals, ϕ_1 and ϕ_2 (we encountered work functions when the photoelectric effect was introduced in Chapter 34), and the Fermi energies of the two metals. When the metals are in contact, the total energy of the system is lowered if electrons near the boundary move from the metal with the higher Fermi energy into the metal with the lower Fermi energy until the Fermi energies of the two metals are the same, as shown in Figure 38-11. When equilibrium is established, the metal with the lower initial Fermi energy is negatively charged and the other is positively charged, so that between them there is a potential difference V_{contact} given by

$$V_{\text{contact}} = \frac{\phi_1 - \phi_2}{e} \qquad\qquad 38\text{-}27$$

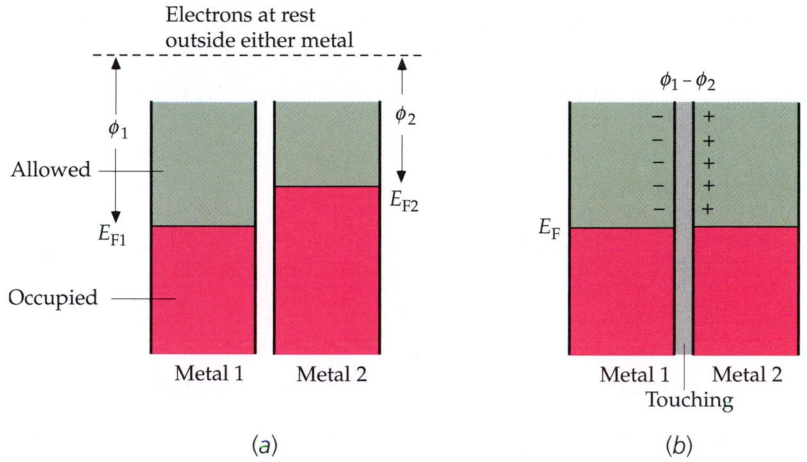

FIGURE 38-11 (*a*) Energy levels for two different metals with different Fermi energies and work functions. (*b*) When the metals are in contact, electrons flow from the metal that initially has the higher Fermi energy to the metal that initially has the lower Fermi energy until the Fermi energies are equal.

Table 38-2 lists the work functions for several metals.

TABLE 38-2

Work Functions for Some Metals

	Metal	ϕ, eV		Metal	ϕ, eV
Ag	Silver	4.7	K	Potassium	2.1
Au	Gold	4.8	Mn	Manganese	3.8
Ca	Calcium	3.2	Na	Sodium	2.3
Cu	Copper	4.1	Ni	Nickel	5.2

The threshold wavelength for the photoelectric effect is 271 nm for tungsten and 262 nm for silver. What is the contact potential developed when silver and tungsten are placed in contact?

PICTURE THE PROBLEM The contact potential is proportional to the difference in the work functions for the two metals. The work function ϕ can be found from the given threshold wavelengths using $\phi = hc/\lambda_t$ (Equation 34-4).

1. The contact potential is given by Equation 38-27:

$$V_{contact} = \frac{\phi_1 - \phi_2}{e}$$

2. The work function is related to the threshold wavelength (Equation 34-4):

$$\phi = \frac{hc}{\lambda_t}$$

3. Substitute $\lambda_t = 271$ nm for tungsten:

$$\phi_W = \frac{hc}{\lambda_t} = \frac{1240 \text{ eV·nm}}{271 \text{ nm}} = 4.58 \text{ eV}$$

4. Substitute $\lambda_t = 262$ nm for silver:

$$\phi_{Ag} = \frac{1240 \text{ eV·nm}}{262 \text{ nm}} = 4.73 \text{ eV}$$

5. The contact potential is thus:

$$V_{contact} = \frac{\phi_{Ag} - \phi_W}{e} = 4.73 \text{ V} - 4.58 \text{ V}$$

$$= \boxed{0.15 \text{ V}}$$

Heat Capacity Due to Electrons in a Metal

The quantum-mechanical modification of the electron distribution in metals allows us to understand why the contribution of the electron gas to the heat capacity of a metal is much less that of the ions. According to the classical equipartition theorem, the energy of the lattice ions in n moles of a solid is $3nRT$, and thus the molar specific heat is $c' = 3R$, where R is the universal gas constant (see Section 18-7). In a metal, there is a free electron gas containing a number of electrons approximately equal to the number of lattice ions. If these electrons obey the classical equipartition theorem, they should have an energy of $\frac{3}{2}nRT$ and contribute an additional $\frac{3}{2}R$ to the molar specific heat. But measured heat capacities of metals are just slightly greater than those of insulators. We can understand this because at some temperature T, only those electrons with energies near the Fermi energy can be excited by random collisions with the lattice ions. The number of these electrons is of the order of $(kT/E_F)N$, where N is the total number of electrons. The energy of these electrons is increased from that at $T = 0$ by an amount that is of the order of kT. So the total increase in thermal energy is of the order of $(kT/E_F)N \times kT$. We can thus express the energy of N electrons at temperature T as

$$E = NE_{av}(0) + \alpha N \frac{kT}{E_F} kT \qquad\qquad 38\text{-}28$$

where α is some constant that we expect to be of the order of 1 if our reasoning is correct. The calculation of α is quite difficult. The result is $\alpha = \pi^2/4$. Using this result and writing E_F in terms of the Fermi temperature, $E_F = kT_F$, we obtain the following for the contribution of the electron gas to the heat capacity at constant volume:

$$C_V = \frac{dE}{dT} = 2\alpha N k \frac{kT}{E_F} = \frac{\pi^2}{2} n R \frac{T}{T_F}$$

where we have written Nk in terms of the gas constant R ($Nk = nR$). The molar specific heat at constant volume is then

$$c_V' = \frac{\pi^2}{2} R \frac{T}{T_F} \qquad\qquad\qquad 38\text{-}29$$

We can see that because of the large value of T_F, the contribution of the electron gas is a small fraction of R at ordinary temperatures. Because $T_F = 81{,}700$ K for copper, the molar specific heat of the electron gas at $T = 300$ K is

$$c_V' = \frac{\pi^2}{2} \left(\frac{300 \text{ K}}{81{,}700} \right) R \approx 0.02\, R$$

which is in good agreement with experiment.

38-4 Quantum Theory of Electrical Conduction

We can use Equation 38-14 for the resistivity if we use the Fermi speed u_F in place of v_{av}:

$$\rho = \frac{m_e u_F}{n_e e^2 \lambda} \qquad\qquad\qquad 38\text{-}30$$

We now have two problems. First, since the Fermi speed u_F is approximately independent of temperature, the resistivity given by Equation 38-30 is independent of temperature unless the mean free path depends on it. The second problem concerns magnitudes. As mentioned earlier, the classical expression for resistivity using v_{av} calculated from the Maxwell–Boltzmann distribution gives values that are about 6 times too large at $T = 300$ K. Since the Fermi speed u_F is about 16 times the Maxwell–Boltzmann value of v_{av}, the magnitude of ρ predicted by Equation 38-30 will be approximately 100 times greater than the experimentally determined value. The resolution of both of these problems lies in the calculation of the mean free path λ.

The Scattering of Electron Waves

In Equation 38-16 for the classical mean free path $\lambda = 1/(n_{ion}A)$, the quantity $A = \pi r^2$ is the area of the lattice ion as seen by an electron. In the quantum calculation, the mean free path is related to the scattering of electron waves by the crystal lattice. Detailed calculations show that, for a *perfectly* ordered crystal, $\lambda = \infty$; that is, there is no scattering of the electron waves. The scattering of electron waves arises because of *imperfections* in the crystal lattice, which have nothing to do with the actual cross-sectional area A of the lattice ions. According to the quantum theory of electron scattering, A depends merely on *deviations* of the lattice ions from a perfectly ordered array and not on the size of the ions. The most common causes of such deviations are thermal vibrations of the lattice ions or impurities.

We can use $\lambda = 1/(n_{ion}A)$ for the mean free path if we reinterpret the area A. Figure 38-12 compares the classical picture and the quantum picture of this area. In the quantum picture, the lattice ions are points that have no size but present an

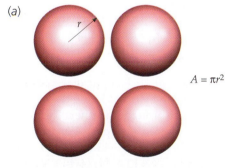

(a)

$A = \pi r^2$

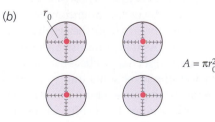

(b) r_0

$A = \pi r_0^2$

FIGURE 38-12 (*a*) Classical picture of the lattice ions as spherical balls of radius r that present an area πr^2 to the electrons. (*b*) Quantum-mechanical picture of the lattice ions as points that are vibrating in three dimensions. The area presented to the electrons is πr_0^2, where r_0 is the amplitude of oscillation of the ions.

area $A = \pi r_0^2$, where r_0 is the amplitude of thermal vibrations. In Chapter 14, we saw that the energy of vibration in simple harmonic motion is proportional to the square of the amplitude, which is πr_0^2. Thus, the effective area A is proportional to the energy of vibration of the lattice ions. From the equipartition theorem,[†] we know that the average energy of vibration is proportional to kT. Thus, A is proportional to T, and λ is proportional to $1/T$. Then the resistivity given by Equation 38-14 is proportional to T, in agreement with experiment.

The effective area A due to thermal vibrations can be calculated, and the results give values for the resistivity that are in agreement with experiment. At $T = 300$ K, for example, the effective area turns out to be about 100 times smaller than the actual area of a lattice ion. We see, therefore, that the free-electron model of metals gives a good account of electrical conduction if the classical mean speed v_{av} is replaced by the Fermi speed u_F and if the collisions between electrons and the lattice ions are interpreted in terms of the scattering of electron waves, for which only deviations from a perfectly ordered lattice are important.

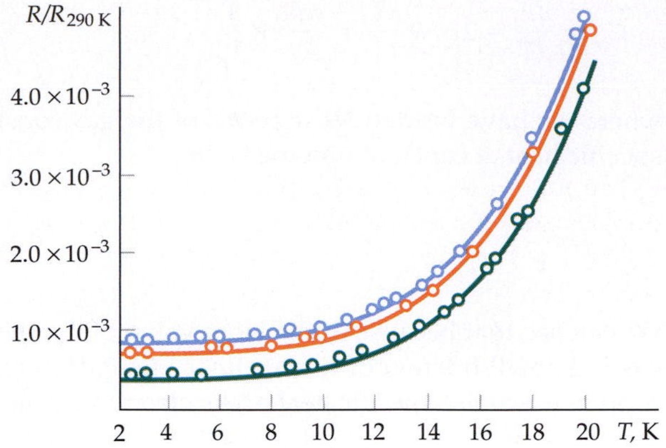

The presence of impurities in a metal also causes deviations from perfect regularity in the crystal lattice. The effects of impurities on resistivity are approximately independent of temperature. The resistivity of a metal containing impurities can be written $\rho = \rho_t + \rho_i$, where ρ_t is the resistivity due to the thermal motion of the lattice ions and ρ_i is the resistivity due to impurities. Figure 38-13 shows typical resistance curves versus temperature curves for metals with impurities. As the absolute temperature approaches zero, ρ_t approaches zero, and the resistivity approaches the constant ρ_i due to impurities.

FIGURE 38-13 Relative resistance versus temperature for three samples of sodium. The three curves have the same temperature dependence but different magnitudes because of differing amounts of impurities in the samples.

38-5 Band Theory of Solids

Resistivities vary enormously between insulators and conductors. For a typical insulator, such as quartz, $\rho \sim 10^{16}$ $\Omega \cdot$m, whereas for a typical conductor, $\rho \sim 10^{-8}$ $\Omega \cdot$m. The reason for this enormous variation is the variation in the number density of free electrons n_e. To understand this variation, we consider the effect of the lattice on the electron energy levels.

We begin by considering the energy levels of the individual atoms as they are brought together. The allowed energy levels in an isolated atom are often far apart. For example, in hydrogen, the lowest allowed energy $E_1 = -13.6$ eV is 10.2 eV below the next lowest allowed energy $E_2 = (-13.6 \text{ eV})/4 = -3.4$ eV.[‡] Let us consider two identical atoms and focus our attention on one particular energy level. When the atoms are far apart, the energy of a particular level is the same for each atom. As the atoms are brought closer together, the energy level for each atom changes because of the influence of the other atom. As a result, the level splits into two levels of slightly different energies for the two-atom system. If we bring three atoms close together, a particular energy level splits into three separate levels of slightly different energies. Figure 38-14 shows the energy splitting of two energy levels for six atoms as a function of the separation of the atoms.

If we have N identical atoms, a particular energy level in the isolated atom splits into N different, closely spaced energy levels when the atoms are close together. In a macroscopic solid, N is very large— of the order of 10^{23}—so each energy level splits into a very large

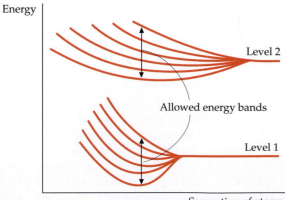

FIGURE 38-14 Energy splitting of two energy levels for six atoms as a function of the separation of the atoms. When there are many atoms, each level splits into a near-continuum of levels called a band.

---

† The equipartition theorem does hold for the lattice ions, which obey the Maxwell–Boltzmann energy distribution.
‡ The energy levels in hydrogen are discussed in Chapter 36.

number of levels called a **band.** The levels are spaced almost continuously within the band. There is a separate band of levels for each particular energy level of the isolated atom. The bands may be widely separated in energy, they may be close together, or they may even overlap, depending on the kind of atom and the type of bonding in the solid.

The lowest energy bands, corresponding to the lowest energy levels of the atoms in the lattice, are filled with electrons that are bound to the individual atoms. The electrons that can take part in conduction occupy the higher energy bands. The highest energy band that contains electrons is called the **valence band.** The valence band may be completely filled with electrons or only partially filled, depending on the kind of atom and the type of bonding in the solid.

We can now understand why some solids are conductors and why others are insulators. If the valence band is only partially full, there are many available empty energy states in the band, and the electrons in the band can easily be raised to a higher energy state by an electric field. Accordingly, this material is a good conductor. If the valence band is full and there is a large energy gap between it and the next available band, a typical applied electric field will be too weak to excite an electron from the upper energy levels of the filled band across the large gap into the energy levels of the empty band, so the material is an insulator. The lowest band in which there are unoccupied states is called the **conduction band.** In a conductor, the valence band is only partially filled, so the valence band is also the conduction band. An energy gap between allowed bands is called a **forbidden energy band.**

The band structure for a conductor, such as copper, is shown in Figure 38-15a. The lower bands (not shown) are filled with the inner electrons of the atoms. The valence band is only about half full. When an electric field is established in the conductor, the electrons in the conduction band are accelerated, which means that their energy is increased. This is consistent with the Pauli exclusion principle because there are many empty energy states just above those occupied by electrons in this band. These electrons are thus the conduction electrons.

Figure 38-15b shows the band structure for magnesium, which is also a conductor. In this case, the highest occupied band is full, but there is an empty band above it that overlaps it. The two bands thus form a combined valence–conduction band that is only partially filled.

Figure 38-15c shows the band structure for a typical insulator. At $T = 0$ K, the valence band is completely full. The next energy band containing empty energy states, the conduction band, is separated from the valence band by a large energy gap. At $T = 0$, the conduction band is empty. At ordinary temperatures, a few electrons can be excited to states in this band, but most cannot be excited to states because the energy gap is large compared with the energy an electron might obtain by thermal excitation. Very few electrons can be thermally excited to the nearly empty conduction band, even at fairly high temperatures. When an electric field of ordinary magnitude is established in the solid, electrons cannot be

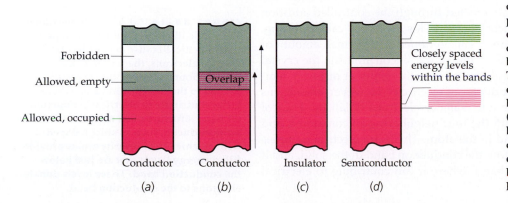

Conductor   Conductor   Insulator   Semiconductor

(a)   (b)   (c)   (d)

Forbidden

Allowed, empty

Allowed, occupied

Overlap

Closely spaced energy levels within the bands

FIGURE 38-15 Four possible band structures for a solid. (*a*) A typical conductor. The valence band is only partially full, so electrons can be easily excited to nearby energy states. (*b*) A conductor in which the allowed energy bands overlap. (*c*) A typical insulator. There is a forbidden band with a large energy gap between the filled valence band and the conduction band. (*d*) A semiconductor. The energy gap between the filled valence band and the conduction band is very small, so some electrons are excited to the conduction band at normal temperatures, leaving holes in the valence band.

accelerated because there are no empty energy states at nearby energies. We describe this by saying that there are no free electrons. The small conductivity that is observed is due to the very few electrons that are thermally excited into the nearly empty conduction band. When an electric field applied to an insulator is sufficiently strong to cause an electron to be excited across the energy gap to the empty band, dielectric breakdown occurs.

In some materials, the energy gap between the filled valence band and the empty conduction band is very small, as shown in Figure 38-15d. At $T = 0$, there are no electrons in the conduction band and the material is an insulator. However, at ordinary temperatures, there are an appreciable number of electrons in the conduction band due to thermal excitation. Such a material is called an **intrinsic semiconductor.** For typical intrinsic semiconductors, such as silicon and germanium, the energy gap is only about 1 eV. In the presence of an electric field, the electrons in the conduction band can be accelerated because there are empty states nearby. Also, for each electron in the conduction band there is a vacancy, or hole, in the nearly filled valence band. In the presence of an electric field, electrons in this band can also be excited to a vacant energy level. This contributes to the electric current and is most easily described as the motion of a hole in the direction of the field and opposite to the motion of the electrons. The hole thus acts like a positive charge. To visualize the conduction of holes, think of a two-lane, one-way road with one lane full of parked cars and the other lane empty. If a car moves out of the filled lane into the empty lane, it can move ahead freely. As the other cars move up to occupy the space left, the empty space propagates backward in the direction opposite the motion of the cars. Both the forward motion of the car in the nearly empty lane and the backward propagation of the empty space contribute to a net forward propagation of the cars.

An interesting characteristic of semiconductors is that the resistivity of the material decreases as the temperature increases, which is contrary to the case for normal conductors. The reason is that as the temperature increases, the number of free electrons increases because there are more electrons in the conduction band. The number of holes in the valence band also increases, of course. In semiconductors, the effect of the increase in the number of charge carriers, both electrons and holes, exceeds the effect of the increase in resistivity due to the increased scattering of the electrons by the lattice ions due to thermal vibrations. Semiconductors therefore have a negative temperature coefficient of resistivity.

38-6 Semiconductors

The semiconducting property of intrinsic semiconductors materials makes them useful as a basis for electronic circuit components whose resistivity can be controlled by application of an external voltage or current. Most such *solid-state devices*, however, such as the semiconductor diode and the transistor, make use of **impurity semiconductors,** which are created through the controlled addition of certain impurities to intrinsic semiconductors. This process is called **doping.** Figure 38-16a is a schematic illustration of silicon doped with a small amount of arsenic so that the arsenic atoms replace a few of the silicon atoms in the crystal lattice. The conduction band of pure silicon is virtually empty at ordinary temperatures, so pure silicon is a poor conductor of electricity. However, arsenic has five valence electrons rather than the four valence electrons of silicon. Four of these electrons take part in bonds with the four neighboring silicon atoms, and the fifth electron is very loosely bound to the atom. This extra electron occupies an energy level that is just slightly below the conduction band in the solid, and it is easily excited into the conduction band, where it can contribute to electrical conduction.

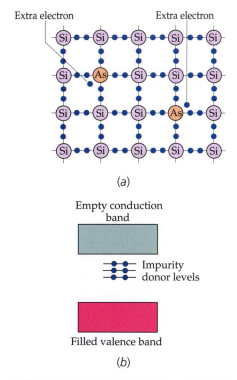

(a)

Empty conduction band

Impurity donor levels

Filled valence band

(b)

FIGURE 38-16 (*a*) A two-dimensional schematic illustration of silicon doped with arsenic. Because arsenic has five valence electrons, there is an extra, weakly bound electron that is easily excited to the conduction band, where it can contribute to electrical conduction. (*b*) Band structure of an *n*-type semiconductor, such as silicon doped with arsenic. The impurity atoms provide filled energy levels that are just below the conduction band. These levels donate electrons to the conduction band.

The effect on the band structure of a silicon crystal achieved by doping it with arsenic is shown in Figure 38-16b. The levels shown just below the conduction band are due to the extra electrons of the arsenic atoms. These levels are called **donor levels** because they donate electrons to the conduction band without leaving holes in the valence band. Such a semiconductor is called an ***n*-type semiconductor** because the major charge carriers are negative electrons. The conductivity of a doped semiconductor can be controlled by controlling the amount of impurity added. The addition of just one part per million can increase the conductivity by several orders of magnitude.

Another type of impurity semiconductor can be made by replacing a silicon atom with a gallium atom, which has three valence electrons (Figure 38-17a). The gallium atom accepts electrons from the valence band to complete its four covalent bonds, thus creating a hole in the valence band. The effect on the band structure of silicon achieved by doping it with gallium is shown in Figure 38-17b. The empty levels shown just above the valence band are due to the holes from the ionized gallium atoms. These levels are called **acceptor levels** because they accept electrons from the filled valence band when these electrons are thermally excited to a higher energy state. This creates holes in the valence band that are free to propagate in the direction of an electric field. Such a semiconductor is called a ***p*-type semiconductor** because the charge carriers are positive holes. The fact that conduction is due to the motion of positive holes can be verified by the Hall effect. (The Hall effect is discussed in chapter 26.)

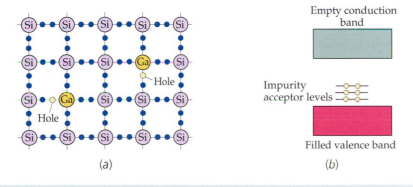

(a)   (b)

FIGURE 38-17 (a) A two-dimensional schematic illustration of silicon doped with gallium. Because gallium has only three valence electrons, there is a hole in one of its bonds. As electrons move into the hole the hole moves about, contributing to the conduction of electrical current. (b) Band structure of a *p*-type semiconductor, such as silicon doped with gallium. The impurity atoms provide empty energy levels just above the filled valence band that accept electrons from the valence band.

Synthetic crystal silicon is produced beginning with a raw material containing silicon (for instance, common beach sand), separating out the silicon, and melting it. From a seed crystal, the molten silicon grows into a cylindrical crystal, such as the one shown here. The crystals (typically about 1.3 m long) are formed under highly controlled conditions to ensure that they are flawless and the crystals are then sliced into thousands of thin wafers onto which the layers of an integrated circuit are etched.

NUMBER DENSITY OF FREE ELECTRONS IN ARSENIC-DOPED SILICON

EXAMPLE 38-7 Try It Yourself

The number of free electrons in pure silicon is approximately 10^{10} electrons/cm³ at ordinary temperatures. If one silicon atom out of every million atoms is replaced by an arsenic atom, how many free electrons per cubic centimeter are there? (The density of silicon is 2.33 g/cm³ and its molar mass is 28.1 g/mol.)

PICTURE THE PROBLEM The number of silicon atoms per cubic centimeter, n_{Si} can be found from $n_{Si} = N_A \rho / M$. Then, since each arsenic atom contributes one free electron, the number of electrons contributed by the arsenic atoms is $10^{-6} n_S$.

Cover the column to the right and try these on your own before looking at the answers.

Steps	Answers

1. Calculate the number of silicon atoms per cubic centimeter.

$$n_{Si} = \frac{\rho N_A}{M}$$

$$= \frac{(2.33 \text{ g/cm}^3)(6.02 \times 10^{23} \text{ atoms/mol})}{28.1 \text{ g/mol}}$$

$$= 4.99 \times 10^{22} \text{ atoms/cm}^3$$

2. Multiply by 10^{-6} to obtain the number of arsenic atoms per cubic centimeter, which equals the added number of free electrons per cubic centimeter.

$$n_e = 10^{-6} n_{Si} = \boxed{4.99 \times 10^{16} \text{ electrons/cm}^3}$$

REMARKS Because silicon has so few free electrons per atom, the number of conduction electrons is increased by a factor of approximately 5 million by doping silicon with just one arsenic atom per million silicon atoms.

EXERCISE How many free electrons are there per silicon atom in pure silicon? (*Answer* 2×10^{-13})

*38-7 Semiconductor Junctions and Devices

Semiconductor devices such as diodes and transistors make use of n-type semiconductors and p-type semiconductors joined together, as shown in Figure 38-18. In practice, the two types of semiconductors are often incorporated into a single silicon crystal doped with donor impurities on one side and acceptor impurities on the other side. The region in which the semiconductor changes from a p-type semiconductor to an n-type semiconductor is called a **junction**.

When an n-type semiconductor and a p-type semiconductor are placed in contact, the initially unequal concentrations of electrons and holes result in the diffusion of electrons across the junction from the n side to the p side and holes from the p side to the n side until equilibrium is established. The result of this diffusion is a net transport of positive charge from the p side to the n side. Unlike the case when two different metals are in contact, the electrons cannot travel very far from the junction region because the semiconductor is not a particularly good conductor. The diffusion of electrons and holes therefore creates a double layer of charge at the junction similar to that on a parallel-plate capacitor. There is, thus, a potential difference V across the junction, which tends to inhibit further diffusion. In equilibrium, the n side with its net positive charge will be at a higher potential than the p side with its net negative charge. In the junction region, between the charge layers, there will be very few charge carriers of either type, so the junction region has a high resistance. Figure 38-19 shows the energy-level diagram for a pn junction. The junction region is also called the **depletion region** because it has been depleted of charge carriers.

*Diodes

In Figure 38-20, an external potential difference has been applied across a pn junction by connecting a battery and a resistor to the semiconductor. When the positive terminal of the battery is connected to the p side of the junction, as shown in Figure 38-20a, the junction is said to be **forward biased**. Forward

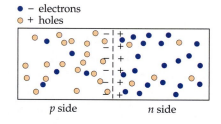

- • − electrons
- ○ + holes

p side n side

FIGURE 38-18 A pn junction. Because of the difference in their concentrations on either side of the pn junction, holes diffuse from the p side to the n side, and electrons diffuse from the n side to the p side. As a result, there is a double layer of charge at the junction, with the p side being negative and the n side being positive.

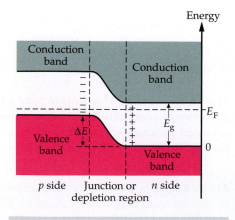

FIGURE 38-19 Electron energy levels for a pn junction.

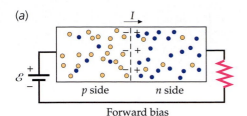

(a) Forward bias

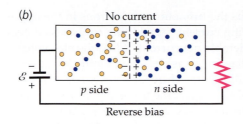

(b) Reverse bias

FIGURE 38-20 A *pn*-junction diode. (*a*) Forward-biased *pn* junction. The applied potential difference enhances the diffusion of holes from the *p* side to the *n* side and of electrons from the *n* side to the *p* side, resulting in a current *I*. (*b*) Reverse-biased *pn* junction. The applied potential difference inhibits the further diffusion of holes and electrons across the junction, so there is no current.

biasing lowers the potential across the junction. The diffusion of electrons and holes is thereby increased as they attempt to reestablish equilibrium, resulting in a current in the circuit.

If the positive terminal of the battery is connected to the *n* side of the junction, as shown in Figure 38-20*b*, the junction is said to be **reverse biased.** Reverse biasing tends to increase the potential difference across the junction, thereby further inhibiting diffusion. Figure 38-21 shows a plot of current versus voltage for a typical semiconductor junction. Essentially, the junction conducts only in one direction. A single-junction semiconductor device is called a **diode.**[†] Diodes have many uses. One is to convert alternating current into direct current, a process called rectification.

Note that the current in Figure 38-21 suddenly increases in magnitude at extreme values of reverse bias. In such large electric fields, electrons are stripped from their atomic bonds and accelerated across the junction. These electrons, in turn, cause others to break loose. This effect is called **avalanche breakdown.** Although such a breakdown can be disastrous in a circuit where it is not intended, the fact that it occurs at a sharply defined voltage makes it of use in a special voltage reference standard known as a **Zener diode.** Zener diodes are also used to protect devices from excessively high voltages.

An interesting effect, one that we discuss only qualitatively, occurs if both the *n* side and the *p* side of a *pn*-junction diode are so heavily doped that the donors on the *n* side provide so many electrons that the lower part of the conduction band is practically filled, and the acceptors on the *p* side accept so many electrons that the upper part of the valence band is nearly empty. Figure 38-22*a* shows the energy-level diagram for this situation. Because the depletion region is now so narrow, electrons can easily penetrate the potential barrier across the junction and tunnel to the other side. The flow of electrons through the barrier is called a **tunneling current,** and such a heavily doped diode is called a **tunnel diode.**

At equilibrium with no bias, there is an equal tunneling current in each direction. When a small bias voltage is applied across the junction, the energy-level diagram is as shown in Figure 38-22*b*, and the tunneling of electrons from the

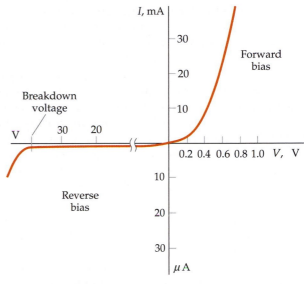

FIGURE 38-21 Plot of current versus applied voltage across a *pn* junction. Note the different scales on both axes for the forward and reverse bias conditions.

FIGURE 38-22 Electron energy levels for a heavily doped *pn*-junction tunnel diode. (*a*) With no bias voltage, some electrons tunnel in each direction. (*b*) With a small bias voltage, the tunneling current is enhanced in one direction, making a sizable contribution to the net current. (*c*) With further increases in the bias voltage, the tunneling current decreases dramatically.

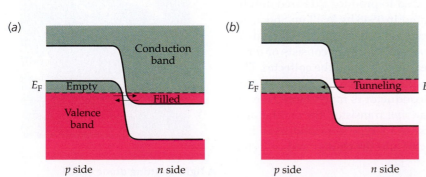

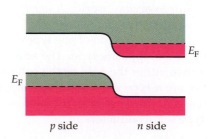

[†] The name *diode* originates from a vacuum tube device consisting of just two electrodes that also conducts electric current in one direction only.

n side to the *p* side is increased, whereas the tunneling of electrons in the opposite direction is decreased. This tunneling current, in addition to the usual current due to diffusion, results in a considerable net current. When the bias voltage is increased slightly, the energy-level diagram is as shown in Figure 38-22*c*, and the tunneling current is decreased. Although the diffusion current is increased, the net current is decreased. At large bias voltages, the tunneling current is completely negligible, and the total current increases with increasing bias voltage due to diffusion, as in an ordinary *pn*-junction diode. Figure 38-23 shows the current curve versus the voltage curve for a tunnel diode. Such diodes are used in electric circuits because of their very fast response time. When operated near the peak in the current curve versus the voltage curve, a small change in bias voltage results in a large change in the current.

Another use for the *pn*-junction semiconductor is the **solar cell,** which is illustrated schematically in Figure 38-24. When a photon of energy greater than the gap energy (1.1 eV in silicon) strikes the *p*-type region, it can excite an electron from the valence band into the conduction band, leaving a hole in the valence band. This region is already rich in holes. Some of the electrons created by the photons will recombine with holes, but some will migrate to the junction. From there, they are accelerated into the *n*-type region by the electric field between the double layer of charge. This creates an excess negative charge in the *n*-type region and an excess positive charge in the *p*-type region. The result is a potential difference between the two regions, which in practice is approximately 0.6 V. If a load resistance is connected across the two regions, a charge flows through the resistance. Some of the incident light energy is thus converted into electrical energy. The current in the resistor is proportional to the number of incident photons, which is in turn proportional to the intensity of the incident light.

There are many other applications of semiconductors with *pn* junctions. Particle detectors, called **surface-barrier detectors,** consist of a *pn*-junction semiconductor with a large reverse bias so that there is ordinarily no current. When a high-energy particle, such as an electron, passes through the semiconductor, it creates many electron–hole pairs as it loses energy. The resulting current pulse signals the passage of the particle. **Light-emitting diodes** (LEDs) are *pn*-junction semiconductors with a large forward bias that produces a large excess concentration of electrons on the *p* side and holes on the *n* side of the junction. Under these conditions, the diode emits light as the electrons and holes recombine. This is essentially the reverse of the process that occurs in a solar cell, in which electron–hole pairs are created by the absorption of light. LEDs are commonly used as warning indicators and as sources of infrared light beams.

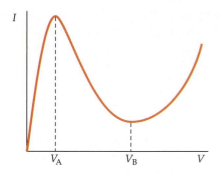

FIGURE 38-23 Current versus applied (bias) voltage *V* for a tunnel diode. For $V < V_A$, an increase in the bias voltage *V* enhances tunneling. For $V_A < V < V_B$, an increase in the bias voltage inhibits tunneling. For $V > V_B$, the tunneling is negligible, and the diode behaves like an ordinary *pn*-junction diode.

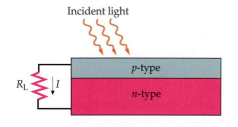

FIGURE 38-24 A *pn*-junction semiconductor as a solar cell. When light strikes the *p*-type region, electron–hole pairs are created, resulting in a current through the load resistance R_L.

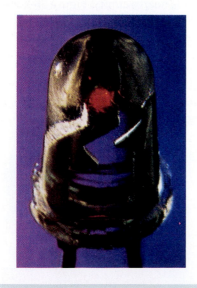

A light-emitting diode (LED).

*Transistors

The transistor, a semiconducting device that is used to produce a desired output signal in response to an input signal, was invented in 1948 by William Shockley, John Bardeen, and Walter Brattain and has revolutionized the electronics industry and our everyday world. A *simple bipolar junction transistor*[†] consists of three distinct semiconductor regions called the **emitter,** the **base,** and the **collector.** The base is a very thin region of one type of semiconductor sandwiched between two regions of the opposite type. The emitter semiconductor is much more heavily doped than either the base or the collector. In an *npn* transistor, the emitter and collector are *n*-type semiconductors and the base is a *p*-type semiconductor; in a *pnp* transistor, the base is an *n*-type semiconductor and the emitter and collector are *p*-type semiconductors.

---

† Besides the bipolar junction transistor, there are other categories of transistors, notably, the field-effect transistor.

Figure 38-25 and Figure 38-26 show, respectively, a *pnp* transistor and an *npn* transistor with the symbols used to represent each transistor in circuit diagrams. We see that either transistor consists of two *pn* junctions. We will discuss the operation of a *pnp* transistor. The operation of an *npn* transistor is similar.

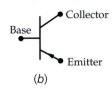

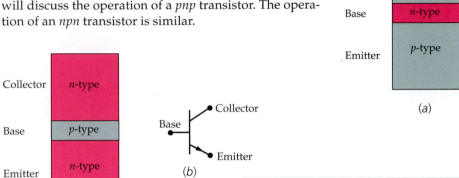

(a)

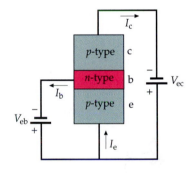

FIGURE 38-25 A *pnp* transistor. (a) The heavily doped emitter emits holes that pass through the thin base to the collector. (b) Symbol for a *pnp* transistor in a circuit. The arrow points in the direction of the conventional current, which is the same as that of the emitted holes.

FIGURE 38-26 An *npn* transistor. (a) The heavily doped emitter emits electrons that pass through the thin base to the collector. (b) Symbol for an *npn* transistor. The arrow points in the direction of the conventional current, which is opposite the direction of the emitted electrons.

In normal operation, the emitter–base junction is forward biased, and the base–collector junction is reverse biased, as shown in Figure 38-27. The heavily doped *p*-type emitter emits holes that flow toward the emitter–base junction. This flow constitutes the emitter current I_e. Because the base is very thin, most of these holes flow across the base into the collector. This flow in the collector constitutes a current I_c. However, some of the holes recombine in the base producing a positive charge that inhibits the further flow of current. To prevent this, some of the holes that do not reach the collector are drawn off the base as a base current I_b in a wire connected to the base. In Figure 38-27, therefore, I_c is almost but not quite equal to I_e, and I_b is much smaller than either I_c or I_e. It is customary to express I_c as

$$I_c = \beta I_b \qquad\qquad 38\text{-}31$$

where β is called the **current gain** of the transistor. Transistors can be designed to have values of β as low as ten or as high as several hundred.

Figure 38-28 shows a simple *pnp* transistor used as an amplifier. A small, time-varying input voltage v_s is connected in series with a bias voltage V_{eb}. The base current is then the sum of a steady current I_b produced by the bias voltage V_{eb} and a varying current i_b due to the signal voltage v_s. Because v_s may at any

FIGURE 38-27 A *pnp* transistor biased for normal operation. Holes from the emitter can easily diffuse across the base, which is only tens of nanometers thick. Most of the holes flow to the collector, producing the current I_c.

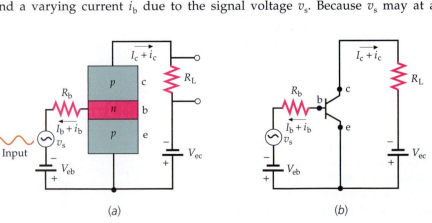

(a)   (b)

FIGURE 38-28 (a) A *pnp* transistor used as an amplifier. A small change i_b in the base current results in a large change i_c in the collector current. Thus, a small signal in the base circuit results in a large signal across the load resistor R_L in the collector circuit. (b) The same circuit as in Figure 38-28a with the conventional symbol for the transistor.

instant be either positive or negative, the bias voltage V_{eb} must be large enough to ensure that there is always a forward bias on the emitter–base junction. The collector current will consist of two parts: a direct current $I_c = \beta I_b$ and an alternating current $i_c = \beta i_b$. We thus have a current amplifier in which the time-varying output current i_c is β times the input current i_b. In such an amplifier, the steady currents I_c and I_b, although essential to the operation of the transistor, are usually not of interest. The input signal voltage v_s is related to the base current by Ohm's law:

$$i_b = \frac{v_s}{R_b + r_b} \qquad\qquad 38\text{-}32$$

where r_b is the internal resistance of the transistor between the base and emitter. Similarly, the collector current i_c produces a voltage v_L across the output or load resistance R_L given by

$$v_L = i_c R_L \qquad\qquad 38\text{-}33$$

Using Equation 38-31 and Equation 38-32, we have

$$i_c = \beta i_b = \beta \frac{v_s}{R_b + r_b} \qquad\qquad 38\text{-}34$$

The output voltage is thus related to the input voltage by

$$v_L = \beta \frac{v_s}{R_b + r_b} R_L = \beta \frac{R_L}{R_b + r_b} v_s \qquad\qquad 38\text{-}35$$

The ratio of the output voltage to the input voltage is the **voltage gain** of the amplifier:

$$\text{Voltage gain} = \frac{v_L}{v_s} = \beta \frac{R_L}{R_b + r_b} \qquad\qquad 38\text{-}36$$

A typical amplifier (e.g., in a tape player) has several transistors, similar to the one shown in Figure 38-28, connected in series so that the output of one transistor serves as the input for the next. Thus, the very small voltage fluctuations produced by the motion of the magnetic tape past the pickup heads controls the large amounts of power required to drive the loudspeakers. The power delivered to the speakers is supplied by the dc sources connected to each transistor.

The technology of semiconductors extends well beyond individual transistors and diodes. Many of the electronic devices we now take for granted, such as laptop computers and the processors that govern the operation of vehicles and appliances, rely on large-scale integration of many transistors and other circuit components on a single chip. Large-scale integration combined with advanced concepts in semiconductor theory has created remarkable new instruments for scientific research.

38-8 Superconductivity

There are some materials for which the resistivity suddenly drops to zero below a certain temperature T_c, which is called the **critical temperature**. This amazing phenomenon, called **superconductivity,** was discovered in 1911 by the Dutch physicist H. Kamerlingh Onnes, who developed a technique for liquefying helium (boiling point 4.2 K) and put his technique to work exploring the properties of materials at temperatures in this range. Figure 38-29 shows Onnes's plot of the resistance of mercury versus temperature. The critical temperature for mercury is approximately the same as the boiling point of helium, which is 4.2 K. Critical temperatures for other superconducting elements range from less than 0.1 K for hafnium and iridium to 9.2 K for niobium. The temperature range for superconductors goes much higher for a number of metallic compounds. For example, the superconducting alloy Nb_3Ge, discovered in 1973, has a critical temperature of 25 K, which was the highest known until 1986, when the discoveries of J. Georg Bednorz and K. Alexander Müller launched the era of high-temperature superconductors, now defined as materials that exhibit superconductivity at temperatures above 77 K (the temperature at which nitrogen boils). To date (April 2003), the highest temperature at which superconductivity has been demonstrated, using thallium doped $HgBa_2Ca_2Cu_3O_8$+delta, is 138 K at atmospheric pressure. At extremely high pressures, some materials exhibit superconductivity at temperatures as high as 164 K.

The resistivity of a superconductor is zero. There can be a current in a superconductor even when there is no emf in the superconducting circuit. Indeed, in superconducting rings in which there was no electric field, steady currents have been observed to persist for years without apparent loss. Despite the cost and inconvenience of refrigeration with expensive liquid helium, many superconducting magnets have been built using superconducting materials, because such magnets require no power expenditure to maintain the large current needed to produce a large magnetic field.

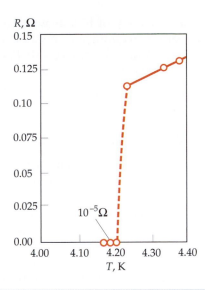

FIGURE 38-29 Plot by H. Kamerlingh Onnes of the resistance of mercury versus temperature, showing the sudden decrease at the critical temperature of $T = 4.2$ K.

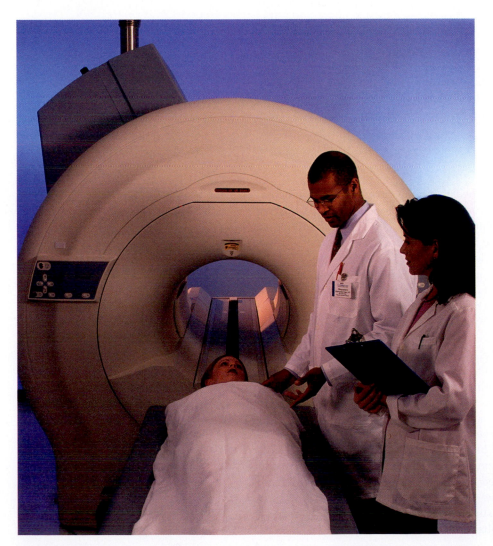

The wires for the magnetic field of a magnetic resonance imaging (MRI) machine carry large currents. To keep the wires from overheating, they are maintained at superconducting temperatures. To accomplish this, they are immersed in liquid helium.

The discovery of high-temperature superconductors has revolutionized the study of superconductivity because relatively inexpensive liquid nitrogen, which boils at 77 K, can be used for a coolant. However, many problems, such as brittleness and the toxicity of the materials, make these new superconductors difficult to use. The search continues for new materials that will superconduct at even higher temperatures.

The BCS Theory

It had been recognized for some time that superconductivity is due to a collective action of the conducting electrons. In 1957, John Bardeen, Leon Cooper, and Robert Schrieffer published a successful theory of superconductivity now known by the initials of the inventors as the **BCS theory.** According to this theory, the electrons in a superconductor are coupled in pairs at low temperatures. The coupling comes about because of the interaction between electrons and the crystal lattice. One electron interacts with the lattice and perturbs it. The perturbed lattice interacts with another electron in such a way that there is an attraction between the two electrons that at low temperatures can exceed the Coulomb repulsion between them. The electrons form a bound state called a **Cooper pair.** The electrons in a Cooper pair have equal and opposite spins, so they form a system with zero spin. Each Cooper pair acts as a *single particle* with zero spin, in other words, as a boson. Bosons do not obey the exclusion principle. Any number of Cooper pairs may be in the same quantum state with the same energy. In the ground state of a superconductor (at $T = 0$), all the electrons are in Cooper pairs and all the Cooper pairs are in the same energy state. In the superconducting state, the Cooper pairs are correlated so that they all act together. An electric current can be produced in a superconductor because all of the electrons in this collective state move together. But energy cannot be dissipated by individual collisions of electron and lattice ions unless the temperature is high enough to break the binding of the Cooper pairs. The required energy is called the **superconducting energy gap** E_g. In the BCS theory, this energy at absolute zero is related to the critical temperature by

$$E_g = 3.5 \, kT_c \tag{38-37}$$

The energy gap can be determined by measuring the current across a junction between a normal metal and a superconductor as a function of voltage. Consider two metals separated by a layer of insulating material, such as aluminum oxide, that is only a few nanometers thick. The insulating material between the metals forms a barrier that prevents most electrons from traversing the junction. However, waves can tunnel through a barrier if the barrier is not too thick, even if the energy of the wave is less than that of the barrier.

When the materials on either side of the gap are normal nonsuperconducting metals, the current resulting from the tunneling of electrons through the insulating layer obeys Ohm's law for low applied voltages (Figure 38-30a). When one of the metals is a normal metal and the other is a superconductor, there is no current (at absolute zero) unless the applied voltage V is greater than a critical voltage $V_c = E_g/(2e)$, where E_g is the superconductor energy gap. Figure 38-30b shows the plot of current versus voltage for this situation. The current escalates rapidly when the energy $2eV$ absorbed by a Cooper pair traversing the barrier approaches $E_g = 2eV_c$, the minimum energy needed to break up the pair. (The small current visible in Figure 38-30b before the critical voltage is reached is present because at any temperature above absolute zero some of the electrons in the superconductor are thermally excited above the energy gap and are therefore not paired.) At voltages slightly above V_c, the current versus voltage curve becomes that for a normal metal. The superconducting energy gap can thus be measured by measuring the average voltage for the transition region.

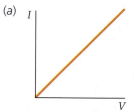

(a)

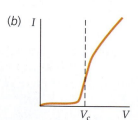

(b)

FIGURE 38-30 Tunneling current versus voltage for a junction of two metals separated by a thin oxide layer. (*a*) When both metals are normal metals, the current is proportional to the voltage, as predicted by Ohm's law. (*b*) When one metal is a normal metal and another metal is a superconductor, the current is approximately zero until the applied voltage V approaches the critical voltage $V_c = E_g/(2e)$.

EXAMPLE 38-8

Calculate the superconducting energy gap for mercury (T_c = 4.2 K) predicted by the BCS theory.

1. The BCS prediction for the energy gap is:

$$E_g = 3.5kT_c$$

2. Substitute T_c = 4.2 K:

$$E_g = 3.5kT_c$$

$$= 3.5(1.38 \times 10^{-23} \text{ J/K})(4.2 \text{ K})\left(\frac{1 \text{ ev}}{1.6 \times 10^{-19} \text{ J}}\right)$$

$$= \boxed{1.27 \times 10^{-3} \text{ eV}}$$

Note that the energy gap for a typical superconductor is much smaller than the energy gap for a typical semiconductor, which is of the order of 1 eV. As the temperature is increased from $T = 0$, some of the Cooper pairs are broken. Then there are fewer pairs available for each pair to interact with, and the energy gap is reduced until at $T = T_c$ the energy gap is zero (Figure 38-31).

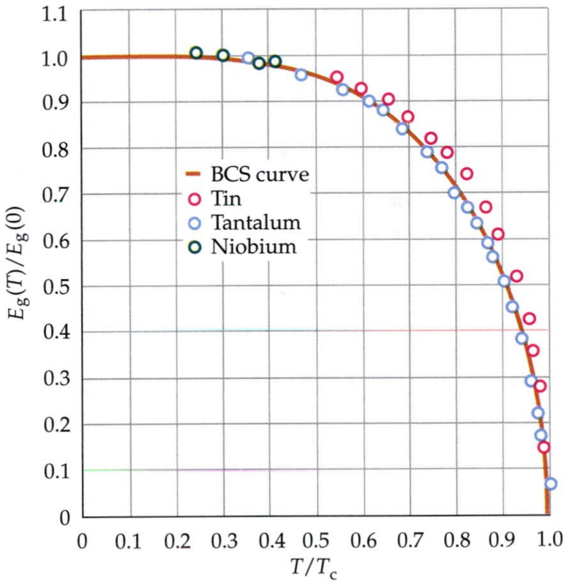

The Josephson Effect

When two superconductors are separated by a thin nonsuperconducting barrier (e.g., a layer of aluminum oxide a few nanometers thick), the junction is called a **Josephson junction,** based on the prediction in 1962 by Brian Josephson that Cooper pairs could tunnel across such a junction from one superconductor to the other with no resistance. The tunneling of Cooper pairs constitutes a current, which does not require a voltage to be applied across the junction. The current depends on the difference in phase of the wave functions that describe the Cooper pairs. Let ϕ_1 be the phase constant for the wave function of a Cooper pair in one superconductor. All the Cooper pairs in a superconductor act coherently and have the same phase constant. If ϕ_2 is the phase constant for the Cooper pairs in the second superconductor, the current across the junction is given by

$$I = I_{max} \sin(\phi_2 - \phi_1) \qquad\qquad 38\text{-}38$$

where I_{max} is the maximum current, which depends on the thickness of the barrier. This result has been observed experimentally and is known as the **dc Josephson effect.**

Josephson also predicted that if a dc voltage V were applied across a Josephson junction, there would be a current that alternates with frequency f given by

$$f = \frac{2e}{h} V \qquad\qquad 38\text{-}39$$

This result, known as the **ac Josephson effect,** has been observed experimentally, and careful measurement of the frequency allows a precise determination of the ratio e/h. Because frequency can be measured very accurately, the ac Josephson effect is also used to establish precise voltage standards. The inverse effect, in which the application of an alternating voltage across a Josephson junction results in a dc current, has also been observed.

FIGURE 38-31 Ratio of the energy gap at temperature T to that at temperature $T = 0$ as a function of the relative temperature T/T_c. The solid curve is that predicted by the BCS theory.

EXAMPLE 38-9

Using $e = 1.602 \times 10^{-19}$ C and $h = 6.626 \times 10^{-34}$ J·s, calculate the frequency of the Josephson current if the applied voltage is 1 μV.

Substitute the given values into Equation 38-39 to calculate f:

$$f = \frac{2e}{h} V = \frac{2(1.602 \times 10^{-19} \text{ C})}{6.626 \times 10^{-34} \text{ J·s}} (10^{-6} \text{ V})$$

$$= 4.835 \times 10^8 \text{ Hz} = \boxed{483.5 \text{ MHz}}$$

38-9 The Fermi–Dirac Distribution†

The classical Maxwell–Boltzmann distribution (Equation 17-39) gives the number of molecules with energy E in the range between E and $E + dE$. It is the product of $g(E) dE$ where $g(E)$ is the **density of states** (number of energy states in the range dE) and the Boltzmann factor $e^{-E/(kT)}$, which is the probability of a state being occupied. The distribution function for free electrons in a metal is called the **Fermi–Dirac distribution.** The Fermi–Dirac distribution can be written in the same form as the Maxwell–Boltzmann distribution, with the density of states calculated from quantum theory and the Boltzmann factor replaced by the Fermi factor. Let $n(E) dE$ be the number of electrons with energies between E and $E + dE$. This number is written

$$n(E) \, dE = g(E) \, dE \, f(E) \qquad\qquad 38\text{-}40$$

ENERGY DISTRIBUTION FUNCTION

where $g(E) dE$ is the number of states between E and $E + dE$ and $f(E)$ is the probability of a state being occupied, which is the Fermi factor. The density of states in three dimensions is somewhat difficult to calculate, so we just give the result. For electrons in a metal of volume V, the density of states is

$$g(E) = \frac{8\sqrt{2}\pi m_e^{3/2} V}{h^3} E^{1/2} \qquad\qquad 38\text{-}41$$

DENSITY OF STATES

As in the classical Maxwell–Boltzmann distribution, the density of states is proportional to $E^{1/2}$.

At $T = 0$, the Fermi factor is given by Equation 38-24:

$$f(E) = 1, \qquad E < E_F$$
$$f(E) = 0, \qquad E > E_F$$

The integral of $n(E) dE$ over all energies gives the total number of electrons N. We can derive Equation 38-22a for the Fermi energy at $T = 0$ by integrating $n(E) dE$ from $E = 0$ to $E = \infty$. We obtain

---

† This material is somewhat complicated. You may wish to read it in two passes, the first to gain an overview of the topic, and the second to understand it in depth.

$$N = \int_0^\infty n(E)\,dE = \int_0^{E_F} n(E)\,dE + \int_{E_F}^\infty n(E)\,dE = \frac{8\sqrt{2}\pi m_e^{3/2}V}{h^3} \int_0^{E_F} E^{1/2}\,dE + 0$$

$$= \frac{16\sqrt{2}\pi m_e^{3/2}V}{3h^3} E_F^{3/2}$$

Note that at $T = 0$, $n(E)$ is zero for $E > E_F$. Solving for E_F gives the Fermi energy at $T = 0$:

$$E_F = \frac{h^2}{8m_e}\left(\frac{3N}{\pi V}\right)^{2/3} \qquad\qquad \text{38-42}$$

which is Equation 38-22a. In terms of the Fermi energy, the density of states (Equation 38-41) is

$$g(E) = \frac{8\sqrt{2}\pi m_e^{3/2}V}{h^3}E^{1/2} = \frac{3}{2}NE_F^{-3/2}E^{1/2} \qquad\qquad \text{38-43}$$

DENSITY OF STATES IN TERMS OF E_F

which is obtained by solving Equation 38-42 for m_e, and then substituting for m_e in Equation 38-41. The average energy at $T = 0$ is calculated from

$$E_{av} = \frac{\displaystyle\int_0^{E_F} Eg(E)\,dE}{\displaystyle\int_0^{E_F} g(E)\,dE} = \frac{1}{N}\int_0^{E_F} Eg(E)\,dE$$

where $N = \int_0^{E_F} g(E)\,dE$ is the total number of electrons. Performing the integration, we obtain Equation 38-23:

$$E_{av} = \tfrac{3}{5}E_F \qquad\qquad \text{38-44}$$

AVERAGE ENERGY AT $T = 0$

At $T > 0$, the Fermi factor is more complicated. It can be shown to be

$$f(E) = \frac{1}{e^{(E-E_F)/(kT)} + 1} \qquad\qquad \text{38-45}$$

FERMI FACTOR

We can see from this equation that for E greater than E_F, $e^{(E-E_F)/(kt)}$ becomes very large as T approaches zero, so at $T = 0$, the Fermi factor is zero for $E > E_F$. On the other hand, for E less than E_F, $e^{(E-E_F)/(kt)}$ approaches 0 as T approaches zero, so at $T = 0$, $f(E) = 1$ for $E < E_F$. Thus, the Fermi factor given by Equation 38-45 holds for all temperatures. Note also that for any nonzero value of T, $f(E) = \tfrac{1}{2}$ at $E = E_F$.
The complete Fermi–Dirac distribution function is thus

$$n(E)\,dE = g(E)f(E)\,dE = \frac{8\sqrt{2}\pi m_e^{3/2}V}{h^3}E^{1/2}\frac{1}{e^{(E-E_F)/(kT)} + 1}\,dE \qquad\qquad \text{38-46}$$

FERMI–DIRAC DISTRIBUTION

We can see that for those few electrons with energies much greater than the Fermi energy, the Fermi factor approaches $1/e^{(E-E_F)/(kt)} = e^{(E_F-E)/(kt)} = e^{E_F/(kT)}e^{-E/(kT)}$, which is proportional to $e^{-E/(kT)}$. Thus, the high-energy tail of the Fermi–Dirac energy distribution decreases as $e^{-E/(kT)}$, just like the classical Maxwell–Boltzmann energy distribution. The reason is that in this high-energy region, there are many unoccupied energy states and few electrons, so the Pauli exclusion principle is not important, and the distribution approaches the classical distribution. This result has practical importance because it applies to the conduction electrons in semiconductors.

FERMI FACTOR FOR COPPER AT 300 K   **EXAMPLE 38-10**

At what energy is the Fermi factor equal to 0.1 for copper at $T = 300$ K?

PICTURE THE PROBLEM We set $f(E) = 0.1$ in Equation 38-45, using $T = 300$ K and $E_F = 7.04$ eV from Table 38-1 and solve for E.

1. Solve Equation 38-45 for $e^{(E-E_F)/(kt)}$:

$$f(E) = \frac{1}{e^{(E-E_F)/(kt)} + 1}$$

so

$$e^{(E-E_F)/(kt)} = \frac{1}{f(E)} - 1 = \frac{1}{0.1} - 1 = 9$$

2. Take the logarithm of both sides:

$$\frac{E - E_F}{kT} = \ln 9$$

3. Solve for E. For E_F, use the value for E_F at $T = 0$ K listed in Table 38-1:

$$E = E_F + (\ln 9)kT$$

$$= 7.04 \text{ eV} + (\ln 9)(8.62 \times 10^{-5} \text{ eV/K})(300 \text{ K})$$

$$= \boxed{7.10 \text{ eV}}$$

REMARKS The Fermi factor drops from about 1 to 0.1 at just 0.06 eV above the Fermi energy of approximately 7 eV.

PROBABILITY OF A HIGHER ENERGY STATE BEING OCCUPIED   **EXAMPLE 38-11**

Find the probability that an energy state in copper 0.1 eV above the Fermi energy is occupied at $T = 300$ K.

PICTURE THE PROBLEM The probability is the Fermi factor given in Equation 38-45, with $E_F = 7.04$ eV, and $E = 7.14$ eV.

1. The probability of a state being occupied equals the Fermi factor:

$$P = f(E) = \frac{1}{e^{(E-E_F)/(kT)} + 1}$$

2. Calculate the exponent in the Fermi factor (exponents are always dimensionless):

$$\frac{E - E_F}{kT} = \frac{7.14 \text{ eV} - 7.04 \text{ eV}}{(8.62 \times 10^{-5} \text{ eV/K})(300 \text{ K})} = 3.87$$

3. Use this result to calculate the Fermi factor:

$$f(E) = \frac{1}{e^{(E-E_F)/(kT)} + 1} = \frac{1}{e^{3.87} + 1}$$

$$= \frac{1}{48 + 1} = \boxed{0.0204}$$

REMARKS The probability of an electron having an energy 0.1 eV above the Fermi energy at 300 K is only about 2 percent.

PROBABILITY OF A LOWER ENERGY STATE BEING OCCUPIED   **EXAMPLE 3 8 - 1 2**   **Try It Yourself**

Find the probability that an energy state in copper 0.1 eV *below* the Fermi energy is occupied at $T = 300$ K.

PICTURE THE PROBLEM   The probability is the Fermi factor given in Equation 38-45, with $E_F = 7.04$ eV, and $E = 6.94$ eV.

Cover the column to the right and try these on your own before looking at the answers.

Steps	Answers
1. Write the Fermi factor.	$f(E) = \dfrac{1}{e^{(E - E_F)/(kT)} + 1}$
2. Calculate the exponent in the Fermi factor.	$\dfrac{E - E_F}{kT} = \dfrac{6.94 \text{ eV} - 7.04 \text{ eV}}{(8.62 \times 10^{-5} \text{ eV/K})(300 \text{ K})} = -3.87$
3. Use your result from step 2 to calculate the Fermi factor.	$f(E) = \dfrac{1}{e^{(E - E_F)/(kT)} + 1} = \dfrac{1}{e^{-3.87} + 1}$
	$= \dfrac{1}{0.021 + 1} = \boxed{0.979}$

REMARKS   The probability of an electron having an energy of 0.1 eV *below* the Fermi energy at 300 K is approximately 98 percent.

EXERCISE   What is the probability of an energy state 0.1 eV below the Fermi energy being unoccupied at 300 K? (*Answer* $1 - 0.98 = 0.02$ or 2 percent. This is the probability of there being a hole at this energy.)

SUMMARY

Topic	Relevant Equations and Remarks
1. The Structure of Solids	Solids are often found in crystalline form in which a small structure, which is called the unit cell, is repeated over and over. A crystal may have a face-centered-cubic, body-centered-cubic, hexagonal close-packed, or other structure depending on the type of bonding between the atoms, ions, or molecules in the crystal and on the relative sizes of the atoms.

Potential energy

$$U = -\alpha \frac{ke^2}{r} + \frac{A}{r^n}$$ 38-3

where r is the separation distance between neighboring ions and α is the Madelung constant, which depends on the geometry of the crystal and is of the order of 1.8, and n is approximately 9.

2. A Microscopic Picture of Conduction

Resistivity

$$\rho = \frac{m_e v_{av}}{n_e e^2 \lambda}$$ 38-14

where v_{av} is the average speed of the electrons and λ is their mean free path between collisions with the lattice ions.

Mean free path	$$\lambda = \frac{vt}{n_{ion}\pi r^2 vt} = \frac{1}{n_{ion}\pi r^2} = \frac{1}{n_{ion}A}$$	**38-16**

where n_{ion} is the number of lattice ions per unit volume, r is their effective radius, and A is their effective cross-sectional area.

3. Classical Interpretation of v_{av} and λ	v_{av} is determined from the Maxwell–Boltzmann distribution, and r is the actual radius of a lattice ion.

4. Quantum Interpretation of v_{av} and λ	v_{av} is determined from the Fermi–Dirac distribution and is approximately constant independent of temperature. The mean free path is determined from the scattering of electron waves, which occurs only because of deviations from a perfectly ordered array. The radius r is the amplitude of vibration of the lattice ion, which is proportional to $\sqrt{T}$, so A is proportional to T.

5. The Fermi Electron Gas

Fermi energy E_F at $T = 0$	E_F is the energy of the last filled (or half-filled) energy state.

E_F at $T > 0$	E_F is the energy at which the probability of being occupied is $\frac{1}{2}$.

Approximate magnitude of E_F	E_F is between 5 eV and 10 eV for most metals.

Dependence of E_F on the number density of free electrons	$$E_F = \frac{h^2}{8m_e}\left(\frac{3N}{\pi V}\right)^{2/3} = (0.365 \text{ eV·nm}^2)\left(\frac{N}{V}\right)^{2/3}$$	**38-22a, 38-22b**

Average energy at $T = 0$	$E_{av} = \frac{3}{5}E_F$	**38-23**

Fermi factor at $T = 0$	The Fermi factor $f(E)$ is the probability of a state being occupied.	
	$f(E) = 1, \quad E < E_F$	
	$f(E) = 0, \quad E > E_F$	**38-24**

Fermi temperature	$$T_F = \frac{E_F}{k}$$	**38-25**

Fermi speed	$$u_F = \sqrt{\frac{2E_F}{m_e}}$$	**38-26**

Contact potential	When two different metals are placed in contact, electrons flow from the metal with the higher Fermi energy to the metal with the lower Fermi energy until the Fermi energies of the two metals are equal. In equilibrium, there is a potential difference between the metals that is equal to the difference in the work function of the two metals divided by the electronic charge e:	
	$$V_{contact} = \frac{\phi_1 - \phi_2}{e}$$	**38-27**

Specific heat due to conduction electrons	$$c_V' = \frac{\pi^2}{2}R\frac{T}{T_F}$$	**38-29**

6. Band Theory of Solids	When many atoms are brought together to form a solid, the individual energy levels are split into bands of allowed energies. The splitting depends on the type of bonding and the lattice separation. The highest energy band that contains electrons is called the valence band. In a conductor, the valence band is only partially full, so there are many available empty energy states for excited electrons. In an insulator, the valence band is completely full and there is a large energy gap between it and the next

allowed band, the conduction band. In a semiconductor, the energy gap between the filled valence band and the empty conduction band is small; so, at ordinary temperatures, an appreciable number of electrons are thermally excited into the conduction band.

7.	**Semiconductors**	The conductivity of a semiconductor can be greatly increased by doping. In an n-type semiconductor, the doping adds electrons at energies just below that of the conduction band. In a p-type semiconductor, holes are added at energies just above that of the valence band.
8.	***Semiconductor Junctions and Devices**	
	*Junctions	Semiconductor devices such as diodes and transistors make use of n-type semiconductors and p-type semiconductors. The two types of semiconductors are typically a single silicon crystal doped with donor impurities on one side and acceptor impurities on the other side. The region in which the semiconductor changes from a p-type semiconductor to an n-type semiconductor is called a junction. Junctions are used in diodes, solar cells, surface barrier detectors, LEDs, and transistors.
	*Diodes	A diode is a single-junction device that carries current in one direction only.
	*Zener diodes	A Zener diode is a diode with a very high reverse bias. It breaks down suddenly at a distinct voltage and can therefore be used as a voltage reference standard.
	*Tunnel diodes	A tunnel diode is a diode that is heavily doped so that electrons tunnel through the depletion barrier. At normal operation, a small change in bias voltage results in a large change in current.
	*Transistors	A transistor consists of a very thin semiconductor of one type sandwiched between two semiconductors of the opposite type. Transistors are used in amplifiers because a small variation in the base current results in a large variation in the collector current.
9.	**Superconductivity**	In a superconductor, the resistance drops suddenly to zero below a critical temperature T_c. Superconductors with critical temperatures as high as 138 K have been discovered.
	The BCS theory	Superconductivity is described by a theory of quantum mechanics called the BCS theory in which the free electrons form Cooper pairs. The energy needed to break up a Cooper pair is called the superconducting energy gap E_g. When all the electrons are paired, individual electrons cannot be scattered by a lattice ion, so the resistance is zero.
	Tunneling	When a normal conductor is separated from a superconductor by a thin layer of oxide, electrons can tunnel through the energy barrier if the applied voltage across the layer is $E_\mathrm{g}/(2e)$, where E_g is the energy needed to break up a Cooper pair. The energy gap E_g can be determined by a measurement of the tunneling current versus the applied voltage.
	Josephson junction	A system of two superconductors separated by a thin layer of nonconducting material is called a Josephson junction.
	dc Josephson effect	A dc current is observed to tunnel through a Josephson junction even in the absence of voltage across the junction.
	ac Josephson effect	When a dc voltage V is applied across a Josephson junction, an ac current is observed with a frequency $$f = \frac{2e}{h} V \qquad \text{38-39}$$ Measurement of the frequency of this current allows a precise determination of the ratio e/h.

10. The Fermi–Dirac Distribution

The number of electrons with energies between E and $E + dE$ is given by

$$n(E)\, dE = g(E)\, dE\, f(E) \qquad\qquad \textbf{38-40}$$

where $g(E)$ is the density of states and $f(E)$ is the Fermi factor.

Density of states	$g(E) = \dfrac{8\sqrt{2}\,\pi m_e^{3/2} V}{h^3} E^{1/2}$	**38-41**
Fermi factor at temperature	$f(E) = \dfrac{1}{e^{(E - E_F)/(kT)} + 1}$	**38-45**

PROBLEMS

- Single-concept, single-step, relatively easy
- •• Intermediate-level, may require synthesis of concepts
- ••• Challenging
- **SSM** Solution is in the *Student Solutions Manual*
- **iSOLVE** Problems available on iSOLVE online homework service
- **iSOLVE ✓** These "Checkpoint" online homework service problems ask students additional questions about their confidence level, and how they arrived at their answer.

In a few problems, you are given more data than you actually need; in a few other problems, you are required to supply data from your general knowledge, outside sources, or informed estimates.

Conceptual Problems

1 • In the classical model of conduction, the electron loses energy on average in a collision because it loses the drift velocity it had picked up since the last collision. Where does this energy appear?

2 • **SSM** When the temperature of pure copper is lowered from 300 K to 4 K, its resistivity drops by a much greater factor than that of brass when it is cooled the same way. Why?

3 • Thomas refuses to believe that a potential difference can be created simply by bringing two different metals into contact with each other. John talks him into making a small wager, and is about to cash in. (a) Which two metals from Table 38-2 would demonstrate his point most effectively? (b) What is the value of that contact potential?

4 • (a) In Problem 3, which choices of different metals would make the least impressive demonstration? (b) What is the value of that contact potential?

5 • A metal is a good conductor because the valence energy band for electrons is (a) completely full. (b) full, but there is only a small gap to a higher empty band. (c) partly full. (d) empty. (e) None of these is correct.

6 • Insulators are poor conductors of electricity because (a) there is a small energy gap between the valence band and the next higher band where electrons can exist. (b) there is a large energy gap between the full valence band and the next higher band where electrons can exist. (c) the valence band has a few vacancies for electrons. (d) the valence band is only partly full. (e) None of these is correct.

7 • True or false:

(a) Solids that are good electrical conductors are usually good heat conductors.
(b) The classical free-electron theory adequately explains the heat capacity of metals.
(c) At $T = 0$, the Fermi factor is either 1 or 0.
(d) The Fermi energy is the average energy of an electron in a solid.
(e) The contact potential between two metals is proportional to the difference in the work functions of the two metals.
(f) At $T = 0$, an intrinsic semiconductor is an insulator.
(g) Semiconductors conduct current in one direction only.

8 • **SSM** How does the change in the resistivity of copper compare with that of silicon when the temperature increases?

9 • Which of the following elements are most likely to act as acceptor impurities in germanium? (a) bromine (b) gallium (c) silicon (d) phosphorus (e) magnesium

10 • Which of the following elements are most likely to serve as donor impurities in germanium? (a) bromine (b) gallium (c) silicon (d) phosphorus (e) magnesium

11 • An electron hole is created when a photon is absorbed by a semiconductor. How does this enable the semiconductor to conduct electricity?

12 • Examine the positions of phosphorus, P; boron, B; thallium, Tl; and antimony, Sb in Table 36-1. (a) Which of these elements can be used to dope silicon to create an n-type semiconductor? (b) Which of these elements can be used to dope silicon to create a p-type semiconductor?

13 • When light strikes the p-type semiconductor in a pn junction solar cell, (a) only free electrons are created. (b) only positive holes are created. (c) both electrons and holes are created. (d) positive protons are created. (e) None of these is correct.

Estimation and Approximation

14 • The ratio of the resistivity of the most resistive (least conductive) material to that of the least resistive material (excluding superconductors) is approximately 10^{24}. You can develop a feeling for how remarkable this range is by considering what the ratio is of the largest to smallest values of other material properties. Choose any three properties of matter, and using tables in this book or some other resource, calculate the ratio of the largest instance of the property to the smallest instance of that property (other than zero) and rank these in decreasing order. Can you find any other property that shows a range as large as that of electrical resistivity?

15 • A device is said to be "ohmic" if a graph of current versus applied voltage is a straight line, and the resistance R of the device is the slope of this line. A pn junction is an example of a nonohmic device, as may be seen from Figure 38-21. For nonohmic devices, it is sometimes convenient to define the *differential resistance* as the reciprocal of the slope of the I versus V curve. Using the curve in Figure 38-21, estimate the differential resistance of the pn junction at applied voltages of –20 V, +0.2 V, +0.4 V, +0.6 V, and +0.8 V.

The Structure of Solids

16 • **iSOLVE** Calculate the distance r_0 between the K^+ and the Cl^- ions in KCl, assuming that each ion occupies a cubic volume of side r_0. The molar mass of KCl is 74.55 g/mol and its density is 1.984 g/cm³.

17 • **iSOLVE** The distance between the Li^+ and Cl^- ions in LiCl is 0.257 nm. Use this and the molar mass of LiCl (42.4 g/mol) to compute the density of LiCl.

18 • **SSM** **iSOLVE** Find the value of n in Equation 38-6 that gives the measured dissociation energy of 741 kJ/mol for LiCl, which has the same structure as NaCl and for which $r_0 =$ 0.257 nm.

19 •• (a) Use Equation 38-6 and calculate $U(r_0)$ for calcium oxide, CaO, where $r_0 = 0.208$ nm. Assume $n = 8$. (b) If $n = 10$, what is the fractional change in $U(r_0)$?

A Microscopic Picture of Conduction

20 • A measure of the density of the free-electron gas in a metal is the distance r_s, which is defined as the radius of the sphere whose volume equals the volume per conduction electron. (a) Show that $r_s = [3/(4\pi n)]^{1/3}$, where n is the free-electron number density. (b) Calculate r_s for copper in nanometers.

21 • **iSOLVE ✓** (a) Given a mean free path $\lambda = 0.4$ nm and a mean speed $v_{av} = 1.17 \times 10^5$ m/s for the current flow in copper at a temperature of 300 K, calculate the classical value for the resistivity ρ of copper. (b) The classical model suggests that the mean free path is temperature independent and that v_{av} depends on temperature. From this model, what would ρ be at 100 K?

22 •• **SSM** Silicon has an atomic weight of 28.09 and a density of 2.41×10^3 kg/m³. Each atom of silicon has two valence electrons and the Fermi energy of the material is 4.88 eV. (a) Given that the electron mean free path at room temperature is $\lambda = 27.0$ nm, estimate the resistivity. (b) The accepted value for the resistivity of silicon is 640 $\Omega \cdot$m (at room temperature). How does this accepted value compare to the value calculated in part (a)?

The Fermi Electron Gas

23 • Calculate the number density of free electrons in (a) Ag ($\rho = 10.5$ g/cm³) and (b) Au ($\rho = 19.3$ g/cm³), assuming one free electron per atom, and compare your results with the values listed in Table 38-1.

24 • The density of aluminum is 2.7 g/cm³. How many free electrons are present per aluminum atom?

25 • The density of tin is 7.3 g/cm³. How many free electrons are present per tin atom?

26 • **SSM** **iSOLVE** Calculate the Fermi temperature for (a) Al, (b) K, and (c) Sn.

27 • **iSOLVE** What is the speed of a conduction electron whose energy is equal to the Fermi energy E_F for (a) Na, (b) Au, and (c) Sn?

28 • **iSOLVE** Calculate the Fermi energy for (a) Al, (b) K, and (c) Sn using the number densities given in Table 38-1.

29 • **iSOLVE** Find the average energy of the conduction electrons at $T = 0$ in (a) copper and (b) lithium.

30 • **iSOLVE** Calculate (a) the Fermi temperature and (b) the Fermi energy at $T = 0$ for iron.

31 •• **SSM** (a) Assuming that gold contributes one free electron per atom to the metal, calculate the electron density in gold knowing that its atomic weight is 196.97 and its mass density is 19.3×10^3 kg/m³. (b) If the Fermi speed for gold is 1.39×10^6 m/s, what is the Fermi energy in electron volts? (c) By what factor is the Fermi energy higher than the kT energy at room temperature? (d) Explain the difference between the Fermi energy and the kT energy.

32 •• **SSM** The pressure of an ideal gas is related to the average energy of the gas particles by $PV = \frac{2}{3}NE_{av}$, where N is the number of particles and E_{av} is the average energy. Use this to calculate the pressure of the Fermi electron gas in copper in newtons per square meter, and compare your result with atmospheric pressure, which is about 10^5 N/m². (Note: The units are most easily handled by using the conversion factors 1 N/m² $= 1$ J/m³ and 1 eV $= 1.6 \times 10^{-19}$ J.)

33 •• The bulk modulus B of a material can be defined by

$$B = -V \frac{\partial P}{\partial V}$$

(a) Use the ideal-gas relation $PV = \frac{2}{3}NE_{av}$ and Equation 38-22 and Equation 38-23 to show that

$$P = \frac{2NE_F}{5V} = CV^{-5/3}$$

where C is a constant independent of V. (b) Show that the bulk modulus of the Fermi electron gas is therefore

$$B = \frac{5}{3}P = \frac{2NE_F}{5V}$$

(c) Compute the bulk modulus in newtons per square meter for the Fermi electron gas in copper and compare your result with the measured value of 140×10^9 N/m^2.

34 • Calculate the contact potential between (a) Ag and Cu, (b) Ag and Ni, and (c) Ca and Cu.

Heat Capacity Due to Electrons in a Metal

35 •• **SSM** Gold has a Fermi energy of 5.53 eV. Determine the molar specific heat at constant volume and room temperature for gold.

Quantum Theory of Electrical Conduction

36 • The resistivities of Na, Au, and Sn at $T = 273$ K are 4.2 $\mu\Omega$·cm, 2.04 $\mu\Omega$·cm, and 10.6 $\mu\Omega$·cm, respectively. Use these values and the Fermi speeds calculated in Problem 27 to find the mean free paths λ for the conduction electrons in these elements.

37 •• **SSM** The resistivity of pure copper is increased approximately 1×10^{-8} Ω·m by the addition of 1 percent (by number of atoms) of an impurity throughout the metal. The mean free path depends on both the impurity and the oscillations of the lattice ions according to the equation $1/\lambda = 1/\lambda_t + 1/\lambda_i$. (a) Estimate λ_i from the data given in Table 38-1. (b) If r is the effective radius of an impurity lattice ion seen by an electron, the scattering cross section is πr^2. Estimate this area, using the fact that r is related to λ_i by Equation 38-16.

Band Theory of Solids

38 • When light of wavelength 380.0 nm falls on a semiconductor, electrons are promoted from the valence band to the conduction band. Calculate the energy gap, in electron volts, for this semiconductor.

39 • **SSM** **SOLVE** You are an electron sitting at the top of the valence band in a silicon atom, longing to jump across the 1.14-eV energy gap that separates you from the bottom of the conduction band and all of the adventures that it may contain. What you need, of course, is a photon. What is the maximum photon wavelength that will get you across the gap?

40 • Repeat Problem 39 for germanium, for which the energy gap is 0.74 eV.

41 • Repeat Problem 39 for diamond, for which the energy gap is 7.0 eV.

42 •• **SOLVE**✓ A photon of wavelength 3.35 μm has just enough energy to raise an electron from the valence band

to the conduction band in a lead sulfide crystal. (a) Find the energy gap between these bands in lead sulfide. (b) Find the temperature T for which kT equals this energy gap.

Semiconductors

43 • The donor energy levels in an n-type semiconductor are 0.01 eV below the conduction band. Find the temperature for which $kT = 0.01$ eV.

44 •• When a thin slab of semiconducting material is illuminated with monochromatic electromagnetic radiation, most of the radiation is transmitted through the slab if the wavelength is greater than 1.85 mm. For wavelengths less than 1.85 mm, most of the incident radiation is absorbed. Determine the energy gap of this semiconductor.

45 •• The relative binding of the extra electron in the arsenic atom that replaces an atom in silicon or germanium can be understood from a calculation of the first Bohr orbit of this electron in these materials. Four of arsenic's outer electrons form covalent bonds, so the fifth electron sees a singly charged center of attraction. This model is a modified hydrogen atom. In the Bohr model of the hydrogen atom, the electron moves in free space at a radius a_0 given by Equation 36-12, which is

$$a_0 = \frac{4\pi\epsilon_0\hbar^2}{m_e e^2}$$

When an electron moves in a crystal, we can approximate the effect of the other atoms by replacing ϵ_0 with $\kappa\epsilon_0$ and m_e with an effective mass for the electron. For silicon, κ is 12 and the effective mass is approximately $0.2m_e$. For germanium, κ is 16 and the effective mass is approximately $0.1m_e$. Estimate the Bohr radii for the outer electron as it orbits the impurity arsenic atom in silicon and germanium.

46 •• **SSM** The ground-state energy of the hydrogen atom is given by

$$E_1 = -\frac{mk^2e^4}{2\hbar^2} = -\frac{e^2 m_e}{8\epsilon_0^2 h^2}.$$

Modify this equation in the spirit of Problem 45 by replacing ϵ_0 by $\kappa\epsilon_0$ and m_e by an effective mass for the electron to estimate the binding energy of the extra electron of an impurity arsenic atom in (a) silicon and (b) germanium.

47 •• A doped n-type silicon sample with 10^{16} electrons per cubic centimeter in the conduction band has a resistivity of 5×10^{-3} Ω·m at 300 K. Find the mean free path of the electrons. Use the effective mass of $0.2m_e$ for the mass of the electrons. (See Problem 45.) Compare this mean free path with that of conduction electrons in copper at 300 K.

48 •• The measured Hall coefficient of a doped silicon sample is 0.04 V·m/A·T at room temperature. If all the doping impurities have contributed to the total charge carriers of the sample, find (a) the type of impurity (donor or acceptor) used to dope the sample and (b) the concentration of these impurities.

*Semiconductor Junctions and Devices

49 •• Simple theory for the current versus the bias voltage across a pn junction yields the equation $I = I_0(e^{eV_b/kT} - 1)$.

Sketch I versus V_b for both positive and negative values of V_b using this equation.

50 • The plate current in an *npn* transistor circuit is 25.0 mA. If 88 percent of the electrons emitted reach the collector, what is the base current?

51 •• **SSM** **SOLVE** In Figure 38-27 for the *pnp*-transistor amplifier, suppose $R_b = 2$ kΩ and $R_L = 10$ kΩ. Suppose further that a 10-μA ac base current generates a 0.5-mA ac collector current. What is the voltage gain of the amplifier?

52 •• **SOLVE** ✓ Germanium can be used to measure the energy of incident particles. Consider a 660-keV gamma ray emitted from ^{137}Cs. (*a*) Given that the band gap in germanium is 0.72 eV, how many electron–hole pairs can be generated as this gamma ray travels through germanium? (*b*) The number of pairs N in Part (*a*) will have statistical fluctuations given by $\pm\sqrt{N}$. What then is the energy resolution of this detector in this photon energy region?

53 •• Make a sketch showing the valence and conduction band edges and Fermi energy of a *pn*-junction diode when biased (*a*) in the forward direction and (*b*) in the reverse direction.

54 •• **SSM** A "good" silicon diode has the current–voltage characteristic given in Problem 49. Let $kT = 0.025$ eV (room temperature) and the saturation current $I_0 = 1$ nA. (*a*) Show that for small reverse-bias voltages, the resistance is 25 MΩ. (*Hint: Do a Taylor expansion of the exponential function or use your calculator and enter small values for V_b.*) (*b*) Find the dc resistance for a reverse bias of 0.5 V. (*c*) Find the dc resistance for a 0.5-V forward bias. What is the current in this case? (*d*) Calculate the ac resistance dV/dI for a 0.5-V forward bias.

55 •• A slab of silicon of thickness $t = 1.0$ mm and width $w = 1.0$ cm is placed in a magnetic field $B = 0.4$ T. The slab is in the xy plane, and the magnetic field points in the positive z direction. When a current of 0.2 A flows through the sample in the positive x direction, a voltage difference of 5 mV develops across the width of the sample with the electric field in the sample pointing in the positive y direction. Determine the semiconductor type (n or p) and the concentration of charge carriers.

The BCS Theory

56 • (*a*) Use Equation 38-37 to calculate the superconducting energy gap for tin, and compare your result with the measured value of 6×10^{-4} eV. (*b*) Use the measured value to calculate the wavelength of a photon having sufficient energy to break up Cooper pairs in tin ($T_c = 3.72$ K) at $T = 0$.

57 • **SSM** Repeat Problem 56 for lead ($T_c = 7.19$ K), which has a measured energy gap of 2.73×10^{-3} eV.

The Fermi–Dirac Distribution

58 •• The number of electrons in the conduction band of an insulator or intrinsic semiconductor is governed chiefly by the Fermi factor. Since the valence band in these materials is nearly filled and the conduction band is nearly empty, the Fermi energy E_F is generally midway between the top of the valence band and the bottom of the conduction band, that is,

at $E_g/2$, where E_g is the band gap between the two bands and the energy is measured from the top of the valence band. (*a*) In silicon, $E_g \approx 1.0$ eV. Show that in this case the Fermi factor for electrons at the bottom of the conduction band is given by $\exp(-E_g/2kT)$ and evaluate this factor. Discuss the significance of this result if there are 10^{22} valence electrons per cubic centimeter and the probability of finding an electron in the conduction band is given by the Fermi factor. (*b*) Repeat the calculation in Part (*a*) for an insulator with a band gap of 6.0 eV.

59 •• Approximately how many energy states, with energies between 2.00 eV and 2.20 eV, are available to electrons in a cube of silver measuring 1.00 mm on a side?

60 •• **SSM** (*a*) Use Equation 38-22*a* to calculate the Fermi energy for silver. (*b*) Determine the average energy of a free electron and (*c*) find the Fermi speed for silver.

61 •• **SOLVE** Show that at $E = E_F$, the Fermi factor is $F = 0.5$.

62 •• **SOLVE** ✓ What is the difference between the energies at which the Fermi factor is 0.9 and 0.1 at 300 K in (*a*) copper, (*b*) potassium, and (*c*) aluminum.

63 •• **SSM** **SOLVE** ✓ What is the probability that a conduction electron in silver will have a kinetic energy of 4.9 eV at $T = 300$ K?

64 •• Show that $g(E) = (3N/2)E_F^{-3/2}E^{1/2}$ (Equation 38-43) follows from Equation 38-41 for $g(E)$, and from Equation 38-22*a* for E_F.

65 •• Carry out the integration $E_{av} = (1/N)\int_0^{E_F} Eg(e)\,dE$ to show that the average energy at $T = 0$ is $\frac{3}{5}E_F$.

66 •• The density of the electron states in a metal can be written $g(E) = AE^{1/2}$, where A is a constant and E is measured from the bottom of the conduction band. (*a*) Show that the total number of states is $\frac{2}{3}AE_F^{3/2}$. (*b*) Approximately what fraction of the conduction electrons are within kT of the Fermi energy? (*c*) Evaluate this fraction for copper at $T = 300$ K.

67 •• **SOLVE** ✓ What is the probability that a conduction electron in silver will have a kinetic energy of 5.49 eV at $T = 300$ K?

68 •• **SOLVE** Use Equation 38-41 for the density of states to estimate the fraction of the conduction electrons in copper that can absorb energy from collisions with the vibrating lattice ions at (*a*) 77 K and (*b*) 300 K.

69 •• In an intrinsic semiconductor, the Fermi energy is about midway between the top of the valence band and the bottom of the conduction band. In germanium, the forbidden energy band has a width of 0.7 eV. Show that at room temperature the distribution function of electrons in the conduction band is given by the Maxwell–Boltzmann distribution function.

70 ••• **SSM** (*a*) Show that for $E \geq 0$, the Fermi factor may be written as $f(E) = 1/(Ce^{E/kT} + 1)$. (*b*) Show that if $C \gg e^{-E/(kT)}$, $f(E) = Ae^{-E/(kT)} \ll 1$; in other words, show that the Fermi factor is a constant times the classical Boltzmann factor if $A \ll 1$. (*c*) Use $\int n(E)\,dE = N$ and Equation 38-41 to determine the constant A. (*d*) Using the result obtained in Part (*c*), show that the classical approximation is applicable when the electron concentration is very small and/or the temperature is very high. (*e*) Most semiconductors have impurities added in a process called doping, which increases the free electron concentration so that it is about $10^{17}/cm^3$ at room temperature. Show that for these systems, the classical distribution function is applicable.

71 ••• Show that the condition for the applicability of the classical distribution function for an electron gas ($A \ll 1$ in Problem 70) is equivalent to the requirement that the average separation between electrons is much greater than their de Broglie wavelength.

72 ••• The root-mean-square (rms) value of a variable is obtained by calculating the average value of the square of that variable and then taking the square root of the result. Use this procedure to determine the rms energy of a Fermi distribution. Express your result in terms of E_F and compare it to the average energy. Why do E_{av} and E_{rms} differ?

General Problems

73 • The density of potassium is 0.851 g/cm³. How many free electrons are there per potassium atom?

74 • Calculate the number density of free electrons for (*a*) Mg ($\rho = 1.74$ g/cm³) and (*b*) Zn ($\rho = 7.1$ g/cm³), assuming two free electrons per atom, and compare your results with the values listed in Table 38-1.

75 •• **ISOLVE** Estimate the fraction of free electrons in copper that are in excited states above the Fermi energy at (*a*) 300 K (about room temperature) and (*b*) at 1000 K.

76 •• **SSM** Determine the energy that has 10 percent free electron occupancy probability for manganese at T = 1300 K.

77 •• The semiconducting compound CdSe is widely used for light-emitting diodes (LEDs). The energy gap in CdSe is 1.8 eV. What is the frequency of the light emitted by a CdSe LED?

78 ••• **SSM** A 2-cm² wafer of pure silicon is irradiated with light having a wavelength of 775 nm. The intensity of the light beam is 4.0 W/m² and every photon that strikes the sample is absorbed and creates an electron–hole pair. (*a*) How many electron–hole pairs are produced in one second? (*b*) If the number of electron–hole pairs in the sample is 6.25×10^{11} in the steady state, at what rate do the electron–hole pairs recombine? (*c*) If every recombination event results in the radiation of one photon, at what rate is energy radiated by the sample?

Relativity

THE ANDROMEDA GALAXY BY MEASURING THE FREQUENCY OF THE LIGHT COMING TO US FROM DISTANT OBJECTS, WE ARE ABLE TO DETERMINE HOW FAST THESE OBJECTS ARE APPROACHING TOWARD US OR RECEDING FROM US.

? **Have you wondered how the frequency of the light enables us to determine the speed of recession of a distant galaxy? This is discussed in Example 39-5.**

The theory of relativity consists of two rather different theories, the special theory and the general theory. The special theory, developed by Albert Einstein and others in 1905, concerns the comparison of measurements made in different inertial reference frames moving with constant velocity relative to one another. Its consequences, which can be derived with a minimum of mathematics, are applicable in a wide variety of situations encountered in physics and in engineering. On the other hand, the general theory, also developed by Einstein and others around 1916, is concerned with accelerated reference frames and gravity. A thorough understanding of the general theory requires sophisticated mathematics, and the applications of this theory are chiefly in the area of gravitation. The general theory is of great importance in cosmology, but it is rarely encountered in other areas of physics or in engineering. The general theory is used, however, in the engineering of the Global Positioning System (GPS).[†]

---

† The satellites used in GPS contain atomic clocks.

> In this chapter, we concentrate on the special theory (often referred to as *special relativity*). General relativity will be discussed briefly near the end of the chapter.

39-1 Newtonian Relativity

Newton's first law does not distinguish between a particle at rest and a particle moving with constant velocity. If there is no net external force acting, the particle will remain in its initial state, either at rest or moving with its initial velocity. A particle at rest relative to you is moving with constant velocity relative to an observer who is moving with constant velocity relative to you. How might we distinguish whether you and the particle are at rest and the second observer is moving with constant velocity, or the second observer is at rest and you and the particle are moving?

Let us consider some simple experiments. Suppose we have a railway boxcar moving along a straight, flat track with a constant velocity v. We note that a ball at rest in the boxcar remains at rest. If we drop the ball, it falls straight down, relative to the boxcar, with an acceleration g due to gravity. Of course, when viewed from the track the ball moves along a parabolic path because it has an initial velocity v to the right. No mechanics experiment that we can do— measuring the period of a pendulum, observing the collisions between two objects, or whatever—will tell us whether the boxcar is moving and the track is at rest or the track is moving and the boxcar is at rest. If we have a coordinate system attached to the track and another attached to the boxcar, Newton's laws hold in either system.

A set of coordinate systems at rest relative to each other is called a *reference frame*. A reference frame in which Newton's laws hold is called an *inertial reference frame*.[†] All reference frames moving at constant velocity relative to an inertial reference frame are also inertial reference frames. If we have two inertial reference frames moving with constant velocity relative to each other, there are no mechanics experiments that can tell us which is at rest and which is moving or if they are both moving. This result is known as the principle of **Newtonian relativity**:

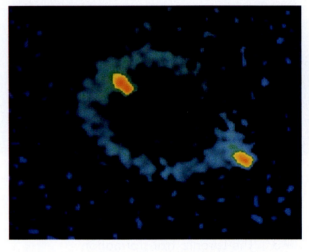

This ring-like structure of the radio source MG1131 + 0456 is thought to be due to *gravitational lensing*, first proposed by Albert Einstein in 1936, in which a source is imaged into a ring by a large, massive object in the foreground.

Absolute motion cannot be detected.

PRINCIPLE OF NEWTONIAN RELATIVITY

This principle was well known by Galileo, Newton, and others in the seventeenth century. By the late nineteenth century, however, this view had changed. It was then generally thought that Newtonian relativity was not valid and that absolute motion could be detected in principle by a measurement of the speed of light.

Ether and the Speed of Light

We saw in Chapter 15 that the velocity of a wave depends on the properties of the medium in which the wave travels and not on the velocity of the source of the waves. For example, the velocity of sound relative to still air depends on the temperature of the air. Light and other electromagnetic waves (radio, X rays, etc.) travel through a vacuum with a speed $c \approx 3 \times 10^8$ m/s that is predicted by James Clerk Maxwell's equations for electricity and magnetism. But what is this speed

---

† Reference frames were first discussed in Section 2-1. Inertial reference frames were also discussed in Section 4-1.

relative to? What is the equivalent of still air for a vacuum? A proposed medium for the propagation of light was called the *ether*; it was thought to pervade all space. The velocity of light relative to the ether was assumed to be c, as predicted by Maxwell's equations. The velocity of any object relative to the ether was considered its absolute velocity.

Albert Michelson, first in 1881 and then again with Edward Morley in 1887, set out to measure the velocity of the earth relative to the ether by an ingenious experiment in which the velocity of light relative to the earth was compared for two light beams, one in the direction of the earth's motion relative to the sun and the other perpendicular to the direction of the earth's motion. Despite painstakingly careful measurements, they could detect no difference. The experiment has since been repeated under various conditions by a number of people, and no difference has ever been found. The absolute motion of the earth relative to the ether cannot be detected.

39-2　Einstein's Postulates

In 1905, at the age of 26, Albert Einstein published a paper on the electrodynamics of moving bodies.[†] In this paper, he postulated that absolute motion cannot be detected by any experiment. That is, there is no ether. The earth can be considered to be at rest and the velocity of light will be the same in any direction.[‡] His theory of special relativity can be derived from two postulates. Simply stated, these postulates are as follows:

Postulate 1: Absolute uniform motion cannot be detected.

Postulate 2: The speed of light is independent of the motion of the source.

EINSTEIN'S POSTULATES

Postulate 1 is merely an extension of the Newtonian principle of relativity to include all types of physical measurements (not just those that are mechanical). Postulate 2 describes a common property of all waves. For example, the speed of sound waves does not depend on the motion of the sound source. The sound waves from a car horn travel through the air with the same velocity independent of whether the car is moving or not. The speed of the waves depends only on the properties of the air, such as its temperature.

Although each postulate seems quite reasonable, many of the implications of the two postulates together are quite surprising and contradict what is often called common sense. For example, one important implication of these postulates is that every observer measures the same value for the speed of light independent of the relative motion of the source and the observer. Consider a light source S and two observers, R_1 at rest relative to S and R_2 moving toward S with speed v, as shown in Figure 39-1a. The speed of light measured by R_1 is $c = 3 \times 10^8$ m/s. What is the speed measured by R_2? The answer is *not* $c + v$. By postulate 1, Figure 39-1a is equivalent to Figure 39-1b, in which R_2 is at rest and the source S and R_1 are moving with speed v. That is, since absolute motion cannot be detected, it is not possible to say which is really moving and which is at rest. By postulate 2, the speed of light from a moving source is independent of the

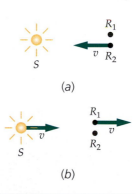

(a)

(b)

FIGURE 39-1　(*a*) A stationary light source S and a stationary observer R_1, with a second observer R_2 moving toward the source with speed v. (*b*) In the reference frame in which the observer R_2 is at rest, the light source S and observer R_1 move to the right with speed v. If absolute motion cannot be detected, the two views are equivalent. Since the speed of light does not depend on the motion of the source, observer R_2 measures the same value for that speed as observer R_1.

† *Annalen der Physik*, vol. 17, 1905, p. 841. For a translation from the original German, see W. Perrett and G. B. Jeffery (trans.), *The Principle of Relativity: A Collection of Original Memoirs on the Special and General Theory of Relativity* by H. A. Lorentz, A. Einstein, H. Minkowski, and W. Weyl, Dover, New York, 1923.

‡ Einstein did not set out to explain the results of the Michelson–Morley experiment. His theory arose from his considerations of the theory of electricity and magnetism and the unusual property of electromagnetic waves that they propagate in a vacuum. In his first paper, which contains the complete theory of special relativity, he made only a passing reference to the Michelson–Morley experiment, and in later years he could not recall whether he was aware of the details of this experiment before he published his theory.

motion of the source. Thus, looking at Figure 39-1*b*, we see that R_2 measures the speed of light to be c, just as R_1 does. This result is often considered as an alternative to Einstein's second postulate:

> Postulate 2 (alternate): Every observer measures the same value c for the speed of light.

This result contradicts our intuitive ideas about relative velocities. If a car moves at 50 km/h away from an observer and another car moves at 80 km/h in the same direction, the velocity of the second car relative to the first car is 30 km/h. This result is easily measured and conforms to our intuition. However, according to Einstein's postulates, if a light beam is moving in the direction of the cars, observers in both cars will measure the same speed for the light beam. Our intuitive ideas about the combination of velocities are approximations that hold only when the speeds are very small compared with the speed of light. Even in an airplane moving with the speed of sound, to measure the speed of light accurately enough to distinguish the difference between the results c and $c + v$, where v is the speed of the plane, would require a measurement with six-digit accuracy.

39-3 The Lorentz Transformation

Einstein's postulates have important consequences for measuring time intervals and space intervals, as well as relative velocities. Throughout this chapter, we will be comparing measurements of the positions and times of events (such as lightning flashes) made by observers who are moving relative to each other. We will use a rectangular coordinate system xyz with origin O, called the S reference frame, and another system $x'y'z'$ with origin O', called the S' frame, that is moving with a constant velocity $\vec{v}$ relative to the S frame. Relative to the S' frame, the S frame is moving with a constant velocity $-\vec{v}$. For simplicity, we will consider the S' frame to be moving along the x axis in the positive x direction relative to S. In each frame, we will assume that there are as many observers as are needed who are equipped with measuring devices, such as clocks and metersticks, that are identical when compared at rest (see Figure 39-2).

We will use Einstein's postulates to find the general relation between the coordinates x, y, and z and the time t of an event as seen in reference frame S and the coordinates x', y', and z' and the time t' of the same event as seen in reference frame S', which is moving with uniform velocity relative to S. We assume that the origins are coincident at time $t = t' = 0$. The classical relation, called the **Galilean transformation,** is

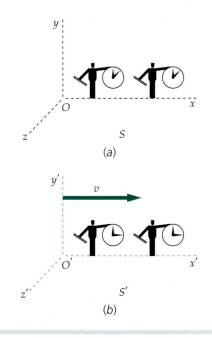

FIGURE 39-2 Coordinate reference frames S and S' moving with relative speed v. In each frame, there are observers with metersticks and clocks that are identical when compared at rest.

$$x = x' + vt', \qquad y = y', \qquad z = z', \qquad t = t' \qquad \text{39-1}a$$

GALILEAN TRANSFORMATION

The inverse transformation is

$$x' = x - vt, \qquad y' = y, \qquad z' = z, \qquad t' = t \qquad \text{39-1}b$$

These equations are consistent with experimental observations as long as v is much less than c. They lead to the familiar classical addition law for velocities. If a particle has velocity $u_x = dx/dt$ in frame S, its velocity in frame S' is

$$u'_x = \frac{dx'}{dt'} = \frac{dx'}{dt} = \frac{dx}{dt} - v = u_x - v \qquad \text{39-2}$$

If we differentiate this equation again, we find that the acceleration of the particle is the same in both frames:

$$a_x = \frac{du_x}{dt} = \frac{du'_x}{dt'} = a'_x$$

It should be clear that the Galilean transformation is not consistent with Einstein's postulates of special relativity. If light moves along the x axis with speed $u'_x = c$ in S', these equations imply that the speed in S' is $u_x = c + v$ rather than $u_x = c$, which is consistent with Einstein's postulates and with experiment. The classical transformation equations must therefore be modified to make them consistent with Einstein's postulates. We will give a brief outline of one method of obtaining the relativistic transformation.

We assume that the relativistic transformation equation for x is the same as the classical equation (Equation 39-1a) except for a constant multiplier on the right side. That is, we assume the equation is of the form

$$x = \gamma(x' + vt') \qquad\qquad 39\text{-}3$$

where γ is a constant that can depend on v and c but not on the coordinates. The inverse transformation must look the same except for the sign of the velocity:

$$x' = \gamma(x - vt) \qquad\qquad 39\text{-}4$$

Let us consider a light pulse that starts at the origin of S at $t = 0$. Since we have assumed that the origins are coincident at $t = t' = 0$, the pulse also starts at the origin of S' at $t' = 0$. Einstein's postulates require that the equation for the x component of the wave front of the light pulse is $x = ct$ in frame S and $x' = ct'$ in frame S'. Substituting ct for x and ct' for x' in Equation 39-3 and Equation 39-4, we obtain

$$ct = \gamma(ct' + vt') = \gamma(c + v)t' \qquad\qquad 39\text{-}5$$

and

$$ct' = \gamma(ct - vt) = \gamma(c - v)t \qquad\qquad 39\text{-}6$$

We can eliminate the ratio t'/t from these two equations and determine γ. Thus,

$$\gamma = \frac{1}{\sqrt{1 - \dfrac{v^2}{c^2}}} \qquad\qquad 39\text{-}7$$

Note that γ is always greater than 1, and that when v is much less than c, $\gamma \approx 1$. The relativistic transformation for x and x' is therefore given by Equation 39-3 and Equation 39-4, with γ given by Equation 39-7. We can obtain equations for t and t' by combining Equation 39-3 with the inverse transformation given by Equation 39-4. Substituting $x = \gamma(x' + vt')$ for x in Equation 39-4, we obtain

$$x' = \gamma\big[\gamma(x' + vt') - vt\big] \qquad\qquad 39\text{-}8$$

which can be solved for t in terms of x' and t'. The complete relativistic transformation is

$$x = \gamma(x' + vt'), \qquad y = y', \qquad z = z' \tag{39-9}$$

$$t = \gamma\left(t' + \frac{vx'}{c^2}\right) \tag{39-10}$$

LORENTZ TRANSFORMATION

The inverse transformation is

$$x' = \gamma(x - vt), \qquad y' = y, \qquad z' = z \tag{39-11}$$

$$t' = \gamma\left(t - \frac{vx}{c^2}\right) \tag{39-12}$$

The transformation described by Equation 39-9 through Equation 39-12 is called the **Lorentz transformation.** It relates the space and time coordinates x, y, z, and t of an event in frame S to the coordinates x', y', z', and t' of the same event as seen in frame S', which is moving along the x axis with speed v relative to frame S.

We will now look at some applications of the Lorentz transformation.

Time Dilation

Consider two events that occur at a single point x'_0 at times t'_1 and t'_2 in frame S'. We can find the times t_1 and t_2 for these events in S from Equation 39-10. We have

$$t_1 = \gamma\left(t'_1 + \frac{vx'_0}{c^2}\right)$$

and

$$t_2 = \gamma\left(t'_2 + \frac{vx'_0}{c^2}\right)$$

so

$$t_2 - t_1 = \gamma(t'_2 - t'_1)$$

The time between events that happen at the *same place* in a reference frame is called **proper time** t_{p}. In this case, the time interval $t'_2 - t'_1$ measured in frame S' is proper time. The time interval Δt measured in any other reference frame is always longer than the proper time. This expansion is called **time dilation:**

$$\Delta t = \gamma \, \Delta t_{\mathrm{p}} \tag{39-13}$$

TIME DILATION

SPATIAL SEPARATION AND TEMPORAL SEPARATION OF TWO EVENTS   **EXAMPLE 39-1**

Two events occur at the same point x'_0 at times t'_1 and t'_2 in frame S', which is traveling at speed v relative to frame S. (*a*) What is the spatial separation of these events in frame S? (*b*) What is the temporal separation of these events in frame S?

PICTURE THE PROBLEM The spatial separation in S is $x_2 - x_1$, where x_2 and x_1 are the coordinates of the events in S, which are found using Equation 39-9.

(a) 1. The position x_1 in S is given by Equation 39-9 with $x_1' = x_0'$:

$$x_1 = \gamma(x_0' + vt_1')$$

2. Similarly, the position x_2 in S is given by:

$$x_2 = \gamma(x_0' + vt_2')$$

3. Subtract to find the spatial separation:

$$\Delta x = x_2 - x_1 = \gamma v(t_2' - t_1') = \boxed{\dfrac{v(t_2' - t_1')}{\sqrt{1 - (v^2/c^2)}}}$$

(b) Using the time dilation formula, relate the two time intervals. The two events occur at the same place in S', so the proper time between the two events is $\Delta t_p = t_2' - t_1'$:

$$\Delta t = t_2 - t_1 = \gamma(t_2' - t_1') = \boxed{\dfrac{(t_2' - t_1')}{\sqrt{1 - (v^2/c^2)}}}$$

REMARKS Dividing the Part (a) result by the Part (b) result gives $\Delta x/\Delta t = v$. The spatial separation of these two events in S is the distance a fixed point, such as x_0' in S', moves in S during the time interval between the events in S.

We can understand time dilation directly from Einstein's postulates without using the Lorentz transformation. Figure 39-3a shows an observer A' a distance D from a mirror. The observer and the mirror are in a spaceship that is at rest in frame S'. The observer explodes a flash gun and measures the time interval $\Delta t'$ between the original flash and his seeing the return flash from the mirror. Because light travels with speed c, this time is

$$\Delta t' = \frac{2D}{c}$$

We now consider these same two events, the original flash of light and the receiving of the return flash, as observed in reference frame S, in which observer A' and the mirror are moving to the right with speed v, as shown in Figure 39-3b.

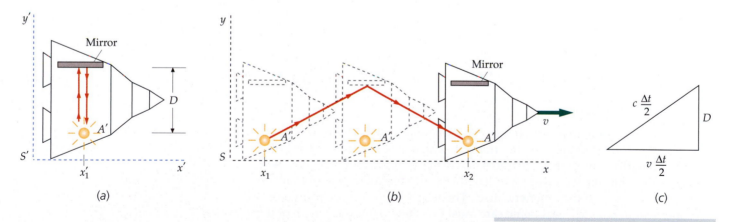

(a)   (b)   (c)

The events happen at two different places x_1 and x_2 in frame S. During the time interval Δt (as measured in S) between the original flash and the return flash, observer A' and his spaceship have moved a horizontal distance $v\,\Delta t$. In Figure 39-3b, we can see that the path traveled by the light is longer in S than in S'. However, by Einstein's postulates, light travels with the same speed c in frame S as it does in frame S'. Because light travels farther in S at the same speed, it takes longer in S to reach the mirror and return. The time interval in S is thus longer than it is in S'. From the triangle in Figure 39-3c, we have

$$\left(\frac{c\,\Delta t}{2}\right)^2 = D^2 + \left(\frac{v\,\Delta t}{2}\right)^2$$

FIGURE 39-3 (a) Observer A' and the mirror are in a spaceship at rest in frame S'. The time it takes for the light pulse to reach the mirror and return is measured by A' to be $2D/c$. (b) In frame S, the spaceship is moving to the right with speed v. If the speed of light is the same in both frames, the time it takes for the light to reach the mirror and return is longer than $2D/c$ in S because the distance traveled is greater than $2D$. (c) A right triangle for computing the time Δt in frame S.

or

$$\Delta t = \frac{2D}{\sqrt{c^2 - v^2}} = \frac{2D}{c} \frac{1}{\sqrt{1 - (v^2/c^2)}}$$

Using $\Delta t' = 2D/c$, we obtain

$$\Delta t = \frac{\Delta t'}{\sqrt{1 - (v^2/c^2)}} = \gamma \Delta t'$$

HOW LONG IS A ONE-HOUR NAP?   **E X A M P L E   3 9 - 2**   **Try It Yourself**

Astronauts in a spaceship traveling at $v = 0.6c$ relative to the earth sign off from space control, saying that they are going to nap for 1 h and then call back. How long does their nap last as measured on the earth?

PICTURE THE PROBLEM Because the astronauts go to sleep and wake up at the same place in their reference frame, the time interval for their nap of 1 h as measured by them is proper time. In the earth's reference frame, they move a considerable distance between these two events. The time interval measured in the earth's frame (using two clocks located at those events) is longer by the factor γ.

Cover the column to the right and try these on your own before looking at the answers.

Steps	Answers
1. Relate the time interval measured on the earth Δt to the proper time Δt_p.	$\Delta t = \gamma \Delta t_p$
2. Calculate γ for $v = 0.6c$.	$\gamma = 1.25$
3. Substitute to calculate the time of the nap in the earth's frame.	$\Delta t = \gamma \Delta t_p = \boxed{1.25 \text{ h}}$

EXERCISE If the spaceship is moving at $v = 0.8c$, how long would a 1 h nap last as measured on the earth? (*Answer* 1.67 h)

Length Contraction

A phenomenon closely related to time dilation is **length contraction**. The length of an object measured in the reference frame in which the object is at rest is called its **proper length** L_p. In a reference frame in which the object is moving, the measured length is shorter than its proper length. Consider a rod at rest in frame S' with one end at x_2' and the other end at x_1'. The length of the rod in this frame is its proper length $L_p = x_2' - x_1'$. Some care must be taken to find the length of the rod in frame S. In this frame, the rod is moving to the right with speed v, the speed of frame S'. The length of the rod in frame S is defined as $L = x_2 - x_1$, where x_2 is the position of one end at some time t_2, and x_1 is the position of the other end *at the same time* $t_1 = t_2$ as measured in frame S. Equation 39-11 is convenient to use to calculate $x_2 - x_1$ at some time t because it relates x and x' to t, whereas Equation 39-9 is not convenient because it relates x and x' to t':

$$x_2' = \gamma(x_2 - vt_2)$$

and

$$x_1' = \gamma(x_1 - vt_1)$$

Since $t_2 = t_1$, we obtain

$$x_2' - x_1' = \gamma(x_2 - x_1)$$

$$x_2 - x_1 = \frac{1}{\gamma}(x_2' - x_1') = (x_2' - x_1')\sqrt{1 - \frac{v^2}{c^2}}$$

or

$$L = \frac{1}{\gamma}L_p = L_p\sqrt{1 - \frac{v^2}{c^2}} \qquad\qquad 39\text{-}14$$

LENGTH CONTRACTION

Thus, the length of a rod is smaller when it is measured in a frame in which it is moving. Before Einstein's paper was published, Hendrik A. Lorentz and George F. FitzGerald tried to explain the null result of the Michelson–Morley experiment by assuming that distances in the direction of motion contracted by the amount given in Equation 39-14. This length contraction is now known as the **Lorentz–FitzGerald contraction.**

THE LENGTH OF A MOVING METERSTICK   **E X A M P L E   3 9 - 3**

A stick that has a proper length of 1 m moves in a direction along its length with speed v relative to you. The length of the stick as measured by you is 0.914 m. What is the speed v?

PICTURE THE PROBLEM Since both L and L_p are given, we can find v directly from Equation 39-14.

1. Equation 39-14 relates the lengths L and L_p and the speed v:

$$L = L_p\sqrt{1 - \frac{v^2}{c^2}}$$

2. Solve for v:

$$v = c\sqrt{1 - \frac{L^2}{L_p^2}} = c\sqrt{1 - \frac{(0.914\ \text{m})^2}{(1\ \text{m})^2}} = \boxed{0.406c}$$

An interesting example of time dilation or length contraction is afforded by the appearance of muons as secondary radiation from cosmic rays. Muons decay according to the statistical law of radioactivity:

$$N(t) = N_0 e^{-t/\tau} \qquad\qquad 39\text{-}15$$

where N_0 is the original number of muons at time $t = 0$, $N(t)$ is the number remaining at time t, and τ is the mean lifetime, which is approximately 2 μs for muons at rest. Since muons are created (from the decay of pions) high in the atmosphere, usually several thousand meters above sea level, few muons should reach sea level. A typical muon moving with speed 0.9978c would travel only about 600 m in 2 μs. However, the lifetime of the muon measured in the earth's reference frame is increased by the factor $1/\sqrt{1 - (v^2/c^2)}$, which is 15 for this particular speed. The mean lifetime measured in the earth's reference frame is therefore 30 μs, and a muon with speed 0.9978c travels approximately 9000 m in this time. From the muon's point of view, it lives only 2 μs, but the atmosphere is rushing past it with a speed of 0.9978c. The distance of 9000 m in the earth's

frame is thus contracted to only 600 m in the muon's frame, as indicated in Figure 39-4.

It is easy to distinguish experimentally between the classical and
relativistic predictions of the observation of muons at sea level. Suppose that we observe 10^8 muons at an altitude of 9000 m in some
time interval with a muon detector. How many would we expect to
observe at sea level in the same time interval? According to the nonrelativistic prediction, the time it takes for these muons to travel
9000 m is $(9000 \text{ m})/(0.998c) \approx 30 \mu s$, which is 15 lifetimes. Substituting $N_0 = 10^8$ and $t = 15\tau$ into Equation 39-15, we obtain

$$N = 10^8 e^{-15} = 30.6$$

We would thus expect all but about 31 of the original 100 million muons to decay
before reaching sea level.

According to the relativistic prediction, the earth must travel only the contracted distance of 600 m in the rest frame of the muon. This takes only $2 \mu s = 1\tau$.
Therefore, the number of muons expected at sea level is

$$N = 10^8 e^{-1} = 3.68 \times 10^7$$

Thus, relativity predicts that we would observe 36.8 million muons in the same
time interval. Experiments of this type have confirmed the relativistic predictions.

The Relativistic Doppler Effect

For light or other electromagnetic waves in a vacuum, a distinction between
motion of source and receiver cannot be made. Therefore, the expressions we
derived in Chapter 15 for the Doppler effect cannot be correct for light. The
reason is that in that derivation, we assumed the time intervals in the reference
frames of the source and receiver to be the same.

Consider a source moving toward a receiver with velocity v, relative to the
receiver. If the source emits N electromagnetic waves in a time Δt_R (measured in
the frame of the receiver), the first wave will travel a distance $c \, \Delta t_R$ and the source
will travel a distance $v \, \Delta t_R$ measured in the frame of the receiver. The wavelength
will be

$$\lambda' = \frac{c \, \Delta t_R - v \, \Delta t_R}{N}$$

The frequency f' observed by the receiver will therefore be

$$f' = \frac{c}{\lambda'} = \frac{c}{c - v} \frac{N}{\Delta t_R} = \frac{1}{1 - (v/c)} \frac{N}{\Delta t_R}$$

If the frequency of the source is f_0, it will emit $N = f_0 \, \Delta t_S$ waves in the time Δt_S
measured by the source. Then

$$f' = \frac{1}{1 - (v/c)} \frac{N}{\Delta t_R} = \frac{1}{1 - (v/c)} \frac{f_0 \, \Delta t_S}{\Delta t_R} = \frac{f_0}{1 - (v/c)} \frac{\Delta t_S}{\Delta t_R}$$

Here Δt_S is the proper time interval (the first wave and the Nth wave are emitted
at the same place in the source's reference frame). Times Δt_S and Δt_R are related
by Equation 39-13 for time dilation:

$$\Delta t_R = \gamma \, \Delta t_S = \frac{\Delta t_S}{\sqrt{1 - (v^2/c^2)}}$$

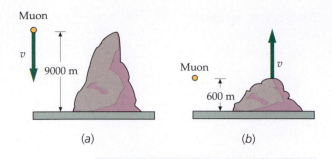

Muon

v

9000 m

Muon

600 m

v

(a)   (b)

FIGURE 39-4 Although muons are
created high above the earth and their
mean lifetime is only about 2 μs when at
rest, many appear at the earth's surface.
(*a*) In the earth's reference frame, a
typical muon moving at 0.998c has a mean
lifetime of 30 μs and travels 9000 m in
this time. (*b*) In the reference frame of
the muon, the distance traveled by the
earth is only 600 m in the muon's
lifetime of 2 μs.

Thus, when the source and the receiver are moving toward one another we obtain

$$f' = \frac{f_0}{1 - (v/c)} \frac{1}{\gamma} = \frac{\sqrt{1 - (v/c)^2}}{1 - (v/c)} f_0 = \sqrt{\frac{1 + (v/c)}{1 - (v/c)}} f_0, \quad \text{approaching} \quad \text{39-16}a$$

This differs from our classical equation only in the time-dilation factor. It is left as a problem (Problem 27) for you to show that the same results are obtained if the calculations are done in the reference frame of the source.

When the source and the receiver are moving away from one another, the same analysis shows that the observed frequency is given by

$$f' = \frac{\sqrt{1 - (v/c)^2}}{1 + (v/c)} f_0 = \sqrt{\frac{1 - (v/c)}{1 + (v/c)}} f_0, \quad \text{receding} \qquad \text{39-16}b$$

An application of the relativistic Doppler effect is the **redshift** observed in the light from distant galaxies. Because the galaxies are moving away from us, the light they emit is shifted toward the longer red wavelengths. The speed of the galaxies relative to us can be determined by measuring this shift.

CONVINCING THE JUDGE   **EXAMPLE 39-4**   **Put It in Context**

As part of a community volunteering option on your campus, you are spending the day shadowing two police officers. You have just had the excitement of pulling over a car that went through a red light. The driver claims that the red light looked green because the car was moving toward the stoplight, which shifted the wavelength of the observed light. You quickly do some calculations to see if the driver has a reasonable case or not.

PICTURE THE PROBLEM We can use the Doppler shift formula for approaching objects in Equation 39-16a. This will tell us the velocity, but we need to know the frequencies of the light. We can make good guesses for the wavelengths of red light and green light and use the definition of the speed of a wave $c = f\lambda$ to determine the frequencies.

1. The observer is approaching the light source, so we use the Doppler formula (Equation 39-16a) for approaching sources:

$$f' = \sqrt{\frac{1 + (v/c)}{1 - (v/c)}} f_0$$

2. Substitute c/λ for f, then simplify:

$$\frac{c}{\lambda'} = \sqrt{\frac{1 + (v/c)}{1 - (v/c)}} \frac{c}{\lambda_0}$$

$$\left(\frac{\lambda_0}{\lambda'}\right)^2 = \frac{1 + (v/c)}{1 - (v/c)}$$

3. Cross multiply and solve for v/c:

$$(\lambda_0)^2\left(1 - \frac{v}{c}\right) = (\lambda')^2\left(1 + \frac{v}{c}\right)$$

$$(\lambda_0)^2 - (\lambda')^2 = \left[(\lambda_0)^2 + (\lambda')^2\right]\left(\frac{v}{c}\right)$$

$$\frac{v}{c} = \frac{(\lambda_0)^2 - (\lambda')^2}{(\lambda_0)^2 + (\lambda')^2} = \frac{1 - (\lambda'/\lambda_0)^2}{1 + (\lambda'/\lambda_0)^2}$$

4. The values for the wavelengths for the colors of the visible spectrum can be found in Table 30-1. The wavelengths for red are 725 nm or longer, and the wavelengths for green are 675 nm or shorter. Solve for the speed needed to shift the wavelength from 725 nm to 675 nm:

$$\frac{\lambda'}{\lambda_0} = \frac{675 \text{ nm}}{725 \text{ nm}} = 0.931$$

$$\frac{v}{c} = \frac{1 - 0.931^2}{1 + 0.931^2} = 0.0713$$

$$v = 0.0713c = 2.14 \times 10^7 \text{ m/s} = 4.79 \times 10^7 \text{ mi/h}$$

5. This speed is beyond any possible speed for a car: | The driver does not have a plausible case.

FINDING SPEED FROM THE DOPPLER SHIFT **EXAMPLE 39-5 Try It Yourself**

The longest wavelength of light emitted by hydrogen in the Balmer series is $\lambda_0 = 656$ nm. In light from a distant galaxy, this wavelength is measured to be $\lambda' = 1458$ nm. Find the speed at which the distant galaxy is receding from the earth.

Cover the column to the right and try these on your own before looking at the answers.

Steps **Answers**

1. Use Equation 39-16b to relate the speed v to the received frequency f' and the emitted frequency f_0.

$$f' = \sqrt{\frac{1 - (v/c)}{1 + (v/c)}} f_0$$

2. Substitute $f' = c/\lambda'$ and $f_0 = c/\lambda_0$ and solve for v/c.

$$\frac{v}{c} = \frac{1 - (\lambda_0/\lambda')^2}{1 + (\lambda_0/\lambda')^2} = 0.664$$

$$v = \boxed{0.664c}$$

39-4 Clock Synchronization and Simultaneity

We saw in Section 39-3 that proper time is the time interval between two events that occur at the same point in some reference frame. It can therefore be measured on a single clock. (Remember, in each frame there is a clock at each point in space, and the time of an event in a given frame is measured by the clock at that point.) However, in another reference frame moving relative to the first, the same two events occur at different places, so two clocks are needed to record the times. The time of each event is measured on a different clock, and the interval is found by subtraction. This procedure requires that the clocks be **synchronized.** We will show in this section that

Two clocks that are synchronized in one reference frame are typically not synchronized in any other frame moving relative to the first frame.

SYNCHRONIZED CLOCKS

Here is a corollary to this result:

Two events that are simultaneous in one reference frame typically are not simultaneous in another frame that is moving relative to the first.[†]

SIMULTANEOUS EVENTS

[†] This is true *unless* the x coordinates of the two events are equal, where the x axis is parallel with the relative velocity of the two frames.

Comprehension of these facts usually resolves all relativity paradoxes. Unfortunately, the intuitive (and incorrect) belief that simultaneity is an absolute relation is difficult to overcome.

Suppose we have two clocks at rest at point A and point B a distance L apart in frame S. How can we synchronize these two clocks? If an observer at A looks at the clock at B and sets her clock to read the same time, the clocks will not be synchronized because of the time L/c it takes light to travel from one clock to another. To synchronize the clocks, the observer at A must set her clock ahead by the time L/c. Then she will see that the clock at B reads a time that is L/c behind the time on her clock, but she will calculate that the clocks are synchronized when she allows for the time L/c for the light to reach her. Any other observers in S (except those equidistant from the clocks) will see the clocks reading different times, but they will also calculate that the clocks are synchronized when they correct for the time it takes the light to reach them. An equivalent method for synchronizing two clocks would be for an observer C at a point midway between the clocks to send a light signal and for the observers at A and B to set their clocks to some prearranged time when they receive the signal.

We now examine the question of **simultaneity.** Suppose A and B agree to explode flashguns at t_0 (having previously synchronized their clocks). Observer C will see the light from the two flashes at the same time, and because he is equidistant from A and B, he will conclude that the flashes were simultaneous. Other observers in frame S will see the light from A or B first, depending on their location, but after correcting for the time the light takes to reach them, they also will conclude that the flashes were simultaneous. We can thus define simultaneity as follows:

> Two events in a reference frame are simultaneous if light signals from the events reach an observer halfway between the events at the same time.

DEFINITION—SIMULTANEITY

To show that two events that are simultaneous in frame S are not simultaneous in another frame S' moving relative to S, we will use an example introduced by Einstein. A train is moving with speed v past a station platform. We will consider the train to be at rest in S' and the platform to be at rest in S. We have observers A', B', and C' at the front, back, and middle of the train. We now suppose that the train and platform are struck by lightning at the front and back of the train and that the lightning bolts are simultaneous in the frame of the platform S (Figure 39-5). That is, an observer C on the platform halfway between the positions A and B, where the lightning strikes, sees the two flashes at the same time. It is convenient to suppose that the lightning scorches the train and platform so that the events can be easily located. Because C' is in the middle of the train, halfway between the places on the train that are scorched, the events are simultaneous in S' only if C' sees the flashes at the same time. However, the flash from the front of the train is seen by C' before the flash from the back of the

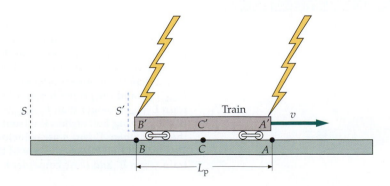

FIGURE 39-5 In frame S attached to the platform, simultaneous lightning bolts strike the ends of a train traveling with speed v. The light from these simultaneous events reaches observer C, standing midway between the events, at the same time. The distance between the bolts is $L_{\text{p,platform}}$.

train. We can understand this by considering the motion of C' as seen in frame S (Figure 39-6). By the time the light from the front flash reaches C', C' has moved some distance toward the front flash and some distance away from the back flash. Thus, the light from the back flash has not yet reached C', as indicated in the figure. Observer C' must therefore conclude that the events are not simultaneous and that the front of the train was struck before the back. Furthermore, all observers in S' on the train will agree with C' when they have corrected for the time it takes the light to reach them.

Figure 39-7 shows the events of the lightning bolts as seen in the reference frame of the train (S'). In this frame the platform is moving, so the distance between the burns on the platform is contracted. The platform is shorter than it is in S, and, since the train is at rest, the train is longer than its contracted length in S. When the lightning bolt strikes the front of the train at A', the front of the train is at point A, and the back of the train has not yet reached point B. Later, when the lightning bolt strikes the back of the train at B', the back has reached point B on the platform.

The time discrepancy of two clocks that are synchronized in frame S as seen in frame S' can be found from the Lorentz transformation equations. Suppose we have clocks at points x_1 and x_2 that are synchronized in S. What are the times t_1 and t_2 on these clocks as observed from frame S' at a time t_0'? From Equation 39-12, we have

$$t_0' = \gamma\left(t_1 - \frac{vx_1}{c^2}\right)$$

and

$$t_0' = \gamma\left(t_2 - \frac{vx_2}{c^2}\right)$$

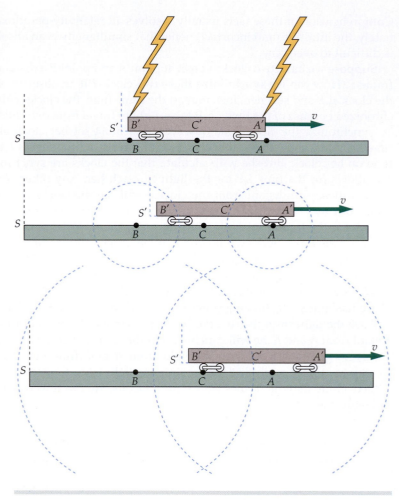

FIGURE 39-6 In frame S attached to the platform, the light from the lightning bolt at the front of the train reaches observer C', standing on the train at its midpoint, before the light from the bolt at the back of the train. Since C' is midway between the events (which occur at the front and rear of the train), these events are not simultaneous for him.

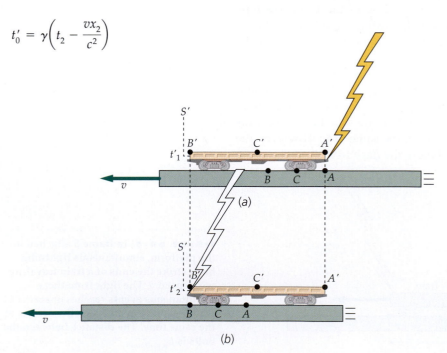

FIGURE 39-7 The lightning bolts of Figure 39-5 as seen in frame S' of the train. In this frame, the distance between A and B on the platform is less than $L_{\text{p,platform}}$, and the proper length of the train $L_{\text{p,train}}$ is longer than $L_{\text{p,platform}}$. The first lightning bolt strikes the front of the train when A' and A are coincident. The second bolt strikes the rear of the train when B' and B are coincident.

Then

$$t_2 - t_1 = \frac{v}{c^2}(x_2 - x_1)$$

Note that the chasing clock (at x_2) leads the other (at x_1) by an amount that is proportional to their proper separation $L_p = x_2 - x_1$.

If two clocks are synchronized in the frame in which they are both at rest, in a frame in which they are moving along the line through both clocks, the chasing clock leads (shows a later time) by an amount

$$\Delta t_S = L_p \frac{v}{c^2} \qquad\qquad 39\text{-}17$$

where L_p is the proper distance between the clocks.

CHASING CLOCK SHOWS LATER TIME

A numerical example should help clarify time dilation, clock synchronization, and the internal consistency of these results.

SYNCHRONIZING CLOCKS   **EXAMPLE   39-6**

An observer in a spaceship has a flashgun and a mirror, as shown in Figure 39-3. The distance from the gun to the mirror is 15 light-minutes (written 15c·min) and the spaceship, at rest in frame S', travels with speed $v = 0.8c$ relative to a very long space platform that is at rest in frame S. The platform has two synchronized clocks, one clock at the position x_1 of the spaceship when the observer explodes the flashgun, and the other clock at the position x_2 of the spaceship when the light returns to the gun from the mirror. Find the time intervals between the events (exploding the flashgun and receiving the return flash from the mirror) (a) in the frame of the spaceship and (b) in the frame of the platform. (c) Find the distance traveled by the spaceship and (d) the amount by which the clocks on the platform are out of synchronization according to observers on the spaceship.

(a) 1. In the spaceship, the light travels from the gun to the mirror and back, a total distance $D = 30$ c·min. The time required is D/c:

$$\Delta t' = \frac{D}{c} = \frac{30\,c\cdot\text{min}}{c} = 30\text{ min}$$

2. Since these events happen at the same place in the spaceship, the time interval is proper time:

$$\Delta t_p = \boxed{30\text{ min}}$$

(b) 1. In frame S, the time between the events is longer by the factor γ:

$$\Delta t = \gamma\,\Delta t_p = \gamma(30\text{ min})$$

2. Calculate γ:

$$\gamma = \frac{1}{\sqrt{1-(v^2/c^2)}} = \frac{1}{\sqrt{1-(0.8)^2}} = \frac{1}{\sqrt{0.36}} = \frac{5}{3}$$

3. Use this value of γ to calculate the time between the events as observed in frame S:

$$\Delta t = \gamma\,\Delta t_p = \tfrac{5}{3}(30\text{ min}) = \boxed{50\text{ min}}$$

(c) In frame S, the distance traveled by the spaceship is $v\,\Delta t$:

$$x_2 - x_1 = v\,\Delta t = (0.8c)(50\text{ min}) = \boxed{40\,c\cdot\text{min}}$$

(d) 1. The amount that the clocks on the platform are out of synchronization is related to the proper distance between the clocks L_p:

$$\Delta t_s = L_p \frac{v}{c^2}$$

2. The Part (c) result is the proper distance between the clocks on the platform:

$$L_p = x_2 - x_1 = 40 \, c \cdot min$$

so

$$\Delta t_s = L_p \frac{v}{c^2} = (40 \, c \cdot min)\frac{(0.8c)}{c^2} = \boxed{32 \, min}$$

REMARKS Observers on the platform would say that the spaceship's clock is running slow because it records a time of only 30 min between the events, whereas the time measured by observers on the platform is 50 min.

Figure 39-8 shows the situation viewed from the spaceship in S'. The platform is traveling past the ship with speed 0.8c. There is a clock at point x_1, which coincides with the ship when the flashgun is exploded, and another at point x_2, which coincides with the ship when the return flash is received from the mirror. We assume that the clock at x_1 reads 12:00 noon at the time of the light flash. The clocks at x_1 and x_2 are synchronized in S but not in S'. In S', the clock at x_2, which is chasing the one at x_1, leads by 32 min; it would thus read 12:32 to an observer in S'. When the spaceship coincides with x_2, the clock there reads 12:50. The time between the events is therefore 50 min in S. Note that according to observers in S', this clock ticks off 50 min − 32 min = 18 min for a trip that takes 30 min in S'. Thus, observers in S' see this clock run slow by the factor 30/18 = 5/3.

Every observer in one frame sees the clocks in the other frame run slow. According to observers in S, who measure 50 min for the time interval, the time interval in S' (30 min) is too small, so they see the single clock in S' run too slow by the factor 5/3. According to the observers in S', the observers in S measure a time that is too *long* despite the fact that their clocks run too slow because the clocks in S are out of synchronization. The clocks tick off only 18 min, but the second clock leads the first clock by 32 min, so the time interval is 50 min.

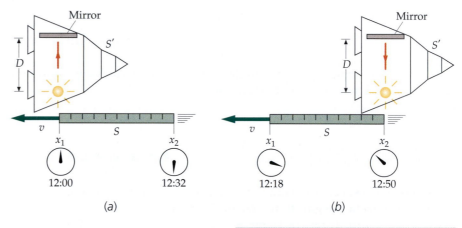

(a)

(b)

FIGURE 39-8 Clocks on a platform as observed from the spaceship's frame of reference S'. During the time $\Delta t' = $ 30 min it takes for the platform to pass the spaceship, the clocks on the platform run slow and tick off (30 min)/γ = 18 min. But the clocks are unsynchronized, with the chasing clock leading by $L_p v/c^2$, which for this case is 32 min. The time it takes for the spaceship to go from x_1 to x_2, as measured on the platform, is therefore 32 min + 18 min = 50 min.

The Twin Paradox

Homer and Ulysses are identical twins. Ulysses travels at high speed to a planet beyond the solar system and returns while Homer remains at home. When they are together again, which twin is older, or are they the same age? The correct answer is that Homer, the twin who stays at home, is older. This problem, with variations, has been the subject of spirited debate for decades, though there are very few who disagree with the answer. The problem appears to be a paradox because of the seemingly symmetric roles played by the twins with the asymmetric result in their aging. The paradox is resolved when the asymmetry of the twins' roles is noted. The relativistic result conflicts with common sense based on our strong but incorrect belief in absolute simultaneity. We will consider a particular case with some numerical magnitudes that, though impractical, make the calculations easy.

Let planet P and Homer on the earth be at rest in reference frame S a distance L_p apart, as illustrated in Figure 39-9. We neglect the motion of the earth. Reference frames S' and S'' are moving with speed v toward and away from the planet, respectively. Ulysses quickly accelerates to speed v, then coasts in S' until he reaches the planet, where he quickly decelerates to a stop and is momentarily at rest in S. To return, Ulysses quickly accelerates to speed v toward the earth and then coasts in S'' until he reaches the earth, where he quickly decelerates to a stop. We can assume that the acceleration (and deceleration) times are negligible compared with the coasting times. We use the following values for illustration: $L_p = 8$ light-years ($8\ c{\cdot}y$) and $v = 0.8c$. Then $\sqrt{1 - (v^2/c^2)} = 3/5$ and $\gamma = 5/3$.

It is easy to analyze the problem from Homer's point of view on the earth. According to Homer's clock, Ulysses coasts in S' for a time $L_p/v = 10$ y and in S'' for an equal time. Thus, Homer is 20 y older when Ulysses returns. The time interval in S' between Ulysses's leaving the earth and his arriving at the planet is shorter because it is proper time. The time it takes to reach the planet by Ulysses's clock is

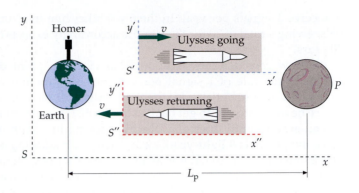

FIGURE 39-9 The twin paradox. The earth and a distant planet are fixed in frame S. Ulysses coasts in frame S' to the planet and then coasts back in frame S''. His twin Homer stays on the earth. When Ulysses returns, he is younger than his twin. The roles played by the twins are **not symmetric. Homer remains in one inertial reference frame, but Ulysses must accelerate if he is to return home.**

$$\Delta t' = \frac{\Delta t}{\gamma} = \frac{10\ \text{y}}{5/3} = 6\ \text{y}$$

Since the same time is required for the return trip, Ulysses will have recorded 12 y for the round trip and will be 8 y younger than Homer upon his return.

From Ulysses's point of view, the distance from the earth to the planet is contracted and is only

$$L' = \frac{L_p}{\gamma} = \frac{8\ c{\cdot}y}{5/3} = 4.8\ c{\cdot}y$$

At $v = 0.8c$, it takes only 6 y each way.

The real difficulty in this problem is for Ulysses to understand why his twin aged 20 y during his absence. If we consider Ulysses as being at rest and Homer as moving away, Homer's clock should run slow and measure only $3/5(6\ \text{y}) = 3.6\ \text{y}$. Then why shouldn't Homer age only 7.2 y during the round trip? This, of course, is the paradox. The difficulty with the analysis from the point of view of Ulysses is that he does not remain in an inertial frame. What happens while Ulysses is stopping and starting? To investigate this problem in detail, we would need to treat accelerated reference frames, a subject dealt with in the study of general relativity and beyond the scope of this book. However, we can get some insight into the problem by having the twins send regular signals to each other so that they can record the other's age continuously. If they arrange to send a signal once a year, each can determine the age of the other merely by counting the signals received. The arrival frequency of the signals will not be 1 per year because of the Doppler shift. The frequency observed will be given by Equation 39-16a and Equation 39-16b. Using $v/c = 0.8$ and $v^2/c^2 = 0.64$, we have for the case in which the twins are receding from each other

$$f' = \frac{\sqrt{1 - (v^2/c^2)}}{1 + (v/c)} f_0 = \frac{\sqrt{1 - 0.64}}{1 + 0.8} f_0 = \frac{1}{3} f_0$$

When they are approaching, Equation 39-16a gives $f' = 3f_0$.

Consider the situation first from the point of view of Ulysses. During the 6 y it takes him to reach the planet (remember that the distance is contracted in his frame), he receives signals at the rate of $\frac{1}{3}$ signal per year, and so he receives 2 signals. As soon as Ulysses turns around and starts back to the earth, he begins to receive 3 signals per year. In the 6 y it takes him to return he receives 18 signals, giving a total of 20 for the trip. He accordingly expects his twin to have aged 20 years.

We now consider the situation from Homer's point of view. He receives signals at the rate of $\frac{1}{3}$ signal per year not only for the 10 y it takes Ulysses to reach the planet but also for the time it takes for the last signal sent by Ulysses before he turns around to get back to the earth. (He cannot know that Ulysses has turned around until the signals begin reaching him with increased frequency.) Since the planet is 8 light-years away, there is an additional 8 y of receiving signals at the rate of $\frac{1}{3}$ signal per year. During the first 18 y, Homer receives 6 signals. In the final 2 y before Ulysses arrives, Homer receives 6 signals, or 3 per year. (The first signal sent after Ulysses turns around takes 8 y to reach the earth, whereas Ulysses, traveling at 0.8c, takes 10 y to return and therefore arrives just 2 y after Homer begins to receive signals at the faster rate.) Thus, Homer expects Ulysses to have aged 12 y. In this analysis, the asymmetry of the twins' roles is apparent. When they are together again, both twins agree that the one who has been accelerated will be younger than the one who stayed home.

The predictions of the special theory of relativity concerning the twin paradox have been tested using small particles that can be accelerated to such large speeds that γ is appreciably greater than 1. Unstable particles can be accelerated and trapped in circular orbits in a magnetic field, for example, and their lifetimes can then be compared with those of identical particles at rest. In all such experiments, the accelerated particles live longer on the average than the particles at rest, as predicted. These predictions have also been confirmed by the results of an experiment in which high-precision atomic clocks were flown around the world in commercial airplanes, but the analysis of this experiment is complicated due to the necessity of including gravitational effects treated in the general theory of relativity.

39-5 The Velocity Transformation

We can find how velocities transform from one reference frame to another by differentiating the Lorentz transformation equations. Suppose a particle has velocity $u'_x = dx'/dt'$ in frame S', which is moving to the right with speed v relative to frame S. The particle's velocity in frame S is

$$u_x = \frac{dx}{dt}$$

From the Lorentz transformation equations (Equation 39-9 and Equation 39-10), we have

$$dx = \gamma(dx' + v\,dt')$$

and

$$dt = \gamma\left(dt' + \frac{v\,dx'}{c^2}\right)$$

The velocity in S is thus

$$u_x = \frac{dx}{dt} = \frac{\gamma(dx' + v\,dt')}{\gamma\left(dt' + \dfrac{v\,dx'}{c^2}\right)} = \frac{\dfrac{dx'}{dt'} + v}{1 + \dfrac{v}{c^2}\dfrac{dx'}{dt'}} = \frac{u'_x + v}{1 + \dfrac{v\,u'_x}{c^2}}$$

If a particle has components of velocity along the y or z axes, we can use the same relation between dt and dt', with $dy = dy'$ and $dz = dz'$, to obtain

$$u_y = \frac{dy}{dt} = \frac{dy'}{\gamma\left(dt' + \dfrac{v\,dx'}{c^2}\right)} = \frac{\dfrac{dy'}{dt'}}{\gamma\left(1 + \dfrac{v}{c^2}\dfrac{dx'}{dt'}\right)} = \frac{u'_y}{\gamma\left(1 + \dfrac{vu'_x}{c^2}\right)}$$

and

$$u_z = \frac{u'_z}{\gamma\left(1 + \dfrac{vu'_x}{c^2}\right)}$$

The complete relativistic velocity transformation is

$$u_x = \frac{u'_x + v}{1 + \dfrac{vu'_x}{c^2}} \qquad\qquad 39\text{-}18a$$

$$u_y = \frac{u'_y}{\gamma\left(1 + \dfrac{vu'_x}{c^2}\right)} \qquad\qquad 39\text{-}18b$$

$$u_z = \frac{u'_z}{\gamma\left(1 + \dfrac{vu'_x}{c^2}\right)} \qquad\qquad 39\text{-}18c$$

RELATIVISTIC VELOCITY TRANSFORMATION

The inverse velocity transformation equations are

$$u'_x = \frac{u_x - v}{1 - \dfrac{vu_x}{c^2}} \qquad\qquad 39\text{-}19a$$

$$u'_y = \frac{u_y}{\gamma\left(1 - \dfrac{vu_x}{c^2}\right)} \qquad\qquad 39\text{-}19b$$

$$u'_z = \frac{u_z}{\gamma\left(1 - \dfrac{vu_x}{c^2}\right)} \qquad\qquad 39\text{-}19c$$

These equations differ from the classical and intuitive result $u_x = u'_x + v$, $u_y = u'_y$, and $u_z = u'_z$ because the denominators in the equations are not equal to 1. When v and u'_x are small compared with the speed of light c, $\gamma \approx 1$ and $vu'_x/c^2 \ll 1$. Then the relativistic and classical expressions are the same.

RELATIVE VELOCITY AT NONRELATIVISTIC SPEEDS　　　　**E X A M P L E   3 9 - 7**

A supersonic plane moves away from you along the x axis with speed 1000 m/s (about 3 times the speed of sound) relative to you. A second plane moves along the x axis away from you, and away from the first plane, at speed 500 m/s relative to the first plane. How fast is the second plane moving relative to you?

PICTURE THE PROBLEM These speeds are so small compared with c that we expect the classical equations for combining velocities to be accurate. We show this by calculating the correction term in the denominator of Equation 39-18a. Let frame S be your rest frame and frame S' be moving with velocity $v = 1000$ m/s. The first plane is then at rest in frame S', and the second plane has velocity $u_x' = 500$ m/s in S'.

1. Let S and S' be the reference frames of you and the first plane, respectively. Also, let u_x and u_x' be the velocities of the second plane relative to S and S', respectively. Equation 39-18a can be used to find u_x. The velocity of the second plane relative to you is v:

$$u_x = \frac{u_x' + v}{1 + \frac{v u_x'}{c^2}}$$

2. If the correction term in the denominator is neglgible, Equation 39-18a gives the classical formula for combining velocities. Calculate the value of this correction term:

$$\frac{v u_x'}{c^2} = \frac{(1000)(500)}{(3 \times 10^8)^2} \approx 5.6 \times 10^{-12}$$

3. This correction term is so small that the classical and relativistic results are essentially the same:

$$u_x \approx u_x' + v$$

$$= 500 \text{ m/s} + 1000 \text{ m/s} = \boxed{1500 \text{ m/s}}$$

RELATIVE VELOCITY AT RELATIVISTIC SPEEDS   **E X A M P L E   3 9 - 8**

Work Example 39-7 if the first plane moves with speed $v = 0.8c$ relative to you and the second plane moves with the same speed $0.8c$ relative to the first plane.

PICTURE THE PROBLEM These speeds are not small compared with c, so we use the relativistic expression (Equation 39-18a). We again assume that you are at rest in frame S and the first plane is at rest in frame S' that is moving at $v = 0.8c$ relative to you. The velocity of the second plane in S' is $u_x' = 0.8c$.

Use Equation 39-18a to calculate the speed of the second plane relative to you:

$$u_x = \frac{u_x' + v}{1 + \frac{v u_x'}{c^2}} = \frac{0.8c + 0.8c}{1 + \frac{(0.8c)(0.8c)}{c^2}} = \frac{1.6c}{1.64} = \boxed{0.98c}$$

The result in Example 39-8 is quite different from the classically expected result of $0.8c + 0.8c = 1.6c$. In fact, it can be shown from Equations 39-18 that if the speed of an object is less than c in one frame, it is less than c in all other frames moving relative to that frame with a speed less than c. (See Problem 23.) We will see in Section 39-7 that it takes an infinite amount of energy to accelerate a particle to the speed of light. The speed of light c is thus an upper, unattainable limit for the speed of a particle with mass. (There are massless particles, such as photons, that always move at the speed of light.)

RELATIVE SPEED OF A PHOTON   **E X A M P L E   3 9 - 9**

A photon moves along the x axis in frame S', with speed $u_x' = c$. What is its speed in frame S?

The speed in S is given by Equation 39-18a:

$$u_x = \frac{u_x' + v}{1 + \frac{v u_x'}{c^2}} = \frac{c + v}{1 + \frac{vc}{c^2}} = \frac{c + v}{1 + \frac{v}{c}} = \frac{c + v}{\frac{1}{c}(c + v)} = \boxed{c}$$

REMARKS The speed in both frames is c, independent of v. This is in accord with Einstein's postulates.

ROCKETS PASSING IN OPPOSITE DIRECTIONS   **EXAMPLE 39-10**

Two spaceships, each 100 m long when measured at rest, travel toward each other with speeds of $0.85c$ relative to the earth. (*a*) How long is each spaceship as measured by someone on the earth? (*b*) How fast is each spaceship traveling as measured by an observer on one of the spaceships? (*c*) How long is one spaceship when measured by an observer on one of the spaceships? (*d*) At time $t = 0$ on the earth, the front ends of the ships are together as they just begin to pass each other. At what time on the earth are their back ends together?

PICTURE THE PROBLEM (*a*) The length of each spaceship as measured on the earth is the contracted length $\sqrt{1 - (v_1^2/c^2)}\ L_p$ (Equation 39-14), where v_1 is the speed of either spaceship. To solve Part (*b*), let the earth be in frame S, and the spaceship on the left be in frame S' moving with velocity $v = 0.85c$ relative to S. Then the spaceship on the right moves with velocity $u_x = -0.85c$, as shown in Figure 39-10. (*c*) The length of one spaceship as seen by the other is $\sqrt{1 - (v_2^2/c^2)}\ L_p$, where v_2 is the speed of one spaceship relative to the other.

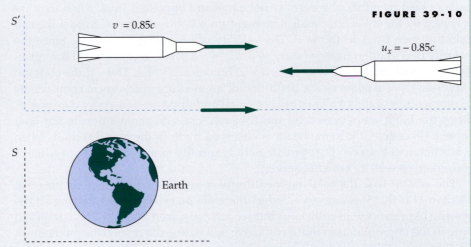

FIGURE 39-10

$v = 0.85c$

$u_x = -0.85c$

S'

S

Earth

(*a*) The length of each spaceship in the earth's frame is the proper length divided by γ:

$$L = \sqrt{1 - \frac{v_1^2}{c^2}}\ L_p = \sqrt{1 - \frac{(0.85c)^2}{c^2}}\ (100\ \text{m}) = \boxed{52.7\ \text{m}}$$

(*b*) Use the velocity transformation formula (Equation 39-19*a*) to find the velocity u_x' of the spaceship on the right as seen in frame S':

$$u_x' = \frac{u_x - v}{1 - \dfrac{vu_x}{c^2}} = \frac{-0.85c - 0.85c}{1 - \dfrac{(0.85c)(-0.85c)}{c^2}} = \frac{-1.70c}{1.7225} = \boxed{-0.987c}$$

(*c*) In the frame of the left spaceship, the right spaceship is moving with speed $v_2 = |u_x'| = 0.987c$. Use this to calculate the contracted length of the spaceship on the right:

$$L = \sqrt{1 - \frac{v_1^2}{c^2}}\ L_p = \sqrt{1 - \frac{(0.987c)^2}{c^2}}\ (100\ \text{m}) = \boxed{16.1\ \text{m}}$$

(*d*) If the front ends of the spaceships are together at $t = 0$ on the earth, their back ends will be together after the time it takes either spaceship to move the length of the spaceship in the earth's frame:

$$t = \frac{L}{v_1} = \frac{52.7\ \text{m}}{0.85c} = \frac{52.7\ \text{m}}{(0.85)(3 \times 10^8\ \text{m/s})} = \boxed{2.07 \times 10^{-7}\ \text{s}}$$

39-6 Relativistic Momentum

We have seen in previous sections that Einstein's postulates require important modifications in our ideas of simultaneity and in our measurements of time and length. Einstein's postulates also require modifications in our concepts of mass, momentum, and energy. In classical mechanics, the momentum of a particle is defined as the product of its mass and its velocity, $m\vec{u}$, where $\vec{u}$ is the velocity. In an isolated system of particles, with no net force acting on the system, the total momentum of the system remains constant.

We can see from a simple thought experiment that the quantity $\Sigma m_i \vec{u}_i$ is not conserved in an isolated system. We consider two observers: observer A in reference frame S and observer B in frame S', which is moving to the right in the x direction with speed v with respect to frame S. Each has a ball of mass m. The two balls are identical when compared at rest. One observer throws his ball up with a speed u_0 relative to him and the other throws his ball down with a speed u_0 relative to him, so that each ball travels a distance L, makes an elastic collision with the other ball, and returns. Figure 39-11 shows how the collision looks in each reference frame. Classically, each ball has vertical momentum of magnitude mu_0. Since the vertical components of the momenta are equal and opposite, the total vertical component of momentum is zero before the collision. The collision merely reverses the momentum of each ball, so the total vertical momentum is zero after the collision.

Relativistically, however, the vertical components of the velocities of the two balls as seen by either observer are not equal and opposite. Thus, when they are reversed by the collision, classical momentum is not conserved. Consider the collision as seen by A in frame S. The velocity of his ball is $u_{Ay} = +u_0$. Since the velocity of B's ball in frame S' is $u'_{Bx} = 0$, $u'_{By} = -u_0$, the y component of the velocity of B's ball in frame S is $u_{By} = -u_0/\gamma$ (Equation 39-18b). Thus, if the classical expression $m\vec{u}$ is taken as the definition of momentum, the vertical components of momentum of the two balls are not equal and opposite as seen by observer A. Since the balls are reversed by the collision, classical momentum is not conserved. Of course, the same result is observed by B. In the classical limit, when u is much less than c, γ is approximately 1, and the momentum of the system is conserved as seen by either observer.

The reason that the total momentum of a system is important in classical mechanics is that it is conserved when there are no external forces acting on the system, as is the case in collisions. But we have just seen that $\Sigma m_i \vec{u}_i$ is conserved only in the approximation that $u \ll c$. We will define the relativistic momentum $\vec{p}$ of a particle to have the following properties:

1. In collisions, $\vec{p}$ is conserved.

2. As u/c approaches zero, $\vec{p}$ approaches $m\vec{u}$.

We will show that the quantity

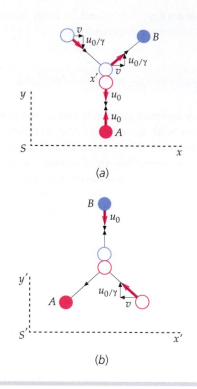

FIGURE 39-11 (*a*) Elastic collision of two identical balls as seen in frame S. The vertical component of the velocity of ball B is u_0/γ in S if it is u_0 in S'. (*b*) The same collision as seen in S'. In this frame, ball A has a vertical component of velocity equal to u_0/γ.

$$\vec{p} = \frac{m\vec{u}}{\sqrt{1 - \dfrac{u^2}{c^2}}}$$

39-20

RELATIVISTIC MOMENTUM

is conserved in the elastic collision shown in Figure 39-11. Since this quantity also approaches $m\vec{u}$ as u/c approaches zero, we take this equation for the definition of the **relativistic momentum** of a particle.

One interpretation of Equation 39-20 is that the mass of an object increases with speed. Then the quantity $m_{rel} = m/\sqrt{1 - (u^2/c^2)}$ is called the *relativistic mass*. The relativistic mass of a particle when it is at rest in some reference frame is then called its *rest mass m*. In this chapter, we will treat the terms mass and rest mass as synonymous, and both terms will be labeled m.

Illustration of Conservation of the Relativistic Momentum

We will compute the y component of the relativistic momentum of each particle in the reference frame S for the collision of Figure 39-11 and show that the y component of the total relativistic momentum is zero. The speed of ball A in S is u_0, so the y component of its relativistic momentum is

$$p_{Ay} = \frac{mu_0}{\sqrt{1-(u_0^2/c^2)}}$$

The speed of ball B in S is more complicated. Its x component is v and its y component is $-u_0/\gamma$. Thus,

$$u_B^2 = u_{Bx}^2 + u_{By}^2 = v^2 + \left[-u_0\sqrt{1-(v^2/c^2)}\,\right]^2 = v^2 + u_0^2 - \frac{u_0^2 v^2}{c^2}$$

Using this result to compute $\sqrt{1-(u_B^2/c^2)}$, we obtain

$$1 - \frac{u_B^2}{c^2} = 1 - \frac{v^2}{c^2} - \frac{u_0^2}{c^2} + \frac{u_0^2 v^2}{c^4} = \left(1 - \frac{v^2}{c^2}\right)\left(1 - \frac{u_0^2}{c^2}\right)$$

and

$$\sqrt{1-(u_B^2/c^2)} = \sqrt{1-(v^2/c^2)}\,\sqrt{1-(u_0^2/c^2)} = (1/\gamma)\,\sqrt{1-(u_0^2/c^2)}$$

The y component of the relativistic momentum of ball B as seen in S is therefore

$$p_{By} = \frac{mu_{By}}{\sqrt{1-(u_B^2/c^2)}} = \frac{-mu_0/\gamma}{(1/\gamma)\,\sqrt{1-(u_0^2/c^2)}} = \frac{-mu_0}{\sqrt{1-(u_0^2/c^2)}}$$

Since $p_{By} = -p_{Ay}$, the y component of the total momentum of the two balls is zero. If the speed of each ball is reversed by the collision, the total momentum will remain zero and momentum will be conserved.

39-7 Relativistic Energy

In classical mechanics, the work done by the net force acting on a particle equals the change in the kinetic energy of the particle. In relativistic mechanics, we equate the net force to the rate of change of the relativistic momentum. The work done by the net force can then be calculated and set equal to the change in kinetic energy.

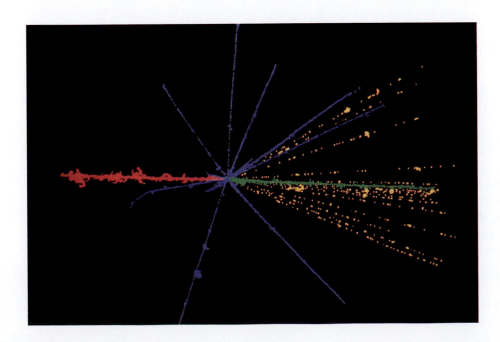

The creation of elementary particles demonstrates the conversion of kinetic energy to rest energy. In this 1950 photograph of a cosmic ray shower, a high-energy sulfur nucleus (red) collides with a nucleus in a photographic emulsion and produces a spray of particles, including a fluorine nucleus (green), other nuclear fragments (blue), and approximately 16 pions (yellow).

As in classical mechanics, we will define kinetic energy as the work done by the net force in accelerating a particle from rest to some final velocity u_f. Considering one dimension only, we have

$$K = \int_{u=0}^{u=u_f} F_{\text{net}} \, ds = \int_0^{u_f} \frac{dp}{dt} \, ds = \int_0^{u_f} u \, dp = \int_0^{u_f} u \, d\left(\frac{mu}{\sqrt{1 - (u^2/c^2)}} \right) \qquad \text{39-21}$$

where we have used $u = ds/dt$. It is left as a problem (Problem 37) for you to show that

$$d\left(\frac{mu}{\sqrt{1 - (u^2/c^2)}} \right) = m\left(1 - \frac{u^2}{c^2} \right)^{-3/2} du$$

If we substitute this expression into the integrand in Equation 39-21, we obtain

$$K = \int_0^{u_f} u \, d\left(\frac{mu}{\sqrt{1 - (u^2/c^2)}} \right) = \int_0^{u_f} m\left(1 - \frac{u^2}{c^2} \right)^{-3/2} u \, du$$

$$= mc^2 \left(\frac{1}{\sqrt{1 - (u_f^2/c^2)}} - 1 \right)$$

or

$$K = \frac{mc^2}{\sqrt{1 - (u^2/c^2)}} - mc^2 \qquad \text{39-22}$$

RELATIVISTIC KINETIC ENERGY

(In this expression the final speed u_f is arbitrary, so the subscript f is not needed.)

The expression for kinetic energy consists of two terms. The first term depends on the speed of the particle. The second, mc^2, is independent of the speed. The quantity mc^2 is called the **rest energy** E_0 of the particle. The rest energy is the product of the mass and c^2:

$$E_0 = mc^2 \qquad \text{39-23}$$

REST ENERGY

The total **relativistic energy** E is then defined to be the sum of the kinetic energy and the rest energy:

$$E = K + mc^2 = \frac{mc^2}{\sqrt{1 - (u^2/c^2)}} \qquad \text{39-24}$$

RELATIVISTIC ENERGY

Thus, the work done by an unbalanced force increases the energy from the rest energy mc^2 to the final energy $mc^2/\sqrt{1 - (u^2/c^2)} = m_{\text{rel}}c^2$, where $m_{\text{rel}} = m/\sqrt{1 - (u^2/c^2)}$ is the relativistic mass. We can obtain a useful expression for the velocity of a particle by multiplying Equation 39-20 for the relativistic momentum by c^2 and comparing the result with Equation 39-24 for the relativistic energy. We have

$$pc^2 = \frac{mc^2 u}{\sqrt{1 - (u^2/c^2)}} = Eu$$

or

$$\frac{u}{c} = \frac{pc}{E}$$

39-25

Energies in atomic and nuclear physics are usually expressed in units of electron volts (eV) or mega-electron volts (MeV):

$$1\ eV = 1.602 \times 10^{-19}\ J$$

A convenient unit for the masses of atomic particles is eV/c^2 or MeV/c^2, which is the rest energy of the particle divided by c^2. The rest energies of some elementary particles and light nuclei are given in Table 39-1.

TABLE 39-1

Rest Energies of Some Elementary Particles and Light Nuclei

Particle	Symbol	Rest energy, MeV
Photon	γ	0
Electron (positron)	e or e^- (e^+)	0.5110
Muon	$\mu^\pm$	105.7
Pion	π^0	135
	$\pi^\pm$	139.6
Proton	p	938.280
Neutron	n	939.573
Deuteron	^{2}H or d	1875.628
Triton	^{3}H or t	2808.944
Helium-3	^{3}He	2808.41
Alpha particle	^{4}He or α	3727.409

Total Energy, Kinetic Energy, and Momentum   **EXAMPLE 39-11**

An electron (rest energy 0.511 MeV) moves with speed $u = 0.8c$. Find (a) its total energy, (b) its kinetic energy, and (c) the magnitude of its momentum.

(a) The total energy is given by Equation 39-24:

$$E = \frac{mc^2}{\sqrt{1-(u^2/c^2)}} = \frac{0.511\ MeV}{\sqrt{1-0.64}} = \frac{0.511\ MeV}{0.6} = \boxed{0.852\ MeV}$$

(b) The kinetic energy is the total energy minus the rest energy:

$$K = E - mc^2 = 0.852\ MeV - 0.511\ MeV = \boxed{0.341 MeV}$$

(c) The magnitude of the momentum is found from Equation 39-20. We can simplify by multiplying both numerator and denominator by c^2 and using the Part (a) result:

$$p = \frac{mu}{\sqrt{1-(u^2/c^2)}}$$

$$= \frac{mc^2}{\sqrt{1-(u^2/c^2)}}\frac{u}{c^2} = (0.852\ MeV)\frac{0.8c}{c^2} = \boxed{0.681\ MeV/c}$$

REMARKS The technique used to solve Part (c) (multiplying numerator and denominator by c^2) is equivalent to using Equation 39-25.

The expression for kinetic energy given by Equation 39-22 does not look much like the classical expression $\frac{1}{2}mu^2$. However, when u is much less than c, we can approximate $1/\sqrt{1 - (u^2/c^2)}$ using the binomial expansion

$$(1 + x)^n = 1 + nx + n(n - 1)\frac{x^2}{2} + \cdots \approx 1 + nx \qquad 39\text{-}26$$

Then

$$\frac{1}{\sqrt{1 - (u^2/c^2)}} = \left(1 - \frac{u^2}{c^2}\right)^{-1/2} \approx 1 + \frac{1}{2}\frac{u^2}{c^2}$$

From this result, when u is much less than c, the expression for relativistic kinetic energy becomes

$$K = mc^2\left[\frac{1}{\sqrt{1 - (u^2/c^2)}} - 1\right] \approx mc^2\left(1 + \frac{1}{2}\frac{u^2}{c^2} - 1\right) = \frac{1}{2}mu^2$$

Thus, at low speeds, the relativistic expression is the same as the classical expression.

We note from Equation 39-24 that as the speed u approaches the speed of light c, the energy of the particle becomes very large because $1/\sqrt{1 - (u^2/c^2)}$ becomes very large. At $u = c$, the energy becomes infinite. For u greater than c, $\sqrt{1 - (u^2/c^2)}$ is the square root of a negative number and is therefore imaginary. A simple interpretation of the result that it takes an infinite amount of energy to accelerate a particle to the speed of light is that no particle that is ever at rest in any inertial reference frame can travel as fast or faster than the speed of light c. As we noted in Example 39-8, if the speed of a particle is less than c in one reference frame, it is less than c in all other reference frames moving relative to that frame at speeds less than c.

In practical applications, the momentum or energy of a particle is often known rather than the speed. Equation 39-20 for the relativistic momentum and Equation 39-24 for the relativistic energy can be combined to eliminate the speed u. The result is

$$E^2 = p^2c^2 + (mc^2)^2 \qquad 39\text{-}27$$

RELATION FOR TOTAL ENERGY, MOMENTUM, AND REST ENERGY

This useful equation can be conveniently remembered from the right triangle shown in Figure 39-12. If the energy of a particle is much greater than its rest energy mc^2, the second term on the right side of Equation 39-27 can be neglected, giving the useful approximation

$$E \approx pc, \quad \text{for } E \gg mc^2 \qquad 39\text{-}28$$

Equation 39-28 is an exact relation between energy and momentum for particles with no mass, such as photons.

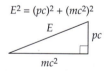

$$E^2 = (pc)^2 + (mc^2)^2$$

FIGURE 39-12 Right triangle to remember Equation 39-27.

EXERCISE A proton (mass 938 MeV/c^2) has a total energy of 1400 MeV. Find (a) $1/\sqrt{1 - (u^2/c^2)}$, (b) the momentum of the proton, and (c) the speed u of the proton. (Answer (a) 1.49, (b) $p = 1.04 \times 10^3$ MeV/c, and (c) $u = 0.74c$)

Mass and Energy

Einstein considered Equation 39-23 relating the energy of a particle to its mass to be the most significant result of the theory of relativity. Energy and inertia, which were formerly two distinct concepts, are related through this famous equation. As discussed in Chapter 7, the conversion of rest energy to kinetic energy with a corresponding decrease in mass is a common occurrence in radioactive decay and nuclear reactions, including nuclear fission and nuclear fusion. We illustrated this in Section 7-3 with the deuteron, whose mass is $2.22 \text{ MeV}/c^2$ less than the mass of its parts, a proton and a neutron. When a neutron and a proton combine to form a deuteron, 2.22 MeV of energy is released. The breaking up of a deuteron into a neutron and a proton requires 2.22 MeV of energy input. The proton and the neutron are thus bound together in a deuteron by a binding energy of 2.22 MeV. Any stable composite particle, such as a deuteron or a helium nucleus (2 neutrons plus 2 protons), that is made up of other particles has a mass and rest energy that are less than the sum of the masses and rest energies of its parts. The difference in rest energy is the binding energy of the composite particle. The binding energies of atoms and molecules are of the order of a few electron volts, which leads to a negligible difference in mass between the composite particle and its parts. The binding energies of nuclei are of the order of several MeV, which leads to a noticeable difference in mass. Some very heavy nuclei, such as radium, are radioactive and decay into a lighter nucleus plus an alpha particle. In this case, the original nucleus has a rest energy greater than that of the decay particles. The excess energy appears as the kinetic energy of the decay products.

To further illustrate the interrelation of mass and energy, we consider a perfectly inelastic collision of two particles. Classically, kinetic energy is lost in such a collision. Relativistically, this loss in kinetic energy shows up as an increase in rest energy of the system; that is, the total energy of the system is conserved. Consider a particle of mass m_1 moving with initial speed u_1 that collides with a particle of mass m_2 moving with initial speed u_2. The particles collide and stick together, forming a particle of mass M that moves with speed u_f, as shown in Figure 39-13. The initial total energy of particle 1 is

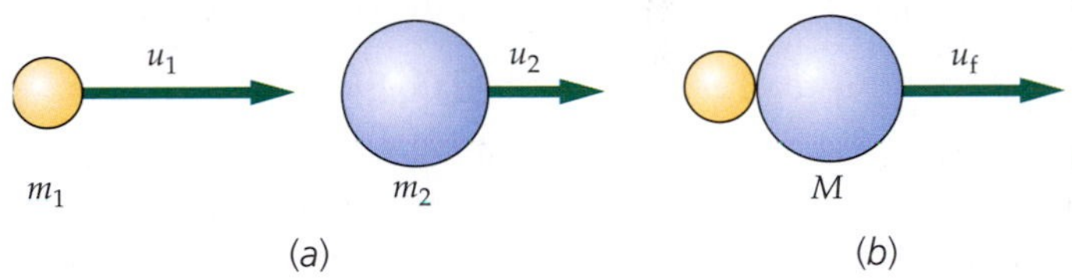

$$E_1 = K_1 + m_1 c^2$$

where K_1 is its initial kinetic energy. Similarly the initial total energy of particle 2 is

$$E_2 = K_2 + m_2 c^2$$

The total initial energy of the system is

$$E_i = E_1 + E_2 = K_1 + m_1 c^2 + K_2 + m_2 c^2 = K_i + M_i c^2$$

where $K_i = K_1 + K_2$ and $M_i = m_1 + m_2$ are the initial kinetic energy and initial mass of the system. The final total energy of the system is

$$E_f = K_f + M_f c^2$$

If we set the final total energy equal to the initial total energy, we obtain

$$K_f + M_f c^2 = K_i + M_i c^2$$

Rearranging gives $K_f - K_i = -(M_f - M_i)c^2$, which can be expressed

$$\Delta K + (\Delta M)c^2 = 0 \qquad\qquad 39\text{-}29$$

where $\Delta M = M_f - M_i$ is the change in mass of the system.

(a)   (b)

FIGURE 39-13 A perfectly inelastic collision between two particles. One particle of mass m_1 collides with another particle of mass m_2. After the collision, the particles stick together, forming a composite particle of mass M that moves with speed u_f so that relativistic momentum is conserved. Kinetic energy is lost in this process. If we assume that the total energy is conserved, the loss in kinetic energy must equal c^2 times the increase in the mass of the system.

TOTALLY INELASTIC COLLISION **E X A M P L E 3 9 - 1 2**

A particle of mass 2 MeV/c^2 and kinetic energy 3 MeV collides with a stationary particle of mass 4 MeV/c^2. After the collision, the two particles stick together. Find (a) the initial momentum of the system, (b) the final velocity of the two-particle system, and (c) the mass of the two-particle system.

PICTURE THE PROBLEM (a) The initial momentum of the system is the initial momentum of the incoming particle, which can be found from the total energy of the particle. (b) The final velocity of the system can be found from its total energy and momentum using $u/c = pc/E$ (Equation 39-25). The energy is found from conservation of energy, and the momentum from conservation of momentum. (c) Since the final energy and momentum are known, the final mass can be found from $E^2 = p^2c^2 + (Mc^2)^2$.

(a) 1. The initial momentum of the system is the initial momentum of the incoming particle. The momentum of a particle is related to its energy and mass (Equation 39-27):

$$E_1^2 = p_1^2c^2 + (m_1c^2)^2$$
$$p_1c = \sqrt{E_1^2 - (m_1c^2)^2}$$

2. The total energy of the moving particle is the sum of its kinetic energy and its rest energy:

$$E_1 = 3\,\text{MeV} + 2\,\text{MeV} = 5\,\text{MeV}$$

3. Use this total energy to calculate the momentum:

$$p_1c = \sqrt{E_1^2 - (m_1c^2)^2} = \sqrt{(5\,\text{MeV})^2 - (2\,\text{MeV})^2} = \sqrt{21}\,\text{MeV}$$

$$p_1 = \boxed{4.58\,\text{MeV}/c}$$

(b) 1. We can find the final velocity of the system from its total energy E_f and its momentum p_f using Equation 39-25:

$$\frac{u_f}{c} = \frac{p_f c}{E_f}$$

2. By the conservation of total energy, the final energy of the system equals the initial total energy of the two particles:

$$E_f = E_i = E_1 + E_2 = 5\,\text{MeV} + 4\,\text{MeV} = 9\,\text{MeV}$$

3. By the conservation of momentum, the final momentum of the two-particle system equals the initial momentum:

$$p_f = 4.58\,\text{MeV}/c$$

4. Calculate the velocity of the two-particle system from its total energy and momentum using $u/c = pc/E$:

$$\frac{u_f}{c} = \frac{p_f c}{E_f} = \frac{4.58\,\text{MeV}}{9\,\text{MeV}} = 0.509$$

$$u_f = \boxed{0.509c}$$

(c) We can find the mass M_f of the final two-particle system from Equation 39-27 using $pc = 4.58$ MeV and $E = 9$ MeV:

$$E_f^2 = (p_f c)^2 + (M_f c^2)^2$$
$$(9\,\text{MeV})^2 = (4.58\,\text{MeV})^2 + (M_f c^2)^2$$

$$M_f = \boxed{7.75\,\text{MeV}/c^2}$$

REMARKS Note that the mass of the system increased from 6 MeV/c^2 to 7.75 MeV/c^2. This increase times c^2 equals the loss in kinetic energy of the system, as you will show in the following exercise.

EXERCISE (a) Find the final kinetic energy of the two-particle system in Example 39-12. (b) Find the loss in kinetic energy, K_{loss}, in the collision. (c) Show that $K_{loss} = (\Delta M)c^2$, where ΔM is the change in mass of the system. [*Answer* (a) $K_f = E_f - M_f c^2 = 9\,\text{MeV} - 7.75\,\text{MeV} = 1.25\,\text{MeV}$, (b) $K_{loss} = K_i - K_f = 3\,\text{MeV} - 1.25\,\text{MeV} = 1.75\,\text{MeV}$, and (c) $(\Delta M)c^2 = (M_f - M_i)c^2 = 7.75\,\text{MeV} - (2\,\text{MeV} + 4\,\text{MeV}) = 1.75\,\text{MeV} = K_{loss}$]

MOMENTUM AND TOTAL-ENERGY CONSERVATION   **EXAMPLE 39-13**

A 1×10^6-kg rocket has 1×10^3 kg of fuel on board. The rocket is parked in space when it suddenly becomes necessary to accelerate. The rocket engines ignite, and the 1×10^3 kg of fuel are consumed. The exhaust (spent fuel) is ejected during a very short time interval at a speed of $c/2$ relative to S—the inertial reference frame in which the rocket is initially at rest. (*a*) Calculate the change in the mass of the rocket-fuel system. (*b*) Calculate the final speed of the rocket u_R relative to S. (*c*) Again, calculate the final speed of the rocket relative to S, this time using classical (newtonian) mechanics.

PICTURE THE PROBLEM The speed of the rocket and the change in the mass of the system can be calculated via conservation of momentum and conservation of energy. In reference frame S, the total momentum of the rocket plus fuel is zero. After the burn, the magnitude of the momentum of the rocket equals that of the ejected fuel. Let $m_R = 1 \times 10^6$ kg be the mass of the rocket, not including the mass of the fuel, let $m_{F,i} = 1 \times 10^3$ kg be the mass of the fuel *before* the burn, and let $m_{F,f}$ be the mass of the fuel *after* the burn. The mass of the rocket, m_R, remains fixed, but during the burn the mass of the fuel decreases. (The fuel has less chemical energy after the burn, and so has less mass as well.)

(*a*) 1. The magnitudes of the momentum of the rocket and the momentum of the ejected fuel are equal. For the reasons stated above, the mass of the rocket, not including the 1×10^3 kg of fuel, does not change during the burn:

$$p_R = p_F$$

$$\frac{m_R u_R}{\sqrt{1 - (u_R^2/c^2)}} = \frac{m_{F,f} u_F}{\sqrt{1 - (u_F^2/c^2)}} = p$$

where

$p = p_R = p_F$, $m_R = 1 \times 10^6$ kg, $u_F = 0.5c$, and u_R is the final speed of the rocket.

2. The total energy of the system does not change:

$$E_f = E_i$$

3. The initial energy is the rest energy of the rocket and fuel before the burn. The final energy is the energy of the rocket plus energy of the fuel. The energy of each is related to its momentum by Equation 39-27:

$$E_i = m_R c^2 + m_{F,i} c^2 = (m_R + m_{F,i})c^2$$

$$E_{R,f}^2 = p^2 c^2 + (m_R c^2)^2$$

$$E_{F,f}^2 = p^2 c^2 + (m_{F,f} c^2)^2$$

so

$$E_f = E_{R,f} + E_{F,f}$$

$$E_f = \sqrt{p^2 c^2 + (m_R c^2)^2} + \sqrt{p^2 c^2 + (m_{F,f} c^2)^2}$$

4. Equate the initial and final energies:

$$\sqrt{p^2 c^2 + (m_R c^2)^2} + \sqrt{p^2 c^2 + (m_{F,f} c^2)^2} = (m_R + m_{F,i})c^2$$

5. The step 4 result and the step 1 result,

$$p = \frac{m_{F,f} u_F}{\sqrt{1 - (u_F^2/c^2)}}$$, constitute two simultaneous equations with unknowns p and $m_{F,f}$. Solving for $m_{F,f}$ gives:

$$m_{F,f} = 866 \text{ kg}$$

so

$$m_{loss} = m_{F,i} - m_{F,f} = 1000 \text{ kg} - 866 \text{ kg} = \boxed{134 \text{ kg}}$$

(*b*) 1. To solve for u_R, we use Equation 39-25:

$$\frac{u_R}{c} = \frac{pc}{E_{R,f}}$$

2. To solve for p, we substitute the value for $m_{F,f}$ into the Part (*a*), step 1 result:

$$p = \frac{m_{F,f} u_F}{\sqrt{1 - (u_F^2/c^2)}} = \frac{(866 \text{ kg})\frac{1}{2}c}{\sqrt{1 - \frac{1}{4}}} = (5.00 \times 10^2 \text{ kg})c$$

3. We use the value for p to solve for $E_{R,f}$:

$$E_{R,f}^2 = p^2c^2 + (m_R c^2)^2 = (5.00 \times 10^2 \text{ kg})^2c^4 + (10^6 \text{ kg})^2c^4 = (1.00 \times 10^{12} \text{ kg}^2)c^4$$

so

$$E_{R,f} = (1.00 \times 10^6 \text{ kg})c^2$$

4. Using our Part (b), step 1 result, we solve for u_R:

$$u_R = \frac{pc^2}{E_{R,f}} = \frac{(5.00 \times 10^2 \text{ kg})c^3}{(1.00 \times 10^6 \text{ kg})c^2} = \boxed{5.00 \times 10^{-4}c = 1.50 \times 10^5 \text{ m/s}}$$

(c) Equate the magnitude of the classical expressions for the momentum of the rocket and burned fuel and solve for u_R:

$$m_R u_R = m_F u_F$$

$$u_R = \frac{m_F}{m_R}u_F = \frac{10^3 \text{ kg}}{10^6 \text{ kg}}0.5c = \boxed{1.5 \times 10^5 \text{ m/s}}$$

REMARK If carried out to five figures, the relativistic calculation gives $u_R = 4.9994 \times 10^4\, c$ for the final speed of the rocket. However, the classical calculation gives $u_R = 5.0000 \times 10^4\, c$. These two values differ by less than one part in 8000.

EXERCISE If the matter being ejected were a 1×10^3-kg rigid block launched by a spring with one end attached to the rocket, would the rest mass of the block change or would the rest mass of the spring change? (*Answer* Only the rest mass of the spring would change.)

39-8 General Relativity

The generalization of the theory of relativity to noninertial reference frames by Einstein in 1916 is known as the general theory of relativity. It is much more difficult mathematically than the special theory of relativity, and there are fewer situations in which it can be tested. Nevertheless, its importance calls for a brief qualitative discussion.

The basis of the general theory of relativity is the **principle of equivalence:**

> A homogeneous gravitational field is completely equivalent to a uniformly accelerated reference frame.

PRINCIPLE OF EQUIVALENCE

This principle arises in Newtonian mechanics because of the apparent identity of gravitational mass and inertial mass. In a uniform gravitational field, all objects fall with the same acceleration $\vec{g}$ independent of their mass because the gravitational force is proportional to the (gravitational) mass, whereas the acceleration varies inversely with the (inertial) mass. Consider a compartment in space undergoing a uniform acceleration $\vec{a}$, as shown in Figure 39-14a. No mechanics experiment can be performed *inside* the compartment that will distinguish whether the compartment is actually accelerating in space or is at rest (or is moving with uniform velocity) in the presence of a uniform gravitational field $\vec{g} = -\vec{a}$, as shown in Figure 39-14b. If objects are dropped in the compartment, they will fall to the floor with an acceleration $\vec{g} = -\vec{a}$. If people stand on a spring scale, it will read their weight of magnitude ma.

Einstein assumed that the principle of equivalence applies to all physics and not just to mechanics. In effect, he assumed that there is no experiment of any kind that can distinguish uniformly accelerated motion from the presence of a gravitational field.

One consequence of the principle of equivalence—the deflection of a light beam in a gravitational field—was one of the first to be tested experimentally. In a region

(a)

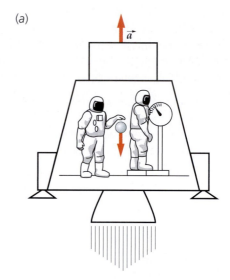

(b)

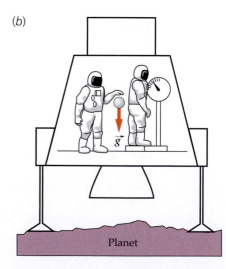

FIGURE 39-14 The results of experiments in a uniformly accelerated reference frame (*a*) cannot be distinguished from those in a uniform gravitational field (*b*) if the acceleration $\vec{a}$ and the gravitational field $\vec{g}$ have the same magnitude.

with no gravitational field, a light beam will travel in a straight line at speed c. The principle of equivalence tells us that a region with no gravitational field exists only in a compartment that is in free fall. Figure 39-15 shows a beam of light entering a compartment that is accelerating relative to a nearby reference frame in free fall. Successive positions of the compartment at equal time intervals are shown in Figure 39-15a. Because the compartment is accelerating, the distance it moves in each time interval increases with time. The path of the beam of light as observed from inside the compartment is therefore a parabola, as shown in Figure 39-15b. But according to the principle of equivalence, there is no way to distinguish between an accelerating compartment and one moving with uniform velocity in a uniform gravitational field. We conclude, therefore, that a beam of light will accelerate in a gravitational field, just like objects that have mass. For example, near the surface of the earth, light will fall with an acceleration of 9.81 m/s^2. This is difficult to observe because of the enormous speed of light. In a distance of 3000 km, which takes light about 0.01 s to traverse, a beam of light should fall approximately 0.5 mm. Einstein pointed out that the deflection of a light beam in a gravitational field might be observed when light from a distant star passes close to the sun, as illustrated in Figure 39-16. Because of the brightness of the sun, this cannot ordinarily be seen. Such a deflection was first observed in 1919 during an eclipse of the sun. This well-publicized observation brought instant worldwide fame to Einstein.

A second prediction from Einstein's theory of general relativity, which we will not discuss in detail, is the excess precession of the perihelion of the orbit of Mercury of about $0.01°$ per century. This effect had been known and unexplained for some time, so, in a sense, explaining it constituted an immediate success of the theory.

A third prediction of general relativity concerns the change in time intervals and frequencies of light in a gravitational field. In Chapter 11, we found that the gravitational potential energy between two masses M and m a distance r apart is

$$U = -\frac{GMm}{r}$$

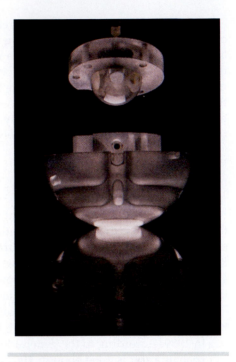

The quartz sphere in the top part of the container is probably the world's most perfectly round object. It is designed to spin as a gyroscope in a satellite orbiting the earth. General relativity predicts that the rotation of the earth will cause the axis of rotation of the gyroscope to precess in a circle at a rate of approximately 1 revolution in 100,000 years.

where G is the universal gravitational constant, and the point of zero potential energy has been chosen to be when the separation of the masses is infinite. The potential energy per unit mass near a mass M is called the *gravitational potential* ϕ:

$$\phi = -\frac{GM}{r} \qquad\qquad 39\text{-}30$$

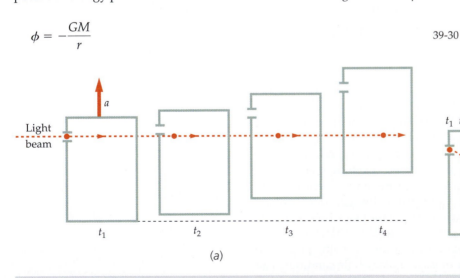

(a)   (b)

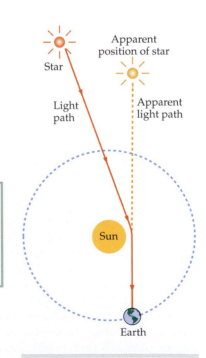

FIGURE 39-15 (*a*) A light beam moving in a straight line through a compartment that is undergoing uniform acceleration relative to a nearby reference frame in free fall. The position of the beam is shown at equally spaced times t_1, t_2, t_3, and t_4. (*b*) In the reference frame of the compartment, the light travels in a parabolic path as a ball would if it were projected horizontally. The vertical displacements are greatly exaggerated in Figure 39-15a and Figure 39-15b for emphasis.

FIGURE 39-16 The deflection (greatly exaggerated) of a beam of light due to the gravitational attraction of the sun.

According to the general theory of relativity, clocks run more slowly in regions of lower gravitational potential. (Since the gravitational potential is negative, as can be seen from Equation 39-30, the nearer the mass the more negative, and therefore the lower the gravitational potential.) If Δt_1 is a time interval between two events measured by a clock where the gravitational potential is ϕ_1 and Δt_2 is the interval between the same events as measured by a clock where the gravitational potential is ϕ_2, general relativity predicts that the fractional difference between these times will be approximately[†]

$$\frac{\Delta t_2 - \Delta t_1}{\Delta t} = \frac{1}{c^2}(\phi_2 - \phi_1)$$

39-31

A clock in a region of low gravitational potential will therefore run slower than a clock in a region of high potential. Since a vibrating atom can be considered to be a clock, the frequency of vibration of an atom in a region of low potential, such as near the sun, will be lower than the frequency of vibration of the same atom on the earth. This shift toward a lower frequency, and therefore a longer wavelength, is called the **gravitational redshift.**

As our final example of the predictions of general relativity, we mention **black holes,** which were first predicted by J. Robert Oppenheimer and Hartland Snyder in 1939. According to the general theory of relativity, if the density of an object such as a star is great enough, its gravitational attraction will be so great that once inside a critical radius, nothing can escape, not even light or other electromagnetic radiation. (The effect of a black hole on objects outside the critical radius is the same as that of any other mass.) A remarkable property of such an object is that nothing that happens inside it can be communicated to the outside. As sometimes occurs in physics, a simple but incorrect calculation gives the correct results for the relation between the mass and the critical radius of a black hole. In Newtonian mechanics, the speed needed for a particle to escape from the surface of a planet or a star of mass M and radius R is given by Equation 11-21:

$$v_e = \sqrt{\frac{2GM}{R}}$$

If we set the escape speed equal to the speed of light and solve for the radius, we obtain the critical radius R_S, called the **Schwarzschild radius:**

$$R_S = \frac{2GM}{c^2}$$

39-32

For an object with a mass equal to five times that of our sun (theoretically the minimum mass for a black hole) to be a black hole, its radius would have to be approximately 15 km. Since no radiation is emitted from a black hole and its radius is expected to be small, the detection of a black hole is not easy. The best chance of detection occurs if a black hole is a close companion to a normal star in a binary star system. Then both stars revolve around their center of mass and the gravitational field of the black hole will pull gas from the normal star into the black hole. However, to conserve angular momentum, the gas does not go straight into the black hole. Instead, the gas orbits around the black hole in a disk, called an accretion disk, while slowly being pulled closer to the black hole. The gas in this disk emits X rays because the temperature of the gas being pulled inward reaches several millions of kelvins. The mass of a black-hole candidate can often be estimated. An estimated mass of at least five solar masses, along with the emission of X rays, establishes a strong inference that the candidate is, in fact, a black hole. In addition to the black holes just described, there are supermassive black holes that exist at the centers of galaxies. At the center of the Milky Way is a supermassive black hole with a mass of about two million solar masses.

This extremely accurate hydrogen maser clock was launched in a satellite in 1976, and its time was compared to that of an identical clock on the earth. In accordance with the prediction of general relativity, the clock on the earth, where the gravitational potential was lower, *lost* approximately 4.3×10^{-10} s each second compared with the clock orbiting the earth at an altitude of approximately 10,000 km.

[†] Since this shift is usually very small, it does not matter by which interval we divide on the left side of the equation.

SUMMARY

Topic	Relevant Equations and Remarks	

1. Einstein's Postulates	The special theory of relativity is based on two postulates of Albert Einstein. All of the results of special relativity can be derived from these postulates.
	    Postulate 1: Absolute uniform motion cannot be detected.
	    Postulate 2: The speed of light is independent of the motion of the source.
	An important implication of these postulates is
	    Postulate 2 (alternate): Every observer measures the same value c for the speed of light.	

2. The Lorentz Transformation	$x = \gamma(x' + vt'), \quad y = y', \quad z = z'$	39-9
	$t = \gamma\left(t' + \dfrac{vx'}{c^2}\right)$	39-10
	$\gamma = \dfrac{1}{\sqrt{1 - (v^2/c^2)}}$	39-7
Inverse transformation	$x' = \gamma(x - vt), \quad y' = y, \quad z' = z$	39-11
	$t' = \gamma\left(t - \dfrac{vx}{c^2}\right)$	39-12
3. Time Dilation	The time interval measured between two events that occur at the same point in space in some reference frame is called the proper time t_p. In another reference frame in which the events occur at different places, the time interval between the events is longer by the factor γ.	
	$\Delta t = \gamma \, \Delta t_p$	39-13
4. Length Contraction	The length of an object measured in the reference frame in which the object is at rest is called its proper length L_p. When measured in another reference frame, the length of the object is	
	$L = \dfrac{L_p}{\gamma}$	39-14
5. The Relativistic Doppler Effect	$f' = \dfrac{\sqrt{1 - (v^2/c^2)}}{1 - (v/c)} f_0, \qquad$ approaching	39-16a
	$f' = \dfrac{\sqrt{1 - (v^2/c^2)}}{1 + (v/c)} f_0, \qquad$ receding	39-16b
6. Clock Synchronization and Simultaneity	Two events that are simultaneous in one reference frame typically are not simultaneous in another frame that is moving relative to the first. If two clocks are synchronized in the frame in which they are at rest, they will be out of synchronization in another frame. In the frame in which they are moving, the chasing clock leads by an amount	
	$\Delta t_S = L_p \dfrac{v}{c^2}$	39-17
	where L_p is the proper distance between the clocks.	
7. The Velocity Transformation	$u_x = \dfrac{u'_x + v}{1 + (vu'_x/c^2)}$	39-18a

$$u_y = \frac{u'_y}{\gamma[1 + (vu'_x/c^2)]}$$ 39-18*b*

$$u_z = \frac{u'_z}{\gamma[1 + (vu'_x/c^2)]}$$ 39-18*c*

Inverse velocity transformation

$$u'_x = \frac{u_x - v}{1 - (vu_x/c^2)}$$ 39-19*a*

$$u'_y = \frac{u_y}{\gamma[1 - (vu_x/c^2)]}$$ 39-19*b*

$$u'_z = \frac{u_z}{\gamma[1 - (vu_x/c^2)]}$$ 39-19*c*

8. Relativistic Momentum

$$\vec{p} = \frac{m\vec{u}}{\sqrt{1 - (u^2/c^2)}}$$ 39-20

where *m* is the mass of the particle.

9. Relativistic Energy

Kinetic energy

$$K = \frac{mc^2}{\sqrt{1 - (u^2/c^2)}} - mc^2 = \frac{mc^2}{\sqrt{1 - (u^2/c^2)}} - E_0$$ 39-22

Rest energy

$$E_0 = mc^2$$ 39-23

Total energy

$$E = K + E_0 = \frac{mc^2}{\sqrt{1 - (u^2/c^2)}}$$ 39-24

10. Useful Formulas for Speed, Energy, and Momentum

$$\frac{u}{c} = \frac{pc}{E}$$ 39-25

$$E^2 = p^2c^2 + (mc^2)^2$$ 39-27

$$E \approx pc, \quad \text{for } E \gg mc^2$$ 39-28

PROBLEMS

- Single-concept, single-step, relatively easy
- •• Intermediate-level, may require synthesis of concepts
- ••• Challenging
- **SSM** Solution is in the *Student Solutions Manual*
- **iSOLVE** Problems available on iSOLVE online homework service
- **iSOLVE✓** These "Checkpoint" online homework service problems ask students additional questions about their confidence level, and how they arrived at their answer.

In a few problems, you are given more data than you actually need; in a few other problems, you are required to supply data from your general knowledge, outside sources, or informed estimates.

Conceptual Problems

1 • **SSM** The approximate total energy of a particle of mass *m* moving at speed $u \ll c$ is (*a*) $mc^2 + \frac{1}{2}mu^2$. (*b*) $\frac{1}{2}mu^2$. (*c*) *cmu*. (*d*) mc^2. (*e*) $\frac{1}{2}cmu$.

2 • **SSM** A set of twins work in an office building. One twin works on the top floor and the other twin works in the basement. Considering general relativity, which twin will age more quickly? (*a*) They will age at the same rate. (*b*) The twin who works on the top floor will age more quickly. (*c*) The twin who works in the basement will age more quickly. (*d*) It depends on the speed of the office building. (*e*) None of these is correct.

3 • True or false:

(*a*) The speed of light is the same in all reference frames.

(b) Proper time is the shortest time interval between two events.

(c) Absolute motion can be determined by means of length contraction.

(d) The light-year is a unit of distance.

(e) Simultaneous events must occur at the same place.

(f) If two events are not simultaneous in one frame, they cannot be simultaneous in any other frame.

(g) If two particles are tightly bound together by strong attractive forces, the mass of the system is less than the sum of the masses of the individual particles when separated.

4 • An observer sees a system consisting of a mass oscillating on the end of a spring moving past at a speed u and notes that the period of the system is T. Another observer, who is moving with the mass–spring system, also measures its period. The second observer will find a period that is (a) equal to T. (b) less than T. (c) greater than T. (d) either (a) or (b) depending on whether the system was approaching or receding from the first observer. (e) There is not sufficient information to answer the question.

5 • The Lorentz transformation for y and z is the same as the classical result: $y = y'$ and $z = z'$. Yet the relativistic velocity transformation does not give the classical result $u_y = u_y'$ and $u_z = u_z'$. Explain.

Estimation and Approximation

6 • • The sun radiates energy at the rate of approximately 4×10^{26} W. Assume that this energy is produced by a reaction whose net result is the fusion of 4 H nuclei to form 1 He nucleus, with the release of 25 MeV for each He nucleus formed. Calculate the sun's loss of mass per day.

7 • • **SSM** The most distant galaxies that can be seen by the Hubble telescope are moving away from us with a red-shift parameter of about $z = 5$. (See Problem 30 for a definition of z.) (a) What is the velocity of these galaxies relative to us (expressed as a fraction of the speed of light)? (b) *Hubble's law* states that the recession velocity is given by the expression $v = Hx$, where v is the velocity of recession, x is the distance, and H is the Hubble constant, $H = 75$ km/s/Mpc. (1 pc = 3.26 $c\cdot$y.) Estimate the distance of such a galaxy using the information given.

Time Dilation and Length Contraction

8 • The proper mean lifetime of a muon is 2 μs. Muons in a beam are traveling through a laboratory at 0.95c. (a) What is their mean lifetime as measured in the laboratory? (b) How far do they travel, on average, before they decay?

9 • • In the Stanford linear collider, small bundles of electrons and positrons are fired at each other. In the laboratory's frame of reference, each bundle is approximately 1 cm long and 10 μm in diameter. In the collision region, each particle has an energy of 50 GeV, and the electrons and the positrons are moving in opposite directions. (a) How long and how wide is each bundle in its own reference frame? (b) What must be the minimum proper length of the accelerator for a bundle to have both its ends simultaneously in the accelerator in its own reference frame? (The ac-

tual length of the accelerator is less than 1000 m.) (c) What is the length of a positron bundle in the reference frame of the electron bundle?

10 • • **SSM** Unobtainium (Un) is an unstable particle that decays into normalium (Nr) and standardium (St) particles. (a) An accelerator produces a beam of Un that travels to a detector located 100 m away from the accelerator. The particles travel with a velocity of $v = 0.866c$. How long do the particles take (in the laboratory frame) to get to the detector? (b) By the time the particles get to the detector, half of the particles have decayed. What is the half-life of Un? (*Note:* Half-life as it would be measured in a frame moving with the particles.) (c) A new detector is going to be used, which is located 1000 m away from the accelerator. How fast should the particles be moving if half of the particles are to make it to the new detector?

11 • • Star A and Star B are at rest relative to the earth. Star A is 27 $c\cdot$y from earth, and Star B is located beyond (behind) Star A as viewed from earth. (a) A spaceship is making a trip from earth to Star A at a speed such that the trip from earth to Star A takes 12 y according to clocks on the spaceship. At what speed, relative to earth, must the ship travel? (Assume that the times for acceleration are very short compared to the overall trip time.) (b) Upon reaching Star A, the ship speeds up and departs for Star B at a speed such that the gamma factor, γ, is twice that of Part (a). The trip from Star A to Star B takes 5 y (ship's time). How far, in $c\cdot$y, is Star B from Star A in the rest frame of the earth and the two stars? (c) Upon reaching Star B, the ship departs for earth at the same speed as in Part (b). It takes it 10 y (ship's time) to return to earth. If you were born on earth the day the ship left earth (and you remain on earth), how old are you on the day the ship returns to earth?

12 • A spaceship travels to a star 35 $c\cdot$y away at a speed of 2.7×10^8 m/s. How long does the spaceship take to get to the star (a) as measured on the earth and (b) as measured by a passenger on the spaceship?

13 • Use the binomial expansion equation

$$(1 + x)^n = 1 + nx + \frac{n(n-1)}{2}x^2 + \ldots \approx 1 + nx, \quad \text{for } x \ll 1$$

to derive the following results for the case when v is much less than c.

(a) $\gamma \approx 1 + \dfrac{1}{2}\dfrac{v^2}{c^2}$

(b) $\dfrac{1}{\gamma} \approx 1 - \dfrac{1}{2}\dfrac{v^2}{c^2}$

(c) $\gamma - 1 \approx 1 - \dfrac{1}{\gamma} \approx \dfrac{1}{2}\dfrac{v^2}{c^2}$

14 • • A clock on Spaceship A measures the time interval between two events, both of which occur at the location of the clock. You are on Spaceship B. According to your careful measurements, the time interval between the two events is 1 percent longer than that measured by the two clocks on Spaceship A. How fast is Ship A moving relative to Ship B. (Use one or more of the results of Problem 13.)

15 • • If a plane flies at a speed of 2000 km/h, how long must the plane fly before its clock loses 1 s because of time dilation? (Use one or more of the results of Problem 13.)

The Lorentz Transformation, Clock Synchronization, and Simultaneity

16 •• Show that when $v << c$ the transformation equations for x, t, and u reduce to the Galilean equations.

17 •• **SSM** **iSOLVE** A spaceship of proper length L_p = 400 m moves past a transmitting station at a speed of $0.76c$. At the instant that the nose of the spaceship passes the transmitter, clocks at the transmitter and in the nose of the spaceship are synchronized to $t = t' = 0$. The instant that the tail of the spaceship passes the transmitter a signal is sent and subsequently detected by the receiver in the nose of the spaceship. (*a*) When, according to the clock in the spaceship, is the signal sent? (*b*) When, according to the clock at the transmitter, is the signal received by the spaceship? (*c*) When, according to the clock in the spaceship, is the signal received? (*d*) Where, according to an observer at the transmitter, is the nose of the spaceship when the signal is received?

18 •• In frame S, event B occurs 2 μs after event A, which occurs at $x = 1.5$ km from event A. How fast must an observer be moving along the $+x$ axis so that events A and B occur simultaneously? Is it possible for event B to precede event A for some observer?

19 •• Observers in reference frame S see an explosion located at $x_1 = 480$ m. A second explosion occurs 5 μs later at $x_2 = 1200$ m. In reference frame S', which is moving along the $+x$ axis at speed v, the explosions occur at the same point in space. What is the separation in time between the two explosions as measured in S'?

20 ••• Two events in S are separated by a distance $D = x_2 - x_1$ and a time $T = t_2 - t_1$. (*a*) Use the Lorentz transformation to show that in frame S', which is moving with speed v relative to S, the time separation is $t'_2 - t'_1 = \gamma(T - vD/c^2)$. (*b*) Show that the events can be simultaneous in frame S' only if D is greater than cT. (*c*) If one of the events is the *cause* of the other, the separation D must be less than cT, since D/c is the smallest time that a signal can take to travel from x_1 to x_2 in frame S. Show that if D is less than cT, t'_2 is greater than t'_1 in all reference frames. This shows that if the cause precedes the effect in one frame, it must precede it in all reference frames. (*d*) Suppose that a signal could be sent with speed $c' > c$ so that in frame S the cause precedes the effect by the time $T = D/c'$. Show that there is then a reference frame moving with speed v less than c in which the effect precedes the cause.

21 ••• A rocket with a proper length of 700 m is moving to the right at a speed of $0.9c$. It has two clocks, one in the nose and one in the tail, that have been synchronized in the frame of the rocket. A clock on the ground and the nose clock on the rocket both read $t = 0$ as they pass. (*a*) At $t = 0$, what does the tail clock on the rocket read as seen by an observer on the ground? When the tail clock on the rocket passes the ground clock, (*b*) what does the tail clock read as seen by an observer on the ground, (*c*) what does the nose clock read as seen by an observer on the ground, and (*d*) what does the nose clock read as seen by an observer on the rocket? (*e*) At $t = 1$ h, as measured on the rocket, a light signal is sent from the nose of the rocket to an observer standing by the ground clock. What does the ground clock read when the observer receives this signal? (*f*) When the observer on the ground receives the signal, he sends a return signal to the nose of the rocket. When is this signal received at the nose of the rocket as seen on the rocket?

22 ••• **SSM** An observer in frame S standing at the origin observes two flashes of colored light separated spatially by $\Delta x = 2400$ m. A blue flash occurs first, followed by a red flash 5 μs later. An observer in S' moving along the x axis at speed v relative to S also observes the flashes 5 μs apart and with a separation of 2400 m, but the red flash is observed first. Find the magnitude and direction of v.

The Velocity Transformation

23 •• Show that if u'_x and v in Equation 39-18a are both positive and less than c, then u_x is positive and less than c. [*Hint: Let $u'_x = (1 - \varepsilon_1)c$ and $v = (1 - \varepsilon_2)c$, where ε_1 and ε_2 are positive numbers that are less than 1.*]

24 •• **SSM** A spaceship, at rest in a certain reference frame S, is given a speed increment of $0.50c$ (call this boost 1). Relative to its new rest frame, the spaceship is given a further $0.50c$ increment 10 seconds later (as measured in its new rest frame; call this boost 2). This process is continued indefinitely, at 10-s intervals, as measured in the rest frame of the ship. (Assume that the boost itself takes a very short time compared to 10 s.) (*a*) Using a spreadsheet program, calculate and graph the velocity of the spaceship in reference frame S as a function of the boost number for boost 1 to boost 10. (*b*) Graph the gamma factor the same way. (*c*) How many boosts does it take until the velocity of the ship in S is greater than $0.999c$? (*d*) How far has the spaceship moved after 5 boosts, as measured in reference frame S? What is the average speed of the spaceship (between boost 1 and boost 5) as measured in S?

The Relativistic Doppler Effect

25 • Sodium light of wavelength 589 nm is emitted by a source that is moving toward the earth with speed v. The wavelength measured in the frame of the earth is 547 nm. Find v.

26 • **iSOLVE** ✓ A distant galaxy is moving away from us at a speed of 1.85×10^7 m/s. Calculate the fractional redshift $(\lambda' - \lambda_0)/\lambda_0$ in the light from this galaxy.

27 •• Derive Equation 39-16a for the frequency received by an observer moving with speed v toward a stationary source of electromagnetic waves.

28 • Show that if v is much less than c, the Doppler shift is given approximately by

$$\Delta f/f \approx \pm v/c$$

29 •• **SSM** **iSOLVE** ✓ A clock is placed in a satellite that orbits the earth with a period of 90 min. By what time interval will this clock differ from an identical clock on the earth after 1 y? (Assume that special relativity applies and neglect general relativity.)

30 •• For light that is Doppler-shifted with respect to an observer, define the redshift parameter

$$z = \frac{f - f'}{f'}$$

where f is the frequency of the light measured in the rest frame of the emitter, and f' is the frequency measured in the rest frame of the observer. If the emitter is moving directly away from the observer, show that the relative velocity between the emitter and the observer is

$$v = c\left(\frac{u^2 - 1}{u^2 + 1}\right)$$

where $u = z + 1$.

31 • A light beam moves along the y' axis with speed c in frame S', which is moving to the right with speed v relative to frame S. (a) Find the x and y components of the velocity of the light beam in frame S. (b) Show that the magnitude of the velocity of the light beam in S is c.

32 • **iSOLVE✔** A spaceship is moving east at speed $0.90c$ relative to the earth. A second spaceship is moving west at speed $0.90c$ relative to the earth. What is the speed of one spaceship relative to the other spaceship?

33 •• **SSM** A particle moves with speed $0.8c$ along the x'' axis of frame S'', which moves with speed $0.8c$ along the x' axis relative to frame S'. Frame S' moves with speed $0.8c$ along the x axis relative to frame S. (a) Find the speed of the particle relative to frame S'. (b) Find the speed of the particle relative to frame S.

Relativistic Momentum and Relativistic Energy

34 • **SSM** A proton (rest energy 938 MeV) has a total energy of 2200 MeV. (a) What is its speed? (b) What is its momentum?

35 • If the kinetic energy of a particle equals twice its rest energy, what percentage error is made by using $p = mu$ for its momentum?

36 •• **iSOLVE** A particle with momentum of 6 MeV/c has total energy of 8 MeV. (a) Determine the mass of the particle. (b) What is the energy of the particle in a reference frame in which its momentum is 4 MeV/c? (c) What are the relative velocities of the two reference frames?

37 •• Show that

$$d\left(\frac{mu}{\sqrt{1 - (u^2/c^2)}}\right) = m\left(1 - \frac{u^2}{c^2}\right)^{-3/2} du$$

38 •• **iSOLVE** The K^0 particle has a mass of 497.7 MeV/c^2. It decays into a π^- and π^+, each with mass 139.6 MeV/c^2. Following the decay of a K^0, one of the pions is at rest in the laboratory. Determine the kinetic energy of the other pion and of the K^0 prior to the decay.

39 •• **SSM** Two protons approach each other head-on at $0.5c$ relative to reference frame S'. (a) Calculate the total kinetic energy of the two protons as seen in frame S'. (b) Calculate the total kinetic energy of the protons as seen in reference frame S, which is moving with speed $0.5c$ relative to S' so that one of the protons is at rest.

40 •• An antiproton has the same rest energy as a proton. It is created in the reaction $p + p \rightarrow p + p + p + \bar{p}$. In an experiment, protons at rest in the laboratory are bombarded with protons of kinetic energy K_L, which must be great enough so that kinetic energy equal to $2mc^2$ can be converted into the rest energy of the two particles. In the frame of the laboratory, the total kinetic energy cannot be converted into rest energy because of conservation of momentum. However, in the zero-momentum reference frame in which the two initial protons are moving toward each other with equal speed u, the total kinetic energy can be converted into rest energy. (a) Find the speed of each proton u so that the total kinetic energy in the zero-momentum frame is $2mc^2$. (b) Transform to the laboratory's frame in which one proton is at rest, and find the speed u' of the other proton. (c) Show that the kinetic energy of the moving proton in the laboratory's frame is $K_L = 6mc^2$.

41 ••• **iSOLVE** A particle of mass 1 MeV/c^2 and kinetic energy 2 MeV collides with a stationary particle of mass 2 MeV/c^2. After the collision, the particles stick together. Find (a) the speed of the first particle before the collision, (b) the total energy of the first particle before the collision, (c) the initial total momentum of the system, (d) the total kinetic energy after the collision, and (e) the mass of the system after the collision.

General Relativity

42 •• **SSM** Light traveling in the direction of increasing gravitational potential undergoes a frequency redshift. Calculate the shift in wavelength if a beam of light of wavelength $\lambda = 632.8$ nm is sent up a vertical shaft of height $L = 100$ m.

43 •• Let us revisit a problem from Chapter 3: Two cannons are pointed directly toward each other, as shown in Figure 39-17. When fired, the cannonballs will follow the trajectories shown. Point P is the point where the trajectories cross each other. Ignore the effects of air resistance. Using the principle of equivalence, show that if the cannons are fired simultaneously, the cannonballs will hit each other at point P.

FIGURE 39-17 Problem 43

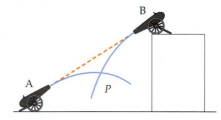

44 ••• A horizontal turntable rotates with angular speed ω. There is a clock at the center of the turntable and one at a distance r from the center. In an inertial reference frame, the clock at distance r is moving with speed $u = r\omega$. (a) Show that from time dilation according to special relativity, time intervals Δt_0 for the clock at rest and Δt_r for the moving clock are related by

$$\frac{\Delta t_r - \Delta t_0}{\Delta t_0} = -\frac{r^2\omega^2}{2c^2}, \quad \text{if } r\omega \ll c$$

(b) In a reference frame rotating with the table, both clocks are at rest. Show that the clock at distance r experiences a pseudo-force $F_r = mr\omega^2$ in this accelerated frame and that this is equivalent to a difference in gravitational potential between r and the origin of $\phi_r - \phi_0 = -\frac{1}{2}r^2\omega^2$. Use this potential difference given in Part (b) to show that in this frame the difference in time intervals is the same as in the inertial frame.

General Problems

45 • **ISOLVE✓** How fast must a muon travel so that its mean lifetime is 46 μs if its mean lifetime at rest is 2 μs?

46 • **ISOLVE** A distant galaxy is moving away from the earth with a speed that results in each wavelength received on the earth being shifted so that $\lambda' = 2\lambda_0$. Find the speed of the galaxy relative to the earth.

47 •• **SSM** Frames S and S' are moving relative to each other along the x and x' axes. Observers in the two frames set their clocks to $t = 0$ when the origins coincide. In frame S, event 1 occurs at $x_1 = 1.0$ $c\cdot$y and $t_1 = 1$ y and event 2 occurs at $x_2 = 2.0$ $c\cdot$y and $t_2 = 0.5$ y. These events occur simultaneously in frame S'. (a) Find the magnitude and direction of the velocity of S' relative to S. (b) At what time do both these events occur as measured in S'?

48 •• An interstellar spaceship travels from the earth to a distant star system 12 light-years away (as measured in the earth's frame). The trip takes 15 y as measured on the spaceship. (a) What is the speed of the spaceship relative to the earth? (b) When the ship arrives, it sends a signal to the earth. How long after the ship leaves the earth will it be before the earth receives the signal?

49 •• The neutral pion π^0 has a mass of 135 MeV/c^2. This particle can be created in a proton–proton collision:

$$p + p \rightarrow p + p + \pi^0$$

Determine the threshold kinetic energy for the creation of a π^0 in a collision of a moving proton and a stationary proton. (See Problem 40.)

50 •• A rocket with a proper length of 1000 m moves in the $+x$ direction at 0.6c with respect to an observer on the ground. An astronaut stands at the rear of the rocket and fires a bullet toward the front of the rocket at 0.8c relative to the rocket. How long does it take the bullet to reach the front of the rocket (a) as measured in the frame of the rocket, (b) as measured in the frame of the ground, and (c) as measured in the frame of the bullet?

51 ••• **SSM** In a simple thought experiment, Einstein showed that there is mass associated with electromagnetic radiation. Consider a box of length L and mass M resting on a frictionless surface. At the left wall of the box is a light source that emits radiation of energy E, which is absorbed at the right wall of the box. According to classical electromagnetic theory, this radiation carries momentum of magnitude $p = E/c$ (Equation 32-13). (a) Find the recoil velocity of the box so that momentum is conserved when the light is emitted. (Since p is small and M is large, you may use classical mechanics.) (b) When the light is absorbed at the right wall of the box the box stops, so the total momentum remains zero. If we neglect the very small velocity of the box, the time it takes for the radiation to travel across the box is $\Delta t = L/c$. Find the distance moved by the box in this time. (c) Show that if the center of

mass of the system is to remain at the same place, the radiation must carry mass $m = E/c^2$.

52 ••• Reference frame S' is moving along the x' axis at 0.6c relative to frame S. A particle that is originally at $x' = 10$ m at $t_1' = 0$ is suddenly accelerated and then moves at a constant speed of $c/3$ in the $-x'$ direction until time $t_2' = 60$ m/c, when it is suddenly brought to rest. As observed in frame S, find (a) the speed of the particle, (b) the distance and the direction that the particle traveled from t_1' to t_2', and (c) the time the particle traveled.

53 ••• In reference frame S, the acceleration of a particle is $\vec{a} = a_x\hat{i} + a_y\hat{j} + a_z\hat{k}$. Derive expressions for the acceleration components a_x', a_y', and a_z' of the particle in reference frame S' that is moving relative to S in the x direction with velocity v.

54 ••• Using the relativistic conservation of momentum and energy and the relation between energy and momentum for a photon $E = pc$, prove that a free electron (i.e., one not bound to an atomic nucleus) cannot absorb or emit a photon.

55 ••• **SSM** When a projectile particle with kinetic energy greater than the threshold kinetic energy K_{th} strikes a stationary target particle, one or more particles may be created in the inelastic collision. Show that the threshold kinetic energy of the projectile is given by

$$K_{th} = \frac{(\Sigma m_{in} + \Sigma m_{fin})(\Sigma m_{fin} - \Sigma m_{in})c^2}{2m_{target}}$$

Here Σm_{in} is the sum of the masses of the projectile and target particles, Σm_{fin} is the sum of the masses of the final particles, and m_{target} is the mass of the target particle. Use this expression to determine the threshold kinetic energy of protons incident on a stationary proton target for the production of a proton–antiproton pair; compare your result with the result of Problem 40.

56 ••• A particle of mass M decays into two identical particles of mass m, where $m = 0.3M$. Prior to the decay, the particle of mass M has an energy of $4Mc^2$ in the laboratory. The velocities of the decay products are along the direction of motion of M. Find the velocities of the decay products in the laboratory.

57 ••• A stick of proper length L_p makes an angle θ with the x axis in frame S. Show that the angle θ' made with the x' axis in frame S', which is moving along the $+x$ axis with speed v, is given by $\tan \theta' = \gamma \tan \theta$ and that the length of the stick in S' is

$$L' = L_p\left(\frac{1}{\gamma^2}\cos^2\theta + \sin^2\theta\right)^{1/2}$$

58 ••• Show that if a particle moves at an angle θ with the x axis with speed u in frame S, it moves at an angle θ' with the x' axis in S' given by

$$\tan \theta' = \frac{\sin \theta}{\gamma[\cos \theta - (v/u)]}$$

59 ••• **SSM** For the special case of a particle moving with speed u along the y axis in frame S, show that its momentum and energy in frame S', a frame that is moving along the x axis with velocity v, are related to its momentum and energy in S by the transformation equations

$$p'_x = \gamma\left(p_x - \frac{vE}{c^2}\right), \qquad p'_y = p_{y'}, \qquad p'_z = p_z$$

$$\frac{E'}{c} = \gamma\left(\frac{E}{c} - \frac{vp_x}{c}\right)$$

Compare these equations with the Lorentz transformation for x', y', z', and t'. These equations show that the quantities $p_{x'}$, $p_{y'}$, $p_{z'}$, and E/c transform in the same way as do x, y, z, and ct.

60 ••• The equation for the spherical wavefront of a light pulse that begins at the origin at time $t = 0$ is $x^2 + y^2 + z^2 - (ct)^2 = 0$. Using the Lorentz transformation, show that such a light pulse also has a spherical wavefront in frame S' by showing that $x'^2 + y'^2 + z'^2 - (ct')^2 = 0$ in S'.

61 ••• In Problem 60, you showed that the quantity $x^2 + y^2 + z^2 - (ct)^2$ has the same value (0) in both S and S'. Such a quantity is called an *invariant*. From the results of Problem 59, the quantity $p_x^2 + p_y^2 + p_z^2 - E^2/c^2$ must also be an invariant. Show that this quantity has the value $-m^2c^2$ in both the S and S' reference frames.

62 ••• **SSM** A long rod that is parallel to the x axis is in free fall with acceleration g parallel to the $-y$ axis. An observer in a rocket moving with speed v parallel to the x axis passes by and watches the rod falling. Using the Lorentz transformations, show that the observer will measure the rod to be bent into a parabolic shape. Is the parabola concave upward or concave downward?

Nuclear Physics

A NUCLEAR POWER PLANT IN GERMANY THE FISSION REACTOR CORE IS HOUSED IN A HEMISPHERICAL CONTAINMENT STRUCTURE (CENTER). TWO LARGE COOLING TOWERS ARE TO ITS LEFT.

? **How much energy is released in the fission of 1 g of ^{235}U? This calculation is the subject of Example 40-6.**

To the chemist, the atomic nucleus is essentially a point charge that contains most of the mass of the atom. It plays a negligible role in the structure of atoms and molecules.

➤ In this chapter, we will look at the nucleus from the physicist's perspective and see how the protons and neutrons that make up the nucleus have played important roles in our everyday life as well as in the history and structure of the universe. The fission of very heavy nuclei, such as uranium, is a major source of power today, while the fusion of very light nuclei is the energy source that powers the stars, including our sun, and may hold the key to our energy needs of the future.

40-1 Properties of Nuclei

The nucleus of an atom contains just two kinds of particles, protons and neutrons,[†] which have approximately the same mass (the neutron is approximately 0.2 percent more massive). The proton has a charge of $+e$ and the neutron is uncharged. The number of protons, Z, is the atomic number of the atom, which

---

† The most prevalent hydrogen nucleus contains a single proton.

also equals the number of electrons in the atom. The number of neutrons, N, is approximately equal to Z for light nuclei, and for heavier nuclei the number of neutrons is increasingly greater than Z. The total number of nucleons[†] $A = N + Z$ is called the mass number of the nucleus. A particular nuclear species is called a **nuclide.** Two or more nuclides with the same atomic number Z but with different N and A numbers are called **isotopes.** A particular nuclide is designated by its atomic symbol (H for hydrogen, He for helium, etc.), with the mass number A as a presuperscript. The lightest element, hydrogen, has three isotopes: protium, ^{1}H, whose nucleus is just a single proton; deuterium, ^{2}H, whose nucleus contains one proton and one neutron; and tritium, ^{3}H, whose nucleus contains one proton and two neutrons. Although the mass of the deuterium atom is about twice the mass of the protium atom and the mass of the tritium atom is about three times the mass of protium, these three atoms have nearly identical chemical properties because they each have one electron. On the average, there are about three stable isotopes for each element, although some atoms have only one stable isotope while others have five or six. The most common isotope of the second lightest element, helium, is ^{4}He. The ^{4}He nucleus is also known as an α particle. Another isotope of helium is ^{3}He.

Inside the nucleus, the nucleons exert a strong attractive force on their nearby neighbors. This force, called the **strong nuclear force** or the **hadronic force,** is much stronger than the electrostatic force of repulsion between the protons and is very much stronger than the gravitational forces between the nucleons. (Gravity is so weak that it can always be neglected in nuclear physics.) The strong nuclear force is roughly the same between two neutrons, two protons, or a neutron and a proton. Two protons, of course, also exert a repulsive electrostatic force on each other due to their charges, which tends to weaken the attraction between them somewhat. The strong nuclear force decreases rapidly with distance, and it is negligible when two nucleons are more than a few femtometers apart.

Size and Shape

The size and shape of the nucleus can be determined by bombarding it with high-energy particles and observing the scattering. The results depend somewhat on the kind of experiment. For example, a scattering experiment using electrons measures the charge distribution of the nucleus, whereas a scattering experiment using neutrons determines the region of influence of the strong nuclear force. A wide variety of experiments suggest that most nuclei are approximately spherical, with radii given approximately by

$$R = R_0 A^{1/3}$$

40-1

NUCLEAR RADIUS

where R_0 is approximately 1.2 fm. The fact that the radius of a spherical nucleus is proportional to $A^{1/3}$ implies that the volume of the nucleus is proportional to A. Since the mass of the nucleus is also approximately proportional to A, the densities of all nuclei are approximately the same. This is analogous to a drop of liquid, which also has constant density independent of its size. The **liquid-drop model** of the nucleus has proved quite successful in explaining nuclear behavior, especially the fission of heavy nuclei.

[†] The word *nucleon* refers to either a neutron or a proton.

N and Z Numbers

For light nuclei, the greatest stability is achieved when the numbers of protons and neutrons are approximately equal, $N \approx Z$. For heavier nuclei, instability caused by the electrostatic repulsion between the protons is minimized when there are more neutrons than protons. We can see this by looking at the N and Z numbers for the most abundant isotopes of some representative elements: for ${}^{16}_{8}O$, $N = 8$ and $Z = 8$; for ${}^{40}_{20}Ca$, $N = 20$ and $Z = 20$; for ${}^{56}_{26}Fe$, $N = 30$ and $Z = 26$; for ${}^{207}_{82}Pb$, $N = 125$ and $Z = 82$; and for ${}^{238}_{92}U$, $N = 146$ and $Z = 92$. (The atomic number Z has been included here as a presubscript of the atomic symbol for emphasis. It is not actually needed because the atomic number is implied by the atomic symbol.)

Figure 40-1 shows a plot of N versus Z for the known stable nuclei. The curve follows the straight line $N = Z$ for small values of N and Z. We can understand this tendency for N and Z to be equal by considering the total energy of A particles in a one-dimensional box. For $A = 8$, Figure 40-2 shows the energy levels for eight neutrons and for four neutrons and four protons. Because of the exclusion principle, only two identical particles (with opposite spins) can be in the same space state. Since protons and neutrons are not identical, we can put two each in a state, as shown in Figure 40-2b. Thus, the total energy for four protons and four neutrons is less than the total energy for eight neutrons (or eight protons), as shown in Figure 40-2a. When the Coulomb energy of repulsion, which is proportional to Z^2, is included, this result changes somewhat. For large values of A and Z, the total energy may be increased less by adding two neutrons than by adding one neutron and one proton because of the electrostatic repulsion involved in the latter case. This explains why $N > Z$ for the larger values of A (for the heavier nuclei).

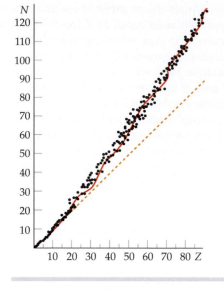

FIGURE 40-1 Plot of number of neutrons N versus number of protons Z for the stable nuclides. The dashed line is $N = Z$.

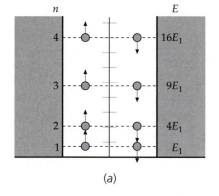

(a)

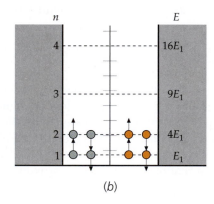

(b)

FIGURE 40-2 (a) Eight neutrons in a one-dimensional box. In accordance with the exclusion principle, only two neutrons (with opposite spins) can be in a given energy level. (b) Four neutrons and four protons in a one-dimensional box. Because protons and neutrons are not identical particles, two of each can be in the same energy level. The total energy is much less for this case than for that in Figure 40-2a.

EXERCISE (a) Calculate the total energy of the eight neutrons in the one-dimensional box shown in Figure 40-2a. (b) Calculate the total energy of the four neutrons and four protons in the one-dimensional box shown in Figure 40-2b. (*Answer* (a) $60E_1$ (b) $20E_1$)

Mass and Binding Energy

The mass of a nucleus is less than the mass of its parts by E_b/c^2, where E_b is the binding energy and c is the speed of light. When two or more nucleons fuse together to form a nucleus, the total mass decreases and energy is given off. Conversely, to break up a nucleus into its parts, energy must be put into the system to produce the increase in mass.

Atomic masses and nuclear masses are often given in unified mass units (u), defined as one-twelfth the mass of the neutral ${}^{12}C$ atom. The rest energy of one unified mass unit is

$$(1 \text{ u})c^2 = 931.5 \text{ MeV} \qquad \qquad 40\text{-}2$$

Consider ^{4}He, for example, which consists of two protons and two neutrons. The mass of an atom can be accurately measured in a mass spectrometer. The mass of the ^{4}He atom is 4.002 603 u. This includes the masses of the two electrons in the atom. The mass of the ^{1}H atom is 1.007 825 u, and the mass of the neutron is 1.008 665 u. The sum of the masses of two ^{1}H atoms plus two neutrons is 2(1.007 825 u) + 2(1.008 665 u) = 4.032 980 u, which is greater than the mass of the ^{4}He atom by 0.030 377 u.[†] We can find the binding energy of the ^{4}He nucleus from this mass difference of 0.030 377 u by using the mass conversion factor $(1\ u)c^2 = 931.5$ MeV from Equation 40-2. Then

$$(0.030\ 377\ u)c^2 = (0.030\ 377\ u)c^2 \times \frac{931.5\ \text{MeV}/c^2}{1\ u}$$
$$= 28.30\ \text{MeV}$$

The total binding energy of ^{4}He is thus 28.30 MeV. In general, the binding energy of a nucleus of an atom of atomic mass M_A containing Z protons and N neutrons is found by calculating the difference between the mass of the parts and the mass of the nucleus and then multiplying by c^2:

$$E_b = (ZM_H + Nm_n - M_A)c^2 \qquad\qquad \text{40-3}$$

TOTAL NUCLEAR BINDING ENERGY

where M_H is the mass of the ^{1}H atom and m_n is the mass of the neutron. (Note that the mass of the Z electrons in the term ZM_H is canceled by the mass of the Z electrons in the term M_A.[‡]) The atomic masses of the neutron and of some selected isotopes are listed in Table 40-1.

[†] Note that by using the masses of two ^{1}H atoms rather than two protons, the masses of the electrons in the atom are accounted for. We do this because it is atomic masses, not nuclear masses, that are measured directly and listed in mass tables.

[‡] The mass associated with the binding energies of the electrons are not accounted for in this calculation.

TABLE 40-1

Atomic Masses of the Neutron and Selected Isotopes[†]

Element	Symbol	Z	Atomic mass, u	Element	Symbol	Z	Atomic mass, u
Neutron	n	0	1.008 665	Oxygen	^{16}O	8	15.994 915
Hydrogen	^{1}H	1	1.007 825	Sodium	^{23}Na	11	22.989 770
Deuterium	^{2}H or D	1	2.013 553	Potassium	^{39}K	19	38.963 707
Tritium	^{3}H or T	1	3.016 049	Iron	^{56}Fe	26	55.934 942
Helium	^{3}He	2	3.016 029	Copper	^{63}Cu	29	62.929 601
	^{4}He	2	4.002 603	Silver	^{107}Ag	47	106.905 093
Lithium	^{6}Li	3	6.015 122	Gold	^{197}Au	79	196.966 552
	^{7}Li	3	7.016 004	Lead	^{208}Pb	82	207.976 636
Boron	^{10}B	5	10.012 937	Polonium	^{212}Po	84	211.988 852
Carbon	^{12}C	6	12.000 000	Radon	^{222}Rn	86	222.017 571
	^{13}C	6	13.003 354	Radium	^{226}Ra	88	226.025 403
	^{14}C	6	14.003 242	Uranium	^{238}U	92	238.050 783
Nitrogen	^{13}N	7	13.005 739	Plutonium	^{242}Pu	94	242.058 737
	^{14}N	7	14.003 074				

[†]Mass values obtained at <http://physics.nist.gov/PhysRefData/Compositions/index.html>.

BINDING ENERGY OF THE LAST NEUTRON   **EXAMPLE   40-1**

Find the binding energy of the last neutron in ^{4}He.

PICTURE THE PROBLEM The binding energy is c^2 times the difference in mass of ^{3}He plus a neutron and ^{4}He. We find these masses from Table 40-1 and convert to energy using Equation 40-3.

1. Add the mass of the neutron to that of ^{3}He:

$$m_{^3\text{He}} + m_n = 3.016\ 029\ \text{u} + 1.008\ 665\ \text{u}$$

$$= 4.024\ 694\ \text{u}$$

2. Subtract the mass of ^{4}He from the result:

$$\Delta m = (m_{^3\text{He}} + m_n) - m_{^4\text{He}}$$

$$= 4.024\ 694\ \text{u} - 4.002\ 603\ \text{u} = 0.022\ 091\ \text{u}$$

3. Multiply this mass difference by c^2 and convert to MeV:

$$E_b = (\Delta m)c^2$$

$$= (0.022\ 091\ \text{u})c^2 \times \frac{931.5\ \text{MeV}/c^2}{1\ \text{u}}$$

$$= \boxed{20.58\ \text{MeV}}$$

Figure 40-3 shows the binding energy per nucleon E_b/A versus A. The mean value is approximately 8.3 MeV. The flatness of this curve for $A > 50$ shows that E_b is approximately proportional to A. This indicates that there is saturation of nuclear forces in the nucleus as would be the case if each nucleon were attracted only to its nearest neighbors. Such a situation also leads to a constant nuclear density consistent with the measurements of the radius. If, for example, there were no saturation and each nucleon bonded to each other nucleon, there would be $A - 1$ bonds for each nucleon and a total of $A(A - 1)$ bonds altogether. The total binding energy, which is a measure of the energy needed to break all these bonds, would then be proportional to $A(A - 1)$, and E_b/A would not be approximately constant. The steep rise in the curve for low A is due to the increase in the number of nearest neighbors and therefore to the increased number of bonds per nucleon. The gradual decrease at high A is due to the Coulomb repulsion of the protons, which increases as Z^2 and decreases the binding energy. Eventually, for very large A, this Coulomb repulsion becomes so great that a nucleus with A greater than approximately 300 is unstable and undergoes spontaneous fission.

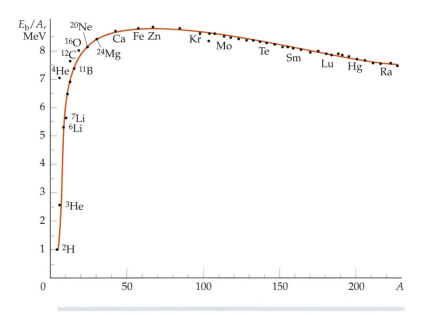

FIGURE 40-3 The binding energy per nucleon versus the mass number A. For nuclei with values of A greater than 50, the curve is approximately constant, indicating that the total binding energy is approximately proportional to A.

40-2 Radioactivity

Many nuclei are radioactive; that is, they decay into other nuclei by the emission of particles such as photons, electrons, neutrons, or α particles. The terms α decay, β decay, and γ decay were used before it was discovered that α particles are

^{4}He nuclei, β particles are either electrons (β^-) or positrons[†] (β^+), and γ-rays are photons. The rate of decay is not constant over time, but decreases exponentially. *This exponential time dependence is characteristic of all radioactivity and indicates that radioactive decay is a statistical process.* Because each nucleus is well shielded from others by the atomic electrons, pressure and temperature changes have little or no effect on the rate of radioactive decay or other nuclear properties.

Let N be the number of radioactive nuclei at some time t. If the decay of an individual nucleus is a random event, we expect the number of nuclei that decay in some time interval dt to be proportional to N and to dt. Because of these decays, the number N will decrease. The change in N is given by

$$dN = -\lambda N\, dt \qquad\qquad 40\text{-}4$$

where λ is a constant of proportionality called the **decay constant.** The rate of change of N, dN/dt, is proportional to N. This is characteristic of exponential decay. To solve Equation 40-4 for N, we first divide each side by N, thus separating the variables N and t:

$$\frac{dN}{N} = -\lambda\, dt$$

Integrating, we obtain

$$\int_{N_0}^{N'} \frac{dN}{N} = -\lambda \int_0^{t'} dt$$

or

$$\ln \frac{N'}{N_0} = -\lambda t' \qquad\qquad 40\text{-}5$$

where N' is the number of nuclei that remain at time t'. For convenience, we drop the primes from N' and t'. This introduces no ambiguity because the parameters N and t have been integrated out of the equation. Taking the exponential of each side, we obtain

$$\frac{N}{N_0} = e^{-\lambda t}$$

or

$$N = N_0 e^{-\lambda t} \qquad\qquad 40\text{-}6$$

The number of radioactive decays per second is called the decay rate R:

$$R = -\frac{dN}{dt} = \lambda N = \lambda N_0 e^{-\lambda t} = R_0 e^{-\lambda t} \qquad\qquad 40\text{-}7$$

DECAY RATE

where

$$R_0 = \lambda N_0 \qquad\qquad 40\text{-}8$$

---

† The positron is identical to an electron except it has a charge of $+e$.

is the rate of decay at time $t = 0$. The decay rate R is the quantity that is determined experimentally.

The average or **mean lifetime** τ is equal to the reciprocal of the decay constant (see Problem 42):

$$\tau = \frac{1}{\lambda} \qquad \qquad \text{40-9}$$

The mean lifetime is analogous to the time constant in the exponential decrease in the charge on a capacitor in an RC circuit that we discussed in Section 25-6. After a time equal to the mean lifetime, the number of radioactive nuclei and the decay rate are each equal to $e^{-1} = 37$ percent of their original values. The **half-life** $t_{1/2}$ is defined as the time it takes for the number of nuclei and the decay rate to decrease by half. Setting $t = t_{1/2}$ and $N = N_0/2$ in Equation 40-6 gives

$$\frac{N_0}{2} = N_0 e^{-\lambda t_{1/2}} \qquad \qquad \text{40-10}$$

or

$$e^{+\lambda t_{1/2}} = 2$$

Solving for $t_{1/2}$ gives

$$t_{1/2} = \frac{\ln 2}{\lambda} = \frac{0.693}{\lambda} = 0.693\tau \qquad \qquad \text{40-11}$$

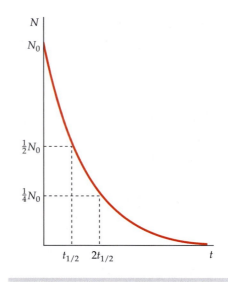

Figure 40-4 shows a plot of N versus t. If we multiply the numbers on the N axis by λ, this graph becomes a plot of R versus t. After each time interval of one half-life, both the number of nuclei left and the decay rate have decreased to half of their previous values. For example, if the decay rate is R_0 initially, it will be $\frac{1}{2} R_0$ after one half-life, $(\frac{1}{2})(\frac{1}{2})R_0$ after two half-lives, and so forth. After n half-lives, the decay rate will be

$$R = (\tfrac{1}{2})^n R_0 \qquad \qquad \text{40-12}$$

FIGURE 40-4 Exponential radioactive decay. After each half-life $t_{1/2}$, the number of nuclei remaining has decreased by one-half. The decay rate $R = \lambda N$ has the same time dependence, as does N.

The half-lives of radioactive nuclei vary from very small times (less than 1 μs) to very large times (up to 10^{16} y).

COUNTING RATE FOR RADIOACTIVE DECAY   **E X A M P L E   4 0 - 2**

A radioactive source has a half-life of 1 min. At time $t = 0$, the radioactive source is placed near a detector, and the counting rate (the number of decay particles detected per unit time) is observed to be 2000 counts/s. Find the counting rate at times $t = 1$ min, $t = 2$ min, $t = 3$ min, and $t = 10$ min.

PICTURE THE PROBLEM The counting rate decreases by a factor of 2 each minute.

1. Since the half-life is 1 min, the counting rate will be half as great at $t = 1$ min as at $t = 0$:
$$r_1 = \tfrac{1}{2}r_0 = \tfrac{1}{2}(2000 \text{ counts/s})$$
$$= \boxed{1000 \text{ counts/s at 1 min}}$$

2. At $t = 2$ min, the rate is half that at 1 min. It decreases by one-half each minute:

$$r_2 = \tfrac{1}{2}r_1 = \tfrac{1}{2}(1000 \text{ counts/s})$$

$$= \boxed{500 \text{ counts/s at 2 min}}$$

$$r_3 = \tfrac{1}{2}r_2 = \tfrac{1}{2}(500 \text{ counts/s})$$

$$= \boxed{250 \text{ counts/s at 3 min}}$$

3. At $t = 10$ min, the rate will be $(\tfrac{1}{2})^{10}$ times the initial rate:

$$r_{10} = (\tfrac{1}{2})^{10}r_0 = (\tfrac{1}{2})^{10}\,(2000 \text{ counts/s})$$

$$= 1.95 \text{ counts/s}$$

$$\approx \boxed{2 \text{ counts/s at 10 min}}$$

DETECTION-EFFICIENCY CONSIDERATIONS   **EXAMPLE 40-3**

If the detection efficiency in Example 40-2 is 20 percent, (*a*) how many radioactive nuclei are there at time $t = 0$ and (*b*) at time $t = 1$ min? (*c*) How many nuclei decay in the first minute?

PICTURE THE PROBLEM The detection efficiency depends on the probability that a radioactive decay particle will enter the detector and the probability that upon entering the detector it will produce a count. If the efficiency is 20 percent, the decay rate must be five times the counting rate.

(*a*) 1. The number of radioactive nuclei is related to the decay rate R, and the decay constant λ:

$$R = \lambda N$$

2. The decay constant is related to the half-life:

$$\lambda = \frac{0.693}{t_{1/2}} = \frac{0.693}{1 \text{ min}} = 0.693 \text{ min}^{-1}$$

3. Because the detection efficiency is 20 percent, the decay rate is five times the counting rate. Calculate the initial decay rate:

$$R_0 = 5 \times 2000 \text{ counts/s}$$

$$= 10{,}000 \text{ s}^{-1}$$

4. Substitute to calculate the initial number of radioactive nuclei N_0 at $t = 0$:

$$N_0 = \frac{R_0}{\lambda} = \frac{10{,}000 \text{ s}^{-1}}{0.693 \text{ min}^{-1}} \times \frac{60 \text{ s}}{1 \text{ min}}$$

$$= \boxed{8.66 \times 10^5}$$

(*b*) At time $t = 1$ min $= t_{1/2}$, there are half as many radioactive nuclei as at $t = 0$:

$$N_1 = \tfrac{1}{2}(8.66 \times 10^5) = \boxed{4.33 \times 10^5}$$

(*c*) The number of nuclei that decay in the first minute is $N_0 - N_1$:

$$\Delta N = N_0 - N_1$$

$$= 8.66 \times 10^5 - 4.33 \times 10^5$$

$$= \boxed{4.33 \times 10^5}$$

The SI unit of radioactive decay is the **becquerel** (Bq), which is defined as one decay per second:

$$1 \text{ Bq} = 1 \text{ decay/s} \qquad\qquad\qquad 40\text{-}13$$

A historical unit that applies to all types of radioactivity is the **curie** (Ci), which is defined as

$$1 \text{ Ci} = 3.7 \times 10^{10} \text{ decays/s} = 3.7 \times 10^{10} \text{ Bq} \qquad \text{40-14}$$

The curie is the rate at which radiation is emitted by 1 g of radium. Since this is a very large unit, the millicurie (mCi) or microcurie (μCi) are often used.

Beta Decay

Beta decay occurs in nuclei that have too many neutrons or too few neutrons for stability. In β decay, A remains the same while Z either increases by 1 (β^- decay) or decreases by 1 (β^+ decay).

The simplest example of β decay is the decay of the free neutron into a proton plus an electron. (The half-life of a free neutron is about 10.8 min.) The energy of decay is 0.782 MeV, which is the difference between the rest energy of the neutron and the rest energy of the proton plus electron. More generally, in β^- decay, a nucleus of mass number A and atomic number Z decays into a nucleus, referred to as the **daughter nucleus,** of mass number A and atomic number $Z' = Z + 1$ with the emission of an electron. If the decay energy were shared by only the daughter nucleus and the emitted electron, the energy of the electron would be uniquely determined by the conservation of energy and momentum. Experimentally, however, the energies of the electrons emitted in the β^- decay of a nucleus are observed to vary from zero to the maximum energy available. A typical energy spectrum for these electrons is shown in Figure 40-5.

To explain the apparent nonconservation of energy in β decay, Wolfgang Pauli in 1930 suggested that a third particle, which he called the **neutrino,** is also emitted. Because the measured maximum energy of the emitted electrons is equal to the total available for the decay, the rest energy and therefore the mass of the neutrino was assumed to be zero. (It is now believed that the mass of the neutrino is very small but not zero.) In 1948, measurements of the momenta of the emitted electron and the recoiling nucleus showed that the neutrino was also needed for the conservation of linear momentum in β decay. The neutrino was first observed experimentally in 1957. It is now known that there are at least three kinds of neutrinos, one (v_e) associated with electrons, one (v_μ) associated with muons, and one (v_τ), which was first observed at Fermi National Laboratory in 2000, associated with the tau particle, τ. Moreover, each neutrino has an antiparticle, written $\bar{v}_e$, $\bar{v}_\mu$, and $\bar{v}_\tau$. It is the electron antineutrino that is emitted in the decay of a neutron, which is written[†]

$$n \rightarrow p + \beta^- + \bar{v}_e \qquad \text{40-15}$$

In β^+ decay, a proton changes into a neutron with the emission of a positron (and a neutrino). A free proton cannot decay by positron emission because of conservation of energy (the mass of the neutron plus the positron is greater than that of the proton); however, because of binding-energy effects, a proton inside a nucleus can decay. A typical β^+ decay is

$$^{13}_{7}\text{N} \rightarrow {}^{13}_{6}\text{C} + \beta^+ + v_e \qquad \text{40-16}$$

The electrons or the positrons emitted in β decay do not exist inside the nucleus. They are created in the process of decay, just as photons are created when an atom makes a transition from a higher energy state to a lower energy state.

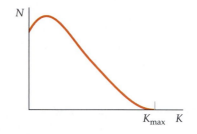

FIGURE 40-5 Number of electrons emitted in β^- decay versus kinetic energy. The fact that all the electrons do not have the same energy K_{max} suggests that another particle, one that shares the energy available for decay, is emitted.

---

† This reaction is also written $n \rightarrow p + e^- + \bar{v}_e$.

An important example of β decay is that of ^{14}C, which is used in radioactive carbon dating:

$$^{14}C \rightarrow {}^{14}N + \beta^- + \bar{v}_e \qquad\qquad 40\text{-}17$$

The half-life for this decay is 5730 y. The radioactive isotope ^{14}C is produced in the upper atmosphere in nuclear reactions caused by cosmic rays. The chemical behavior of carbon atoms with ^{14}C nuclei is the same as those with ordinary ^{12}C nuclei. For example, atoms with these nuclei combine with oxygen to form CO_2 molecules. Since living organisms continually exchange CO_2 with the atmosphere, the ratio of ^{14}C to ^{12}C in a living organism is the same as the equilibrium ratio in the atmosphere, which is about 1.3×10^{-12}. After an organism dies, it no longer absorbs ^{14}C from the atmosphere, so the ratio of ^{14}C to ^{12}C continually decreases due to the radioactive decay of ^{14}C. The number of ^{14}C decays per minute per gram of carbon in a living organism can be calculated from the known half-life of ^{14}C and the number of ^{14}C nuclei in a gram of carbon. The result is that there are approximately 15.0 decays per minute per gram of carbon in a living organism. Using this result and the measured number of decays per minute per gram of carbon in a nonliving sample of bone, wood, or other object containing carbon, we can determine the age of the sample. For example, if the measured rate were 7.5 decays per minute per gram, the age of the sample would be one half-life = 5730 years.

How Old Is the Artifact?　　　　　　　　　**EXAMPLE　40-4**　　**Put It in Context**

You have a summer job working in an archeological research lab. Your supervisor calls to tell you that they found a new bone at their current dig and asks you to determine the age of the bone from a sample that is in the mail. When the bone arrives, you take a sample that contains 200 grams of carbon and you find a beta decay rate of 400 decays/min.

PICTURE THE PROBLEM We first obtain a rough estimate of the age of the bone. If the bone were from a living organism, we would expect the decay rate to be [(15 decays/min)/g](200 g) = 3000 decays/min. Since 400/3000 is roughly 1/8 (actually 1/7.5), the sample must be approximately three half-lives old, which is about 3(5730 y) = 17,190 y. To find the age of the bone more accurately, we need to determine the number of half-lives of the bone. We can do this by using the equality $R_n = (1/2)^n R_0$ where R_n is the current decay rate, R_0 is the initial decay rate, and n is the number of half-lives. We can determine the initial decay rate by multiplying the decay rate per gram times the mass of the carbon of the sample.

1. Write the decay rate after n half-lives in terms of the initial decay rate:

$$R_n = \left(\tfrac{1}{2}\right)^n R_0$$

2. Calculate the initial decay rate (the decay for 200 g of carbon when the organism stopped breathing):

$$R_0 = [(15 \text{ decays/min})/g](200 \text{ g})$$
$$= 3000 \text{ decays/min}$$

3. Substitute the values for R_0 and R_n into the step 1 equation and solve for n:

$$R_n = \left(\tfrac{1}{2}\right)^n R_0$$

$$400 \frac{\text{decays}}{\text{min}} = \left(\tfrac{1}{2}\right)^n 3000 \frac{\text{decays}}{\text{min}}$$

$$\left(\tfrac{1}{2}\right)^n = \frac{400}{3000}$$

$$2^n = \frac{3000}{400} = 7.5$$

4. We solve for n by taking the logarithm of each side:

$$n \ln 2 = \ln 7.5$$

$$n = \frac{\ln 7.5}{\ln 2} = 2.91$$

5. The age of the bone is $nt_{1/2}$:

$$t = nt_{1/2} = 2.91(5730 \text{ y}) = \boxed{1.67 \times 10^4 \text{ y}}$$

EXERCISE Picture the Problem of Example 40-4 states "Since 400/3000 is roughly 1/8 (actually 1/7.5), the sample must be approximately three half-lives old, ..." Explain. [*Answer* It is because $\frac{1}{7.5} \approx \frac{1}{8} = (\frac{1}{2})^3$, so $n = 3$.]

Gamma Decay

In γ decay, a nucleus in an excited state decays to a lower-energy state by the emission of a photon. This is the nuclear counterpart of spontaneous emission of photons by atoms and molecules. Unlike β decay or α decay, neither the mass number A nor the atomic number Z change during γ decay. Since the spacing of the nuclear energy levels is of the order of 1 MeV (as compared with spacing of the order of 1 eV in atoms), the wavelengths of the emitted photons are of the order of 1 pm (1 pm = 10^{-12} m):

$$\lambda = \frac{hc}{E} \approx \frac{1240 \text{ eV·nm}}{1 \text{ MeV}} = 0.00124 \text{ nm} = 1.24 \text{ pm}$$

The mean lifetime for γ decay is often very short. Usually it is observed only because it follows either α decay or β decay. For example, if a radioactive parent nucleus decays by β decay to an excited state of the daughter nucleus, the daughter nucleus then decays to its ground state by γ emission. Direct measurements of mean lifetimes as short as approximately 10^{-11} s are possible. Measurements of mean lifetimes shorter than 10^{-11} s are difficult, but they can sometimes be made by indirect methods.

A few γ emitters have very long lifetimes, of the order of hours. Nuclear energy states that have such long lifetimes are called **metastable states.**

Alpha Decay

All very heavy nuclei ($Z > 83$) are theoretically unstable via α decay because the mass of the original radioactive nucleus is greater than the sum of the masses of the decay products—an α particle and the daughter nucleus. Consider the decay of ^{232}Th ($Z = 90$) into ^{228}Ra ($Z = 88$) plus an α particle. This is written as

$$^{232}\text{Th} \rightarrow {}^{228}\text{Ra} + \alpha = {}^{228}\text{Ra} + {}^4\text{He} \qquad \qquad 40\text{-}18$$

The mass of the ^{232}Th atom is 232.038 050 u. The mass of the daughter atom ^{228}Ra is 228.031 064 u. Adding 4.002 603 u to this for the mass of ^{4}He, we get 232.033 667 u for the total mass of the decay products. This is less than the mass of ^{232}Th by 0.004 382 u, which multiplied by 931.5 MeV/c^2 gives 4.08 MeV/c^2 for the excess mass of ^{232}Th over that of the decay products. The isotope ^{232}Th is therefore theoretically unstable to α decay. This decay does in fact occur in nature with the emission of an α particle of kinetic energy 4.08 MeV. (The kinetic energy of the α particle is actually somewhat less than 4.08 MeV because some of the decay energy is shared by the recoiling ^{228}Ra nucleus.)

When a nucleus emits an α particle, both N and Z decrease by 2 and A decreases by 4. The daughter of a radioactive nucleus is often itself radioactive and decays by either α decay or β decay or both. If the original nucleus has a mass number A that is 4 times an integer, the daughter nucleus and all those in the chain will also have mass numbers equal to 4 times an integer. Similarly, if the mass number of the original nucleus is $4n + 1$, where n is an integer, all the nuclei in the decay chain will have mass numbers given by $4n + 1$, with n decreasing by one at each decay. We can see, therefore, that there are four possible α-decay chains, depending on whether A equals $4n$, $4n + 1$, $4n + 2$, or $4n + 3$, where n is an integer. All but one of these decay chains are found on the earth. The $4n + 1$ series is not found because its longest-lived member (other than the stable end product ^{209}Bi) is ^{237}Np, which has a half-life of only 2×10^6 y. Because this is much less than the age of the earth, this series has disappeared.

Figure 40-6 shows the thorium series, for which $A = 4n$. It begins with an α decay from ^{232}Th to ^{228}Ra. The daughter nuclide of an α decay is on the left or neutron-rich side of the stability curve (the dashed line in the figure), so it often decays by β^- decay. In the thorium series, ^{228}Ra decays by β^- decay to ^{228}Ac, which in turn decays by β^- decay to ^{228}Th. There are then four α decays to ^{212}Pb, which decays by β^- decay to ^{212}Bi. The series branches at ^{212}Bi, which decays either by α decay to ^{208}Tl or by β^- decay to ^{212}Po. The branches meet at the stable lead isotope ^{208}Pb.

The energies of α particles from natural radioactive sources range from approximately 4 MeV to 7 MeV, and the half-lives of the sources range from approximately 10^{-5} s to 10^{10} y. In general, the smaller the energy of the emitted α particle, the longer the half-life. As we discussed in Section 35-4, the enormous variation in half-lives was explained by George Gamow in 1928. He considered α decay to be a process in which an α particle is first formed inside a nucleus and then tunnels through the Coulomb barrier (Figure 40-7). A slight increase in the energy of the α particle reduces the relative height $U - E$ of the barrier and also the thickness. Because the probability of penetration is so sensitive to the relative height and thickness of the barrier, a small increase in E leads to a large increase in the probability of barrier penetration and therefore to a shorter lifetime. Gamow was able to derive an expression for the half-life as a function of E that is in excellent agreement with experimental results.

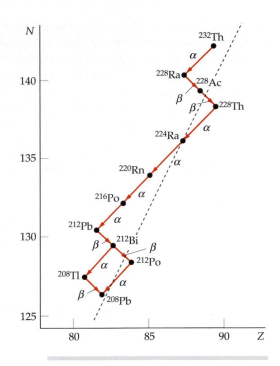

FIGURE 40-6 The thorium $(4n)$ α decay series. The dashed line is the curve of stability.

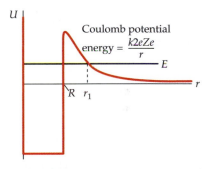

FIGURE 40-7 A model of the potential energy for an α particle and a nucleus. The strong attractive nuclear force that exists for values of r less than the nuclear radius R is indicated by the potential well. Outside the nucleus, the nuclear force is negligible, and the potential energy is given by Coulomb's law $U = +k2eZe/r$, where Ze is the nuclear charge and $2e$ is the charge of the α particle. The energy E is the kinetic energy of the α particle when it is far away from the nucleus. A small increase in E reduces the relative height $U - E$ of the barrier and also reduces its thickness, leading to a much greater chance of penetration. An increase in the energy of the emitted α particles by a factor of 2 results in a reduction of the half-life by a factor of more than 10^{20}.

40-3 Nuclear Reactions

Information about nuclei is typically obtained by bombarding the nuclei with various particles and observing the results. Although the first experiments of this type were limited by the need to use naturally occurring radiation, they produced many important discoveries. In 1932, J. D. Cockcroft and E. T. S. Walton succeeded in producing the reaction

$$p + {}^{7}\text{Li} \rightarrow {}^{8}\text{Be} \rightarrow {}^{4}\text{He} + {}^{4}\text{He}$$

using artifically accelerated protons. At about the same time, the Van de Graaff electrostatic generator was built (by R. Van de Graaff in 1931) as was the first cyclotron (by E. O. Lawrence and M. S. Livingston in 1932). Since then, enormous advances in the technology for accelerating and detecting particles have been made, and many nuclear reactions have been studied.

When a particle is incident on a nucleus, several different things can happen. The incident particle may be scattered elastically or inelastically, or the incident particle may be absorbed by the nucleus, and another particle or particles may be emitted. In inelastic scattering, the nucleus is left in an excited state and subsequently decays by emitting photons (or other particles).

The amount of energy released or absorbed in a reaction (in the center of mass reference frame) is called the **Q value** of the reaction. The Q value equals c^2 times this mass difference. When energy is released by a nuclear reaction, the reaction is said to be an **exothermic reaction.** In an exothermic reaction, the total mass of the incoming particles is greater than the total mass of the outgoing particles, and the Q value is positive. If the total mass of the incoming particles is less than that of the outgoing particles, energy is required for the reaction to take place, and the reaction is said to be an **endothermic reaction.** The Q value of an endothermic reaction is negative. In general, if Δm is the increase in mass, the Q value is

$$Q = -(\Delta m)c^2 \qquad \qquad \text{40-19}$$

Q VALUE

An endothermic reaction cannot take place below a specific threshold energy. In the laboratory reference frame in which stationary particles are bombarded by incoming particles, the threshold energy is somewhat greater than $|Q|$ because the outgoing particles must have some kinetic energy to conserve momentum.

A measure of the effective size of a nucleus for a particular nuclear reaction is the **cross section** σ. If I is the number of incident particles per unit time per unit area (the incident intensity) and R is the number of reactions per unit time per nucleus, the cross section is

$$\sigma = \frac{R}{I} \qquad \qquad \text{40-20}$$

The cross section σ has the dimensions of area. Since nuclear cross sections are of the order of the square of the nuclear radius, a convenient unit for them is the **barn,** which is defined as

$$1 \text{ barn} = 10^{-28} \text{ m}^2 \qquad \qquad \text{40-21}$$

The cross section for a particular reaction is a function of energy. For an endothermic reaction, it is zero for energies below the threshold energy.

EXOTHERMIC OR ENDOTHERMIC? **EXAMPLE 40-5**

Find the Q value of the reaction p + ^{7}Li → ^{4}He + ^{4}He and state whether the reaction is exothermic or endothermic.

PICTURE THE PROBLEM We find the masses of the atoms from Table 40-1 and calculate the difference in the total mass of the outgoing particles and the incoming particles. The Q value is $-(\Delta m)c^2$. If we use the mass of hydrogen rather than the mass of the proton, there will be four electrons on each side of the reaction, so the electron masses will cancel.

1. Find the mass of each atom from Table 40-1:

^{1}H   1.007 825 u

^{7}Li   7.016 004 u

^{4}He   4.002 603 u

2. Calculate the initial mass m_i of the incoming particles: $m_i = 1.007\ 825\ \text{u} + 7.016\ 004\ \text{u} = 8.023\ 829\ \text{u}$

3. Calculate the final mass m_f: $m_f = 2(4.002\ 603\ \text{u}) = 8.005\ 206\ \text{u}$

4. Calculate the increase in mass: $\Delta m = m_f - m_i = 8.005\ 206\ \text{u} - 8.023\ 829\ \text{u}$

$$= -0.018\ 623\ \text{u}$$

5. Calculate the Q value:

$$Q = -(\Delta m)c^2 = (+0.018\,623\text{ u})c^2 \times \left[931.5\text{ MeV}/(\text{u·}c^2)\right]$$

$$= \boxed{17.35\text{ MeV}}$$

$\boxed{Q \text{ is positive, so the reaction is exothermic.}}$

REMARKS   Since the initial mass is greater than the final mass, the initial energy is greater than the final energy and the reaction is exothermic, yielding 17.35 MeV.

Reactions With Neutrons

Nuclear reactions that involve neutrons are important for understanding nuclear reactors. The most likely reaction between a nucleus and a neutron having an energy of more than about 1 MeV is scattering. However, even if the scattering is elastic, the neutron loses some energy to the nucleus because the nucleus recoils. If a neutron is scattered many times in a material, its energy decreases until the neutron is of the order of the energy of thermal motion kT, where k is Boltzmann's constant and T is the absolute temperature. (At ordinary room temperatures, kT is approximately 0.025 eV.) The neutron is then equally likely to gain or lose energy from a nucleus when it is elastically scattered. A neutron with energy of the order of kT is called a **thermal neutron.**

At low energies, a neutron is likely to be captured, with the emission of a γ ray from the excited nucleus. Figure 40-8 shows the neutron-capture cross section for silver as a function of the energy of the neutron. The large peak in this curve is called a **resonance.** Except for the resonance, the cross section varies fairly smoothly with energy, decreasing with increasing energy roughly as $1/v$, where v is the speed of the neutron. We can understand this energy dependence as follows: Consider a neutron moving with speed v near a nucleus of diameter $2R$. The time it takes the neutron to pass the nucleus is $2R/v$. Thus, the neutron-capture cross section is proportional to the time spent by the neutron in the vicinity of the silver nucleus. The dashed line in Figure 40-8 indicates this $1/v$ dependence. At the maximum of the resonance, the value of the cross section is very large ($\sigma > 5000$ barns) compared with a value of only about 10 barns just past the resonance. Many elements show similar resonances in their neutron-capture cross sections. For example, the maximum cross section for ^{113}Cd is approximately 57,000 barns. This material is thus very useful for shielding against low-energy neutrons.

An important nuclear reaction that involves neutrons is fission, which is discussed in the next section.

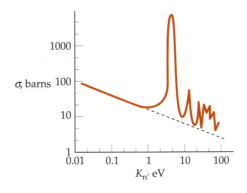

FIGURE 40-8 Neutron-capture cross section for silver as a function of the energy of the neutron. The straight line indicates the $1/v$ dependence of the cross section, which is proportional to the time spent by the neutron in the vicinity of the silver nucleus. Superimposed on this dependence are a large resonance and several smaller resonances.

40-4 Fission and Fusion

Figure 40-9 shows a plot of the nuclear mass difference per nucleon $(M - Zm_p - Nm_n)/A$ in units of MeV/c^2 versus A. This is just the negative of the binding-energy curve shown in Figure 40-3. From Figure 40-9, we can see that the mass per nucleon for both very heavy ($A \approx 200$) and very light ($A \leq 20$) nuclides is more than that for nuclides of intermediate mass. Thus, energy is released when a very heavy nucleus, such as ^{235}U, breaks up into two lighter nuclei—a process called **fission**—or when two very light nuclei, such as ^{2}H and ^{3}H, fuse together to form a nucleus of greater mass—a process called **fusion.**

The application of both fission and fusion to the development of nuclear weapons has had a profound effect on our lives during the past 58 years. The peaceful application of these reactions to the development of energy resources

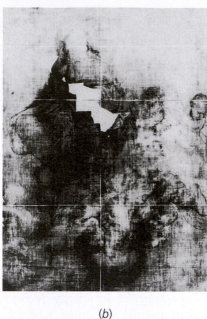

(a) (b) (c)

Hidden layers in paintings are analyzed by bombarding the painting with neutrons and observing the radiative emissions from nuclei that have captured a neutron. Different elements used in the painting have different half-lives. (a) Van Dyck's painting *Saint Rosalie Interceding for the Plague-Stricken of Palermo*. The black-and-white images in (b) and (c) were formed using a special film sensitive to electrons emitted by the radioactively decaying elements. Image (b), taken a few hours after the neutron irradiation, reveals the presence of manganese, found in umber, which is a dark earth pigment used for the painting's base layer. (Blank areas show where modern repairs, free of manganese, have been made.) The image in (c) was taken 4 days later, after the umber emissions had died away and when phosphorus, found in charcoal and boneblack, was the main radiating element. Upside down is revealed a sketch of Van Dyck himself. The self-portrait, executed in charcoal, had been overpainted by the artist.

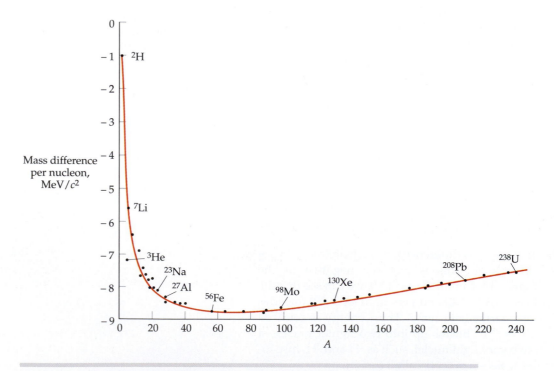

FIGURE 40-9 Plot of mass difference per nucleon $(M - Zm_p - Nm_n)/A$ in units of MeV/c^2 versus A. The rest mass per nucleon is less for nuclei of intermediate mass than for very light nuclei or very heavy nuclei.

may have an even greater effect in the future. We will look at some of the features of fission and fusion that are important for their application in reactors to generate power.

Fission

Very heavy nuclei (Z > 92) are subject to spontaneous fission. They break apart into two nuclei even if the nuclei are left to themselves with no outside disturbance. We can understand this by considering the analogy of a charged liquid drop. If the drop is not too large, surface tension can overcome the repulsive forces of the charges and hold the drop together. There is, however, a certain maximum size beyond which the drop will be unstable and will spontaneously break apart. Spontaneous fission puts an upper limit on the size of a nucleus and therefore on the number of elements that are possible.

Some heavy nuclei—uranium and plutonium, in particular—can be induced to fission by the capture of a neutron. In the fission of ^{235}U, for example, the uranium nucleus is excited by the capture of a neutron, causing it to split into two nuclei and emit several neutrons. The Coulomb force of repulsion drives the fission fragments apart, with the energy eventually showing up as thermal energy. Consider, for example, the fission of a nucleus of mass number $A = 200$ into two nuclei of mass number $A = 100$. Since the rest energy for $A = 200$ is about 1 MeV per nucleon greater than that for $A = 100$, approximately 200 MeV per nucleus is released in such a fission. This is a large amount of energy. By contrast, in the chemical reaction of combustion, only about 4 eV of energy is released per molecule of oxygen consumed.

ENERGY RELEASED IN THE FISSION OF ^{235}U   **E X A M P L E   4 0 - 6**

Calculate the total energy in kilowatt-hours released in the fission of 1 g of ^{235}U, assuming that 200 MeV is released per fission.

PICTURE THE PROBLEM We need to find the number of uranium nuclei in one gram of ^{235}U, which we find using the fact that there are Avogadro's number ($N_A = 6.02 \times 10^{23}$) of nuclei in 235 grams.

1. The total energy is the number of nuclei times the energy per nucleus:

$$E = NE_{nucleus} = N(200 \text{ MeV/nucleus})$$

2. Calculate N:

$$N = \frac{6.02 \times 10^{23} \text{ nuclei/mol}}{235 \text{ g/mol}} \times 1 \text{ g}$$

$$= 2.56 \times 10^{21} \text{ nuclei}$$

3. Calculate the energy per gram in eV and convert to kW·h:

$$E = \frac{200 \times 10^6 \text{ eV}}{1 \text{ nucleus}} \times 2.56 \times 10^{21} \text{ nuclei}$$

$$= 5.12 \times 10^{29} \text{ eV} = 8.19 \times 10^{10} \text{ J}$$

$$= 8.19 \times 10^7 \text{ kW·s} = \boxed{2.28 \times 10^4 \text{ kW·h}}$$

The fission of uranium was discovered in 1938 by Otto Hahn and Fritz Strassmann, who found, by careful chemical analysis, that medium-mass elements (e.g., barium and lanthanum) were produced in the bombardment of uranium with neutrons. The discovery that several neutrons were emitted in the fission process led to speculation concerning the possibility of using these neutrons to cause further fissions, thereby producing a chain reaction. When ^{235}U captures a neutron, the resulting ^{236}U nucleus emits γ-rays as it deexcites to the ground state approximately 15 percent of the time and undergoes fission approximately 85 percent of the time. The fission process is somewhat analogous to the oscillation of a liquid drop, as shown in Figure 40-10. If the oscillations are violent enough, the drop splits in two. Using the liquid-drop model, Niels Bohr and John Wheeler calculated the critical energy E_c needed by the ^{236}U nucleus to undergo fission. (^{236}U is the nucleus formed momentarily by the capture of a neutron by ^{235}U.) For this nucleus, the critical energy is 5.3 MeV, which is less than the 6.4 MeV of excitation energy produced when ^{235}U captures a neutron. The capture of a neutron by ^{235}U therefore produces an excited state of the ^{236}U nucleus that has more than enough energy to break apart. On the other hand, the critical energy for fission of the ^{239}U nucleus is 5.9 MeV. The capture of a neutron by a ^{238}U nucleus produces an excitation energy of only 5.2 MeV. Therefore, when a neutron is captured by ^{238}U to form ^{239}U, the excitation energy is not great enough for fission to occur. In this case, the excited ^{239}U nucleus deexcites by γ emission and then decays to $^{239}N_p$ by β decay, and then again to ^{239}Pu by β decay.

A fissioning nucleus can break into two medium-mass fragments in many different ways, as shown in Figure 40-11. Depending on the particular reaction, 1, 2, or 3 neutrons may be emitted. The average number of neutrons emitted in the fission of ^{235}U is approximately 2.5. A typical fission reaction is

$$n + {}^{235}U \rightarrow {}^{141}Ba + {}^{92}Kr + 3n$$

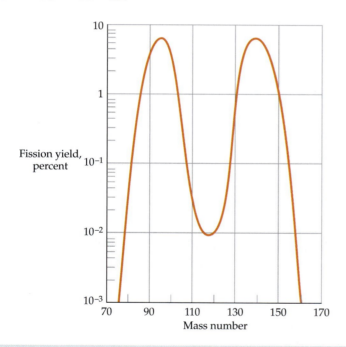

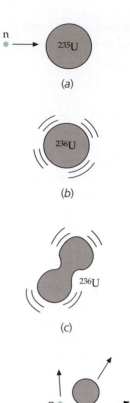

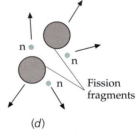

FIGURE 40-10 Schematic illustration of nuclear fission. (*a*) The absorption of a neutron by ^{235}U leads to (*b*) ^{236}U in an excited state. (*c*) The oscillation of ^{236}U has become unstable. (*d*) The nucleus splits apart into two nuclei of medium mass and emits several neutrons that can produce fission in other nuclei.

FIGURE 40-11 Distribution of the possible fission fragments of ^{235}U. The splitting of ^{235}U into two fragments of unequal mass is more likely than its splitting into fragments of equal mass.

Nuclear Fission Reactors

To sustain a chain reaction in a fission reactor, one of the neutrons (on the average) emitted in the fission of ^{235}U must be captured by another ^{235}U nucleus and cause it to fission. The **reproduction constant** k of a reactor is defined as the average number of neutrons from each fission that cause a subsequent fission. The maximum possible value of k is 2.5, but it is normally less than this for two important reasons: (1) Some of the neutrons may escape from the region containing fissionable nuclei and (2) some of the neutrons may be captured by nonfissioning nuclei in the reactor. If k is exactly 1, the reaction will be self-sustaining. If k is less than 1, the reaction will die out. If k is significantly greater than 1, the reaction

The inside of a nuclear power plant in Kent, England. A technician is standing on the reactor charge transfer plate, into which uranium fuel rods fit.

rate will increase rapidly and *run away.* In the design of nuclear bombs, such a runaway reaction is desired. In power reactors, the value of k must be kept very nearly equal to 1.

Since the neutrons emitted in fission have energies of the order of 1 MeV, whereas the chance for neutron capture leading to fission in ^{235}U is largest at small energies, the chain reaction can be sustained only if the neutrons are slowed down before they escape from the reactor. At high energies (1 MeV to 2 MeV), neutrons lose energy rapidly by inelastic scattering from ^{238}U, the principal constituent of natural uranium. (Natural uranium contains 99.3 percent ^{238}U and only 0.7 percent fissionable ^{235}U.) Once the neutron energy is below the excitation energies of the nuclei in the reactor (about 1 MeV), the main process of energy loss is by elastic scattering, in which a fast neutron collides with a nucleus at rest and transfers some of its kinetic energy to that nucleus. Such energy transfers are efficient only if the masses of the two bodies are comparable. A neutron will not transfer much energy in an elastic collision with a heavy uranium nucleus. Such a collision is like one between a marble and a billiard ball. The marble will be deflected by the much more massive billiard ball, and very little of its kinetic energy will be transferred to the billiard ball. A **moderator** consisting of material, such as water or carbon, that contains light nuclei is therefore placed around the fissionable material in the core of the reactor to slow down the neutrons. The neutrons are slowed down by elastic collisions with the nuclei of the moderator until they are in thermal equilibrium with the moderator. Because of the relatively large neutron-capture cross section of the hydrogen nucleus in water, reactors that use ordinary water as a moderator cannot easily achieve $k \approx 1$ unless they use enriched uranium, in which the ^{235}U content has been increased from 0.7 percent to between 1 percent and 4 percent. Natural uranium can be used if heavy water (D_2O) is used instead of ordinary (light) water (H_2O) as the moderator. Although heavy water is expensive, most Canadian reactors use heavy water for a moderator to avoid the cost of constructing uranium-enrichment facilities.

Figure 40-12 shows some of the features of a pressurized-water reactor commonly used in the United States to generate electricity. Fission in the core heats the water to a high temperature in the primary loop, which is closed. This water, which also serves as the moderator, is under high pressure to prevent the water from boiling. The hot water is pumped to a heat exchanger, where it heats the water in the secondary loop and converts the water to steam, which is then used to drive the turbines that produce electrical power. Note that the water in the secondary loop is isolated from the water in the primary loop to prevent its contamination by the radioactive nuclei in the reactor core.

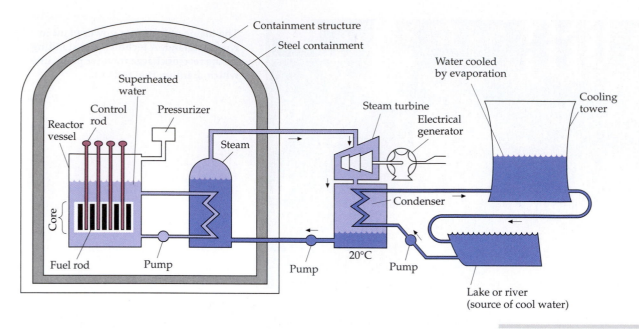

FIGURE 40-12 Simplified drawing of a pressurized-water reactor. The water in contact with the reactor core serves as both the moderator and the heat-transfer material. It is isolated from the water used to produce the steam that drives the turbines. Many features, such as the backup cooling mechanisms, are not shown here.

The ability to control the reproduction factor k precisely is important if a power reactor is to be operated safely. Both natural negative-feedback mechanisms and mechanical methods of control are used. If k is greater than 1, the reaction rate increases and the temperature of the reactor increases. If water is used as a moderator, its density decreases with increasing temperature and the water becomes a less effective moderator. A second important control method is the use of control rods made of a material, such as cadmium, that has a very large neutron-capture cross section. When the reactor is started up, the control rods are inserted so that k is less than 1. As the rods are gradually withdrawn from the reactor, fewer neutrons are captured by the control rods and k increases to 1. If k becomes greater than 1, the rods are inserted further.

Mechanical control of the reaction rate of a nuclear reactor using control rods is possible only because some of the neutrons emitted in the fission process are **delayed neutrons.** The time needed for a neutron to slow down from 1 MeV or 2 MeV to the thermal-energy level and then be captured is only of the order of a millisecond. If all the neutrons emitted in fission were prompt neutrons, that is, emitted immediately in the fission process, mechanical control would not be possible because the reactor would run away before the rods could be further inserted. However, approximately 0.65 percent of the neutrons emitted are delayed by an average time of about 14 s. These neutrons are emitted not in the fission process itself but in the decay of the fission fragments. The effect of the delayed neutrons can be seen in the following examples.

DOUBLING TIME   **EXAMPLE 40-7**

If the average time between fission generations (the time it takes for a neutron emitted in one fission to cause another) is $t_1 = 1 \text{ ms} = 0.001 \text{ s}$ and if the average number of neutrons from each fission that cause a subsequent fission is 1.001, how long will it take for the reaction rate to double?

PICTURE THE PROBLEM The reaction rate is the number of nuclei that fission per unit time. The time to double the reaction rate is the product of the number of generations N needed to double the reaction rate and the generation time. If $k = 1.001$, the reaction rate after N generations is 1.001^N. We find the number of generations by setting 1.001^N equal to 2 and solving for N.

1. Set 1.001^N equal to 2 and solve for N:

$$(1.001)^N = 2$$

$$N \ln 1.001 = \ln 2$$

$$N = \frac{\ln 2}{\ln 1.001} = 693$$

2. Multiply the number of generations by the generation time: $t = Nt_1 = 693(0.001 \text{ s}) = \boxed{0.693 \text{ s}}$

REMARKS The doubling time of about 0.7 s is not enough time for insertion of control rods.

DELAYED NEUTRONS AND CONTROL-ROD INSERTION   **EXAMPLE 40-8**   **Try It Yourself**

Assuming that 0.65 percent of the neutrons emitted are delayed by 14 s, find the average generation time and the doubling time if $k = 1.001$.

PICTURE THE PROBLEM The doubling time is Nt_{av}, where t_{av} is the average time between generations. Since 99.35 percent of the generation times are 0.001 s and 0.65 percent are 14 s, the average generation time is 0.9935(0.001 s) + 0.0065(14 s).

Cover the column to the right and try these on your own before looking at the answers.

Steps	Answers
1. Compute the average generation time.	$t_{av} = 0.9935(0.001 \text{ s}) + 0.0065(14 \text{ s})$
	$= 0.092 \text{ s}$
2. Use your result to find the time for 693 generations.	$t = \boxed{63.8 \text{ s}}$

REMARKS Even though the number of delayed neutrons is less than 1 percent, they have a large effect on the doubling time. Here they increase the generation time by a factor of 92, resulting in a doubling time of about 64 s, which is plenty of time for mechanical insertion of control rods.

Because of the limited supply of natural uranium, the small fraction of ^{235}U in natural uranium, and the limited capacity of enrichment facilities, reactors based on the fission of ^{235}U cannot meet our energy needs for very long. A promising alternative is the **breeder reactor.** When the relatively plentiful but nonfissionable ^{238}U nucleus captures a neutron, it decays by β decay (with a half-life of 20 min) to ^{239}Np, which in turn decays by β decay (with a half-life of 2.35 days) to the fissionable nuclide ^{239}Pu. Since ^{239}Pu fissions with fast neutrons, no moderator is needed. A reactor initially fueled with a mixture of ^{238}U and ^{239}Pu will breed as much fuel as it uses or more if one or more of the neutrons emitted in the fission of ^{239}Pu is captured by ^{238}U. Practical studies indicate that a typical breeder reactor can be expected to double its fuel supply in 7 to 10 years.

There are two major safety problems inherent with breeder reactors. The fraction of delayed neutrons is only 0.3 percent for the fission of ^{239}Pu, so the time between generations is much less than that for ordinary reactors. Mechanical control is therefore much more difficult. Also, because the operating temperature of a breeder reactor is relatively high and a moderator is not desired, a heat-transfer material, such as liquid sodium metal, is used rather than water (which is the moderator as well as the heat-transfer material in an ordinary reactor). If the temperature of the reactor increases, the resulting decrease in the density of the heat-transfer material leads to positive feedback, since it will absorb fewer neutrons than before. Because of these safety considerations, breeder reactors are not yet in commercial use in the United States. There are, however, several in operation in France, Great Britain, and the former Soviet Union.

Fusion

In fusion, two light nuclei, such as deuterium (^{2}H) and tritium (^{3}H), fuse together to form a heavier nucleus. A typical fusion reaction is

$$^2\text{H} + {}^3\text{H} \rightarrow {}^4\text{He} + \text{n} + 17.6 \text{ MeV}$$

The energy released in fusion depends on the particular reaction. For the ^{2}H + ^{3}H reaction, the energy released is 17.6 MeV. Although this is less than the energy released in a fission reaction, it is a greater amount of energy per unit mass. The energy released in this fusion reaction is (17.6 MeV)/(5 nucleons) = 3.52 MeV per nucleon. This is approximately 3.5 times as great as the 1 MeV per nucleon released in fission.

The production of power from the fusion of light nuclei holds great promise because of the relative abundance of the fuel and the absence of some of the dangers inherent in fission reactors. Unfortunately, the technology necessary to make fusion a practical source of energy has not yet been developed. We will consider the ^{2}H + ^{3}H reaction; other reactions present similar problems.

Because of the Coulomb repulsion between the ^{2}H and ^{3}H nuclei, very large kinetic energies, of the order of 1 MeV, are needed to get the nuclei close enough together for the attractive nuclear forces to become effective and to cause fusion. Such energies can be obtained in an accelerator, but since the scattering of one nucleus by the other is much more probable than fusion, the bombardment of one nucleus by another in an accelerator requires the input of more energy than is recovered. To obtain energy from fusion, the particles must be heated to a temperature great enough for the fusion reaction to occur as the result of random thermal collisions. Because a significant number of particles have kinetic energies greater than the mean kinetic energy, $\frac{3}{2}kT$, and because some particles can tunnel through the Coulomb barrier, a temperature T corresponding to $kT \approx 10$ keV is adequate to ensure that a reasonable number of fusion reactions will occur if the density of the particles is sufficiently high. The temperature corresponding to $kT = 10$ keV is of the order of 10^8 K. These temperatures occur in the interiors of stars, where such reactions are common. At these temperatures, a gas consists of positive ions and negative electrons and is called a **plasma**. One of the problems arising in attempts to produce controlled fusion reactions is the problem of confining the plasma long enough for the reactions to take place. In the interior of the sun, the plasma is confined by the enormous gravitational field of the sun. In a laboratory on the earth, confinement is a difficult problem.

The energy required to heat a plasma is proportional to the number density of its ions, n, whereas the collision rate is proportional to n^2, the square of the number density. If τ is the confinement time, the output energy is proportional to $n^2\tau$. If the output energy is to exceed the input energy, we must have

$$C_1 n^2 \tau > C_2 n$$

where C_1 and C_2 are constants. In 1957, the British physicist J. D. Lawson evaluated these constants from estimates of the efficiencies of various hypothetical fusion reactors and derived the following relation between density and confinement time, known as **Lawson's criterion:**

$$n\tau > 10^{20} \text{ s·particles/m}^3 \qquad \text{40-22}$$

LAWSON'S CRITERION

If Lawson's criterion is met and the thermal energy of the ions is great enough ($kT \sim 10$ keV), the energy released by a fusion reactor will just equal the energy input; that is, the reactor will just break even. For the reactor to be practical, much more energy must be released.

Two schemes for achieving Lawson's criterion are currently under investigation. In one scheme, **magnetic confinement,** a magnetic field is used to confine the plasma (see Section 26-2). In the most common arrangement, first developed in the former USSR and called the Tokamak, the plasma is confined in a large toroid. The magnetic field is a combination of the doughnut-shaped magnetic field due to the windings of the toroid and the self-field due to the current of the circulating plasma. The break-even point has been achieved recently using magnetic confinement, but we are still a long way from building a practical fusion reactor.

In a second scheme, called **inertial confinement,** a pellet of solid deuterium and tritium is bombarded from all sides by intense pulsed laser beams of energies of the order of 10^4 J lasting about 10^{-8} s. (Intense beams of ions are also used.) Computer simulation studies indicate that the pellet should be compressed to approximately 10^4 times its normal density and heated to a temperature greater than 10^8 K. This should produce approximately 10^6 J of fusion energy in 10^{-11} s, which is so brief that confinement is achieved by inertia alone.

Because the break-even point is just barely being achieved in magnetic-confinement fusion, and because the building of a fusion reactor involves many practical problems that have not yet been solved, the availability of fusion to meet our energy needs is not expected for at least several decades. However, fusion holds great promise as an energy source for the future.

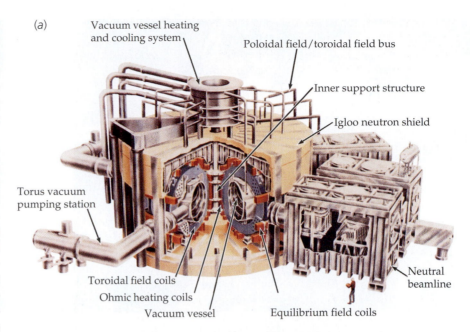

(a)

Vacuum vessel heating and cooling system

Poloidal field/toroidal field bus

Inner support structure

Igloo neutron shield

Torus vacuum pumping station

Toroidal field coils

Ohmic heating coils

Vacuum vessel

Equilibrium field coils

Neutral beamline

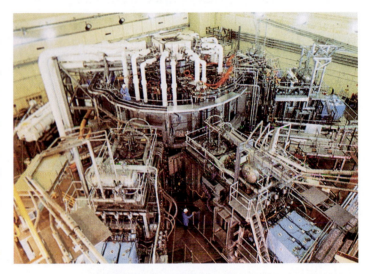

(b)

(a) Schematic of the Tokamak Fusion Test Reactor (TFTR). The toroidal coils, surrounding the doughnut-shaped vacuum vessel, are designed to conduct current for 3-s pulses, separated by waiting times of 5 min. Pulses peak at 73,000 A, producing a magnetic field of 5.2 T. This magnetic field is the principal means of confining the deuterium–tritium plasma that circulates within the vacuum vessel. Current for the pulses is delivered by converting the rotational energy of two 600-ton flywheels. Sets of poloidal coils, perpendicular to the toroidal coils, carry an oscillating current that generates a current through the confined plasma itself, heating it ohmically. Additional poloidal fields help stabilize the confined plasma. Between four and six neutral-beam injection systems (only one of which is shown in the schematic) are used to inject high-energy deuterium atoms into the deuterium–tritium plasma, heating beyond what could be obtained ohmically, ultimately to the point of fusion. (b) The TFTR itself. The diameter of the vacuum vessel is 7.7 m. (c) An 800-kA plasma, lasting 1.6 s, as it discharges within the vacuum vessel.

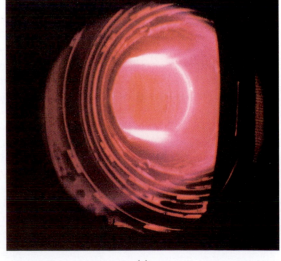

(c)

(a)

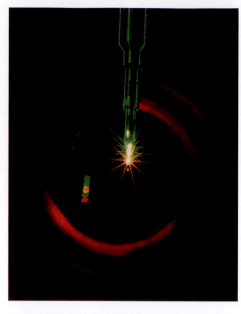

(b)

(a) The Nova target chamber, an aluminum sphere approximately 5 m in diameter, inside which 10 beams from the world's most powerful laser converge onto a hydrogen-containing pellet 0.5 mm in diameter. (b) The resulting fusion reaction is visible as a tiny star, lasting 10^{-10} s, releasing 10^{13} neutrons.

SUMMARY

Topic	Relevant Equations and Remarks
1. **Properties of Nuclei**	Nuclei have N neutrons, Z protons, and a mass number $A = N + Z$. For light nuclei, N and Z are approximately equal, whereas for heavy nuclei, N is greater than Z.
Isotopes	Isotopes consist of two or more nuclei with the same atomic number Z but with different values of N and A.
Size and shape	Most nuclei are approximately spherical in shape and have a volume that is proportional to A. Because the mass is proportional to A, nuclear density is independent of A.
Radius	$R = R_0 A^{1/3} \approx (1.2 \text{ fm}) A^{1/3}$ **40-1**
Mass and binding energy	The mass of a stable nucleus is less than the sum of the masses of its nucleons. The mass difference Δm times c^2 equals the binding energy E_b of the nucleus. The binding energy is approximately proportional to the mass number A.
2. **Radioactivity**	Unstable nuclei are radioactive and decay by emitting α particles (^{4}He nuclei), β particles (electrons or positrons), or γ-rays (photons). All radioactivity is statistical in nature and follows an exponential decay law: $N = N_0 e^{-\lambda t}$ **40-6**

| Decay law | $R = \lambda N = R_0 e^{-\lambda t}$ | **40-7** |

| Mean life | $\tau = \dfrac{1}{\lambda}$ | **40-9** |

| Half-life | $t_{1/2} = 0.693\tau$ | **40-11** |

The half-lives of α decay range from a fraction of a second to millions of years. For β decay, the half-lives range up to hours or days. For γ decay, the half-lives are usually less than a microsecond.

| Decay-rate units | The number of decays per second of 1 g of radium is the curie (Ci). |

$$1\ \text{Ci} = 3.7 \times 10^{10}\ \text{decays/s} = 3.7 \times 10^{10}\ \text{Bq}$$

$$(1\ \text{Bq} = 1\ \text{decay/s})$$

3. Nuclear Reactions

| Q value | The Q value equals c^2 times the total mass of the incoming particles less the total mass of the outgoing particles in the center of mass reference frame. If the net mass change is Δm, the Q value is |

$$Q = -(\Delta m)c^2 \qquad\qquad \textbf{40-19}$$

| Exothermic reaction | The mass decreases, Q is positive and measures the energy released. |

| Endothermic reaction | The mass increases, Q is negative. Then $|Q|$ is the threshold energy for the reaction in the center of mass reference frame. |

4. Fission

Fission occurs when some heavy elements, such as ^{235}U or ^{239}Pu, capture a neutron and split apart into two medium-mass nuclei. The two nuclei then fly apart because of electrostatic repulsion, releasing a large amount of energy. A chain reaction is possible because several neutrons are emitted by a nucleus when it undergoes fission. A chain reaction can be sustained in a reactor if, on the average, one of the emitted neutrons is slowed down by scattering in the reactor and is then captured by another fissionable nucleus. Very heavy nuclei ($Z > 92$) are subject to spontaneous fission.

5. Fusion

A large amount of energy is released when two light nuclei, such as ^{2}H and ^{3}H, fuse together. Fusion takes place spontaneously inside the sun and other stars, where the temperature is great enough (about 10^8 K) for thermal motion to bring the charged hydrogen ions close enough together to fuse. Although controlled fusion holds great promise as a future energy source, practical difficulties have thus far prevented its development.

| Lawson criterion | The minimum product of particle density n and confinement time τ to get more energy out of a fusion reactor than is put in is $n\tau > 10^{20}$ s·particles/m^3. |

PROBLEMS

- Single-concept, single-step, relatively easy
- • Intermediate-level, may require synthesis of concepts
- • • Challenging
- SSM Solution is in the *Student Solutions Manual*
- iSOLVE Problems available on iSOLVE online homework service
- iSOLVE✓ These "Checkpoint" online homework service problems ask students additional questions about their confidence level, and how they arrived at their answer.

In a few problems, you are given more data than you actually need; in a few other problems, you are required to supply data from your general knowledge, outside sources, or informed estimates.

Conceptual Problems

1 • Give the symbols for two other isotopes of (*a*) ^{14}N, (*b*) ^{56}Fe, and (*c*) ^{118}Sn.

2 • Why is the decay series $A = 4n + 1$ not found in nature?

3 • A decay by α emission is often followed by β decay. When this occurs, it is by β^- and not β^+ decay. Why?

4 • SSM The half-life of ^{14}C is much less than the age of the universe, yet ^{14}C is found in nature. Why?

5 • What effect would a long-term variation in cosmic-ray activity have on the accuracy of ^{14}C dating?

6 • Why is there not an element with $Z = 130$?

7 • Why is a moderator needed in an ordinary nuclear fission reactor?

8 • Explain why water is more effective than lead in slowing down fast neutrons.

9 • The stable isotope of sodium is ^{23}Na. What kind of radioactivity would you expect of (*a*) ^{22}Na and (*b*) ^{24}Na?

10 • What is the advantage of a breeder reactor over an ordinary reactor? What are the disadvantages?

11 • True or false:

(*a*) The atomic nucleus contains protons, neutrons, and electrons.

(*b*) The mass of ^{2}H is less than the mass of a proton plus a neutron.

(*c*) After two half-lives, all the radioactive nuclei in a given sample have decayed.

(*d*) In a breeder reactor, fuel can be produced as fast as it is consumed.

12 • Why do extreme changes in the temperature or the pressure of a radioactive sample have little or no effect on the radioactivity?

13 • SSM Write balanced reactions for each of the following nuclear decays: (*a*) beta decay of ^{16}N, (*b*) alpha decay of ^{248}Fm, (*c*) positron decay of ^{12}N, (*d*) beta decay of ^{81}Se, (*e*) positron decay of ^{61}Cu, and (*f*) alpha decay of ^{228}Th.

14 • SSM Write and balance reaction equations for each of the following: (*a*) ^{240}Pu undergoes spontaneous fission to form two fission fragments and three neutrons. One of the fission fragments is a ^{90}Sr nucleus. (*b*) A ^{72}Ge nucleus absorbs an alpha particle and ejects a photon. (*c*) An ^{127}I nucleus absorbs a deuteron and ejects a neutron. (*d*) A ^{235}U nucleus absorbs a slow neutron and fissions forming a ^{113}Ag nucleus, two neutrons, and another fission fragment. (*e*) A ^{55}Mn nucleus is struck with a high-energy ^{7}Li nucleus resulting in a triton, ^{3}H, and a new nucleus. (*f*) ^{238}U absorbs a slow neutron, resulting in a compound nucleus that emits a beta particle, followed a short time later by a second beta particle. What is the resulting nucleus?

Estimation and Approximation

15 • We found in Chapter 25 that the ratio of the resistivity of the most insulating material to that of the least resistive material (excluding superconductors) is approximately 10^{22}. There are very few properties of materials that show such a wide range of values. Using information in the textbook or other resources, find the ratio of largest to smallest for some nuclear related properties of matter. Some examples might be the range of mass densities found in an atom, the half-life of radioactive nuclei, or the range of nuclear masses.

16 • • According to the United States Department of Energy, the U.S. population consumes approximately 10^{20} joules of energy each year. Estimate the mass (in kg) of (*a*) uranium that would be needed to produce this much energy using nuclear fission and (*b*) deuterium and tritium that would be needed to produce this much energy using nuclear fusion.

Properties of Nuclei

17 • SSM Calculate the binding energy and the binding energy per nucleon from the masses given in Table 40-1 for (*a*) ^{12}C, (*b*) ^{56}Fe, and (*c*) ^{238}U.

18 • Repeat Problem 17 for (*a*) ^{6}Li, (*b*) ^{39}K, and (*c*) ^{208}Pb.

19 • Use Equation 40-1 to compute the radii of the following nuclei: (*a*) ^{16}O, (*b*) ^{56}Fe, and (*c*) ^{197}Au.

20 • In a fission process, a ^{239}Pu nucleus splits into two nuclei whose mass number ratio is 3 to 1. Calculate the radii of the nuclei formed in this process.

21 • • SSM The neutron, when isolated from an atomic nucleus, decays into a proton, an electron, and an antineutrino as follows: $^1_0$n $\rightarrow$ $^1_1$H $+$ $^{\ 0}_{-1}$e $+$ $^0_0\bar{\nu}$. The thermal energy of a neutron is of the order of kT, where k is the Boltzmann constant. (a) In both joules and electron volts, calculate the energy of a thermal neutron at 25°C. (b) What is the speed of this thermal neutron? (c) A beam of monoenergetic thermal neutrons is produced at 25°C with an intensity I. After traveling 1350 km, the beam has an intensity of $I/2$. Using this information, estimate the half-life of the neutron. Express your answer in minutes.

22 • Use Equation 40-1 for the radius of a spherical nucleus and the approximation that the mass of a nucleus of mass number A is $A \times (1u)$ to calculate the density of nuclear matter in grams per cubic centimeter.

23 •• Consider the following fission process: $^{235}_{92}\text{U} + ^1_0\text{n} \rightarrow ^{95}_{42}\text{Mo} + ^{139}_{57}\text{La} + 2^1_0\text{n}$. Determine the electrostatic potential energy, in MeV, of the reaction products when the ^{95}Mo nucleus and the ^{139}La nucleus are just touching immediately after being formed in the fission process.

24 •• **SSM** In 1920, 12 years before the discovery of the neutron, Ernest Rutherford argued that proton–electron pairs might exist in the confines of the nucleus in order to explain the mass number, A, being greater than the nuclear charge, Z. He also used this argument to account for the source of beta particles in radioactive decay. Rutherford's scattering experiments in 1910 showed that the nucleus had a diameter of approximately 10 fm. Using this nuclear diameter, the uncertainty principle, and given that beta particles have an energy range of 0.02 MeV to 3.40 MeV, show why electrons cannot be contained within the nucleus.

Radioactivity

25 • **iSOLVE** Homer enters the visitors' chambers, and his Geiger beeper goes off. He shuts off the beep, removes the device from his shoulder patch, and holds it near the only new object in the room—an orb that is to be presented as a gift from the visiting Cartesians. Pushing a button marked "monitor," Homer reads that the orb is a radioactive source with a counting rate of 4000 counts/s. After 10 min, the counting rate has dropped to 1000 counts/s. The source's half-life appears on the Geiger-beeper display. (a) What is the half-life of the source? (b) What will the counting rate be 20 min after the monitoring device was switched on?

26 • **iSOLVE**✔ A certain source gives 2000 counts/s at time $t = 0$. Its half-life is 2 min. What is the counting rate after (a) 4 min, (b) 6 min, and (c) 8 min?

27 • **iSOLVE**✔ The counting rate from a radioactive source is 8000 counts/s at time $t = 0$, and 10 min later the rate is 1000 counts/s. (a) What is the half-life? (b) What is the decay constant? (c) What is the counting rate after 20 min?

28 • **iSOLVE** The half-life of radium is 1620 y. Calculate the number of disintegrations per second of 1 g of radium, and show that the disintegration rate is approximately 1 Ci.

29 • **iSOLVE** A radioactive silver foil ($t_{1/2} = 2.4$ min) is placed near a Geiger counter and 1000 counts/s are observed at time $t = 0$. (a) What is the counting rate at $t = 2.4$ min and at $t = 4.8$ min? (b) If the counting efficiency is 20 percent, how many radioactive nuclei are there at time $t = 0$? At time

$t = 2.4$ min? (c) At what time will the counting rate be about 30 counts/s?

30 • Use Table 40-1 to calculate the energy in MeV for the α decay of (a) ^{226}Ra and (b) ^{242}Pu.

31 •• **SSM** Plutonium is a highly hazardous and toxic material to the human body. Once it enters the body it collects primarily in the bones, although it also can be found in other organs. Red blood cells are synthesized within the marrow of the bones. The isotope ^{239}Pu is an alpha emitter with a half-life of 24,360 years. Since alpha particles are an ionizing radiation, the blood-making ability of the marrow is, in time, destroyed by the presence of ^{239}Pu. In addition, many kinds of cancers will also be initiated in the surrounding tissues by the ionizing effects of the alpha particles. (a) If a person accidentally ingested 2.0 μg of ^{239}Pu and it is absorbed by the bones of the victim, how many alpha particles are produced per second within the skeleton? (b) When, in years, will the activity be 1000 alpha particles per second?

32 •• Consider a parent, $^A_Z X$, alpha-emitting nucleus initially at rest. The nucleus decays into a daughter nucleus, $^{A-4}_{Z-2} Y$, and an alpha particle as follows: $^A_Z X \rightarrow ^{A-4}_{Z-2} Y + ^4_2 \alpha + Q$. (a) Show that the alpha particle has a kinetic energy of $(A - 4)Q/A$. (b) Show that the kinetic energy of the recoiling daughter nucleus is given by $K_Y = 4Q/A$.

33 •• **SSM** The fissile material ^{239}Pu is an alpha emitter. Write the equation of this reaction. Given that ^{239}Pu, ^{235}U, and an alpha particle have respective masses of 239.052 156 u, 235.043 923 u, and 4.002 603 u, use the relations appearing in Problem 32 to calculate the kinetic energies of the alpha particle and the recoiling daughter nucleus.

34 • Through a friend in the security department at the museum, Angela obtained a sample with 175 g of carbon. The decay rate of ^{14}C was 8.1 Bq. How old is the sample?

35 • **iSOLVE** A sample of a radioactive isotope is found to have an activity of 115.0 Bq immediately after it is pulled from the reactor that formed the isotope. Its activity 2 h 15 min later is measured to be 85.2 Bq. (a) Calculate the decay constant and the half-life of the sample. (b) How many radioactive nuclei were there in the sample initially?

36 •• **SSM** Radiation has long been used in medical therapy to control the development and growth of cancer cells. Cobalt-60, a gamma emitter of 1.17 MeV and 1.33 MeV energies, is used to irradiate and destroy deep-seated cancers. Small needles made of ^{60}Co of a specified activity are encased in gold and used as body implants in tumors for time periods that are related to tumor size, tumor cell reproductive rate, and the activity of the needle. (a) A 1.00 μg sample of ^{60}Co, with a half-life of 5.27 y, is prepared in the cyclotron of a medical center to irradiate a small internal tumor with gamma rays. In curies, determine the initial activity of the sample. (b) What is the activity of the sample after 1.75 y?

37 •• **iSOLVE** Measurements of the activity of a radioactive sample have yielded the results shown in the following table. Plot the activity as a function of time, using semilogarithmic paper, and determine the decay constant and the half-life of the radioisotope.

Time, min	Activity	Time, min	Activity
0	4287	20	880
5	2800	30	412
10	1960	40	188
15	1326	60	42

38 •• (a) Show that if the decay rate is R_0 at time $t = 0$ and R_1 at some later time t_1, the decay constant is given by $\lambda = t_1^{-1} \ln(R_0/R_1)$ and the half-life is given by $t_{1/2} = 0.693t_1/\ln(R_0/R_1)$. (b) Use these results to find the decay constant and the half-life if the decay rate is 1200 Bq at $t = 0$ and 800 Bq at $t_1 = 60$ s.

39 •• **ISOLVE** A wooden casket is thought to be 18,000 years old. How much carbon would have to be recovered from this object to yield a ^{14}C counting rate of no less than 5 counts/min?

40 •• **ISOLVE** A 1.00-mg sample of substance of atomic mass 59.934 u emits β particles with an activity of 1.131 Ci. Find the decay constant for this substance in s^{-1} and its half-life in years.

41 •• **SSM** The counting rate from a radioactive source is measured every minute. The resulting counts per second are 1000, 820, 673, 552, 453, 371, 305, and 250. Plot the counting rate versus time on semilogarithmic graph paper, and use your graph to find the half-life of the source.

42 •• **ISOLVE** A sample of radioactive material is initially found to have an activity of 115.0 decays/min. After 4 d 5 h, its activity is measured to be 73.5 decays/min. (a) Calculate the half-life of the material. (b) How long (from the initial time) will it take for the sample to reach an activity level of 10.0 decays/min?

43 •• **ISOLVE** ✓ The rubidium isotope ^{87}Rb is a β emitter with a half-life of 4.9×10^{10} y that decays into ^{87}Sr. It is used to determine the age of rocks and fossils. Rocks containing the fossils of early animals contain a ratio of ^{87}Sr to ^{87}Rb of 0.0100. Assuming that there was no ^{87}Sr present when the rocks were formed, calculate the age of these fossils.

44 ••• If there are N_0 radioactive nuclei at time $t = 0$, the number that decay in some time interval dt at time t is $-dN = \lambda N_0 e^{-\lambda t} dt$. If we multiply this number by the lifetime t of these nuclei, sum over all the possible lifetimes from $t = 0$ to $t = \infty$, and divide by the total number of nuclei, we get the mean lifetime τ.

$$\tau = \frac{1}{N_0} \int_0^\infty t|dN| = \int_0^\infty t\lambda e^{-\lambda t} dt$$

Show that $\tau = 1/\lambda$.

Nuclear Reactions

45 • Using Table 40-1, find the Q values for the following reactions: (a) ^{1}H + ^{3}H → ^{3}He + n + Q and (b) ^{2}H + ^{2}H → ^{3}He + n + Q.

46 • Using Table 40-1, find the Q values for the following reactions: (a) ^{2}H + ^{2}H → ^{3}H + ^{1}H + Q, (b) ^{2}H + ^{3}He → ^{4}He + ^{1}H + Q, and (c) ^{6}Li + n → ^{3}H + ^{4}He + Q.

47 •• **SSM** (a) Use the atomic masses $m = 14.003\ 242$ u for $^{14}_6$C and $m = 14.003\ 074$ u for $^{14}_7$N to calculate the Q value (in MeV) for the β decay

$$^{14}_6\text{C} \rightarrow ^{14}_7\text{N} + \beta^- + \bar{v}_e$$

(b) Explain why you do not need to add the mass of the β^- to that of atomic $^{14}_7$N for this calculation.

48 •• (a) Use the atomic masses $m = 13.005\ 738$ u for ^{13}N and $m = 13.003\ 354$ u for $^{13}_6$C to calculate the Q value (in MeV) for the β decay

$$^{13}_7\text{N} \rightarrow ^{13}_6\text{C} + \beta^+ + v_e$$

(b) Explain why you need to add two electron masses to the mass of $^{13}_6$C in the calculation of the Q value for this reaction.

Fission and Fusion

49 • **SSM** **ISOLVE** ✓ Assuming an average energy of 200 MeV per fission, calculate the number of fissions per second needed for a 500-MW reactor.

50 • **ISOLVE** If the reproduction factor in a reactor is $k = 1.1$, find the number of generations needed for the power level to (a) double, (b) increase by a factor of 10, and (c) increase by a factor of 100. Find the time needed in each case if (d) there are no delayed neutrons, so that the time between generations is 1 ms and (e) there are delayed neutrons that make the average time between generations 100 ms.

51 •• **SSM** Consider the following fission reaction: $^{235}_{92}$U + $^1_0$n → $^{95}_{42}$Mo + $^{139}_{57}$La + 2^1_0n + Q. The masses of the neutron, U, Mo, and La are 1.008 665 u, 235.043 923 u, 94.905 842 u, and 138.906 348 u, respectively. Calculate the Q value, in MeV, for this fission reaction. Compare the result to the result obtained in Problem 23.

52 •• **ISOLVE** In 1989, researchers claimed to have achieved fusion in an electrochemical cell at room temperature. They claimed a power output of 4 W from deuterium fusion reactions in the palladium electrode of their apparatus. If the two most likely reactions are

$$^2\text{H} + ^2\text{H} \rightarrow ^3\text{He} + \text{n} + 3.27 \text{ MeV}$$

and

$$^2\text{H} + ^2\text{H} \rightarrow ^3\text{He} + ^1\text{H} + 4.03 \text{ MeV}$$

with 50 percent of the reactions going by each branch, how many neutrons per second would we expect to be emitted in the generation of 4 W of power?

53 •• **ISOLVE** A fusion reactor that uses only deuterium for fuel would have the two reactions in Problem 52 taking place in the reactor. The ^{3}H produced in the second reaction reacts immediately with another ^{2}H to produce

$$^3\text{H} + ^2\text{H} \rightarrow ^4\text{He} + \text{n} + 17.6 \text{ MeV}$$

The ratio of ^{2}H to ^{1}H atoms in naturally occurring hydrogen is 1.5×10^{-4}. How much energy would be produced from 4 L of water if all of the ^{2}H nuclei undergo fusion?

54 ••• **SSM** **iSOLVE** The fusion reaction between ^{2}H and ^{3}H is

$$^3H + {}^2H \to {}^4He + n + 17.6 \text{ MeV}$$

Using the conservation of momentum and the given Q value, find the final energies of both the ^{4}He nucleus and the neutron, assuming the initial kinetic energy of the system is 1.00 MeV and the initial momentum of the system is zero.

55 ••• Energy is generated in the sun and other stars by fusion. One of the fusion cycles, the proton–proton cycle, consists of the following reactions:

$$^1H + {}^1H \to {}^2H + \beta^+ + v_e$$

$$^1H + {}^2H \to {}^3He + \gamma$$

followed by

$$^1H + {}^3H \to {}^4He + \beta^+ + v_e$$

(a) Show that the net effect of these reactions is

$$4^1H \to {}^4He + 2\beta^+ + 2v_e + \gamma$$

(b) Show that rest energy of 24.7 MeV is released in this cycle (not counting the energy of 1.02 MeV released when each positron meets an electron and the two annihilate). (c) The sun radiates energy at the rate of approximately 4×10^{26} W. Assuming that this is due to the conversion of four protons into helium plus γ-rays and neutrinos, which releases 26.7 MeV, what is the rate of proton consumption in the sun? How long will the sun last if it continues to radiate at its present level? (Assume that protons constitute about half of the total mass, 2×10^{30} kg, of the sun.)

General Problems

56 • (a) Show that $ke^2 = 1.44$ MeV·fm, where k is the Coulomb constant and e is the electron charge. (b) Show that $hc = 1240$ MeV·fm.

57 • **SSM** **iSOLVE✔** The counting rate from a radioactive source is 6400 counts/s. The half-life of the source is 10 s. Make a plot of the counting rate as a function of time for times up to 1 min. What is the decay constant for this source?

58 • Find the energy needed to remove a neutron from (a) ^{4}He and (b) ^{7}Li.

59 • The isotope ^{14}C decays according to $^{14}C \to {}^{14}N + e^- + \bar{v}_e$. The atomic mass of ^{14}N is 14.003 074 u. Determine the maximum kinetic energy of the electron. (Neglect recoil of the nitrogen atom.)

60 • **iSOLVE** A neutron star is an object of nuclear density. If our sun were to collapse to a neutron star, what would be the radius of that object?

61 •• **SSM** Show that the ^{109}Ag nucleus is stable against alpha decay, $^{109}_{47}Ag \to {}^4_2He + {}^{105}_{45}Rh + Q$. The mass of the ^{109}Ag nucleus is 108.904 756 u, and the products of the decay are 4.002 603 u and 104.905 250 u, respectively.

62 •• Gamma rays can be used to induce photofission (fission triggered by the absorption of a photon) in nuclei. Calculate the threshold photon wavelength for the following nuclear reaction: $^2H + \gamma \to {}^1H + {}^1n$. Use Table 40-1 for the masses of the interacting particles.

63 • The relative abundance of ^{40}K (molecular mass 40.0 g/mol) is 1.2×10^{-4}. The isotope ^{40}K is radioactive with a half-life of 1.3×10^9 y. Potassium is an essential element of every living cell. In the human body the mass of potassium constitutes approximately 0.36 percent of the total mass. Determine the activity of this radioactive source in a student whose mass is 60 kg.

64 •• When a positron makes contact with an electron, the electron–positron pair annihilate via the reaction $\beta^+ + \beta^- \to 2\gamma$. Calculate the minimum total energy, in MeV, of the two photons created when a positron–electron pair annihilate.

65 •• The isotope ^{24}Na is a β emitter with a half-life of 15 h. A saline solution containing this radioactive isotope with an activity of 600 kBq is injected into the bloodstream of a patient. Ten hours later, the activity of 1 mL of blood from this individual yields a counting rate of 60 Bq. Determine the volume of blood in this patient.

66 •• **SSM** (a) Determine the closest distance of approach of an 8-MeV α particle in a head-on collision with a nucleus of ^{197}Au and a nucleus of ^{10}B, neglecting the recoil of the struck nuclei. (b) Repeat the calculation taking into account the recoil of the struck nuclei.

67 •• Twelve nucleons are in a one-dimensional infinite square well of length $L = 3$ fm. (a) Using the approximation that the mass of a nucleon is 1 u, find the lowest energy of a nucleon in the well. Express your answer in MeV. What is the ground-state energy of the system of 12 nucleons in the well if (b) all the nucleons are neutrons so that there can be only 2 in each state and (c) 6 of the nucleons are neutrons and 6 are protons so that there can be 4 nucleons in each state? (Neglect the energy of Coulomb repulsion of the protons.)

68 •• The helium nucleus or α particle is a very tightly bound system. Nuclei with $N = Z = 2n$, where n is an integer (e.g., ^{12}C, ^{16}O, ^{20}Ne, and ^{24}Mg), may be thought of as agglomerates of α particles. (a) Use this model to estimate the binding energy of a pair of α particles from the atomic masses of ^{4}He and ^{16}O. Assume that the four α particles in ^{16}O form a regular tetrahedron with one α particle at each vertex. (b) From the result obtained in Part (a) determine, on the basis of this model, the binding energy of ^{12}C and compare your result with the result obtained from the atomic mass of ^{12}C.

69 •• Radioactive nuclei with a decay constant of λ are produced in an accelerator at a constant rate R_p. The number of radioactive nuclei N then obeys the equation $dN/dt = R_p - \lambda N$. (a) If N is zero at $t = 0$, sketch N versus t for this situation. (b) The isotope ^{62}Cu is produced at a rate of 100 per second by placing ordinary copper (^{63}Cu) in a beam of high-energy photons. The reaction is

$$\lambda + {}^{63}Cu \to {}^{62}Cu + n$$

^{62}Cu decays by β decay with a half-life of 10 min. After a time long enough so that $dN/dt \approx 0$, how many ^{62}Cu nuclei are there?

70 •• **SSM** **iSOLVE** The total energy consumed in the United States in 1 y is approximately 7.0×10^{19} J. How many kilograms of ^{235}U would be needed to provide this amount of energy if we assume that 200 MeV of energy is released by each fissioning uranium nucleus, that all of the uranium atoms undergo fission, and that all of the energy-conversion mechanisms used are 100 percent efficient?

71 •• (a) Find the wavelength of a particle in the ground state of a one-dimensional infinite square well of length $L = 2$ fm. (b) Find the momentum in units of MeV/c for a particle with this wavelength. (c) Show that the total energy of an electron with this wavelength is approximately $E \approx pc$. (d) What is the kinetic energy of an electron in the ground state of this well? This calculation shows that if an electron were confined in a region of space as small as a nucleus, it would have a very large kinetic energy.

72 •• If ^{12}C, ^{11}B, and ^{1}H have respective masses of 12.000 000 u, 11.009 306 u, and 1.007 825 u, determine the minimum energy, Q, in MeV, required to remove a proton from a ^{12}C nucleus.

73 ••• [SSM] Assume that a neutron decays into a proton plus an electron without the emission of a neutrino. The energy shared by the proton and the electron is then 0.782 MeV. In the rest frame of the neutron, the total momentum is zero, so the momentum of the proton must be equal and opposite the momentum of the electron. This determines the relative energies of the two particles, but because the electron is relativistic, the exact calculation of these relative energies is somewhat difficult. (a) Assume that the kinetic energy of the electron is 0.782 MeV and calculate the momentum p of the electron in units of MeV/c. (Hint: Use Equation 39-28.) (b) Using your result from Part (a), calculate the kinetic energy $p^2/2m_p$ of the proton. (c) Since the total energy of the electron plus the proton is 0.782 MeV, the calculation in Part (b) gives a correction to the assumption that the energy of the electron is 0.782 MeV. What percentage of 0.782 MeV is this correction?

74 ••• Consider a neutron of mass m moving with speed v_L and making an elastic head-on collision with a nucleus of mass M that is at rest in the laboratory frame of reference. (a) Show that the speed of the center of mass in the lab frame is $V = mv_L/(m + M)$. (b) What is the speed of the nucleus in the center-of-mass frame before the collision and after the collision? (c) What is the speed of the nucleus in the lab frame after the collision? (d) Show that the energy of the nucleus after the collision in the lab frame is

$$\frac{1}{2}M(2V)^2 = \frac{4mM}{(m + M)^2}\left(\frac{1}{2}mv_L^2\right)$$

(e) Show that the fraction of the energy lost by the neutron in this elastic collision is

$$\frac{-\Delta E}{E} = \frac{4mM}{(m + M)^2} = \frac{4(m/M)}{(1 + m/M)^2}$$

75 ••• (a) Use the result from the Part (e) equation of Problem 74 to show that after N head-on collisions of a neutron with carbon nuclei at rest, the energy of the neutron is approximately $(0.714)^N E_0$, where E_0 is its original energy. (b) How many head-on collisions are required to reduce the energy of the neutron from 2 MeV to 0.02 eV, assuming stationary carbon nuclei?

76 ••• On the average, a neutron loses 63 percent of its energy in a collision with a hydrogen atom and 11 percent of its energy in a collision with a carbon atom. Calculate the number of collisions needed to reduce the energy of a neutron from 2 MeV to 0.02 eV if the neutron collides with (a) hydrogen atoms and (b) carbon atoms. (See Problem 75.)

77 ••• [SSM] Frequently, the daughter of a radioactive parent is itself radioactive. Suppose the parent, designated by A, has a decay constant λ_A; while the daughter, designated by B, has a decay constant λ_B. The number of nuclei of B are then given by the solution to the differential equation

$$dN_B/dt = \lambda_A N_A - \lambda_B N_B$$

(a) Justify this differential equation. (b) Show that the solution for this equation is

$$N_B(t) = \frac{N_{A0}\lambda_A}{\lambda_B - \lambda_A}(e^{-\lambda_A t} - e^{-\lambda_B t})$$

where N_{A0} is the number of A nuclei present at $t = 0$ when there are no B nuclei. (c) Show that $N_B(t) > 0$ whether $\lambda_A > \lambda_B$ or $\lambda_B > \lambda_A$. (d) Make a plot of $N_A(t)$ and $N_B(t)$ as a function of time when $\tau_B = 3\tau_A$.

78 ••• Suppose isotope A decays to isotope B with a decay constant λ_A, and isotope B in turn decays with a decay constant λ_B. Suppose a sample contains, at $t = 0$, only isotope A. Derive an expression for the time at which the number of isotope B nuclei will be a maximum. (See Problem 77.)

79 ••• An example of the situation discussed in Problem 77 is the radioactive isotope ^{229}Th, an α emitter with a half-life of 7300 y. Its daughter, ^{225}Ra, is a β emitter with a half-life of 14.8 d. In this instance, as in many instances, the half-life of the parent is much longer than the half-life of the daughter. Using the expression given in Problem 77, Part (b), starting with a sample of pure ^{229}Th containing N_{A0} nuclei, show that the number, N_B, of ^{225}Ra nuclei will, after several years, be a constant given by

$$N_B = \frac{\lambda_A}{\lambda_B}N_A$$

The number of daughter nuclei are said to be in *secular equilibrium*.

Elementary Particles and the Beginning of the Universe

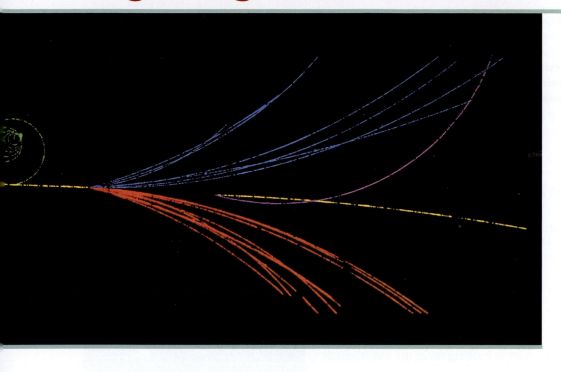

TRACKS IN A BUBBLE CHAMBER PRODUCED BY AN INCOMING HIGH-ENERGY PROTON (YELLOW) INCIDENT FROM THE LEFT, COLLIDING WITH A PROTON AT REST. THE SMALL GREEN SPIRAL IS AN ELECTRON KNOCKED OUT OF AN ATOM. IT CURVES TO THE LEFT BECAUSE OF AN EXTERNAL MAGNETIC FIELD IN THE CHAMBER. THE COLLISION PRODUCES SEVEN NEGATIVE PARTICLES (π^-)(BLUE); A NEUTRAL PARTICLE Λ^0 THAT LEAVES NO TRACK; AND NINE POSITIVE PARTICLES (RED) INCLUDING SEVEN π^+, A K^+, AND A PROTON. THE Λ^0 TRAVELS IN THE ORIGINAL DIRECTION OF THE INCOMING PROTON BEFORE DECAYING INTO A PROTON (YELLOW) AND A π^- (PURPLE).

I n John Dalton's atomic theory of matter (1808), the atom was considered to be the smallest indivisible constituent of matter, that is, an elementary particle. Then, with the discovery of the electron by J. J. Thomson (1897), the Bohr theory of the nuclear atom (1913), and the discovery of the neutron (1932), it became clear that atoms and even nuclei have considerable structure. For a time, it was thought that there were just four "elementary" particles: proton, neutron, electron, and photon. However, the positron or antielectron was discovered in 1932, and shortly thereafter the muon, the pion, and many other particles were predicted and discovered.

Since the 1950s, enormous sums of money have been spent constructing particle accelerators of greater and greater energies in hopes of finding particles predicted by various theories. At present, we know of several hundred particles that at one time or another have been considered to be elementary, and research teams at the giant accelerator laboratories around the world are searching for and finding new particles. Some of these particles have such short lifetimes (of the order of 10^{-23} s) that they can be detected only indirectly. Many particles are observed only in nuclear reactions with high-energy accelerators. In addition to the usual particle properties of mass, charge, and spin, new properties have been found and given whimsical names such as strangeness, charm, color, topness, and bottomness.

➤ In this chapter, we will first look at the various ways of classifying the multitude of particles that have been found. We will then describe the current theory of elementary particles, called the *standard model,* in which all matter in nature—from the exotic particles produced in the giant accelerator laboratories to ordinary grains of sand—is considered to be constructed from just two families of elementary particles, leptons and quarks. In the final section, we will use our knowledge of elementary particles to discuss the big bang theory of the origin of the universe.

41-1 Hadrons and Leptons

All the different forces observed in nature, from ordinary friction to the tremendous forces involved in supernova explosions, can be understood in terms of the four basic interactions: (1) the strong nuclear interaction (also called the hadronic interaction), (2) the electromagnetic interaction, (3) the weak (nuclear) interaction, and (4) the gravitational interaction. The four basic interactions provide a convenient structure for the classification of particles. Some particles participate in all four interactions, whereas other particles participate in only some of the interactions. For example, all particles participate in gravity, the weakest of the interactions. All particles that carry electric charge participate in the electromagnetic interaction.

Particles that interact via the strong interaction are called **hadrons.** There are two kinds of hadrons: **baryons,** which have spin $\frac{1}{2}$ (or $\frac{3}{2}$, $\frac{5}{2}$, etc.), and **mesons,** which have zero or integral spin. Baryons, which include nucleons, are the most massive of the elementary particles. Mesons have intermediate masses between the mass of the electron and the mass of the proton. Particles that decay via the strong interaction have very short lifetimes of the order of 10^{-23} s, which is about the time it takes light to travel a distance equal to the diameter of a nucleus. On the other hand, particles that decay via the weak interaction have much longer lifetimes of the order of 10^{-10} s. Table 41-1 lists some of the properties of those hadrons that are stable against decay via the strong interaction.

Hadrons are rather complicated entities with complex structures. If we use the term *elementary particle* to mean a point particle without structure that is not constructed from some more elementary entities, hadrons do not fit the bill. It is now believed that all hadrons are composed of more fundamental entities called *quarks,* which are truly elementary particles.

Particles that participate in the weak interaction but not in the strong interaction are called **leptons.** These include electrons, muons, and neutrinos, which are all less massive than the lightest hadron. The word *lepton,* meaning "light particle," was chosen to reflect the relatively small mass of these particles. However, the most recently discovered lepton, the *tau,* found by Martin Lewis Perl in 1975, has a mass of 1784 MeV/c^2, nearly twice the mass of the proton (938 MeV/c^2), so we now have a "heavy lepton." As far as we know, leptons are point particles with no structure and can be considered to be truly elementary in the sense that they are not composed of other particles.

There are six leptons. They are the electron and the electron neutrino, the muon and the muon neutrino, and the tau and the tau neutrino. (Each of these leptons has an antiparticle.) The masses of the electron, the muon, and the tau are quite different. The mass of the electron is 0.511 MeV/c^2, the mass of the muon is 106 MeV/c^2, and the mass of the tau is 1784 MeV/c^2. The standard model predicts that neutrinos, like photons, are without mass. Neutrinos were originally

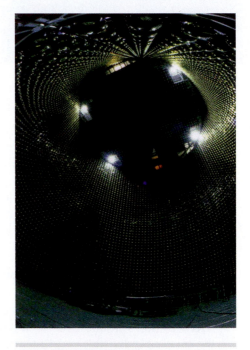

The Super-Kamiokande detector, built in Japan in 1996 as a joint Japanese–American experiment, is essentially a water tank the size of a large cathedral installed in a deep zinc mine 1 mile inside a mountain. When neutrinos pass through the tank, one of the nutrinos occasionally collides with an atom, sending blue light through the water to an array of detectors. This photograph shows the detector wall and top with approximately 9000 photomultiplier tubes that help detect the neutrinos. Experimental results reported in June 1998 indicate that the mass of the neutrino cannot be zero.

TABLE 41-1

Hadrons That Are Stable Against Decay via the Strong Nuclear Interaction

Name	Symbol	Mass, MeV/c^2	Spin, $\hbar$	Charge, e	Antiparticle	Mean Lifetime, s	Typical Decay Products[†]
Baryons							
Nucleon	p (proton)	938.3	$\frac{1}{2}$	+1	p^-	Infinite	
	n (neutron)	939.6	$\frac{1}{2}$	0	$\bar{\mathrm{n}}$	930	$p + e^- + \bar{\nu}_e$
Lambda	Λ^0	1116	$\frac{1}{2}$	0	$\bar{\Lambda}^0$	2.5×10^{-10}	$p + \pi^-$
Sigma[‡]	Σ^+	1189	$\frac{1}{2}$	+1	$\bar{\Sigma}^-$	0.8×10^{-10}	$n + \pi^+$
	Σ^0	1193	$\frac{1}{2}$	0	$\bar{\Sigma}^0$	10^{-20}	$\Lambda^0 + \gamma$
	Σ^-	1197	$\frac{1}{2}$	−1	$\bar{\Sigma}^+$	1.7×10^{-10}	$n + \pi^-$
Xi	Ξ^0	1315	$\frac{1}{2}$	0	$\bar{\Xi}^0$	3.0×10^{-10}	$\Lambda^0 + \pi^0$
	Ξ^-	1321	$\frac{1}{2}$	−1	Ξ^+	1.7×10^{-10}	$\Lambda^0 + \pi^-$
Omega	Ω^-	1672	$\frac{3}{2}$	−1	Ω^+	1.3×10^{-10}	$\Xi^0 + \pi^-$
Mesons							
Pion	π^+	139.6	0	+1	π^-	2.6×10^{-8}	$\mu^+ + \nu_\mu$
	π^0	135	0	0	π^0	0.8×10^{-16}	$\gamma + \gamma$
	π^-	139.6	0	−1	π^+	2.6×10^{-8}	$\mu^- + \bar{\nu}_\mu$
Kaon[§]	K^+	493.7	0	+1	K^-	1.24×10^{-8}	$\pi^+ + \pi^0$
	K^0	497.7	0	0	$\bar{K}^0$	0.88×10^{-10}	$\pi^+ + \pi^-$
						and	
						5.2×10^{-8}	$\pi^+ + e^- + \bar{\nu}_e$
Eta	η^0	549	0	0		2×10^{-19}	$\gamma + \gamma$

† Other decay modes also occur for most particles.
‡ The Σ^0 is included here for completeness even though it does decay via the strong interaction.
§ The K^0 has two distinct lifetimes, sometimes referred to as K^0_{short} and K^0_{long}. All other particles have a unique lifetime.

thought to be massless. However, there is now strong evidence that their mass, though very small, is greater than zero. In the late 1990s, experiments using a detector in Japan called the Super-Kamiokande (Super-K) found that neutrinos emitted from the sun arrived on the earth in much smaller numbers than the numbers predicted from the fusion processes in the sun. This can be explained if the mass of the neutrino were not zero.[†] In addition, a neutrino mass as small as a few eV/c^2 would have great cosmological significance. The answer to the question of whether the universe will continue to expand indefinitely or will reach a maximum size and begin to contract depends on the total mass in the universe. Thus, the answer could depend on whether the mass of the neutrino is actually zero, or is merely small, since the cosmic density of each species of neutrino is ~ 100 per cm^3. The observation of electron neutrinos from the supernova 1987A puts an upper limit on the mass of these neutrinos. Since the velocity of a particle with mass depends on its energy, the arrival time of a burst of neutrinos with mass from a supernova would be spread out in time. The fact that the electron neutrinos from the 1987 supernova all arrived at the earth within 13 s of one another results in an upper limit of about 16 eV/c^2 for their mass. Note that

† The connection between the shortfall of solar-neutrino detections and the mass of the neutrino is elucidated in "On Morphing Neutrinos and Why They Must Have Mass" by Eugene Hecht, *The Physics Teacher* 41 (2003): 164–168.

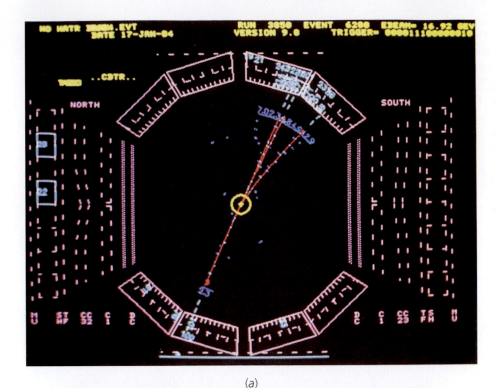

(a)

(b)

(*a*) A computer display of the production and decay of a τ^+ and τ^- pair. An electron and a positron annihilate at the center marked by the yellow cross, producing a τ^+ and τ^- pair, which travel in opposite directions, but quickly decay while still inside the beam pipe (yellow circle). The τ^+ decays into two invisible neutrinos and a μ^+, which travels toward the bottom left. Its track in the drift chamber is calculated by a computer and indicated in red. It penetrates the lead–argon counters outlined in purple and is detected at the blue dot near the bottom blue line that marks the end of a muon detector. The τ^- decays into three charged pions (red tracks moving upward) plus invisible neutrinos. (*b*) The Mark I detector, built by a team from the Stanford Linear Accelerator Center (SLAC) and the Lawrence Berkeley Laboratory, became famous for many discoveries, including the ψ/ J meson and the τ lepton. Tracks of particles are recorded by wire spark chambers wrapped in concentric cylinders around the beam pipe extending out to the ring where physicist Carl Friedberg has his right foot. Beyond this are two rings of protruding tubes, housing photomultipliers that view various scintillation counters. The rectangular magnets at the left guide the counterrotating beams that collide in the center of the detector.

an upper limit does not imply that the mass is not zero. Measurements of the relative number of muon neutrinos and electron neutrinos entering the huge, underground Super-K detector suggest that at least one type of neutrino can oscillate between types (e.g., between a mu neutrino and a tau neutrino). Further measurements of antineutrinos from nuclear reactors strongly shows that all three types of neutrinos oscillate between types and thus have mass. Measurements made in Japan, using the *Kam*ioka *L*iquid *S*cintillator *A*nti-*N*eutrino *D*etector (KamLAND), show that oscillations from one species of neutrino to another species of neutrino can be observed over path lengths as short as 180 km (Figure 41-1).

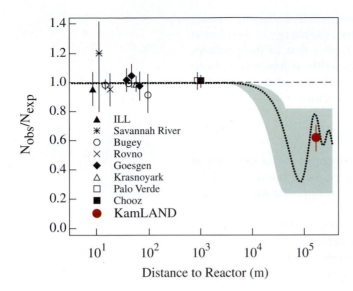

FIGURE 41-1 First evidence for antineutrino disappearance. The ratio of the number of antineutrinos observed N_{obs} to the number that one would expect to observe N_{exp} (assuming no neutrino oscillations) is plotted versus distance to the nearest antineutrino sources. The KamLAND site is 180,000 m (180 km) from nearby antineutrino sources (nuclear reactors), while the other eight detector sites are less than 1000 m from nearby nuclear reactors. For these eight sites, $N_{obs}/N_{exp} = 1.0$, which is what is expected assuming no neutrino oscillations. However, the KamLAND detector found $N_{obs}/N_{exp} = 0.6$. This result is strong evidence that while neutrinos do not oscillate in significant numbers while traveling over path lengths of less than 1.0 km, they do oscillate in significant numbers while traveling over path lengths only a few orders of magnitude longer than that.

41-2 Spin and Antiparticles

One important characteristic of a particle is its intrinsic spin angular momentum. We have already discussed the fact that the electron has a quantum number m_s that corresponds to the z component of its intrinsic spin characterized by the quantum number $s = \frac{1}{2}$. Protons, neutrons, neutrinos, and the various other particles that also have an intrinsic spin characterized by the quantum number $s = \frac{1}{2}$ are called **spin-$\frac{1}{2}$ particles.** Particles that have spin $\frac{1}{2}$ (or $\frac{3}{2}$, $\frac{5}{2}$, etc.) are called fermions and obey the Pauli exclusion principle. Particles such as pions and other mesons have zero spin or integral spin ($s = 0, 1, 2$, etc.). These particles are called bosons and do not obey the Pauli exclusion principle. That is, any number of these particles can be in the same quantum state.

Spin-$\frac{1}{2}$ particles are described by the Dirac equation, which is an extension of the Schrödinger equation that includes special relativity. One feature of Paul Dirac's theory, proposed in 1927, is the prediction of the existence of antiparticles. In special relativity, the energy of a particle is related to the mass and the momentum of the particle by $E = \pm\sqrt{p^2c^2 + m^2c^4}$ (Equation 39-28). We usually choose the positive solution and dismiss the negative-energy solution with a physical argument. However, the Dirac equation requires the existence of wave functions that correspond to the negative-energy states. Dirac got around this difficulty by postulating that all the negative-energy states were filled and would therefore not be observable. Only holes in the "infinite sea" of negative-energy states would be observed. For example, a hole in the negative sea of electron energy states would appear as a particle identical to the electron except with positive charge. When such a particle came in the vicinity of an electron the two particles would annihilate, releasing two photons with a total energy of $2m_ec^2$, where m_e is the mass of the electron. This interpretation received little attention until a particle with just these properties, called the positron, was discovered in 1932 by Carl Anderson.

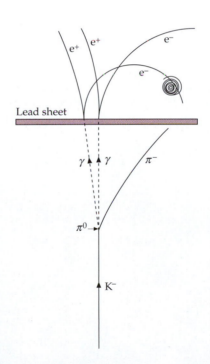

A negative kaon (K^-) enters a bubble chamber from the bottom and decays into a π^-, which moves off to the right, and a π^0, which immediately decays into two photons whose paths are indicated by the dashed lines in the drawing. Each photon interacts in the lead sheet, producing an electron–positron pair. The spiral at the right is another electron that has been knocked out of an atom in the chamber. (Other extraneous tracks have been removed from the photograph.)

Antiparticles are never created alone but always in particle–antiparticle pairs. In the creation of an electron–positron pair by a photon, the energy of the photon must be greater than the rest energy of the electron plus that of the positron, which is $2m_e c^2 \approx 1.02$ MeV. Although the positron is stable, it has only a short-term existence in our universe because of the large supply of electrons in matter. The fate of a positron is annihilation according to the reaction

$$e^+ + e^- \rightarrow \gamma + \gamma \qquad\qquad 41\text{-}1$$

The probability of this reaction is large only if the positron is at rest or nearly at rest. In the center-of-mass reference frame, the momentum of the two particles prior to annihilation is zero, so two photons moving in opposite directions are needed to conserve linear momentum.

The fact that we call electrons *particles* and positrons *antiparticles* does not imply that positrons are less fundamental than electrons. It merely reflects the nature of our universe. If our matter were made up of negative protons and positive electrons, then positive protons and negative electrons would suffer quick annihilation and would be called antiparticles.

The antiproton (p⁻) was discovered in 1955 by Emilio Segrè and Owen Chamberlain using a beam of protons in the Bevatron at Berkeley to produce the reaction[†]

$$p^+ + p^+ \rightarrow p^+ + p^+ + p^+ + p^- \qquad\qquad 41\text{-}2$$

[†] The antiproton is sometimes denoted by $\bar{p}$ rather than p⁻. For neutral particles, such as the neutron, the bar must be used to denote the antiparticle. Thus, the antineutron is denoted by $\bar{n}$. The normal electron and proton are often denoted by e and p without the minus or plus superscripts.

An aerial view of the European Laboratory for Particle Physics (CERN) just outside of Geneva, Switzerland. The large circle shows the Large Electron–Positron collider (LEP) tunnel, which is 27 km in circumference. The irregular dashed line is the border between France and Switzerland.

The creation of a proton–antiproton pair (Figure 41-2) requires kinetic energy of at least $2m_p c^2 = 1877$ MeV $= 1.877$ GeV in the zero-momentum reference frame in which the two protons approach each other with equal and opposite momenta. In the laboratory frame in which one of the protons is initially at rest, the kinetic energy of the incoming proton must be at least $6m_p c^2 = 5.63$ GeV (see Problem 40 of Chapter 39). This energy was not available in laboratories before the development of high-energy accelerators in the 1950s. Antiprotons annihilate with protons to produce two gamma rays in a reaction similar to the reaction in Equation 41-1.

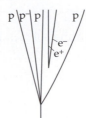

FIGURE 41-2 Bubble-chamber tracks that show the creation of a proton–antiproton pair in the collision of an incident 25-GeV proton, with a stationary proton in liquid hydrogen.

The tunnel of the proton–antiproton collider at CERN. The same bending magnets and focusing magnets can be used for protons or antiprotons moving in opposite directions. The rectangular box in the foreground is a focusing magnet, and the next four boxes are the bending magnets.

PROTON–ANTIPROTON ANNIHILATION   **EXAMPLE 41-1**   **Put It in Context**

You have been reading about nuclear physics and particle interactions. In particular, you have been looking at the reaction $p^+ + p^- \rightarrow \gamma + \gamma$ (proton–antiproton annihilation). You wonder if the photons produced are visible to the human eye if the two protons are initially at rest. Are the photons visible to the human eye?

PICTURE THE PROBLEM If the photons are visible, they should have wavelengths in the visible range (400 nm to 800 nm). Because the proton and the antiproton are at rest, conservation of momentum requires that the two photons created in their annihilation have equal and opposite momenta and therefore equal energies, frequencies, and wavelengths. Conservation of energy implies that the photons have a combined energy equal to the rest energy of the proton plus the rest energy of the antiproton (approximately 938 MeV each).

1. Set the total energy of the two photons, $2E_\gamma$, equal to the rest energy of the proton plus antiproton and solve for E_γ:

$$2E_\gamma = 2m_p c^2$$

so

$$E_\gamma = m_p c^2 = 938 \text{ MeV}$$

2. Set the energy of the photon equal to $hf = hc/\lambda$ and solve for the wavelength λ:

$$E_\gamma = hf = \frac{hc}{\lambda}$$

$$\lambda = \frac{hc}{E_\gamma} = \frac{1240 \text{ eV·nm}}{938 \text{ MeV}}$$

$$= 1.32 \times 10^{-6} \text{ nm} = 1.32 \text{ fm}$$

3. Compare this wavelength with the wavelengths of visible light:

The photons are *not* in the visible spectrum.

REMARKS The wavelength of the photons is more than eight orders of magnitude less than 400 nm—the shortest wavelength in the visible spectrum.

41-3 The Conservation Laws

One of the maxims of nature is "anything that can happen does." If a conceivable decay or reaction does not occur, there must be a reason. The reason is usually expressed in terms of a conservation law. The conservation of energy rules out the decay of any particle for which the total mass of the decay products would be greater than the initial mass of the particle before decay. The conservation of linear momentum requires that when an electron and a positron at rest annihilate, two photons must be emitted. Angular momentum must also be conserved in a reaction or a decay. A fourth conservation law that restricts the possible particle decays and reactions is that of electric charge. The net electric charge before a decay or a reaction must equal the net charge after the decay or the reaction.

There are two additional conservation laws that are important in the reactions and the decays of elementary particles: the conservation of baryon number and the conservation of lepton number. Consider the proposed decay

$$p \rightarrow \pi^0 + e^+$$

This decay would conserve charge, energy, angular momentum, and linear momentum, but it does not occur. It does not conserve either lepton number or baryon number. The conservation of lepton number and baryon number implies that whenever a lepton or a baryon is created, an antiparticle of the same type is also created. We assign the **lepton number** $L = +1$ to all leptons, $L = -1$ to all antileptons, and $L = 0$ to all other particles. Similarly, the **baryon number** $B = +1$ is assigned to all baryons, $B = -1$ to all antibaryons, and $B = 0$ to all other particles. The baryon numbers and the lepton numbers cannot change in a reaction or a decay. The conservation of baryon number along with the conservation of energy implies that the least massive baryon, the proton, must be stable.

The conservation of lepton number implies that the neutrino emitted in the β decay of the free neutron is an antineutrino:

$$n \rightarrow p^+ + e^- + \bar{\nu}_e \qquad \qquad 41\text{-}3$$

The fact that neutrinos and antineutrinos are different is illustrated by an experiment in which ^{37}Cl is bombarded with an intense antineutrino beam from the decay of reactor neutrons. If neutrinos and antineutrinos were the same, we would expect the following reaction:

$$^{37}\text{Cl} + \bar{\nu}_e \rightarrow {}^{37}\text{Ar} + e^- \qquad \qquad 41\text{-}4$$

This reaction is not observed. However, if protons are bombarded with anti-neutrinos, the reaction

$$p + \bar{\nu}_e \rightarrow n + e^+ \qquad 41\text{-}5$$

is observed. Note that the lepton number is -1 on the left side of the reaction in Equation 41-4 and $+1$ on the right side of the reaction. But the lepton number is -1 on both sides of the reaction in Equation 41-5.

Not only are neutrinos and antineutrinos distinct particles, but the neutrinos associated with electrons are distinct from the neutrinos associated with muons. Electron-like leptons (e and ν_e), muon-like leptons (μ and ν_μ), and tau-like leptons (τ and ν_τ) are each separately conserved, so we assign separate lepton numbers L_e, L_μ, and L_τ to the particles. For e and ν_e, $L_e = +1$; for their anti-particles, $L_e = -1$; and for all other particles, $L_e = 0$. The lepton numbers L_μ and L_τ are similarly assigned.

WHAT LAWS ARE BEING VIOLATED?   **EXAMPLE 41-2**

What conservation laws (if any) are violated by the following proposed decays: (a) $n \rightarrow p + \pi$, (b) $\Lambda^0 \rightarrow p^- + \pi^+$, and (c) $\mu^- \rightarrow e^- + \gamma$?

(a) There are no leptons in this decay, so there is no problem with the conservation of lepton number. The net charge is zero before the decay and after the decay, so charge is conserved. Also, the baryon number is $+1$ before the decay and after the decay. However, the rest energy of the proton (938.3 MeV) plus the rest energy of the pion (139.6 MeV) is greater than the rest energy of the neutron (939.6 MeV). Thus, this decay violates the conservation of energy.

(b) Again, there are no leptons involved, and the net charge is zero before the decay and after the decay. Also, the rest energy of the Λ^0 (1116 MeV) is greater than the rest energy of the antiproton (938.3 MeV) plus the rest energy of the pion (139.6 MeV), so energy is conserved, with the loss in rest energy equaling the gain in kinetic energy of the decay products. However, this decay does not conserve baryon number, which is $+1$ for the Λ^0 and -1 for the antiproton.

(c) This reaction does not conserve muon lepton number or electron lepton number. The muon does decay via $\mu^- \rightarrow e^- + \bar{\nu}_e + \nu_\mu$, which does conserve both muon lepton numbers and electron lepton numbers.

There are some conservation laws that are not universal but apply only to certain kinds of interactions. In particular, there are quantities that are conserved in decays and reactions that occur via the strong interaction but not in decays or reactions that occur via the weak interaction. One of these quantities that is particularly important is **strangeness,** introduced by M. Gell-Mann and K. Nishijima in 1952 to explain the strange behavior of some of the heavy baryons and mesons. Consider the reaction

$$p + \pi^- \rightarrow \Lambda^0 + K^0 \qquad 41\text{-}6$$

The proton and the pion interact via the strong interaction. Both the Λ^0 and K^0 decay into hadrons

$$\Lambda^0 \rightarrow p + \pi^- \qquad 41\text{-}7$$

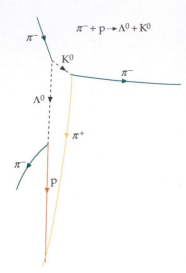

$$\pi^- + p \rightarrow \Lambda^0 + K^0$$

An early photograph of bubble-chamber tracks at the Lawrence Berkeley Laboratory, showing the production and the decay of two strange particles, the K^0 and the Λ^0. These neutral particles are identified by the tracks of their decay particles. The lambda particle was named because of the similarity of the tracks of its decay particles to the Greek letter Λ. (The blue tracks are particles not involved in the reaction of Equation 41-6.)

and

$$K^0 \rightarrow \pi^+ + \pi^- \qquad\qquad 41\text{-}8$$

However, the decay times for both the Λ^0 and K^0 are of the order of 10^{-10} s, which is characteristic of the weak interaction, rather than 10^{-23} s, which would be expected for the strong interaction. Other particles showing similar behavior were called **strange particles.** These particles are always produced in pairs and never singly, even when all other conservation laws are met. This behavior is described by assigning a new property called strangeness to these particles. In reactions and decays that occur via the strong interaction, strangeness is conserved. In reactions and decays that occur via the weak interaction, the strangeness can change by ±1. The strangeness of the ordinary hadrons—the nucleons and pions—was arbitrarily taken to be zero. The strangeness of the K^0 was arbitrarily chosen to be $+1$. The strangeness of the Λ^0 particle must then be -1 so that strangeness is conserved in the reaction of Equation 41-6. The strangeness of other particles could then be assigned by looking at their various reactions and decays. In reactions and decays that occur via the weak interaction, the strangeness can change by ±1.

Figure 41-3 shows the masses of the baryons and the mesons that are stable against decay via the strong interaction versus strangeness. We can see from this figure that these particles cluster in multiplets of one, two, or three particles of approximately equal mass, and that the strangeness of a multiplet of particles is related to the *center of charge* of the multiplet.

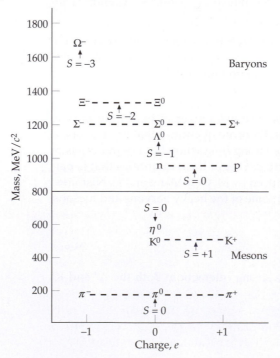

FIGURE 41-3 The strangeness of hadrons shown on a plot of mass versus charge. The strangeness of a baryon-charge multiplet is related to the number of places the center of charge of the multiplet is displaced from that of the nucleon doublet. For each displacement of e, the strangeness changes by ±1. For mesons, the strangeness is related to the number of places the center of charge is displaced from that of the pion triplet. Because of the unfortunate original assignment of $+1$ for the strangeness of kaons, all of the baryons that are stable against decay via the strong interaction have negative or zero strangeness.

EXAMPLE   41 - 3

State whether the following decays can occur via the strong interaction, via the weak interaction, or not at all: (*a*) $\Sigma^+ \rightarrow p + \pi^0$, (*b*) $\Sigma^0 \rightarrow \Lambda^0 + \gamma$, and (*c*) $\Xi^0 \rightarrow n + \pi^0$.

PICTURE THE PROBLEM We first note that the mass of each decaying particle is greater than the mass of the decay products, so there is no problem with energy conservation in any of the decays. In addition, there are no leptons involved in any of the decays, and charge and baryon number are both conserved in all the decays. The decay will occur via the strong interaction if strangeness is conserved. If $\Delta S = \pm 1$, the decay will occur via the weak interaction. If $|S|$ changes by more than 1, the decay will not occur.

(*a*) From Figure 41-3, we can see that the strangeness of the Σ^+ is -1, whereas the strangeness of both the proton and the pion is zero. This decay is possible via the weak interaction but not the strong interaction. It is, in fact, one of the decay modes of the Σ^+ particle with a lifetime of the order of 10^{-10} s.

(*b*) Since the strangeness of both the Σ^0 and Λ^0 is -1, this decay can proceed via the strong interaction. It is, in fact, the dominant mode of decay of the Σ^0 particle with a lifetime of approximately 10^{-20} s.

(*c*) The strangeness of the Ξ^0 is -2, whereas the strangeness of both the neutron and the pion is zero. Since strangeness cannot change by 2 in a decay or in a reaction, this decay cannot occur.

41-4 Quarks

Leptons appear to be truly elementary particles in that they do not break down into smaller entities and they seem to have no measurable size or structure. Hadrons, on the other hand, are complex particles with size and structure, and they decay into other hadrons. Furthermore, at the present time, there are only six known leptons, whereas there are many more hadrons. Except for the Σ^0 particle, Table 41-1 includes only hadrons that are stable against decay via the strong interaction. Hundreds of other hadrons have been discovered; their properties, such as charge, spin, mass, strangeness, and decay schemes, have been measured.

The most important advance in our understanding of elementary particles was the quark model proposed by M. Gell-Mann and G. Zweig in 1963 in which all hadrons consist of combinations of two or three truly elementary particles called **quarks.**[†] In the original model, quarks came in three types, called **flavors,** labeled *u*, *d*, and *s* (for *up*, *down*, and *strange*). An unusual property of quarks is that they carry fractional electron charges. The charge of the *u* quark is $+\frac{2}{3}e$ and the charge of the *d* and *s* quarks is $\frac{1}{3}e$. Each quark has spin $\frac{1}{2}$ and a baryon number of $\frac{1}{3}$. The strangeness of the *u* and *d* quark is 0, and the strangeness of the *s* quark is -1. Each quark has an antiquark with the opposite electric charge, baryon number, and strangeness. Baryons consist of three quarks (or three antiquarks for antiparticles), whereas mesons consist of a quark and an antiquark, giving mesons a baryon number $B = 0$, as required. The proton consists of the combination *uud* and the neutron consists of the combination *udd*. Baryons with a strangeness $S = -1$ contain one *s* quark. All the particles listed in Table 41-1 can be constructed from these three quarks and three antiquarks.[‡] The great strength of the quark model is that all the allowed combinations of three quarks or

---

[†] The name *quark* was chosen by M. Gell-Mann from a quotation from *Finnegan's Wake* by James Joyce.

[‡] The correct quark combinations of hadrons are not always obvious, because of the symmetry requirements on the total wave function. For example, the π^0 meson is represented by a linear combination of $u\bar{u}$ and $d\bar{d}$.

quark–antiquark pairs result in known hadrons. Strong evidence for the existence of quarks inside a nucleon is provided by high-energy scattering experiments called *deep inelastic scattering*. In these experiments, a nucleon is bombarded with electrons, muons, or neutrinos of energies from 15 GeV to 200 GeV. Analyses of particles scattered at large angles indicate that inside the nucleon are three spin-$\frac{1}{2}$ particles of sizes much smaller than that of the nucleon. These experiments are analogous to Rutherford's scattering of α particles by atoms in which the presence of a tiny nucleus in the atom was inferred from the large-angle scattering of the α particles.

GIVEN THE CONSTITUENT QUARK SPECIES, IDENTIFY THE PARTICLE

EXAMPLE 41-4

What are the properties of the particles made up of the following quarks: (a) $u\bar{d}$, (b) $\bar{u}d$, (c) dds, and (d) uss?

PICTURE THE PROBLEM Baryons are made up of three quarks, whereas mesons consist of a quark and an antiquark. We add the electric charges of the quarks to find the total charge of the hadron. We also find the strangeness of the hadron by adding the strangeness of the quarks.

(a) Because $u\bar{d}$ is a quark–antiquark combination, it has baryon number 0 and is therefore a meson. There is no strange quark here, so the strangeness of the meson is zero. The charge of the up quark is $+\frac{2}{3}e$ and the charge of the antidown quark is $+\frac{1}{3}e$, so the charge of the meson is $+1e$. This is the quark combination of the π^{+} meson.

(b) The particle $\bar{u}d$ is also a meson with zero strangeness. Its electric charge is $-\frac{2}{3}e + (-\frac{1}{3}e) = -1e$. This is the quark combination of the π^{-} meson.

(c) The particle dds is a baryon with strangeness -1 because it contains one strange quark. Its electric charge is $-\frac{1}{3}e - \frac{1}{3}e - \frac{1}{3}e = -1e$. This is the quark combination for the Σ^{-} particle.

(d) The particle uss is a baryon with strangeness -2. Its electric charge is $+\frac{2}{3}e - \frac{1}{3}e - \frac{1}{3}e = 0$. This is the quark combination for the Ξ^{0} particle.

MASTER the CONCEPT WEB

In 1967, a fourth quark was proposed to explain some discrepancies between experimental determinations of certain decay rates and calculations based on the quark model. The fourth quark is labeled c for a new property called **charm.** Like strangeness, charm is conserved in strong interactions but changes by ± 1 in weak interactions. In 1975, a new heavy meson called the ψ/J particle (or simply the ψ particle) was discovered that has the properties expected of a $c\bar{c}$ combination. Since then, other mesons with combinations such as $c\bar{d}$ and $\bar{c}d$ as well as baryons containing the charmed quark, have been discovered. Two more quarks labeled t and b (for *top* and *bottom*) were proposed in the 1970s. In 1977, a massive new meson called the Y meson or **bottomonium,** which is considered to have the quark combination $b\bar{b}$, was discovered. The top quark was observed in 1995. The properties of the six quarks are listed in Table 41-2.

The six quarks and six leptons (and their antiparticles) are thought to be the fundamental elementary particles of which all matter is composed. Table 41-3 lists the masses of the fundamental particles. In this table, the masses given for neutrinos are upper limits. The masses given for quarks are educated guesses. There is experimental evidence for the existence of each of these particles.

TABLE 41-2

Properties of Quarks and Antiquarks

Flavor	Spin	Charge	Baryon Number	Strangeness	Charm	Topness	Bottomness
Quarks							
u (up)	$\frac{1}{2}\hbar$	$+\frac{2}{3}e$	$+\frac{1}{3}$	0	0	0	0
d (down)	$\frac{1}{2}\hbar$	$-\frac{1}{3}e$	$+\frac{1}{3}$	0	0	0	0
s (strange)	$\frac{1}{2}\hbar$	$-\frac{1}{3}e$	$+\frac{1}{3}$	-1	0	0	0
c (charmed)	$\frac{1}{2}\hbar$	$+\frac{2}{3}e$	$+\frac{1}{3}$	0	$+1$	0	0
t (top)	$\frac{1}{2}\hbar$	$+\frac{2}{3}e$	$+\frac{1}{3}$	0	0	$+1$	0
b (bottom)	$\frac{1}{2}\hbar$	$-\frac{1}{3}e$	$+\frac{1}{3}$	0	0	0	$+1$
Antiquarks							
$\bar{u}$	$\frac{1}{2}\hbar$	$-\frac{2}{3}e$	$-\frac{1}{3}$	0	0	0	0
$\bar{d}$	$\frac{1}{2}\hbar$	$+\frac{1}{3}e$	$-\frac{1}{3}$	0	0	0	0
$\bar{s}$	$\frac{1}{2}\hbar$	$+\frac{1}{3}e$	$-\frac{1}{3}$	$+1$	0	0	0
$\bar{c}$	$\frac{1}{2}\hbar$	$-\frac{2}{3}e$	$-\frac{1}{3}$	0	-1	0	0
$\bar{t}$	$\frac{1}{2}\hbar$	$-\frac{2}{3}e$	$-\frac{1}{3}$	0	0	-1	0
$\bar{b}$	$\frac{1}{2}\hbar$	$+\frac{1}{3}e$	$-\frac{1}{3}$	0	0	0	-1

Quark Confinement

Despite considerable experimental effort, no isolated quark has ever been observed. It is now believed that it is impossible to obtain an isolated quark. Although the force between quarks is not known, it is believed that the potential energy of two quarks increases with increasing separation distance so that an infinite amount of energy would be needed to separate the quarks completely. This would be true, for example, if the force of attraction between two quarks remains constant or increases with separation distance, rather than decreasing with increasing separation distance as is the case for other fundamental forces, such as the electric force between two charges, the gravitational force between two masses, and the strong nuclear force between two hadrons.

When a large amount of energy is added to a quark system, such as a nucleon, a quark–antiquark pair is created and the original quarks remain confined within the original system. Because quarks cannot be isolated, but are always bound in a baryon or a meson, the mass of a quark cannot be accurately known, which is why the masses listed in Table 41-3 are merely educated guesses.

41-5 Field Particles

In addition to the six fundamental leptons and six fundamental quarks, there are other particles, called *field particles* or *field quanta*, that are associated with the forces exerted by one elementary particle on another. In **quantum electrodynamics,** the electromagnetic field of a single charged particle is described by **virtual photons** that are continuously being emitted and reabsorbed by the particle. If we put energy into the system by accelerating the charge, some of these virtual photons are shaken off and

TABLE 41-3

Masses of Fundamental Particles

Particle	Mass
Quarks	
u (up)	336 MeV/c^2
d (down)	338 MeV/c^2
s (strange)	540 MeV/c^2
c (charmed)	1,500 MeV/c^2
t (top)	174,000 MeV/c^2
b (bottom)	4,500 MeV/c^2
Leptons	
e^- (electron)	0.511 MeV/c^2
ν_e (electron neutrino)	< 7 eV/c^2
μ^- (muon)	105.659 MeV/c^2
ν_μ (muon neutrino)	< 0.27 MeV/c^2
τ^- (tau)	1,784 MeV/c^2
ν_τ (tau neutrino)	< 31 MeV/c^2

become real, observable photons. The photon is said to mediate the electromagnetic interaction. Each of the four basic interactions can be described via mediating field particles.

The field quantum associated with the gravitational interaction, called the **graviton,** has not yet been observed. The gravitational *charge* analogous to electric charge is mass.

The weak interaction is thought to be mediated by three field quanta called **vector bosons:** W^+, W^-, and Z^0. These particles were predicted by Sheldon Glashow, Abdus Salam, and Steven Weinberg in a theory called the *electroweak theory*, which we discuss in the next section. The W and Z particles were first observed in 1983 by a group of over a hundred scientists led by Carlo Rubbia using the high-energy accelerator at CERN in Geneva, Switzerland. The masses of the $W^\pm$ particles (about 80 GeV/c^2) and the Z particle (about 91 GeV/c^2) measured in this experiment were in excellent agreement with those predicted by the electroweak theory. (The W^- particle is the antiparticle of the W^+ particle, so they must have identical masses.)

The field quanta associated with the strong force between quarks are called **gluons.** Isolated gluons have not been observed experimentally. The *charge* responsible for the strong interactions comes in three varieties, labeled *red, green,* and *blue* (analogous with the three primary colors), and the strong charge is called the **color charge.** The field theory for strong interactions, analogous to quantum electrodynamics for electromagnetic interactions, is called **quantum chromodynamics (QCD).**

Table 41-4 lists the bosons responsible for mediating the basic interactions.

TABLE 41-4

Bosons That Mediate the Basic Interactions

Interaction	Boson	Spin	Mass	Electric Charge
Strong	g (gluon)†	1	0	0
Weak	$W^\pm$	1	80.22 GeV/c^2	$\pm 1e$
	Z^0	1	91.19 GeV/c^2	0
Electromagnetic	γ (photon)	1	0	0
Gravitational	Graviton†	2	0	0

† Not yet observed.

41-6 The Electroweak Theory

In the **electroweak theory,** the electromagnetic and weak interactions are considered to be two different manifestations of a more fundamental electroweak interaction. At very high energies (>> 100 GeV), the electroweak interaction would be mediated by four bosons. From symmetry considerations, these would be a triplet consisting of W^+, W^0, and W^-, all of equal mass, and a singlet B^0 of some other mass. Neither the W^0 nor the B^0 would be observed directly, but one linear combination of the W^0 and the B^0 would be the Z^0 and another would be the photon. At ordinary energies, the symmetry is broken. This leads to the separation of the electromagnetic interaction mediated by the massless photon and the weak interaction mediated by the W^+, W^-, and Z^0 particles. The fact that the photon is massless and that the W and Z particles have masses of the order of 100 GeV/c^2 shows that the symmetry assumed in the electroweak theory does not exist at lower energies.

The symmetry-breaking mechanism is called a **Higgs field,** which requires a new boson, the **Higgs boson,** whose rest energy is expected to be of the order of 1 TeV (1 TeV = 10^{12} eV). The Higgs boson has not yet been observed. Calculations show that Higgs bosons (if they exist) should be produced in head-on collisions between protons of energies of the order of 20 TeV. Such energies are not presently available.

41-7 The Standard Model

The combination of the quark model, electroweak theory, and quantum chromodynamics is called the **standard model.** In this model, the fundamental particles are the leptons and quarks, each of which comes in six flavors, as shown in Table 41-3; the force carriers are the photon, the $W^\pm$ and Z particles, and the gluons (of which there are eight types). The leptons and quarks are all spin-$\frac{1}{2}$ fermions, which obey the Pauli exclusion principle, and the force carriers are integral-spin bosons, which do not obey the Pauli exclusion principle. Every force in nature is due to one of the four basic interactions: strong, electromagnetic, weak, and gravitational. A particle experiences one of the basic interactions if it carries a charge associated with that interaction. Electric charge is the familiar charge that we have studied previously. Weak charge, also called flavor charge, is carried by leptons and quarks. The charge associated with the strong interaction is called color charge and is carried by quarks and gluons but not by leptons. The charge associated with the gravitational force is mass. It is important to note that the photon, which mediates the electromagnetic interaction, does not carry electric charge. Similarly, the $W^\pm$ and Z particles, which mediate the weak interaction, do not carry weak charge. However, the gluons, which mediate the strong interaction, do carry color charge. This fact is related to the confinement of quarks as discussed in Section 41-4.

All matter is made up of leptons or quarks. There are no known composite particles consisting of leptons bound together by the weak force. Leptons exist only as isolated particles. Hadrons (baryons and mesons) are composite particles consisting of quarks bound together by the color charge. A result of QCD theory is that only color-neutral combinations of quarks are allowed. Three quarks of different colors can combine to form color-neutral baryons, such as the neutron and the proton. Mesons contain a quark and an antiquark and are also color-neutral. Excited states of hadrons are considered to be different particles. For example, the Δ^+ particle is an excited state of the proton. Both are made up of the *uud* quarks, but the proton is in the ground state with spin $\frac{1}{2}$ and a rest energy of 938 MeV, whereas the Δ^+ particle is in the first excited state with spin $\frac{3}{2}$ and a rest energy of 1232 MeV. The two *u* quarks can be in the same spin state in the Δ^+ without violating the exclusion principle, because they have different color. All baryons eventually decay to the lightest baryon, the proton. The proton cannot decay because of conservation of energy and conservation of baryon number.

The strong interaction has two parts, the fundamental interaction or color interaction and what is called the *residual strong interaction.* The fundamental interaction is responsible for the force exerted by one quark on another quark and is mediated by gluons. The residual strong interaction is responsible for the force between color-neutral nucleons, such as the neutron and the proton. This force is due to the residual strong interactions between the color-charged quarks that make up the nucleons and can be viewed as being mediated by the exchange of mesons. The residual strong interaction between color-neutral nucleons can be thought of as analogous to the residual electromagnetic interaction between neutral atoms that bind them together to form molecules. Table 41-5 lists some of the properties of the basic interactions.

TABLE 41-5

Properties of the Basic Interactions

| | Gravitational | Weak | Electromagnetic | Strong | |
				Fundamental	Residual
Acts on	Mass	Flavor	Electric charge	Color charge	
Particles experiencing	All	Quarks, leptons	Electrically charged	Quarks, gluons	Hadrons
Particles mediating	Graviton	$W^\pm, Z$	γ	Gluons	Mesons
Strength for two quarks at 10^{-18} m[†]	10^{-41}	0.8	1	25	(not applicable)
Strength for two protons in nucleus[†]	10^{-36}	10^{-7}	1	(not applicable)	20

† Strengths are relative to electromagnetic strength.

For each particle there is an antiparticle. A particle and its antiparticle have identical mass and spin but opposite electric charge. For leptons, the lepton numbers L_e, L_μ, and L_τ of the antiparticles are the negatives of the corresponding numbers for the particles. For example, the lepton number for the electron is $L_e = +1$, and the lepton number for the positron is $L_e = -1$. For hadrons, the baryon number, strangeness, charm, topness, and bottomness are the sums of those quantities for the quarks that make up the hadron. The number of each antiparticle is the negative of the number for the corresponding particle. For example, the lambda particle Λ^0, which is made up of the *uds* quarks, has $B = 1$ and $S = -1$, whereas its antiparticle $\overline{\Lambda}^0$, which is made up of the $\overline{u}\overline{d}\overline{s}$ quarks, has $B = -1$ and $S = +1$. A particle such as the photon γ or the Z^0 particle that has zero electric charge, $B = 0$, $L = 0$, $S = 0$; and zero charm, topness, and bottomness is its own antiparticle. Note that the K^0 meson ($d\overline{s}$) has a zero value for all of these quantities except strangeness, which is +1. Its antiparticle, the $\overline{K}^0$ meson ($\overline{d}s$), has strangeness -1, which makes it distinct from the K^0. The π^+ ($u\overline{d}$) and π^- ($\overline{u}d$) are somewhat special in that they have electric charge but zero values for L, B, and S. They are antiparticles of each other, but since there is no conservation law for mesons, it is impossible to say which is the particle and which is the antiparticle. Similarly, the W^+ and W^- are antiparticles of each other.

Grand Unification Theories

With the success of the electroweak theory, attempts have been made to combine the strong, electromagnetic, and weak interactions in various **grand unification theories (GUTs).** In one of these theories, leptons and quarks are considered to be two aspects of a single class of particles. Under certain conditions, a quark could change into a lepton and vice versa, even though this would appear to violate the conservation of lepton number and baryon number. One of the exciting predictions of this theory is that the proton is not stable but merely has a very long lifetime of the order of 10^{32} y. Such a long lifetime makes proton decay difficult to observe. However, projects are ongoing in which detectors monitor very large numbers of protons in search of an event indicating the decay of a proton.

41-8 The Evolution of the Universe

In the presently accepted model, the universe began with a singular cataclysmic event called the **big bang** and is expanding. The first evidence that the universe is expanding was the astronomer Edwin Powell Hubble's discovery of the relation between the redshifts in the spectra of galaxies and their distances from us. This relation is illustrated in Figure 41-4 for a group of spiral galaxies used by astronomers for calibrating distances. Provided that the redshift is due to the Doppler effect, the recession velocity v of a galaxy is related to its distance r from us by Hubble's law,

$$v = Hr \qquad\qquad 41\text{-}9$$

where H is the **Hubble constant.** In principle, the value of H is easy to obtain since it relies on the direct calculation of v from redshift measurements. However, astronomical distances are very difficult to obtain, and they have been computed for only a fraction of the 10^{10} or so galaxies in the observable universe. Thus, the value of H changes as distance calibration data are refined. The currently accepted value of the Hubble constant is

$$H = \frac{23 \text{ km/s}}{10^6 \; c \cdot y} \qquad\qquad 41\text{-}10$$

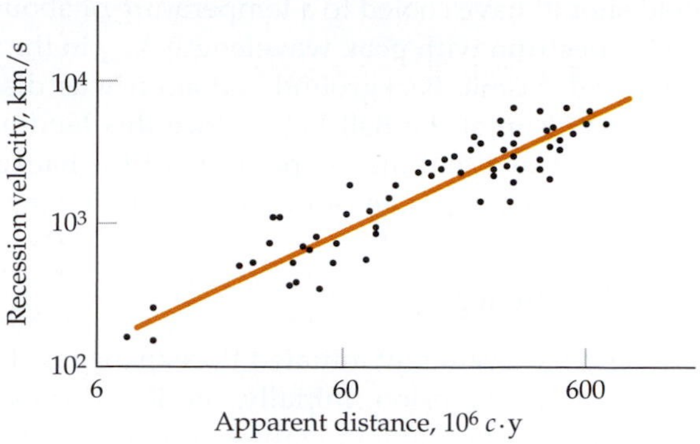

FIGURE 41-4 A plot of the recession velocities of individual galaxies versus apparent distance.

Hubble's law tells us that the galaxies are all rushing away from us, with those the farthest away moving the fastest. However, there is no reason why our location should be special. An observer in any galaxy would make the same observations and compute the same Hubble constant. Thus, Hubble's law suggests that all of the galaxies are receding from each other at an average speed of 23 km/s per $10^6 \; c \cdot y$ of separation. In other words, the universe is expanding. Notice that the basic dimension of H is reciprocal time. The quantity $1/H$ is called the **Hubble age** and equals about 1.3×10^{10} y. This would correspond to the age of the universe if the gravitational pull on the receding galaxies were ignored.

Using Hubble's Law   **EXAMPLE 41-5**

Redshift measurements of a galaxy in the constellation Virgo yield a recession velocity of 1200 km/s. How far is it to that galaxy?

PICTURE THE PROBLEM We calculate the distance from Hubble's law.

Use Hubble's law to find r:
$$r = \frac{v}{H} = (1200 \text{ km/s})\frac{10^6 \; c \cdot y}{23 \text{ km/s}} = \boxed{52 \times 10^6 \; c \cdot y}$$

EXERCISE Show that $1/H = 1.3 \times 10^{10}$ y.

The 2.7-K Background Radiation

In investigating ways of accounting for the cosmic abundance of elements heavier than hydrogen, cosmologists recognized that nucleosynthesis in stars could explain the abundance of elements heavier than helium but could not by itself explain that of helium. Helium must therefore have been formed during the big bang. To synthesize an amount of helium sufficient to account for its present abundance, the big bang would have to have occurred at an extremely high initial temperature to provide the necessary reaction rate before fusion was shut down by the decreasing density of the very rapid initial expansion. The high temperature implies a corresponding thermal (blackbody) radiation field that would cool as the expansion progressed. Theoretical analysis predicted that from the estimated time of the big bang to the present, the remnants of the radiation field should have cooled to a temperature of about 3 K, corresponding to a black-body spectrum with peak wavelength λ_{max} in the microwave region. In 1965, the predicted cosmic background radiation was discovered by Arno Penzias and Robert Wilson at the Bell Labs. Since this landmark discovery, careful analysis has established that the temperature of the background field is 2.7 ± 0.1 K and has shown that it has an isotropic distribution in space.

The Big Bang

The singular event that initiated the expansion of the universe is thought to have been a huge explosion. Initially, the four forces of nature (strong, electromagnetic, weak, and gravitational) were unified into a single force. Physicists have been successful in developing theoretical descriptions that unify the first three forces, but a theory of quantum gravity, needed for the extreme densities of the single-force period, does not yet exist. Consequently, until the cooling universe "froze" or "condensed out" the gravitational force at approximately 10^{-43} s after the big bang, when the temperature was still 10^{32} K, we have no means of describing what was occurring. At this point, the average energy of the particles created would have been about 10^{19} GeV. As the universe continued to cool below 10^{32} K, the three forces other than gravity remained unified and are described by the grand unification theories (GUTs). Quarks and leptons were indistinguishable and particle quantum numbers were not conserved. It was during this period that a slight excess of quarks over antiquarks occurred, roughly 1 part in 10^9, that ultimately resulted in the predominance of matter over antimatter that we now observe in the universe.

At 10^{-35} s, the universe had expanded sufficiently to cool to approximately 10^{27} K, at which point another phase transition occurred as the strong force condensed out of the GUTs group, leaving only the electromagnetic and weak forces still unified as the **electroweak force.** During this period, the previously free quarks in the dense mixture of roughly equal numbers of quarks, leptons, their antiparticles, and photons began to combine into hadrons and their antiparticles, including the nucleons. By the time the universe had cooled to approximately 10^{13} K, at about $t = 10^{-6}$ s, the hadrons had mostly disappeared. This is because 10^{13} K corresponds to $kT \sim 1$ GeV, which is the minimum energy needed to create nucleons and antinucleons from the photons present via the reactions

$$\gamma \rightarrow p^+ + p^- \tag{41-11a}$$

and

$$\gamma \rightarrow n^+ + \bar{n} \tag{41-11b}$$

The particle–antiparticle pairs annihilated and there was no new production to replace them. Only the slight earlier excess of quarks over antiquarks led to a slight excess of protons and neutrons over their antiparticles. The annihilations resulted in photons and leptons, and after about $t = 10^{-4}$ s, those particles in roughly equal numbers dominated the universe. This was the **lepton era.** At about $t = 10$ s, the temperature had fallen to 10^{10} K ($kT \sim 1$ MeV). Further expansion and cooling dropped the average photon energy below the energy needed to form an electron–positron pair. Annihilation then removed all of the positrons as it had the antiprotons and antineutrons earlier, leaving only the small excess of electrons arising from charge conservation, and the **radiation era** began. The particles present were primarily photons and neutrinos.

Within a few more minutes, the temperature dropped sufficiently to enable fusing protons and neutrons to form nuclei that were not immediately photodis-integrated. The nuclei of deuterium, helium, and lithium were produced in this **nucleosynthesis period,** but the rapid expansion soon dropped the temperature too low for the fusion to continue and the formation of heavier elements had to await the birth of stars.

A long time later, when the temperature had dropped to about 3000 K as the universe grew to about 1/1000 of its present size, kT dropped below typical atomic ionization energies and atoms were formed. By then, the expansion had redshifted the radiation field so that the total radiation energy was about equal to the energy represented by the remaining mass. As expansion and cooling continued, the energy of the steadily redshifting radiation declined at a steady rate until, at $t = 10^{10}$ y (now), matter came to dominate the universe, with its energy density exceeding that of the 2.7-K radiation remaining from the big bang by a factor of about 1000.

SUMMARY

Topic	Relevant Equations and Remarks
1. Basic Interactions	There are four basic interactions: strong, electromagnetic, weak, and gravitational.
Strong	The *charge* associated with the strong interaction is called color. Quarks and gluons have color and experience the strong interaction. Hadrons (baryons and mesons) experience a residual strong interaction resulting from the fundamental strong interaction between the quarks that make up the hadrons. Decay times via strong interaction are typically 10^{-23} s.
Electromagnetic	All particles with electric charge experience the force due to the electromagnetic interaction.
Weak	The *charge* associated with the weak interaction is called flavor. Quarks and leptons have flavor and experience the weak interaction. Decay times via weak interaction are typically 10^{-10} s.
Gravitational	All particles with mass experience the force due to the gravitational interaction.
2. Fundamental Particles	There are two families of fundamental particles, leptons and quarks, each containing six members. It is thought that these particles have no size and no internal structure.
Leptons	Leptons are spin-$\frac{1}{2}$ fermions: the electron e and its neutrino ν_e, the muon μ and its neutrino ν_μ, and the tau τ and its neutrino ν_τ. The electron, muon, and tau have mass, electric charge, and flavor, but not color; so they participate in the gravitational, electromagnetic, and weak interactions, but not the strong interaction. The neutrinos have flavor but no electric charge and no color. They appear to have a very small mass.
Quarks	There are six quarks, called up u, down d, strange s, charmed c, top t, and bottom b. Each is a spin-$\frac{1}{2}$ fermion. The quarks participate in all of the basic interactions. Because they are always confined in mesons or baryons, their masses can only be estimated.
3. Hadrons	Hadrons are composite particles that are made up of quarks. There are two types of hadrons, baryons and mesons. Baryons, which include the neutron and proton, are fermions of half-integral spin consisting of three quarks. Mesons, which include pions and kaons, have zero or integral spin. Hadrons interact with each other via the residual strong interaction.

4. Field Particles

In addition to the six fundamental leptons and six fundamental quarks, there are field particles that are associated with the basic interactions.

Interaction	Field Particle
Gravitational	Graviton
Electromagnetic	Photon
Weak	W^+, W^-, Z^0
Strong	Gluons

5. The Conservation Laws

Some quantities, such as energy, linear momentum, electric charge, angular momentum, baryon number, and each of the three lepton numbers, are strictly conserved in all reactions and decays. Others, such as strangeness and charm, are conserved in reactions and decays that proceed via the strong interaction but not in those that proceed via the weak interaction.

6. Particles and Antiparticles

Particles and their antiparticles have identical masses but opposite values for their other properties, such as charge, lepton number, baryon number, and strangeness. Particle–antiparticle pairs can be produced in various nuclear reactions if the energy available is greater than $2mc^2$, where m is the mass of the particle.

7. Hubble's Law

Hubble's law relates the recession velocity of a galaxy, determined from the redshift of its spectrum, to the distance of the galaxy from us:

$$v = Hr \tag{41-9}$$

where the Hubble constant $H = 23$ km/s per million light-years. From Hubble's law, we conclude that the universe is expanding and that the expansion began approximately $1/H$ years ago.

8. The Big Bang

According to the model currently used to describe the evolution of the universe, the universe began with a big bang approximately 10^{10} years ago. The big bang model is supported by substantial experimental observations, including the isotropic, 2.7-K, background blackbody radiation spectrum.

PROBLEMS

- • Single-concept, single-step, relatively easy
- •• Intermediate-level, may require synthesis of concepts
- ••• Challenging
- SSM Solution is in the *Student Solutions Manual*
- iSOLVE Problems available on iSOLVE online homework service
- iSOLVE✓ These "Checkpoint" online homework service problems ask students additional questions about their confidence level, and how they arrived at their answer.

In a few problems, you are given more data than you actually need; in a few other problems, you are required to supply data from your general knowledge, outside sources, or informed estimates.

Conceptual Problems

1 • How are baryons and mesons similar? How are they different?

2 • The muon and the pion have nearly the same mass. How do these particles differ?

3 • SSM How can you tell whether a decay proceeds via the strong interaction or via the weak interaction?

4 • True or false:
(a) All baryons are hadrons.
(b) All hadrons are baryons.

5 • True or false: Mesons are spin-$\frac{1}{2}$ particles.

6 • How can you tell whether a particle is a meson or a baryon by looking at its quark content?

7 • Are there any quark–antiquark combinations that result in a nonintegral electric charge?

8 • True or false:

(a) Leptons consist of three quarks.
(b) The times for decays via the weak interaction are typically longer than the times for decays via the strong interaction.
(c) Electrons interact with protons via the strong interaction.
(d) Strangeness is not conserved in weak interactions.
(e) Neutrons have no charm.

9 • **SSM** Based on the assumption that a pion, π^+, interacts with an antiproton, $\bar{p}$, is it possible that a proton, p, could be produced by such an interaction?

Estimation and Approximation

10 •• Grand unification theories predict that the proton has a long but finite lifetime. Current experiments based on detecting the decay of protons in water infer that this lifetime is at least 10^{32} years. Assume 10^{32} years is, in fact, the lifetime of the proton. Estimate the expected time between proton-decays that occur in the water of a filled Olympic-size swimming pool. An Olympic-size swimming pool is 100 m × 25 m × 2 m. Give your answer in days.

11 • **iSOLVE** Table 41-5 lists some properties of the four fundamental interactions. To gain a better appreciation for the significance of this table, confirm the numerical entries in the second and fourth column of the last row of the table by estimating the ratio of the electromagnetic force to the gravitational force between two protons located in a nucleus.

Spin and Antiparticles

12 • **SSM** Two pions at rest annihilate according to the reaction $\pi^+ + \pi^- \to \gamma + \gamma$. (a) Why must the energies of the two γ-rays be equal? (b) Find the energy of each γ-ray. (c) Find the wavelength of each γ-ray.

13 • Find the minimum energy of the photon needed for the following pair-production reactions: (a) $\gamma \to \pi^+ + \pi^-$, (b) $\gamma \to p + p^-$, and (c) $\gamma \to \mu^- + \mu^+$.

The Conservation Laws

14 • State which of the following decays or reactions violate one or more of the conservation laws, and give the law or laws violated in each case: (a) $p^+ \to n + e^+ + \bar{\nu}_e$, (b) $n \to p^+ + \pi^-$, (c) $e^+ + e^- \to \gamma$, (d) $p + p^- \to \gamma + \gamma$, and (e) $\bar{\nu}_e + p \to n + e^+$.

15 • Determine the change in strangeness in each reaction that follows, and state whether the reaction can proceed via the strong interaction, via the weak interaction, or not at all: (a) $\Omega^- \to \Xi^0 + \pi^-$, (b) $\Xi^0 \to p + \pi^- + \pi^0$, and (c) $\Lambda^0 \to p^+ + \pi^-$.

16 • Determine the change in strangeness for each decay, and state whether the decay can proceed via the strong interaction, via the weak interaction, or not at all: (a) $\Omega^- \to \Lambda^0 + K^-$ and (b) $\Xi^0 \to p + \pi^-$.

17 • Determine the change in strangeness for each decay, and state whether the decay can proceed via the strong interaction, via the weak interaction, or not at all: (a) $\Omega^- \to \Lambda^0 + \bar{\nu}_e + e^-$ and (b) $\Sigma^+ \to p + \pi^0$.

18 • (a) Which of the following decays of the τ particle is possible?

$$\tau \to \mu^- + \bar{\nu}_\mu + \nu_\tau$$
$$\tau \to \mu^- + \nu_\mu + \bar{\nu}_\tau$$

(b) Explain why the other decay is not possible. (c) Calculate the kinetic energy of the decay products for the decay that is possible.

19 •• **iSOLVE** ✓ Consider the following decay chain:

$$\Omega^- \to \Xi^0 + \pi^-$$
$$\Xi^0 \to \Sigma^+ + e^- + \bar{\nu}_e$$
$$\pi^- \to \mu^- + \bar{\nu}_\mu$$
$$\Sigma^+ \to n + \pi^+$$
$$\pi^+ \to \mu^+ + \nu_\mu$$
$$\mu^+ \to e^+ + \bar{\nu}_\mu + \nu_e$$
$$\mu^- \to e^- + \bar{\nu}_e + \nu_\mu$$

(a) Are all the final products shown stable? If not, finish the decay chain. (b) Write the overall decay reaction for Ω^- to the final products. (c) Check the overall decay reaction for the conservation of electric charge, baryon number, lepton number, and strangeness.

20 •• **SSM** **iSOLVE** Test the following decays for violation of the conservation of energy, electric charge, baryon number, and lepton number: (a) $n \to \pi^+ + \pi^- + \mu^+ + \mu^-$ and (b) $\pi^0 \to e^+ + e^- + \gamma$. Assume that linear momentum and angular momentum are conserved. State which conservation laws (if any) are violated in each decay.

Quarks

21 • Find the baryon number, charge, and strangeness for the following quark combinations and identify the hadron: (a) uud, (b) udd, (c) uus, (d) dds, (e) uss, and (f) dss.

22 • Repeat Problem 21 for the following quark combinations: (a) $u\bar{d}$, (b) $\bar{u}d$, (c) $u\bar{s}$, and (d) $\bar{u}s$.

23 • The Δ^{++} particle is a baryon that decays via the strong interaction. Its strangeness, charm, topness, and bottomness are all zero. What combination of quarks gives a particle with these properties?

24 • Find a possible combination of quarks that gives the correct values for electric charge, baryon number, and strangeness for (a) K^+ and (b) K^0.

25 • The D^+ meson has no strangeness, but it has charm of +1. (a) What is a possible quark combination that will give the correct properties for this particle? (b) Repeat Part (a) for the D^- meson, which is the antiparticle of the D^+ meson.

26 • Find a possible combination of quarks that gives the correct values for electric charge, baryon number, and strangeness for (a) K^- (the K^- is the antiparticle of the K^+) and (b) $\bar{K}^0$.

27 •• **SSM** Find a possible quark combination for the following particles: (a) Λ^0, (b) p^-, and (c) Σ^-.

28 •• Find a possible quark combination for the following particles: (a) $\bar{n}$, (b) Ξ^0, and (c) Σ^+.

29 •• Find a possible quark combination for the following particles: (a) Ω^- and (b) Ξ^-.

30 •• State the properties of the particles made up of the following quarks: (a) ddd, (b) $u\bar{c}$, (c) $u\bar{b}$, and (d) $\overline{sss}$.

The Evolution of the Universe

31 • **SSM** A galaxy is receding from the earth at 2.5 percent the speed of light. Estimate the distance from the earth to this galaxy.

32 • Estimate the speed of a galaxy that is 12×10^9 $c\cdot$y away from us.

33 •• Equation 39-16b deals with the relativistic Doppler frequency shift for a light from a source that is receding from an observer. Show that the relativistic Doppler wavelength shift is

$$\lambda' = \lambda_0 \sqrt{\frac{1 + v/c}{1 - v/c}}$$

34 •• **SSM** The red line in the spectrum of atomic hydrogen is frequently referred to as the $H\alpha$ line, and it has a wavelength of 656.3 nm. Using Hubble's law and the relativistic Doppler equation from Problem 33, determine the wavelength of the $H\alpha$ line in the spectrum emitted from galaxies at distances of (a) 5×10^6 $c\cdot$y, (b) 50×10^6 $c\cdot$y, (c) 500×10^6 $c\cdot$y, and (d) 5×10^9 $c\cdot$y from the earth.

General Problems

35 • (a) What conditions are necessary for a particle and its antiparticle to be identical? Find the antiparticle for (b) π^0 and (c) Ξ^0.

36 •• Consider the following decay chain:

$\Xi^0 \rightarrow \Lambda^0 + \pi^0$

$\Lambda^0 \rightarrow p + \pi^-$

$\pi^0 \rightarrow \gamma + \gamma$

$\pi^- \rightarrow \mu^- + \bar{\nu}_\mu$

$\mu^- \rightarrow e^- + \bar{\nu}_e + \nu_\mu$

(a) Are all the final products shown stable? If not, finish the decay chain. (b) Write the overall decay reaction for Ξ^0 to the final products. (c) Check the overall decay reaction for the conservation of electric charge, baryon number, lepton number, and strangeness. (d) In the first step of the chain, could the Λ^0 have been a Σ^0?

37 •• **SSM** In Problem 36, one of the reactions is $\pi^0 \rightarrow \gamma + \gamma$. (a) In terms of the quark model, show how this reaction can take place. (b) Why is it that the number of photons produced must be at least two?

38 •• Test the following decays for violation of the conservation of energy, electric charge, baryon number, and lepton number: (a) $\Lambda^0 \rightarrow p + \pi^-$, (b) $\Sigma^- \rightarrow n + p^-$, and (c) $\mu^- \rightarrow e^- + \bar{\nu}_e + \nu_\mu$. Assume that linear momentum and angular momentum are conserved. State which conservation laws (if any) are violated in each decay.

39 •• **SSM** Using Figure 41-2 and the laws of conservation of charge number, baryon number, strangeness, and spin, identify the unknown particle in each of the following strong reactions: (a) $p + \pi^- \rightarrow \Sigma^0 + ?$, (b) $p + p \rightarrow \pi^+ + n + K^+ + ?$, and (c) $p + \overline{K}^- \rightarrow \Xi^- + ?$

40 •• **I SOLVE ✓** Consider the following high-energy particle reaction: $p + p \rightarrow \Lambda^0 + K^0 + p + ?$ In this reaction, stationary protons are bombarded with a beam of high-energy protons. (a) Use the laws of conservation of charge number, baryon number, strangeness (Figure 41-2), and spin to determine the unknown particle. (b) Calculate the Q value for the reaction. (c) The threshold energy K_{th} for this reaction is given by

$$K_{th} = -\frac{Q}{2m_p}(m_p + m_p + M_1 + M_2 + M_3 + M_4)$$

where $M_1, M_2, M_3,$ and M_4 are the masses of the reaction products. Find K_{th}.

41 •• **I SOLVE ✓** Light from a distant galaxy shows a redshift of the $H\alpha$ line of hydrogen from 656.3 nm to 1458 nm. (a) What is the recessional velocity of the galaxy? (b) Estimate the distance to this galaxy.

42 ••• **I SOLVE** (a) Calculate the total kinetic energy of the decay products for the decay $\Lambda^0 \rightarrow p + \pi^-$. Assume the Λ^0 is initially at rest. (b) Find the ratio of the kinetic energy of the pion to the kinetic energy of the proton. (c) Find the kinetic energies of the proton and the pion for this decay.

43 ••• **SSM** A Σ^0 particle at rest decays into a Λ^0 plus a photon. (a) What is the total energy of the decay products? (b) Assuming that the kinetic energy of the Λ^0 is negligible compared with the energy of the photon, calculate the approximate momentum of the photon. (c) Use your result from Part (b) to calculate the kinetic energy of the Λ^0. (d) Use your result from Part (c) to obtain a better estimate of the momentum and the energy of the photon.

44 ••• In this problem, you will calculate the difference in the time of arrival of two neutrinos of different energy from a supernova that is 170,000 light-years away. Let the energies of the neutrinos be $E_1 = 20$ MeV and $E_2 = 5$ MeV, and assume that the mass of a neutrino is 20 eV$/c^2$. Because the total energy of the neutrinos is so much greater than their rest energy, the neutrinos have speeds that are very nearly equal to c and energies that are approximately $E \approx pc$. (a) If t_1 and t_2 are the times it takes for neutrinos of speeds u_1 and u_2 to travel a distance x, show that

$$\Delta t = t_2 - t_1 = x \frac{u_1 - u_2}{u_1 u_2} \approx \frac{x \, \Delta u}{c^2}$$

(b) The speed of a neutrino of mass m and total energy E can be found from Equation 39-25. Show that when $E >> mc^2$, the speed u is given approximately by

$$\frac{u}{c} \approx 1 - \frac{1}{2}\left(\frac{mc^2}{E}\right)^2$$

(c) Use the results from Part (b) to calculate $u_1 - u_2$ for the energies and mass given, and calculate Δt from the result from Part (a) for $x = 170,000$ $c\cdot$y. (d) Repeat the calculation in Part (c) using $mc^2 = 40$ eV for the rest energy of a neutrino.

APPENDIX A

SI Units and Conversion Factors

Basic Units

Length	The *meter* (m) is the distance traveled by light in a vacuum in $1/299{,}792{,}458$ s.
Time	The *second* (s) is the duration of 9,192,631,770 periods of the radiation corresponding to the transition between the two hyperfine levels of the ground state of the ^{133}Cs atom.
Mass	The *kilogram* (kg) is the mass of the international standard body preserved at Sèvres, France.
Current	The *ampere* (A) is that current in two very long parallel wires 1 m apart that gives rise to a magnetic force per unit length of 2×10^{-7} N/m.
Temperature	The *kelvin* (K) is 1/273.16 of the thermodynamic temperature of the triple point of water.
Luminous intensity	The *candela* (cd) is the luminous intensity, in the perpendicular direction, of a surface of area $1/600{,}000$ m^2 of a blackbody at the temperature of freezing platinum at a pressure of 1 atm.

Derived Units

Force	newton (N)	$1\text{ N} = 1\text{ kg·m/s}^2$
Work, energy	joule (J)	$1\text{ J} = 1\text{ N·m}$
Power	watt (W)	$1\text{ W} = 1\text{ J/s}$
Frequency	hertz (Hz)	$1\text{ Hz} = \text{cy/s}$
Charge	coulomb (C)	$1\text{ C} = 1\text{ A·s}$
Potential	volt (V)	$1\text{ V} = 1\text{ J/C}$
Resistance	ohm (Ω)	$1\text{ }\Omega = 1\text{ V/A}$
Capacitance	farad (F)	$1\text{ F} = 1\text{ C/V}$
Magnetic field	tesla (T)	$1\text{ T} = 1\text{ N/(A·m)}$
Magnetic flux	weber (Wb)	$1\text{ Wb} = 1\text{ T·m}^2$
Inductance	henry (H)	$1\text{ H} = 1\text{ J/A}^2$

Conversion Factors

Conversion factors are written as equations for simplicity;
relations marked with an asterisk are exact.

Length

1 km = 0.6215 mi

1 mi = 1.609 km

1 m = 1.0936 yd = 3.281 ft = 39.37 in.

*1 in. = 2.54 cm

*1 ft = 12 in. = 30.48 cm

*1 yd = 3 ft = 91.44 cm

1 lightyear = 1 $c \cdot$y = 9.461×10^{15} m

*1 Å = 0.1 nm

Area

*1 m^2 = 10^4 cm^2

1 km^2 = 0.3861 mi^2 = 247.1 acres

*1 $in.^2$ = 6.4516 cm^2

1 ft^2 = 9.29×10^{-2} m^2

1 m^2 = 10.76 ft^2

*1 acre = 43,560 ft^2

1 mi^2 = 640 acres = 2.590 km^2

Volume

*1 m^3 = 10^6 cm^3

*1 L = 1000 cm^3 = 10^{-3} m^3

1 gal = 3.786 L

1 gal = 4 qt = 8 pt = 128 oz = 231 in^3

1 in^3 = 16.39 cm^3

1 ft^3 = 1728 $in.^3$ = 28.32 L
       = 2.832×10^4 cm^3

Time

*1 h = 60 min = 3.6 ks

*1 d = 24 h = 1440 min = 86.4 ks

1 y = 365.24 d = 3.156×10^7 s

Speed

*1 m/s = 3.6 km/h

1 km/h = 0.2778 m/s = 0.6215 mi/h

1 mi/h = 0.4470 m/s = 1.609 km/h

1 mi/h = 1.467 ft/s

Angle and Angular Speed

*π rad = 180°

1 rad = 57.30°

1° = 1.745×10^{-2} rad

1 rev/min = 0.1047 rad/s

1 rad/s = 9.549 rev/min

Mass

*1 kg = 1000 g

*1 tonne = 1000 kg = 1 Mg

1 u = 1.6606×10^{-27} kg

1 kg = 6.022×10^{26} u

1 slug = 14.59 kg

1 kg = 6.852×10^{-2} slug

1 u = 931.50 MeV/c^2

Density

*1 g/cm^3 = 1000 kg/m^3 = 1 kg/L

(1 g/cm^3)g = 62.4 lb/ft^3

Force

1 N = 0.2248 lb = 10^5 dyn

*1 lb = 4.448222 N

(1 kg)g = 2.2046 lb

Pressure

*1 Pa = 1 N/m^2

*1 atm = 101.325 kPa = 1.01325 bars

1 atm = 14.7 $lb/in.^2$ = 760 mmHg
       = 29.9 in.Hg = 33.8 ftH$_2$O

1 $lb/in.^2$ = 6.895 kPa

1 torr = 1 mmHg = 133.32 Pa

1 bar = 100 kPa

Energy

*1 kW$\cdot$h = 3.6 MJ

*1 cal = 4.1840 J

1 ft$\cdot$lb = 1.356 J = 1.286×10^{-3} Btu

*1 L$\cdot$atm = 101.325 J

1 L$\cdot$atm = 24.217 cal

1 Btu = 778 ft$\cdot$lb = 252 cal = 1054.35 J

1 eV = 1.602×10^{-19} J

1 u$\cdot c^2$ = 931.50 MeV

*1 erg = 10^{-7} J

Power

1 horsepower = 550 ft$\cdot$lb/s = 745.7 W

1 Btu/h = 1.055 kW

1 W = 1.341×10^{-3} horsepower
    = 0.7376 ft$\cdot$lb/s

Magnetic Field

*1 T = 10^4 G

Thermal Conductivity

1 W/(m$\cdot$K) = 6.938 Btu$\cdot$in./(h$\cdot ft^2 \cdot$F°)

1 Btu$\cdot$in./(h$\cdot ft^2 \cdot$F°) = 0.1441 W/(m$\cdot$K)

APPENDIX B

Numerical Data

Terrestrial Data

Free-fall acceleration g	9.80665 m/s^2; 32.1740 ft/s^2
(Standard value at sea level at 45° latitude)[†]	
Standard value	
At sea level, at equator[†]	9.7804 m/s^2
At sea level, at poles[†]	9.8322 m/s^2
Mass of earth M_E	5.98×10^{24} kg
Radius of earth R_E, mean	6.37×10^6 m; 3960 mi
Escape speed $\sqrt{2R_E g}$	1.12×10^4 m/s; 6.95 mi/s
Solar constant[‡]	1.35 kW/m^2
Standard temperature and pressure (STP):	
Temperature	273.15 K
Pressure	101.325 kPa (1.00 atm)
Molar mass of air	28.97 g/mol
Density of air (STP), ρ_{air}	1.293 kg/m^3
Speed of sound (STP)	331 m/s
Heat of fusion of H_2O (0°C, 1 atm)	333.5 kJ/kg
Heat of vaporization of H_2O (100°C, 1 atm)	2.257 MJ/kg.

† Measured relative to the earth's surface.
‡ Average power incident normally on 1 m^2 outside the earth's atmosphere at the mean distance from the earth to the sun.

Astronomical Data[†]

Earth	
Distance to moon[‡]	3.844×10^8 m; 2.389×10^5 mi
Distance to sun, mean[‡]	1.496×10^{11} m; 9.30×10^7 mi; 1.00 AU
Orbital speed, mean	2.98×10^4 m/s
Moon	
Mass	7.35×10^{22} kg
Radius	1.738×10^6 m
Period	27.32 d
Acceleration of gravity at surface	1.62 m/s^2
Sun	
Mass	1.99×10^{30} kg
Radius	6.96×10^8 m

† Additional solar-system data is available from NASA at <http://nssdc.gsfc.nasa.gov/planetary/planetfact.html>.
‡ Center to center.

Physical Constants[†]

Gravitational constant	G	$6.673(10) \times 10^{-11}$ N·m^2/kg^2
Speed of light	c	$2.997\ 924\ 58 \times 10^8$ m/s
Fundamental charge	e	$1.602\ 1764\ 62(63) \times 10^{-19}$ C
Avogadro's number	N_A	$6.022\ 141\ 99(47) \times 10^{23}$ particles/mol
Gas constant	R	$8.314\ 472(15)$ J/(mol·K)
		$1.987\ 2065(36)$ cal/(mol·K)
		$8.205\ 746(15) \times 10^{-2}$ L·atm/(mol·K)
Boltzmann constant	$k = R/N_A$	$1.380\ 6503(24) \times 10^{-23}$ J/K
		$8.617\ 342(15) \times 10^{-5}$ eV/K
Stefan-Boltzmann constant	$\sigma = (\pi^2/60)k^4/(\hbar^3 c^2)$	$5.670\ 400(40) \times 10^{-8}$ W/(m^{2}k^4)
Atomic mass constant	$m_u = \frac{1}{12}m(^{12}C)$	$1.660\ 538\ 73(13) \times 10^{-27}$ kg = 1u
Coulomb constant	$k = 1/(4\pi\epsilon_0)$	$8.987\ 551\ 788 \ldots \times 10^9$ N·m^2/C^2
Permittivity of free space	ϵ_0	$8.854\ 187\ 817 \ldots \times 10^{-12}$ C^2/(N·m^2)
Permeability of free space	μ_0	$4\pi \times 10^{-7}$ N/A^2
		$1.256\ 637 \times 10^{-6}$ N/A^2
Planck's constant	h	$6.626\ 068\ 76(52) \times 10^{-34}$ J·s
		$4.135\ 667\ 27(16) \times 10^{-15}$ eV·s
	$\hbar = h/2\pi$	$1.054\ 571\ 596(82) \times 10^{-34}$ J·s
		$6.582\ 118\ 89(26) \times 10^{-16}$ eV·s
Mass of electron	m_e	$9.109\ 381\ 88(72) \times 10^{-31}$ kg
		$0.510\ 998\ 902(21)$ MeV/c^2
Mass of proton	m_p	$1.672\ 621\ 58(13) \times 10^{-27}$ kg
		$938.271\ 998(38) \times$ MeV/c^2
Mass of neutron	m_n	$1.674\ 927\ 16(13) \times 10^{-27}$ kg
		$939.565\ 330(38)$ MeV/c^2
Bohr magneton	$m_B = eh/2m_e$	$9.274\ 0008\ 99(37) \times 10^{-24}$ J/T
		$5.788\ 381\ 749(43) \times 10^{-5}$ eV/T
Nuclear magneton	$m_n = eh/2m_p$	$5.050\ 783\ 17(20) \times 10^{-27}$ J/T
		$3.152\ 451\ 238(24) \times 10^{-8}$ eV/T
Magnetic flux quantum	$\phi_0 = h/2e$	$2.067\ 833\ 636(81) \times 10^{-15}$ T·m^2
Quantized Hall resistance	$R_K = h/e^2$	$2.581\ 280\ 7572(95) \times 10^4\ \Omega$
Rydberg constant	R_H	$1.097\ 373\ 156\ 8549(83) \times 10^7$ m^{-1}
Josephson frequency-voltage quotient	$K_J = 2e/h$	$4.835\ 978\ 98(19) \times 10^{14}$ Hz/V
Compton wavelength	$\lambda_C = h/m_e c$	$2.426\ 310\ 215(18) \times 10^{-12}$ m

[†] The values for these and other constants may be found on the Internet at http://physics.nist.gov/cuu/Constants/index.html. The numbers in parentheses represent the uncertainties in the last two digits. (For example, 2.044 43(13) stands for 2.044 43 ± 0.000 13.) Values with without uncertainties are exact, including those values with ellipses (like the value of pi is exactly 3.1415. . .).

For additional data, see the following tables in the text.

Geometry and Trigonometry

$C = \pi d = 2\pi r$ definition of π

$A = \pi r^2$ area of circle

$V = \frac{4}{3}\pi r^3$ spherical volume

$A = dV/dr = 4\pi r^2$ spherical surface area

$V = A_{\text{base}}L = \pi r^2 L$ cylindrical volume

$A = dV/dr = 2\pi rL$ cylindrical surface area

$o = h\sin\theta$

$a = h\cos\theta$

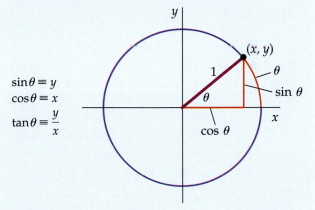

$\sin^2\theta + \cos^2\theta = 1$

$\sin(A \pm B) = \sin A \cos B \pm \cos A \sin B$

$\cos(A \pm B) = \cos A \cos B \mp \sin A \sin B$

$\sin A \pm \sin B = 2\sin[\frac{1}{2}(A \pm B)]\cos[\frac{1}{2}(A \mp B)]$

$\sin\theta \equiv y$

$\cos\theta \equiv x$

$\tan\theta \equiv \dfrac{y}{x}$

IF $|\theta| \ll 1$, THEN

    $\cos\theta \approx 1$ AND $\tan\theta \approx \sin\theta \approx \theta$ (θ in radians)

Quadratic Formula

If $ax^2 + bx + c = 0$, then $x = \dfrac{-b \pm \sqrt{b^2 - 4ac}}{2a}$

Binomial Expansion

If $|x| < 1$, then $(1 + x)^n =$

$$1 + nx + \frac{n(n-1)}{2!}x^2 + \frac{n(n-1)(n-2)}{3!}x^3 + \dots$$

If $|x| \ll 1$, then $(1 + x)^n \approx 1 + nx$

If $|\Delta x|$ is small, then $\Delta F \approx \dfrac{dF}{dx}\Delta x$

Periodic Table of Elements

1																	18
1 **H** 1.00797	2											13	14	15	16	17	2 **He** 4.003
3 **Li** 6.941	4 **Be** 9.012											5 **B** 10.81	6 **C** 12.011	7 **N** 14.007	8 **O** 15.9994	9 **F** 19.00	10 **Ne** 20.179
11 **Na** 22.990	12 **Mg** 24.31	3	4	5	6	7	8	9	10	11	12	13 **Al** 26.98	14 **Si** 28.09	15 **P** 30.974	16 **S** 32.064	17 **Cl** 35.453	18 **Ar** 39.948
19 **K** 39.102	20 **Ca** 40.08	21 **Sc** 44.96	22 **Ti** 47.88	23 **V** 50.94	24 **Cr** 52.00	25 **Mn** 54.94	26 **Fe** 55.85	27 **Co** 58.93	28 **Ni** 58.69	29 **Cu** 63.55	30 **Zn** 65.38	31 **Ga** 69.72	32 **Ge** 72.59	33 **As** 74.92	34 **Se** 78.96	35 **Br** 79.90	36 **Kr** 83.80
37 **Rb** 85.47	38 **Sr** 87.62	39 **Y** 88.906	40 **Zr** 91.22	41 **Nb** 92.91	42 **Mo** 95.94	43 **Tc** (98)	44 **Ru** 101.1	45 **Rh** 102.905	46 **Pd** 106.4	47 **Ag** 107.870	48 **Cd** 112.41	49 **In** 114.82	50 **Sn** 118.69	51 **Sb** 121.75	52 **Te** 127.60	53 **I** 126.90	54 **Xe** 131.29
55 **Cs** 132.905	56 **Ba** 137.33	57–71 **Rare Earths**	72 **Hf** 178.49	73 **Ta** 180.95	74 **W** 183.85	75 **Re** 186.2	76 **Os** 190.2	77 **Ir** 192.2	78 **Pt** 195.09	79 **Au** 196.97	80 **Hg** 200.59	81 **Tl** 204.37	82 **Pb** 207.19	83 **Bi** 208.98	84 **Po** (210)	85 **At** (210)	86 **Rn** (222)
87 **Fr** (223)	88 **Ra** (226)	89–103 Actinides	104 **Rf** (261)	105 **Ha** (260)	106 (263)	107 (262)	108 (265)	109 (266)									

Rare Earths (Lanthanides)	57 **La** 138.91	58 **Ce** 140.12	59 **Pr** 140.91	60 **Nd** 144.24	61 **Pm** (147)	62 **Sm** 150.36	63 **Eu** 152.0	64 **Gd** 157.25	65 **Tb** 158.92	66 **Dy** 162.50	67 **Ho** 164.93	68 **Er** 167.26	69 **Tm** 168.93	70 **Yb** 173.04	71 **Lu** 174.97
Actinides	89 **Ac** 227.03	90 **Th** 232.04	91 **Pa** 231.04	92 **U** 238.03	93 **Np** 237.05	94 **Pu** (244)	95 **Am** (243)	96 **Cm** (247)	97 **Bk** (247)	98 **Cf** (251)	99 **Es** (252)	100 **Fm** (257)	101 **Md** (258)	102 **No** (259)	103 **Lr** (260)

The 1–18 group designation has been recommended by the International Union of Pure and Applied Chemistry (IUPAC).

Atomic Numbers and Atomic Masses[†]

Name	Symbol	Atomic Number	Mass	Name	Symbol	Atomic Number	Mass
Actinium	Ac	89	227.03	Mercury	Hg	80	200.59
Aluminum	Al	13	26.98	Molybdenum	Mo	42	95.94
Americium	Am	95	(243)	Neodymium	Nd	60	144.24
Antimony	Sb	51	121.75	Neon	Ne	10	20.179
Argon	Ar	18	39.948	Neptunium	Np	93	237.05
Arsenic	As	33	74.92	Nickel	Ni	28	58.69
Astatine	At	85	(210)	Niobium	Nb	41	92.91
Barium	Ba	56	137.3	Nitrogen	N	7	14.007
Berkelium	Bk	97	(247)	Nobelium	No	102	(259)
Beryllium	Be	4	9.012	Osmium	Os	76	190.2
Bismuth	Bi	83	208.98	Oxygen	O	8	15.9994
Boron	B	5	10.81	Palladium	Pd	46	106.4
Bromine	Br	35	79.90	Phosphorus	P	15	30.974
Cadmium	Cd	48	112.41	Platinum	Pt	78	195.09
Calcium	Ca	20	40.08	Plutonium	Pu	94	(244)
Californium	Cf	98	(251)	Polonium	Po	84	(210)
Carbon	C	6	12.011	Potassium	K	19	39.098
Cerium	Ce	58	140.12	Praseodymium	Pr	59	140.91
Cesium	Cs	55	132.905	Promethium	Pm	61	(147)
Chlorine	Cl	17	35.453	Protactinium	Pa	91	231.04
Chromium	Cr	24	52.00	Radium	Ra	88	(226)
Cobalt	Co	27	58.93	Radon	Rn	86	(222)
Copper	Cu	29	63.55	Rhenium	Re	75	186.2
Curium	Cm	96	(247)	Rhodium	Rh	45	102.905
Dysprosium	Dy	66	162.50	Rubidium	Rb	37	85.47
Einsteinium	Es	99	(252)	Ruthenium	Ru	44	101.1
Erbium	Er	68	167.26	Rutherfordium	Rf	104	(261)
Europium	Eu	63	152.0	Samarium	Sm	62	150.36
Fermium	Fm	100	(257)	Scandium	Sc	21	44.96
Fluorine	F	9	19.00	Selenium	Se	34	78.96
Francium	Fr	87	(223)	Silicon	Si	14	28.09
Gadolinium	Gd	64	157.25	Silver	Ag	47	107.870
Gallium	Ga	31	69.72	Sodium	Na	11	22.990
Germanium	Ge	32	72.59	Strontium	Sr	38	87.62
Gold	Au	79	196.97	Sulfur	S	16	32.064
Hafnium	Hf	72	178.49	Tantalum	Ta	73	180.95
Hahnium	Ha	105	(260)	Technetium	Tc	43	(98)
Helium	He	2	4.003	Tellurium	Te	52	127.60
Holmium	Ho	67	164.93	Terbium	Tb	65	158.92
Hydrogen	H	1	1.0079	Thallium	Tl	81	204.37
Indium	In	49	114.82	Thorium	Th	90	232.04
Iodine	I	53	126.90	Thulium	Tm	69	168.93
Iridium	Ir	77	192.2	Tin	Sn	50	118.69
Iron	Fe	26	55.85	Titanium	Ti	22	47.88
Krypton	Kr	36	83.80	Tungsten	W	74	183.85
Lanthanum	La	57	138.91	Uranium	U	92	238.03
Lawrencium	Lr	103	(260)	Vanadium	V	23	50.94
Lead	Pb	82	207.2	Xenon	Xe	54	131.29
Lithium	Li	3	6.941	Ytterbium	Yb	70	173.04
Lutetium	Lu	71	174.97	Yttrium	Y	39	88.906
Magnesium	Mg	12	24.31	Zinc	Zn	30	65.38
Manganese	Mn	25	54.94	Zirconium	Zr	40	91.22
Mendelevium	Md	101	(258)				

[†] More precise values for the atomic masses, along with the uncertainties in the masses, can be found at http://physics.nist.gov/PhysRefData/.

ILLUSTRATION CREDITS

American; **(e)** © 1988 by David Scharf. All rights reserved; **p. 1012 Figure 31-19 (a)** © 1987 Ken Kay/Fundamental Photographs; **(b)** Courtesy Battelle-Northwest Laboratories; **(bottom)** © 1990 Richard Magna/Fundamental Photographs; **p. 1013 Figure 31-20 (b)** © Macduff Everton/CORBIS; **p. 1014** © 1987 Pete Saloutos/The Stock Market; **p. 1015 Figure 31-23 (b)** © 1987 Ken Kay/Fundamental Photographs; **p. 1017 Figure 31-27 (c)** © Ted Horowitz/The Stock Market; **(bottom a)** © Dan Boyd/Courtesy Naval Research Laboratory; **(bottom b)** Courtesy AT&T Archives; **p. 1018 Figure 31-28 (c)** © Robert Greenler; **p. 1019** © David Parker/Science Photo Library/Photo Researchers; **p. 1020 (a)** © Robert Greenler; **(b)** Giovanni DeAmici, NSF, Lawrence Berkeley Laboratory; **p. 1021 (a and b)** Larry Langrill; **p. 1023 (a)** © 1970 Fundamental Photographs; **(b)** 1990 PAR/NYC, Inc./Photo by Elizabeth Algieri; **p. 1025 Figure 31-43 (b)** © 1987 Paul Silverman Photographs; **p. 1026 (a)** Glen A. Izett, U.S. Geological Survey, Denver, Colorado; **(b)** Glen A. Izett, U.S. Geological Survey, Denver, Colorado; **(c)** Dr. Anthony J. Gow/Cold Regions Research and Engineering Laboratory, Hanover, New Hampshire; **(d)** Dr. Anthony J. Gow/Cold Regions Research and Engineering Laboratory, Hanover, New Hampshire; **(e)** © Sepp Seitz/Woodfin Camp and Associates.

Chapter 32

Opener p. 1038 Photo by Gene Mosca; **p. 1039 Figure 32-2** Photo by Demetrios Zangos; **p. 1048 (a and b)** © 1990 Richard Megna/Fundamental Photographs; **p. 1054 Figure 32-29 (a, bottom)** Nils Abramson; **(b, bottom)** © 1974 Fundamental Photographs; **Figure 32.30 (b)** © Fundamental Photographs; **p. 1057** © Bohdan Hrynewych/Stock Boston; **p. 1068 (a, b, and c)** Lennart Nilsson, **(d)** Courtesy IMEC and University of Pennsylvania Department of Electrical Engineering; **p. 1072 (a)** © Scala/Art Resource; **(b)** © Royal Astronomical Society Library; **(c)** Lick Observatory, courtesy of the University of California Regents; **(d)** California Institute of Technology; **(e)** © 1980 Gary Ladd; **p. 1073 (a, b and c)** © California Association for Research in Astronomy; **(bottom)** Courtesy of NASA.

Chapter 33

Opener p. 1084 © Aaron Haupt/Photo Researchers, NY; **p. 1087 Figure 33-3 (a)** Courtesy of Bausch & Lomb; **p. 1088 Figure 33-5 (a and b)** Courtesy T. A. Wiggins; **Figure 33-6** From *PSSC Physics*, 2nd Edition, 1965. D.C. Heath & Co. and Education Development Center, Newton, MA; **p. 1091 Figure 33-9 (a)** Courtesy of Michael Cagnet; **p. 1092 Figure 33-11 (a)** Courtesy of Michael Cagnet; **p. 1093 Figure 33-14 (a)** Courtesy of Michael Cagnet; **p. 1097 Figure 33-21** Courtesy Michael Cagnet; **p. 1102 Figure 33-30 (a and b)** M. Cagnet, M. Façon, J.C. Thrierr, *Atlas of Optical Phenomena*; **Figure 33-31 (a)** Courtesy Battelle-Northwest Laboratories; **p. 1103 Figures 33-32, 33-33,** and **33-35 (a and b)** Courtesy of Michael Cagnet; **p. 1104 Figure 33-36** Courtesy of National Radio Astronomy Observatory/Associated Universities, Inc./National Science Foundation. Photographer: Kelly Gatlin. Digital composite: Patricia Smiley; **p. 1105 (bottom)** © Kevin R. Morris/CORBIS; **p. 1106 Figure 33-38 (a)** Clarence Bennett/Oakland University, Rochester, Michigan; **(b)** NRAO/AUI/Science Photo Library/Photo Researchers; **p. 1108 (a and b)** © 1981 by Ronald R. Erickson, Hologram by Nicklaus Phillips, 1978, for Digital Equipment Corporation.

Chapter 34

Opener p. 1117 Courtesy of Akira Tononmura, Advanced Research Laboratory, Hitachi, Ltd.; **p. 1128 (a & b)** *PSSC Physics*, 2nd ed., 1965. D.C. Heath & Co., and Education Development Center, Inc., Newton, MA; **(c)** C. G. Shull; **(d)** Claus Jönsson; **p. 1128 (bottom)** Jack Griffith/University of North Carolina

Chapter 35

Opener p. 1149 Courtesy of IBM and the IBM Almadin Laboratories; **p. 1161** Education Development Center

Chapter 36

Opener p. 1171 Courtesy of NASA and the Harvard-Smithsonian Center for Astrophysics; **p. 1172** Adapted from Eastman Kodak and Wabash Instrument Corporation; **p. 1192 (a & b)** A Jayaraman/AT&T Bell Labs; **p. 1193** © David Parker/Photo Researchers; **p. 1198** © Robert Landau/Westlight.

Chapter 37

Opener p. 1208 (a) Norman Collection for the History of Molecular Biology, **(b)** © A. Barrington Brown/Photo Researchers, NY; **p. 1214** © Will and Demi McIntire/Photo Researchers; **p. 1221** Courtesy of Dr. J.A. Marquissee.

Chapter 38

Opener p. 1228 © 2003 John Alves/Mystic Wanderer Images; **p. 1230 (a)** Richard Walters, 2/89, p. 52/*Discover*; **(b)** © Dr. Jeremy Burgess/Science Photo Library/Photo Researchers; **(c)** © Thomas R. Taylor/Photo Researchers; **(d)** Courtesy the AT&T Archives; **p. 1233 (a)** Chris Kovach, 3/91, p. 69/*Discover*; **(b)** Srinivas Manne, University of California, Santa Barbara; **(c)** Dr. F.A. Quiocho and J.S. Spurlino/Howard Hughes Medical Institute, Baylor College of Medicine; **(d)** W. Krätschmer/Max-Planck Institute for Nuclear Physics; **(e)** © Kenneth Weard/BioGrafx/Science Source/Photo Researchers; **p. 1247** Museum of Modern Art; **p. 1250** © C. Falco/Photo Researchers; **p. 1253** © Royalty Free/CORBIS

Chapter 39

Opener p. 1267 Courtesy of NASA; **p. 1268** Courtesy NRAO/AUI; **p. 1289** © C. Powell, P. Fowler, & D. Perkins/Science Photo Library/Photo Researchers; **p. 1297** © Michael Freeman

Chapter 40

Opener p. 1306 © Hans Wolf/The Image Bank; **p. 1320(a)** © 1991 by the Metropolitan Museum of Art, **(b & c)** Courtesy of Paintings Conservation Dept., Metropolitan Museum of Art; **p. 1323** © Jerry Mason/Photo Researchers; **p. 1327 (all)** Courtesy of the Princeton Plasma Physics Laboratory; **p. 1328 (a & b)** Courtesy of the Lawrence Livermore National Laboratory/U.S. Department of Energy

Chapter 41

Opener p. 1335 © Lawrence Livermore Lab/Science Photo Library/Photo Researchers; **p. 1336** ICCR (Institute for Cosmic Ray Research), The University of Tokyo; **p. 1338 (top)** © Science Photo Library/Photo Researchers, (bot), © Lawrence Berkeley Laboratory/Science Photo Library/Photo Researchers; **p. 1339 (top)** © Lawrence Berkeley Laboratory/Science Photo Library/Photo Researchers, **(bot)** Fig. 4 from "First Results from KamLAND: Evidence for Reactor Antineutrino Disappearance" by the KamLAND Collaboration, *Physical Review Letters*, Vol. 90, No. 2, December 17, 2003. Copyright © 2003 The American Physical Society. Reprinted with permission; **p. 1340** Richard Ehrlich; **p. 1341 (both)** CERN; **p. 1344** © Lawrence Berkeley Laboratory/Science Photo Library/Photo Researchers.

ANSWERS

Problem answers are calculated using $g = 9.81 \text{ m/s}^2$ unless otherwise specified in the Problem. Differences in the last figure can easily result from differences in rounding the input data and are not important.

Chapter 21

1. *Similarities:* The force between charges and masses vary as $1/r^2$.

   Differences: There are positive and negative charges but only positive masses. Like charges repel; like masses attract. The gravitational constant G is many orders of magnitude smaller than the Coulomb constant k.

3. *(c)*

5. *(a)*

 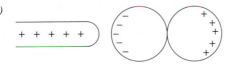

   (b)

7. *(d)*

9. *(d)*

11.

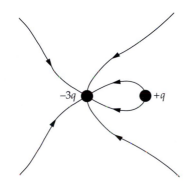

13. *(a)*

15. Because $\theta \neq 0$, a dipole in a uniform electric field will experience a restoring torque whose magnitude is $pE \sin \theta$. Hence, it will oscillate about its equilibrium orientation, $\theta = 0$. If the field is nonuniform and $dE/dx > 0$, the dipole will accelerate in the x direction as it oscillates about $\theta = 0$.

17. *(a)* The force between the balls is diminished because the field produced by the two charges creates a dipolar field that opposes that of the two charges when they are out of the water.

    (b) The force is again reduced because a dipole is induced on the third metal ball.

19. Assume that the wand has a negative charge. When the charged wand is brought near the tinfoil, the side nearer the wand becomes positively charged by induction, and so it swings toward the wand. When it touches the wand, some of the negative charge is transferred to the tinfoil, which thus has a net negative charge, and is now repelled by the wand.

21. *(a)* 3.46×10^{10} N

    (b) $32.0\ \mu$C

23. 141 nC

25. 5.00×10^{12}

27. 4.82×10^7 C

29. $(1.50 \times 10^{-2}\ \text{N})\,\hat{\imath}$

31. $(-8.64\ \text{N})\hat{\jmath}$

33. $\vec{F}_1 = (0.899\ \text{N})\,\hat{\imath} + (1.80\ \text{N})\hat{\jmath}$
    $\vec{F}_2 = (-1.28\ \text{N})\,\hat{\imath} - (1.16\ \text{N})\hat{\jmath}$
    $\vec{F}_3 = (0.391\ \text{N})\,\hat{\imath} - (0.640\ \text{N})\hat{\jmath}$

35. $\vec{F}_q = \dfrac{kqQ}{R^2}\left(1 + \dfrac{\sqrt{2}}{2}\right)\hat{\imath}$

37. *(a)* $\vec{E}(6\ \text{m}) = (999\ \text{N/C})\,\hat{\imath}$

    (b) $\vec{E}(-10\ \text{m}) = (-360\ \text{N/C})\,\hat{\imath}$

    (c)

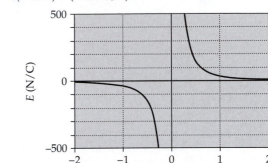

39. *(a)* $(400\ \text{kN/C})\hat{\jmath}$

    (b) $(-1.60\ \text{mN})\vec{\jmath}$

    (c) -40.0 nC

41. *(a)* $\vec{E}_x = (34.5\ \text{kN/C})\,\hat{\imath}$

    (b) $\vec{F} = (69.0\ \mu\text{N})\,\hat{\imath}$

43. *(a)* 12.9 kN/C, 231°

    (b) 2.08×10^{-15} N, 51.3°

45. *(a)* 1.90 kN/C, 235°

    (b) 3.04×10^{-16} N, 235°

47. *(a)* Because E_x is in the x direction, a positive test charge that is displaced from $(0, 0)$ in the x direction will experience a force in the x direction and accelerate in the x direction. Consequently, the equilibrium at $(0, 0)$ is unstable for a small displacement along the x axis. If the positive test charge is displaced in the y direction, the charge at $+a$ will exert a greater force than the charge at $-a$, and the net force is then in the $-y$ direction; that is, it is a restoring force. Consequently, the equilibrium at $(0, 0)$ is stable for small displacements along the y direction.

    (b) Following the same arguments as in Part *(a)*, one finds that, for a negative test charge, the equilibrium is stable at $(0, 0)$ for displacements along the x direction and the equilibrium is unstable for displacements along the y direction.

    (c) Because the two $+q$ charges repel, the charge Q at $(0, 0)$ must be a negative charge. Because the force between charges varies as $1/r^2$, and the negative charge is midway between the two positive charges, $Q = -q/4$.

    (d) If the charge Q is displaced, the equilibrium is the same as discussed in Part *(b)*. If either of the $+q$ charges are displaced, the system is unstable.

49. *(a)* 1.76×10^{11} C/kg

    (b) 1.76×10^{13} m/s²; The direction of the acceleration is opposite the electric field.

    (c) $0.170\ \mu$s

    (d) 25.5 cm

51. (a) $(-7.03 \times 10^{13} \text{ m/s}^2)\hat{j}$

     (b) 50.0 ns

     (c) $(-8.79 \text{ cm})\hat{j}$

53. 800 μC

55. 4.07 cm

57. (a) 8.00×10^{-18} C·m

     (b)

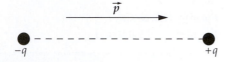

$-q$ $+q$

61. (a) $\vec{F}_{\text{net}} = Cp\hat{i}$

     (b) $\vec{F}_{\text{net}} = p_x \dfrac{dE_x}{dx} \hat{i}$

63. (a) 1.86×10^{-9} kg

     (b) 1.24×10^{36}

65. $\vec{E}_{P_2} = (1.73 \times 10^6 \text{ N/C})\hat{i}$. While the separation of the two charges of the dipole is more than 10 percent of the distance to the point of interest, that is, x is not much greater than a, this result is in excellent agreement with the result of Problem 64.

67. (a) $q_1 = 3.99 \ \mu\text{C}, q_2 = 2.01 \ \mu\text{C}$; or $q_1 = 2.01 \ \mu\text{C}, q_2 = 3.99 \ \mu\text{C}$

     (b) $q_1 = 7.12 \ \mu\text{C}; q_2 = -1.12 \ \mu\text{C}$

71. (a) $q_1 = 17.5 \ \mu\text{C}, q_2 = 183 \ \mu\text{C}$; or $q_1 = 183 \ \mu\text{C}, q_2 = 17.5 \ \mu\text{C}$

     (b) $q_1 = -15.0 \ \mu\text{C}; q_2 = 215 \ \mu\text{C}$

73. (a) 0.225 N

     (b) 0.113 N·m; counterclockwise

     (c) 0.0461 kg

     (d) 5.03×10^{-7} C

75. (a) $q_1 = 28.0 \ \mu\text{C}, q_2 = 172 \ \mu\text{C}$; or $q_1 = 172 \ \mu\text{C}, q_2 = 28.0 \ \mu\text{C}$

     (b) 250 N

77. (a) $-97.2 \ \mu$C

     (b) $x_1 = 0.0508$ m; $x_2 = 0.169$ m

79. (a) $10.3°$

     (b) $9.86°$

81. $\dfrac{d^2 \theta}{dt^2} = -\dfrac{2qE}{ma} \theta; T = 2\pi \sqrt{\dfrac{ma}{2qE}}$

83. $v = \sqrt{\dfrac{ke^2}{2mr}}$

87. (a) $8.48°$

     (b) $\theta_1 = 9.42°; \theta_1 = 6.98°$

89. $\vec{E} = (-1.10 \times 10^4 \text{ N/C})\hat{j}$

91. (a) $\vec{E}_P = \dfrac{2kQy}{(a^2 + y^2)^{3/2}} \hat{j}$

     (b) $\vec{F}_y = \dfrac{2kqQy}{(a^2 + y^2)^{3/2}} \hat{j}$

     (c) The differential equation of motion is $\dfrac{d^2 y}{dt^2} + \dfrac{16kqQ}{mL^3} y = 0$

     (d) 9.37 Hz

93. (b) 5.15×10^{-5} m/s

Chapter 22

1. (a) False. Gauss's law states that the net flux through any surface is given by $\phi_{\text{net}} = \oint_S E_n \, dA = 4\pi k Q_{\text{inside}}$. While it is true that Gauss's law is easiest to apply to symmetric charge distributions, it holds for *any* surface.

    (b) True

3. The electric field is that due to all the charges, inside and outside the surface. Gauss's law states that the net flux through any surface is given by $\phi_{\text{net}} = \oint_S E_n \, dA = 4\pi k Q_{\text{inside}}$. The lines of flux through a Gaussian surface begin on charges on one side of the surface and terminate on charges on the other side of the surface.

5. (a) False. Consider a spherical shell, in which there is no charge, in the vicinity of an infinite sheet of charge. The electric field due to the infinite sheet would be non-zero everywhere on the spherical surface.

    (b) True (assuming there are no charges inside the shell).

    (c) True

    (d) False. Consider a spherical conducting shell. Such a surface will have equal charges on its inner and outer surfaces but, because their areas differ, so will their charge densities.

7. (a)

9. (b)

11. (c)

13. False. A physical quantity is discontinuous if its value on one side of a boundary differs from that on the other. We can show that this statement is false by citing a counterexample. Consider the field of a uniformly charged sphere. ρ is discontinuous at the surface, E is not.

15. 3×10^6 V/m; 5.31×10^{-5} C/m²

17. (a) 17.5 nC

    (b) 26.2 N/C

    (c) 4.37 N/C

    (d) 2.57 mN/C

    (e) 2.52 mN/C

19. (a) 4.69×10^5 N/C

    (b) 1.13×10^6 N/C

    (c) 1.54×10^3 N/C

    (d) 1.55×10^3 N/C; This result is greater than the result for Part (c) by less than 0.1%.

21. (a) 0.300 nC

    (b) 1.43 kN/C

    (c) 183 N/C

    (d) 0.133149 N/C

    (e) 0.133147 N/C

23. (a) $E_x(0.2a) = 0.189\dfrac{kQ}{a^2}$

    (b) $E_x(0.5a) = 0.358\dfrac{kQ}{a^2}$

    (c) $E_x(0.7a) = 0.385\dfrac{kQ}{a^2}$

    (d) $E_x(a) = 0.354\dfrac{kQ}{a^2}$

(e) $E_x(2a) = 0.179\dfrac{kQ}{a^2}$

(f)

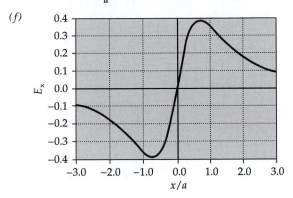

25.

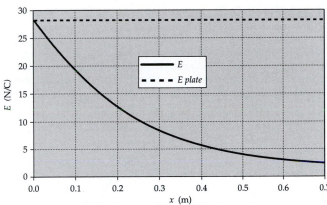

The magnitudes of the electric fields differ by more than 10% for $x = 0.03$ m.

31. (a) 20.0 N·m²/C

(b) 17.3 N·m²/C

33. (a) $\phi_{\text{right}} = 1.51$ N·m²/C; $\phi_{\text{left}} = 1.51$ N·m²/C

(b) $\phi_{\text{curved}} = 0$

(c) $\phi_{\text{net}} = 3.02$ N·m²/C

(d) $Q_{\text{inside}} = 2.67 \times 10^{-11}$ C

35. (a) 3.14 m²

(b) 7.19×10^4 N/C

(c) 2.26×10^5 N·m²/C

(d) No. The flux through the surface is independent of where the charge is located inside the sphere.

(e) 2.26×10^5 N·m²/C

37. 3.77×10^4 N·m²/C

39. (a) $E_{r<R_1} = 0$; $E_{R_1<r<R_1} = \dfrac{kq_1}{r^2}$; $E_{r>R_2} = \dfrac{k(q_1 + q_2)}{r^2}$

(b) -1

(c)

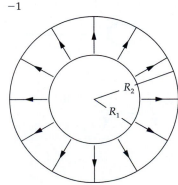

41. (a) 0.407 nC

(b) 339 N/C

(c) 999 N/C

(d) 983 N/C

(e) 366 N/C

43. 3.77 N/C

45. (a) $2\pi BR^2$

(b) $E_r(r > R) = \dfrac{BR^2}{2\,\epsilon_0 r^2}$; $E_r(r < R) = \dfrac{B}{2\epsilon_0}$

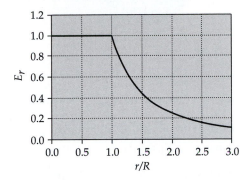

47. (a) $E_r(r < a) = 0$; $E_r(a < r < b) = \dfrac{\rho}{3\epsilon_0 r^2}(r^3 - a^3)$;

$E_r(r > b) = \dfrac{\rho}{3\epsilon_0 r^2}(b^3 - a^3)$

49. (a) 679 nC

(b) $E(2\text{ cm}) = 0$

(c) $E(5.9\text{ cm}) = 0$

(d) $E(6.1\text{ cm}) = 1.00$ kN/C

(e) $E(10\text{ cm}) = 610$ N/C

51. (a) 679 nC

(b) $E_r(2\text{ cm}) = 339$ N/C

(c) $E_r(5.9\text{ cm}) = 1.00$ kN/C

(d) $E_r(6.1\text{ cm}) = 1.00$ kN/C

(e) $E_r(10\text{ cm}) = 610$ N/C

53. (a) $E_n(r < 1.5\text{ cm}) = 0$; $E_n(1.5\text{ cm} < r < 4.5\text{ cm}) = \dfrac{(108\text{ N·m/C})}{r}$;

$E_n(4.5\text{ cm} < r < 6.5\text{ cm}) = 0$

(b) $\sigma_1 = 21.2$ nC/m²; $\sigma_2 = -14.7$ nC/m²

55. (b) $E_n(r < R) = \dfrac{b}{4\epsilon_0}r^3$; $E_n(r > R) = \dfrac{bR^4}{4r\epsilon_0}$

57. (a) $\lambda_{\text{inner}} = 18.8$ nC/m

(b) $E_n(r < 1.5\text{ cm}) = 22.6$ kN/C;

$E_n(1.5\text{ cm} < r < 4.5\text{ cm}) = \dfrac{339\text{ N·m/C}}{r}$;

$E_n(4.5\text{ cm} < r < 6.5\text{ cm}) = 0$;

$E_n(r > 6.5\text{ cm}) = \dfrac{339\text{ N·m/C}}{r}$

59. 9.42 kN/C

61. (a) $E_n(r < a) = \dfrac{kq}{r^2}$; $E_n(a < r < b) = 0$; $E_n(r > b) = \dfrac{kq}{r^2}$

    (b)

    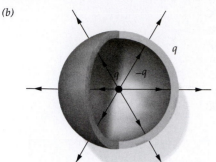

    (c) $\sigma_{inner} = -\dfrac{q}{4\pi a^2}$; $\sigma_{outer} = \dfrac{q}{4\pi b^2}$

63. (a) $\sigma_{inner} = -0.553\ \mu C/m^2$; $\sigma_{outer} = 0.246\ \mu C/m^2$

    (b) $E_n(r < a) = (2.25 \times 10^4\ N{\cdot}m^2/C)\left(\dfrac{1}{r^2}\right)$;

    $E_n(0.6\ m < r < 0.9\ m) = 0$;

    $E_n(r > 0.9\ m) = (2.25 \times 10^4\ N{\cdot}m^2/C)\left(\dfrac{1}{r^2}\right)$

    (c) $\sigma_{inner} = -0.553\ \mu C/m^2$; $\sigma_{outer} = 0.589\ \mu C/m^2$;

    $E_n(r < a) = (2.25 \times 10^4\ N{\cdot}m^2/C)\left(\dfrac{1}{r^2}\right)$;

    $E_n(0.6\ m < r < 0.9\ m) = 0$;

    $E_n(r > 0.9\ m) = (5.39 \times 10^4\ N{\cdot}m^2/C)\left(\dfrac{1}{r^2}\right)$

65. (a) $\sigma_{face} = 1.60\ \mu C/m^2$; $E_{slab} = 1.81 \times 10^5\ N/C$

    (b) $\vec{E}_{near} = (-0.680 \times 10^5\ N/C)\hat{r}$; $\vec{E}_{far} = (2.94 \times 10^5\ N/C)\hat{r}$;

    $\sigma_{near} = 0.602\ \mu C/m^2$; $\sigma_{near} = 2.60\ \mu C/m^2$

67. $1.15 \times 10^5\ N/C$

69. (a) $\dfrac{Q}{8\pi\epsilon_0 r^2}$

    (b) $\dfrac{Q^2 a^2}{32\pi\epsilon_0 r^4}$

    (c) $\dfrac{Q^2}{32\pi^2\epsilon_0 r^4}$

71. $1.11 \times 10^6\ N/C$

73. (a) $\vec{E}(0, 0) = (339\ kN/C)\hat{i}$

    (b) $\vec{E}(0.2\ m, 0.1\ m) = (1310\ kN/C)\hat{i} + (-268\ kN/C)\hat{j}$

    (c) $\vec{E}(0.5\ m, 0.2\ m) = (203\ kN/C)\hat{i}$

75. (a) $\dfrac{e}{\pi a_0^3}$

    (b) $E(r) = \dfrac{ke}{r^2} e^{-2r/a_0}\left(1 + \dfrac{2r}{a_0} + \dfrac{2r^2}{a_0^2}\right)$

77. (a) $\dfrac{q_1}{q_2} = \dfrac{r_1}{r_2}$; $E_1 > E_2$

    (b) Because $E_1 > E_2$, the resultant field points toward s_2.

    (c) $E_1 = E_2$

    (d) $\vec{E} = 0$; If $E \propto 1/r$, then s_2 would produce the stronger field at P and $\vec{E}$ would point toward s_1.

79. (a) If Q is positive, the field at the origin points radially outward.

    (b) $E_{center} = \dfrac{kQ\ell}{2\pi R^3}$

81. (a) $E(0.4\ m, 0) = 203\ kN/C$; $\theta = 56.2°$

    (b) $E(2.5\ m, 0) = 263\ kN/C$; $\theta = 153°$

83. $T = 2\pi R \sqrt{\dfrac{m}{2k\lambda q}}$

85. $7.42\ rad/s$

87. (b) $\vec{E}_1 = \dfrac{\rho b}{3\epsilon_0}\hat{r}$; $\vec{E}_2 = \dfrac{\rho b}{3\epsilon_0}\hat{r}$

89. (b) $\vec{E}_1 = \dfrac{(2\rho + \rho')b}{3\epsilon_0}\hat{r}$; $\vec{E}_2 = \dfrac{\rho'}{3\epsilon_0}b\hat{r}$

91. $200\ N/C$

95. $0.5R$

97. $4.49 \times 10^{14}\ s^{-1}$

Chapter 23

1. A positive charge will move in whatever direction reduces its potential energy. The positive charge will reduce its potential energy if it moves toward a region of lower electric potential.

3. If V is constant, its gradient is zero; consequently $\vec{E} = 0$.

5. Because the field lines are always perpendicular to equipotential surfaces, you always move perpendicular to the field.

7.

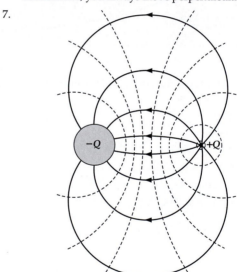

9.

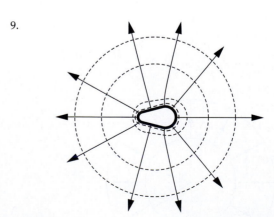

11. (b)

13. (c)

15. (a) No. The potential at the surface of a conductor also depends on the local radius of the surface. Hence, r and σ can vary in such a way that V is constant.

    (b) Yes; Yes.

17. 3.00×10^9 V

19. (a) $K = 0.719$ MeV

    (b) 0.0767%

21. (a) -8.00 kV

    (b) -24.0 mJ

    (c) 24.0 mJ

    (d) $-(2 \text{ kV/m})x$

    (e) $4 \text{ kV} - (2 \text{ kV/m})x$

    (f) $2 \text{ kV} - (2 \text{ kV/m})x$

23. (a) 4.50 kV

    (b) 13.5 mJ

    (c) 13.5 mJ

25. (a) 3.10×10^7 m/s

    (b) 2.50 MV/m

27. (a) $r = \dfrac{2kZe^2}{E}$

    (b) 45.4 fm; 25.3 fm

29. (a) 12.9 kV

    (b) 7.55 kV

    (c) 4.44 kV

31. (a) 270 kV

    (b) 191 kV

33. (a) $V = kq\left(\dfrac{1}{|x - a|} + \dfrac{1}{|x + a|}\right)$

    (b)

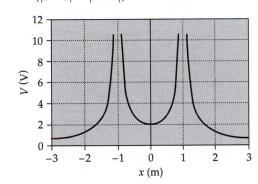

    (c) 0; 0

35. (a) $V_b - V_a$ is positive.

    (b) 25.0 kV/m

37. (a) 8.99 kV; 8.96 kV

    (b) 3.00 kV/m

    (c) 3.00 kV/m

    (d) 8.99 kV

39. (a) $V_b - V_a$ is negative.

    (b) 5.00 kV/m

41. (a) $V(x) = k\left(\dfrac{q}{|x|} + \dfrac{3q}{|x - 1|}\right)$

    (b) -0.500 m; 0.250 m

43. (a) $V(x) = kq\left(\dfrac{2}{\sqrt{x^2 + a^2}} + \dfrac{1}{|x - a|}\right)$

    (b) $E_x(x > a) = \dfrac{2kqx}{(x^2 + a^2)^{3/2}} + \dfrac{kq}{(x - a)^2}$

45. (a) 6.02 kV

    (b) -12.7 kV

    (c) -42.3 kV

47. (a) $V(x, 0) = \dfrac{kQ}{L} \ln\left(\dfrac{\sqrt{x^2 + L^2/4} + L/2}{\sqrt{x^2 + L^2/4} - L/2}\right)$

49. (a) $Q = \frac{1}{2}\pi\sigma_0 R^2$

    (b) $V = \dfrac{2\pi k\sigma_0}{R^2}\left(\dfrac{R^2 - 2x^2}{3}\sqrt{x^2 + R^2} + \dfrac{2x^3}{3}\right)$

51. (a) $V(x) = 2\pi k\sigma_0\left(2\sqrt{x^2 + \dfrac{R^2}{2}} - \sqrt{x^2 + R^2} - x\right)$

    (b) $V(x) = \dfrac{\pi k\sigma_0 R^4}{8x^3}$

53. (a) $V(x) = \dfrac{kQ}{L} \ln\left(\dfrac{x + \dfrac{L}{2}}{x - \dfrac{L}{2}}\right)$

    (b) $V(x) = \dfrac{kQ}{x}$

55. $V_b - V_a = -\dfrac{2kq}{L} \ln\left(\dfrac{b}{a}\right)$

57. (a) $V_\text{I} = \dfrac{\sigma}{\epsilon_0}x; \quad V_\text{II} = 0; \quad V_\text{III} = \dfrac{\sigma}{\epsilon_0}(a - x)$

    (b) $V_\text{I} = 0; \quad V_\text{II} = -\dfrac{\sigma}{\epsilon_0}x; \quad V_\text{III} = -\dfrac{\sigma}{\epsilon_0}a$

59. (a) $V_1 = \dfrac{kQ}{R^3}r^2$

    (b) $dV_2 = \dfrac{3kQ}{R^3}r'\,dr'$

    (c) $V_2 = \dfrac{3kQ}{2R^3}(R^2 - r^2)$

    (d) $V = \dfrac{kQ}{2R}\left(3 - \dfrac{r^2}{R^2}\right)$

(c) $E_x(0.25 \text{ m}) = (21.3 \text{ m}^{-2})kq; \quad E_x(-0.5 \text{ m}) = (-2.67 \text{ m}^{-2})kq$

(d)

61. $r_{20\,V} = 0.499$ m; $r_{40\,V} = 0.250$ m; $r_{60\,V} = 0.166$ m; $r_{80\,V} = 0.125$ m; $r_{100\,V} = 0.0999$ m

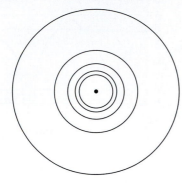

The equipotential surfaces are not equally spaced.

63. $26.6\ \mu C/m^2$

65. $V_a - V_b = V_a = kq\left(\dfrac{1}{a} - \dfrac{1}{b}\right)$

67. (a)

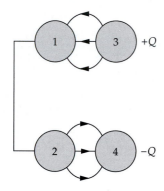

+Q

−Q

(b) $V_1 = V_2$ because the spheres are connected. From the direction of the electric field lines, it follows that $V_3 > V_1$.

(c) If sphere 3 and sphere 4 are connected, $V_3 = V_4$. The conditions of Part (b) can only be satisfied if all potentials are zero. Consequently the charge on each sphere is zero.

69. (a) $V(x) = \dfrac{2kq}{\sqrt{x^2 + a^2}}$

(b) $\vec{E}(x) = \dfrac{2kqx}{(x^2 + a^2)^{3/2}}\ \hat{i}$

71. (a) $V(x, y) = \dfrac{\lambda}{2\pi\epsilon_0} \ln\left[\dfrac{\sqrt{(x - a)^2 + y^2}}{\sqrt{(x + a)^2 + y^2}}\right]$; $V(0, y) = 0$

(b)

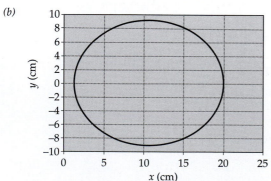

73. $\rho_0 = \dfrac{e}{\pi a^3}$

75. $-20\ \mu C$

77. (a) $W_{+Q \to +a} = \dfrac{kQ^2}{2a}$

(b) $W_{-Q \to 0} = \dfrac{-2kQ^2}{a}$

(c) $W_{-Q \to 2a} = \dfrac{2kQ^2}{3a}$

79. 1.38×10^5 m/s

81. $R_2 = \frac{2}{3} R_1$

83. 7.12 nC

85. (a) $\Delta V = \dfrac{\sigma d}{\epsilon_0}$

(b) $\Delta V' = \dfrac{\sigma}{\epsilon_0}(d - a)$

87. (b) $\sigma = \dfrac{qd}{4\pi(d^2 + r^2)^{3/2}}$

(c) $W = \dfrac{kq^2}{6d}$

89. (a) $V(c) = 0$; $V(b) = kQ\left(\dfrac{1}{b} - \dfrac{1}{c}\right)$; $V(a) = kQ\left(\dfrac{1}{b} - \dfrac{1}{c}\right)$

(b) $Q_b = Q$; $V(a) = V(c) = 0$; $Q_a = -Q\dfrac{a(c - b)}{b(c - a)}$;

$Q_c = -Q\dfrac{c(b - a)}{b(c - a)}$; $V(b) = kQ\dfrac{(c - b)(b - a)}{b^2(c - a)}$

91. $W = \dfrac{3Q^2}{20\pi\epsilon_0 R}$

93. (a) $R' = 0.794R$

(b) $\Delta E = 0.370E$

95. (a) $V_{av} = \dfrac{q}{4\pi\epsilon_0 R}$

(b) The superposition principle tells us that the potential at any point is the sum of the potentials due to any charge distributions in space. Because this result is independent of any properties of the sphere, this result must hold for any sphere and for any configuration of charges outside of the sphere.

Chapter 24

1. (c)

3. True

5. (d)

7. Both statements are true.

9. True

11. (a) False

(b) False

(c) False

13. 0.104 nF/m $\leq C/L \leq 0.173$ nF/m

15. 9.03×10^{10} J

17. (a) 30 mJ

(b) -5.99 mJ

(c) -18.0 mJ

19. $22.2\ \mu J$

21. $v = q\sqrt{\dfrac{6\sqrt{2}k}{ma}}$

23. 75.0 nF

25. (a) 15.0 mJ

   (b) 45.0 mJ

27. (a) 0.625 J

   (b) 1.88 J

29. (a) 100 kV/m

   (b) 44.3 mJ/m³

   (c) 88.6 μJ

   (d) 17.7 nF

   (e) 88.5 μJ; in agreement with Part (c)

31. (a) 11.1 nC

   (b) 0.553 μJ

33. (a) 100

   (b) 10 V

   (c) charge: 1.00 kV; difference: 10.0 μC

35. $C_{eq} = C_2 + \dfrac{C_1 C_3}{C_1 + C_3}$

37. (a) 40.0 μC

   (b) $V_{10} = 4.00$ V; $V_{20} = 2.00$ V

39. (a) 15.2 μF

   (b) 2.40 mC; 0.632 mC

   (c) 0.304 J

41. (a) 0.242 μF

   (b) 2.42 μC; $Q_1 = 1.93$ μC; $Q_{0.25} = 0.483$ μC

   (c) 12.1 μJ

43. Place four of the capacitors in series. Then the potential across each capacitor is 100 V when the potential across the combination is 400 V. The equivalent capacitance of the series is 2/4 μF = 0.5 μF. If we place four such series combinations in parallel, as shown in the circuit diagram, the total capacitance between the terminals is 2 μF.

45. (a) $C_{eq} = 0.618$ μF

   (b) 1.618 μF

47. (a) 40.0 V

   (b) 4.24 m

49. (a) 0.333 mm

   (b) 3.76 m³

51. (a) $E_{r<R_1} = 0$; $u_{r<R_1} = 0$; $E_{r>R_2} = 0$; $u_{r>R_2} = 0$

   (b) $\dfrac{kQ^2}{L} \ln\left(\dfrac{R_2}{R_1}\right)$

   (c) $\dfrac{kQ^2}{L} \ln\left(\dfrac{R_2}{R_1}\right)$

53. $C = \dfrac{\epsilon_0 (R_2^2 - R_1^2)}{2d}(\theta - \Delta\theta)$

57. $2R$

59. (a) 2.00 kV

   (b) 0

61. (a) 2.40 μF

   (b) 360 μJ

63. (a) 6.00 V

   (b) 1.15 mJ

   (c) 0.288 mJ

65. (a) 200 V

   (b) $q_1 = -254$ μC; $q_2 = 146$ μC; $q_3 = 546$ μC

   (c) $V_1 = -127$ V; $V_2 = 36.5$ V; $V_3 = 91.0$ V

67. 2.71 nF

69. (a) $E_{r<R_1} = 0$; $u_{r<R_1} = 0$; $E_{r>R_2} = 0$; $u_{r>R_2} = 0$

   (b) $dU = \dfrac{kQ^2}{2\kappa r^2} dr$

   (c) $U = \dfrac{1}{2} Q^2 \left(\dfrac{R_2 - R_1}{4\pi\kappa\epsilon_0 R_1 R_2}\right)$

71. $C_{eq} = \left(\dfrac{4\kappa_1\kappa_2}{3\kappa_1 + \kappa_2}\right) C_0$

73. $C_{eq} = \left(\dfrac{\kappa d}{\kappa(d - t) + t}\right) C_0$

75. (a) 5.00

   (b) 1.25

   (c) 50.0

77. $C = \left(\kappa_3 + \dfrac{2\kappa_1\kappa_2}{\kappa_1 + \kappa_2}\right)\left(\dfrac{\epsilon_0 A}{2d}\right)$

79. (a) $C = \dfrac{3\epsilon_0 A}{y_0 \ln(4)}$

   (b) $\left.\dfrac{\sigma_b}{\sigma_f}\right|_{y=0} = 0$; $\left.\dfrac{\sigma_b}{\sigma_f}\right|_{y=y_0} = 0.750$

   (c) $\rho(y) = \dfrac{3\sigma}{[y_0(1 + 3y/y_0)^2]}$

   (d) $\rho = -\frac{3}{4}\sigma$, which is the charge per unit area in the dielectric, and just cancels out the induced surface charge density.

81. (a) 14.0 μF

   (b) 1.14 μF

83. 1.00 mm

85. $C_2 C_3 = C_1 C_4$

87. (a) $C_{new} = \dfrac{\epsilon_0 A}{2d}$

   (b) $V_{new} = 2V$

   (c) $U_{new} = \dfrac{\epsilon_0 A V^2}{d}$

   (d) $W = \dfrac{\epsilon_0 A V^2}{2d}$

89. (a) 2.22 nF

   (b) 66.6 μC

91. $Q_1 = 267$ μC; $Q_2 = 133$ μC

95. (a) $U = \dfrac{Q^2}{2\epsilon_0 A} x$

   (b) $dU = \dfrac{Q^2}{2\epsilon_0 A} dx$

   (c) $F = \dfrac{Q^2}{2\epsilon_0 A}$

   (d) $F = \frac{1}{2} QE$

97. (a) $U = \dfrac{Q^2 d}{2\epsilon_0 a[(\kappa - 1)x + a]}$

   (b) $F = \dfrac{(\kappa - 1)Q^2 d}{2a\epsilon_0[(\kappa - 1)x + a]^2}$

   (c) $F = \dfrac{(\kappa - 1)a\epsilon_0 V^2}{2d}$

   (d) This force originates from the fringing fields around the edges of the capacitor. The effect of the force is to pull the dielectric into the space between the capacitor plates.

99. $2.55 \, \mu\text{J}$

101. (a) Because F increases as ℓ decreases, a decrease in plate separation will unbalance the system and the balance is unstable.

(b) $V = \ell \sqrt{\dfrac{2Mg}{\epsilon_0 A}}$

103. (a) $Q_1 = (200 \text{ V})C_1; Q_2 = (200 \text{ V})\kappa C_1$

(b) $U = (2 \times 10^4 \text{ V}^2)(1 + \kappa)C_1$

(c) $U_f = (10^4 \text{ V}^2)C_1(1 + \kappa)^2$

(d) $V_f = 100(1 + \kappa)\text{V}$

105. (a) $0.225 \, \text{J}$

(b) $3.50 \text{ mC}; 1.00 \text{ mC}$

(c) 2.25 mC

(d) $0.506 \, \text{J}$

107. $0.100 \, \mu\text{F}; 16.0 \, \mu\text{C}$

109. (a) 1.00 mJ

(b) $Q_1' = 47.6 \, \mu\text{C}; Q_2' = 152 \, \mu\text{C}$

(c) 0.476 mJ

111. (a) $C(V) = C_0 \left(1 + \dfrac{\kappa \epsilon_0 V^2}{2Yd^2}\right)$

(b) 7.97 kV

(c) $0.209 \text{ percent}; 99.8 \text{ percent}$

Chapter 25

1. When current flows, the charges are not in equilibrium. In that case, the electric field provides the force needed for the charge flow.

3. (e)

5. (c)

7. (d)

9. You should decrease the resistance. Because the voltage across the resistor is constant, the heat out is given by $P = V^2/R$. Hence, decreasing the resistance will increase P.

11. (a)

13. (a)

15. (b)

17. (b)

19. (b)

21. (e)

23. A small resistance, because $P = \mathcal{E}^2/R$.

25. Yes. Kirchhoff's rules are statements of the conservation of energy and charge and therefore apply to all circuits.

27. (a) 3.12 V

(b) 78.0 mV/m

(c) 18.7 W

29. 2.03 m

31. 2.08 mm

33. 0.281 mm/s

35. (a) $5.93 \times 10^7 \text{ m/s}$

(b) $37.3 \, \mu\text{A}$

37. $0.210 \text{ mm/s}; 0.531 \text{ mm/s}$

39. (a) $1.04 \times 10^8 \text{ m}^{-1}$

(b) $1.04 \times 10^{14} \text{ m}^{-3}$

41. (a) $33.3 \, \Omega$

(b) 0.751 A

43. 8.98 mm

45. 1.95 V

47. $62.2 \, c{\cdot}y$

49. (a) 0.170

(b) E is greater in the iron wire.

51. $1.20 \, \Omega$

53. $0.0314 \, \Omega$

55. (b) 90.0 mA

57. $\dfrac{\rho L}{\pi a b}$

59. (a) $\dfrac{\rho}{2\pi L} \ln \dfrac{b}{a}$

(b) 2.05 A

61. $45.6°\text{C}$

63. (a) 15.0 A

(b) $11.1 \, \Omega$

(c) 1.30 kW

65. (a) $\dfrac{1}{A}[\rho_1 L_1 + \rho_2 L_2]$

(b) 264

67. (a) 636 K

(b) As the filament heats up its resistance increases, leading to more power being dissipated, leading to further heat, leading to a higher temperature, and so on. This thermal runaway can burn out the filament if not controlled.

69. (a) 5.00 mA

(b) 50.0 V

71. 180 J

73. (a) 240 W

(b) 228 W

(c) 43.2 kJ

(d) 2.16 kJ

75. (a) 6.91 MJ

(b) 12.8 h

77. (a) 26.7 kW

(b) 5.76 MC

(c) 69.1 MJ

(d) 57.6 km

(e) $\$0.03/\text{km}$

79. (a) $1.33 \, \Omega$

(b) $I_4 = 3.00 \text{ A}; I_3 = 4.00 \text{ A}; I_6 = 2.00 \text{ A}$

81. (b) Because the potential difference between points c and d is zero, no current would flow through the resistor connected between these two points, and the addition of that resistor would not change the network.

83. $450 \, \Omega$

85. (b)

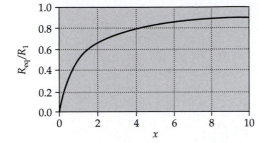

87. (a) 6.00 Ω

(b) 0.667 A; 1.33 A; 0.667 A

89. 8

91. (a) $R_3 = \dfrac{R_1^2}{R_1 + R_2}$

(b) $R_2 = 0$

(c) $R_1 = \dfrac{R_3 + \sqrt{R_3^2 + 4R_2 R_3}}{2}$

93. (a) 4.00 A

(b) 2.00 V

(c) 1.00 Ω

95. (a)

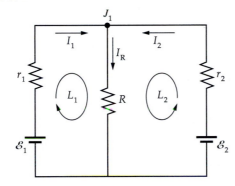

(b) $I_1 = -57.0$ A; $I_2 = 63.0$ A; $I_R = 6.00$ A

(c) $P_2 = 794$ W; $P_1 = 650$ W; $P_{r_1} = 32.5$ W; $P_{r_2} = 39.7$ W; $P_R = 72.0$ W

97. (a) $I_1 = 0.667$ A; $I_2 = 0.889$ A; $I_3 = 1.56$ A

(b) 9.36 V

(c) 8.00 W; 10.7 W

99. If $r = R$, both arrangements provide the same power to the load. Examination of the second derivative of P_p at $R = \frac{1}{2}r$ shows that $R = \frac{1}{2}r$ corresponds to a maximum value of P_p and hence, for the parallel combination, the power delivered to the load is greater if $R < r$ and is at a maximum when $R = \frac{1}{2}r$.

103. 2.40 V

105. (a) 3.33 V

(b) 3.33 V

(c) 3.13 V

(d) 2.00 V

(e) 0.435 V

(f) 1.67 MΩ

107. 2.50 Ω

109. 195 kΩ

111. (a) 600 μC

(b) 0.200 A

(c) 3.00 ms

(d) 81.2 μC

113. 2.18 MΩ

115. (a) 8.00 μC

(b) 73.7 ms

117. (a) 5.69 μC

(b) 1.10 μA

(c) 1.10 μA

(d) 6.60 μW

(e) 2.42 μW

(f) 4.17 μW

121. (a) 0.250 A

(b) 62.5 mA

(c) $I_2(t) = (62.5 \text{ mA})\left(1 - e^{-t/0.750 \text{ ms}}\right)$

123. (a) 48.0 μA

(b) 0.866 s

125. (a)

127. (b)

129. (a) 30.0 A

(b) 4.00 V

131. $R = 14.0$ Ω; $\mathcal{E} = -7.00$ V

133. (a) 43.9 Ω

(b) 300 Ω

(c) 3.80 kΩ

135. (a) $2.19 \times 10^{13}/\text{s}$

(b) 210 J/s

(c) 27.6 s

137. 0.164 L/s

139. This result holds independently of the geometries of the capacitor and the resistor.

141. (a) 10.0 ms

(b) $V(t) = \dfrac{\mathcal{E}}{\tau}t$

(c) 1.00 GΩ

(d) 60.9 ps

(e) $P_1 = 6.17$ nW; $P_2 = 2.89$ kW

145. (a) $I(t) = \dfrac{V_0}{R}e^{-t/\tau}$

(b) $P(t) = \dfrac{V_0^2}{R}e^{-2t/\tau}$

(c) $E = \frac{1}{2}V_0^2 C_{eq}$; This is exactly the difference between the initial and final stored energies found in the preceding problem, which confirms the statement at the end of that problem that the difference in the stored energies equals the energy dissipated in the resistor.

149. (a) 10^{12}

(b) 0.160 mA

(c) 64.0 kW

(d) 640 MW

(e) 10^{-4}

151. $R_{eq} = \dfrac{R_1 + \sqrt{R_1^2 + 4R_1 R_2}}{2}$

Chapter 26

1. *(b)*

3. False

5. The alternating current running through the filament is changing direction every 1/60 s, so in a magnetic field the filament experiences a force that alternates in direction at that frequency.

7. *(a)* True

   (b) True

   (c) True

   (d) False

   (e) True

9. Upward

11. *(a)*

13. From relativity; This is equivalent to the electron moving from right to left at velocity v with the magnet stationary. When the electron is directly over the magnet, the field points directly up, so there is a force directed out of the page on the electron.

15. If only $\vec{F}$ and I are known, one can only conclude that the magnetic field $\vec{B}$ is in the plane perpendicular to $\vec{F}$. The specific direction of $\vec{B}$ is undetermined.

17. *(a)* 177 C/kg

    (b) 53.1 nC

19. *(a)* $-(3.80 \text{ mN})\hat{k}$

    (b) $-(7.51 \text{ mN})\hat{k}$

    (c) 0

    (d) $(7.51 \text{ mN})\hat{j}$

21. 0.962 N

23. 0.621 pN; $\theta_x = 108°$; $\theta_y = 102°$; $\theta_z = 158°$

25. 1.48 A

29. $\vec{B} = (10 \text{ T})\hat{i} + (10 \text{ T})\hat{j} - (15 \text{ T})\hat{k}$

31. *(a)* 87.4 ns

    (b) 6.47×10^7 m/s

    (c) 11.4 MeV

33. *(a)* 142 m

    (b) 2.84 m

35. *(a)* $\dfrac{v_\alpha}{v_p} = \dfrac{1}{2}$

    (b) $\dfrac{K_\alpha}{K_p} = 1$

    (c) $\dfrac{L_\alpha}{L_p} = 2$

39. *(a)* $\phi = 24.0°$; $r_p = 0.492$ m; $v_p = 2.83 \times 10^7$ m/s

    (b) $v_d = 1.41 \times 10^7$ m/s

41. *(a)* 1.64×10^6 m/s

    (b) 14.1 keV

    (c) 7.67 eV

43. *(a)* 7.35 mm

    (b) 66.3 μT

45. *(a)* 63.3 cm

    (b) 2.60 cm

47. $\Delta t_{58} = 15.8 \ \mu$s; $\Delta t_{60} = 16.3 \ \mu$s

49. *(a)* 21.3 MHz

    (b) 46.0 MeV

    (c) 10.7 MHz; 23.0 MeV

53. *(a)* 0.302 A·m²

    (b) 0.131 N·m

55. *(a)* 0

    (b) $\pm(2.70 \times 10^{-3} \text{ N·m})\hat{j}$

57. $B = \dfrac{mg}{I\pi R}$

59. *(a)*

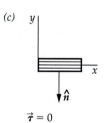

$\vec{\tau} = (0.840 \text{ N·m})\hat{k}$

(b)

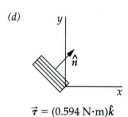

$\vec{\tau} = 0$

(c)

$\vec{\tau} = 0$

(d)

$\vec{\tau} = (0.594 \text{ N·m})\hat{k}$

61. 0.377 A·m²; Into the page.

67. $\mu = \frac{4}{3} \pi \sigma R^4 \omega$

69. *(a)* $\tau = \frac{1}{4} \pi \sigma r^4 \omega B \sin \theta$

    (b) $\Omega = \dfrac{\pi \sigma r^2 B}{2m} \sin \theta$

71. *(a)* 3.69×10^{-5} m/s

    (b) 1.48 μV

73. 1.02 mV

75. 3.46

77. *(a)* 0.131 μs

    (b) 2.41×10^7 m/s

    (c) 12.0 MeV

81. *(a)* $\vec{B} = -\dfrac{mg}{I\ell} \tan \theta \, \hat{u}_v$

    (b) $g \sin \theta$

83. (a) $\vec{F} = (1.60 \times 10^{-18} \text{ N})\hat{j}$

    (b) $\vec{E} = (10.0 \text{ V/m})\hat{j}$

    (c) $\Delta V = 20.0 \text{ V}$

85. 5.10 m

Chapter 27

1. (a) The electric forces are repulsive; The magnetic forces are attractive (the two charges moving in the same direction act like two currents in the same direction).

   (b) The electric forces are again repulsive; The magnetic forces are also repulsive.

3.

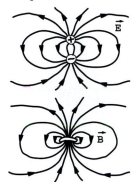

5. (a)

7. (e)

9.

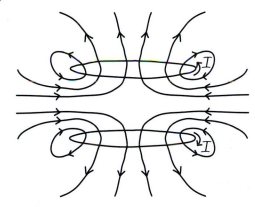

11. (a) True

    (b) True

13. (a) False

    (b) True

    (c) False

    (d) False

    (e) True

15. No. The classical relation between magnetic moment and angular momentum is $\vec{u} = \dfrac{q}{2m}\vec{L}$. Thus, if the angular momentum of the particle is zero, its magnetic moment will also be zero.

17. From Ampère's law, the current enclosed by a closed path within the tube is zero, and from the cylindrical symmetry it follows that $B = 0$ everywhere within the tube.

19. H_2, CO_2, and N_2 are diamagnetic ($\chi_m < 0$); O_2 is paramagnetic ($\chi_m > 0$).

21. 60.0 μT

23. (a) $\vec{B}(0, 0) = -(9.00 \text{ pT})\hat{k}$

    (b) $\vec{B}(0, 1 \text{ m}) = -(36.0 \text{ pT})\hat{k}$

    (c) $\vec{B}(0, 3 \text{ m}) = (36.0 \text{ pT})\hat{k}$

    (d) $\vec{B}(0, 4 \text{ m}) = (9.00 \text{ pT})\hat{k}$

25. (a) $\vec{B}(1 \text{ m}, 3 \text{ m}) = 0$

    (b) $\vec{B}(6 \text{ m}, 4 \text{ m}) = -(3.56 \times 10^{-23} \text{ T})\hat{k}$

    (c) $\vec{B}(3 \text{ m}, 6 \text{ m}) = (4.00 \times 10^{-23} \text{ T})\hat{k}$

27. $\dfrac{F_B}{F_E} = \dfrac{v^2}{c^2}$

29. $d\vec{B}(3 \text{ m}, 0, 0) = -(9.60 \text{ pT})\hat{i}$

31. (a) $B(0) = 54.5 \ \mu\text{T}$

    (b) $B(0.01 \text{ m}) = 46.5 \ \mu\text{T}$

    (c) $B(0.02 \text{ m}) = 31.4 \ \mu\text{T}$

    (d) $B(0.35 \text{ m}) = 33.9 \text{ nT}$

33. (a) 19.1 cm

    (b) 45.3 cm

    (c) 99.5 cm

35. (a)

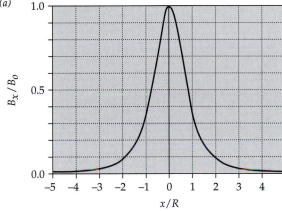

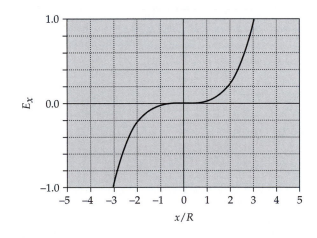

(b)

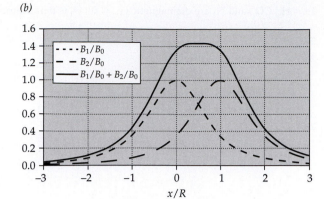

39.

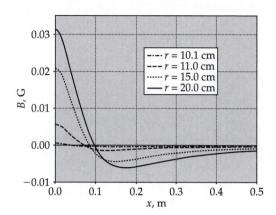

43. *(a)* $\vec{B}(-3\ \text{cm}) = -(88.6\ \mu\text{T})\hat{k}$
 (b) $\vec{B}(0) = 0$
 (c) $\vec{B}(3\ \text{cm}) = -(88.6\ \mu\text{T})\hat{k}$
 (d) $\vec{B}(9\ \text{cm}) = -(160\ \mu\text{T})\hat{k}$

45. *(a)* $\vec{B}(-3\ \text{cm}) = -(177\ \mu\text{T})\hat{k}$
 (b) $\vec{B}(0) = -(133\ \mu\text{T})\hat{k}$
 (c) $\vec{B}(3\ \text{cm}) = -(177\ \mu\text{T})\hat{k}$
 (d) $\vec{B}(9\ \text{cm}) = (106\ \mu\text{T})\hat{k}$

47. *(a)* $\vec{B}(z = 8\ \text{cm}) = (64.0\ \mu\text{T})\hat{j}$
 (b) $\vec{B}(z = 8\ \text{cm}) = -(48.0\ \mu\text{T})\hat{k}$

49. *(a)* Because the currents repel, they are antiparallel.
 (b) 39.3 mA

51. 80.2 A

53. *(a)* $4.50 \times 10^{-4}\ \text{N/m}$
 (b) $30.0\ \mu\text{T}$

55. *(a)* 80.0 A
 (b) $-(0.240\ \text{mT})\hat{j}$

57. *(a)* $\dfrac{3\sqrt{2}\,\mu_0 I^2}{4\pi a}$
 (b) $\dfrac{\sqrt{2}\,\mu_0 I^2}{4\pi a}$

59. *(a)* 3.25 mT
 (b) 3.25 mT
 (c) 1.63 mT

63. $B_{\text{inside}} = 0;\ B_{\text{outside}} = \dfrac{\mu_0 I}{2\pi R}$

65. *(a)* $B_{r<R} = \dfrac{\mu_0 I}{2\pi R}$
 (b) $B_{r>R} = 0$

69. *(a)* $B_{r<a} = 0$
 (b) $B_{a<r<b} = \dfrac{\mu_0 I}{2\pi R}\dfrac{r^2 - a^2}{b^2 - a^2}$
 (c) $B_{r>b} = \dfrac{\mu_0 I}{2\pi R}$

71. *(a)* 27.3 mT
 (b) 20.0 mT

73. *(a)* 10.1 mT
 (b) 10.1 mT; 1.52 T

75. *(a)* 0.544 A/m
 (b) 10.054 A/m

77. -4.00×10^{-5}

79.

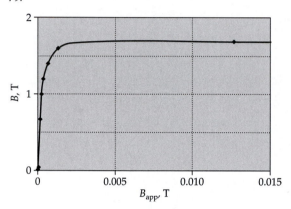

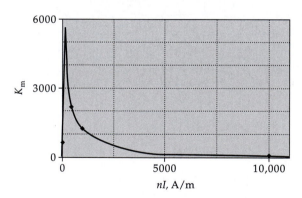

83. $1.69\ \mu_{\text{B}}$

85. *(b)* 7.46×10^{-4}

87. $B_{\text{app}} = \dfrac{\mu_0 NI}{2\pi a};\ B = \dfrac{\mu_0 NI}{2\pi a} + \mu_0 M$

89. *(a)* 30.2 mT
 (b) 6.96 A/m
 (c) 30.2 mT

91. $K_m = 11.75; \mu = 1.48 \times 10^{-5}$

93. (a) 12.6 mT

    (b) 1.36×10^6 A/m

    (c) 137

95. (a) 60.3 mT

    (b) 24.0 A

97. (a) 1.42×10^6 A/m

    (b) $K_m = 90.0; \mu = 1.13 \times 10^{-4}$ T·m/A; $\chi_m = 89.0$

99. (a) $(8.00 \text{ T/m})r$

    (b) $(3.20 \times 10^{-3} \text{ T·m})\dfrac{1}{r}$

    (c) $(8.00 \times 10^{-6} \text{ T·m})\dfrac{1}{r}$

    (d) Note that the field in the ferromagnetic region is the field that would be produced in a nonmagnetic region by a current of $400I = 1600$ A. The amperian current on the inside of the surface of the ferromagnetic material must therefore be $1600 \text{ A} - 40 \text{ A} = 1560$ A in the direction of I. On the outside surface, there must then be an amperian current of 1560 A in the opposite direction.

101. $\dfrac{\mu_0 I}{4}\left(\dfrac{1}{R_1} - \dfrac{1}{R_2}\right)\hat{\imath}$

103. $\dfrac{\mu_0}{2\pi}\dfrac{I}{a}(1 + \sqrt{2})$

105. (a) $\vec{F}_2 = (1.00 \times 10^{-4} \text{ N})\hat{\imath}; \vec{F}_4 = (-0.286 \times 10^{-4} \text{ N})\hat{\imath}$

    (b) $\vec{F}_{\text{net}} = (0.714 \times 10^{-4} \text{ N})\hat{\imath}$

107. $(7.07 \ \mu\text{T})\hat{\imath}$

109.

Graph: B (G) on vertical axis (0 to 4), r (mm) on horizontal axis (0 to 26). Curve rises sharply to a peak near r = 2–3 mm at about B = 4 G, then decreases gradually toward about 0.4 G at r = 26 mm.

111. $\kappa = 0.246$ N·m/rad; $T = 0.523$ s

115. (a) 5.24×10^{-2} A·m^2

    (b) 7.70×10^5 A/m

    (c) 2.31×10^4 A

117. (a) 70.6 A·m^2

    (b) 17.7 N·m

119. 3.18 cm

121. (a) 10.0 μT

    (b) 10.0 μT

    (c) 5.00 μT

123. 2.24 A

125. (c) $\dfrac{\mu_0 \omega \sigma}{2}\left(\dfrac{R^2 + 2x^2}{\sqrt{R^2 + x^2}} - 2x\right)$

Chapter 28

1. (d)

3. (a) If the current in B is clockwise, the loops repel one another.

    (b) If the current in B is counterclockwise, the loops attract one another.

5. (a) and (b)

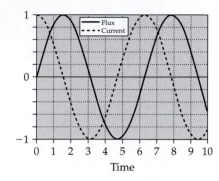

7. The protection is needed because if the current is suddenly interrupted, the resulting emf generated across the inductor due to the large flux change can blow out the inductor. The diode allows the current to flow (in a loop) even when the switch is opened.

9. (a) False

    (b) True

    (c) True

    (d) False

    (e) True

11. The time-varying magnetic field of the magnet sets up eddy currents in the metal tube. The eddy currents establish a magnetic field with a magnetic moment opposite to that of the moving magnet; thus, the magnet is slowed down. If the tube is made of a nonconducting material, there are no eddy currents.

13. (a) 3.14 rad/s

    (b) 1.94 mV

    (c) No. To generate an emf of 1 V, the students would have to rotate the jump rope about 500 times faster.

    (d) The use of multiple strands of lighter wire (so that the composite wire could be rotated at the same angular speed) looped several times around would increase the induced emf.

15. 2.00 kV

17. (a) 0

    (b) 1.37×10^{-5} Wb

    (c) 0

    (d) 1.19×10^{-5} Wb

19. $\pi r^2 B$

21. 6.74×10^{-3} Wb

23. (a) $\mu_0 n I N \pi R_1^2$

    (b) $\mu_0 n I N \pi R_3^2$

25. $\dfrac{\mu_0 I}{4\pi}$

27. (a) 0.314 mV

    (b) 0.785 mA

    (c) 0.247 μW

29. (a)

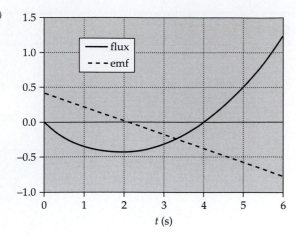

(b) Referring to the graph, we see that the flux is a minimum at $t = 2$ s and that $V = 0$ at this instant.

(c) The flux is zero at $t = 0$ and $t = 4$ s. At these times, $\mathcal{E} = 0.4$ V and -0.4 V, respectively.

31. (a) -1.26 mC

(b) 12.6 mA

(c) 630 mV

33. 79.8 μT

35. (a) $0.693 \dfrac{\mu_0 a}{\pi}$

(b) $4.16 \ \mu\Omega$.

(c) Because the magnetic flux due to I is increasing into the page, the induced current will be in such a direction that its magnetic field will oppose this increase; that is, it will be out of the page. Thus, the induced current is counter-clockwise.

37. 400 m/s

39. (a)

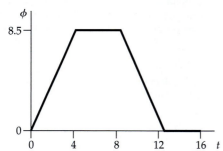

(b)

41. (a) $\dfrac{dv}{dt} = \dfrac{B\ell}{mR}(\mathcal{E} - B\ell v)$

(b) $v_t = \dfrac{\mathcal{E}}{B\ell}$

(c) 0

43. $x = \dfrac{mv_0 R}{B^2\ell^2}$

47. (d)

49. (a) $(15.6 \text{ T·m}) \, v$

(b) 1.61 cm/s

51. (a) $\dfrac{\mu_0 Ivb}{2\pi}\left(\dfrac{1}{d + vt} - \dfrac{1}{d + a + vt}\right)$

(b) $\dfrac{\mu_0 Ivb}{2\pi}\left(\dfrac{1}{d + vt} - \dfrac{1}{d + a + vt}\right)$

53. (a) $24 \text{ Wb} + (1600 \text{ H·A/s})t$

(b) -1.60 kV

55. (a) 6.03 mT

(b) 7.58×10^{-4} Wb

(c) 0.253 mH

(d) 37.9 mV

57. $0; 162 \ \Omega$

63. $\dfrac{\mu_0 I^2}{16\pi}$

65. $0.157 \ \mu$H

67. (a) $I = 0; dI/dt = 25.0$ A/s

(b) $I = 2.27$ A; $dI/dt = 0.5$ A/s

(c) $I = 7.90$ A; $dI/dt = 9.20$ A/s

(d) $I = 10.8$ A; $dI/dt = 3.38$ A/s

69. (a) 44.1 W

(b) 40.4 W

(c) 3.62 W

71. (a) 5.77 s

(b) 28.9 H

73. (a) 3.00 kA/s

(b) 1.50 kA/s

(c) 80.0 mA

(d) 0.123 ms

75. (a) 1.00 A; 0

(b) 100 V; 100 V

(c)

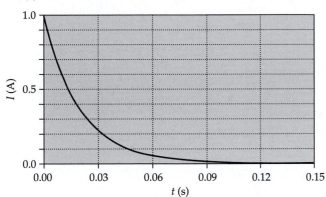

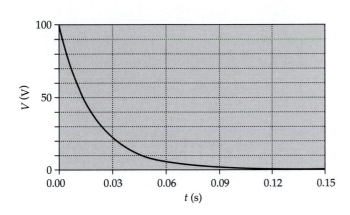

77. (a) 88.1 ms

(b) 35.2 mH

79. (a) $\dfrac{dI_R}{dt}\Big|_{t=0} = 9.00 \text{ kA/s}; \dfrac{dI_{8\,mH}}{dt} = 3.00 \text{ kA/s};$

$\dfrac{dI_{4\,mH}}{dt} = 6.00 \text{ kA/s}$

(b) 1.60 A

81. (a) 3.53 J

(b) 1.61 J

(c) 1.92 J

83. (a) 7.07 mV

(b) 6.64 mV

85. (b) 275 rad/s

89. (a) As the magnet passes through the coil, it induces an emf because of the changing flux through the coil. This allows the coil to sense when the magnet is passing through it.

(b) One cannot use a cylinder made of conductive material because eddy currents induced in the material by a falling magnet would slow the magnet.

(c) As the magnet approaches the loop, the flux increases, resulting in the increasing voltage signal. When the magnet is passing the coil, the flux goes from increasing to decreasing, so the induced emf becomes zero and then negative. The time at which the induced emf is zero is the time at which the magnet is at the center of the coil.

(d)

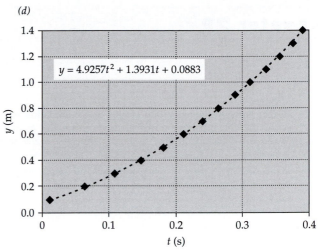

9.85 m/s²

91. $I(t) = (0.350 \text{ A}) \sin(2 \text{ rad/s})t$

93. (a) $-\frac{1}{2} r \mu_0 n I_0 \omega \cos \omega t$

(b) $-\dfrac{\mu_0 n R^2 I_0 \omega}{2r} \cos \omega t$

97. (a) 30.8 N/m

(b) $\dfrac{B y_0 \omega w}{R} \cos \omega t$

(c) $\beta = \dfrac{B^2 w^2}{R}$

(d)

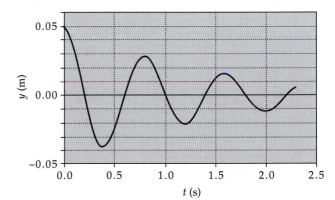

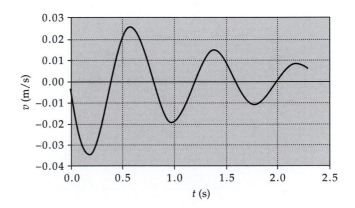

Chapter 29

1. *(b)*

3. *(b)*

5. *(c)*

7. Yes to both questions. While charge is accumulating on the capacitor, the capacitor absorbs power from the generator. When the capacitor is discharging, it supplies power to the generator.

9. To make an *LC* circuit with a small resonance frequency requires a large inductance and large capacitance. Neither is easy to construct.

11. Yes. The power factor is defined to be cos δ = R/Z and, because Z is frequency dependent, so is cos δ.

13. 0

15. True

17. *(a)* False
    (b) True

19. *(a)* 39.8 Hz
    (b) 15.1 V

21. *(a)* 13.6 V
    (b) 486 Hz

23. *(a)* 0.833 A
    (b) 1.18 A
    (c) 200 W

25. *(a)* 0.377 Ω
    (b) 3.77 Ω
    (c) 37.7 Ω

27. 1.59 kHz

29. *(a)* 25.1 mA
    (b) 17.8 mA

31. *(a)* (0.346 A) cos ωt
    (b) (0.346 A) cos ωt
    (c) (0.344 A) cos(ωt + 0.165 rad)

33. *(a)* 1.26 ms
    (b) 88.0 mH

35. *(a)* 2.25 mJ
    (b) 712 Hz
    (c) 0.671 A

37. *(a)*

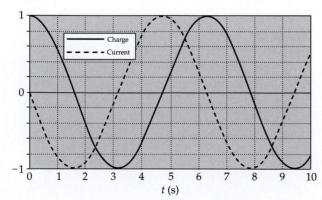

39. 29.2 mH

41. *(a)* 0.333
    (b) 26.7 Ω
    (c) 0.200 H
    (d) I lags $\mathcal{E}$ by 70.5°

43. 0.397

45. *(a)* $I_{R,\text{rms}}$ = 6.20 A; $I_{R_L,\text{rms}}$ = 2.79 A; $I_{L,\text{rms}}$ = 5.52 A
    (b) $I_{R,\text{rms}}$ = 3.29 A; $I_{R_L,\text{rms}}$ = 2.95 A; $I_{L,\text{rms}}$ = 1.47 A
    (c) 50.3 percent; 80.5 percent

47. 60.0 V

49. *(b)* −90°
    (c) 0

57.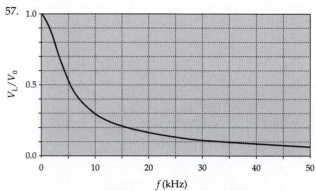

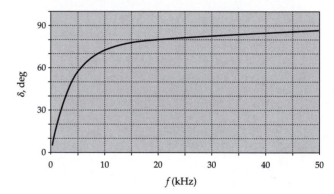

59. $\Delta\omega = R/2L$

61. 170 μF

63. *(a)* $I = -(18.75 \text{ mA}) \sin\left(1250t + \dfrac{\pi}{4}\right)$
    (b) 22.86 μF
    (c) $U_e = (4.92 \ \mu\text{J}) \cos^2\left(1250t + \dfrac{\pi}{4}\right)$

    $U_m = (4.92 \ \mu\text{J}) \sin^2\left(1250t + \dfrac{\pi}{4}\right)$

    $U = U_m + U_e = 4.92 \ \mu\text{J}$

65. *(a)* 5.39×10^{-16} F
    (b) $f(x) = \dfrac{70.0 \text{ MHz}}{\sqrt{1 - (3.96 \text{ m}^{-1})x}}$

67. *(a)* 0.0444
    (b) 491 rad/s or 509 rad/s

71. *(a)* 1.13 kHz
    (b) X_C = 79.6 Ω; X_L = 62.8 Ω
    (c) Z = 17.5 Ω; 4.04 A
    (d) δ = −73.4°

73. (a) 14.1
    (b) 79.8 Hz
    (c) 0.275

75. (a) 10.0 A
    (b) 53.1°
    (c) 332 μF
    (d) 133 V

77. (b) $\delta = -\dfrac{\pi}{2} + \omega RC$

    (c) $\delta = \dfrac{\pi}{2} - \dfrac{R}{\omega L}$

79. (a) 80.5 V
    (b) 78.0 V
    (c) 166 V
    (d) 112 V
    (e) 183 V

81. 0.935 μF

83. (a)

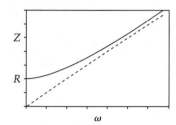

    (b)

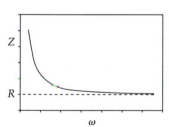

    (c)

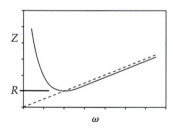

85. (a) $L = 0.800$ mH; $C = 12.5$ μF
    (b) 1.60
    (c) 2.00 A

87. (a) 12.0 Ω
    (b) 7.20 Ω; 9.60 Ω
    (c) If the current leads the emf, the reactance is capacitive.

91. (a) 1.64×10^3 rad/s
    (b) $I_{C,\text{rms}} = 1.39$ A; $\delta_C = -84.4°$
        $I_{L,\text{rms}} = 1.41$ A; $\delta_L = 78.5°$
        $I_{\text{rms}} = 0.417$ A

93. (a) 0.396 μF
    (b) 840 Ω

95. (a) 13.26 kHz
    (b) 200 mA; 600 V
    (c) $V_L = 433$ V; $V_C = 419$ V

103. (a) 1.67 rad/s
     (b) 3.96 W; 7.69 W

105. (a) 1/5
     (b) 50.0 A

107. (a) 1.50 A
     (b) 19

109. 3333

113. (a) 265 Ω
     (b) 2.65 Ω
     (c) 2.65 mΩ

115. (a) 12.0 V
     (b) 8.49 V

117. (a) $Q_1 = (60\ \mu\text{F})\cos(120\pi t) + 72\ \mu\text{F}$;
         $Q_2 = (30\ \mu\text{F})\cos(120\pi t) + 36\ \mu\text{F}$
     (b) $I = -(33.9\ \text{mA})\sin(120\pi t)$
     (c) 4.36 mJ
     (d) 36.0 μJ

119. $I_{\text{max}} = 1.06$ A; $I_{\text{min}} = -0.0560$ A; $I_{\text{av}} = 0.500$ A; $I_{\text{rms}} = 0.636$ A

121. The inductance acts as a short circuit to the constant voltage source. The current is infinite at all times. Consequently, $I_{\text{max}} = I_{\text{rms}} = \infty$. There is no minimum current.

Chapter 30

1. (a) False
   (b) True
   (c) True
   (d) True
   (e) False
   (f) True

3. X rays have greater frequencies, whereas light waves have longer wavelengths (see Table 30-1).

5. Consulting Table 30-1, we see that FM radio waves and television waves have wavelengths of the order of a few meters.

7. The dipole antenna should be in the horizontal plane and normal to the line from the transmitter to the receiver.

9. $I = 2.94 \times 10^7$ W/m²; $P = 5.20$ mW

11. (a) $E_{\text{rms}} = 719$ V/m; $B_{\text{rms}} = 2.40$ μT
    (b) $P_{\text{av}} = 3.87 \times 10^{26}$ W
    (c) $I = 6.36 \times 10^7$ W/m²; $P_r = 0.212$ Pa

13. $F_r = 7.09 \times 10^7$ N; Because the ratio of these forces is 1.65×10^{-14} for the earth and 4.26×10^{-14} for Mars, Mars has the larger ratio. The reason that the ratio is higher for Mars is that the dependence of the radiation pressure on the distance from the sun is the same for both forces (r^{-2}), whereas the dependence on the radii of the planets is different. Radiation pressure varies as R^2, whereas the gravitational force varies as R^3 (assuming that the two planets have the same density, an assumption that is nearly true). Consequently, the ratio of the forces goes as $R^2/R^3 = R^{-1}$. Because Mars is small than the earth, the ratio is larger.

15. (a) 3.40×10^{14} V/m·s

    (b) 5.00 A

19. (a) 10.0 A

    (b) 2.26×10^{12} V/m·s

    (c) 7.90×10^{-7} T·m

21. (a) $I = \dfrac{A(0.01 \text{ V/s})}{\rho d} t$

    (b) $I_d = \dfrac{(0.01 \text{ V/s})\epsilon_0 A}{d}$

    (c) $t = \epsilon_0 \rho$

25. (a) 300 m

    (b) 3.00 m

27. 3.00×10^{18} Hz

29. (a) $30.0°$

    (b) 7.07 m

31. $4.14 \ \mu\text{W/m}^2$; 5.21×10^{17} W/cm²·s

33. $0.151 \ \mu\text{W/m}^2$

35. (a) 283 V/m

    (b) $0.943 \ \mu$T

    (c) 212 W/m²

    (d) $0.707 \ \mu$Pa

39. (a) 40.0 nN

    (b) 80.0 nN

41. (a) 3.46 V/m; 11.5 nT

    (b) 0.346 V/m; 1.15 nT

    (c) 0.0346 V/m; 0.115 nT

43. $E_{rms} = 75.2$ kV/m; $B_{rms} = 0.251$ mT

45. (a) Positive x direction

    (b) 0.628 m; 477 MHz

    (c) $\vec{E}(x,t) = (194 \text{ V/m}) \cos[10x - (3 \times 10^9)t]\hat{j}$
$\vec{B}(x,t) = (0.647 \ \mu\text{T}) \cos[10x - (3 \times 10^9)t]\hat{k}$

47. 6.10×10^{-3} degrees

49. 3.42 MW/m²

55. The current induced in a loop antenna is proportional to the time-varying magnetic field. For a maximum signal, the antenna's plane should make an angle $\theta = 0°$ with the line from the antenna to the transmitter. For any other angle, the induced current is proportional to $\cos \theta$. The intensity of the signal is therefore proportional to $\cos \theta$.

57. 72.6 nV

59. (a) $I = V_0 \left(\dfrac{1}{R} \sin \omega t + \dfrac{\omega \epsilon_0 \pi a^2}{d} \cos \omega t \right)$

    (b) $B(r) = \dfrac{\mu_0 V_0}{2 \pi r} \left(\dfrac{1}{R} \sin \omega t + \omega \dfrac{r^2}{a^2} \cos \omega t \right)$

    (c) $\delta = \tan^{-1} \left(\dfrac{R \omega \epsilon_0 \pi a^2}{d} \right)$

61. (a) $\vec{S} = \dfrac{1}{\mu_0 c} [E_{1,0}^2 \cos^2(k_1 x - \omega_1 t) + 2E_{1,0}E_{2,0} \cos(k_1 x - \omega_1 t)$
$\times \cos(k_2 x - \omega_2 t + \delta) + E_{2,0}^2 \cos^2(k_2 x - \omega_2 t + \delta)]\hat{i}$

    (b) $\vec{S}_{av} = \dfrac{1}{2\mu_0 c} [E_{1,0}^2 + E_{2,0}^2]\hat{i}$

    (c) $\vec{S} = \dfrac{1}{\mu_0 c} [E_{1,0}^2 \cos^2(k_1 x - \omega_1 t) - E_{2,0}^2 \cos^2(k_2 x + \omega_2 t + \delta)]\hat{i}$

    (d) $\vec{S}_{av} = \dfrac{1}{2\mu_0 c} [E_{1,0}^2 - E_{2,0}^2]\hat{i}$

63. (a) 9.15×10^{-15} T

    (b) $(1.01 \ \mu\text{V}) \cos(8.80 \times 10^5 \text{ s}^{-1})t$

    (c) $(5.49 \ \mu\text{V}) \sin(8.80 \times 10^5 \text{ s}^{-1})t$

65. (a) $E = \dfrac{I\rho}{\pi a^2}$

    (b) $B = \dfrac{\mu_0 I}{2\pi a}$

    (c) $\vec{S} = -\dfrac{I^2 \rho}{2\pi^2 a^3}\hat{r}$

67. $0.574 \ \mu$m

69. 3.33 mN

Chapter 31

1. The population inversion between the state $E_{2,Ne}$ and the state 1.96 eV below it (see Figure 31-9) is achieved by inelastic collisions between neon atoms and helium atoms excited to the state $E_{2,He}$.

3. The layer of water greatly reduces the light reflected back from the car's headlights, but increases the light reflected by the road of light from the headlights of oncoming cars.

5. The change in atmospheric density results in refraction of the light from the sun, bending it toward the earth. Consequently, the sun can be seen even after it is just below the horizon. Also, the light from the lower portion of the sun is refracted more than the light from the upper portion, so the lower part appears to be slightly higher in the sky. The effect is an apparent flattening of the disk into an ellipse.

7. He takes the path LES because the time required to reach the swimmer is the least for this path.

9. (d)

11.

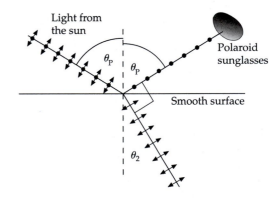

13. (c)

15. In resonance absorption, the molecules respond to the frequency of the light through the Einstein photon relation $E = hf$. Thus, the color appears to be the same in spite of the fact that the wavelength has changed.

17. (a) 2:00 A.M., September 1

    (b) 2:08 A.M., September 1

19. (a) 15.0 mJ

    (b) 5.25×10^{16}

21. (a) 435 nm

    (b) 1210 nm

23. (a) 387.5 nm; $\lambda_{21} = 1138$ nm; $\lambda_{10} = 587.7$ nm

(b) $\lambda_{03} = 285.1$ nm; $\lambda_{32} = 1078$ nm; $\lambda_{21} = 1138$ nm; $\lambda_{10} = 587.7$ nm; $\lambda_{31} = 553.6$ nm; $\lambda_{20} = 387.5$ nm

25. (a)

27. 5 min, 23 s

29. (a) 20.0 μs

(b) $\Delta t_{\text{reaction}} \approx 10^4 \, \Delta t$

33. $v_{\text{water}} = 2.25 \times 10^8$ m/s; $v_{\text{glass}} = 2.00 \times 10^8$ m/s

35. (a) 50.2°

(b) 38.8°

(c) 26.3°

37. 92.2 percent

41. 62.5°

43. 102 m²

45. 1.30

47. 5.43°

49. (a) 48.7°

(b) Note that θ_2 equals the critical angle for a water–air interface. Therefore, the ray will not leave the water for $\theta_1 \geq 41.8°$.

51. 1.02°

53. (a) 53.1°

(b) 56.3°

55. (a) $\frac{1}{8}I_0$

(b) $\frac{3}{32}I_0$

57. (a) 30.0°

(b) 1.73

59. $I_3 = \frac{1}{8}I_0 \sin^2 2\omega t$

61. 13

63. $I_4 = 0.211I_0$

65. Right circularly polarized; $\vec{E} = E_0 \sin(kx + \omega t)\hat{j} - E_0 \cos(kx + \omega t)\hat{k}$

67. 35.3°

69. 1.45 m

71. 3.42 m

73. (a) 36.8°

(b) 38.7°

75. (a) −1.00 m

(b) $\theta_i = 26.6°$; $\theta_r = 26.6°$

77. For silicate flint glass: $\theta_p = 58.3°$; for borate flint glass: $\theta_p = 57.5°$; for quartz glass: $\theta_p = 57.0°$; for silicate crown glass: $\theta_p = 56.5°$

79. (b) $\theta_p > \theta_c$

81. (a) 1.33

(b) $\theta_c = 37.2°$

(c) $\theta_2 = 48.8°$

83. (a) $\dfrac{I_t}{I_0} = \left[\dfrac{4n}{(n+1)^2}\right]^{2N}$

(b) 0.783

(c) ≈ 28

85. (a) 24.0°

(b) 4.45 km

(c) $\theta = \tan^{-1}\left(\dfrac{v_{\text{earth}}}{c}\right)$

(d) 2.99×10^8 m/s

Chapter 32

1. Yes. Note that a virtual image is seen because the eye focuses the diverging rays to form a real image on the retina. Similarly, the camera lens can focus the diverging rays onto the film.

3. (a) False

(b) False

(c) True

(d) False

5. A convex mirror always produces a virtual erect image that is smaller than the object. It never produces an enlarged image.

7. (b)

9. (a) The lens will be positive if its index of refraction is greater than that of the surrounding medium, and the lens is thicker in the middle than at the edges. Conversely, if the index of refraction of the lens is less than that of the surrounding medium, the lens will be positive if it is thinner at its center than at the edges.

(b) The lens will be negative if its index of refraction is greater than that of the surrounding medium, and the lens is thinner at the center than at the edges. Conversely, if the index of refraction of the lens is less than that of the surrounding medium, the lens will be negative if it is thicker at the center than at the edges.

11. (d)

13. The eye accommodates by varying the power of a lens located a fixed distance away from the retina. A camera, on the other hand, has a fixed power lens that can move with respect to the film location.

15. (b)

17. (d)

19. (a)

21. Microscopes ordinarily produce images (either the intermediate one produced by the objective or the one viewed through the eyepiece) that are larger than the object being viewed. A telescope, on the other hand, ordinarily produces images that are much reduced compared to the object. The object is normally viewed from a great distance, and the telescope magnifies the angle subtended by the object.

23. Plano-convex lens: $r_1 = -16.2$ cm; $r_2 = \infty$

Biconvex with equal curvature: $r_1 = -32.4$ cm; $r_2 = 32.4$ cm

Biconvex with unequal curvature: $r_1 = 16.2$ cm; $r_2 = 8.10$ cm

25.

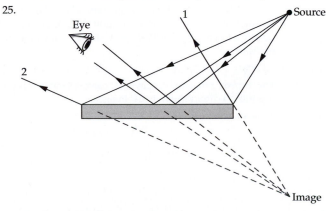

27.

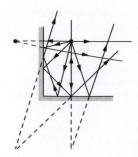

29. (a) The first image in the mirror on the left is 10 cm behind the mirror. The mirror on the right forms an image 20 cm behind that mirror or 50 cm from the left mirror. This image will result in a second image 50 cm behind the left mirror. The first image in the left mirror is 40 cm from the right mirror and forms an image 40 cm behind the right mirror or 70 cm from the left mirror. That image gives an image 70 cm behind the left mirror. The fourth image behind the left mirror is 110 cm behind that mirror.

(b) Proceeding as in Part (a) for the mirror on the right, one finds the location of the images to be 20 cm, 40 cm, 80 cm, and 100 cm behind the right-hand mirror.

31. (a) $s' = 15.8$ cm; $m = -0.316$; Because the image distance is positive and the lateral magnification is less than one and negative, the image is real, inverted, and reduced.

(b) $s' = 24.0$ cm; $m = -1$; Because the image distance is positive and the lateral magnification is one and negative, the image is real, inverted, and the same size as the object.

(c) $s' = \infty$ and there is no image.

(d) $s' = -24.0$ cm; $m = 3$; Because the image distance is negative and the lateral magnification is three and positive, the image is virtual, erect, and three times the size of the object.

33. (a) $s' = -9.85$ cm; $m = 0.179$; Because the image distance is negative and the lateral magnification is less than one in magnitude and positive, the image is virtual, erect, and reduced.

(b) $s' = -8.00$ cm; $m = 0.333$; Because the image distance is negative and the lateral magnification is less than one in magnitude and positive, the image is virtual, erect, and reduced.

(c) $s' = -6.00$ cm; $m = 0.5$; Because the image distance is negative and the lateral magnification is one-half in magnitude and positive, the image is virtual, erect, and half the size of the object.

(d) $s' = -4.80$ cm; $m = 0.600$; Because the image distance is negative and the lateral magnification is less than one and positive, the image is virtual, erect, and six-tenths the size of the object.

35. (a) 5.13 cm

(b) The mirror must be concave. A convex mirror always produces a diminished virtual image.

37. $s' = -4.00$ m; $y' = 3.68$ cm

39. (a)

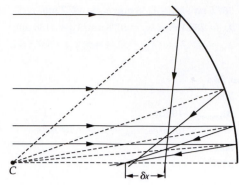

$\delta x = 1.02$ cm

(b) By blocking off the edges of the mirror so that only paraxial rays within 2 cm of the mirror axis are reflected, the spread is reduced by about 83 percent.

41. (a) -1.33 m

(b) Because $f_{small} < 0$, the small mirror is convex.

43. (a) $s' = -8.58$ cm, where the minus sign tells us that the image is 5.58 cm from the front surface of the bowl.

(b) $s' = -35.9$ cm, where the minus sign tells us that the image is 35.9 cm from the front surface of the bowl.

45. 14.4 cm

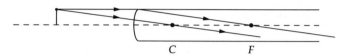

47. (a) $s' = -104$ cm, where the negative image position tells us that the image is 104 cm in front of the surface and is virtual.

(b) $s' = -8.29$ cm, where the minus sign tells us that the image is 8.29 cm in front of the surface and is virtual.

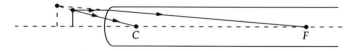

(c) $s' = 63.5$ cm, that is, the image is 63.5 cm behind (to the right of) the surface (at the focal point) and is real.

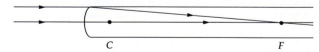

49. (a) $s' = 64.0$ cm

(b) $s' = -80.0$ cm

(c) The final image is 96 cm − 80 cm = 16 cm from the surface, the radius of the surface is 8 cm and is virtual.

51. (a)   19.0 cm

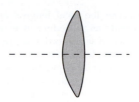

(b)   30.0 cm

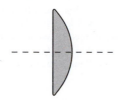

(c)   −15.0 cm

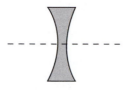

(d)   −52.0 cm

53. (a)   −30.3 cm

(b)   −22.0 cm

(c)   −0.275

(d)   Because $s' < 0$ and $m < 0$, the image is virtual and upright.

55.

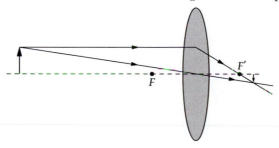

$s' = 16.7$ cm; $y' = -2.00$ cm; The image is real, inverted, and diminished. Because $s' > 0$ and $y' = -2.00$ cm, the image is real, inverted, and diminished in agreement with the ray diagram.

57.

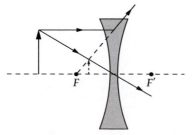

$s' = -6.67$ cm; $y' = 0.500$ cm; The image is virtual, erect, and diminished. Because $s' < 0$ and $y' = 0.500$ cm, the image is virtual, erect, and about one-third the size of the object in agreement with the ray diagram.

59. (a)

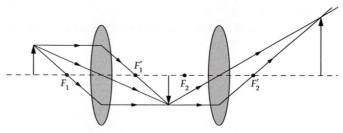

$s'_2 = 30$ cm and the final image is 85.0 cm from the object; $m_2 = -2$

(b)   Because $s'_2 > 0$ and $m = m_1 m_2 = 2$, the image is real and upright.

(c)   The image magnification is twice the size of the object.

61. (b)   3.70 m

63. (a) and (b)

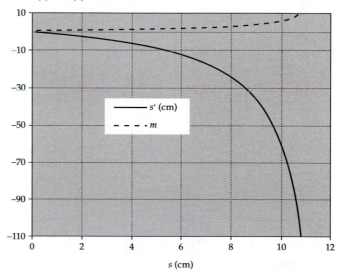

(c)   The images are virtual and erect for this range of object distances.

(d)   The asymptote of the graph of s' versus s corresponds to the image approaching infinity as the object distance approaches the focal length of the lens. The horizontal asymptote of the graph of m versus s indicates that, as the object moves toward the lens, the height of the image formed by the lens approaches the height of the object.

65. 15.0 cm; The final image is 50 cm from the object, real, inverted, and the same size as the object.

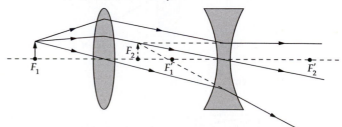

69. (a) 41.2 cm

    (b) −1.53

    (c) Because $m < 0$, the image is inverted. Because $s_2' > 0$, the image is real.

71. (a) False

    (b) True

73. (c) 40.0 D; 4.00 D

75. (c) −7.33 D

77. 1.72 mm

79. (a) 80.0 μrad

    (b) 1.60 mm

81. 0.444 D

83. 3.07 D

85. 6.00

87. 5.00

89. (a) 3.00

    (b) 4.00

93. (a) −1.88

    (b) −18.8

95. −232

97. (a) 9.00 mm

    (b) −20.0; −0.180 rad

99. (a) $P_{\text{Palomar}} = (25.0)P_{\text{Yerkes}}$

    (b) −134

101. (b)

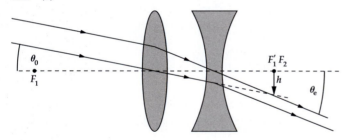

103. −1/150

105. 1.34 mm

107. (a) $s = 5.00$ cm; $s' = -10.0$ cm

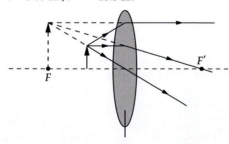

    (b) $s = 15.0$ cm; $s' = 30.0$ cm

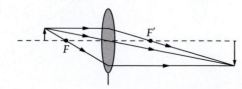

109. (a) Because the focal lengths appear in the magnification formula as a product, it would appear that it does not matter in which order we use them. The usual arrangement would be to use the shorter focal length lens as the objective, but we get the same magnification in the reverse order. What difference does it make then? It makes no difference in this problem. However, it is generally true that the smaller the focal length of a lens, the smaller its diameter. This condition makes it harder to use the shorter focal length lens, with its smaller diameter, as the eyepiece lens. If we separate the objective and eyepiece lenses by $L + f_e + f_o = 16$ cm + 7.5 cm + 2.5 cm = 26.0 cm, the overall magnification will be −21.3.

    (b)

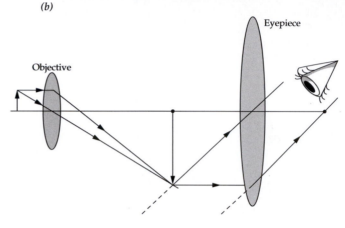

111. 3.70 m

113. (a) 9.52 cm

    (b) −1.19

    (c)

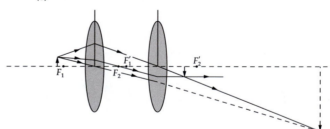

115. (a) 21.3 cm

    (b) 79.2 cm

117. 0.0971 m/s

119. (a) 22.5 cm

    (b) 18.0 cm

    (c)

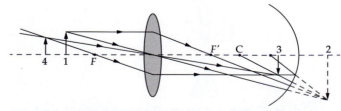

    To see this image, the eye must be to the left of the image 4.

121. 36.8 cm

123. (a) The final image is 0.8 cm behind the mirror surface.

    (b) The final image is at the mirror surface.

Chapter 33

1. The energy is distributed nonuniformly in space; in some regions the energy is below average (destructive interference), in others it is higher than average (constructive interference).

3. The thickness of the air space between the flat glass and the lens is approximately proportional to the square of d, the diameter of the ring. Consequently, the separation between adjacent rings is proportional to $1/d$.

5. If the film is thick, the various colors (i.e., different wavelengths) will give constructive and destructive interference at that thickness. Consequently, what one observes is the reflected intensity of white light.

7. The first zeros in the intensity occur at angles given by $\sin \theta = \lambda/a$. Hence, decreasing a increases θ, and the diffraction pattern becomes wider.

9. (a)

11. (a) False
    (b) True
    (c) True
    (d) True
    (e) True

13. $3.44 \ \mu$m

15. $\theta_{\text{red}} \approx 1.80°$; light, $\theta_{\text{blue}} \approx 1.25°$.

17. (a) 300 nm
    (b) 135°

19. 164°

21. $5.46 \ \mu$m $< d < 5.75 \ \mu$m

23. (a) 600 nm
    (b) From the table, we see that the only wavelengths in the visible spectrum are 720 nm, 514 nm, and 400 nm.
    (c) From the table, we see that the missing wavelengths in the visible spectrum are 720 nm, 514 nm, and 400 nm.

25. 476 nm

27. (c) The transmitted pattern is complementary to the reflected pattern.
    (d) 68 bright fringes
    (e) 1.14 cm
    (f) The wavelength of the light in the film becomes $\lambda_{\text{air}}/n = 444$ nm. The separation between fringes is reduced and the number of fringes that will be seen is increased by the factor $n = 1.33$.

29. 0.535 nm; 0.926 nm

31. 4.95 mm

33. (a) $50.0 \ \mu$m
    (b) Not with the unaided eye. The separation is too small to be observed with the naked eye.
    (c) 0.500 mm

35. 625 nm and 417 nm

37. (a) 0.600 mrad
    (b) 6.00 mrad
    (c) 60.0 mrad

39. (a) 1.25 km
    (b) 376 m
    (c) 2.6×10^{-14}

41. (a) $20.0 \ \mu$m
    (b) 9

43. There are eight interference fringes on each side of the central maximum. The secondary diffraction maximum is half as wide as the central one. It follows that it will contain eight interference maxima.

45. $3.61 \sin(\omega t - 56.3°)$

47. $0.0162 I_0$

49. (b) 6.00 mm
    (c) The width for four sources is half the width for two sources.

51. (a)

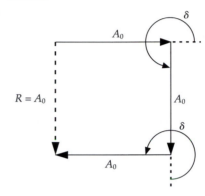

    (b) $5.56 \ \text{mW/m}^2$

53. (a) 8.54 mrad
    (b) 6.83 cm

55. 7.00 mm

57. 484 m

59. 5.00×10^9 m

61. 86.9 mrad; 82.1 mrad

63. 40.6°; 38.0°

65. 30.0°

67. Because $m_{\text{max}} = 2.98$, one can see the complete spectrum only for $m = 1$ and 2. Because 700 nm $< 2 \times 400$ nm, there is no overlap of the second-order spectrum into the first-order spectrum; however, there is overlap of long wavelengths in the second order with short wavelengths in the third-order spectrum.

69. (a) $y_1 = 0.353$ m; $y_2 = 0.706$ m
    (b) $\Delta y = 88.4 \ \mu$m
    (c) 8000

71. $R = 3.09 \times 10^5$; $n = 5.14 \times 10^4 \ \text{cm}^{-1}$

73. (a) $n = 750 \ \text{cm}^{-1}$
    (b) 4.20 cm; 12.6 cm

75. (a) $\phi = \sin^{-1}\left(\dfrac{m\lambda}{a}\right)$
    (b) 64.2°

77. (a) 0.150 mm
    (b) $3.33 \times 10^3 \ \text{m}^{-1}$

79. 1.68 cm

81. 0.130 mrad

85. (a) 97.8 nm
    (b) No, because 180 nm is not in the visible portion of the spectrum.
    (c) 0.273; 0.124

87. 12.3 m

Chapter 34

1. (c)

3. (a) True

   (b) False

   (c) True

   (d) True

5. (a)

7. (a) True

   (b) True

   (c) True

   (d) False

9. (c)

11. In the photoelectric effect, an electron absorbs the energy of a single photon. Therefore, $K_{max} = hf - \phi$, independent of the number of photons incident on the surface. However, the number of photons incident on the surface determines the number of electrons that are emitted.

13. According to quantum theory, the average value of many measurements of the same quantity will yield the expectation value of that quantity. However, any single measurement may differ from the expectation value.

15. (a)

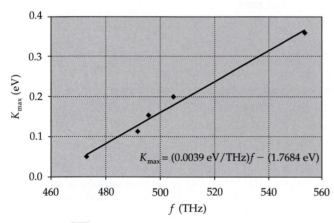

$$K_{max} = (0.0039\ \text{eV/THz})f - (1.7684\ \text{eV})$$

   (b) 1.77 eV

   (c) Cesium

17. Soccer ball

19. (a) 2.42×10^{14} Hz

   (b) 2.42×10^{17} Hz

   (c) 2.42×10^{20} Hz

21. (a) 12.4 keV

   (b) 1.24 GeV

23. $1.95 \times 10^{16}\ \text{s}^{-1}$

25. (a) 4.13 eV

   (b) 2.10 eV

   (c) 0.784 eV

   (d) 590 nm

27. (a) 653 nm; 4.59×10^{14} Hz

   (b) 3.06 eV

   (c) 1.64 eV

29. 1.21 pm

31. 180 pm

33. $p_1 = 9.32 \times 10^{-24}\ \text{kg·m/s};\ p_e = 1.80 \times 10^{-23}\ \text{kg·m/s}$

35. 42

37. 2.91 nm

39. (a) $p_e = 2.09 \times 10^{-22}\ \text{N·s};\ p_p = 8.95 \times 10^{-21}\ \text{N·s};$
   $p_\alpha = 1.79 \times 10^{-20}\ \text{N·s}$

   (b) $\lambda_p = 7.41 \times 10^{-14}\ \text{m};\ \lambda_e = 3.17 \times 10^{-12}\ \text{m};\ \lambda_\alpha = 3.70 \times 10^{-14}\ \text{m}$

41. 20.2 fm

43. (a) 0.820 meV

   (b) 820 MeV

45. 0.167 nm

47. 4.65 pm

49. 1.66×10^{-33} m; This is many orders of magnitude smaller than even the diameter of a proton.

51. 0.0872 nm; This distance is of the order of the size of an atom.

53. (a) $E_1 = 206$ MeV; $E_2 = 824$ MeV; $E_3 = 1.85$ GeV

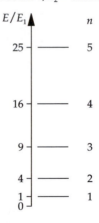

   (b) 2.01 fm

   (c) 1.20 fm

   (d) 0.752 fm

55. (a) 0.004

   (b) 0.003

   (c) 0

57. (a) $\dfrac{L}{2}$

   (b) $0.321L^2$

59. (a) $\dfrac{1}{\sqrt{a}}$

   (b) 0.865

61. (a) 0.500

   (b) 0.402

   (c) 0.750

65. $<x> = 0;\ <x^2> = L^2\left[\dfrac{1}{12} - \dfrac{1}{2\pi^2}\right]$

67. (a) 3.10 eV

   (b) 6.25×10^{16} eV

   (c) 2.02×10^{16}

69. (a) 1.00 μm; 10^{-16} kg·m/s

   (b) 0.949×10^{12}

71. 121 eV

73. 6.80×10^3 km

75. (a) 3.18 W/m²

    (b) 1.04 × 10¹⁵

79. 1.28 MeV

81. 1.04 eV; 554 nm

83. (b) 0.2 percent

    (c) Classically, the energy is continuous. For very large values of n, the energy difference between adjacent levels is infinitesimal.

85. (a) 6.25 × 10⁻⁴ eV/s

    (b) 53.3 min

Chapter 35

1. True

3. (a)   (b)

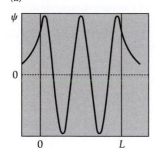 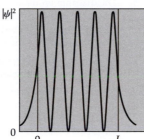

11. (a) 9.49 × 10⁻⁹ m

    (b) 4.19 MeV

13. $\Delta x \, \Delta p = \dfrac{\hbar}{2}$

15. (a) $\sqrt{\dfrac{3}{2}} k_1$

    (b) 0.0102

    (c) 0.990

    (d) 9.90 × 10⁵

17. 0.341

19. (a) $r_{1,4\,\text{MeV}} = 6.62 \times 10^{-14}$ m; $r_{1,7\,\text{MeV}} = 3.78 \times 10^{-14}$ m

    (b) $T_{4\,\text{MeV}} = 3.27 \times 10^{-51}$; $T_{7\,\text{MeV}} = 1.55 \times 10^{-4}$

21. $\psi = A \sin\left(\dfrac{n_1 \pi}{L_1} x\right) \sin\left(\dfrac{n_2 \pi}{2L_1} y\right) \sin\left(\dfrac{n_3 \pi}{3L_1} z\right)$

23. $\psi = A \sin\left(\dfrac{n_1 \pi}{L_1} x\right) \sin\left(\dfrac{n_2 \pi}{2L_1} y\right) \sin\left(\dfrac{n_3 \pi}{4L_1} z\right)$

25. (a) $\psi(x, y) = A \sin\dfrac{n\pi}{L} x \sin\dfrac{m\pi}{L} y$

    (b) $E_{n,m} = \dfrac{h^2}{8mL^2}(n^2 + m^2)$

    (c) $E_{1,2} = E_{2,1} = \dfrac{5h^2}{8mL^2}$

    (d) (1, 7), (7, 1), (5, 5); $E = \dfrac{25h^2}{4mL^2}$

27. $E_{0,10\,\text{bosons}} = \dfrac{5h^2}{4mL^2}$

33. $E_0 = \dfrac{5h^2}{mL^2}$; $E_1 = E_2 = \dfrac{21h^2}{4mL^2}$

39. $A_2 = 4\sqrt{\dfrac{8m\omega_0}{h}}$

Chapter 36

1. Examination of Figure 36-4 indicates that as n increases, the spacing of adjacent energy levels decreases.

3. (a)

5. (d)

7. (a)

9. (c)

11. In conformity with the exclusion principle, the total number of electrons that can be accommodated in states of quantum number n is n^2 (see Problem 37). The fact that closed shells correspond to $2n^2$ electrons indicates that there is another quantum number that can have two possible values.

13. (a) phosphorus

    (b) chromium

15. (d)

17. The optical spectrum of any atom is due to the configuration of its outer-shell electrons. Ionizing the next atom in the periodic table gives you an ion with the same number of outer-shell electrons and almost the same nuclear charge. Hence, the spectra should be very similar.

19. (a) allowed

    (b) not allowed

    (c) not allowed

    (d) allowed

    (e) allowed

21. (b) 75.2 nK

23. (a) 103 nm

    (b) 97.3 nm

25. (a) 1.51 eV; 821 nm

    (b) 0.661 eV; 1876 nm

    0.967 eV; 1282 nm

    1.13 eV; 1097 nm

$6 \rightarrow 3$	$5 \rightarrow 3$		$4 \rightarrow 3$
1097 nm	1282 nm		1876 nm

27. (a) 657.8 nm

    (b) 1.0945 × 10⁷ m⁻¹

29. (b) 1.096776 × 10⁷ m⁻¹; 1.097374 × 10⁷ m⁻¹; 0.0546 percent

31. (a) 1.49 × 10⁻³⁴ J·s

    (b) −1, 0, +1

    (c)

```
z
        ↗ m = 1
↗
—→ m = 0
        ↘ m = −1
```

33. (a) 0, 1, 2

    (b) 0, −1, 0, +1; −2, −1, 0, +1, +2

    (c) 18

35. (a) 45.0°

    (b) 26.6°

    (c) 8.05°

37. (a) $6\hbar^2$

    (b) $4\hbar^2$

    (c) $2\hbar^2$

39. (a) 4

    (b)

n	ℓ	m	(n, ℓ, m)
2	0	0	$(2, 0, 0)$
2	1	−1	$(2, 1, -1)$
2	1	0	$(2, 1, 0)$
2	1	1	$(2, 1, 1)$

41. (a) $\psi_{2,0,0}(a_0) = \dfrac{0.0605}{a_0^{3/2}}$

    (b) $[\psi_{2,0,0}(a_0)]^2 = \dfrac{0.00366}{a_0^3}$

    (c) $P(a_0) = \dfrac{0.0460}{a_0}$

43. (a) 9.20×10^{-4}

    (b) 0

49. 0.323

51. ℓ must equal to 0 or 1

53.

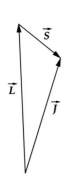

$j = \dfrac{5}{2}$

$j = \dfrac{7}{2}$

55. (c)

57. (a) $1s^2 2s^2 2p^6 3s^2 3p^1$

    (b) $1s^2 2s^2 2p^6 3s^2 3p^6 3d^4 4s$

59. (a) 2s or 2p

    (b) $1s^2 2s^2 2p^6 3p$

    (c) 1s2s

61. (a) 0.0610 nm; 0.0678 nm

    (b) 0.0542 nm

63. (a) 1.00 nm

    (b) 0.155 nm

65. $n_i = 4$ to $n_f = 1$

67. $n_i = 3$ to $n_f = 2$; $n_i = 9$ to $n_f = 3$; $n_i = 7$ to $n_f = 4$

69. (a) 1.6179 eV; 1.6106 eV

    (b) 0.00730 eV

    (c) 63.0 T

73. (a) 1.06 GHz

    (b) 28.4 cm; microwave

75. (a) $R_H = 1.096776 \times 10^7 \text{ m}^{-1}$; $R_D = 1.097075 \times 10^7 \text{ m}^{-1}$

    (b) 0.179 nm

77. (a) $R_T = 1.097175 \times 10^7 \text{ m}^{-1}$

    (b) 0.0600 nm; 0.238 nm

Chapter 37

1. Yes. Because the center of charge of the positive Na ion does not coincide with the center of charge for the negative Cl ion, the NaCl molecule has a permanent dipole moment. Hence, it is a polar molecule.

3. No. Neon occurs naturally as Ne, not Ne_2. Neon is a rare gas atom with a closed shell electron configuration.

5. The diagram would consist of a nonbonding ground state with no vibrational or rotational states for ArF (similar to the upper curve in Figure 37-4) but for ArF* there should be a bonding excited state as in the lower part of the figure.

7. A stiff spring will have a force constant of about 2×10^3 N/m, about the same as the force constant obtained in Example 37-4.

9. For H_2, the concentration of negative charge between the two protons holds the protons together. In the H_2^+ ion, there is only one electron that is shared by the two positive charges so that most of the electronic charge is again between the two protons. However, the negative charge in the H_2^+ ion is not as effective as the larger charge in the H_2 molecule, and the protons should be farther apart. The experimental values support this argument. For H_2, $r_0 = 0.074$ nm, while for H_2^+, $r_0 = 0.106$ nm.

11. With more than two atoms in the molecule, there will be more than just one frequency of vibration because there are more possible relative motions. In advanced mechanics these are known as normal modes of vibration.

13. (a) $\ell = 2.15 \times 10^{30}$

    (b) 5.10×10^{-65}

15. 0.946 nm

17. (a) 23.0 kcal/mol

    (b) 98.2 kcal/mol

19. 0.44 eV

21.

![Graph of U(r) versus r showing a potential energy curve with a minimum, with horizontal dashed lines at energy levels E_2 and E_1, and corresponding values $r_{1,av}$ and $r_{2,av}$ marked on the horizontal axis.]

23. (a) −6.64 eV

    (b) 5.70 eV; 0.63 eV

25. (a) 0.31 eV

    (b) $n = 19.7$; $C = 1.37 \times 10^{-13}$ eV·nm$^{19.7}$

27. 0.121 nm

29. 41

33. (a) 0.179 eV

(b) 2.80×10^{-47} kg·m^2

(c) 0.132 nm

35.

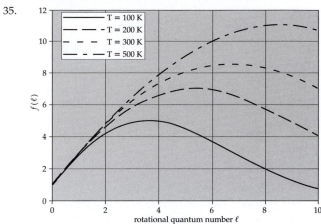

37. 0.9722 u; 0.9737 u; 0.00150; 0.00119

39. 0.955 meV

41. 1.55 kN/m

43. $r_0 = 0.074$ nm; $U_0 = 4.52$ eV

45. $1/x^4$

47. (a) 1.45×10^{-46} kg·m^2; 0.239 meV

(b) 0.239 eV

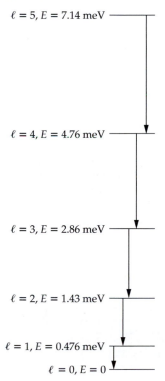

(c) $\lambda_{1,0} = 2596$ μm; $\lambda_{2,1} = 1298$ μm; $\lambda_{3,2} = 865$ μm; $\lambda_{4,3} = 649$ μm; $\lambda_{5,4} = 519$ μm; microwave

Chapter 38

1. The energy lost by the electrons in collision with the ions of the crystal lattice appears as Joule heat (I^2R).

3. (a) Examining Table 38-2, we see that the greatest difference between the work functions will occur when potassium and nickel are joined.

(b) 3.10 V

5. (c)

7. (a) True

(b) False

(c) True

(d) False

(e) True

(f) True

(g) False

9. (b)

11. The excited electron is in the conduction band and can conduct electricity. A hole is left in the valence band allowing the positive hole to move through the band also contributing to the current.

13. (c)

15. ∞, 40 Ω, 20 Ω, 10 Ω, 5 Ω

17. 2.07 g/cm^3

19. (a) -10.6 eV

(b) -2.83 percent

21. (a) 0.123 Ω·m

(b) 0.0707 Ω·m

23. (a) $n_{Ag} = 5.86 \times 10^{22}$ electrons/cm^3

(b) $n_{Ag} = 5.90 \times 10^{22}$ electrons/cm^3; Both these results agree with the values in Table 38-1.

25. 4.00

27. (a) 1.07×10^6 m/s

(b) 1.39×10^6 m/s

(c) 1.89×10^6 m/s

29. (a) 4.22 eV

(b) 2.85 eV

31. (a) 5.90×10^{28} e/m^3

(b) 5.49 eV

(c) 211

(d) E_F is 211 times kT at room temperature. There are so many free electrons present that most of them are crowded, as described by the Pauli exclusion principle, up to energies far higher than they would be according to the classical model.

33. (c) 63.7×10^9 N/m^2; $B = 0.455B_{Cu}$

35. 0.192 J/mol·K

37. (a) 65.6 nm

(b) 0.0180 nm^2

39. 1.09 μm

41. 177 nm

43. 116 K

45. 3.17 nm; 8.46 nm

47. 37.2 nm; $\lambda_{Cu} = 38.8$ nm; The mean free paths agree to within 4.02 percent.

49.

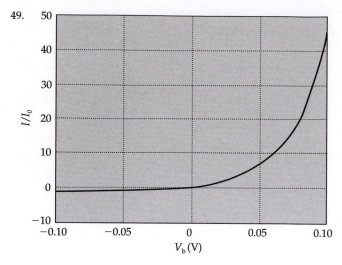

51. 250

53. (a) (b)

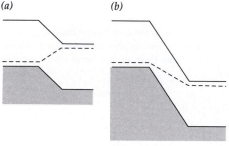

55. $n = 1.00 \times 10^{23}$ m^{-3}; the semiconductor is p-type.

57. (a) 2.17 meV; $0.8E_{g,measured}$

 (b) 0.454 mm

59. 1.97×10^{18}

63. 1

67. 0.596

73. 1.07

75. (a) 5.51×10^{-3}

 (b) 1.84×10^{-2}

77. 689 nm

Chapter 39

1. (a)

3. (a) True

 (b) True

 (c) False

 (d) True

 (e) False

 (f) False

 (g) True

5. Although $\Delta y = \Delta y'$; $\Delta t \neq \Delta t'$. Consequently, $u_y = \Delta y/\Delta t' \neq \Delta y'/\Delta t' = u_y'$.

7. (a) 0.946

 (b) 12.3 Gc·y

9. (a) $L_p = 978.5$ m; the width of the beam is unchanged

 (b) 9.57×10^7 m

 (c) 0.102 μm

11. (a) $0.914c$

 (b) $22.5c$·y

 (c) 101 y

15. 1.85×10^4 y

17. (a) 1.76 μs

 (b) 6.32 μs

 (c) 3.09 μs

 (d) 1.70 km

19. 4.39 μs

21. (a) 2.10 μs

 (b) 2.59 μs

 (c) 0.49 μs

 (d) 2.59 μs

 (e) 4.36 h

 (f) 18.8 h

25. 2.22×10^7 m/s

29. 10.5 ms

31. (a) $u_x = v; u_y = \dfrac{c}{\gamma}$

33. (a) $0.976c$

 (b) $0.997c$

35. 66.7 percent

39. (a) 290 MeV

 (b) 629 MeV

41. (a) $0.943c$

 (b) 3.00 MeV

 (c) 2.83 MeV/c

 (d) 0.878 MeV

 (e) 4.12 MeV/c^2

43. In a freely falling reference frame, both cannonballs travel along straight lines, so they must hit each other, as they were pointed at each other when they were fired.

45. $0.999c$

47. (a) S' moves in the negative x direction

 (b) 1.73 y

49. 281 MeV

51. (a) $v = -\dfrac{E}{Mc}$

 (b) $d = -\dfrac{LE}{Mc^2}$

53. $a_x' = \dfrac{a_x}{\gamma^3 \delta^3}; a_y' = \dfrac{a_y}{\gamma^2 \delta^2} + \dfrac{v u_y}{\gamma^3 \delta^3 c^2} a_x$

Chapter 40

1. (a) ^{15}N, ^{16}N

 (b) ^{54}Fe, ^{55}Fe

 (c) ^{54}Fe, ^{55}Fe

3. Generally, β-decay leaves the daughter nucleus neutron rich, that is, above the line of stability. The daughter nucleus therefore tends to decay via β^- emission, which converts a nuclear neutron to a proton.

5. It would make the dating unreliable because the current concentration of ^{14}C is not equal to that at some earlier time.

7. The probability for neutron capture by the fissionable nucleus is large only for slow (thermal) neutrons. The neutrons emitted in the fission process are fast (high energy) neutrons and must be slowed to thermal neutrons before they are likely to be captured by another fissionable nucleus.

9. (a) β^-
   (b) β^+

11. (a) False
    (b) True
    (c) False
    (d) True

13. (a) $^{16}_{7}\text{N} \rightarrow {}^{16}_{8}\text{O} + {}^{0}_{-1}\beta + {}^{0}_{0}\overline{v} + Q$
    (b) $^{248}_{100}\text{Fm} \rightarrow {}^{244}_{98}\text{Cf} + {}^{4}_{2}\text{He} + Q$
    (c) $^{12}_{7}\text{N} \rightarrow {}^{12}_{6}\text{C} + {}^{0}_{+1}\beta + {}^{0}_{0}v + Q$
    (d) $^{81}_{34}\text{Se} \rightarrow {}^{81}_{35}\text{Br} + {}^{0}_{-1}\beta + {}^{0}_{0}\overline{v} + Q$
    (e) $^{61}_{29}\text{Cu} \rightarrow {}^{61}_{28}\text{Ni} + {}^{0}_{+1}\beta + {}^{0}_{0}v + Q$
    (f) $^{228}_{90}\text{Th} \rightarrow {}^{224}_{88}\text{Ra} + {}^{4}_{2}\text{He} + Q$

15. Mass density = 10^{15}, half life = 10^{15}, and nuclear masses = 2

17. (a) 92.2 MeV; 7.68 MeV
    (b) 492 MeV; 8.79 MeV
    (c) 1802 MeV; 7.57 MeV

19. (a) 3.02 fm
    (b) 4.59 fm
    (c) 6.98 fm

21. (a) 4.11×10^{-21} J; 25.7 MeV
    (b) 2.22 km/s
    (c) 10.1 min

23. 295 MeV

25. (a) 5 min
    (b) 250 Bq

27. (a) 200 s
    (b) 3.47×10^{-3} s^{-1}
    (c) 125 Bq

29. (a) 500 Bq; 250 Bq
    (b) 1.04×10^6; 5.19×10^5
    (c) 12.1 min

31. (a) 4.55×10^3 α/s
    (b) 5.32×10^4 y

33. $^{239}_{94}\text{Pu} \rightarrow {}^{235}_{92}\text{U} + {}^{4}_{2}\alpha + Q$; $Q = 5.25$ MeV; $K_\alpha = 5.16$ MeV; $K_U = 87.9$ keV

35. (a) 0.133 h^{-1}; 5.20 h
    (b) 3.11×10^6

37.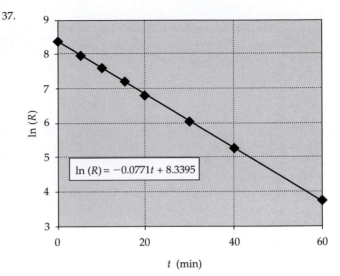

$\lambda = 0.0771$ min^{-1}

$t_{1/2} = 8.99$ min

39. 2.94 g

41. 3.50 min

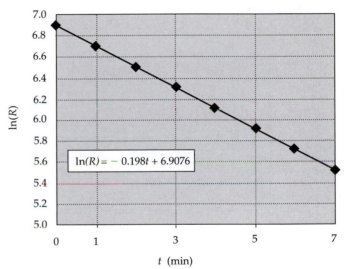

43. 7.03×10^8 y

45. (a) 0.156 MeV
    (b) 3.27 MeV

47. (a) 0.158 MeV
    (b) The masses given are for atoms, not nuclei, so for nuclear masses the masses are too large by the atomic number times the mass of an electron. For the given nuclear reaction, the mass of the carbon atom is too large by $6m_e$ and the mass of the nitrogen atom is too large by $7m_e$. Subtracting $6m_e$ from both sides of the reaction equation leaves an extra electron mass on the right. Not including the mass of the beta particle (electron) is mathematically equivalent to explicitly subtracting $1m_e$ from the right side of the equation.

49. 1.56×10^{19} s^{-1}

51. 208 MeV; 88.1 percent

53. 3.20×10^{10} J

55. (a) $4^1\text{H} \rightarrow {}^4\text{He} + 2\beta^+ + 2v_e + \gamma$
    (b) 24.7 eV
    (c) 3.74×10^{38} s^{-1}; 5.04×10^{10} y

57. 0.693 s^{-1}

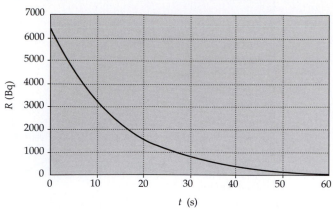

59. 156 keV

61. −2.88 MeV

63. 6.75×10^3 Bq

65. 6.30 L

67. (a) 22.9 MeV

   (b) 4.17 GeV

   (c) 1.28 GeV

69. (a)

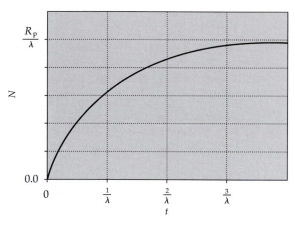

   (b) 8.66×10^4

71. (a) 4.00 fm

   (b) 310 MeV/c

   (d) 310 MeV

75. (a) 1.188 MeV/c

   (b) 752 eV

   (c) 0.0962 percent

77. (b) 55

79. (d)

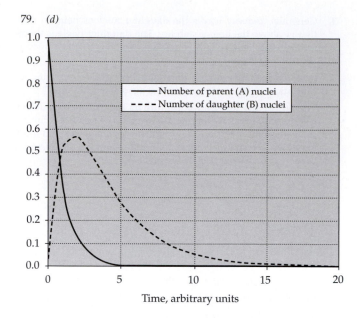

Chapter 41

1. *Similarities:* Baryons and mesons are hadrons; that is, they participate in the strong interaction. Both are composed of quarks.

   Differences: Baryons consist of three quarks and are fermions. Mesons consist of two quarks and are bosons. Baryons have baryon number +1 or −1. Mesons have baryon number 0.

3. A decay process involving the strong interaction has a very short lifetime ($\sim 10^{-23}$ s), whereas decay processes that proceed via the weak interaction have lifetimes of order 10^{-10} s.

5. False

7. No. From Table 41-2, it is evident that any quark–antiquark combination always results in an integral or zero charge.

9. No. Such a reaction is impossible. A proton requires three quarks. Three quarks are not available because a pion is made of a quark and an antiquark and the antiproton consists of three antiquarks.

11. (a) 279.2 MeV

   (b) 1877 MeV

   (c) 211.3 MeV

13. (a) +1; Because $\Delta S = +1$, the reaction can proceed via the weak interaction.

   (b) +2; Because $\Delta S = +2$, the reaction is not allowed.

   (c) +1; Because $\Delta S = +1$, the reaction can proceed via the weak interaction.

15. (a) +2; Because $\Delta S = +2$, the reaction is not allowed.

   (b) +1; Because $\Delta S = +1$, the reaction can proceed via the weak interaction.

17. (a) No. The neutron is not stable. $n \rightarrow p^+ + e^- + \bar{v}_e$

    (b) $\Omega^- \rightarrow p^+ + e^+ + 3e^- + v_e + 3\bar{v}_e + 2\bar{v}_\mu + 2v_\mu$

    (c) Because $Q = -1$ before and after the decay, charge is conserved. Because $B = 1$ before and after the decay, the baryon number is conserved. Because $L_e = 0$ before and after the decay, the lepton number for electrons is conserved. Strangeness is not conserved. However, in each baryon decay $\Delta S = +1$, and each decay is allowed via the weak interaction.

19.

Combination		B	Q	S	Hadron
(a)	uud	1	+1	0	p^+
(b)	udd	1	0	0	n
(c)	uus	1	+1	−1	Σ^+
(d)	dds	1	−1	−1	Σ^-
(e)	uss	1	0	−2	Ξ^0
(f)	dss	1	−1	−2	Ξ^-

21. From Table 41-2, we see that to satisfy the conditions of charge $= +2$ and zero strangeness, charm, topness, and bottomness, the quark combination must be uuu.

23. (a) $c\bar{d}$

    (b) $\bar{c}d$

25. (a) uds

    (b) $\bar{u}\bar{u}\bar{d}$

    (c) dds

27. (a) sss

    (b) ssd

29. 3.26×10^8 $c \cdot y$

33. (a) It must be a meson, and it must consist of a quark and its antiquark.

    (b) The π^0 is its own antiparticle.

    (c) The Ξ^0 is a baryon; it cannot be its own antiparticle. The antiparticle is the $\bar{\Xi}^0 = \overline{uss}$.

35. (a) The u and $\bar{u}$ annihilate, resulting in the photons.

    (b) Two photons are created to conserve linear momentum.

37. (a) These properties indicate that the particle is the kaon, K^0.

    (b) These properties indicate that the particle is either the Σ^0 or the Λ^0 baryon.

    (c) These properties indicate that the particle is the kaon, K^+.

39. (a) 1.99×10^5 km/s

    (b) 8.65×10^9 $c \cdot y$

41. (a) 1193 MeV

    (b) 77.0 MeV/c

    (c) 2.66 MeV

    (d) 74.3 MeV; 74.3 MeV/c

INDEX

Physical Constants[†]

Atomic mass constant	$m_u = \frac{1}{12}m(^{12}C)$	$1\,u = 1.660\,538\,73(13) \times 10^{-27}\,\text{kg}$
Avogadro's number	N_A	$6.022\,141\,99(47) \times 10^{23}\,\text{particles/mol}$
Boltzmann constant	$k = R/N_A$	$1.380\,6503(24) \times 10^{-23}\,\text{J/K}$ $8.617\,342(15) \times 10^{-5}\,\text{eV/K}$
Bohr magneton	$m_B = e\hbar/(2m_e)$	$9.274\,008\,99(37) \times 10^{-24}\,\text{J/T} =$ $5.788\,381\,749(43) \times 10^{-5}\,\text{eV/T}$
Coulomb constant	$k = 1/(4\pi\epsilon_0)$	$8.987\,551\,788\ldots \times 10^{9}\,\text{N·m}^2/\text{C}^2$
Compton wavelength	$\lambda_C = h/(m_e c)$	$2.426\,310\,215(18) \times 10^{-12}\,\text{m}$
Fundamental charge	e	$1.602\,176\,462(63) \times 10^{-19}\,\text{C}$
Gas constant	R	$8.314\,472(15)\,\text{J/(mol·K)} =$ $1.987\,2065(36)\,\text{cal/(mol·K)} =$ $8.205\,746(15) \times 10^{-2}\,\text{L·atm/(mol·K)}$
Gravitational constant	G	$6.673(10) \times 10^{-11}\,\text{N·m}^2/\text{kg}^2$
Mass of electron	m_e	$9.109\,381\,88(72) \times 10^{-31}\,\text{kg} =$ $0.510\,998\,902(21)\,\text{MeV}/c^2$
Mass of proton	m_p	$1.672\,621\,58(13) \times 10^{-27}\,\text{kg} =$ $938.271\,998(38)\,\text{MeV}/c^2$
Mass of neutron	m_n	$1.674\,927\,16(13) \times 10^{-27}\,\text{kg} =$ $939.565\,330(38)\,\text{MeV}/c^2$
Permittivity of free space	ϵ_0	$8.854\,187\,817\ldots \times 10^{-12}\,\text{C}^2/(\text{N·m}^2)$
Permeability of free space	μ_0	$4\pi \times 10^{-7}\,\text{N/A}^2$
Planck's constant	h	$6.626\,068\,76(52) \times 10^{-34}\,\text{J·s} =$ $4.135\,667\,27(16) \times 10^{-15}\,\text{eV·s}$
	$\hbar = h/(2\pi)$	$1.054\,571\,596(82) \times 10^{-34}\,\text{J·s} =$ $6.582\,118\,89(26) \times 10^{-16}\,\text{eV·s}$
Speed of light	c	$2.997\,924\,58 \times 10^{8}\,\text{m/s}$
Stefan-Boltzmann constant	σ	$5.670\,400(40) \times 10^{-8}\,\text{W/(m}^2\text{·K}^4)$

† The values for these and other constants can be found in Appendix B as well as on the Internet at http://physics.nist.gov/cuu/Constants/index.html. The numbers in parentheses represent the uncertainties in the last two digits. (For example, 2.044 43(13) stands for 2.044 43 ± 0.000 13.) Values without uncertainties are exact. Values with ellipses are exact (like the number $\pi = 3.1415\ldots$).

Derivatives and Definite Integrals

$$\frac{d}{dx}\sin ax = a\cos ax \qquad \int_0^\infty e^{-ax}\,dx = \frac{1}{a} \qquad \int_0^\infty x^2 e^{-ax^2}\,dx = \frac{1}{4}\sqrt{\frac{\pi}{a^3}}$$

$$\frac{d}{dx}\cos ax = -a\sin ax \qquad \int_0^\infty e^{-ax^2}\,dx = \frac{1}{2}\sqrt{\frac{\pi}{a}} \qquad \int_0^\infty x^3 e^{-ax^2}\,dx = \frac{4}{a^2}$$

$$\frac{d}{dx}e^{ax} = ae^{ax} \qquad \int_0^\infty xe^{-ax^2}\,dx = \frac{2}{a} \qquad \int_0^\infty x^4 e^{-ax^2}\,dx = \frac{3}{8}\sqrt{\frac{\pi}{a^5}}$$

The a in the six integrals is a positive constant.

Vector Products

$$\vec{A} \cdot \vec{B} = AB\cos\theta \qquad \vec{A} \times \vec{B} = AB\sin\theta\,\hat{n} \quad (\hat{n} \text{ obtained using right-hand rule})$$

For additional data, see the following tables in the text.

Geometry and Trigonometry

$C = \pi d = 2\pi r$ definition of π

$A = \pi r^2$ area of circle

$V = \frac{4}{3}\pi r^3$ spherical volume

$A = dV/dr = 4\pi r^2$ spherical surface area

$V = A_{\text{base}}L = \pi r^2 L$ cylindrical volume

$A = dV/dr = 2\pi rL$ cylindrical surface area

$o = h\sin\theta$
$a = h\cos\theta$

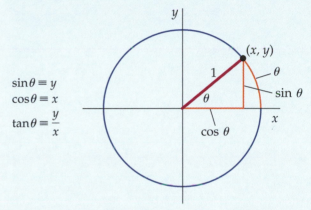

$\sin^2\theta + \cos^2\theta = 1$

$\sin(A \pm B) = \sin A \cos B \pm \cos A \sin B$

$\cos(A \pm B) = \cos A \cos B \mp \sin A \sin B$

$\sin A \pm \sin B = 2\sin[\frac{1}{2}(A \pm B)]\cos[\frac{1}{2}(A \mp B)]$

$\sin\theta \equiv y$
$\cos\theta \equiv x$
$\tan\theta \equiv \dfrac{y}{x}$

IF $|\theta| \ll 1$, THEN
$\cos\theta \approx 1$ AND $\tan\theta \approx \sin\theta \approx \theta$ (θ in radians)

Quadratic Formula

If $ax^2 + bx + c = 0$, then $x = \dfrac{-b \pm \sqrt{b^2 - 4ac}}{2a}$

Binomial Expansion

If $|x| < 1$, then $(1 + x)^n =$
$$1 + nx + \frac{n(n-1)}{2!}x^2 + \frac{n(n-1)(n-2)}{3!}x^3 + \dots$$

If $|x| \ll 1$, then $(1 + x)^n \approx 1 + nx$

If $|\Delta x|$ is small, then $\Delta F \approx \dfrac{dF}{dx}\Delta x$